For the Student: 10 Ways to Succeed with Algebra

1. Attend class regularly. Pay attention in class and take careful notes. In particular, note the problems your teacher works and copy the complete solutions. Keep these notes separate from your homework.

2. Ask questions. Don't hesitate to ask questions in class. Other students may have the same questions but be reluctant to ask them, and everyone will benefit from the answers.

3. Read your text carefully. Many students go directly to the exercise sets without taking time to read the text and examples. *Read the complete section,* with pencil, paper, and calculator handy. Pay special attention to the colored boxes, which contain key definitions, rules, and procedures, as well as the Caution, Note, and Problem-Solving Hint boxes. Doing so will pay off when you tackle the homework problems.

4. Reread your class notes. Before starting your homework, rework the problems your teacher did in class. This will reinforce what you have learned. Teachers often hear the comment, *"I understand it perfectly when you do it, but I get stuck when I try to work the problem myself."*

5. Practice by working problems. Do your homework only *after* reading the text and reviewing your class notes. Check your work against the answer section or the *Student's Solutions Manual.* If you make an error and are unable to determine what went wrong, mark that problem and ask your instructor about it. Then work more problems of the same type to reinforce what you have learned.

6. Work neatly. Write symbols neatly. Skip lines between steps. Write large enough so that others can read your work. Use pencil. Make sure that problems are clearly separated from each other.

7. Review the material. After completing each section, look over the text again. Decide on the main objectives, and don't be content until you feel that you have mastered them. (In this book, objectives are clearly stated both at the beginning and within each section.) Write a summary of the section or make an outline for future reference.

8. Prepare for tests. The chapter summaries in the text are an excellent way for you to review key terms, new symbols, and important concepts from the chapter. After working through the chapter review exercises, use the chapter test as a practice test. Work the problems under test conditions, without looking at the text or answers until you are finished. Time yourself. When you have finished, check your answers against the answer section and rework any that you missed.

9. Learn from your mistakes. Keep all graded assignments, quizzes, and tests that are returned to you. Be sure to correct any errors on them and use them to study for future tests and the final exam.

10. Be diligent and don't give up. The authors of this text can tell you that they did not always understand a topic the first time they saw it. Don't worry if you also find this to be true. As you read more about a topic and work through the problems, you will gain understanding; the thrill of finally "getting it" is a great feeling. Listen to the words of the late Jim Valvano: *Never give up!*

Triangles and Angles

Right Triangle

Triangle has one 90°
(right) angle.

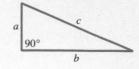

Pythagorean Formula
(for right triangles)

$a^2 + b^2 = c^2$

Right Angle

Measure is 90°.

Isosceles Triangle

Two sides are equal.

$AB = BC$

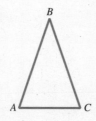

Straight Angle

Measure is 180°.

Equilateral Triangle

All sides are equal.

$AB = BC = CA$

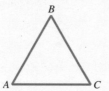

Complementary Angles

The sum of the measures of
two complementary
angles is 90°.

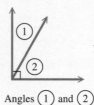

Angles ① and ②
are complementary.

Sum of the Angles of Any Triangle

$A + B + C = 180°$

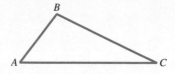

Supplementary Angles

The sum of the
measures of two
supplementary
angles is 180°.

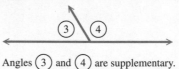

Angles ③ and ④ are supplementary.

Similar Triangles

Corresponding angles are
equal; corresponding sides
are proportional.

$A = D, B = E, C = F$

$\dfrac{AB}{DE} = \dfrac{AC}{DF} = \dfrac{BC}{EF}$

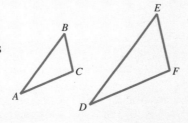

Vertical Angles

Vertical angles have
equal measures.

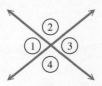

Angle ① = Angle ③

Angle ② = Angle ④

Intermediate Algebra

10th EDITION

MARGARET L. LIAL
American River College

JOHN HORNSBY
University of New Orleans

TERRY McGINNIS

PEARSON

Addison
Wesley

Boston   San Francisco   New York
London   Toronto   Sydney   Tokyo   Singapore   Madrid
Mexico City   Munich   Paris   Cape Town   Hong Kong   Montreal

Publisher: Greg Tobin
Editor in Chief: Maureen O'Connor
Senior Project Editor: Lauren Morse
Assistant Editor: Caroline Case
Senior Managing Editor: Karen Wernholm
Senior Production Supervisor: Kathleen A. Manley
Senior Designer: Dennis Schaefer
Photo Researcher: Beth Anderson
Digital Assets Manager: Marianne Groth
Media Producer: Sharon Tomasulo
Software Development: John O'Brien, MathXL; Ted Hartman, TestGen
Marketing Manager: Michelle Renda
Marketing Assistant: Alexandra Waibel
Senior Author Support/Technology Specialist: Joe Vetere
Senior Prepress Supervisor: Caroline Fell
Rights and Permissions Advisor: Dana Weightman
Manufacturing Manager: Evelyn Beaton
Media Buyer: Ginny Michaud
Text Design: IKO Ink
Production Coordination and Composition: WestWords/PMG
Illustrations: Network Graphics

Cover image: Cattail Marsh © Copyright Lorraine Cota Manley

Photo Credits: see page I-11

Many of the designations used by manufacturers and sellers to distinguish their products are claimed as trademarks. Where those designations appear in this book, and Addison-Wesley was aware of a trademark claim, the designations have been printed in initial caps or all caps.

Library of Congress Cataloging-in-Publication Data
Lial, Margaret L.
    Intermediate algebra.— 10th ed. / Margaret L. Lial, John Hornsby, Terry McGinnis.
      p. cm.
    Includes index.
    ISBN 0-321-55764-6
    1. Algebra—Textbooks.   I. Hornsby, E. John.   II. McGinnis, Terry.   III. Title.

    QA152.3.L534   2007
    512—dc22

                          2006052486
ISBN-13: 978-0-321-55764-3   ISBN-10: 0-321-55764-6

5 6 7 8 9 10—DOW—11 10 09

Contents

List of Applications

Preface

The tenth edition of *Intermediate Algebra* continues our ongoing commitment to provide the best possible text and supplements package to help instructors teach and students succeed. To that end, we have addressed the diverse needs of today's students through a more open design, updated figures and graphs, helpful features, careful explanations of topics, and a comprehensive package of supplements and study aids. We have also taken special care to respond to the suggestions of users and reviewers and have added many new examples and exercises based on their feedback. Students who have never studied algebra—as well as those who require further review of basic algebraic concepts before taking additional courses in mathematics, business, science, nursing, or other fields—will benefit from the text's student-oriented approach.

This text is part of a series that includes the following books:

- *Beginning Algebra,* Tenth Edition, by Lial, Hornsby, and McGinnis
- *Beginning and Intermediate Algebra,* Fourth Edition, by Lial, Hornsby, and McGinnis
- *Algebra for College Students,* Sixth Edition, by Lial, Hornsby, and McGinnis

Key Features

We believe students and instructors will welcome the following helpful features.

Enhanced Annotated Instructor's Edition   For easier reference, margin answers in the *Annotated Instructor's Edition* are now given in a single- or double-column format whenever possible. In addition, the authors have added approximately 45 new Teaching Tips and more than 100 new and updated Classroom Examples.

NEW   *Tab Your Way to Success!*   A "Tab Your Way to Success!" guide provides students with color-coded Post-It® tabs to mark important pages of the text that they may need to return to for review work, test preparation, or instructor help.

Chapter Openers   New and updated chapter openers feature real-world applications of mathematics that are relevant to students and tied to specific material within the chapters. Examples of topics include Americans' spending on pets, television ownership, and the relationship between temperature scales. (See pages 1, 53, and 147—Chapters 1, 2, and 3.)

Real-Life Applications   We are always on the lookout for interesting data to use in real-life applications. As a result, we have included many new or updated examples and exercises from fields such as business, pop culture, sports, the life sciences, and technology that show the relevance of algebra to daily life. (See pages 148, 182, and 265.) A comprehensive List of Applications appears at the beginning of the text. (See pages vii–x.)

Figures, Photos, and **NEW** *Hand-Drawn Graphs*   Today's students are more visually oriented than ever. Thus, we have made a concerted effort to include mathematical figures, diagrams, tables, and graphs, including the new "hand-drawn" style of graphs, whenever possible. (See pages 150, 194, and 314.) Many of the graphs also use a style similar to that seen by students in today's print and electronic media. Photos have been incorporated to enhance applications in examples and exercises. (See pages 48, 227, and 292.)

Emphasis on Problem Solving   Introduced in Chapter 2, our six-step problem-solving method is integrated throughout the text. The six steps, *Read, Assign a Variable, Write an Equation, Solve, State the Answer,* and *Check,* are emphasized in boldface type and repeated in examples and exercises to reinforce the problem-solving process for students. (See pages 75, 254, and 432.) **PROBLEM-SOLVING HINT** boxes provide students with helpful problem-solving tips and strategies. (See pages 66, 73, and 89.)

Learning Objectives   Each section begins with clearly stated, numbered objectives, and the included material is directly keyed to these objectives so that students know exactly what is covered in each section. (See pages 35, 286, and 498.)

Cautions and Notes   One of the most popular features of previous editions, **CAUTION** and **NOTE** boxes warn students about common errors and emphasize important ideas throughout the exposition. (See pages 7, 168, and 476.) Highlighted in bright yellow, the text design makes them easy to spot.

NEW   *Pointers*   Pointers from the authors have been added to examples and provide students with important on-the-spot reminders and warnings about common pitfalls. (See pages 17, 104, and 183.)

Connections   Connections boxes provide connections to the real world or to other mathematical concepts, historical background, and thought-provoking questions for writing, class discussion, or group work. (See pages 80, 376, and 487.)

Now Try Exercises   To actively engage students in the learning process, each example concludes with a reference to one or more parallel exercises from the corresponding exercise set. In this way, students are able to immediately apply and reinforce the concepts and skills presented in the examples. These Now Try exercises are now marked with gray screens in the exercise sets so they can be easily spotted. **NEW** Using the new Video Lectures on CD or DVD with Solution Clips, students can watch an instructor work through the complete solution to one Now Try problem for every example in the text. Exercises with a solution on video are marked with a CD icon ⊙ in the exercise sets. (See pages 18, 166, and 259.)

Ample and Varied Exercise Sets   One of the most commonly mentioned strengths of this text is its exercise sets. The text contains a wealth of exercises to provide students with opportunities to practice, apply, connect, and extend the algebraic concepts and skills they are learning. Numerous illustrations, tables, graphs, and photos have been added to the exercise sets to help students visualize the problems they are solving. Problem types include writing ✎, estimation, graphing calculator ▦, and challenging "brain buster" exercises that go beyond the examples as well as applications and multiple-choice, matching, true/false, and fill-in-the-blank problems.

NEW • *Concept Check* exercises facilitate mathematical thinking and conceptual understanding. (See pages 41, 212, and 351.)

NEW • *WHAT WENT WRONG?* exercises ask students to identify typical errors in solutions and work the problems correctly. (See pages 60, 296, and 351.)

NEW • **PREVIEW EXERCISES**, brought back from earlier editions by popular request, *review* previously-studied concepts and *preview* skills needed for the upcoming section. (See pages 62, 199, and 301.)

• *Relating Concepts Exercises* These sets of exercises help students tie together topics and develop problem-solving skills as they compare and contrast ideas, identify and describe patterns, and extend concepts to new situations. (See pages 87, 191, and 365.) These exercises make great collaborative activities for pairs or small groups of students.

• *Summary Exercises* These special exercise sets provide students with the all-important *mixed* review problems they need to master topics. Summaries of solution methods or additional examples are often included. (See pages 97, 132, and 426.)

• *Technology Insights Exercises* We assume that all students who use this text have access to scientific calculators. *While graphing calculators are not required for this text,* some students may go on to courses that use them. For this reason, we have included Technology Insights exercises in selected exercise sets. These exercises provide an opportunity for students to interpret typical results seen on graphing calculator screens. Actual calculator screens from the Texas Instruments TI-83/84 Plus graphing calculator are featured. (See pages 159, 301, and 379.)

Group Activities Appearing at the end of each chapter, these real-data activities allow students to apply the mathematical content of the chapter in a collaborative setting. (See pages 133, 274, and 524.)

Ample Opportunity for Review Each chapter concludes with a Chapter Summary that features Key Terms, New Symbols, Test Your Word Power, and a Quick Review of each section's content with additional examples. A comprehensive set of Chapter Review Exercises, keyed to individual sections, is included, as are Mixed Review Exercises and a Chapter Test. Beginning with Chapter 2, each chapter concludes with a set of Cumulative Review Exercises that cover material going back to Chapter 1. **NEW** The new Pass the Test: Chapter Test Solutions on Video with Interactive Chapter Summaries CD includes many helpful review resources based on the Chapter Summary. (See pages 216, 278, 341, and 387.)

Test Your Word Power To help students understand and master mathematical vocabulary, this feature is found in each Chapter Summary. Key terms from the chapter are presented along with four possible definitions in a multiple-choice format. Answers and examples illustrating each term are provided. **NEW** Interactive versions of the Key Terms and Test Your Word Power are available on the Pass the Test CD. (See pages 44, 134, and 216.)

Glossary A comprehensive glossary of key terms from throughout the text is included at the back of the book. (See pages G-1 to G-7.)

What content changes have been made?

A primary focus of this revision of the text was to polish and enhance individual presentations of topics and exercise sets, based on user and reviewer feedback, and we have worked hard to do this throughout the book. Some of the specific content changes include the following:

- There are approximately 1200 new and updated exercises, including many problems that focus on drill, skill development, and review. These include new Concept Check exercises, *WHAT WENT WRONG?* problems, and Preview Exercises.

- When a new type of graph is introduced (Sections 3.1, 3.4, 5.3, 7.4, 8.1, 9.5, 9.6, 10.2, 10.3, and 11.1–11.3), a new "hand-drawn" graph style is used to simulate what a student might actually sketch on graph paper.

- Real-world data in over 300 applications have been updated.

- Chapter 3 includes a new set of summary exercises on slopes and equations of lines.

- Variation, formerly covered in Chapter 3, now appears in Section 7.6.

- Composition of functions is now covered in Section 5.3.

- The presentation of the following topics are among those that have been expanded:

    Review of fractions (Section 1.1)
    Compound inequalities (Section 2.6)
    Midpoint formula (Section 3.1)
    Graphs of linear and constant functions (Section 3.5)
    Composition of functions (Section 5.3)
    Summary Exercises on Operations with Radicals and Rational Exponents (Chapter 8)

What supplements are available?

For a comprehensive list of the supplements and study aids that accompany *Intermediate Algebra,* Tenth Edition, see pages xvi and xviii.

Acknowledgments

The comments, criticisms, and suggestions of users, nonusers, instructors, and students have positively shaped this textbook over the years, and we are most grateful for the many responses we have received. Thanks to the following people for their review work, feedback, assistance at various meetings, and additional media contributions:

Barbara Aaker, *Community College of Denver*
Viola Lee Bean, *Boise State University*
Kim Bennekin, *Georgia Perimeter College*
Dixie Blackinton, *Weber State University*
Tim Caldwell, *Meridian Community College*
Sally Casey, *Shawnee Community College*
Callie Daniels, *St. Charles Community College*
Cheryl Davids, *Central Carolina Technical College*
Chris Diorietes, *Fayetteville Technical Community College*
Sylvia Dreyfus, *Meridian Community College*
Lucy Edwards, *Las Positas College*
LaTonya Ellis, *Bishop State Community College*
Beverly Hall, *Fayetteville Technical Community College*
Sandee House, *Georgia Perimeter College*
Lynette King, *Gadsden State Community College*
Linda Kodama, *Kapi´olani Community College*

Ted Koukounas, *Suffolk Community College*
Karen McKarnin, *Allen County Community College*
James Metz, *Kapi´olani Community College*
Jean Millen, *Georgia Perimeter College*
Molly Misko, *Gadsden State Community College*
Jane Roads, *Moberly Area Community College*
Melanie Smith, *Bishop State Community College*
Erik Stubsten, *Chattanooga State Technical Community College*
Tong Wagner, *Greenville Technical College*
Sessia Wyche, *University of Texas at Brownsville*

Special thanks are due all those instructors at Broward Community College for their insightful comments.

Over the years, we have come to rely on an extensive team of experienced professionals. Our sincere thanks go to these dedicated individuals at Addison-Wesley, who worked long and hard to make this revision a success: Greg Tobin, Maureen O'Connor, Lauren Morse, Michelle Renda, Caroline Case, Alexandra Waibel, Kathy Manley, and Sharon Smith.

Abby Tanenbaum did an outstanding job helping us revise traditional and real-data applications. Melena Fenn provided excellent production work. Thanks are due Jeff Cole, who supplied accurate, helpful solutions manuals, and Jim Ball, who provided the comprehensive Printed Test Bank. We are most grateful to Becky Troutman for preparing the comprehensive List of Applications; and Janis Cimperman, Steve Ouellette, and Cathy Ferrer for accuracy checking page proofs.

As an author team, we are committed to the goal stated earlier in this Preface—to provide the best possible text and supplements package to help instructors teach and students succeed. We are most grateful to all those over the years who have aspired to this goal with us. As we continue to work toward it, we would welcome any comments or suggestions you might have. Please feel free to send your comments via e-mail to math@aw.com.

Margaret L. Lial
John Hornsby
Terry McGinnis

STUDENT SUPPLEMENTS

Student's Solutions Manual
- Provides detailed solutions to the odd-numbered section-level exercises and summary exercises and to all Relating Concepts, Chapter Review, Chapter Test, and Cumulative Review Exercises

ISBNs: 0-321-44115-X and 978-0-321-44115-7

NEW Video Lectures on CD with Solution Clips
NEW Video Lectures on DVD with Solution Clips
- Complete set of digitized videos on CD-ROM (or DVD) for students to use at home or on campus
- Includes a full lecture for each section of the text
- Students can also choose to watch an instructor work through the solutions to exercises that have been correlated to all examples from the text (one exercise for each example)
- Each exercise that has a video solution available is denoted in the exercise sets by a CD icon ⊙
- Optional captioning in English and Spanish is available for the lecture portion of this product (Video Lectures on CD only)

CD ISBNs: 0-321-44959-2 and 978-0-321-44959-7
DVD ISBNs: 0-321-44482-5 and 978-0-321-44482-0

NEW Pass the Test: Chapter Test Solutions on Video with Interactive Chapter Summaries on CD
Included with each Student Edition of the book, this CD-ROM contains:
- Interactive "Key Terms" with definitions
- Interactive "Test Your Word Power"
- Summary lectures for each key concept from the "Quick Review" for each chapter
- Video footage of an instructor working through the complete solutions for all chapter test problems

Additional Skill and Drill Manual
- Provides additional practice and test preparation for students

ISBNs: 0-321-44637-2 and 978-0-321-44637-4

NEW MathXL® Tutorials on CD
- Provides algorithmically generated practice exercises that correlate at the objective level to the content of the text
- Every exercise is accompanied by an example and a guided solution, and selected exercises may also include a video clip
- The software provides helpful feedback and can generate printed summaries of students' progress

ISBNs: 0-321-44961-4 and 978-0-321-44961-0

INSTRUCTOR SUPPLEMENTS

Annotated Instructor's Edition
- Provides answers to all text exercises in color in the margin next to the corresponding problems, along with teaching tips and extra examples for the classroom
- Icons identify writing ✎ and calculator ▦ exercises

ISBNs: 0-321-44783-2 and 978-0-321-44783-8

Instructor's Solutions Manual
- Provides complete solutions to all text exercises
- **NEW** Now includes solutions to all Classroom Examples

ISBNs: 0-321-44799-9 and 978-0-321-44799-9

Instructor and Adjunct Support Manual
- Includes resources designed to help both new and adjunct faculty with course preparation and classroom management
- Offers helpful teaching tips correlated to the sections of the text

ISBNs: 0-321-44636-4 and 978-0-321-44636-7

NEW Online Lesson Plans
- Lesson plans for each section of the book
- Worksheets covering additional topics
- Correlation to California Mathematics Content Standards for Algebra II

Printed Test Bank and Instructor's Resource Guide
- The test bank contains two diagnostic pretests, four free-response and two multiple-choice test forms per chapter, and two final exams
- The resource guide contains additional practice exercises for most objectives of every section and a conversion guide from the ninth to the tenth edition

ISBNs: 0-321-44480-9 and 978-0-321-44480-6

Online Answer Book
- Provides answers to all the exercises in the text

PowerPoint Lecture Slides
- Presents key concepts and definitions from the text
- Provides complete solutions to all Classroom Examples from the Annotated Instructor's Edition

NEW Active Learning Lecture Slides
- Multiple choice questions are available for each section of the book, allowing instructors to quickly assess mastery of material in class
- Available in PowerPoint, these slides can be used with classroom response systems

STUDENT SUPPLEMENTS

Addison-Wesley Math Tutor Center

- Staffed by qualified mathematics instructors
- Provides tutoring on examples and odd-numbered exercises from the textbook through a registration number with a new textbook or purchased separately
- Accessible via toll-free telephone, toll-free fax, e-mail, or the Internet
  www.aw-bc/tutorcenter

InterAct Math Tutorial Website www.interactmath.com

- Get practice and tutorial help online!
- Retry an exercise as many times as you like with new values each time for unlimited practice and mastery
- Every exercise is accompanied by an interactive guided solution that gives you helpful feedback when an incorrect answer is entered
- View the steps of a worked-out sample problem similar to the one you're working on

NEW Worksheets for Classroom or Lab Practice

These lab- and classroom-friendly workbooks offer the following resources for every section of the text:

- A list of learning objectives
- Vocabulary practice problems
- Extra practice exercises with ample space for students to show their work

INSTRUCTOR SUPPLEMENTS

TestGen®

- **NEW** Now includes a pre-made-test for each chapter that has been correlated problem-by-problem to the chapter tests in the book
- Enables instructors to build, edit, print, and administer tests using a computerized bank of questions developed to cover all text objectives
- Algorithmically based, TestGen® allows instructors to create multiple but equivalent versions of the same question or test with the click of a button
- Instructors can also modify test bank questions or add new questions
- Tests can be printed or administered online
- Available on a dual-platform Windows/Macintosh CD-ROM

ISBNs: 0-321-44481-7 and 978-0-321-44481-3

Adjunct Support Center

The Math Adjunct Support Center is staffed by qualified mathematics instructors with over 50 years combined experience at both the community college and university level. Assistance is provided for faculty in the following areas:

- Suggested syllabus consultation
- Tips on using materials packaged with your book
- Book-specific content assistance
- Teaching suggestions including advice on classroom strategies

www.aw-bc.com/tutorcenter/math-adjunct.html

Available for Students and Instructors

MathXL®   MathXL® is a powerful online homework, tutorial, and assessment system that accompanies your Addison-Wesley textbook in mathematics or statistics. With MathXL, instructors can create, edit, and assign online homework and tests using algorithmically generated exercises correlated at the objective level to the textbook. They can also create and assign their own online exercises and import TestGen tests for added flexibility. All student work is tracked in MathXL's online gradebook. Students can take chapter tests in MathXL and receive personalized study plans based on their test results. The study plan diagnoses weaknesses and links students directly to tutorial exercises for the objectives they need to study and retest. Students can also access supplemental video clips directly from selected exercises. MathXL is available to qualified adopters. For more information, visit our Web site at www.mathxl.com, or contact your Addison-Wesley sales representative.

MyMathLab®   MyMathLab® is a series of text-specific, easily customizable online courses for Addison-Wesley textbooks in mathematics and statistics. MyMathLab is powered by CourseCompass™—Pearson Education's online teaching and learning environment—and by MathXL®—our online homework, tutorial, and assessment system. MyMathLab gives instructors the tools they need to deliver all or a portion of their course online, whether students are in a lab setting or working from home. MyMathLab provides a rich and flexible set of course materials, featuring free-response exercises that are algorithmically generated for unlimited practice and mastery. Students can also use online tools, such as video lectures, animations, and a multimedia textbook, to independently improve their understanding and performance. Instructors can use MyMathLab's homework and test managers to select and assign online exercises correlated directly to the textbook, and they can also create and assign their own online exercises and import TestGen tests for added flexibility. MyMathLab's online gradebook—designed specifically for mathematics and statistics—automatically tracks students' homework and test results and gives the instructor control over how to calculate final grades. Instructors can also add offline (paper-and-pencil) grades to the MathXL gradebook. MyMathLab is available to qualified adopters. For more information, visit our Web site at www.mymathlab.com or contact your Addison-Wesley sales representative.

Feature Walkthrough

3

Graphs, Linear Equations, and Functions

The two most common measures of temperature are Fahrenheit (F) and Celsius (C). We often see signs displaying one or both of these types of temperature. It is fairly common knowledge that water freezes at 32°F, or 0°C, and boils at 212°F, or 100°C. Because there is a *linear* relationship between the Fahrenheit and Celsius temperature scales, using these two equivalences, we can derive the familiar formulas for converting from one scale to the other, as seen in Exercises 91–96 of Section 3.3.

Graphs are widely used in the media because they present a great deal of information in a concise form. In this chapter, we see how information such as the relationship between the two temperature scales can be depicted by graphs.

3.1 The Rectangular Coordinate System

3.2 The Slope of a Line

3.3 Linear Equations in Two Variables

Summary Exercises on Slopes and Equations of Lines

3.4 Linear Inequalities in Two Variables

3.5 Introduction to Functions

Chapter Opener

Each chapter opens with an application and section outline. The application in the opener is tied to material presented later in the chapter.

---

148 **CHAPTER 3** Graphs, Linear Equations, and Functions

3.1 The Rectangular Coordinate System

OBJECTIVES

1 Interpret a line graph.
2 Plot ordered pairs.
3 Find ordered pairs that satisfy a given equation.
4 Graph lines.
5 Find *x*- and *y*-intercepts.
6 Recognize equations of horizontal and vertical lines and lines passing through the origin.
7 Use the midpoint formula.
8 Use a graphing calculator to graph an equation.

OBJECTIVE 1 Interpret a line graph. The line graph in Figure 1 shows personal spending (in billions of dollars) on medical care in the United States from 1997 through 2003. About how much was spent on medical care in 2002? (We will answer this question shortly.)

Personal Spending on Medical Care

Source: U.S. Centers for Medicare and Medicaid Services.

FIGURE 1

The line graph in Figure 1 presents information based on a method for locating a point in a plane developed by René Descartes, a 17th-century French mathematician. Legend has it that Descartes, who was lying in bed ill, was watching a fly crawl about on the ceiling near a corner of the room. It occurred to him that the location of the fly on the ceiling could be described by determining its distances from the two adjacent walls. See the figure in the margin. We use this insight to plot points and graph linear equations in two variables whose graphs are straight lines.

OBJECTIVE 2 Plot ordered pairs. Each of the pairs of numbers (3, 1), (−5, 6), and (4, −1) is an example of an **ordered pair**—that is, a pair of numbers written within parentheses in which the order of the numbers is important. We graph an ordered pair by using two perpendicular number lines that intersect at their 0 points, as shown in the plane in Figure 2. The common 0 point is called the **origin**. The position of any point in this plane is determined by referring to the horizontal number line, or **x-axis**, and the vertical number line, or **y-axis**. The first number in the ordered pair indicates the position relative to the *x*-axis, and the second number indicates the position relative to the *y*-axis. The *x*-axis and the *y*-axis make up a **rectangular** (or **Cartesian**, for Descartes) **coordinate system**.

To locate, or **plot**, the point on the graph that corresponds to the ordered pair (3, 1), we move three units from 0 to the right along the *x*-axis and then one unit up parallel to the *y*-axis. The point corresponding to the ordered pair (3, 1) is labeled *A* in Figure 3 on the next page. Additional points are labeled *B–E*. The phrase "the point corresponding to the ordered pair (3, 1)" is often abbreviated as "the point (3, 1)." The numbers in an ordered pair are called the **coordinates** of the corresponding point.

Locating a fly on a ceiling

Origin *y*-axis

x-axis

FIGURE 2

Learning Objectives

Each section opens with a highlighted list of clearly stated, numbered *learning objectives.* These learning objectives are restated throughout the section where appropriate for reinforcement.

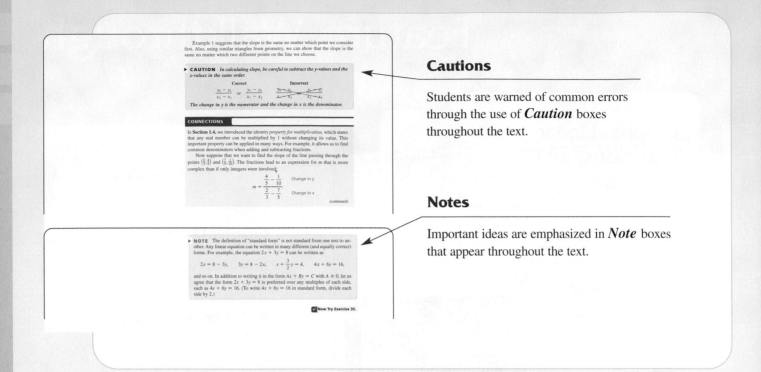

Cautions

Students are warned of common errors through the use of **Caution** boxes throughout the text.

Notes

Important ideas are emphasized in **Note** boxes that appear throughout the text.

Now Try Exercises

Now Try Exercises are found after each example to encourage students to work exercises in the exercise sets that parallel the example just studied. **NEW**—Students can watch an instructor working through the complete solution to one *Now Try* problem for every example in the text on the new Video Lectures on CD or DVD with Solution Clips.

Classroom Examples and Teaching Tips

The *Annotated Instructor's Edition* provides answers to all text exercises and Group Activities in color in the margin or next to the corresponding exercise. ***Classroom Examples*** are also included to provide instructors with examples that are different from those that students have in their textbooks. Solutions to the Classroom Examples are found in the *Instructor's Solutions Manual* or in the PowerPoint Lecture Slides. ***Teaching Tips*** offer guidance on presenting the material at hand.

NEW Pointers

Examples have been made even more student-friendly with ***pointers*** from the authors that provide on-the-spot reminders and warnings about common pitfalls.

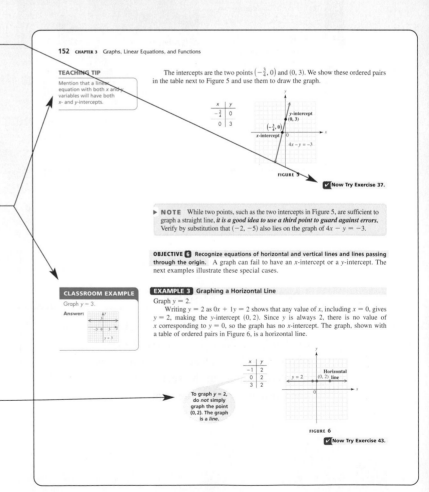

Connections

Connections boxes provide connections to the real world or to other mathematical concepts, historical background, and offer thought-provoking questions for writing or class discussion.

Writing Exercises

Writing exercises abound in the Lial series through the Connections boxes and also in the exercise sets (as marked with a pencil icon ✏).

NEW Preview Exercises

Preview Exercises have been added to the end of each section exercise set to help students transition from one section to the next.

PREVIEW EXERCISES

Solve each equation for y. See Section 2.2.

99. $3x + 2y = 8$ **100.** $4x + 3y = 0$ **101.** $y - 2 = 4(x + 3)$

Write each equation in the form $Ax + By = C$. See Section 2.1.

102. $y - (-2) = \frac{3}{2}(x - 5)$ **103.** $y - (-1) = \frac{5}{3}[x - (-4)]$

104. $y - 7 = -\frac{1}{4}[x - (-3)]$ **105.** $y - (-1) = -\frac{1}{2}[x - (-2)]$

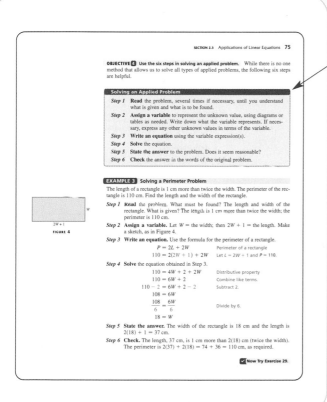

Problem Solving

The Lial *six-step problem-solving method* is introduced in Chapter 2 and is then continually reinforced in examples, exercises, and problem-solving hint boxes throughout the text.

Summary Exercises

Summary Exercises appear in selected chapters to provide students with *mixed* practice problems needed to master topics.

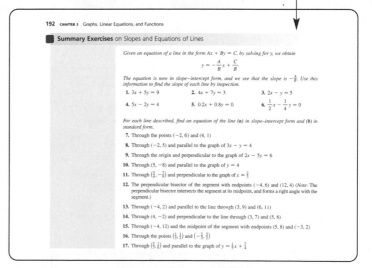

Ample and Varied Exercise Sets

This text contains over 5100 exercises, including over 1400 review exercises, plus numerous conceptual and writing exercises that go beyond the examples. Multiple-choice, matching, true/false, and completion exercises help to provide variety. Exercises suitable for graphing calculator use are marked with an icon. **NEW** Students can watch an instructor work through the complete solution to all exercises marked with a CD icon on the Video Lectures on CD or DVD with Solution Clips.

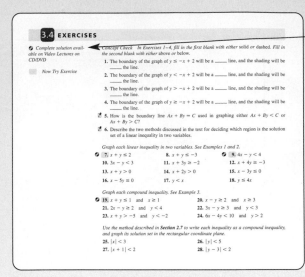

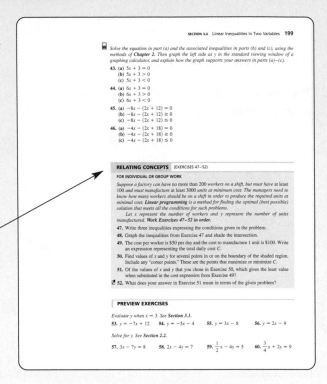

Relating Concepts

Found in selected exercise sets, these exercises tie together topics and highlight the relationships among various concepts and skills.

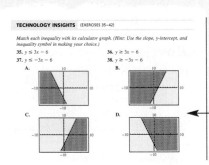

Technology Insights

Technology Insights exercises are found in selected exercise sets throughout the text. These exercises illustrate the power of graphing calculators and provide an opportunity for students to interpret typical results seen on graphing calculator screens. (A graphing calculator is *not* required to complete these exercises).

Group Activities

Appearing at the end of each chapter, these activities allow students to work collaboratively to solve a problem related to the chapter material.

Ample Opportunity for Review

One of the most popular features of the Lial textbooks is the extensive and well thought-out end-of-chapter material. At the end of each chapter, students will find:

Key Terms and New Symbols that are keyed back to the appropriate section for easy reference and study. **NEW** Look for interactive *Key Terms* with definitions on the Pass the Test CD.

Test Your Word Power to help students understand and master mathematical vocabulary; key terms from the chapter are presented with four possible definitions in multiple-choice format. **NEW** Look for an interactive version of *Test Your Word Power* on the Pass the Test CD.

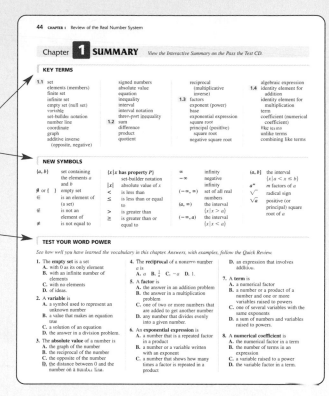

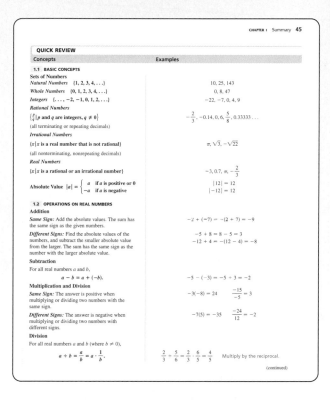

Quick Review sections give students main concepts from the chapter (referenced back to the appropriate section) and an adjacent example of each concept. **NEW** Quick Review Summary Lectures are available for each concept on the Pass The Test CD.

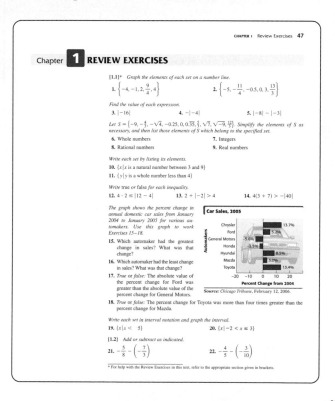

Review Exercises are keyed to the appropriate sections so that students can refer to examples of that type of problem if they need help.

Mixed Review Exercises require students to solve problems without the help of section references.

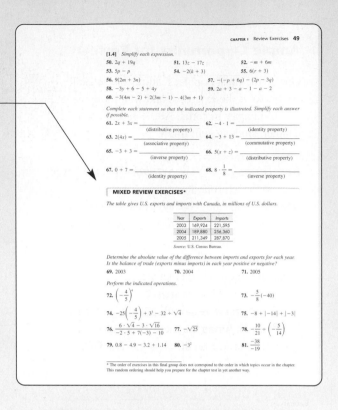

[1.4] *Simplify each expression.*

50. $2q + 19q$ **51.** $13z - 17z$ **52.** $-m + 6m$
53. $5p - p$ **54.** $-2(k + 3)$ **55.** $6(r + 3)$
56. $9(2m + 3n)$ **57.** $-(-p + 6q) - (2p - 3q)$
58. $-3y + 6 - 5 + 4y$ **59.** $2a + 3 - a - 1 - a - 2$
60. $-3(4m - 2) + 2(3m - 1) - 4(3m + 1)$

Complete each statement so that the indicated property is illustrated. Simplify each answer if possible.

61. $2x + 3x =$ _____ **62.** $-4 \cdot 1 =$ _____
(distributive property) (identity property)

63. $2(4x) =$ _____ **64.** $-3 + 13 =$ _____
(associative property) (commutative property)

65. $-3 + 3 =$ _____ **66.** $5(x + z) =$ _____
(inverse property) (distributive property)

67. $0 + 7 =$ _____ **68.** $8 \cdot \frac{1}{8} =$ _____
(identity property) (inverse property)

MIXED REVIEW EXERCISES*

The table gives U.S. exports and imports with Canada, in millions of U.S. dollars.

Year	Exports	Imports
2003	169,924	221,595
2004	189,880	256,360
2005	211,349	287,870

Source: U.S. Census Bureau.

Determine the absolute value of the difference between imports and exports for each year. Is the balance of trade (exports minus imports) in each year positive or negative?

69. 2003 **70.** 2004 **71.** 2005

Perform the indicated operations.

72. $\left(-\frac{4}{5}\right)^4$ **73.** $-\frac{5}{8}(-40)$

74. $-25\left(-\frac{4}{5}\right) + 3^3 - 32 \div \sqrt{4}$ **75.** $-8 + |-14| + |-3|$

76. $\frac{6 \cdot \sqrt{4} - 3 \cdot \sqrt{16}}{-2 \cdot 5 + 7(-3) - 10}$ **77.** $-\sqrt{25}$ **78.** $-\frac{10}{21} \div \left(-\frac{5}{14}\right)$

79. $0.8 - 4.9 - 3.2 + 1.14$ **80.** -3^2 **81.** $\frac{-38}{-19}$

* The order of exercises in this final group does not correspond to the order in which topics occur in the chapter. This random ordering should help you prepare for the chapter test in yet another way.

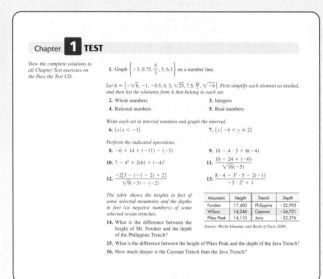

Chapter **1** TEST

View the complete Chapter Test solutions to all Chapter Test exercises on the Pass the Test CD.

1. Graph $\left\{-3, 0.75, \frac{5}{3}, 5, 6.3\right\}$ on a number line.

Let $A = \left\{-\sqrt{6}, -1, -0.5, 0, 3, \sqrt{25}, 7.5, \frac{24}{2}, \sqrt{-4}\right\}$. *First simplify each element as needed, and then list the elements from A that belong to each set.*

2. Whole numbers **3.** Integers
4. Rational numbers **5.** Real numbers

Write each set in interval notation and graph the interval.

6. $\{x \mid x < -3\}$ **7.** $\{y \mid -4 < y \le 2\}$

Perform the indicated operations.

8. $-6 + 14 + (-11) - (-3)$ **9.** $10 - 4 \cdot 3 + 6(-4)$
10. $7 - 4^2 + 2(6) + (-4)^2$ **11.** $\frac{10 - 24 + (-6)}{\sqrt{16}(-5)}$
12. $\frac{-2[3 - (-1 - 2) + 2]}{\sqrt{9}(-3) - (-2)}$ **13.** $\frac{8 \cdot 4 - 3^2 \cdot 5 - 2(-1)}{-3 \cdot 2^3 + 1}$

The table shows the heights in feet of some selected mountains and the depths in feet (as negative numbers) of some selected ocean trenches.

Mountain	Height	Trench	Depth
Foraker	17,400	Philippine	−32,995
Wilson	14,246	Cayman	−24,721
Pikes Peak	14,110	Java	−23,376

Source: World Almanac and Book of Facts 2006.

14. What is the difference between the height of Mt. Foraker and the depth of the Philippine Trench?
15. What is the difference between the height of Pikes Peak and the depth of the Java Trench?
16. How much deeper is the Cayman Trench than the Java Trench?

Chapter Tests help students practice for the real thing. **NEW** The Pass the Test CD in the back of the text offers video of an instructor working through the complete solution for every exercise from the chapter tests.

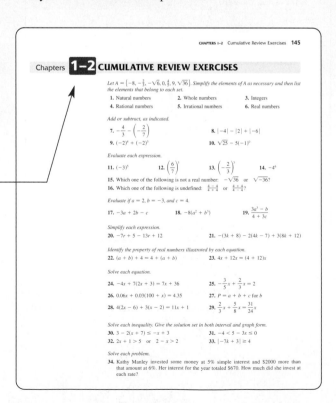

Chapters **1–2** CUMULATIVE REVIEW EXERCISES

Let $A = \left\{-8, -\frac{7}{3}, -\sqrt{6}, 0, \frac{4}{3}, 9, \sqrt{36}\right\}$. *Simplify the elements of A as necessary and then list the elements that belong to each set.*

1. Natural numbers **2.** Whole numbers **3.** Integers
4. Rational numbers **5.** Irrational numbers **6.** Real numbers

Add or subtract, as indicated.

7. $-\frac{4}{3} - \left(-\frac{2}{7}\right)$ **8.** $|-4| - |2| + |-6|$
9. $(-2)^4 + (-2)^5$ **10.** $\sqrt{25} - 5(-1)^0$

Evaluate each expression.

11. $(-3)^5$ **12.** $\left(\frac{6}{7}\right)^3$ **13.** $\left(-\frac{2}{3}\right)^3$ **14.** -4^6
15. Which one of the following is not a real number: $-\sqrt{36}$ or $\sqrt{-36}$?
16. Which one of the following is undefined: $\frac{4 - 4}{4 + 4}$ or $\frac{4 + 4}{4 - 4}$?

Evaluate if $a = 2$, $b = -3$, *and* $c = 4$.

17. $-3a + 2b - c$ **18.** $-8(a^2 + b^3)$ **19.** $\frac{3a^3 - b}{4 + 3c}$

Simplify each expression.

20. $-7r + 5 - 13r + 12$ **21.** $-(3k + 8) - 2(4k - 7) + 3(8k + 12)$

Identify the property of real numbers illustrated by each equation.

22. $(a + b) + 4 = 4 + (a + b)$ **23.** $4x + 12x = (4 + 12)x$

Solve each equation.

24. $-4x + 7(2x + 3) = 7x + 36$ **25.** $-\frac{3}{5}x + \frac{2}{3}x = 2$
26. $0.06x + 0.03(100 + x) = 4.35$ **27.** $P = a + b + c$ for b
28. $4(2x - 6) + 3(x - 2) = 11x + 1$ **29.** $\frac{2}{3}x + \frac{5}{8}x = \frac{31}{24}x$

Solve each inequality. Give the solution set in both interval and graph form.

30. $3 - 2(x + 7) \le -x + 3$ **31.** $-4 < 5 - 3x \le 0$
32. $2x + 1 > 5$ or $2 - x > 2$ **33.** $|-7k + 3| \ge 4$

Solve each problem.

34. Kathy Manley invested some money at 5% simple interest and $2000 more than that amount at 6%. Her interest for the year totaled $670. How much did she invest at each rate?

Cumulative Review Exercises gather various types of exercises from preceding chapters to help students remember and retain what they are learning throughout the course.

Tab Your Way to Success

Use these tabs to mark important sections and pages.

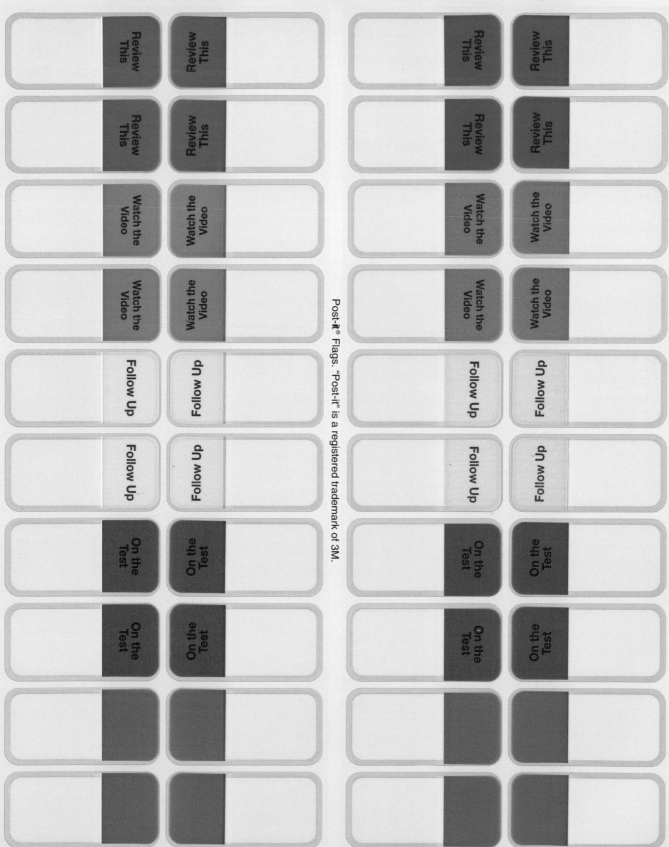

Post-it® Flags. "Post-it" is a registered trademark of 3M.

TABS

TAKE CHARGE OF YOUR LEARNING

These tabs offer several ways to be successful in your mathematics class by taking charge of your own learning.

Review This
Use this tab to identify the most important information in each chapter so that you can refer to these pages easily.

Watch the Video
Tab an exercise for which you would like to see the complete solution worked out on video.

Follow Up
Tab any material you need extra help on or wish to discuss with your instructor.

On the Test
Mark material that your instructor emphasizes or material that you expect will be on the test.

Blank Tab
Use this tab to mark any information you might like to refer to again easily.

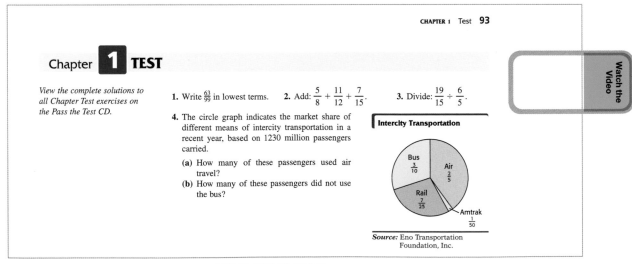

Chapter 1 TEST

View the complete solutions to all Chapter Test exercises on the Pass the Test CD.

1. Write $\frac{63}{99}$ in lowest terms.   2. Add: $\frac{5}{8} + \frac{11}{12} + \frac{7}{15}$.   3. Divide: $\frac{19}{15} \div \frac{6}{5}$.

4. The circle graph indicates the market share of different means of intercity transportation in a recent year, based on 1230 million passengers carried.

   (a) How many of these passengers used air travel?
   (b) How many of these passengers did not use the bus?

Intercity Transportation

Bus $\frac{3}{10}$

Air $\frac{2}{5}$

Rail $\frac{7}{25}$

Amtrak $\frac{1}{50}$

Source: Eno Transportation Foundation, Inc.

Watch the Video

ISBN-13: 978-0-321-50135-6
ISBN-10:   0-321-50135-7

EAN

9 780321 501356

90000

NEW *Pass the Test: Chapter Test Solutions on Video with Interactive Chapter Summaries on CD*, included with each new copy of the book, is based on the end-of-chapter material from the book and offers the following tools for each chapter:

- Interactive *Key Terms* with definitions
- Interactive *Test Your Word Power* vocabulary exercises
- Summary lectures for each key concept from the *Quick Review*
- Video footage of an instructor working through the complete solutions for all chapter test problems.

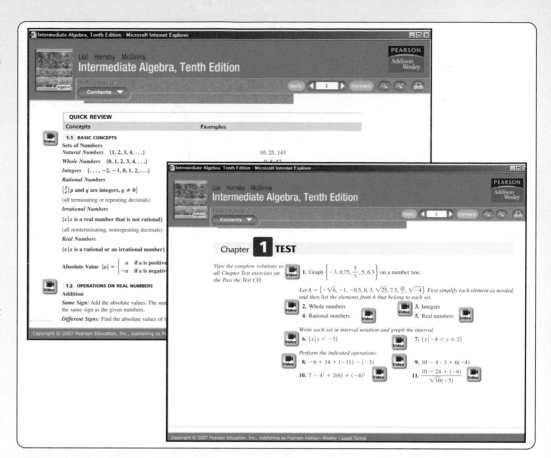

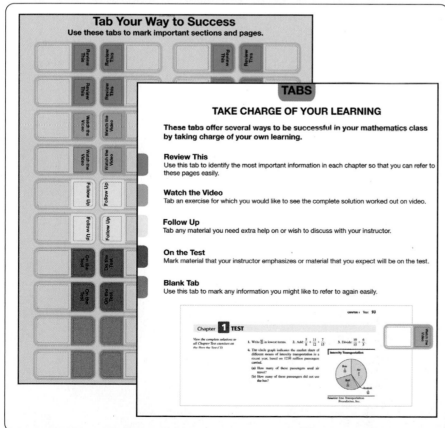

NEW *Tab Your Way to Success* appears at the front of the text. This page of reusable, color-coded Post-It® tabs makes it easy for students to flag pages they want to return to for review, test preparation, or instructor help.

Review of the Real Number System

Americans are crazy about their pets. Over 64 million U.S. households owned pets in 2005. Combined, these households spent more than \$35 billion pampering their animal friends. The fastest-growing segment of the pet industry is the high-end luxury area, which includes everything from gourmet pet foods, designer toys, and specialty furniture to groomers, dog walkers, boarding in posh pet hotels, and even pet therapists. (*Source:* American Pet Products Manufacturers Association.)

In Exercise 101 of Section 1.3, we use an *algebraic expression*, one of the topics of this chapter, to determine how much Americans have spent annually on their pets in recent years.

1.1 Basic Concepts

In this chapter, we review some of the basic symbols and rules of algebra.

OBJECTIVE 1 Write sets using set notation. A **set** is a collection of objects called the **elements** or **members** of the set. In algebra, the elements of a set are usually numbers. Set braces, { }, are used to enclose the elements. For example, 2 is an element of the set $\{1, 2, 3\}$. Since we can count the number of elements in the set $\{1, 2, 3\}$, it is a **finite set.**

In our study of algebra, we refer to certain sets of numbers by name. The set

$$N = \{1, 2, 3, 4, 5, 6, \ldots\} \qquad \text{Natural (counting) numbers}$$

is called the **natural numbers,** or the **counting numbers.** The three dots (*ellipsis* points) show that the list continues in the same pattern indefinitely. We cannot list all of the elements of the set of natural numbers, so it is an **infinite set.**

When 0 is included with the set of natural numbers, we have the set of **whole numbers,** written

$$W = \{0, 1, 2, 3, 4, 5, 6, \ldots\}. \qquad \text{Whole numbers}$$

The set containing no elements, such as the set of whole numbers less than 0, is called the **empty set,** or **null set,** usually written $\emptyset$ or { }.

▶ **CAUTION** Do not write $\{\emptyset\}$ for the empty set; $\{\emptyset\}$ is a set with one element: $\emptyset$. Use the notation $\emptyset$ or { } for the empty set.

To write the fact that 2 is an element of the set $\{1, 2, 3\}$, we use the symbol $\in$ (read "is an element of").

$$2 \in \{1, 2, 3\}$$

The number 2 is also an element of the set of natural numbers N, so we may write

$$2 \in N.$$

To show that 0 is *not* an element of set N, we draw a slash through the symbol $\in$.

$$0 \notin N$$

Two sets are equal if they contain exactly the same elements. For example, $\{1, 2\} = \{2, 1\}$. (Order doesn't matter.) However, $\{1, 2\} \neq \{0, 1, 2\}$ ($\neq$ means "is not equal to"), since one set contains the element 0 while the other does not.

In algebra, letters called **variables** are often used to represent numbers or to define sets of numbers. For example,

$$\{x \mid x \text{ is a natural number between 3 and 15}\}$$

(read "the set of all elements x such that x is a natural number between 3 and 15") defines the set

$$\{4, 5, 6, 7, \ldots, 14\}.$$

The notation $\{x \mid x$ is a natural number between 3 and 15$\}$ is an example of **set-builder notation.**

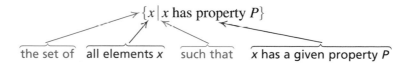

$\{x \mid x$ has property $P\}$

the set of   all elements x such that x has a given property P

EXAMPLE 1 Listing the Elements in Sets

List the elements in each set.

(a) $\{x \mid x$ is a natural number less than 4$\}$

The natural numbers less than 4 are 1, 2, and 3. This set is $\{1, 2, 3\}$.

(b) $\{y \mid y$ is one of the first five even natural numbers$\}$ is $\{2, 4, 6, 8, 10\}$.

(c) $\{z \mid z$ is a natural number greater than or equal to 7$\}$

The set of natural numbers greater than or equal to 7 is an infinite set, written with ellipsis points as

$$\{7, 8, 9, 10, \ldots\}.$$

✔ **Now Try Exercise 1.**

EXAMPLE 2 Using Set-Builder Notation to Describe Sets

Use set-builder notation to describe each set.

(a) $\{1, 3, 5, 7, 9\}$

There are often several ways to describe a set in set-builder notation. One way to describe the given set is

$$\{y \mid y \text{ is one of the first five odd natural numbers}\}.$$

(b) $\{5, 10, 15, \ldots\}$

This set can be described as $\{x \mid x$ is a multiple of 5 greater than 0$\}$.

✔ **Now Try Exercises 13 and 15.**

OBJECTIVE 2 Use number lines. A good way to get a picture of a set of numbers is to use a **number line.** To construct a number line, choose any point on a horizontal line and label it 0. Next, choose a point to the right of 0 and label it 1. The distance from 0 to 1 establishes a scale that can be used to locate more points, with positive numbers to the right of 0 and negative numbers to the left of 0. See Figure 1.

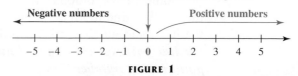

The number 0 is neither positive nor negative.

Negative numbers   Positive numbers

$-5 \quad -4 \quad -3 \quad -2 \quad -1 \quad 0 \quad 1 \quad 2 \quad 3 \quad 4 \quad 5$

FIGURE 1

The set of numbers identified on the number line in Figure 1, including positive and negative numbers and 0, is part of the set of **integers,** written

$$I = \{\ldots, -3, -2, -1, 0, 1, 2, 3, \ldots\}. \quad \text{Integers}$$

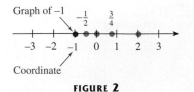

Graph of –1

Coordinate

FIGURE 2

Each number on a number line is called the **coordinate** of the point that it labels, while the point is the **graph** of the number. Figure 2 shows a number line with several points graphed on it.

The fractions $-\frac{1}{2}$ and $\frac{3}{4}$, graphed on the number line in Figure 2, are examples of *rational numbers*. A **rational number** can be expressed as the quotient of two integers, with denominator not 0. The set of all rational numbers is written

$$\left\{ \frac{p}{q} \,\middle|\, p \text{ and } q \text{ are integers, } q \neq 0 \right\}. \quad \text{Rational numbers}$$

The set of rational numbers includes the natural numbers, whole numbers, and integers, since these numbers can be written as fractions. For example, $14 = \frac{14}{1}$, $-3 = \frac{-3}{1}$, and $0 = \frac{0}{1}$. A rational number written as a fraction, such as $\frac{1}{8}$ or $\frac{2}{3}$, can also be expressed as a decimal by dividing the numerator by the denominator as follows.

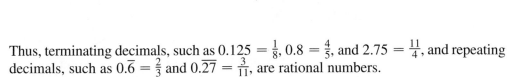

$$\frac{1}{8} = 0.125$$

$$\frac{2}{3} = 0.\overline{6}$$

A bar is written over the repeating digit(s).

Thus, terminating decimals, such as $0.125 = \frac{1}{8}$, $0.8 = \frac{4}{5}$, and $2.75 = \frac{11}{4}$, and repeating decimals, such as $0.\overline{6} = \frac{2}{3}$ and $0.\overline{27} = \frac{3}{11}$, are rational numbers.

Decimal numbers that neither terminate nor repeat are *not* rational and thus are called **irrational numbers.** Many square roots are irrational numbers; for example, $\sqrt{2} = 1.4142136\ldots$ and $-\sqrt{7} = -2.6457513\ldots$ repeat indefinitely without pattern. $\left(\text{Some square roots } are \text{ rational: } \sqrt{16} = 4, \sqrt{100} = 10, \text{ and so on.}\right)$ Another irrational number is π, the ratio of the distance around, or circumference of, a circle to its diameter.

Some of the rational and irrational numbers just discussed are graphed on the number line in Figure 3 on the next page. The rational numbers together with the irrational numbers make up the set of **real numbers.** *Every point on a number line corresponds to a real number, and every real number corresponds to a point on the number line.*

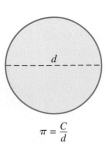

$\pi = \dfrac{C}{d}$

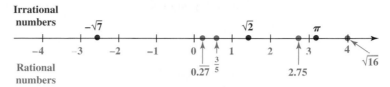

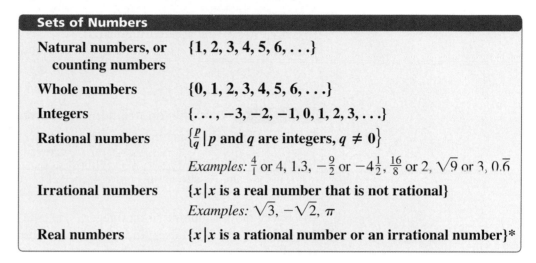

FIGURE 3

OBJECTIVE 3 Know the common sets of numbers.

Sets of Numbers	
Natural numbers, or counting numbers	$\{1, 2, 3, 4, 5, 6, \ldots\}$
Whole numbers	$\{0, 1, 2, 3, 4, 5, 6, \ldots\}$
Integers	$\{\ldots, -3, -2, -1, 0, 1, 2, 3, \ldots\}$
Rational numbers	$\left\{ \frac{p}{q} \mid p \text{ and } q \text{ are integers, } q \neq 0 \right\}$
	Examples: $\frac{4}{1}$ or 4, 1.3, $-\frac{9}{2}$ or $-4\frac{1}{2}$, $\frac{16}{8}$ or 2, $\sqrt{9}$ or 3, $0.\overline{6}$
Irrational numbers	$\{x \mid x \text{ is a real number that is not rational}\}$
	Examples: $\sqrt{3}, -\sqrt{2}, \pi$
Real numbers	$\{x \mid x \text{ is a rational number or an irrational number}\}$*

Figure 4 shows that the set of real numbers includes both the rational and irrational numbers. Every real number is either rational or irrational. Also, notice that the integers are elements of the set of rational numbers and that the whole numbers and natural numbers are elements of the set of integers.

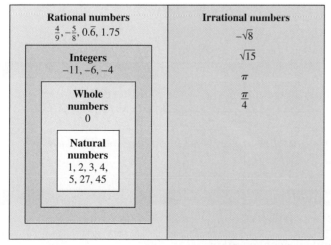

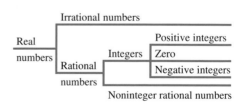

FIGURE 4 The Real Numbers

*An example of a number that is not real is $\sqrt{-1}$. This number, part of the *complex number system*, is discussed in **Chapter 8**.

EXAMPLE 3 Identifying Examples of Number Sets

Which numbers in

$$\left\{-8, -\sqrt{5}, -\frac{9}{64}, 0, 0.5, \frac{1}{3}, 1.\overline{12}, \sqrt{3}, 2, \pi\right\}$$

are elements of each set?

(a) Integers

   −8, 0, and 2 are integers.

(b) Rational numbers

   $-8, -\frac{9}{64}, 0, 0.5, \frac{1}{3}, 1.\overline{12}$, and 2 are rational numbers.

(c) Irrational numbers

   $-\sqrt{5}, \sqrt{3}$, and π are irrational numbers.

(d) Real numbers

   All the numbers in the given set are real numbers.

✔ Now Try Exercise 23.

EXAMPLE 4 Determining Relationships between Sets of Numbers

Decide whether each statement is *true* or *false.*

(a) All irrational numbers are real numbers.

   This is true. As shown in Figure 4, the set of real numbers includes all irrational numbers.

(b) Every rational number is an integer.

   This statement is false. Although some rational numbers are integers, other rational numbers, such as $\frac{2}{3}$ and $-\frac{1}{4}$, are not.

✔ Now Try Exercise 25.

OBJECTIVE 4 Find additive inverses.   Look at the number line in Figure 5. For each positive number, there is a negative number on the opposite side of 0 that lies the same distance from 0. These pairs of numbers are called *additive inverses, opposites,* or *negatives* of each other. For example, 3 is the additive inverse of −3, and −3 is the additive inverse of 3.

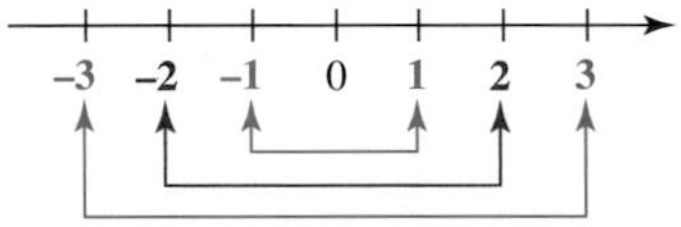

Additive inverses (opposites)

FIGURE 5

Additive Inverse

For any real number a, the number $-a$ is the **additive inverse** of a.

Change the sign of a number to get its additive inverse. The sum of a number and its additive inverse is always 0.

Uses of the Symbol −

The symbol "−" is used to indicate any of the following:

1. a negative number, such as −9 or −15;

2. the additive inverse of a number, as in "−4 is the additive inverse of 4";

3. subtraction, as in 12 − 3.

In the expression $-(-5)$, the symbol "$-$" is being used in two ways: the first $-$ indicates the additive inverse (or opposite) of -5, and the second indicates a negative number, -5. Since the additive inverse of -5 is 5, then

$$-(-5) = 5.$$

This example suggests the following property.

$-(-a)$

For any real number a, $\quad -(-a) = a.$

Number	Additive Inverse
6	-6
-4	4
$\frac{2}{3}$	$-\frac{2}{3}$
-8.7	8.7
0	0

Numbers written with positive or negative signs, such as $+4$, $+8$, -9, and -5, are called **signed numbers.** A positive number can be called a signed number even though the positive sign is usually left off. The table in the margin shows the additive inverses of several signed numbers. The number 0 is its own additive inverse.

OBJECTIVE 5 Use absolute value. Geometrically, the **absolute value** of a number a, written $|a|$, is the distance on the number line from 0 to a. For example, the absolute value of 5 is the same as the absolute value of -5 because each number lies five units from 0. See Figure 6. That is,

$$|5| = 5 \quad \text{and} \quad |-5| = 5.$$

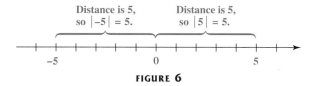

Distance is 5, so $|-5| = 5$.   Distance is 5, so $|5| = 5$.

FIGURE 6

▶ **CAUTION** *Because absolute value represents distance, and distance is never negative, the absolute value of a number is always positive or 0.*

The formal definition of absolute value follows.

Absolute Value

For any real number a, $\quad |a| = \begin{cases} a & \text{if } a \text{ is positive or 0} \\ -a & \text{if } a \text{ is negative.} \end{cases}$

The second part of this definition, $|a| = -a$ if a is negative, requires careful thought. If a is a *negative* number, then $-a$, the additive inverse or opposite of a, is a positive number. Thus, $|a|$ is positive. For example, if $a = -3$, then

$$|a| = |-3| = -(-3) = 3. \qquad |a| = -a \text{ if } a \text{ is negative.}$$

EXAMPLE 5 Finding Absolute Value

Simplify by finding each absolute value.

(a) $|13| = 13$ **(b)** $|-2| = -(-2) = 2$ **(c)** $|0| = 0$

(d) $-|8|$

Evaluate the absolute value first. Then find the additive inverse.

$$-|8| = -(8) = -8$$

(e) $-|-8|$

Work as in part (d): $|-8| = 8$,   so $-|-8| = -(8) = -8$.

(f) $|5| + |-2| = 5 + 2 = 7$

(g) $|5 - 2| = |3| = 3$

> ✔ **Now Try Exercises 43, 47, 49, and 53.**

Absolute value is useful in applications comparing size without regard to sign.

EXAMPLE 6 Comparing Rates of Change in Industries

The projected annual rates of change in employment (in percent) in some of the fastest-growing and in some of the most rapidly-declining industries from 2002 through 2012 are shown in the table.

Industry (2002–2012)	Annual Rate of Change (in percent)
Software publishers	5.3
Care services for the elderly	4.5
Child day-care services	3.6
Cut-and-sew apparel manufacturing	−12.2
Fabric mills	−5.9
Metal ore mining	−4.8

Source: U.S. Bureau of Labor Statistics.

What industry in the list is expected to see the greatest change? the least change?

We want the greatest *change,* without regard to whether the change is an increase or a decrease. Look for the number in the list with the largest absolute value. That number is found in cut-and-sew apparel manufacturing, since $|-12.2| = 12.2$. Similarly, the least change is in the child day-care services industry: $|3.6| = 3.6$.

> ✔ **Now Try Exercise 59.**

OBJECTIVE 6 Use inequality symbols. The statement $4 + 2 = 6$ is an **equation**—a statement that two quantities are equal. The statement $4 \neq 6$ (read "4 is not equal to 6") is an **inequality**—a statement that two quantities are *not* equal. When two numbers are not equal, one must be less than the other. The symbol $<$ means "is less than." For example,

$$8 < 9, \quad -6 < 15, \quad -6 < -1, \quad \text{and} \quad 0 < \frac{4}{3}.$$

The symbol $>$ means "is greater than." For example,

$$12 > 5, \quad 9 > -2, \quad -4 > -6, \quad \text{and} \quad \frac{6}{5} > 0.$$

In each case, the symbol "points" toward the smaller number.

The number line in Figure 7 shows the graphs of the numbers 4 and 9. We know that $4 < 9$. On the graph, 4 is to the left of 9. *The smaller of two numbers is always to the left of the other on a number line.*

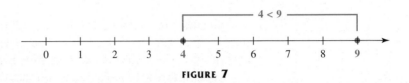

FIGURE 7

Inequalities on a Number Line

On a number line,

$a < b$ if a is to the left of b; $\quad a > b$ if a is to the right of b.

We can use a number line to determine order. As shown on the number line in Figure 8, -6 is located to the left of 1. For this reason, $-6 < 1$. Also, $1 > -6$. From the same number line,

$$-5 < -2, \quad \text{or} \quad -2 > -5.$$

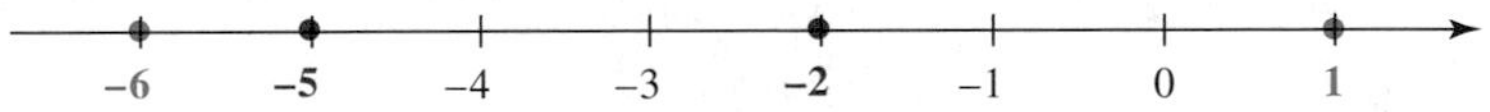

FIGURE 8

▶ **CAUTION** *Be careful when ordering negative numbers.* Since -5 is to the left of -2 on the number line in Figure 8, $-5 < -2$, or $-2 > -5$. In each case, the symbol points to -5, the smaller number.

✔ **Now Try Exercises 65 and 73.**

The following table summarizes results about positive and negative numbers in both words and symbols.

Words	Symbols
Every negative number is less than 0.	If a is negative, then $a < 0$.
Every positive number is greater than 0.	If a is positive, then $a > 0$.
0 is neither positive nor negative.	

In addition to the symbols $\neq$, $<$, and $>$, the symbols $\leq$ and $\geq$ are often used.

INEQUALITY SYMBOLS

Symbol	Meaning	Example
$\neq$	is not equal to	$3 \neq 7$
$<$	is less than	$-4 < -1$
$>$	is greater than	$3 > -2$
$\leq$	is less than or equal to	$6 \leq 6$
$\geq$	is greater than or equal to	$-8 \geq -10$

Inequality	Why It Is True
$6 \leq 8$	$6 < 8$
$-2 \leq -2$	$-2 = -2$
$-9 \geq -12$	$-9 > -12$
$-3 \geq -3$	$-3 = -3$
$6 \cdot 4 \leq 5(5)$	$24 < 25$

The table in the margin shows several inequalities and why each is true. Notice the reason that $-2 \leq -2$ is true: **With the symbol $\leq$, if either the $<$ part or the $=$ part is true, then the inequality is true. This is also the case with the $\geq$ symbol.**

In the last row of the table, recall that the dot in $6 \cdot 4$ indicates the product 6×4, or 24, and $5(5)$ means 5×5, or 25. Thus, the inequality $6 \cdot 4 \leq 5(5)$ becomes $24 \leq 25$, which is true.

✔ **Now Try Exercise 95.**

OBJECTIVE 7 Graph sets of real numbers. Inequality symbols and variables are used to write sets of real numbers. For example, the set $\{x \mid x > -2\}$ consists of all the real numbers greater than -2. On a number line, we graph the elements of this set by drawing an arrow from -2 to the right. We use a parenthesis at -2 to indicate that -2 is *not* an element of the given set. See Figure 9.

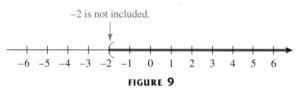

FIGURE 9

The set of numbers greater than -2 is an example of an **interval** on the number line. To write intervals, we use **interval notation.** We write the interval of all numbers greater than -2 as $(-2, \infty)$. The **infinity symbol** ∞ does not indicate a number; it shows that the interval includes all real numbers greater than -2. The left parenthesis indicates that -2 is not included. *A parenthesis is always used next to the infinity symbol.* The set of all real numbers is written in interval notation as $(-\infty, \infty)$.

EXAMPLE 7 Graphing an Inequality Written in Interval Notation

Write $\{x \mid x < 4\}$ in interval notation and graph the interval.

The interval is written $(-\infty, 4)$. The graph is shown in Figure 10. Since the elements of the set are all real numbers *less than* 4, the graph extends to the left.

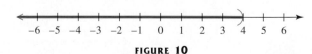

FIGURE 10

✔ **Now Try Exercise 101.**

The set $\{x \mid x \leq -6\}$ includes all real numbers less than or equal to -6. To show that -6 is part of the set, a square bracket is used at -6, as shown in Figure 11. In interval notation, this set is written $(-\infty, -6]$.

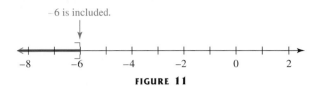

FIGURE 11

EXAMPLE 8 Graphing an Inequality Written in Interval Notation

Write $\{x \mid x \geq -4\}$ in interval notation and graph the interval.

This set is written in interval notation as $[-4, \infty)$. The graph is shown in Figure 12. We use a square bracket at -4, since -4 is part of the set.

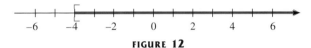

FIGURE 12

✔ Now Try Exercise 103.

We sometimes graph sets of numbers that are *between* two given numbers. For example, the set $\{x \mid -2 < x < 4\}$ includes all real numbers between -2 and 4, but *not* the numbers -2 and 4 themselves. This set is written in interval notation as $(-2, 4)$. The graph has a heavy line between -2 and 4, with parentheses at -2 and 4. See Figure 13. The inequality $-2 < x < 4$, called a **three-part inequality,** is read "-2 is less than x and x is less than 4," or "x is between -2 and 4."

FIGURE 13

EXAMPLE 9 Graphing a Three-Part Inequality

Write $\{x \mid 3 < x \leq 10\}$ in interval notation and graph the interval.

Use a parenthesis at 3 and a square bracket at 10 to get $(3, 10]$ in interval notation. The graph is shown in Figure 14. Read the inequality $3 < x \leq 10$ as "3 is less than x and x is less than or equal to 10," or "x is between 3 and 10, excluding 3 and including 10."

FIGURE 14

✔ Now Try Exercise 109.

▶ **NOTE** Some books use open circles and solid dots instead of parentheses and brackets when graphing inequalities. For example, the graph of $\{x \mid x > -2\}$ would have an open circle at -2. The graph of $\{x \mid x \geq -4\}$ would have a solid dot at -4. The graph of $\{x \mid -3 < x \leq 10\}$ would have an open circle at -3 and a solid dot at 10.

1.1 EXERCISES

Complete solution available on Video Lecture on CD/DVD

Now Try Exercise

Write each set by listing its elements. See Example 1.

1. $\{x \mid x$ is a natural number less than 6$\}$ **2.** $\{m \mid m$ is a natural number less than 9$\}$

3. $\{z \mid z$ is an integer greater than 4$\}$ **4.** $\{y \mid y$ is an integer greater than 8$\}$

5. $\{z \mid z$ is an integer less than or equal to 4$\}$ **6.** $\{p \mid p$ is an integer less than 3$\}$

7. $\{a \mid a$ is an even integer greater than 8$\}$ **8.** $\{k \mid k$ is an odd integer less than 1$\}$

9. $\{x \mid x$ is an irrational number that is also rational$\}$

10. $\{r \mid r$ is a number that is both positive and negative$\}$

11. $\{p \mid p$ is a number whose absolute value is 4$\}$

12. $\{w \mid w$ is a number whose absolute value is 7$\}$

Write each set using set-builder notation. See Example 2. (More than one description is possible.)

13. $\{2, 4, 6, 8\}$ **14.** $\{11, 12, 13, 14\}$

15. $\{4, 8, 12, 16, \ldots\}$ **16.** $\{\ldots, -6, -3, 0, 3, 6, \ldots\}$

17. *Concept Check*   A student claimed that $\{x \mid x$ is a natural number greater than 3$\}$ and $\{y \mid y$ is a natural number greater than 3$\}$ actually name the same set, even though different variables are used. Was this student correct?

Graph the elements of each set on a number line. See Objective 2.

18. $\{-3, -1, 0, 4, 6\}$ **19.** $\{-4, -2, 0, 3, 5\}$

20. $\left\{-\dfrac{2}{3}, 0, \dfrac{4}{5}, \dfrac{12}{5}, \dfrac{9}{2}, 4.8\right\}$ **21.** $\left\{-\dfrac{6}{5}, -\dfrac{1}{4}, 0, \dfrac{5}{6}, \dfrac{13}{4}, 5.2, \dfrac{11}{2}\right\}$

*Which elements of each set are (**a**) natural numbers, (**b**) whole numbers, (**c**) integers, (**d**) rational numbers, (**e**) irrational numbers, (**f**) real numbers? See Example 3.*

22. $\left\{-8, -\sqrt{5}, -0.6, 0, \dfrac{3}{4}, \sqrt{3}, \pi, 5, \dfrac{13}{2}, 17, \dfrac{40}{2}\right\}$

23. $\left\{-9, -\sqrt{6}, -0.7, 0, \dfrac{6}{7}, \sqrt{7}, 4.\overline{6}, 8, \dfrac{21}{2}, 13, \dfrac{75}{5}\right\}$

24. *Concept Check*   Give a real number that statisfies each condition.

(a) An integer between 6.75 and 7.75   **(b)** A rational number between $\frac{1}{4}$ and $\frac{3}{4}$

(c) A whole number that is not a natural number   **(d)** An integer that is not a whole number

(e) An irrational number between $\sqrt{4}$ and $\sqrt{9}$

Decide whether each statement is true *or* false. *If it is false, tell why. See Example 4.*

25. Every integer is a whole number.   **26.** Every natural number is an integer.

27. Every irrational number is an integer.   **28.** Every integer is a rational number.

29. Every natural number is a whole number.   **30.** Some rational numbers are irrational.

31. Some rational numbers are whole numbers.   **32.** Some real numbers are integers.

33. The absolute value of any number is the same as the absolute value of its additive inverse.

34. The absolute value of any nonzero number is positive.

35. *Concept Check*   Match each expression in Column I with its value in Column II. Choices in Column II may be used once, more than once, or not at all.

I	II
(a) $-(-4)$	**A.** 4
(b) $|-4|$	**B.** -4
(c) $-|-4|$	**C.** Both A and B
(d) $-|-(-4)|$	**D.** Neither A nor B

36. *Concept Check*   For what value(s) of x is $|x| = 4$ true?

*Give **(a)** the additive inverse and **(b)** the absolute value of each number. See the discussion of additive inverses and Example 5.*

37. 6   **38.** 8   **39.** -12 **40.** -15 **41.** $\dfrac{6}{5}$ **42.** 0.13

Simplify by finding each absolute value. See Example 5.

43. $|-8|$ **44.** $|-11|$ **45.** $\left|\dfrac{3}{2}\right|$ **46.** $\left|\dfrac{7}{4}\right|$

47. $-|5|$ **48.** $-|17|$ **49.** $-|-2|$ **50.** $-|-8|$

51. $-|4.5|$ **52.** $-|12.6|$ **53.** $|-2| + |3|$ **54.** $|-16| + |12|$

55. $|-9| - |-3|$ **56.** $|-10| - |-5|$

57. $|-1| + |-2| - |-3|$ **58.** $|-6| + |-4| - |-10|$

Solve each problem. See Example 6.

59. The table shows the percent change in population from 2000 through 2004 for selected cities in the United States.

City	Percent Change
Las Vegas	11.5
Los Angeles	4.1
Chicago	-1.2
Philadelphia	-3.1
Phoenix	7.3
Detroit	-5.4

Source: U.S. Census Bureau.

(a) Which city had the greatest change in population? What was this change? Was it an increase or a decline?

(b) Which city had the least change in population? What was this change? Was it an increase or a decline?

60. The table gives the net trade balance, in millions of U.S. dollars, for selected U.S. trade partners for January 2006.

Country	Trade Balance (in millions of dollars)
India	-1257
China	$-17,911$
Netherlands	756
France	-85
Turkey	-78
Australia	925

Source: U.S. Census Bureau.

A negative balance means that imports to the U.S. exceeded exports from the U.S., while a positive balance means that exports exceeded imports.

(a) Which country had the greatest discrepancy between exports and imports? Explain.

(b) Which country had the least discrepancy between exports and imports? Explain.

Sea level refers to the surface of the ocean. The depth of a body of water such as an ocean or sea can be expressed as a negative number, representing average depth in feet below sea level. By contrast, the altitude of a mountain can be expressed as a positive number, indicating its height in feet above sea level. The table gives selected depths and heights.

Body of Water	Average Depth in Feet (as a negative number)	Mountain	Altitude in Feet (as a positive number)
Pacific Ocean	−12,925	McKinley	20,320
South China Sea	−4,802	Point Success	14,158
Gulf of California	−2,375	Matlalcueyetl	14,636
Caribbean Sea	−8,448	Rainier	14,410
Indian Ocean	−12,598	Steele	16,644

Source: World Almanac and Book of Facts 2006.

61. List the bodies of water in order, starting with the deepest and ending with the shallowest.

62. List the mountains in order, starting with the shortest and ending with the tallest.

63. *True* or *false:* The absolute value of the depth of the Pacific Ocean is greater than the absolute value of the depth of the Indian Ocean.

64. *True* or *false:* The absolute value of the depth of the Gulf of California is greater than the absolute value of the depth of the Caribbean Sea.

Use the number line to answer true *or* false *to each statement. See Objective 6.*

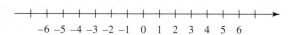

65. $-6 < -2$ **66.** $-4 < -3$ **67.** $-4 > -3$ **68.** $-2 > -1$

69. $3 > -2$ **70.** $5 > -3$ **71.** $-3 \geq -3$ **72.** $-4 \leq -4$

Rewrite each statement with $>$ so that it uses $<$ instead; rewrite each statement with $<$ so that it uses $>$. See Objective 6.

73. $6 > 2$ **74.** $4 > 1$ **75.** $-9 < 4$ **76.** $-5 < 1$

77. $-5 > -10$ **78.** $-8 > -12$ **79.** $0 < x$ **80.** $-2 < x$

Use an inequality symbol to write each statement.

81. 7 is greater than y.

82. -4 is less than 12.

83. 5 is greater than or equal to 5.

84. -3 is less than or equal to -3.

85. $3t - 4$ is less than or equal to 10.

86. $5x + 4$ is greater than or equal to 19.

87. $5x + 3$ is not equal to 0.

88. $6x + 7$ is not equal to -3.

89. t is between -3 and 5.

90. r is between -4 and 12.

91. $3x$ is between -3 and 4, including -3 and excluding 4.

92. $5y$ is between -2 and 6, excluding -2 and including 6.

Simplify. Then tell whether the resulting statement is true *or* false. *See Objective 6.*

93. $-6 < 7 + 3$

94. $-7 < 4 + 2$

95. $2 \cdot 5 \geq 4 + 6$

96. $8 + 7 \leq 3 \cdot 5$

97. $-|-3| \geq -3$

98. $-|-5| \leq -5$

99. $-8 > -|-6|$

100. $-9 > -|-4|$

Write each set in interval notation and graph the interval. See Examples 7–9.

101. $\{x \mid x > -1\}$

102. $\{x \mid x < 5\}$

103. $\{x \mid x \leq 6\}$

104. $\{x \mid x \geq -3\}$

105. $\{x \mid 0 < x < 3.5\}$

106. $\{x \mid -4 < x < 6.1\}$

107. $\{x \mid 2 \leq x \leq 7\}$

108. $\{x \mid -3 \leq x \leq -2\}$

109. $\{x \mid -4 < x \leq 3\}$

110. $\{x \mid 3 \leq x < 6\}$

111. $\{x \mid 0 < x \leq 3\}$

112. $\{x \mid -1 \leq x < 6\}$

The graph shows egg production in millions of eggs in selected states for 2004 and 2005. Use this graph to work Exercises 113–116.

113. In 2005, which states had production greater than 6000 million eggs?

114. In which states was 2005 egg production less than 2004 egg production?

115. If x represents 2005 egg production for Texas (TX) and y represents 2005 egg production for Ohio (OH), which is true, $x < y$ or $x > y$?

116. If x represents 2005 egg production for Iowa (IA) and y represents 2005 egg production for Pennsylvania (PA), write two inequalities that compare the production in these two states.

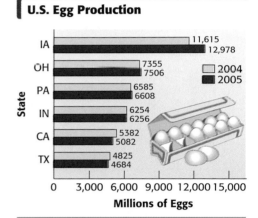

U.S. Egg Production

Source: Iowa Agricultural Statistics.

117. List the sets of numbers introduced in this section. Give a short explanation, including three examples, for each set.

118. List at least five symbols introduced in this section, and give a true statement involving each one.

1.2 Operations on Real Numbers

We review the rules for adding, subtracting, multiplying, and dividing real numbers.

OBJECTIVE 1 Add real numbers. Recall that the answer to an addition problem is called the **sum.** The rules for adding real numbers follow.

Adding Real Numbers

Same sign To add two numbers with the *same* sign, add their absolute values. The sum has the same sign as the given numbers.

Different signs To add two numbers with *different* signs, find the absolute values of the numbers, and subtract the smaller absolute value from the larger. The sum has the same sign as the number with the larger absolute value.

EXAMPLE 1 Adding Two Negative Real Numbers

Find each sum.

(a) $-12 + (-8)$

First find the absolute values.

$$|-12| = 12 \quad \text{and} \quad |-8| = 8$$

Because -12 and -8 have the *same* sign, add their absolute values.

> Both numbers are negative, so the answer will be negative.

$$-12 + (-8) = -(12 + 8) \qquad \text{Add the absolute values.}$$
$$= -(20)$$
$$= -20$$

(b) $-6 + (-3) = -(|-6| + |-3|)$ Add the absolute values.
$$= -(6 + 3) = -9$$

(c) $-1.2 + (-0.4) = -(1.2 + 0.4) = -1.6$

(d) $-\dfrac{5}{6} + \left(-\dfrac{1}{3}\right) = -\left(\dfrac{5}{6} + \dfrac{1}{3}\right)$ Add the absolute values. Both numbers are negative, so the answer will be negative.

$$= -\left(\dfrac{5}{6} + \dfrac{2}{6}\right) \qquad \text{The least common denominator is 6; } \dfrac{1 \cdot 2}{3 \cdot 2} = \dfrac{2}{6}$$

$$= -\dfrac{7}{6} \qquad \text{Add numerators; keep the same denominator.}$$

✔ **Now Try Exercise 11.**

EXAMPLE 2 Adding Real Numbers with Different Signs

Find each sum.

(a) $-17 + 11$

First find the absolute values.

$$|-17| = 17 \quad \text{and} \quad |11| = 11$$

Because -17 and 11 have *different* signs, subtract their absolute values.
$$17 - 11 = 6$$
The number -17 has a larger absolute value than 11, so the answer is negative.

$$-17 + 11 = -6$$

> The sum is negative because $|-17| > |11|$.

(b) $4 + (-1)$

Subtract the absolute values, 4 and 1. Because 4 has the larger absolute value, the sum must be positive.
$$4 + (-1) = 4 - 1 = 3$$

> The sum is positive because $|4| > |-1|$.

(c) $-9 + 17 = 17 - 9 = 8$

(d) $-2.3 + 5.6 = 5.6 - 2.3 = 3.3$

(e) $-16 + 12$

The absolute values are 16 and 12. Subtract the absolute values.

$$-16 + 12 = -(16 - 12) = -4$$

> The sum is negative because $|-16| > |12|$.

(f) $-\dfrac{4}{5} + \dfrac{2}{3}$

The least common denominator is 15. Write each fraction in terms of the common denominator.

$$-\frac{4}{5} = -\frac{4 \cdot 3}{5 \cdot 3} = -\frac{12}{15} \quad \text{and} \quad \frac{2}{3} = \frac{2 \cdot 5}{3 \cdot 5} = \frac{10}{15}$$

$$-\frac{4}{5} + \frac{2}{3} = -\frac{12}{15} + \frac{10}{15}$$

$$= -\left(\frac{12}{15} - \frac{10}{15}\right)$$

Subtract the absolute values. $-\frac{12}{15}$ has the larger absolute value, so the answer will be negative.

$$= -\frac{2}{15}$$

Subtract numerators; keep the same denominator.

✔ **Now Try Exercises 13, 15, and 17.**

OBJECTIVE 2 Subtract real numbers. Recall that the answer to a subtraction problem is called the **difference.** Compare the following two statements.

$$6 - 4 = 2$$
$$6 + (-4) = 2$$

Thus, $6 - 4 = 6 + (-4)$. To subtract 4 from 6, we add the additive inverse of 4 to 6. This example suggests the following definition of subtraction.

Subtraction

For all real numbers a and b,
$$a - b = a + (-b).$$
That is, to subtract b from a, add the additive inverse (or opposite) of b to a.

EXAMPLE 3 Subtracting Real Numbers

Find each difference.

Change to addition.

The additive inverse of 8 is −8.

(a) $6 - 8 = 6 + (-8) = -2$

Change to additon.

The additive inverse of 4 is −4.

(b) $-12 - 4 = -12 + (-4) = -16$

(c) $-10 - (-7) = -10 + 7$ The additive inverse of −7 is 7.

$= -3$

(d) $-2.4 - (-8.1) = -2.4 + 8.1 = 5.7$

(e) $\dfrac{5}{6} - \left(-\dfrac{3}{8}\right) = \dfrac{5}{6} + \dfrac{3}{8}$ To subtract, add the additive inverse (opposite).

$= \dfrac{5 \cdot 4}{6 \cdot 4} + \dfrac{3 \cdot 3}{8 \cdot 3}$ Write each fraction with the least common denominator, 24.

$= \dfrac{20}{24} + \dfrac{9}{24}$

$= \dfrac{29}{24}$ Add numerators; keep the same denominator.

✔ **Now Try Exercises 19, 25, and 27.**

When working a problem that involves both addition and subtraction, add and subtract in order from left to right. Work inside brackets or parentheses first.

EXAMPLE 4 Adding and Subtracting Real Numbers

Perform the indicated operations.

(a) $15 - (-3) - 5 - 12$

$= (15 + 3) - 5 - 12$ Work from left to right.

$= 18 - 5 - 12$

$= 13 - 12$

$= 1$

(b) $-9 - [-8 - (-4)] + 6$

$= -9 - [-8 + 4] + 6$ Work inside brackets.

$= -9 - [-4] + 6$

$= -9 + 4 + 6$ Add the additive inverse.

$= -5 + 6$ Work from left to right.

$= 1$

✔ **Now Try Exercises 39 and 41.**

OBJECTIVE 3 Find the distance between two points on a number line. The number line in Figure 15 shows several points.

FIGURE 15

To find the distance between the points 4 and 7, we subtract $7 - 4 = 3$. Since distance is always positive (or 0), we must be careful to subtract in such a way that the answer is positive (or 0). Or, to avoid this problem altogether, we can find the absolute value of the difference. Then the distance between 4 and 7 is either

$$|7 - 4| = |3| = 3 \qquad \text{or} \qquad |4 - 7| = |-3| = 3.$$

This discussion can be summarized as follows.

Distance

The **distance** between two points on a number line is the absolute value of the difference between their coordinates.

EXAMPLE 5 Finding Distance between Points on the Number Line

Find the distance between each pair of points listed from Figure 15.

(a) 8 and -4

Find the absolute value of the difference of the numbers, taken in either order.

$$|8 - (-4)| = 12 \qquad \text{or} \qquad |-4 - 8| = 12$$

(b) -4 and -6

$$|-4 - (-6)| = 2 \qquad \text{or} \qquad |-6 - (-4)| = 2$$

✔ **Now Try Exercise 51.**

OBJECTIVE 4 Multiply real numbers. The answer to a multiplication problem is called the **product.** For example, 24 is the product of 8 and 3.

Multiplying Real Numbers

Same sign The product of two numbers with the *same* sign is positive.

Different signs The product of two numbers with *different* signs is negative.

EXAMPLE 6 Multiplying Real Numbers

Find each product.

(a) $-3(-9) = 27$ Same sign; product is positive.

(b) $-0.5(-0.4) = 0.2$

(c) $-\dfrac{3}{4}\left(-\dfrac{5}{6}\right) = \dfrac{15}{24}$ Multiply numerators; multiply denominators.

$\phantom{-\dfrac{3}{4}\left(-\dfrac{5}{6}\right)} = \dfrac{5 \cdot 3}{8 \cdot 3}$ Factor to write in lowest terms.

$\phantom{-\dfrac{3}{4}\left(-\dfrac{5}{6}\right)} = \dfrac{5}{8}$ Divide out the common factor, 3.

(d) $6(-9) = -54$ Different signs; product is negative.

$-6 = -\dfrac{6}{1}$

(e) $-0.05(0.3) = -0.015$ **(f)** $-\dfrac{5}{8}\left(\dfrac{12}{13}\right) = -\dfrac{15}{26}$ **(g)** $\dfrac{2}{3}(-6) = -4$

✔ **Now Try Exercises 59, 63, 65, and 71.**

OBJECTIVE 5 Divide real numbers. The result of dividing one number by another is called the **quotient.** The quotient of two real numbers a and b ($b \neq 0$) is the real number q such that $q \cdot b = a$. That is,

$$a \div b = q \quad \text{only if} \quad q \cdot b = a.$$

For example, $36 \div 9 = 4$, since $4 \cdot 9 = 36$. Similarly, $35 \div (-5) = -7$, since $-7(-5) = 35$.

The quotient $a \div b$ can also be denoted $\dfrac{a}{b}$. Thus, $35 \div (-5)$ can be written $\dfrac{35}{-5}$. Then, $\dfrac{35}{-5} = -7$, because -7 answers the question, "What number multiplied by -5 gives the product 35?"

Now consider $\dfrac{5}{0}$. On the one hand, there is *no* number whose product with 0 gives 5. On the other hand, $\dfrac{0}{0}$ would be satisfied by *every* real number, because any number multiplied by 0 gives 0. When dividing, we always want a *unique* quotient. ***Therefore, division by 0 is undefined.***

▶ **CAUTION** Division by 0 is undefined. However, dividing 0 by a nonzero number gives the quotient 0. For example,

$$\dfrac{6}{0} \text{ is undefined,} \quad \text{but} \quad \dfrac{0}{6} = 0 \quad (\text{because } 0 \cdot 6 = 0).$$

Be careful when 0 is involved in a division problem.

✔ **Now Try Exercises 79 and 81.**

Recall that $\dfrac{a}{b} = a \cdot \dfrac{1}{b}$. ***Thus, dividing by b is the same as multiplying by $\dfrac{1}{b}$.*** If $b \neq 0$, then $\dfrac{1}{b}$ is the **reciprocal,** or **multiplicative inverse,** of b. When multiplied, reciprocals have a product of 1. The table on the next page gives several numbers and their reciprocals. There is no reciprocal for 0 because there is no number that can be multiplied by 0 to give the product 1.

Number	Reciprocal
$-\frac{2}{5}$	$-\frac{5}{2}$
-6	$-\frac{1}{6}$
$\frac{7}{11}$	$\frac{11}{7}$
0.05	20
0	None

$$\left.\begin{array}{l} -\frac{2}{5}\left(-\frac{5}{2}\right) = 1 \\ -6\left(-\frac{1}{6}\right) = 1 \\ \frac{7}{11}\left(\frac{11}{7}\right) = 1 \\ 0.05(20) = 1 \end{array}\right\} \begin{array}{l} \text{Reciprocals have} \\ \text{a product of 1.} \end{array}$$

▶ **CAUTION** *A number and its additive inverse have opposite signs; however, a number and its reciprocal always have the same sign.*

The preceding discussion suggests the following definition of division.

Division

For all real numbers a and b (where $b \neq 0$),

$$a \div b = \frac{a}{b} = a \cdot \frac{1}{b}.$$

That is, multiply the first number by the reciprocal of the second number.

Since division is defined as multiplication by the reciprocal, the rules for signs of quotients are the same as those for signs of products.

Dividing Real Numbers

Same sign   The quotient of two nonzero real numbers with the *same* sign is positive.

Different signs   The quotient of two nonzero real numbers with *different* signs is negative.

EXAMPLE 7 Dividing Real Numbers

Find each quotient.

(a) $\dfrac{-12}{4} = -12 \cdot \dfrac{1}{4} = -3 \qquad \frac{a}{b} = a \cdot \frac{1}{b}$

(b) $\dfrac{6}{-3} = 6\left(-\dfrac{1}{3}\right) = -2 \qquad$ The reciprocal of -3 is $-\frac{1}{3}$.

(c) $\dfrac{-\dfrac{2}{3}}{-\dfrac{5}{9}} = -\dfrac{2}{3} \cdot \left(-\dfrac{9}{5}\right) = \dfrac{6}{5} \qquad$ The reciprocal of $-\frac{5}{9}$ is $-\frac{9}{5}$.

This is a *complex fraction* **(Section 7.3)**—a fraction that has a fraction in the numerator, the denominator, or both.

(d) $-\dfrac{9}{14} \div \dfrac{3}{7} = -\dfrac{9}{14} \cdot \dfrac{7}{3}$　　Multiply by the reciprocal.

$\qquad\qquad = -\dfrac{63}{42}$　　Multiply numerators; multiply denominators.

$\qquad\qquad = -\dfrac{7 \cdot 3 \cdot 3}{7 \cdot 3 \cdot 2}$　　Factor.

$\qquad\qquad = -\dfrac{3}{2}$　　Lowest terms

✔ **Now Try Exercises 73, 75, 83, and 85.**

The rules for multiplication and division suggest the following results.

Equivalent Forms of a Fraction

The fractions $\dfrac{-x}{y}$, $\dfrac{x}{-y}$, and $-\dfrac{x}{y}$ are equivalent ($y \neq 0$).

Example: $\dfrac{-4}{7} = \dfrac{4}{-7} = -\dfrac{4}{7}$

The fractions $\dfrac{x}{y}$ and $\dfrac{-x}{-y}$ are equivalent ($y \neq 0$).

Example: $\dfrac{4}{7} = \dfrac{-4}{-7}$

Every fraction has three signs: the sign of the numerator, the sign of the denominator, and the sign of the fraction itself. Changing any two of these three signs does not change the value of the fraction. Changing only one sign, or changing all three, *does* change the value.

1.2 EXERCISES

Concept Check   Complete each statement and give an example.

1. The sum of a positive number and a negative number is 0 if _____.

2. The sum of two positive numbers is a _____ number.

3. The sum of two negative numbers is a _____ number.

4. The sum of a positive number and a negative number is negative if _____.

5. The sum of a positive number and a negative number is positive if _____.

6. The difference between two positive numbers is negative if _____.

7. The difference between two negative numbers is negative if _____.

8. The product of two numbers with the same sign is _____.

9. The product of two numbers with different signs is _____.

10. The quotient formed by any nonzero number divided by 0 is _____, and the quotient formed by 0 divided by any nonzero number is _____.

Add or subtract as indicated. See Examples 1–3.

11. $-6 + (-13)$ **12.** $-8 + (-15)$ **13.** $13 + (-4)$

14. $19 + (-13)$ **15.** $-\dfrac{7}{3} + \dfrac{3}{4}$ **16.** $-\dfrac{5}{6} + \dfrac{3}{8}$

17. $-2.3 + 0.45$ **18.** $-0.238 + 4.55$ **19.** $-6 - 5$

20. $-8 - 13$ **21.** $8 - (-13)$ **22.** $13 - (-22)$

23. $-16 - (-3)$ **24.** $-21 - (-8)$ **25.** $-12.31 - (-2.13)$

26. $-15.88 - (-9.22)$ **27.** $\dfrac{9}{10} - \left(-\dfrac{4}{3}\right)$ **28.** $\dfrac{3}{14} - \left(-\dfrac{1}{4}\right)$

29. $|-8 - 6|$ **30.** $|-7 - 9|$ **31.** $-|-4 + 9|$

32. $-|-5 + 7|$ **33.** $-2 - |-4|$ **34.** $9 - |-13|$

Perform the indicated operations. See Example 4.

35. $-7 + 5 - 9$ **36.** $-12 + 13 - 19$ **37.** $6 - (-2) + 8$

38. $7 - (-3) + 12$ **39.** $-9 \quad 4 - (-3) + 6$ **40.** $-10 - 5 - (-12) + 8$

41. $-8 - (-12) - (2 - 6)$ **42.** $-3 + (-14) + (-5 + 3)$ **43.** $-0.382 + 4 - 0.6$

44. $3 - 2.94 - (-0.63)$ **45.** $\left(-\dfrac{5}{4} - \dfrac{2}{3}\right) + \dfrac{1}{6}$ **46.** $\left(-\dfrac{5}{8} + \dfrac{1}{4}\right) - \left(-\dfrac{1}{4}\right)$

47. $-\dfrac{3}{4} - \left(\dfrac{1}{2} - \dfrac{3}{8}\right)$ **48.** $\dfrac{7}{5} - \left(\dfrac{9}{10} - \dfrac{3}{2}\right)$

49. $|-11| - |-5| - |7| + |-2|$ **50.** $|-6| + |-3| - |4| - |-8|$

The number line has several points labeled. Find the distance between each pair of points. See Example 5.

51. *A* and *B*   **52.** *A* and *C*   **53.** *D* and *F*   **54.** *E* and *C*

55. A statement that is often heard is "Two negatives give a positive." When is this true? When is it not true? Give a more precise statement that conveys this message.

56. Explain why the reciprocal of a nonzero number must have the same sign as the number.

Multiply. See Example 6.

57. $5(-7)$ **58.** $6(-6)$ **59.** $-8(-5)$ **60.** $-10(-4)$

61. $-10\left(-\dfrac{1}{5}\right)$ **62.** $-\dfrac{1}{2}(-12)$ **63.** $\dfrac{3}{4}(-16)$ **64.** $\dfrac{4}{5}(-35)$

65. $-\dfrac{5}{2}\left(-\dfrac{12}{25}\right)$ **66.** $-\dfrac{9}{7}\left(-\dfrac{35}{36}\right)$ **67.** $-\dfrac{3}{8}\left(-\dfrac{24}{9}\right)$ **68.** $-\dfrac{2}{11}\left(-\dfrac{99}{4}\right)$

69. $-2.4(-2.45)$ **70.** $-3.45(-2.14)$ **71.** $3.4(-3.14)$ **72.** $5.66(-2.1)$

Divide where possible. See Example 7.

73. $\dfrac{-14}{2}$ **74.** $\dfrac{-26}{13}$ **75.** $\dfrac{-24}{-4}$ **76.** $\dfrac{-36}{-9}$ **77.** $\dfrac{100}{-25}$

78. $\dfrac{300}{-60}$ **79.** $\dfrac{0}{-8}$ **80.** $\dfrac{0}{-10}$ **81.** $\dfrac{5}{0}$ **82.** $\dfrac{12}{0}$

83. $-\dfrac{10}{17} \div \left(-\dfrac{12}{5}\right)$ **84.** $-\dfrac{22}{23} \div \left(-\dfrac{33}{4}\right)$ **85.** $\dfrac{\dfrac{12}{13}}{-\dfrac{4}{3}}$ **86.** $\dfrac{\dfrac{5}{6}}{-\dfrac{1}{30}}$

87. $\dfrac{-27.72}{13.2}$ **88.** $\dfrac{-126.7}{36.2}$ **89.** $\dfrac{-100}{-0.01}$ **90.** $\dfrac{-50}{-0.05}$

Exercises 91–116 provide more practice on operations with fractions and decimals. Perform the indicated operations.

91. $\dfrac{1}{6} - \left(-\dfrac{7}{9}\right)$ **92.** $\dfrac{7}{10} - \left(-\dfrac{5}{6}\right)$ **93.** $-\dfrac{1}{9} + \dfrac{7}{12}$

94. $-\dfrac{1}{12} + \dfrac{11}{16}$ **95.** $-\dfrac{3}{8} - \dfrac{5}{12}$ **96.** $-\dfrac{11}{15} - \dfrac{2}{9}$

97. $-\dfrac{7}{30} + \dfrac{2}{45} - \dfrac{3}{10}$ **98.** $-\dfrac{8}{15} - \dfrac{3}{20} + \dfrac{5}{6}$ **99.** $\dfrac{8}{25}\left(-\dfrac{5}{12}\right)$

100. $\dfrac{9}{20}\left(-\dfrac{4}{15}\right)$ **101.** $\dfrac{5}{6}\left(-\dfrac{9}{10}\right)\left(-\dfrac{4}{5}\right)$ **102.** $\dfrac{2}{3}\left(-\dfrac{9}{20}\right)\left(-\dfrac{5}{12}\right)$

103. $\dfrac{7}{6} \div \left(-\dfrac{9}{10}\right)$ **104.** $\dfrac{8}{5} \div \left(-\dfrac{18}{25}\right)$ **105.** $\dfrac{-\dfrac{8}{9}}{2}$

106. $\dfrac{-\dfrac{15}{16}}{3}$ **107.** $-8.6 - 3.751$ **108.** $-27.8 - 13.582$

109. $(-4.2)(1.4)(2.7)$ **110.** $(1.9)(-10.3)(0.04)$ **111.** $-24.84 \div 6$

112. $-32.84 \div 4$ **113.** $-2496 \div (-0.52)$ **114.** $-161.7 \div (-0.75)$

115. $-14.23 + 9.81 + 74.63 - 18.715$

116. $-89.416 + 21.32 - 478.91 + 298.213$

Solve each problem.

117. The highest temperature ever recorded in Juneau, Alaska, was 90°F. The lowest temperature ever recorded there was −22°F. What is the difference between these two temperatures? (*Source: World Almanac and Book of Facts 2006.*)

118. On August 10, 1936, a temperature of 120°F was recorded in Ponds, Arkansas. On February 13, 1905, Ozark, Arkansas, recorded a temperature of −29°F. What is the difference between these two temperatures? (*Source: World Almanac and Book of Facts 2006.*)

119. Andrew McGinnis has $48.35 in his checking account. He uses his debit card to make purchases of $35.99 and $20.00, which overdraws his account. His bank charges his account an overdraft fee of $28.50. He then deposits his paycheck for $66.27 from his part-time job at Arby's. What is the balance in his account?

120. Kayla Koolbeck has $37.50 in her checking account. She uses her debit card to make purchases of $25.99 and $19.34, which overdraws her account. Her bank charges her account an overdraft fee of $25.00. She then deposits her paycheck for $58.66 from her part-time job at Subway. What is the balance in her account?

121. Eric Holub owes $382.45 on his Visa account. He returns two items costing $25.10 and $34.50 for credit. Then he makes purchases of $45.00 and $98.17.

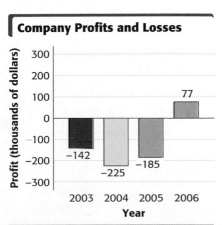

   (a) How much should his payment be if he wants to pay off the balance on the account?

   (b) Instead of paying off the balance, he makes a payment of $300 and then incurs a finance charge of $24.66. What is the balance on his account?

122. Kate Hilby owes $237.59 on her MasterCard account. She returns one item costing $47.25 for credit and then makes two purchases of $12.39 and $20.00.

   (a) How much should her payment be if she wants to pay off the balance on the account?

   (b) Instead of paying off the balance, she makes a payment of $75.00 and incurs a finance charge of $32.06. What is the balance on her account?

123. The graph shows profits and losses in thousands of dollars for a private company for the years 2003 through 2006.

   (a) What was the total profit or loss for the years 2003 through 2006?

   (b) Find the difference between the profit or loss in 2006 and that in 2005.

   (c) Find the difference between the profit or loss in 2004 and that in 2003.

Company Profits and Losses

124. The graph shows annual returns in percent for Class A shares of the AIM Charter Fund for the years 2000 through 2005.

(a) Find the sum of the percents for the years shown in the graph.

(b) Find the difference between the returns in 2003 and 2002.

(c) Find the difference between the returns in 2001 and 2000.

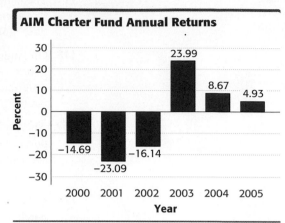

AIM Charter Fund Annual Returns

Source: AIM.

125. The table shows Social Security finances (in billions of dollars).

Year	Tax Revenue	Cost of Benefits
2000	538	409
2010*	916	710
2020*	1479	1405
2030*	2041	2542

*Projected

Source: Social Security Board of Trustees.

(a) Find the difference between Social Security tax revenue and cost of benefits for each year shown in the table.

(b) Interpret your answer for 2030.

1.3 Exponents, Roots, and Order of Operations

OBJECTIVES

1 Use exponents.

2 Find square roots.

3 Use the order of operations.

4 Evaluate algebraic expressions for given values of variables.

Two or more numbers whose product is a third number are **factors** of that third number. For example, 2 and 6 are factors of 12, since $2 \cdot 6 = 12$. Other integer factors of 12 are 1, 3, 4, 12, -1, -2, -3, -4, -6, and -12.

OBJECTIVE 1 Use exponents.   In algebra, we use *exponents* as a way of writing products of repeated factors. For example, the product $2 \cdot 2 \cdot 2 \cdot 2 \cdot 2$ is written

$$\underbrace{2 \cdot 2 \cdot 2 \cdot 2 \cdot 2}_{\text{5 factors of 2}} = 2^5.$$

The number 5 shows that 2 is used as a factor 5 times. The number 5 is the *exponent,* and 2 is the *base.*

$$2^5 \longleftarrow \text{Exponent}$$
$$\text{Base}$$

Read 2^5 as "2 to the fifth power," or "2 to the fifth." Multiplying the five 2s gives

$$2^5 = 2 \cdot 2 \cdot 2 \cdot 2 \cdot 2 = 32.$$

Exponential Expression

If a is a real number and n is a natural number, then

$$a^n = \underbrace{a \cdot a \cdot a \cdot \ldots \cdot a,}_{n \text{ factors of } a}$$

where n is the **exponent,** a is the **base,** and a^n is an **exponential expression.** Exponents are also called **powers.**

EXAMPLE 1 Using Exponential Notation

Write using exponents.

(a) $4 \cdot 4 \cdot 4$
Here, 4 is used as a factor 3 times.

$$\underbrace{4 \cdot 4 \cdot 4}_{3 \text{ factors of } 4} = 4^3$$

Read 4^3 as "4 cubed."

(b) $\dfrac{3}{5} \cdot \dfrac{3}{5} = \left(\dfrac{3}{5}\right)^2$ 2 factors of $\frac{3}{5}$

Read $\left(\frac{3}{5}\right)^2$ as "$\frac{3}{5}$ squared."

(c) $(-6)(-6)(-6)(-6) = (-6)^4$
Read $(-6)^4$ as "-6 to the fourth power," or "-6 to the fourth."

(d) $(0.3)(0.3)(0.3)(0.3)(0.3) = (0.3)^5$ **(e)** $x \cdot x \cdot x \cdot x \cdot x \cdot x = x^6$

✔ **Now Try Exercises 13, 15, 17, and 19.**

In parts (a) and (b) of Example 1, we used the terms *squared* and *cubed* to refer to powers of 2 and 3, respectively. The term *squared* comes from the figure of a square, which has the same measure for both length and width, as shown in Figure 16(a). Similarly, the term *cubed* comes from the figure of a cube. As shown in Figure 16(b), the length, width, and height of a cube have the same measure.

(a) $3 \cdot 3 = 3$ squared, or 3^2

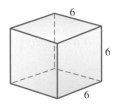

(b) $6 \cdot 6 \cdot 6 = 6$ cubed, or 6^3

FIGURE 16

EXAMPLE 2 Evaluating Exponential Expressions

Evaluate.

(a) $5^2 = 5 \cdot 5 = 25$ 5 is used as a factor 2 times.

$5^2 = 5 \cdot 5$, NOT $5 \cdot 2$.

(b) $\left(\dfrac{2}{3}\right)^3 = \dfrac{2}{3} \cdot \dfrac{2}{3} \cdot \dfrac{2}{3} = \dfrac{8}{27}$ $\dfrac{2}{3}$ is used as a factor 3 times.

(c) $2^6 = 2 \cdot 2 \cdot 2 \cdot 2 \cdot 2 \cdot 2 = 64$

✔ **Now Try Exercises 21 and 27.**

EXAMPLE 3   **Evaluating Exponential Expressions with Negative Signs**

Evaluate.

(a) $(-3)^5 = (-3)(-3)(-3)(-3)(-3) = -243$ The base is -3.

(b) $(-2)^6 = (-2)(-2)(-2)(-2)(-2)(-2) = 64$ The base is -2.

(c) -2^6

There are no parentheses. The exponent 6 applies *only* to the number 2, not to -2.

$$-2^6 = -(2 \cdot 2 \cdot 2 \cdot 2 \cdot 2 \cdot 2) = -64$$ The base is 2.

✔ **Now Try Exercises 29, 31, and 33.**

Examples 3(a) and (b) suggest the following generalizations.

> The product of an *odd* number of negative factors is negative.
>
> The product of an *even* number of negative factors is positive.

▶ **CAUTION**   As shown in Examples 3(b) and (c), it is important to distinguish between $-a^n$ and $(-a)^n$.

$$-a^n = -1\underbrace{(a \cdot a \cdot a \cdot \ldots \cdot a)}_{n \text{ factors of } a}$$ The base is a.

$$(-a)^n = \underbrace{(-a)(-a) \cdot \ldots \cdot (-a)}_{n \text{ factors of } -a}$$ The base is $-a$.

Be careful when evaluating an exponential expression with a negative sign.

OBJECTIVE 2   Find square roots.   As we saw in Example 2(a), $5^2 = 5 \cdot 5 = 25$, so 5 squared is 25. The opposite (inverse) of squaring a number is called taking its **square root.** For example, a square root of 25 is 5. Another square root of 25 is -5, since $(-5)^2 = 25$. Thus, 25 has two square roots: 5 and -5.

We write the **positive** or **principal square root** of a number with the symbol $\sqrt{}$, called a **radical sign.** For example, the positive or principal square root of 25 is written $\sqrt{25} = 5$. The **negative square root** of 25 is written $-\sqrt{25} = -5$. *Since the square of any nonzero real number is positive, the square root of a negative number, such as $\sqrt{-25}$, is not a real number.*

EXAMPLE 4 Finding Square Roots

Find each square root that is a real number.

(a) $\sqrt{36} = 6$, since 6 is positive and $6^2 = 36$.

(b) $\sqrt{0} = 0$, since $0^2 = 0$.

(c) $\sqrt{\dfrac{9}{16}} = \dfrac{3}{4}$, since $\left(\dfrac{3}{4}\right)^2 = \dfrac{9}{16}$.

(d) $\sqrt{0.16} = 0.4$, since $(0.4)^2 = 0.16$.

(e) $\sqrt{100} = 10$, since $10^2 = 100$.

(f) $-\sqrt{100} = -10$, since the negative sign is outside the radical sign.

(g) $\sqrt{-100}$ is not a real number, because the negative sign is inside the radical sign. No *real number* squared equals -100.

Notice the difference among the square roots in parts (e), (f), and (g). Part (e) is the positive or principal square root of 100, part (f) is the negative square root of 100, and part (g) is the square root of -100, which is not a real number.

✔ **Now Try Exercises 37, 41, 43, and 47.**

▶ **CAUTION** The symbol $\sqrt{}$ is used only for the *positive* square root, except that $\sqrt{0} = 0$. The symbol $-\sqrt{}$ is used for the negative square root.

OBJECTIVE 3 Use the order of operations. To simplify $5 + 2 \cdot 3$, what should we do first—add 5 and 2 or multiply 2 and 3? When an expression involves more than one operation symbol, we use the following **order of operations.**

Order of Operations

1. Work separately above and below any **fraction bar.**
2. If **grouping symbols** such as **parentheses ( )**, **brackets [ ]**, or **absolute value bars | |** are present, start with the innermost set and work outward.
3. Evaluate all **powers, roots,** and **absolute values.**
4. **Multiply** or **divide** in order from left to right.
5. **Add** or **subtract** in order from left to right.

▶ **NOTE** Some students like to use the mnemonic "Please Excuse My Dear Aunt Sally" to help remember the rules for order of operations.

Please	Excuse	My	Dear	Aunt	Sally
↑	↑	↑	↑	↑	↑
Parentheses	Exponents	Multiply	Divide	Add	Subtract

Be sure to multiply or divide *in order from left to right;* then add or subtract *in order from left to right.*

EXAMPLE 5 Using the Order of Operations

Simplify.

(a) $5 + 2 \cdot 3$

$= 5 + 6$ Multiply.

$= 11$ Add.

(b) $24 \div 3 \cdot 2 + 6$

Multiplications and divisions are done in the order in which they appear from left to right, so divide first.

$$24 \div 3 \cdot 2 + 6$$

$= 8 \cdot 2 + 6$ Divide.

$= 16 + 6$ Multiply.

$= 22$ Add.

✔ **Now Try Exercises 53 and 57.**

EXAMPLE 6 Using the Order of Operations

Simplify.

(a) $10 \div 5 + 2|3 - 4|$

$= 10 \div 5 + 2|-1|$ Subtract inside the absolute value bars.

$= 10 \div 5 + 2 \cdot 1$ Take the absolute value.

$= 2 + 2$ Divide; multiply.

$= 4$ Add.

(b) $4 \cdot 3^2 + 7 - (2 + 8)$

$= 4 \cdot 3^2 + 7 - 10$ Add inside parentheses.

$= 4 \cdot 9 + 7 - 10$ Evaluate the power.

$3^2 = 3 \cdot 3,$ **NOT 3 · 2.** $= 36 + 7 - 10$ Multiply.

$= 43 - 10$ Add.

$= 33$ Subtract.

(c) $\dfrac{1}{2} \cdot 4 + (6 \div 3 - 7)$

$= \dfrac{1}{2} \cdot 4 + (2 - 7)$ Divide inside parentheses.

$= \dfrac{1}{2} \cdot 4 + (-5)$ Subtract inside parentheses.

$= 2 + (-5)$ Multiply.

$= -3$ Add.

✔ **Now Try Exercises 65 and 71.**

EXAMPLE 7 Using the Order of Operations

Simplify $\dfrac{5 + (-2^3)(2)}{6 \cdot \sqrt{9} - 9 \cdot 2}$.

Work separately above and below the fraction bar.

$$\frac{5 + (-2^3)(2)}{6 \cdot \sqrt{9} - 9 \cdot 2}$$

$$= \frac{5 + (-8)(2)}{6 \cdot 3 - 9 \cdot 2} \qquad \text{Evaluate the power and the root.}$$

$$= \frac{5 - 16}{18 - 18} \qquad \text{Multiply.}$$

$$= \frac{-11}{0} \qquad \text{Subtract.}$$

Since division by 0 is undefined, the given expression is undefined.

✔ **Now Try Exercise 75.**

OBJECTIVE 4 Evaluate algebraic expressions for given values of variables. Any sequence of numbers, variables, operation symbols, and/or grouping symbols formed in accordance with the rules of algebra is called an **algebraic expression.**

$$6ab, \qquad 5m - 9n, \qquad \text{and} \qquad -2(x^2 + 4y) \qquad \text{Algebraic expressions}$$

Algebraic expressions have different numerical values for different values of the variables. We evaluate such expressions by *substituting* given values for the variables.

For example, if movie tickets cost \$8 each, the amount in dollars you pay for x tickets can be represented by the algebraic expression $8x$. We can substitute different numbers of tickets to get the costs of purchasing those tickets.

EXAMPLE 8 Evaluating Algebraic Expressions

Evaluate each expression if $m = -4$, $n = 5$, $p = -6$, and $q = 25$.

> Use parentheses around substituted values to avoid errors.

(a) $5m - 9n$

$$= 5(-4) - 9(5) \qquad \text{Substitute; let } m = -4 \text{ and } n = 5.$$

$$= -20 - 45 \qquad \text{Multiply.}$$

$$= -65 \qquad \text{Subtract.}$$

(b) $\dfrac{m + 2n}{4p}$

$$= \frac{-4 + 2(5)}{4(-6)} \qquad \text{Substitute; let } m = -4, n = 5, \text{ and } p = -6.$$

$$= \frac{-4 + 10}{-24} \qquad \text{Work separately above and below the fraction bar.}$$

$$= \frac{6}{-24} = -\frac{1}{4} \qquad \text{Write in lowest terms; also, } \frac{a}{-b} = -\frac{a}{b}.$$

(c) $-3m^3 - n^2\left(\sqrt{q}\right)$

$= -3(-4)^3 - (5)^2\left(\sqrt{25}\right)$ Substitute; let $m = -4$, $n = 5$, and $q = 25$.

$= -3(-64) - 25(5)$ Evaluate the powers and the root.

$= 192 - 125$ Multiply.

$= 67$ Subtract.

Notice the careful use of parentheses around substituted values.

✔ **Now Try Exercises 79 and 85.**

1.3 EXERCISES

Concept Check   *Decide whether each statement is* true *or* false. *If it is false, correct the statement so that it is true.*

1. $-4^6 = (-4)^6$

2. $-4^7 = (-4)^7$

3. $\sqrt{16}$ is a positive number.

4. $3 + 5 \cdot 6 = 3 + (5 \cdot 6)$

5. $(-2)^7$ is a negative number.

6. $(-2)^8$ is a positive number.

7. The product of 8 positive factors and 8 negative factors is positive.

8. The product of 3 positive factors and 3 negative factors is positive.

9. In the exponential expression -3^5, -3 is the base.

10. $\sqrt{a}$ is positive for all positive numbers a.

Concept Check   *In Exercises 11 and 12, evaluate each exponential expression.*

11. (a) 8^2 **(b)** -8^2

     (c) $(-8)^2$ **(d)** $-(-8)^2$

12. (a) 4^3 **(b)** -4^3

     (c) $(-4)^3$ **(d)** $-(-4)^3$

Write each expression by using exponents. See Example 1.

13. $10 \cdot 10 \cdot 10 \cdot 10$

14. $8 \cdot 8 \cdot 8$

15. $\dfrac{3}{4} \cdot \dfrac{3}{4} \cdot \dfrac{3}{4} \cdot \dfrac{3}{4} \cdot \dfrac{3}{4}$

16. $\dfrac{1}{2} \cdot \dfrac{1}{2}$

17. $(-9)(-9)(-9)$

18. $(-4)(-4)(-4)(-4)$

19. $z \cdot z \cdot z \cdot z \cdot z \cdot z \cdot z$

20. $a \cdot a \cdot a \cdot a \cdot a$

Evaluate each expression. See Examples 2 and 3.

21. 4^2

22. 2^4

23. 0.28^3

24. 0.91^3

25. $\left(\dfrac{1}{5}\right)^3$

26. $\left(\dfrac{1}{6}\right)^4$

27. $\left(\dfrac{4}{5}\right)^4$

28. $\left(\dfrac{7}{10}\right)^3$

29. $(-5)^3$

30. $(-2)^5$

31. $(-2)^8$

32. $(-3)^6$

33. -3^6

34. -4^6

35. -8^4

36. -10^3

Find each square root. If it is not a real number, say so. See Example 4.

37. $\sqrt{81}$

38. $\sqrt{64}$

39. $\sqrt{169}$

40. $\sqrt{225}$

41. $-\sqrt{400}$

42. $-\sqrt{900}$

43. $\sqrt{\dfrac{100}{121}}$

44. $\sqrt{\dfrac{225}{169}}$

45. $-\sqrt{0.49}$

46. $-\sqrt{0.64}$

47. $\sqrt{-36}$

48. $\sqrt{-121}$

49. *Concept Check* Match each square root with the appropriate value or description.

    (a) $\sqrt{144}$ **(b)** $\sqrt{-144}$ **(c)** $-\sqrt{144}$

    A. -12 **B.** 12 **C.** Not a real number

50. Explain why $\sqrt{-900}$ is not a real number.

Concept Check   *In Exercises 51 and 52, a represents a positive number.*

51. Is $-\sqrt{-a}$ positive, negative, or not a real number?

52. Is $-\sqrt{a}$ positive, negative, or not a real number?

Simplify each expression. Use the order of operations. See Examples 5–7.

53. $12 + 3 \cdot 4$

54. $15 + 5 \cdot 2$

55. $6 \cdot 3 - 12 \div 4$

56. $9 \cdot 4 - 8 \div 2$

57. $10 + 30 \div 2 \cdot 3$

58. $12 + 24 \div 3 \cdot 2$

59. $-3(5)^2 - (-2)(-8)$

60. $-9(2)^2 - (-3)(-2)$

61. $5 - 7 \cdot 3 - (-2)^3$

62. $-4 - 3 \cdot 5 + 6^2$

63. $-7(\sqrt{36}) - (-2)(-3)$

64. $-8(\sqrt{64}) - (-3)(-7)$

65. $6|4 - 5| - 24 \div 3$

66. $-4|2 - 4| + 8 \cdot 2$

67. $|-6 - 5|(-8) + 3^2$

68. $(-6 - 3)|-2 - 3| \div 9$

69. $6 + \dfrac{2}{3}(-9) - \dfrac{5}{8} \cdot 16$

70. $7 - \dfrac{3}{4}(-8) + 12 \cdot \dfrac{5}{6}$

71. $-14\left(-\dfrac{2}{7}\right) \div (2 \cdot 6 - 10)$

72. $-12\left(-\dfrac{3}{4}\right) - (6 \cdot 5 \div 3)$

73. $\dfrac{(-5 + \sqrt{4})(-2^2)}{-5 - 1}$

74. $\dfrac{(-9 + \sqrt{16})(-3^2)}{-4 - 1}$

75. $\dfrac{2(-5) + (-3)(-2)}{-8 + 3^2 - 1}$

76. $\dfrac{3(-4) + (-5)(-8)}{2^3 - 2 - 6}$

77. $\dfrac{5 - 3\left(\dfrac{-5 - 9}{-7}\right) - 6}{-9 - 11 + 3 \cdot 7}$

78. $\dfrac{-4\left(\dfrac{12 - (-8)}{3 \cdot 2 + 4}\right) - 5(-1 - 7)}{-9 - (-7) - [-5 - (-8)]}$

Evaluate each expression if $a = -3$, $b = 64$, and $c = 6$. See Example 8.

79. $3a + \sqrt{b}$

80. $-2a - \sqrt{b}$

81. $\sqrt{b} + c - a$

82. $\sqrt{b} - c + a$

83. $4a^3 + 2c$

84. $-3a^4 - 3c$

85. $\dfrac{2c + a^3}{4b + 6a}$

86. $\dfrac{3c + a^2}{2b - 6c}$

Evaluate each expression if $w = 4$, $x = -\frac{3}{4}$, $y = \frac{1}{2}$, and $z = 1.25$. See Example 8.

87. $wy - 8x$

88. $wz - 12y$

89. $xy + y^4$

90. $xy - x^2$

91. $-w + 2x + 3y + z$

92. $w - 6x + 5y - 3z$

93. $\dfrac{7x + 9y}{w}$

94. $\dfrac{7y - 5x}{2w}$

Solve each problem.

Residents of Linn County, Iowa, in the Cedar Rapids Community School District can use the expression

$$(v \times 0.5485 - 4850) \div 1000 \times 31.44$$

to determine their property taxes, where v is assessed home value. (*Source: The Gazette,* Cedar Rapids, Iowa, August 19, 2000.) Use the expression to calculate the amount of property taxes to the nearest dollar that the owner of a home with each of the following values would pay. Follow the order of operations.

95. $100,000   **96.** $150,000   **97.** $200,000

The Blood Alcohol Concentration (BAC) of a person who has been drinking is given by the expression

number of oz $\times$ % alcohol $\times$ 0.075 $\div$ body weight in lb $-$ hr of drinking $\times$ 0.015.

(*Source:* Lawlor, J., *Auto Math Handbook: Mathematical Calculations, Theory, and Formulas for Automotive Enthusiasts,* HP Books, 1991.)

98. Suppose a policeman stops a 190-lb man who, in 2 hr, has ingested four 12-oz beers (48 oz), each having a 3.2% alcohol content.

   (a) Substitute the values into the formula, and write the expression for the man's BAC.
   (b) Calculate the man's BAC to the nearest thousandth. Follow the order of operations.

99. Find the BAC to the nearest thousandth for a 135-lb woman who, in 3 hr, has drunk three 12-oz beers (36 oz), each having a 4.0% alcohol content.

100. (a) Calculate the BACs in Exercises 98 and 99 if each person weighs 25 lb more and the rest of the variables stay the same. How does increased weight affect a person's BAC?
   (b) Predict how decreased weight would affect the BAC of each person in Exercises 98 and 99. Calculate the BACs if each person weighs 25 lb less and the rest of the variables stay the same.

101. An approximation of the amount in billions of dollars that Americans have spent on their pets from 1994 to 2005 can be obtained by substituting a given year for x in the expression

$$1.718x - 3409.$$

(*Source:* American Pet Products Manufacturers Association.) Approximate the amount spent in each year. Round answers to the nearest tenth.

   (a) 1994   **(b)** 2000   **(c)** 2005
   (d) How has the amount Americans have spent on their pets changed from 1994 to 2005?

102. An approximation of federal spending on education in billions of dollars from 2001 through 2005 can be obtained using the expression

$$9.0499x - 18,071.87,$$

where x represents the year. (*Source*: U.S. Department of the Treasury.)

(a) Use this expression to complete the table. Round answers to the nearest tenth.

Year	Education Spending (in billions of dollars)
2001	37.0
2002	46.0
2003	
2004	
2005	

(b) How has the amount of federal spending on education changed from 2001 to 2005?

1.4 Properties of Real Numbers

OBJECTIVES

1 Use the distributive property.

2 Use the inverse properties.

3 Use the identity properties.

4 Use the commutative and associative properties.

5 Use the multiplication property of 0.

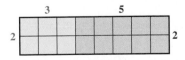

Area of left part is $2 \cdot 3 = 6$.
Area of right part is $2 \cdot 5 = 10$.
Area of total rectangle is $2(3 + 5) = 16$.

FIGURE 17

The study of any object is simplified when we know the properties of the object. For example, a property of water is that it freezes when cooled to 0°C. Knowing this helps us to predict the behavior of water.

The study of numbers is no different. The basic properties of real numbers studied in this section reflect results that occur consistently in work with numbers, so they have been generalized to apply to expressions with variables as well.

OBJECTIVE 1 Use the distributive property. Notice that

$$2(3 + 5) = 2 \cdot 8 = 16$$

and $\qquad 2 \cdot 3 + 2 \cdot 5 = 6 + 10 = 16,$

so $\qquad 2(3 + 5) = 2 \cdot 3 + 2 \cdot 5.$

This idea is illustrated by the divided rectangle in Figure 17. Similarly,

$$-4[5 + (-3)] = -4(2) = -8$$

and $\qquad -4(5) + (-4)(-3) = -20 + 12 = -8,$

so $\qquad -4[5 + (-3)] = -4(5) + (-4)(-3).$

These examples are generalized to *all* real numbers as the **distributive property of multiplication with respect to addition,** or simply the **distributive property.**

Distributive Property

For any real numbers a, b, and c,

$$a(b + c) = ab + ac \qquad \text{and} \qquad (b + c)a = ba + ca.$$

The distributive property can also be written

$$ab + ac = a(b + c) \qquad \text{and} \qquad ba + ca = (b + c)a$$

and can be extended to more than two numbers as well.

$$a(b + c + d) = ab + ac + ad$$

The distributive property provides a way to rewrite a product $a(b + c)$ as a sum $ab + ac$ or a sum as a product.

> ▶ **NOTE** When we rewrite $a(b + c)$ as $ab + ac$, we sometimes refer to the process as "removing" or "clearing" parentheses.

EXAMPLE 1 Using the Distributive Property

Use the distributive property to rewrite each expression.

(a) $3(x + y)$ Use the first form of the property to
   $= 3x + 3y$ rewrite the given product as a sum.

(b) $-2(5 + k)$
   $= -2(5) + (-2)(k)$
   $= -10 - 2k$

(c) $4x + 8x$ Use the second form of the property to
   $= (4 + 8)x$ rewrite the given sum as a product.
   $= 12x$

(d) $3r - 7r$
   $= 3r + (-7r)$ Definition of subtraction
   $= [3 + (-7)]r$ Distributive property
   $= -4r$

(e) $5p + 7q$

Because there is no common number or variable here, we cannot use the distributive property to rewrite the expression.

(f) $6(x + 2y - 3z)$
   $= 6x + 6(2y) + 6(-3z)$
   $= 6x + 12y - 18z$

✔ **Now Try Exercises 11, 13, 15, and 19.**

The distributive property can also be used for subtraction (Example 1(d)), so

$$a(b - c) = ab - ac.$$

OBJECTIVE ② Use the inverse properties. In **Section 1.1**, we saw that the *additive inverse* (or *opposite*) of a number a is $-a$ and that additive inverses have a sum of 0.

$$5 \text{ and } -5, \quad -\frac{1}{2} \text{ and } \frac{1}{2}, \quad -34 \text{ and } 34 \qquad \text{Additive inverses (sum of 0)}$$

In **Section 1.2,** we saw that the *multiplicative inverse (or reciprocal)* of a number a is $\frac{1}{a}$ (where $a \neq 0$) and that multiplicative inverses have a product of 1.

$$5 \text{ and } \frac{1}{5}, \quad -\frac{1}{2} \text{ and } -2, \quad \frac{3}{4} \text{ and } \frac{4}{3}$$
Multiplicative inverses (product of 1)

This discussion leads to the **inverse properties** of addition and multiplication, which can be extended to the real numbers of algebra.

Inverse Properties

For any real number a,

$$a + (-a) = 0 \quad \text{ and } \quad -a + a = 0$$

$$a \cdot \frac{1}{a} = 1 \quad \text{ and } \quad \frac{1}{a} \cdot a = 1 \quad (a \neq 0).$$

The inverse properties "undo" addition or multiplication. Think of putting on your shoes when you get up in the morning and then taking them off before you go to bed at night. These are inverse operations that undo each other.

OBJECTIVE 3 Use the identity properties.   The numbers 0 and 1 each have a special property. Zero is the only number that can be added to any number to get that number. Adding 0 to any number leaves the identity of the number unchanged. For this reason, 0 is called the **identity element for addition,** or the **additive identity.** In a similar way, multiplying any number by 1 leaves the identity of the number unchanged, so 1 is the **identity element for multiplication,** or the **multiplicative identity.** The **identity properties** summarize this discussion and extend these properties from arithmetic to algebra.

Identity Properties

For any real number a,

$$a + 0 = 0 + a = a$$

$$a \cdot 1 = 1 \cdot a = a.$$

The identity properties leave the identity of a real number unchanged. Think of a child wearing a costume on Halloween. The child's appearance is changed, but his or her identity is unchanged.

EXAMPLE 2 Using the Identity Property $1 \cdot a = a$

Simplify each expression.

(a) $12m + m$

$\quad = 12m + 1m \qquad$ Identity property; $m = 1 \cdot m$, or $1m$

$\quad = (12 + 1)m \qquad$ Distributive property

$\quad = 13m \qquad$ Add inside parentheses.

(b) $y + y$

$= 1y + 1y$ Identity property

$= (1 + 1)y$ Distributive property

$= 2y$ Add inside parentheses.

(c) $-(m - 5n)$

$= -1(m - 5n)$ Identity property

$= -1(m) + (-1)(-5n)$ Distributive property

$= -m + 5n$ Multiply.

> Multiply *each* term by −1. Be careful with signs.

✔ **Now Try Exercises 21 and 23.**

Term	Numerical Coefficient
$-7y$	-7
$34r^3$	34
$-26x^5yz^4$	-26
$-k = -1k$	-1
$r = 1r$	1
$\dfrac{3x}{8} = \dfrac{3}{8}x$	$\dfrac{3}{8}$
$\dfrac{x}{3} = \dfrac{1x}{3} = \dfrac{1}{3}x$	$\dfrac{1}{3}$

Expressions such as $12m$ and $5n$ from Example 2 are examples of *terms*. A **term** is a number or the product of a number and one or more variables raised to powers. The numerical factor in a term is called the **numerical coefficient,** or just the **coefficient.** Some examples of terms and their coefficients are shown in the table in the margin.

Terms with exactly the same variables raised to exactly the same powers are called **like terms.** Some examples of like terms are

$$5p \text{ and } -21p \qquad -6x^2 \text{ and } 9x^2. \quad \text{Like terms}$$

Some examples of **unlike terms** are

$$3m \text{ and } 16x \qquad 7y^3 \text{ and } -3y^2. \quad \text{Unlike terms}$$

OBJECTIVE 4 Use the commutative and associative properties. Simplifying expressions as in parts (a) and (b) of Example 2 is called **combining like terms.** *Only like terms may be combined.* To combine like terms in an expression such as

$$-2m + 5m + 3 - 6m + 8,$$

we need two more properties. From arithmetic, we know that

$$3 + 9 = 12 \qquad \text{and} \qquad 9 + 3 = 12$$
$$3 \cdot 9 = 27 \qquad \text{and} \qquad 9 \cdot 3 = 27.$$

The order of the numbers being added or multiplied does not matter. The same answers result. Also,

$$(5 + 7) + 2 = 12 + 2 = 14$$
$$5 + (7 + 2) = 5 + 9 = 14,$$

and

$$(5 \cdot 7) \cdot 2 = 35 \cdot 2 = 70$$
$$5(7 \cdot 2) = 5 \cdot 14 = 70.$$

The way in which the numbers being added or multiplied are grouped does not matter. The same answers result.

These arithmetic examples can be extended to algebra.

Commutative and Associative Properties

For any real numbers a, b, and c,

$$a + b = b + a$$

and

$$ab = ba.$$

$\left.\right\}$ Commutative properties

Interchange the order of the two terms or factors.

Also,

$$a + (b + c) = (a + b) + c$$

and

$$a(bc) = (ab)c.$$

$\left.\right\}$ Associative properties

Shift parentheses among the three terms or factors; the order stays the same.

The commutative properties are used to change the *order* of the terms or factors in an expression. Think of commuting from home to work and then from work to home. The associative properties are used to *regroup* the terms or factors of an expression. Remember, to *associate* is to be part of a group.

EXAMPLE 3 Using the Commutative and Associative Properties

Simplify $-2m + 5m + 3 - 6m + 8$.

$$-2m + 5m + 3 - 6m + 8$$

$= (-2m + 5m) + 3 - 6m + 8$	Order of operations
$= (-2 + 5)m + 3 - 6m + 8$	Distributive property
$= 3m + 3 - 6m + 8$	Add inside parentheses.

By the order of operations, the next step would be to add $3m$ and 3, but they are unlike terms. To get $3m$ and $-6m$ together, use the associative and commutative properties. Begin by inserting parentheses and brackets according to the order of operations.

$= [(3m + 3) - 6m] + 8$	
$= [3m + (3 - 6m)] + 8$	Associative property
$= [3m + (-6m + 3)] + 8$	Commutative property
$= [(3m + [-6m]) + 3] + 8$	Associative property
$= (-3m + 3) + 8$	Combine like terms.
$= -3m + (3 + 8)$	Associative property
$= -3m + 11$	Add.

In practice, many of these steps are not written down, but you should realize that the commutative and associative properties are used whenever the terms in an expression are rearranged to combine like terms.

✔ **Now Try Exercise 27.**

> **EXAMPLE 4** Using the Properties of Real Numbers

Simplify each expression.

(a) $5y - 8y - 6y + 11y$

$\quad = (5 - 8 - 6 + 11)y \qquad$ Distributive property

$\quad = 2y \qquad\qquad\qquad$ Combine like terms.

(b) $\qquad\quad 3x + 4 - 5(x + 1) - 8$

$\qquad = 3x + 4 - 5x - 5 - 8 \qquad$ Distributive property

$\qquad = 3x - 5x + 4 - 5 - 8 \qquad$ Commutative property

$\qquad = -2x - 9 \qquad\qquad\qquad$ Combine like terms.

Be careful with signs.

(c) $8 - (3m + 2)$

$\quad = 8 - 1(3m + 2) \qquad$ Identity property

$\quad = 8 - 3m - 2 \qquad\quad$ Distributive property

$\quad = 6 - 3m \qquad\qquad\quad$ Combine like terms.

(d) $3x(5)(y)$

$\quad = [3x(5)]y \qquad$ Order of operations

$\quad = [3(x \cdot 5)]y \qquad$ Associative property

$\quad = [3(5x)]y \qquad$ Commutative property

$\quad = [(3 \cdot 5)x]y \qquad$ Associative property

$\quad = (15x)y \qquad$ Multiply.

$\quad = 15(xy) \qquad$ Associative property

$\quad = 15xy$

As previously mentioned, many of these steps are not usually written out.

✔ **Now Try Exercises 29 and 31.**

▶ **CAUTION** Be careful. The distributive property does not apply in Example 4(d), because there is no addition involved.

$$(3x)(5)(y) \neq (3x)(5) \cdot (3x)(y)$$

OBJECTIVE 5 Use the multiplication property of 0. The additive identity property gives a special property of 0, namely, that $a + 0 = a$ for any real number a. The **multiplication property of 0** gives a special property of 0 that involves multiplication. The product of any real number and 0 is 0.

> **Multiplication Property of 0**
>
> For any real number a,
>
> $$a \cdot 0 = 0 \qquad \text{and} \qquad 0 \cdot a = 0.$$

1.4 EXERCISES

⊙ *Complete solution available on Video Lectures on CD/DVD*

Now Try Exercise

Concept Check   *Choose the correct response in Exercises 1–4.*

1. The identity element for addition is

    A. $-a$ **B.** 0   **C.** 1   **D.** $\dfrac{1}{a}$.

2. The identity element for multiplication is

    A. $-a$ **B.** 0   **C.** 1   **D.** $\dfrac{1}{a}$.

3. The additive inverse of a is

    A. $-a$ **B.** 0   **C.** 1   **D.** $\dfrac{1}{a}$.

4. The multiplicative inverse of a, where $a \neq 0$, is

    A. $-a$ **B.** 0   **C.** 1   **D.** $\dfrac{1}{a}$.

Concept Check   *Complete each statement.*

5. The multiplication property of 0 says that the _____ of 0 and any real number is _____.

6. The commutative property is used to change the _____ of two terms or factors.

7. The associative property is used to change the _____ of three terms or factors.

8. Like terms are terms with the _____ variables raised to the _____ powers.

9. When simplifying an expression, only _____ terms can be combined.

10. The coefficient in the term $-8yz^2$ is _____.

Simplify each expression. See Examples 1 and 2.

⊙ **11.** $2(m + p)$ **12.** $3(a + b)$ **13.** $-12(x - y)$ **14.** $-10(p - q)$

15. $5k + 3k$ **16.** $6a + 5a$ **17.** $7r - 9r$ **18.** $4n - 6n$

19. $-8z + 4w$ **20.** $-12k + 3r$ ⊙ **21.** $a + 7a$ **22.** $s + 9s$

23. $-(2d - f)$ **24.** $-(3m - n)$

Simplify each expression. See Examples 1–4.

25. $-12y + 4y + 3 + 2y$ **26.** $-5r - 9r + 8r - 5$

⊙ **27.** $-6p + 5 - 4p + 6 + 11p$ **28.** $-8x - 12 + 3x - 5x + 9$

⊙ **29.** $3(k + 2) - 5k + 6 + 3$ **30.** $5(r - 3) + 6r - 2r + 4$

31. $-2(m + 1) - (m - 4)$ **32.** $6(a - 5) - (a + 6)$

33. $0.25(8 + 4p) - 0.5(6 + 2p)$ **34.** $0.4(10 - 5x) - 0.8(5 + 10x)$

35. $-(2p + 5) + 3(2p + 4) - 2p$ **36.** $-(7m - 12) - 2(4m + 7) - 8m$

37. $2 + 3(2z - 5) - 3(4z + 6) - 8$ **38.** $-4 + 4(4k - 3) - 6(2k + 8) + 7$

Concept Check   *Complete each statement so that the indicated property is illustrated. Simplify each answer if possible.*

39. $5x + 8x =$ _____
                 (distributive property)

40. $9y - 6y =$ _____
                 (distributive property)

41. $5(9r) =$ _____
                 (associative property)

42. $-4 + (12 + 8) =$ _____
                 (associative property)

43. $5x + 9y =$ _____
                 (commutative property)

44. $-5 \cdot 7 =$ _____
                 (commutative property)

45. $1 \cdot 7 =$ _____
(identity property)

46. $-12x + 0 =$ _____
(identity property)

47. $-\dfrac{1}{4}ty + \dfrac{1}{4}ty =$ _____
(inverse property)

48. $-\dfrac{9}{8}\left(-\dfrac{8}{9}\right) =$ _____
(inverse property)

49. $8(-4 + x) =$ _____
(distributive property)

50. $3(x - y + z) =$ _____
(distributive property)

51. $0(0.875x + 9y - 88z) =$ _____
(multiplication property of 0)

52. $0(35t^2 - 8t + 12) =$ _____
(multiplication property of 0)

53. *Concept Check*　Give an "everyday" example of a commutative operation and of an operation that is not commutative.

54. *Concept Check*　Give an "everyday" example of inverse operations.

The distributive property can be used to mentally perform calculations. For example, calculate $38 \cdot 17 + 38 \cdot 3$ as follows:

$$
\begin{aligned}
38 \cdot 17 + 38 \cdot 3 &= 38(17 + 3) &&\text{Distributive property} \\
&= 38(20) &&\text{Add inside parentheses.} \\
&= 760. &&\text{Multiply.}
\end{aligned}
$$

Use the distributive property to calculate each value mentally.

55. $96 \cdot 19 + 4 \cdot 19$

56. $27 \cdot 60 + 27 \cdot 40$

57. $58 \cdot \dfrac{3}{2} - 8 \cdot \dfrac{3}{2}$

58. $8.75(15) - 8.75(5)$

59. $4.31(69) + 4.31(31)$

60. $\dfrac{8}{5}(17) + \dfrac{8}{5}(13)$

RELATING CONCEPTS　(EXERCISES 61–66)

FOR INDIVIDUAL OR GROUP WORK

When simplifying the expression $3x + 4 + 2x + 7$ to $5x + 11$, some steps are usually done mentally. **Work Exercises 61–66 in order,** *providing the property that justifies each statement in the given simplification. (These steps could be done in other orders.)*

$$3x + 4 + 2x + 7$$

61.　$= (3x + 4) + (2x + 7)$ _____

62.　$= 3x + (4 + 2x) + 7$ _____

63.　$= 3x + (2x + 4) + 7$ _____

64.　$= (3x + 2x) + (4 + 7)$ _____

65.　$= (3 + 2)x + (4 + 7)$ _____

66.　$= 5x + 11$ _____

---

67. By the distributive property, $a(b + c) = ab + ac$. This property is more completely named the distributive property of multiplication with respect to addition. Is there a distributive property of addition with respect to multiplication? That is, does

$$a + (b \cdot c) = (a + b)(a + c)$$

for all real numbers, a, b, and c? To find out, try some sample values of a, b, and c.

68. Explain how the distributive property is used to combine like terms. Give an example.

Chapter **1**	**Group Activity**

HOW AMERICANS SPEND THEIR MONEY

Objective Construct and read bar graphs and circle graphs.

Graphs and tables are a great way of presenting information. They allow you to make comparisons and approximations, as well as draw conclusions more quickly than you would by reading a page of text with the same information.

Listed in the table are common personal consumption expenditures of Americans during the years 1997 through 2003.

Personal Consumption Expenditures (in billions of dollars)

Category	1997	1998	1999	2000	2001	2002	2003
Food and Tobacco	$832.3	$900.2	$963.8	$1003.7	$1047.8	$1095.0	$1152.6
Clothing, Accessories, Jewelry	353.3	368.3	397.2	397.0	396.8	404.4	412.3
Personal Care	79.4	80.5	86.0	93.4	94.3	95.8	96.9
Housing	829.8	858.2	906.2	1006.5	1073.7	1144.8	1188.4
Household Operation	620.7	643.8	682.5	719.3	738.4	746.0	779.6
Medical Care	957.3	1040.9	1102.6	1218.3	1322.8	1444.9	1557.2
Personal Business	459.1	533.7	586.2	539.1	539.0	552.1	577.7
Transportation	636.4	648.6	705.5	853.4	874.0	877.5	925.5
Recreation	462.9	489.8	534.9	585.7	603.4	628.3	660.7
Education and Research	129.4	139.4	148.9	163.6	176.3	190.7	201.7
Religious and Welfare Activities	157.6	162.6	170.2	172.3	186.1	202.9	211.2
TOTAL	$5518.2	$5866.0	$6284.0	$6752.3	$7052.6	$7382.4	$7763.8

Source: U.S. Bureau of Economic Analysis.

A. Have each person in the group construct a bar graph for one category of personal consumption expenditures over the 7-yr period shown in the table.

B. Have each person in the group construct a circle graph for one year shown in the table. Include all 11 categories of personal consumption expenditures. Use approximate values as needed.

C. As a group, examine the graphs you have constructed.

1. Discuss any conclusions you can draw from them.

2. What do the graphs tell you about how personal consumption of Americans changed over the 7 yr?

3. Write a paragraph that summarizes the group's conclusions.

Chapter **1** SUMMARY *View the Interactive Summary on the Pass the Test CD.*

KEY TERMS

1.1 set
elements (members)
finite set
infinite set
empty set (null set)
variable
set-builder notation
number line
coordinate
graph
additive inverse
(opposite, negative)

signed numbers
absolute value
equation
inequality
interval
interval notation
three-part inequality
1.2 sum
difference
product
quotient

reciprocal
(multiplicative
inverse)
1.3 factors
exponent (power)
base
exponential expression
square root
principal (positive)
square root
negative square root

algebraic expression
1.4 identity element for
addition
identity element for
multiplication
term
coefficient (numerical
coefficient)
like terms
unlike terms
combining like terms

NEW SYMBOLS

$\{a, b\}$ set containing
the elements a
and b

$\emptyset$ or $\{\ \}$ empty set

$\in$ is an element of
(a set)

$\notin$ is not an
element of

$\neq$ is not equal to

$\{x\,|\,x \text{ has property } P\}$
set-builder notation

$|x|$ absolute value of x

$<$ is less than

$\leq$ is less than or equal
to

$>$ is greater than

$\geq$ is greater than or
equal to

∞ infinity

$-\infty$ negative
infinity

$(-\infty, \infty)$ set of all real
numbers

(a, ∞) the interval
$\{x\,|\,x > a\}$

$(-\infty, a)$ the interval
$\{x\,|\,x < a\}$

$(a, b]$ the interval
$\{x\,|\,a < x \leq b\}$

a^m m factors of a

$\sqrt{\ }$ radical sign

$\sqrt{a}$ positive (or
principal) square
root of a

TEST YOUR WORD POWER

See how well you have learned the vocabulary in this chapter. Answers, with examples, follow the Quick Review.

1. The **empty set** is a set
   A. with 0 as its only element
   B. with an infinite number of
      elements
   C. with no elements
   D. of ideas.

2. A **variable** is
   A. a symbol used to represent an
      unknown number
   B. a value that makes an equation
      true
   C. a solution of an equation
   D. the answer in a division problem.

3. The **absolute value** of a number is
   A. the graph of the number
   B. the reciprocal of the number
   C. the opposite of the number
   D. the distance between 0 and the
      number on a number line.

4. The **reciprocal** of a nonzero number
   a is
   A. a B. $\frac{1}{a}$ C. $-a$ D. 1.

5. A **factor** is
   A. the answer in an addition problem
   B. the answer in a multiplication
      problem
   C. one of two or more numbers that
      are added to get another number
   D. any number that divides evenly
      into a given number.

6. An **exponential expression** is
   A. a number that is a repeated factor
      in a product
   B. a number or a variable written
      with an exponent
   C. a number that shows how many
      times a factor is repeated in a
      product

D. an expression that involves
      addition.

7. A **term** is
   A. a numerical factor
   B. a number or a product of a
      number and one or more
      variables raised to powers
   C. one of several variables with the
      same exponents
   D. a sum of numbers and variables
      raised to powers.

8. A **numerical coefficient** is
   A. the numerical factor in a term
   B. the number of terms in an
      expression
   C. a variable raised to a power
   D. the variable factor in a term.

QUICK REVIEW

Concepts	Examples

1.1 BASIC CONCEPTS

Sets of Numbers

Natural Numbers $\{1, 2, 3, 4, \ldots\}$

Whole Numbers $\{0, 1, 2, 3, 4, \ldots\}$

Integers $\{\ldots, -2, -1, 0, 1, 2, \ldots\}$

Rational Numbers

$\left\{\frac{p}{q}\,\middle|\,p \text{ and } q \text{ are integers, } q \neq 0\right\}$

(all terminating or repeating decimals)

Irrational Numbers

$\{x \mid x \text{ is a real number that is not rational}\}$

(all nonterminating, nonrepeating decimals)

Real Numbers

$\{x \mid x \text{ is a rational or an irrational number}\}$

Absolute Value $|a| = \begin{cases} a & \text{if } a \text{ is positive or } 0 \\ -a & \text{if } a \text{ is negative} \end{cases}$

Examples:

$10, 25, 143$

$0, 8, 47$

$-22, -7, 0, 4, 9$

$-\dfrac{2}{3}, -0.14, 0, 6, \dfrac{5}{8}, 0.33333\ldots$

$\pi, \sqrt{3}, -\sqrt{22}$

$-3, 0.7, \pi, -\dfrac{2}{3}$

$|12| = 12$
$|-12| = 12$

1.2 OPERATIONS ON REAL NUMBERS

Addition

Same Sign: Add the absolute values. The sum has the same sign as the given numbers.

Different Signs: Find the absolute values of the numbers, and subtract the smaller absolute value from the larger. The sum has the same sign as the number with the larger absolute value.

$-2 + (-7) = -(2 + 7) = -9$

$-5 + 8 = 8 - 5 = 3$
$-12 + 4 = -(12 - 4) = -8$

Subtraction

For all real numbers a and b,

$$a - b = a + (-b).$$

$-5 - (-3) = -5 + 3 = -2$

Multiplication and Division

Same Sign: The answer is positive when multiplying or dividing two numbers with the same sign.

$-3(-8) = 24 \qquad \dfrac{-15}{-5} = 3$

Different Signs: The answer is negative when multiplying or dividing two numbers with different signs.

$-7(5) = -35 \qquad \dfrac{-24}{12} = -2$

Division

For all real numbers a and b (where $b \neq 0$),

$$a \div b = \frac{a}{b} = a \cdot \frac{1}{b}.$$

$\dfrac{2}{3} \div \dfrac{5}{6} = \dfrac{2}{3} \cdot \dfrac{6}{5} = \dfrac{4}{5}$ Multiply by the reciprocal.

(continued)

Concepts	Examples

1.3 EXPONENTS, ROOTS, AND ORDER OF OPERATIONS

The product of an even number of negative factors is positive. The product of an odd number of negative factors is negative.

$(-5)^2$ is positive: $(-5)^2 = (-5)(-5) = 25$

$(-5)^3$ is negative: $(-5)^3 = (-5)(-5)(-5) = -125$

Order of Operations

1. Work separately above and below any fraction bar.
2. If parentheses, brackets, or absolute value bars are present, start with the innermost set and work outward.
3. Evaluate all exponents, roots, and absolute values.
4. Multiply or divide in order from left to right.
5. Add or subtract in order from left to right.

$$\frac{12 + 3}{5 \cdot 2} = \frac{15}{10} = \frac{3}{2}$$

$$(-6)[2^2 - (3 + 4)] + 3$$
$$= (-6)[2^2 - 7] + 3$$
$$= (-6)[4 - 7] + 3$$
$$= (-6)[-3] + 3$$
$$= 18 + 3$$
$$= 21$$

1.4 PROPERTIES OF REAL NUMBERS

For real numbers a, b, and c,

Distributive Property

$a(b + c) = ab + ac$

$12(4 + 2) = 12 \cdot 4 + 12 \cdot 2$

Inverse Properties

$a + (-a) = 0$ and $-a + a = 0$

$a \cdot \dfrac{1}{a} = 1$ and $\dfrac{1}{a} \cdot a = 1$

$5 + (-5) = 0$ $-12 + 12 = 0$

$5 \cdot \dfrac{1}{5} = 1$ $-\dfrac{1}{3}(-3) = 1$

Identity Properties

$a + 0 = 0 + a = a$ and $a \cdot 1 = 1 \cdot a = a$

$-32 + 0 = -32$ $17.5 \cdot 1 = 17.5$

Commutative Properties

$a + b = b + a$ and $ab = ba$

$9 + (-3) = -3 + 9$ $6(-4) = (-4)6$

Associative Properties

$a + (b + c) = (a + b) + c$ and $a(bc) = (ab)c$

$7 + (5 + 3) = (7 + 5) + 3$ $-4(6 \cdot 3) = (-4 \cdot 6)3$

Multiplication Property of 0

$a \cdot 0 = 0$ and $0 \cdot a = 0.$

$4 \cdot 0 = 0$ $0(-3) = 0$

Answers to Test Your Word Power

1. C; *Example:* The set of whole numbers less than 0 is the empty set, written $\emptyset$.

2. A; *Examples: a, b, c*

3. D; *Examples:* $|2| = 2$ and $|-2| = 2$

4. B; *Examples:* 3 is the reciprocal of $\frac{1}{3}$; $-\frac{5}{2}$ is the reciprocal of $-\frac{2}{5}$.

5. D; *Examples:* 2 and 5 are factors of 10, since both divide evenly (without remainder) into 10; other integer factors of 10 are $-10, -5, -2, -1, 1,$ and 10.

6. B; *Examples:* 3^4 and x^{10}

7. B; *Examples:* $6, \frac{x}{2}, -4ab^2$

8. A; *Examples:* The term $8z$ has numerical coefficient 8, and $-10x^3y$ has numerical coefficient -10.

Chapter **1** REVIEW EXERCISES

[1.1] * *Graph the elements of each set on a number line.*

1. $\left\{-4, -1, 2, \dfrac{9}{4}, 4\right\}$

2. $\left\{-5, -\dfrac{11}{4}, -0.5, 0, 3, \dfrac{13}{3}\right\}$

Find the value of each expression.

3. $|-16|$

4. $-|-4|$

5. $|-8| - |-3|$

Let $S = \left\{-9, -\dfrac{4}{3}, -\sqrt{4}, -0.25, 0, 0.\overline{35}, \dfrac{5}{3}, \sqrt{7}, \sqrt{-9}, \dfrac{12}{3}\right\}$. *Simplify the elements of S as necessary, and then list those elements of S which belong to the specified set.*

6. Whole numbers

7. Integers

8. Rational numbers

9. Real numbers

Write each set by listing its elements.

10. $\{x \mid x$ is a natural number between 3 and 9$\}$

11. $\{y \mid y$ is a whole number less than 4$\}$

Write true *or* false *for each inequality.*

12. $4 \cdot 2 \le |12 - 4|$

13. $2 + |-2| > 4$

14. $4(3 + 7) > -|40|$

The graph shows the percent change in annual domestic car sales from January 2004 to January 2005 for various automakers. Use this graph to work Exercises 15–18.

15. Which automaker had the greatest change in sales? What was that change?

16. Which automaker had the least change in sales? What was that change?

17. *True* or *false:* The absolute value of the percent change for Ford was greater than the absolute value of the percent change for General Motors.

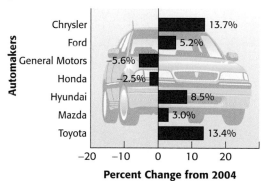

Car Sales, 2005

Automakers

Chrysler	13.7%
Ford	5.2%
General Motors	−5.6%
Honda	−2.5%
Hyundai	8.5%
Mazda	3.0%
Toyota	13.4%

−20   −10   0   10   20

Percent Change from 2004

Source: *Chicago Tribune,* February 12, 2006.

18. *True* or *false:* The percent change for Toyota was more than four times greater than the percent change for Mazda.

Write each set in interval notation and graph the interval.

19. $\{x \mid x < -5\}$

20. $\{x \mid -2 < x \le 3\}$

[1.2] *Add or subtract as indicated.*

21. $-\dfrac{5}{8} - \left(-\dfrac{7}{3}\right)$

22. $-\dfrac{4}{5} - \left(-\dfrac{3}{10}\right)$

* For help with the Review Exercises in this text, refer to the appropriate section given in brackets.

23. $-5 + (-11) + 20 - 7$

24. $-9.42 + 1.83 - 7.6 - 1.9$

25. $-15 + (-13) + (-11)$

26. $-1 - 3 - (-10) + (-7)$

27. $\dfrac{3}{4} - \left(\dfrac{1}{2} - \dfrac{9}{10}\right)$

28. $-|-12| - |-9| + (-4) - |10|$

29. In 2003, Krispy Kreme Doughnuts reported a profit of $13.1 million. In 2004, the low-carb diet craze was responsible for a first-quarter loss in doughnut sales of $24.4 million. Find the difference between these two amounts. (*Source*: Krispy Kreme Doughnuts.)

Multiply or divide as indicated.

30. $2(-5)(-3)(-3)$

31. $-\dfrac{3}{7}\left(-\dfrac{14}{9}\right)$

32. $\dfrac{75}{-5}$

33. $\dfrac{-2.3754}{-0.74}$

34. *Concept Check* Which one of the following is undefined: $\dfrac{5}{7 - 7}$ or $\dfrac{7 - 7}{5}$?

[1.3] *Evaluate each expression.*

35. 10^4

36. $\left(\dfrac{3}{7}\right)^3$

37. $(-5)^3$

38. -5^3

Find each square root. If it is not a real number, say so.

39. $\sqrt{400}$

40. $\sqrt{\dfrac{64}{121}}$

41. $-\sqrt{0.81}$

42. $\sqrt{-64}$

Simplify each expression.

43. $-14\left(\dfrac{3}{7}\right) + 6 \div 3$

44. $-\dfrac{2}{3}[5(-2) + 8 - 4^3]$

45. $\dfrac{-5(3^2) + 9\left(\sqrt{4}\right) - 5}{6 - 5(-2)}$

Evaluate each expression if $k = -4$, $m = 2$, and $n = 16$.

46. $4k - 7m$

47. $-3\sqrt{n} + m + 5k$

48. $\dfrac{4m^3 - 3n}{7k^2 - 10}$

49. The following expression for *body mass index* (BMI) can help determine ideal body weight.

$$704 \times \text{(weight in pounds)} \div \text{(height in inches)}^2$$

A BMI of 19 to 25 corresponds to a healthy weight. (*Source: Washington Post.*)

(a) Baseball player Carlos Beltran is 6 ft, 1 in., tall and weighs 190 lb. (*Source: Street & Smith's Baseball 2004 Yearbook.*) Find his BMI (to the nearest whole number).

(b) Calculate your BMI.

[1.4]   *Simplify each expression.*

50. $2q + 19q$ **51.** $13z - 17z$ **52.** $-m + 6m$

53. $5p - p$ **54.** $-2(k + 3)$ **55.** $6(r + 3)$

56. $9(2m + 3n)$ **57.** $-(-p + 6q) - (2p - 3q)$

58. $-3y + 6 - 5 + 4y$ **59.** $2a + 3 - a - 1 - a - 2$

60. $-3(4m - 2) + 2(3m - 1) - 4(3m + 1)$

Complete each statement so that the indicated property is illustrated. Simplify each answer if possible.

61. $2x + 3x =$ _____   **62.** $-4 \cdot 1 =$ _____
(distributive property)   (identity property)

63. $2(4x) =$ _____   **64.** $-3 + 13 =$ _____
(associative property)   (commutative property)

65. $-3 + 3 =$ _____   **66.** $5(x + z) =$ _____
(inverse property)   (distributive property)

67. $0 + 7 =$ _____   **68.** $8 \cdot \dfrac{1}{8} =$ _____
(identity property)   (inverse property)

MIXED REVIEW EXERCISES*

The table gives U.S. exports and imports with Canada, in millions of U.S. dollars.

Year	Exports	Imports
2003	169,924	221,595
2004	189,880	256,360
2005	211,349	287,870

Source: U.S. Census Bureau.

Determine the absolute value of the difference between imports and exports for each year. Is the balance of trade (exports minus imports) in each year positive or negative?

69. 2003   **70.** 2004   **71.** 2005

Perform the indicated operations.

72. $\left(-\dfrac{4}{5}\right)^4$

73. $-\dfrac{5}{8}(-40)$

74. $-25\left(-\dfrac{4}{5}\right) + 3^3 - 32 \div \sqrt{4}$

75. $-8 + |-14| + |-3|$

76. $\dfrac{6 \cdot \sqrt{4} - 3 \cdot \sqrt{16}}{-2 \cdot 5 + 7(-3) - 10}$ **77.** $-\sqrt{25}$

78. $-\dfrac{10}{21} \div \left(-\dfrac{5}{14}\right)$

79. $0.8 - 4.9 - 3.2 + 1.14$ **80.** -3^2

81. $\dfrac{-38}{-19}$

---

* The order of exercises in this final group does not correspond to the order in which topics occur in the chapter. This random ordering should help you prepare for the chapter test in yet another way.

82. $-2(k - 1) + 3k - k$ **83.** $-\sqrt{-100}$ **84.** $-(3k - 4h)$

85. $-4.6(2.48)$ **86.** $-\dfrac{2}{3}(-15) + (2^4 - 8 \div 4)$

87. $-2x + 5 - 4x - 1$ **88.** $-\dfrac{2}{3} - \left(\dfrac{1}{6} - \dfrac{5}{9}\right)$

89. Evaluate $-m(3k^2 + 5m)$ if **(a)** $k = -4$ and $m = 2$ and **(b)** $k = \frac{1}{2}$ and $m = -\frac{3}{4}$.

90. *Concept Check*   To evaluate $(3 + 2)^2$, should you work within the parentheses first, or should you square 3 and square 2 and then add?

Chapter **1** TEST

View the complete solutions to all Chapter Test exercises on the Pass the Test CD.

1. Graph $\left\{-3, 0.75, \dfrac{5}{3}, 5, 6.3\right\}$ on a number line.

Let $A = \left\{-\sqrt{6}, -1, -0.5, 0, 3, \sqrt{25}, 7.5, \frac{24}{2}, \sqrt{-4}\right\}$. First simplify each element as needed, and then list the elements from A that belong to each set.

2. Whole numbers   **3.** Integers

4. Rational numbers   **5.** Real numbers

Write each set in interval notation and graph the interval.

6. $\{x \mid x < -3\}$ **7.** $\{y \mid -4 < y \le 2\}$

Perform the indicated operations.

8. $-6 + 14 + (-11) - (-3)$ **9.** $10 - 4 \cdot 3 + 6(-4)$

10. $7 - 4^2 + 2(6) + (-4)^2$ **11.** $\dfrac{10 - 24 + (-6)}{\sqrt{16}(-5)}$

12. $\dfrac{-2[3 - (-1 - 2) + 2]}{\sqrt{9}(-3) - (-2)}$ **13.** $\dfrac{8 \cdot 4 - 3^2 \cdot 5 - 2(-1)}{-3 \cdot 2^3 + 1}$

The table shows the heights in feet of some selected mountains and the depths in feet (as negative numbers) of some selected ocean trenches.

Mountain	Height	Trench	Depth
Foraker	17,400	Philippine	−32,995
Wilson	14,246	Cayman	−24,721
Pikes Peak	14,110	Java	−23,376

Source: World Almanac and Book of Facts 2006.

14. What is the difference between the height of Mt. Foraker and the depth of the Philippine Trench?

15. What is the difference between the height of Pikes Peak and the depth of the Java Trench?

16. How much deeper is the Cayman Trench than the Java Trench?

Find each square root. If the number is not real, say so.

17. $\sqrt{196}$ **18.** $-\sqrt{225}$ **19.** $\sqrt{-16}$

20. *Concept Check*   For the expression $\sqrt{a}$, under what conditions will its value be **(a)** positive,   **(b)** not real,   **(c)** 0?

21. Evaluate $\dfrac{8k + 2m^2}{r - 2}$ if $k = -3$, $m = -3$, and $r = 25$.

22. Simplify $-3(2k - 4) + 4(3k - 5) - 2 + 4k$.

23. How does the subtraction sign affect the terms $-4r$ and 6 when $(3r + 8) - (-4r + 6)$ is simplified? What is the simplified form?

Match each statement in Column I with the appropriate property in Column II. Answers may be used more than once.

I	II
24. $6 + (-6) = 0$	**A.** Distributive property
25. $-2 + (3 + 6) = (-2 + 3) + 6$	**B.** Inverse property
26. $5x + 15x = (5 + 15)x$	**C.** Identity property
27. $13 \cdot 0 = 0$	**D.** Associative property
28. $-9 + 0 = -9$	**E.** Commutative property
29. $4 \cdot 1 = 4$	**F.** Multiplication property of 0
30. $(a + b) + c = (b + a) + c$	

Linear Equations, Inequalities, and Applications

Television, first operational in the 1940s, has become the most widespread form of communication in the world. In 2005, 106.9 million homes—98% of all U.S. households—owned at least one TV set, and average viewing time among all viewers exceeded 31 hours per week. Favorite prime-time television programs were *CSI* and *American Idol*. (*Source:* Nielsen Media Research; *Microsoft Encarta Encyclopedia.*)

In Section 2.2, we discuss the concept of *percent*—one of the most common everyday applications of mathematics—and use it in Exercises 51–54 to determine additional information about television viewing in U.S. households.

2.1 Linear Equations in One Variable

OBJECTIVES

1 Decide whether a number is a solution of a linear equation.

2 Solve linear equations by using the addition and multiplication properties of equality.

3 Solve linear equations by using the distributive property.

4 Solve linear equations with fractions or decimals.

5 Identify conditional equations, contradictions, and identities.

In **Chapter 1,** we reviewed *algebraic expressions.* Examples include

$$8x + 9, \quad y - 4, \quad \text{and} \quad \frac{x^3 y^8}{z}. \qquad \text{Algebraic expressions}$$

Equations and inequalities compare algebraic expressions, just as a balance scale compares the weights of two quantities. Recall from **Section 1.1** that an *equation* is a statement that two algebraic expressions are equal. It is important to be able to distinguish between algebraic expressions and equations. *An equation always contains an equals sign, while an expression does not.*

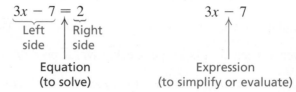

Equation
(to solve)

Expression
(to simplify or evaluate)

✔ **Now Try Exercise 5.**

A *linear equation in one variable* involves only real numbers and one variable raised to the first power. Examples include

$$x + 1 = -2, \qquad x - 3 = 5, \qquad \text{and} \qquad 2k + 5 = 10. \qquad \text{Linear equations}$$

> **Linear Equation in One Variable**
>
> A **linear equation in one variable** can be written in the form
> $$Ax + B = C,$$
> where A, B, and C are real numbers, with $A \neq 0$.

A linear equation is a **first-degree equation,** since the greatest power on the variable is 1. Some equations that are not linear (that is, *nonlinear*) are

$$x^2 + 3y = 5, \quad \frac{8}{x} = -22, \quad \text{and} \quad \sqrt{x} = 6. \qquad \text{Nonlinear equations}$$

OBJECTIVE 1 Decide whether a number is a solution of a linear equation. If the variable in an equation can be replaced by a real number that makes the statement true, then that number is a **solution** of the equation. For example, 8 is a solution of the equation $x - 3 = 5$, since replacing x with 8 gives a true statement. An equation is *solved* by finding its **solution set,** the set of all solutions. The solution set of the equation $x - 3 = 5$ is $\{8\}$.

Equivalent equations are related equations that have the same solution set. To solve an equation, we usually start with the given equation and replace it with a series of simpler equivalent equations. For example,

$$5x + 2 = 17, \quad 5x = 15, \quad \text{and} \quad x = 3 \qquad \text{Equivalent equations}$$

are all equivalent, since each has the solution set $\{3\}$.

OBJECTIVE ② Solve linear equations by using the addition and multiplication properties of equality. We use two important properties to produce equivalent equations.

Addition and Multiplication Properties of Equality

Addition Property of Equality

For all real numbers A, B, and C, the equations

$$A = B \quad \text{and} \quad A + C = B + C$$

are equivalent.

   That is, the same number may be added to each side of an equation without changing the solution set.

Multiplication Property of Equality

For all real numbers A and B, and for $C \neq 0$, the equations

$$A = B \quad \text{and} \quad AC = BC$$

are equivalent.

   That is, each side of an equation may be multiplied by the same nonzero number without changing the solution set.

Because subtraction and division are defined in terms of addition and multiplication, respectively, the preceding properties can be extended:

   The same number may be subtracted from each side of an equation, and each side of an equation may be divided by the same nonzero number, without changing the solution set.

EXAMPLE 1 Using the Properties of Equality to Solve a Linear Equation

Solve $4x - 2x - 5 = 4 + 6x + 3$.

   The goal is to isolate x on one side of the equation.

$4x - 2x - 5 = 4 + 6x + 3$	
$2x - 5 = 7 + 6x$	Combine like terms.
$2x - 5 + 5 = 7 + 6x + 5$	Add 5 to each side.
$2x = 12 + 6x$	Combine like terms.
$2x - 6x = 12 + 6x - 6x$	Subtract 6x from each side.
$-4x = 12$	Combine like terms.
$\dfrac{-4x}{-4} = \dfrac{12}{-4}$	Divide each side by −4.
$x = -3$	

To be sure that -3 is the solution, check by substituting for x in the *original* equation.

Check: $4x - 2x - 5 = 4 + 6x + 3$ Original equation

$4(-3) - 2(-3) - 5 = 4 + 6(-3) + 3$?   Let $x = -3$.

$-12 + 6 - 5 = 4 - 18 + 3$?   Multiply.

Use parentheses around substituted values to avoid errors.

$-11 = -11$ True

This is *not* the solution.

The true statement indicates that $\{-3\}$ is the solution set.

✔ **Now Try Exercise 15.**

▶ **CAUTION**   Notice in Example 1 that the equality symbols are aligned in a column. *Do not use more than one equality symbol in a horizontal line of work when solving an equation.*

We use the following steps to solve a linear equation in one variable.

Solving a Linear Equation in One Variable

Step 1   **Clear fractions.** Eliminate any fractions by multiplying each side by the least common denominator.

Step 2   **Simplify each side separately.** Use the distributive property to clear parentheses and combine like terms as needed.

Step 3   **Isolate the variable terms on one side.** Use the addition property to get all terms with variables on one side of the equation and all numbers on the other.

Step 4   **Isolate the variable.** Use the multiplication property to get an equation with just the variable (with coefficient 1) on one side.

Step 5   **Check.** Substitute the proposed solution into the original equation.

OBJECTIVE 3   Solve linear equations by using the distributive property.   In Example 1, we did not use Step 1 or the distributive property in Step 2 as given in the box. Many equations, however, will require one or both of these steps.

EXAMPLE 2   Using the Distributive Property to Solve a Linear Equation

Solve $2(k - 5) + 3k = k + 6$.

Step 1   Since there are no fractions in this equation, Step 1 does not apply.

Step 2   Use the distributive property to simplify and combine terms on the left.

Be sure to distribute over *all* terms within the parentheses.

$2(k - 5) + 3k = k + 6$

$2k + 2(-5) + 3k = k + 6$ Distributive property

$2k - 10 + 3k = k + 6$ Multiply.

$5k - 10 = k + 6$ Combine like terms.

Step 3 Next, use the addition property of equality.

$$5k - 10 + 10 = k + 6 + 10 \qquad \text{Add 10.}$$
$$5k = k + 16 \qquad \text{Combine like terms.}$$
$$5k - k = k + 16 - k \qquad \text{Subtract } k.$$
$$4k = 16 \qquad \text{Combine like terms.}$$

Step 4 Use the multiplication property of equality to isolate k on the left.

$$\frac{4k}{4} = \frac{16}{4} \qquad \text{Divide by 4.}$$
$$k = 4$$

Step 5 Check by substituting 4 for k in the original equation.

$$\begin{aligned}
\textit{Check:} \quad 2(k - 5) + 3k &= k + 6 && \text{Original equation} \\
2(4 - 5) + 3(4) &= 4 + 6 \quad ? && \text{Let } k = 4. \\
2(-1) + 12 &= 10 \quad ? \\
10 &= 10 && \text{True}
\end{aligned}$$

Always check your work.

The solution checks, so $\{4\}$ is the solution set.

✔ **Now Try Exercise 19.**

OBJECTIVE ④ Solve linear equations with fractions or decimals. When fractions or decimals appear as coefficients in equations, our work can be made easier if we multiply each side of the equation by the least common denominator (LCD) of all the fractions. This is an application of the multiplication property of equality, and it produces an equivalent equation with integer coefficients.

EXAMPLE 3 Solving a Linear Equation with Fractions

Solve $\dfrac{x + 7}{6} + \dfrac{2x - 8}{2} = -4$.

Start by eliminating the fractions. Multiply both sides by the LCD, 6.

Step 1
$$6\left(\frac{x + 7}{6} + \frac{2x - 8}{2}\right) = 6(-4) \qquad \text{The LCD is 6.}$$

Step 2
$$6\left(\frac{x + 7}{6}\right) + 6\left(\frac{2x - 8}{2}\right) = 6(-4) \qquad \text{Distributive property}$$

$$\frac{6(x + 7)}{6} + \frac{6(2x - 8)}{2} = -24 \qquad \text{Multiply; } 6 = \tfrac{6}{1}$$

$$x + 7 + 3(2x - 8) = -24$$
$$x + 7 + 3(2x) + 3(-8) = -24 \qquad \text{Distributive property}$$
$$x + 7 + 6x - 24 = -24 \qquad \text{Multiply.}$$
$$7x - 17 = -24 \qquad \text{Combine like terms.}$$

Step 3
$$7x - 17 + 17 = -24 + 17 \qquad \text{Add 17.}$$
$$7x = -7 \qquad \text{Combine like terms.}$$

Step 4
$$\frac{7x}{7} = \frac{-7}{7}$$ Divide by 7.

$$x = -1$$

Step 5   *Check* by substituting -1 for x in the original equation.

$$\frac{x+7}{6} + \frac{2x-8}{2} = -4$$

$$\frac{-1+7}{6} + \frac{2(-1)-8}{2} = -4 \quad ? \quad \text{Let } x = -1.$$

$$\frac{6}{6} + \frac{-10}{2} = -4 \quad ?$$

$$1 - 5 = -4 \quad ?$$

$$-4 = -4 \qquad \text{True}$$

The solution checks, so the solution set is $\{-1\}$.

✔ **Now Try Exercise 59.**

In **Sections 2.2 and 2.3,** we solve problems dealing with interest rates and concentrations of solutions. These problems involve percents that are converted to decimals. The equations that are used to solve such problems have decimal coefficients. We can clear these decimals by multiplying by a power of 10, such as

$$10^1 = 10, \quad 10^2 = 100, \quad \text{and so on.}$$

This allows us to obtain integer coefficients.

EXAMPLE 4   **Solving a Linear Equation with Decimals**

Solve $0.06x + 0.09(15 - x) = 0.07(15)$.

Since each decimal number is given in hundredths, multiply both sides of the equation by 100. A number can be multiplied by 100 by moving the decimal point two places to the right. To multiply the second term, $0.09(15 - x)$, by 100—that is, $100(0.09)(15 - x)$—remember the associative property: To multiply three terms, first multiply any two of them. Here we multiply $100(0.09)$ first to get 9, so the product $100(0.09)(15 - x)$ becomes $9(15 - x)$.

$$0.06x + 0.09(15 - x) = 0.07(15)$$

$$0.06x + 0.09(15 - x) = 0.07(15) \qquad \text{Multiply each term by 100.}$$

Move decimal points 2 places to the right.

$$6x + 9(15 - x) = 7(15)$$

$$6x + 9(15) - 9x = 105 \qquad \text{Distributive property; multiply.}$$

$$-3x + 135 = 105 \qquad \text{Combine like terms; multiply.}$$

$$-3x + 135 - 135 = 105 - 135 \qquad \text{Subtract 135.}$$

$$-3x = -30 \qquad \text{Combine like terms.}$$

$$\frac{-3x}{-3} = \frac{-30}{-3} \qquad \text{Divide by } -3.$$

$$x = 10$$

Check:
$$0.06x + 0.09(15 - x) = 0.07(15)$$
$$0.06(10) + 0.09(15 - 10) = 0.07(15) \qquad ? \qquad \text{Let } x = 10.$$
$$0.6 + 0.09(5) = 1.05 \qquad ?$$
$$0.6 + 0.45 = 1.05 \qquad ?$$
$$1.05 = 1.05 \qquad \text{True}$$

A true statement results, so the solution set is $\{10\}$.

✔ **Now Try Exercise 63.**

OBJECTIVE 5 Identify conditional equations, contradictions, and identities. In Examples 1–4, all of the equations had solution sets containing *one* element, such as $\{10\}$ in Example 4. Some equations, however, have no solutions, while others have an infinite number of solutions. The table gives the names of these types of equations.

Type of Linear Equation	Number of Solutions	Indication when Solving
Conditional	One	Final line is $x = $ a number. (See Example 5(a).)
Identity	Infinite; solution set {all real numbers}	Final line is true, such as $0 = 0$. (See Example 5(b).)
Contradiction	None; solution set ∅	Final line is false, such as $-15 = -20$. (See Example 5(c).)

EXAMPLE 5 Recognizing Conditional Equations, Identities, and Contradictions

Solve each equation. Decide whether it is a *conditional equation*, an *identity*, or a *contradiction*.

(a)
$$5x - 9 = 4(x - 3)$$
$$5x - 9 = 4x - 12 \qquad \text{Distributive property}$$
$$5x - 9 - 4x = 4x - 12 - 4x \qquad \text{Subtract } 4x.$$
$$x - 9 = -12 \qquad \text{Combine like terms.}$$
$$x - 9 + 9 = -12 + 9 \qquad \text{Add 9.}$$
$$x = -3$$

The solution set, $\{-3\}$, has only one element, so $5x - 9 = 4(x - 3)$ is a conditional equation.

(b)
$$5x - 15 = 5(x - 3)$$
$$5x - 15 = 5x - 15 \qquad \text{Distributive property}$$
$$5x - 15 - 5x + 15 = 5x - 15 - 5x + 15 \qquad \text{Subtract } 5x; \text{ add 15.}$$
$$0 = 0 \qquad \text{True}$$

The final line, $0 = 0$, indicates that the solution set is {all real numbers}, and the equation $5x - 15 = 5(x - 3)$ is an identity. (The first step yielded $5x - 15 = 5x - 15$, which is *true* for all values of x. We could have identified the equation as an identity at that point.)

(c)
$$5x - 15 = 5(x - 4)$$
$$5x - 15 = 5x - 20 \qquad \text{Distributive property}$$
$$5x - 15 - 5x = 5x - 20 - 5x \qquad \text{Subtract } 5x.$$
$$-15 = -20 \qquad \text{False}$$

Since the result, $-15 = -20$, is *false,* the equation has no solution. The solution set is $\emptyset$, so the equation $5x - 15 = 5(x - 4)$ is a contradiction.

✔ **Now Try Exercises 17, 23, and 33.**

2.1 EXERCISES

⊕ *Complete solution available on Video Lectures on CD/DVD*

▨ *Now Try Exercise*

1. *Concept Check*   Which equations are linear equations in x?

   A. $3x + x - 1 = 0$ **B.** $8 = x^2$ **C.** $6x + 2 = 9$ **D.** $\dfrac{1}{2}x - \dfrac{1}{x} = 0$

▨ 2. Which of the equations in Exercise 1 are nonlinear equations in x? Explain why.

▨ 3. Decide whether 6 is a solution of $3(x + 4) = 5x$ by substituting 6 for x. If it is not a solution, explain why.

▨ 4. Use substitution to decide whether -2 is a solution of $5(x + 4) - 3(x + 6) = 9(x + 1)$. If it is not a solution, explain why.

⊕ 5. *Concept Check*   Identify each of the following as an *expression* or an *equation.*

   (a) $3x = 6$ **(b)** $3x + 6$
   (c) $5x + 6(x - 3) = 12x + 6$ **(d)** $5x + 6(x - 3) - (12x + 6)$

▨ 6. In Example 1, a student looked at the check and thought that $\{-11\}$ should be given as the solution set. Explain why this is not correct.

7. *Concept Check*   The following "solution" contains a common student error.

   $$8x - 2(2x - 3) = 3x + 7$$
   $$8x - 4x - 6 = 3x + 7 \qquad \text{Distributive property}$$
   $$4x - 6 = 3x + 7 \qquad \text{Combine like terms.}$$
   $$x = 13 \qquad \text{Subtract } 3x; \text{ add 6.}$$

   WHAT WENT WRONG? Find the correct solution.

8. *Concept Check*   When clearing parentheses in the expression

   $$-5m - (2m - 4) + 5$$

   on the right side of the equation in Exercise 39, the $-$ sign before the parentheses acts like a factor representing what number? Clear parentheses and simplify this expression.

▨ 9. Explain the distinction between a conditional equation, an identity, and a contradiction.

10. *Concept Check*   A student tried to solve the equation $8x = 7x$ by dividing each side by x, obtaining $8 = 7$. He gave the solution set as $\emptyset$. **WHAT WENT WRONG?**

Solve each equation, and check your solution. If applicable, tell whether the equation is an identity or a contradiction. See Examples 1, 2, and 5.

11. $7x + 8 = 1$ **12.** $5x - 4 = 21$ **13.** $5x + 2 = 3x - 6$

14. $9p + 1 = 7p - 9$ ◐ **15.** $7x - 5x + 15 = x + 8$ **16.** $2x + 4 - x = 4x - 5$

◐ **17.** $12w + 15w - 9 + 5 = -3w + 5 - 9$ **18.** $-4t + 5t - 8 + 4 = 6t - 4$

◐ **19.** $3(2t - 4) = 20 - 2t$ **20.** $2(3 - 2x) = (x - 4)$

21. $-5(x + 1) + 3x + 2 = 6x + 4$ **22.** $5(x + 3) + 4x - 5 = 4 - 2x$

23. $-2x + 5x - 9 = 3(x - 4) - 5$ **24.** $-6x + 2x - 11 = -2(2x - 3) + 4$

25. $2(x + 3) = -4(x + 1)$ **26.** $4(t - 9) = 8(t + 3)$

27. $3(2w + 1) - 2(w - 2) = 5$ **28.** $4(x - 2) + 2(x + 3) = 6$

29. $2x + 3(x - 4) = 2(x - 3)$ **30.** $6x - 3(5x + 2) = 4(1 - x)$

31. $6p - 4(3 - 2p) = 5(p - 4) - 10$ **32.** $-2k - 3(4 - 2k) = 2(k - 3) + 2$

33. $-2(t + 3) - t - 4 = -3(t + 4) + 2$ **34.** $4(2d + 7) = 2d + 25 + 3(2d + 1)$

35. $2[w - (2w + 4) + 3] = 2(w + 1)$ **36.** $4[2t - (3 - t) + 5] = -(2 + 7t)$

37. $-[2z - (5z + 2)] = 2 + (2z + 7)$ **38.** $-[6x - (4x + 8)] = 9 + (6x + 3)$

39. $-3m + 6 - 5(m - 1) = -5m - (2m - 4) + 5$

40. $4(k + 2) - 8k - 5 = -3k + 9 - 2(k + 6)$

41. $7[2 - (3 + 4x)] - 2x = -9 + 2(1 - 15x)$

42. $4[6 - (1 + 2x)] + 10x = 2(10 - 3x) + 8x$

43. $-[3x - (2x + 5)] = -4 - [3(2x - 4) - 3x]$

44. $2[-(x - 1) + 4] = 5 + [-(6x - 7) + 9x]$

45. *Concept Check*   To solve the linear equation

$$\frac{8x}{3} - \frac{5x}{4} = -13,$$

we multiply each side by the least common denominator of all the fractions in the equation. What is this least common denominator?

✐ **46.** Suppose that in solving the equation

$$\frac{1}{3}x + \frac{1}{2}x = \frac{1}{6}x,$$

we begin by multiplying each side by 12, rather than the *least* common denominator, 6. Would we get the correct solution? Explain.

47. *Concept Check*   To solve a linear equation with decimals, we usually begin by multiplying by a power of 10 so that all coefficients are integers. What is the least power of 10 that will accomplish this goal in each equation?

    (a) $0.05x + 0.12(x + 5000) = 940$ (Exercise 63)
    (b) $0.006(x + 2) = 0.007x + 0.009$ (Exercise 69)

48. *Concept Check* The expression $0.06(10 - x)(100)$ is equivalent to which of the following?

    A. $0.06 - 0.06x$ **B.** $60 - 6x$ **C.** $6 - 6x$ **D.** $6 - 0.06x$

Solve each equation, and check your solution. See Examples 3 and 4.

49. $-\dfrac{5}{9}k = 2$

50. $\dfrac{3}{11}z = -5$

51. $\dfrac{6}{5}x = -1$

52. $-\dfrac{7}{8}r = 6$

53. $\dfrac{m}{2} + \dfrac{m}{3} = 5$

54. $\dfrac{x}{5} - \dfrac{x}{4} = 1$

55. $\dfrac{3x}{4} + \dfrac{5x}{2} = 13$

56. $\dfrac{8x}{3} - \dfrac{2x}{4} = -13$

57. $\dfrac{x - 10}{5} + \dfrac{2}{5} = -\dfrac{x}{3}$

58. $\dfrac{2r - 3}{7} + \dfrac{3}{7} = -\dfrac{r}{3}$

59. $\dfrac{3x - 1}{4} + \dfrac{x + 3}{6} = 3$

60. $\dfrac{3x + 2}{7} - \dfrac{x + 4}{5} = 2$

61. $\dfrac{4t + 1}{3} = \dfrac{t + 5}{6} + \dfrac{t - 3}{6}$

62. $\dfrac{2x + 5}{5} = \dfrac{3x + 1}{2} + \dfrac{-x + 7}{2}$

63. $0.05x + 0.12(x + 5000) = 940$

64. $0.09k + 0.13(k + 300) = 61$

65. $0.02(50) + 0.08r = 0.04(50 + r)$

66. $0.20(14{,}000) + 0.14t = 0.18(14{,}000 + t)$

67. $0.05x + 0.10(200 - x) = 0.45x$

68. $0.08x + 0.12(260 - x) = 0.48x$

69. $0.006(x + 2) = 0.007x + 0.009$

70. $0.004x + 0.006(50 - x) = 0.004(68)$

Most of the exercise sets in the rest of the book end with brief sets of "Preview Exercises." These exercises are designed to help you review *ideas introduced earlier, as well as* preview *ideas needed for the next section. If you need help with these Preview Exercises, look in the section or sections indicated.*

PREVIEW EXERCISES

*Use the given value(s) to evaluate each expression. See **Section 1.3.***

71. $2L + 2W$; $L = 10, W = 8$

72. rt; $r = 0.15, t = 3$

73. $\dfrac{1}{3}Bh$; $B = 27, h = 8$

74. prt; $p = 8000, r = 0.06, t = 2$

75. $\dfrac{5}{9}(F - 32)$; $F = 122$

76. $\dfrac{9}{5}C + 32$; $C = 60$

77. $\dfrac{1}{2}h(b + B)$; $B = 9, b = 4, h = 3$

78. $\dfrac{1}{2}bh$; $b = 21, h = 7$

2.2   Formulas

A **mathematical model** is an equation or inequality that describes a real situation. Models for many applied problems already exist; they are called *formulas*. A **formula** is an equation in which variables are used to describe a relationship. Some formulas that we will be using are

$$d = rt, \qquad I = prt, \qquad \text{and} \qquad P = 2L + 2W. \qquad \text{Formulas}$$

Some common formulas used in algebra are given inside the covers of this book.

OBJECTIVE ❶ Solve a formula for a specified variable.   The formula $I = prt$ says that interest on a loan or investment equals principal (amount borrowed or invested), times rate (annually, in percent), times time (in years) at the given interest rate. To determine how long it will take for an investment at a stated interest rate to earn a predetermined amount of interest, it would help to first solve the formula for t. This process is called **solving for a specified variable** or **solving a literal equation.**

The steps used in the examples that follow are similar to those used in solving linear equations in **Section 2.1**. *When you are solving for a specified variable, the key is to treat that variable as if it were the only one; treat all other variables like numbers (constants).*

The following additional suggestions may be helpful.

Solving for a Specified Variable

Step 1   Transform so that all terms containing the specified variable are on one side of the equation and all terms without that variable are on the other side.

Step 2   If necessary, use the distributive property to combine the terms with the specified variable.* The result should be the product of a sum or difference and the variable.

Step 3   Divide both sides by the factor that is the coefficient of the specified variable.

EXAMPLE 1   Solving for a Specified Variable

Solve the formula $P = 2L + 2W$ for W.

This formula gives the relationship between the perimeter P (distance around) a rectangle, the length L of the rectangle, and the width W of the rectangle. See Figure 1.

Solve the formula for W by isolating W on one side of the equals sign.

$$P = 2L + 2W$$

Step 1 $\quad P - 2L = 2L + 2W - 2L \qquad$ Subtract $2L$ from both sides.

$\qquad\qquad\quad P - 2L = 2W \qquad\qquad\qquad$ Combine like terms.

Perimeter P, the distance around a rectangle, is given by
$$P = 2L + 2W.$$

FIGURE 1

---

*Using the distributive property to write $ab + ac$ as $a(b + c)$ is called *factoring*. See **Chapter 6.**

Step 2 is not needed here.

Step 3
$$\frac{P - 2L}{2} = \frac{2W}{2} \qquad \text{Divide both sides by 2.}$$

$$\frac{P - 2L}{2} = W, \quad \text{or} \quad W = \frac{P - 2L}{2}$$

✔ **Now Try Exercise 9.**

▶ **CAUTION** In Step 3 of Example 1, do not simplify the fraction by dividing 2 into the term $2L$. The subtraction in the numerator must be done before the division.

$$\frac{P - 2L}{2} \neq P - L$$

EXAMPLE 2 Solving a Formula Having Parentheses

The formula for the perimeter of a rectangle is sometimes written in the equivalent form $P = 2(L + W)$. Solve this form for W.

One way to begin is to use the distributive property on the right side of the equation to get $P = 2L + 2W$, which we would then solve as in Example 1. Another way to begin is to divide by the coefficient 2.

$$P = 2(L + W)$$

$$\frac{P}{2} = L + W \qquad \text{Divide both sides by 2.}$$

$$\frac{P}{2} - L = W, \quad \text{or} \quad W = \frac{P}{2} - L \qquad \text{Subtract } L \text{ from both sides.}$$

We can show that this result is equivalent to our result in Example 1 by multiplying L by $\frac{2}{2}$.

$$\frac{P}{2} - \frac{2}{2}(L) = W \qquad \tfrac{2}{2} = 1, \text{ so } L = \tfrac{2}{2}(L).$$

$$\frac{P}{2} - \frac{2L}{2} = W$$

$$\frac{P - 2L}{2} = W \qquad \text{Subtract fractions: } \tfrac{a}{c} - \tfrac{b}{c} = \tfrac{a - b}{c}.$$

The final line agrees with the result in Example 1.

✔ **Now Try Exercise 15.**

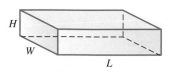

Rectangular solid

$A = 2HW + 2LW + 2LH$

FIGURE 2

A rectangular solid has the shape of a box, but is solid. See Figure 2. The labels H, W, and L represent the height, width, and length of the figure, respectively. The surface area of any solid three-dimensional figure is the total area of its surface. For a rectangular solid, the surface area A is

$$A = 2HW + 2LW + 2LH.$$

> **EXAMPLE 3** Using the Distributive Property to Solve for a Specified Variable

Solve the formula $A = 2HW + 2LW + 2LH$ for L.

To solve for the length L, treat L as the only variable and treat all other variables as constants.

We must isolate the L-terms.

$$A = 2HW + 2LW + 2LH$$

$$A - 2HW = 2LW + 2LH \qquad \text{Subtract } 2HW.$$

$$A - 2HW = L(2W + 2H) \qquad \text{Distributive property}$$

This is a key step.

$$\frac{A - 2HW}{2W + 2H} = L, \quad \text{or} \quad L = \frac{A - 2HW}{2W + 2H} \qquad \text{Divide by } 2W + 2H.$$

✔ **Now Try Exercise 21.**

▶ **CAUTION** The most common error in working a problem like Example 3 is not using the distributive property correctly. We must write the expression so that the specified variable is a *factor*; then we can divide by its coefficient in the final step.

OBJECTIVE 2 Solve applied problems by using formulas. The distance formula $d = rt$ relates d, the distance traveled, r, the rate or speed, and t, the travel time.

> **EXAMPLE 4** Finding Average Speed

Janet Branson found that usually it took her $\frac{3}{4}$ hr each day to drive a distance of 15 mi to work. What was her speed?

Find the formula for speed (rate) r by solving $d = rt$ for r.

$$d = rt$$

$$\frac{d}{t} = \frac{rt}{t} \qquad \text{Divide by } t.$$

$$\frac{d}{t} = r, \quad \text{or} \quad r = \frac{d}{t}$$

Notice that only Step 3 was needed to solve for r in this example. Now find Janet's speed by substituting the given values of d and t into the last formula.

$$r = \frac{d}{t}$$

$$r = \frac{15}{\frac{3}{4}} \qquad \text{Let } d = 15, t = \frac{3}{4}.$$

$$r = 15 \cdot \frac{4}{3} \qquad \text{Multiply by the reciprocal of } \frac{3}{4}.$$

$$r = 20$$

Her speed averaged 20 mph. (That is, at times she may have traveled a little faster or a little slower than 20 mph, but overall, her speed was 20 mph.)

✔ **Now Try Exercise 25.**

▶ **PROBLEM-SOLVING HINT**   As seen in Example 4, it may be convenient to first solve for a specific unknown variable *before* substituting the given values. This is particularly useful when we wish to substitute several different values for the same variable. For example, an economics class might need to solve the equation $I = prt$ for r to find rates that produce specific amounts of interest for various principals and times.

OBJECTIVE 3 Solve percent problems.   An important everyday use of mathematics involves the concept of percent. Percent is written with the symbol %. The word **percent** means "per one hundred." One percent means "one per one hundred" or "one one-hundredth."

$$1\% = 0.01 \qquad \text{or} \qquad 1\% = \frac{1}{100}$$

The following formula can be used to solve a percent problem:

$$\frac{\textbf{partial amount}}{\textbf{whole amount}} = \textbf{percent (represented as a decimal).}$$

For example, if a class consists of 50 students, and 32 are males, then the percent of males in the class is

$$\frac{\text{partial amount}}{\text{whole amount}} = \frac{32}{50} = 0.64, \quad \text{or} \quad 64\%.$$

EXAMPLE 5 Solving Percent Problems

(a) A 50-L mixture of acid and water contains 10 L of acid. What is the percent of acid in the mixture?

   The amount of the mixture is 50 L, and the part that is acid is 10 L. Let x represent the percent of acid. Then the percent of acid in the mixture is

$$x = \frac{10}{50} \begin{matrix} \leftarrow \text{partial amount} \\ \leftarrow \text{whole amount} \end{matrix}$$

$$x = 0.20, \quad \text{or} \quad 20\%.$$

(b) If a savings account balance of $3550 earns 6% interest in one year, how much interest is earned?

   Let x represent the amount of interest earned (that is, the part of the whole amount invested). Since $6\% = 6 \cdot 0.01 = 0.06$, the equation is

$$\frac{x}{3550} = 0.06 \qquad \tfrac{\text{partial amount}}{\text{whole amount}} = \text{percent}$$

$$x = 0.06(3550) \qquad \text{Multiply by 3550.}$$

$$x = 213.$$

The interest earned is $213.

✔ **Now Try Exercises 37 and 39.**

EXAMPLE 6 Interpreting Percents from a Graph

In 2005, people in the United States spent an estimated $35.9 billion on their pets. Use the graph in Figure 3 to determine how much of this amount was spent on pet food.

Spending On Kitty And Rover

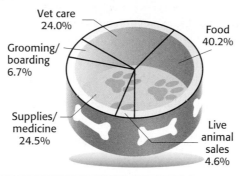

Vet care
24.0%

Food
40.2%

Grooming/
boarding
6.7%

Supplies/
medicine
24.5%

Live
animal
sales
4.6%

Source: American Pet Products Manufacturers
Association, Inc.

FIGURE 3

According to the graph, 40.2% was spent on food. Let x represent this amount in billions of dollars.

$$\frac{x}{35.9} = 0.402 \qquad 40.2\% = 0.402$$

$$x = 0.402(35.9) \qquad \text{Multiply by 35.9.}$$

$$x = 14.4 \qquad \text{Nearest tenth}$$

About $14.4 billion was spent on pet food.

✔ **Now Try Exercise 47.**

2.2 EXERCISES

🌐 *Complete solution available on Video Lectures on CD/DVD*

Now Try Exercise

RELATING CONCEPTS (EXERCISES 1–6)

FOR INDIVIDUAL OR GROUP WORK

Consider the following equations:

First Equation	*Second Equation*
$x = \dfrac{5x + 8}{3}$	$t = \dfrac{bt + k}{c} \quad (c \neq 0).$

Solving the second equation for t requires the same logic as solving the first equation for x. When solving for t, we treat all other variables as though they were constants. **Work Exercises 1–6 in order,** *to see the "parallel logic" of solving for x and solving for t.*

1. (a) Clear the first equation of fractions by multiplying each side by 3.
   (b) Clear the second equation of fractions by multiplying each side by c.

(continued)

2. (a) Get the terms involving x on the left side of the first equation by subtracting $5x$ from each side.
   (b) Get the terms involving t on the left side of the second equation by subtracting bt from each side.

3. (a) Combine like terms on the left side of the first equation. What property allows us to write $3x - 5x$ as $(3 - 5)x = -2x$?
   (b) Write the expression on the left side of the second equation so that t is a factor. What property allows us to do this?

4. (a) Divide each side of the first equation by the coefficient of x.
   (b) Divide each side of the second equation by the coefficient of t.

5. Look at your answer for the second equation. What restriction must be placed on the variables? Why is this necessary?

✎ **6.** Write a short paragraph summarizing what you have learned in this group of exercises.

Solve each formula for the specified variable. See Examples 1 and 2.

7. $I = prt$ for r (simple interest)

8. $d = rt$ for t (distance)

🌐 **9.** $P = 2L + 2W$ for L
(perimeter of a rectangle)

10. $A = bh$ for b (area of a parallelogram)

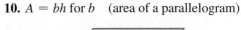

11. $V = LWH$
(volume of a rectangular solid)
   (a) for W **(b)** for H

12. $P = a + b + c$
(perimeter of a triangle)
   (a) for b **(b)** for c

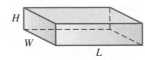

13. $C = 2\pi r$ for r
(circumference of a circle)

14. $A = \dfrac{1}{2}bh$ for h
(area of a triangle)

15. $A = \dfrac{1}{2}h(b + B)$ (area of a trapezoid)
   (a) for h **(b)** for B

16. $S = 2\pi rh + 2\pi r^2$ for h
(surface area of a right circular cylinder)

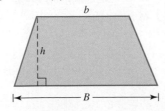

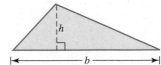

17. $F = \dfrac{9}{5}C + 32$ for C

(Celsius to Fahrenheit)

18. $C = \dfrac{5}{9}(F - 32)$ for F

(Fahrenheit to Celsius)

19. *Concept Check* When a formula is solved for a particular variable, several different equivalent forms may be possible. If we solve $A = \frac{1}{2}bh$ for h, one possible correct answer is

$$h = \frac{2A}{b}.$$

Which one of the following is *not* equivalent to this?

A. $h = 2\left(\dfrac{A}{b}\right)$ **B.** $h = 2A\left(\dfrac{1}{b}\right)$ **C.** $h = \dfrac{A}{\frac{1}{2}b}$ **D.** $h = \dfrac{\frac{1}{2}A}{b}$

20. *Concept Check* Suppose a student solved the formula $A = 2HW + 2LW + 2LH$ for L as follows:

$$A = 2HW + 2LW + 2LH$$
$$A - 2LW - 2HW = 2LH$$
$$\frac{A - 2LW - 2HW}{2H} = L.$$

WHAT WENT WRONG?

Solve each equation for the specified variable. See Example 3.

21. $2k + ar = r - 3y$ for r

22. $4s + 7p = tp - 7$ for p

23. $w = \dfrac{3y - x}{y}$ for y

24. $c = \dfrac{-2t + 4}{t}$ for t

Solve each problem. See Example 4.

25. In 2005, Jeff Gordon won the Daytona 500 (mile) race with a speed of 135.173 mph. Find his time to the nearest thousandth. (*Source: World Almanac and Book of Facts 2006.*)

26. In 2004, rain shortened the Indianapolis 500 race to 450 mi. It was won by Buddy Rice, who averaged 138.518 mph. What was his time to the nearest thousandth? (*Source:* indy500.com)

27. Faye Korn traveled from Kansas City to Louisville, a distance of 520 mi, in 10 hr. Find her rate in miles per hour.

28. The distance from Melbourne to London is 10,500 mi. If a jet averages 500 mph between the two cities, what is its travel time in hours?

29. As of 2005, the highest temperature ever recorded in Chicago was 40°C. Find the corresponding Fahrenheit temperature. (*Source: World Almanac and Book of Facts 2006.*)

30. As of 2005, the lowest temperature ever recorded in Memphis was $-13°$F. Find the corresponding Celsius temperature. (*Source: World Almanac and Book of Facts 2006.*)

31. The base of the Great Pyramid of Cheops is a square whose perimeter is 920 m. What is the length of each side of this square? (*Source: Atlas of Ancient Archaeology.*)

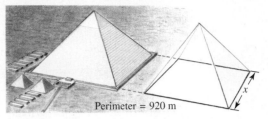

Perimeter = 920 m

32. Marina City in Chicago is a complex of two residential towers that resemble corncobs. Each tower has a concrete cylindrical core with a 35-ft diameter and is 588 ft tall. Find the volume of the core of one of the towers to the nearest whole number. (*Hint:* Use the π key on your calculator.) (*Source:* www.architechgallery.com; www.aviewoncities.com)

33. The circumference of a circle is 480π in. What is the radius of the circle? What is its diameter?

34. The radius of a circle is 2.5 in. What is the diameter of the circle? What is its circumference?

35. A sheet of standard-size copy paper measures 8.5 in. by 11 in. If a ream (500 sheets) of this paper has a volume of 187 in.^{3}, how thick is the ream?

36. Copy paper (Exercise 35) also comes in legal size, which has the same width, but is longer than standard size. If a ream of legal-size copy paper has the same thickness as the standard-size paper and a volume of 238 in.^{3}, what is the length of a sheet of legal paper?

Solve each problem. See Example 5.

🌐 **37.** A mixture of alcohol and water contains a total of 36 oz of liquid. There are 9 oz of pure alcohol in the mixture. What percent of the mixture is water? What percent is alcohol?

38. A mixture of acid and water is 35% acid. If the mixture contains a total of 40 L, how many liters of pure acid are in the mixture? How many liters of pure water are in the mixture?

39. A real-estate agent earned $6300 commission on a property sale of $210,000. What is her rate of commission?

40. A certificate of deposit for 1 yr pays $221 simple interest on a principal of $3400. What is the interest rate being paid on this deposit?

*When a consumer loan is paid off ahead of schedule, the finance charge is less than if the loan were paid off over its scheduled life. By one method, called the **rule of 78,** the amount of unearned interest (the finance charge that need not be paid) is given by*

$$u = f \cdot \frac{k(k+1)}{n(n+1)},$$

where u is the amount of unearned interest (money saved) when a loan scheduled to run for n payments is paid off k payments ahead of schedule. The total scheduled finance charge is f.

Use the formula for the rule of 78 to solve Exercises 41–44.

41. Rhonda Alessi bought a new Ford and agreed to pay it off in 36 monthly payments. The total finance charge was $700. Find the unearned interest if she paid the loan off 4 payments ahead of schedule.

42. Charles Vosburg bought a car and agreed to pay it off in 36 monthly payments. The total finance charge on the loan was $600. With 12 payments remaining, Charles decided to pay the loan in full. Find the amount of unearned interest.

43. The finance charge on a loan taken out by Vic Denicola is $380.50. If 24 equal monthly installments were needed to repay the loan, and the loan is paid in full with 8 months remaining, find the amount of unearned interest.

44. Adrian Ortega is scheduled to repay a loan in 24 equal monthly installments. The total finance charge on the loan is $450. With 9 payments remaining, he decides to repay the loan in full. Find the amount of unearned interest.

Exercises 45 and 46 deal with winning percentage in the standings of baseball teams. Winning percentage (Pct.) is commonly expressed as a decimal rounded to the nearest thousandth. To find the winning percentage of a team, divide the number of wins (W) by the total number of games played (W + L).

45. At the start of play on May 4, 2006, the standings of the Central Division of the American League were as shown. Find the winning percentages of the following teams.

    (a) Detroit   **(b)** Cleveland
    (c) Minnesota   **(d)** Kansas City

	W	L	Pct.
Chicago	19	8	.704
Detroit	19	9	
Cleveland	14	13	
Minnesota	11	16	
Kansas City	5	20	

46. Repeat Exercise 45 for the following standings of the Central Division of the National League.

    (a) Cincinnati   **(b)** Houston
    (c) Chicago   **(d)** Pittsburgh

	W	L	Pct.
Cincinnati	19	9	
Houston	18	9	
St. Louis	17	11	.607
Chicago	14	12	
Milwaukee	15	13	.536
Pittsburgh	8	21	

An average middle-income family will spend $242,070 to raise a child born in 2004 from birth to age 17. The graph shows the percents spent for various categories. Use the graph to answer Exercises 47–50. See Example 6.

◉ 47. To the nearest dollar, how much will be spent to provide housing for the child?

48. To the nearest dollar, how much will be spent for health care?

49. Use your answer from Exercise 48 to find how much will be spent for transportation.

The Cost of Parenthood

Housing 34%
Miscellaneous 11%
Child care/ education 11%
Health care 7%
Clothing 6%
Transportation 14%
Food 17%

Source: U.S. Department of Agriculture.

50. About $41,000 will be spent for food. To the nearest percent, what percent of the cost of raising a child from birth to age 17 is this? Does your answer agree with the percent shown in the graph?

Television networks have been losing viewers to cable programming since 1982, as the two graphs show. Use these graphs to answer Exercises 51–54. See Example 6.

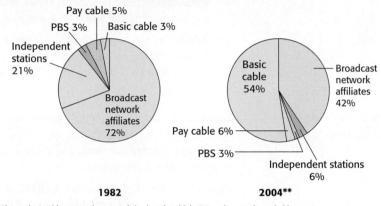

Shifting Share of the Television Audience as More Homes Receive Cable Programming*

1982

2004**

*Shares don't add to 100% because of viewing of multiple TV sets in some households.
**Independent stations include all superstations except TBS; broadcast affiliates include Fox; cable includes TBS.

Source: Nielsen Media Research.

51. In a typical group of 50,000 television viewers, how many would have watched basic cable in 1982?

52. In 1982, how many of a typical group of 110,000 viewers watched independent stations?

53. How many of a typical group of 35,000 viewers watched basic cable in 2004?

54. In a typical group of 65,000 viewers, how many watched independent stations in 2004?

PREVIEW EXERCISES

*Solve each equation. See **Section 2.1**.*

55. $4x + 4(x + 7) = 124$

56. $x + 0.20x = 66$

57. $\dfrac{5}{3} + \dfrac{2}{3}x = 2$

58. $0.07x + 0.05(9000 - x) = 510$

59. $2.4 + 0.4x = 0.25(6 + x)$

60. $5(x - 2) + 8x = 16$

*Evaluate. See **Section 1.2**.*

61. The product of -3 and 5, divided by 1 less than 6

62. Half of -18, added to the reciprocal of $\frac{1}{5}$

63. The sum of 6 and -9, multiplied by the additive inverse of 2

64. The product of -2 and 4, added to the product of -9 and -3

2.3 Applications of Linear Equations

OBJECTIVE 1 Translate from words to mathematical expressions. Producing a mathematical model of a real situation often involves translating verbal statements into mathematical statements.

▶ **PROBLEM-SOLVING HINT** Usually, there are key words and phrases in a verbal problem that translate into mathematical expressions involving addition, subtraction, multiplication, and division. Translations of some commonly used expressions follow.

TRANSLATING FROM WORDS TO MATHEMATICAL EXPRESSIONS

Verbal Expression	Mathematical Expression (where x and y are numbers)
Addition	
The **sum** of a number and 7	$x + 7$
6 **more than** a number	$x + 6$
3 **plus** a number	$3 + x$
24 **added to** a number	$x + 24$
A number **increased by** 5	$x + 5$
The **sum** of two numbers	$x + y$
Subtraction	
2 **less than** a number	$x - 2$
2 **less** a number	$2 - x$
12 **minus** a number	$12 - x$
A number **decreased by** 12	$x - 12$
A number **subtracted from** 10	$10 - x$
From a number, **subtract** 10	$x - 10$
The **difference between** two numbers	$x - y$
Multiplication	
16 **times** a number	$16x$
A number **multiplied by** 6	$6x$
$\frac{2}{3}$ **of** a number (used with fractions and percent)	$\frac{2}{3}x$
$\frac{3}{4}$ **as much as** a number	$\frac{3}{4}x$
Twice (2 times) a number	$2x$
The **product** of two numbers	xy
Division	
The **quotient** of 8 and a number	$\frac{8}{x}$ $(x \neq 0)$
A number **divided by** 13	$\frac{x}{13}$
The **ratio** of two numbers or the **quotient** of two numbers	$\frac{x}{y}$ $(y \neq 0)$

▶ **CAUTION** Because subtraction and division are not commutative operations, it is important to correctly translate expressions involving them. For example, "2 less than a number" is translated as $x - 2$, *not* $2 - x$. "A number subtracted from 10" is expressed as $10 - x$, *not* $x - 10$.

For division, the number *by which* we are dividing is the denominator, and the number *into which* we are dividing is the numerator. For example, "a number divided by 13" and "13 divided into x" both translate as $\frac{x}{13}$. Similarly, "the quotient of x and y" is translated as $\frac{x}{y}$.

OBJECTIVE 2 **Write equations from given information.** The symbol for equality, $=$, is often indicated by the word *is*. In fact, because equal mathematical expressions represent names for the same number, any words that indicate the idea of "sameness" translate to $=$.

EXAMPLE 1 **Translating Words into Equations**

Translate each verbal sentence into an equation.

Verbal Sentence	Equation
Twice a number, decreased by 3, is 42.	$2x - 3 = 42$
The product of a number and 12, decreased by 7, is 105.	$12x - 7 = 105$
The quotient of a number and the number plus 4 is 28.	$\dfrac{x}{x + 4} = 28$
The quotient of a number and 4, plus the number, is 10.	$\dfrac{x}{4} + x = 10$

✔ **Now Try Exercises 7 and 17.**

OBJECTIVE 3 **Distinguish between expressions and equations.** An expression translates as a phrase. An equation includes the $=$ symbol, with something on either side of it, and translates as a sentence.

EXAMPLE 2 **Distinguishing between Expressions and Equations**

Decide whether each is an *expression* or an *equation*.

(a) $2(3 + x) - 4x + 7$

There is no equals sign, so this is an expression.

(b) $2(3 + x) - 4x + 7 = -1$

Because there is an equals sign with something on either side of it, this is an equation.

Note that the expression in part (a) simplifies to the expression $-2x + 13$, and the equation in part (b) has solution 7.

✔ **Now Try Exercises 21 and 23.**

OBJECTIVE 4 Use the six steps in solving an applied problem.   While there is no one method that allows us to solve all types of applied problems, the following six steps are helpful.

Solving an Applied Problem

Step 1   **Read** the problem, several times if necessary, until you understand what is given and what is to be found.

Step 2   **Assign a variable** to represent the unknown value, using diagrams or tables as needed. Write down what the variable represents. If necessary, express any other unknown values in terms of the variable.

Step 3   **Write an equation** using the variable expression(s).

Step 4   **Solve** the equation.

Step 5   **State the answer** to the problem. Does it seem reasonable?

Step 6   **Check** the answer in the words of the original problem.

EXAMPLE 3   **Solving a Perimeter Problem**

The length of a rectangle is 1 cm more than twice the width. The perimeter of the rectangle is 110 cm. Find the length and the width of the rectangle.

$2W + 1$

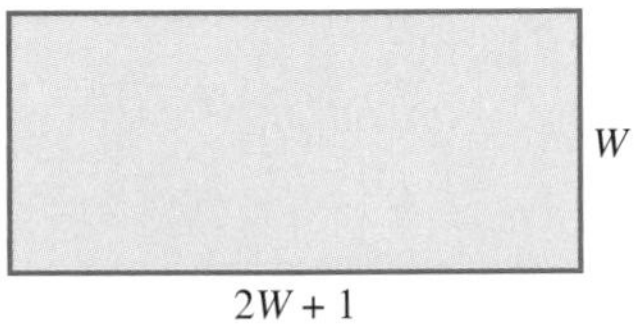

FIGURE 4

Step 1   **Read** the problem. What must be found? The length and width of the rectangle. What is given? The length is 1 cm more than twice the width; the perimeter is 110 cm.

Step 2   **Assign a variable.** Let W = the width; then $2W + 1$ = the length. Make a sketch, as in Figure 4.

Step 3   **Write an equation.** Use the formula for the perimeter of a rectangle.

$$P = 2L + 2W \qquad \text{Perimeter of a rectangle}$$
$$110 = 2(2W + 1) + 2W \qquad \text{Let } L = 2W + 1 \text{ and } P = 110.$$

Step 4   **Solve** the equation obtained in Step 3.

$$110 = 4W + 2 + 2W \qquad \text{Distributive property}$$
$$110 = 6W + 2 \qquad \text{Combine like terms.}$$
$$110 - 2 = 6W + 2 - 2 \qquad \text{Subtract 2.}$$
$$108 = 6W$$
$$\frac{108}{6} = \frac{6W}{6} \qquad \text{Divide by 6.}$$
$$18 = W$$

Step 5   **State the answer.** The width of the rectangle is 18 cm and the length is $2(18) + 1 = 37$ cm.

Step 6   **Check.** The length, 37 cm, is 1 cm more than $2(18)$ cm (twice the width). The perimeter is $2(37) + 2(18) = 74 + 36 = 110$ cm, as required.

✔ **Now Try Exercise 29.**

EXAMPLE 4 Finding Unknown Numerical Quantities

Two outstanding major league pitchers in recent years are Randy Johnson and Johan Santana. In 2004, the two pitchers had a combined total of 555 strikeouts. Johnson had 25 more strikeouts than Santana. How many strikeouts did each pitcher have? (*Source: World Almanac and Book of Facts 2006.*)

Step 1 **Read** the problem. We are asked to find the number of strikeouts each pitcher had.

Step 2 **Assign a variable** to represent the number of strikeouts for one of the men.

Let s = the number of strikeouts for Johan Santana.

We must also find the number of strikeouts for Randy Johnson. Since he had 25 more strikeouts than Santana,

$s + 25$ = the number of strikeouts for Johnson.

Step 3 **Write an equation.** The sum of the numbers of strikeouts is 555, so

$$\underset{\displaystyle s}{\text{Santana's strikeouts}} \;+\; \underset{\displaystyle (s+25)}{\text{Johnson's strikeouts}} \;=\; \underset{\displaystyle 555.}{\text{Total}}$$

Step 4 **Solve** the equation.

$$s + (s + 25) = 555$$
$$2s + 25 = 555 \qquad \text{Combine like terms.}$$
$$2s + 25 - 25 = 555 - 25 \qquad \text{Subtract 25.}$$
$$2s = 530$$
$$\frac{2s}{2} = \frac{530}{2} \qquad \text{Divide by 2.}$$

Don't stop here. $\qquad s = 265$

Step 5 **State the answer.** We let s represent the number of strikeouts for Santana, so Santana had 265. Then Johnson had

$$s + 25 = 265 + 25 = 290 \text{ strikeouts.}$$

Step 6 **Check.** 290 is 25 more than 265, and $265 + 290 = 555$. The conditions of the problem are satisfied, and our answer checks.

✔ **Now Try Exercise 35.**

▶ **CAUTION**   A common error in solving applied problems is forgetting to answer all the questions asked in the problem. In Example 4, we were asked for the number of strikeouts for *each* player, so there was extra work in Step 5 in order to find Johnson's number.

OBJECTIVE 5 **Solve percent problems.**   Recall from **Section 2.2** that percent means "per one hundred," so 5% means 0.05, 14% means 0.14, and so on.

EXAMPLE 5 Solving a Percent Problem

In 2002, there were 301 long-distance area codes in the United States, an increase of 250% over the number when the area code plan originated in 1947. How many area codes were there in 1947? (*Source:* SBC Telephone Directory.)

Step 1 **Read** the problem. We are given that the number of area codes increased by 250% from 1947 to 2002, and there were 301 area codes in 2002. We must find the original number of area codes.

Step 2 **Assign a variable.**

Let x = the number of area codes in 1947.

Since $250\% = 250(0.01) = 2.5$,

$2.5x$ = the number of codes added since then.

Step 3 **Write an equation** from the given information.

the number in 1947 + the increase = 301

$$x \quad + \quad 2.5x \quad = 301$$

Note the x in $2.5x$.

Step 4 **Solve** the equation.

$$1x + 2.5x = 301 \qquad \text{Identity property}$$
$$3.5x = 301 \qquad \text{Combine like terms.}$$
$$x = 86 \qquad \text{Divide by 3.5.}$$

Step 5 **State the answer.** There were 86 area codes in 1947.

Step 6 **Check** that the increase, $301 - 86 = 215$, is 250% of 86.

$$250\% \cdot 86 = 250(0.01)(86) = 215, \quad \text{as required.}$$

✔ **Now Try Exercise 45.**

▶ **CAUTION** Watch for two common errors that occur in solving problems like the one in Example 5.

1. Do not try to find 250% of 301 and subtract that amount from 301. The 250% should be applied to *the amount in 1947, not the amount in 2002.*

2. Do not write the equation as

$$x + 2.5 = 301. \qquad \text{Incorrect}$$

The percent must be multiplied by some amount; in this case, the amount is the number of area codes in 1947, giving $2.5x$.

OBJECTIVE 6 Solve investment problems. The investment problems in this chapter deal with *simple interest.* In most real-world applications, *compound interest* (covered in a later chapter) is used.

EXAMPLE 6 Solving an Investment Problem

After winning the state lottery, Mark LeBeau has $40,000 to invest. He will put part of the money in an account paying 4% interest and the remainder into stocks paying 6% interest. His accountant tells him that the total annual income from these investments should be $2040. How much should he invest at each rate?

Step 1 **Read** the problem again. We must find the two amounts.

Step 2 **Assign a variable.**

Let x = the amount to invest at 4%;

then 40,000 − x = the amount to invest at 6%.

The formula for interest is $I = prt$. Here the time t is 1 yr. Use a table to organize the given information.

Principal	Rate (as a decimal)	Interest
x	0.04	0.04x
40,000 − x	0.06	0.06(40,000 − x)
40,000	✕✕✕✕	2040

Multiply principal, rate, and time (here, 1 yr) to get interest.

← Total

Step 3 **Write an equation.** The last column of the table gives the equation.

$$\underset{\downarrow}{\text{interest at 4\%}} + \underset{\downarrow}{\text{interest at 6\%}} = \underset{\downarrow}{\text{total interest}}$$
$$0.04x + 0.06(40,000 - x) = 2040$$

Step 4 **Solve** the equation. We do so without clearing decimals.

$$0.04x + 0.06(40,000) - 0.06x = 2040 \qquad \text{Distributive property}$$
$$-0.02x + 2400 = 2040 \qquad \text{Combine like terms; multiply.}$$
$$-0.02x = -360 \qquad \text{Subtract 2400.}$$
$$x = 18,000 \qquad \text{Divide by } -0.02.$$

Step 5 **State the answer.** Mark should invest $18,000 of the money at 4% and $40,000 − $18,000 = $22,000 at 6%.

Step 6 **Check** by finding the annual interest at each rate. The sum of these two amounts should total $2040.

$$0.04(\$18,000) = \$720 \quad \text{and} \quad 0.06(\$22,000) = \$1320$$
$$\$720 + \$1320 = \$2040, \quad \text{as required.}$$

✔ **Now Try Exercise 49.**

▶ **PROBLEM-SOLVING HINT** In Example 6, we chose to let the variable represent the amount invested at 4%. Students often ask, "Can I let the variable represent the other unknown?" The answer is yes. The equation will be different, but in the end the answers will be the same.

OBJECTIVE 7 Solve mixture problems. Mixture problems involving rates of concentration can be solved with linear equations.

EXAMPLE 7 Solving a Mixture Problem

A chemist must mix 8 L of a 40% acid solution with some 70% solution to get a 50% solution. How much of the 70% solution should be used?

Step 1 **Read** the problem. The problem asks for the amount of 70% solution to be used.

Step 2 **Assign a variable.** Let x = the number of liters of 70% solution to be used. The information in the problem is illustrated in Figure 5.

FIGURE 5

Use the given information to complete a table.

Number of Liters	Percent (as a decimal)	Liters of Pure Acid
8	0.40	0.40(8) = 3.2
x	0.70	0.70x
8 + x	0.50	0.50(8 + x)

Sum must equal

The numbers in the last column were found by multiplying the strengths by the numbers of liters. The number of liters of pure acid in the 40% solution plus the number of liters in the 70% solution must equal the number of liters in the 50% solution.

Step 3 **Write an equation.**

$$3.2 + 0.70x = 0.50(8 + x)$$

Step 4 **Solve.**

$$3.2 + 0.70x = 4 + 0.50x \qquad \text{Distributive property}$$
$$0.20x = 0.8 \qquad \text{Subtract 3.2 and 0.50}x.$$
$$x = 4 \qquad \text{Divide by 0.20.}$$

Step 5 **State the answer.** The chemist should use 4 L of the 70% solution.

Step 6 **Check.** 8 L of 40% solution plus 4 L of 70% solution is

$$8(0.40) + 4(0.70) = 6\,\text{L}$$

of acid. Similarly, 8 + 4 or 12 L of 50% solution has

$$12(0.50) = 6\,\text{L}$$

of acid in the mixture. The total amount of pure acid is 6 L both before and after mixing, so the answer checks.

✔ **Now Try Exercise 55.**

> ▶ **PROBLEM-SOLVING HINT** Remember that when pure water is added to a solution, water is 0% of the chemical (acid, alcohol, etc.). Similarly, pure chemical is 100% chemical.

EXAMPLE 8 Solving a Mixture Problem when One Ingredient Is Pure

The octane rating of gasoline is a measure of its antiknock qualities. For a standard fuel, the octane rating is the percent of isooctane. How many liters of pure isooctane should be mixed with 200 L of 94% isooctane, referred to as 94 octane, to get a mixture that is 98% isooctane?

Step 1 **Read** the problem. The problem asks for the amount of pure isooctane.

Step 2 **Assign a variable.** Let $x =$ the number of liters of pure (100%) isooctane. Complete a table. Recall that $100\% = 100(0.01) = 1$.

Number of Liters	Percent (as a decimal)	Liters of Pure Isooctane
x	1	x
200	0.94	0.94(200)
$x + 200$	0.98	0.98(x + 200)

Step 3 **Write an equation.** The equation comes from the last column of the table.

$$x + 0.94(200) = 0.98(x + 200)$$

Step 4 **Solve.**

$$x + 0.94(200) = 0.98x + 0.98(200) \qquad \text{Distributive property}$$
$$x + 188 = 0.98x + 196 \qquad \text{Multiply.}$$
$$0.02x = 8 \qquad \text{Subtract } 0.98x \text{ and } 188.$$
$$x = 400 \qquad \text{Divide by } 0.02.$$

Step 5 **State the answer.** 400 L of isooctane is needed.

Step 6 **Check** by showing that $400 + 0.94(200) = 0.98(400 + 200)$ is true.

✔ **Now Try Exercise 59.**

CONNECTIONS

Probably the most famous study of problem-solving techniques was developed by George Polya. Among his many publications was the modern classic *How to Solve It*. In this book, Polya proposed a four-step process for problem solving.

Polya's Four-Step Process for Problem Solving

1. **Understand the problem.** You must first decide what you are to find.
2. **Devise a plan.** Some strategies that may prove useful follow.

(continued)

Problem-Solving Strategies

If a formula applies, use it.   Look for a pattern.

Write an equation and solve it.   Use trial and error.

Draw a sketch.   Work backward.

Make a table or a chart.

We used the first of these strategies in **Section 2.2.** In this section, we used the other three strategies on the left.

3. Carry out the plan. This is where the algebraic techniques you are learning in this book can be helpful.

4. Look back and check. Is your answer reasonable? Does it answer the question that was asked?

For Discussion or Writing

Compare Polya's four steps with the six steps for problem solving given earlier. Which of our steps correspond with each of Polya's steps?

2.3 EXERCISES

⊕ *Complete solution available on Video Lectures on CD/DVD*

Now Try Exercise

Concept Check   In each of the following, (a) translate as an expression and (b) translate as an equation or inequality. Use x to represent the number.

1. (a) 12 more than a number
   (b) 12 is more than a number.

2. (a) 3 less than a number
   (b) 3 is less than a number.

3. (a) 4 less than a number
   (b) 4 is less than a number.

4. (a) 6 greater than a number
   (b) 6 is greater than a number.

5. *Concept Check*   Which one of the following is *not* a valid translation of "20% of a number," where x represents the number?

   A. $0.20x$ **B.** $0.2x$ **C.** $\dfrac{x}{5}$ **D.** $20x$

✎ **6.** Explain why $13 - x$ is *not* a correct translation of "13 less than a number."

Translate each verbal phrase into a mathematical expression. Use x to represent the unknown number. See Example 1.

⊕ **7.** Twice a number, decreased by 13

8. The product of 6 and a number, decreased by 12

9. 12 increased by three times a number

10. 12 more than one-half of a number

11. The product of 8 and 12 less than a number

12. The product of 9 more than a number and 6 less than the number

13. The quotient of three times a number and 7

14. The quotient of 6 and five times a nonzero number

Use the variable x for the unknown, and write an equation representing the verbal sentence. Then solve the problem. See Example 1.

15. The sum of a number and 6 is -31. Find the number.

16. The sum of a number and -4 is 12. Find the number.

17. If the product of a number and -4 is subtracted from the number, the result is 9 more than the number. Find the number.

18. If the quotient of a number and 6 is added to twice the number, the result is 8 less than the number. Find the number.

19. When $\frac{2}{3}$ of a number is subtracted from 12, the result is 10. Find the number.

20. When 75% of a number is added to 6, the result is 3 more than the number. Find the number.

Decide whether each is an expression *or an* equation. *See Example 2.*

21. $5(x + 3) - 8(2x - 6)$

22. $-7(z + 4) + 13(z - 6)$

23. $5(x + 3) - 8(2x - 6) = 12$

24. $-7(z + 4) + 13(z - 6) = 18$

25. $\dfrac{r}{2} - \dfrac{r + 9}{6} - 8$

26. $\dfrac{r}{2} - \dfrac{r + 9}{6} = 8$

Concept Check *In Exercises 27 and 28, complete the six suggested problem-solving steps to solve each problem.*

27. Two of the leading U.S. research universities are the Massachusetts Institute of Technology (MIT) and Stanford University. In 2002, these two universities secured 230 patents on various inventions. Stanford secured 38 fewer patents than MIT. How many patents did each university secure? (*Source:* Association of University Technology Managers.)

Step 1 **Read** the problem carefully. We are asked to find _____.

Step 2 **Assign a variable.** Let $x =$ the number of patents that MIT secured. Then $x - 38 =$ the number of _____.

Step 3 **Write an equation.** _____ + _____ = 230

Step 4 **Solve** the equation. $x =$ _____

Step 5 **State the answer.** MIT secured _____ patents, and Stanford secured _____ patents.

Step 6 **Check.** The number of Stanford patents was _____ fewer than the number of _____, and the total number of patents was 134 + _____ = _____.

28. In a recent sample of book buyers, 70 more shopped at large-chain bookstores than at small-chain/independent bookstores. A total of 442 book buyers shopped at these two types of stores. How many buyers shopped at each type of bookstore? (*Source:* Book Industry Study Group.)

Step 1 **Read** the problem carefully. We are asked to find _____.

Step 2 **Assign a variable.** Let $x =$ the number of book buyers at large-chain bookstores. Then $x - 70 =$ the number of _____.

Step 3 **Write an equation.** _____ + _____ = 442

Step 4 **Solve** the equation. $x =$ _____

Step 5 **State the answer.** There were _____ large-chain bookstore shoppers and _____ small-chain/independent shoppers.

Step 6 **Check.** The number of _____ was _____ more than the number of _____, and the total number of these shoppers was 256 + _____ = _____.

Solve each problem. See Examples 3 and 4.

29. The John Hancock Center in Chicago has a rectangular base. The length of the base measures 65 ft less than twice the width. The perimeter of the base is 860 ft. What are the dimensions of the base?

30. The John Hancock Center (Exercise 29) tapers as it rises. The top floor is rectangular and has perimeter 520 ft. The width of the top floor measures 20 ft more than one-half its length. What are the dimensions of the top floor?

The perimeter of the top floor is 520 ft.

$\frac{1}{2}L + 20$

L

$2W - 65$ W

The perimeter of the base is 860 ft.

31. Grant Wood painted his most famous work, *American Gothic,* in 1930 on composition board with perimeter 108.44 in. If the painting is 5.54 in. taller than it is wide, find the dimensions of the painting. (*Source: The Gazette,* Cedar Rapids, Iowa, March 12, 2004.)

32. The perimeter of a certain rectangle is 16 times the width. The length is 12 cm more than the width. Find the length and width of the rectangle.

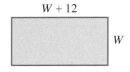

$W + 12$

W

33. The Bermuda Triangle supposedly causes trouble for aircraft pilots. It has a perimeter of 3075 mi. The shortest side measures 75 mi less than the middle side, and the longest side measures 375 mi more than the middle side. Find the lengths of the three sides.

34. The Vietnam Veterans Memorial in Washington, DC, is in the shape of two sides of an isosceles triangle. If the two walls of equal length were joined by a straight line of 438 ft, the perimeter of the resulting triangle would be 931.5 ft. Find the lengths of the two walls. (*Source:* Pamphlet obtained at Vietnam Veterans Memorial.)

438 ft

35. The two companies with top revenues in the Fortune 500 list for 2005 were Exxon Mobil and Wal-Mart. Their revenues together totaled $656 billion. Wal-Mart revenues were $24 billion less than Exxon Mobil revenues. What were the revenues of each corporation? (*Source:* www.money.cnn.com)

36. Two of the longest-running Broadway shows were *Cats,* which played from 1982 through 2000, and *Les Misérables,* which played from 1987 through 2005. Together, there were 14,165 performances of these two shows during their Broadway runs. There were 805 fewer performances of *Les Misérables* than of *Cats.* How many performances were there of each show? (*Source:* The League of American Theatres and Producers.)

37. Galileo Galilei conducted experiments involving Italy's famous Leaning Tower of Pisa to investigate the relationship between an object's speed of fall and its weight. The Leaning Tower is 804 ft shorter than the Eiffel Tower in Paris, France. The two towers have a total height of 1164 ft. How tall is each tower? (*Source: Microsoft Encarta Encyclopedia.*)

38. In 2003, the New York Yankees and the New York Mets had the highest payrolls in Major League Baseball. The Mets' payroll was $32.8 million less than the Yankees' payroll, and the two payrolls totaled $266.6 million. What was the payroll for each team? (*Source:* Associated Press.)

39. In the 2004 presidential election, George W. Bush and John Kerry together received 537 electoral votes. Bush received 35 more votes than Kerry. How many votes did each candidate receive? (*Source:* Congressional Quarterly, Inc.)

40. Ted Williams and Rogers Hornsby were two great hitters. Together, they got 5584 hits in their careers. Hornsby got 276 more hits than Williams. How many base hits did each get? (*Source:* Neft, D. S., and R. M. Cohen, *The Sports Encyclopedia: Baseball,* St. Martins Griffin; New York, 1997.)

41. In 2005, the number of participants in the ACT exam was 1,186,251. In 1990, a total of 817,000 took the exam. By what percent did the number increase over this period of time, to the nearest tenth of a percent? (*Source:* ACT.)

42. Composite scores on the ACT exam fell from 21.0 in 2001 to 20.8 in 2002. What percent decrease was the drop? (*Source:* ACT.)

43. In 1995, the average cost of tuition and fees at public four-year universities in the United States was $2811 for full-time students. By 2005, it had risen approximately 95%. To the nearest dollar, what was the approximate cost in 2005? (*Source:* The College Board.)

44. In 1995, the average cost of tuition and fees at private four-year universities in the United States was $12,216 for full-time students. By 2005, it had risen approximately 73.8%. To the nearest dollar, what was the approximate cost in 2005? (*Source:* The College Board.)

45. In 2005, the average cost of a traditional Thanksgiving dinner for 10, featuring turkey, stuffing, cranberries, pumpkin pie, and trimmings, was $36.78, an increase of 3.1% over the cost in 2004. What was the cost, to the nearest cent, in 2004? (*Source:* American Farm Bureau.)

46. Refer to Exercise 45. The first year that information on the cost of a traditional Thanksgiving dinner was collected was 1987. The 2005 cost of $36.78 was an increase of 37.5% over the cost in 1987. What was the cost, to the nearest cent, in 1987? (*Source:* American Farm Bureau.)

47. At the end of a day, Jeff Hornsby found that the total cash register receipts at the motel where he works amounted to $2725. This included the 9% sales tax charged. Find the amount of the tax.

48. Fino Roverato sold his house for $159,000. He got this amount knowing that he would have to pay a 6% commission to his agent. What amount did he have after the agent was paid?

Solve each investment problem. See Example 6.

49. Carter Fenton earned $12,000 last year by giving tennis lessons. He invested part of the money at 3% simple interest and the rest at 4%. In one year, he earned a total of $440 in interest. How much did he invest at each rate?

Principal	Rate (as a decimal)	Interest
x	0.03	
	0.04	

50. Caroline Case won $60,000 on a slot machine in Las Vegas. She invested part of the money at 2% simple interest and the rest at 3%. In one year, she earned a total of $1600 in interest. How much was invested at each rate?

Principal	Rate (as a decimal)	Interest
x	0.02	

51. Ashley O'Shaughnessy won $5000 in a contest. She invested some of the money at 5% simple interest and $400 less than twice that amount at 6.5%. In one year, she earned $298 in interest. How much did she invest at each rate?

52. Toshiro Hashimoto invested some money at 4.5% simple interest and $1000 more than four times that amount at 6%. His total annual income for one year from interest on the two investments was $801. How much did he invest at each rate?

53. Vincente and Ricarda Pérez have invested $27,000 in bonds paying 7%. How much additional money should they invest in a certificate of deposit paying 4% simple interest so that the total return on the two investments will be 6%?

54. Rebecca Herst received a year-end bonus of $17,000 from her company and invested the money in an account paying 6.5%. How much additional money should she deposit in an account paying 5% so that the return on the two investments will be 6%?

Solve each problem involving rates of concentration and mixtures. See Examples 7 and 8.

55. Ten liters of a 4% acid solution must be mixed with a 10% solution to get a 6% solution. How many liters of the 10% solution are needed?

Liters of Solution	Percent (as a decimal)	Liters of Pure Acid
10	0.04	
x	0.10	
	0.06	

56. How many liters of a 14% alcohol solution must be mixed with 20 L of a 50% solution to get a 30% solution?

Liters of Solution	Percent (as a decimal)	Liters of Pure Alcohol
x	0.14	
	0.50	

57. In a chemistry class, 12 L of a 12% alcohol solution must be mixed with a 20% solution to get a 14% solution. How many liters of the 20% solution are needed?

58. How many liters of a 10% alcohol solution must be mixed with 40 L of a 50% solution to get a 40% solution?

59. How much pure dye must be added to 4 gal of a 25% dye solution to increase the solution to 40%? (*Hint:* Pure dye is 100% dye.)

60. How much water must be added to 6 gal of a 4% insecticide solution to reduce the concentration to 3%? (*Hint:* Water is 0% insecticide.)

61. Randall Albritton wants to mix 50 lb of nuts worth $2 per lb with some nuts worth $6 per lb to make a mixture worth $5 per lb. How many pounds of $6 nuts must he use?

Pounds of Nuts	Cost per Pound	Total Cost

62. Lee Ann Spahr wants to mix tea worth 2¢ per oz with 100 oz of tea worth 5¢ per oz to make a mixture worth 3¢ per oz. How much 2¢ tea should be used?

Ounces of Tea	Cost per Ounce	Total Cost

63. Why is it impossible to mix candy worth $4 per lb and candy worth $5 per lb to obtain a final mixture worth $6 per lb?

64. Write an equation based on the following problem, solve the equation, and explain why the problem has no solution:

How much 30% acid should be mixed with 15 L of 50% acid to obtain a mixture that is 60% acid?

RELATING CONCEPTS (EXERCISES 65–69)

FOR INDIVIDUAL OR GROUP WORK

Consider each problem.

Problem A   Jack has $800 invested in two accounts. One pays 5% interest per year and the other pays 10% interest per year. The amount of yearly interest is the same as he would get if the entire $800 was invested at 8.75%. How much does he have invested at each rate?

Problem B   Jill has 800 L of acid solution. She obtained it by mixing some 5% acid with some 10% acid. Her final mixture of 800 L is 8.75% acid. How much of each of the 5% and 10% solutions did she use to get her final mixture?

In Problem A, let x represent the amount invested at 5% interest, and in Problem B, let y represent the amount of 5% acid used. **Work Exercises 65–69 in order.**

65. (a) Write an expression in x that represents the amount of money Jack invested at 10% in Problem A.
   (b) Write an expression in y that represents the amount of 10% acid solution Jill used in Problem B.

66. (a) Write expressions that represent the amount of interest Jack earns per year at 5% and at 10%.
   (b) Write expressions that represent the amount of pure acid in Jill's 5% and 10% acid solutions.

67. (a) The sum of the two expressions in part (a) of Exercise 66 must equal the total amount of interest earned in one year. Write an equation representing this fact.
   (b) The sum of the two expressions in part (b) of Exercise 66 must equal the amount of pure acid in the final mixture. Write an equation representing this fact.

68. (a) Solve Problem A.   **(b)** Solve Problem B.

69. Explain the similarities between the processes used in solving Problems A and B.

PREVIEW EXERCISES

*Solve each problem. See **Section 2.2**.*

70. Use $d = rt$ to find d if $r = 50$ and $t = 4$.

71. Use $P = 2L + 2W$ to find P if $L = 10$ and $W = 6$.

72. Use $P = 2L + 2W$ to find W if $P = 80$ and $L = 34$.

73. Use $P = a + b + c$ to find a if $b = 13$, $c = 14$, and $P = 46$.

74. Use $A = \frac{1}{2}h(b + B)$ to find h if $A - 156$, $b - 12$, and $B = 14$.

75. Use $d = rt$ to find r if $d = 75$ and $t = 15$.

2.4 Further Applications of Linear Equations

There are three common applications of linear equations that we did not discuss in **Section 2.3:** money problems, uniform motion problems, and problems involving the angles of a triangle.

OBJECTIVE 1 Solve problems about different denominations of money. These problems are similar to the simple interest problems in **Section 2.3.**

▶ **PROBLEM-SOLVING HINT** In problems involving money, use the basic fact that

$$\begin{array}{c}\textbf{number of monetary}\\\textbf{units of the same kind}\end{array} \times \textbf{denomination} = \begin{array}{c}\textbf{total monetary}\\\textbf{value}\end{array}.$$

For example, 30 dimes have a monetary value of $30(\$0.10) = \$3.00$. Fifteen 5-dollar bills have a value of $15(\$5) = \$75$.

EXAMPLE 1 Solving a Money Denomination Problem

For a bill totaling $5.65, a cashier received 25 coins consisting of nickels and quarters. How many of each denomination of coin did the cashier receive?

Step 1 **Read** the problem. The problem asks that we find the number of nickels and the number of quarters the cashier received.

Step 2 **Assign a variable.**

Let x = the number of nickels;

then $25 - x$ = the number of quarters.

We can organize the information in a table.

	Number of Coins	Denomination	Value
Nickels	x	0.05	$0.05x$
Quarters	$25 - x$	0.25	$0.25(25 - x)$
			5.65 ← Total

Step 3 **Write an equation.** From the last column of the table,

$$0.05x + 0.25(25 - x) = 5.65.$$

Step 4 **Solve.**

$$0.05x + 0.25(25 - x) = 5.65$$
$$5x + 25(25 - x) = 565 \quad \text{Multiply by 100.}$$
$$5x + 625 - 25x = 565 \quad \text{Distributive property}$$
$$-20x = -60 \quad \text{Subtract 625; combine terms.}$$
$$x = 3 \quad \text{Divide by } -20.$$

Move decimal points 2 places to the right.

Step 5 **State the answer.** The cashier has 3 nickels and $25 - 3 = 22$ quarters.

Step 6 **Check.** The cashier has $3 + 22 = 25$ coins, and the value of the coins is $\$0.05(3) + \$0.25(22) = \$5.65$, as required.

> ✔ **Now Try Exercise 11.**

▶ **CAUTION**   *Be sure that your answer is reasonable* when you are working problems like Example 1. Because you are dealing with a number of coins, the correct answer can be neither negative nor a fraction.

OBJECTIVE 2   Solve problems about uniform motion.

▶ **PROBLEM-SOLVING HINT**   Uniform motion problems use the distance formula $d = rt$. In this formula, *when rate (or speed) is given in miles per hour, time must be given in hours.* To solve such problems, *draw a sketch* to illustrate what is happening in the problem, and *make a table* to summarize the given information.

EXAMPLE 2   **Solving a Motion Problem (Motion in Opposite Directions)**

Two cars leave the same place at the same time, one going east and the other west. The eastbound car averages 40 mph, while the westbound car averages 50 mph. In how many hours will they be 300 mi apart?

Step 1 **Read** the problem. We are looking for the time it takes for the two cars to be 300 mi apart.

Step 2 **Assign a variable.** A sketch shows what is happening in the problem: The cars are going in *opposite* directions. See Figure 6.

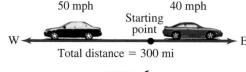

FIGURE 6

Let $x =$ the time traveled by each car.

Summarize the information of the problem in a table. *Fill in each distance by multiplying rate by time,* using the formula $d = rt$.

	Rate	Time	Distance
Eastbound Car	40	x	$40x$
Westbound Car	50	x	$50x$
			300	← Total

Step 3 **Write an equation.** The sum of the two distances is 300.

$$40x + 50x = 300$$

Step 4 **Solve.** $\qquad 90x = 300 \qquad\qquad$ Combine like terms.

$$x = \frac{300}{90} = \frac{10}{3} \qquad \text{Divide by 90; lowest terms}$$

Step 5 **State the answer.** The cars travel $\frac{10}{3} = 3\frac{1}{3}$ hr, or 3 hr, 20 min.

Step 6 **Check.** The eastbound car traveled $40\left(\frac{10}{3}\right) = \frac{400}{3}$ mi, and the westbound car traveled $50\left(\frac{10}{3}\right) = \frac{500}{3}$ mi, for a total of $\frac{400}{3} + \frac{500}{3} = \frac{900}{3} = 300$ mi, as required.

✔ **Now Try Exercise 21.**

▶ **CAUTION** It is a common error to write 300 as the distance traveled by each car in Example 2. Three hundred miles is the *total* distance traveled.

As in Example 2, in general, the equation for a problem involving motion in *opposite* directions is of the form

partial distance + partial distance = total distance.

EXAMPLE 3 Solving a Motion Problem (Motion in the Same Direction)

Jeff can bike to work in $\frac{3}{4}$ hr. When he takes the bus, the trip takes $\frac{1}{4}$ hr. If the bus travels 20 mph faster than Jeff rides his bike, how far is it to his workplace?

Step 1 **Read** the problem. We must find the distance between Jeff's home and his workplace.

Step 2 **Assign a variable.** Although the problem asks for a distance, it is easier here to let x be Jeff's speed when he rides his bike to work. Then the speed of the bus is $x + 20$. For the trip by bike,

$$d = rt = x \cdot \frac{3}{4} = \frac{3}{4}x,$$

and by bus, $\qquad d = rt = (x + 20) \cdot \frac{1}{4} = \frac{1}{4}(x + 20).$

Summarize this information in a table.

	Rate	Time	Distance	
Bike	x	$\frac{3}{4}$	$\frac{3}{4}x$	⎤
Bus	$x + 20$	$\frac{1}{4}$	$\frac{1}{4}(x + 20)$	⎦ Same

Step 3 **Write an equation.** The key to setting up the correct equation is to understand that the distance in each case is the same. See Figure 7.

Home   Workplace

FIGURE 7

$$\frac{3}{4}x = \frac{1}{4}(x + 20)$$ The distance is the same
in each case.

Step 4 **Solve.** $4\left(\dfrac{3}{4}x\right) = 4\left(\dfrac{1}{4}\right)(x + 20)$ Multiply by 4.

$$3x = x + 20$$ Multiply.

$$2x = 20$$ Subtract x.

$$x = 10$$ Divide by 2.

Step 5 **State the answer.** The required distance is

$$d = \frac{3}{4}x = \frac{3}{4}(10) = \frac{30}{4} = 7.5 \text{ mi.} \longleftarrow$$

Step 6 **Check** by finding the distance using

$$d = \frac{1}{4}(x + 20) = \frac{1}{4}(10 + 20) = \frac{30}{4} = 7.5 \text{ mi.} \longleftarrow$$

The same
result

✔ **Now Try Exercise 25.**

As in Example 3, the equation for a problem involving motion in the *same* direction is often of the form

one distance = other distance.

▶ **PROBLEM-SOLVING HINT**   In Example 3, it was easier to let the variable represent a quantity other than the one that we were asked to find. This is the case in some problems. It takes practice to learn when this approach is the best, and practice means working lots of problems.

OBJECTIVE 3 **Solve problems about angles.**   An important result of Euclidean geometry (the geometry of the Greek mathematician Euclid) is that *the sum of the angle measures of any triangle is* **180°.** This property is used in the next example.

EXAMPLE 4 **Finding Angle Measures**

Find the value of x, and determine the measure of each angle in Figure 8.

Step 1 **Read** the problem. We are asked to find the measure
of each angle.

Step 2 **Assign a variable.**

Let x = the measure of one angle.

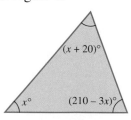

FIGURE 8

Step 3 **Write an equation.** The sum of the three measures shown in the figure must be 180°.

$$x + (x + 20) + (210 - 3x) = 180$$

Step 4 **Solve.**

$$-x + 230 = 180 \qquad \text{Combine like terms.}$$
$$-x = -50 \qquad \text{Subtract 230.}$$
$$x = 50 \qquad \text{Divide by } -1.$$

Step 5 **State the answer.** One angle measures 50°. The other two angles measure

$$x + 20 = 50 + 20 = 70°$$

and $\qquad 210 - 3x = 210 - 3(50) = 60°.$

Step 6 **Check.** Since $50° + 70° + 60° = 180°$, the answers are correct.

✔ **Now Try Exercise 31.**

2.4 EXERCISES

⊙ *Complete solution available on Video Lectures on CD/DVD*

▨ *Now Try Exercise*

Concept Check   Solve each problem.

1. What amount of money is found in a coin hoard containing 38 nickels and 26 dimes?

2. The distance between Cape Town, South Africa, and Miami is 7700 mi. If a jet averages 480 mph between the two cities, what is its travel time in hours?

3. Tri Phong traveled from Chicago to Des Moines, a distance of 300 mi, in 5 hr. What was his rate in miles per hour?

4. A square has perimeter 40 in. What would be the perimeter of an equilateral triangle whose sides each measure the same length as the side of the square?

✐ *Write a short explanation in Exercises 5–8.*

5. Read over Example 3 in this section. The solution of the equation is 10. Why is *10 mph* not the answer to the problem?

6. Suppose that you know that two angles of a triangle have equal measures and the third angle measures 36°. Explain in a few words the strategy you would use to find the measures of the equal angles without actually writing an equation.

7. In a problem about the number of coins of different denominations, would an answer that is a fraction be reasonable? What about a negative number?

8. In a motion problem the rate is given as *x* mph and the time is given as 30 min. What variable expression represents the distance in miles?

Solve each problem. See Example 1.

9. Otis Taylor has a box of coins that he uses when he plays poker with his friends. The box currently contains 44 coins, consisting of pennies, dimes, and quarters. The number of pennies is equal to the number of dimes, and the total value is $4.37. How many of each denomination of coin does he have in the box?

Number of Coins	Denomination	Value
x	0.01	0.01*x*
x		
	0.25	
✕✕✕✕✕✕✕✕✕		4.37 ← Total

10. Nana Nantambu found some coins while looking under her sofa pillows. There were equal numbers of nickels and quarters and twice as many half-dollars as quarters. If she found $2.60 in all, how many of each denomination of coin did she find?

Number of Coins	Denomination	Value
x	0.05	0.05x
x		
2x	0.50	
		2.60 ← Total

11. In Canada, $1 and $2 bills have been replaced by coins. The $1 coins are called "loonies" because they have a picture of a loon (a well-known Canadian bird) on the reverse, and the $2 coins are called "toonies." When Marissa returned home to San Francisco from a trip to Vancouver, she found that she had acquired 37 of these coins, with a total value of 51 Canadian dollars. How many coins of each denomination did she have?

12. Luke Corey works at an ice cream shop. At the end of his shift, he counted the bills in his cash drawer and found 119 bills with a total value of $347. If all of the bills are $5 bills and $1 bills, how many of each denomination were in his cash drawer?

13. Dave Bowers collects U.S. gold coins. He has a collection of 41 coins. Some are $10 coins, and the rest are $20 coins. If the face value of the coins is $540, how many of each denomination does he have?

14. In the 19th century, the United States minted two-cent and three-cent pieces. Frances Steib has three times as many three-cent pieces as two-cent pieces, and the face value of these coins is $2.42. How many of each denomination does she have?

15. In 2006, general admission to the Field Museum in Chicago cost $12 for adults and $7 for children and seniors. If $18,430 was collected from the sale of 2010 general admission tickets, how many adult tickets were sold? (*Source:* www.fieldmuseum.org)

16. For a high school production of *West Side Story*, student tickets cost $5 each while non-student tickets cost $8. If 480 tickets were sold for the Saturday night show and a total of $2895 was collected, how many tickets of each type were sold?

In Exercises 17–20, find the rate on the basis of the information provided. Use a calculator and round your answers to the nearest hundredth. All events were at the 2004 Summer Olympics in Athens, Greece. (Source: World Almanac and Book of Facts 2006.)

	Event	Participant	Distance	Time
17.	100-m hurdles, women	Joanna Hayes, USA	100 m	12.37 sec
18.	400-m hurdles, women	Fani Halkia, Greece	400 m	52.82 sec
19.	400-m hurdles, men	Felix Sánchez, Dominican Republic	400 m	47.63 sec
20.	400-m run, men	Jeremy Wariner, USA	400 m	44.00 sec

Solve each problem. See Examples 2 and 3.

21. Two steamers leave a port on a river at the same time, traveling in opposite directions. Each is traveling 22 mph. How long will it take for them to be 110 mi apart?

	Rate	Time	Distance
First Steamer		t	
Second Steamer	22		
			110

22. A train leaves Kansas City, Kansas, and travels north at 85 km per hr. Another train leaves at the same time and travels south at 95 km per hr. How long will it take before they are 315 km apart?

	Rate	Time	Distance
First Train	85	t	
Second Train			
			315

23. Agents Mulder and Scully are driving to Georgia to investigate "Big Blue," a giant aquatic reptile reported to inhabit one of the local lakes. Mulder leaves Washington at 8:30 A.M. and averages 65 mph. His partner, Scully, leaves at 9:00 A.M., following the same path and averaging 68 mph. At what time will Scully catch up with Mulder?

	Rate	Time	Distance
Mulder			
Scully			

24. Lois and Clark are covering separate stories and have to travel in opposite directions. Lois leaves the *Daily Planet* at 8:00 A.M. and travels at 35 mph. Clark leaves at 8:15 A.M. and travels at 40 mph. At what time will they be 140 mi apart?

	Rate	Time	Distance
Lois			
Clark			

25. It took Charmaine 3.6 hr to drive to her mother's house on Saturday morning for a weekend visit. On her return trip on Sunday night, traffic was heavier, so the trip took her 4 hr. Her average speed on Sunday was 5 mph slower than on Saturday. What was her average speed on Sunday?

	Rate	Time	Distance
Saturday			
Sunday			

26. Sara Schneider commutes to her office in Redwood City, California, by train. When she walks to the train station, it takes her 40 min. When she rides her bike, it takes her 12 min. Her average walking speed is 7 mph less than her average biking speed. Find the distance from Sara's house to the train station.

	Rate	Time	Distance
Walking			
Biking			

27. Johnny leaves Memphis to visit his cousin, Anne Hoffman, who lives in the town of Hornsby, Tennessee, 80 mi away. He travels at an average speed of 50 mph. One-half hour later, Anne leaves to visit Johnny, traveling at an average speed of 60 mph. How long after Anne leaves will it be before they meet?

28. On an automobile trip, Leanne Moen maintained a steady speed for the first two hours. Rush-hour traffic slowed her speed by 25 mph for the last part of the trip. The entire trip, a distance of 125 mi, took $2\frac{1}{2}$ hr. What was her speed during the first part of the trip?

Find the measure of each angle in the triangles shown. See Example 4.

29.

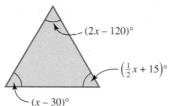

$(2x - 120)°$

$\left(\frac{1}{2}x + 15\right)°$

$(x - 30)°$

30.

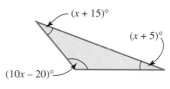

$(x + 15)°$

$(x + 5)°$

$(10x - 20)°$

31.

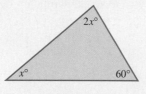

$(9x - 4)°$

$(3x + 7)°$

$(4x + 1)°$

32.

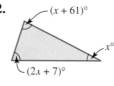

$(x + 61)°$

$x°$

$(2x + 7)°$

RELATING CONCEPTS (EXERCISES 33–36)

FOR INDIVIDUAL OR GROUP WORK

Consider the following two figures. **Work Exercises 33–36 in order.**

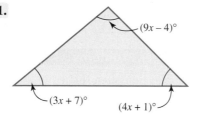

$2x°$

$x°$ $60°$

FIGURE A

$60°$ $y°$

FIGURE B

33. Solve for the measures of the unknown angles in Figure A.

34. Solve for the measure of the unknown angle marked $y°$ in Figure B.

35. Add the measures of the two angles you found in Exercise 33. How does the sum compare to the measure of the angle you found in Exercise 34?

36. From Exercises 33–35, make a conjecture (an educated guess) about the relationship among the angles marked ①, ②, and ③ in the figure shown at the right.

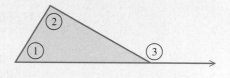

In Exercises 37 and 38, the angles marked with variable expressions are called **vertical angles.** *It is shown in geometry that vertical angles have equal measures. Find the measure of each angle.*

37.

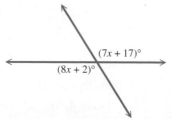

$(7x + 17)°$

$(8x + 2)°$

38.

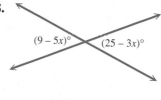

$(9 - 5x)°$ $(25 - 3x)°$

39. Two angles whose sum is 90° are called **complementary angles.** Find the measures of the complementary angles shown in the figure.

$(5x - 1)°$
$(2x)°$

40. Two angles whose sum is 180° are called **supplementary angles.** Find the measures of the supplementary angles shown in the figure.

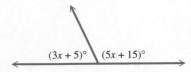

$(3x + 5)°$ $(5x + 15)°$

Consecutive Integer Problems

Another type of application often studied in algebra courses involves *consecutive integers.* **Consecutive integers** are integers that follow each other in counting order, such as 8, 9, and 10. Suppose we wish to solve the following problem:

Find three consecutive integers such that the sum of the first and third, increased by 3, is 50 more than the second.

Let x = the first of the unknown integers, $x + 1$ = the second, and $x + 2$ = the third. The equation to solve would be

Sum of the first and third	increased by 3	is	50 more than the second.
↓	↓	↓	↓
$x + (x + 2)$	$+ 3$	$=$	$(x + 1) + 50.$

$$2x + 5 = x + 51$$
$$x = 46$$

The solution of this equation is 46, so the first integer is $x = 46$, the second is $x + 1 = 47$, and the third is $x + 2 = 48$. The three integers are 46, 47, and 48. Check by substituting these numbers back into the words of the original problem.

Solve each problem involving consecutive integers.

41. Find three consecutive integers such that the sum of the first and twice the second is 17 more than twice the third.

42. Find four consecutive integers such that the sum of the first three is 54 more than the fourth.

43. If I add my current age to the age I will be next year on this date, the sum is 103 years. How old will I be 10 years from today?

44. Two pages facing each other in this book have 193 as the sum of their page numbers. What are the two page numbers?

PREVIEW EXERCISES

Graph each interval. See Section 1.1.

45. $(4, \infty)$

46. $(-\infty, -2]$

47. $(-2, 6)$

48. $[-1, 6]$

49. $[-4, 9)$

50. $(-4, 9]$

Summary Exercises on Solving Applied Problems

The applications that follow are of the various types introduced in this chapter. Use the strategies you have developed to solve each problem.

1. The length of a rectangle is 3 in. more than its width. If the length were decreased by 2 in. and the width were increased by 1 in., the perimeter of the resulting rectangle would be 24 in. Find the dimensions of the original rectangle.

$x + 3$

x

2. A farmer wishes to enclose a rectangular region with 210 m of fencing in such a way that the length is twice the width and the region is divided into two equal parts, as shown in the figure. What length and width should be used?

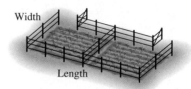

Width

Length

3. After a discount of 37%, the sale price for a *Harry Potter* Paperback Boxed Set (Books 1–6) by J. K. Rowling was $32.09. What was the regular price of the set of books to the nearest cent? (*Source:* amazon.com)

4. An electronics store offered a DVD recorder for $255, the sale price after the regular price was discounted 40%. What was the regular price?

5. An amount of money is invested at 4% annual simple interest, and twice that amount is invested at 5%. The total annual interest is $112. How much is invested at each rate?

6. An amount of money is invested at 3% annual simple interest, and $2000 more than that amount is invested at 4%. The total annual interest is $920. How much is invested at each rate?

7. The popular television program *Frasier,* which concluded an 11-yr run on May 13, 2004, won 9 fewer than twice as many Emmy awards as *The Simpsons.* As of early 2004, the two series had won a total of 51 Emmys. How many Emmys had each series won? (*Source*: Academy of Television Arts and Sciences.)

8. As of May 2004, the two all-time top-grossing American movies were *Titanic* and *Star Wars: Episode IV—A New Hope. Titanic* grossed $139.8 million more than *Star Wars.* Together, the two films brought in $1061.8 million. How much did each movie gross? (*Source: Variety.*)

9. Atlanta and Cincinnati are 440 mi apart. John leaves Cincinnati, driving toward Atlanta at an average speed of 60 mph. Pat leaves Atlanta at the same time, driving toward Cincinnati in her antique auto, averaging 28 mph. How long will it take them to meet?

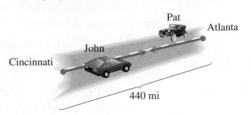

10. Hailu Negussie from Ethiopia won the 2005 men's Boston Marathon with a winning time of 2 hr, 11 min, 45 sec, or 2.196 hr. The women's race was won by Catherine Ndereba from Kenya, whose winning time was 2 hr, 25 min, 13 sec, or 2.420 hr. Ndereba's average rate was 1.1 mph slower than Negussie's. Find the average rate for each runner, to the nearest hundredth. (*Source: World Almanac and Book of Facts 2006.*)

11. A pharmacist has 20 L of a 10% drug solution. How many liters of 5% solution must be added to get a mixture that is 8%?

12. A certain metal is 20% tin. How many kilograms of this metal must be mixed with 80 kg of a metal that is 70% tin to get a metal that is 50% tin?

13. A cashier has a total of 126 bills in fives and tens. The total value of the money is $840. How many of each type of bill does he have?

14. The top-grossing domestic movie in 2004 was *Shrek 2*. On the opening weekend, one theater showing this movie took in $18,060 by selling a total of 2460 tickets, some at $8 and the rest at $6. How many tickets were sold at each price? (*Source: Variety.*)

15. Find the measure of each angle.

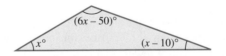

16. Find the measure of each marked angle.

17. The sum of the least and greatest of three consecutive integers is 32 more than the middle integer. What are the three integers?

18. If the lesser of two consecutive odd integers is doubled, the result is 7 more than the greater of the two integers. Find the two integers.

19. The perimeter of a triangle is 34 in. The middle side is twice as long as the shortest side. The longest side is 2 in. less than three times the shortest side. Find the lengths of the three sides.

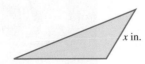

20. The perimeter of a rectangle is 43 in. more than the length. The width is 10 in. Find the length of the rectangle.

2.5 Linear Inequalities in One Variable

In **Section 1.1,** we used interval notation to write solution sets of inequalities, with a parenthesis to indicate that an endpoint is not included and a square bracket to indicate that an endpoint is included. We summarize the various types of intervals here.

INTERVAL NOTATION

Type of Interval	Set	Interval Notation	Graph
Open interval	$\{x \mid a < x\}$	(a, ∞)	
	$\{x \mid a < x < b\}$	(a, b)	
	$\{x \mid x < b\}$	$(-\infty, b)$	
	$\{x \mid x \text{ is a real number}\}$	$(-\infty, \infty)$	
Half-open interval	$\{x \mid a \le x\}$	$[a, \infty)$	
	$\{x \mid a < x \le b\}$	$(a, b]$	
	$\{x \mid a \le x < b\}$	$[a, b)$	
	$\{x \mid x \le b\}$	$(-\infty, b]$	
Closed interval	$\{x \mid a \le x \le b\}$	$[a, b]$	

An **inequality** says that two expressions are *not* equal. Solving inequalities is similar to solving equations.

Linear Inequality in One Variable

A **linear inequality in one variable** can be written in the form

$$Ax + B < C,$$

where A, B, and C are real numbers, with $A \ne 0$.

(Throughout this section, we give definitions and rules only for $<$, but they are also valid for $>$, $\le$, and $\ge$.) Examples of linear inequalities include

$$x + 5 < 2, \quad x - 3 \ge 5, \quad \text{and} \quad 2k + 5 \le 10. \qquad \text{Linear inequalities}$$

OBJECTIVE 1 Solve linear inequalities by using the addition property. We solve an inequality by finding all numbers that make the inequality true. Usually, an inequality has an infinite number of solutions. These solutions, like solutions of equations, are found by producing a series of simpler related equivalent inequalities. **Equivalent inequalities** are inequalities with the same solution set. We use two important properties to produce equivalent inequalities.

Addition Property of Inequality

For all real numbers A, B, and C, the inequalities

$$A < B \qquad \text{and} \qquad A + C < B + C$$

are equivalent.

That is, adding the same number to each side of an inequality does not change the solution set.

EXAMPLE 1 Using the Addition Property of Inequality

Solve $x - 7 < -12$ and graph the solution set.

$$x - 7 < -12$$
$$x - 7 + 7 < -12 + 7 \qquad \text{Add 7.}$$
$$x < -5$$

Check: Substitute -5 for x in the *equation* $x - 7 = -12$.

$$x - 7 = -12$$
$$-5 - 7 = -12 \qquad ? \qquad \text{Let } x = -5.$$
$$-12 = -12 \qquad \text{True}$$

This shows that -5 is the boundary point. Now test a number on each side of -5 to verify that numbers *less than* -5 make the inequality true. Choose -4 and -6.

$$x - 7 < -12$$

$-4 - 7 < -12$? Let $x = -4$.	$-6 - 7 < -12$? Let $x = -6$.
$-11 < -12$ False	$-13 < -12$ True
-4 is not in the solution set.	-6 is in the solution set.

The check confirms that $(-\infty, -5)$, graphed in Figure 9, is the correct solution set.

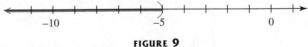

FIGURE 9

✔ **Now Try Exercise 11.**

As with equations, the addition property can be used to *subtract* the same number from each side of an inequality.

EXAMPLE 2 Using the Addition Property of Inequality

Solve $14 + 2m \leq 3m$ and graph the solution set.

$$14 + 2m \leq 3m$$
$$14 + 2m - 2m \leq 3m - 2m \qquad \text{Subtract } 2m.$$
$$14 \leq m \qquad \text{Combine like terms.}$$

Be careful. $\qquad m \geq 14 \qquad \text{Rewrite.}$

The inequality $14 \leq m$ (14 is less than or equal to m) can also be written $m \geq 14$ (m is greater than or equal to 14). Notice that in each case the inequality symbol points to the lesser number, 14.

Check: $\qquad 14 + 2m = 3m$
$$14 + 2(14) = 3(14) \qquad ? \qquad \text{Let } m = 14.$$
$$42 = 42 \qquad \text{True}$$

So 14 satisfies the equality part of $\leq$. Choose 10 and 15 as test points.

$$14 + 2m < 3m$$

$14 + 2(10) < 3(10)$? Let $m = 10$.	$14 + 2(15) < 3(15)$? Let $m = 15$.
$34 < 30$ False	$44 < 45$ True
10 is not in the solution set.	15 is in the solution set.

The check confirms that $[14, \infty)$ is the correct solution set. See Figure 10.

FIGURE 10

✔ **Now Try Exercise 27.**

OBJECTIVE ❷ Solve linear inequalities by using the multiplication property. Solving an inequality such as $3x \leq 15$ requires dividing each side by 3, using the *multiplication property of inequality*. To see how this property works, start with the true statement

$$-2 < 5.$$

Multiply each side by, say, 8.

$$-2(8) < 5(8) \qquad \text{Multiply by 8.}$$
$$-16 < 40 \qquad \text{True}$$

This gives a true statement. Start again with $-2 < 5$, and multiply each side by -8.

$$-2(-8) < 5(-8) \qquad \text{Multiply by } -8.$$
$$16 < -40 \qquad \text{False}$$

The result, $16 < -40$, is false. To make it true, we must change the direction of the inequality symbol to get

$$16 > -40. \qquad \text{True}$$

As these examples suggest, multiplying each side of an inequality by a *negative* number requires reversing the direction of the inequality symbol. The same is true for dividing by a negative number, since division is defined in terms of multiplication.

Multiplication Property of Inequality

For all real numbers A, B, and C, with $C \neq 0$,

(a) the inequalities

$$A < B \qquad \text{and} \qquad AC < BC \qquad \text{are equivalent if } C > 0;$$

(b) the inequalities

$$A < B \qquad \text{and} \qquad AC > BC \qquad \text{are equivalent if } C < 0.$$

That is, each side of an inequality may be multiplied (or divided) by a *positive* number without changing the direction of the inequality symbol. ***Multiplying (or dividing) by a negative number requires that we reverse the inequality symbol.***

EXAMPLE 3 Using the Multiplication Property of Inequality

Solve each inequality and graph the solution set.

(a) $5m \leq -30$

Divide each side by 5. ***Since 5 > 0, do not reverse the inequality symbol.***

$$5m \leq -30$$
$$\frac{5m}{5} \leq \frac{-30}{5} \qquad \text{Divide by 5.}$$
$$m \leq -6$$

Check that the solution set is the interval $(-\infty, -6]$, graphed in Figure 11.

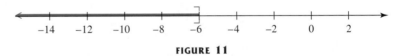

FIGURE 11

(b) $-4k \leq 32$

Divide each side by -4. ***Since $-4 < 0$, reverse the inequality symbol.***

$$-4k \leq 32$$
$$\frac{-4k}{-4} \geq \frac{32}{-4} \qquad \text{Divide by } -4; \text{ reverse the symbol.}$$

Reverse the inequality symbol when dividing by a *negative* number.

$$k \geq -8$$

Check the solution set. Figure 12 shows the graph of the solution set, $[-8, \infty)$.

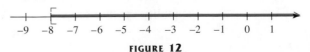

FIGURE 12

✔ **Now Try Exercises 15 and 19.**

The steps used in solving a linear inequality are given here.

Solving a Linear Inequality

Step 1 **Simplify each side separately.** Use the distributive property to clear parentheses and combine like terms as needed.

Step 2 **Isolate the variable terms on one side.** Use the addition property of inequality to get all terms with variables on one side of the inequality and all numbers on the other side.

Step 3 **Isolate the variable.** Use the multiplication property of inequality to change the inequality to the form

$$x < k \quad \text{or} \quad x > k.$$

▶ **CAUTION** *Reverse the direction of the inequality symbol when multiplying or dividing each side of an inequality by a negative number.*

EXAMPLE 4 **Solving a Linear Inequality by Using the Distributive Property**

Solve $-3(x + 4) + 2 \geq 7 - x$ and graph the solution set.

Step 1 $-3(x + 4) + 2 \geq 7 - x$

$\qquad -3x - 12 + 2 \geq 7 - x$ Distributive property

$\qquad -3x - 10 \geq 7 - x$

Step 2 $-3x - 10 + x \geq 7 - x + x$ Add x.

$\qquad -2x - 10 \geq 7$

$\qquad -2x - 10 + 10 \geq 7 + 10$ Add 10.

$\qquad -2x \geq 17$

Step 3 $\dfrac{-2x}{-2} \leq \dfrac{17}{-2}$ Divide by -2; change $\geq$ to $\leq$.

Be sure to reverse the inequality symbol. $x \leq -\dfrac{17}{2}$

Figure 13 shows the graph of the solution set, $\left(-\infty, -\frac{17}{2}\right]$.

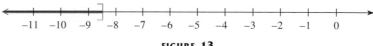

FIGURE 13

✔ **Now Try Exercise 31.**

▶ **NOTE** In Step 2 of Example 4, if we add $3x$ (instead of x) to both sides of the inequality, then we have

$$-3x - 10 + 3x \geq 7 - x + 3x \qquad \text{Add } 3x.$$
$$-10 \geq 2x + 7$$
$$-10 - 7 \geq 2x + 7 - 7 \qquad \text{Subtract 7.}$$
$$-17 \geq 2x$$
$$-\frac{17}{2} \geq x. \qquad\qquad \text{Divide by 2.}$$

The result, "$-\frac{17}{2}$ is greater than or equal to x," means the same thing as "x is less than or equal to $-\frac{17}{2}$." Thus, the solution set is the same.

EXAMPLE 5 **Solving a Linear Inequality with Fractions**

Solve $-\frac{2}{3}(r - 3) - \frac{1}{2} < \frac{1}{2}(5 - r)$ and graph the solution set.

To clear fractions, multiply each side by the least common denominator, 6.

$$-\frac{2}{3}(r - 3) - \frac{1}{2} < \frac{1}{2}(5 - r)$$

$$6\left[-\frac{2}{3}(r - 3) - \frac{1}{2}\right] < 6\left[\frac{1}{2}(5 - r)\right] \qquad \text{Multiply by 6.}$$

Be careful here.

$$6\left[-\frac{2}{3}(r - 3)\right] - 6\left(\frac{1}{2}\right) < 6\left[\frac{1}{2}(5 - r)\right] \qquad \text{Distributive property}$$

$$-4(r - 3) - 3 < 3(5 - r) \qquad \text{Multiply.}$$

Step 1
$$-4r + 12 - 3 < 15 - 3r \qquad \text{Distributive property}$$
$$-4r + 9 < 15 - 3r$$

Step 2
$$-4r + 9 + 3r < 15 - 3r + 3r \qquad \text{Add } 3r.$$
$$-r + 9 < 15$$
$$-r + 9 - 9 < 15 - 9 \qquad \text{Subtract 9.}$$
$$-r < 6$$

Step 3
$$-1(-r) > -1(6) \qquad \begin{array}{l}\text{Multiply by } -1;\\ \text{change} < \text{to} >.\end{array}$$

Remember to reverse the inequality symbol when multiplying by a negative *number.*

$$r > -6$$

Check that the solution set is $(-6, \infty)$. See the graph in Figure 14.

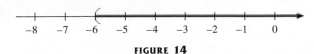

FIGURE 14

✔ **Now Try Exercise 37.**

OBJECTIVE 3 **Solve linear inequalities with three parts.** For some applications, it is necessary to work with a **three-part inequality** such as

$$3 < x + 2 < 8,$$

where $x + 2$ is *between* 3 and 8. To solve this inequality, we subtract 2 from each of the three parts of the inequality, giving

$$3 - 2 < x + 2 - 2 < 8 - 2$$
$$1 < x < 6.$$

Thus, x must be between 1 and 6, so that $x + 2$ will be between 3 and 8. The solution set, $(1, 6)$, is graphed in Figure 15.

FIGURE 15

▶ **CAUTION** *In three-part inequalities, the order of the parts is important.* It would be *wrong* to write an inequality as $8 < x + 2 < 3$, since this would imply that $8 < 3$, a false statement. *In general, three-part inequalities are written so that the symbols point in the same direction and both point toward the lesser number.*

EXAMPLE 6 Solving a Three-Part Inequality

Solve $-2 \le -3k - 1 \le 5$ and graph the solution set.

$$-2 \le -3k - 1 \le 5$$
$$-2 + 1 \le -3k - 1 + 1 \le 5 + 1 \qquad \text{Add 1 to each part.}$$
$$-1 \le -3k \le 6$$
$$\frac{-1}{-3} \ge \frac{-3k}{-3} \ge \frac{6}{-3} \qquad \text{Divide each part by } -3; \\ \text{reverse the inequality symbols.}$$
$$\frac{1}{3} \ge k \ge -2 \qquad \text{Rewrite in the order on the number line.}$$
$$-2 \le k \le \frac{1}{3}$$

Check that the solution set is $\left[-2, \frac{1}{3}\right]$, as shown in Figure 16.

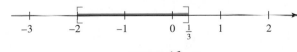

FIGURE 16

✔ Now Try Exercise 59.

The following table gives examples of the types of solution sets to be expected from solving linear equations or linear inequalities.

SOLUTION SETS OF LINEAR EQUATIONS AND INEQUALITIES

Equation or Inequality	Typical Solution Set	Graph of Solution Set
Linear equation $5x + 4 = 14$	$\{2\}$	
Linear inequality $5x + 4 < 14$ or $5x + 4 > 14$	$(-\infty, 2)$ $(2, \infty)$	
Three-part inequality $-1 \le 5x + 4 \le 14$	$[-1, 2]$	

OBJECTIVE **4** **Solve applied problems by using linear inequalities.** In addition to the familiar "is less than" and "is greater than," other expressions such as "is no more than" and "is at least" indicate inequalities, as shown in the table.

Word Expression	Interpretation
a is at least b	$a \ge b$
a is no less than b	$a \ge b$
a is at most b	$a \le b$
a is no more than b	$a \le b$

EXAMPLE 7 **Using a Linear Inequality to Solve a Rental Problem**

A rental company charges $15 to rent a chain saw, plus $2 per hr. Tom Ruhberg can spend no more than $35 to clear some logs from his yard. What is the *maximum* amount of time he can use the rented saw?

Step 1 **Read** the problem again.

Step 2 **Assign a variable.** Let $h =$ the number of hours he can rent the saw.

Step 3 **Write an inequality.** He must pay $15, plus $2h$, to rent the saw for h hours, and this amount must be *no more than* $35.

$$\underbrace{\text{Cost of renting}}_{15 + 2h} \quad \underbrace{\text{is no more than}}_{\le} \quad \underbrace{\text{35 dollars.}}_{35}$$

Step 4 **Solve.**
$$2h \le 20 \qquad \text{Subtract 15.}$$
$$h \le 10 \qquad \text{Divide by 2.}$$

Step 5 **State the answer.** He can use the saw for a maximum of 10 hr. (Of course, he may use it for less time, as indicated by the inequality $h \le 10$.)

Step 6 **Check.** If Tom uses the saw for 10 hr, he will spend $15 + 2(10) = 35$ dollars, the maximum amount.

✔ **Now Try Exercise 73.**

EXAMPLE 8 **Finding an Average Test Score**

Martha has scores of 88, 86, and 90 on her first three algebra tests. An average score of at least 90 will earn an A in the class. What possible scores on her fourth test will earn her an A average?

Let $x =$ the score on the fourth test. Her average score must be at least 90. To find the average of four numbers, add them and then divide by 4.

$$\underbrace{\frac{88 + 86 + 90 + x}{4}}_{\text{Average}} \quad \underbrace{\ge}_{\substack{\text{is at} \\ \text{least}}} \quad \underbrace{90}_{90.}$$

$$\frac{264 + x}{4} \ge 90 \qquad \text{Add the scores.}$$

$$264 + x \ge 360 \qquad \text{Multiply by 4.}$$

$$x \ge 96 \qquad \text{Subtract 264.}$$

She must score 96 or more on her fourth test.

Check: $\dfrac{88 + 86 + 90 + 96}{4} = \dfrac{360}{4} = 90,$ the minimum score.

A score of 96 or more will give an average of at least 90, as required.

✔ **Now Try Exercise 71.**

2.5 EXERCISES

● *Complete solution available on Video Lectures on CD/DVD*

▨ *Now Try Exercise*

Concept Check *Match each inequality in Column I with the correct graph or interval in Column II.*

I

1. $x \le 3$

2. $x > 3$

3. $x < 3$

4. $x \ge 3$

5. $-3 \le x \le 3$

6. $-3 < x < 3$

II

A. ![number line with bracket at 3 going right]

B. ![number line with parenthesis at 3 going left]

C. $(3, \infty)$

D. $(-\infty, 3]$

E. $(-3, 3)$

F. $[-3, 3]$

7. *Concept Check* Refer to the graph, and write an inequality or a three-part inequality for each description.

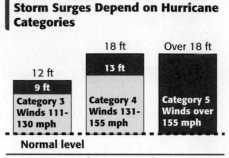

Storm Surges Depend on Hurricane Categories

Source: National Oceanic and Atmospheric Administration.

(a) The wind speed *s* (in miles per hour) of a Category 4 hurricane

(b) The wind speed *s* (in miles per hour) of a Category 5 hurricane

(c) The storm surge *x* (in feet) from a Category 3 hurricane

(d) The storm surge *x* (in feet) from a Category 5 hurricane

8. *Concept Check* Dr. Paul Donohue writes a syndicated column in which readers question him on a variety of health topics. Reader C. J. wrote, "Many people say they can weigh more because they have a large frame. How is frame size determined?" Here is Dr. Donohue's response:

> *"For a man, a wrist circumference between 6.75 and 7.25 in. [inclusive] indicates a medium frame. Anything above is a large frame and anything below, a small frame."*

Using *x* to represent wrist circumference in inches, write an inequality or a three-part inequality that represents wrist circumference for a male with a

(a) small frame   **(b)** medium frame   **(c)** large frame.

(*Source: The Gazette,* Cedar Rapids, Iowa, October 4, 2004.)

9. *Concept Check* A student solved the following inequality as shown.

$$4x \geq -64$$
$$\frac{4x}{4} \leq \frac{-64}{4}$$
$$x \leq -16$$

Solution set: $(-\infty, -16]$

WHAT WENT WRONG? Give the correct solution set.

10. Explain how to determine whether to use parentheses or brackets when graphing the solution set of an inequality.

Solve each inequality. Give the solution set in both interval and graph form. See Examples 1–5.

11. $x - 4 \geq 12$

12. $t - 3 \geq 7$

13. $3k + 1 > 22$

14. $5z + 6 < 76$

15. $4x < -16$

16. $2m > -10$

17. $-\dfrac{3}{4}r \geq 30$

18. $-\dfrac{2}{3}x \leq 12$

19. $-1.3m \geq -5.2$

20. $-2.5y \leq -1.25$

21. $5t + 2 \leq -48$

22. $4x + 1 \leq -31$

23. $\dfrac{5z - 6}{8} < 8$

24. $\dfrac{3k - 1}{4} > 5$

25. $\dfrac{2k - 5}{-4} > 5$

26. $\dfrac{3z - 2}{-5} < 6$

27. $6x - 4 \geq -2x$

28. $-2m + 8 \leq 2m$

29. $m - 2(m - 4) \leq 3m$

30. $x + 4(2x - 1) \geq x$

31. $-(4 + r) + 2 - 3r < -14$

32. $-(9 + k) - 5 + 4k \geq 4$

33. $-3(z - 6) > 2z - 2$

34. $-2(x + 4) \leq 6x + 16$

35. $\dfrac{2}{3}(3k - 1) \geq \dfrac{3}{2}(2k - 3)$

36. $\dfrac{7}{5}(10m - 1) < \dfrac{2}{3}(6m + 5)$

37. $-\dfrac{1}{4}(p + 6) + \dfrac{3}{2}(2p - 5) < 10$

38. $\dfrac{3}{5}(k - 2) - \dfrac{1}{4}(2k - 7) \leq 3$

39. $3(2x - 4) - 4x < 2x + 3$

40. $7(4 - x) + 5x < 2(16 - x)$

41. $8\left(\dfrac{1}{2}x + 3\right) < 8\left(\dfrac{1}{2}x - 1\right)$

42. $10x + 2(x - 4) < 12x - 10$

RELATING CONCEPTS (EXERCISES 43–47)

FOR INDIVIDUAL OR GROUP WORK

Work Exercises 43–47 in order.

43. Solve the linear equation $5(x + 3) - 2(x - 4) = 2(x + 7)$, and graph the solution set on a number line.

44. Solve the linear inequality $5(x + 3) - 2(x - 4) > 2(x + 7)$, and graph the solution set on a number line.

45. Solve the linear inequality $5(x + 3) - 2(x - 4) < 2(x + 7)$, and graph the solution set on a number line.

46. Graph all the solution sets of the equation and inequalities in Exercises 43–45 on the same number line. What set do you obtain?

47. Based on the results of Exercises 43–45, complete the following, using a conjecture (educated guess): The solution set of $-3(x + 2) = 3x + 12$ is $\{-3\}$, and the solution set of $-3(x + 2) < 3x + 12$ is $(-3, \infty)$. Therefore the solution set of $-3(x + 2) > 3x + 12$ is _____.

48. *Concept Check* Which is the graph of $-2 < x$?

A. ⟵————⟨——→
      -2

B. ⟵——⟩————→
         -2

C. ⟵————⟦——→
      -2

D. ⟵——⟧————→
         -2

Solve each inequality. Give the solution set in both interval and graph form. See Example 6.

49. $-4 < x - 5 < 6$

50. $-1 < x + 1 < 8$

51. $-9 \le k + 5 \le 15$

52. $-4 \le m + 3 \le 10$

53. $-6 \le 2z + 4 \le 16$

54. $-15 < 3p + 6 < -12$

55. $-19 \le 3x - 5 \le 1$

56. $-16 < 3t + 2 < -10$

57. $-1 \le \dfrac{2x - 5}{6} \le 5$

58. $-3 \le \dfrac{3m + 1}{4} \le 3$

59. $4 \le -9x + 5 < 8$

60. $4 \le -2x + 3 < 8$

Find the unknown numbers in each description.

61. Six times a number is between -12 and 12.

62. Half a number is between -3 and 2.

63. When 1 is added to twice a number, the result is greater than or equal to 7.

64. If 8 is subtracted from a number, then the result is at least 5.

65. One third of a number is added to 6, giving a result of at least 3.

66. Three times a number, minus 5, is no more than 7.

The weather forecast by time of day for the U.S. Olympic Track and Field Trials in Sacramento, California, is shown in the figure. Use this graph to work Exercises 67–70.

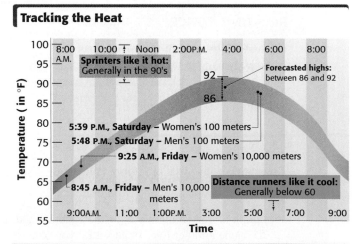

Source: Accuweather, Bee research.

67. Sprinters prefer Fahrenheit temperatures in the 90s. Using the upper boundary of the forecast, during what period is the temperature expected to be at least 90°F?

68. Distance runners prefer cool temperatures. During what period are temperatures predicted to be no more than 70°F? Use the lower forecast boundary.

69. What range of temperatures is predicted for the Women's 100-m event?

70. What range of temperatures is forecast for the Men's 10,000-m event?

Solve each problem. See Examples 7 and 8.

71. Finley Westmoreland earned scores of 90 and 82 on his first two tests in English literature. What score must he make on his third test to keep an average of 84 or greater?

72. Jack Hornsby scored 92 and 96 on his first two tests in "Methods in Teaching Mathematics." What score must he make on his third test to keep an average of 90 or greater?

73. Amber is signing up for cell phone service. She is trying to decide between Plan A, which costs $54.99 a month with a free phone included, and Plan B, which costs $49.99 a month, but would require her to buy a phone for $129. Under either plan, Amber does not expect to go over the included number of monthly minutes. After how many months would Plan B be a better deal?

74. Craig and Julie Phillips need to rent a truck to move their belongings to their new apartment. They can rent a truck of the size they need from U-Haul for $29.95 a day plus 28 cents per mile or from Budget Truck Rentals for $34.95 a day plus 25 cents per mile. After how many miles would the Budget rental be a better deal than the U-Haul one?

A product will produce a profit only when the revenue R from selling the product exceeds the cost C of producing it. In Exercises 75 and 76, find the least whole number of units x that must be sold for the business to show a profit for the item described.

75. Peripheral Visions, Inc., finds that the cost of producing x studio-quality DVDs is $C = 20x + 100$, while the revenue produced from them is $R = 24x$ (C and R in dollars).

76. Speedy Delivery finds that the cost of making x deliveries is $C = 3x + 2300$, while the revenue produced from them is $R = 5.50x$ (C and R in dollars).

77. A body mass index (BMI) between 19 and 25 is considered healthy. Use the formula

$$\text{BMI} = \frac{704 \times (\text{weight in pounds})}{(\text{height in inches})^2}$$

to find the weight range w, to the nearest pound, that gives a healthy BMI for each height. (*Source: Washington Post.*)

(a) 72 in. **(b)** Your height in inches

78. To achieve the maximum benefit from exercising, the heart rate, in beats per minute, should be in the target heart rate (THR) zone. For a person aged A, the formula is

$$0.7(220 - A) \leq \text{THR} \leq 0.85(220 - A).$$

Find the THR to the nearest whole number for each age. (*Source:* Hockey, Robert V., *Physical Fitness: The Pathway to Healthful Living,* Times Mirror/Mosby College Publishing, 1989.)

(a) 35   **(b)** Your age

PREVIEW EXERCISES

*Each exercise requires the graph of two inequalities. Graph them and respond to the statement that follows. See **Section 1.1.***

79. (a) Graph $x > 4$.   **(b)** Graph $x < 5$.
(c) Describe the set of numbers belonging to *both* of these sets.

80. (a) Graph $m < 7$.   **(b)** Graph $m < 9$.
(c) Describe the set of numbers belonging to *both* of these sets.

81. (a) Graph $t < 5$.   **(b)** Graph $t > 4$.
(c) Describe the set of numbers belonging to *either one or both* of these sets.

82. (a) Graph $s < -3$.   **(b)** Graph $s > -1$.
(c) Describe the set of numbers belonging to *neither* of these sets.

2.6 Set Operations and Compound Inequalities

The table shows symptoms of an underactive thyroid and an overactive thyroid.

Underactive Thyroid	Overactive Thyroid
Sleepiness, s	Insomnia, i
Dry hands, d	Moist hands, m
Intolerance of cold, c	Intolerance of heat, h
Goiter, g	Goiter, g

Source: The Merck Manual of Diagnosis and Therapy, 16th Edition, Merck Research Laboratories, 1992.

Let N be the set of symptoms of an underactive thyroid, and let O be the set of symptoms of an overactive thyroid. Suppose we are interested in the set of symptoms that are found in *both* sets N and O. In this section, we discuss the use of the words *and* and *or* as they relate to sets and inequalities.

OBJECTIVE 1 Find the intersection of two sets. The intersection of two sets is defined with the word *and*.

Intersection of Sets

For any two sets A and B, the **intersection** of A and B, symbolized $A \cap B$, is defined as follows:
$$A \cap B = \{x \mid x \text{ is an element of } A \text{ and } x \text{ is an element of } B\}.$$

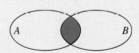

EXAMPLE 1 Finding the Intersection of Two Sets

Let $A = \{1, 2, 3, 4\}$ and $B = \{2, 4, 6\}$. Find $A \cap B$.

The set $A \cap B$ contains those elements that belong to both A *and* B: the numbers 2 and 4. Therefore,
$$A \cap B = \{1, 2, 3, 4\} \cap \{2, 4, 6\}$$
$$= \{2, 4\}.$$

✔ **Now Try Exercise 7.**

A **compound inequality** consists of two inequalities linked by a connective word such as *and* or *or*. Examples of compound inequalities are
$$x + 1 \le 9 \quad \text{and} \quad x - 2 \ge 3$$
$$2x > 4 \quad \text{or} \quad 3x - 6 < 5. \qquad \text{Compound inequalities}$$

OBJECTIVE 2 Solve compound inequalities with the word *and*. Use the following steps.

Solving a Compound Inequality with *and*

Step 1   Solve each inequality individually.

Step 2   Since the inequalities are joined with *and,* the solution set of the compound inequality will include all numbers that satisfy both inequalities in Step 1 (the intersection of the solution sets).

EXAMPLE 2   **Solving a Compound Inequality with *and***

Solve the compound inequality

$$x + 1 \leq 9 \quad \text{and} \quad x - 2 \geq 3.$$

Step 1   Solve each inequality individually.

$$
\begin{array}{ccc}
x + 1 \leq 9 & \text{and} & x - 2 \geq 3 \\
x + 1 - 1 \leq 9 - 1 & \text{and} & x - 2 + 2 \geq 3 + 2 \\
x \leq 8 & \text{and} & x \geq 5
\end{array}
$$

Step 2   Because the inequalities are joined with the word *and,* the solution set will include all numbers that satisfy both inequalities in Step 1 at the same time. Thus, the compound inequality is true whenever $x \leq 8$ and $x \geq 5$ are both true. The top graph in Figure 17 shows $x \leq 8$, and the bottom graph shows $x \geq 5$.

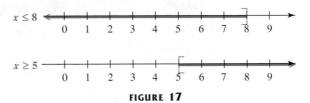

FIGURE 17

Find the intersection of the two graphs in Figure 17 to get the solution set of the compound inequality. Figure 18 shows that the solution set, in interval notation, is $[5, 8]$.

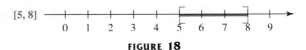

FIGURE 18

✔ **Now Try Exercise 25.**

EXAMPLE 3   **Solving a Compound Inequality with *and***

Solve the compound inequality

$$-3x - 2 > 5 \quad \text{and} \quad 5x - 1 \leq -21.$$

Step 1   Solve each inequality individually.

$$
\begin{array}{ccc}
-3x - 2 > 5 & \text{and} & 5x - 1 \leq -21 \\
-3x > 7 & \text{and} & 5x \leq -20 \\
x < -\dfrac{7}{3} & \text{and} & x \leq -4
\end{array}
$$

Remember to reverse the inequality symbol.

The graphs of $x < -\frac{7}{3}$ and $x \le -4$ are shown in Figure 19.

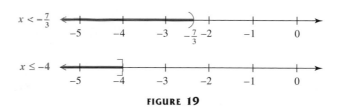

FIGURE 19

Step 2 Now find all values of x that satisfy both conditions; that is, find the real numbers that are less than $-\frac{7}{3}$ and also less than or equal to -4. As shown by the graph in Figure 20, the solution set is $(-\infty, -4]$.

FIGURE 20

> ✔ **Now Try Exercise 29.**

EXAMPLE 4 Solving a Compound Inequality with *and*

Solve $x + 2 < 5$ and $x - 10 > 2$.

First solve each inequality individually.

$$x + 2 < 5 \quad \text{and} \quad x - 10 > 2$$
$$x < 3 \quad \text{and} \quad x > 12$$

The graphs of $x < 3$ and $x > 12$ are shown in Figure 21.

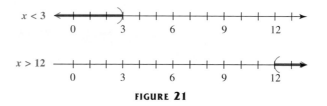

FIGURE 21

There is no number that is both less than 3 *and* greater than 12, so the given compound inequality has no solution. The solution set is $\emptyset$. See Figure 22.

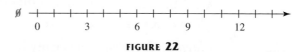

FIGURE 22

> ✔ **Now Try Exercise 23.**

OBJECTIVE 3 **Find the union of two sets.** The union of two sets is defined with the word *or*.

Union of Sets

For any two sets A and B, the **union** of A and B, symbolized $A \cup B$, is defined as follows:

$$A \cup B = \{x \mid x \text{ is an element of } A \textbf{ or } x \text{ is an element of } B\}.$$

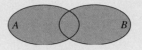

EXAMPLE 5 Finding the Union of Two Sets

Let $A = \{1, 2, 3, 4\}$ and $B = \{2, 4, 6\}$. Find $A \cup B$.

Begin by listing all the elements of set A: 1, 2, 3, 4. Then list any additional elements from set B. In this case the elements 2 and 4 are already listed, so the only additional element is 6. Therefore,

$$\begin{aligned} A \cup B &= \{1, 2, 3, 4\} \cup \{2, 4, 6\} \\ &= \{1, 2, 3, 4, 6\}. \end{aligned}$$

The union consists of all elements in either A *or* B (or both).

✔ **Now Try Exercise 13.**

▶ **NOTE** In Example 5, notice that although the elements 2 and 4 appeared in both sets A and B, they are written only once in $A \cup B$.

OBJECTIVE 4 Solve compound inequalities with the word *or*. Use the following steps.

Solving a Compound Inequality with *or*

Step 1 Solve each inequality individually.

Step 2 Since the inequalities are joined with *or*, the solution set of the compound inequality includes all numbers that satisfy either one of the two inequalities in Step 1 (the union of the solution sets).

EXAMPLE 6 Solving a Compound Inequality with *or*

Solve $6x - 4 < 2x$ or $-3x \leq -9$.

Step 1 Solve each inequality individually.

$$6x - 4 < 2x \quad \text{or} \quad -3x \leq -9$$
$$4x < 4$$
$$x < 1 \quad \text{or} \quad x \geq 3$$

> Remember to reverse the inequality symbol.

The graphs of these two inequalities are shown in Figure 23 on the next page.

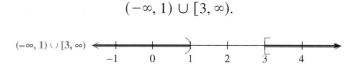

FIGURE 23

Step 2 Since the inequalities are joined with *or,* find the union of the two solution sets. The union is shown in Figure 24 and is written

$$(-\infty, 1) \cup [3, \infty).$$

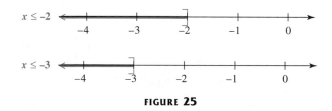

FIGURE 24

✔**Now Try Exercise 41.**

▶ **CAUTION** When inequalities are used to write the solution set in Example 6, it *must* be written as

$$x < 1 \quad \text{or} \quad x \geq 3,$$

which keeps the numbers 1 and 3 in their order on the number line. Writing $3 \leq x < 1$ would imply that $3 \leq 1$, which is ***FALSE.*** There is no other way to write the solution set of such a union.

EXAMPLE 7 **Solving a Compound Inequality with *or***

Solve $-4x + 1 \geq 9$ or $5x + 3 \leq -12$.

First we solve each inequality individually.

$$-4x + 1 \geq 9 \quad \text{or} \quad 5x + 3 \leq -12$$
$$-4x \geq 8 \quad \text{or} \quad 5x \leq -15$$
$$x \leq -2 \quad \text{or} \quad x \leq -3$$

The graphs of these two inequalities are shown in Figure 25.

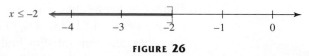

FIGURE 25

By taking the union, we obtain the interval $(-\infty, -2]$. See Figure 26.

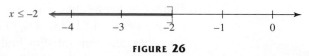

FIGURE 26

✔**Now Try Exercise 35.**

EXAMPLE 8 Solving a Compound Inequality with *or*

Solve $-2x + 5 \geq 11$ or $4x - 7 \geq -27$.

$$-2x + 5 \geq 11 \quad \text{or} \quad 4x - 7 \geq -27$$
$$-2x \geq 6 \quad \text{or} \quad 4x \geq -20$$
$$x \leq -3 \quad \text{or} \quad x \geq -5$$

The graphs of these two inequalities are shown in Figure 27.

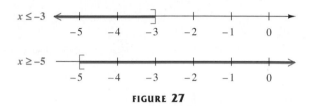

FIGURE 27

By taking the union, we obtain every real number as a solution, since every real number satisfies at least one of the two inequalities. The set of all real numbers is written in interval notation as $(-\infty, \infty)$ and graphed as in Figure 28.

FIGURE 28

✔ Now Try Exercise 45.

EXAMPLE 9 Applying Intersection and Union

The five highest-grossing domestic films (adjusted for inflation) as of July, 2005, are listed in the table.

FIVE ALL-TIME HIGHEST-GROSSING DOMESTIC FILMS

Film	Admissions	Gross Income
Gone with the Wind	202,044,569	$1,293,085,000
Star Wars	178,119,595	$1,139,965,000
The Sound of Music	142,415,376	$911,458,000
E.T.	141,925,359	$908,322,298
The Ten Commandments	131,000,000	$838,400,000

Source: Exhibitor Relations Co., Inc.

List the elements of the following sets.

(a) The set of the top five films with admissions greater than 180,000,000 *and* gross income greater than $1,000,000,000

The only film that satisfies both conditions is *Gone with the Wind,* so the set is

{*Gone with the Wind*}.

(b) The set of the top five films with admissions less than 170,000,000 *or* gross income greater than $1,000,000,000

Here, any film that satisfies at least one of the conditions is in the set. This set includes all five films:

{*Gone with the Wind, Star Wars, The Sound of Music, E.T., The Ten Commandments*}.

✔ **Now Try Exercise 63.**

2.6 EXERCISES

⊕ *Complete solution available on Video Lectures on CD/DVD*

▢ *Now Try Exercise*

Concept Check *Decide whether each statement is* true *or* false. *If it is false, explain why.*

1. The union of the solution sets of $x + 1 = 5$, $x + 1 < 5$, and $x + 1 > 5$ is $(-\infty, \infty)$.

2. The intersection of the sets $\{x \mid x \geq 7\}$ and $\{x \mid x \leq 7\}$ is $\emptyset$.

3. The union of the sets $(-\infty, 8)$ and $(8, \infty)$ is $\{8\}$.

4. The intersection of the sets $(-\infty, 8]$ and $[8, \infty)$ is $\{8\}$.

5. The intersection of the set of rational numbers and the set of irrational numbers is $\{0\}$.

6. The union of the set of rational numbers and the set of irrational numbers is the set of real numbers.

Let $A = \{1, 2, 3, 4, 5, 6\}$, $B = \{1, 3, 5\}$, $C = \{1, 6\}$, and $D = \{4\}$. Specify each set. See Examples 1 and 5.

⊕ **7.** $B \cap A$ **8.** $A \cap B$ **9.** $A \cap D$ **10.** $B \cap C$

11. $B \cap \emptyset$ **12.** $A \cap \emptyset$ ⊕ **13.** $A \cup B$ **14.** $B \cup D$

Concept Check *Two sets are specified by graphs. Graph the intersection of the two sets.*

15. 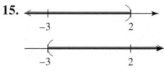   **16.**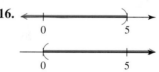

17. 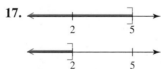   **18.**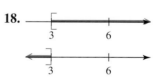

For each compound inequality, give the solution set in both interval and graph form. See Examples 2–4.

19. $x < 2$ and $x > -3$ **20.** $x < 5$ and $x > 0$ **21.** $x \leq 2$ and $x \leq 5$

22. $x \geq 3$ and $x \geq 6$ ⊕ **23.** $x \leq 3$ and $x \geq 6$ **24.** $x \leq -1$ and $x \geq 3$

⊕ **25.** $x - 3 \leq 6$ and $x + 2 \geq 7$ **26.** $x + 5 \leq 11$ and $x - 3 \geq -1$

27. $-3x > 3$ and $x + 3 > 0$ **28.** $-3x < 3$ and $x + 2 < 6$

⊕ **29.** $3x - 4 \leq 8$ and $-4x + 1 \geq -15$ **30.** $7x + 6 \leq 48$ and $-4x \geq -24$

Concept Check *Two sets are specified by graphs. Graph the union of the two sets.*

31.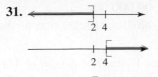

32.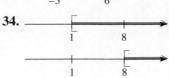

33.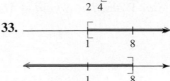

34.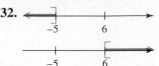

For each compound inequality, give the solution set in both interval and graph form. See Examples 6–8.

35. $x \leq 1$ or $x \leq 8$ **36.** $x \geq 1$ or $x \geq 8$ **37.** $x \geq -2$ or $x \geq 5$

38. $x \leq -2$ or $x \leq 6$ **39.** $x \geq -2$ or $x \leq 4$ **40.** $x \geq 5$ or $x \leq 7$

41. $x + 2 > 7$ or $1 - x > 6$ **42.** $x + 1 > 3$ or $x + 4 < 2$

43. $x + 1 > 3$ or $-4x + 1 > 5$ **44.** $3x < x + 12$ or $x + 1 > 10$

45. $4x + 1 \geq -7$ or $-2x + 3 \geq 5$ **46.** $3x + 2 \leq -7$ or $-2x + 1 \leq 9$

Concept Check *Express each set in the simplest interval form. (Hint: Graph each set and look for the intersection or union.)*

47. $(-\infty, -1] \cap [-4, \infty)$ **48.** $[-1, \infty) \cap (-\infty, 9]$

49. $(-\infty, -6] \cap [-9, \infty)$ **50.** $(5, 11] \cap [6, \infty)$

51. $(-\infty, 3) \cup (-\infty, -2)$ **52.** $[-9, 1] \cup (-\infty, -3)$

53. $[3, 6] \cup (4, 9)$ **54.** $[-1, 2] \cup (0, 5)$

For each compound inequality, decide whether intersection *or* union *should be used. Then give the solution set in both interval and graph form. See Examples 2–4 and 6–8.*

55. $x < -1$ and $x > -5$ **56.** $x > -1$ and $x < 7$

57. $x < 4$ or $x < -2$ **58.** $x < 5$ or $x < -3$

59. $-3x \leq -6$ or $-3x \geq 0$ **60.** $2x - 6 \leq -18$ and $2x \geq -18$

61. $x + 1 \geq 5$ and $x - 2 \leq 10$ **62.** $-8x \leq -24$ or $-5x \geq 15$

Average expenses for full-time college students at all two- and four-year institutions during a recent academic year are shown in the table.

COLLEGE EXPENSES (IN DOLLARS)

Type of Expense	Public Schools	Private Schools
Tuition and fees	2928	16,517
Board rates	2702	3236
Dormitory charges	2925	3750

Source: U.S. National Center for Education Statistics.

Refer to the table on college expenses on the preceding page. List the elements of each set.
See Example 9.

63. The set of expenses that are less than $3000 for public schools *and* are greater than $5000 for private schools

64. The set of expenses that are less than $2800 for public schools *and* are less than $4000 for private schools

65. The set of expenses that are greater than $2900 for public schools *or* are greater than $5000 for private schools

66. The set of expenses that are greater than $4,000 *or* are less than $2700

RELATING CONCEPTS (EXERCISES 67–72)

FOR INDIVIDUAL OR GROUP WORK

The figures represent the backyards of neighbors Luigi, Maria, Than, and Joe. Find the area and the perimeter of each yard. Suppose that each resident has 150 ft of fencing and enough sod to cover 1400 ft² of lawn. Give the name or names of the residents whose yards satisfy each description. **Work Exercises 67–72 in order.**

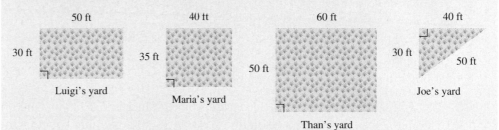

67. The yard can be fenced *and* the yard can be sodded.

68. The yard can be fenced *and* the yard cannot be sodded.

69. The yard cannot be fenced *and* the yard can be sodded.

70. The yard cannot be fenced *and* the yard cannot be sodded.

71. The yard can be fenced *or* the yard can be sodded.

72. The yard cannot be fenced *or* the yard can be sodded.

PREVIEW EXERCISES

*Solve each inequality. See **Section 2.5.***

73. $2y - 4 \le 3y + 2$

74. $5t - 8 < 6t - 7$

75. $-5 < 2r + 1 < 5$

76. $-7 \le 3w - 2 < 7$

*Evaluate. See **Sections 1.1 and 1.2.***

77. $-|6| - |-11| + (-4)$

78. $(-5) - |-9| + |5 - 4|$

2.7   Absolute Value Equations and Inequalities

In a production line, quality is controlled by randomly choosing items from the line and checking to see how selected measurements vary from the optimum measure. The differences are sometimes positive and sometimes negative, so they are expressed as absolute values. For example, a machine that fills quart milk cartons might be set to release 1 qt (32 oz), plus or minus 2 oz per carton. Then the number of ounces in each carton should satisfy the *absolute value inequality* $|x - 32| \leq 2$, where x is the number of ounces.

OBJECTIVE 1   Use the distance definition of absolute value.   In **Section 1.1,** we saw that the absolute value of a number x, written $|x|$, represents the distance from x to 0 on the number line. For example, the solutions of $|x| = 4$ are 4 and -4, as shown in Figure 29.

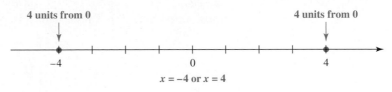

$x = -4$ or $x = 4$

FIGURE 29

Because absolute value represents distance from 0, it is reasonable to interpret the solutions of $|x| > 4$ to be all numbers that are *more* than four units from 0. The set $(-\infty, -4) \cup (4, \infty)$ fits this description. Figure 30 shows the graph of the solution set of $|x| > 4$. Because the graph consists of two separate intervals, the solution set is described with *or*: $x < -4$ or $x > 4$.

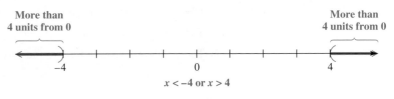

$x < -4$ or $x > 4$

FIGURE 30

The solution set of $|x| < 4$ consists of all numbers that are *less* than 4 units from 0 on the number line. Another way of thinking about this is to think of all numbers *between* -4 and 4. This set of numbers is given by $(-4, 4)$, as shown in Figure 31. Here, the graph shows that $-4 < x < 4$, which means $x > -4$ *and* $x < 4$.

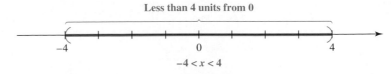

$-4 < x < 4$

FIGURE 31

The equation and inequalities just described are examples of **absolute value equations and inequalities.** They involve the absolute value of a variable expression and generally take the form

$$|ax + b| = k, \qquad |ax + b| > k, \qquad \text{or} \qquad |ax + b| < k,$$

where k is a positive number. From Figures 29–31, we see that

$|x| = 4$ has the same solution set as $x = -4$ or $x = 4$,

$|x| > 4$ has the same solution set as $x < -4$ or $x > 4$,

$|x| < 4$ has the same solution set as $x > -4$ and $x < 4$.

Thus, we solve an absolute value equation or inequality by solving the appropriate compound equation or inequality.

Solving Absolute Value Equations and Inequalities

Let k be a positive real number and p and q be real numbers.

1. To solve $|ax + b| = k,$ solve the compound equation

$$ax + b = k \quad \text{or} \quad ax + b = -k.$$

The solution set is usually of the form $\{p, q\}$, which includes two numbers.

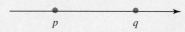

2. To solve $|ax + b| > k,$ solve the compound inequality

$$ax + b > k \quad \text{or} \quad ax + b < -k.$$

The solution set is of the form $(-\infty, p) \cup (q, \infty)$, which consists of two separate intervals.

3. To solve $|ax + b| < k,$ solve the three-part inequality

$$-k < ax + b < k.$$

The solution set is of the form (p, q), a single interval.

▶ **NOTE** Some people prefer to write the compound statements in parts 1 and 2 of the preceding box as

$$ax + b = k \quad \text{or} \quad -(ax + b) = k$$

and

$$ax + b > k \quad \text{or} \quad -(ax + b) > k.$$

These forms produce the same results.

OBJECTIVE 2 Solve equations of the form $|ax + b| = k$, for $k > 0$. *Remember that because absolute value refers to distance from the origin, an absolute value equation will have two parts.*

EXAMPLE 1 Solving an Absolute Value Equation

Solve $|2x + 1| = 7$.

For $|2x + 1|$ to equal 7, $2x + 1$ must be 7 units from 0 on the number line. This can happen only when $2x + 1 = 7$ or $2x + 1 = -7$. This is the first case in the preceding box. Solve this compound equation as follows:

$$2x + 1 = 7 \quad \text{or} \quad 2x + 1 = -7$$
$$2x = 6 \quad \text{or} \quad 2x = -8$$
$$x = 3 \quad \text{or} \quad x = -4.$$

Check by substituting 3 and then -4 into the original absolute value equation to verify that the solution set is $\{-4, 3\}$. The graph is shown in Figure 32.

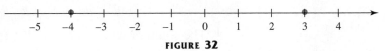

FIGURE 32

✔ **Now Try Exercise 11.**

OBJECTIVE 3 Solve inequalities of the form $|ax + b| < k$ and of the form $|ax + b| > k$, for $k > 0$.

EXAMPLE 2 Solving an Absolute Value Inequality with >

Solve $|2x + 1| > 7$.

By Case 2 of the previous box, this absolute value inequality is rewritten as

$$2x + 1 > 7 \quad \text{or} \quad 2x + 1 < -7,$$

because $2x + 1$ must represent a number that is *more* than 7 units from 0 on either side of the number line. Now, solve the compound inequality.

$$2x + 1 > 7 \quad \text{or} \quad 2x + 1 < -7$$
$$2x > 6 \quad \text{or} \quad 2x < -8$$
$$x > 3 \quad \text{or} \quad x < -4$$

Check these solutions. The solution set is $(-\infty, -4) \cup (3, \infty)$. See Figure 33. Notice that the graph consists of two intervals.

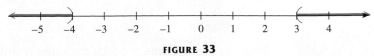

FIGURE 33

✔ **Now Try Exercise 25.**

EXAMPLE 3 Solving an Absolute Value Inequality with <

Solve $|2x + 1| < 7$.

The expression $2x + 1$ must represent a number that is less than 7 units from 0 on either side of the number line. Another way of thinking about this is to realize that $2x + 1$ must be between -7 and 7. As Case 3 of the previous box shows, that relationship is written as a three-part inequality.

$$-7 < 2x + 1 < 7$$
$$-8 < 2x < 6 \qquad \text{Subtract 1 from each part.}$$
$$-4 < x < 3 \qquad \text{Divide each part by 2.}$$

Check that the solution set is $(-4, 3)$. The graph consists of the single interval shown in Figure 34.

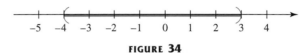

FIGURE 34

✔ **Now Try Exercise 39.**

Look back at Figures 32, 33, and 34, with the graphs of $|2x + 1| = 7$, $|2x + 1| > 7$, and $|2x + 1| < 7$, respectively. If we find the union of the three sets, we get the set of all real numbers. This is because, for any value of x, $|2x + 1|$ will satisfy one and only one of the following: It is equal to 7, greater than 7, or less than 7.

▶ **CAUTION** When solving absolute value equations and inequalities of the types in Examples 1, 2, and 3, remember the following.

1. The methods described apply when the constant is alone on one side of the equation or inequality and is *positive.*

2. Absolute value equations and absolute value inequalities of the form $|ax + b| > k$ translate into "or" compound statements.

3. Absolute value inequalities of the form $|ax + b| < k$ translate into "and" compound statements, which may be written as three-part inequalities.

4. An "or" statement *cannot* be written in three parts. It would be incorrect to write $-7 > 2x + 1 > 7$ in Example 2, because this would imply that $-7 > 7$, which is *false.*

OBJECTIVE 4 Solve absolute value equations that involve rewriting. Sometimes an absolute value equation or inequality requires some rewriting before it can be set up as a compound statement, as shown in the next example.

EXAMPLE 4 Solving an Absolute Value Equation That Requires Rewriting

Solve $|x + 3| + 5 = 12$.

First get the absolute value alone on one side of the equals sign.

$$|x + 3| + 5 - 5 = 12 - 5 \qquad \text{Subtract 5.}$$
$$|x + 3| = 7$$

Now use the method shown in Example 1 to solve $|x + 3| = 7$.

$$x + 3 = 7 \quad \text{or} \quad x + 3 = -7$$
$$x = 4 \quad \text{or} \quad x = -10$$

Check that the solution set is $\{-10, 4\}$ by substituting into the original equation.

> We write solutions in the order in which they appear on a number line.

✔ **Now Try Exercise 65.**

We use a similar method to solve an absolute value *inequality* that requires rewriting:

$\|x + 3\| + 5 \geq 12$	$\|x + 3\| + 5 \leq 12$
$\|x + 3\| \geq 7$	$\|x + 3\| \leq 7$
$x + 3 \geq 7 \quad \text{or} \quad x + 3 \leq -7$	$-7 \leq x + 3 \leq 7$
$x \geq 4 \quad \text{or} \quad x \leq -10.$	$-10 \leq x \leq 4.$
Solution set: $(-\infty, -10] \cup [4, \infty)$	Solution set: $[-10, 4]$

OBJECTIVE 5 **Solve equations of the form $|ax + b| = |cx + d|$.** By definition, for two expressions to have the same absolute value, they must either be equal or be negatives of each other.

Solving $|ax + b| = |cx + d|$

To solve an absolute value equation of the form
$$|ax + b| = |cx + d|,$$
solve the compound equation
$$ax + b = cx + d \quad \text{or} \quad ax + b = -(cx + d).$$

EXAMPLE 5 **Solving an Equation with Two Absolute Values**

Solve $|z + 6| = |2z - 3|$.

This equation is satisfied either if $z + 6$ and $2z - 3$ are equal to each other or if $z + 6$ and $2z - 3$ are negatives of each other.

$$z + 6 = 2z - 3 \quad \text{or} \quad z + 6 = -(2z - 3)$$
$$z + 9 = 2z \quad \text{or} \quad z + 6 = -2z + 3$$
$$9 = z \quad \text{or} \quad 3z = -3$$
$$z = -1$$

Check that the solution set is $\{-1, 9\}$.

✔ **Now Try Exercise 71.**

OBJECTIVE 6 Solve special cases of absolute value equations and inequalities. When an absolute value equation or inequality involves a *negative constant or 0* alone on one side, use the properties of absolute value to solve the equation or inequality. Keep the following in mind.

Special Cases of Absolute Value

1. The absolute value of an expression can never be negative; that is, $|a| \geq 0$ for all real numbers a.

2. The absolute value of an expression equals 0 only when the expression is equal to 0.

EXAMPLE 6 Solving Special Cases of Absolute Value Equations

Solve each equation.

(a) $|5r - 3| = -4$

See Case 1 in the preceding box. *The absolute value of an expression can never be negative,* so there are no solutions for this equation. The solution set is $\emptyset$.

(b) $|7x - 3| = 0$

See Case 2 in the preceding box. The expression $7x - 3$ will equal 0 *only* if

$$7x - 3 = 0.$$

The solution of this equation is $\frac{3}{7}$. Thus, the solution set is $\left\{\frac{3}{7}\right\}$, with just one element. Check by substitution into the original equation.

✔ **Now Try Exercises 81 and 83.**

EXAMPLE 7 Solving Special Cases of Absolute Value Inequalities

Solve each inequality.

(a) $|x| \geq -4$

The absolute value of a number is always greater than or equal to 0. Thus, $|x| \geq -4$ is true for *all* real numbers. The solution set is $(-\infty, \infty)$.

(b)
$$|k + 6| - 3 < -5$$
$$|k + 6| < -2 \qquad \text{Add 3 to each side.}$$

There is no number whose absolute value is less than -2, so this inequality has no solution. The solution set is $\emptyset$.

(c)
$$|m - 7| + 4 \leq 4$$
$$|m - 7| \leq 0 \qquad \text{Subtract 4 from each side.}$$

The value of $|m - 7|$ will never be less than 0. However, $|m - 7|$ will equal 0 when $m = 7$. Therefore, the solution set is $\{7\}$.

✔ **Now Try Exercises 79, 89, and 95.**

CONNECTIONS

Absolute value is used to find the relative error of a measurement in science, engineering, manufacturing, and other fields. If x_t represents the expected value of a measurement and x represents the actual measurement, then the relative error in x equals the absolute value of the difference between x_t and x, divided by x_t. That is,

$$\text{relative error in } x = \left| \frac{x_t - x}{x_t} \right|.$$

In many situations in the work world, the relative error must be less than some predetermined amount. For example, suppose a machine filling *quart* milk cartons is set for a relative error no greater than 0.05. Here $x_t = 32$ oz, the relative error $= 0.05$ oz, and we must find x, given that

$$\left| \frac{32 - x}{32} \right| \le 0.05.$$

For Discussion or Writing

With this tolerance level, how many ounces may a carton contain?

2.7　EXERCISES

Concept Check　*Match each absolute value equation or inequality in Column I with the graph of its solution set in Column II.*

I	II	I	II
1. $|x| = 5$　**A.**		**2.** $|x| = 9$　**A.**	
$|x| < 5$　**B.**		$|x| > 9$　**B.**	
$|x| > 5$　**C.**		$|x| \ge 9$　**C.**	
$|x| \le 5$　**D.**		$|x| < 9$　**D.**	
$|x| \ge 5$　**E.**		$|x| \le 9$　**E.**	

3. *Concept Check*　How many solutions will $|ax + b| = k$ have if

(a) $k = 0$　(b) $k > 0$　(c) $k < 0$?

4. Explain when to use *and* and when to use *or* if you are solving an absolute value equation or inequality of the form $|ax + b| = k$, $|ax + b| < k$, or $|ax + b| > k$, where k is a positive number.

Solve each equation. See Example 1.

5. $|x| = 12$

6. $|k| = 14$

7. $|4x| = 20$

8. $|5x| = 30$

9. $|y - 3| = 9$

10. $|p - 5| = 13$

11. $|2x - 1| = 11$

12. $|2y + 3| = 19$

13. $|4r - 5| = 17$

14. $|5t - 1| = 21$

15. $|2y + 5| = 14$

16. $|2x - 9| = 18$

17. $\left| \dfrac{1}{2}x + 3 \right| = 2$

18. $\left| \dfrac{2}{3}q - 1 \right| = 5$

19. $\left| 1 + \dfrac{3}{4}k \right| = 7$

20. $\left| 2 - \dfrac{5}{2}m \right| = 14$

Solve each inequality and graph the solution set. See Example 2.

21. $|x| > 3$

22. $|y| > 5$

23. $|k| \geq 4$

24. $|r| \geq 6$

25. $|r + 5| \geq 20$

26. $|3x - 1| \geq 8$

27. $|t + 2| > 10$

28. $|4x + 1| > 21$

29. $|3 - x| > 5$

30. $|5 - x| > 3$

31. $|-5x + 3| \geq 12$

32. $|-2x - 4| \geq 5$

33. *Concept Check* The graph of the solution set of $|2x + 1| = 9$ is given here.

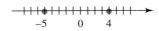

Without actually doing the algebraic work, graph the solution set of each inequality, referring to the graph shown.

(a) $|2x + 1| < 9$ **(b)** $|2x + 1| > 9$

34. *Concept Check* The graph of the solution set of $|3x - 4| < 5$ is given here.

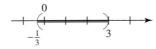

Without actually doing the algebraic work, graph the solution set of the following, referring to the graph shown.

(a) $|3x - 4| = 5$ **(b)** $|3x - 4| > 5$

Solve each inequality and graph the solution set. See Example 3. (Hint: Compare your answers with those in Exercises 21–32.)

35. $|x| \leq 3$

36. $|y| \leq 5$

37. $|k| < 4$

38. $|r| < 6$

39. $|r + 5| \leq 20$

40. $|3x - 1| < 8$

41. $|t + 2| \leq 10$

42. $|4x + 1| < 21$

43. $|3 - x| \leq 5$

44. $|5 - x| \leq 3$

45. $|-5x + 3| \leq 12$

46. $|-2x - 4| \leq 5$

Exercises 47–62 represent a sampling of the various types of absolute value equations and inequalities covered in Exercises 1–46. Decide which method of solution applies, and find the solution set. In Exercises 47–58, graph the solution set. See Examples 1–3.

47. $|-4 + k| > 9$ **48.** $|-3 + t| > 8$ **49.** $|r + 5| > 20$

50. $|2x - 1| < 7$ **51.** $|7 + 2z| = 5$ **52.** $|9 - 3p| = 3$

53. $|3r - 1| \leq 11$ **54.** $|2s - 6| \leq 6$ **55.** $|-6x - 6| \leq 1$

56. $|-2x - 6| \leq 5$ **57.** $|2x - 1| \geq 7$ **58.** $|-4 + k| \leq 9$

59. $|x + 2| = 3$ **60.** $|x + 3| = 10$

61. $|x - 6| = 3$ **62.** $|x - 4| = 1$

Solve each equation or inequality. See Example 4.

63. $|x| - 1 = 4$ **64.** $|x| + 3 = 10$ 💿 **65.** $|x + 4| + 1 = 2$

66. $|x + 5| - 2 = 12$ **67.** $|2x + 1| + 3 > 8$ **68.** $|6x - 1| - 2 > 6$

69. $|x + 5| - 6 \leq -1$ **70.** $|r - 2| - 3 \leq 4$

Solve each equation. See Example 5.

💿 **71.** $|3x + 1| = |2x + 4|$ **72.** $|7x + 12| = |x - 8|$

73. $\left| m - \dfrac{1}{2} \right| = \left| \dfrac{1}{2} m - 2 \right|$ **74.** $\left| \dfrac{2}{3} r - 2 \right| = \left| \dfrac{1}{3} r + 3 \right|$

75. $|6x| = |9x + 1|$ **76.** $|13x| = |2x + 1|$

77. $|2p - 6| = |2p + 11|$ **78.** $|3x - 1| = |3x + 9|$

Solve each equation or inequality. See Examples 6 and 7.

💿 **79.** $|x| \geq -10$ **80.** $|x| \geq -15$ 💿 **81.** $|12t - 3| = -8$

82. $|13w + 1| = -3$ **83.** $|4x + 1| = 0$ **84.** $|6r - 2| = 0$

85. $|2q - 1| = -6$ **86.** $|8n + 4| = -4$

87. $|x + 5| > -9$ **88.** $|x + 9| > -3$

89. $|7x + 3| \leq 0$ **90.** $|4x - 1| \leq 0$

91. $|5x - 2| = 0$ **92.** $|4 + 7x| = 0$

93. $|x - 2| + 3 \geq 2$ **94.** $|k - 4| + 5 \geq 4$

95. $|10z + 7| + 3 < 1$ **96.** $|4x + 1| - 2 < -5$

97. The 2005 recommended daily intake (RDI) of calcium for females aged 19–50 is 1000 mg. Actual mineral needs vary from person to person. Write an absolute value inequality, with x representing the RDI, to express the RDI plus or minus 100 mg, and solve the inequality. (*Source:* Food and Nutrition Board, National Academy of Sciences—Institute of Medicine.)

98. The average clotting time of blood is 7.45 sec, with a variation of plus or minus 3.6 sec. Write this statement as an absolute value inequality with x representing the time, and solve the inequality.

RELATING CONCEPTS (EXERCISES 99–102)

FOR INDIVIDUAL OR GROUP WORK

The 10 tallest buildings in Kansas City, Missouri, are listed along with their heights.

Building	Height (in feet)
One Kansas City Place	632
Town Pavilion	591
Hyatt Regency Crown Center	504
Kansas City Power and Light	481
Fidelity Bank and Trust Building	454
City Hall	443
1201 Walnut	427
Federal Office Building	413
Commerce Tower	407
City Center Square	404

Source: World Almanac and Book of Facts 2006.

Use this information to **work Exercises 99–102 in order.**

99. To find the average of a group of numbers, we add the numbers and then divide by the number of numbers added. Use a calculator to find the average of the heights.

100. Let k represent the average height of these buildings. If a height x satisfies the inequality

$$|x - k| < t,$$

then the height is said to be within t feet of the average. Using your result from Exercise 99, list the buildings that are within 50 ft of the average.

101. Repeat Exercise 100, but list the buildings that are within 75 ft of the average.

102. **(a)** Write an absolute value inequality that describes the height of a building that is *not* within 75 ft of the average.

(b) Solve the inequality you wrote in part (a).

(c) Use the result of part (b) to list the buildings that are not within 75 ft of the average.

(d) Confirm that your answer to part (c) makes sense by comparing it with your answer to Exercise 101.

PREVIEW EXERCISES

*For the equations (**a**)* $3x + 2y = 24$, *and (**b**)* $-2x + 5y = 20$, *find y for the given value of x. See **Section 1.3.***

103. $x = 0$ **104.** $x = -2$ **105.** $x = 8$ **106.** $x = 1.5$

■ **Summary Exercises** on Solving Linear and Absolute Value Equations and Inequalities

This section of miscellaneous equations and inequalities provides practice in solving all such types introduced in this chapter. You might wish to refer to the boxes in the chapter that summarize the various methods of solution.

Solve each equation or inequality. Give the solution set in set notation for equations and in interval notation for inequalities.

1. $4z + 1 = 49$

2. $|m - 1| = 6$

3. $6q - 9 = 12 + 3q$

4. $3p + 7 = 9 + 8p$

5. $|a + 3| = -4$

6. $2m + 1 \leq m$

7. $8r + 2 \geq 5r$

8. $4(a - 11) + 3a = 20a - 31$

9. $2q - 1 = -7$

10. $|3q - 7| - 4 = 0$

11. $6z - 5 \leq 3z + 10$

12. $|5z - 8| + 9 \geq 7$

13. $9x - 3(x + 1) = 8x - 7$

14. $|m| \geq 8$

15. $9x - 5 \geq 9x + 3$

16. $13p - 5 > 13p - 8$

17. $|q| < 5.5$

18. $4z - 1 = 12 + z$

19. $\dfrac{2}{3}x + 8 = \dfrac{1}{4}x$

20. $-\dfrac{5}{8}z \geq -20$

21. $\dfrac{1}{4}p < -6$

22. $7z - 3 + 2z = 9z - 8z$

23. $\dfrac{3}{5}q - \dfrac{1}{10} = 2$

24. $|r - 1| < 7$

25. $r + 9 + 7r = 4(3 + 2r) - 3$

26. $6 - 3(2 - p) < 2(1 + p) + 3$

27. $|2p - 3| > 11$

28. $\dfrac{x}{4} - \dfrac{2x}{3} = -10$

29. $|5a + 1| \leq 0$

30. $5z - (3 + z) \geq 2(3z + 1)$

31. $-2 \leq 3x - 1 \leq 8$

32. $-1 \leq 6 - x \leq 5$

33. $|7z - 1| = |5z + 3|$

34. $|p + 2| = |p + 4|$

35. $|1 - 3x| \geq 4$

36. $\dfrac{1}{2} \leq \dfrac{2}{3}r \leq \dfrac{5}{4}$

37. $-(m + 4) + 2 = 3m + 8$

38. $\dfrac{p}{6} - \dfrac{3p}{5} = p - 86$

39. $-6 \leq \dfrac{3}{2} - x \leq 6$

40. $|5 - x| < 4$

41. $|x - 1| \geq -6$

42. $|2r - 5| = |r + 4|$

43. $8q - (1 - q) = 3(1 + 3q) - 4$

44. $8t - (t + 3) = -(2t + 1) - 12$

45. $|r - 5| = |r + 9|$

46. $|r + 2| < -3$

47. $2x + 1 > 5$ or $3x + 4 < 1$

48. $1 - 2x \geq 5$ and $7 + 3x \geq -2$

Chapter **2** **Group Activity**

COMPARING LONG-DISTANCE COSTS

Objective Write an inequality to solve an applied problem.

Cellular phones are popular tools for both local and long-distance phone calls. Frequently, rate plans include long-distance telephoning as an option. The plans vary among different companies and often offer a limited number of "anytime" minutes.

Consider the following pricing schemes for regular and cellular phones:

- The long-distance plan for an *in-home* phone costs $6.95 per month plus $0.05 per min for long-distance calls both within your state or between states, with no limit to the number of minutes of call time.

- One option for a *cellular* phone is a flat monthly fee of $59.99 that includes 450 min of "anytime" local or long-distance calls.
  Note: Basic phone rates are *not* included in the in-home plan, but since you intend to have an in-home phone anyway, you can disregard those costs. Also, calls in excess of the limits for the cellular plan are expensive: $0.35 per minute over the maximum. You do *not* expect to exceed the number of minutes included in the basic cellular rate plan, so do not worry about those extra charges.

We might ask, "Which plan is more economical?"

A. To answer this question, let x represent the number of minutes of long-distance calls in a month.

   1. Write an expression that represents the monthly costs for the in-home rate plan.

   2. Write the expression that represents the monthly cost for the cellular rate plan.

   3. How many minutes of long-distance calls would you have to make in one month with the in-home phone to exceed the cost of the cellular phone plan? Write a linear inequality which states that the in-home rate plan costs more than the cellular rate plan.

   4. Solve the linear inequality and answer the question posed in Problem 3. What does your answer mean in terms of comparing phone costs?

B. Analyze your answers.

   1. Compare phone costs for the two plans.

   2. Suppose you use the cellular phone plan for 450 min (the maximum number of minutes without incurring excess charges). How much more money would you pay compared with the in-home plan's charges?

C. There were actually two different rate plans for *in-home* long distance. Another plan costs $3.95 per month, plus $0.07 per minute. How many minutes of long distance would you have to call in one month before the $0.05 plan costs less than the $0.07 plan?

Chapter **2** SUMMARY   *View the Interactive Summary on the Pass the Test CD.*

KEY TERMS

2.1 linear (first-degree)
equation in one
variable
solution
solution set
equivalent equations

conditional equation
contradiction
identity
2.2 mathematical model
formula
percent

2.5 inequality
linear inequality in
one variable
equivalent inequalities
three-part inequality
2.6 intersection

compound inequality
union
2.7 absolute value
equation
absolute value
inequality

NEW SYMBOLS

$1°$ one degree

∞ infinity
$-\infty$ negative infinity

$(-\infty, \infty)$ the set of real
numbers

$\cap$ set intersection
$\cup$ set union

TEST YOUR WORD POWER

See how well you have learned the vocabulary in this chapter. Answers, with examples, follow the Quick Review.

1. An **algebraic expression** is
   A. an expression that uses any of the
   four basic operations or the
   operation of raising to powers or
   taking roots on any collection of
   variables and numbers formed
   according to the rules of algebra
   B. an equation that uses any of the
   four basic operations or the
   operation of raising to powers or
   taking roots on any collection of
   variables and numbers formed
   according to the rules of algebra
   C. an equation in algebra

   D. an expression that contains
   fractions.
2. An **equation** is
   A. an algebraic expression
   B. an expression that contains
   fractions
   C. an expression that uses any of
   the four basic operations or the
   operation of raising to powers or
   taking roots on any collection of
   variables and numbers formed
   according to the rules of algebra
   D. a statement that two algebraic
   expressions are equal.

3. The **intersection** of two sets A and
   B is the set of elements that belong
   A. to both A and B
   B. to either A or B, or both
   C. to either A or B, but not both
   D. to just A.

4. The **union** of two sets A and B is the
   set of elements that belong
   A. to both A and B
   B. to either A or B, or both
   C. to either A or B, but not both
   D. to just B.

QUICK REVIEW

Concepts	Examples

2.1 LINEAR EQUATIONS IN ONE VARIABLE

**Addition and Multiplication Properties
of Equality**
The same number may be added to (or subtracted
from) each side of an equation to obtain an
equivalent equation. Similarly, the same nonzero
number may be multiplied by or divided into each
side of an equation to obtain an equivalent equation.

Solve.

$$x - 5 = 10$$

$$x - 5 + 5 = 10 + 5$$

$$x = 15$$

The solution set is $\{15\}$.

$$\frac{1}{2}x = 10$$

$$2\left(\frac{1}{2}x\right) = 2(10)$$

$$x = 20$$

The solution set is $\{20\}$.

(continued)

Concepts	Examples

Solving a Linear Equation in One Variable

Step 1 Clear fractions.

Step 2 Simplify each side separately.

Step 3 Isolate the variable terms on one side.

Solve $4(8 - 3t) = 32 - 8(t + 2)$.

$$32 - 12t = 32 - 8t - 16 \qquad \text{Distributive property}$$
$$32 - 12t = 16 - 8t$$

$$32 - 12t + 12t = 16 - 8t + 12t \qquad \text{Add 12t.}$$
$$32 = 16 + 4t$$
$$32 - 16 = 16 + 4t - 16 \qquad \text{Subtract 16.}$$
$$16 = 4t$$

Step 4 Isolate the variable.

$$\frac{16}{4} = \frac{4t}{4} \qquad \text{Divide by 4.}$$
$$4 = t$$

Step 5 Check.

The solution set is $\{4\}$. This can be checked by substituting 4 for t in the original equation.

2.2 FORMULAS

Solving a Formula for a Specified Variable

Step 1 Get all terms with the specified variable on one side and all terms without that variable on the other side.

Step 2 If necessary, use the distributive property to combine terms with the specified variable.

Step 3 Divide both sides by the factor that is the coefficient of the specified variable.

Solve $A = \dfrac{1}{2}bh$ for h.

$$A = \frac{1}{2}bh$$

$$2A = 2\left(\frac{1}{2}bh\right) \qquad \text{Multiply by 2.}$$

$$2A = bh$$

$$\frac{2A}{b} = h, \quad \text{or} \quad h = \frac{2A}{b} \qquad \text{Divide by } b.$$

2.3 APPLICATIONS OF LINEAR EQUATIONS

Solving an Applied Problem

Step 1 Read the problem.

Step 2 Assign a variable.

How many liters of 30% alcohol solution and 80% alcohol solution must be mixed to obtain 100 L of 50% alcohol solution?

Let $\quad x =$ number of liters of 30% solution needed;

then $100 - x =$ number of liters of 80% solution needed.

Liters of Solution	Percent (as a decimal)	Liters of Pure Alcohol
x	0.30	$0.30x$
$100 - x$	0.80	$0.80(100 - x)$
100	0.50	$0.50(100)$

Step 3 Write an equation.

Step 4 Solve the equation.

Step 5 State the answer.

Step 6 Check.

The equation is $\quad 0.30x + 0.80(100 - x) = 0.50(100)$.

The solution of the equation is 60. Thus, 60 L of 30% solution and $100 - 60 = 40$ L of 80% solution are needed.

$$0.30(60) + 0.80(100 - 60) = 50 \text{ is true.}$$

(continued)

Concepts	Examples

2.4 FURTHER APPLICATIONS OF LINEAR EQUATIONS

To solve a uniform motion problem, draw a sketch and make a table. Use the formula **d = rt.**

Two cars start from towns 400 mi apart and travel toward each other. They meet after 4 hr. Find the speed of each car if one travels 20 mph faster than the other.

Let x = speed of the slower car in miles per hour;

then $x + 20$ = speed of the faster car.

Use the information in the problem and $d = rt$ to complete a table.

	Rate	Time	Distance
Slower Car	x	4	$4x$
Faster Car	$x + 20$	4	$4(x + 20)$
			400 ←Total

A sketch shows that the sum of the distances, $4x$ and $4(x + 20)$, must be 400.

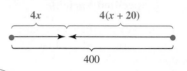

The equation is

$$4x + 4(x + 20) = 400.$$

Problems involving denominations of money and mixture problems are solved by methods similar to the one used for the mixture problem shown in the example for **Section 2.3.**

Solving this equation gives $x = 40$. The slower car travels 40 mph, and the faster car travels $40 + 20 = 60$ mph.

2.5 LINEAR INEQUALITIES IN ONE VARIABLE

Solving a Linear Inequality in One Variable

Step 1 Simplify each side of the inequality by clearing parentheses and combining like terms.

Step 2 Use the addition property of inequality to get all terms with variables on one side and all terms without variables on the other side.

Step 3 Use the multiplication property of inequality to write the inequality in the form $x < k$ or $x > k$.

If an inequality is multiplied or divided by a negative number, the inequality symbol must be reversed.

Solve $3(x + 2) - 5x \leq 12$.

$$3x + 6 - 5x \leq 12 \qquad \text{Distributive property}$$
$$-2x + 6 \leq 12$$
$$-2x + 6 - 6 \leq 12 - 6 \qquad \text{Subtract 6.}$$
$$-2x \leq 6$$
$$\frac{-2x}{-2} \geq \frac{6}{-2} \qquad \begin{array}{l}\text{Divide by } -2;\\ \text{change } \leq \text{ to } \geq.\end{array}$$
$$x \geq -3$$

The solution set $[-3, \infty)$ is graphed here.

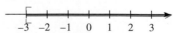

(continued)

Concepts	Examples

To solve a three-part inequality, work with all three parts at the same time.

Solve $-4 < 2x + 3 \le 7$.

$$-4 - 3 < 2x + 3 - 3 \le 7 - 3 \quad \text{Subtract 3.}$$
$$-7 < 2x \le 4$$
$$\frac{-7}{2} < \frac{2x}{2} \le \frac{4}{2} \quad \text{Divide by 2.}$$
$$-\frac{7}{2} < x \le 2$$

The solution set, $\left(-\frac{7}{2}, 2\right]$, is graphed here.

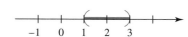

2.6 SET OPERATIONS AND COMPOUND INEQUALITIES

Solving a Compound Inequality

Step 1 Solve each inequality in the compound inequality individually.

Step 2 If the inequalities are joined with *and,* then the solution set is the intersection of the two individual solution sets.

If the inequalities are joined with *or,* then the solution set is the union of the two individual solution sets.

Solve $x + 1 > 2$ and $2x < 6$.

$$x + 1 > 2 \quad \text{and} \quad 2x < 6$$
$$x > 1 \quad \text{and} \quad x < 3$$

The solution set is $(1, 3)$.

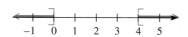

Solve $x \ge 4$ or $x \le 0$.

The solution set is $(-\infty, 0] \cup [4, \infty)$.

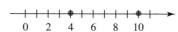

2.7 ABSOLUTE VALUE EQUATIONS AND INEQUALITIES

Solving Absolute Value Equations and Inequalities

Let k be a positive number.

To solve $|ax + b| = k,$ solve the compound equation

$$ax + b = k \quad \text{or} \quad ax + b = -k.$$

To solve $|ax + b| > k,$ solve the compound inequality

$$ax + b > k \quad \text{or} \quad ax + b < -k.$$

Solve $|x - 7| = 3$.

$$x - 7 = 3 \quad \text{or} \quad x - 7 = -3$$
$$x = 10 \quad \text{or} \quad x = 4$$

The solution set is $\{4, 10\}$.

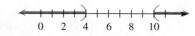

Solve $|x - 7| > 3$.

$$x - 7 > 3 \quad \text{or} \quad x - 7 < -3$$
$$x > 10 \quad \text{or} \quad x < 4$$

The solution set is $(-\infty, 4) \cup (10, \infty)$.

(continued)

Concepts	Examples
To solve $\lvert ax + b\rvert < k,$ solve the compound inequality $$-k < ax + b < k.$$	Solve $\lvert x - 7\rvert < 3.$ $$-3 < x - 7 < 3$$ $$4 < x < 10 \qquad \text{Add 7.}$$ The solution set is $(4, 10)$.

Concepts	Examples
To solve an absolute value equation of the form $$\lvert ax + b\rvert = \lvert cx + d\rvert,$$ solve the compound equation $$ax + b = cx + d \quad \text{or} \quad ax + b = -(cx + d).$$	Solve $\lvert x + 2\rvert = \lvert 2x - 6\rvert.$ $$x + 2 = 2x - 6 \quad \text{or} \quad x + 2 = -(2x - 6)$$ $$x = 8 \qquad\qquad x + 2 = -2x + 6$$ $$3x = 4$$ $$x = \frac{4}{3}$$ The solution set is $\left\{\frac{4}{3}, 8\right\}$.

Answers to Test Your Word Power

1. A; *Examples:* $\dfrac{3y - 1}{2}, 6 + \sqrt{2x}, 4a^3b - c$ **2.** D; *Examples:* $2a + 3 = 7, 3y = -8,$ $x^2 = 4$ **3.** A; *Example:* If $A = \{2, 4, 6, 8\}$ and $B = \{1, 2, 3\}$, then $A \cap B = \{2\}$.

4. B; *Example:* Using the sets A and B from Answer 3, $A \cup B = \{1, 2, 3, 4, 6, 8\}$.

Chapter 2 REVIEW EXERCISES

[2.1] *Solve each equation.*

1. $-(8 + 3z) + 5 = 2z + 6$

2. $-\dfrac{3}{4}x = -12$

3. $\dfrac{2q + 1}{3} - \dfrac{q - 1}{4} = 0$

4. $5(2x - 3) = 6(x - 1) + 4x$

Solve each equation. Then tell whether the equation is conditional, *an* identity, *or a* contradiction.

5. $7r - 3(2r - 5) + 5 + 3r = 4r + 20$

6. $8p - 4p - (p - 7) + 9p + 6 = 12p - 7$

7. $-2r + 6(r - 1) + 3r - (4 - r) = -(r + 5) - 5$

[2.2] *Solve each formula for the specified variable.*

8. $V = LWH$ for L

9. $A = \dfrac{1}{2}h(b + B)$ for b

Solve each equation for x.

10. $M = -\dfrac{1}{4}(x + 3y)$

11. $P = \dfrac{3}{4}x - 12$

12. Give the steps you would use to solve $-2x + 5 = 7$.

[2.2, 2.3] *Solve each problem.*

13. A rectangular solid has a volume of 180 ft³. Its length is 6 ft and its width is 5 ft. Find its height.

14. The total number of deaths from AIDS in the United States in 2003 was 17,849. In 2004, this figure had decreased to 15,798. What approximate percent decrease did this represent? (*Source:* U.S. Centers for Disease Control.)

15. Find the simple-interest rate that Francesco Castellucio is earning if his principal of $30,000 earns $7800 interest in 4 yr.

16. If the Fahrenheit temperature is 77°, what is the corresponding Celsius temperature?

For 2005, total U.S. government spending was about $2500 billion (or $2.5 trillion). The circle graph shows how the spending was divided.

17. About how much was spent on Social Security?

18. About how much did the U.S. government spend on education and social services in 2005?

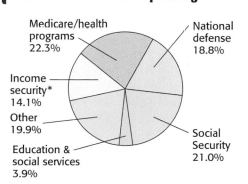

2005 U.S. Government Spending

Medicare/health programs 22.3%

National defense 18.8%

Income security* 14.1%

Other 19.9%

Education & social services 3.9%

Social Security 21.0%

*Includes pensions for government workers, unemployment compensation, food stamps, and other such programs.

Source: U.S. Office of Management and Budget.

Write each phrase as a mathematical expression, using x as the variable.

19. One-third of a number, subtracted from 9

20. The product of 4 and a number, divided by 9 more than the number

Solve each problem.

21. The length of a rectangle is 3 m less than twice the width. The perimeter of the rectangle is 42 m. Find the length and width of the rectangle.

22. In a triangle with two sides of equal length, the third side measures 15 in. less than the sum of the two equal sides. The perimeter of the triangle is 53 in. Find the lengths of the three sides.

23. A candy clerk has three times as many kilograms of chocolate creams as peanut clusters. The clerk has 48 kg of the two candies altogether. How many kilograms of peanut clusters does the clerk have?

24. How many liters of a 20% solution of a chemical should be mixed with 15 L of a 50% solution to get a 30% mixture?

25. How much water should be added to 30 L of a 40% acid solution to reduce it to a 30% solution?

Liters of Solution	Percent (as a decimal)	Liters of Pure Acid
	0.40	
x		
	0.30	

26. Jay Jenkins invested some money at 6% and $4000 less than that amount at 4%. Find the amount invested at each rate if his total annual interest income is $840.

Principal	Rate (as a decimal)	Interest
x	0.06	
	0.04	

[2.4]

27. A grocery store clerk has $3.50 in dimes and quarters in her cash drawer. The number of dimes is 1 less than twice the number of quarters. How many of each denomination are there?

28. When Jim emptied his pockets one evening, he found he had 19 nickels and dimes with a total value of $1.55. How many of each denomination did he have?

29. *Concept Check*   Which choice is the best *estimate* for the average speed of a trip of 405 mi that lasted 8.2 hr?

A. 50 mph   **B.** 30 mph   **C.** 60 mph   **D.** 40 mph

30. (a)   A driver averaged 53 mph and took 10 hr to travel from Memphis to Chicago. What is the distance between Memphis and Chicago?

(b)   A small plane traveled from Warsaw to Rome, averaging 164 mph. The trip took 2 hr. What is the distance from Warsaw to Rome?

31. A passenger train and a freight train leave a town at the same time and go in opposite directions. They travel at speeds of 60 mph and 75 mph, respectively. How long will it take for them to be 297 mi apart?

	Rate	Time	Distance
Passenger Train	60	x	
Freight Train	75	x	

32. Two cars leave towns 230 km apart at the same time, traveling directly toward one another. One car travels 15 km per hr slower than the other. They pass one another 2 hr later. What are their speeds?

	Rate	Time	Distance
Faster car	x	2	
Slower car	x − 15	2	

33. An automobile averaged 45 mph for the first part of a trip and 50 mph for the second part. If the entire trip took 4 hr and covered 195 mi, for how long was the rate 45 mph?

34. An 85-mi trip to the beach took the Valenzuela family 2 hr. During the second hour, a rainstorm caused them to average 7 mph less than they traveled during the first hour. Find their average rate for the first hour.

35. Find the measure of each angle in the triangle.

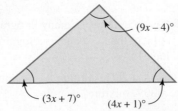

36. Find the measure of each marked angle.

$(15x + 15)°$ $(3x + 3)°$

[2.5] *Solve each inequality. Express the solution set in interval form.*

37. $-\dfrac{2}{3}k < 6$

38. $-5x - 4 \geq 11$

39. $\dfrac{6a + 3}{-4} < -3$

40. $5 - (6 - 4k) \geq 2k - 7$

41. $8 \leq 3z - 1 < 14$

42. $\dfrac{5}{3}(m - 2) + \dfrac{2}{5}(m + 1) > 1$

Solve each problem.

43. The perimeter of a rectangular playground must be no greater than 120 m. One dimension of the playground must be 22 m. Find the possible lengths of the other dimension of the playground.

22 m

44. A group of college students wants to buy tickets to attend a performance of Monty Python's *Spamalot* at the Cadillac Palace Theatre in Chicago. The best price they can find is a group rate of $89 per ticket if 10 or more tickets are purchased at the same time. If they have $2000 available to spend on tickets and they qualify for a $50 group discount, how many tickets can they purchase?

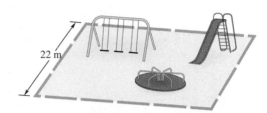

45. To pass algebra, a student must have an average of at least 70 on five tests. On the first four tests, a student has scores of 75, 79, 64, and 71. What possible scores on the fifth test would guarantee the student a passing grade in the class?

46. While solving the inequality

$$10x + 2(x - 4) < 12x - 13,$$

a student did all the work correctly and obtained the statement $-8 < -13$. The student did not know what to do at this point, because the variable "disappeared." How would you explain to the student the interpretation of this result?

[2.6] *Let $A = \{a, b, c, d\}$, $B = \{a, c, e, f\}$, and $C = \{a, e, f, g\}$. Find each set.*

47. $A \cap B$

48. $A \cap C$

49. $B \cup C$

50. $A \cup C$

Solve each compound inequality. Give the solution set in both interval and graph form.

51. $x > 6$ and $x < 9$

52. $x + 4 > 12$ and $x - 2 < 12$

53. $x > 5$ or $x \leq -3$

54. $x \geq -2$ or $x < 2$

55. $x - 4 > 6$ and $x + 3 \leq 10$

56. $-5x + 1 \geq 11$ or $3x + 5 \geq 26$

Express each union or intersection in simplest interval form.

57. $(-3, \infty) \cap (-\infty, 4)$

58. $(-\infty, 6) \cap (-\infty, 2)$

59. $(4, \infty) \cup (9, \infty)$

60. $(1, 2) \cup (1, \infty)$

[2.7] *Solve each absolute value equation.*

61. $|x| = 7$ **62.** $|x + 2| = 9$ **63.** $|3k - 7| = 8$

64. $|z - 4| = -12$ **65.** $|2k - 7| + 4 = 11$ **66.** $|4a + 2| - 7 = -3$

67. $|3p + 1| = |p + 2|$ **68.** $|2m - 1| = |2m + 3|$

Solve each absolute value inequality. Give the solution set in interval form.

69. $|p| < 14$ **70.** $|-t + 6| \leq 7$

71. $|2p + 5| \leq 1$ **72.** $|x + 1| \geq -3$

MIXED REVIEW EXERCISES*

Solve.

73. $5 - (6 - 4k) > 2k - 5$ **74.** $ak + bt = 6t - sk$ for k

75. $x < 3$ and $x \geq -2$ **76.** $\dfrac{4x + 2}{4} + \dfrac{3x - 1}{8} = \dfrac{x + 6}{16}$

77. $|3k + 6| \geq 0$ **78.** $-5r \geq -10$

79. A newspaper recycling collection bin is in the shape of a box 1.5 ft wide and 5 ft long. If the volume of the bin is 75 ft³, find the height.

80. The sum of the first and third of three consecutive integers is 47 more than the second integer. What are the integers?

81. $|3x + 2| + 4 = 9$ **82.** $0.05x + 0.03(1200 - x) = 42$

83. $|m + 3| \leq 13$ **84.** $\dfrac{3}{4}(a - 2) - \dfrac{1}{3}(5 - 2a) < -2$

85. $-4 < 3 - 2k < 9$ **86.** $-0.3x + 2.1(x - 4) \leq -6.6$

87. The complement of an angle measures 10° less than one-fifth of its supplement. Find the measure of the angle.

88. To qualify for a company pension plan, an employee must average at least $1000 per month in earnings. During the first four months of the year, an employee made $900, $1200, $1040, and $760. What possible amounts earned during the fifth month will qualify the employee?

89. $|5r - 1| > 14$ **90.** $x \geq -2$ or $x < 4$

91. How many liters of a 20% solution of a chemical should be mixed with 10 L of a 50% solution to get a 40% mixture?

92. $|m - 1| = |2m + 3|$ **93.** $\dfrac{3x}{5} - \dfrac{x}{2} = 3$ **94.** $|m + 3| \leq 1$

95. $|3k - 7| = 4$ **96.** $5(2x - 7) = 2(5x + 3)$

97. *Concept Check* If $k < 0$, what is the solution set of

     (a) $|5x + 3| < k$ **(b)** $|5x + 3| > k$ **(c)** $|5x + 3| = k$?

---

*The order of exercises in this final group does not correspond to the order in which topics occur in the chapter. This random ordering should help you prepare for the chapter test.

In Exercises 98 and 99, sketch the graph of each solution set.

98. $x > 6$ and $x < 8$

99. $-5x + 1 \geq 11$ or $3x + 5 \geq 26$

100. The numbers of civilian workers (to the nearest thousand) for several states in 2004 are shown in the table.

NUMBER OF WORKERS

State	Female	Male
Illinois	2,979,000	3,407,000
Maine	334,000	362,000
North Carolina	1,973,000	2,270,000
Oregon	840,000	1,010,000
Utah	538,000	668,000
Wisconsin	1,456,000	1,615,000

Source: U.S. Bureau of Labor Statistics.

List the elements of each set.

(a) The set of states with less than 1 million female workers *and* more than 1 million male workers

(b) The set of states with less than 1 million female workers *or* more than 2 million male workers

(c) The set of states with a total of more than 7 million civilian workers

Chapter 2 TEST

Solve each equation.

1. $3(2x - 2) - 4(x + 6) = 3x + 8 + x$

2. $0.08x + 0.06(x + 9) = 1.24$

3. $\dfrac{x + 6}{10} + \dfrac{x - 4}{15} = \dfrac{x + 2}{6}$

4. Solve each equation. Then tell whether the equation is a *conditional equation,* an *identity,* or a *contradiction.*

(a) $3x - (2 - x) + 4x + 2 = 8x + 3$

(b) $\dfrac{x}{3} + 7 = \dfrac{5x}{6} - 2 - \dfrac{x}{2} + 9$

(c) $-4(2x - 6) = 5x + 24 - 7x$

5. Solve $-16t^2 + vt - S = 0$ for v.

6. Solve $ar + 2 = 3r - 6t$ for r.

Solve each problem.

7. The 2005 Indianapolis 500 (mile) race was won by Dan Wheldon, who averaged 157.603 mph. What was Wheldon's time to the nearest thousandth of an hour? (*Source: World Almanac and Book of Facts 2006.*)

8. A certificate of deposit pays $2281.25 in simple interest for 1 yr on a principal of $36,500. What is the rate of interest?

9. In 2005, there were 37,142 offices, stations, and branches of the U.S. Postal Service, of which 27,385 were actually classified as post offices. What percent, to the nearest tenth, were classified as post offices? (*Source:* U.S. Postal Service.)

10. Tyler McGinnis invested some money at 3% simple interest and some at 5% simple interest. The total amount of his investments was $28,000, and the interest he earned during the first year was $1240. How much did he invest at each rate?

11. Two cars leave from the same point at the same time, traveling in opposite directions. One travels 15 mph slower than the other. After 6 hr, they are 630 mi apart. Find the rate of each car.

12. Find the measure of each angle.

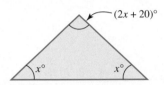

Solve each inequality. Give the solution set in both interval and graph form.

13. $4 - 6(x + 3) \leq -2 - 3(x + 6) + 3x$

14. $-\dfrac{4}{7}x > -16$

15. $-6 \leq \dfrac{4}{3}x - 2 \leq 2$

16. *Concept Check* Which one of the following inequalities is equivalent to $x < -3$?

A. $-3x < 9$ **B.** $-3x > -9$ **C.** $-3x > 9$ **D.** $-3x < -9$

Solve each problem.

17. A student must have an average of at least 80 on the four tests in a course to get a B. The student had scores of 83, 76, and 79 on the first three tests. What minimum score on the fourth test would guarantee the student a B in the course?

18. A product will break even or produce a profit only if the revenue R (in dollars) from selling the product is at least equal to the cost C (in dollars) of producing it. Suppose that the cost to produce x units of carpet is $C = 50x + 5000$, while the revenue is $R = 60x$. For what values of x is R at least equal to C?

19. Let $A = \{1, 2, 5, 7\}$ and $B = \{1, 5, 9, 12\}$. Find each of the following sets.

(a) $A \cap B$ **(b)** $A \cup B$

Solve each compound or absolute value inequality.

20. $3k \geq 6$ and $k - 4 < 5$

21. $-4x \leq -24$ or $4x - 2 < 10$

22. $|4x + 3| \leq 7$

23. $|5 - 6x| > 12$

24. $|7 - x| \leq -1$

25. $|-3x + 4| - 4 < -1$

Solve each absolute value equation.

26. $|3k - 2| + 1 = 8$

27. $|3 - 5x| = |2x + 8|$

28. *Concept Check* If $k < 0$, what is the solution set of

(a) $|8x - 5| < k$ **(b)** $|8x - 5| > k$ **(c)** $|8x - 5| = k$?

Chapters 1–2 CUMULATIVE REVIEW EXERCISES

Let $A = \left\{-8, -\frac{2}{3}, -\sqrt{6}, 0, \frac{4}{5}, 9, \sqrt{36}\right\}$. Simplify the elements of A as necessary and then list the elements that belong to each set.

1. Natural numbers
2. Whole numbers
3. Integers
4. Rational numbers
5. Irrational numbers
6. Real numbers

Add or subtract, as indicated.

7. $-\frac{4}{3} - \left(-\frac{2}{7}\right)$

8. $|-4| - |2| + |-6|$

9. $(-2)^4 + (-2)^3$

10. $\sqrt{25} - 5(-1)^0$

Evaluate each expression.

11. $(-3)^5$

12. $\left(\frac{6}{7}\right)^3$

13. $\left(-\frac{2}{3}\right)^3$

14. -4^6

15. Which one of the following is not a real number: $-\sqrt{36}$ or $\sqrt{-36}$?

16. Which one of the following is undefined: $\frac{4-4}{4+4}$ or $\frac{4+4}{4-4}$?

Evaluate if $a = 2$, $b = -3$, and $c = 4$.

17. $-3a + 2b - c$

18. $-8(a^2 + b^3)$

19. $\dfrac{3a^3 - b}{4 + 3c}$

Simplify each expression.

20. $-7r + 5 - 13r + 12$

21. $-(3k + 8) - 2(4k - 7) + 3(8k + 12)$

Identify the property of real numbers illustrated by each equation.

22. $(a + b) + 4 = 4 + (a + b)$

23. $4x + 12x = (4 + 12)x$

Solve each equation.

24. $-4x + 7(2x + 3) = 7x + 36$

25. $-\frac{3}{5}x + \frac{2}{3}x = 2$

26. $0.06x + 0.03(100 + x) = 4.35$

27. $P = a + b + c$ for b

28. $4(2x - 6) + 3(x - 2) = 11x + 1$

29. $\frac{2}{3}x + \frac{5}{8}x = \frac{31}{24}x$

Solve each inequality. Give the solution set in both interval and graph form.

30. $3 - 2(x + 7) \le -x + 3$

31. $-4 < 5 - 3x \le 0$

32. $2x + 1 > 5$ or $2 - x > 2$

33. $|-7k + 3| \ge 4$

Solve each problem.

34. Kathy Manley invested some money at 5% simple interest and $2000 more than that amount at 6%. Her interest for the year totaled $670. How much did she invest at each rate?

35. A dietician must use three foods—A, B, and C—in a diet. He must include twice as many grams of food A as food C, and 5 g of food B. The three foods must total at most 24 g. What is the largest amount of food C that the dietician can use?

36. Michelle Renda got scores of 88 and 78 on her first two tests. What score must she make on her third test to keep an average of 80 or greater?

37. How much pure alcohol should be added to 7 L of 10% alcohol to increase the concentration to 30% alcohol?

38. A coin collection contains 29 coins and consists of pennies, nickels, and quarters. The number of quarters is 4 less than the number of nickels, and the face value of the collection is $2.69. How many of each denomination are there in the collection?

39. Clark's rule, a formula used in reducing drug dosage according to weight from the recommended adult dosage to a child dosage, is

$$\frac{\text{weight of child in pounds}}{150} \times \text{adult dose} = \text{child's dose}.$$

Find a child's dosage if the child weighs 55 lb and the recommended adult dosage is 120 mg.

40. Since 1975, the number of daily newspapers has steadily declined. According to the table,

    (a) by how many did the number of daily newspapers decrease between 1990 and 2003?

    (b) by what *percent* did the number of daily newspapers decrease from 1990 to 2003?

Year	Number of Daily Newspapers
1975	1756
1980	1745
1985	1676
1990	1611
1995	1533
2000	1480
2001	1468
2002	1457
2003	1456

Source: Editorial & Publisher Co.

Graphs, Linear Equations, and Functions

The two most common measures of temperature are Fahrenheit (F) and Celsius (C). We often see signs displaying one or both of these types of temperature. It is fairly common knowledge that water freezes at 32°F, or 0°C, and boils at 212°F, or 100°C. Because there is a *linear* relationship between the Fahrenheit and Celsius temperature scales, using these two equivalences, we can derive the familiar formulas for converting from one scale to the other, as seen in Exercises 91–96 of Section 3.3.

Graphs are widely used in the media because they present a great deal of information in a concise form. In this chapter, we see how information such as the relationship between the two temperature scales can be depicted by graphs.

3.1 The Rectangular Coordinate System

OBJECTIVE ❶ Interpret a line graph. The line graph in Figure 1 shows personal spending (in billions of dollars) on medical care in the United States from 1997 through 2003. About how much was spent on medical care in 2002? (We will answer this question shortly.)

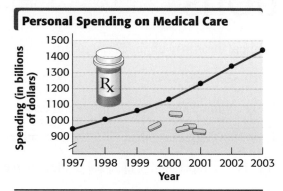

Personal Spending on Medical Care

Source: U.S. Centers for Medicare and Medicaid Services.

FIGURE 1

The line graph in Figure 1 presents information based on a method for locating a point in a plane developed by René Descartes, a 17th-century French mathematician. Legend has it that Descartes, who was lying in bed ill, was watching a fly crawl about on the ceiling near a corner of the room. It occurred to him that the location of the fly on the ceiling could be described by determining its distances from the two adjacent walls. See the figure in the margin. We use this insight to plot points and graph linear equations in two variables whose graphs are straight lines.

Locating a fly
on a ceiling

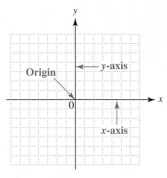

FIGURE 2

OBJECTIVE ❷ Plot ordered pairs. Each of the pairs of numbers $(3, 1)$, $(-5, 6)$, and $(4, -1)$ is an example of an **ordered pair**—that is, a pair of numbers written within parentheses in which the order of the numbers is important. We graph an ordered pair by using two perpendicular number lines that intersect at their 0 points, as shown in the plane in Figure 2. The common 0 point is called the **origin.** The position of any point in this plane is determined by referring to the horizontal number line, or ***x*-axis,** and the vertical number line, or ***y*-axis.** The first number in the ordered pair indicates the position relative to the *x*-axis, and the second number indicates the position relative to the *y*-axis. The *x*-axis and the *y*-axis make up a **rectangular** (or **Cartesian,** for Descartes) **coordinate system.**

To locate, or **plot,** the point on the graph that corresponds to the ordered pair $(3, 1)$, we move three units from 0 to the right along the *x*-axis and then one unit up parallel to the *y*-axis. The point corresponding to the ordered pair $(3, 1)$ is labeled *A* in Figure 3 on the next page. Additional points are labeled *B–E.* The phrase "the point corresponding to the ordered pair $(3, 1)$" is often abbreviated as "the point $(3, 1)$." The numbers in an ordered pair are called the **coordinates** of the corresponding point.

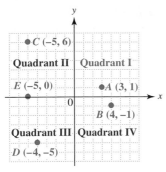

FIGURE 3

We can relate this method of locating ordered pairs to the line graph in Figure 1. We move along the horizontal axis to a year and then up parallel to the vertical axis to approximate medical spending for that year. Thus, we can write the ordered pair (2002, 1340) to indicate that in 2002 personal spending on medical care was about $1340 billion.

▶ **CAUTION** The parentheses used to represent an ordered pair are also used to represent an open interval (introduced in **Section 1.1**). The context of the discussion tells whether ordered pairs or open intervals are being represented.

The four regions of the graph, shown in Figure 3, are called **quadrants I, II, III, and IV,** reading counterclockwise from the upper right quadrant. The points on the x-axis and y-axis do not belong to any quadrant. For example, point E in Figure 3 belongs to no quadrant.

✔ **Now Try Exercises 3, 9, 13, 15, and 21.**

OBJECTIVE 3 Find ordered pairs that satisfy a given equation. Each solution of an equation with two variables, such as

$$2x + 3y = 6,$$

includes two numbers, one for each variable. To keep track of which number goes with which variable, we write the solutions as ordered pairs. (If x and y are used as the variables, the x-value is given first.) For example, we can show that $(6, -2)$ is a solution of $2x + 3y = 6$ by substitution.

$$2x + 3y = 6$$
$$2(6) + 3(-2) = 6 \quad ? \quad \text{Let } x = 6, y = -2.$$
$$12 - 6 = 6 \quad ?$$
$$6 = 6 \quad \text{True}$$

Use parentheses to avoid errors.

Because the ordered pair $(6, -2)$ makes the equation true, it is a solution. On the other hand, $(5, 1)$ is *not* a solution of the equation $2x + 3y = 6$, because

$$2x + 3y = 2(5) + 3(1)$$
$$= 10 + 3$$
$$= 13, \quad \textbf{not} \quad 6.$$

To find ordered pairs that satisfy an equation, select any number for one of the variables, substitute it into the equation for that variable, and then solve for the other variable. Two other ordered pairs satisfying $2x + 3y = 6$ are $(0, 2)$ and $(3, 0)$.

Since any real number could be selected for one variable and would lead to a real number for the other variable, linear equations in two variables have an infinite number of solutions.

EXAMPLE 1 Completing Ordered Pairs and Making a Table

In parts (a) and (b), complete each ordered pair for $2x + 3y = 6$. Then, in part (c), write the results as a table of ordered pairs.

(a) $(-3, \quad)$

Replace x with -3 in the equation to find y.

$$2x + 3y = 6$$
$$2(-3) + 3y = 6 \qquad \text{Let } x = -3.$$
$$-6 + 3y = 6$$
$$3y = 12$$
$$y = 4$$

The ordered pair is $(-3, 4)$.

(b) $(\quad, -4)$

Replace y with -4 in the equation to find x.

$$2x + 3y = 6$$
$$2x + 3(-4) = 6 \qquad \text{Let } y = -4.$$
$$2x - 12 = 6$$
$$2x = 18$$
$$x = 9$$

The ordered pair is $(9, -4)$.

(c) We write a table of these ordered pairs as shown.

x	y	
-3	4	$\leftarrow$ Represents the ordered pair $(-3, 4)$
9	-4	$\leftarrow$ Represents the ordered pair $(9, -4)$

✔ **Now Try Exercise 25(a).**

OBJECTIVE 4 Graph lines. The **graph of an equation** is the set of points corresponding to *all* ordered pairs that satisfy the equation. It gives a "picture" of the equation.

To graph an equation, we plot several ordered pairs that satisfy the equation until we have enough points to suggest the shape of the graph. For example, to graph $2x + 3y = 6$, we plot all ordered pairs found in Objective 3 and Example 1. These points, shown in a table of values and plotted in Figure 4(a), appear to lie on a straight line. If all the ordered pairs that satisfy the equation $2x + 3y = 6$ were graphed, they would form the straight line shown in Figure 4(b).

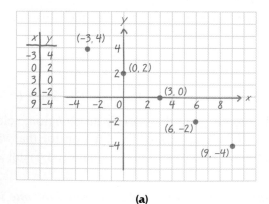

(a)

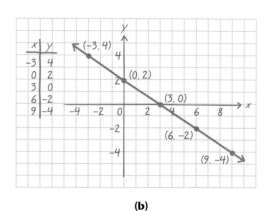

(b)

FIGURE 4

✔ **Now Try Exercise 25(b).**

The equation $2x + 3y = 6$ is called a **first-degree equation,** because it has no term with a variable to a power greater than 1.

The graph of any first-degree equation in two variables is a straight line.

Since first-degree equations with two variables have straight-line graphs, they are called *linear equations in two variables.* (We discussed linear equations in one variable in **Chapter 2.**)

Linear Equation in Two Variables

A **linear equation in two variables** can be written in the form

$$Ax + By = C,$$

where A, B, and C are real numbers (A and B not both 0). This form is called **standard form.**

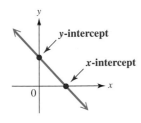

OBJECTIVE 5 **Find *x*- and *y*-intercepts.** A straight line is determined if any two different points on the line are known, so finding two different points is enough to graph the line. Two useful points for graphing are the *x*- and *y*-intercepts. The ***x*-intercept** is the point (if any) where the line intersects the *x*-axis; likewise, the ***y*-intercept** is the point (if any) where the line intersects the *y*-axis.* In Figure 4(b), the *y*-value of the point where the line intersects the *x*-axis is 0. Similarly, the *x*-value of the point where the line intersects the *y*-axis is 0. This suggests a method for finding the *x*- and *y*-intercepts.

Finding Intercepts

When graphing the equation of a line,

$$\text{let } y = 0 \text{ to find the } x\text{-intercept;}$$

$$\text{let } x = 0 \text{ to find the } y\text{-intercept.}$$

EXAMPLE 2 Finding Intercepts

Find the *x*- and *y*-intercepts of $4x - y = -3$ and graph the equation.

We find the *x*-intercept by letting $y = 0$.

$$4x - 0 = -3 \qquad \text{Let } y = 0.$$

$$4x = -3$$

$$x = -\frac{3}{4} \qquad x\text{-intercept is } \left(-\frac{3}{4}, 0\right).$$

For the *y*-intercept, we let $x = 0$.

$$4(0) - y = -3 \qquad \text{Let } x = 0.$$

$$-y = -3$$

$$y = 3 \qquad y\text{-intercept is } (0, 3).$$

---

*Some texts define an intercept as a number, not a point. For example, "*y*-intercept $(0, 4)$" would be given as "*y*-intercept 4."

The intercepts are the two points $\left(-\frac{3}{4}, 0\right)$ and $(0, 3)$. We show these ordered pairs in the table next to Figure 5 and use them to draw the graph.

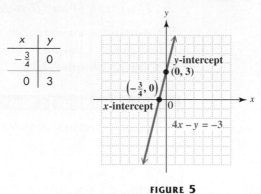

FIGURE 5

✔ **Now Try Exercise 37.**

▶ **NOTE** While two points, such as the two intercepts in Figure 5, are sufficient to graph a straight line, *it is a good idea to use a third point to guard against errors.* Verify by substitution that $(-2, -5)$ also lies on the graph of $4x - y = -3$.

OBJECTIVE 6 **Recognize equations of horizontal and vertical lines and lines passing through the origin.** A graph can fail to have an x-intercept or a y-intercept. The next examples illustrate these special cases.

EXAMPLE 3 **Graphing a Horizontal Line**

Graph $y = 2$.

Writing $y = 2$ as $0x + 1y = 2$ shows that any value of x, including $x = 0$, gives $y = 2$, making the y-intercept $(0, 2)$. Since y is always 2, there is no value of x corresponding to $y = 0$, so the graph has no x-intercept. The graph, shown with a table of ordered pairs in Figure 6, is a horizontal line.

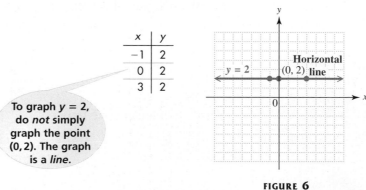

To graph $y = 2$, do *not* simply graph the point $(0, 2)$. The graph is a *line*.

FIGURE 6

✔ **Now Try Exercise 43.**

EXAMPLE 4 Graphing a Vertical Line

Graph $x + 1 = 0$.

The form $1x + 0y = -1$ shows that every value of y leads to $x = -1$, making the x-intercept $(-1, 0)$. No value of y makes $x = 0$, so the graph has no y-intercept. The only way a straight line can have no y-intercept is to be vertical, as shown in Figure 7.

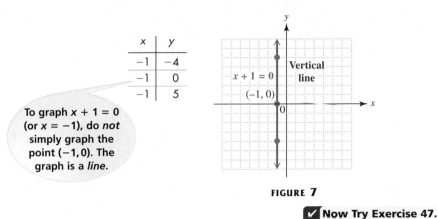

x	y
-1	-4
-1	0
-1	5

To graph $x + 1 = 0$ (or $x = -1$), do *not* simply graph the point $(-1, 0)$. The graph is a *line*.

FIGURE 7

✔ **Now Try Exercise 47.**

Some lines have their x- and y-intercepts at the origin.

EXAMPLE 5 Graphing a Line That Passes through the Origin

Graph $x + 2y = 0$.

Find the intercepts.

x-intercept	**y-intercept**
$x + 2y = 0$	$x + 2y = 0$
$x + 2(0) = 0$ Let $y = 0$.	$0 + 2y = 0$ Let $x = 0$.
$x + 0 = 0$	$y = 0$ y-intercept is $(0, 0)$.
$x = 0$ x-intercept is $(0, 0)$.	

Both intercepts are the same point, $(0, 0)$, which means that the graph passes through the origin. To find another point to graph the line, choose any nonzero number for x, say, $x = 4$, and solve for y.

$$x + 2y = 0$$
$$4 + 2y = 0 \qquad \text{Let } x = 4.$$
$$2y = -4$$
$$y = -2$$

This gives the ordered pair $(4, -2)$. To find the additional point, we could have chosen any number (except 0) for y instead of x.

The points $(0, 0)$ and $(4, -2)$ lead to the graph shown in Figure 8 on the next page. As a check, verify that $(-2, 1)$ also lies on the line.

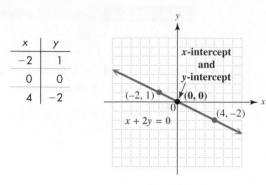

FIGURE 8

✔ **Now Try Exercise 49.**

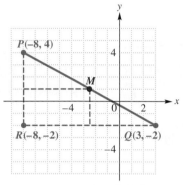

FIGURE 9

OBJECTIVE 7 **Use the midpoint formula.** If the coordinates of the endpoints of a line segment are known, then the coordinates of the *midpoint* of the segment can be found. Figure 9 shows that segment PQ has endpoints $P(-8, 4)$ and $Q(3, -2)$. R is the point with the same x-coordinate as P and the same y-coordinate as Q. So the coordinates of R are $(-8, -2)$.

The x-coordinate of the midpoint M of PQ is the same as the x-coordinate of the midpoint of RQ. Since RQ is horizontal, the x-coordinate of its midpoint is the *average* of the x-coordinates of its endpoints:

$$\frac{1}{2}(-8 + 3) = -2.5.$$

The y-coordinate of M is the average of the y-coordinates of the midpoint of PR:

$$\frac{1}{2}(4 + (-2)) = 1.$$

The midpoint of PQ is $M(-2.5, 1)$. This discussion leads to the *midpoint formula*.

Midpoint Formula

If the endpoints of a line segment PQ are (x_1, y_1) and (x_2, y_2), its midpoint M is

$$\left(\frac{x_1 + x_2}{2}, \frac{y_1 + y_2}{2}\right).$$

The small numbers 1 and 2 in these ordered pairs are called *subscripts*. Read (x_1, y_1) as "x-sub-one, y-sub-one."

EXAMPLE 6 **Finding the Coordinates of a Midpoint**

Find the coordinates of the midpoint of line segment PQ with endpoints $P(4, -3)$ and $Q(6, -1)$.

Use the midpoint formula with $x_1 = 4$, $x_2 = 6$, $y_1 = -3$, and $y_2 = -1$:

$$\left(\frac{4 + 6}{2}, \frac{-3 + (-1)}{2}\right) = \left(\frac{10}{2}, \frac{-4}{2}\right) = (5, -2). \leftarrow \text{Midpoint}$$

✔ **Now Try Exercise 55.**

▶ **NOTE** When finding the coordinates of the midpoint of a line segment, remember that you are finding the *average* of the *x*-coordinates and the *average* of the *y*-coordinates of the endpoints of the segment. In both cases, add the corresponding coordinates and divide the sum by 2.

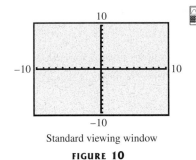

Standard viewing window

FIGURE 10

OBJECTIVE 8 Use a graphing calculator to graph an equation. When graphing by hand, we first set up a rectangular coordinate system, then plot points, and, finally, draw the graph. Similarly, when graphing with a graphing calculator, we first tell the calculator how to set up a rectangular coordinate system. This involves choosing the minimum and maximum *x*- and *y*-values that will determine the viewing screen. In the screen shown in Figure 10, we chose minimum *x*- and *y*-values of −10 and maximum *x*- and *y*-values of 10. The *scale* on each axis determines the distance between the tick marks; in the screen shown, the scale is 1 for both axes. We refer to this screen as the *standard viewing window*.

To graph an equation, we usually need to solve the equation for *y* in order to enter it into the calculator. Once the equation is graphed, we can use the calculator to find the intercepts or any other point on the graph.

EXAMPLE 7 Graphing a Linear Equation with a Graphing Calculator and Finding the Intercepts

Use a graphing calculator to graph $4x - y = 3$.

Because we want to be able to see the intercepts on the screen, we use them to determine an appropriate window. Here, the *x*-intercept is $(0.75, 0)$ and the *y*-intercept is $(0, -3)$. Although many choices are possible, we choose the standard viewing window. We must solve the equation for *y* to enter it into the calculator.

$$4x - y = 3$$
$$-y = -4x + 3 \qquad \text{Subtract } 4x.$$
$$y = 4x - 3 \qquad \text{Multiply by } -1.$$

The graph in Figure 11 also gives the intercepts at the bottoms of the screens.

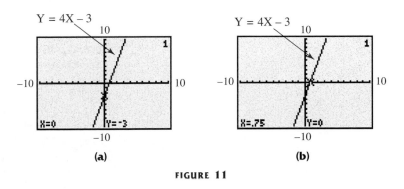

        (a)   (b)

FIGURE 11

Some calculators have the capability of locating the *x*-intercept (called "Root" or "Zero"). Consult your owner's manual.

✔ **Now Try Exercise 77.**

3.1 EXERCISES

Complete solution available on Video Lectures on CD/DVD

Now Try Exercise

In Exercises 1 and 2, answer each question by locating ordered pairs on the graphs. See Objective 1.

1. The graph indicates U.S. federal government tax revenues in billions of dollars.

   (a) If the ordered pair (x, y) represents a point on the graph, what does x represent? What does y represent?

   (b) Estimate revenue in 2002.

   (c) Write an ordered pair (x, y) that gives approximate federal tax revenues in 2002.

   (d) What does the ordered pair (2000, 2030) mean in the context of this graph?

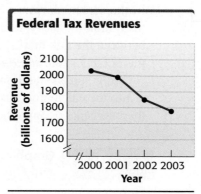

Federal Tax Revenues

Source: U.S. Office of Management and Budget.

2. The graph shows the percent of women in mathematics or computer science professions.

   (a) If the ordered pair (x, y) represents a point on the graph, what does x represent? What does y represent?

   (b) In what decade (10-yr period) did the percent of women in mathematics or computer science professions decrease?

   (c) Write an ordered pair (x, y) that gives the approximate percent of women in mathematics or computer science professions in 1990.

   (d) What does the ordered pair (2000, 30) mean in the context of this graph?

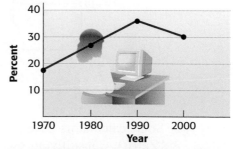

Women in Mathematics or Computer Science Professions

Source: U.S. Census Bureau and Bureau of Labor Statistics.

Concept Check *Fill in each blank with the correct response.*

3. The point with coordinates $(0, 0)$ is called the _____ of a rectangular coordinate system.

4. For any value of x, the point $(x, 0)$ lies on the _____-axis.

5. To find the x-intercept of a line, we let _____ equal 0 and solve for _____; to find the y-intercept, we let _____ equal 0 and solve for _____.

6. The equation _____ $= 4$ has a horizontal line as its graph.
        $(x$ or $y)$

7. To graph a straight line, we must find a minimum of _____ points.

8. The point (_____, 4) is on the graph of $2x - 3y = 0$.

Name the quadrant, if any, in which each point is located.

9. (a) $(1, 6)$ **(b)** $(-4, -2)$ **10. (a)** $(-2, -10)$ **(b)** $(4, 8)$
    (c) $(-3, 6)$ **(d)** $(7, -5)$ **(c)** $(-9, 12)$ **(d)** $(3, -9)$
    (e) $(-3, 0)$ **(f)** $(0, -0.5)$ **(e)** $(0, -8)$ **(f)** $(2.3, 0)$

11. *Concept Check*   Use the given information to determine the quadrants in which the point (x, y) may lie.
    (a) $xy > 0$ **(b)** $xy < 0$ **(c)** $\dfrac{x}{y} < 0$ **(d)** $\dfrac{x}{y} > 0$

12. *Concept Check*   What must be true about the coordinates of any point that lies along an axis?

Plot each point in a rectangular coordinate system. See Objective 2.

13. $(2, 3)$ **14.** $(-1, 2)$ **15.** $(-3, -2)$ **16.** $(1, -4)$ **17.** $(0, 5)$

18. $(-2, -4)$ **19.** $(-2, 4)$ **20.** $(3, 0)$ **21.** $(-2, 0)$ **22.** $(3, -3)$

*In Exercises 23–32, **(a)** complete the given table for each equation and then **(b)** graph the equation. See Example 1 and Figure 4.*

23. $y = x - 4$ **24.** $y = x + 3$ **25.** $x - y = 3$ **26.** $x - y = 5$

x	y
0	
1	
2	
3	
4	

x	y
0	
1	
2	
3	
4	

x	y
0	
	0
5	
2	

x	y
0	
	0
1	
3	

27. $x + 2y = 5$ **28.** $x + 3y = -5$ **29.** $4x - 5y = 20$

x	y
0	
	0
2	
	2

x	y
0	
	0
1	
	-1

x	y
0	
	0
2	
	-3

30. $6x - 5y = 30$

x	y
0	
	0
3	
	−2

31. $y = -2x + 3$

x	y
0	
1	
2	
3	

32. $y = -3x + 1$

x	y
0	
1	
2	
3	

33. *Concept Check* Consider the patterns formed in the tables for Exercises 23 and 31. Fill in each blank with the appropriate number.

(a) In Exercise 23, for every increase in x by 1 unit, y increases by _____ unit(s).

(b) In Exercise 31, for every increase in x by 1 unit, y decreases by _____ unit(s).

(c) On the basis of your observations in parts (a) and (b), make a conjecture about a similar pattern for $y = 2x + 4$. Then test your conjecture.

34. Explain why the graph of $x + y = k$ cannot pass through quadrant III if $k > 0$.

35. A student attempted to graph $4x + 5y = 0$ by finding intercepts. First she let $x = 0$ and found y; then she let $y = 0$ and found x. In both cases, the resulting point was $(0, 0)$. She knew that she needed at least two points to graph the line, but was unsure what to do next because finding intercepts gave her only one point. Explain to her what to do next.

36. *Concept Check* What is the equation of the x-axis? What is the equation of the y-axis?

Find the x- and y-intercepts. Then graph each equation. See Examples 2–5.

37. $2x + 3y = 12$

38. $5x + 2y = 10$

39. $x - 3y = 6$

40. $x - 2y = -4$

41. $\dfrac{2}{3}x - 3y = 7$

42. $\dfrac{5}{7}x + \dfrac{6}{7}y = -2$

43. $y = 5$

44. $y = -3$

45. $x = 2$

46. $x = -3$

47. $x + 4 = 0$

48. $y + 2 = 0$

49. $x + 5y = 0$

50. $x - 3y = 0$

51. $2x = 3y$

52. $4y = 3x$

53. $-\dfrac{2}{3}y = x$

54. $3y = -\dfrac{4}{3}x$

Find the midpoint of each segment with the given endpoints. See Example 6.

55. $(-8, 4)$ and $(-2, -6)$

56. $(5, 2)$ and $(-1, 8)$

57. $(3, -6)$ and $(6, 3)$

58. $(-10, 4)$ and $(7, 1)$

59. $(-9, 3)$ and $(9, 8)$

60. $(4, -3)$ and $(-1, 3)$

61. $(2.5, 3.1)$ and $(1.7, -1.3)$

62. $(6.2, 5.8)$ and $(1.4, -0.6)$

63. $\left(\dfrac{1}{2}, \dfrac{1}{3}\right)$ and $\left(\dfrac{3}{2}, \dfrac{5}{3}\right)$

64. $\left(\dfrac{21}{4}, \dfrac{2}{5}\right)$ and $\left(\dfrac{7}{4}, \dfrac{3}{5}\right)$

65. $\left(-\dfrac{1}{3}, \dfrac{2}{7}\right)$ and $\left(-\dfrac{1}{2}, \dfrac{1}{14}\right)$

66. $\left(\dfrac{3}{5}, -\dfrac{1}{3}\right)$ and $\left(\dfrac{1}{2}, -\dfrac{7}{2}\right)$

Suppose that segment PQ has the given coordinates for one endpoint P and for its midpoint M. Find the coordinates of the other endpoint Q. (Hint: Represent Q by (x, y) and write two equations using the midpoint formula: one involving x and the other involving y. Then solve for x and y.)

67. $P(5, 8)$, $M(8, 2)$

68. $P(7, 10)$, $M(5, 3)$

69. $P\left(\frac{1}{3}, \frac{1}{5}\right)$, $M\left(\frac{3}{2}, 1\right)$

70. $P(2.5, 1.75)$, $M(3, 2)$

A linear equation can be used as a model to describe real data in some cases. Exercises 71 and 72 are based on this idea.

71. The number of U.S. travelers to other countries for 2000–2003 is approximated by the linear equation

$$y = -1237x + 60{,}936,$$

where y is the number of travelers, in thousands, in year x. In the equation, $x = 0$ corresponds to 2000, $x = 1$ corresponds to 2001, and so on. Use the equation to approximate the number of U.S. travelers to other countries in 2003. (*Source:* U.S. Department of Commerce.)

72. The total expenditures for dental services in the United States from 1990 through 2003 can be approximated by the linear equation

$$y = 3.32x + 28.7,$$

where y is the expenditure in billions of dollars. In the equation, $x = 0$ corresponds to 1990, $x = 1$ corresponds to 1991, and so on. On the basis of this equation, find the total amount spent on dental services in the United States in 2002. (*Source:* U.S. Centers for Medicare and Medicaid Services.)

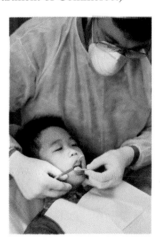

TECHNOLOGY INSIGHTS (EXERCISES 73–76)

73. The screens show the graph of one of the equations in A–D. Which equation is it?

    A. $3x + 2y = 6$ **B.** $-3x + 2y = 6$ **C.** $-3x - 2y = 6$ **D.** $3x - 2y = 6$

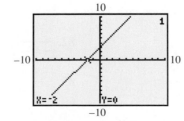 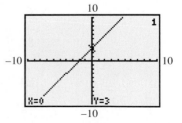

(continued)

74. The table of ordered pairs was generated by a graphing calculator with a TABLE feature.

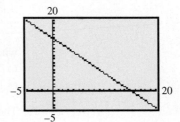

(a) What is the x-intercept?

(b) What is the y-intercept?

(c) Which equation corresponds to this table of values?

   A. $Y_1 = 2X - 3$ **B.** $Y_1 = -2X - 3$

   C. $Y_1 = 2X + 3$ **D.** $Y_1 = -2X + 3$

75. The screens each show the graph of $x + y = 15$ (which was entered as $y = -x + 15$). However, different viewing windows are used. Which window would be more useful for this graph? Why?

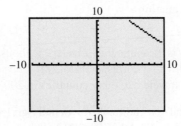

76. The screen shows the graph of $x + 2y = 0$ from Example 5. In what form should you enter the equation into the calculator?

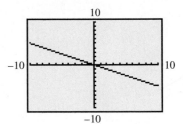

Graph each equation with a graphing calculator. Use the standard window. See Example 7.

77. $5x + 2y = -10$

78. $4x - y = 10$

79. $3.6x - y = -5.8$

80. $0.3x + 0.4y = -0.6$

PREVIEW EXERCISES

*Find each quotient. See **Section 1.2**.*

81. $\dfrac{6 - 2}{5 - 3}$

82. $\dfrac{5 - 7}{-4 - 2}$

83. $\dfrac{4 - (-1)}{-3 - (-5)}$

84. $\dfrac{-6 - 0}{0 - (-3)}$

85. $\dfrac{-5 - (-5)}{3 - 2}$

86. $\dfrac{7 - (-2)}{-3 - (-3)}$

3.2 The Slope of a Line

Slope (steepness) is used in many practical ways. The slope of a highway (sometimes called the *grade*) is often given as a percent. For example, a 10% $\left(\text{or } \frac{10}{100} = \frac{1}{10}\right)$ slope means that the highway rises 1 unit for every 10 horizontal units. Stairs and roofs have slopes too, as shown in Figure 12.

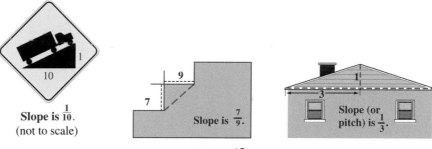

Slope is $\frac{1}{10}$.
(not to scale)

Slope is $\frac{7}{9}$.

Slope (or pitch) is $\frac{1}{3}$.

FIGURE 12

In each example mentioned, slope is the ratio of vertical change, or **rise,** to horizontal change, or **run.** A simple way to remember this is to think, *"Slope is rise over run."*

OBJECTIVE ❶ Find the slope of a line, given two points on the line. To get a formal definition of the slope of a line, we designate two different points on the line. To differentiate between the points, we write them as (x_1, y_1) and (x_2, y_2). See Figure 13.

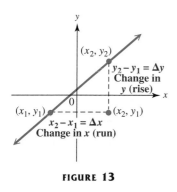

FIGURE 13

As we move along the line in Figure 13 from (x_1, y_1) to (x_2, y_2), the y-value changes (vertically) from y_1 to y_2, an amount equal to $y_2 - y_1$. As y changes from y_1 to y_2, the value of x changes (horizontally) from x_1 to x_2 by the amount $x_2 - x_1$. (The Greek letter **delta, Δ,** is used in mathematics to denote "change in," so Δy and Δx represent the change in y and the change in x, respectively.) The ratio of the change in y to the change in x (the rise over the run) is called the *slope* of the line, with the letter m traditionally used for slope.

Slope Formula

The **slope** of the line through the distinct points (x_1, y_1) and (x_2, y_2) is

$$m = \frac{\text{rise}}{\text{run}} = \frac{\text{change in } y}{\text{change in } x} = \frac{\Delta y}{\Delta x} = \frac{y_2 - y_1}{x_2 - x_1} \quad (x_1 \neq x_2).$$

EXAMPLE 1 Finding the Slope of a Line

Find the slope of the line through the points $(2, -1)$ and $(-5, 3)$.

If $(2, -1) = (x_1, y_1)$ and $(-5, 3) = (x_2, y_2)$, then

$$m = \frac{y_2 - y_1}{x_2 - x_1} = \frac{3 - (-1)}{-5 - 2} = \frac{4}{-7} = -\frac{4}{7}.$$

Thus, the slope is $-\frac{4}{7}$. See Figure 14 on the next page.

If the ordered pairs are interchanged so that $(2, -1) = (x_2, y_2)$ and $(-5, 3) = (x_1, y_1)$ in the slope formula, the slope is the same.

$$m = \frac{-1 - 3}{2 - (-5)} = \frac{-4}{7} = -\frac{4}{7}$$

y-values are in the *numerator*, *x*-values in the *denominator*.

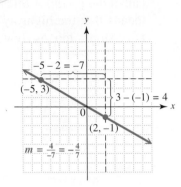

FIGURE 14

✔ **Now Try Exercise 23.**

Example 1 suggests that the slope is the same no matter which point we consider first. Also, using similar triangles from geometry, we can show that the slope is the same no matter which two different points on the line we choose.

▶ **CAUTION** *In calculating slope, be careful to subtract the y-values and the x-values in the same order.*

Correct	Incorrect

$$\frac{y_2 - y_1}{x_2 - x_1} \quad \text{or} \quad \frac{y_1 - y_2}{x_1 - x_2} \qquad \frac{y_2 - y_1}{x_1 - x_2} \quad \text{or} \quad \frac{y_1 - y_2}{x_2 - x_1}$$

The change in y is the numerator and the change in x is the denominator.

CONNECTIONS

In **Section 1.4,** we introduced the *identity property for multiplication,* which states that any real number can be multiplied by 1 without changing its value. This important property can be applied in many ways. For example, it allows us to find common denominators when adding and subtracting fractions.

Now suppose that we want to find the slope of the line passing through the points $\left(\frac{2}{3}, \frac{4}{5}\right)$ and $\left(\frac{7}{5}, \frac{1}{10}\right)$. The fractions lead to an expression for m that is more complex than if only integers were involved:

$$m = \frac{\dfrac{4}{5} - \dfrac{1}{10}}{\dfrac{2}{3} - \dfrac{7}{5}}.$$

Change in *y*

Change in *x*

(continued)

This expression is called a *complex fraction* (**Section 7.3**). One method to simplify a complex fraction involves *multiplying the complex fraction by 1 in the form of a number divided by itself.* We choose the least common denominator of all the fractions in the complex fraction and then multiply both the numerator and denominator by that number. Here, the least common denominator is 30:

$$m = \dfrac{\dfrac{4}{5} - \dfrac{1}{10}}{\dfrac{2}{3} - \dfrac{7}{5}} = \dfrac{\dfrac{4}{5} - \dfrac{1}{10}}{\dfrac{2}{3} - \dfrac{7}{5}} \cdot \dfrac{30}{30} = \dfrac{\dfrac{4}{5}(30) - \dfrac{1}{10}(30)}{\dfrac{2}{3}(30) - \dfrac{7}{5}(30)} = \dfrac{24 - 3}{20 - 42} = \dfrac{21}{-22} = -\dfrac{21}{22}.$$

$\underbrace{}$ Multiply by 1. $\qquad$ $\underbrace{}$ Distributive property

Thus, the slope has been computed and a lot of messy arithmetic has been avoided, all because of the identity property for multiplication.

For Discussion or Writing

1. With the method just described, what number would you use in simplifying the following complex fraction? (*Hint:* Find the least common denominator of all the fractions within the complex fraction.)

$$\dfrac{\dfrac{1}{6} - \dfrac{3}{20}}{\dfrac{2}{15} - \dfrac{7}{12}}$$

2. Find the slope of the line containing each pair of points.

(a) $\left(\dfrac{1}{5}, \dfrac{1}{3}\right)$ and $\left(\dfrac{2}{3}, \dfrac{6}{5}\right)$ $\qquad$ **(b)** $\left(-\dfrac{5}{2}, \dfrac{1}{6}\right)$ and $\left(\dfrac{5}{3}, \dfrac{3}{8}\right)$

(c) $\left(-\dfrac{5}{6}, -\dfrac{1}{2}\right)$ and $\left(-\dfrac{1}{3}, -\dfrac{3}{2}\right)$

OBJECTIVE **2** **Find the slope of a line, given an equation of the line.** When an equation of a line is given, one way to find the slope is to use the definition of slope by first finding two different points on the line.

EXAMPLE 2 **Finding the Slope of a Line**

Find the slope of the line $4x - y = -8$.

The intercepts can be used as the two different points needed to find the slope. Let $y = 0$ to find that the *x*-intercept is $(-2, 0)$. Then let $x = 0$ to find that the *y*-intercept is $(0, 8)$. Use these two points in the slope formula. The slope is

$$m = \dfrac{\text{rise}}{\text{run}} = \dfrac{8 - 0}{0 - (-2)} = \dfrac{8}{2} = 4.$$

✔ **Now Try Exercise 41.**

EXAMPLE 3 Finding Slopes of Horizontal and Vertical Lines

Find the slope of each line.

(a) $y = 2$

The graph of $y = 2$ is a horizontal line. To find the slope, select two different points on the line, such as $(3, 2)$ and $(-1, 2)$, and use the slope formula.

$$m = \frac{\text{rise}}{\text{run}} = \frac{2 - 2}{3 - (-1)} = \frac{0}{4} = 0$$

In this case, the *rise* is 0, so the slope is 0.

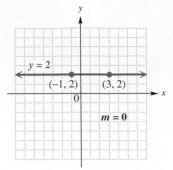

(b) $x + 1 = 0$

The graph of $x + 1 = 0$, or $x = -1$, is a vertical line. Two points that satisfy the equation $x = -1$ are $(-1, 5)$ and $(-1, -4)$. If we use these two points to try to find the slope, we obtain

$$m = \frac{\text{rise}}{\text{run}} = \frac{-4 - 5}{-1 - (-1)} = \frac{-9}{0}.$$

Since division by 0 is undefined, the slope is undefined. This is why the definition of slope includes the restriction that $x_1 \neq x_2$.

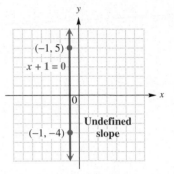

✔ **Now Try Exercises 47 and 49.**

Example 3 illustrates the following important concepts.

Horizontal and Vertical Lines

- An equation of the form $y = b$ always intersects the y-axis; thus, the line with that equation is horizontal and has slope 0.

- An equation of the form $x = a$ always intersects the x-axis; thus, the line with that equation is vertical and has undefined slope.

The slope of a line can also be found directly from its equation. Look again at the equation $4x - y = -8$ from Example 2. Solve this equation for y.

$$4x - y = -8 \qquad \text{Equation from Example 2}$$
$$-y = -4x - 8 \qquad \text{Subtract } 4x.$$
$$y = 4x + 8 \qquad \text{Multiply by } -1.$$

Notice that the slope, 4, found with the slope formula in Example 2 is the same number as the coefficient of x in the equation $y = 4x + 8$. We will see in the next section that this always happens, *as long as the equation is solved for y.*

EXAMPLE 4 Finding the Slope from an Equation

Find the slope of the graph of $3x - 5y = 8$.

Solve the equation for y.

$$3x - 5y = 8$$
$$-5y = -3x + 8 \qquad \text{Subtract } 3x.$$
$$y = \frac{3}{5}x - \frac{8}{5} \qquad \text{Divide by } -5.$$

The slope is given by the coefficient of x, so the slope is $\frac{3}{5}$.

✔ **Now Try Exercise 43.**

OBJECTIVE 3 Graph a line, given its slope and a point on the line. Example 5 shows how to graph a straight line by using the geometric interpretation of slope and one point on the line.

EXAMPLE 5 Using the Slope and a Point to Graph Lines

Graph each line.

(a) With slope $\frac{2}{3}$ and y-intercept $(0, -4)$

Begin by plotting the point $P(0, -4)$, as shown in Figure 15. Then use the slope to find a second point. From the slope formula,

$$m = \frac{\text{change in } y}{\text{change in } x} = \frac{2}{3},$$

so move 2 units *up* and then 3 units to the *right* to locate another point on the graph, $R(3, -2)$. The line through $P(0, -4)$ and R is the required graph.

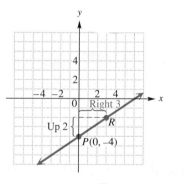

FIGURE 15

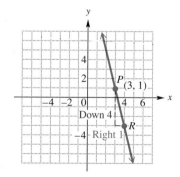

FIGURE 16

(b) Through $(3, 1)$ with slope -4

Start by locating the point $P(3, 1)$ on a graph. Find a second point R on the line by writing the slope -4 as $\frac{-4}{1}$ and using the slope formula.

$$m = \frac{\text{change in } y}{\text{change in } x} = \frac{-4}{1}$$

Move 4 units *down* from $(3, 1)$, and then move 1 unit to the *right*. Draw a line through this second point R and $(3, 1)$, as shown in Figure 16.

The slope also could be written as

$$m = \frac{\text{change in } y}{\text{change in } x} = \frac{4}{-1}.$$

In this case, the second point R is located 4 units *up* and 1 unit to the *left*. Verify that this approach also produces the line in Figure 16.

✔ **Now Try Exercises 55 and 57.**

In Example 5(a), the slope of the line is the *positive* number $\frac{2}{3}$. The graph of the line in Figure 15 goes up (rises) from left to right. The line in Example 5(b) has *negative* slope -4. As Figure 16 shows, its graph goes down (falls) from left to right. These facts suggest the following generalization.

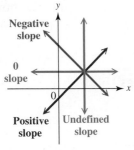

FIGURE 17

> ### Orientation of a Line in the Plane
>
> A positive slope indicates that the line goes *up* (rises) from left to right.
>
> A negative slope indicates that the line goes *down* (falls) from left to right.

Figure 17 shows lines of positive, 0, negative, and undefined slopes.

OBJECTIVE 4 **Use slopes to determine whether two lines are parallel, perpendicular, or neither.** The slopes of a pair of parallel or perpendicular lines are related in a special way. Recall that the slope of a line measures the steepness of the line. Since parallel lines have equal steepness, their slopes must be equal; also, lines with the same slope are parallel.

> ### Slopes of Parallel Lines
>
> Two nonvertical lines with the same slope are parallel.
>
> Two nonvertical parallel lines have the same slope.

EXAMPLE 6 **Determining whether Two Lines Are Parallel**

Determine whether the lines L_1, through $(-2, 1)$ and $(4, 5)$, and L_2, through $(3, 0)$ and $(0, -2)$, are parallel.

The slope of L_1 is

$$m_1 = \frac{5 - 1}{4 - (-2)} = \frac{4}{6} = \frac{2}{3}.$$

The slope of L_2 is

$$m_2 = \frac{-2 - 0}{0 - 3} = \frac{-2}{-3} = \frac{2}{3}.$$

Because the slopes are equal, the two lines are parallel.

✔ **Now Try Exercise 65.**

To see how the slopes of perpendicular lines are related, consider a nonvertical line with slope $\frac{a}{b}$. If this line is rotated 90°, the vertical change and the horizontal change are interchanged and the slope is $-\frac{b}{a}$, since the horizontal change is now negative. See Figure 18. Thus, the slopes of perpendicular lines have product -1 and are negative reciprocals of each other. For example, if the slopes of two lines are $\frac{3}{4}$ and $-\frac{4}{3}$, then the lines are perpendicular because

$$\frac{3}{4}\left(-\frac{4}{3}\right) = -1.$$

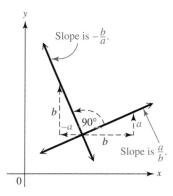

FIGURE 18

Slopes of Perpendicular Lines

Two perpendicular lines, neither of which is parallel to an axis, have slopes that are negative reciprocals; that is, their product is -1. Also, lines with slopes that are negative reciprocals are perpendicular.

EXAMPLE 7 **Determining whether Two Lines Are Perpendicular**

Determine whether the lines with equations $2y = 3x - 6$ and $2x + 3y = -6$ are perpendicular.

Find the slope of each line by first solving each equation for y.

$$2y = 3x - 6$$
$$y = \frac{3}{2}x - 3$$
$$\uparrow$$
Slope

$$2x + 3y = -6$$
$$3y = -2x - 6$$
$$y = -\frac{2}{3}x - 2$$
$$\uparrow$$
Slope

Since the product of the slopes of the two lines is $\frac{3}{2}\left(-\frac{2}{3}\right) = -1$, the lines are perpendicular.

✔ **Now Try Exercise 67.**

▶ **NOTE** In Example 7, alternatively, we could have found the slope of each line by using intercepts and the slope formula. For the graph of the equation $2y = 3x - 6$, the x-intercept is $(2, 0)$ and the y-intercept is $(0, -3)$. Thus,

$$m = \frac{0 - (-3)}{2 - 0} = \frac{3}{2}, \quad \text{as in Example 7.}$$

Find the intercepts of the graph of $2x + 3y = -6$ and use them to confirm the slope $-\frac{2}{3}$. Since the slopes are negative reciprocals, the lines are perpendicular.

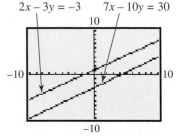

The graphs are *not* parallel, although they may appear to be.

FIGURE 19

We must be careful when interpreting calculator graphs of parallel and perpendicular lines. For example, the graphs of the equations in Figure 19 appear to be parallel. However, checking their slopes algebraically, we find that

$$2x - 3y = -3 \qquad\qquad 7x - 10y = 30$$
$$-3y = -2x - 3 \qquad\qquad -10y = -7x + 30$$
$$y = \frac{2}{3}x + 1 \qquad\qquad y = \frac{7}{10}x - 3.$$

Since the slopes $\frac{2}{3}$ and $\frac{7}{10}$ are not equal, the lines are *not* parallel.

Figure 20(a) shows graphs of the perpendicular lines from Example 7. As graphed in the standard viewing window, the lines do not appear to be perpendicular. However, if we use a *square viewing window* as in Figure 20(b), we get a more realistic view. (Many graphing calculators can set a square window automatically. See your owner's manual.) *We cannot rely completely on what we see on a calculator screen—we must understand the mathematical concepts as well.*

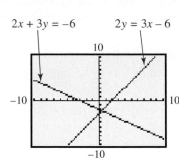

In the standard window, the lines *do not* appear to be perpendicular.

(a)

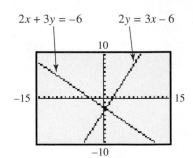

In the square window, the lines *do* appear to be perpendicular.

(b)

FIGURE 20

OBJECTIVE 5 Solve problems involving average rate of change. We know that the slope of a line is the ratio of the vertical change in y to the horizontal change in x. Thus, slope gives the *average rate of change* in y per unit change in x, where the value of y depends on the value of x. The next examples illustrate this idea. We assume a linear relationship between x and y.

EXAMPLE 8 Interpreting Slope as Average Rate of Change

The graph in Figure 21 approximates the average number of hours per year spent watching cable and satellite TV for each person in the United States during the years 2000 through 2004. Find the average rate of change in number of hours per year.

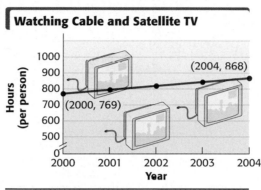

Watching Cable and Satellite TV

Source: Veronis Suhler Stevenson.

FIGURE 21

To determine the average rate of change, we need two pairs of data. From the graph, if $x = 2000$, then $y = 769$ and if $x = 2004$, then $y = 868$. Thus, we have the ordered pairs $(2000, 769)$ and $(2004, 868)$. By the slope formula,

$$\text{average rate of change} = \frac{868 - 769}{2004 - 2000} = \frac{99}{4} = 24.75.$$

> A positive slope indicates an increase.

This means that the average time per person spent watching cable and satellite TV *increased* by about 25 hr per year from 2000 through 2004.

✔ **Now Try Exercise 85.**

EXAMPLE 9 Interpreting Slope as Average Rate of Change

During the year 2000, the average person in the U.S. spent 866 hr watching broadcast TV. In 2004, the average number of hours per person spent watching broadcast TV was 678. Find the average rate of change in number of hours per year. (*Source:* Veronis Suhler Stevenson.)

To use the slope formula, we need two ordered pairs. Here, we let one ordered pair be $(2000, 866)$ and the other be $(2004, 678)$.

$$\text{average rate of change} = \frac{678 - 866}{2004 - 2000} = \frac{-188}{4} = -47$$

> A negative slope indicates a decrease.

The graph in Figure 22 confirms that the line through the ordered pairs falls from left to right and therefore has negative slope. Thus, the average time per person spent watching broadcast TV *decreased* by 47 hr per year from 2000 through 2004.

✔ **Now Try Exercise 87.**

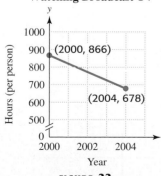

Watching Broadcast TV

FIGURE 22

3.2 EXERCISES

⊕ *Complete solution available on Video Lectures on CD/DVD*

Now Try Exercise

1. *Concept Check* A ski slope drops 30 ft for every horizontal 100 ft. Which of the following express its slope? (There are several correct choices.)

   A. 0.3   **B.** $\dfrac{3}{10}$ **C.** $3\dfrac{1}{3}$

   D. $\dfrac{30}{100}$ **E.** $\dfrac{10}{3}$ **F.** 30

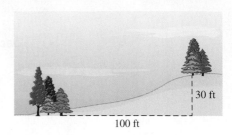

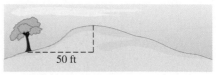
100 ft   30 ft

2. *Concept Check* A hill has slope 0.05. How many feet in the vertical direction correspond to a run of 50 ft?

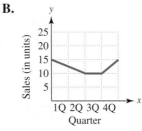
50 ft

NOT TO SCALE

3. *Concept Check* Match each situation in (a)–(d) with the most appropriate graph in A–D.

   (a) Sales rose sharply during the first quarter, leveled off during the second quarter, and then rose slowly for the rest of the year.
   (b) Sales fell sharply during the first quarter and then rose slowly during the second and third quarters before leveling off for the rest of the year.
   (c) Sales rose sharply during the first quarter and then fell to the original level during the second quarter before rising steadily for the rest of the year.
   (d) Sales fell during the first two quarters of the year, leveled off during the third quarter, and rose during the fourth quarter.

   A.

   Sales (in units) vs *Quarter* (1Q 2Q 3Q 4Q)

   B.

   Sales (in units) vs *Quarter* (1Q 2Q 3Q 4Q)

   C.

   Sales (in units) vs *Quarter* (1Q 2Q 3Q 4Q)

   D.

   Sales (in units) vs *Quarter* (1Q 2Q 3Q 4Q)

Determine the slope of each line segment in the given figure.

4. *AB*   **5.** *BC*   **6.** *CD*

7. *DE*   **8.** *EF*   **9.** *FG*

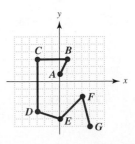

10. *Concept Check*   If B and D were joined by a line segment in the figure for Exercises 4–9, what would be the slope of the segment?

Calculate the value of each slope m, if possible, by using the slope formula. See Example 1.

11. $m = \dfrac{6 - 2}{5 - 3}$

12. $m = \dfrac{5 - 7}{-4 - 2}$

13. $m = \dfrac{4 - (-1)}{-3 - (-5)}$

14. $m = \dfrac{-6 - 0}{0 - (-3)}$

15. $m = \dfrac{-5 - (-5)}{3 - 2}$

16. $m = \dfrac{-2 - (-2)}{4 - (-3)}$

17. $m = \dfrac{3 - 8}{-2 - (-2)}$

18. $m = \dfrac{5 - 6}{-8 - (-8)}$

19. $m = \dfrac{\dfrac{4}{3} + \dfrac{1}{2}}{\dfrac{1}{6} - \dfrac{1}{6}}$

20. Which of the following forms of the slope formula are correct? Explain.

   A. $m = \dfrac{y_1 - y_2}{x_2 - x_1}$ **B.** $m = \dfrac{y_1 - y_2}{x_1 - x_2}$ **C.** $m = \dfrac{x_2 - x_1}{y_2 - y_1}$ **D.** $m = \dfrac{y_2 - y_1}{x_2 - x_1}$

Find the slope of the line through each pair of points. See Example 1 and the Connections that follow Example 1.

21. $(-2, -3)$ and $(-1, 5)$ **22.** $(-4, 3)$ and $(-3, 4)$ **23.** $(-4, 1)$ and $(2, 6)$

24. $(-3, -3)$ and $(5, 6)$ **25.** $(2, 4)$ and $(-4, 4)$ **26.** $(-6, 3)$ and $(2, 3)$

27. $(1.5, 2.6)$ and $(0.5, 3.6)$ **28.** $(3.4, 4.2)$ and $(1.4, 10.2)$

29. $\left(\dfrac{1}{6}, \dfrac{1}{2}\right)$ and $\left(\dfrac{5}{6}, \dfrac{9}{2}\right)$ **30.** $\left(\dfrac{3}{4}, \dfrac{1}{3}\right)$ and $\left(\dfrac{5}{4}, \dfrac{10}{3}\right)$

31. $\left(-\dfrac{2}{9}, \dfrac{5}{18}\right)$ and $\left(\dfrac{1}{18}, -\dfrac{5}{9}\right)$ **32.** $\left(-\dfrac{4}{5}, \dfrac{9}{10}\right)$ and $\left(-\dfrac{3}{10}, \dfrac{1}{5}\right)$

Find the slope of each line.

33.

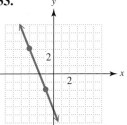

34.

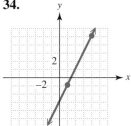

35.

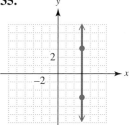

36. *Concept Check*   Let k be the number of letters in your last name. Sketch the graph of $y = k$. What is the slope of this line?

Concept Check   *On the basis of the figure shown here, determine which line satisfies the given description.*

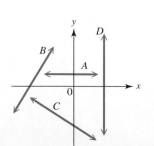

37. The line has positive slope.

38. The line has negative slope.

39. The line has slope 0.

40. The line has undefined slope.

Find the slope of the line and sketch the graph. See Examples 1–4.

41. $x + 2y = 4$

42. $x + 3y = -6$

43. $5x - 2y = 10$

44. $4x - y = 4$

45. $y = 4x$

46. $y = -3x$

47. $x - 3 = 0$

48. $x + 2 = 0$

49. $y = -5$

50. $y = -4$

51. $2y = 3$

52. $3x = 4$

Graph the line described. See Example 5.

53. Through $(-4, 2)$; $m = \dfrac{1}{2}$

54. Through $(-2, -3)$; $m = \dfrac{5}{4}$

55. y-intercept $(0, -2)$; $m = -\dfrac{2}{3}$

56. y-intercept $(0, -4)$; $m = -\dfrac{3}{2}$

57. Through $(-1, -2)$; $m = 3$

58. Through $(-2, -4)$; $m = 4$

59. $m = 0$; through $(2, -5)$

60. $m = 0$; through $(5, 3)$

61. Undefined slope; through $(-3, 1)$

62. Undefined slope; through $(-4, 1)$

63. *Concept Check* If a line has slope $-\frac{4}{9}$, then any line parallel to it has slope _____, and any line perpendicular to it has slope _____.

64. *Concept Check* If a line has slope 0.2, then any line parallel to it has slope _____, and any line perpendicular to it has slope _____.

Decide whether each pair of lines is parallel, perpendicular, *or* neither. *See Examples 6 and 7.*

65. The line through $(15, 9)$ and $(12, -7)$ and the line through $(8, -4)$ and $(5, -20)$

66. The line through $(4, 6)$ and $(-8, 7)$ and the line through $(-5, 5)$ and $(7, 4)$

67. $x + 4y = 7$ and $4x - y = 3$

68. $2x + 5y = -7$ and $5x - 2y = 1$

69. $4x - 3y = 6$ and $3x - 4y = 2$

70. $2x + y = 6$ and $x - y = 4$

71. $x = 6$ and $6 - x = 8$

72. $3x = y$ and $2y - 6x = 5$

73. $4x + y = 0$ and $5x - 8 = 2y$

74. $2x + 5y = -8$ and $6 + 2x = 5y$

75. $2x = y + 3$ and $2y + x = 3$

76. $4x - 3y = 8$ and $4y + 3x = 12$

Find and interpret the average rate of change illustrated in each graph.

77.

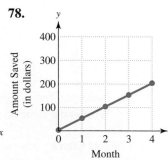

78.

79.

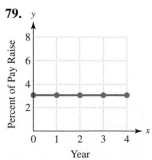

80. *Concept Check* If the graph of a linear equation rises from left to right, then the average rate of change is _____. If the graph of a linear equation falls from
(positive/negative)
left to right, then the average rate of change is _____.
(positive/negative)

Concept Check   *Solve each problem.*

81. When designing the arena now known as TD Banknorth Garden in Boston, architects designed the ramps leading up to the entrances so that circus elephants would be able to walk up the ramps. The maximum grade (or slope) that an elephant will walk on is 13%. Suppose that such a ramp was constructed with a horizontal run of 150 ft. What would be the maximum vertical rise the architects could use?

82. The upper deck at U.S. Cellular Field (formerly Comiskey Park) in Chicago has produced, among other complaints, displeasure with its steepness. It is 160 ft from home plate to the front of the upper deck and 250 ft from home plate to the back. The top of the upper deck is 63 ft above the bottom. What is its slope? (Consider the slope as a positive number here.)

150 ft

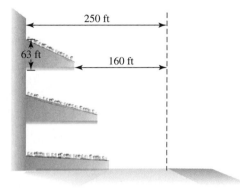

250 ft

63 ft

160 ft

Solve each problem. See Examples 8 and 9.

83. The table gives the number of cellular telephone subscribers (in thousands) from 1999 through 2004.

    (a) Find the average rate of change in subscribers for 1999–2000, 2000–2001, and so on.

    (b) Is the average rate of change in subscribers in successive years approximately the same? If the ordered pairs in the table were plotted, could an approximately straight line be drawn through them?

CELLULAR TELEPHONE SUBSCRIBERS

Year	Subscribers (in thousands)
1999	86,047
2000	109,478
2001	128,375
2002	140,767
2003	158,722
2004	182,140

Source: CTIA: The Wireless Association.

84. The table gives book publishers' approximate net dollar sales (in millions) from 1995 through 2000.

    (a) Find the average rate of change for 1995–1996, 1995–1999, and 1998–2000.

    (b) What do you notice about your answers in part (a)? What does this tell you?

BOOK PUBLISHERS' SALES

Year	Sales (in millions)
1995	19,000
1996	20,000
1997	21,000
1998	22,000
1999	23,000
2000	24,000

Source: Book Industry Study Group.

85. Personal spending on recreation in the United States (in billions of dollars) in recent years is closely approximated by the graph.

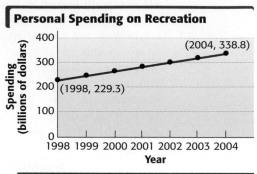

Source: U.S. Department of Commerce.

(a) Use the given ordered pairs to determine the average rate of change in these expenditures per year.

(b) Explain how a positive slope is interpreted in this situation.

86. The graph provides a good approximation of the number of drive-in theaters in the United States from 2000 through 2004.

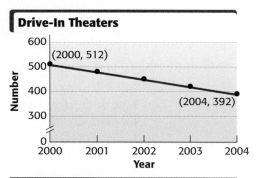

Source: Neilsen EDI.

(a) Use the given ordered pairs to find the average rate of change in the number of drive-in theaters per year during this period.

(b) Explain how a negative slope is interpreted in this situation.

87. When introduced in 1997, a DVD player sold for about $500. Five years later, the average price was $155. Find and interpret the average rate of change in price per year. (*Source: The Gazette,* Cedar Rapids, Iowa, June 22, 2002.)

88. In 1997, when DVD players entered the market, 0.349 million (that is, 349,000) were sold. Five years later, sales of DVD players reached 15.5 million (estimated). Find and interpret the average rate of change in sales, in millions per year. Round your answer to the nearest hundredth. (*Source: The Gazette,* Cedar Rapids, Iowa, June 22, 2002.)

The next problems are "brain busters." Use your knowledge of the slopes of parallel and perpendicular lines.

89. Show that $(-13, -9)$, $(-11, -1)$, $(2, -2)$, and $(4, 6)$ are the vertices of a parallelogram. (*Hint:* A parallelogram is a four-sided figure with opposite sides parallel.)

90. Is the figure with vertices at $(-11, -5)$, $(-2, -19)$, $(12, -10)$, and $(3, 4)$ a parallelogram? Is it a rectangle? (*Hint:* A rectangle is a parallelogram with a right angle.)

TECHNOLOGY INSIGHTS (EXERCISES 91 AND 92)

91. The graphing calculator screen shows two lines. One is the graph of $y_1 = -2x + 3$ and the other is the graph of $y_2 = 3x - 4$. Which is which?

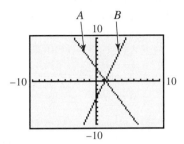

92. The graphing calculator screen shows two lines. One is the graph of $y_1 = 2x - 5$ and the other is the graph of $y_2 = 4x - 5$. Which is which?

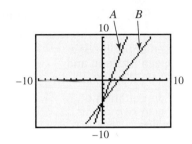

RELATING CONCEPTS (EXERCISES 93–98)

FOR INDIVIDUAL OR GROUP WORK

Three points that lie on the same straight line are said to be **collinear.** *Consider the points $A(3, 1)$, $B(6, 2)$, and $C(9, 3)$.* **Work Exercises 93–98 in order.**

93. Find the slope of segment AB.

94. Find the slope of segment BC.

95. Find the slope of segment AC.

96. If slope of segment AB = slope of segment BC = slope of segment AC, then A, B, and C are collinear. Use the results of Exercises 93–95 to show that this statement is satisfied.

97. Use the slope formula to determine whether the points $(1, -2)$, $(3, -1)$, and $(5, 0)$ are collinear.

98. Repeat Exercise 97 for the points $(0, 6)$, $(4, -5)$, and $(-2, 12)$.

PREVIEW EXERCISES

Solve each equation for y. See **Section 2.2.**

99. $3x + 2y = 8$ **100.** $4x + 3y = 0$ **101.** $y - 2 = 4(x + 3)$

Write each equation in the form $Ax + By = C$. See **Section 2.1.**

102. $y - (-2) = \dfrac{3}{2}(x - 5)$ **103.** $y - (-1) = \dfrac{5}{3}[x - (-4)]$

104. $y - 7 = -\dfrac{1}{4}[x - (-3)]$ **105.** $y - (-1) = -\dfrac{1}{2}[x - (-2)]$

3.3 Linear Equations in Two Variables

OBJECTIVES

1. Write an equation of a line, given its slope and y-intercept.
2. Graph a line, using its slope and y-intercept.
3. Write an equation of a line, given its slope and a point on the line.
4. Write an equation of a line, given two points on the line.
5. Write an equation of a line parallel or perpendicular to a given line.
6. Write an equation of a line that models real data.
7. Use a graphing calculator to solve linear equations in one variable.

OBJECTIVE 1 Write an equation of a line, given its slope and y-intercept. In the previous section, we found the slope of a line from the equation of the line by solving the equation for y. For example, we found that the slope of the line with equation $y = 4x + 8$ is 4, the coefficient of x. What does the number 8 represent?

To find out, suppose a line has slope m and y-intercept $(0, b)$. We can find an equation of this line by choosing another point (x, y) on the line, as shown in Figure 23. Using the slope formula gives

$$m = \frac{y - b}{x - 0}$$

$$m = \frac{y - b}{x}$$

$$mx = y - b \qquad \text{Multiply by } x.$$

$$mx + b = y \qquad \text{Add } b.$$

$$y = mx + b. \qquad \text{Rewrite.}$$

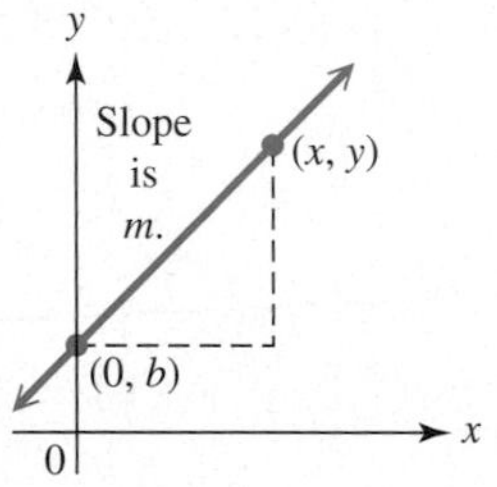

FIGURE 23

This last equation is called the *slope–intercept form* of the equation of a line, because we can identify the slope and y-intercept at a glance. Thus, in the line with equation $y = 4x + 8$, the number 8 indicates that the y-intercept is $(0, 8)$.

Slope–Intercept Form

The **slope–intercept form** of the equation of a line with slope m and y-intercept $(0, b)$ is

$$\mathbf{y = mx + b.}$$

↑ Slope ↑ y-intercept is $(0, b)$.

EXAMPLE 1 Using the Slope–Intercept Form to Find an Equation of a Line

Find an equation of the line with slope $-\frac{4}{5}$ and y-intercept $(0, -2)$.

Here, $m = -\frac{4}{5}$ and $b = -2$. Substitute these values into the slope–intercept form.

$$y = mx + b \qquad \text{Slope–intercept form}$$

$$y = -\frac{4}{5}x - 2 \qquad m = -\tfrac{4}{5};\ b = -2$$

✔ **Now Try Exercise 19.**

OBJECTIVE 2 Graph a line, using its slope and y-intercept. If the equation of a line is written in slope–intercept form, we can use the slope and y-intercept to obtain the graph of the equation. (We first saw this approach in Example 5(a) in **Section 3.2.**)

EXAMPLE 2 Graphing Lines Using Slope and *y*-Intercept

Graph each line, using the slope and *y*-intercept.

(a) $y = 3x - 6$

Here, $m = 3$ and $b = -6$. Plot the *y*-intercept $(0, -6)$. The slope 3 can be interpreted as

$$m = \frac{\text{rise}}{\text{run}} = \frac{\text{change in } y}{\text{change in } x} = \frac{3}{1}.$$

From $(0, -6)$, move 3 units *up* and 1 unit to the *right,* and plot a second point at $(1, -3)$. Join the two points with a straight line to obtain the graph in Figure 24.

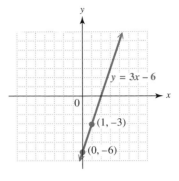

FIGURE 24

(b) $3y + 2x = 9$

Write the equation in slope–intercept form by solving for *y*.

$$3y + 2x = 9$$
$$3y = -2x + 9 \qquad \text{Subtract } 2x.$$
$$y = -\frac{2}{3}x + 3 \qquad \text{Slope–intercept form}$$

Slope ⎯⎯⎯↑   ↑⎯⎯ *y*-intercept is $(0, 3)$.

To graph this equation, plot the *y*-intercept $(0, 3)$. The slope can be interpreted as either $\frac{-2}{3}$ or $\frac{2}{-3}$. Using $\frac{-2}{3}$, begin at $(0, 3)$ and move 2 units *down* and 3 units to the *right* to locate the point $(3, 1)$. The line through these two points is the required graph. See Figure 25. $\left(\text{Verify that the point obtained with } \frac{2}{-3} \text{ as the slope is also on this line.}\right)$

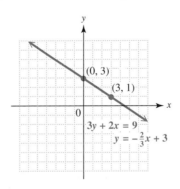

FIGURE 25

✔ **Now Try Exercise 25.**

▶ **NOTE** The slope–intercept form of a linear equation is the most useful for several reasons. Every linear equation (of a nonvertical line) has a *unique* (one and only one) slope–intercept form. In **Section 3.5,** we study *linear functions,* which are defined using slope–intercept form. Also, this is the form we use when graphing a line with a graphing calculator. (See **Section 3.1,** Example 7.)

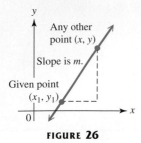

FIGURE 26

OBJECTIVE 3 Write an equation of a line, given its slope and a point on the line. Let m represent the slope of a line and (x_1, y_1) represent a given point on the line. Let (x, y) represent any other point on the line. See Figure 26. Then, by the slope formula,

$$m = \frac{y - y_1}{x - x_1}$$

$$m(x - x_1) = y - y_1 \qquad \text{Multiply each side by } x - x_1.$$

$$y - y_1 = m(x - x_1). \qquad \text{Rewrite.}$$

This last equation is the *point–slope form* of the equation of a line.

Point–Slope Form

The **point–slope form** of the equation of a line with slope m passing through the point (x_1, y_1) is

$$\overset{\text{Slope}}{\underset{\text{Given point}}{y - y_1 = m(x - x_1).}}$$

EXAMPLE 3 Finding an Equation of a Line, Given the Slope and a Point

Find an equation of the line with slope $\frac{1}{3}$ and passing through the point $(-2, 5)$.

Method 1: Use the point–slope form of the equation of a line, with $(x_1, y_1) = (-2, 5)$ and $m = \frac{1}{3}$.

$$y - y_1 = m(x - x_1) \qquad \text{Point–slope form}$$

$$y - 5 = \frac{1}{3}[x - (-2)] \qquad y_1 = 5, \; m = \frac{1}{3}, \; x_1 = -2$$

$$y - 5 = \frac{1}{3}(x + 2)$$

$$3y - 15 = x + 2 \qquad \text{Multiply by 3.}$$

$$-x + 3y = 17 \qquad (*) \qquad \text{Subtract } x; \text{ add 15.}$$

By solving for y, we find the slope–intercept form to be $y = \frac{1}{3}x + \frac{17}{3}$.

Method 2: An alternative method for finding this equation uses the slope–intercept form, with $(x, y) = (-2, 5)$ and $m = \frac{1}{3}$.

$$y = mx + b \qquad \text{Slope–intercept form}$$

$$5 = \frac{1}{3}(-2) + b \qquad \text{Substitute for } y, m, \text{ and } x.$$

$$5 = -\frac{2}{3} + b$$

$$b = \frac{17}{3} \qquad \text{Solve for } b.$$

Knowing that $m = \frac{1}{3}$ and $b = \frac{17}{3}$ gives the equation $y = \frac{1}{3}x + \frac{17}{3}$.

In **Section 3.1,** we defined *standard form* for a linear equation as

$$Ax + By = C,$$

where A, B, and C are real numbers. Most often, however, A, B, and C are integers. In this case, let us agree that integers A, B, and C have no common factor (except 1) and $A \geq 0$. For example, equation (*) in Example 3, $-x + 3y = 17$, is written in standard form by multiplying each side by -1 to obtain $x - 3y = -17$.

▶ **NOTE** The definition of "standard form" is not standard from one text to another. Any linear equation can be written in many different (and equally correct) forms. For example, the equation $2x + 3y = 8$ can be written as

$$2x = 8 - 3y, \qquad 3y = 8 - 2x, \qquad x + \frac{3}{2}y = 4, \qquad 4x + 6y = 16,$$

and so on. In addition to writing it in the form $Ax + By = C$ with $A \geq 0$, let us agree that the form $2x + 3y = 8$ is preferred over any multiples of each side, such as $4x + 6y = 16$. (To write $4x + 6y = 16$ in standard form, divide each side by 2.)

✔ **Now Try Exercise 35.**

OBJECTIVE 4 Write an equation of a line, given two points on the line. To find an equation of a line when two points on the line are known, first use the slope formula to find the slope of the line. Then use the slope with either of the given points and the point–slope form of the equation of a line.

EXAMPLE 4 Finding an Equation of a Line, Given Two Points

Find an equation of the line passing through the points $(-4, 3)$ and $(5, -7)$. Write the equation in standard form.

First find the slope by the slope formula.

$$m = \frac{-7 - 3}{5 - (-4)} = -\frac{10}{9}$$

Use either $(-4, 3)$ or $(5, -7)$ as (x_1, y_1) in the point–slope form of the equation of a line. If you choose $(-4, 3)$, then $-4 = x_1$ and $3 = y_1$.

$$y - y_1 = m(x - x_1) \qquad \text{Point–slope form}$$

$$y - 3 = -\frac{10}{9}[x - (-4)] \qquad y_1 = 3, m = -\tfrac{10}{9}, x_1 = -4$$

$$y - 3 = -\frac{10}{9}(x + 4)$$

$$9y - 27 = -10x - 40 \qquad \text{Multiply by 9; distributive property.}$$

$$10x + 9y = -13 \qquad \text{Standard form}$$

Verify that if $(5, -7)$ were used, the same equation would result.

✔ **Now Try Exercise 51.**

A horizontal line has slope 0. Using point–slope form, we find that the equation of a horizontal line through the point (a, b) is

$$y - y_1 = m(x - x_1) \qquad \text{Point–slope form}$$
$$y - b = 0(x - a) \qquad y_1 = b,\ m = 0,\ x_1 = a$$
$$y - b = 0 \qquad \text{Multiplication property of 0}$$
$$y = b. \qquad \text{Add } b.$$

Notice that point–slope form does not apply to a vertical line, since the slope of a vertical line is undefined. A vertical line through the point (a, b) has equation $x = a$.

In summary, horizontal and vertical lines have the following special equations.

Equations of Horizontal and Vertical Lines

The horizontal line through the point (a, b) has equation $y = b.$

The vertical line through the point (a, b) has equation $x = a.$

✔ **Now Try Exercises 43 and 45.**

OBJECTIVE 5 Write an equation of a line parallel or perpendicular to a given line. As mentioned in **Section 3.2**, parallel lines have the same slope and perpendicular lines have slopes that are negative reciprocals of each other.

EXAMPLE 5 Finding Equations of Parallel or Perpendicular Lines

Find an equation of the line passing through the point $(-4, 5)$ and **(a)** parallel to the line $2x + 3y = 6$; **(b)** perpendicular to the line $2x + 3y = 6$. Write each equation in slope–intercept form.

(a) We find the slope of the line $2x + 3y = 6$ by solving for y.

$$2x + 3y = 6$$
$$3y = -2x + 6 \qquad \text{Subtract } 2x.$$
$$y = -\frac{2}{3}x + 2 \qquad \text{Divide by 3.}$$
$$\underset{\text{Slope}}{\uparrow}$$

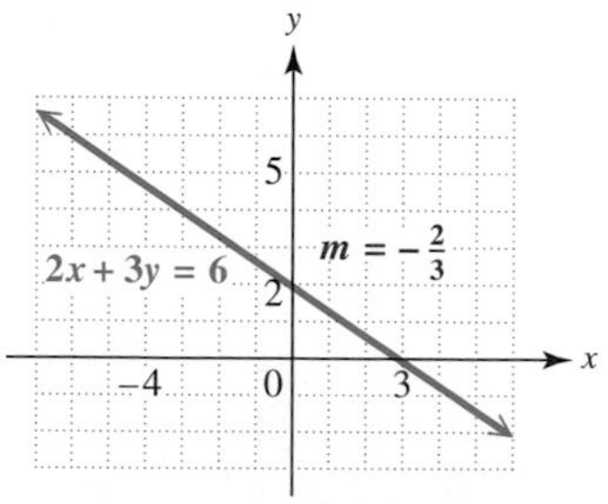

FIGURE 27

The slope of the line is given by the coefficient of x, so $m = -\frac{2}{3}$. See Figure 27.

The required equation of the line through $(-4, 5)$ and parallel to $2x + 3y = 6$ must also have slope $-\frac{2}{3}$. To find this equation, we use the point–slope form, with $(x_1, y_1) = (-4, 5)$ and $m = -\frac{2}{3}$.

$$y - 5 = -\frac{2}{3}[x - (-4)] \qquad y_1 = 5,\ m = -\frac{2}{3},\ x_1 = -4$$

$$y - 5 = -\frac{2}{3}(x + 4)$$

$$y - 5 = -\frac{2}{3}x - \frac{8}{3}$$ Distributive property

$$y = -\frac{2}{3}x - \frac{8}{3} + \frac{15}{3}$$ Add $5 = \frac{15}{3}$.

$$y = -\frac{2}{3}x + \frac{7}{3}$$ Combine like terms.

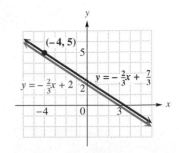

FIGURE 28

We did not clear fractions here because we want the equation in slope–intercept form—that is, solved for y. Both lines are shown in Figure 28.

(b) To be perpendicular to the line $2x + 3y = 6$, a line must have a slope that is the negative reciprocal of $-\frac{2}{3}$, which is $\frac{3}{2}$. We use $(-4, 5)$ and slope $\frac{3}{2}$ in the point–slope form to get the equation of the perpendicular line shown in Figure 29.

$$y - 5 = \frac{3}{2}[x - (-4)] \qquad y_1 = 5, \ m = \frac{3}{2}, \ x_1 = -4$$

$$y - 5 = \frac{3}{2}(x + 4)$$

$$y - 5 = \frac{3}{2}x + 6 \qquad \text{Distributive property}$$

$$y = \frac{3}{2}x + 11 \qquad \text{Add 5.}$$

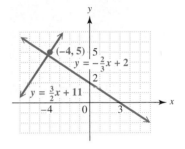

FIGURE 29

✔ **Now Try Exercises 63 and 67.**

A summary of the various forms of linear equations follows.

Forms of Linear Equations

Equation	Description	When to Use
$y = mx + b$	**Slope–Intercept Form** Slope is m. y-intercept is $(0, b)$.	The slope and y-intercept can be easily identified and used to quickly graph the equation.
$y - y_1 = m(x - x_1)$	**Point–Slope Form** Slope is m. Line passes through (x_1, y_1).	This form is ideal for finding the equation of a line if the slope and a point on the line or two points on the line are known.
$Ax + By = C$	**Standard Form** (A, B, and C integers, $A \geq 0$) Slope is $-\frac{A}{B}$ $(B \neq 0)$. x-intercept is $\left(\frac{C}{A}, 0\right)$ $(A \neq 0)$. y-intercept is $\left(0, \frac{C}{B}\right)$ $(B \neq 0)$.	The x- and y-intercepts can be found quickly and used to graph the equation. The slope must be calculated.
$y = b$	**Horizontal Line** Slope is 0. y-intercept is $(0, b)$.	If the graph intersects only the y-axis, then y is the only variable in the equation.
$x = a$	**Vertical Line** Slope is undefined. x-intercept is $(a, 0)$.	If the graph intersects only the x-axis, then x is the only variable in the equation.

OBJECTIVE 6   Write an equation of a line that models real data.   We can use the information presented in this section to write equations of lines that mathematically describe, or *model,* real data if the given set of data changes at a fairly constant rate. In this case, the data fit a linear pattern, and the rate of change is the slope of the line.

EXAMPLE 6   **Determining a Linear Equation to Describe Real Data**

Suppose it is time to fill your car with gasoline. At your local station, 89-octane gas is selling for $3.20 per gal.

(a) Write an equation that describes the cost y to buy x gallons of gas.

Experience has taught you that the total price you pay is determined by the number of gallons you buy multiplied by the price per gallon (in this case, $3.20). As you pump the gas, two sets of numbers spin by: the number of gallons pumped and the price for that number of gallons.

The table uses ordered pairs to illustrate this situation.

Number of Gallons Pumped	Price of This Number of Gallons
0	0($3.20) = $ 0.00
1	1($3.20) = $ 3.20
2	2($3.20) = $ 6.40
3	3($3.20) = $ 9.60
4	4($3.20) = $12.80

If we let x denote the number of gallons pumped, then the total price y in dollars can be found by the linear equation

Total price ⎯⎯⎯⌄   ⌄⎯⎯⎯ Number of gallons

$$y = 3.20x.$$

Theoretically, there are infinitely many ordered pairs (x, y) that satisfy this equation, but here we are limited to nonnegative values for x, since we cannot have a negative number of gallons. In this situation, there is also a practical maximum value for x that varies from one car to another. What determines this maximum value?

(b) You can also get a car wash at the gas station if you pay an additional $3.00. Write an equation that defines the price for gas and a car wash.

Since an additional $3.00 will be charged, you pay $3.20x + 3.00$ dollars for x gallons of gas and a car wash, or

$$y = 3.2x + 3. \qquad \text{Delete unnecessary zeros.}$$

(c) Interpret the ordered pairs $(5, 19)$ and $(10, 35)$ in relation to the equation from part (b).

The ordered pair $(5, 19)$ indicates that the price of 5 gal of gas and a car wash is $19.00. Similarly, $(10, 35)$ indicates that the price of 10 gal of gas and a car wash is $35.00.

✔ **Now Try Exercises 71 and 75.**

▶ **NOTE** In Example 6(a), the ordered pair $(0, 0)$ satisfied the equation, so the linear equation has the form $y = mx$, where $b = 0$. If a realistic situation involves an initial charge plus a charge per unit, as in Example 6(b), the equation has the form $y = mx + b$, where $b \neq 0$.

EXAMPLE 7 **Finding an Equation of a Line That Models Data**

Average annual tuition and fees for in-state students at public four-year colleges are shown in the table for selected years and graphed as ordered pairs of points in the *scatter diagram* in Figure 30, where $x = 0$ represents 1990, $x = 4$ represents 1994, and so on, and y represents the cost in dollars.

Year	Cost (in dollars)
1990	2035
1994	2820
1996	3151
1998	3486
2000	3774
2002	4273

Source: U.S. National Center for Education Statistics.

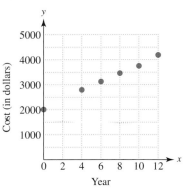

FIGURE 30

(a) Find an equation that models the data.

Since the points in Figure 30 lie approximately on a straight line, we can write a linear equation that models the relationship between year x and cost y. We choose two data points, $(0, 2035)$ and $(12, 4273)$, to find the slope of the line.

$$m = \frac{4273 - 2035}{12 - 0} = \frac{2238}{12} = 186.5$$

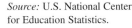

Start with the x- and y-values of the same point.

The slope 186.5 indicates that the cost of tuition and fees increased by about $186.50 per year from 1990 to 2002. We use this slope, the y-intercept $(0, 2035)$, and the slope–intercept form to write an equation of the line. Thus,

$$y = 186.5x + 2035.$$

(b) Use the equation from part (a) to approximate the cost of tuition and fees at public four-year colleges in 2004.

The value $x = 14$ corresponds to the year 2004.

$$y = 186.5x + 2035$$
$$y = 186.5(14) + 2035 \qquad \text{Substitute 14 for } x.$$
$$y = 4646$$

According to the model, average tuition and fees for in-state students at public four-year colleges in 2004 were about $4646.

✔ **Now Try Exercise 81.**

▶ **NOTE** In Example 7, if we had chosen different data points, we would have gotten a slightly different equation. However, all such equations should be similar.

EXAMPLE 8 Finding an Equation of a Line That Models Data

Retail spending (in billions of dollars) on prescription drugs in the United States is shown in the graph in Figure 31.

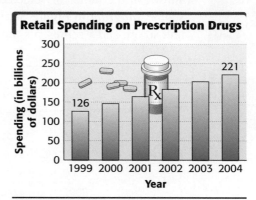

FIGURE 31

(a) Write an equation that models the data.

The data shown in the bar graph increase linearly; that is, we could draw a straight line through the tops of any two bars that would be close to the top of each bar. We can use the data and the point–slope form of the equation of a line to get an equation that models the relationship between year x and spending on prescription drugs, y. If we let $x = 9$ represent 1999, $x = 10$ represent 2000, and so on, the given data for 1999 and 2004 can be written as the ordered pairs $(9, 126)$ and $(14, 221)$. The slope of the line through these two points is

$$m = \frac{221 - 126}{14 - 9} = \frac{95}{5} = 19.$$

Thus, retail spending on prescription drugs increased by about \$19 billion per year. Using this slope, one of the points, say, $(9, 126)$, and the point–slope form, we obtain the following:

$$y - y_1 = m(x - x_1) \qquad \text{Point–slope form}$$
$$y - 126 = 19(x - 9) \qquad (x_1, y_1) = (9, 126); \ m = 19$$

Either point can be used here. **(14, 221)** provides the same answer.

$$y - 126 = 19x - 171 \qquad \text{Distributive property}$$
$$y = 19x - 45. \qquad \text{Slope–intercept form}$$

Thus, retail spending y (in billions of dollars) on prescription drugs in the United States in year x can be approximated by the equation $y = 19x - 45$.

(b) Use the equation from part (a) to predict retail spending on prescription drugs in the United States in 2007. (Assume a constant rate of change.)

Since $x = 9$ represents 1999 and 2007 is 8 yr after 1999, $x = 17$ represents 2007.

$$y = 19x - 45$$
$$y = 19(17) - 45 \qquad \text{Substitute 17 for } x.$$
$$y = 278$$

According to the model, \$278 billion will be spent on prescription drugs in 2007.

✔ **Now Try Exercise 83.**

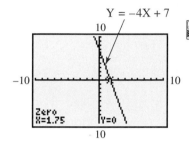

$Y = -4X + 7$

FIGURE 32

OBJECTIVE 7 Use a graphing calculator to solve linear equations in one variable. Figure 32 shows the graph of

$$Y = -4X + 7.$$

From the values at the bottom of the screen, we see that when X = 1.75, Y = 0. This means that X = 1.75 satisfies the equation

$$-4X + 7 = 0,$$

a linear equation in one variable. Therefore, the solution set of $-4X + 7 = 0$ is {1.75}. We can verify this algebraically by substitution. (The word "Zero" indicates that the *x*-intercept has been located.)

EXAMPLE 9 Solving an Equation with a Graphing Calculator

Use a graphing calculator to solve $-2x - 4(2 - x) = 3x + 4$.

We must write the equation as an equivalent equation with 0 on one side.

$$-2x - 4(2 - x) - 3x - 4 = 0 \qquad \text{Subtract } 3x \text{ and 4.}$$

Then we graph

$$Y = -2X - 4(2 - X) - 3X - 4$$

to find the *x*-intercept. The standard viewing window *cannot* be used because the *x*-intercept does not lie in the interval $[-10, 10]$. As seen in Figure 33, the *x*-intercept of the graph is the point $(-12, 0)$; thus, the solution of the equation is -12. The solution set is $\{-12\}$.

$Y = -2X - 4(2 - X) - 3X - 4$

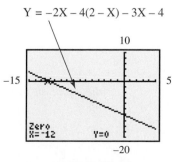

FIGURE 33

✔ **Now Try Exercise 85.**

3.3 EXERCISES

● *Complete solution available on Video Lectures on CD/DVD*

▢ *Now Try Exercise*

Concept Check *In Exercises 1–6, provide the appropriate response.*

1. The following equations all represent the same line. Which one is in standard form as defined in the text?

A. $3x - 2y = 5$ **B.** $2y = 3x - 5$ **C.** $\dfrac{3}{5}x - \dfrac{2}{5}y = 1$ **D.** $3x = 2y + 5$

2. Which equation is in point–slope form?

A. $y = 6x + 2$ **B.** $4x + y = 9$ **C.** $y - 3 = 2(x - 1)$ **D.** $2y = 3x - 7$

3. Which equation in Exercise 2 is in slope–intercept form?

4. Write the equation $y + 2 = -3(x - 4)$ in slope–intercept form.

5. Write the equation from Exercise 4 in standard form.

6. Write the equation $10x - 7y = 70$ in slope–intercept form.

Concept Check *Match each equation with the graph that it most closely resembles. (Hint: Determine the signs of m and b to help you make your decision.)*

7. $y = 2x + 3$

8. $y = -2x + 3$

9. $y = -2x - 3$

10. $y = 2x - 3$

11. $y = 2x$

12. $y = -2x$

13. $y = 3$

14. $y = -3$

A.

B.

C.

D.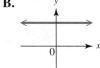

E.

F.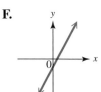

G.

H.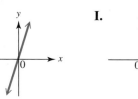

I.

Find the equation in slope–intercept form of the line satisfying the given conditions. See Example 1.

15. $m = 5;\ b = 15$

16. $m = -2;\ b = 12$

17. $m = -\dfrac{2}{3};\ b = \dfrac{4}{5}$

18. $m = -\dfrac{5}{8};\ b = -\dfrac{1}{3}$

● **19.** Slope $\dfrac{2}{5}$; y-intercept $(0, 5)$

20. Slope $-\dfrac{3}{4}$; y-intercept $(0, 7)$

Concept Check Write an equation in slope–intercept form of the line shown in each graph. *(Hint: Use the indicated points to find the slope.)*

21.

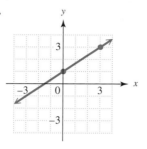

22.

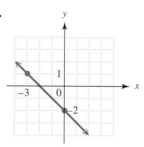

For each equation, (a) write it in slope–intercept form, (b) give the slope of the line, (c) give the y-intercept, and (d) graph the line. See Example 2.

23. $-x + y = 4$ **24.** $-x + y = 6$ **25.** $6x + 5y = 30$

26. $3x + 4y = 12$ **27.** $4x - 5y = 20$ **28.** $7x - 3y = 3$

29. $x + 2y = -4$ **30.** $x + 3y = -9$ **31.** $-4x + 3y = 12$

32. *Concept Check* Express the slope *m* of the graph of $Ax + By = C$ $(A \neq 0)$ in terms of *A* and *B*.

Find an equation of the line that satisfies the given conditions. (a) Write the equation in standard form. (b) Write the equation in slope–intercept form. See Example 3.

33. Through $(5, 8)$; slope -2

34. Through $(12, 10)$; slope 1

35. Through $(-2, 4)$; slope $-\dfrac{3}{4}$

36. Through $(-1, 6)$; slope $-\dfrac{5}{6}$

37. Through $(-5, 4)$; slope $\dfrac{1}{2}$

38. Through $(7, -2)$; slope $\dfrac{1}{4}$

39. *x*-intercept $(3, 0)$; slope 4

40. *x*-intercept $(-2, 0)$; slope -5

41. Through $(2, 6.8)$; slope 1.4

42. Through $(6, -1.2)$; slope 0.8

Find an equation of the line that satisfies the given conditions.

43. Through $(9, 5)$; slope 0

44. Through $(-4, -2)$; slope 0

45. Through $(9, 10)$; undefined slope

46. Through $(-2, 8)$; undefined slope

47. Through $(0.5, 0.2)$; vertical

48. Through $\left(\dfrac{5}{8}, \dfrac{2}{9}\right)$; vertical

49. Through $(-7, 8)$; horizontal

50. Through $(2, 7)$; horizontal

Find an equation of the line passing through the given points. (a) Write the equation in standard form. (b) Write the equation in slope–intercept form if possible. See Example 4.

51. $(3, 4)$ and $(5, 8)$

52. $(5, -2)$ and $(-3, 14)$

53. $(6, 1)$ and $(-2, 5)$

54. $(-2, 5)$ and $(-8, 1)$

55. $\left(-\dfrac{2}{5}, \dfrac{2}{5}\right)$ and $\left(\dfrac{4}{3}, \dfrac{2}{3}\right)$

56. $\left(\dfrac{3}{4}, \dfrac{8}{3}\right)$ and $\left(\dfrac{2}{5}, \dfrac{2}{3}\right)$

57. $(2, 5)$ and $(1, 5)$

58. $(-2, 2)$ and $(4, 2)$

59. $(7, 6)$ and $(7, -8)$ **60.** $(13, 5)$ and $(13, -1)$

61. $\left(\dfrac{1}{2}, -3\right)$ and $\left(-\dfrac{2}{3}, -3\right)$ **62.** $\left(-\dfrac{4}{9}, -6\right)$ and $\left(\dfrac{12}{7}, -6\right)$

Find an equation of the line that satisfies the given conditions. **(a)** *Write the equation in slope–intercept form.* **(b)** *Write the eqution in standard form. See Example 5.*

63. Through $(7, 2)$; parallel to $3x - y = 8$

64. Through $(4, 1)$; parallel to $2x + 5y = 10$

65. Through $(-2, -2)$; parallel to $-x + 2y = 10$

66. Through $(-1, 3)$; parallel to $-x + 3y = 12$

67. Through $(8, 5)$; perpendicular to $2x - y = 7$

68. Through $(2, -7)$; perpendicular to $5x + 2y = 18$

69. Through $(-2, 7)$; perpendicular to $x = 9$

70. Through $(8, 4)$; perpendicular to $x = -3$

Write an equation in the form $y = mx$ for each situation. Then give the three ordered pairs associated with the equation for x-values 0, 5, and 10. See Example 6(a).

71. x represents the number of hours traveling at 45 mph, and y represents the distance traveled (in miles).

72. x represents the number of compact discs sold at \$16 each, and y represents the total cost of the discs (in dollars).

73. x represents the number of gallons of gas sold at \$3.01 per gal, and y represents the total cost of the gasoline (in dollars).

74. x represents the number of days a DVD movie is rented at \$3.50 per day, and y represents the total charge for the rental (in dollars).

For each situation, **(a)** *write an equation in the form $y = mx + b$,* **(b)** *find and interpret the ordered pair associated with the equation for $x = 5$, and* **(c)** *answer the question. See Examples 6(b) and 6(c).*

75. A membership in the Midwest Athletic Club costs \$99, plus \$39 per month. (*Source:* Midwest Athletic Club.) Let x represent the number of months selected. How much does the first year's membership cost?

76. For a family membership, the athletic club in Exercise 75 charges a membership fee of \$159, plus \$60 for each additional family member after the first. Let x represent the number of additional family members. What is the membership fee for a four-person family?

77. A cell phone plan includes 450 anytime minutes for \$35 per month, plus \$19.95 for a Nokia 5165 cell phone and \$25 for a one-time activation fee. (*Source:* U.S. Cellular.) Let x represent the number of months of service. If you sign a 1-yr contract, how much will this cell phone package cost? (Assume that you never use more than the allotted number of minutes.)

78. Another cell phone plan includes 900 anytime minutes for $50 per month, plus a one-time activation fee of $25. A Nokia 5165 cell phone is included at no additional charge. (*Source:* U.S. Cellular.) Let *x* represent the number of months of service. If you sign a 2-yr contract, how much will this cell phone plan cost? (Assume that you never use more than the allotted number of minutes.)

79. There is a $30 fee to rent a chain saw, plus $6 per day. Let *x* represent the number of days the saw is rented and *y* represent the charge to the user in dollars. If the total charge is $138, for how many days is the saw rented?

80. A rental car costs $50 plus $0.20 per mile. Let *x* represent the number of miles driven and *y* represent the total charge to the renter. How many miles was the car driven if the renter paid $84.60?

Solve each problem. In part (a), give equations in slope–intercept form. See Examples 7 and 8. (Source for Exercises 81 and 82: Jupiter Media Metrix.)

81. The percent of U.S. households that access the Internet by dial-up is shown in the graph, where the year 2000 corresponds to *x* = 0.

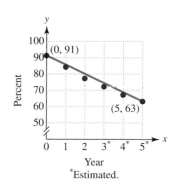

   (a) Use the ordered pairs from the graph to write an equation that models the data. What does the slope tell us in the context of this problem?
   (b) Use the equation from part (a) to predict the percent of U.S. households that were expected to access the Internet by dial-up in 2006. Round your answer to the nearest percent.

82. The percent of households that access the Internet by high-speed broadband is shown in the graph, where the year 2000 corresponds to *x* = 0.

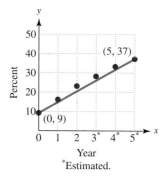

   (a) Use the ordered pairs from the graph to write an equation that models the data. What does the slope tell us in the context of this problem?
   (b) Use the equation from part (a) to predict the percent of U.S. households that were expected to access the Internet by broadband in 2006. Round your answer to the nearest percent.

83. Median household income of African-Americans is shown in the bar graph.

   (a) Use the information given for the years 1995 and 2003, letting *x* = 5 represent 1995, *x* = 13 represent 2003, and *y* represent the median income, to write an equation that models median household income.
   (b) Use the equation to approximate the median income for 1999. How does your result compare with the actual value, $27,910?

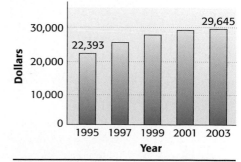

Median Household Income for African-Americans

Source: U.S. Census Bureau.

84. The number of post offices in the United States is shown in the bar graph.

(a) Use the information given for the years 1995 and 2000, letting $x = 5$ represent 1995, $x = 10$ represent 2000, and y represent the number of post offices, to write an equation that models the data.

(b) Use the equation to approximate the number of post offices in 1998. How does this result compare with the actual value, 27,952?

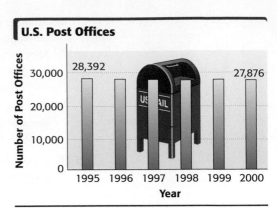

U.S. Post Offices

Source: U.S. Postal Service, *Annual Report of the Postmaster General.*

TECHNOLOGY INSIGHTS   (EXERCISES 85–90)

In Exercises 85–90, do the following. See Example 9.

(a) Simplify and write the given equation so that the right side is 0. Then replace 0 with y.

(b) The graph of the equation for y is shown with each exercise. Use the graph to determine the solution of the equation.

*(c) Solve the equation with the methods of **Chapter 2.***

85. $2x + 7 - x = 4x - 2$

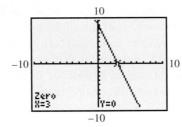

86. $7x - 2x + 4 - 5 = 3x + 1$

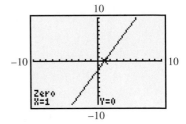

87. $3(2x + 1) - 2(x - 2) = 5$

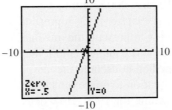

88. $4x - 3(4 - 2x) = 2(x - 3) + 6x + 2$

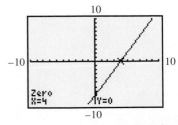

(continued)

89. The graph of y_1 is shown in the standard viewing window. Which is the only choice that can be the solution of the equation $y_1 = 0$?

    A. -15 **B.** 0 **C.** 5 **D.** 15

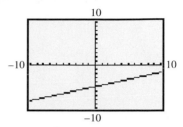

90. **(a)** Solve $-2(x - 5) = -x - 2$, using the methods of **Chapter 2.**

    (b) Explain why the standard viewing window of a graphing calculator cannot graphically support the solution found in part (a). What minimum and maximum x-values would make it possible for the solution to be seen?

RELATING CONCEPTS (EXERCISES 91–96)

FOR INDIVIDUAL OR GROUP WORK

*In **Section 2.2**, we learned how formulas can be applied to problem solving. **Work Exercises 91–96 in order,** to see how the formula that relates Celsius and Fahrenheit temperatures is derived.*

91. There is a linear relationship between Celsius and Fahrenheit temperatures. When $C = 0°$, $F = $ _____°, and when $C = 100°$, $F = $ _____°.

92. Think of ordered pairs of temperatures (C, F), where C and F represent corresponding Celsius and Fahrenheit temperatures. The equation that relates the two scales has a straight-line graph that contains the two points determined in Exercise 91. What are these two points?

93. Find the slope of the line described in Exercise 92.

94. Now think of the point–slope form of the equation in terms of C and F, where C replaces x and F replaces y. Use the slope you found in Exercise 93 and one of the two points determined earlier, and find the equation that gives F in terms of C.

95. To obtain another form of the formula, use the equation you found in Exercise 94 and solve for C in terms of F.

96. The equation found in Exercise 94 is graphed on the graphing calculator screen shown here. Interpret the display at the bottom in the context of this group of exercises.

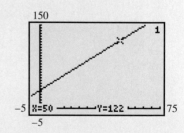

PREVIEW EXERCISES

*Solve each inequality. See **Section 2.5**.*

97. $2x + 5 < 9$ **98.** $-x + 4 > 3$ **99.** $5 - 3x \geq 9$ **100.** $-x \leq 0$

Summary Exercises on Slopes and Equations of Lines

Given an equation of a line in the form $Ax + By = C$, by solving for y, we obtain

$$y = -\frac{A}{B}x + \frac{C}{B}.$$

The equation is now in slope–intercept form, and we see that the slope is $-\frac{A}{B}$. Use this information to find the slope of each line by inspection.

1. $3x + 5y = 9$ **2.** $4x + 7y = 3$ **3.** $2x - y = 5$

4. $5x - 2y = 4$ **5.** $0.2x + 0.8y = 0$ **6.** $\frac{1}{2}x - \frac{1}{4}y = 0$

For each line described, find an equation of the line (a) in slope–intercept form and (b) in standard form.

7. Through the points $(-2, 6)$ and $(4, 1)$

8. Through $(-2, 5)$ and parallel to the graph of $3x - y = 4$

9. Through the origin and perpendicular to the graph of $2x - 5y = 6$

10. Through $(5, -8)$ and parallel to the graph of $y = 4$

11. Through $\left(\frac{3}{4}, -\frac{7}{9}\right)$ and perpendicular to the graph of $x = \frac{2}{3}$

12. The perpendicular bisector of the segment with endpoints $(-4, 6)$ and $(12, 4)$ (*Note:* The perpendicular bisector intersects the segment at its midpoint, and forms a right angle with the segment.)

13. Through $(-4, 2)$ and parallel to the line through $(3, 9)$ and $(6, 11)$

14. Through $(4, -2)$ and perpendicular to the line through $(3, 7)$ and $(5, 6)$

15. Through $(-4, 12)$ and the midpoint of the segment with endpoints $(5, 8)$ and $(-3, 2)$

16. Through the points $\left(\frac{1}{3}, \frac{1}{2}\right)$ and $\left(-\frac{2}{3}, \frac{5}{2}\right)$

17. Through $\left(\frac{3}{7}, \frac{1}{6}\right)$ and parallel to the graph of $y = \frac{1}{5}x + \frac{7}{4}$

18. Through $\left(\frac{4}{7}, \frac{5}{6}\right)$ and perpendicular to the graph of $y = \frac{4}{3}x + \frac{3}{8}$

19. Through the points $(0.3, 1.5)$ and $(0.4, 1.7)$

20. Through $(2.5, 1.75)$ and parallel to the graph of $y = 0.5x + 3.25$

Concept Check   *Match the description in Column 1 with the correct equation in Column II.*

I	II
21. Slope -0.5, $b = -2$	**A.** $y = -\frac{1}{2}x$
22. x-intercept $(4, 0)$, y-intercept $(0, 2)$	**B.** $y = -\frac{1}{2}x - 2$
23. Passes through $(4, -2)$ and $(0, 0)$	**C.** $x - 2y = 2$
24. $m = \frac{1}{2}$, passes through $(-2, -2)$	**D.** $x + 2y = 4$
25. $m = \frac{1}{2}$, passes through the origin	**E.** $x = 2y$

3.4 Linear Inequalities in Two Variables

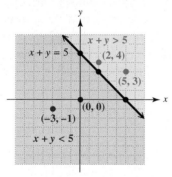

FIGURE 34

OBJECTIVE 1 Graph linear inequalities in two variables. In **Chapter 2,** we graphed linear inequalities in one variable on the number line. In this section, we graph linear inequalities in two variables on a rectangular coordinate system.

Linear Inequality in Two Variables

An inequality that can be written as

$$Ax + By < C \qquad \text{or} \qquad Ax + By > C,$$

where A, B, and C are real numbers and A and B are not both 0, is a **linear inequality in two variables.**

The symbols $\leq$ and $\geq$ may replace $<$ and $>$ in the definition.

Consider the graph in Figure 34. The graph of the line $x + y = 5$ divides the points in the rectangular coordinate system into three sets:

1. Those points that lie on the line itself and satisfy the equation $x + y = 5$ [like $(0, 5)$, $(2, 3)$, and $(5, 0)$];

2. Those that lie in the half-plane above the line and satisfy the inequality $x + y > 5$ [like $(5, 3)$ and $(2, 4)$];

3. Those that lie in the half-plane below the line and satisfy the inequality $x + y < 5$ [like $(0, 0)$ and $(-3, -1)$].

The graph of the line $x + y = 5$ is called the **boundary line** for the inequalities $x + y > 5$ and $x + y < 5$. Graphs of linear inequalities in two variables are *regions* in the real number plane that may or may not include boundary lines.

To graph a linear inequality in two variables, follow these steps.

Graphing a Linear Inequality

Step 1 **Draw the graph of the straight line that is the boundary.** Make the line solid if the inequality involves $\leq$ or $\geq$. Make the line dashed if the inequality involves $<$ or $>$.

Step 2 **Choose a test point.** Choose any point not on the line, and substitute the coordinates of that point in the inequality.

Step 3 **Shade the appropriate region.** Shade the region that includes the test point if it satisfies the original inequality. Otherwise, shade the region on the other side of the boundary line.

▶ **CAUTION** When drawing the boundary line in Step 1, be careful to draw a solid line if the inequality includes equality ($\leq, \geq$) or a dashed line if equality is not included ($<, >$). Students often make errors in this step.

EXAMPLE 1 Graphing a Linear Inequality

Graph $3x + 2y \geq 6$.

Step 1 First graph the line $3x + 2y = 6$. The graph of this line, the boundary of the graph of the inequality, is shown in Figure 35.

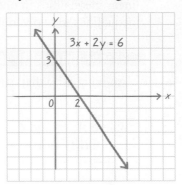

FIGURE 35

Step 2 The graph of the inequality $3x + 2y \geq 6$ includes the points of the line $3x + 2y = 6$ and either the points *above* the line $3x + 2y = 6$ or the points *below* that line. To decide which, select any point not on the boundary line $3x + 2y = 6$ as a test point. The origin, $(0, 0)$, is often a good choice because the substitution is easy. Substitute the values from the test point $(0, 0)$ for x and y in the inequality $3x + 2y > 6$.

$$3(0) + 2(0) > 6 \quad ?$$
$$0 > 6 \qquad \text{False}$$

Step 3 Because the result is false, $(0, 0)$ does *not* satisfy the inequality, so the solution set includes all points on the other side of the line. This region is shaded in Figure 36.

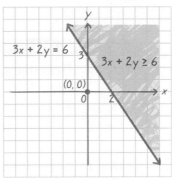

FIGURE 36

✔ **Now Try Exercise 7.**

If the inequality is written in the form $y > mx + b$ or $y < mx + b$, then the inequality symbol indicates which half-plane to shade.

If $y > mx + b$, then shade above the boundary line;

if $y < mx + b$, then shade below the boundary line.

This method works only if the inequality is solved for y.

▶ **CAUTION** A common error in using the method just described is to use the original inequality symbol when deciding which half-plane to shade. Be sure to use the inequality symbol found in the inequality *after* it is solved for *y*.

EXAMPLE 2 Graphing a Linear Inequality

Graph $x - 3y < 4$.

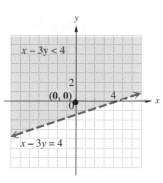

FIGURE 37

First graph the boundary line, shown in Figure 37. The points of the boundary line do not belong to the inequality $x - 3y < 4$ (because the inequality symbol is $<$, not $\leq$). For this reason, the line is dashed. Now solve the inequality for *y*.

$$x - 3y < 4$$
$$-3y < -x + 4 \qquad \text{Subtract } x.$$
$$y > \frac{1}{3}x - \frac{4}{3} \qquad \text{Multiply by } -\frac{1}{3}\text{; change} < \text{to} >.$$

Because of the *is greater than* symbol that occurs **when the inequality is solved for *y*,** shade *above* the line. As a check, choose a test point not on the line, say, $(0, 0)$, and substitute for *x* and *y* in the original inequality.

$$0 - 3(0) < 4 \qquad ?$$
$$0 < 4 \qquad \text{True}$$

This result agrees with the decision to shade above the line. The solution set, graphed in Figure 37, includes only those points in the shaded half-plane (not those on the line).

✔ **Now Try Exercise 9.**

OBJECTIVE 2 Graph the intersection of two linear inequalities. A pair of inequalities joined with the word *and* is interpreted as the intersection of the solution sets of the inequalities. *The graph of the intersection of two or more inequalities is the region of the plane where all points satisfy all of the inequalities at the same time.*

EXAMPLE 3 Graphing the Intersection of Two Inequalities

Graph $2x + 4y \geq 5$ and $x \geq 1$.

To begin, we graph each of the two inequalities $2x + 4y \geq 5$ and $x \geq 1$ separately, as shown in Figures 38(a) and (b). Then we use heavy shading to identify the intersection of the graphs, as shown in Figure 38(c).

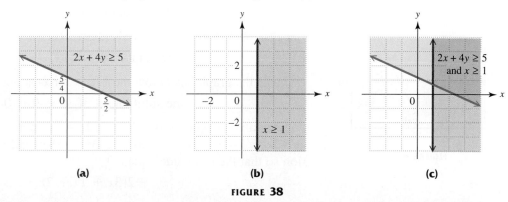

(a)   **(b)**   **(c)**

FIGURE 38

In practice, the graphs in Figures 38(a) and (b) are graphed on the same axes. To check the graph in Figure 38(c), we use a test point from each of the four regions formed by the intersection of the boundary lines. Verify that only ordered pairs in the heavily shaded region satisfy both inequalities.

> ✔ **Now Try Exercise 19.**

OBJECTIVE 3 Graph the union of two linear inequalities. When two inequalities are joined by the word *or,* we must find the union of the graphs of the inequalities. *The graph of the union of two inequalities includes all of the points that satisfy either inequality.*

EXAMPLE 4 Graphing the Union of Two Inequalities

Graph $2x + 4y \geq 5$ or $x \geq 1$.

The graphs of the two inequalities are shown in Figures 38(a) and (b) in Example 3. The graph of the union is shown in Figure 39.

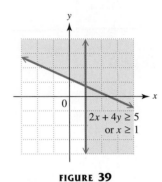

$2x + 4y \geq 5$
or $x \geq 1$

FIGURE 39

> ✔ **Now Try Exercise 29.**

OBJECTIVE 4 Use a graphing calculator to solve linear inequalities in one variable. Recall from **Section 3.3** that the x-intercept of the graph of the line $y = mx + b$ indicates the solution of the equation $mx + b = 0$. We can extend this observation to find solutions of the associated inequalities $mx + b > 0$ and $mx + b < 0$. The solution set of $mx + b > 0$ is the set of all x-values for which the graph of $y = mx + b$ is *above* the x-axis. (We consider points above because the symbol is $>$.) The solution set of $mx + b < 0$ is the set of all x-values for which the graph of $y = mx + b$ is *below* the x-axis. (We consider points below because the symbol is $<$.)

For example, in Figure 40, the x-intercept of $y = 3x - 9$ is $(3, 0)$. Therefore,

the solution set of $3x - 9 = 0$ is $\{3\}$.

Because the graph of y lies above the x-axis for x-values greater than 3,

the solution set of $3x - 9 > 0$ is $(3, \infty)$.

Because the graph lies below the x-axis for x-values less than 3,

the solution set of $3x - 9 < 0$ is $(-\infty, 3)$.

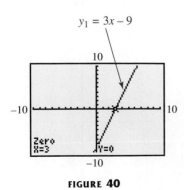

$y_1 = 3x - 9$

FIGURE 40

To solve the equation $-2(3x + 1) = -2x + 18$ and the associated inequalities $-2(3x + 1) > -2x + 18$ and $-2(3x + 1) < -2x + 18$, we must rewrite the equation so that the right side equals 0:

$$-2(3x + 1) + 2x - 18 = 0.$$

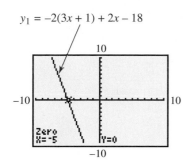

$y_1 = -2(3x + 1) + 2x - 18$

FIGURE 41

Graphing

$$y = -2(3x + 1) + 2x - 18$$

yields the *x*-intercept $(-5, 0)$, as shown in Figure 41. Because the graph of *y* lies *above* the *x*-axis for *x*-values less than -5,

the solution set of $-2(3x + 1) > -2x + 18$ is $(-\infty, -5)$.

Because the graph of *y* lies *below* the *x*-axis for *x*-values greater than -5,

the solution set of $-2(3x + 1) < -2x + 18$ is $(-5, \infty)$.

✔ **Now Try Exercise 39.**

3.4 EXERCISES

Complete solution available on Video Lectures on CD/DVD

Now Try Exercise

Concept Check In Exercises 1–4, fill in the first blank with either solid or dashed. Fill in the second blank with either above or below.

1. The boundary of the graph of $y \le -x + 2$ will be a _____ line, and the shading will be _____ the line.

2. The boundary of the graph of $y < -x + 2$ will be a _____ line, and the shading will be _____ the line.

3. The boundary of the graph of $y > -x + 2$ will be a _____ line, and the shading will be _____ the line.

4. The boundary of the graph of $y \ge -x + 2$ will be a _____ line, and the shading will be _____ the line.

5. How is the boundary line $Ax + By = C$ used in graphing either $Ax + By < C$ or $Ax + By > C$?

6. Describe the two methods discussed in the text for deciding which region is the solution set of a linear inequality in two variables.

Graph each linear inequality in two variables. See Examples 1 and 2.

7. $x + y \le 2$ **8.** $x + y \le -3$ **9.** $4x - y < 4$

10. $3x - y < 3$ **11.** $x + 3y \ge -2$ **12.** $x + 4y \ge -3$

13. $x + y > 0$ **14.** $x + 2y > 0$ **15.** $x - 3y \le 0$

16. $x - 5y \le 0$ **17.** $y < x$ **18.** $y \le 4x$

Graph each compound inequality. See Example 3.

19. $x + y \le 1$ and $x \ge 1$ **20.** $x - y \ge 2$ and $x \ge 3$

21. $2x - y \ge 2$ and $y < 4$ **22.** $3x - y \ge 3$ and $y < 3$

23. $x + y > -5$ and $y < -2$ **24.** $6x - 4y < 10$ and $y > 2$

*Use the method described in **Section 2.7** to write each inequality as a compound inequality, and graph its solution set in the rectangular coordinate plane.*

25. $|x| < 3$ **26.** $|y| < 5$

27. $|x + 1| < 2$ **28.** $|y - 3| < 2$

Graph each compound inequality. See Example 4.

 29. $x - y \geq 1$ or $y \geq 2$ **30.** $x + y \leq 2$ or $y \geq 3$

31. $x - 2 > y$ or $x < 1$ **32.** $x + 3 < y$ or $x > 3$

33. $3x + 2y < 6$ or $x - 2y > 2$ **34.** $x - y \geq 1$ or $x + y \leq 4$

TECHNOLOGY INSIGHTS (EXERCISES 35–42)

Match each inequality with its calculator graph. (Hint: Use the slope, y-intercept, and inequality symbol in making your choice.)

35. $y \leq 3x - 6$ **36.** $y \geq 3x - 6$

37. $y \leq -3x - 6$ **38.** $y \geq -3x - 6$

A.

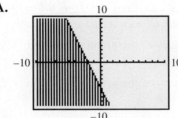

B.

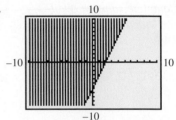

C.

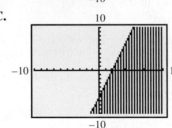

D.

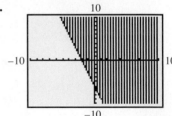

The graph of a linear equation $y = mx + b$ is shown on a graphing calculator screen, along with the x-value of the x-intercept of the line. Use the screen to solve (a) $y = 0$, (b) $y < 0$, and (c) $y > 0$. See Objective 4.

39.

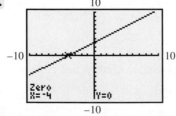

40.

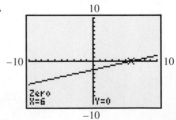

41.

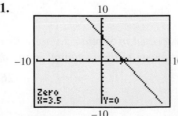

42.

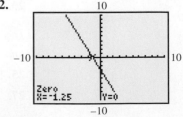

*Solve the equation in part (a) and the associated inequalities in parts (b) and (c), using the methods of **Chapter 2**. Then graph the left side as y in the standard viewing window of a graphing calculator, and explain how the graph supports your answers in parts (a)–(c).*

43. (a) $5x + 3 = 0$
　　(b) $5x + 3 > 0$
　　(c) $5x + 3 < 0$

44. (a) $6x + 3 = 0$
　　(b) $6x + 3 > 0$
　　(c) $6x + 3 < 0$

45. (a) $-8x - (2x + 12) = 0$
　　(b) $-8x - (2x + 12) \geq 0$
　　(c) $-8x - (2x + 12) \leq 0$

46. (a) $-4x - (2x + 18) = 0$
　　(b) $-4x - (2x + 18) \geq 0$
　　(c) $-4x - (2x + 18) \leq 0$

RELATING CONCEPTS (EXERCISES 47–52)

FOR INDIVIDUAL OR GROUP WORK

Suppose a factory can have no more than 200 *workers on a shift, but must have* at least 100 *and must manufacture* at least 3000 *units at minimum cost. The managers need to know how many workers should be on a shift in order to produce the required units at minimal cost.* **Linear programming** *is a method for finding the optimal (best possible) solution that meets all the conditions for such problems.*

Let x represent the number of workers and y represent the number of units manufactured. **Work Exercises 47–52 in order.**

47. Write three inequalities expressing the conditions given in the problem.

48. Graph the inequalities from Exercise 47 and shade the intersection.

49. The cost per worker is $50 per day and the cost to manufacture 1 unit is $100. Write an expression representing the total daily cost C.

50. Find values of x and y for several points in or on the boundary of the shaded region. Include any "corner points." These are the points that maximize or minimize C.

51. Of the values of x and y that you chose in Exercise 50, which gives the least value when substituted in the cost expression from Exercise 49?

52. What does your answer in Exercise 51 mean in terms of the given problem?

PREVIEW EXERCISES

*Evaluate y when x = 3. See **Section 3.1**.*

53. $y = -7x + 12$　　　**54.** $y = -5x - 4$　　　**55.** $y = 3x - 8$　　　**56.** $y = 2x - 9$

*Solve for y. See **Section 2.2**.*

57. $3x - 7y = 8$　　　**58.** $2x - 4y = 7$　　　**59.** $\dfrac{1}{2}x - 4y = 5$　　　**60.** $\dfrac{3}{4}x + 2y = 9$

3.5 Introduction to Functions

OBJECTIVE 1 Distinguish between independent and dependent variables. We often describe one quantity in terms of another. Consider the following:

- The amount of your paycheck if you are paid hourly depends on the number of hours you worked.

- The cost at the gas station depends on the number of gallons of gas you pumped into your car.

- The distance traveled by a car moving at a constant speed depends on the time traveled.

We can use ordered pairs to represent these corresponding quantities. For example, we indicate the relationship between the amount of your paycheck and your hours worked by writing ordered pairs in which the first number represents hours worked and the second number represents the paycheck amount in dollars. Then the ordered pair $(5, 40)$ indicates that when you work 5 hr, your paycheck is $40. Similarly, the ordered pairs $(10, 80)$ and $(20, 160)$ show that working 10 hr results in an $80 paycheck and working 20 hr results in a $160 paycheck. In this example, what would the ordered pair $(40, 320)$ indicate?

Since the amount of your paycheck *depends* on the number of hours worked, your paycheck amount is called the *dependent variable,* and the number of hours worked is called the *independent variable.* Generalizing, if the value of the variable y depends on the value of the variable x, then y is the **dependent variable** and x is the **independent variable.**

Independent variable ⌐↓ ↓⌐ Dependent variable

$$(x, y)$$

OBJECTIVE 2 Define and identify relations and functions. Since we can write related quantities as ordered pairs, a set of ordered pairs such as

$$\{(5, 40), (10, 80), (20, 160), (40, 320)\}$$

is called a *relation.*

Relation
A **relation** is a set of ordered pairs.

A special kind of relation called a *function* is very important in mathematics and its applications.

Function
A **function** is a relation in which, for each value of the first component of the ordered pairs, there is *exactly one value* of the second component.

EXAMPLE 1 Determining whether Relations Are Functions

Tell whether each relation defines a function.

$$F = \{(1, 2), (-2, 4), (3, -1)\}$$
$$G = \{(-2, -1), (-1, 0), (0, 1), (1, 2), (2, 2)\}$$
$$H = \{(-4, 1), (-2, 1), (-2, 0)\}$$

Relations F and G are functions, because, for each different x-value, there is exactly one y-value. Notice that in G, the last two ordered pairs have the same y-value. (1 is paired with 2, and 2 is paired with 2.) This does not violate the definition of a function, since the first components (x-values) are different and each is paired with only one second component (y-value).

In relation H, however, the last two ordered pairs have the *same x*-value paired with *two different y*-values (-2 is paired with both 1 and 0), so H is a relation, but not a function. ***In a function, no two ordered pairs can have the same first component and different second components.***

Different y-values

$$H = \{(-4, 1), (-2, 1), (-2, 0)\} \qquad \text{Not a function}$$

Same x-value

✔ **Now Try Exercises 5 and 7.**

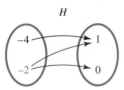

F is a function.

H is not a function.

FIGURE 42

Relations and functions can also be expressed as a correspondence or *mapping* from one set to another, as shown in Figure 42 for function F and relation H from Example 1. The arrow from 1 to 2 indicates that the ordered pair (1, 2) belongs to F. Each first component is paired with exactly one second component. In the mapping for set H, which is not a function, the first component -2 is paired with two different second components, 1 and 0.

Since relations and functions are sets of ordered pairs, we can represent them by tables and graphs. A table and graph for function F is shown in Figure 43.

Finally, we can describe a relation or function by using a rule that tells how to determine the dependent variable for a specific value of the independent variable. The rule may be given in words: the dependent variable is twice the independent variable. Usually the rule is given as an equation. For example, if the value of y is twice the value of x, the rule is

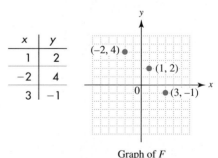

x	y
1	2
-2	4
3	-1

Graph of F

FIGURE 43

$$y = 2x.$$

Dependent variable

Independent variable

▶ **NOTE** Another way to think of a function relationship is to think of the independent variable as an input and the dependent variable as an output. This is illustrated by the input–output (function) machine for the function defined by $y = 2x$.

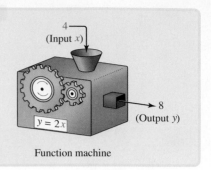
4 (Input x)
$y = 2x$
8 (Output y)
Function machine

In a function, there is exactly one value of the dependent variable, the second component, for each value of the independent variable, the first component. This is what makes functions so important in applications.

▶ **NOTE** The relation from the beginning of this section representing hours worked and corresponding paycheck amount is a function, since each x-value is paired with exactly one y-value.

OBJECTIVE 3 Find the domain and range.

Domain and Range

In a relation, the set of all values of the independent variable (x) is the **domain.** The set of all values of the dependent variable (y) is the **range.**

EXAMPLE 2 Finding Domains and Ranges of Relations

Give the domain and range of each relation. Tell whether the relation defines a function.

(a) $\{(3, -1), (4, 2), (4, 5), (6, 8)\}$

The domain, the set of x-values, is $\{3, 4, 6\}$; the range, the set of y-values, is $\{-1, 2, 5, 8\}$. This relation is not a function because the same x-value 4 is paired with two different y-values, 2 and 5.

(b)

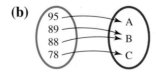
95
89
88
78
A
B
C

The domain of this relation is

$$\{95, 89, 88, 78\};$$

the range is

$$\{A, B, C\}.$$

This mapping defines a function: Each x-value corresponds to exactly one y-value.

(c)

x	y
-5	2
0	2
5	2

This is a table of ordered pairs, so the domain is the set of x-values $\{-5, 0, 5\}$ and the range is the set of y-values $\{2\}$. The table defines a function because each different x-value corresponds to exactly one y-value (even though it is the same y-value).

✔ **Now Try Exercises 11, 13, and 15.**

A graph gives a picture of a relation and can be used to determine its domain and range.

EXAMPLE 3 **Finding Domains and Ranges from Graphs**

Give the domain and range of each relation.

(a)

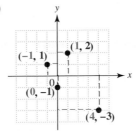

This relation includes the four ordered pairs that are graphed. The domain is the set of *x*-values,

$$\{-1, 0, 1, 4\}.$$

The range is the set of *y*-values,

$$\{-3, -1, 1, 2\}.$$

(b)

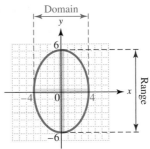

The *x*-values of the points on the graph include all numbers between -4 and 4, inclusive. The *y*-values include all numbers between -6 and 6, inclusive. In interval notation,

the domain is $[-4, 4]$;
the range is $[-6, 6]$.

(c)

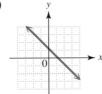

The arrowheads indicate that the line extends indefinitely left and right, as well as up and down. Therefore, both the domain and the range include all real numbers, written $(-\infty, \infty)$.

(d)

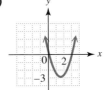

The arrowheads indicate that the graph extends indefinitely left and right, as well as upward. The domain is $(-\infty, \infty)$. Because there is a least *y*-value, -3, the range includes all numbers greater than or equal to -3, written $[-3, \infty)$.

✔ **Now Try Exercises 17 and 19.**

Since relations are often defined by equations, such as $y = 2x + 3$ and $y^2 = x$, we must sometimes determine the domain of a relation from its equation. In this book, we assume the following agreement on the domain of a relation.

Agreement on Domain

Unless specified otherwise, the domain of a relation is assumed to be all real numbers that produce real numbers when substituted for the independent variable.

To illustrate this agreement, since any real number can be used as a replacement for x in $y = 2x + 3$, its domain is the set of all real numbers. As another example, the function defined by $y = \frac{1}{x}$ has all real numbers except 0 as domain, since y is undefined if $x = 0$. Also, the domain of $y = \sqrt{x}$ is all nonnegative real numbers. *In general, the domain of a function defined by an algebraic expression is all real numbers, except those numbers that lead to division by 0 or an even root of a negative number.*

OBJECTIVE 4 **Identify functions defined by graphs and equations.** Since each value of x leads to only one value of y in a function, any vertical line drawn through the graph of a function must intersect the graph in at most one point. This is the *vertical line test* for a function.

Vertical Line Test

If every vertical line intersects the graph of a relation in no more than one point, then the relation is a function.

For example, the graph shown in Figure 44(a) is not the graph of a function, since a vertical line intersects the graph in more than one point. The graph in Figure 44(b) does represent a function.

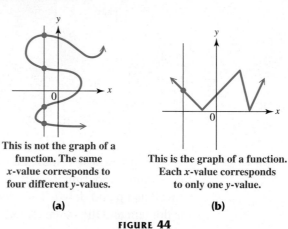

This is not the graph of a
function. The same
x-value corresponds to
four different y-values.

(a)

This is the graph of a function.
Each x-value corresponds
to only one y-value.

(b)

FIGURE 44

EXAMPLE 4 **Using the Vertical Line Test**

Use the vertical line test to determine whether each relation graphed in Example 3 is a function.

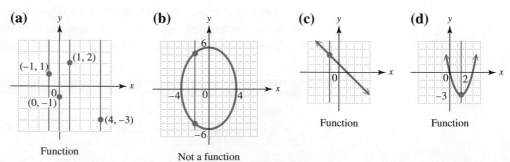

The graphs in (a), (c), and (d) represent functions. The graph of the relation in (b) fails the vertical line test, since the same *x*-value corresponds to two different *y*-values; therefore, it is not the graph of a function.

✔ **Now Try Exercise 21.**

▶ **NOTE** Graphs that do not represent functions are still relations. *Remember that all equations and graphs represent relations and that all relations have a domain and range.*

The vertical line test is a simple method for identifying a function defined by a graph. It is more difficult to decide whether a relation defined by an equation is a function. The next example gives some hints that may help.

EXAMPLE 5 Identifying Functions from Their Equations

Decide whether each relation defines *y* as a function of *x*, and give the domain.

(a) $y = x + 4$

In the defining equation (or rule) $y - x + 4$, *y* is always found by adding 4 to *x*. Thus, each value of *x* corresponds to just one value of *y*, and the relation defines a function; *x* can be any real number, so the domain is $\{x \mid x \text{ is a real number}\}$, or $(-\infty, \infty)$.

(b) $y = \sqrt{2x - 1}$

For any choice of *x* in the domain, there is exactly one corresponding value for *y* (the radical is a nonnegative number), so this equation defines a function. Refer to the agreement on domain stated previously. Since the equation involves a square root, the quantity under the radical sign cannot be negative. Thus, numbers in the domain must satisfy

$$2x - 1 \geq 0$$
$$2x \geq 1$$
$$x \geq \frac{1}{2},$$

and the domain of the function is $\left[\frac{1}{2}, \infty\right)$.

(c) $y^2 = x$

The ordered pairs $(16, 4)$ and $(16, -4)$ both satisfy this equation. Since one value of *x*, 16, corresponds to two values of *y*, 4 and -4, this equation does not define a function. (When solved for *y*, *two* equations are obtained: $y = \sqrt{x}$ and $y = -\sqrt{x}$. See **Section 9.1.**) Because *x* is equal to the square of *y*, the values of *x* must always be nonnegative. The domain of the relation is $[0, \infty)$.

(d) $y \leq x - 1$

By definition, *y* is a function of *x* if every value of *x* leads to exactly one value of *y*. In this example, a particular value of *x*, say, 1, corresponds to many values of *y*. The ordered pairs $(1, 0)$, $(1, -1)$, $(1, -2)$, $(1, -3)$, and so on all satisfy the inequality. Thus, this relation does not define a function. Any number can be used for *x*, so the domain is the set of real numbers, $(-\infty, \infty)$.

(e) $y = \dfrac{5}{x - 1}$

Given any value of x in the domain, we find y by subtracting 1 and then dividing the result into 5. This process produces exactly one value of y for each value in the domain, so the given equation defines a function. The domain includes all real numbers except those which make the denominator 0. We find these numbers by setting the denominator equal to 0 and solving for x.

$$x - 1 = 0$$
$$x = 1$$

The domain includes all real numbers *except* 1, written as $(-\infty, 1) \cup (1, \infty)$.

✔ **Now Try Exercises 27, 29, and 35.**

In summary, three variations of the definition of a function are given here.

Variations of the Definition of a Function

1. A **function** is a relation in which, for each value of the first component of the ordered pairs, there is exactly one value of the second component.

2. A **function** is a set of distinct ordered pairs in which no first component is repeated.

3. A **function** is a rule or correspondence that assigns exactly one range value to each domain value.

OBJECTIVE 5 Use function notation. When a function f is defined with a rule or an equation using x and y for the independent and dependent variables, we say, "y is a function of x" to emphasize that y *depends on* x. We use the notation

$$y = f(x),$$

called **function notation,** to express this and read $f(x)$ as "f of x." (In this special notation, the parentheses do not indicate multiplication.) The letter f is a name for this particular function. For example, if $y = 9x - 5$, we can name this function f and write

$$f(x) = 9x - 5.$$

Note that $f(x)$ *is just another name for the dependent variable y.* For example, if $y = f(x) = 9x - 5$ and $x = 2$, then we find y, or $f(2)$, by replacing x with 2.

$$y = f(2)$$
$$= 9 \cdot 2 - 5$$
$$= 18 - 5$$
$$= 13.$$

The statement "If $x = 2$, then $y = 13$" is represented by the ordered pair $(2, 13)$ and is abbreviated with function notation as

$$f(2) = 13.$$

Read $f(2)$ as "f of 2" or "f at 2." Also,

$$f(0) = 9 \cdot 0 - 5 = -5 \qquad \text{and} \qquad f(-3) = 9(-3) - 5 = -32.$$

These ideas and the symbols used to represent them can be illustrated as follows.

Name of the function

Defining expression

$$y = f(x) = 9x - 5$$

Value of the function   Name of the independent variable

▶ **CAUTION**   The symbol $f(x)$ *does not* indicate "f times x," but represents the y-value associated with the indicated x-value. As just shown, $f(2)$ is the y-value that corresponds to the x-value 2.

EXAMPLE 6   Using Function Notation

Let $f(x) = -x^2 + 5x - 3$. Find the following.

(a) $f(2)$

Do *not* read this as "f times 2." Read it as "f of 2."

$f(x) = -x^2 + 5x - 3$	The base in $-x^2$ is x, not $(-x)$.
$f(2) = -2^2 + 5 \cdot 2 - 3$	Replace x with 2.
$= -4 + 10 - 3$	Apply the exponent; multiply.
$= 3$	Add and subtract.

Thus, $f(2) = 3$, and the ordered pair $(2, 3)$ belongs to f.

(b) $f(q)$

Replace x with q.

$$f(x) = -x^2 + 5x - 3$$
$$f(q) = -q^2 + 5q - 3$$

The replacement of one variable with another expression (such as q) is important in later courses.

✔ **Now Try Exercises 41 and 49.**

Sometimes letters other than f, such as g, h, or capital letters F, G, and H are used to name functions.

EXAMPLE 7   Using Function Notation

Let $g(x) = 2x + 3$. Find and simplify $g(a + 1)$.

$$g(x) = 2x + 3$$
$$g(a + 1) = 2(a + 1) + 3 \qquad \text{Replace } x \text{ with } a + 1.$$
$$= 2a + 2 + 3$$
$$= 2a + 5$$

✔ **Now Try Exercise 53.**

EXAMPLE 8 Using Function Notation

For each function, find $f(3)$.

(a) $f(x) = 3x - 7$

$f(3) = 3(3) - 7$ Replace x with 3.

$f(3) = 2$

(b) $f = \{(-3, 5), (0, 3), (3, 1), (6, -1)\}$

We want $f(3)$, the y-value of the ordered pair whose first component is $x = 3$. As indicated by the ordered pair $(3, 1)$, when $x = 3$, $y = 1$, so $f(3) = 1$.

(c)

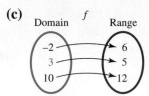

The domain element 3 is paired with 5 in the range, so $f(3) = 5$.

(d) The function graphed in Figure 44

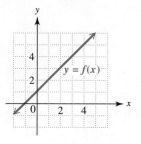

FIGURE 44   FIGURE 45

To evaluate $f(3)$, find 3 on the x-axis. See Figure 45. Then move up until the graph of f is reached. Moving horizontally to the y-axis gives 4 for the corresponding y-value. Thus, $f(3) = 4$.

✔ **Now Try Exercises 61, 63, and 65.**

If a function f is defined by an equation with x and y, and y is not solved for x, use the following steps to find $f(x)$.

Finding an Expression for $f(x)$

Step 1   Solve the equation for y.

Step 2   Replace y with $f(x)$.

EXAMPLE 9 Writing Equations Using Function Notation

Rewrite each equation using function notation $f(x)$. Then find $f(-2)$ and $f(a)$.

(a) $y = x^2 + 1$

This equation is already solved for y. Since $y = f(x)$,

$$f(x) = x^2 + 1.$$

To find $f(-2)$, let $x = -2$.

$$f(-2) = (-2)^2 + 1$$
$$= 4 + 1$$
$$= 5$$

Find $f(a)$ by letting $x = a$: $f(a) = a^2 + 1$.

(b) $x - 4y = 5$

First solve $x - 4y = 5$ for y. Then replace y with $f(x)$.

$$x - 4y = 5$$
$$x - 5 = 4y$$
$$y = \frac{x - 5}{4} \quad \text{so} \quad f(x) = \frac{1}{4}x - \frac{5}{4}$$

Now find $f(-2)$ and $f(a)$.

$$f(-2) = \frac{1}{4}(-2) - \frac{5}{4} = -\frac{7}{4} \qquad \text{Let } x = -2.$$

$$f(a) = \frac{1}{4}a - \frac{5}{4} \qquad \text{Let } x = a.$$

✔ **Now Try Exercise 67.**

OBJECTIVE 6 **Graph linear and constant functions.** Linear equations (except for vertical lines with equations $x = a$) define *linear functions*.

> **Linear Function**
>
> A function that can be defined by
>
> $$f(x) = ax + b$$
>
> for real numbers a and b is a **linear function.** The value of a is the slope m of the graph of the function.

A linear function defined by $f(x) = b$ (whose graph is a horizontal line) is sometimes called a **constant function.** The domain of any linear function is $(-\infty, \infty)$. The range of a nonconstant linear function is $(-\infty, \infty)$, while the range of the constant function defined by $f(x) = b$ is $\{b\}$.

EXAMPLE 10 **Graphing Linear and Constant Functions**

Graph each function. Give the domain and range.

(a) $f(x) = \frac{1}{4}x - \frac{5}{4}$

Recall from **Section 3.3** that the graph of $y = \frac{1}{4}x - \frac{5}{4}$ has slope $m = \frac{1}{4}$ and y-intercept $\left(0, -\frac{5}{4}\right)$. So here we have

$$f(x) = \frac{1}{4}x - \frac{5}{4}.$$

Slope ⬏   ⬏ y-intercept is $\left(0, -\frac{5}{4}\right)$.

To graph this function, plot the *y*-intercept $\left(0, -\frac{5}{4}\right)$ and use the definition of slope as $\frac{\text{rise}}{\text{run}}$ to find a second point on the line. Since the slope is $\frac{1}{4}$, move 1 unit up from $\left(0, -\frac{5}{4}\right)$ and 4 units to the right. Draw the straight line through the points to obtain the graph shown in Figure 46. The domain and range are both $(-\infty, \infty)$.

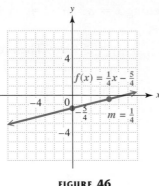

FIGURE 46

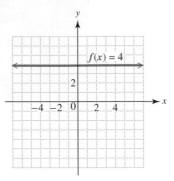

FIGURE 47

(b) $f(x) = 4$

This is a constant function. Its graph is the horizontal line containing all points with *y*-coordinate equal to 4. See Figure 47. The domain is $(-\infty, \infty)$ and the range is $\{4\}$.

✔ **Now Try Exercises 75 and 81.**

3.5 EXERCISES

✎ **1.** In your own words, define a function and give an example.

✎ **2.** In your own words, define the domain of a function and give an example.

3. *Concept Check* In an ordered pair of a relation, is the first element the independent or the dependent variable?

4. *Concept Check* Give an example of a relation that is not a function and that has domain $\{-3, 2, 6\}$ and range $\{4, 6\}$. (There are many possible correct answers.)

Tell whether each relation defines a function. See Example 1.

🌐 **5.** $\{(5, 1), (3, 2), (4, 9), (7, 6)\}$ **6.** $\{(8, 0), (5, 4), (9, 3), (3, 8)\}$

7. $\{(2, 4), (0, 2), (2, 5)\}$ **8.** $\{(9, -2), (-3, 5), (9, 2)\}$

9. $\{(-3, 1), (4, 1), (-2, 7)\}$ **10.** $\{(-12, 5), (-10, 3), (8, 3)\}$

Decide whether each relation defines a function, and give the domain and range. See Examples 1–4.

🌐 **11.** $\{(1, 1), (1, -1), (0, 0), (2, 4), (2, -4)\}$ **12.** $\{(2, 5), (3, 7), (4, 9), (5, 11)\}$

13.

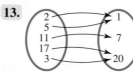

14.

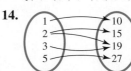

15.

x	y
1	5
1	2
1	−1
1	−4

16.

x	y
4	−3
2	−3
0	−3
−2	−3

17.

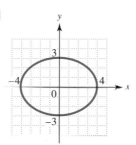

18.

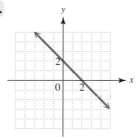

19.

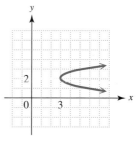

20.

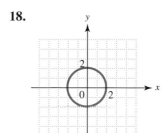

21.

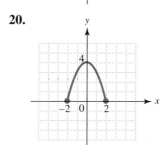

22.

Decide whether each relation defines y as a function of x. (Solve for y first if necessary.) Give the domain. See Example 5.

23. $y = x^2$ **24.** $y = x^3$ **25.** $x = y^6$ **26.** $x = y^4$

27. $y = 2x - 6$ **28.** $y = -6x + 8$ **29.** $x + y < 4$ **30.** $x - y < 3$

31. $y = \sqrt{x}$ **32.** $y = -\sqrt{x}$ **33.** $xy = 1$ **34.** $xy = -3$

35. $y = \sqrt{4x + 2}$ **36.** $y = \sqrt{9 - 2x}$ **37.** $y = \dfrac{2}{x - 4}$ **38.** $y = \dfrac{-7}{x + 2}$

39. *Concept Check*   Choose the correct response: The notation $f(3)$ means

  A. the variable f times 3, or $3f$.
  B. the value of the dependent variable when the independent variable is 3.
  C. the value of the independent variable when the dependent variable is 3.
  D. f equals 3.

40. *Concept Check*   Give an example of a function from everyday life. (*Hint:* Fill in the blanks: _____ depends on _____, so _____ is a function of _____.)

Let $f(x) = -3x + 4$ and $g(x) = -x^2 + 4x + 1$. Find the following. See Examples 6 and 7.

41. $f(0)$ | **42.** $f(-3)$ | **43.** $g(-2)$ | **44.** $g(10)$

45. $f\left(\dfrac{1}{3}\right)$ | **46.** $f\left(\dfrac{7}{3}\right)$ | **47.** $g(0.5)$ | **48.** $g(1.5)$

49. $f(p)$ | **50.** $g(k)$ | **51.** $f(-x)$ | **52.** $g(-x)$

53. $f(x + 2)$ | **54.** $f(x - 2)$ | **55.** $g(\pi)$ | **56.** $g(e)$

57. $f(x + h)$ | **58.** $f(x + h) - f(x)$ | **59.** $f(4) - g(4)$ | **60.** $f(10) - g(10)$

For each function, find (a) $f(2)$ and (b) $f(-1)$. See Example 8.

61. $f = \{(-1, 3), (4, 7), (0, 6), (2, 2)\}$ **62.** $f = \{(2, 5), (3, 9), (-1, 11), (5, 3)\}$

63.

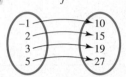

64.

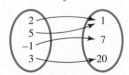

65.

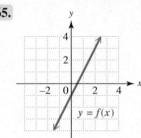

66.

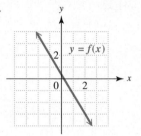

An equation that defines y as a function f of x is given. (a) Solve for y in terms of x, and replace y with the function notation $f(x)$. (b) Find $f(3)$. See Example 9.

67. $x + 3y = 12$ | **68.** $x - 4y = 8$ | **69.** $y + 2x^2 = 3$

70. $y - 3x^2 = 2$ | **71.** $4x - 3y = 8$ | **72.** $-2x + 5y = 9$

73. *Concept Check* Fill in each blank with the correct response.

The equation $2x + y = 4$ has a straight _____ as its graph. One point that lies on the graph is $(3, \underline{\quad})$. If we solve the equation for y and use function notation, we obtain $f(x) = \underline{\quad}$. For this function, $f(3) = \underline{\quad}$, meaning that the point $(\underline{\quad}, \underline{\quad})$ lies on the graph of the function.

74. *Concept Check* Which of the following defines y as a linear function of x?

A. $y = \dfrac{1}{4}x - \dfrac{5}{4}$ **B.** $y = \dfrac{1}{x}$ **C.** $y = x^2$ **D.** $y = \sqrt{x}$

Graph each linear function. Give the domain and range. See Example 10.

75. $f(x) = -2x + 5$ | **76.** $g(x) = 4x - 1$ | **77.** $h(x) = \dfrac{1}{2}x + 2$

78. $F(x) = -\dfrac{1}{4}x + 1$ | **79.** $G(x) = 2x$ | **80.** $H(x) = -3x$

81. $g(x) = -4$ | **82.** $f(x) = 5$ | **83.** $f(x) = 0$

84. *Concept Check* What is the name that is usually given to the graph in Exercise 83?

Solve each problem.

85. A package weighing x pounds costs $f(x)$ dollars to mail to a given location, where

$$f(x) = 2.75x.$$

(a) Evaluate $f(3)$.

(b) In your own words, describe what 3 and the value $f(3)$ mean in part (a), using the terminology *independent variable* and *dependent variable*.

(c) How much would it cost to mail a 5-lb package? Interpret this question and its answer, using function notation.

86. Suppose that a taxicab driver charges $2.50 per mile.

(a) Fill in the table with the correct response for the price $f(x)$ he charges for a trip of x miles.

x	$f(x)$
0	
1	
2	
3	

(b) The linear function that gives a rule for the amount charged is $f(x) = $ _____.

(c) Graph this function for the domain $\{0, 1, 2, 3\}$.

87. Forensic scientists use the lengths of certain bones to calculate the height of a person. Two bones often used are the tibia (t), the bone from the ankle to the knee, and the femur (r), the bone from the knee to the hip socket. A person's height (h) is determined from the lengths of these bones by using functions defined by the following formulas. All measurements are in centimeters.

For men: $h(r) = 69.09 + 2.24r$ or $h(t) = 81.69 + 2.39t$
For women: $h(r) = 61.41 + 2.32r$ or $h(t) = 72.57 + 2.53t$

(a) Find the height of a man with a femur measuring 56 cm.

(b) Find the height of a man with a tibia measuring 40 cm.

(c) Find the height of a woman with a femur measuring 50 cm.

(d) Find the height of a woman with a tibia measuring 36 cm.

Femur

Tibia

88. Federal regulations set standards for the size of the quarters of marine mammals. A pool to house sea otters must have a volume of "the square of the sea otter's average adult length (in meters) multiplied by 3.14 and by 0.91 meter." If x represents the sea otter's average adult length and $f(x)$ represents the volume (in cubic meters) of the corresponding pool size, this formula can be written as

$$f(x) = 0.91(3.14)x^2.$$

Find the volume of the pool for each adult sea otter length (in meters). Round answers to the nearest hundredth.

(a) 0.8   (b) 1.0   (c) 1.2   (d) 1.5

89. Refer to the graph to answer the questions.

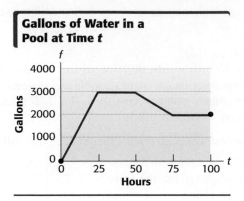

Gallons of Water in a Pool at Time t

(a) What numbers are possible values of the independent variable? the dependent variable?
(b) For how long is the water level increasing? decreasing?
(c) How many gallons of water are in the pool after 90 hr?
(d) Call this function f. What is $f(0)$? What does it mean?
(e) What is $f(25)$? What does it mean?

90. The graph shows the daily number of megawatts of electricity used on a record-breaking summer day in Sacramento, California.

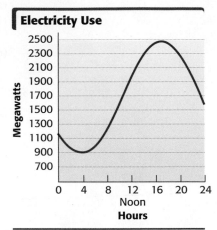

Electricity Use

Source: Sacramento Municipal Utility District.

(a) Why is this the graph of a function?
(b) What is the domain?
(c) Estimate the number of megawatts used at 8 A.M.
(d) At what time was the most electricity used? the least electricity?
(e) Call this function f. What is $f(12)$? What does it mean?

TECHNOLOGY INSIGHTS (EXERCISES 91 AND 92)

91. The calculator screen shows the graph of a linear function $y = f(x)$, along with the display of coordinates of a point on the graph. Use function notation to write what the display indicates.

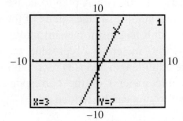

92. The table was generated by a graphing calculator for a linear function $Y_1 = f(X)$. Use the table to work parts (a)–(e).

(a) What is $f(2)$?
(b) If $f(X) = -3.7$, what is the value of X?
(c) What is the slope of the line?
(d) What is the y-intercept of the line?
(e) Find the expression for $f(X)$.

X	Y₁
0	3.5
1	2.3
2	1.1
3	-.1
4	-1.3
5	-2.5
6	-3.7

X=0

PREVIEW EXERCISES

Solve for y if x = −3. See **Section 2.1.**

93. $6x + 5y = 2$ **94.** $\dfrac{4}{3}x + y = 9$ **95.** $1.5x + 2.5y = 5.5$

Solve each equation. See **Section 2.1.**

96. $2\left(\dfrac{3x - 1}{2}\right) - 2x = -4$ **97.** $-4\left(\dfrac{x - 3}{3}\right) + 2x = 8$ **98.** $-2\left(\dfrac{3x + 1}{5}\right) + x = -4$

Chapter **3** Group Activity

CHOOSING AN ENERGY SOURCE (OR HOW TO GET INTO HOT WATER)

Objective Write and graph linear functions that model given data.

There are many different ways to heat water. In this activity, you will look at three different energy sources that may be used to provide heat for a 40-gal home water tank.

Have each student in your group choose one of the three types of water heaters listed in the table.

Type of Hot Water Heater	Size (in gallons)	Price	Operating Cost per Month (manufacturer's estimate)	Hot Water Temperature
Kenmore Economizer 6 —Electric	40	$139.99	$35.00	120°–130°
Kenmore Economizer 6 —Natural Gas	40	$139.99	$13.25	120°–130°
Sunbather Water Heater —Solar	40	$950.00	$0.00	*

Source: Jade Mountain.

A. Using data for the water heater you selected, find a linear equation that represents the total cost y of heating water with respect to time. Let x represent number of months. Write the equation in slope-intercept form.

B. Graph your equation, using domain [0, 60] and range [0, 1000].

C. As a group, compare the graphs of your equations.

   1. What are the y-intercepts?

   2. How do the slopes compare? Which is the steepest? Which has 0 slope?

D. Discuss other factors to consider in choosing each type of water heater. Which of these water heaters would you choose to heat your home?

---

*No temperature is listed, but the ad says, "Best for warm climates, preheating water, summer only in cold places, or when hot water needed only in afternoons and evenings."

Chapter **3** **SUMMARY** *View the Interactive Summary on the Pass the Test CD.*

KEY TERMS

3.1 ordered pair
origin
x-axis
y-axis
rectangular (Cartesian)
coordinate system
plot
coordinate
quadrant

graph of an equation
first-degree equation
linear equation in two
variables
standard form
x-intercept
y-intercept
3.2 rise
run

slope
3.3 slope–intercept form
point–slope form
3.4 linear inequality in
two variables
boundary line
3.5 dependent variable
independent variable
relation

function
domain
range
function notation
linear function
constant function

NEW SYMBOLS

(a, b) ordered pair

x_1 a specific value of the
variable x (read
"x-sub-one"), usually
used to distinguish that
value from a different
value of x

Δ Greek letter delta
m slope

$f(x)$ function of x (read
"f of x")

TEST YOUR WORD POWER

See how well you have learned the vocabulary in this chapter. Answers, with examples, follow the Quick Review.

1. An **ordered pair** is a pair of
   numbers written
   A. in numerical order between
   brackets
   B. between parentheses or brackets
   C. between parentheses in which
   order is important
   D. between parentheses in which
   order does not matter.

2. A **linear equation in two variables**
   is an equation that can be written in
   the form
   A. $Ax + By < C$
   B. $ax = b$
   C. $y = x^2$
   D. $Ax + By = C$.

3. An **intercept** is
   A. the point where the x-axis and
   y-axis intersect
   B. a pair of numbers written
   between parentheses in which
   order matters

   C. one of the four regions
   determined by a rectangular
   coordinate system
   D. the point where a graph intersects
   the x-axis or the y-axis.

4. The **slope** of a line is
   A. the measure of the run over the
   rise of the line
   B. the distance between two points
   on the line
   C. the ratio of the change in y to the
   change in x along the line
   D. the horizontal change compared
   with the vertical change between
   two points on the line.

5. In a relationship between two
   variables x and y, the **independent
   variable** is
   A. x if x depends on y
   B. x if y depends on x
   C. either x or y
   D. the larger of x and y.

6. In a relationship between two
   variables x and y, the **dependent
   variable** is
   A. y if y depends on x
   B. y if x depends on y
   C. either x or y
   D. the smaller of x and y.

7. A **relation** is
   A. a set of ordered pairs
   B. the ratio of the change in y to the
   change in x along a line
   C. the set of all possible values of
   the independent variable
   D. all the second components of a
   set of ordered pairs.

8. A **function** is
   A. the pair of numbers in an ordered
   pair
   B. a set of ordered pairs in which
   each x-value corresponds to
   exactly one y-value

(continued)

C. a pair of numbers written between parentheses in which order matters
D. the set of all ordered pairs that satisfy an equation.

9. The **domain** of a function is
   A. the set of all possible values of the dependent variable y

B. a set of ordered pairs
C. the difference between the x-values
D. the set of all possible values of the independent variable x.

10. The **range** of a function is
   A. the set of all possible values of the dependent variable y

B. a set of ordered pairs
C. the difference between the y-values
D. the set of all possible values of the independent variable x.

QUICK REVIEW

Concepts	Examples

3.1 THE RECTANGULAR COORDINATE SYSTEM

Finding Intercepts

To find the x-intercept, let $y = 0$ and solve for x.

To find the y-intercept, let $x = 0$ and solve for y.

Find the intercepts of the graph of $2x + 3y = 12$.

x-intercept
$$2x + 3(0) = 12$$
$$2x = 12$$
$$x = 6$$
The x-intercept is $(6, 0)$.

y-intercept
$$2(0) + 3y = 12$$
$$3y = 12$$
$$y = 4$$
The y-intercept is $(0, 4)$.

3.2 THE SLOPE OF A LINE

If $x_2 \neq x_1$, then

$$m = \frac{\text{rise}}{\text{run}} = \frac{\text{change in } y}{\text{change in } x} = \frac{\Delta y}{\Delta x} = \frac{y_2 - y_1}{x_2 - x_1}.$$

Find the slope of the graph of $2x + 3y = 12$.

   Use the intercepts $(6, 0)$ and $(0, 4)$ and the slope formula.

$$m = \frac{4 - 0}{0 - 6} = \frac{4}{-6} = -\frac{2}{3} \qquad x_1 = 6,\, y_1 = 0,\, x_2 = 0,\, y_2 = 4$$

A vertical line has undefined slope.
A horizontal line has 0 slope.

The graph of the line $x = 3$ has undefined slope.
The graph of the line $y = -5$ has slope $m = 0$.

Parallel lines have equal slopes.

The lines $y = 2x + 3$ and $4x - 2y = 6$ are parallel; both have $m = 2$.

$$y = 2x + 3$$
$$m = 2$$

$$4x - 2y = 6$$
$$-2y = -4x + 6$$
$$y = 2x - 3$$
$$m = 2$$

The slopes of perpendicular lines are negative reciprocals with a product of -1.

The lines $y = 3x - 1$ and $x + 3y = 4$ are perpendicular; their slopes are negative reciprocals.

$$y = 3x - 1$$
$$m = 3$$

$$x + 3y = 4$$
$$3y = -x + 4$$
$$y = -\frac{1}{3}x + \frac{4}{3}$$
$$m = -\frac{1}{3}$$

(continued)

Concepts	Examples

3.3 LINEAR EQUATIONS IN TWO VARIABLES

Slope–Intercept Form
$y = mx + b$

$y = 2x + 3$ $m = 2$, *y*-intercept is (0, 3).

Point–Slope Form
$y - y_1 = m(x - x_1)$

$y - 3 = 4(x - 5)$ (5, 3) is on the line, $m = 4$.

Standard Form
$Ax + By = C$ (*A*, *B*, *C* integers, $A \geq 0$)

$2x - 5y = 8$ Standard form

Horizontal Line
$y = b$

$y = 4$ Horizontal line

Vertical Line
$x = a$

$x = -1$ Vertical line

3.4 LINEAR INEQUALITIES IN TWO VARIABLES

Graphing a Linear Inequality

Graph $2x - 3y \leq 6$.

Step 1   Draw the graph of the line that is the boundary. Make the line solid if the inequality involves $\leq$ or $\geq$. Make the line dashed if the inequality involves $<$ or $>$.

Draw the graph of $2x - 3y = 6$. Use a solid line because of the inclusion of equality in the symbol $\leq$.

Step 2   Choose any point not on the line as a test point. Substitute the coordinates into the inequality.

Choose (0, 0), for example.

$2(0) - 3(0) = 0$, and $0 \leq 6$ True

Step 3   Shade the region that includes the test point if the test point satisfies the original inequality. Otherwise, shade the region on the other side of the boundary line.

Shade the side of the line that includes (0, 0).

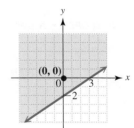

3.5 INTRODUCTION TO FUNCTIONS

A **function** is a set of ordered pairs such that, for each first component, there is one and only one second component. The set of first components is called the **domain,** and the set of second components is called the **range.**

$y = f(x) = x^2$ defines a function *f* with domain $(-\infty, \infty)$ and range $[0, \infty)$.

To evaluate a function *f*, where $f(x)$ defines the range value for a given value of *x* in the domain, substitute the value wherever *x* appears.

If $f(x) = x^2 - 7x + 12$, then

$$f(1) = 1^2 - 7(1) + 12 = 6.$$

To write an equation that defines a function *f* in function notation, do the following:

Write $2x + 3y = 12$ using notation for a function *f*.

Step 1   Solve the equation for *y*.

$3y = -2x + 12$ Subtract 2*x*.

$y = -\dfrac{2}{3}x + 4$ Divide by 3.

Step 2   Replace *y* with $f(x)$.

$f(x) = -\dfrac{2}{3}x + 4$ Function notation

(continued)

Answers to Test Your Word Power

1. C; *Examples:* (0, 3), (3, 8), (4, 0)

2. D; *Examples:* $3x + 2y = 6$, $x = y - 7, 4x = y$

3. D; *Example:* In Figure 4(b) of **Section 3.1,** the *x*-intercept is (3, 0) and the *y*-intercept is (0, 2).

4. C; *Example:* The line through (3, 6) and (5, 4) has slope $\frac{4-6}{5-3} = \frac{-2}{2} = -1$.

5. B; *Example:* See Answer 6, which follows.

6. A; *Example:* When you borrow money, the amount you borrow (independent variable) determines the size of your payments (dependent variable).

7. A; *Example:* The set {(2, 0), (4, 3), (6, 6), (8, 9)} defines a relation.

8. B; The relation given in Answer 7 is a function, since the *x*-value of each ordered pair corresponds to exactly one *y*-value.

9. D; *Example:* In the function in Answer 7, the domain is the set of *x*-values, {2, 4, 6, 8}.

10. A; *Example:* In the function in Answer 7, the range is the set of *y*-values, {0, 3, 6, 9}.

Chapter **3** REVIEW EXERCISES

[3.1] *Complete the table of ordered pairs for each equation. Then graph the equation.*

1. $3x + 2y = 10$

x	y
0	
	0
2	
	-2

2. $x - y = 8$

x	y
2	
	-3
3	
	-2

Find the x- and y-intercepts and then graph each equation.

3. $4x - 3y = 12$

4. $5x + 7y = 28$

5. $2x + 5y = 20$

6. $x - 4y = 8$

7. Explain how the signs of the *x*- and *y*-coordinates of a point determine the quadrant in which the point lies.

[3.2] *Find the slope of each line.*

8. Through $(-1, 2)$ and $(4, -5)$

9. Through $(0, 3)$ and $(-2, 4)$

10. $y = 2x + 3$

11. $3x - 4y = 5$

12. $x = 5$

13. Parallel to $3y = 2x + 5$

14. Perpendicular to $3x - y = 4$

15. Through $(-1, 5)$ and $(-1, -4)$

16.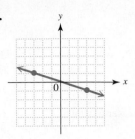

17.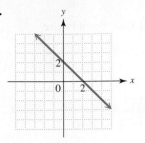

Tell whether each line has positive, negative, 0, *or* undefined *slope.*

18.   **19.**   **20.**   **21.**

22. *Concept Check*   If a walkway rises 2 ft for every 10 ft on the horizontal, which of the following express its slope (or grade)? (There are several correct choices.)

A. 0.2   **B.** $\dfrac{2}{10}$ **C.** $\dfrac{1}{5}$

D. 20%   **E.** 5   **F.** $\dfrac{20}{100}$

G. 500%   **H.** $\dfrac{10}{2}$ **I.** -5

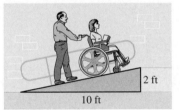

23. *Concept Check*   If the pitch of a roof is $\frac{1}{4}$, how many feet in the horizontal direction correspond to a rise of 3 ft?

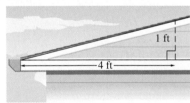

24. Family income in the United States has increased steadily for many years (primarily due to inflation). In 1980, the median family income was about $21,000 per year. In 2003, it was about $52,700 per year. Find the average rate of change of median family income to the nearest dollar over that period. (*Source:* U.S. Census Bureau.)

[3.3]   *Find an equation for each line.* **(a)** *Write the equation in slope–intercept form.* **(b)** *Write the equation in standard form.*

25. Slope $-\dfrac{1}{3}$; *y*-intercept $(0, -1)$

26. Slope 0; *y*-intercept $(0, -2)$

27. Slope $-\dfrac{4}{3}$; through $(2, 7)$

28. Slope 3; through $(-1, 4)$

29. Vertical; through $(2, 5)$

30. Through $(2, -5)$ and $(1, 4)$

31. Through $(-3, -1)$ and $(2, 6)$

32. The line pictured in Exercise 17

33. Parallel to $4x - y = 3$ and through $(7, -1)$

34. Perpendicular to $2x - 5y = 7$ and through $(4, 3)$

35. The Midwest Athletic Club (**Section 3.3,** Exercises 75 and 76) offers two special membership plans. (*Source:* Midwest Athletic Club.) For each plan, write a linear equation in slope–intercept form and give the cost y in dollars of a 1-yr membership. Let x represent the number of months.

   (a) Executive VIP/Gold membership: $159 fee, plus $57 per month
   (b) Executive Regular/Silver membership: $159 fee, plus $47 per month

36. The percent of tax returns filed electronically for the years 1996–2002 is shown in the graph.

   (a) Use the information given for the years 1996 and 2002, letting $x = 6$ represent 1996, $x = 12$ represent 2002, and y represent the percent of returns filed electronically to find a linear equation that models the data. Write the equation in slope intercept form. Interpret the slope of the graph.
   (b) Use your equation from part (a) to predict the percent of tax returns that were filed electronically in 2005.

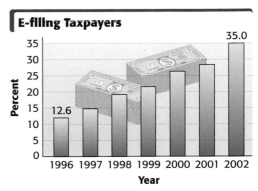

Source: Internal Revenue Service.

[3.4]   *Graph the solution set of each inequality or compound inequality.*

37. $3x - 2y \le 12$

38. $5x - y > 6$

39. $2x + y \le 1$ and $x \ge 2y$

40. $x \ge 2$ or $y \ge 2$

41. *Concept Check*   Which one of the following has as its graph a dashed boundary line and shading below the line?

   A. $y \ge 4x + 3$ **B.** $y > 4x + 3$ **C.** $y \le 4x + 3$ **D.** $y < 4x + 3$

[3.5]   *In Exercises 42–45, give the domain and range of each relation. Identify any functions.*

42. $\{(-4, 2), (-4, -2), (1, 5), (1, -5)\}$

43.

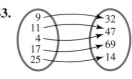

44.

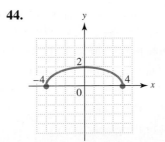

45.

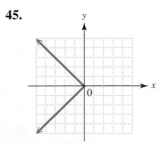

Determine whether each equation or inequality defines y as a function of x. Give the domain in each case. Identify any linear functions.

46. $y = 3x - 3$

47. $y < x + 2$

48. $y = |x|$

49. $y = \sqrt{4x + 7}$

50. $x = y^2$

51. $y = \dfrac{7}{x - 6}$

52. Explain the test that allows us to determine whether a graph is that of a function.

Given $f(x) = -2x^2 + 3x - 6$, find each function value or expression.

53. $f(0)$

54. $f(2.1)$

55. $f\left(-\dfrac{1}{2}\right)$

56. $f(k)$

57. The table shows life expectancy at birth in the United States for selected years.

(a) Does the table define a function?
(b) What are the domain and range?
(c) Call this function f. Give two ordered pairs that belong to f.
(d) Find $f(1973)$. What does this mean?
(e) If $f(x) = 75.5$, what does x equal?

Year	Life Expectancy at Birth (years)
1943	63.3
1953	68.8
1963	69.9
1973	71.4
1983	74.6
1993	75.5
2003	77.6

Source: Centers for Disease Control and Prevention.

58. The equation $2x^2 - y = 0$ defines y as a function f of x. Write it using function notation, and find $f(3)$.

59. *Concept Check* Suppose that $2x - 5y = 7$ defines y as a function f of x. If $y = f(x)$, which one of the following defines the same function?

A. $f(x) = -\dfrac{2}{5}x + \dfrac{7}{5}$

B. $f(x) = -\dfrac{2}{5}x - \dfrac{7}{5}$

C. $f(x) = \dfrac{2}{5}x - \dfrac{7}{5}$

D. $f(x) = \dfrac{2}{5}x + \dfrac{7}{5}$

60. Can the graph of a linear function have an undefined slope? Explain.

RELATING CONCEPTS (EXERCISES 61–72)

FOR INDIVIDUAL OR GROUP WORK

Refer to the straight-line graph and **work Exercises 61–72 in order.**

61. By just looking at the graph, how can you tell whether the slope is positive, negative, 0, or undefined?

62. Use the slope formula to find the slope of the line.

63. What is the slope of any line parallel to the line shown? perpendicular to the line shown?

64. Find the x-intercept of the graph.

65. Find the y-intercept of the graph.

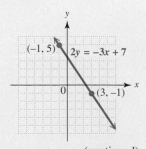

(continued)

66. Use function notation to write the equation of the line. Use f to designate the function.

67. Find $f(8)$.

68. If $f(x) = -8$, what is the value of x?

69. Graph the solution set of $f(x) \geq 0$.

70. What is the solution set of $f(x) = 0$?

71. What is the solution set of $f(x) < 0$? (Use the graph and the result of Exercise 70.)

72. What is the solution set of $f(x) > 0$? (Use the graph and the result of Exercise 70.)

Chapter 3 TEST

 *View the complete solutions to all Chapter Test exercises on the Pass the Test CD.*

1. Complete the table of ordered pairs for the equation $2x - 3y = 12$.

x	y
1	
3	
	-4

Find the x- and y-intercepts, and graph each equation.

2. $3x - 2y = 20$ **3.** $y = 5$ **4.** $x = 2$

5. Find the slope of the line through the points $(6, 4)$ and $(-4, -1)$.

6. *Concept Check*   Describe how the graph of a line with undefined slope is situated in a rectangular coordinate system.

Determine whether each pair of lines is parallel, perpendicular, *or* neither.

7. $5x - y = 8$ and $5y = -x + 3$ **8.** $2y = 3x + 12$ and $3y = 2x - 5$

9. In 1980, there were 119,000 farms in Iowa. As of 2005, there were 89,000. Find and interpret the average rate of change in the number of farms per year. (*Source:* U.S. Department of Agriculture.)

*Find an equation of each line, and write it in **(a)** slope–intercept form if possible and **(b)** standard form.*

10. Through $(4, -1)$; $m = -5$

11. Through $(-3, 14)$; horizontal

12. Through $(-7, 2)$ and parallel to $3x + 5y = 6$

13. Through $(-7, 2)$ and perpendicular to $y = 2x$

14. Through $(-2, 3)$ and $(6, -1)$

15. Through $(5, -6)$; vertical

16. *Concept Check* Which one of the following has positive slope and negative y-coordinate for its y-intercept?

A.
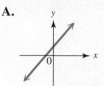

B.
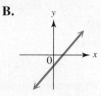

C.
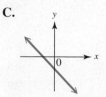

D.

17. The bar graph shows median household income for Hispanics.

(a) Use the information for the years 1995 and 2001 to find an equation that models the data. Let $x = 5$ represent 1995, $x = 11$ represent 2001, and y represent the median income. Write the equation in slope–intercept form.

(b) Use the equation from part (a) to approximate median household income for 1999 to the nearest dollar. How does your result compare against the actual value, $30,735?

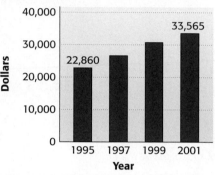

Source: U.S. Census Bureau.

Graph each inequality or compound inequality.

18. $3x - 2y > 6$

19. $y < 2x - 1$ and $x - y < 3$

20. Which one of the following is the graph of a function?

A.
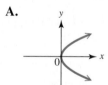

B.
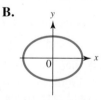

C.
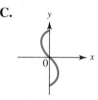

D.

21. Which of the following does not define y as a function of x?

A. $\{(0, 1), (-2, 3), (4, 8)\}$ **B.** $y = 2x - 6$ **C.** $y = \sqrt{x + 2}$ **D.**

x	y
0	1
3	2
0	2
6	3

Give the domain and range of the relation shown in each of the following.

22. Choice A of Exercise 20

23. Choice A of Exercise 21

24. For $f(x) = -x^2 + 2x - 1$,

(a) find $f(1)$. (b) find $f(a)$.

25. Graph the linear function defined by $f(x) = \frac{2}{3}x - 1$. What is its domain and range?

Chapters **1–3** CUMULATIVE REVIEW EXERCISES

Decide whether each statement is always true, sometimes true, *or* never true. *If the statement is* sometimes true, *give examples in which it is true and in which it is false.*

1. The absolute value of a negative number equals the additive inverse of the number.

2. The quotient of two integers with nonzero denominator is a rational number.

3. The sum of two negative numbers is positive.

4. The sum of a positive number and a negative number is 0.

Perform each operation.

5. $-|-2| - 4 + |-3| + 7$

6. $(-0.8)^2$

7. $\sqrt{-64}$

8. $-\dfrac{2}{3}\left(-\dfrac{12}{5}\right)$

Simplify.

9. $-(-4m + 3)$

10. $3x^2 - 4x + 4 + 9x - x^2$

11. $\dfrac{(4^2 - 4) - (-1)7}{4 + (-6)}$

12. Write $-3 < x \le 5$ in interval notation.

13. Is $\sqrt{\dfrac{-2 + 4}{-5}}$ a real number?

Evaluate each expression if $p = -4$, $q = \frac{1}{2}$, and $r = 16$.

14. $-3(2q - 3p)$

15. $|p|^3 - |16q^3|$

16. $\dfrac{\sqrt{r}}{8p + 2r}$

Solve.

17. $2z - 5 + 3z = 2 - z$

18. $\dfrac{3a - 1}{5} + \dfrac{a + 2}{2} = -\dfrac{3}{10}$

19. $V = \dfrac{1}{3}\pi r^2 h$ for h

Solve each problem.

20. As of 2003, the lowest temperature ever recorded in South Dakota was $-50°C$ at McIntosh on January 21, 1985. What was the corresponding Fahrenheit temperature? (*Source: World Almanac and Book of Facts 2006.*)

21. If each side of a square were increased by 4 in., the perimeter would be 8 in. less than twice the perimeter of the original square. Find the length of a side of the original square.

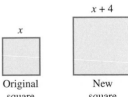

Original square   New square

22. Two planes leave the Dallas–Fort Worth airport at the same time. One travels east at 550 mph, and the other travels west at 500 mph. Assuming no wind, how long will it take for the planes to be 2100 mi apart?

West ← Airport → East

Solve. Write each solution set in interval notation and graph it.

23. $-4 < 3 - 2k < 9$

24. $-0.3x + 2.1(x - 4) \leq -6.6$

25. $\dfrac{1}{2}x > 3$ and $\dfrac{1}{3}x < \dfrac{8}{3}$

26. $-5x + 1 \geq 11$ or $3x + 5 > 26$

Solve.

27. $|2k - 7| + 4 = 11$

28. $|3m + 6| \geq 0$

29. How are the solution sets of a linear equation and the two associated inequalities related?

30. Complete the table of ordered pairs at the right for the linear equation $3x - 4y = 12$.

x	y
0	
	0
2	

31. Find the *x*- and *y*-intercepts of the line with equation $3x + 5y = 12$, and graph the line.

32. Consider the points $A(-2, 1)$ and $B(3, -5)$.

   (a) Find the slope of the line *AB*.

   (b) Find the slope of a line perpendicular to line *AB*.

33. Graph the inequality $-2x + y < -6$.

*Write an equation for each line. Express the equation **(a)** in slope–intercept form if possible and **(b)** in standard form.*

34. Slope $-\dfrac{3}{4}$; *y*-intercept $(0, -1)$

35. Horizontal; through $(2, -2)$

36. Through $(4, -3)$ and $(1, 1)$

37. Vertical; through $(4, -6)$

38. Give the domain and range of the relation. Does it define a function? Explain.

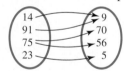

39. Consider the function defined by
$$f(x) = -4x + 10.$$

   (a) Find the domain and range.

   (b) Evaluate $f(-3)$.

40. Use the information in the graph to find and interpret the average rate of change in the number of motor scooters sold in the United States from 1997 to 2004.

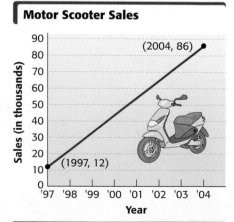

Motor Scooter Sales

Source: Motorcycle Industry Council.

Systems of Linear Equations

In the early 1970s, the NBC television network aired *The Bill Cosby Show,* in which the popular comedian played Chet Kincaid, a Los Angeles high school physical education teacher. In the episode "Let *x* Equal a Lousy Weekend," Chet must substitute for the algebra teacher. He and the entire class are stumped by the following problem:

How many pounds of candy that sells for $0.75 per lb must be mixed with candy that sells for $1.25 per lb to obtain 9 lb of a mixture that should sell for $0.96 per lb?

The smartest student in the class eventually helps Chet solve this problem. In Exercise 31 of Section 4.3, we ask you to use a *system of linear equations,* the topic of this chapter, to do so.

4.1 Systems of Linear Equations in Two Variables

In recent years, the sale of digital cameras has increased, while that of conventional cameras has decreased. These trends can be seen in the graph in Figure 1. The two straight-line graphs intersect at the point in time when the two types of cameras had the *same* sales.

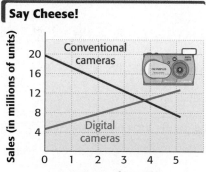

Say Cheese!

Sales (in millions of units)

Conventional cameras

Digital cameras

Years since 2000

Source: Consumer Electronics Association.

FIGURE 1

$$2.5x + y = 19.4$$
$$-1.7x + y = 4.4$$

Linear system of equations

(Here, $x = 0$ represents 2000, $x = 1$ represents 2001, and so on; y represents sales in millions of units.)

As shown beside Figure 1, we can use a linear equation to model the graph of digital camera sales (the blue equation) and another linear equation to model the graph of conventional camera sales (the red equation). Such a set of equations is called a **system of equations**—in this case, a **linear system of equations.** The point where the graphs in Figure 1 intersect is a solution of each of the individual equations. It is also the solution of the linear system of equations.

OBJECTIVE 1 Decide whether an ordered pair is a solution of a linear system. The **solution set of a linear system** of equations contains all ordered pairs that satisfy all the equations of the system *at the same time.*

EXAMPLE 1 Deciding whether an Ordered Pair Is a Solution

Decide whether the given ordered pair is a solution of the given system.

(a) $x + y = 6$
$4x - y = 14$; $(4, 2)$

Replace x with 4 and y with 2 in each equation of the system.

$x + y = 6$	$4x - y = 14$
$4 + 2 = 6$?	$4(4) - 2 = 14$?
$6 = 6$ True	$14 = 14$ True

Since $(4, 2)$ makes *both* equations true, $(4, 2)$ is a solution of the system.

(b) $3x + 2y = 11$
$x + 5y = 36$; $(-1, 7)$

$$3x + 2y = 11$$
$$3(-1) + 2(7) = 11 \quad ?$$
$$-3 + 14 = 11 \quad ?$$
$$11 = 11 \quad \text{True}$$

$$x + 5y = 36$$
$$-1 + 5(7) = 36 \quad ?$$
$$-1 + 35 = 36 \quad ?$$
$$34 = 36 \quad \text{False}$$

> Use parentheses when substituting to avoid errors.

The ordered pair $(-1, 7)$ is not a solution of the system, since it does not make *both* equations true.

✔ **Now Try Exercises 11 and 13.**

OBJECTIVE 2 Solve linear systems by graphing. One way to find the solution set of a linear system of equations is to graph each equation and find the point where the graphs intersect.

EXAMPLE 2 Solving a System by Graphing

Solve the system of equations by graphing.

$$x + y = 5 \quad (1)$$
$$2x - y = 4 \quad (2)$$

When we graph these linear equations as shown in Figure 2, the graph suggests that the point of intersection is the ordered pair $(3, 2)$.

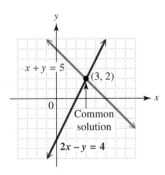

FIGURE 2

To be sure that $(3, 2)$ is a solution of *both* equations, we check by substituting 3 for x and 2 for y in each equation.

Check: $\quad x + y = 5 \quad (1)$
$$3 + 2 = 5 \quad ?$$
$$5 = 5 \quad \text{True}$$

$\quad 2x - y = 4 \quad (2)$
$$2(3) - 2 = 4 \quad ?$$
$$6 - 2 = 4 \quad ?$$
$$4 = 4 \quad \text{True}$$

Since $(3, 2)$ makes both equations true, $\{(3, 2)\}$ is the solution set of the system.

✔ **Now Try Exercise 15.**

There are three possibilities for the solution set of a linear system in two variables.

> ### Graphs of Linear Systems in Two Variables
>
> 1. **The two graphs intersect in a single point.** The coordinates of this point give the only solution of the system. Since the system has a solution, it is **consistent.** The equations are *not* equivalent, so they are **independent.** See Figure 3(a).
> 2. **The graphs are parallel lines.** There is no solution common to both equations, so the solution set is ∅ and the system is **inconsistent.** Since the equations are *not* equivalent, they are **independent.** See Figure 3(b).
> 3. **The graphs are the same line.** Since any solution of one equation of the system is a solution of the other, the solution set is an infinite set of ordered pairs representing the points on the line. This type of system is **consistent** because there is a solution. The equations are equivalent, so they are **dependent.** See Figure 3(c).
>
>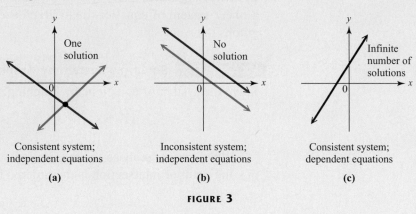
>
> | Consistent system; independent equations | Inconsistent system; independent equations | Consistent system; dependent equations |
> | :---: | :---: | :---: |
> | **(a)** | **(b)** | **(c)** |
>
> **FIGURE 3**

OBJECTIVE **3** Solve linear systems (with two equations and two variables) by substitution. Since it can be difficult to read exact coordinates, especially if they are not integers, from a graph, we usually use algebraic methods to solve systems. One such method, the **substitution method,** is most useful for solving linear systems in which one equation is solved or can be easily solved for one variable in terms of the other.

EXAMPLE 3 Solving a System by Substitution

Solve the system

$$2x - y = 6 \qquad (1)$$
$$x = y + 2. \qquad (2)$$

Since equation (2) is solved for x, substitute $y + 2$ for x in equation (1).

$$2x - y = 6 \qquad \text{(1)}$$
$$2(y + 2) - y = 6 \qquad \text{Let } x = y + 2.$$
$$2y + 4 - y = 6 \qquad \text{Distributive property}$$

Be sure to use parentheses here.

$$y + 4 = 6 \qquad \text{Combine like terms.}$$
$$y = 2 \qquad \text{Subtract 4.}$$

We found y. Now find x by substituting 2 for y in equation (2).

$$x = y + 2 = 2 + 2 = 4$$

> Write the *x*-value first in the ordered pair.

Thus, $x = 4$ and $y = 2$, giving the ordered pair $(4, 2)$. Check this solution in both equations of the original system.

Check:

$2x - y = 6$	(1)	$x = y + 2$	(2)
$2(4) - 2 = 6$?		$4 = 2 + 2$?	
$8 - 2 = 6$?		$4 = 4$	True
$6 = 6$	True		

Since $(4, 2)$ makes both equations true, the solution set is $\{(4, 2)\}$.

✔ **Now Try Exercise 19.**

The substitution method is summarized as follows.

Solving a Linear System by Substitution

Step 1 **Solve one of the equations for either variable.** If one of the equations has a variable term with coefficient 1 or -1, choose that equation since the substitution method is usually easier this way.

Step 2 **Substitute** for that variable in the other equation. The result should be an equation with just one variable.

Step 3 **Solve** the equation from Step 2.

Step 4 **Find the other value.** Substitute the result from Step 3 into the equation from Step 1 to find the value of the other variable.

Step 5 **Check** the solution in both of the original equations. Then write the solution set.

EXAMPLE 4 Solving a System by Substitution

Solve the system

$$3x + 2y = 13 \qquad (1)$$
$$4x - y = -1. \qquad (2)$$

Step 1 Solve one of the equations for either x or y. Since the coefficient of y in equation (2) is -1, it is easiest to solve for y in equation (2).

$4x - y = -1$	(2)
$-y = -1 - 4x$	Subtract 4*x*.
$y = 1 + 4x$	Multiply by -1.

Step 2 Substitute $1 + 4x$ for y in equation (1).

$3x + 2y = 13$	(1)
$3x + 2(1 + 4x) = 13$	Let *y* = 1 + 4*x*.

Step 3 Solve for x.

$$3x + 2(1 + 4x) = 13 \qquad \text{Equation from Step 2}$$
$$3x + 2 + 8x = 13 \qquad \text{Distributive property}$$
$$11x = 11 \qquad \text{Combine like terms; subtract 2.}$$
$$x = 1 \qquad \text{Divide by 11.}$$

Step 4 Now find y. Since $y = 1 + 4x$ and $x = 1$,

$$y = 1 + 4(1) = 5. \qquad \text{Let } x = 1.$$

Step 5 Check the solution $(1, 5)$ in both equations (1) and (2).

$3x + 2y = 13$ (1)		$4x - y = -1$ (2)	
$3(1) + 2(5) = 13$?		$4(1) - 5 = -1$?	
$3 + 10 = 13$?		$4 - 5 = -1$?	
$13 = 13$ True		$-1 = -1$ True	

The solution set is $\{(1, 5)\}$.

✔ **Now Try Exercise 25.**

EXAMPLE 5 **Solving a System with Fractional Coefficients**

Solve the system

$$\frac{2}{3}x - \frac{1}{2}y = \frac{7}{6} \qquad (1)$$
$$3x - y = 6. \qquad (2)$$

This system will be easier to solve if we clear the fractions in equation (1).

$$6\left(\frac{2}{3}x - \frac{1}{2}y\right) = 6\left(\frac{7}{6}\right) \qquad \text{Multiply (1) by the LCD, 6.}$$
$$6 \cdot \frac{2}{3}x - 6 \cdot \frac{1}{2}y = 6 \cdot \frac{7}{6} \qquad \text{Distributive property}$$

Remember to multiply each term by 6.

$$4x - 3y = 7 \qquad (3)$$

Now the system consists of equations (2) and (3). Solve equation (2) for y.

$$3x - y = 6 \qquad (2)$$
$$-y = 6 - 3x \qquad \text{Subtract } 3x.$$
$$y = 3x - 6 \qquad \text{Multiply by } -1; \text{ rewrite.}$$

Substitute $3x - 6$ for y in equation (3).

$$4x - 3y = 7 \qquad (3)$$
$$4x - 3(3x - 6) = 7 \qquad \text{Let } y = 3x - 6.$$
$$4x - 9x + 18 = 7 \qquad \text{Distributive property}$$

Be careful with signs.

$$-5x + 18 = 7 \qquad \text{Combine like terms.}$$
$$-5x = -11 \qquad \text{Subtract 18.}$$
$$x = \frac{11}{5} \qquad \text{Divide by } -5.$$

Since $y = 3x - 6$ and $x = \frac{11}{5}$,

$$y = 3\left(\frac{11}{5}\right) - 6 = \frac{33}{5} - \frac{30}{5} = \frac{3}{5}.$$

A check verifies that the solution set is $\left\{\left(\frac{11}{5}, \frac{3}{5}\right)\right\}$.

✔ **Now Try Exercise 29.**

▶ **NOTE** If an equation in a system contains decimal coefficients, it is best to first clear the decimals by multiplying by an appropriate power of 10, depending on the number of decimal places. Then solve the system. For example, we multiply *each side* of the equation

$$0.5x + 0.75y = 3.25$$

by 10^2, or 100, to get the equivalent equation

$$50x + 75y = 325.$$

OBJECTIVE 4 Solve linear systems (with two equations and two variables) by elimination. Another algebraic method, the **elimination method,** involves combining the two equations in a system so that one variable is eliminated. This is done using the following logic:

If $a = b$ and $c = d$, then $a + c = b + d$.

EXAMPLE 6 Solving a System by Elimination

Solve the system

$$2x + 3y = -6 \quad \text{(1)}$$
$$4x - 3y = 6. \quad \text{(2)}$$

Notice that adding the equations together will eliminate the variable y.

$$
\begin{array}{rl}
2x + 3y = -6 & \text{(1)} \\
\underline{4x - 3y = 6} & \text{(2)} \\
6x = 0 & \text{Add.} \\
x = 0 & \text{Solve for } x.
\end{array}
$$

To find y, substitute 0 for x in either equation (1) or equation (2).

$$
\begin{array}{rl}
2x + 3y = -6 & \text{(1)} \\
2(0) + 3y = -6 & \text{Let } x = 0. \\
0 + 3y = -6 & \text{Multiply.} \\
3y = -6 & \text{Add.} \\
y = -2 & \text{Divide by 3.}
\end{array}
$$

The solution is $(0, -2)$. Check mentally by substituting 0 for x and -2 for y in both equations of the original system. The solution set is $\{(0, -2)\}$.

✔ **Now Try Exercise 35.**

By adding the equations in Example 6, we eliminated the variable y because the coefficients of the y-terms were opposites. In many cases the coefficients will *not* be opposites, and we must transform one or both equations so that the coefficients of one pair of variable terms are opposites.

The elimination method is summarized as follows.

Solving a Linear System by Elimination

Step 1 **Write both equations in standard form $Ax + By = C$.**

Step 2 **Make the coefficients of one pair of variable terms opposites.** Multiply one or both equations by appropriate numbers so that the sum of the coefficients of either the x- or y-terms is 0.

Step 3 **Add the new equations to eliminate a variable.** The sum should be an equation with just one variable.

Step 4 **Solve the equation from Step 3 for the remaining variable.**

Step 5 **Find the other value.** Substitute the result of Step 4 into either of the original equations and solve for the other variable.

Step 6 **Check the solution in both of the original equations. Then write the solution set.**

EXAMPLE 7 Solving a System by Elimination

Solve the system

$$5x - 2y = 4 \qquad (1)$$
$$2x + 3y = 13. \qquad (2)$$

Step 1 Both equations are in standard form.

Step 2 Suppose that you wish to eliminate the variable x. One way to do this is to multiply equation (1) by 2 and equation (2) by -5.

$$10x - 4y = 8 \qquad \text{2 times each side of equation (1)}$$
$$-10x - 15y = -65 \qquad \text{-5 times each side of equation (2)}$$

Step 3 Now add.

$$\begin{array}{r} 10x - 4y = 8 \\ -10x - 15y = -65 \\ \hline -19y = -57 \end{array} \qquad \text{Add.}$$

Step 4 Solve for y. $\qquad\qquad y = 3 \qquad$ Divide by -19.

Step 5 To find x, substitute 3 for y in either equation (1) or equation (2).

$$2x + 3y = 13 \qquad (2)$$
$$2x + 3(3) = 13 \qquad \text{Let } y = 3.$$
$$2x + 9 = 13 \qquad \text{Multiply.}$$
$$2x = 4 \qquad \text{Subtract 9.}$$
$$x = 2 \qquad \text{Divide by 2.}$$

Step 6 The solution is $(2, 3)$. To check, substitute 2 for x and 3 for y in both equations (1) and (2).

$5x - 2y = 4$	(1)	$2x + 3y = 13$	(2)
$5(2) - 2(3) = 4$	?	$2(2) + 3(3) = 13$	?
$10 - 6 = 4$	?	$4 + 9 = 13$	?
$4 = 4$	True	$13 = 13$	True

The solution set is $\{(2, 3)\}$.

✔ **Now Try Exercise 39.**

OBJECTIVE 5 Solve special systems. As we saw in Figures 3(b) and (c), some systems of linear equations have no solution or an infinite number of solutions.

EXAMPLE 8 Solving a System of Dependent Equations

Solve the system

$$2x - y = 3 \qquad (1)$$
$$6x - 3y = 9. \qquad (2)$$

We multiply equation (1) by -3 and then add the result to equation (2).

$-6x + 3y = -9$	-3 times each side of equation (1)
$\underline{6x - 3y = 9}$	(2)
$0 = 0$	True

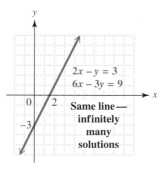

Adding these equations gives the true statement $0 = 0$. In the original system, we could get equation (2) from equation (1) by multiplying equation (1) by 3. Because of this, equations (1) and (2) are equivalent and have the same graph, as shown in Figure 4. The equations are dependent. The solution set is the set of all points on the line with equation $2x - y = 3$, written in set-builder notation **(Section 1.1)** as

$$\{(x, y) \mid 2x - y = 3\}$$

and read "the set of all ordered pairs (x, y), such that $2x - y = 3$."

FIGURE 4

✔ **Now Try Exercise 41.**

▶ **NOTE** When a system has dependent equations and an infinite number of solutions, as in Example 8, either equation of the system could be used to write the solution set. We prefer to use the equation (in standard form) with coefficients that are integers having no common factor (except 1).

EXAMPLE 9 Solving an Inconsistent System

Solve the system

$$x + 3y = 4 \qquad (1)$$
$$-2x - 6y = 3. \qquad (2)$$

Multiply equation (1) by 2, and then add the result to equation (2).

$$
\begin{array}{ll}
2x + 6y = 8 & \text{Equation (1) multiplied by 2} \\
\underline{-2x - 6y = 3} & \text{(2)} \\
0 = 11 & \text{False}
\end{array}
$$

The result of the addition step is a false statement, which indicates that the system is inconsistent. As shown in Figure 5, the graphs of the equations of the system are parallel lines. There are no ordered pairs that satisfy both equations, so there is no solution for the system; the solution set is ∅.

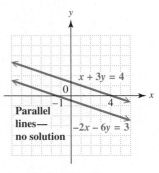

FIGURE 5

✔ **Now Try Exercise 45.**

The results of Examples 8 and 9 are generalized as follows.

> **Special Cases of Linear Systems**
>
> If both variables are eliminated when a system of linear equations is solved,
>
> **1.** there are infinitely many solutions if the resulting statement is *true;*
>
> **2.** there is no solution if the resulting statement is *false.*

    Slopes and y-intercepts can be used to decide whether the graphs of a system of equations are parallel lines or whether they coincide. In Example 8, writing each equation in slope–intercept form shows that both lines have slope 2 and y-intercept $(0, -3)$, so the graphs are the same line and the system has an infinite solution set.

    In Example 9, both equations have slope $-\frac{1}{3}$, but y-intercepts $\left(0, \frac{4}{3}\right)$ and $\left(0, -\frac{1}{2}\right)$, showing that the graphs are two distinct parallel lines. Thus, the system has no solution.

✔ **Now Try Exercises 53 and 55.**

 OBJECTIVE 6 Recognize how a graphing calculator is used to solve a linear system. In Example 2, we showed how to solve the system

$$x + y = 5$$
$$2x - y = 4$$

by graphing the two lines and finding their point of intersection. We can also do this with a graphing calculator.

EXAMPLE 10 **Finding the Solution Set of a System from a Graphing Calculator Screen**

Solve the system

$$x + y = 5$$
$$2x - y = 4.$$

See Figure 6. The two lines were graphed by solving the first equation to get

$$y = 5 - x$$

and the second to get

$$y = 2x - 4.$$

The coordinates of their point of intersection are displayed at the bottom of the screen, indicating that the solution set is $\{(3, 2)\}$. (Compare this graph with the one in Figure 2.)

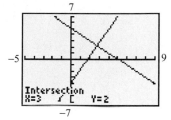

FIGURE 6

✔ **Now Try Exercise 71.**

4.1 EXERCISES

● *Complete solution available on Video Lectures on CD/DVD*

▢ *Now Try Exercise*

Concept Check   Fill in the blanks with the correct responses.

1. If $(3, -6)$ is a solution of a linear system in two variables, then substituting _____ for x and _____ for y leads to true statements in *both* equations.

2. A solution of a system of independent linear equations in two variables is a(n) _____.

3. If solving a system leads to a false statement such as $0 = 5$, the solution set is _____.

4. If solving a system leads to a true statement such as $0 = 0$, the system has _____ equations.

5. If the two lines forming a system have the same slope and different y-intercepts, the system has _____ solution(s).
    (how many?)

6. If the two lines forming a system have different slopes, the system has _____ solution(s).
    (how many?)

7. *Concept Check* Which ordered pair could be a solution of the graphed system of equations? Why?

    A. $(3, 3)$
    B. $(-3, 3)$
    C. $(-3, -3)$
    D. $(3, -3)$

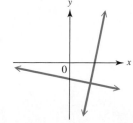

8. *Concept Check* Which ordered pair could be a solution of the graphed system of equations? Why?

    A. $(3, 0)$
    B. $(-3, 0)$
    C. $(0, 3)$
    D. $(0, -3)$

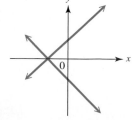

9. *Concept Check* Match each system with the correct graph.

(a) $x + y = 6$
  $x - y = 0$

(b) $x + y = -6$
  $x - y = 0$

(c) $x + y = 0$
  $x - y = -6$

(d) $x + y = 0$
  $x - y = 6$

A.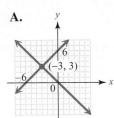

B.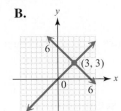

C.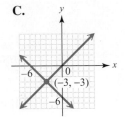

D.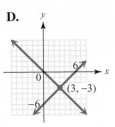

Decide whether the given ordered pair is a solution of the given system. See Example 1.

10. $x + y = 6$
  $x - y = 4$; $(5, 1)$

11. $x - y = 17$
  $x + y = -1$; $(8, -9)$

12. $2x - y = 8$
  $3x + 2y = 20$; $(5, 2)$

13. $3x - 5y = -12$
  $x - y = 1$; $(-1, 2)$

Solve each system by graphing. See Example 2.

14. $x + y = 4$
  $2x - y = 2$

15. $x + y = -5$
  $-2x + y = 1$

16. $x - 4y = -4$
  $3x + y = 1$

Solve each system by substitution. If the system is inconsistent or has dependent equations, say so. See Examples 3–5, 8, and 9.

17. $4x + y = 6$
  $y = 2x$

18. $2x - y = 6$
  $y = 5x$

19. $-x - 4y = -14$
  $y = 2x - 1$

20. $-3x - 5y = -17$
  $y = 4x + 8$

21. $3x - 4y = -22$
  $-3x + y = 0$

22. $-3x + y = -5$
  $x + 2y = 0$

23. $5x - 4y = 9$
  $3 - 2y = -x$

24. $6x - y = -9$
  $4 + 7x = -y$

25. $4x - 5y = -11$
  $x + 2y = 7$

26. $3x - y = 10$
  $2x + 5y = 1$

27. $x = 3y + 5$
  $x = \dfrac{3}{2} y$

28. $x = 6y - 2$
  $x = \dfrac{3}{4} y$

29. $\dfrac{1}{2} x + \dfrac{1}{3} y = 3$
  $-3x + y = 0$

30. $\dfrac{1}{4} x - \dfrac{1}{5} y = 9$
  $5x - y = 0$

31. $y = 2x$
  $4x - 2y = 0$

32. $x = 3y$
  $3x - 9y = 0$

33. $x = 5y$
  $5x - 25y = 5$

34. $y = -4x$
  $8x + 2y = 4$

Solve each system by elimination. If the system is inconsistent or has dependent equations, say so. See Examples 5–9.

35. $-2x + 3y = -16$
$2x - 5y = 24$

36. $6x + 5y = -7$
$-6x - 11y = 1$

37. $2x - 5y = 11$
$3x + y = 8$

38. $-2x + 3y = 1$
$-4x + y = -3$

39. $3x + 4y = -6$
$5x + 3y = 1$

40. $4x + 3y = 1$
$3x + 2y = 2$

41. $7x + 2y = 6$
$-14x - 4y = -12$

42. $x - 4y = 2$
$4x - 16y = 8$

43. $3x + 3y = 0$
$4x + 2y = 3$

44. $8x + 4y = 0$
$4x \quad 2y = 2$

45. $5x - 5y = 3$
$x - y = 12$

46. $2x - 3y = 7$
$-4x + 6y = 14$

47. $x + y = 0$
$2x - 2y = 0$

48. $3x + 3y = 0$
$-2x - y = 0$

49. $x - \dfrac{1}{2}y = 2$
$-x + \dfrac{2}{5}y = -\dfrac{8}{5}$

50. $\dfrac{3}{2}x + y = 3$
$\dfrac{2}{3}x + \dfrac{1}{3}y = 1$

51. $\dfrac{1}{2}x + \dfrac{1}{3}y = -\dfrac{1}{3}$
$\dfrac{1}{2}x + 2y = -7$

52. $\dfrac{1}{5}x + y = \dfrac{6}{5}$
$\dfrac{1}{10}x + \dfrac{1}{3}y = \dfrac{5}{6}$

Concept Check *Write each equation in slope–intercept form and then tell how many solutions the system has. Do not actually solve.*

53. $3x + 7y = 4$
$6x + 14y = 3$

54. $-x + 2y = 8$
$4x - 8y = 1$

55. $2x = -3y + 1$
$6x = -9y + 3$

56. $5x = -2y + 1$
$10x = -4y + 2$

57. Assuming that you want to minimize the amount of work required, tell whether you would use the substitution or elimination method to solve each system. Explain your answers. *Do not actually solve.*

(a) $6x - y = 5$
$y = 11x$

(b) $3x + y = -7$
$x - y = -5$

(c) $3x - 2y = 0$
$9x + 8y = 7$

Solve each system by the method of your choice. See Examples 3–9. (For Exercises 58–60, see your answers to Exercise 57.)

58. $6x - y = 5$
$y = 11x$

59. $3x + y = -7$
$x - y = -5$

60. $3x - 2y = 0$
$9x + 8y = 7$

61. $2x + 3y = 10$
$-3x + y = 18$

62. $3x - 5y = 7$
$2x + 3y = 30$

63. $\dfrac{1}{2}x - \dfrac{1}{8}y = -\dfrac{1}{4}$
$4x - y = -2$

64. $\dfrac{1}{6}x + \dfrac{1}{3}y = 8$
$\dfrac{1}{4}x + \dfrac{1}{2}y = 12$

65. $0.3x + 0.2y = 0.4$
$0.5x + 0.4y = 0.7$

66. $0.2x + 0.5y = 6$
$0.4x + y = 9$

TECHNOLOGY INSIGHTS (EXERCISES 67–70)

67. The table shown was generated by a graphing calculator. The functions defined by Y_1 and Y_2 are linear. Based on the table, what are the coordinates of the point of intersection of the graphs?

X	Y₁	Y₂
0	-7	-1
1	-6	-2
2	-5	-3
3	-4	-4
4	-3	-5
5	-2	-6
6	-1	-7

X=0

68. The functions defined by Y_1 and Y_2 in the table are linear.

(a) Use the methods of **Chapter 3** to find the equation for Y_1.

(b) Use the methods of **Chapter 3** to find the equation for Y_2.

(c) Solve the system of equations formed by Y_1 and Y_2.

X	Y₁	Y₂
0	4	7
1	8	5
2	12	3
3	16	1
4	20	-1
5	24	-3
6	28	-5

X=0

69. The solution set of the system

$$y_1 = 3x - 5$$
$$y_2 = -4x + 2$$

is $\{(1, -2)\}$. Using slopes and y-intercepts, determine which one of the two calculator graphs is the appropriate one for this system.

A.

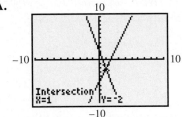

B.

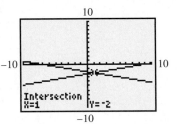

70. Which one of the ordered pairs listed is the only possible solution of the system whose graphs are shown in the standard viewing window of a graphing calculator?

A. $(15, -15)$ **B.** $(15, 15)$
C. $(-15, 15)$ **D.** $(-15, -15)$

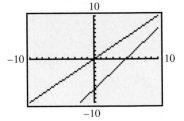

*For each system, **(a)** solve by elimination or substitution and **(b)** use a graphing calculator to support your result. In part (b), be sure to solve each equation for y first. See Example 10.*

71. $x + y = 10$
$2x - y = 5$

72. $6x + y = 5$
$-x + y = -9$

73. $3x - 2y = 4$
$3x + y = -2$

74. $2x - 3y = 3$
$2x + 2y = 8$

Answer the questions in Exercises 75–78 by observing the graphs provided.

75. The figure shows graphs that represent supply and demand for a certain brand of low-fat frozen yogurt at various prices per half-gallon (in dollars).

(a) At what price does supply equal demand?

(b) For how many half-gallons does supply equal demand?

(c) What are the supply and demand at a price of $2 per half-gallon?

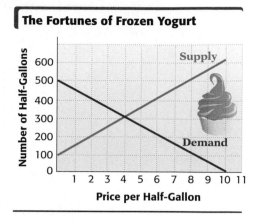

The Fortunes of Frozen Yogurt

76. La Bronda Jones compared the monthly payments she would incur for two types of mortgages: fixed rate and variable rate. Her observations led to the following graphs.

(a) For which years would the monthly payment be more for the fixed-rate mortgage than for the variable-rate mortgage?

(b) In what year would the payments be the same, and what would those payments be?

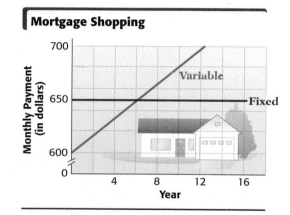

Mortgage Shopping

77. The graph shows network share (the percentage of TV sets in use) for the early evening news programs for three major broadcast networks from 1994 through 2004.

(a) Between what years did the ABC early evening news dominate?

(b) During what year did ABC's dominance end? Which network equaled ABC's share that year? What was that share?

(c) During what years did ABC and NBC have equal network share? What was the share for each of those years?

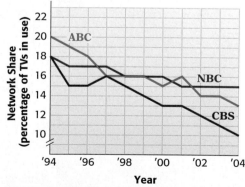

Who's Watching the Evening News?

Source: Nielsen Media Research.

(d) Find the first year on the graph in which two networks had equal share. Which networks were these? Write their share as an ordered pair of the form (year, share).

(e) Describe the general trend in viewership for the three major networks during the years shown.

78. The graph shows how the production of vinyl LPs, audiocassettes, and compact discs (CDs) changed over the final years of the 20th century.

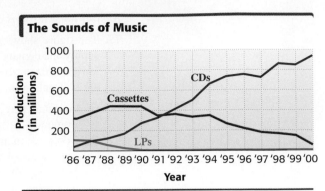

The Sounds of Music

Source: Recording Industry Association of America.

(a) In what year did cassette production and CD production reach an equal level? What was that level?

(b) Express the point of intersection of the graphs of LP production and CD production as an ordered pair of the form (year, production level).

(c) Between what years did cassette production first stabilize and remain fairly constant?

(d) Describe the trend in CD production from 1986 through 2000. If a straight line were used to approximate its graph, would the line have positive, negative, or 0 slope?

(e) If a straight line were used to approximate the graph of cassette production from 1990 through 2000, would the line have positive, negative, or 0 slope? Explain.

Use the graph in Figure 1 at the beginning of this section (repeated here) to work Exercises 79–82.

79. For which years during the period 2000–2004 were sales of digital cameras less than sales of conventional cameras?

80. Estimate the year in which sales for the two types of cameras were the same. About what was this sales figure?

81. If $x = 0$ represents 2000 and $x = 4$ represents 2004, sales (y) in millions of units can be modeled by the linear equations in the following system.

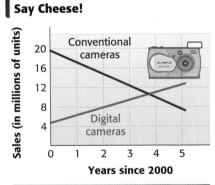

Say Cheese!

Source: Consumer Electronics Association.

$$2.5x + y = 19.4 \quad \text{Conventional cameras}$$
$$-1.7x + y = 4.4 \quad \text{Digital cameras}$$

Solve this system. Express values to the nearest tenth. Write the solution as an ordered pair of the form (year, sales).

82. Interpret your answer for Exercise 81. How does that answer compare with your estimate from Exercise 80?

Systems such as those in Exercises 83–86 can be solved by elimination. One way to do this is to let $p = \frac{1}{x}$ and $q = \frac{1}{y}$. Substitute, solve for p and q, and then find x and y. (For example, in Exercise 83, $\frac{3}{x} = 3 \cdot \frac{1}{x} = 3p$.) Use this method to solve each system.

83.
$$\frac{3}{x} + \frac{4}{y} = \frac{5}{2}$$
$$\frac{5}{x} - \frac{3}{y} = \frac{7}{4}$$

84.
$$\frac{2}{x} - \frac{5}{y} = \frac{3}{2}$$
$$\frac{4}{x} + \frac{1}{y} = \frac{4}{5}$$

85. $\dfrac{2}{x} + \dfrac{3}{y} = \dfrac{11}{2}$

    $-\dfrac{1}{x} + \dfrac{2}{y} = -1$

86. $\dfrac{4}{x} - \dfrac{9}{y} = -1$

    $-\dfrac{7}{x} + \dfrac{6}{y} = -\dfrac{3}{2}$

The next problems are "brain busters." Solve by any method. Assume that a and b represent nonzero constants.

87. $ax + by = c$

    $ax - 2by = c$

88. $ax + by = 2$

    $-ax + 2by = 1$

89. $2ax - y = 3$

    $y = 5ax$

90. $3ax + 2y = 1$

    $-ax + y = 2$

RELATING CONCEPTS (EXERCISES 91–94)

FOR INDIVIDUAL OR GROUP WORK

Work Exercises 91–94 in order, *to see the connections between systems of linear equations and the graphs of linear functions.*

91. Use elimination or substitution to solve the system

$$3x + y = 6 \quad (1)$$
$$-2x + 3y = 7. \quad (2)$$

92. For equation (1) in the system of Exercise 91, solve for y and rename it $f(x)$. What special kind of function is f?

93. For equation (2) in the system of Exercise 91, solve for y and rename it $g(x)$. What special kind of function is g?

94. Use the result of Exercise 91 to fill in the blanks with the appropriate responses:

Because the graphs of f and g are straight lines that are neither parallel nor coincide, they intersect in exactly _____ point. The coordinates of the point are (_____, _____). In function notation, this intersection is given by $f($_____$) = $ _____ and $g($_____$) = $ _____.

PREVIEW EXERCISES

Multiply both sides of each equation by the given number. See **Section 2.1.**

95. $2x - 3y + z = 5$ by 4

96. $-3x + 8y - z = 0$ by -3

Solve for z if x = 1 and y = −2. See **Section 2.1.**

97. $x + 2y + 3z = 9$

98. $-3x - y + z = 1$

By what number must the first equation be multiplied so that x is eliminated when the two equations are added? See **Section 4.1.**

99. $x + 2y - z = 0$

    $3x - 4y + 2z = 6$

100. $x - 2y + 5z = -7$

    $-2x - 3y + 4z = -14$

4.2 Systems of Linear Equations in Three Variables

A solution of an equation in three variables, such as

$$2x + 3y - z = 4, \qquad \text{Linear equation in three variables}$$

is called an **ordered triple** and is written (x, y, z). For example, the ordered triple $(0, 1, -1)$ is a solution of the preceding equation, because

$$2(0) + 3(1) - (-1) = 4$$

is a true statement. Verify that another solution of this equation is $(10, -3, 7)$.

We now extend the term *linear equation* to equations of the form

$$Ax + By + Cz + \cdots + Dw = K,$$

where not all the coefficients $A, B, C, \ldots, D$ equal 0. For example,

$$2x + 3y - 5z = 7 \quad \text{and} \quad x - 2y - z + 3u - 2w = 8$$

are linear equations, the first with three variables and the second with five.

OBJECTIVE 1 Understand the geometry of systems of three equations in three variables.
Consider the solution of a system such as

$$\begin{aligned} 4x + 8y + \ z &= 2 \\ x + 7y - 3z &= -14 \qquad \text{System of linear equations} \\ 2x - 3y + 2z &= 3. \qquad \text{in three variables} \end{aligned}$$

Theoretically, a system of this type can be solved by graphing. However, the graph of a linear equation with three variables is a *plane,* not a line. Since visualizing a plane requires three-dimensional graphing, the method of graphing is not practical with these systems. However, it does illustrate the number of solutions possible for such systems, as shown in Figure 7.

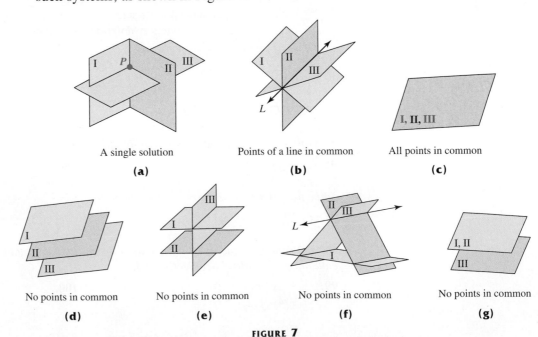

A single solution
(a)

Points of a line in common
(b)

All points in common
(c)

No points in common
(d)

No points in common
(e)

No points in common
(f)

No points in common
(g)

FIGURE 7

Figure 7 illustrates the following cases.

Graphs of Linear Systems in Three Variables

1. **The three planes may meet at a single, common point** that is the solution of the system. See Figure 7(a).

2. **The three planes may have the points of a line in common,** so that the infinite set of points that satisfy the equation of the line is the solution of the system. See Figure 7(b).

3. **The three planes may coincide,** so that the solution of the system is the set of all points on a plane. See Figure 7(c).

4. **The planes may have no points common to all three,** so that there is no solution of the system. See Figures 7(d)–(g).

OBJECTIVE 2 Solve linear systems (with three equations and three variables) by elimination. Since graphing to find the solution set of a system of three equations in three variables is impractical, these systems are solved with an extension of the elimination method from **Section 4.1,** summarized as follows.

Solving a Linear System in Three Variables

Step 1 **Eliminate a variable.** Use the elimination method to eliminate any variable from any two of the original equations. The result is an equation in two variables.

Step 2 **Eliminate the same variable again.** Eliminate the *same* variable from any *other* two equations. The result is an equation in the same two variables as in Step 1.

Step 3 **Eliminate a different variable and solve.** Use the elimination method to eliminate a second variable from the two equations in two variables that result from Steps 1 and 2. The result is an equation in one variable which gives the value of that variable.

Step 4 **Find a second value.** Substitute the value of the variable found in Step 3 into either of the equations in two variables to find the value of the second variable.

Step 5 **Find a third value.** Use the values of the two variables from Steps 3 and 4 to find the value of the third variable by substituting into an appropriate equation.

Step 6 **Check** the solution in all of the original equations. Then write the solution set.

EXAMPLE 1 Solving a System in Three Variables

Solve the system

$$
\begin{aligned}
4x + 8y + z &= 2 & (1) \\
x + 7y - 3z &= -14 & (2) \\
2x - 3y + 2z &= 3. & (3)
\end{aligned}
$$

$$4x + 8y + z = 2 \quad (1)$$
$$x + 7y - 3z = -14 \quad (2)$$
$$2x - 3y + 2z = 3 \quad (3)$$

The original system
of Example 1

Step 1 We must eliminate a variable from the sum of two equations. **The choice of which variable to eliminate is arbitrary.** Suppose we decide to eliminate z. We multiply equation (1) by 3 and then add the result to equation (2).

$$12x + 24y + 3z = 6 \qquad \text{Multiply each side of (1) by 3.}$$
$$\underline{x + 7y - 3z = -14} \qquad (2)$$
$$13x + 31y = -8 \qquad \text{Add.} \quad (4)$$

Step 2 Equation (4) has only two variables. To get another equation without z, we multiply equation (1) by -2 and add the result to equation (3). **It is essential at this point to eliminate the same variable, z.**

> Make sure equation (5) has the same variables as equation (4).

$$-8x - 16y - 2z = -4 \qquad \text{Multiply each side of (1) by } -2.$$
$$\underline{2x - 3y + 2z = 3} \qquad (3)$$
$$-6x - 19y = -1 \qquad \text{Add.} \quad (5)$$

Step 3 Now we solve the system of equations (4) and (5) for x and y. This step is possible only if the *same* variable was eliminated in Steps 1 and 2.

$$78x + 186y = -48 \qquad \text{Multiply each side of (4) by 6.}$$
$$\underline{-78x - 247y = -13} \qquad \text{Multiply each side of (5) by 13.}$$
$$-61y = -61 \qquad \text{Add.}$$
$$y = 1$$

Step 4 We substitute 1 for y in either equation (4) or equation (5) to find x.

$$-6x - 19y = -1 \qquad (5)$$
$$-6x - 19(1) = -1 \qquad \text{Let } y = 1.$$
$$-6x - 19 = -1 \qquad \text{Multiply.}$$
$$-6x = 18 \qquad \text{Add 19.}$$
$$x = -3 \qquad \text{Divide by } -6.$$

Step 5 We substitute -3 for x and 1 for y in any one of the three original equations to find z. Choosing (1) gives

$$4x + 8y + z = 2 \qquad (1)$$
$$4(-3) + 8(1) + z = 2 \qquad \text{Let } x = -3 \text{ and } y = 1.$$
$$-4 + z = 2$$
$$z = 6.$$

> Write the values of x, y, and z in the correct order.

Step 6 It appears that the ordered triple $(-3, 1, 6)$ is the only solution of the system. We must check that this solution satisfies all three equations of the system.

Check:
$$4x + 8y + z = 2 \qquad (1)$$
$$4(-3) + 8(1) + 6 = 2 \qquad ?$$
$$-12 + 8 + 6 = 2 \qquad ?$$
$$2 = 2 \qquad \text{True}$$

Because $(-3, 1, 6)$ also satisfies equations (2) and (3), the solution set is $\{(-3, 1, 6)\}$.

✔ **Now Try Exercise 3.**

OBJECTIVE 3 Solve linear systems (with three equations and three variables) in which some of the equations have missing terms. When this situation occurs, one elimination step can be omitted.

EXAMPLE 2 Solving a System of Equations with Missing Terms

Solve the system

$$
\begin{aligned}
6x - 12y &= -5 \quad &(1)\\
8y + z &= 0 \quad &(2)\\
9x - z &= 12. \quad &(3)
\end{aligned}
$$

Since equation (3) is missing the variable y, use equations (1) and (2) to eliminate y.

$$
\begin{array}{ll}
12x - 24y = -10 & \text{Multiply each side of (1) by 2.}\\
\underline{ 24y + 3z = 0} & \text{Multiply each side of (2) by 3.}\\
12x + 3z = -10 & \text{Add. \quad (4)}
\end{array}
$$

Use this result, together with equation (3), to eliminate z. Multiply equation (3) by 3.

$$
\begin{array}{ll}
27x - 3z = 36 & \text{Multiply each side of (3) by 3.}\\
\underline{12x + 3z = -10} & (4)\\
39x = 26 & \text{Add.}\\
x = \dfrac{26}{39} & \text{Divide by 39.}\\
x = \dfrac{2}{3} & \text{Lowest terms}
\end{array}
$$

Substituting this value for x in equation (3) gives

$$
\begin{array}{ll}
9x - z = 12 & (3)\\
9\left(\dfrac{2}{3}\right) - z = 12 & \text{Let } x = \tfrac{2}{3}.\\
6 - z = 12 & \text{Multiply.}\\
z = -6. & \text{Subtract 6; multiply by } -1.
\end{array}
$$

Substituting -6 for z in equation (2) gives

$$
\begin{array}{ll}
8y + z = 0 & (2)\\
8y - 6 = 0 & \text{Let } z = -6.\\
8y = 6 & \text{Add 6.}\\
y = \dfrac{6}{8} & \text{Divide by 8.}\\
y = \dfrac{3}{4}. & \text{Lowest terms}
\end{array}
$$

Check in each of the original equations of the system to verify that the solution set is $\left\{\left(\tfrac{2}{3}, \tfrac{3}{4}, -6\right)\right\}$.

✔ **Now Try Exercise 21.**

OBJECTIVE 4 Solve special systems. Linear systems with three variables may be inconsistent or may include dependent equations.

EXAMPLE 3 Solving an Inconsistent System with Three Variables

Solve the system

$$2x - 4y + 6z = 5 \qquad (1)$$
$$-x + 3y - 2z = -1 \qquad (2)$$
$$x - 2y + 3z = 1. \qquad (3)$$

Eliminate x by adding equations (2) and (3) to get the equation

$$y + z = 0.$$

Eliminate x again, using equations (1) and (3).

$$\begin{array}{ll} -2x + 4y - 6z = -2 & \text{Multiply each side of (3) by } -2. \\ \underline{2x - 4y + 6z = 5} & (1) \\ 0 = 3 & \text{False} \end{array}$$

The resulting false statement indicates that equations (1) and (3) have no common solution. Thus, the system is inconsistent and the solution set is $\emptyset$. The graph of this system would show the two planes parallel to one another.

✔ **Now Try Exercise 27.**

▶ **NOTE** If you get a false statement when adding as in Example 3, you do not need to go any further with the solution. Since two of the three planes are parallel, it is not possible for the three planes to have any points in common.

EXAMPLE 4 Solving a System of Dependent Equations with Three Variables

Solve the system

$$2x - 3y + 4z = 8 \qquad (1)$$
$$-x + \frac{3}{2}y - 2z = -4 \qquad (2)$$
$$6x - 9y + 12z = 24. \qquad (3)$$

Multiplying each side of equation (1) by 3 gives equation (3). Multiplying each side of equation (2) by -6 also gives equation (3). Because of this, the equations are dependent. All three equations have the same graph, as illustrated in Figure 7(c). The solution set is written

$$\{(x, y, z) \mid 2x - 3y + 4z = 8\}. \qquad \text{Set-builder notation}$$

Although any one of the three equations could be used to write the solution set, we use the equation in standard form with coefficients that are integers with no common factor (except 1), as we did in **Section 4.1.**

✔ **Now Try Exercise 33.**

EXAMPLE 5 Solving Another Special System

Solve the system

$$2x - y + 3z = 6 \qquad (1)$$
$$x - \frac{1}{2}y + \frac{3}{2}z = 3 \qquad (2)$$
$$4x - 2y + 6z = 1. \qquad (3)$$

Multiplying each side of equation (2) by 2 gives equation (1), so these two equations are dependent. Equations (1) and (3) are not equivalent, however. Multiplying equation (3) by $\frac{1}{2}$ does *not* give equation (1). Instead, we obtain two equations with the same coefficients, but with different constant terms. The graphs of these two equations have no points in common (that is, the planes are parallel). Thus, the system is inconsistent and the solution set is $\emptyset$, as illustrated in Figure 7(g).

✔ **Now Try Exercise 37.**

4.2 EXERCISES

Complete solution available on Video Lectures on CD/DVD

Now Try Exercise

1. *Concept Check* The two equations $\begin{array}{l} x + y + z = 6 \\ 2x - y + z = 3 \end{array}$ have a common solution of $(1, 2, 3)$. Which equation would complete a system of three linear equations in three variables having solution set $\{(1, 2, 3)\}$?

A. $3x + 2y - z = 1$ **B.** $3x + 2y - z = 4$

C. $3x + 2y - z = 5$ **D.** $3x + 2y - z = 6$

2. Explain what the following statement means: $\{(-1, 2, 3)\}$ is the solution set of the system

$$2x + y + z = 3$$
$$3x - y + z = -2$$
$$4x - y + 2z = 0.$$

Solve each system of equations. See Example 1.

3. $2x - 5y + 3z = -1$
$x + 4y - 2z = 9$
$x - 2y - 4z = -5$

4. $x + 3y - 6z = 7$
$2x - y + z = 1$
$x + 2y + 2z = -1$

5. $3x + 2y + z = 8$
$2x - 3y + 2z = -16$
$x + 4y - z = 20$

6. $-3x + y - z = -10$
$-4x + 2y + 3z = -1$
$2x + 3y - 2z = -5$

7. $2x + 5y + 2z = 0$
$4x - 7y - 3z = 1$
$3x - 8y - 2z = -6$

8. $5x - 2y + 3z = -9$
$4x + 3y + 5z = 4$
$2x + 4y - 2z = 14$

9. $x + 2y + z = 4$
$2x + y - z = -1$
$x - y - z = -2$

10. $x - 2y + 5z = -7$
$-2x - 3y + 4z = -14$
$-3x + 5y - z = -7$

11. $\frac{1}{3}x + \frac{1}{6}y - \frac{2}{3}z = -1$
$-\frac{3}{4}x - \frac{1}{3}y - \frac{1}{4}z = 3$
$\frac{1}{2}x + \frac{3}{2}y + \frac{3}{4}z = 21$

12. $\frac{2}{3}x - \frac{1}{4}y + \frac{5}{8}z = 0$
$\frac{1}{5}x + \frac{2}{3}y - \frac{1}{4}z = -7$
$-\frac{3}{5}x + \frac{4}{3}y - \frac{7}{8}z = -5$

13. $-x + 2y + 6z = 2$
$3x + 2y + 6z = 6$
$x + 4y - 3z = 1$

14. $2x + y + 2z = 1$
$x + 2y + z = 2$
$x - y - z = 0$

15. $x + y - z = -2$
$2x - y + z = -5$
$-x + 2y - 3z = -4$

16. $x + 2y + 3z = 1$
$-x - y + 3z = 2$
$-6x + y + z = -2$

Solve each system of equations. See Example 2.

17. $2x - 3y + 2z = -1$
$x + 2y + z = 17$
$2y - z = 7$

18. $2x - y + 3z = 6$
$x + 2y - z = 8$
$2y + z = 1$

19. $4x + 2y - 3z = 6$
$x - 4y + z = -4$
$-x + 2z = 2$

20. $2x + 3y - 4z = 4$
$x - 6y + z = -16$
$-x + 3z = 8$

21. $2x + y = 6$
$3y - 2z = -4$
$3x - 5z = -7$

22. $4x - 8y = -7$
$4y + z = 7$
$-8x + z = -4$

23. $-5x + 2y + z = 5$
$-3x - 2y - z = 3$
$-x + 6y = 1$

24. $x + y - z = 0$
$2y - z = 1$
$2x + 3y - 4z = -4$

25. $4x - z = -6$
$\dfrac{3}{5}y + \dfrac{1}{2}z = 0$
$\dfrac{1}{3}x + \dfrac{2}{3}z = -5$

26. $5x - 2z = 8$
$4y + 3z = -9$
$\dfrac{1}{2}x + \dfrac{2}{3}y = -1$

Solve each system of equations. If the system is inconsistent or has dependent equations, say so. See Examples 1, 3, 4, and 5.

27. $2x + 2y - 6z = 5$
$-3x + y - z = -2$
$-x - y + 3z = 4$

28. $-2x + 5y + z = -3$
$5x + 14y - z = -11$
$7x + 9y - 2z = -5$

29. $-5x + 5y - 20z = -40$
$x - y + 4z = 8$
$3x - 3y + 12z = 24$

30. $x + 4y - z = 3$
$-2x - 8y + 2z = -6$
$3x + 12y - 3z = 9$

31. $x + 5y - 2z = -1$
$-2x + 8y + z = -4$
$3x - y + 5z = 19$

32. $x + 3y + z = 2$
$4x + y + 2z = -4$
$5x + 2y + 3z = -2$

33. $2x + y - z = 6$
$4x + 2y - 2z = 12$
$-x - \dfrac{1}{2}y + \dfrac{1}{2}z = -3$

34. $2x - 8y + 2z = -10$
$-x + 4y - z = 5$
$\dfrac{1}{8}x - \dfrac{1}{2}y + \dfrac{1}{8}z = -\dfrac{5}{8}$

35. $x + y - 2z = 0$
$3x - y + z = 0$
$4x + 2y - z = 0$

36. $2x + 3y - z = 0$
$x - 4y + 2z = 0$
$3x - 5y - z = 0$

37. $x - 2y + \dfrac{1}{3}z = 4$
$3x - 6y + z = 12$
$-6x + 12y - 2z = -3$

38. $4x + y - 2z = 3$
$x + \dfrac{1}{4}y - \dfrac{1}{2}z = \dfrac{3}{4}$
$2x + \dfrac{1}{2}y - z = 1$

The next problems are "brain busters." Extend the method of this section to solve each system. Express the solution in the form (x, y, z, w).

39.
$$x + y + z - w = 5$$
$$2x + y - z + w = 3$$
$$x - 2y + 3z + w = 18$$
$$-x - y + z + 2w = 8$$

40.
$$3x + y - z + 2w = 9$$
$$x + y + 2z - w = 10$$
$$x - y - z + 3w = -2$$
$$-x + y - z + w = -6$$

41.
$$3x + y - z + w = -3$$
$$2x + 4y + z - w = -7$$
$$-2x + 3y - 5z + w = 3$$
$$5x + 4y - 5z + 2w = -7$$

42.
$$x - 3y + 7z + w = 11$$
$$2x + 4y + 6z - 3w = -3$$
$$3x + 2y + z + 2w = 19$$
$$4x + y - 3z + w = 22$$

PREVIEW EXERCISES

Solve each problem. See ***Sections 2.3 and 2.4.***

43. The perimeter of a triangle is 323 in. The shortest side measures five-sixths the length of the longest side, and the medium side measures 17 in. less than the longest side. Find the lengths of the sides of the triangle.

44. The sum of the three angles of a triangle is 180°. The largest angle is twice the measure of the smallest, and the third angle measures 10° less than the largest. Find the measures of the three angles.

45. The sum of three numbers is 16. The greatest number is −3 times the least, while the middle number is four less than the greatest. Find the three numbers.

46. Margaret Maggio has a collection of pennies, dimes, and quarters. The number of dimes is one less than twice the number of pennies. If there are 27 coins in all worth a total of $4.20, how many of each denomination of coin is in the collection?

4.3 Applications of Systems of Linear Equations

OBJECTIVES

1 Solve geometry problems by using two variables.

2 Solve money problems by using two variables.

3 Solve mixture problems by using two variables.

4 Solve distance–rate–time problems by using two variables.

5 Solve problems with three variables by using a system of three equations.

Many applied problems involve more than one unknown quantity. Although some problems with two unknowns can be solved by using just one variable, it is often easier to use two variables and a system of equations. Problems that can be solved by writing a system of equations have been of interest historically. The following problem appeared in a Hindu work that dates back to about 850 A.D. (See Exercise 35.)

The mixed price of 9 citrons (a lemonlike fruit shown in the photo) and 7 fragrant wood apples is 107; again, the mixed price of 7 citrons and 9 fragrant wood apples is 101. O you arithmetician, tell me quickly the price of a citron and the price of a wood apple here, having distinctly separated those prices well.

The following steps, based on the six-step problem-solving method of **Section 2.3,** give a strategy for solving applied problems by using more than one variable.

Solving an Applied Problem by Writing a System of Equations

Step 1 **Read** the problem carefully until you understand what is given and what is to be found.

Step 2 **Assign variables** to represent the unknown values, using diagrams or tables as needed. *Write down* what each variable represents.

Step 3 **Write a system of equations** that relates the unknowns.

Step 4 **Solve** the system of equations.

Step 5 **State the answer** to the problem. Does it seem reasonable?

Step 6 **Check** the answer in the words of the original problem.

OBJECTIVE 1 Solve geometry problems by using two variables.

EXAMPLE 1 Finding the Dimensions of a Soccer Field

Unlike a football field, whose dimensions cannot vary, a rectangular soccer field may have a width between 50 and 100 yd and a length between 50 and 100 yd. Suppose that one particular soccer field has a perimeter of 320 yd. Its length measures 40 yd more than its width. What are the dimensions of this field? (*Source: Microsoft Encarta Encyclopedia.*)

Step 1 **Read** the problem again. We are asked to find the dimensions of the field.

Step 2 **Assign variables.** Let L = the length and W = the width. See Figure 8.

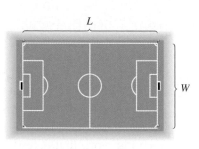

FIGURE 8

Step 3 **Write a system of equations.** Because the perimeter is 320 yd, we find one equation by using the perimeter formula:

$$2L + 2W = 320.$$

Because the length is 40 yd more than the width, we have

$$L = W + 40.$$

The system is

$$2L + 2W = 320 \qquad (1)$$
$$L = W + 40. \qquad (2)$$

Step 4 **Solve** the system of equations. Since equation (2), $L = W + 40$, is solved for L, we use the substitution method. We substitute $W + 40$ for L in equation (1) and solve for W.

$$2L + 2W = 320 \quad \text{(1)}$$
$$2(W + 40) + 2W = 320 \quad \text{Let } L = W + 40.$$
$$2W + 80 + 2W = 320 \quad \text{Distributive property}$$
$$4W + 80 = 320 \quad \text{Combine like terms.}$$
$$4W = 240 \quad \text{Subtract 80.}$$
$$W = 60 \quad \text{Divide by 4.}$$

> Be sure to use parentheses around $W + 40$.

Let $W = 60$ in the equation $L = W + 40$ to find L.

$$L = 60 + 40 = 100$$

Step 5 **State the answer.** The length is 100 yd, and the width is 60 yd. This is reasonable, since both dimensions are within the ranges given in the problem.

Step 6 **Check.** The perimeter of this soccer field is

$$2(100) + 2(60) = 320 \text{ yd,}$$

and the length, 100 yd, is 40 yd more than the width, since

$$100 - 40 = 60.$$

The answer is correct.

✔ **Now Try Exercise 3.**

OBJECTIVE 2 Solve money problems by using two variables.

EXAMPLE 2 **Solving a Problem about Ticket Prices**

It was reported in March 2004 that during the National Hockey League and National Basketball Association seasons, two hockey tickets and one basketball ticket purchased at their average prices would have cost $126.77. One hockey ticket and two basketball tickets would have cost $128.86. What were the average ticket prices for the two sports? (*Source:* Team Marketing Report, Chicago.)

Step 1 **Read** the problem again. There are two unknowns.

Step 2 **Assign variables.**

Let h = the average price for a hockey ticket

and b = the average price for a basketball ticket.

Step 3 **Write a system of equations.** Because two hockey tickets and one basketball ticket cost a total of $126.77, one equation for the system is

$$2h + b = 126.77.$$

By similar reasoning, the second equation is

$$h + 2b = 128.86.$$

Therefore, the system is

$$2h + \ b = 126.77 \quad \text{(1)}$$
$$h + 2b = 128.86. \quad \text{(2)}$$

Step 4 **Solve** the system. To eliminate h, multiply equation (2), $h + 2b = 128.86$, by -2 and add.

$$
\begin{array}{ll}
2h + b = 126.77 & (1) \\
\underline{-2h - 4b = -257.72} & \text{Multiply each side of (2) by } -2. \\
 -3b = -130.95 & \text{Add.} \\
b = 43.65 & \text{Divide by } -3.
\end{array}
$$

To find the value of h, let $b = 43.65$ in equation (2).

$$
\begin{array}{ll}
h + 2b = 128.86 & (2) \\
h + 2(43.65) = 128.86 & \text{Let } b = 43.65. \\
h + 87.30 = 128.86 & \text{Multiply.} \\
h = 41.56 & \text{Subtract 87.30.}
\end{array}
$$

Step 5 **State the answer.** The average price for one basketball ticket was $43.65. For one hockey ticket, the average price was $41.56.

Step 6 **Check** that these values satisfy the conditions stated in the problem.

✔ **Now Try Exercise 11.**

OBJECTIVE 3 **Solve mixture problems by using two variables.** We solved mixture problems in **Section 2.3** by using one variable. Many mixture problems can also be solved by using more than one variable and a system of equations.

EXAMPLE 3 **Solving a Mixture Problem**

How many ounces each of 5% hydrochloric acid and 20% hydrochloric acid must be combined to get 10 oz of solution that is 12.5% hydrochloric acid?

Step 1 **Read** the problem. Two solutions of different strengths are being mixed together to get a specific amount of a solution with an "in-between" strength.

Step 2 **Assign variables.**

Let x = the number of ounces of 5% solution

and y = the number of ounces of 20% solution.

Use a table to summarize the information from the problem. We multiply the amount of each solution (given in the first column) by its concentration of acid (given in the second column) to get the amount of acid in that solution (given in the third column).

Ounces of Solution	Percent (as a decimal)	Ounces of Pure Acid
x	5% = 0.05	0.05x
y	20% = 0.20	0.20y
10	12.5% = 0.125	(0.125)10

Gives equation (1)   Gives equation (2)

Figure 9 illustrates what is happening in the problem.

FIGURE 9

Step 3 **Write a system of equations.** When the x ounces of 5% solution and the y ounces of 20% solution are combined, the total number of ounces is 10, so

$$x + y = 10. \quad (1)$$

The number of ounces of acid in the 5% solution $(0.05x)$ plus the number of ounces of acid in the 20% solution $(0.20y)$ should equal the total number of ounces of acid in the mixture, which is $(0.125)10$, or 1.25.

$$0.05x + 0.20y = 1.25 \quad (2)$$

Notice that these equations can be quickly determined by reading down the table or using the labels in Figure 9.

Step 4 **Solve** the system of equations (1) and (2). Eliminate x by first multiplying equation (2) by 100 to clear it of decimals and then multiplying equation (1) by -5.

$$
\begin{array}{ll}
5x + 20y = 125 & \text{Multiply each side of (2) by 100.} \\
\underline{-5x - 5y = -50} & \text{Multiply each side of (1) by } -5. \\
15y = 75 & \text{Add.} \\
y = 5 & \text{Divide by 15.}
\end{array}
$$

Because $y = 5$ and $x + y = 10$, x is also 5.

Step 5 **State the answer.** The desired mixture will require 5 oz of the 5% solution and 5 oz of the 20% solution.

Step 6 **Check.**

Total amount of solution: $x + y = 5 \text{ oz} + 5 \text{ oz} = 10 \text{ oz},$ as required.

Total amount of acid: 5% of 5 oz + 20% of 5 oz $= 0.05(5) + 0.20(5)$
$$= 1.25 \text{ oz}$$

Percent of acid in solution:

$$
\begin{array}{l}
\text{Total acid} \longrightarrow \dfrac{1.25}{10} = 0.125, \quad \text{or} \quad 12.5\%, \quad \text{as required.} \\
\text{Total solution} \longrightarrow
\end{array}
$$

✔ **Now Try Exercise 17.**

OBJECTIVE 4 Solve distance–rate–time problems by using two variables. Motion problems require the distance formula $d = rt,$ where d is distance, r is rate (or speed), and t is time. These applications often lead to systems of equations.

EXAMPLE 4 Solving a Motion Problem

A car travels 250 km in the same time that a truck travels 225 km. If the speed of the car is 8 km per hr faster than the speed of the truck, find both speeds.

Step 1 **Read** the problem again. Given the distances traveled, we need to find the speed of each vehicle.

Step 2 **Assign variables.**

$$\text{Let } x = \text{the speed of the car,}$$
$$\text{and } y = \text{the speed of the truck.}$$

As in Example 3, a table helps organize the information. Fill in the distance for each vehicle, and use the assigned variables for the unknown speeds (rates).

	d	r	t
Car	250	x	$\frac{250}{x}$
Truck	225	y	$\frac{225}{y}$

The times must be equal.

To get the expressions for time, we solved the distance formula $d = rt$ for t. Since $\frac{d}{r} = t$, the two times can be written as $\frac{250}{x}$ and $\frac{225}{y}$, respectively.

Step 3 **Write a system of equations.** The problem states that the car travels 8 km per hr faster than the truck. Since the two speeds are x and y,

$$x = y + 8. \quad (1)$$

Both vehicles travel for the same time, so from the table,

$$\text{Time for car} \longrightarrow \frac{250}{x} = \frac{225}{y}. \longleftarrow \text{Time for truck}$$

This is not a linear equation. However, multiplying each side by xy gives

$$250y = 225x, \quad (2)$$

which is linear. The system is

$$x = y + 8$$
$$250y = 225x.$$

Step 4 **Solve** the system by substitution. Replace x with $y + 8$ in equation (2).

$250y = 225x$	(2)
$250y = 225(y + 8)$	Let $x = y + 8$.
$250y = 225y + 1800$	Distributive property
$25y = 1800$	Subtract $225y$.
$y = 72$	Divide by 25.

Be sure to use parentheses around $y + 8$.

Because $x = y + 8$, the value of x is $72 + 8 = 80$.

Step 5 **State the answer.** The car's speed is 80 km per hr, and the truck's speed is 72 km per hr.

Step 6 **Check.**

$$Car: \quad t = \frac{d}{r} = \frac{250}{80} = 3.125$$

$$Truck: \quad t = \frac{d}{r} = \frac{225}{72} = 3.125$$

Times are equal.

Since $80 - 72 = 8$, the conditions of the problem are satisfied.

✔ **Now Try Exercise 27.**

OBJECTIVE 5 Solve problems with three variables by using a system of three equations. To solve such problems, we extend the method used for two unknowns. Since three variables are used, three equations are necessary to find a solution.

EXAMPLE 5 Solving a Problem Involving Prices

At Panera Bread, a loaf of honey wheat bread costs $2.59, a loaf of sunflower bread costs $2.99, and a loaf of French bread costs $3.29. On a recent day, three times as many loaves of honey wheat were sold as sunflower. The number of loaves of French bread sold was 5 less than the number of loaves of honey wheat sold. Total receipts for these breads were $66.07. How many loaves of each type of bread were sold? (*Source:* Panera Bread menu.)

Step 1 **Read** the problem again. There are three unknowns in this problem.

Step 2 **Assign variables** to represent the three unknowns.

Let $x =$ the number of loaves of honey wheat bread,

$y =$ the number of loaves of sunflower bread,

and $z =$ the number of loaves of French bread.

Step 3 **Write a system of three equations,** using the information in the problem. Since three times as many loaves of honey wheat were sold as sunflower,

$$x = 3y, \quad \text{or} \quad x - 3y = 0. \quad (1)$$

Also,

Number of loaves of French bread	equals	5 less than the number of loaves of honey wheat.
↓	↓	↓
z	$=$	$x - 5,$

so $\qquad\qquad x - z = 5. \quad (2)$

Multiplying the cost of a loaf of each kind of bread by the number of loaves of that kind sold and adding gives the total receipts.

$$2.59x + 2.99y + 3.29z = 66.07$$

Multiply each side of this equation by 100 to clear it of decimals.

$$259x + 299y + 329z = 6607 \quad (3)$$

Step 4 **Solve** the system of three equations,

$$x - 3y = 0 \qquad (1)$$
$$x - z = 5 \qquad (2)$$
$$259x + 299y + 329z = 6607, \qquad (3)$$

using the method shown in **Section 4.2** to find

$$x = 12, \quad y = 4, \quad \text{and} \quad z = 7.$$

Step 5 **State the answer.** The solution is $(12, 4, 7)$, so 12 loaves of honey wheat, 4 loaves of sunflower, and 7 loaves of French bread were sold.

Step 6 **Check.** Since $12 = 3 \cdot 4$, the number of loaves of honey wheat is three times the number of loaves of sunflower. Also, $12 - 7 = 5$, so the number of loaves of French bread is 5 less than the number of loaves of honey wheat. Multiply the appropriate cost per loaf by the number of loaves sold and add the results to check that total receipts were \$66.07.

✔ **Now Try Exercise 45.**

EXAMPLE 6 Solving a Business Production Problem

A company produces three color television sets: models X, Y, and Z. Each model X set requires 2 hr of electronics work, 2 hr of assembly time, and 1 hr of finishing time. Each model Y requires 1, 3, and 1 hr of electronics, assembly, and finishing time, respectively. Each model Z requires 3, 2, and 2 hr of the same work, respectively. There are 100 hr available for electronics, 100 hr available for assembly, and 65 hr available for finishing per week. How many of each model should be produced each week if all available time must be used?

Step 1 **Read** the problem again. There are three unknowns.

Step 2 **Assign variables.**

Let $x =$ the number of model X produced per week,

$y =$ the number of model Y produced per week,

and $z =$ the number of model Z produced per week.

We organize the information in a table.

	Each Model X	Each Model Y	Each Model Z	Totals
Hours of Electronics Work	2	1	3	100
Hours of Assembly Time	2	3	2	100
Hours of Finishing Time	1	1	2	65

Step 3 **Write a system of three equations.** The x model X sets require $2x$ hours of electronics, the y model Y sets require $1y$ (or y) hours of electronics, and the

z model Z sets require $3z$ hours of electronics. Since 100 hr are available for electronics,

$$2x + y + 3z = 100. \quad (1)$$

Similarly, from the fact that 100 hr are available for assembly,

$$2x + 3y + 2z = 100, \quad (2)$$

and the fact that 65 hr are available for finishing leads to the equation

$$x + y + 2z = 65. \quad (3)$$

Notice that by reading across the table, we can easily determine the coefficients and constants in the equations of the system.

Step 4 **Solve** the system of equations (1), (2), and (3), namely,

$$2x + y + 3z = 100$$
$$2x + 3y + 2z = 100$$
$$x + y + 2z = 65,$$

to find $x = 15$, $y = 10$, and $z = 20$.

Step 5 **State the answer.** The company should produce 15 model X, 10 model Y, and 20 model Z sets per week.

Step 6 **Check** that these values satisfy the conditions of the problem.

✔ **Now Try Exercise 47.**

4.3 EXERCISES

🌐 *Complete solution available on Video Lectures on CD/DVD*

Now Try Exercise

Solve each problem. See Example 1.

1. During the 2005 Major League Baseball season, the Chicago White Sox played 162 games. They won 36 more games than they lost. What was their win–loss record that year?

2. Refer to Exercise 1. During the same 162-game season, the Kansas City Royals lost 50 more games than they won. What was the team's win–loss record?

2005 MLB Final Standings American League Central

Team	W	L
Chicago	—	—
Cleveland	93	69
Minnesota	83	79
Detroit	71	91
Kansas City	—	—

Source: www.mlb.com

🌐 3. Venus and Serena measured a tennis court and found that it was 42 ft longer than it was wide and had a perimeter of 228 ft. What were the length and the width of the tennis court?

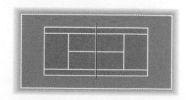

4. Jason and Vince found that the width of their basketball court was 44 ft less than the length. If the perimeter was 288 ft, what were the length and the width of their court?

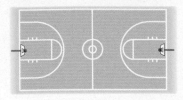

5. The two biggest Fortune 500 companies in 2005 were Wal-Mart and ExxonMobil. ExxonMobil's revenue was $24 billion more than that of Wal-Mart. Total revenue for the two companies was $656 billion. What was the revenue for each company? (*Source:* Fortune 500.)

6. In 2004, U.S. exports to Canada were $79,388 million more than exports to Mexico. Together, exports to these two countries totaled $300,938 million. How much were exports to each country? (*Source:* U.S. Census Bureau.)

In Exercises 7 and 8, find the measures of the angles marked x and y. Remember that (1) the sum of the measures of the angles of a triangle is 180°, (2) supplementary angles have a sum of 180°, and (3) vertical angles have equal measures.

7.

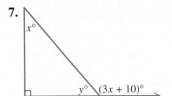

8.

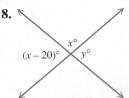

The Fan Cost Index (FCI) represents the cost of four average-price tickets, four small soft drinks, two small beers, four hot dogs, parking for one car, two game programs, and two souvenir caps to a sporting event. (Source: www.teammarketing.com)

Use the concept of FCI in Exercises 9 and 10. See Example 2.

9. For the 2005–2006 season, the FCI prices for the National Hockey League and the National Basketball Association totaled $514.69. The hockey FCI was $20.05 less than that of basketball. What were the FCIs for these sports?

10. In 2005, the FCI prices for Major League Baseball and the National Football League totaled $501.01. The football FCI was $158.63 more than that of baseball. What were the FCIs for these sports?

Solve each problem. See Example 2.

11. Andrew McGinnis works at Arby's. During one particular day, he sold 15 Regular Roast Beef sandwiches and 10 Large Roast Beef sandwiches, totaling $77.75. Another day, he sold 30 Regular Roast Beef sandwiches and 5 Large Roast Beef sandwiches, totaling $92.65. How much did each type of sandwich cost? (*Source:* Arby's menu.)

12. London and New York are among the most expensive cities worldwide for business travelers. On the basis of average costs per day for each city (which include the costs of business-class lodging and three meals), 2 days in London and 3 days in New York costs $2099, while 4 days in London and 2 days in New York costs $2586. What is the average cost per day for each city? (*Source:* Runzheimer International.)

Concept Check *The formulas p = br (percentage = base × rate) and I = prt (simple interest = principal × rate × time) are used in the applications in Exercises 17–24. To prepare to use these formulas, answer the questions in Exercises 13 and 14.*

13. If a container of liquid contains 60 oz of solution, what is the number of ounces of pure acid if the given solution contains the following acid concentrations?

   (a) 10%   **(b)** 25%   **(c)** 40%   **(d)** 50%

14. If $5000 is invested in an account paying simple annual interest, how much interest will be earned during the first year at the following rates?

   (a) 2%   **(b)** 3%   **(c)** 4%   **(d)** 3.5%

15. *Concept Check*   If a pound of turkey costs $1.29, give an expression for the cost of *x* pounds.

16. *Concept Check*   If a ticket to the movie *Madagascar* costs $8 and *y* tickets are sold, give an expression for the amount collected from the sale.

Solve each problem. See Example 3.

17. How many gallons each of 25% alcohol and 35% alcohol should be mixed to get 20 gal of 32% alcohol?

Gallons of Solution	Percent (as a decimal)	Gallons of Pure Alcohol
x	25% = 0.25	
y	35% = 0.35	
20	32% =	

18. How many liters each of 15% acid and 33% acid should be mixed to get 120 L of 21% acid?

Liters of Solution	Percent (as a decimal)	Liters of Pure Acid
x	15% = 0.15	
y	33% =	
120	21% =	

19. Pure acid is to be added to a 10% acid solution to obtain 54 L of a 20% acid solution. What amounts of each should be used?

20. A truck radiator holds 36 L of fluid. How much pure antifreeze must be added to a mixture that is 4% antifreeze to fill the radiator with a mixture that is 20% antifreeze?

21. A party mix is made by adding nuts that sell for $2.50 per kg to a cereal mixture that sells for $1 per kg. How much of each should be added to get 30 kg of a mix that will sell for $1.70 per kg?

	Number of Kilograms	Price per Kilogram	Value
Nuts	x	2.50	
Cereal	y	1.00	
Mixture		1.70	

22. A popular fruit drink is made by mixing fruit juices. Such a drink with 50% juice is to be mixed with another drink that is 30% juice to get 200 L of a drink that is 45% juice. How much of each should be used?

	Liters of Drink	Percent (as a decimal)	Liters of Pure Juice
50% Juice	x	0.50	
30% Juice	y	0.30	
Mixture		0.45	

23. A total of $3000 is invested, part at 2% simple interest and part at 4%. If the total annual return from the two investments is $100, how much is invested at each rate?

Principal	Rate (as a decimal)	Interest
x	0.02	0.02x
y	0.04	0.04y
3000		100

24. An investor will invest a total of $15,000 in two accounts, one paying 4% annual simple interest and the other 3%. If he wants to earn $550 annual interest, how much should he invest at each rate?

Principal	Rate (as a decimal)	Interest
x	0.04	
y	0.03	
15,000		

Concept Check *The formula d = rt (distance = rate × time) is used in the applications in Exercises 27–30. To prepare to use this formula, work Exercises 25 and 26.*

25. If the speed of a boat in still water is 10 mph, and the speed of the current of a river is x mph, what is the speed of the boat

Upstream (against the current)

Downstream (with the current)

(a) going upstream (that is, against the current, which slows the boat down);

(b) going downstream (that is, with the current, which speeds the boat up)?

26. If the speed of a killer whale is 25 mph and the whale swims for y hours, give an expression for the number of miles the whale travels.

Solve each problem. See Example 4.

27. A train travels 150 km in the same time that a plane covers 400 km. If the speed of the plane is 20 km per hr less than 3 times the speed of the train, find both speeds.

	r	t	d
Train	x		150
Plane	y		400

28. A freight train and an express train leave towns 390 km apart, traveling toward one another. The freight train travels 30 km per hr slower than the express train. They pass one another 3 hr later. What are their speeds?

	r	t	d
Freight Train	x	3	
Express Train	y	3	

29. In his motorboat, Bill Ruhberg travels upstream at top speed to his favorite fishing spot, a distance of 36 mi, in 2 hr. Returning, he finds that the trip downstream, still at top speed, takes only 1.5 hr. Find the speed of Bill's boat and the speed of the current. Let x = the speed of the boat and y = the speed of the current.

	r	t	d
Upstream	x − y	2	
Downstream	x + y		

30. Traveling for 3 hr into a steady head wind, a plane flies 1650 mi. The pilot determines that flying *with* the same wind for 2 hr, he could make a trip of 1300 mi. Find the speed of the plane and the speed of the wind.

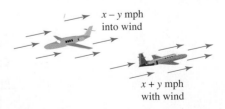

x − y mph into wind

x + y mph with wind

Solve each problem by using two variables. See Examples 1–4.

31. (See the Chapter Introduction.) How many pounds of candy that sells for $0.75 per lb must be mixed with candy that sells for $1.25 per lb to obtain 9 lb of a mixture that should sell for $0.96 per lb?

32. At age 54, rock icon Bruce Springsteen generated the most revenue on the North American concert circuit in 2003. Springsteen and second-place Céline Dion together took in $196.4 million from ticket sales. If Dion took in $35.4 million less than Springsteen, how much did each generate? (*Source:* Pollstar.)

33. Tickets to a production of *Cats* at Broward Community College cost $5 for general admission or $4 with a student ID. If 184 people paid to see a performance and $812 was collected, how many of each type of ticket were sold?

34. At a business meeting at Panera Bread, the bill for two cappuccinos and three house lattes was $14.55. At another table, the bill for one cappuccino and two house lattes was $8.77. How much did each type of beverage cost? (*Source:* Panera Bread menu.)

35. The mixed price of 9 citrons and 7 fragrant wood apples is 107; again, the mixed price of 7 citrons and 9 fragrant wood apples is 101. O you arithmetician, tell me quickly the price of a citron and the price of a wood apple here, having distinctly separated those prices well. (*Source:* Hindu work, A.D. 850.)

36. Braving blizzard conditions on the planet Hoth, Luke Skywalker sets out at top speed in his snow speeder for a rebel base 4800 mi away. He travels into a steady head wind and makes the trip in 3 hr. Returning, he finds that the trip back, still at top speed but now with a tailwind, takes only 2 hr. Find the top speed of Luke's snow speeder and the speed of the wind.

	r	t	d
Into Head Wind			
With Tailwind			

Solve each problem by using three variables. See Examples 5 and 6. (In Exercises 37–40, remember that the sum of the measures of the angles of a triangle is $180°$.)

37. In the figure, $z = x + 10$ and $x + y = 100$. Determine a third equation involving x, y, and z, and then find the measures of the three angles.

38. In the figure, x is 10 less than y and 20 less than z. Write a system of equations and find the measures of the three angles.

39. In a certain triangle, the measure of the second angle is 10° more than three times the first. The third angle measure is equal to the sum of the measures of the other two. Find the measures of the three angles.

40. The measure of the largest angle of a triangle is 12° less than the sum of the measures of the other two. The smallest angle measures 58° less than the largest. Find the measures of the angles.

41. The perimeter of a triangle is 70 cm. The longest side is 4 cm less than the sum of the other two sides. Twice the shortest side is 9 cm less than the longest side. Find the length of each side of the triangle.

42. The perimeter of a triangle is 56 in. The longest side measures 4 in. less than the sum of the other two sides. Three times the shortest side is 4 in. more than the longest side. Find the lengths of the three sides.

43. In the 2004 Summer Olympics in Athens, Greece, the United States earned 6 more gold medals than bronze. The number of silver medals earned was 19 less than twice the number of bronze medals. The United States earned a total of 103 medals. How many of each kind of medal did the United States earn? (*Source: World Almanac and Book of Facts 2006.*)

44. In a random sample of 100 Americans of voting age, 10 more Americans identify themselves as Independents than Republicans. Six fewer Americans identify themselves as Republicans than Democrats. Assuming that all of those sampled are Republican, Democrat, or Independent, how many of those in the sample identify themselves with each political affiliation? (*Source:* The Gallup Organization.)

45. Tickets for one show on the Harlem Globetrotters' 2006 "Unstoppable" Tour cost $14, $20, or, for VIP seats, $50. So far, five times as many $14 tickets have been sold as VIP tickets. The number of $14 tickets is 15 more than the sum of the number of $20 tickets and the number of VIP tickets. Sales of all three kinds of tickets total $11,700. How many of each kind of ticket have been sold? (*Source:* www.ticketmaster.com)

46. Three kinds of tickets are available for a *Third Day* concert: "up close," "in the middle," and "far out." "Up close" tickets cost $10 more than "in the middle" tickets, while "in the middle" tickets cost $10 more than "far out" tickets. Twice the cost of an "up close" ticket is $20 more than 3 times the cost of a "far out" ticket. Find the price of each kind of ticket.

47. A wholesaler supplies college T-shirts to three college bookstores: A, B, and C. The wholesaler recently shipped a total of 800 T-shirts to the three bookstores. In order to meet student demand at the three colleges, twice as many T-shirts were shipped to bookstore B as to bookstore A, and the number shipped to bookstore C was 40 less than the sum of the numbers shipped to the other two bookstores. How many T-shirts were shipped to each bookstore?

48. An office supply store sells three models of computer desks: A, B, and C. In January, the store sold a total of 85 computer desks. The number of model B desks was five more than the number of model C desks, and the number of model A desks was four more than twice the number of model C desks. How many of each model did the store sell in January?

49. A plant food is to be made from three chemicals. The mix must include 60% of the first and second chemicals. The second and third chemicals must be in the ratio of 4 to 3 by weight. How much of each chemical is needed to make 750 kg of the plant food?

50. How many ounces of 5% hydrochloric acid, 20% hydrochloric acid, and water must be combined to get 10 oz of solution that is 8.5% hydrochloric acid if the amount of water used must equal the total amount of the other two solutions?

51. During the 2005–2006 National Hockey League regular season, the Calgary Flames played 82 games. Together, their wins and losses totaled 71. They tied 14 fewer games than they lost. How many wins, losses, and ties did they have that year?

Team	GP	W	L	T	Pts	GF	GA
Calgary	82	—	—	—	103	218	200
Colorado	82	43	30	9	95	283	257
Edmonton	82	41	28	13	95	256	251
Vancouver	82	42	32	8	92	256	255
Minnesota	82	30	36	8	84	231	215

Source: www.sportzdomain.com

52. During the 2005–2006 National Hockey League season, the Boston Bruins played 82 games. Their losses and ties totaled 53, and they had 8 fewer wins than losses. How many wins, losses, and ties did they have that year?

Team	GP	W	L	T	Pts	GF	GA
Ottawa	82	52	21	9	113	314	211
Buffalo	82	52	24	6	110	281	239
Montreal	82	42	31	9	93	243	247
Toronto	82	41	33	8	90	257	270
Boston	82	—	—	—	74	230	266

Source: www.sportzdomain.com

PREVIEW EXERCISES

Give **(a)** *the additive inverse and* **(b)** *the multiplicative inverse (reciprocal) of each number. See Sections 1.1 and 1.2.*

53. -6 **54.** 0.2 **55.** $\dfrac{7}{8}$ **56.** 2.25

4.4 Solving Systems of Linear Equations by Matrix Methods

OBJECTIVE 1 Define a matrix. A **matrix** is an ordered array of numbers, such as

$$\text{Rows} \begin{bmatrix} 2 & 3 & 5 \\ 7 & 1 & 2 \end{bmatrix}. \quad \text{Matrix}$$

Columns

The numbers are called **elements** of the matrix. Matrices (the plural of *matrix*) are named according to the number of **rows** and **columns** they contain. The rows are read horizontally, and the columns are read vertically. For example, the first row in the preceding matrix is 2   3   5 and the first column is $\frac{2}{7}$. This matrix is a 2×3 (read "two by three") matrix, because it has 2 rows and 3 columns. The number of rows followed by the number of columns gives the **dimensions** of the matrix.

$$\begin{bmatrix} -1 & 0 \\ 1 & -2 \end{bmatrix} \quad \begin{array}{c} 2 \times 2 \\ \text{matrix} \end{array} \qquad \begin{bmatrix} 8 & -1 & -3 \\ 2 & 1 & 6 \\ 0 & 5 & -3 \\ 5 & 9 & 7 \end{bmatrix} \quad \begin{array}{c} 4 \times 3 \\ \text{matrix} \end{array}$$

A **square matrix** is a matrix that has the same number of rows as columns. The 2×2 matrix above is a square matrix.

Figure 10 shows how a graphing calculator displays the preceding two matrices. Work with matrices is made much easier by the use of technology when it is available. Consult your owner's manual for details.

In this section, we discuss a matrix method of solving linear systems that is really just a very structured way of using the elimination method. The advantage of this new method is that it can be done by a graphing calculator or a computer.

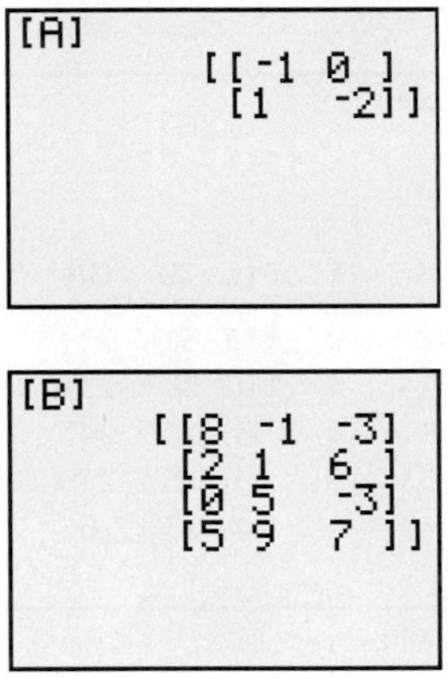

```
[A]
      [[-1  0 ]
       [1  -2]]
```

```
[B]
      [[8 -1 -3]
       [2  1  6 ]
       [0  5 -3]
       [5  9  7 ]]
```

FIGURE 10

OBJECTIVE 2 Write the augmented matrix of a system. To solve a linear system using matrices, we begin by writing an *augmented matrix* of the system. An **augmented matrix** has a vertical bar that separates the columns of the matrix into two groups. For example, to solve the system

$$\begin{aligned} x - 3y &= 1 \\ 2x + y &= -5, \end{aligned} \qquad \text{start with the augmented matrix} \qquad \left[\begin{array}{cc|c} 1 & -3 & 1 \\ 2 & 1 & -5 \end{array} \right].$$

Place the coefficients of the variables to the left of the bar, and the constants to the right. The bar separates the coefficients from the constants. *The matrix is just a shorthand way of writing the system of equations, so the rows of the augmented matrix can be treated the same as the equations of a system of equations.*

We know that exchanging the positions of two equations in a system does not change the system. Also, multiplying any equation in a system by a nonzero number does not change the system. Comparable changes to the augmented matrix of a system of equations produce new matrices that correspond to systems with the same solutions as the original system.

The following **row operations** produce new matrices that lead to systems having the same solutions as the original system.

Matrix Row Operations

1. Any two rows of the matrix may be interchanged.

2. The elements of any row may be multiplied by any nonzero real number.

3. Any row may be changed by adding to the elements of the row the product of a real number and the corresponding elements of another row.

Examples of these row operations follow.

Row operation 1:

$$\begin{bmatrix} 2 & 3 & 9 \\ 4 & 8 & -3 \\ 1 & 0 & 7 \end{bmatrix} \text{ becomes } \begin{bmatrix} 1 & 0 & 7 \\ 4 & 8 & -3 \\ 2 & 3 & 9 \end{bmatrix}.$$
Interchange row 1 and row 3.

Row operation 2:

$$\begin{bmatrix} 2 & 3 & 9 \\ 4 & 8 & -3 \\ 1 & 0 & 7 \end{bmatrix} \text{ becomes } \begin{bmatrix} 6 & 9 & 27 \\ 4 & 8 & -3 \\ 1 & 0 & 7 \end{bmatrix}.$$
Multiply the numbers in row 1 by 3.

Row operation 3:

$$\begin{bmatrix} 2 & 3 & 9 \\ 4 & 8 & -3 \\ 1 & 0 & 7 \end{bmatrix} \text{ becomes } \begin{bmatrix} 0 & 3 & -5 \\ 4 & 8 & -3 \\ 1 & 0 & 7 \end{bmatrix}.$$
Multiply the numbers in row 3 by -2; add them to the corresponding numbers in row 1.

The third row operation corresponds to the way we eliminated a variable from a pair of equations in previous sections.

OBJECTIVE 3 Use row operations to solve a system with two equations. Row operations can be used to rewrite a matrix until it is the matrix of a system whose solution is easy to find. The goal is a matrix in the form

$$\left[\begin{array}{cc|c} 1 & a & b \\ 0 & 1 & c \end{array}\right] \quad \text{or} \quad \left[\begin{array}{ccc|c} 1 & a & b & c \\ 0 & 1 & d & e \\ 0 & 0 & 1 & f \end{array}\right]$$

for systems with two and three equations, respectively. Notice that there are 1's down the diagonal from upper left to lower right and 0's below the 1's. A matrix written this way is said to be in **row echelon form.** When these matrices are rewritten as systems of equations, the value of one variable is known and the rest can be found by substitution. The following examples illustrate this method.

> **EXAMPLE 1** **Using Row Operations to Solve a System with Two Variables**

Use row operations to solve the system

$$x - 3y = 1$$
$$2x + y = -5.$$

We start with the augmented matrix of the system.

$$\begin{bmatrix} 1 & -3 & | & 1 \\ 2 & 1 & | & -5 \end{bmatrix}$$ Augmented matrix

Now we use the various row operations to change this matrix into one that leads to a system that is easier to solve.

It is best to work by columns. We start with the first column and make sure that there is a 1 in the first row, first column, position. There already is a 1 in this position. Next, we get 0 in every position below the first. To get a 0 in row two, column one, we use the third row operation and add to the numbers in row two the result of multiplying each number in row one by -2. (We abbreviate this as $-2R_1 + R_2$.) Row one remains unchanged.

$$\begin{bmatrix} 1 & -3 & | & 1 \\ 2 + 1(-2) & 1 + (-3)(-2) & | & -5 + 1(-2) \end{bmatrix}$$

Original number -2 times number
from row two   from row one

$$\begin{bmatrix} 1 & -3 & | & 1 \\ 0 & 7 & | & -7 \end{bmatrix}$$ $-2R_1 + R_2$

The matrix now has a 1 in the first position of column one, with 0 in every position below the first.

Now we go to column two. The number 1 is needed in row two, column two. We get this 1 by using the second row operation, multiplying each number of row two by $\frac{1}{7}$.

Stop here—this matrix is in row echelon form.
$$\begin{bmatrix} 1 & -3 & | & 1 \\ 0 & 1 & | & -1 \end{bmatrix}$$ $\frac{1}{7}R_2$

This augmented matrix leads to the system of equations

$$\begin{align} 1x - 3y &= 1 \\ 0x + 1y &= -1, \end{align} \quad \text{or} \quad \begin{align} x - 3y &= 1 \\ y &= -1. \end{align}$$

From the second equation, $y = -1$, we substitute -1 for y in the first equation to get

$$\begin{align} x - 3y &= 1 \\ x - 3(-1) &= 1 \qquad \text{Let } y = -1. \\ x + 3 &= 1 \qquad \text{Multiply.} \\ x &= -2. \qquad \text{Subtract 3.} \end{align}$$

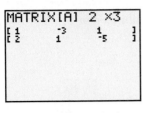

(a)

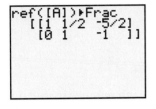

(b)

FIGURE 11

The solution set of the system is $\{(-2, -1)\}$. Check this solution by substitution in both equations of the system.

> Write the values of x and y in the correct order.

✔ **Now Try Exercise 3.**

If the augmented matrix of the system in Example 1 is entered as matrix [A] in a graphing calculator (Figure 11(a)) and the row echelon form of the matrix is found (Figure 11(b)), then the system becomes

$$x + \frac{1}{2}y = -\frac{5}{2}$$
$$y = -1.$$

While this system looks different from the one we obtained in Example 1, it is equivalent, since its solution set is also $\{(-2, -1)\}$.

OBJECTIVE 4 Use row operations to solve a system with three equations. A linear system with three equations is solved in a similar way. We use row operations to get 1's down the diagonal from left to right and all 0's below each 1.

EXAMPLE 2 Using Row Operations to Solve a System with Three Variables

Use row operations to solve the system

$$x - y + 5z = -6$$
$$3x + 3y - z = 10$$
$$x + 3y + 2z = 5.$$

Start by writing the augmented matrix of the system.

$$\begin{bmatrix} 1 & -1 & 5 & | & -6 \\ 3 & 3 & -1 & | & 10 \\ 1 & 3 & 2 & | & 5 \end{bmatrix} \quad \text{Augmented matrix}$$

This matrix already has 1 in row one, column one. Next get 0's in the rest of column one. First, add to row two the results of multiplying each number of row one by -3. This gives the matrix

$$\begin{bmatrix} 1 & -1 & 5 & | & -6 \\ 0 & 6 & -16 & | & 28 \\ 1 & 3 & 2 & | & 5 \end{bmatrix}. \quad -3R_1 + R_2$$

Now add to the numbers in row three the results of multiplying each number of row one by -1.

$$\begin{bmatrix} 1 & -1 & 5 & | & -6 \\ 0 & 6 & -16 & | & 28 \\ 0 & 4 & -3 & | & 11 \end{bmatrix} \quad -1R_1 + R_3$$

Introduce 1 in row two, column two, by multiplying each number in row two by $\frac{1}{6}$.

$$\begin{bmatrix} 1 & -1 & 5 & | & -6 \\ 0 & 1 & -\frac{8}{3} & | & \frac{14}{3} \\ 0 & 4 & -3 & | & 11 \end{bmatrix} \quad \frac{1}{6}R_2$$

To obtain 0 in row three, column two, add to row three the results of multiplying each number in row two by -4.

$$\begin{bmatrix} 1 & -1 & 5 & | & -6 \\ 0 & 1 & -\frac{8}{3} & | & \frac{14}{3} \\ 0 & 0 & \frac{23}{3} & | & -\frac{23}{3} \end{bmatrix} \quad -4R_2 + R_3$$

Finally, obtain 1 in row three, column three, by multiplying each number in row three by $\frac{3}{23}$.

$$\begin{bmatrix} 1 & -1 & 5 & | & -6 \\ 0 & 1 & -\frac{8}{3} & | & \frac{14}{3} \\ 0 & 0 & 1 & | & -1 \end{bmatrix} \quad \frac{3}{23}R_3$$

This final matrix gives the system of equations

$$x - y + 5z = -6$$
$$y - \frac{8}{3}z = \frac{14}{3}$$
$$z = -1.$$

Substitute -1 for z in the second equation, $y - \frac{8}{3}z = \frac{14}{3}$, to find that $y = 2$. Finally, substitute 2 for y and -1 for z in the first equation, $x - y + 5z = -6$, to determine that $x = 1$. The solution set of the original system is $\{(1, 2, -1)\}$. Check by substitution.

✔ **Now Try Exercise 15.**

OBJECTIVE 5 **Use row operations to solve special systems.** In the final example, we show how to recognize inconsistent systems or systems with dependent equations when solving these systems with row operations.

EXAMPLE 3 **Recognizing Inconsistent Systems or Dependent Equations**

Use row operations to solve each system.

(a) $2x - 3y = 8$
$-6x + 9y = 4$

$$\begin{bmatrix} 2 & -3 & | & 8 \\ -6 & 9 & | & 4 \end{bmatrix} \quad \text{Write the augmented matrix.}$$

$$\begin{bmatrix} 1 & -\frac{3}{2} & | & 4 \\ -6 & 9 & | & 4 \end{bmatrix} \quad \frac{1}{2}R_1$$

$$\begin{bmatrix} 1 & -\frac{3}{2} & | & 4 \\ 0 & 0 & | & 28 \end{bmatrix} \quad 6R_1 + R_2$$

The corresponding system of equations is

$$x - \frac{3}{2}y = 4$$

$$0 = 28, \quad \text{False}$$

which has no solution and is inconsistent. The solution set is $\emptyset$.

(b) $-10x + 12y = 30$

$5x - 6y = -15$

$$\begin{bmatrix} -10 & 12 & \big| & 30 \\ 5 & -6 & \big| & -15 \end{bmatrix} \quad \begin{array}{l}\text{Write the}\\ \text{augmented matrix.}\end{array}$$

$$\begin{bmatrix} 1 & -\frac{6}{5} & \big| & -3 \\ 5 & -6 & \big| & -15 \end{bmatrix} \quad -\tfrac{1}{10}R_1$$

$$\begin{bmatrix} 1 & -\frac{6}{5} & \big| & -3 \\ 0 & 0 & \big| & 0 \end{bmatrix} \quad -5R_1 + R_2$$

The corresponding system is

$$x - \frac{6}{5}y = -3$$

$$0 = 0, \quad \text{True}$$

which has dependent equations. Using the second equation of the given system, which is in standard form, we write the solution set as

$$\{(x, y) \mid 5x - 6y = -15\}.$$

✔ **Now Try Exercises 11 and 13.**

4.4 EXERCISES

1. *Concept Check*   Consider the matrix $\begin{bmatrix} -2 & 3 & 1 \\ 0 & 5 & -3 \\ 1 & 4 & 8 \end{bmatrix}$ and answer the following.

(a) What are the elements of the second row?

(b) What are the elements of the third column?

✍ (c) Is this a square matrix? Explain why or why not.

(d) Give the matrix obtained by interchanging the first and third rows.

(e) Give the matrix obtained by multiplying the first row by $-\frac{1}{2}$.

(f) Give the matrix obtained by multiplying the third row by 3 and adding to the first row.

2. *Concept Check*   Give the dimensions of each matrix.

(a) $\begin{bmatrix} 3 & -7 \\ 4 & 5 \\ -1 & 0 \end{bmatrix}$

(b) $\begin{bmatrix} 4 & 9 & 0 \\ -1 & 2 & -4 \end{bmatrix}$

(c) $\begin{bmatrix} 6 & 3 \\ -2 & 5 \\ 4 & 10 \\ 1 & -1 \end{bmatrix}$

Complete the steps in the matrix solution of each system by filling in the boxes. Give the final system and the solution set. See Example 1.

3. $\begin{aligned} 4x + 8y &= 44 \\ 2x - y &= -3 \end{aligned}$ $\rightarrow$ $\left[\begin{array}{cc|c} 4 & 8 & 44 \\ 2 & -1 & -3 \end{array}\right]$

$\left[\begin{array}{cc|c} 1 & \blacksquare & \blacksquare \\ 2 & -1 & -3 \end{array}\right]$ $\frac{1}{4}R_1$

$\left[\begin{array}{cc|c} 1 & 2 & 11 \\ 0 & \blacksquare & \blacksquare \end{array}\right]$ $-2R_1 + R_2$

$\left[\begin{array}{cc|c} 1 & 2 & 11 \\ 0 & 1 & \blacksquare \end{array}\right]$ $-\frac{1}{5}R_2$

4. $\begin{aligned} 2x - 5y &= -1 \\ 3x + y &= 7 \end{aligned}$ $\rightarrow$ $\left[\begin{array}{cc|c} 2 & -5 & -1 \\ 3 & 1 & 7 \end{array}\right]$

$\left[\begin{array}{cc|c} 1 & -\frac{5}{2} & \blacksquare \\ 3 & 1 & 7 \end{array}\right]$ $\frac{1}{2}R_1$

$\left[\begin{array}{cc|c} 1 & -\frac{5}{2} & -\frac{1}{2} \\ 0 & \blacksquare & \blacksquare \end{array}\right]$ $-3R_1 + R_2$

$\left[\begin{array}{cc|c} 1 & -\frac{5}{2} & -\frac{1}{2} \\ 0 & 1 & \blacksquare \end{array}\right]$ $\frac{2}{17}R_2$

Use row operations to solve each system. See Examples 1 and 3.

5. $\begin{aligned} x + y &= 5 \\ x - y &= 3 \end{aligned}$

6. $\begin{aligned} x + 2y &= 7 \\ x - y &= -2 \end{aligned}$

7. $\begin{aligned} 2x + 4y &= 6 \\ 3x - y &= 2 \end{aligned}$

8. $\begin{aligned} 4x + 5y &= -7 \\ x - y &= 5 \end{aligned}$

9. $\begin{aligned} 3x + 4y &= 13 \\ 2x - 3y &= -14 \end{aligned}$

10. $\begin{aligned} 5x + 2y &= 8 \\ 3x - y &= 7 \end{aligned}$

11. $\begin{aligned} -4x + 12y &= 36 \\ x - 3y &= 9 \end{aligned}$

12. $\begin{aligned} 2x - 4y &= 8 \\ -3x + 6y &= 5 \end{aligned}$

13. $\begin{aligned} 2x + y &= 4 \\ 4x + 2y &= 8 \end{aligned}$

14. $\begin{aligned} -3x - 4y &= 1 \\ 6x + 8y &= -2 \end{aligned}$

Complete the steps in the matrix solution of each system by filling in the boxes. Give the final system and the solution set. See Example 2.

15. $\begin{aligned} x + y - z &= -3 \\ 2x + y + z &= 4 \\ 5x - y + 2z &= 23 \end{aligned}$

$\left[\begin{array}{ccc|c} 1 & 1 & -1 & -3 \\ 2 & 1 & 1 & 4 \\ 5 & -1 & 2 & 23 \end{array}\right]$

$\left[\begin{array}{ccc|c} 1 & 1 & -1 & -3 \\ 0 & \blacksquare & \blacksquare & \blacksquare \\ 0 & \blacksquare & \blacksquare & \blacksquare \end{array}\right]$ $\begin{aligned} -2R_1 + R_2 \\ -5R_1 + R_3 \end{aligned}$

$\left[\begin{array}{ccc|c} 1 & 1 & -1 & -3 \\ 0 & 1 & \blacksquare & \blacksquare \\ 0 & -6 & 7 & 38 \end{array}\right]$ $-1R_2$

$\left[\begin{array}{ccc|c} 1 & 1 & -1 & -3 \\ 0 & 1 & -3 & -10 \\ 0 & 0 & \blacksquare & \blacksquare \end{array}\right]$ $6R_2 + R_3$

$\left[\begin{array}{ccc|c} 1 & 1 & -1 & -3 \\ 0 & 1 & -3 & -10 \\ 0 & 0 & 1 & \blacksquare \end{array}\right]$ $-\frac{1}{11}R_3$

16. $\begin{aligned} 2x + y + 2z &= 11 \\ 2x - y - z &= -3 \\ 3x + 2y + z &= 9 \end{aligned}$

$\left[\begin{array}{ccc|c} 2 & 1 & 2 & 11 \\ 2 & -1 & -1 & -3 \\ 3 & 2 & 1 & 9 \end{array}\right]$

$\left[\begin{array}{ccc|c} 1 & \blacksquare & \blacksquare & \blacksquare \\ 2 & -1 & -1 & -3 \\ 3 & 2 & 1 & 9 \end{array}\right]$ $\frac{1}{2}R_1$

$\left[\begin{array}{ccc|c} 1 & \frac{1}{2} & 1 & \frac{11}{2} \\ 0 & \blacksquare & \blacksquare & \blacksquare \\ 0 & \blacksquare & \blacksquare & \blacksquare \end{array}\right]$ $\begin{aligned} -2R_1 + R_2 \\ -3R_1 + R_3 \end{aligned}$

$\left[\begin{array}{ccc|c} 1 & \frac{1}{2} & 1 & \frac{11}{2} \\ 0 & 1 & \blacksquare & \blacksquare \\ 0 & \frac{1}{2} & -2 & -\frac{15}{2} \end{array}\right]$ $-\frac{1}{2}R_2$

$\left[\begin{array}{ccc|c} 1 & \frac{1}{2} & 1 & \frac{11}{2} \\ 0 & 1 & \frac{3}{2} & 7 \\ 0 & 0 & \blacksquare & \blacksquare \end{array}\right]$ $-\frac{1}{2}R_2 + R_3$

$\left[\begin{array}{ccc|c} 1 & \frac{1}{2} & 1 & \frac{11}{2} \\ 0 & 1 & \frac{3}{2} & 7 \\ 0 & 0 & 1 & \blacksquare \end{array}\right]$ $-\frac{4}{11}R_3$

Use row operations to solve each system. See Examples 2 and 3.

17. $x + y - 3z = 1$
$2x - y + z = 9$
$3x + y - 4z = 8$

18. $2x + 4y - 3z = -18$
$3x + y - z = -5$
$x - 2y + 4z = 14$

19. $x + y - z = 6$
$2x - y + z = -9$
$x - 2y + 3z = 1$

20. $x + 3y - 6z = 7$
$2x - y + 2z = 0$
$x + y + 2z = -1$

21. $x - y = 1$
$y - z = 6$
$x + z = -1$

22. $x + y = 1$
$2x - z = 0$
$y + 2z = -2$

23. $x - 2y + z = 4$
$3x - 6y + 3z = 12$
$-2x + 4y - 2z = -8$

24. $4x + 8y + 4z = 9$
$x + 3y + 4z = 10$
$5x + 10y + 5z = 12$

25. $x + 2y + 3z = -2$
$2x + 4y + 6z = -5$
$x - y + 2z = 6$

26. $x + 3y + z = 1$
$2x + 6y + 2z = 2$
$3x + 9y + 3z = 3$

The augmented matrix of the system in Exercise 3 is shown in the graphing calculator screen on the left as matrix [A]. *The screen in the middle shows the row echelon form for* [A]. *Compare it with the matrix shown in the answer section for Exercise 3. The screen on the right shows the "reduced" row echelon form, and from this it can be determined by inspection that the solution set of the system is* $\{(1, 5)\}$.

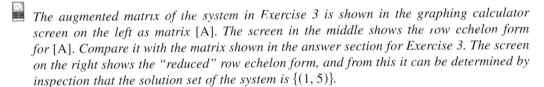

Use a graphing calculator and either one of the two matrix methods illustrated to solve each system.

27. $4x + y = 5$
$2x + y = 3$

28. $5x + 3y = 7$
$7x - 3y = -19$

29. $5x + y - 3z = -6$
$2x + 3y + z = 5$
$-3x - 2y + 4z = 3$

30. $x + y + z = 3$
$3x - 3y - 4z = -1$
$x + y + 3z = 11$

31. $x + z = -3$
$y + z = 3$
$x + y = 8$

32. $x - y = -1$
$-y + z = -2$
$x + z = -2$

PREVIEW EXERCISES

*Evaluate each exponential expression. See **Section 1.3**.*

33. 2^6

34. $(-4)^5$

35. $(-5)^4$

36. -5^4

37. $\left(\dfrac{3}{4}\right)^4$

38. $\left(-\dfrac{3}{2}\right)^5$

| Chapter **4** | **Group Activity** |

CREATING LINEAR MODELS TO ANALYZE THE NUMBER OF AIDS PATIENTS

Objective Create linear models to describe data.

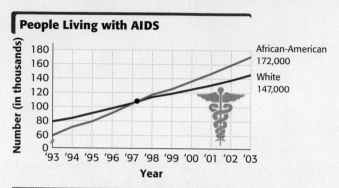

People Living with AIDS

Source: U.S. Centers for Disease Control and Prevention.

A. The graph shows the number of African-Americans and Whites living with AIDS in the United States. The two lines were obtained by joining data points that are of the form (year, number of people in thousands). Let $x = 0$ represent the year 1993, $x = 1$ represent 1994, and so on, and estimate the value of y (in thousands) from the graph for each year for African-Americans and for Whites. Fill in the table.

Year, x	Number of African-Americans with AIDS (y, in thousands)	Number of Whites with AIDS (y, in thousands)
0 (1993)	60	80
1 (1994)		
2 (1995)		
3 (1996)		
4 (1997)		
5 (1998)		
6 (1999)		
7 (2000)		
8 (2001)		
9 (2002)		
10 (2003)	172	147

Use two data points to find an equation of the line describing the data for African-Americans. Then find an equation describing the data for Whites.

B. The two equations from part A form a system of two linear equations in two variables. Solve this system, using any method you wish.

C. The x-value of the solution of the system in part B should correspond to the year in which the two lines intersect. Look at the graph again. Does your x-value correspond the way it should?

Chapter **4** SUMMARY
View the Interactive Summary on the Pass the Test CD.

KEY TERMS

4.1 system of equations
system of linear
equations
solution set of a linear
system

consistent system
independent equations
inconsistent system
dependent equations
elimination method

substitution method
4.2 ordered triple
4.4 matrix
element of a matrix
row

column
square matrix
augmented matrix
row operations
row echelon form

NEW SYMBOLS

(x, y, z) ordered triple

$\begin{bmatrix} a & b & c \\ d & e & f \end{bmatrix}$ matrix with two rows, three columns

TEST YOUR WORD POWER

See how well you have learned the vocabulary in this chapter. Answers, with examples, follow the Quick Review.

1. A **system of equations** consists of
   A. at least two equations with different variables
   B. two or more equations that have an infinite number of solutions
   C. two or more equations that are to be solved at the same time
   D. two or more inequalities that are to be solved.

2. The **solution set of a system of equations** is
   A. all ordered pairs that satisfy one equation of the system
   B. all ordered pairs that satisfy all the equations of the system at the same time

C. any ordered pair that satisfies one or more equations of the system
D. the set of values that make all the equations of the system false.

3. An **inconsistent system** is a system of equations
   A. with one solution
   B. with no solution
   C. with an infinite number of solutions
   D. that have the same graph.

4. **Dependent equations**
   A. have different graphs
   B. have no solution

C. have one solution
D. are different forms of the same equation.

5. A **matrix** is
   A. an ordered pair of numbers
   B. an array of numbers with the same number of rows and columns
   C. a pair of numbers written between brackets
   D. a rectangular array of numbers.

QUICK REVIEW

Concepts	Examples

4.1 SYSTEMS OF LINEAR EQUATIONS IN TWO VARIABLES

Solving a Linear System by Substitution

Step 1 Solve one of the equations for either variable.

Solve by substitution: $4x - y = 7$ (1)
$3x + 2y = 30.$ (2)

Solve for y in equation (1): $y = 4x - 7$.

(continued)

Concepts	Examples

Step 2 Substitute for that variable in the other equation. The result should be an equation with just one variable.

Substitute $4x - 7$ for y in equation (2), and solve for x.

$$3x + 2y = 30 \quad (2)$$
$$3x + 2(4x - 7) = 30 \quad \text{Let } y = 4x - 7.$$
$$3x + 8x - 14 = 30 \quad \text{Distributive property}$$
$$11x = 44 \quad \text{Combine terms; add 14.}$$
$$x = 4 \quad \text{Divide by 11.}$$

Step 3 Solve the equation from Step 2.

Step 4 Find the value of the other variable by substituting the result from Step 3 into the equation from Step 1.

Substitute 4 for x in the equation $y = 4x - 7$ to find that $y = 4(4) - 7 = 9$.

Step 5 Check the solution in both of the original equations. Then write the solution set.

Check to see that $\{(4, 9)\}$ is the solution set.

Solving a Linear System by Elimination

Step 1 Write both equations in standard form.

Solve by elimination:
$$5x + y = 2 \quad (1)$$
$$2x - 3y = 11. \quad (2)$$

Step 2 Make the coefficients of one pair of variable terms opposites.

To eliminate y, multiply equation (1) by 3 and add the result to equation (2).

Step 3 Add the new equations. The sum should be an equation with just one variable.

$$15x + 3y = 6 \quad \text{3 times equation (1)}$$
$$\underline{2x - 3y = 11} \quad (2)$$
$$17x = 17 \quad \text{Add.}$$
$$x = 1 \quad \text{Divide by 17.}$$

Step 4 Solve the equation from Step 3.

Step 5 Find the value of the other variable by substituting the result of Step 4 into either of the original equations.

Let $x = 1$ in equation (1), and solve for y.
$$5(1) + y = 2$$
$$y = -3$$

Step 6 Check the solution in both of the original equations. Then write the solution set.

Check to verify that $\{(1, -3)\}$ is the solution set.

4.2   SYSTEMS OF LINEAR EQUATIONS IN THREE VARIABLES

Solving a Linear System in Three Variables

Step 1 Use the elimination method to eliminate any variable from any two of the original equations.

Solve the system
$$x + 2y - z = 6 \quad (1)$$
$$x + y + z = 6 \quad (2)$$
$$2x + y - z = 7. \quad (3)$$

Add equations (1) and (2); z is eliminated and the result is $2x + 3y = 12$.

Step 2 Eliminate the *same* variable from any *other* two equations.

Eliminate z again by adding equations (2) and (3) to get $3x + 2y = 13$. Now solve the system
$$2x + 3y = 12 \quad (4)$$
$$3x + 2y = 13. \quad (5)$$

(continued)

Concepts	Examples

Step 3 Eliminate a second variable from the two equations in two variables that result from Steps 1 and 2. The result is an equation in one variable that gives the value of that variable.

To eliminate x, multiply equation (4) by -3 and equation (5) by 2.

$$\begin{array}{r} -6x - 9y = -36 \\ 6x + 4y = 26 \\ \hline -5y = -10 \\ y = 2 \end{array}$$

Step 4 Substitute the value of the variable found in Step 3 into either of the equations in two variables to find the value of the second variable.

Let $y = 2$ in equation (4).

$$2x + 3(2) = 12$$
$$2x + 6 = 12$$
$$2x = 6$$
$$x = 3$$

Step 5 Use the values of the two variables from Steps 3 and 4 to find the value of the third variable by substituting into an appropriate equation.

Let $y = 2$ and $x = 3$ in any of the original equations to find that $z = 1$.

Step 6 Check the solution in all of the original equations. Then write the solution set.

Check. The solution set is $\{(3, 2, 1)\}$.

4.3 APPLICATIONS OF SYSTEMS OF LINEAR EQUATIONS

Step 1 Read the problem carefully.

The perimeter of a rectangle is 18 ft. The length is 3 ft more than twice the width. What are the dimensions of the rectangle?

Step 2 Assign variables.

Let $x =$ the length and $y =$ the width.

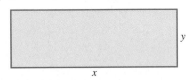

Step 3 Write a system of equations that relates the unknowns.

From the perimeter formula, one equation is $2x + 2y = 18$. From the problem, another equation is $x = 2y + 3$.

Step 4 Solve the system.

Solve the system to find that $x = 7$ and $y = 2$.

Steps 5 and 6 State the answer and check.

The length is 7 ft and the width is 2 ft. Since the perimeter is

$2(7) + 2(2) = 18$, and $3 + 2(2) = 7$, the answer checks.

4.4 SOLVING SYSTEMS OF LINEAR EQUATIONS BY MATRIX METHODS

Matrix Row Operations
1. Any two rows of the matrix may be interchanged.

$$\begin{bmatrix} 1 & 5 & 7 \\ 3 & 9 & -2 \\ 0 & 6 & 4 \end{bmatrix} \text{ becomes } \begin{bmatrix} 3 & 9 & -2 \\ 1 & 5 & 7 \\ 0 & 6 & 4 \end{bmatrix}$$

Interchange R_1 and R_2.

2. The elements of any row may be multiplied by any nonzero real number.

$$\begin{bmatrix} 1 & 5 & 7 \\ 3 & 9 & -2 \\ 0 & 6 & 4 \end{bmatrix} \text{ becomes } \begin{bmatrix} 1 & 5 & 7 \\ 1 & 3 & -\frac{2}{3} \\ 0 & 6 & 4 \end{bmatrix}$$

$\frac{1}{3} R_2$

(continued)

Concepts	Examples

3. Any row may be changed by adding to the elements of the row the product of a real number and the elements of another row.

$$\begin{bmatrix} 1 & 5 & 7 \\ 3 & 9 & -2 \\ 0 & 6 & 4 \end{bmatrix} \text{ becomes } \begin{bmatrix} 1 & 5 & 7 \\ 0 & -6 & -23 \\ 0 & 6 & 4 \end{bmatrix} \quad -3R_1 + R_2$$

A system can be solved by matrix methods. Write the augmented matrix and use row operations to obtain a matrix in row echelon form.

Solve using row operations: $\begin{aligned} x + 3y &= 7 \\ 2x + y &= 4. \end{aligned}$

$$\begin{bmatrix} 1 & 3 & | & 7 \\ 2 & 1 & | & 4 \end{bmatrix} \quad \text{Augmented matrix}$$

$$\begin{bmatrix} 1 & 3 & | & 7 \\ 0 & -5 & | & -10 \end{bmatrix} \quad -2R_1 + R_2$$

$$\begin{bmatrix} 1 & 3 & | & 7 \\ 0 & 1 & | & 2 \end{bmatrix} \quad -\tfrac{1}{5}R_2 \xrightarrow{\text{implies}} \begin{aligned} x + 3y &= 7 \\ y &= 2 \end{aligned}$$

When $y = 2$, $x + 3(2) = 7$, so $x = 1$. The solution set is $\{(1, 2)\}$.

Answers to Test Your Word Power

1. C;

Example: $\begin{aligned} 3x - y &= 3 \\ 2x + y &= 7 \end{aligned}$

2. B; *Example:* The ordered pair $(2, 3)$ satisfies both equations of the system in Answer 1, so $\{(2, 3)\}$ is the solution set of the system.

3. B; *Example:* The equations of two parallel lines form an inconsistent system; their graphs never intver intersect, so the system has no solution.

4. D; *Example:* The equations $4x - y = 8$ and $8x - 2y = 16$ are dependent because their graphs are the same line.

5. D;

Examples: $\begin{bmatrix} 3 & -1 & 0 \\ 4 & 2 & 1 \end{bmatrix}, \begin{bmatrix} 1 & 2 \\ 4 & 3 \end{bmatrix}$

Chapter **4** REVIEW EXERCISES

[4.1]

1. The graph shows the trends during the years 1975 through 2002 relating to bachelor's degrees awarded in the United States.

(a) Between what years shown on the horizontal axis did the number of degrees for men equal that for women?

(b) When the number of degrees for men was equal to that for women, what was that number (approximately)?

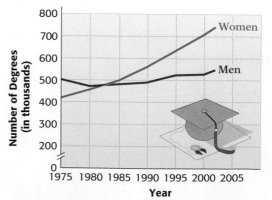

Bachelor's Degrees in the United States

Source: U.S. National Center for Education Statistics, *Digest of Education Statistics*, annual.

2. Solve the system by graphing.

$$x + 3y = 8$$
$$2x - y = 2$$

3. *Concept Check* Which one of the following ordered pairs is not a solution of the equation $3x + 2y = 6$?

A. $(2, 0)$ **B.** $(0, 3)$ **C.** $(4, -3)$ **D.** $(3, -2)$

4. *Concept Check* Suppose that two linear equations are graphed on the same set of coordinate axes. Sketch what the graph might look like if the system has the given description.

(a) The system has a single solution. **(b)** The system has no solution.

(c) The system has infinitely many solutions.

Solve each system by the substitution method. If a system is inconsistent or has dependent equations, say so.

5. $3x + y = -4$

$x = \dfrac{2}{3}y$

6. $9x - y = -4$

$y = x + 4$

7. $-5x + 2y = -2$

$x + 6y = 26$

Solve each system of equations by the elimination method. If a system is inconsistent or has dependent equations, say so.

8. $5x + y = 12$

$2x - 2y = 0$

9. $x - 4y = -4$

$3x + y = 1$

10. $6x + 5y = 4$

$-4x + 2y = 8$

11. $\dfrac{1}{6}x + \dfrac{1}{6}y = -\dfrac{1}{2}$

$x - y = -9$

12. $-3x + y = 6$

$y = 6 + 3x$

13. $5x - 4y = 2$

$-10x + 8y = 7$

14. Without doing any algebraic work, but answering on the basis of your knowledge of the graphs of the two lines, explain why the system

$$y = 3x + 2$$
$$y = 3x - 4$$

has $\emptyset$ as its solution set.

[4.2] *Solve each system. If a system is inconsistent or has dependent equations, say so.*

15. $2x + 3y - z = -16$

$x + 2y + 2z = -3$

$-3x + y + z = -5$

16. $4x - y = 2$

$3y + z = 9$

$x + 2z = 7$

17. $3x - y - z = -8$

$4x + 2y + 3z = 15$

$-6x + 2y + 2z = 10$

[4.3] *Solve each problem by using a system of equations.*

18. A regulation National Hockey League ice rink has perimeter 570 ft. The length of the rink is 30 ft longer than twice the width. What are the dimensions of an NHL ice rink? (*Source: Microsoft Encarta Encyclopedia.*)

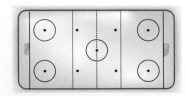

19. In 2004, the Boston Red Sox and the Chicago Cubs had the most expensive average ticket prices in Major League Baseball. Four Red Sox tickets and four Cubs tickets purchased at their average prices cost $276.88, while two Red Sox tickets and six Cubs tickets cost $252.24. Find the average ticket price for a Red Sox ticket and a Cubs ticket. (*Source:* AP.)

20. A plane flies 560 mi in 1.75 hr traveling with the wind. The return trip later against the same wind takes the plane 2 hr. Find the speed of the plane and the speed of the wind. Let x = the speed of the plane and y = the speed of the wind.

	r	t	d
With Wind	$x + y$	1.75	
Against Wind		2	

21. Sweet's Candy Store is offering a special mix for Valentine's Day. Ms. Sweet will mix some $2-per-lb nuts with some $1-per-lb chocolate candy to get 100 lb of mix, which she will sell at $1.30 per lb. How many pounds of each should she use?

	Number of Pounds	Price per Pound	Value
Nuts	x		
Chocolate	y		
Mixture	100		

22. The sum of the measures of the angles of a triangle is 180°. The largest angle measures 10° less than the sum of the other two. The measure of the middle-sized angle is the average of the other two. Find the measures of the three angles.

23. Maria Gonzales sells real estate. On three recent sales, she made 10% commission, 6% commission, and 5% commission. Her total commissions on these sales were $17,000, and she sold property worth $280,000. If the 5% sale amounted to the sum of the other two, what were the three sales prices?

24. How many liters each of 8%, 10%, and 20% hydrogen peroxide should be mixed together to get 8 L of 12.5% solution if the amount of 8% solution used must be 2 L more than the amount of 20% solution used?

25. In the great baseball year of 1961, Yankee teammates Mickey Mantle, Roger Maris, and John Blanchard combined for 136 home runs. Mantle hit 7 fewer than Maris. Maris hit 40 more than Blanchard. What were the home run totals for each player? (*Source:* Neft, David S., and Richard M. Cohen, *The Sports Encyclopedia: Baseball 1997.*)

[4.4] *Solve each system of equations by using row operations.*

26. $2x + 5y = -4$
$4x - y = 14$

27. $6x + 3y = 9$
$-7x + 2y = 17$

28. $x + 2y - z = 1$
$3x + 4y + 2z = -2$
$-2x - y + z = -1$

29. $x + 3y = 7$
$3x + z = 2$
$y - 2z = 4$

MIXED REVIEW EXERCISES

30. *Concept Check* Which system, A or B, would be easier to solve using the substitution method? Why?

A. $5x - 3y = 7$
$2x + 8y = 3$

B. $7x + 2y = 4$
$y = -3x + 1$

Solve by any method.

31. $\frac{2}{3}x + \frac{1}{6}y = \frac{19}{2}$
$\frac{1}{3}x - \frac{2}{9}y = 2$

32. $2x + 5y - z = 12$
$-x + y - 4z = -10$
$-8x - 20y + 4z = 31$

33. $x = 7y + 10$
$2x + 3y = 3$

34. $x + 4y = 17$
$-3x + 2y = -9$

35. $-7x + 3y = 12$
$5x + 2y = 8$

36. $2x - 5y = 8$
$3x + 4y = 10$

37. To make a 10% acid solution for chemistry class, Xavier wants to mix some 5% solution with 10 L of 20% solution. How many liters of 5% solution should he use?

38. In the 2006 Winter Olympics in Turino, Italy, the top medal-winning countries were Germany, the United States, and Canada, with a combined total of 78 medals. Germany won four more medals than the United States, while Canada won one fewer medal than the United States. How many medals did each country win? (*Source:* www.nbcolympics.com)

Chapter 4 TEST

View the complete solutions to all Chapter Test exercises on the Pass the Test CD.

If the rates of growth between 1990 and 2000 continue, the populations of Houston, Phoenix, Dallas, and Philadelphia will follow the trends indicated in the graph. Use the graph to work Exercises 1 and 2.

1. (a) Which of these cities will experience population growth?
(b) Which city will experience population decline?
(c) Rank the city populations from least to greatest for the year 2000.

2. (a) In which year will the population of Dallas equal that of Philadelphia? About what will this population be?
(b) Write as an ordered pair (year, population in millions) the point at which Houston and Phoenix will have the same population.

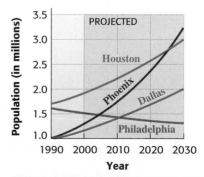

The Growth Game
Size of cities if the rate of population growth from 1990 to 2000 continues:

Source: U.S. Census Bureau, *Chronicle* research.

3. Use a graph to solve the system $\quad\begin{aligned}x + y &= 7\\ x - y &= 5.\end{aligned}$

Solve each system by substitution or elimination. If a system is inconsistent or has dependent equations, say so.

4. $2x - 3y = 24$
$y = -\dfrac{2}{3}x$

5. $3x - y = -8$
$2x + 6y = 3$

6. $12x - 5y = 8$
$3x = \dfrac{5}{4}y + 2$

7. $3x + y = 12$
$2x - y = 3$

8. $-5x + 2y = -4$
$6x + 3y = -6$

9. $3x + 4y = 8$
$8y = 7 - 6x$

10. $3x + 5y + 3z = 2$
$6x + 5y + z = 0$
$3x + 10y - 2z = 6$

11. $4x + y + z = 11$
$x - y - z = 4$
$y + 2z = 0$

Solve each problem using a system of equations.

12. Julia Roberts is one of the biggest box-office stars in Hollywood. As of June 2004, her two top-grossing domestic films, *Ocean's Eleven* and *Runaway Bride,* earned $335.5 million together. If *Runaway Bride* grossed $31.3 million less than *Ocean's Eleven,* how much did each film gross? (*Source:* www.the-numbers.com)

13. Two cars start from points 420 mi apart and travel toward each other. They meet after 3.5 hr. Find the average speed of each car if one travels 30 mph slower than the other.

14. A chemist needs 12 L of a 40% alcohol solution. She must mix a 20% solution and a 50% solution. How many liters of each will be required to obtain what she needs?

15. A local electronics store will sell seven AC adaptors and two rechargeable flashlights for $86, or three AC adaptors and four rechargeable flashlights for $84. What is the price of a single AC adaptor and a single rechargeable flashlight?

16. The owner of a tea shop wants to mix three kinds of tea to make 100 oz of a mixture that will sell for $0.83 per oz. He uses Orange Pekoe, which sells for $0.80 per oz, Irish Breakfast, for $0.85 per oz, and Earl Grey, for $0.95 per oz. If he wants to use twice as much Orange Pekoe as Irish Breakfast, how much of each kind of tea should he use?

Solve each system using row operations.

17. $3x + 2y = 4$
$5x + 5y = 9$

18. $x + 3y + 2z = 11$
$3x + 7y + 4z = 23$
$5x + 3y - 5z = -14$

Chapters **1–4** CUMULATIVE REVIEW EXERCISES

Perform each operation.

1. $-\dfrac{3}{4} - \dfrac{2}{5}$

2. $\dfrac{8}{15} \div \left(-\dfrac{12}{5}\right)$

Evaluate each expression if possible.

3. $(-3)^4$

4. -3^4

5. $-(-3)^4$

6. $\sqrt{0.49}$

7. $-\sqrt{0.49}$

8. $\sqrt{-0.49}$

Evaluate if $x = -4$, $y = 3$, and $z = 6$.

9. $|2x| + 3y - z^3$

10. $-5(x^3 - y^3)$

11. Which property of real numbers justifies the statement $5 + (3 \cdot 6) = 5 + (6 \cdot 3)$?

Solve each equation.

12. $7(2x + 3) - 4(2x + 1) - 2(x + 1)$

13. $|6x - 8| = 4$

14. $ax + by = cx + d$ for x

15. $0.04x + 0.06(x - 1) = 1.04$

Solve each inequality.

16. $\dfrac{2}{3}x + \dfrac{5}{12}x \le 20$

17. $|3x + 2| \le 4$

18. $|12t + 7| \ge 0$

19. A survey measured public recognition of the most popular contemporary advertising slogans. Complete the results shown in the table if 2500 people were surveyed.

Slogan (product or company)	Percent Recognition (nearest tenth of a percent)	Actual Number Who Recognized Slogan (nearest whole number)
Please Don't Squeeze the . . . (Charmin)	80.4%	
The Breakfast of Champions (Wheaties)	72.5%	
The King of Beers (Budweiser)		1570
Like a Good Neighbor (State Farm)		1430

[Other slogans included "You're in Good Hands" (Allstate), "Snap, Crackle, Pop" (Rice Krispies), and "The Un-Cola" (7-Up).]
Source: Department of Integrated Marketing Communications, Northwestern University.

Solve each problem.

20. A triangle has an area of 42 m². The base is 14 m long. Find the height of the triangle.

21. A jar contains only pennies, nickels, and dimes. The number of dimes is one more than the number of nickels, and the number of pennies is six more than the number of nickels. How many of each denomination can be found in the jar, if the total value is $4.80?

22. Two angles of a triangle have the same measure. The measure of the third angle is 4° less than twice the measure of each of the equal angles. Find the measures of the three angles.

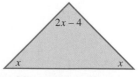

Measures are in degrees.

In Exercises 23–27, point A has coordinates $(-2, 6)$ *and point B has coordinates* $(4, -2)$.

23. What is the equation of the horizontal line through A?

24. What is the equation of the vertical line through B?

25. What is the slope of line AB?

26. What is the slope of a line perpendicular to line AB?

27. What is the standard form of the equation of line AB?

28. Graph the line having slope $\frac{2}{3}$ and passing through the point $(-1, -3)$.

29. Graph the inequality $-3x - 2y \leq 6$.

30. Given that $f(x) = x^2 + 3x - 6$, find **(a)** $f(-3)$ and   **(b)** $f(a)$.

Solve by any method.

31. $\begin{aligned} -2x + 3y &= -15 \\ 4x - y &= 15 \end{aligned}$

32. $\begin{aligned} x - 3y &= 7 \\ 2x - 6y &= 14 \end{aligned}$

33. $\begin{aligned} x + y + z &= 10 \\ x - y - z &= 0 \\ -x + y - z &= -4 \end{aligned}$

In Exercises 34–37, solve each problem by using a system of equations.

34. Mabel Johnston bought apples and oranges at De Ville's Grocery. She bought 6 lb of fruit. Oranges cost $0.90 per lb, while apples cost $0.70 per lb. If she spent a total of $5.20, how many pounds of each kind of fruit did she buy?

35. Ten years after the original Tickle Me Elmo became a must-have toy, a new version, called T.M.X., was released in the fall of 2006. The original Tickle Me Elmo's average cost was $12.37 less than the recommended cost of T.M.X., and one of each cost $67.63. Find the average cost of Tickle Me Elmo and the recommended cost of T.M.X. (*Source:* NPD Group, Inc.; *USA Today.*)

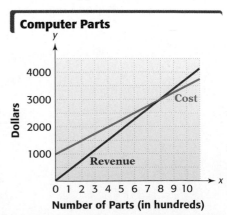

36. Butch and Peggy need some cardboard boxes. They can buy 10 small and 20 large boxes for $65, or 6 small and 10 large boxes for $34. Find the cost of each size of box.

37. At the New Roads Nut Shop, 6 lb of peanuts and 12 lb of cashews cost $60, while 3 lb of peanuts and 4 lb of cashews cost $22. Find the cost of each type of nut.

38. The graph shows a company's costs to produce computer parts and the revenue from the sale of computer parts.

   (a) At what production level does the cost equal the revenue? What is the revenue at that point?

   (b) Profit is revenue less cost. Estimate the profit on the sale of 1100 parts.

Computer Parts

Exponents, Polynomials, and Polynomial Functions

In 1980, MasterCard International Incorporated first began offering debit cards in an effort to challenge Visa, the leader in credit card transactions at that time. Now immensely popular, debit cards draw money from consumers' bank accounts rather than from established lines of credit. It is estimated that nearly 300 million debit cards are now in use. (*Source: Microsoft Encarta Encyclopedia*; HSN Consultants, Inc.)

We introduced the concept of a function in Section 3.5 and extend our work to include *polynomial functions* in this chapter. In Exercise 11 of Section 5.3, we use a polynomial function to model the number of bank debit cards issued.

5.1 Integer Exponents and Scientific Notation

Recall that we use exponents to write products of repeated factors. For example,

$$2^5 \quad \text{is defined as} \quad 2 \cdot 2 \cdot 2 \cdot 2 \cdot 2 = 32.$$

The number 5, the *exponent,* shows that the *base* 2 appears as a factor five times. The quantity 2^5 is called an *exponential* or a *power.* We read 2^5 as "2 to the fifth power" or "2 to the fifth."

OBJECTIVE 1 **Use the product rule for exponents.** There are rules that simplify work with exponents. For example, the product $2^5 \cdot 2^3$ can be simplified as follows.

$$\overbrace{2^5 \cdot 2^3 = (2 \cdot 2 \cdot 2 \cdot 2 \cdot 2)(2 \cdot 2 \cdot 2) = 2^8}^{5 + 3 = 8}$$

This result, that products of exponential expressions with the *same base* are found by adding exponents, is generalized as the **product rule for exponents.**

Product Rule for Exponents

If m and n are natural numbers and a is any real number, then

$$a^m \cdot a^n = a^{m+n}.$$

That is, when multiplying powers of like bases, keep the same base and add the exponents.

To see that the product rule is true, use the definition of an exponent as follows.

$$a^m = \underbrace{a \cdot a \cdot a \cdot \ldots \cdot a}_{a \text{ appears as a factor } m \text{ times.}} \qquad a^n = \underbrace{a \cdot a \cdot a \cdot \ldots \cdot a}_{a \text{ appears as a factor } n \text{ times.}}$$

From this, $\quad a^m \cdot a^n = \underbrace{a \cdot a \cdot a \cdot \ldots \cdot a}_{m \text{ factors}} \cdot \underbrace{a \cdot a \cdot a \cdot \ldots \cdot a}_{n \text{ factors}}$

$$= \underbrace{a \cdot a \cdot a \cdot \ldots \cdot a}_{(m+n) \text{ factors}}$$

$$a^m \cdot a^n = a^{m+n}.$$

EXAMPLE 1 **Using the Product Rule for Exponents**

Apply the product rule for exponents, if possible, in each case.

(a) $3^4 \cdot 3^7 = 3^{4+7} = 3^{11}$

> Do *not* multiply the bases! Keep the same base.

(b) $5^3 \cdot 5 = 5^3 \cdot 5^1 = 5^{3+1} = 5^4$

(c) $y^3 \cdot y^8 \cdot y^2 = y^{3+8+2} = y^{13}$

(d) $(5y^2)(-3y^4) = 5(-3)y^2y^4$ Associative and commutative properties

$$= -15y^{2+4} \qquad \text{Product rule}$$

$$= -15y^6$$

(e) $(7p^3q)(2p^5q^2) = 7(2)p^3p^5q^1q^2 = 14p^8q^3$

(f) $x^2 \cdot y^4$ The product rule does not apply because the bases are not the same.

> ✔ **Now Try Exercises 7, 11, 13, and 15.**

▶ **CAUTION**   Be careful in problems like Example 1(a) not to multiply the bases. Notice that $3^4 \cdot 3^7 = 3^{11}$, *not* 9^{11}. *Remember to keep the same base and add the exponents.*

OBJECTIVE 2   Define 0 and negative exponents.   Suppose we multiply 4^2 by 4^0. Extending the product rule, we should have

$$4^2 \cdot 4^0 = 4^{2+0} = 4^2.$$

For the product rule to hold, 4^0 must equal 1, so we define a^0 that way for any nonzero real number a.

Zero Exponent

If a is any nonzero real number, then

$$a^0 = 1.$$

---

*The expression 0^0 is undefined.**

EXAMPLE 2   Using 0 as an Exponent

Evaluate.

(a) $6^0 = 1$

> The base is 6, *not* −6.

(b) $(-6)^0 = 1$

> Here the base is −6.

(c) $-6^0 = -(6^0) = -(1) = -1$

(d) $-(-6)^0 = -1$

(e) $5^0 + 12^0 = 1 + 1 = 2$

(f) $(8k)^0 = 1, \quad k \neq 0$

> ✔ **Now Try Exercises 17 and 25.**

To define a negative exponent, we extend the product rule. For example,

$$8^2 \cdot 8^{-2} = 8^{2+(-2)} = 8^0 = 1.$$

Here 8^{-2} is the reciprocal of 8^2. But $\frac{1}{8^2}$ is the reciprocal of 8^2, and a number can have only one reciprocal. Therefore, $8^{-2} = \frac{1}{8^2}$. We generalize this result as follows.

Negative Exponent

For any natural number n and any nonzero real number a,

$$a^{-n} = \frac{1}{a^n}.$$

---

*In advanced treatments, 0^0 is called an *indeterminate form.*

With this definition, the expression a^n is meaningful for any integer exponent n and any nonzero real number a. The product rule is valid for integers m and n.

▶ **CAUTION** *A negative exponent does not indicate a negative number;* negative exponents lead to reciprocals. For example,

$$3^{-2} = \frac{1}{3^2} = \frac{1}{9} \quad \text{Not negative} \quad \bigg| \quad -3^{-2} = -\frac{1}{3^2} = -\frac{1}{9}. \quad \text{Negative}$$

EXAMPLE 3 Using Negative Exponents

In parts (a)–(f), write each exponential with only positive exponents.

(a) $2^{-3} = \dfrac{1}{2^3}$

(b) $6^{-1} = \dfrac{1}{6^1} = \dfrac{1}{6}$

(c) $(5z)^{-3} = \dfrac{1}{(5z)^3}, \quad z \neq 0$
↑
Base is 5z.

(d) $5z^{-3} = 5\left(\dfrac{1}{z^3}\right) = \dfrac{5}{z^3}, \quad z \neq 0$
↑
Base is z.

(e) $-m^{-2} = -\dfrac{1}{m^2}, \quad m \neq 0$

(What is the base here?)

(f) $(-m)^{-2} = \dfrac{1}{(-m)^2}, \quad m \neq 0$

(What is the base here?)

In parts (g) and (h), evaluate.

(g) $3^{-1} + 4^{-1} = \dfrac{1}{3} + \dfrac{1}{4} = \dfrac{4}{12} + \dfrac{3}{12} = \dfrac{7}{12} \qquad \frac{1}{3} \cdot \frac{4}{4} = \frac{4}{12}; \frac{1}{4} \cdot \frac{3}{3} = \frac{3}{12}$

> The product rule does *not* apply here.

(h) $5^{-1} - 2^{-1} = \dfrac{1}{5} - \dfrac{1}{2} = \dfrac{2}{10} - \dfrac{5}{10} = -\dfrac{3}{10}$

✔ **Now Try Exercises 33, 37, 39, 41, and 45.**

▶ **CAUTION** In Example 3(g), note that $3^{-1} + 4^{-1} \neq (3 + 4)^{-1}$. The expression on the left is equal to $\frac{7}{12}$, as shown in the example, while the expression on the right is $7^{-1} = \frac{1}{7}$. Similar reasoning can be applied to part (h).

EXAMPLE 4 Using Negative Exponents

Evaluate.

(a) $\dfrac{1}{2^{-3}} = \dfrac{1}{\frac{1}{2^3}} = 1 \div \dfrac{1}{2^3} = 1 \cdot \dfrac{2^3}{1} = 2^3 = 8$

Multiply by the reciprocal of the divisor.

(b) $\dfrac{2^{-3}}{3^{-2}} = \dfrac{\frac{1}{2^3}}{\frac{1}{3^2}} = \dfrac{1}{2^3} \div \dfrac{1}{3^2} = \dfrac{1}{2^3} \cdot \dfrac{3^2}{1} = \dfrac{3^2}{2^3} = \dfrac{9}{8}$

✔ **Now Try Exercises 51 and 53.**

Example 4 suggests the following generalizations.

Special Rules for Negative Exponents

If $a \neq 0$ and $b \neq 0$, then $\quad \dfrac{1}{a^{-n}} = a^n \quad$ and $\quad \dfrac{a^{-n}}{b^{-m}} = \dfrac{b^m}{a^n}.$

OBJECTIVE 3 Use the quotient rule for exponents. A quotient, such as $\dfrac{a^8}{a^3}$, can be simplified in much the same way as a product. (In all quotients of this type, assume that the denominator is not 0.) Using the definition of an exponent gives

$$\frac{a^8}{a^3} = \frac{a \cdot a \cdot a \cdot a \cdot a \cdot a \cdot a \cdot a}{a \cdot a \cdot a} = a \cdot a \cdot a \cdot a \cdot a = a^5.$$

Notice that $8 - 3 = 5$. In the same way,

$$\frac{a^3}{a^8} = \frac{a \cdot a \cdot a}{a \cdot a \cdot a \cdot a \cdot a \cdot a \cdot a \cdot a} = \frac{1}{a^5} = a^{-5}.$$

Here $3 - 8 = -5$. These examples suggest the **quotient rule for exponents.**

Quotient Rule for Exponents

If a is any nonzero real number and m and n are integers, then

$$\frac{a^m}{a^n} = a^{m-n}.$$

That is, when dividing powers of like bases, keep the same base and subtract the exponent of the denominator from the exponent of the numerator.

EXAMPLE 5 Using the Quotient Rule for Exponents

Apply the quotient rule for exponents, if possible, and write each result with only positive exponents.

Numerator exponent
Denominator exponent

(a) $\dfrac{3^7}{3^2} = 3^{7-2} = 3^5$

Minus sign

(b) $\dfrac{p^6}{p^2} = p^{6-2} = p^4, \quad p \neq 0$

(c) $\dfrac{k^7}{k^{12}} = k^{7-12} = k^{-5} = \dfrac{1}{k^5}, \quad k \neq 0$

(d) $\dfrac{2^7}{2^{-3}} = 2^{7-(-3)} = 2^{7+3} = 2^{10}$

Use parentheses to avoid errors.

(e) $\dfrac{8^{-2}}{8^5} = 8^{-2-5} = 8^{-7} = \dfrac{1}{8^7}$

(f) $\dfrac{6}{6^{-1}} = \dfrac{6^1}{6^{-1}} = 6^{1-(-1)} = 6^2$

(g) $\dfrac{z^{-5}}{z^{-8}} = z^{-5-(-8)} = z^3, \quad z \neq 0$

(h) $\dfrac{a^3}{b^4}, \quad b \neq 0$

This expression cannot be simplified further.

Be careful with signs.

The quotient rule does not apply because the bases are different.

✔ **Now Try Exercises 63, 67, 75, and 77.**

▶ **CAUTION** Be careful when working with quotients that involve negative exponents in the denominator. Be sure to write the numerator exponent, then a minus sign, and then the denominator exponent. Use parentheses.

OBJECTIVE 4 Use the power rules for exponents. We can simplify $(3^4)^2$ as

$$(3^4)^2 = 3^4 \cdot 3^4 = 3^{4+4} = 3^8,$$

where $4 \cdot 2 = 8$. This example suggests the first **power rule for exponents.** The other two power rules can be demonstrated with similar examples.

Power Rules for Exponents

If a and b are real numbers and m and n are integers, then

(a) $(a^m)^n = a^{mn}$,   **(b)** $(ab)^m = a^m b^m$,   and   **(c)** $\left(\dfrac{a}{b}\right)^m = \dfrac{a^m}{b^m}$ $(b \neq 0)$.

That is,

(a) To raise a power to a power, multiply exponents.

(b) To raise a product to a power, raise each factor to that power.

(c) To raise a quotient to a power, raise the numerator and the denominator to that power.

EXAMPLE 6 Using the Power Rules for Exponents

Simplify, using the power rules.

(a) $(p^8)^3 = p^{8 \cdot 3} = p^{24}$

(b) $\left(\dfrac{2}{3}\right)^4 = \dfrac{2^4}{3^4} = \dfrac{16}{81}$

(c) $(3y)^4 = 3^4 y^4 = 81y^4$

(d) $(6p^7)^2 = 6^2 p^{7 \cdot 2} = 6^2 p^{14} = 36p^{14}$

(e) $\left(\dfrac{-2m^5}{z}\right)^3 = \dfrac{(-2)^3 m^{5 \cdot 3}}{z^3} = \dfrac{(-2)^3 m^{15}}{z^3} = \dfrac{-8m^{15}}{z^3}, \quad z \neq 0$

✔ **Now Try Exercises 79, 81, 83, and 87.**

The reciprocal of a^n is $\dfrac{1}{a^n} = \left(\dfrac{1}{a}\right)^n$. Also, by definition, a^n and a^{-n} are reciprocals, since

$$a^n \cdot a^{-n} = a^n \cdot \frac{1}{a^n} = 1.$$

Thus, since both are reciprocals of a^n,

$$a^{-n} = \left(\frac{1}{a}\right)^n.$$

Some examples of this result are

$$6^{-3} = \left(\frac{1}{6}\right)^3 \qquad \text{and} \qquad \left(\frac{1}{3}\right)^{-2} = 3^2.$$

This discussion can be generalized as follows.

More Special Rules for Negative Exponents

If $a \neq 0$ and $b \neq 0$ and n is an integer, then

$$a^{-n} = \left(\frac{1}{a}\right)^n \qquad \text{and} \qquad \left(\frac{a}{b}\right)^{-n} = \left(\frac{b}{a}\right)^n.$$

That is, any nonzero number raised to the negative nth power is equal to the reciprocal of that number raised to the nth power.

EXAMPLE 7 Using Negative Exponents with Fractions

Write with only positive exponents and then evaluate.

(a) $\left(\dfrac{3}{7}\right)^{-2} = \left(\dfrac{7}{3}\right)^2 = \dfrac{49}{9}$

(b) $\left(\dfrac{4}{5}\right)^{-3} = \left(\dfrac{5}{4}\right)^3 = \dfrac{125}{64}$

> Change the fraction to its reciprocal and change the sign of the exponent.

✔ **Now Try Exercise 55.**

The definitions and rules of this section are summarized here.

Definitions and Rules for Exponents

For all integers m and n and all real numbers a and b, the following rules apply.

Product Rule $a^m \cdot a^n = a^{m+n}$

Quotient Rule $\dfrac{a^m}{a^n} = a^{m-n} \quad (a \neq 0)$

Zero Exponent $a^0 = 1 \quad (a \neq 0)$

(continued)

Definitions and Rules for Exponents (continued)

Negative Exponent $\quad a^{-n} = \dfrac{1}{a^n} \quad (a \neq 0)$

Power Rules $\quad (a^m)^n = a^{mn} \qquad\qquad\qquad (ab)^m = a^m b^m$

$$\left(\dfrac{a}{b}\right)^m = \dfrac{a^m}{b^m} \quad (b \neq 0)$$

Special Rules $\quad \dfrac{1}{a^{-n}} = a^n \quad (a \neq 0) \qquad \dfrac{a^{-n}}{b^{-m}} = \dfrac{b^m}{a^n} \quad (a, b \neq 0)$

$$a^{-n} = \left(\dfrac{1}{a}\right)^n \quad (a \neq 0) \quad \left(\dfrac{a}{b}\right)^{-n} = \left(\dfrac{b}{a}\right)^n \quad (a, b \neq 0)$$

OBJECTIVE 5 Simplify exponential expressions.

EXAMPLE 8 Using the Definitions and Rules for Exponents

Simplify so that no negative exponents appear in the final result. Assume that all variables represent nonzero real numbers.

(a) $3^2 \cdot 3^{-5} = 3^{2+(-5)} = 3^{-3} = \dfrac{1}{3^3}, \quad \text{or} \quad \dfrac{1}{27}$

(b) $x^{-3}x^{-4}x^2 = x^{-3+(-4)+2} = x^{-5} = \dfrac{1}{x^5}$

(c) $(4^{-2})^{-5} = 4^{(-2)(-5)} = 4^{10}$

(d) $(x^{-4})^6 = x^{(-4)6} = x^{-24} = \dfrac{1}{x^{24}}$

(e) $\dfrac{x^{-4}y^2}{x^2y^{-5}} = \dfrac{x^{-4}}{x^2} \cdot \dfrac{y^2}{y^{-5}}$

$\qquad = x^{-4-2} \cdot y^{2-(-5)}$

$\qquad = x^{-6}y^7$

$\qquad = \dfrac{y^7}{x^6}$

(f) $(2^3 x^{-2})^{-2} = (2^3)^{-2} \cdot (x^{-2})^{-2}$

$\qquad\qquad\quad = 2^{-6}x^4$

$\qquad\qquad\quad = \dfrac{x^4}{2^6}, \quad \text{or} \quad \dfrac{x^4}{64}$

(g) $\left(\dfrac{3x^2}{y}\right)^2 \left(\dfrac{4x^3}{y^{-2}}\right)^{-1} = \dfrac{3^2(x^2)^2}{y^2} \cdot \dfrac{y^{-2}}{4x^3}$ $\qquad$ Combination of rules

$\qquad\qquad = \dfrac{9x^4}{y^2} \cdot \dfrac{y^{-2}}{4x^3}$ $\qquad$ Power rule

$\qquad\qquad = \dfrac{9}{4}x^{4-3}y^{-2-2}$ $\qquad$ Quotient rule

$\qquad\qquad = \dfrac{9x}{4y^4}$

✔ **Now Try Exercises 91, 93, 107, and 117.**

▶ **NOTE** There is often more than one way to simplify expressions like those in Example 8. For instance, we could simplify Example 8(e) as follows.

$$\frac{x^{-4}y^2}{x^2y^{-5}} = \frac{y^5y^2}{x^4x^2} \qquad \text{Use } \frac{a^{-n}}{b^{-m}} = \frac{b^m}{a^n}.$$

$$= \frac{y^7}{x^6} \qquad \text{Product rule}$$

OBJECTIVE 6 Use the rules for exponents with scientific notation. Scientists often use numbers that are very large or very small. For example, the number of one-celled organisms that will sustain a whale for a few hours is 400,000,000,000,000, and the shortest wavelength of visible light is approximately 0.0000004 m. It is simpler to write these numbers in *scientific notation*.

In scientific notation, a number is written with the decimal point after the first nonzero digit and multiplied by a power of 10, as indicated in the following definition.

Scientific Notation

A number is written in **scientific notation** when it is expressed in the form

$$a \times 10^n$$

where $1 \le |a| < 10$ and n is an integer.

For example, in scientific notation,

$$8000 = 8 \times 1000 = 8 \times 10^3.$$

In scientific notation it is customary to use a times sign ($\times$) instead of a multiplication dot. The following numbers are *not* in scientific notation.

$$0.230 \times 10^4 \qquad\qquad 46.5 \times 10^{-3}$$

0.230 is less than 1.　　　　　46.5 is greater than 10.

To write a number in scientific notation, use the following steps. (If the number is negative, go through these steps, and then attach a negative sign to the result.)

Converting to Scientific Notation

Step 1 **Position the decimal point.** Place a caret, ^, to the right of the first nonzero digit, where the decimal point will be placed.

Step 2 **Determine the numeral for the exponent.** Count the number of digits from the decimal point to the caret. This number gives the absolute value of the exponent on 10.

Step 3 **Determine the sign for the exponent.** Decide whether multiplying by 10^n should make the result of Step 1 greater or less. The exponent should be positive to make the result greater; it should be negative to make the result less.

It is helpful to remember that, for $n \ge 1$, $10^{-n} < 1$ and $10^n \ge 10$.

EXAMPLE 9 Writing Numbers in Scientific Notation

Write each number in scientific notation.

(a) 820,000

> **Step 1** Place a caret to the right of the 8 (the first nonzero digit) to mark the new location of the decimal point.
>
> $$8_\wedge 20,000$$
>
> **Step 2** Count from the decimal point, which is understood to be after the last 0, to the caret.
>
> $$8.20,000. \leftarrow \text{Decimal point}$$
> Count 5 places.
>
> **Step 3** Since 8.2 is to be made greater, the exponent on 10 is positive.
>
> $$820,000 = 8.2 \times 10^5$$

(b) 0.0000072

Count from left to right.

$$0.000007.2$$
6 places

Since the number 7.2 is to be made less, the exponent on 10 is negative.

$$0.0000072 = 7.2 \times 10^{-6}$$

(c) $-0.0000462 = -4.62 \times 10^{-5}$
Count 5 places.

✔ **Now Try Exercises 127, 131, and 133.**

To convert scientific notation to standard notation, just work in reverse.

Converting from Scientific Notation

Multiplying a number by a positive power of 10 makes the number greater, so move the decimal point to the right if n is positive in 10^n.

Multiplying by a negative power of 10 makes a number less, so move the decimal point to the left if n is negative.

If n is 0, leave the decimal point where it is.

EXAMPLE 10 Converting from Scientific Notation to Standard Notation

Write each number in standard notation.

(a) 6.93×10^7

$$6.9300000 \qquad \text{Attach 0's as necessary.}$$
7 places

We moved the decimal point 7 places to the right. (We had to attach five 0's.)

$$6.93 \times 10^7 = 69,300,000$$

(b) 4.7×10^{-6}

$$000004.7 \quad \text{Attach 0's as necessary.}$$
6 places

> Add a leading 0 since the decimal is between 0 and 1.

We moved the decimal point 6 places to the left.

$$4.7 \times 10^{-6} = .0000047, \quad \text{or} \quad 0.0000047$$

(c) $-1.083 \times 10^{0} = -1.083 \times 1 = -1.083$

> ✔ **Now Try Exercises 135, 137, and 139.**

When problems require operations with numbers that are very large or very small, it is often advantageous to write the numbers in scientific notation first and then perform the calculations, using the rules for exponents.

EXAMPLE 11 Using Scientific Notation in Computation

Evaluate $\dfrac{1{,}920{,}000 \times 0.0015}{0.000032 \times 45{,}000}$.

First, express all numbers in scientific notation.

$$\frac{1{,}920{,}000 \times 0.0015}{0.000032 \times 45{,}000} = \frac{1.92 \times 10^{6} \times 1.5 \times 10^{-3}}{3.2 \times 10^{-5} \times 4.5 \times 10^{4}}$$

$$= \frac{1.92 \times 1.5 \times 10^{6} \times 10^{-3}}{3.2 \times 4.5 \times 10^{-5} \times 10^{4}} \quad \text{Commutative and associative properties}$$

$$= \frac{1.92 \times 1.5 \times 10^{3}}{3.2 \times 4.5 \times 10^{-1}} \quad \text{Product rule}$$

$$= \frac{1.92 \times 1.5}{3.2 \times 4.5} \times 10^{4} \quad \text{Quotient rule}$$

$$= 0.2 \times 10^{4}$$

> Don't stop here!

$$= (2 \times 10^{-1}) \times 10^{4} \quad \text{Write 0.2 in scientific notation.}$$

$$= 2 \times 10^{3} \quad \text{Product rule}$$

$$= 2000 \quad \text{Standard notation}$$

> ✔ **Now Try Exercise 149.**

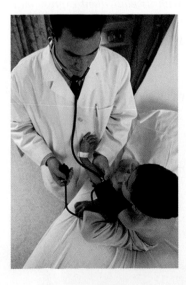

EXAMPLE 12 Using Scientific Notation to Solve Problems

In 1990, the national health care expenditure was $696 billion. By 2003, this figure had risen by a factor of 2.4. The expenditure had risen by almost $2\frac{1}{2}$ times in only 13 yr. (*Source:* U.S. Centers for Medicare & Medicaid Services.)

(a) Write the 1990 health care expenditure in scientific notation.

$$696 \text{ billion} = 696 \times 10^{9} \quad \text{1 billion} = 1{,}000{,}000{,}000 = 10^{9}$$

$$= (6.96 \times 10^{2}) \times 10^{9} \quad \text{Write 696 in scientific notation.}$$

$$= 6.96 \times 10^{11} \quad \text{Product rule}$$

In 1990, the expenditure was 6.96×10^{11}.

(b) What was the expenditure in 2003?

Multiply the result in part (a) by 2.4.

$(6.96 \times 10^{11}) \times 2.4 = (2.4 \times 6.96) \times 10^{11}$ Commutative and associative properties

$= 16.7 \times 10^{11}$ Round to the nearest tenth.

$= (1.67 \times 10^{1}) \times 10^{11}$ Write 16.7 in scientific notation.

$= 1.67 \times 10^{12}$ Product rule

The 2003 expenditure was about \$1,670,000,000,000, almost \$1.7 trillion.

✔ **Now Try Exercise 153.**

5.1 EXERCISES

⊕ *Complete solution available on Video Lectures on CD/DVD*

▨ *Now Try Exercise*

Concept Check *Decide whether each expression has been simplified correctly. If not, correct it.*

1. $(ab)^2 = ab^2$

2. $y^2 \cdot y^6 = y^{12}$

3. $\left(\dfrac{4}{a}\right)^3 = \dfrac{4^3}{a}$ $(a \neq 0)$

4. $xy^0 = 0$ $(y \neq 0)$

▨ **5.** State the product rule for exponents in your own words. Give an example.

6. *Concept Check* Your friend evaluated $4^5 \cdot 4^2$ as 16^7. **WHAT WENT WRONG?** Give the correct answer.

Apply the product rule for exponents, if possible, in each case. See Example 1.

7. $13^4 \cdot 13^8$ **8.** $9^6 \cdot 9^4$ **9.** $x^3 \cdot x^5 \cdot x^9$ **10.** $y^4 \cdot y^5 \cdot y^6$

11. $(-3w^5)(9w^3)$ **12.** $(-5x^2)(3x^4)$ **13.** $(2x^2y^5)(9xy^3)$ **14.** $(8s^4t)(3s^3t^5)$

15. $r^2 \cdot s^4$ **16.** $p^3 \cdot q^2$

In Exercises 17 and 18, match the expression in Column I with its equivalent expression in Column II. Choices may be used once, more than once, or not at all.* See Example 2.

	I	II		I	II
17.	**(a)** 9^0	**A.** 0	**18.**	**(a)** $2x^0$	**A.** 0
	(b) -9^0	**B.** 1		**(b)** $-2x^0$	**B.** 1
	(c) $(-9)^0$	**C.** -1		**(c)** $(2x)^0$	**C.** -1
	(d) $-(-9)^0$	**D.** 9		**(d)** $(-2x)^0$	**D.** 2
		E. -9			**E.** -2

Evaluate. Assume that all variables represent nonzero real numbers.* See Example 2.

19. 25^0 **20.** 14^0 **21.** -7^0 **22.** -10^0

23. $(-15)^0$ **24.** $(-20)^0$ **25.** $3^0 + (-3)^0$ **26.** $5^0 + (-5)^0$

27. $-3^0 + 3^0$ **28.** $-5^0 + 5^0$ **29.** $-4^0 - m^0$ **30.** $-8^0 - k^0$

---

*The authors thank Mitchel Levy of Broward Community College for his suggestions for these exercises.

In Exercises 31 and 32, match the expression in Column I with its equivalent expression in Column II. Choices may be used once, more than once, or not at all.* See Example 3.

	I	II		I	II
31.	**(a)** 4^{-2}	**A.** 16	**32.**	**(a)** 5^{-3}	**A.** 125
	(b) -4^{-2}	**B.** $\dfrac{1}{16}$		**(b)** -5^{-3}	**B.** -125
	(c) $(-4)^{-2}$	**C.** -16		**(c)** $(-5)^{-3}$	**C.** $\dfrac{1}{125}$
	(d) $-(-4)^{-2}$	**D.** $-\dfrac{1}{16}$		**(d)** $-(-5)^{-3}$	**D.** $-\dfrac{1}{125}$

Write each expression with only positive exponents. Assume that all variables represent nonzero real numbers. In Exercises 45–48, simplify each expression. See Example 3.

33. 5^{-4} **34.** 7^{-2} **35.** 8^{-1} **36.** 12^{-1}

37. $(4x)^{-2}$ **38.** $(5t)^{-3}$ **39.** $4x^{-2}$ **40.** $5t^{-3}$

41. $-a^{-3}$ **42.** $-b^{-4}$ **43.** $(-a)^{-4}$ **44.** $(-b)^{-6}$

45. $5^{-1} + 6^{-1}$ **46.** $2^{-1} + 8^{-1}$ **47.** $8^{-1} - 3^{-1}$ **48.** $6^{-1} - 4^{-1}$

49. *Concept Check* Consider the expressions $-a^n$ and $(-a)^n$. In some cases they are equal, and in some cases they are not. Using $n = 2, 3, 4, 5,$ and 6 and $a = 2$, draw a conclusion as to when they are equal and when they are opposites.

50. *Concept Check* Your friend thinks that $(-3)^{-2}$ is a negative number. Why is she incorrect?

Evaluate each expression. See Examples 4 and 7.

51. $\dfrac{1}{4^{-2}}$ **52.** $\dfrac{1}{3^{-3}}$ **53.** $\dfrac{2^{-2}}{3^{-3}}$ **54.** $\dfrac{3^{-3}}{2^{-2}}$

55. $\left(\dfrac{2}{3}\right)^{-3}$ **56.** $\left(\dfrac{3}{2}\right)^{-3}$ **57.** $\left(\dfrac{4}{5}\right)^{-2}$ **58.** $\left(\dfrac{5}{4}\right)^{-2}$

In Exercises 59 and 60, match the expression in Column I with its equivalent expression in Column II. Choices may be used once, more than once, or not at all.*

	I	II		I	II
59.	**(a)** $\left(\dfrac{1}{3}\right)^{-1}$	**A.** $\dfrac{1}{3}$	**60.**	**(a)** $\left(\dfrac{2}{5}\right)^{-2}$	**A.** $\dfrac{25}{4}$
	(b) $\left(-\dfrac{1}{3}\right)^{-1}$	**B.** 3		**(b)** $\left(-\dfrac{2}{5}\right)^{-2}$	**B.** $-\dfrac{25}{4}$
	(c) $-\left(\dfrac{1}{3}\right)^{-1}$	**C.** $-\dfrac{1}{3}$		**(c)** $-\left(\dfrac{2}{5}\right)^{-2}$	**C.** $\dfrac{4}{25}$
	(d) $-\left(-\dfrac{1}{3}\right)^{-1}$	**D.** -3		**(d)** $-\left(-\dfrac{2}{5}\right)^{-2}$	**D.** $-\dfrac{4}{25}$

*The authors thank Mitchel Levy of Broward Community College for his suggestions for these exercises.

✒ **61.** State the quotient rule for exponents in your own words. Give an example.

✒ **62.** State the three power rules for exponents in your own words. Give examples.

Write each result with only positive exponents. Assume that all variables represent nonzero real numbers. See Example 5.

🌐 **63.** $\dfrac{4^8}{4^6}$ **64.** $\dfrac{5^9}{5^7}$ **65.** $\dfrac{x^{12}}{x^8}$ **66.** $\dfrac{y^{14}}{y^{10}}$

67. $\dfrac{r^7}{r^{10}}$ **68.** $\dfrac{y^8}{y^{12}}$ **69.** $\dfrac{6^4}{6^{-2}}$ **70.** $\dfrac{7^5}{7^{-3}}$

71. $\dfrac{6^{-3}}{6^7}$ **72.** $\dfrac{5^{-4}}{5^2}$ **73.** $\dfrac{7}{7^{-1}}$ **74.** $\dfrac{8}{8^{-1}}$

75. $\dfrac{r^{-3}}{r^{-6}}$ **76.** $\dfrac{s^{-4}}{s^{-8}}$ **77.** $\dfrac{x^3}{y^2}$ **78.** $\dfrac{y^5}{t^3}$

Simplify each expression. Assume that all variables represent nonzero real numbers. See Example 6.

🌐 **79.** $(x^3)^6$ **80.** $(y^5)^4$ **81.** $\left(\dfrac{3}{5}\right)^3$ **82.** $\left(\dfrac{4}{3}\right)^2$

83. $(4t)^3$ **84.** $(5t)^4$ **85.** $(-6x^2)^3$ **86.** $(-2x^5)^5$

87. $\left(\dfrac{-4m^2}{t}\right)^3$ **88.** $\left(\dfrac{-5n^4}{r^2}\right)^3$

Simplify each expression so that no negative exponents appear in the final result. Assume that all variables represent nonzero real numbers. See Examples 1–8.

89. $3^5 \cdot 3^{-6}$ **90.** $4^4 \cdot 4^{-6}$ 🌐 **91.** $a^{-3}a^2a^{-4}$

92. $k^{-5}k^{-3}k^4$ **93.** $(k^2)^{-3}k^4$ **94.** $(x^3)^{-4}x^5$

95. $-4r^{-2}(r^4)^2$ **96.** $-2m^{-1}(m^3)^2$ **97.** $(5a^{-1})^4(a^2)^{-3}$

98. $(3p^{-4})^2(p^3)^{-1}$ **99.** $(z^{-4}x^3)^{-1}$ **100.** $(y^{-2}z^4)^{-3}$

101. $7k^2(-2k)(4k^{-5})^0$ **102.** $3a^2(-5a^{-6})(-2a)^0$ **103.** $\dfrac{(p^{-2})^0}{5p^{-4}}$

104. $\dfrac{(m^4)^0}{9m^{-3}}$ **105.** $\dfrac{(3pq)q^2}{6p^2q^4}$ **106.** $\dfrac{(-8xy)y^3}{4x^5y^4}$

107. $\dfrac{4a^5(a^{-1})^3}{(a^{-2})^{-2}}$ **108.** $\dfrac{12k^{-2}(k^{-3})^{-4}}{6k^5}$ **109.** $\dfrac{(-y^{-4})^2}{6(y^{-5})^{-1}}$

110. $\dfrac{2(-m^{-1})^{-4}}{9(m^{-3})^2}$ **111.** $\dfrac{(2k)^2m^{-5}}{(km)^{-3}}$ **112.** $\dfrac{(3rs)^{-2}}{3^2r^2s^{-4}}$

113. $\dfrac{(2k)^2k^3}{k^{-1}k^{-5}}(5k^{-2})^{-3}$ **114.** $\dfrac{(3r^2)^2r^{-5}}{r^{-2}r^3}(2r^{-6})^2$ **115.** $\left(\dfrac{3k^{-2}}{k^4}\right)^{-1} \cdot \dfrac{2}{k}$

116. $\left(\dfrac{7m^{-2}}{m^{-3}}\right)^{-2} \cdot \dfrac{m^3}{4}$ **117.** $\left(\dfrac{2p}{q^2}\right)^3\left(\dfrac{3p^4}{q^{-4}}\right)^{-1}$ **118.** $\left(\dfrac{5z^3}{2a^2}\right)^{-3}\left(\dfrac{8a^{-1}}{15z^{-2}}\right)^{-3}$

119. $\dfrac{2^2y^4(y^{-3})^{-1}}{2^5y^{-2}}$ **120.** $\dfrac{3^{-1}m^4(m^2)^{-1}}{3^2m^{-2}}$

The next problems are "brain busters." Simplify each expression. Assume that all variables represent nonzero real numbers.

121. $\dfrac{(2m^2p^3)^2(4m^2p)^{-2}}{(-3mp^4)^{-1}(2m^3p^4)^3}$

122. $\dfrac{(-5y^3z^4)^2(2yz^5)^{-2}}{10(y^4z)^3(3y^3z^2)^{-1}}$

123. $\dfrac{(-3y^3x^3)(-4y^4x^2)(x^2)^{-4}}{18x^3y^2(y^3)^3(x^3)^{-2}}$

124. $\dfrac{(2m^3x^2)^{-1}(3m^4x)^{-3}}{(m^2x^3)^3(m^2x)^{-5}}$

125. $\left(\dfrac{p^2q^{-1}}{2p^{-2}}\right)^2 \cdot \left(\dfrac{p^3 \cdot 4q^{-2}}{3q^{-5}}\right)^{-1} \cdot \left(\dfrac{pq^{-5}}{q^{-2}}\right)^3$

126. $\left(\dfrac{a^6b^{-2}}{2a^{-2}}\right)^{-1} \cdot \left(\dfrac{6a^{-2}}{5b^{-4}}\right)^2 \cdot \left(\dfrac{2b^{-1}a^2}{3b^{-2}}\right)^{-1}$

Write each number in scientific notation. See Example 9.

127. 530   **128.** 1600   **129.** 0.830   **130.** 0.0072

131. 0.00000692   **132.** 0.875   **133.** $-38{,}500$ **134.** $-976{,}000{,}000$

Write each number in standard notation. See Example 10.

135. 7.2×10^4 **136.** 8.91×10^2 **137.** 2.54×10^{-3}

138. 5.42×10^{-4} **139.** -6×10^4 **140.** -9×10^3

141. 1.2×10^{-5} **142.** 2.7×10^{-6}

Evaluate. See Example 11.

143. $\dfrac{12 \times 10^4}{2 \times 10^6}$

144. $\dfrac{16 \times 10^5}{4 \times 10^8}$

145. $\dfrac{3 \times 10^{-2}}{12 \times 10^3}$

146. $\dfrac{5 \times 10^{-3}}{25 \times 10^2}$

147. $\dfrac{0.05 \times 1600}{0.0004}$

148. $\dfrac{0.003 \times 40{,}000}{0.00012}$

149. $\dfrac{20{,}000 \times 0.018}{300 \times 0.0004}$

150. $\dfrac{840{,}000 \times 0.03}{0.00021 \times 600}$

Solve each problem. See Example 12.

151. The U.S. budget first passed **$1,000,000,000** in 1917. Seventy years later, in 1987, it exceeded **$1,000,000,000,000** for the first time. President George W. Bush's budget request for fiscal-year 2003 was **$2,128,000,000,000.** Stacked in dollar bills, this amount would stretch **144,419** mi, almost two-thirds of the distance to the moon. Write the four bold-faced numbers in scientific notation. (*Source: The Gazette,* Cedar Rapids, Iowa, February 5, 2002.)

152. In 1970, Wal-Mart had **1500** employees. In 1997, Wal-Mart became the largest private employer in the United States, with **680,000** employees. In 1999, Wal-Mart became the largest private employer in the world, with **1,100,000** employees. By 2007, the company is expected to have **2,200,000** employees. Write these four numbers in scientific notation. (*Source:* Wal-Mart.)

153. In 2000, the population of the United States was 281.4 million. (*Source:* U.S. Census Bureau.)

  (a) Write the 2000 population in scientific notation.

  (b) Write $1 trillion—that is, $1,000,000,000,000—in scientific notation.

  (c) Using your answers from parts (a) and (b), calculate how much each person in the United States in the year 2000 would have had to contribute in order to make one person in the nation a trillionaire. Write this amount in standard notation to the nearest dollar.

154. In the early years of the Powerball Lottery, a player would choose five numbers from 1 through 49 and one number from 1 through 42. It can be shown that there are about 8.009×10^7

different ways to do this. Suppose that a group of 2000 persons decided to purchase tickets for all these numbers and each ticket cost $1.00. How much should each person have expected to pay? (*Source:* www.powerball.com)

155. The speed of light is approximately 3×10^{10} cm per sec. How long will it take light to travel 9×10^{12} cm?

156. The average distance from Earth to the sun is 9.3×10^7 mi. How long would it take a rocket, traveling at 2.9×10^3 mph, to reach the sun?

157. A *light-year* is the distance that light travels in one year. Find the number of miles in a light-year if light travels 1.86×10^5 mi per sec.

158. Use the information given in the previous two exercises to find the number of minutes necessary for light from the sun to reach Earth.

159. (a) The planet Mercury has an average distance from the sun of 3.6×10^7 mi, while the average distance of Venus to the sun is 6.7×10^7 mi. How long would it take a spacecraft traveling at 1.55×10^3 mph to travel from Venus to Mercury? (Give your answer in hours, in standard notation.)

  (b) Use the information from part (a) to find the number of days it would take the spacecraft to travel from Venus to Mercury. Round your answer to the nearest whole number of days.

160. When the distance between the centers of the moon and Earth is 4.60×10^8 m, an object on the line joining the centers exerts the same gravitational force on each of those bodies when it is 4.14×10^8 m from the center of Earth. How far is the object from the center of the moon at that point?

TECHNOLOGY INSIGHTS (EXERCISES 161–164)

The screen on the left shows how a graphing calculator displays 250,000 and 0.000000034 in scientific notation. When put in scientific mode, it will calculate and display results as shown in the screen on the right.

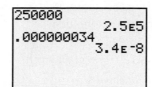

Predict the result the calculator will give for each screen. (Use the usual scientific notation to write your answers.)

161. `(1.5E12)*(5E⁻3)`

162. `(3.2E⁻5)*(3E12)`

163. `(8.4E14)/(2.1E⁻3)`

164. `(2.5E10)/(2E⁻3)`

PREVIEW EXERCISES

*Simplify each expression. See **Section 1.4.***

165. $9x + 5x - x + 8x - 12x$

166. $3p + 2p - 8p - 13p$

167. $6 - 4(3 - z) + 5(2 - 3z)$

168. $2 + 3(5 + q) - 2(6 - q)$

169. $7x - (5 + 5x) + 3$

170. $12q - (6 - 8q) - 4$

5.2 Adding and Subtracting Polynomials

OBJECTIVE 1 Know the basic definitions for polynomials. Just as whole numbers are the basis of arithmetic, *polynomials* are fundamental in algebra. To understand polynomials, we review several words from **Section 1.4.** A **term** is a number, a variable, or the product or quotient of a number and one or more variables raised to powers. Examples of terms include

$$4x, \quad \frac{1}{2}m^5 \left(\text{or } \frac{m^5}{2} \right), \quad -7z^9, \quad 6x^2z, \quad \frac{5}{3x^2}, \quad \text{and} \quad 9. \quad \text{Terms}$$

The number in the product is called the **numerical coefficient,** or just the **coefficient.** In the term $8x^3$, the coefficient is 8. In the term $-4p^5$, it is -4. The coefficient of the term k is understood to be 1 (since k can be written as $1k$). The coefficient of $-r$ is -1. In the term $\frac{x}{3}$, the coefficient is $\frac{1}{3}$, since $\frac{x}{3} = \frac{1x}{3} = \frac{1}{3}x$. More generally, any factor in a term is the coefficient of the product of the remaining factors. For example, $3x^2$ is the coefficient of y in the term $3x^2y$, and $3y$ is the coefficient of x^2 in $3x^2y$.

Any combination of variables or constants (numerical values) joined by the basic operations of addition, subtraction, multiplication, and division (except by 0), or raising to powers or taking roots, formed according to the rules of algebra, is called an **algebraic expression.** The simplest kind of algebraic expression is a *polynomial*.

Polynomial

A **polynomial** is a term or a finite sum of terms in which all variables have whole number exponents and no variables appear in denominators.

Examples of polynomials include

$$3x - 5, \quad 4m^3 - 5m^2p + 8, \quad \text{and} \quad -5t^2s^3. \quad \text{Polynomials}$$

Even though the expression $3x - 5$ involves subtraction, it is a sum of terms, since it could be written as $3x + (-5)$.

Some examples of expressions that are not polynomials are

$$x^{-1} + 3x^{-2}, \quad \sqrt{9 - x}, \quad \text{and} \quad \frac{1}{x}. \quad \text{Not polynomials}$$

The first of these is not a polynomial because it has a negative integer exponent, the second because it involves a variable under a radical, and the third because it contains a variable in the denominator.

Most of the polynomials used in this book contain only one variable. A polynomial containing only the variable x is called a **polynomial in x.** A polynomial in one variable is written in **descending powers** of the variable if the exponents on the variable decrease from left to right. For example,

$$x^5 - 6x^2 + 12x - 5$$

is a polynomial in descending powers of x. The term -5 in this polynomial can be thought of as $-5x^0$, since $-5x^0 = -5(1) = -5$.

EXAMPLE 1 Writing Polynomials in Descending Powers

Write each polynomial in descending powers of the variable.

(a) $y - 6y^3 + 8y^5 - 9y^4 + 12$ is written as $8y^5 - 9y^4 - 6y^3 + y + 12$.

(b) $-2 + m + 6m^2 - 4m^3$ is written as $-4m^3 + 6m^2 + m - 2$.

✔ **Now Try Exercise 1.**

Some polynomials with a specific number of terms are so common that they are given special names. A polynomial with exactly three terms is a **trinomial,** and a polynomial with exactly two terms is a **binomial.** A single-term polynomial is a **monomial.** The table gives examples.

Type of Polynomial	Examples
Monomial	$5x$, $7m^9$, -8, x^2y^2
Binomial	$3x^2 - 6$, $11y + 8$, $5a^2b + 3a$
Trinomial	$y^2 + 11y + 6$, $8p^3 - 7p + 2m$, $-3 + 2k^5 + 9z^4$
None of these	$p^3 - 5p^2 + 2p - 5$, $-9z^3 + 5c^3 + 2m^5 + 11r^2 - 7r$

OBJECTIVE 2 **Find the degree of a polynomial.**   The **degree of a term** with one variable is the exponent on the variable. For example, the degree of $2x^3$ is 3, the degree of $-x^4$ is 4, and the degree of $17x$ (that is, $17x^1$) is 1. The degree of a term with more than one variable is defined to be the sum of the exponents on the variables. For example, the degree of $5x^3y^7$ is 10, because $3 + 7 = 10$.

The greatest degree of any term in a polynomial is called the **degree of the polynomial.** In most cases, we will be interested in finding the degree of a polynomial in one variable. For example, $4x^3 - 2x^2 - 3x + 7$ has degree 3, because the greatest degree of any term is 3 (the degree of $4x^3$).

The table shows several polynomials and their degrees.

Polynomial	Degree
$9x^2 - 5x + 8$	2
$17m^9 + 18m^{14} - 9m^3$	14
$5x$	1, because $5x = 5x^1$
-2	0, because $-2 = -2x^0$ (Any nonzero constant has degree 0.)
$5a^2b^5$	7, because $2 + 5 = 7$
$x^3y^9 + 12xy^4 + 7xy$	12, because the degrees of the terms are 12, 5, and 2, and 12 is the greatest.

▶ **NOTE**   The number 0 has no degree, since 0 times a variable to any power is 0.

✔ **Now Try Exercises 21, 25, and 27.**

OBJECTIVE 3 **Add and subtract polynomials.** We use the distributive property to simplify polynomials by combining terms. For example,

$$x^3 + 4x^2 + 5x^2 - 1 = x^3 + (4 + 5)x^2 - 1 \quad \text{Distributive property}$$
$$= x^3 + 9x^2 - 1.$$

Notice that the terms in the polynomial $4x + 5x^2$ cannot be combined. As these examples suggest, only terms containing exactly the same variables to the same powers may be combined. Recall that such terms are called **like terms.**

> ▶ **CAUTION** *Remember that only like terms can be combined.*

EXAMPLE 2 Combining Like Terms

Combine like terms.

(a) $-5y^3 + 8y^3 - y^3 = (-5 + 8 - 1)y^3$ Distributive property
$$= 2y^3$$

(b) $6x + 5y - 9x + 2y = 6x - 9x + 5y + 2y$ Associative and commutative properties

$$= -3x + 7y \qquad \text{Combine like terms.}$$

Since $-3x$ and $7y$ are unlike terms, no further simplification is possible.

(c) $5x^2y - 6xy^2 + 9x^2y + 13xy^2 = 5x^2y + 9x^2y - 6xy^2 + 13xy^2$
$$= 14x^2y + 7xy^2$$

✔ **Now Try Exercises 31, 37, and 43.**

We use the following rule to add two polynomials.

Adding Polynomials

To add two polynomials, combine like terms.

EXAMPLE 3 Adding Polynomials

Add $(3a^5 - 9a^3 + 4a^2) + (-8a^5 + 8a^3 + 2)$.

Use the commutative and associative properties to rearrange the polynomials so that like terms are together. Then use the distributive property to combine like terms.

$$(3a^5 - 9a^3 + 4a^2) + (-8a^5 + 8a^3 + 2)$$
$$= 3a^5 - 8a^5 - 9a^3 + 8a^3 + 4a^2 + 2$$
$$= -5a^5 - a^3 + 4a^2 + 2 \qquad \text{Combine like terms.}$$

We can add these same two polynomials vertically by placing like terms in columns.

$$3a^5 - 9a^3 + 4a^2$$
$$\underline{-8a^5 + 8a^3 \qquad\quad + 2}$$
$$-5a^5 - \quad a^3 + 4a^2 + 2$$

✔ **Now Try Exercises 51 and 65.**

In **Section 1.2,** we defined subtraction of real numbers as
$$a - b = a + (-b).$$
That is, we add the first number and the negative (or opposite) of the second. We can give a similar definition for subtraction of polynomials by defining the **negative of a polynomial** as that polynomial with the sign of every coefficient changed.

Subtracting Polynomials

To subtract two polynomials, add the first polynomial and the negative of the *second* polynomial.

EXAMPLE 4 Subtracting Polynomials

Subtract $(-6m^2 - 8m + 5) - (-5m^2 + 7m - 8)$.

Change every sign in the second polynomial and add.

$$(-6m^2 - 8m + 5) - (-5m^2 + 7m - 8)$$
$$= -6m^2 - 8m + 5 + 5m^2 - 7m + 8$$
$$= -6m^2 + 5m^2 - 8m - 7m + 5 + 8 \qquad \text{Rearrange terms.}$$
$$= -m^2 - 15m + 13 \qquad \text{Combine like terms.}$$

Check by adding the sum, $-m^2 - 15m + 13$, to the second polynomial. The result should be the first polynomial.

To subtract these two polynomials vertically, write the first polynomial above the second, lining up like terms in columns.

$$-6m^2 - 8m + 5$$
$$\underline{-5m^2 + 7m - 8}$$

Change all the signs in the second polynomial and add.

$$\begin{array}{r} -6m^2 - 8m + 5 \\ \underline{+5m^2 - 7m + 8} \\ -m^2 - 15m + 13 \end{array} \quad \begin{array}{l} \text{Change all signs.} \\[1em] \text{Add in columns.} \end{array}$$

✔ **Now Try Exercises 61 and 69.**

5.2 EXERCISES

◉ *Complete solution available on Video Lectures on CD/DVD*

▨ *Now Try Exercise*

Write each polynomial in descending powers of the variable. See Example 1.

◉ **1.** $2x^3 + x - 3x^2 + 4$ **2.** $3y^2 + y^4 - 2y^3 + y$ **3.** $4p^3 - 8p^5 + p^7$

4. $q^2 + 3q^4 - 2q + 1$ **5.** $-m^3 + 5m^2 + 3m^4 + 10$ **6.** $4 - x + 3x^2$

Give the numerical coefficient and the degree of each term.

7. $7z$ **8.** $3r$ **9.** $-15p^2$ **10.** $-27k^3$ **11.** x^4

12. y^6 **13.** $\dfrac{t}{6}$ **14.** $\dfrac{m}{4}$ **15.** $-mn^5$ **16.** $-a^5b$

Identify each polynomial as a monomial, binomial, trinomial, or none of these. Also, give the degree.

17. 25

18. 5

19. $7m - 22$

20. $6x + 15$

🌐 **21.** $-7y^6 + 11y^8$

22. $12k^2 - 9k^5$

23. $-5m^3 + 6m - 9m^2$

24. $4z^2 - 11z + 2$

25. $-6p^4q - 3p^3q^2 + 2pq^3 - q^4$

26. $8s^3t - 4s^2t^2 + 2st^3 + 9$

27. *Concept Check* Which one of the following is a trinomial in descending powers, having degree 6?

A. $5x^6 - 4x^5 + 12$ **B.** $6x^5 - x^6 + 4$
C. $2x + 4x^2 - x^6$ **D.** $4x^6 - 6x^4 + 9x^2 - 8$

28. *Concept Check* Give an example of a polynomial of four terms in the variable *x*, having degree 5, written in descending powers, and lacking a fourth-degree term.

Combine like terms. See Example 2.

29. $5z^4 + 3z^4$

30. $8r^5 - 2r^5$

🌐 **31.** $-m^3 + 2m^3 + 6m^3$

32. $3p^4 + 5p^4 - 2p^4$

33. $x + x + x + x + x$

34. $z - z - z + z$

35. $m^4 - 3m^2 + m$

36. $5a^5 + 2a^4 - 9a^3$

37. $5t + 4s - 6t + 9s$

38. $8p - 9q - 3p + q$

39. $2k + 3k^2 + 5k^2 - 7$

40. $4x^2 + 2x - 6x^2 - 6$

41. $n^4 - 2n^3 + n^2 - 3n^4 + n^3$

42. $2q^3 + 3q^2 - 4q - q^3 + 5q^2$

43. $3ab^2 + 7a^2b - 5ab^2 + 13a^2b$

44. $6m^2n - 8mn^2 + 3mn^2 - 7m^2n$

45. $4 - (2 + 3m) + 6m + 9$

46. $8a - (3a + 4) - (5a - 3)$

47. $6 + 3p - (2p + 1) - (2p + 9)$

48. $4x - 8 - (-1 + x) - (11x + 5)$

✒ **49.** Define *polynomial* in your own words. Give examples. Include the words *term, monomial, binomial,* and *trinomial* in your explanation.

✒ **50.** Write a paragraph explaining how to add and subtract polynomials. Give examples.

Add or subtract as indicated. See Examples 3 and 4.

🌐 **51.** $(5x^2 + 7x - 4) + (3x^2 - 6x + 2)$

52. $(4k^3 + k^2 + k) + (2k^3 - 4k^2 - 3k)$

53. $(6t^2 - 4t^4 - t) + (3t^4 - 4t^2 + 5)$

54. $(3p^2 + 2p - 5) + (7p^2 - 4p^3 + 3p)$

55. $(y^3 + 3y + 2) + (4y^3 - 3y^2 + 2y - 1)$

56. $(2x^5 - 2x^4 + x^3 - 1) + (x^4 - 3x^3 + 2)$

57. $(3r + 8) - (2r - 5)$

58. $(2d + 7) - (3d - 1)$

59. $(2a^2 + 3a - 1) - (4a^2 + 5a + 6)$

60. $(q^4 - 2q^2 + 10) - (3q^4 + 5q^2 - 5)$

🌐 **61.** $(z^5 + 3z^2 + 2z) - (4z^5 + 2z^2 - 5z)$

62. $(5t^3 - 3t^2 + 2t) - (4t^3 + 2t^2 + 3t)$

63. Add.
$21p - 8$
$\underline{-9p + 4}$

64. Add.
$15m - 9$
$\underline{4m + 12}$

65. Add.
$-12p^2 + 4p - 1$
$\underline{3p^2 + 7p - 8}$

66. Add.
$-6y^3 + 8y + 5$
$\underline{9y^3 + 4y - 6}$

67. Subtract.
$12a + 15$
$\underline{7a - 3}$

68. Subtract.
$-3b + 6$
$\underline{2b - 8}$

69. Subtract.
$$6m^2 - 11m + 5$$
$$\underline{-8m^2 + 2m - 1}$$

70. Subtract.
$$-4z^2 + 2z - 1$$
$$\underline{3z^2 - 5z + 2}$$

71. Add.
$$12z^2 - 11z + 8$$
$$5z^2 + 16z - 2$$
$$\underline{-4z^2 + 5z - 9}$$

72. Add.
$$-6m^3 + 2m^2 + 5m$$
$$8m^3 + 4m^2 - 6m$$
$$\underline{-3m^3 + 2m^2 - 7m}$$

73. Add.
$$6y^3 - 9y^2 + 8$$
$$\underline{4y^3 + 2y^2 + 5y}$$

74. Add.
$$-7r^8 + 2r^6 - r^5$$
$$\underline{ 3r^6 + 5}$$

75. Subtract.
$$-5a^4 + 8a^2 - 9$$
$$\underline{ 6a^3 - a^2 + 2}$$

76. Subtract.
$$ - 2m^3 + 8m^2$$
$$\underline{m^4 - m^3 + 2m}$$

Perform the indicated operations. See Examples 2–4.

77. Subtract $4y^2 - 2y + 3$ from $7y^2 - 6y + 5$.

78. Subtract $-(-4x + 2z^2 + 3m)$ from $[(2z^2 - 3x + m) + (z^2 - 2m)]$.

79. $(-4m^2 + 3n^2 - 5n) - [(3m^2 - 5n^2 + 2n) + (-3m^2) + 4n^2]$

80. $[-(4m^2 - 8m + 4m^3) - (3m^2 + 2m + 5m^3)] + m^2$

81. $[-(y^4 - y^2 + 1) - (y^4 + 2y^2 + 1)] + (3y^4 - 3y^2 - 2)$

82. $[2p - (3p - 6)] - [(5p - (8 - 9p)) + 4p]$

83. $-[3z^2 + 5z - (2z^2 - 6z)] + [(8z^2 - [5z - z^2]) + 2z^2]$

84. $5k - (5k - [2k - (4k - 8k)]) + 11k - (9k - 12k)$

PREVIEW EXERCISES

*Evaluate (**a**) $f(-1)$ and (**b**) $f(2)$ for each function. See **Section 3.5**.*

85. $f(x) = 3x + 1$ **86.** $f(x) = x^2 + 2$ **87.** $f(x) = x^3 - 8$

*Decide whether each graph is the graph of a function. See **Section 3.5**.*

88. 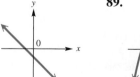   **89.**   **90.**   **91.**

92. If $f(x) = 3x + 1$ and $g(x) = x^2 + 2$, find $f(2) + g(2)$. See **Section 3.5**.

5.3 Polynomial Functions, Graphs, and Composition

OBJECTIVE **1** **Recognize and evaluate polynomial functions.** In **Chapter 3,** we studied linear (first-degree polynomial) functions, defined as $f(x) = ax + b$. Now we consider more general polynomial functions.

> **Polynomial Function**
>
> A **polynomial function of degree n** is defined by
>
> $$f(x) = a_n x^n + a_{n-1}x^{n-1} + \cdots + a_1 x + a_0,$$
>
> for real numbers $a_n, a_{n-1}, \ldots, a_1$, and a_0, where $a_n \neq 0$ and n is a whole number.

Another way of describing a polynomial function is to say that it is a function defined by a polynomial in one variable, consisting of one or more terms. It is usually written in descending powers of the variable, and its degree is the degree of the polynomial that defines it. Suppose we consider the polynomial $3x^2 - 5x + 7$, so that

$$f(x) = 3x^2 - 5x + 7.$$

If $x = -2$, then $f(x) = 3x^2 - 5x + 7$ takes on the value

$$
\begin{aligned}
f(-2) &= 3(-2)^2 - 5(-2) + 7 \quad &\text{Let } x = -2. \\
&= 3(4) + 10 + 7 \\
&= 29.
\end{aligned}
$$

Thus, $f(-2) = 29$, and the ordered pair $(-2, 29)$ belongs to f.

EXAMPLE 1 **Evaluating Polynomial Functions**

Let $f(x) = 4x^3 - x^2 + 5$. Find each value.

(a) $f(3)$

Read this as "*f* of 3," not "*f* times 3."

$$
\begin{aligned}
f(x) &= 4x^3 - x^2 + 5 \\
f(3) &= 4(3)^3 - 3^2 + 5 \quad &\text{Substitute 3 for } x. \\
&= 4(27) - 9 + 5 \quad &\text{Order of operations} \\
&= 108 - 9 + 5 \quad &\text{Multiply.} \\
&= 104 \quad &\text{Subtract; add.}
\end{aligned}
$$

(b) $f(-4) = 4(-4)^3 - (-4)^2 + 5$ Let $x = -4$.

$$
\begin{aligned}
&= 4(-64) - 16 + 5 \\
&= -267
\end{aligned}
$$

Be careful with signs.

✔ **Now Try Exercise 7.**

While f is the most common letter used to represent functions, recall that other letters, such as g and h, are also used. The capital letter P is often used for polynomial functions. Note that the function defined as $P(x) = 4x^3 - x^2 + 5$ yields the same ordered pairs as the function f in Example 1.

OBJECTIVE ② Use a polynomial function to model data.   Polynomial functions can be used to approximate data. Such functions are usually valid for small intervals, and they allow us to predict (with caution) what might happen for values just outside the intervals. These intervals are often periods of years, as shown in Example 2.

EXAMPLE 2 **Using a Polynomial Model to Approximate Data**

The number of U.S. households estimated to see and pay at least one bill on-line each month during the years 2000 through 2006 can be modeled by the polynomial function defined by

$$P(x) = 0.808x^2 + 2.625x + 0.502,$$

where $x = 0$ corresponds to the year 2000, $x = 1$ corresponds to 2001, and so on, and $P(x)$ is in millions. Use this function to approximate the number of households that paid at least one bill on-line each month in 2005.

Since $x = 5$ corresponds to 2005, we must find $P(5)$.

$$P(x) = 0.808x^2 + 2.625x + 0.502$$
$$P(5) = 0.808(5)^2 + 2.625(5) + 0.502 \qquad \text{Let } x = 5.$$
$$= 33.827 \qquad \text{Evaluate.}$$

According to this model, in 2005 about 33.83 million households were expected to pay at least one bill on-line each month.

✔ **Now Try Exercise 9.**

OBJECTIVE ③ Add and subtract polynomial functions.   The operations of addition, subtraction, multiplication, and division are also defined for functions. For example, businesses use the equation "profit equals revenue minus cost," written in function notation as

$$P(x) \ = \ R(x) \ - \ C(x),$$

Profit   Revenue   Cost
function   function   function

where x is the number of items produced and sold. Thus, the profit function is found by subtracting the cost function from the revenue function.

We define the following **operations on functions.**

Adding and Subtracting Functions

If $f(x)$ and $g(x)$ define functions, then

$$(f + g)(x) = f(x) + g(x) \qquad \text{Sum function}$$

and

$$(f - g)(x) = f(x) - g(x). \qquad \text{Difference function}$$

In each case, the domain of the new function is the intersection of the domains of $f(x)$ and $g(x)$.

EXAMPLE 3 Adding and Subtracting Functions

For the polynomial functions defined by

$$f(x) = x^2 - 3x + 7 \quad \text{and} \quad g(x) = -3x^2 - 7x + 7,$$

find **(a)** the sum and **(b)** the difference.

(a) $(f + g)(x) = f(x) + g(x)$ Use the definition.

$$= (x^2 - 3x + 7) + (-3x^2 - 7x + 7) \quad \text{Substitute.}$$

This notation does *not* indicate the distributive property.

$$= -2x^2 - 10x + 14 \quad \text{Add the polynomials.}$$

(b) $(f - g)(x) = f(x) - g(x)$ Use the definition.

$$= (x^2 - 3x + 7) - (-3x^2 - 7x + 7) \quad \text{Substitute.}$$

$$= (x^2 - 3x + 7) + (3x^2 + 7x - 7) \quad \begin{array}{l} \text{Change subtraction} \\ \text{to addition.} \end{array}$$

$$= 4x^2 + 4x \quad \text{Add.}$$

✔ **Now Try Exercise 15.**

EXAMPLE 4 Adding and Subtracting Functions

Find each of the following for the functions defined by

$$f(x) = 10x^2 - 2x \quad \text{and} \quad g(x) = 2x.$$

(a) $(f + g)(2)$

$$(f + g)(2) = f(2) + g(2) \quad \text{Use the definition.}$$

$$\overbrace{f(x) = 10x^2 - 2x} \qquad \overbrace{g(x) = 2x}$$

$$= [10(2)^2 - 2(2)] \quad + \quad 2(2) \quad \text{Substitute.}$$

This is a key step.

$$= [40 - 4] + 4 \quad \text{Order of operations}$$

$$= 40$$

Alternatively, we could first find $(f + g)(x)$.

$$(f + g)(x) = f(x) + g(x) \quad \text{Use the definition.}$$

$$= (10x^2 - 2x) + 2x \quad \text{Substitute.}$$

$$= 10x^2 \quad \text{Combine like terms.}$$

Then,

$$(f + g)(2) = 10(2)^2 = 40. \quad \text{The result is the same.}$$

(b) $(f - g)(x)$ and $(f - g)(1)$

$$(f - g)(x) = f(x) - g(x) \quad \text{Use the definition.}$$

$$= (10x^2 - 2x) - 2x \quad \text{Substitute.}$$

$$= 10x^2 - 4x \quad \text{Combine like terms.}$$

Then,

$$(f - g)(1) = 10(1)^2 - 4(1) = 6. \quad \text{Substitute.}$$

Confirm that $f(1) - g(1)$ gives the same result.

✔ **Now Try Exercises 17 and 19.**

OBJECTIVE 4 Find the composition of functions. The diagram in Figure 1 shows a function f that assigns, to each element x of set X, some element y of set Y. Suppose that a function g takes each element of set Y and assigns a value z of set Z. Then f and g together assign an element x in X to an element z in Z. The result of this process is a new function h that takes an element x in X and assigns it an element z in Z.

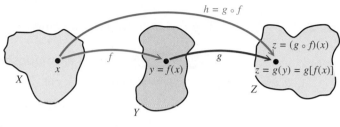

FIGURE 1

This function h is called the *composition* of functions g and f, written $\boldsymbol{g \circ f}$.

Composition of Functions

If f and g are functions, then the **composite function,** or **composition,** of g and f is defined by

$$(g \circ f)(x) = g(f(x))$$

for all x in the domain of f such that $f(x)$ is in the domain of g.

Read $g \circ f$ as "g of f".

As a real-life example of how composite functions occur, consider the following retail situation.

A \$40 pair of blue jeans is on sale for 25% off. If you purchase the jeans before noon, the retailer offers an additional 10% off. What is the final sale price of the blue jeans?

You might be tempted to say that the blue jeans are $25\% + 10\% = 35\%$ off and calculate $\$40(0.35) = \$14$, giving a final sale price of $\$40 - \$14 = \$26$ for the jeans. **This is not correct.** To find the final sale price, we must first find the price after taking 25% off, and then take an additional 10% off that price.

$\$40(0.25) = \$10$, giving a sale price of $\$40 - \$10 = \$30$.   Take 25% off original price.

$\$30(0.10) = \$3$, giving a ***final sale price*** of $\$30 - \$3 = \$27$.   Take additional 10% off.

This is the idea behind composition of functions.

FIGURE 2

As another example of composition, suppose an oil well off the Louisiana coast is leaking, with the leak spreading oil in a circular layer over the surface. See Figure 2. At any time t, in minutes, after the beginning of the leak, the radius of the circular oil slick is given by $r(t) = 5t$ feet. Since $A(r) = \pi r^2$ gives the area of a circle of radius r, the area can be expressed as a function of time by substituting $5t$ for r in $A(r) = \pi r^2$ to get

$$A(r(t)) = \pi(5t)^2 = 25\pi t^2.$$

The function $A(r(t))$ is a composite function of the functions A and r.

EXAMPLE 5 Finding a Composite Function

Let $f(x) = x^2$ and $g(x) = x + 3$. Find $(f \circ g)(4)$.

$$(f \circ g)(4) = f(g(4)) \qquad \text{Definition}$$

Evaluate the "inside" function value first.

$$= f(4 + 3) \qquad \text{Use the rule for } g(x); g(4) = 4 + 3.$$
$$= f(7) \qquad \text{Add.}$$

Now evaluate the "outside" function.

$$= 7^2 \qquad \text{Use the rule for } f(x); f(7) = 7^2.$$
$$= 49$$

✔ **Now Try Exercise 35.**

Notice in Example 5 that if we interchange the order of the functions, the composition of g and f is defined by $g(f(x))$. Once again, letting $x = 4$, we have

$$(g \circ f)(4) = g(f(4)) \qquad \text{Definition}$$
$$= g(4^2) \qquad \text{Use the rule for } f(x); f(4) = 4^2.$$
$$= g(16) \qquad \text{Square 4.}$$
$$= 16 + 3 \qquad \text{Use the rule for } g(x); g(16) = 16 + 3.$$
$$= 19.$$

Here we see that $(f \circ g)(4) \neq (g \circ f)(4)$ because $49 \neq 19$. In general,

$$(f \circ g)(x) \neq (g \circ f)(x).$$

EXAMPLE 6 Finding Composite Functions

Let $f(x) = 4x - 1$ and $g(x) = x^2 + 5$. Find the following.

(a) $(f \circ g)(2)$

$$(f \circ g)(2) = f(g(2))$$
$$= f(2^2 + 5) \qquad g(x) = x^2 + 5$$
$$= f(9) \qquad \text{Work inside the parentheses.}$$
$$= 4(9) - 1 \qquad f(x) = 4x - 1$$
$$= 35 \qquad \text{Multiply; subtract.}$$

(b) $(f \circ g)(x)$

Here, use $g(x)$ as the input for the function f.

$$(f \circ g)(x) = f(g(x))$$
$$= 4(g(x)) - 1 \qquad \text{Use the rule for } f(x); f(x) = 4x - 1.$$
$$= 4(x^2 + 5) - 1 \qquad g(x) = x^2 + 5$$
$$= 4x^2 + 20 - 1 \qquad \text{Distributive property}$$
$$= 4x^2 + 19 \qquad \text{Combine terms.}$$

(c) Find $(f \circ g)(2)$ again, this time using the rule obtained in part (b).

$$(f \circ g)(x) = 4x^2 + 19 \qquad \text{From part (b)}$$
$$(f \circ g)(2) = 4(2)^2 + 19 \qquad \text{Let } x = 2.$$
$$= 4(4) + 19$$
$$= 16 + 19$$
$$= 35$$

The result, 35, is the same as the result in part (a).

✔ **Now Try Exercises 37 and 41.**

OBJECTIVE 5 Graph basic polynomial functions. Functions were introduced in **Section 3.5.** Recall that each input (or x-value) of a function results in one output (or y-value). The simplest polynomial function is the **identity function,** defined by $f(x) = x$. The domain (set of x-values) of this function is all real numbers, $(-\infty, \infty)$, and the function pairs each real number with itself. Therefore, the range (set of y-values) is also $(-\infty, \infty)$. The graph of the function is a straight line, as first seen in **Chapter 3.** (Notice that a *linear function* is a specific kind of polynomial function.) Figure 3 shows the graph of $f(x) = x$ and a table of selected ordered pairs.

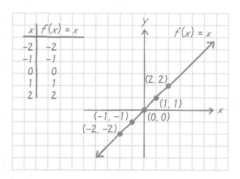

FIGURE 3

Another polynomial function, defined by $f(x) = x^2$, is the **squaring function.** For this function, every real number is paired with its square. The input can be any real number, so the domain is $(-\infty, \infty)$. Since the square of any real number is non-negative, the range is $[0, \infty)$. The graph of the squaring function is a *parabola.* Figure 4 shows the graph and a table of selected ordered pairs.

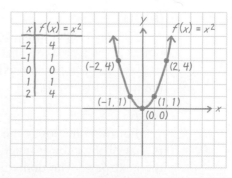

FIGURE 4

The **cubing function** is defined by $f(x) = x^3$. This function pairs every real number with its cube. The domain and the range are both $(-\infty, \infty)$. The graph of the cubing function is neither a line nor a parabola. See Figure 5 and the table of ordered pairs.

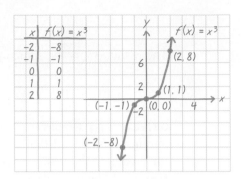

FIGURE 5

EXAMPLE 7	Graphing Variations of the Identity, Squaring, and Cubing Functions

Graph each function by creating a table of ordered pairs. Give the domain and range of each function by observing its graph.

(a) $f(x) = 2x$

To find each range value, multiply the domain value by 2. Plot the points and join them with a straight line. See Figure 6. Both the domain and the range are $(-\infty, \infty)$.

x	$f(x) = 2x$
-2	-4
-1	-2
0	0
1	2
2	4

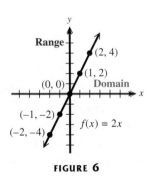

FIGURE 6

x	$f(x) = -x^2$
-2	-4
-1	-1
0	0
1	-1
2	-4

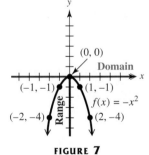

FIGURE 7

(b) $f(x) = -x^2$

For each input x, square it and then take its opposite. Plotting and joining the points gives a parabola that opens down. It is a *reflection* of the graph of the squaring function across the *x*-axis. See the table and Figure 7. The domain is $(-\infty, \infty)$ and the range is $(-\infty, 0]$.

(c) $f(x) = x^3 - 2$

For this function, cube the input and then subtract 2 from the result. The graph is that of the cubing function *shifted* 2 units down. See the table and Figure 8 on the next page. The domain and range are both $(-\infty, \infty)$.

x	$f(x) = x^3 - 2$
-2	-10
-1	-3
0	-2
1	-1
2	6

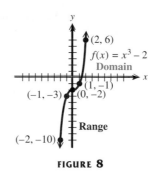

FIGURE 8

✔ **Now Try Exercises 55, 57, and 59.**

5.3 EXERCISES

*For each polynomial function, find (**a**) $f(-1)$ and (**b**) $f(2)$. See Example 1.*

1. $f(x) = 6x - 4$

2. $f(x) = -2x + 5$

3. $f(x) = x^2 - 3x + 4$

4. $f(x) = 3x^2 + x - 5$

5. $f(x) = 5x^4 - 3x^2 + 6$

6. $f(x) = -4x^4 + 2x^2 - 1$

🌐 **7.** $f(x) = -x^2 + 2x^3 - 8$

8. $f(x) = -x^2 - x^3 + 11x$

Solve each problem. See Example 2.

🌐 **9.** The number of airports in the United States during the period from 1980 through 2003 can be approximated by the polynomial function defined by

$$f(x) = -2.19x^2 + 245.7x + 15{,}163,$$

where $x = 0$ represents 1980, $x = 1$ represents 1981, and so on. Use this function to approximate the number of airports in each given year. (*Source:* U.S. Federal Aviation Administration.)

(a) 1980   **(b)** 1990   **(c)** 2003

10. The percent of births to unmarried women during the period from 1990 through 2002 can be approximated by the polynomial function defined by

$$f(x) = 0.010x^3 - 0.245x^2 + 2.09x + 26.6,$$

where $x = 0$ represents 1990, $x = 1$ represents 1991, and so on. Use this function to approximate the percent (to the nearest tenth) of births to unmarried women in each given year. (*Source:* National Center for Health Statistics.)

(a) 1990   **(b)** 1995   **(c)** 2002

11. The number of bank debit cards issued during the period from 1990 through 2000 can be modeled by the polynomial function defined by

$$P(x) = -0.31x^3 + 5.8x^2 - 15x + 9,$$

where $x = 0$ corresponds to the year 1990, $x = 1$ corresponds to 1991, and so on, and $P(x)$ is in millions. Use this function to approximate the number of bank debit cards issued in each given year. Round answers to the nearest million. (*Source: Statistical Abstract of the United States.*)

(a) 1990   **(b)** 1996   **(c)** 1999

12. The total number of cases on the docket of the U.S. Supreme Court during the period from 1980 to 2000 can be modeled by the polynomial function defined by

$$P(x) = -0.0325x^3 + 4.524x^2 + 102.5x + 5128,$$

where $x = 0$ represents 1980, $x = 1$ represents 1981, and so on. Use this function to approximate the number of cases on the docket of the U.S. Supreme Court in each given year. (*Source:* Office of the Clerk, Supreme Court of the United States.)

(a) 1980   **(b)** 1990   **(c)** 2000

*For each pair of functions, find **(a)** $(f + g)(x)$ and **(b)** $(f - g)(x)$. See Example 3.*

13. $f(x) = 5x - 10, g(x) = 3x + 7$ **14.** $f(x) = -4x + 1, g(x) = 6x + 2$

15. $f(x) = 4x^2 + 8x - 3, g(x) = -5x^2 + 4x - 9$

16. $f(x) = 3x^2 - 9x + 10, g(x) = -4x^2 + 2x + 12$

Let $f(x) = x^2 - 9$, $g(x) = 2x$, and $h(x) = x - 3$. Find each of the following. See Example 4.

17. $(f + g)(x)$ **18.** $(f - g)(x)$ **19.** $(f + g)(3)$

20. $(f - g)(-3)$ **21.** $(f - h)(x)$ **22.** $(f + h)(x)$

23. $(f - h)(-3)$ **24.** $(f + h)(-2)$ **25.** $(g + h)(-10)$

26. $(g - h)(10)$ **27.** $(g - h)(-3)$ **28.** $(g + h)\left(\dfrac{1}{3}\right)$

29. $(g + h)\left(\dfrac{1}{4}\right)$ **30.** $(g + h)\left(-\dfrac{1}{4}\right)$ **31.** $(g + h)\left(-\dfrac{1}{2}\right)$

32. Is the equation defining the sum of two functions, $(f + g)(x) = f(x) + g(x)$, an application of the distributive property? Explain.

33. Construct two polynomial functions defined by $f(x)$, a polynomial of degree 3, and $g(x)$, a polyomial of degree 4. Find $(f - g)(x)$ and $(g - f)(x)$. Use your answers to decide whether subtraction of polynomial functions is a commutative operation. Explain.

34. *Concept Check*   Find two polynomial functions defined by $f(x)$ and $g(x)$ such that

$$(f + g)(x) = 3x^3 - x + 3.$$

Let $f(x) = x^2 + 4$, $g(x) = 2x + 3$, and $h(x) = x + 5$. Find each value or expression. See Examples 5 and 6.

35. $(h \circ g)(4)$ **36.** $(f \circ g)(4)$ **37.** $(g \circ f)(6)$ **38.** $(h \circ f)(6)$

39. $(f \circ h)(-2)$ **40.** $(h \circ g)(-2)$ **41.** $(f \circ g)(x)$ **42.** $(g \circ h)(x)$

43. $(f \circ h)(x)$ **44.** $(g \circ f)(x)$ **45.** $(h \circ g)(x)$ **46.** $(h \circ f)(x)$

47. $(f \circ h)\left(\dfrac{1}{2}\right)$ **48.** $(h \circ f)\left(\dfrac{1}{2}\right)$ **49.** $(f \circ g)\left(-\dfrac{1}{2}\right)$ **50.** $(g \circ f)\left(-\dfrac{1}{2}\right)$

Solve each problem.

51. The function defined by $f(x) = 12x$ computes the number of inches in x feet, and the function defined by $g(x) = 5280x$ computes the number of feet in x miles. What is $(f \circ g)(x)$ and what does it compute?

52. The perimeter x of a square with sides of length s is given by the formula $x = 4s$.

   (a) Solve for s in terms of x.
   (b) If y represents the area of this square, write y as a function of the perimeter x.
   (c) Use the composite function of part (b) to find the area of a square with perimeter 6.

53. When a thermal inversion layer is over a city (as happens often in Los Angeles), pollutants cannot rise vertically, but are trapped below the layer and must disperse horizontally. Assume that a factory smokestack begins emitting a pollutant at 8 A.M. Assume that the pollutant disperses horizontally over a circular area. Suppose that t represents the time, in hours, since the factory began emitting pollutants ($t = 0$ represents 8 A.M.), and assume that the radius of the circle of pollution is $r(t) = 2t$ miles. Let $A(r) = \pi r^2$ represent the area of a circle of radius r. Find and interpret $(A \circ r)(t)$.

54. An oil well off the Gulf Coast is leaking, with the leak spreading oil over the surface as a circle. At any time t, in minutes, after the beginning of the leak, the radius of the circular oil slick on the surface is $r(t) = 4t$ feet. Let $A(r) = \pi r^2$ represent the area of a circle of radius r. Find and interpret $(A \circ r)(t)$.

Graph each function. Give the domain and range. See Example 7.

55. $f(x) = -2x + 1$ **56.** $f(x) = 3x + 2$ **57.** $f(x) = -3x^2$

58. $f(x) = \dfrac{1}{2}x^2$ **59.** $f(x) = x^3 + 1$ **60.** $f(x) = -x^3 + 2$

PREVIEW EXERCISES

*Find each product. See **Section 5.1.***

61. $3m^3(4m^2)$ **62.** $5z^2(7z^4)$ **63.** $-3b^5(2a^3b^4)$

64. $-4k^2(-3k^5t^3)$ **65.** $12x^2y(5xy^3)$ **66.** $6mn^3(3m^4n)$

5.4 Multiplying Polynomials

OBJECTIVE 1 Multiply terms. Recall that the product of $3x^4$ and $5x^3$ is found by using the commutative and associative properties, along with the rules for exponents.

$$(3x^4)(5x^3) = 3 \cdot 5 \cdot x^4 \cdot x^3$$
$$= 15x^{4+3}$$
$$= 15x^7$$

EXAMPLE 1 Multiplying Monomials

Find each product.

(a) $-4a^3(3a^5) = -4(3)a^3 \cdot a^5 = -12a^8$

(b) $2m^2z^4(8m^3z^2) = 2(8)m^2 \cdot m^3 \cdot z^4 \cdot z^2 = 16m^5z^6$

✔ **Now Try Exercises 5 and 7.**

OBJECTIVE 2 Multiply any two polynomials. We use the distributive property to extend this process to find the product of any two polynomials.

EXAMPLE 2 Multiplying Polynomials

Find each product.

(a) $-2(8x^3 - 9x^2) = -2(8x^3) - 2(-9x^2)$ Distributive property
$$= -16x^3 + 18x^2$$

(b) $5x^2(-4x^2 + 3x - 2) = 5x^2(-4x^2) + 5x^2(3x) + 5x^2(-2)$
$$= -20x^4 + 15x^3 - 10x^2$$

(c) $(3x - 4)(2x^2 + x)$

Use the distributive property to multiply each term of $2x^2 + x$ by $3x - 4$.

$$(3x - 4)(2x^2 + x) = (3x - 4)(2x^2) + (3x - 4)(x)$$

Here, $3x - 4$ was treated as a single expression so that the distributive property could be used. Now use the distributive property two more times.

$$= 3x(2x^2) + (-4)(2x^2) + (3x)(x) + (-4)(x)$$
$$= 6x^3 - 8x^2 + 3x^2 - 4x$$
$$= 6x^3 - 5x^2 - 4x \quad \text{Combine like terms.}$$

(d) $2x^2(x + 1)(x - 3) = 2x^2[(x + 1)(x) + (x + 1)(-3)]$ Distributive property
$$= 2x^2[x^2 + x - 3x - 3] \quad \text{Distributive property}$$
$$= 2x^2(x^2 - 2x - 3) \quad \text{Combine like terms.}$$
$$= 2x^4 - 4x^3 - 6x^2 \quad \text{Distributive property}$$

✔ **Now Try Exercises 11, 13, and 19.**

It is often easier to multiply polynomials by writing them vertically.

EXAMPLE 3 Multiplying Polynomials Vertically

Find each product.

(a) $(5a - 2b)(3a + b)$

$$
\begin{array}{r}
5a \;\; - 2b \\
3a \;\; + \;\; b \\
\hline
5ab - 2b^2 \quad \leftarrow \text{Multiply } b(5a - 2b). \\
15a^2 - 6ab \qquad\qquad \leftarrow \text{Multiply } 3a(5a - 2b). \\
\hline
15a^2 - \;\; ab - 2b^2 \quad \text{Combine like terms.}
\end{array}
$$

(b) $(3m^3 - 2m^2 + 4)(3m - 5)$

$$
\begin{array}{r}
3m^3 - 2m^2 + \;\; 4 \\
3m \;\; - \;\; 5 \\
\hline
-15m^3 + 10m^2 \qquad\quad - 20 \qquad -5(3m^3 - 2m^2 + 4) \\
9m^4 - \;\; 6m^3 \qquad\qquad + 12m \qquad\quad 3m(3m^3 - 2m^2 + 4) \\
\hline
9m^4 - 21m^3 + 10m^2 + 12m \; - 20 \quad \text{Combine like terms.}
\end{array}
$$

Be sure to write like terms in columns.

✔ **Now Try Exercises 23 and 25.**

▶ **NOTE** We can also use a rectangle to model polynomial multiplication. For example, to find the product

$$(5a - 2b)(3a + b)$$

from Example 3(a), label a rectangle with each term as shown here. Then put the product of each pair of monomials in the appropriate box.

	$3a$	b
$5a$		
$-2b$		

	$3a$	b
$5a$	$15a^2$	$5ab$
$-2b$	$-6ab$	$-2b^2$

The product of the original binomials is the sum of these four monomial products.

$$
\begin{aligned}
(5a - 2b)(3a + b) &= 15a^2 + 5ab - 6ab - 2b^2 \\
&= 15a^2 - ab - 2b^2
\end{aligned}
$$

OBJECTIVE 3 Multiply binomials. When working with polynomials, the product of two binomials occurs repeatedly. There is a shortcut method for finding these products. Recall that a binomial has just two terms, such as $3x - 4$ or $2x + 3$. We can find the product of these binomials by using the distributive property as follows:

$$
\begin{aligned}
(3x - 4)(2x + 3) &= 3x(2x + 3) - 4(2x + 3) \\
&= 3x(2x) + 3x(3) - 4(2x) - 4(3) \\
&= 6x^2 + 9x - 8x - 12.
\end{aligned}
$$

Before combining like terms to find the simplest form of the answer, we check the origin of each of the four terms in the sum $6x^2 + 9x - 8x - 12$. First, $6x^2$ is the product of the two *first* terms.

$$(3x - 4)(2x + 3) \qquad 3x(2x) = 6x^2 \qquad \text{First terms}$$

To get $9x$, the *outer* terms are multiplied.

$$(3x - 4)(2x + 3) \qquad 3x(3) = 9x \qquad \text{Outer terms}$$

The term $-8x$ comes from the *inner* terms.

$$(3x - 4)(2x + 3) \qquad -4(2x) = -8x \qquad \text{Inner terms}$$

Finally, -12 comes from the *last* terms.

$$(3x - 4)(2x + 3) \qquad -4(3) = -12 \qquad \text{Last terms}$$

The product is found by combining these four results.

$$(3x - 4)(2x + 3) = 6x^2 + 9x + (-8x) + (-12)$$
$$= 6x^2 + x - 12$$

To keep track of the order of multiplying terms, we use the initials FOIL (First, Outer, Inner, Last). All the steps of the FOIL method can be done as follows. Try to do as many of these steps as possible mentally.

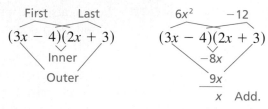

▶ **CAUTION** The FOIL method is an extension of the distributive property, and the acronym *"FOIL" applies only to multiplying two binomials.*

EXAMPLE 4 **Using the FOIL Method**

Use the FOIL method to find each product.

(a) $(4m - 5)(3m + 1)$

First terms $\qquad (4m - 5)(3m + 1) \qquad 4m(3m) = 12m^2$

Outer terms $\qquad (4m - 5)(3m + 1) \qquad 4m(1) = 4m$

Inner terms $\qquad (4m - 5)(3m + 1) \qquad -5(3m) = -15m$

Last terms $\qquad (4m - 5)(3m + 1) \qquad -5(1) = -5$

F　O　I　L

Thus,　$(4m - 5)(3m + 1) = 12m^2 + 4m - 15m - 5$

$= 12m^2 - 11m - 5.$　　Combine like terms.

The procedure can be written in compact form as follows.

$$12m^2 \qquad -5$$
$$(4m - 5)(3m + 1)$$
$$-15m$$
$$4m$$
$$\overline{-11m} \quad \text{Add.}$$

Combine these four results to get $12m^2 - 11m - 5$.

First　Outer　Inner　Last

(b) $(6a - 5b)(3a + 4b) = 18a^2 + 24ab - 15ab - 20b^2$

$= 18a^2 + 9ab - 20b^2$　　Combine like terms.

(c) $(2k + 3z)(5k - 3z) = 10k^2 - 6kz + 15kz - 9z^2$　　FOIL

$= 10k^2 + 9kz - 9z^2$

✔ **Now Try Exercises 35 and 39.**

OBJECTIVE 4 **Find the product of the sum and difference of two terms.** Some types of binomial products occur frequently. The product of the sum and difference of the same two terms, x and y, is

$(x + y)(x - y) = x^2 - xy + xy - y^2$　　FOIL

$= x^2 - y^2.$　　Combine like terms.

Product of the Sum and Difference of Two Terms

The **product of the sum and difference of the two terms x and y** is the difference of the squares of the terms.

$$(x + y)(x - y) = x^2 - y^2$$

EXAMPLE 5 **Multiplying the Sum and Difference of Two Terms**

Find each product.

(a) $(p + 7)(p - 7) = p^2 - 7^2$
$= p^2 - 49$

(b) $(2r + 5)(2r - 5) = (2r)^2 - 5^2$
$= 2^2 r^2 - 25$
$= 4r^2 - 25$

(c) $(6m + 5n)(6m - 5n) = (6m)^2 - (5n)^2$
$= 36m^2 - 25n^2$

(d) $2x^3(x + 3)(x - 3) = 2x^3(x^2 - 9)$
$= 2x^5 - 18x^3$

✔ **Now Try Exercises 47, 51, and 57.**

OBJECTIVE 5 Find the square of a binomial. Another special binomial product is the *square of a binomial*. To find the square of $x + y$, or $(x + y)^2$, multiply $x + y$ by itself.

$$(x + y)(x + y) = x^2 + xy + xy + y^2$$
$$= x^2 + 2xy + y^2$$

A similar result is true for the square of a difference.

Square of a Binomial

The **square of a binomial** is the sum of the square of the first term, twice the product of the two terms, and the square of the last term.

$$(x + y)^2 = x^2 + 2xy + y^2$$
$$(x - y)^2 = x^2 - 2xy + y^2$$

EXAMPLE 6   Squaring Binomials

Find each product.

(a) $(m + 7)^2 = m^2 + 2 \cdot m \cdot 7 + 7^2$ $(x + y)^2 = x^2 + 2xy + y^2$
$$= m^2 + 14m + 49$$

(b) $(p - 5)^2 = p^2 - 2 \cdot p \cdot 5 + 5^2$ $(x - y)^2 = x^2 - 2xy + y^2$
$$= p^2 - 10p + 25$$

(c) $(2p + 3v)^2 = (2p)^2 + 2(2p)(3v) + (3v)^2$
$$= 4p^2 + 12pv + 9v^2$$

(d) $(3r - 5s)^2 = (3r)^2 - 2(3r)(5s) + (5s)^2$
$$= 9r^2 - 30rs + 25s^2$$

✔ **Now Try Exercises 59 and 63.**

▶ **CAUTION**   As the products in the formula for the square of a binomial show,
$$(x + y)^2 \neq x^2 + y^2.$$

More generally,

$$(x + y)^n \neq x^n + y^n \quad (n \neq 1).$$

EXAMPLE 7   Multiplying More Complicated Binomials

Find each product.

(a) $[(3p - 2) + 5q][(3p - 2) - 5q]$
$$= (3p - 2)^2 - (5q)^2 \qquad \text{Product of sum and difference of terms}$$
$$= 9p^2 - 12p + 4 - 25q^2 \qquad \text{Square both quantities.}$$

(b) $[(2z + r) + 1]^2 = (2z + r)^2 + 2(2z + r)(1) + 1^2$ Square of a binomial

$$= 4z^2 + 4zr + r^2 + 4z + 2r + 1$$ Square again; use the distributive property.

(c) $(x + y)^3 = (x + y)^2(x + y)$

This does not equal $x^3 + y^3$.

$$= (x^2 + 2xy + y^2)(x + y)$$ Square $x + y$.

$$= x^3 + 2x^2y + xy^2 + x^2y + 2xy^2 + y^3$$ Distributive property

$$= x^3 + 3x^2y + 3xy^2 + y^3$$ Combine like terms.

(d) $(2a + b)^4 = (2a + b)^2(2a + b)^2$

$$= (4a^2 + 4ab + b^2)(4a^2 + 4ab + b^2)$$ Square $2a + b$.

$$= 16a^4 + 16a^3b + 4a^2b^2 + 16a^3b + 16a^2b^2$$
$$+ 4ab^3 + 4a^2b^2 + 4ab^3 + b^4$$

$$= 16a^4 + 32a^3b + 24a^2b^2 + 8ab^3 + b^4$$

✔ **Now Try Exercises 71, 75, 79, and 83.**

OBJECTIVE 6 Multiply polynomial functions. In **Section 5.3,** we introduced operations on functions and saw how functions can be added and subtracted. Functions can also be multiplied.

Multiplying Functions

If $f(x)$ and $g(x)$ define functions, then

$$(fg)(x) = f(x) \cdot g(x).$$ Product function

The domain of the product function is the intersection of the domains of $f(x)$ and $g(x)$.

EXAMPLE 8 Multiplying Polynomial Functions

For $f(x) = 3x + 4$ and $g(x) = 2x^2 + x$, find $(fg)(x)$ and $(fg)(-1)$.

$$(fg)(x) = f(x) \cdot g(x)$$ Use the definition.

$$= (3x + 4)(2x^2 + x)$$

$$= 6x^3 + 3x^2 + 8x^2 + 4x$$ FOIL

$$= 6x^3 + 11x^2 + 4x$$ Combine like terms.

Then

$$(fg)(-1) = 6(-1)^3 + 11(-1)^2 + 4(-1)$$ Let $x = -1$.

Be careful with signs.

$$= -6 + 11 - 4$$

$$= 1.$$

Confirm that $f(-1) \cdot g(-1)$ is equal to $(fg)(-1)$.

✔ **Now Try Exercises 115 and 117.**

5.4 EXERCISES

Concept Check *Match each product in Column I with the correct polynomial in Column II.*

I	II
1. $(2x - 5)(3x + 4)$	**A.** $6x^2 + 23x + 20$
2. $(2x + 5)(3x + 4)$	**B.** $6x^2 + 7x - 20$
3. $(2x - 5)(3x - 4)$	**C.** $6x^2 - 7x - 20$
4. $(2x + 5)(3x - 4)$	**D.** $6x^2 - 23x + 20$

Find each product. See Examples 1–3.

5. $-8m^3(3m^2)$ **6.** $4p^2(-5p^4)$ **7.** $14x^2y^3(-2x^5y)$

8. $-5m^3n^4(4m^2n^5)$ **9.** $3x(-2x + 5)$ **10.** $5y(-6y - 1)$

11. $-q^3(2 + 3q)$ **12.** $-3a^4(4 - a)$ **13.** $6k^2(3k^2 + 2k + 1)$

14. $5r^3(2r^2 - 3r - 4)$ **15.** $(2m + 3)(3m^2 - 4m - 1)$

16. $(4z - 2)(z^2 + 3z + 5)$ **17.** $m(m + 5)(m - 8)$

18. $p(p - 6)(p + 4)$ **19.** $4z(2z + 1)(3z - 4)$ **20.** $2y(8y - 3)(2y + 1)$

21. $4x^3(x - 3)(x + 2)$ **22.** $2y^5(y - 8)(y + 2)$ **23.** $(2y + 3)(3y - 4)$

24. $(5m - 3)(2m + 6)$

25. $\begin{array}{r} -b^2 + 3b + 3 \\ \underline{2b + 4} \end{array}$ **26.** $\begin{array}{r} -r^2 - 4r + 8 \\ \underline{3r - 2} \end{array}$

27. $\begin{array}{r} 5m - 3n \\ \underline{5m + 3n} \end{array}$ **28.** $\begin{array}{r} 2k + 6q \\ \underline{2k - 6q} \end{array}$ **29.** $\begin{array}{r} 2z^3 - 5z^2 + 8z - 1 \\ \underline{4z + 3} \end{array}$

30. $\begin{array}{r} 3z^4 - 2z^3 + z - 5 \\ \underline{2z - 5} \end{array}$ **31.** $\begin{array}{r} 2p^2 + 3p + 6 \\ \underline{3p^2 - 4p - 1} \end{array}$ **32.** $\begin{array}{r} 5y^2 - 2y + 4 \\ \underline{2y^2 + y + 3} \end{array}$

Use the FOIL method to find each product. See Example 4.

33. $(m + 5)(m - 8)$ **34.** $(p - 6)(p + 4)$ **35.** $(4k + 3)(3k - 2)$

36. $(5w + 2)(2w + 5)$ **37.** $(z - w)(3z + 4w)$ **38.** $(s + t)(2s - 5t)$

39. $(6c - d)(2c + 3d)$ **40.** $(2m - n)(3m + 5n)$ **41.** $(0.2x + 1.3)(0.5x - 0.1)$

42. $(0.5y - 0.4)(0.1y + 2.1)$ **43.** $\left(3w + \dfrac{1}{4}z\right)(w - 2z)$ **44.** $\left(5r - \dfrac{2}{3}y\right)(r + 5y)$

✎ **45.** Describe the FOIL method in your own words.

✎ **46.** Explain why the product of the sum and difference of two terms is not a trinomial.

Find each product. See Example 5.

47. $(2p - 3)(2p + 3)$ **48.** $(3x - 8)(3x + 8)$ **49.** $(5m - 1)(5m + 1)$

50. $(6y + 3)(6y - 3)$ **51.** $(3a + 2c)(3a - 2c)$ **52.** $(5r - 4s)(5r + 4s)$

53. $\left(4x - \dfrac{2}{3}\right)\left(4x + \dfrac{2}{3}\right)$ **54.** $\left(3t + \dfrac{5}{4}\right)\left(3t - \dfrac{5}{4}\right)$ **55.** $(4m + 7n^2)(4m - 7n^2)$

56. $(2k^2 + 6h)(2k^2 - 6h)$ **57.** $3y(5y^3 + 2)(5y^3 - 2)$ **58.** $4x(3x^3 + 4)(3x^3 - 4)$

Find each product. See Example 6.

59. $(y - 5)^2$

60. $(a - 3)^2$

61. $(2p + 7)^2$

62. $(3z + 8)^2$

63. $(4n + 3m)^2$

64. $(5r + 7s)^2$

65. $\left(k - \dfrac{5}{7}p\right)^2$

66. $\left(q - \dfrac{3}{4}r\right)^2$

67. $(0.2x - 1.4y)^2$

68. Explain why $(x + y)^2$ and $x^2 + y^2$ are not equivalent.

69. *Concept Check* Find the product $101 \cdot 99$, using the special product rule $(x + y)(x - y) = x^2 - y^2$.

70. *Concept Check* Repeat Exercise 69 for the product $202 \cdot 198$.

Find each product. See Example 7.

71. $[(5x + 1) + 6y]^2$

72. $[(3m - 2) + p]^2$

73. $[(2a + b) - 3]^2$

74. $[(4k + h) - 4]^2$

75. $[(2a + b) - 3][(2a + b) + 3]$

76. $[(m + p) + 5][(m + p) - 5]$

77. $[(2h - k) + j][(2h - k) - j]$

78. $[(3m - y) + z][(3m - y) - z]$

79. $(y + 2)^3$

80. $(z - 3)^3$

81. $(5r - s)^3$

82. $(x + 3y)^3$

83. $(q - 2)^4$

84. $(r + 3)^4$

The next problems are "brain busters." Find each product.

85. $(2a + b)(3a^2 + 2ab + b^2)$

86. $(m - 5p)(m^2 - 2mp + 3p^2)$

87. $(4z - x)(z^3 - 4z^2x + 2zx^2 - x^3)$

88. $(3r + 2s)(r^3 + 2r^2s - rs^2 + 2s^3)$

89. $(m^2 - 2mp + p^2)(m^2 + 2mp - p^2)$

90. $(3 + x + y)(-3 + x - y)$

91. $ab(a + b)(a + 2b)(a - 3b)$

92. $mp(m - p)(m - 2p)(2m + p)$

In Exercises 93–96, two expressions are given. Replace x with 3 and y with 4 to show that, in general, the two expressions do not equal each other.

93. $(x + y)^2$; $x^2 + y^2$

94. $(x + y)^3$; $x^3 + y^3$

95. $(x + y)^4$; $x^4 + y^4$

96. $(x + y)^5$; $x^5 + y^5$

Find the area of each figure. Express it as a polynomial in descending powers of the variable x. Refer to the formulas on the inside covers of this book if necessary.

97.

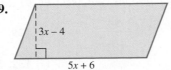

98.

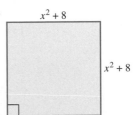

99.

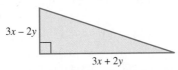

100.

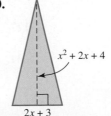

RELATING CONCEPTS (EXERCISES 101–108)

FOR INDIVIDUAL OR GROUP WORK

Consider the figure. **Work Exercises 101–108 in order.**

101. What is the length of each side of the blue square in terms of a and b?

102. What is the formula for the area of a square? Use the formula to write an expression, in the form of a product, for the area of the blue square.

103. Each green rectangle has an area of _____. Therefore, the total area in green is represented by the polynomial _____.

104. The yellow square has an area of _____.

105. The area of the entire colored region is represented by _____, because each side of the entire colored region has length _____.

106. The area of the blue square is equal to the area of the entire colored region, minus the total area of the green squares, minus the area of the yellow square. Write this as a simplified polynomial in a and b.

107. **(a)** What must be true about the expressions for the area of the blue square you found in Exercises 102 and 106?

   (b) Write an equation based on your answer in part (a). How does your equation reinforce one of the main ideas of this section?

108. Draw a figure and give a similar proof for $(a + b)^2 = a^2 + 2ab + b^2$.

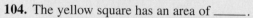

For each pair of functions, find the product $(fg)(x)$. *See Example 8.*

109. $f(x) = 2x$, $g(x) = 5x - 1$

110. $f(x) = 3x$, $g(x) = 6x - 8$

111. $f(x) = x + 1$, $g(x) = 2x - 3$

112. $f(x) = x - 7$, $g(x) = 4x + 5$

113. $f(x) = 2x - 3$, $g(x) = 4x^2 + 6x + 9$

114. $f(x) = 3x + 4$, $g(x) = 9x^2 - 12x + 16$

Let $f(x) = x^2 - 9$, $g(x) = 2x$, *and* $h(x) = x - 3$. *Find each of the following. See Example 8.*

115. $(fg)(x)$

116. $(fh)(x)$

117. $(fg)(2)$

118. $(fh)(1)$

119. $(gh)(x)$

120. $(fh)(-1)$

121. $(gh)(-3)$

122. $(fg)(-2)$

123. $(fg)\left(-\dfrac{1}{2}\right)$

124. $(fg)\left(-\dfrac{1}{3}\right)$

125. $(fh)\left(-\dfrac{1}{4}\right)$

126. $(fh)\left(-\dfrac{1}{5}\right)$

PREVIEW EXERCISES

Perform the indicated operations. See Sections 5.1 and 5.2.

127. $\dfrac{12p^7}{6p^3}$

128. $\dfrac{-9y^{11}}{3y}$

129. $\dfrac{-8a^3b^7}{6a^5b}$

130. $\dfrac{-20r^3s^5}{15rs^9}$

131. Subtract.
$$-3a^2 + 4a - 5$$
$$\underline{5a^2 + 3a - 9}$$

132. Subtract.
$$-4p^2 - 8p + 5$$
$$\underline{3p^2 + 2p + 9}$$

5.5 Dividing Polynomials

OBJECTIVES

1 Divide a polynomial by a monomial.

2 Divide a polynomial by a polynomial of two or more terms.

3 Divide polynomial functions.

OBJECTIVE 1 Divide a polynomial by a monomial. We now discuss polynomial division, beginning with division by a monomial. (Recall that a monomial is a single term, such as $8x$, $-9m^4$, or $11y^2$.)

Dividing a Polynomial by a Monomial

To divide a polynomial by a monomial, divide each term in the polynomial by the monomial, and then write each quotient in lowest terms.

EXAMPLE 1 Dividing a Polynomial by a Monomial

Divide.

(a) $\dfrac{15x^2 - 12x + 6}{3} = \dfrac{15x^2}{3} - \dfrac{12x}{3} + \dfrac{6}{3}$ Divide each term by 3.

$= 5x^2 - 4x + 2$ Write in lowest terms.

Check this answer by multiplying it by the divisor, 3.

$$3\underbrace{(5x^2 - 4x + 2)}_{} = \underbrace{15x^2 - 12x + 6}_{}$$

Divisor Quotient   Original polynomial

(b) $\dfrac{5m^3 - 9m^2 + 10m}{5m^2} = \dfrac{5m^3}{5m^2} - \dfrac{9m^2}{5m^2} + \dfrac{10m}{5m^2}$ Divide each term by $5m^2$.

$= m - \dfrac{9}{5} + \dfrac{2}{m}$ Write in lowest terms.

Check: $5m^2\left(m - \dfrac{9}{5} + \dfrac{2}{m}\right) = 5m^3 - 9m^2 + 10m$,   the original polynomial.

The result $m - \frac{9}{5} + \frac{2}{m}$ is not a polynomial. (Why?) The quotient of two polynomials need not be a polynomial.

(c) $\dfrac{8xy^2 - 9x^2y + 6x^2y^2}{x^2y^2} = \dfrac{8xy^2}{x^2y^2} - \dfrac{9x^2y}{x^2y^2} + \dfrac{6x^2y^2}{x^2y^2}$

$= \dfrac{8}{x} - \dfrac{9}{y} + 6$

✔ **Now Try Exercises 5, 9, and 11.**

OBJECTIVE 2 Divide a polynomial by a polynomial of two or more terms. The process for dividing one polynomial by another polynomial that is not a monomial is similar to that for dividing whole numbers.

EXAMPLE 2 Dividing a Polynomial by a Polynomial

Divide $\dfrac{2m^2 + m - 10}{m - 2}$.

Write the problem as if dividing whole numbers, making sure that both polynomials are written in descending powers of the variables.

$$m - 2 \overline{)2m^2 + m - 10}$$

Divide the first term of $2m^2 + m - 10$ by the first term of $m - 2$. Since $\dfrac{2m^2}{m} = 2m$, place this result above the division line.

$$\begin{array}{r} 2m \quad \longleftarrow \text{Result of } \frac{2m^2}{m} \\ m - 2 \overline{)2m^2 + m - 10} \end{array}$$

Multiply $m - 2$ and $2m$, and write the result below $2m^2 + m - 10$.

$$\begin{array}{r} 2m \\ m - 2 \overline{)2m^2 + m - 10} \\ \underline{2m^2 - 4m} \quad \longleftarrow 2m(m - 2) = 2m^2 - 4m \end{array}$$

Now subtract $2m^2 - 4m$ from $2m^2 + m$. Do this by mentally changing the signs on $2m^2 - 4m$ and *adding*.

$$\begin{array}{r} 2m \\ m - 2 \overline{)2m^2 + m - 10} \\ \underline{2m^2 - 4m} \\ 5m \quad \longleftarrow \text{Subtract. The difference is } 5m. \end{array}$$

To subtract, add the opposite.

Bring down -10 and continue by dividing $5m$ by m.

$$\begin{array}{r} 2m + 5 \longleftarrow \frac{5m}{m} = 5 \\ m - 2 \overline{)2m^2 + m - 10} \\ \underline{2m^2 - 4m} \\ 5m - 10 \longleftarrow \text{Bring down } -10. \\ \underline{5m - 10} \longleftarrow 5(m - 2) = 5m - 10 \\ 0 \longleftarrow \text{Subtract. The difference is } 0. \end{array}$$

Finally, $(2m^2 + m - 10) \div (m - 2) = 2m + 5$. Check by multiplying $m - 2$ and $2m + 5$. The result should be $2m^2 + m - 10$.

✔ Now Try Exercise 17.

EXAMPLE 3 Dividing a Polynomial with a Missing Term

Divide $3x^3 - 2x + 5$ by $x - 3$.

Make sure that $3x^3 - 2x + 5$ is in descending powers of the variable. Add a term with 0 coefficient as a placeholder for the missing x^2-term.

Missing term

$$x - 3 \overline{)3x^3 + 0x^2 - 2x + 5}$$

Start with $\dfrac{3x^3}{x} = 3x^2$.

$$
\begin{array}{r}
3x^2 \quad\quad\quad \longleftarrow \frac{3x^3}{x} = 3x^2 \\
x - 3\overline{)3x^3 + 0x^2 - 2x + 5} \\
\underline{3x^3 - 9x^2} \quad\quad \longleftarrow 3x^2(x-3)
\end{array}
$$

Subtract by mentally changing the signs on $3x^3 - 9x^2$ and adding.

$$
\begin{array}{r}
3x^2 \quad\quad\quad\quad \\
x - 3\overline{)3x^3 + 0x^2 - 2x + 5} \\
\underline{3x^3 - 9x^2} \quad\quad\quad\quad \\
9x^2 \quad\quad\quad \longleftarrow \text{Subtract.}
\end{array}
$$

Bring down the next term.

$$
\begin{array}{r}
3x^2 \quad\quad\quad\quad \\
x - 3\overline{)3x^3 + 0x^2 - 2x + 5} \\
\underline{3x^3 - 9x^2} \quad\quad\quad\quad \\
9x^2 - 2x \quad\quad \longleftarrow \text{Bring down } -2x.
\end{array}
$$

In the next step, $\dfrac{9x^2}{x} = 9x$.

$$
\begin{array}{r}
3x^2 + 9x \quad\quad \longleftarrow \frac{9x^2}{x} = 9x \\
x - 3\overline{)3x^3 + 0x^2 - 2x + 5} \\
\underline{3x^3 - 9x^2} \quad\quad\quad\quad \\
9x^2 - 2x \quad\quad\quad \\
\underline{9x^2 - 27x} \quad\quad \longleftarrow 9x(x-3) \\
25x + 5 \longleftarrow \text{Subtract; bring down 5.}
\end{array}
$$

Finally, $\frac{25x}{x} = 25$.

$$
\begin{array}{r}
3x^2 + 9x + 25 \longleftarrow \frac{25x}{x} = 25 \\
x - 3\overline{)3x^3 + 0x^2 - 2x + 5} \\
\underline{3x^3 - 9x^2} \quad\quad\quad\quad\quad \\
9x^2 - 2x \quad\quad\quad\quad \\
\underline{9x^2 - 27x} \quad\quad\quad\quad \\
25x + 5 \quad\quad \\
\underline{25x - 75} \longleftarrow 25(x-3) \\
80 \longleftarrow \text{Remainder}
\end{array}
$$

Write the remainder, 80, as the numerator of the fraction $\frac{80}{x-3}$. In summary,

$$
\frac{3x^3 - 2x + 5}{x - 3} = 3x^2 + 9x + 25 + \frac{80}{x - 3}.
$$

> Be sure to add $\frac{\text{remainder}}{\text{divisor}}$ here. Don't forget the + sign.

Check by multiplying $x - 3$ and $3x^2 + 9x + 25$ and adding 80 to the result. You should get $3x^3 - 2x + 5$.

✔ **Now Try Exercise 37.**

▶ **CAUTION** Remember to include $\frac{\text{remainder}}{\text{divisor}}$ as part of the answer. *Don't forget to insert a plus sign between the polynomial quotient and this fraction.*

EXAMPLE 4 Dividing by a Polynomial with a Missing Term

Divide $6r^4 + 9r^3 + 2r^2 - 8r + 7$ by $3r^2 - 2$.

The polynomial $3r^2 - 2$ has a missing term. Write it as $3r^2 + 0r - 2$.

$$
\begin{array}{r}
2r^2 + 3r + 2 \\
3r^2 + 0r - 2 \overline{) 6r^4 + 9r^3 + 2r^2 - 8r + 7} \\
\underline{6r^4 + 0r^3 - 4r^2} \\
9r^3 + 6r^2 - 8r \\
\underline{9r^3 + 0r^2 - 6r} \\
6r^2 - 2r + 7 \\
\underline{6r^2 + 0r - 4} \\
-2r + 11
\end{array}
$$

Missing term⟶

Stop when the degree of the remainder is less than the degree of the divisor.

$\leftarrow$ Remainder

The degree of the remainder, $-2r + 11$, is less than the degree of the divisor, $3r^2 - 2$, so the division process is now finished. The result is written

$$2r^2 + 3r + 2 + \frac{-2r + 11}{3r^2 - 2}.$$

✔ **Now Try Exercise 43.**

▶ **CAUTION** Remember the following steps when dividing a polynomial by a polynomial of two or more terms.

1. Be sure the terms in both polynomials are in descending powers.

2. Write any missing terms with 0 placeholders.

EXAMPLE 5 Performing a Division with a Fractional Coefficient in the Quotient

Divide $2p^3 + 5p^2 + p - 2$ by $2p + 2$.

$$\frac{3p^2}{2p} = \frac{3}{2}p$$

$$
\begin{array}{r}
p^2 + \dfrac{3}{2}p - 1 \\
2p + 2 \overline{) 2p^3 + 5p^2 + p - 2} \\
\underline{2p^3 + 2p^2} \\
3p^2 + p \\
\underline{3p^2 + 3p} \\
-2p - 2 \\
\underline{-2p - 2} \\
0
\end{array}
$$

Since the remainder is 0, the quotient is $p^2 + \frac{3}{2}p - 1$.

✔ **Now Try Exercise 45.**

OBJECTIVE 3 Divide polynomial functions. In the preceding sections, we used operations on functions to add, subtract, and multiply polynomial functions. We now define the quotient of two functions.

Dividing Functions

If $f(x)$ and $g(x)$ define functions, then

$$\left(\frac{f}{g}\right)(x) = \frac{f(x)}{g(x)}. \qquad \text{Quotient function}$$

The domain of the quotient function is the intersection of the domains of $f(x)$ and $g(x)$, excluding any values of x for which $g(x) = 0$.

EXAMPLE 6 Dividing Polynomial Functions

For $f(x) = 2x^2 + x - 10$ and $g(x) = x - 2$, find $\left(\frac{f}{g}\right)(x)$ and $\left(\frac{f}{g}\right)(-3)$. What value of x is not in the domain of the quotient function?

$$\left(\frac{f}{g}\right)(x) = \frac{f(x)}{g(x)} = \frac{2x^2 + x - 10}{x - 2}$$

This quotient was found in Example 2, with m replacing x. The result here is $2x + 5$, so

$$\left(\frac{f}{g}\right)(x) = 2x + 5, \qquad x \neq 2.$$

The number 2 is not in the domain because it causes the denominator $g(x) = x - 2$ to equal 0. Then

$$\left(\frac{f}{g}\right)(-3) = 2(-3) + 5 - -1. \qquad \text{Let } x - -3.$$

Verify that the same value is found by evaluating $\frac{f(-3)}{g(-3)}$.

✔ **Now Try Exercises 65 and 67.**

5.5 EXERCISES

⊘ *Complete solution available on Video Lectures on CD/DVD*

Now Try Exercise

Concept Check *Complete each statement with the correct word(s).*

1. We find the quotient of two monomials by using the _____ rule for _____.

2. To divide a polynomial by a monomial, divide each _____ of the polynomial by the _____.

3. When dividing polynomials that are not monomials, first write them in _____ powers.

4. If a polynomial in a division problem has a missing term, insert a term with coefficient equal to _____ as a placeholder.

Divide. See Example 1.

⊘ **5.** $\dfrac{15x^3 - 10x^2 + 5}{5}$

6. $\dfrac{27m^4 - 18m^3 + 9m}{9}$

7. $\dfrac{9y^2 + 12y - 15}{3y}$

8. $\dfrac{80r^2 - 40r + 10}{10r}$

9. $\dfrac{15m^3 + 25m^2 + 30m}{5m^2}$

10. $\dfrac{64x^3 - 72x^2 + 12x}{8x^3}$

11. $\dfrac{14m^2n^2 - 21mn^3 + 28m^2n}{14m^2n}$

12. $\dfrac{24h^2k + 56hk^2 - 28hk}{16h^2k^2}$

13. $\dfrac{8wxy^2 + 3wx^2y + 12w^2xy}{4wx^2y}$

14. $\dfrac{12ab^2c + 10a^2bc + 18abc^2}{6a^2bc}$

Complete the division.

15.
$$
\begin{array}{r}
r^2 \\
3r - 1\overline{)3r^3 - 22r^2 + 25r - 6} \\
\underline{3r^3 - r^2} \\
-21r^2
\end{array}
$$

16.
$$
\begin{array}{r}
3b^2 \\
2b - 5\overline{)6b^3 - 7b^2 - 4b - 40} \\
\underline{6b^3 - 15b^2} \\
8b^2
\end{array}
$$

Divide. See Examples 2–5.

17. $\dfrac{y^2 + y - 20}{y + 5}$

18. $\dfrac{y^2 + 3y - 18}{y + 6}$

19. $\dfrac{q^2 + 4q - 32}{q - 4}$

20. $\dfrac{q^2 + 2q - 35}{q - 5}$

21. $\dfrac{3t^2 + 17t + 10}{3t + 2}$

22. $\dfrac{2k^2 - 3k - 20}{2k + 5}$

23. $\dfrac{p^2 + 2p + 20}{p + 6}$

24. $\dfrac{x^2 + 11x + 16}{x + 8}$

25. $\dfrac{3m^3 + 5m^2 - 5m + 1}{3m - 1}$

26. $\dfrac{8z^3 - 6z^2 - 5z + 3}{4z + 3}$

27. $\dfrac{m^3 - 2m^2 - 9}{m - 3}$

28. $\dfrac{p^3 + 3p^2 - 4}{p + 2}$

29. $(2z^3 - 5z^2 + 6z - 15) \div (2z - 5)$

30. $(3p^3 + p^2 + 18p + 6) \div (3p + 1)$

31. $(4x^3 + 9x^2 - 10x + 3) \div (4x + 1)$

32. $(10z^3 - 26z^2 + 17z - 13) \div (5z - 3)$

33. $\dfrac{6x^3 - 19x^2 + 14x - 15}{3x^2 - 2x + 4}$

34. $\dfrac{8m^3 - 18m^2 + 37m - 13}{2m^2 - 3m + 6}$

35. $(x^3 + 2x - 3) \div (x - 1)$

36. $(2x^3 - 11x^2 + 25) \div (x - 5)$

37. $(3x^3 - x + 4) \div (x - 2)$

38. $(3k^3 + 9k - 14) \div (k - 2)$

39. $\dfrac{4k^4 + 6k^3 + 3k - 1}{2k^2 + 1}$

40. $\dfrac{9k^4 + 12k^3 - 4k - 1}{3k^2 - 1}$

41. $\dfrac{6y^4 + 4y^3 + 4y - 6}{3y^2 + 2y - 3}$

42. $\dfrac{8t^4 + 6t^3 + 12t - 32}{4t^2 + 3t - 8}$

43. $(x^4 - 4x^3 + 5x^2 - 3x + 2) \div (x^2 + 3)$

44. $(3t^4 + 5t^3 - 8t^2 - 13t + 2) \div (t^2 - 5)$

45. $(2p^3 + 7p^2 + 9p + 3) \div (2p + 2)$

46. $(3x^3 + 4x^2 + 7x + 4) \div (3x + 3)$

47. $(3a^2 - 11a + 17) \div (2a + 6)$

48. $(5t^2 + 19t + 7) \div (4t + 12)$

49. $\dfrac{p^3 - 1}{p - 1}$

50. $\dfrac{8a^3 + 1}{2a + 1}$

The next problems are "brain busters". Divide.

51. $\left(2x^2 - \dfrac{7}{3}x - 1\right) \div (3x + 1)$

52. $\left(m^2 + \dfrac{7}{2}m + 3\right) \div (2m + 3)$

53. $\left(3a^2 - \dfrac{23}{4}a - 5\right) \div (4a + 3)$

54. $\left(3q^2 + \dfrac{19}{5}q - 3\right) \div (5q - 2)$

Solve each problem.

55. The volume of a box is $(2p^3 + 15p^2 + 28p)$. The height is p and the length is $(p + 4)$. Give an expression in p that represents the width.

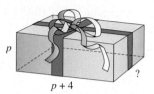

$p + 4$

56. Suppose that a car travels a distance of $(2m^3 + 15m^2 + 35m + 36)$ miles in $(2m + 9)$ hours. Give an expression in m that represents the rate of the car in mph.

57. For $P(x) = x^3 - 4x^2 + 3x - 5$, find $P(-1)$. Then divide $P(x)$ by $D(x) = x + 1$. Compare the remainder with $P(-1)$. What do these results suggest?

58. *Concept Check*   Let $P(x) = 4x^3 - 8x^2 + 13x - 2$ and $D(x) = 2x - 1$. Use division to find polynomials $Q(x)$ and $R(x)$ such that $P(x) = Q(x) \cdot D(x) + R(x)$.

For each pair of functions, find the quotient $\left(\frac{f}{g}\right)(x)$ and give any x-values that are not in the domain of the quotient function. See Example 6.

59. $f(x) = 10x^2 - 2x$, $g(x) = 2x$

60. $f(x) = 18x^2 - 24x$, $g(x) = 3x$

61. $f(x) = 2x^2 - x - 3$, $g(x) = x + 1$

62. $f(x) = 4x^2 - 23x - 35$, $g(x) = x - 7$

63. $f(x) = 8x^3 - 27$, $g(x) = 2x - 3$

64. $f(x) = 27x^3 + 64$, $g(x) = 3x + 4$

Let $f(x) = x^2 - 9$, $g(x) = 2x$, and $h(x) = x - 3$. Find each of the following. See Example 6.

65. $\left(\dfrac{f}{g}\right)(x)$

66. $\left(\dfrac{f}{h}\right)(x)$

67. $\left(\dfrac{f}{g}\right)(2)$

68. $\left(\dfrac{f}{h}\right)(1)$

69. $\left(\dfrac{h}{g}\right)(x)$

70. $\left(\dfrac{g}{h}\right)(x)$

71. $\left(\dfrac{h}{g}\right)(3)$

72. $\left(\dfrac{g}{h}\right)(-1)$

73. $\left(\dfrac{f}{g}\right)\left(\dfrac{1}{2}\right)$

74. $\left(\dfrac{f}{g}\right)\left(\dfrac{3}{2}\right)$

75. $\left(\dfrac{h}{g}\right)\left(-\dfrac{1}{2}\right)$

76. $\left(\dfrac{h}{g}\right)\left(-\dfrac{3}{2}\right)$

PREVIEW EXERCISES

*Use the distributive property to rewrite each expression. See **Section 1.4**.*

77. $9 \cdot 6 + 9 \cdot r^2$

78. $8 \cdot y - 8 \cdot 5t$

79. $7(2x) - 7(3z)$

80. $4(8p) + 4(9y)$

81. $3x(x + 1) + 4(x + 1)$

82. $4p(2p - 3) + 5(2p - 3)$

*Simplify. Write answers with only positive exponents. Assume that all variables represent nonzero real numbers. See **Section 5.1**.*

83. $18z^3w(zw^2)^4$

84. $8z^4(a^2b)^5$

85. $12p^4q^{-2}(5pq)^{-1}$

86. $(5m^2n)^3(-mn^{-1})^2$

Chapter 5 | Group Activity

COMPARING MATHEMATICAL MODELS

Objective Use polynomial models to approximate and compare data.

The number of individuals (in millions) covered by private or governmental health insurance in the United States from 1990 through 2002 can be modeled by the polynomial function defined by

$$P(x) = 3.054x + 248.9 \qquad \text{Linear model}$$

where x represents years since 1990. The number of individuals (in millions) can also be modeled by

$$Q(x) = 0.0006x^2 + 3.047x + 248.9 \qquad \text{Quadratic model}$$

and $\qquad R(x) = 0.0039x^3 - 0.0678x^2 + 3.327x + 248.9,$ $\qquad$ Cubic model

again where x represents years since 1990.

A. Use each model to determine the number of individuals (in millions) covered by private or governmental health insurance in the United States for each year from 1990 through 2002. Complete the table.

	Number of Individuals Insured (in millions)		
Year	Using P(x)	Using Q(x)	Using R(x)
1990			
1995			
2000			
2001			
2002			

B. The actual data are given in the table. Compare the data you entered into the table in part A with the actual data. Which model provides the best approximation of (that is, *best fits*) the actual data? Explain your answer.

Year	Number of Individuals Insured (in millions)
1990	248.9
1995	264.3
2000	279.5
2001	282.1
2002	285.9

Source: U.S. Census Bureau.

C. Use the models to make predictions.

1. Use P, Q, and R to estimate the number of individuals (in millions) who were covered by private or governmental health insurance in 2003.

2. The actual number of individuals insured in 2003 was 288.3 million. How do your estimations compare with the actual number? Which model gave the best approximation?

3. Use each model to predict the number of individuals (in millions) insured in 2010.

4. Discuss the validity of using these models to predict the number of individuals (in millions) covered by private or governmental insurance in years beyond 2002.

Chapter **5** SUMMARY *View the Interactive Summary on the Pass the Test CD.*

KEY TERMS

5.2 term
numerical coefficient
(coefficient)
algebraic expression
polynomial
polynomial in x

descending powers
trinomial
binomial
monomial
degree of a term

degree of a
polynomial
negative of a
polynomial
5.3 polynomial function

composition of
functions
identity function
squaring function
cubing function

NEW SYMBOLS

$(f \circ g)(x) = f(g(x))$ composite function

TEST YOUR WORD POWER

See how well you have learned the vocabulary in this chapter. Answers, with examples, follow the Quick Review.

1. A **polynomial** is an algebraic
expression made up of
   A. a term or a finite product of terms
   with positive coefficients and
   exponents
   B. the sum of two or more terms
   with whole number coefficients
   and exponents
   C. the product of two or more terms
   D. a term or a finite sum of terms
   with real coefficients and whole
   number exponents.

2. A **monomial** is a polynomial with
   A. only one term
   B. exactly two terms
   C. exactly three terms
   D. more than three terms.

3. A **binomial** is a polynomial with
   A. only one term
   B. exactly two terms
   C. exactly three terms
   D. more than three terms.

4. A **trinomial** is a polynomial with
   A. only one term
   B. exactly two terms
   C. exactly three terms
   D. more than three terms.

5. The **FOIL** method is used for
   A. adding two binomials
   B. adding two trinomials
   C. multiplying two binomials
   D. multiplying two trinomials.

QUICK REVIEW

Concepts	Examples

5.1 INTEGER EXPONENTS AND SCIENTIFIC NOTATION

Definitions and Rules for Exponents
For all integers m and n and all real numbers a and b,

Product Rule: $a^m \cdot a^n = a^{m+n}$

Quotient Rule: $\dfrac{a^m}{a^n} = a^{m-n}$ $(a \neq 0)$

Zero Exponent: $a^0 = 1$ $(a \neq 0)$

Apply the rules for exponents.

$$3^4 \cdot 3^2 = 3^6$$

$$\frac{2^5}{2^3} = 2^2$$

$$27^0 = 1, \quad (-5)^0 = 1$$

(continued)

Concepts	Examples

Negative Exponent: $a^{-n} = \dfrac{1}{a^n}$ $(a \neq 0)$

$5^{-2} = \dfrac{1}{5^2}$

Power Rules: $(a^m)^n = a^{mn}$

$(ab)^m = a^m b^m$

$\left(\dfrac{a}{b}\right)^n = \dfrac{a^n}{b^n}$ $(b \neq 0)$.

$(6^3)^4 = 6^{12}$

$(5p)^4 = 5^4 p^4$

$\left(\dfrac{2}{3}\right)^5 = \dfrac{2^5}{3^5}$

Special Rules for Negative Exponents:

$\dfrac{1}{a^{-n}} = a^n$ $(a \neq 0)$

$\dfrac{a^{-n}}{b^{-m}} = \dfrac{b^m}{a^n}$ $(a, b \neq 0)$

$a^{-n} = \left(\dfrac{1}{a}\right)^n$ $(a \neq 0)$

$\left(\dfrac{a}{b}\right)^{-n} = \left(\dfrac{b}{a}\right)^n$ $(a, b \neq 0)$

$\dfrac{1}{3^{-2}} = 3^2$

$\dfrac{5^{-3}}{4^{-6}} = \dfrac{4^6}{5^3}$

$4^{-3} = \left(\dfrac{1}{4}\right)^3$

$\left(\dfrac{4}{7}\right)^{-2} = \left(\dfrac{7}{4}\right)^2$

Scientific Notation

A number is in scientific notation when it is written as a product of a number between 1 and 10 (inclusive of 1) and an integer power of 10.

Write 23,500,000,000 in scientific notation.

$$23{,}500{,}000{,}000 = 2.35 \times 10^{10}$$

Write 4.3×10^{-6} in standard notation.

$$4.3 \times 10^{-6} = 0.0000043$$

5.2 ADDING AND SUBTRACTING POLYNOMIALS

Add or subtract polynomials by combining like terms.

Add. $(x^2 - 2x + 3) + (2x^2 - 8) = 3x^2 - 2x - 5$

Subtract. $(5x^4 + 3x^2) - (7x^4 + x^2 - x) = -2x^4 + 2x^2 + x$

5.3 POLYNOMIAL FUNCTIONS, GRAPHS, AND COMPOSITION

Compostion of f and g

$$(f \circ g)(x) = f(g(x))$$

If $f(x) = x^2$ and $g(x) = 2x + 1$, then

$$(f \circ g)(x) = f(g(x)) = f(2x + 1)$$
$$= (2x + 1)^2$$

and $(g \circ f)(x) = g(f(x)) = g(x^2)$
$$= 2x^2 + 1.$$

The graph of $f(x) = x$ is a line, and the graph of $f(x) = x^2$ is a parabola. These graphs define the identity and squaring functions, respectively.

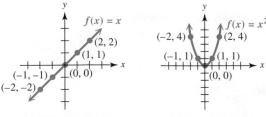

Identity Function   **Squaring Function**

(continued)

Concepts	Examples

The graph of $f(x) = x^3$ defines the cubing function.

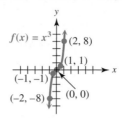

Cubing Function

5.4   MULTIPLYING POLYNOMIALS

To multiply two polynomials, multiply each term of one by each term of the other.

Multiply. $(x^3 + 3x)(4x^2 - 5x + 2)$
$$= 4x^5 + 12x^3 - 5x^4 - 15x^2 + 2x^3 + 6x$$
$$= 4x^5 - 5x^4 + 14x^3 - 15x^2 + 6x$$

To multiply two binomials, use the **FOIL method.** Multiply the **First** terms, the **Outer** terms, the **Inner** terms, and the **Last** terms. Then add these products.

Multiply. $(2x + 3)(x - 7) = 2x(x) + 2x(-7) + 3x + 3(-7)$
$$= 2x^2 - 14x + 3x - 21$$
$$= 2x^2 - 11x - 21$$

Special Products
$$(x + y)(x - y) = x^2 - y^2$$
$$(x + y)^2 = x^2 + 2xy + y^2$$
$$(x - y)^2 = x^2 - 2xy + y^2$$

Multiply.
$$(3m + 8)(3m - 8) = 9m^2 - 64$$
$$(5a + 3b)^2 = 25a^2 + 30ab + 9b^2$$
$$(2k - 1)^2 = 4k^2 - 4k + 1$$

5.5   DIVIDING POLYNOMIALS

Dividing by a Monomial
To divide a polynomial by a monomial, divide each term in the polynomial by the monomial, and then write each fraction in lowest terms.

Divide. $\dfrac{2x^3 - 4x^2 + 6x - 8}{2x} = \dfrac{2x^3}{2x} - \dfrac{4x^2}{2x} + \dfrac{6x}{2x} - \dfrac{8}{2x}$
$$= x^2 - 2x + 3 - \dfrac{4}{x}$$

Dividing by a Polynomial
Use the long division process. The process ends when the remainder is 0 or when the degree of the remainder is less than the degree of the divisor.

Divide. $\dfrac{m^3 - m^2 + 2m + 5}{m + 1}$

$$\begin{array}{r} m^2 - 2m + 4 \\ m + 1\overline{\smash{)}m^3 - m^2 + 2m + 5} \\ \underline{m^3 + m^2 } \\ -2m^2 + 2m \\ \underline{-2m^2 - 2m } \\ 4m + 5 \\ \underline{4m + 4} \\ 1 \leftarrow \text{Remainder} \end{array}$$

The answer is $m^2 - 2m + 4 + \dfrac{1}{m + 1}$.

Answers to Test Your Word Power

1. D; *Example:* $5x^3 + 2x^2 - 7$ **2.** A; *Examples:* $-4, 2x^3, 15a^2b$ **3.** B; *Example:* $3t^3 + 5t$

$$\qquad\qquad\qquad\qquad\qquad\qquad\qquad\; \overset{\text{F}}{} \quad \overset{\text{O}}{} \quad \overset{\text{I}}{} \quad \overset{\text{L}}{}$$

4. C; *Example:* $2a^2 - 3ab + b^2$ **5.** C; *Example:* $(m + 4)(m - 3) = m(m) - 3m + 4m + 4(-3) = m^2 + m - 12$

Chapter **5** REVIEW EXERCISES

[5.1] *Simplify. Write answers with only positive exponents. Assume that all variables represent nonzero real numbers.*

1. 4^3

2. $\left(\dfrac{1}{3}\right)^4$

3. $(-5)^3$

4. $\dfrac{2}{(-3)^{-2}}$

5. $\left(\dfrac{2}{3}\right)^{-4}$

6. $\left(\dfrac{5}{4}\right)^{-2}$

7. $5^{-1} + 6^{-1}$

8. $(5 + 6)^{-1}$

9. $-3^0 + 3^0$

10. $(3^{-4})^2$

11. $(x^{-4})^{-2}$

12. $(xy^{-3})^{-2}$

13. $(z^{-3})^3 z^{-6}$

14. $(5m^{-3})^2 (m^4)^{-3}$

15. $\dfrac{(3r)^2 r^4}{r^{-2} r^{-3}} (9r^{-3})^{-2}$

16. $\left(\dfrac{5z^{-3}}{z^{-1}}\right) \dfrac{5}{z^2}$

17. $\left(\dfrac{6m^{-4}}{m^{-9}}\right)^{-1} \left(\dfrac{m^{-2}}{16}\right)$

18. $\left(\dfrac{3r^5}{5r^{-3}}\right)^{-2} \left(\dfrac{9r^{-1}}{2r^{-5}}\right)^3$

19. $(-3x^4 y^3)(4x^{-2} y^5)$

20. $\dfrac{6m^{-4} n^3}{-3mn^2}$

21. $\dfrac{(5p^{-2}q)(4p^5 q^{-3})}{2p^{-5} q^5}$

22. $\left(\dfrac{a^{-2} b^{-1}}{3a^2}\right)^{-2} \left(\dfrac{b^{-2} \cdot 3a^4}{2b^{-3}}\right)^{-2} \left(\dfrac{a^{-4} b^5}{a^3}\right)^{-2}$

23. Explain the difference between the expressions $(-6)^0$ and -6^0.

24. By choosing a specific value for a, give an example to show that, in general, $(2a)^{-3}$ is not equal to $\dfrac{2}{a^3}$.

25. *Concept Check* Is $\left(\dfrac{a}{b}\right)^{-1} = \dfrac{a^{-1}}{b^{-1}}$ true for all a and $b \neq 0$? If not, explain.

26. *Concept Check* Is $(ab)^{-1} = ab^{-1}$ true for all a and $b \neq 0$? If not, explain.

27. By choosing specific values for x and y, give an example to show that, in general, $(x^2 + y^2)^2 \neq x^4 + y^4$.

Write in scientific notation.

28. 13,450

29. 0.0000000765

30. 0.138

31. In 2000, the total population of the United States was **281,400,000**. Of this amount, **50,454** Americans were centenarians—that is, age **100** or older. Write the three boldfaced numbers in scientific notation. (*Source:* U.S. Census Bureau.)

Write each number in standard notation.

32. 1.21×10^6

33. 5.8×10^{-3}

Find each value. Give answers in both scientific notation and standard notation.

34. $\dfrac{16 \times 10^4}{8 \times 10^8}$

35. $\dfrac{6 \times 10^{-2}}{4 \times 10^{-5}}$

36. $\dfrac{0.0000000164}{0.0004}$

37. $\dfrac{0.0009 \times 12{,}000{,}000}{400{,}000}$

38. In 2005, the estimated population of Luxembourg was 4.69×10^5. The population density was 470 people per mi^2. Based on this information, what is the area of Luxembourg to the nearest square mile? (*Source:* www.worldatlas.com)

39. The population of Fresno, California, is approximately 3.45×10^5. The population density is 5449 per mi^2.

(a) Write the population density in scientific notation.

(b) To the nearest square mile, what is the area of Fresno?

[5.2] *Give the numerical coefficient of each term.*

40. $14p^5$

41. $-z$

42. $\dfrac{x}{10}$

43. $504p^3r^5$

For each polynomial, (a) write it in descending powers, (b) identify it as a monomial, binomial, trinomial, or none of these, and (c) give its degree.

44. $9k + 11k^3 - 3k^2$

45. $14m^6 + 9m^7$

46. $-5y^4 + 3y^3 + 7y^2 - 2y$

47. $-7q^5r^3$

48. *Concept Check*   Give an example of a polynomial in the variable x such that the polynomial has degree 5, is lacking a third-degree term, and is in descending powers of the variable.

Add or subtract as indicated.

49. Add.
$$\begin{array}{r} 3x^2 - 5x + 6 \\ -4x^2 + 2x - 5 \end{array}$$

50. Subtract.
$$\begin{array}{r} -5y^3 \qquad + 8y - 3 \\ 4y^2 + 2y + 9 \end{array}$$

51. $(4a^3 - 9a + 15) - (-2a^3 + 4a^2 + 7a)$

52. $(3y^2 + 2y - 1) + (5y^2 - 11y + 6)$

53. Find the perimeter of the triangle.

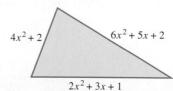

[5.3]

54. Find each of the following for the polynomial function defined by

$$f(x) = -2x^2 + 5x + 7.$$

(a) $f(-2)$ (b) $f(3)$

55. Find each of the following for the polynomial functions defined by

$$f(x) = 2x + 3 \quad \text{and} \quad g(x) = 5x^2 - 3x + 2.$$

(a) $(f + g)(x)$ (b) $(f - g)(x)$ (c) $(f + g)(-1)$ (d) $(f - g)(-1)$

56. Find each of the following for the polynomial functions defined by

$$f(x) = 3x^2 + 2x - 1 \quad \text{and} \quad g(x) = 5x + 7.$$

(a) $(g \circ f)(3)$ (b) $(f \circ g)(3)$ (c) $(f \circ g)(-2)$

(d) $(g \circ f)(-2)$ (e) $(f \circ g)(x)$ (f) $(g \circ f)(x)$

57. The number of people, in millions, enrolled in health maintenance organizations (HMOs) during the period from 1995 through 2003 can be modeled by the polynomial function defined by

$$f(x) = -0.574x^2 + 6.01x + 15.9,$$

where $x = 0$ corresponds to 1995, $x = 1$ corresponds to 1996, and so on. Use this model to approximate the number of people enrolled in each given year. (*Source:* Interstudy; U.S. National Center for Health Statistics.)

(a) 1995   (b) 2000   (c) 2003

Graph each polynomial function defined as follows.

58. $f(x) = -2x + 5$ **59.** $f(x) = x^2 - 6$ **60.** $f(x) = -x^3 + 1$

[5.4] *Find each product.*

61. $-6k(2k^2 + 7)$ **62.** $(3m - 2)(5m + 1)$

63. $(3w - 2t)(2w - 3t)$ **64.** $(2p^2 + 6p)(5p^2 - 4)$

65. $(3q^2 + 2q - 4)(q - 5)$ **66.** $(3z^3 - 2z^2 + 4z - 1)(3z - 2)$

67. $(6r^2 - 1)(6r^2 + 1)$ **68.** $\left(z + \dfrac{3}{5}\right)\left(z - \dfrac{3}{5}\right)$

69. $(4m + 3)^2$ **70.** $t(3t + 2)^2$

[5.5] *Divide.*

71. $\dfrac{4y^3 - 12y^2 + 5y}{4y}$ **72.** $\dfrac{x^3 - 9x^2 + 26x - 30}{x - 5}$

73. $\dfrac{2p^3 + 9p^2 + 27}{2p - 3}$ **74.** $\dfrac{5p^4 + 15p^3 - 33p^2 - 9p + 18}{5p^2 - 3}$

MIXED REVIEW EXERCISES

75. *Concept Check* Match each expression (a)–(i) in Column I with its equivalent expression A–I in Column II. Choices may be used once, more than once, or not at all.

	I			**II**	
(a) 4^{-2}	**(b)** -4^2		**A.** $\dfrac{1}{16}$		**B.** 0
(c) 4^0	**(d)** $(-4)^0$		**C.** 1		**D.** $-\dfrac{1}{16}$
(e) $(-4)^{-2}$	**(f)** -4^0		**E.** -1		**F.** $\dfrac{5}{16}$
(g) $-4^0 + 4^0$	**(h)** $-4^0 - 4^0$		**G.** -16		**H.** -2
(i) $4^{-2} + 4^{-1}$			**I.** none of these		

In Exercises 76–87, perform the indicated operations and then simplify. Write answers with only positive exponents. Assume that all variables represent nonzero real numbers.

76. $\dfrac{6^{-1}y^3(y^2)^{-2}}{6y^{-4}(y^{-1})}$　　**77.** 5^{-3}　　**78.** $(y^6)^{-5}(2y^{-3})^{-4}$

79. $7p^5(3p^4 + p^3 + 2p^2)$　　**80.** $(2x - 9)^2$　　**81.** $\dfrac{(-z^{-2})^3}{5(z^{-3})^{-1}}$

82. $-(-3)^2$　　**83.** $\dfrac{8x^2 - 23x + 2}{x - 3}$

84. $\dfrac{(5z^2x^3)^2(2zx^2)^{-1}}{(-10zx^{-3})^{-2}(3z^{-1}x^{-4})^2}$　　**85.** $[(3m - 5n) + p][(3m - 5n) - p]$

86. $\dfrac{20y^3x^3 + 15y^4x + 25yx^4}{10yx^2}$　　**87.** $(2k - 1) - (3k^2 - 2k + 6)$

88. In 2005, the estimated population of New Zealand was 4.0355×10^6. The population density was 38.9 people per mi^2. Based on this information, what is the area of New Zealand, to the nearest square mile? (*Source:* www.worldatlas.com)

Chapter 5 TEST

View the complete solutions to all Chapter Test exercises on the Pass the Test CD.

1. Match each expression (a)–(i) in Column I with its equivalent expression A–I in Column II. Choices may be used once, more than once, or not at all.

	I			**II**	
(a) 7^{-2}	**(b)** 7^0		**A.** 1		**B.** $\dfrac{1}{9}$
(c) -7^0	**(d)** $(-7)^0$		**C.** $\dfrac{1}{49}$		**D.** -1
(e) -7^2	**(f)** $7^{-1} + 2^{-1}$		**E.** -49		**F.** $\dfrac{9}{14}$
(g) $(7 + 2)^{-1}$	**(h)** $\dfrac{7^{-1}}{2^{-1}}$		**G.** $\dfrac{2}{7}$		**H.** 0
(i) $(-7)^{-2}$			**I.** none of these		

Simplify. Write answers with only positive exponents. Assume that all variables represent nonzero real numbers.

2. $(3x^{-2}y^3)^{-2}(4x^3y^{-4})$

3. $\dfrac{36r^{-4}(r^2)^{-3}}{6r^4}$

4. $\left(\dfrac{4p^2}{q^4}\right)^3\left(\dfrac{6p^8}{q^{-8}}\right)^{-2}$

5. $(-2x^4y^{-3})^0(-4x^{-3}y^{-8})^2$

6. Write 9.1×10^{-7} in standard form.

7. Use scientific notation to simplify $\dfrac{2,500,000 \times 0.00003}{0.05 \times 5,000,000}$. Write the answer in both scientific notation and standard notation.

8. Find each of the following for the functions defined by
$$f(x) = -2x^2 + 5x - 6 \quad \text{and} \quad g(x) = 7x - 3.$$
(a) $f(4)$ **(b)** $(f + g)(x)$ **(c)** $(f - g)(x)$ **(d)** $(f - g)(-2)$

9. Find each of the following for the functions defined by
$$f(x) = 3x + 5 \quad \text{and} \quad g(x) = x^2 + 2.$$
(a) $(f \circ g)(-2)$ **(b)** $(f \circ g)(x)$ **(c)** $(g \circ f)(x)$

Graph each polynomial function.

10. $f(x) = -2x^2 + 3$

11. $f(x) = -x^3 + 3$

12. The number of medical doctors, in thousands, in the United States during the period from 1990 through 2002 can be modeled by the polynomial function defined by
$$f(x) = -0.141x^2 + 21.5x + 616,$$
where $x = 0$ corresponds to 1990, $x = 1$ corresponds to 1991, and so on. Use this model to approximate the number of doctors to the nearest thousand in each given year. (*Source:* American Medical Association.)

(a) 1990   **(b)** 1996   **(c)** 2002

Perform the indicated operations.

13. $(4x^3 - 3x^2 + 2x - 5) - (3x^3 + 11x + 8) + (x^2 - x)$

14. $(5x - 3)(2x + 1)$

15. $(2m - 5)(3m^2 + 4m - 5)$

16. $(6x + y)(6x - y)$

17. $(3k + q)^2$

18. $[2y + (3z - x)][2y - (3z - x)]$

19. $\dfrac{16p^3 - 32p^2 + 24p}{4p^2}$

20. $(x^3 + 3x^2 - 4) \div (x - 1)$

21. If $f(x) = x^2 + 3x + 2$ and $g(x) = x + 1$, find each of the following.
(a) $(fg)(x)$ **(b)** $(fg)(-2)$

22. Use $f(x)$ and $g(x)$ from Exercise 21 to find each of the following.
(a) $\left(\dfrac{f}{g}\right)(x)$ **(b)** $\left(\dfrac{f}{g}\right)(-2)$

Chapters 1–5 CUMULATIVE REVIEW EXERCISES

Match each number in Column I with the choice or choices of sets of numbers in Column II to which the number belongs.

I		II
1. 34		**A.** Natural numbers
2. 0		**B.** Whole numbers
3. 2.16		**C.** Integers
4. $-\sqrt{36}$		**D.** Rational numbers
5. $\sqrt{13}$		**E.** Irrational numbers
6. $-\dfrac{4}{5}$		**F.** Real numbers

Evaluate.

7. $9 \cdot 4 - 16 \div 4$ **8.** $\left(\dfrac{1}{3}\right)^2 - \left(\dfrac{1}{2}\right)^3$ **9.** $-|8 - 13| - |-4| + |-9|$

Solve.

10. $-5(8 - 2z) + 4(7 - z) = 7(8 + z) - 3$ **11.** $3(x + 2) - 5(x + 2) = -2x - 4$

12. $A = p + prt$ for t **13.** $2(m + 5) - 3m + 1 > 5$

14. $|3x - 1| = 2$ **15.** $|3z + 1| \geq 7$

16. *Concept Check* Without showing work, give the solution set of $|x + 1| < -3$.

17. A recent survey polled teens about the most important inventions of the 20th century. Complete the results shown in the table if 1500 teens were surveyed.

Most Important Invention	Percent	Actual Number
Personal computer		480
Pacemaker	26%	
Wireless communication	18%	
Television		150

Source: Lemelson–MIT Program.

18. Find the measure of each angle of the triangle.

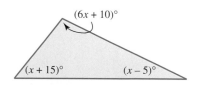

Find the slope of each line described.

19. Through $(-4, 5)$ and $(2, -3)$ **20.** Horizontal, through $(4, 5)$

Find an equation of each line. Write the equation in **(a)** *slope–intercept form and* **(b)** *standard form.*

21. Through $(4, -1)$, $m = -4$ **22.** Through $(0, 0)$ and $(1, 4)$

Graph each equation or inequality.

23. $-3x + 4y = 12$ **24.** $y \leq 2x - 6$ **25.** $3x + 2y < 0$

26. The graph shows the annual number of twin births in the United States for selected years.

✍ **(a)** Use the information given on the graph to find and interpret the average rate of change in the number of twin births per year.

(b) If $x = 0$ represents 1993, use your answer from part (a) to write an equation of the line in slope–intercept form that models the annual number of twin births for the years 1993 through 2003.

(c) Use the equation from part (b) to estimate the approximate number of twin births in 2006.

Seeing Double

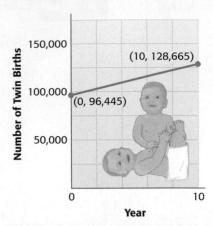

Number of Twin Births

150,000 — (10, 128,665)

100,000 — (0, 96,445)

50,000

0   10

Year

Source: National Center for Health Statistics.

27. Give the domain and range of the relation $\{(-4, -2), (-1, 0), (2, 0), (5, 2)\}$. Does this relation define a function?

28. If $g(x) = -x^2 - 2x + 6$, find $g(3)$.

Solve each system.

29. $3x - 4y = 1$
     $2x + 3y = 12$

30. $3x - 2y = 4$
     $-6x + 4y = 7$

31. $x + 3y - 6z = 7$
     $2x - y + z = 1$
     $x + 2y + 2z = -1$

Use a system of equations to solve each problem.

32. The Star-Spangled Banner that flew over Fort McHenry during the War of 1812 had a perimeter of 144 ft. Its length measured 12 ft more than its width. Find the dimensions of this flag, which is displayed in the Smithsonian Institution's Museum of American History in Washington, DC. (*Source:* National Park Service brochure.)

33. A chemist needs 9 L of a 20% solution of alcohol. She has a 15% solution on hand, as well as a 30% solution. How many liters of the 15% solution and the 30% solution should she mix to get the 20% solution she needs?

Simplify. Write answers with only positive exponents. Assume that all variables represent positive real numbers.

34. $\left(\dfrac{2m^3n}{p^2}\right)^3$

35. $\dfrac{x^{-6}y^3z^{-1}}{x^7y^{-4}z}$

36. $(2m^{-2}n^3)^{-3}$

37. $2^{-1} - 5^{-1}$

Perform the indicated operations.

38. $2(3x^2 - 8x + 1) - 4(x^2 - 3x - 9)$

39. $(3x + 2y)(5x - y)$

40. $(8m + 5n)(8m - 5n)$

41. $(x + 2y)(x^2 - 2xy + 4y^2)$

42. $\dfrac{16x^3y^5 - 8x^2y^2 + 4}{4x^2y}$

43. $\dfrac{m^3 - 3m^2 + 5m - 3}{m - 1}$

44. If $f(x) = x^2 + 6x + 1$ and $g(x) = 2x$, find $(f \circ g)(-2)$.

Factoring

Factoring is used to solve *quadratic equations,* which have many useful applications. An important one is to express the distance a falling or projected object travels in a specific time. Such equations are used in astronomy and the space program to describe the motion of objects in space.

In Section 6.5, we use the concepts of this chapter to explore how to find the heights of objects after they are projected or dropped.

6.1 Greatest Common Factors; Factoring by Grouping

OBJECTIVES

1. Factor out the greatest common factor.
2. Factor by grouping.

Writing a polynomial as the product of two or more simpler polynomials is called **factoring** the polynomial. For example, the product of $3x$ and $5x - 2$ is $15x^2 - 6x$, and $15x^2 - 6x$ can be factored as the product $3x(5x - 2)$.

$$3x(5x - 2) = 15x^2 - 6x \qquad \text{Multiplying}$$
$$15x^2 - 6x = 3x(5x - 2) \qquad \text{Factoring}$$

Notice that both multiplying and factoring use the distributive property, but in opposite directions. *Factoring "undoes," or reverses, multiplying.*

OBJECTIVE 1 Factor out the greatest common factor. The first step in factoring a polynomial is to find the *greatest common factor* for the terms of the polynomial. The **greatest common factor (GCF)** is the largest term that is a factor of all terms in the polynomial. For example, the greatest common factor for $8x + 12$ is 4, since 4 is the largest term that is a factor of (divides into) both $8x$ and 12. Using the distributive property gives

$$8x + 12 = 4(2x) + 4(3)$$
$$= 4(2x + 3).$$

As a check, multiply 4 and $2x + 3$. The result should be $8x + 12$. Using the distributive property this way is called **factoring out the greatest common factor.**

EXAMPLE 1 Factoring Out the Greatest Common Factor

Factor out the greatest common factor.

(a) $9z - 18$

Since 9 is the GCF, factor 9 from each term.
$$9z - 18 = 9 \cdot z - 9 \cdot 2$$
$$= 9(z - 2)$$

Check: $\qquad 9(z - 2) = 9z - 18 \qquad$ Original polynomial

> Always check by multiplying.

(b) $56m + 35p = 7(8m + 5p)$

(c) $2y + 5 \qquad$ There is no common factor other than 1.

(d) $12 + 24z = 12 \cdot 1 + 12 \cdot 2z \qquad$ Identity property
$$= 12(1 + 2z) \qquad \text{12 is the GCF.}$$

> Remember to write the 1.

Check: $\qquad 12(1 + 2z) = 12(1) + 12(2z) \qquad$ Distributive property
$$= 12 + 24z \qquad \text{Original polynomial}$$

✔ **Now Try Exercises 1, 3, and 5.**

EXAMPLE 2 Factoring Out the Greatest Common Factor

Factor out the greatest common factor.

(a) $9x^2 + 12x^3$

The numerical part of the GCF is 3. For the variable parts, x^2 and x^3, use the least exponent that appears on x; here, the least exponent is 2. The GCF is $3x^2$.

$$9x^2 + 12x^3 = 3x^2(3) + 3x^2(4x) \qquad \text{GCF} = 3x^2$$
$$= 3x^2(3 + 4x)$$

(b) $32p^4 - 24p^3 + 40p^5 = 8p^3(4p) + 8p^3(-3) + 8p^3(5p^2) \qquad \text{GCF} = 8p^3$
$$= 8p^3(4p - 3 + 5p^2)$$

(c) $3k^4 - 15k^7 + 24k^9 = 3k^4(1 - 5k^3 + 8k^5)$

> **Remember the 1.**

(d) $24m^3n^2 - 18m^2n + 6m^4n^3$

The numerical part of the GCF is 6. Here, 2 is the least exponent that appears on m, while 1 is the least exponent on n. The GCF is $6m^2n$.

$$24m^3n^2 - 18m^2n + 6m^4n^3 = 6m^2n(4mn) + 6m^2n(-3) + 6m^2n(m^2n^2)$$
$$= 6m^2n(4mn - 3 + m^2n^2)$$

(e) $25x^2y^3 + 30y^5 - 15x^4y^7 = 5y^3(5x^2 + 6y^2 - 3x^4y^4)$

In each case, remember to check the factored form by multiplying.

✔ **Now Try Exercises 7, 11, and 15.**

EXAMPLE 3 Factoring Out a Binomial Factor

Factor out the greatest common factor.

(a) $(x - 5)(x + 6) + (x - 5)(2x + 5)$

The greatest common factor here is the binomial $x - 5$.

$$(x - 5)(x + 6) + (x - 5)(2x + 5) = (x - 5)[(x + 6) + (2x + 5)]$$
$$= (x - 5)(x + 6 + 2x + 5)$$
$$= (x - 5)(3x + 11)$$

(b) $z^2(m + n)^2 + x^2(m + n)^2 = (m + n)^2(z^2 + x^2)$

(c) $p(r + 2s)^2 - q(r + 2s)^3 = (r + 2s)^2[p - q(r + 2s)]$
$$= (r + 2s)^2(p - qr - 2qs)$$

> **Be careful with signs.**

(d) $(p - 5)(p + 2) - (p - 5)(3p + 4)$

$= (p - 5)[(p + 2) - (3p + 4)]$	Factor out the common factor.
$= (p - 5)[p + 2 - 3p - 4]$	Distributive property
$= (p - 5)[-2p - 2]$	Combine like terms.
$= (p - 5)[-2(p + 1)]$	Look for a common factor.
$= -2(p - 5)(p + 1)$	Commutative property

✔ **Now Try Exercises 21 and 25.**

EXAMPLE 4 Factoring Out a Negative Common Factor

Factor $-a^3 + 3a^2 - 5a$ in two ways.

First, a could be used as the common factor, giving

$$-a^3 + 3a^2 - 5a = a(-a^2) + a(3a) + a(-5) \qquad \text{Factor out } a.$$
$$= a(-a^2 + 3a - 5).$$

Because of the leading negative sign, $-a$ could also be used as the common factor.

$$-a^3 + 3a^2 - 5a = -a(a^2) + (-a)(-3a) + (-a)(5) \qquad \text{Factor out } -a.$$
$$= -a(a^2 - 3a + 5)$$

Sometimes there may be a reason to prefer one of these forms, but either is correct.

✔ **Now Try Exercise 33.**

OBJECTIVE 2 **Factor by grouping.** Sometimes the *individual terms* of a polynomial have a greatest common factor of 1, but it still may be possible to factor the polynomial by using a process called *factoring by grouping*. **We usually factor by grouping when a polynomial has more than three terms.**

EXAMPLE 5 Factoring by Grouping

Factor $ax - ay + bx - by$.

Group the terms as follows:

Terms with common factor a Terms with common factor b

$$(ax - ay) + (bx - by).$$

Then factor $ax - ay$ as $a(x - y)$ and factor $bx - by$ as $b(x - y)$.

$$ax - ay + bx - by = (ax - ay) + (bx - by) \qquad \text{Group the terms.}$$
$$= a(x - y) + b(x - y) \qquad \text{Factor each group.}$$
$$= (x - y)(a + b) \qquad \text{The common factor is } x - y.$$

Check by multiplying.

✔ **Now Try Exercise 39.**

EXAMPLE 6 Factoring by Grouping

Factor $3x - 3y - ax + ay$.

Grouping terms gives

$$(3x - 3y) + (-ax + ay) = 3(x - y) + a(-x + y).$$

The factors $(x - y)$ and $(-x + y)$ are opposites, so if we factor out $-a$ instead of a in the second group of terms, we get

$$(3x - 3y) + (-ax + ay) = 3(x - y) - a(x - y) \qquad \text{Be careful with signs.}$$
$$= (x - y)(3 - a).$$

Check: $(x - y)(3 - a) = 3x - ax - 3y + ay \qquad \text{FOIL}$
$$= 3x - 3y - ax + ay \qquad \text{Original polynomial}$$

✔ **Now Try Exercise 43.**

▶ **NOTE** In Example 6, a different grouping would lead to the product
$$(a - 3)(y - x).$$
Verify by multiplying that this is also correct.

Use the following steps to factor by grouping.

Factoring by Grouping

Step 1 **Group terms.** Collect the terms into groups so that each group has a common factor.

Step 2 **Factor within the groups.** Factor out the common factor in each group.

Step 3 **Factor the entire polynomial.** If each group now has a common factor, factor it out. If not, try a different grouping.

Always check the factored form by multiplying.

EXAMPLE 7 **Factoring by Grouping**

Factor $6ax + 12bx + a + 2b$.

$$6ax + 12bx + a + 2b = (6ax + 12bx) + (a + 2b) \qquad \text{Group terms.}$$

Now factor $6x$ from the first group, and use the identity property of multiplication to introduce the factor 1 in the second group.

Remember to write the 1.

$$(6ax + 12bx) + (a + 2b) = 6x(a + 2b) + 1(a + 2b)$$
$$= (a + 2b)(6x + 1) \qquad \text{Factor out } a + 2b.$$

Check: $(a + 2b)(6x + 1) = 6ax + a + 12bx + 2b \qquad \text{FOIL}$
$$= 6ax + 12bx + a + 2b \qquad \text{Original polynomial}$$

✔ **Now Try Exercise 47.**

EXAMPLE 8 **Rearranging Terms before Factoring by Grouping**

Factor $p^2q^2 - 10 - 2q^2 + 5p^2$.

Neither the first two terms nor the last two terms have a common factor except the identity element 1. Rearrange and group the terms as follows:

$$p^2q^2 - 10 - 2q^2 + 5p^2$$
$$= (p^2q^2 - 2q^2) + (5p^2 - 10) \qquad \text{Rearrange and group the terms.}$$
$$= q^2(p^2 - 2) + 5(p^2 - 2) \qquad \text{Factor out the common factors.}$$

Don't stop here.

$$= (p^2 - 2)(q^2 + 5). \qquad \text{Factor out } p^2 - 2.$$

Check: $(p^2 - 2)(q^2 + 5) = p^2q^2 + 5p^2 - 2q^2 - 10 \qquad \text{FOIL}$
$$= p^2q^2 - 10 - 2q^2 + 5p^2 \qquad \text{Original polynomial}$$

✔ **Now Try Exercise 53.**

▶ **CAUTION** In Example 8, do not stop at the step

$$q^2(p^2 - 2) + 5(p^2 - 2).$$

This expression is *not in factored form,* because it is a *sum* of two terms, $q^2(p^2 - 2)$ and $5(p^2 - 2)$, not a *product.*

6.1 EXERCISES

⊙ *Complete solution available on Video Lectures on CD/DVD*

▢ *Now Try Exercise*

Factor out the greatest common factor. Simplify the factors, if possible. See Examples 1–4.

⊙ **1.** $12m - 60$

2. $15r - 45$

3. $4 + 20z$

4. $9 + 27x$

5. $8y - 15$

6. $7x - 40$

⊙ **7.** $8k^3 + 24k$

8. $9z^4 + 81z$

9. $-4p^3q^4 - 2p^2q^5$

10. $-3z^5w^2 - 18z^3w^4$

11. $21x^5 + 35x^4 - 14x^3$

12. $6k^3 - 36k^4 - 48k^5$

13. $10t^5 - 8t^4 - 4t^3$

14. $6p^3 - 3p^2 - 9p^4$

15. $15a^2c^3 - 25ac^2 + 5ac$

16. $15y^3z^3 + 27y^2z^4 - 36yz^5$

17. $16z^2n^6 + 64zn^7 - 32z^3n^3$

18. $5r^3s^5 + 10r^2s^2 - 15r^4s^2$

19. $14a^3b^2 + 7a^2b - 21a^5b^3 + 42ab^4$

20. $12km^3 - 24k^3m^2 + 36k^2m^4 - 60k^4m^3$

⊙ **21.** $(m - 4)(m + 2) + (m - 4)(m + 3)$

22. $(z - 5)(z + 7) + (z - 5)(z + 9)$

23. $(2z - 1)(z + 6) - (2z - 1)(z - 5)$

24. $(3x + 2)(x - 4) - (3x + 2)(x + 8)$

25. $5(2 - x)^2 - 2(2 - x)^3$

26. $2(5 - x)^3 - 3(5 - x)^2$

27. $4(3 - x)^2 - (3 - x)^3 + 3(3 - x)$

28. $2(t - s) + 4(t - s)^2 - (t - s)^3$

29. $15(2z + 1)^3 + 10(2z + 1)^2 - 25(2z + 1)$

30. $6(a + 2b)^2 - 4(a + 2b)^3 + 12(a + 2b)^4$

31. $5(m + p)^3 - 10(m + p)^2 - 15(m + p)^4$

32. $-9a^2(p + q) - 3a^3(p + q)^2 + 6a(p + q)^3$

Factor each polynomial twice. First use a common factor with a positive coefficient, and then use a common factor with a negative coefficient. See Example 4.

⊙ **33.** $-r^3 + 3r^2 + 5r$

34. $-t^4 + 8t^3 - 12t$

35. $-12s^5 + 48s^4$

36. $-16y^4 + 64y^3$

37. $-2x^5 + 6x^3 + 4x^2$

38. $-5a^3 + 10a^4 - 15a^5$

Factor by grouping. See Examples 5–8.

⊙ **39.** $mx + qx + my + qy$

40. $2k + 2h + jk + jh$

41. $10m + 2n + 5mk + nk$

42. $3ma + 3mb + 2ab + 2b^2$

43. $4 - 2q - 6p + 3pq$

44. $20 + 5m + 12n + 3mn$

45. $p^2 - 4zq + pq - 4pz$

46. $r^2 - 9tw + 3rw - 3rt$

47. $2xy + 3y + 2x + 3$

48. $7ab + 35bc + a + 5c$

49. $m^3 + 4m^2 - 6m - 24$

50. $2a^3 + a^2 - 14a - 7$

51. $-3a^3 - 3ab^2 + 2a^2b + 2b^3$

52. $-16m^3 + 4m^2p^2 - 4mp + p^3$

53. $4 + xy - 2y - 2x$

54. $10ab - 21 - 6b + 35a$

55. $8 + 9y^4 - 6y^3 - 12y$

56. $x^3y^2 - 3 - 3y^2 + x^3$

57. $1 - a + ab - b$

58. $2ab^2 - 8b^2 + a - 4$

The next problems are "brain busters." Factor out the variable that is raised to the smaller exponent. (For example, in Exercise 59, factor out m^{-5}.)

59. $3m^{-5} + m^{-3}$

60. $k^{-2} + 2k^{-4}$

61. $3p^{-3} + 2p^{-2}$

62. $-5q^{-3} + 8q^{-2}$

63. *Concept Check*　When directed to completely factor the polynomial

$$4x^2y^5 - 8xy^3,$$

a student wrote

$$2xy^3(2xy^2 - 4).$$

When the teacher did not give him full credit, he complained because when his answer is multiplied out, the result is the original polynomial. ***WHAT WENT WRONG?*** Give the correct answer.

64. *Concept Check*　Refer to Exercise 58. One form of the answer is $(2b^2 + 1)(a - 4)$. Give two other acceptable factored forms of $2ab^2 - 8b^2 + a - 4$.

65. *Concept Check*　Which choice is an example of a polynomial in factored form?

A. $3x^2y^3 + 6x^2(2x + y)$

B. $5(x + y)^2 - 10(x + y)^3$

C. $(-2 + 3x)(5y^2 + 4y + 3)$

D. $(3x + 4)(5x - y) - (3x + 4)(2x - 1)$

PREVIEW EXERCISES

*Find each product. See **Section 5.4**.*

66. $(m - 3)(m + 2)$

67. $(k + 7)(k - 1)$

68. $(2m + 3)(5m - 9)$

69. $(4y - 2)(5y + 6)$

70. $(8z - 3a)(8z - 3a)$

71. $(5x - 2t)(5x + 2t)$

72. $(2x^2 - 5)(x^2 - 6)$

73. $(3y^3 - 4)(2y^3 + 3)$

74. $x(2x + 3)(x - 7)$

75. $5t(3t + 2)(t - 8)$

6.2 Factoring Trinomials

OBJECTIVE 1 Factor trinomials when the coefficient of the squared term is 1. We begin by finding the product of $x + 3$ and $x - 5$.

$$(x + 3)(x - 5) = x^2 - 5x + 3x - 15$$
$$= x^2 - 2x - 15$$

We see by this result that the factored form of $x^2 - 2x - 15$ is $(x + 3)(x - 5)$.

$$\text{Factored form} \longrightarrow \underset{\text{Factoring}}{\overset{\text{Multiplication}}{(x + 3)(x - 5) = x^2 - 2x - 15}} \longleftarrow \text{Product}$$

Since multiplying and factoring are operations that "undo" each other, factoring trinomials involves using FOIL backwards. As shown here, the x^2-term came from multiplying x and x, and -15 came from multiplying 3 and -5.

$$\underset{\text{Product of 3 and }-5 \qquad\qquad \text{is }-15.}{\overset{\text{Product of }x \quad \text{and }x \qquad \text{is }x^2.}{(x + 3)(x - 5) = x^2 - 2x - 15}}$$

We find the $-2x$ in $x^2 - 2x - 15$ by multiplying the outer terms, then the inner terms, and adding.

$$\underset{\text{Inner terms: } 3 \cdot x = 3x}{\overset{\text{Outer terms: } x(-5) = -5x}{(x + 3)(x - 5)}} \qquad \text{Add to get } -2x.$$

Based on this example, use the following steps to factor a trinomial $x^2 + bx + c$, where 1 is the coefficient of the squared term. (A procedure for factoring a trinomial when the coefficient of the squared term is *not* 1 follows later in this section.)

Factoring $x^2 + bx + c$

Step 1 **Find pairs whose product is *c*.** Find all pairs of integers whose product is c, the third term of the trinomial.

Step 2 **Find the pair whose sum is *b*.** Choose the pair whose sum is b, the coefficient of the middle term.

If there are no such integers, the polynomial cannot be factored.

A polynomial that cannot be factored with integer coefficients is a **prime polynomial.**

EXAMPLE 1 Factoring Trinomials in $x^2 + bx + c$ Form

Factor each polynomial.

(a) $y^2 + 2y - 35$

Step 1 Find pairs of numbers whose product is -35.	*Step 2* Write sums of those numbers.
$-35(1)$	$-35 + 1 = -34$
$35(-1)$	$35 + (-1) = 34$
$7(-5)$	$7 + (-5) = 2 \longleftarrow$ Coefficient of the middle term
$5(-7)$	$5 + (-7) = -2$

The required numbers are 7 and -5, so

$$y^2 + 2y - 35 = (y + 7)(y - 5).$$

Check by finding the product of $y + 7$ and $y - 5$.

(b) $r^2 + 8r + 12$

Look for two numbers with a product of 12 and a sum of 8. Of all pairs of numbers having a product of 12, only the pair 6 and 2 has a sum of 8. Therefore,

$$r^2 + 8r + 12 = (r + 6)(r + 2).$$

Because of the commutative property, it would be equally correct to write $(r + 2)(r + 6)$. ***Check by using FOIL to multiply the factored form.***

✔ **Now Try Exercises 5 and 7.**

EXAMPLE 2 Recognizing a Prime Polynomial

Factor $m^2 + 6m + 7$.

Look for two numbers whose product is 7 and whose sum is 6. Only two pairs of integers, 7 and 1 and -7 and -1, give a product of 7. Neither of these pairs has a sum of 6, so $m^2 + 6m + 7$ cannot be factored with integer coefficients and is prime.

✔ **Now Try Exercise 9.**

We use a similar process to factor a trinomial that has more than one variable.

EXAMPLE 3 Factoring a Trinomial in Two Variables

Factor $p^2 + 6ap - 16a^2$.

Look for two expressions whose product is $-16a^2$ and whose sum is $6a$. The quantities $8a$ and $-2a$ have the necessary product and sum, so

$$p^2 + 6ap - 16a^2 = (p + 8a)(p - 2a).$$

Check: $(p + 8a)(p - 2a) = p^2 - 2ap + 8ap - 16a^2$ FOIL

$$= p^2 + 6ap - 16a^2 \qquad \text{Original polynomial}$$

✔ **Now Try Exercise 11.**

A trinomial may have a common factor that should be factored out first.

EXAMPLE 4 Factoring a Trinomial with a Common Factor

Factor $16y^3 - 32y^2 - 48y$.

Start by factoring out the greatest common factor, $16y$.

$$16y^3 - 32y^2 - 48y = 16y(y^2 - 2y - 3)$$

To factor $y^2 - 2y - 3$, look for two integers whose product is -3 and whose sum is -2. The necessary integers are -3 and 1, so

$$16y^3 - 32y^2 - 48y = 16y(y - 3)(y + 1).$$

Remember to include the GCF.

✔ **Now Try Exercise 39.**

▶ **CAUTION** When factoring, always look for a common factor first. Remember to write the common factor as part of the answer.

OBJECTIVE 2 Factor trinomials when the coefficient of the squared term is not 1. We can use a generalization of the method shown in Objective 1 to factor a trinomial of the form $ax^2 + bx + c$, where $a \neq 1$. To factor $3x^2 + 7x + 2$, for example, we first identify the values a, b, and c:

$$ax^2 + bx + c$$
$$3x^2 + 7x + 2, \quad \text{so} \quad a = 3, \quad b = 7, \quad c = 2.$$

The product ac is $3 \cdot 2 = 6$, so we must find integers having a product of 6 and a sum of 7 (since the middle term has coefficient 7). The necessary integers are 1 and 6, so we write $7x$ as $1x + 6x$, or $x + 6x$, giving

$$3x^2 + 7x + 2 = 3x^2 + \underline{x + 6x} + 2.$$

$$x + 6x = 7x$$

$$= (3x^2 + x) + (6x + 2) \qquad \text{Group terms.}$$
$$= x(3x + 1) + 2(3x + 1) \qquad \text{Factor by grouping.}$$
$$= (3x + 1)(x + 2) \qquad \text{Factor out the common factor.}$$

Check by multiplying.

EXAMPLE 5 Factoring a Trinomial in $ax^2 + bx + c$ Form

Factor $12r^2 - 5r - 2$.

Since $a = 12$, $b = -5$, and $c = -2$, the product ac is -24. The two integers whose product is -24 and whose sum is -5 are -8 and 3.

$$12r^2 - 5r - 2 = 12r^2 + 3r - 8r - 2 \qquad \text{Write } -5r \text{ as } 3r - 8r.$$
$$= 3r(4r + 1) - 2(4r + 1) \qquad \text{Factor by grouping.}$$
$$= (4r + 1)(3r - 2) \qquad \text{Factor out the common factor.}$$

Check by multiplying.

✔ **Now Try Exercise 19.**

OBJECTIVE 3 Use an alternative method for factoring trinomials. When the product ac is large, trying repeated combinations and using FOIL is helpful.

EXAMPLE 6 Factoring Trinomials in $ax^2 + bx + c$ Form

Factor each trinomial.

(a) $3x^2 + 7x + 2$

To factor this polynomial, we must find the correct numbers to put in the blanks.

$$3x^2 + 7x + 2 = (\underline{\quad} x + \underline{\quad})(\underline{\quad} x + \underline{\quad})$$

Addition signs are used, since all the signs in the polynomial indicate addition. The first two expressions have a product of $3x^2$, so they must be $3x$ and $1x$, or x.

$$3x^2 + 7x + 2 = (3x + \underline{\quad})(x + \underline{\quad})$$

The product of the two last terms must be 2, so the numbers must be 2 and 1. There is a choice. The 2 could be used with the $3x$ or with the x. Only one of these choices can give the correct middle term, $7x$. We use FOIL to try each one.

$$\overset{\overset{\displaystyle 3x}{\overbrace{\qquad\qquad}}}{(3x + 2)(x + 1)}$$
$$\underset{\underset{\displaystyle 2x}{\underbrace{\qquad\qquad}}}{}$$

$3x + 2x = 5x$

Wrong middle term

$$\overset{\overset{\displaystyle 6x}{\overbrace{\qquad\qquad}}}{(3x + 1)(x + 2)}$$
$$\underset{\underset{\displaystyle x}{\underbrace{\qquad\qquad}}}{}$$

$6x + x = 7x$

Correct middle term

Therefore, $3x^2 + 7x + 2 = (3x + 1)(x + 2)$. (Compare with the solution obtained by factoring by grouping on the preceding page.)

(b) $12r^2 - 5r - 2$

To reduce the number of trials, we note that the trinomial has no common factor (except 1). This means that neither of its factors can have a common factor. We should keep this in mind as we choose factors. We try 4 and 3 for the two first terms.

$$12r^2 - 5r - 2 = (4r\underline{\quad})(3r\underline{\quad})$$

The factors of -2 are -2 and 1 or -1 and 2. We try both possibilities to see if we obtain the correct middle term, $-5r$.

$(4r - 2)(3r + 1)$

Wrong: $4r - 2$ has a common factor of 2, which cannot be correct, since 2 is not a factor of $12r^2 - 5r - 2$.

$$\overset{\overset{\displaystyle 8r}{\overbrace{\qquad\qquad}}}{(4r - 1)(3r + 2)}$$
$$\underset{\underset{\displaystyle -3r}{\underbrace{\qquad}}}{}$$

$8r - 3r = 5r$

Wrong middle term

The middle term on the right is $5r$, instead of the $-5r$ that is needed. We get $-5r$ by interchanging the signs of the second terms in the factors.

$$\overset{\overset{\displaystyle -8r}{\overbrace{\qquad\qquad}}}{(4r + 1)(3r - 2)}$$
$$\underset{\underset{\displaystyle 3r}{\underbrace{\qquad}}}{}$$

$-8r + 3r = -5r$

Correct middle term

Thus, $12r^2 - 5r - 2 = (4r + 1)(3r - 2)$. (Compare with Example 5.)

✔ **Now Try Exercise 21.**

▶ **NOTE** As shown in Example 6(b), if the terms of a polynomial have no common factor (except 1), then none of the terms of its factors can have a common factor. Remembering this will eliminate some potential factors.

We summarize this alternative method of factoring a trinomial in the form

$$ax^2 + bx + c, \quad \text{where } a \neq 1.$$

Factoring $ax^2 + bx + c$

Step 1 **Find pairs whose product is a.** Write all pairs of integer factors of a, the coefficient of the squared term.

Step 2 **Find pairs whose product is c.** Write all pairs of integer factors of c, the last term.

Step 3 **Choose inner and outer terms.** Use FOIL and various combinations of the factors from Steps 1 and 2 until the necessary middle term is found.

If no such combinations exist, the trinomial is prime.

EXAMPLE 7 Factoring a Trinomial in Two Variables

Factor $18m^2 - 19mx - 12x^2$.

There is no common factor (except 1). Follow the steps for factoring a trinomial. There are many possible factors of both 18 and -12. Try 6 and 3 for 18 and -3 and 4 for -12.

$$(6m - 3x)(3m + 4x) \qquad \qquad (6m + 4x)(3m - 3x)$$
Wrong: common factor $\qquad$ Wrong: common factors

Since 6 and 3 do not work as factors of 18, try 9 and 2 instead, with 3 and -4 as factors of -12.

$$(9m + 3x)(2m - 4x) \qquad \qquad (9m - 4x)(2m + 3x)$$

Wrong: common factors

$$27mx$$

$$-8mx$$

$$27mx + (-8mx) = 19mx$$
Wrong middle term

The result on the right differs from the correct middle term only in sign, so interchange the signs of the second terms in the factors.

$$18m^2 - 19mx - 12x^2 = (9m + 4x)(2m - 3x)$$

Check by using FOIL to multiply the factors.

✔ **Now Try Exercise 23.**

EXAMPLE 8 Factoring $ax^2 + bx + c, a < 0$

Factor $-3x^2 + 16x + 12$.

While we could factor directly, it is helpful to first factor out -1 so that the coefficient of the x^2-term is positive.

$$-3x^2 + 16x + 12 = -1(3x^2 - 16x - 12) \qquad \text{Factor out } -1.$$
$$= -1(3x + 2)(x - 6) \qquad \text{Factor the trinomial.}$$
$$= -(3x + 2)(x - 6)$$

This factored form can be written in other ways. Two of them are

$$(-3x - 2)(x - 6) \quad \text{and} \quad (3x + 2)(-x + 6).$$

Verify that these both give the original trinomial when multiplied.

✔ **Now Try Exercise 33.**

EXAMPLE 9 Factoring a Trinomial with a Common Factor

Factor $16y^3 + 24y^2 - 16y$.

$$16y^3 + 24y^2 - 16y = 8y(2y^2 + 3y - 2) \qquad \text{GCF} = 8y$$
$$= 8y(2y - 1)(y + 2) \qquad \text{Factor the trinomial.}$$

Remember the common factor.

✔ **Now Try Exercise 31.**

OBJECTIVE 4 Factor by substitution. Sometimes we can factor a more complicated polynomial by substituting a variable for an expression.

EXAMPLE 10 Factoring a Polynomial by Substitution

Factor $2(x + 3)^2 + 5(x + 3) - 12$.

Since the binomial $x + 3$ appears to powers 2 and 1, we let the substitution variable represent $x + 3$. We may choose any letter we wish except x. We choose t to represent $x + 3$.

$$2(x + 3)^2 + 5(x + 3) - 12 = 2t^2 + 5t - 12 \qquad \text{Let } t = x + 3.$$
$$= (2t - 3)(t + 4) \qquad \text{Factor.}$$
$$= [2(x + 3) - 3][(x + 3) + 4] \qquad \text{Replace } t \text{ with } x + 3.$$
$$= (2x + 6 - 3)(x + 7) \qquad \text{Simplify.}$$
$$= (2x + 3)(x + 7)$$

✔ **Now Try Exercise 49.**

▶ **CAUTION** *Remember to make the final substitution* of $x + 3$ for t in Example 10.

EXAMPLE 11 Factoring a Trinomial in $ax^4 + bx^2 + c$ Form

Factor $6y^4 + 7y^2 - 20$.

The variable y appears to powers in which the larger exponent is twice the smaller exponent. We can let a substitution variable equal the smaller power. Here, we let $t = y^2$.

$$6y^4 + 7y^2 - 20 = 6(y^2)^2 + 7y^2 - 20$$
$$= 6t^2 + 7t - 20 \qquad \text{Substitute.}$$
$$= (3t - 4)(2t + 5) \qquad \text{Factor.}$$
$$= (3y^2 - 4)(2y^2 + 5) \qquad t = y^2$$

Don't stop here.
Replace t with y^2.

✔ **Now Try Exercise 59.**

▶ **N O T E**   Some students feel comfortable factoring polynomials like the one in Example 11 directly, without using the substitution method.

6.2 EXERCISES

⊕ *Complete solution available on Video Lectures on CD/DVD*

▢ *Now Try Exercise*

1. *Concept Check*   Which is *not* a valid way of starting the process of factoring $12x^2 + 29x + 10$?

A. $(12x \quad)(x \quad)$ **B.** $(4x \quad)(3x \quad)$

C. $(6x \quad)(2x \quad)$ **D.** $(8x \quad)(4x \quad)$

2. *Concept Check*   Which is the completely factored form of $2x^6 - 5x^5 - 3x^4$?

A. $x^4(2x + 1)(x - 3)$ **B.** $x^4(2x - 1)(x + 3)$

C. $(2x^5 + x^4)(x - 3)$ **D.** $x^3(2x^2 + x)(x - 3)$

3. *Concept Check*   Which is *not* a factored form of $-x^2 + 16x - 60$?

A. $(x - 10)(-x + 6)$ **B.** $(-x - 10)(x + 6)$

C. $(-x + 10)(x - 6)$ **D.** $-(x - 10)(x - 6)$

4. *Concept Check*   Which is the completely factored form of $4x^2 - 4x - 24$?

A. $4(x - 2)(x + 3)$ **B.** $4(x + 2)(x + 3)$

C. $4(x + 2)(x - 3)$ **D.** $4(x - 2)(x - 3)$

Factor each trinomial. See Examples 1–9.

⊕ **5.** $y^2 + 7y - 30$ **6.** $z^2 + 2z - 24$ **7.** $p^2 + 15p + 56$ **8.** $k^2 - 11k + 30$

⊕ **9.** $m^2 - 11m + 60$ **10.** $p^2 - 12p - 27$ ⊕ **11.** $a^2 - 2ab - 35b^2$

12. $z^2 + 8zw + 15w^2$ **13.** $y^2 - 3yq - 15q^2$ **14.** $k^2 - 11hk + 28h^2$

15. $x^2y^2 + 11xy + 18$ **16.** $p^2q^2 - 5pq - 18$ **17.** $-6m^2 - 13m + 15$

18. $-15y^2 + 17y + 18$ ⊕ **19.** $10x^2 + 3x - 18$ **20.** $8k^2 + 34k + 35$

⊕ **21.** $20k^2 + 47k + 24$ **22.** $27z^2 + 42z - 5$ ⊕ **23.** $15a^2 - 22ab + 8b^2$

24. $15p^2 + 24pq + 8q^2$ **25.** $36m^2 - 60m + 25$ **26.** $25r^2 - 90r + 81$

27. $40x^2 + xy + 6y^2$ **28.** $14c^2 - 17cd - 6d^2$ **29.** $6x^2z^2 + 5xz - 4$

30. $8m^2n^2 - 10mn + 3$ **31.** $24x^2 + 42x + 15$ **32.** $36x^2 + 18x - 4$

33. $-15a^2 - 70a + 120$ **34.** $-12a^2 - 10a + 42$

35. $-11x^3 + 110x^2 - 264x$ **36.** $-9k^3 - 36k^2 + 189k$

37. $2x^3y^3 - 48x^2y^4 + 288xy^5$ **38.** $6m^3n^2 - 24m^2n^3 - 30mn^4$

39. $6a^3 + 12a^2 - 90a$ **40.** $3m^4 + 6m^3 - 72m^2$

41. $13y^3 + 39y^2 - 52y$ **42.** $4p^3 + 24p^2 - 64p$

43. $12p^3 - 12p^2 + 3p$ **44.** $45t^3 + 60t^2 + 20t$

45. *Concept Check*   When a student was given the polynomial $4x^2 + 2x - 20$ to factor completely on a test, the student lost some credit when her answer was $(4x + 10)(x - 2)$. She complained to her teacher that when we multiply $(4x + 10)(x - 2)$, we get the original polynomial. ***WHAT WENT WRONG?*** Give the correct answer.

46. When factoring the polynomial $-4x^2 - 29x + 24$, Terry obtained $(-4x + 3)(x + 8)$, while John got $(4x - 3)(-x - 8)$. Who is correct? Explain your answer.

Factor each trinomial. See Example 10.

47. $12p^6 - 32p^3r + 5r^2$ **48.** $2y^6 + 7xy^3 + 6x^2$

49. $10(k + 1)^2 - 7(k + 1) + 1$ **50.** $4(m - 5)^2 - 4(m - 5) - 15$

51. $3(m + p)^2 - 7(m + p) - 20$ **52.** $4(x - y)^2 - 23(x - y) - 6$

Factor each trinomial. (Hint: Factor out the GCF first.)

53. $a^2(a + b)^2 - ab(a + b)^2 - 6b^2(a + b)^2$

54. $m^2(m - p) + mp(m - p) - 2p^2(m - p)$

55. $p^2(p + q) + 4pq(p + q) + 3q^2(p + q)$

56. $2k^2(5 - y) - 7k(5 - y) + 5(5 - y)$

57. $z^2(z - x) - zx(x - z) - 2x^2(z - x)$

58. $r^2(r - s) - 5rs(s - r) - 6s^2(r - s)$

Factor each trinomial. See Example 11.

59. $p^4 - 10p^2 + 16$ **60.** $k^4 + 10k^2 + 9$ **61.** $2x^4 - 9x^2 - 18$

62. $6z^4 + z^2 - 1$ **63.** $16x^4 + 16x^2 + 3$ **64.** $9r^4 + 9r^2 + 2$

PREVIEW EXERCISES

*Find each product. See **Section 5.4**.*

65. $(3x - 5)(3x + 5)$ **66.** $(8m + 3)(8m - 3)$

67. $(p + 3q)^2$ **68.** $(2z - 7)^2$

69. $(y + 3)(y^2 - 3y + 9)$ **70.** $(3m - 1)(9m^2 + 3m + 1)$

6.3   Special Factoring

OBJECTIVE 1   Factor a difference of squares.   The special products introduced in **Section 5.4** are used in reverse when factoring. Recall that the product of the sum and difference of two terms leads to a **difference of squares.**

Difference of Squares

$$x^2 - y^2 = (x + y)(x - y)$$

EXAMPLE 1   Factoring Differences of Squares

Factor each polynomial.

(a) $t^2 - 36 = t^2 - 6^2$ $36 = 6^2$

$\qquad\qquad\quad = (t + 6)(t - 6)$ Factor the difference of squares.

(b) $4a^2 - 64$

There is a common factor of 4.

$\qquad 4a^2 - 64 = 4(a^2 - 16)$ Factor out the common factor.

$\qquad\qquad\quad = 4(a + 4)(a - 4)$ Factor the difference of squares.

$$A^2 \;-\; B^2 \;=\; (A \;+\; B)\,(A \;-\; B)$$

(c) $16m^2 - 49p^2 = (4m)^2 - (7p)^2 = (4m + 7p)(4m - 7p)$

$$A^2 \;-\; B^2 \;=\; (A \;+\; B)\;(A \;-\; B)$$

(d) $81k^2 - (a + 2)^2 = (9k)^2 - (a + 2)^2 = (9k + a + 2)(9k - (a + 2))$

$\qquad\qquad\qquad\qquad\qquad\qquad\qquad\quad = (9k + a + 2)(9k - a - 2)$

We could have used the method of substitution here.

(e) $x^4 - 81 = (x^2 + 9)(x^2 - 9)$ Factor the difference of squares.

$\qquad\qquad\quad = (x^2 + 9)(x + 3)(x - 3)$ Factor $x^2 - 9$.

✔ **Now Try Exercises 7, 9, 13, 15, and 19.**

▶ **CAUTION**   *Assuming no greatest common factor except 1, it is not possible to factor (with real numbers) a sum of squares such as* $x^2 + 9$ *in Example 1(e). In particular,* $x^2 + y^2 \neq (x + y)^2$, *as shown next.*

OBJECTIVE 2   Factor a perfect square trinomial.   Two other special products from **Section 5.4** lead to the following rules for factoring.

Perfect Square Trinomial

$$x^2 + 2xy + y^2 = (x + y)^2$$
$$x^2 - 2xy + y^2 = (x - y)^2$$

Because the trinomial $x^2 + 2xy + y^2$ is the square of $x + y$, it is called a **perfect square trinomial.** In this pattern, both the first and the last terms of the trinomial must be perfect squares. In the factored form $(x + y)^2$, twice the product of the first and the last terms must give the middle term of the trinomial. You should understand these patterns in words, since they occur with different symbols (other than x and y).

$$4m^2 + 20m + 25 \qquad\qquad p^2 - 8p + 64$$

Perfect square trinomial; $\qquad$ Not a perfect square trinomial;
$4m^2 = (2m)^2$, $25 = 5^2$, $\qquad$ middle term would have to be
and $2(2m)(5) = 20m$. $\qquad\qquad$ $16p$ or $-16p$.

EXAMPLE 2 Factoring Perfect Square Trinomials

Factor each polynomial.

(a) $144p^2 - 120p + 25$

Here, $144p^2 = (12p)^2$ and $25 = 5^2$. The sign on the middle term is $-$, so if $144p^2 - 120p + 25$ is a perfect square trinomial, the factored form will have to be

$$(12p - 5)^2.$$

Take twice the product of the two terms to see if this is correct.

$$2(12p)(-5) = -120p$$

This is the middle term of the given trinomial, so

$$144p^2 - 120p + 25 = (12p - 5)^2.$$

(b) $4m^2 + 20mn + 49n^2$

If this is a perfect square trinomial, it will equal $(2m + 7n)^2$. By the pattern in the box, if multiplied out, this squared binomial has a middle term of

$$2(2m)(7n) = 28mn,$$

which *does not equal* 20mn. Verify that this trinomial cannot be factored by the methods of the previous section either. It is prime.

(c) $(r + 5)^2 + 6(r + 5) + 9 = [(r + 5) + 3]^2$
$$= (r + 8)^2 \qquad\qquad 2(r + 5)(3) = 6(r + 5), \text{ the}$$
$$\text{middle term}$$

(d) $m^2 - 8m + 16 - p^2$

Since there are four terms, use factoring by grouping. The first three terms here form a perfect square trinomial. Group them together, and factor as follows.

$$m^2 - 8m + 16 - p^2 = (m^2 - 8m + 16) - p^2$$
$$= (m - 4)^2 - p^2 \qquad\qquad \text{Factor the perfect square}$$
$$\text{trinomial.}$$
$$= (m - 4 + p)(m - 4 - p) \qquad \text{Factor the difference of}$$
$$\text{squares.}$$

✔ **Now Try Exercises 23, 25, and 33.**

Perfect square trinomials, of course, can be factored by the general methods shown earlier for other trinomials. The patterns given here provide "shortcuts."

OBJECTIVE 3 Factor a difference of cubes. A **difference of cubes**, $x^3 - y^3$, can be factored as follows.

Difference of Cubes

$$x^3 - y^3 = (x - y)(x^2 + xy + y^2)$$

We could check this pattern by finding the product of $x - y$ and $x^2 + xy + y^2$.

EXAMPLE 3 Factoring Differences of Cubes

Factor each polynomial.

$$A^3 - B^3 = (A - B)(A^2 + A \cdot B + B^2)$$

(a) $m^3 - 8 = m^3 - 2^3 = (m - 2)(m^2 + m \cdot 2 + 2^2)$

$$= (m - 2)(m^2 + 2m + 4)$$

Check:

$$(m - 2)(m^2 + 2m + 4)$$

Opposite of the product of the cube roots gives the middle term.

(b) $27x^3 - 8y^3 = (3x)^3 - (2y)^3$

$$= (3x - 2y)[(3x)^2 + (3x)(2y) + (2y)^2]$$

$$= (3x - 2y)(9x^2 + 6xy + 4y^2)$$

(c) $1000k^3 - 27n^3 = (10k)^3 - (3n)^3$

$$= (10k - 3n)[(10k)^2 + (10k)(3n) + (3n)^2]$$

$$= (10k - 3n)(100k^2 + 30kn + 9n^2)$$

✔ **Now Try Exercises 37 and 51.**

OBJECTIVE 4 Factor a sum of cubes. While an expression of the form $x^2 + y^2$ cannot be factored with real numbers, a **sum of cubes** is factored as follows.

Sum of Cubes

$$x^3 + y^3 = (x + y)(x^2 - xy + y^2)$$

To verify this result, find the product of $x + y$ and $x^2 - xy + y^2$. Compare this pattern with the pattern for factoring a difference of cubes.

▶ **NOTE** The sign of the second term in the binomial factor of a sum or difference of cubes is *always the same* as the sign in the original polynomial. In the trinomial factor, the first and last terms are *always positive;* the sign of the middle term is *the opposite of* the sign of the second term in the binomial factor.

EXAMPLE 4 Factoring Sums of Cubes

Factor each polynomial.

(a) $r^3 + 27 = r^3 + 3^3$
$$= (r + 3)(r^2 - 3r + 3^2)$$
$$= (r + 3)(r^2 - 3r + 9)$$

(b) $27z^3 + 125 = (3z)^3 + 5^3$
$$= (3z + 5)[(3z)^2 - (3z)(5) + 5^2]$$
$$= (3z + 5)(9z^2 - 15z + 25)$$

(c) $125t^3 + 216s^6 = (5t)^3 + (6s^2)^3$
$$= (5t + 6s^2)[(5t)^2 - (5t)(6s^2) + (6s^2)^2]$$
$$= (5t + 6s^2)(25t^2 - 30ts^2 + 36s^4)$$

(d) $3x^2 + 192 = 3(x^3 + 64)$ Factor out the common factor.

Remember the common factor.
$$= 3(x + 4)(x^2 - 4x + 16) \qquad \text{Factor the sum of cubes.}$$

(e) $(x + 2)^3 + t^3 = [(x + 2) + t][(x + 2)^2 - (x + 2)t + t^2]$
$$= (x + 2 + t)(x^2 + 4x + 4 - xt - 2t + t^2)$$

✔ **Now Try Exercises 41, 53, 55, and 57.**

▶ **CAUTION** A common error when factoring $x^3 + y^3$ or $x^3 - y^3$ is to think that the xy-term has a coefficient of 2. Since there is no coefficient of 2, expressions of the form $x^2 + xy + y^2$ and $x^2 - xy + y^2$ usually cannot be factored further.

The special types of factoring are summarized here. ***These should be memorized.***

Special Types of Factoring	
Difference of Squares	$x^2 - y^2 = (x + y)(x - y)$
Perfect Square Trinomial	$x^2 + 2xy + y^2 = (x + y)^2$
	$x^2 - 2xy + y^2 = (x - y)^2$
Difference of Cubes	$x^3 - y^3 = (x - y)(x^2 + xy + y^2)$
Sum of Cubes	$x^3 + y^3 = (x + y)(x^2 - xy + y^2)$

6.3 EXERCISES

Concept Check   Work each problem.

1. Which of the following binomials are differences of squares?

   A. $64 - m^2$ **B.** $2x^2 - 25$ **C.** $k^2 + 9$ **D.** $4z^4 - 49$

2. Which of the following binomials are sums or differences of cubes?

   A. $64 + y^3$ **B.** $125 - p^6$ **C.** $9x^3 + 125$ **D.** $(x + y)^3 - 1$

3. Which of the following trinomials are perfect squares?

   A. $x^2 - 8x - 16$ **B.** $4m^2 + 20m + 25$

   C. $9z^4 + 30z^2 + 25$ **D.** $25a^2 - 45a + 81$

4. Of the 12 polynomials listed in Exercises 1–3, which ones can be factored by the methods of this section?

5. The binomial $9x^2 + 81$ is an example of a sum of two squares that can be factored. Under what conditions can the sum of two squares be factored?

6. Insert the correct signs in the blanks.

   (a) $8 + t^3 = (2 \underline{\quad} t)(4 \underline{\quad} 2t \underline{\quad} t^2)$ **(b)** $z^3 - 1 = (z \underline{\quad} 1)(z^2 \underline{\quad} z \underline{\quad} 1)$

Factor each polynomial. See Examples 1–4.

● **7.** $p^2 - 16$ **8.** $k^2 - 9$ **9.** $25x^2 - 4$

10. $36m^2 - 25$ **11.** $18a^2 - 98b^2$ **12.** $32c^2 - 98d^2$

13. $64m^4 - 4y^4$ **14.** $243x^4 - 3t^4$ **15.** $(y + z)^2 - 81$

16. $(h + k)^2 - 9$ **17.** $16 - (x + 3y)^2$ **18.** $64 - (r + 2t)^2$

19. $p^4 - 256$ **20.** $a^4 - 625$ **21.** $k^2 - 6k + 9$

22. $x^2 + 10x + 25$ ● **23.** $4z^2 + 4zw + w^2$ **24.** $9y^2 + 6yz + z^2$

25. $16m^2 - 8m + 1 - n^2$ **26.** $25c^2 - 20c + 4 - d^2$ **27.** $4r^2 - 12r + 9 - s^2$

28. $9a^2 - 24a + 16 - b^2$ **29.** $x^2 - y^2 + 2y - 1$ **30.** $-k^2 - h^2 + 2kh + 4$

31. $98m^2 + 84mn + 18n^2$ **32.** $80z^2 - 40zw + 5w^2$

33. $(p + q)^2 + 2(p + q) + 1$ **34.** $(x + y)^2 + 6(x + y) + 9$

35. $(a - b)^2 + 8(a - b) + 16$ **36.** $(m - n)^2 + 4(m - n) + 4$

● **37.** $x^3 - 27$ **38.** $y^3 - 64$ **39.** $t^3 - 216$

40. $m^3 - 512$ ● **41.** $x^3 + 64$ **42.** $r^3 + 343$

43. $1000 + y^3$ **44.** $729 + x^3$ **45.** $8x^3 + 1$

46. $27y^3 + 1$ **47.** $125x^3 - 216$ **48.** $8w^3 - 125$

49. $x^3 - 8y^3$ **50.** $z^3 - 125p^3$ **51.** $64g^3 - 27h^3$

52. $27a^3 - 8b^3$ **53.** $343p^3 + 125q^3$ **54.** $512t^3 + 27s^3$

55. $24n^3 + 81p^3$ **56.** $250x^3 + 16y^3$ **57.** $(y + z)^3 + 64$

58. $(p - q)^3 + 125$ **59.** $m^6 - 125$ **60.** $27r^6 + 1$

61. $1000x^9 - 27$ **62.** $729p^9 - 64$ **63.** $125y^6 + z^3$

RELATING CONCEPTS (EXERCISES 64–69)

FOR INDIVIDUAL OR GROUP WORK

The binomial $x^6 - y^6$ may be considered either as a difference of squares or a difference of cubes. **Work Exercises 64–69 in order.**

64. Factor $x^6 - y^6$ by first factoring as a difference of squares. Then factor further by considering one of the factors as a sum of cubes and the other factor as a difference of cubes.

65. Based on your answer in Exercise 64, fill in the blank with the correct factors so that $x^6 - y^6$ is factored completely:

$$x^6 - y^6 = (x - y)(x + y) \underline{\hspace{4cm}}.$$

66. Factor $x^6 - y^6$ by first factoring as a difference of cubes. Then factor further by considering one of the factors as a difference of squares.

67. Based on your answer in Exercise 66, fill in the blank with the correct factor so that $x^6 - y^6$ is factored:

$$x^6 - y^6 = (x - y)(x + y) \underline{\hspace{4cm}}.$$

68. Notice that the factor you wrote in the blank in Exercise 67 is a fourth-degree polynomial, while the two factors you wrote in the blank in Exercise 65 are both second-degree polynomials. What must be true about the product of the two factors you wrote in the blank in Exercise 65? Verify this.

69. If you have a choice of factoring as a difference of squares or a difference of cubes, how should you start to more easily obtain the completely factored form of the polynomial? Base the answer on your results in Exercises 64–68 and the methods of factoring explained in this section.

In some cases, the method of factoring by grouping can be combined with the methods of special factoring discussed in this section. For example, to factor $8x^3 + 4x^2 + 27y^3 - 9y^2$, we proceed as follows.

$$8x^3 + 4x^2 + 27y^3 - 9y^2 = (8x^3 + 27y^3) + (4x^2 - 9y^2) \qquad \text{Associative and commutative properties}$$

$$= (2x + 3y)(4x^2 - 6xy + 9y^2) + (2x + 3y)(2x - 3y) \qquad \text{Factor within groups.}$$

$$= (2x + 3y)[(4x^2 - 6xy + 9y^2) + (2x - 3y)] \qquad \text{Factor out the greatest common factor, } 2x + 3y.$$

$$= (2x + 3y)(4x^2 - 6xy + 9y^2 + 2x - 3y) \qquad \text{Combine terms.}$$

In problems such as this, how we choose to group in the first step is essential to factoring correctly. If we reach a "dead end," then we should group differently and try again.

Use the method just described to factor each polynomial.

70. $27x^3 + 9x^2 + y^3 - y^2$

71. $125p^3 + 25p^2 + 8q^3 - 4q^2$

72. $1000k^3 + 20k - m^3 - 2m$

73. $27a^3 + 15a - 64b^3 - 20b$

74. $y^4 + y^3 + y + 1$

75. $8t^4 - 24t^3 + t - 3$

76. $10x^2 + 5x^3 - 10y^2 + 5y^3$

77. $64m^2 - 512m^3 - 81n^2 + 729n^3$

PREVIEW EXERCISES

*Factor completely. See **Sections 6.1 and 6.2.***

78. $24y^3 - 16y^5 + 64y^7$

79. $2ax + ay - 2bx - by$

80. $y^2 - y - 2$

81. $p^2 + 4p - 21$

82. $6t^2 + 19ts - 7s^2$

6.4   A General Approach to Factoring

OBJECTIVES

1 Factor out any common factor.

2 Factor binomials.

3 Factor trinomials.

4 Factor polynomials of more than three terms.

In this section, we summarize and apply the factoring methods presented in the preceding sections. A polynomial is completely factored when it is in the following form:

1. The polynomial is written as a product of prime polynomials with integer coefficients.

2. None of the polynomial factors can be factored further, except that a monomial factor need not be factored completely.

Factoring a Polynomial

Step 1   **Factor out any common factor.**

Step 2   **If the polynomial is a binomial,** check to see if it is the difference of squares, the difference of cubes, or the sum of cubes.

If the polynomial is a trinomial, check to see if it is a perfect square trinomial. If it is not, factor as in **Section 6.2.**

If the polynomial has more than three terms, try to factor by grouping.

Step 3   *Check the factored form by multiplying.*

OBJECTIVE 1   **Factor out any common factor.**   This step is always the same, regardless of the number of terms in the polynomial.

EXAMPLE 1   Factoring Out a Common Factor

Factor each polynomial.

(a) $9p + 45 = 9(p + 5)$

(b) $8m^2p^2 + 4mp = 4mp(2mp + 1)$

(c) $5x(a + b) - y(a + b) = (a + b)(5x - y)$

✔ **Now Try Exercises 13 and 23.**

OBJECTIVE 2   **Factor binomials.**   Use one of the following rules.

Factoring a Binomial

For a **binomial** (two terms), check for the following:

Difference of squares	$x^2 - y^2 = (x + y)(x - y)$
Difference of cubes	$x^3 - y^3 = (x - y)(x^2 + xy + y^2)$
Sum of cubes	$x^3 + y^3 = (x + y)(x^2 - xy + y^2).$

EXAMPLE 2   **Factoring Binomials**

Factor each binomial if possible.

(a) $64m^2 - 9n^2 = (8m)^2 - (3n)^2$ Difference of squares
$$= (8m + 3n)(8m - 3n)$$

(b) $8p^3 - 27 = (2p)^3 - 3^3$ Difference of cubes
$$= (2p - 3)[(2p)^2 + (2p)(3) + 3^2]$$
$$= (2p - 3)(4p^2 + 6p + 9)$$

(c) $1000m^3 + 1 = (10m)^3 + 1^3$ Sum of cubes
$$= (10m + 1)[(10m)^2 - (10m)(1) + 1^2]$$
$$= (10m + 1)(100m^2 - 10m + 1)$$

(d) $25m^2 + 121$ is prime. It is the sum of squares.

✔ **Now Try Exercises 7, 11, 29, and 31.**

▶ **NOTE**   The binomial $25m^2 + 625$ is a sum of squares. It can be factored, however, because it has a common factor.
$$25m^2 + 625 = 25\underbrace{(m^2 + 25)}_{}$$ Factor out the common factor.

This sum of squares cannot
be factored further.

OBJECTIVE 3   **Factor trinomials.**   Consider the following when factoring trinomials.

Factoring a Trinomial

For a **trinomial** (three terms), decide whether it is a perfect square trinomial of the form
$$x^2 + 2xy + y^2 = (x + y)^2 \quad \text{or} \quad x^2 - 2xy + y^2 = (x - y)^2.$$
If not, use the methods of **Section 6.2.**

EXAMPLE 3   **Factoring Trinomials**

Factor each trinomial.

(a) $p^2 + 10p + 25 = (p + 5)^2$ Perfect square trinomial

(b) $49z^2 - 42z + 9 = (7z - 3)^2$ Perfect square trinomial

(c) $y^2 - 5y - 6 = (y - 6)(y + 1)$

The numbers -6 and 1 have a product of -6 and a sum of -5.

(d) $r^2 + 18r + 72 = (r + 6)(r + 12)$

(e) $2k^2 - k - 6 = (2k + 3)(k - 2)$ Use either method from **Section 6.2.**

(f) $28z^2 + 6z - 10 = 2(14z^2 + 3z - 5)$ Factor out the common factor.
$$= 2(7z + 5)(2z - 1)$$

Remember the common factor.

✔ **Now Try Exercises 9, 19, and 41.**

OBJECTIVE 4 Factor polynomials of more than three terms. Try factoring by grouping.

EXAMPLE 4 Factoring Polynomials with More than Three Terms

Factor each polynomial.

(a) $xy^2 - y^3 + x^3 - x^2y = (xy^2 - y^3) + (x^3 - x^2y)$
$$= y^2(x - y) + x^2(x - y)$$
$$= (x - y)(y^2 + x^2)$$

(b) $20k^3 + 4k^2 - 45k - 9 = (20k^3 + 4k^2) - (45k + 9)$ *Be careful with signs.*

$$= 4k^2(5k + 1) - 9(5k + 1)$$
$$= (5k + 1)(4k^2 - 9)$$ *5k + 1 is a common factor.*
$$= (5k + 1)(2k + 3)(2k - 3)$$ Difference of squares

(c) $4a^2 + 4a + 1 - b^2 = (4a^2 + 4a + 1) - b^2$ Associative property
$$= (2a + 1)^2 - b^2$$ Perfect square trinomial
$$= (2a + 1 + b)(2a + 1 - b)$$ Difference of squares

(d) $8m^3 + 4m^2 - n^3 - n^2$

Notice that the terms must be rearranged before grouping, since

$$(8m^3 + 4m^2) - (n^3 + n^2) = 4m^2(2m + 1) - n^2(n + 1),$$

which cannot be factored further. Factor the polynomial as follows:

$$8m^3 + 4m^2 - n^3 - n^2 = (8m^3 - n^3) + (4m^2 - n^2)$$ Group the cubes and squares.

$$= (2m - n)(4m^2 + 2mn + n^2) + (2m - n)(2m + n)$$ Factor each group.

$$= (2m - n)(4m^2 + 2mn + n^2 + 2m + n).$$ Factor out the common factor $2m - n$.

✔ **Now Try Exercises 21 and 45.**

6.4 EXERCISES

▨ *Now Try Exercise*

Factor each polynomial. See Examples 1–4.

1. $100a^2 - 9b^2$

2. $10r^2 + 13r - 3$

3. $3p^4 - 3p^3 - 90p^2$

4. $k^4 - 16$

5. $3a^2pq + 3abpq - 90b^2pq$

6. $49z^2 - 16$

◐ **7.** $225p^2 + 256$

8. $18m^3n + 3m^2n^2 - 6mn^3$

◐ **9.** $6b^2 - 17b - 3$

10. $k^2 - 6k - 16$

11. $x^3 - 1000$

12. $6t^2 + 19tu - 77u^2$

◐ **13.** $4(p + 2) + m(p + 2)$

14. $40p - 32r$

15. $9m^2 - 45m + 18m^3$

16. $4k^2 + 28kr + 49r^2$

17. $54m^3 - 2000$

18. $mn - 2n + 5m - 10$

19. $9m^2 - 30mn + 25n^2$

20. $2a^2 - 7a - 4$

🌐 **21.** $kq - 9q + kr - 9r$

22. $56k^3 - 875$

23. $16z^3x^2 - 32z^2x$

24. $9r^2 + 100$

25. $x^2 + 2x - 35$

26. $9 - a^2 + 2ab - b^2$

27. $x^4 - 625$

28. $2m^2 - mn - 15n^2$

29. $p^3 + 1$

30. $48y^2z^3 - 28y^3z^4$

31. $64m^2 - 625$

32. $14z^2 - 3zk - 2k^2$

33. $12z^3 - 6z^2 + 18z$

34. $225k^2 - 36r^2$

35. $256b^2 - 400c^2$

36. $z^2 - zp - 20p^2$

37. $1000z^3 + 512$

38. $64m^2 - 25n^2$

39. $10r^2 + 23rs - 5s^2$

40. $12k^2 - 17kq - 5q^2$

41. $24p^3q + 52p^2q^2 + 20pq^3$

42. $32x^2 + 16x^3 - 24x^5$

43. $48k^4 - 243$

44. $14x^2 - 25xq - 25q^2$

45. $m^3 + m^2 - n^3 - n^2$

46. $64x^3 + y^3 - 16x^2 + y^2$

47. $x^2 - 4m^2 - 4mn - n^2$

48. $4r^2 - s^2 - 2st - t^2$

49. $18p^5 - 24p^3 + 12p^6$

50. $k^2 - 6k + 16$

51. $2x^2 - 2x - 40$

52. $27x^3 - 3y^3$

53. $(2m + n)^2 - (2m - n)^2$

54. $(3k + 5)^2 - 4(3k + 5) + 4$

55. $50p^2 - 162$

56. $y^2 + 3y - 10$

57. $12m^2rx + 4mnrx + 40n^2rx$

58. $18p^2 + 53pr - 35r^2$

59. $21a^2 - 5ab - 4b^2$

60. $x^2 - 2xy + y^2 - 4$

61. $x^2 - y^2 - 4$

62. $(5r + 2s)^2 - 6(5r + 2s) + 9$

63. $(p + 8q)^2 - 10(p + 8q) + 25$

64. $z^4 - 9z^2 + 20$

65. $21m^4 - 32m^2 - 5$

66. $(x - y)^3 - (27 - y)^3$

67. $(r + 2t)^3 + (r - 3t)^3$

68. $16x^3 + 32x^2 - 9x - 18$

69. $x^5 + 3x^4 - x - 3$

70. $x^{16} - 1$

71. $m^2 - 4m + 4 - n^2 + 6n - 9$

72. $x^2 + 4 + x^2y + 4y$

PREVIEW EXERCISES

Solve each equation. See **Section 2.1.**

73. $3x + 2 = 0$

74. $-2x + 7 = 0$

75. $5x = 0$

76. $-8x = 0$

77. $\frac{1}{2}t + 5 = 0$

78. $-\frac{3}{4}x - 6 = 0$

6.5 Solving Equations by Factoring

OBJECTIVES

1 Learn and use the zero-factor property.

2 Solve applied problems that require the zero-factor property.

In **Chapter 2,** we developed methods for solving linear, or first-degree, equations. Solving higher degree polynomial equations requires other methods, one of which involves factoring.

OBJECTIVE 1 Learn and use the zero-factor property. Solving equations by factoring depends on a special property of the number 0, called the **zero-factor property.**

Zero-Factor Property

If two numbers have a product of 0, then at least one of the numbers must be 0. That is,

$$\text{if } ab = 0, \quad \text{then either} \quad a = 0 \quad \text{or} \quad b = 0.$$

To prove the zero-factor property, we first assume that $a \neq 0$. (If a does equal 0, then the property is proved already.) If $a \neq 0$, then $\frac{1}{a}$ exists, and both sides of $ab = 0$ can be multiplied by $\frac{1}{a}$ to get

$$\frac{1}{a} \cdot ab = \frac{1}{a} \cdot 0$$
$$b = 0.$$

Thus, if $a \neq 0$, then $b = 0$, and the property is proved.

▶ **CAUTION** If $ab = 0$, then $a = 0$ or $b = 0$. However, if $ab = 6$, for example, it is not necessarily true that $a = 6$ or $b = 6$; in fact, it is very likely that *neither $a = 6$ nor $b = 6$. **The zero-factor property works only for a product equal to 0.***

EXAMPLE 1 Using the Zero-Factor Property to Solve an Equation

Solve $(x + 6)(2x - 3) = 0$.

Here, the product of $x + 6$ and $2x - 3$ is 0. By the zero-factor property, this can be true only if

$$x + 6 = 0 \quad \text{or} \quad 2x - 3 = 0. \qquad \text{Zero-factor property}$$

Solve each of these equations.

$$x + 6 = 0 \quad \text{or} \quad 2x - 3 = 0$$
$$x = -6 \quad \text{or} \quad 2x = 3$$
$$x = \frac{3}{2}$$

Check the two solutions -6 and $\frac{3}{2}$ by substitution into the *original* equation.

$$\text{If } x = -6, \text{ then} \qquad\qquad\qquad \text{If } x = \frac{3}{2}, \text{ then}$$

$$
\begin{array}{ll}
(x + 6)(2x - 3) = 0 & \qquad\qquad (x + 6)(2x - 3) = 0 \\
(-6 + 6)[2(-6) - 3] = 0 \quad ? & \qquad \left(\frac{3}{2} + 6\right)\left(2 \cdot \frac{3}{2} - 3\right) = 0 \quad ? \\
0(-15) = 0 \quad ? & \\
0 = 0. \quad \text{True} & \qquad\qquad\qquad \frac{15}{2}(0) = 0 \quad ? \\
& \qquad\qquad\qquad\qquad 0 = 0. \quad \text{True}
\end{array}
$$

Both solutions check; the solution set is $\left\{-6, \frac{3}{2}\right\}$.

✔ **Now Try Exercise 5.**

Since the product $(x + 6)(2x - 3)$ equals $2x^2 + 9x - 18$, the equation of Example 1 has a term with a squared variable and is an example of a *quadratic equation*. *A quadratic equation has degree 2.*

Quadratic Equation

An equation that can be written in the form

$$ax^2 + bx + c = 0,$$

where a, b, and c are real numbers, with $a \neq 0$, is a **quadratic equation.** This form is called **standard form.**

Quadratic equations are discussed in more detail in **Chapter 9.**

The steps for solving a quadratic equation by factoring are summarized here.

Solving a Quadratic Equation by Factoring

Step 1 **Write in standard form.** Rewrite the equation if necessary so that one side is 0.

Step 2 **Factor** the polynomial.

Step 3 **Use the zero-factor property.** Set each variable factor equal to 0.

Step 4 **Find the solution(s).** Solve each equation formed in Step 3.

Step 5 **Check** each solution in the *original* equation.

EXAMPLE 2 Solving Quadratic Equations by Factoring

Solve each equation.

(a) $2x^2 + 3x = 2$

Step 1
$$2x^2 + 3x = 2$$
$$2x^2 + 3x - 2 = 0 \qquad \text{Standard form}$$

Step 2
$$(2x - 1)(x + 2) = 0 \qquad \text{Factor.}$$

Step 3 $\qquad 2x - 1 = 0 \quad \text{or} \quad x + 2 = 0 \qquad$ Zero-factor property

Step 4 $\qquad\qquad 2x = 1 \quad \text{or} \qquad x = -2 \qquad$ Solve each equation.

$$x = \frac{1}{2}$$

Step 5 *Check* each solution in the original equation.

If $x = \frac{1}{2}$, then

$$2x^2 + 3x = 2$$

$$2\left(\frac{1}{2}\right)^2 + 3\left(\frac{1}{2}\right) = 2 \qquad ?$$

$$2\left(\frac{1}{4}\right) + \frac{3}{2} = 2 \qquad ?$$

$$\frac{1}{2} + \frac{3}{2} = 2 \qquad ?$$

$$2 = 2. \qquad \text{True}$$

If $x = -2$, then

$$2x^2 + 3x = 2$$

$$2(-2)^2 + 3(-2) = 2 \qquad ?$$

$$2(4) - 6 = 2 \qquad ?$$

$$8 - 6 = 2 \qquad ?$$

$$2 = 2. \qquad \text{True}$$

> We write solutions in the order they appear on a number line.

Because both solutions check, the solution set is $\left\{-2, \frac{1}{2}\right\}$.

(b) $\qquad\qquad 4x^2 - 4x + 1 = 0 \qquad$ Standard form

$\qquad\qquad\qquad (2x - 1)^2 = 0 \qquad$ Factor.

$\qquad\qquad\qquad 2x - 1 = 0 \qquad$ Zero-factor property

$\qquad\qquad\qquad 2x = 1 \qquad$ Add 1.

$$x = \frac{1}{2} \qquad \text{Divide by 2.}$$

There is only one solution, because the trinomial is a perfect square. The solution set is $\left\{\frac{1}{2}\right\}$.

✔ **Now Try Exercises 11 and 29.**

EXAMPLE 3 Solving a Quadratic Equation with a Missing Constant Term

Solve $4z^2 - 20z = 0$.

This quadratic equation has a missing term. Comparing it with the standard form $ax^2 + bx + c = 0$ shows that $c = 0$. The zero-factor property can still be used.

$$4z^2 - 20z = 0$$

$$4z(z - 5) = 0 \qquad \text{Factor.}$$

> Set each *variable* factor equal to 0.

$\qquad 4z = 0 \quad \text{or} \quad z - 5 = 0 \qquad$ Zero-factor property

$\qquad\; z = 0 \quad \text{or} \qquad z = 5 \qquad$ Solve each equation.

Check: If $z = 0$, then

$$4z^2 - 20z = 0$$

$$4(0)^2 - 20(0) = 0 \qquad ?$$

$$0 - 0 = 0. \qquad \text{True}$$

If $z = 5$, then

$$4z^2 - 20z = 0$$

$$4(5)^2 - 20(5) = 0 \qquad ?$$

$$100 - 100 = 0. \qquad \text{True}$$

The solution set is $\{0, 5\}$.

✔ **Now Try Exercise 19.**

▶ **CAUTION** Remember to include 0 as a solution in Example 3.

EXAMPLE 4 Solving a Quadratic Equation with a Missing Linear Term

Solve $3m^2 - 108 = 0$.

$$3m^2 - 108 = 0$$

$$3(m^2 - 36) = 0 \qquad \text{Factor out 3.}$$

The factor 3 does *not* lead to a solution.

$$3(m + 6)(m - 6) = 0 \qquad \text{Factor } m^2 - 36.$$

$$m + 6 = 0 \quad \text{or} \quad m - 6 = 0 \qquad \text{Zero-factor property}$$

$$m = -6 \quad \text{or} \qquad m = 6$$

Check that the solution set is $\{-6, 6\}$.

✔ **Now Try Exercise 23.**

▶ **CAUTION** The factor 3 in Example 4 is not a *variable* factor, so it does *not* lead to a solution of the equation. In Example 3, however, the factor $4z$ is a variable factor and leads to the solution 0.

EXAMPLE 5 Solving an Equation That Requires Rewriting

Solve $(2q + 1)(q + 1) = 2(1 - q) + 6$.

$$(2q + 1)(q + 1) = 2(1 - q) + 6$$

$$2q^2 + 3q + 1 = 2 - 2q + 6 \qquad \text{Multiply on each side.}$$

$$2q^2 + 5q - 7 = 0 \qquad \text{Standard form}$$

$$(2q + 7)(q - 1) = 0 \qquad \text{Factor.}$$

$$2q + 7 = 0 \quad \text{or} \quad q - 1 = 0 \qquad \text{Zero-factor property}$$

$$q = -\frac{7}{2} \quad \text{or} \qquad q = 1 \qquad \text{Solve each equation.}$$

Check: $\qquad (2q + 1)(q + 1) = 2(1 - q) + 6$

$$\left[2\left(-\frac{7}{2}\right) + 1\right]\left(-\frac{7}{2} + 1\right) = 2\left[1 - \left(-\frac{7}{2}\right)\right] + 6 \quad ? \qquad \text{Let } q = -\tfrac{7}{2}.$$

$$(-7 + 1)\left(-\frac{5}{2}\right) = 2\left(\frac{9}{2}\right) + 6 \qquad\qquad ? \qquad \text{Simplify; } 1 = \tfrac{2}{2}.$$

$$(-6)\left(-\frac{5}{2}\right) = 9 + 6 \qquad\qquad\qquad ?$$

$$15 = 15 \qquad\qquad\qquad\qquad \text{True}$$

Check that 1 is a solution. The solution set is $\left\{-\frac{7}{2}, 1\right\}$.

✔ **Now Try Exercise 35.**

The zero-factor property can be extended to solve certain polynomial equations of degree 3 or higher, as shown in the next example.

EXAMPLE 6 Solving an Equation of Degree 3

Solve $-x^3 + x^2 = -6x$.

Start by adding $6x$ to each side to get 0 on the right side.

$$-x^3 + x^2 + 6x = 0$$

$$x^3 - x^2 - 6x = 0 \qquad \text{Multiply each side by } -1.$$

$$x(x^2 - x - 6) = 0 \qquad \text{Factor out } x.$$

$$x(x - 3)(x + 2) = 0 \qquad \text{Factor the trinomial.}$$

Use the zero-factor property, extended to include the three *variable* factors.

> Remember to set x equal to 0.

$$x = 0 \quad \text{or} \quad x - 3 = 0 \quad \text{or} \quad x + 2 = 0$$

$$x = 3 \quad \text{or} \qquad x = -2$$

Check that the solution set is $\{-2, 0, 3\}$.

✔ **Now Try Exercise 39.**

OBJECTIVE 2 Solve applied problems that require the zero-factor property. The next example shows an application that leads to a quadratic equation. We continue to use the six-step problem-solving method introduced in **Section 2.3.**

EXAMPLE 7 Using a Quadratic Equation in an Application

A piece of sheet metal is in the shape of a parallelogram. The longer sides of the parallelogram are each 8 m longer than the distance between them. The area of the piece is 48 m². Find the length of the longer sides and the distance between them.

Step 1 **Read** the problem again. There will be two answers.

Step 2 **Assign a variable.**

Let $x = $ the distance between the longer sides;

$x + 8 = $ the length of each longer side. (See Figure 1.)

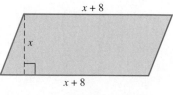

FIGURE 1

Step 3 **Write an equation.** The area of a parallelogram is given by $A = bh$, where b is the length of the longer side and h is the distance between the longer sides. Here, $b = x + 8$ and $h = x$.

$$A = bh$$

$$48 = (x + 8)x \qquad \text{Let } A = 48, \ b = x + 8, \ h = x.$$

Step 4 **Solve.**

$$48 = (x + 8)x$$
$$48 = x^2 + 8x \qquad \text{Distributive property}$$
$$x^2 + 8x - 48 = 0 \qquad \text{Standard form}$$
$$(x + 12)(x - 4) = 0 \qquad \text{Factor.}$$
$$x + 12 = 0 \quad \text{or} \quad x - 4 = 0 \qquad \text{Zero-factor property}$$
$$x = -12 \quad \text{or} \qquad x = 4$$

Step 5 **State the answer.** *A distance cannot be negative, so reject* −12 *as a solution.* The only possible solution is 4, so the distance between the longer sides is 4 m. The length of the longer sides is 4 + 8 = 12 m.

Step 6 **Check.** The length of the longer sides is 8 m more than the distance between them, and the area is 4 · 12 = 48 m², as required, so the answer checks.

✔ **Now Try Exercise 57.**

▶ **CAUTION** When applications lead to quadratic equations, a solution of the equation may not satisfy the physical requirements of the problem, as in Example 7. Reject such solutions.

A function defined by a quadratic polynomial is called a *quadratic function*. (See **Chapter 9.**) The next example uses such a function.

EXAMPLE 8 **Using a Quadratic Function in an Application**

Quadratic functions are used to describe the height a falling object or a projected object reaches in a specific time. For example, if a small rocket is launched vertically upward from ground level with an initial velocity of 128 ft per sec, then its height in feet after t seconds is a function defined by

$$h(t) = -16t^2 + 128t$$

if air resistance is neglected. After how many seconds will the rocket be 220 ft above the ground?

We must let $h(t) = 220$ and solve for t.

$$220 = -16t^2 + 128t \qquad \text{Let } h(t) = 220.$$
$$16t^2 - 128t + 220 = 0 \qquad \text{Standard form}$$
$$4t^2 - 32t + 55 = 0 \qquad \text{Divide by 4.}$$
$$(2t - 5)(2t - 11) = 0 \qquad \text{Factor.}$$
$$2t - 5 = 0 \quad \text{or} \quad 2t - 11 = 0 \qquad \text{Zero-factor property}$$
$$t = 2.5 \quad \text{or} \qquad t = 5.5$$

The rocket will reach a height of 220 ft twice: on its way up at 2.5 sec and again on its way down at 5.5 sec.

✔ **Now Try Exercise 65.**

CONNECTIONS

In **Section 5.3,** we saw that the graph of $f(x) = x^2$ is a parabola. In general, the graph of $f(x) = ax^2 + bx + c$, $a \neq 0$, is a parabola, and the x-intercepts of its graph give the real number solutions of the equation $ax^2 + bx + c = 0$.

A graphing calculator can locate these x-intercepts (called **zeros** of the function) for $Y_1 = f(X) = 2X^2 + 3X - 2$. Notice that this quadratic expression was found on the left side of the equation in Example 2(a) earlier in this section, where the equation was written in standard form.

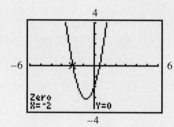 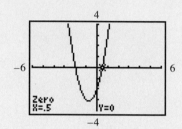

The x-intercepts (zeros) given with the graphs are the same as the solutions found in Example 2(a). This method of graphical solution can be used for any type of equation.

6.5 EXERCISES

Complete solution available on Video Lectures on CD/DVD

Now Try Exercise

1. Explain in your own words how the zero-factor property is used in solving a quadratic equation.

2. *Concept Check*   One of the following equations is *not* in proper form for using the zero-factor property. Which one is it? Tell why it is not in proper form.

A. $(x + 2)(x - 6) = 0$ **B.** $x(3x - 7) = 0$

C. $3t(t + 8)(t - 9) = 0$ **D.** $y(y - 3) + 6(y - 3) = 0$

Solve each equation. See Examples 1–5.

3. $(x + 10)(x - 5) = 0$ **4.** $(x + 7)(x + 3) = 0$ **5.** $(2k - 5)(3k + 8) = 0$

6. $(3q - 4)(2q + 5) = 0$ **7.** $m^2 - 3m - 10 = 0$ **8.** $x^2 + x - 12 = 0$

9. $z^2 + 9z + 18 = 0$ **10.** $x^2 - 18x + 80 = 0$ **11.** $2x^2 = 7x + 4$

12. $2x^2 = 3 - x$ **13.** $15k^2 - 7k = 4$ **14.** $3c^2 + 3 = -10c$

15. $2x^2 - 12 - 4x = x^2 - 3x$

16. $3p^2 + 9p + 30 = 2p^2 - 2p$

17. $(5z + 1)(z + 3) = -2(5z + 1)$

18. $(3x + 1)(x - 3) = 2 + 3(x + 5)$

19. $4p^2 + 16p = 0$

20. $2a^2 - 8a = 0$

21. $6m^2 - 36m = 0$

22. $-3m^2 + 27m = 0$

23. $4p^2 - 16 = 0$

24. $9x^2 - 81 = 0$

25. $-3m^2 + 27 = 0$

26. $-2a^2 + 8 = 0$

27. $-x^2 = 9 - 6x$

28. $-m^2 - 8m = 16$

29. $9k^2 + 24k + 16 = 0$

30. $4m^2 - 20m + 25 = 0$

31. $(x - 3)(x + 5) = -7$

32. $(x + 8)(x - 2) = -21$

33. $(2x + 1)(x - 3) = 6x + 3$

34. $(3x + 2)(x - 3) = 7x - 1$

35. $(x + 3)(x - 6) = (2x + 2)(x - 6)$

36. $(2x + 1)(x + 5) = (x + 11)(x + 3)$

Solve each equation. See Example 6.

37. $2x^3 - 9x^2 - 5x = 0$

38. $6x^3 - 13x^2 - 5x = 0$

39. $x^3 - 2x^2 = 3x$

40. $y^3 - 6y^2 = -8y$

41. $9t^3 = 16t$

42. $25x^3 = 64x$

43. $2r^3 + 5r^2 - 2r - 5 = 0$

44. $2p^3 + p^2 - 98p - 49 = 0$

45. $x^3 - 6x^2 - 9x + 54 = 0$

46. $x^3 - 3x^2 - 4x + 12 = 0$

47. *Concept Check*   A student tried to solve the equation in Exercise 41 by first dividing each side by t, obtaining $9t^2 = 16$. She then solved the resulting equation by the zero-factor property to get the solution set $\left\{-\frac{4}{3}, \frac{4}{3}\right\}$. **WHAT WENT WRONG?** Give the correct solution set.

48. *Concept Check*   Without actually solving each equation, determine which one of the following has 0 in its solution set.

   A. $4x^2 - 25 = 0$ **B.** $x^2 + 2x - 3 = 0$ **C.** $6x^2 + 9x + 1 = 0$ **D.** $x^3 + 4x^2 = 3x$

The next problems are "brain busters." Solve each equation.

49. $2(x - 1)^2 - 7(x - 1) - 15 = 0$

50. $4(2k + 3)^2 - (2k + 3) - 3 = 0$

51. $5(3a - 1)^2 + 3 = -16(3a - 1)$

52. $2(m + 3)^2 = 5(m + 3) - 2$

53. $(2k - 3)^2 = 16k^2$

54. $9p^2 = (5p + 2)^2$

Solve each problem. See Examples 7 and 8.

55. A garden has an area of 320 ft². Its length is 4 ft more than its width. What are the dimensions of the garden? (*Hint:* $320 = 16 \cdot 20$)

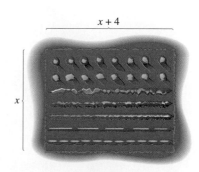

$x + 4$

x

56. A square mirror has sides measuring 2 ft less than the sides of a square painting. If the difference between their areas is 32 ft², find the lengths of the sides of the mirror and the painting.

 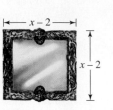

57. The base of a parallelogram is 7 ft more than the height. If the area of the parallelogram is 60 ft², what are the measures of the base and the height?

58. A sign has the shape of a triangle. The length of the base is 3 m less than the height. What are the measures of the base and the height if the area is 44 m²?

59. A farmer has 300 ft of fencing and wants to enclose a rectangular area of 5000 ft². What dimensions should she use? (*Hint:* $5000 = 50 \cdot 100$)

60. A rectangular landfill has an area of 30,000 ft². Its length is 200 ft more than its width. What are the dimensions of the landfill? (*Hint:* $30,000 = 300 \cdot 100$)

61. Find two consecutive integers such that the sum of their squares is 61.

62. Find two consecutive integers such that their product is 72.

63. A box with no top is to be constructed from a piece of cardboard whose length measures 6 in. more than its width. The box is to be formed by cutting squares that measure 2 in. on each side from the four corners and then folding up the sides. If the volume of the box will be 110 in.³, what are the dimensions of the piece of cardboard?

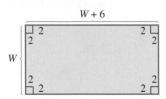

64. The surface area of the box with open top shown in the figure is 161 in.². Find the dimensions of the base. (*Hint:* The surface area of the box is a function defined by $S(x) = x^2 + 16x$.)

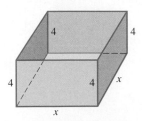

65. If an object is projected upward with an initial velocity of 64 ft per sec from a height of 80 ft, then its height in feet t seconds after it is projected is a function defined by

$$f(t) = -16t^2 + 64t + 80.$$

How long after it is projected will it hit the ground? (*Hint:* When it hits the ground, its height is 0 ft.)

66. Refer to Example 8. After how many seconds will the rocket be 240 ft above the ground? 112 ft above the ground?

67. If a baseball is dropped from a helicopter 625 ft above the ground, then its distance in feet from the ground t seconds later is a function defined by

$$f(t) = -16t^2 + 625.$$

How long after it is dropped will it hit the ground?

68. If a rock is dropped from a building 576 ft high, then its distance in feet from the ground t seconds later is a function defined by

$$f(t) = -16t^2 + 576.$$

How long after it is dropped will it hit the ground?

TECHNOLOGY INSIGHTS (EXERCISES 69–72)

As shown in the Connections box following Example 8, the solutions of the quadratic equation $ax^2 + bx + c = 0$ $(a \neq 0)$ are represented on the graph of the quadratic function $f(x) = ax^2 + bx + c$ by the x-intercepts.

Use the zero-factor property to solve each equation, and confirm that your solutions correspond to the x-intercepts (zeros) shown on the accompanying graphing calculator screens.

69. $2x^2 - 7x - 4 = 0$

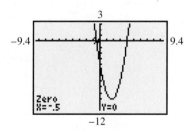 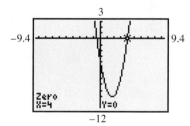

70. $2x^2 + 7x - 15 = 0$

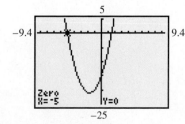 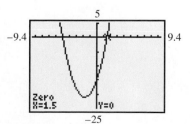

(continued)

71. $-x^2 + 3x = -10$

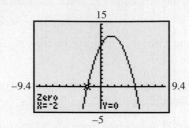 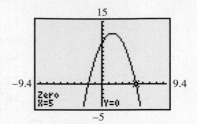

72. $-x^2 + x = -12$

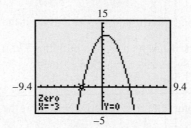 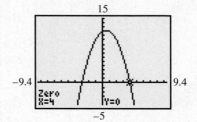

PREVIEW EXERCISES

*Simplify. See **Section 5.1.***

73. $\dfrac{12p^2}{3p}$

74. $\dfrac{-50a^4b^5}{150a^6b^4}$

75. $\dfrac{-27m^2n^5}{36m^6n^8}$

Write each fraction with the indicated denominator.

76. $\dfrac{5}{8} = \dfrac{?}{24}$

77. $\dfrac{12}{25} = \dfrac{?}{75}$

78. $\dfrac{8}{3} = \dfrac{?}{15}$

Chapter **6** Group Activity

FINDING THE HEIGHT OF A PROJECTED OBJECT

Objective Use a quadratic function to model the height of a projected object.

An important application of quadratic functions deals with the height of a projected object as a function of time elapsed after it is projected. We saw instances of this in Example 8 of **Section 6.5** and the corresponding exercises. Now we consider the general case. If air resistance is neglected, the height $f(x)$ (in feet) of an object projected directly upward from an initial height s_0 feet with initial velocity (speed) v_0 feet per second is

$$f(x) = -16x^2 + v_0x + s_0,$$

where x is the number of seconds after the object is projected. The coefficient of x^2, -16, is a constant based on the gravitational force of Earth. This constant varies on other surfaces, such as the moon and other planets.

A. Create a quadratic model for the following situation:

A ball is projected directly upward from an initial height of 100 ft with an initial velocity of 80 ft per sec.

  1. Define a function that describes the height of the ball in terms of time x.

  2. Use a graphing calculator, or calculate several ordered pairs and plot them to graph your function. Use domain $[0, 10]$ and range $[-100, 300]$.*

B. Use your model to get information about the flight of the ball.

  1. Use your graph to estimate the maximum height of the ball and when it reaches that height.

  2. After how many seconds will the ball reach the ground (height 0 ft)? Estimate this answer from the graph and check it in the equation.

  3. Use the graph to estimate the time interval during which the height of the ball is greater than 150 ft. Check your estimate by substituting it into your function.

C. Answer the following questions.

  1. Why must x be a nonnegative number?

  2. What does it mean if $s_0 = 0$?

---

*It is easy to misinterpret the graph of this function. It does not define the *path* followed by the ball; it defines height as a function of time.

Chapter **6 SUMMARY** *View the Interactive Summary on the Pass the Test CD.*

KEY TERMS

6.1 factoring
greatest common
factor (GCF)

6.2 prime polynomial
6.3 difference of squares
perfect square
trinomial

difference of cubes
sum of cubes
6.5 quadratic equation

standard form of a
quadratic equation

TEST YOUR WORD POWER

See how well you have learned the vocabulary in this chapter. Answers, with examples, follow the Quick Review.

1. **Factoring** is
   A. a method of multiplying
      polynomials
   B. the process of writing a
      polynomial as a product
   C. the answer to a multiplication
      problem
   D. a way to add the terms of a
      polynomial.

2. **Factoring by grouping** is used to
   factor
   A. out the GCF
   B. a sum or difference of squares
   C. by the substitution method
   D. a polynomial that has more than
      three terms.

3. A **quadratic equation** is a
   polynomial equation of
   A. degree 1
   B. degree 2

C. degree 3
D. degree 4.

4. The **zero-factor property** is used to
   A. factor a perfect square trinomial
   B. factor by grouping
   C. solve a polynomial equation of
      degree 2 or more
   D. solve a linear equation.

QUICK REVIEW

Concepts	Examples

6.1   GREATEST COMMON FACTORS; FACTORING BY GROUPING

The Greatest Common Factor
The product of the largest common numerical factor and each common variable raised to the least exponent that appears on that variable in any term is the greatest common factor of the terms of the polynomial.

Factor $4x^2y - 50xy^2$.
The greatest common factor is $2xy$.

$$4x^2y - 50xy^2 = 2xy(2x - 25y)$$

Factoring by Grouping

Step 1 Group the terms so that each group has a common factor.

Step 2 Factor out the common factor in each group.

Step 3 If the groups now have a common factor, factor it out. If not, try a different grouping.

Always check the factored form by multiplying.

Factor by grouping.

$$5a - 5b - ax + bx = (5a - 5b) + (-ax + bx)$$
$$= 5(a - b) - x(a - b)$$
$$= (a - b)(5 - x)$$

Be careful
with signs.

(continued)

Concepts	Examples

6.2 FACTORING TRINOMIALS

To factor a trinomial, choose factors of the first term and factors of the last term. Then place them within a pair of parentheses of this form:

$$(\qquad)(\qquad).$$

Try various combinations of the factors until the correct middle term of the trinomial is found.

Factor $15x^2 + 14x - 8$.
The factors of 15 are 5 and 3, and 15 and 1.

The factors of -8 are -4 and 2, 4 and -2, -1 and 8, and 1 and -8.

Various combinations lead to the correct factorization

$$15x^2 + 14x - 8 = (5x - 2)(3x + 4).$$

Check by multiplying.

6.3 SPECIAL FACTORING

Difference of Squares

$$x^2 - y^2 = (x + y)(x - y)$$

Factor.

$$4m^2 - 25n^2 = (2m)^2 - (5n)^2$$
$$= (2m + 5n)(2m - 5n)$$

Perfect Square Trinomials

$$x^2 + 2xy + y^2 = (x + y)^2$$
$$x^2 - 2xy + y^2 = (x - y)^2$$

$$9y^2 + 6y + 1 = (3y + 1)^2$$
$$16p^2 - 56p + 49 = (4p - 7)^2$$

Difference of Cubes

$$x^3 - y^3 = (x - y)(x^2 + xy + y^2)$$

$$8 - 27a^3 = (2 - 3a)(4 + 6a + 9a^2)$$

Sum of Cubes

$$x^3 + y^3 = (x + y)(x^2 - xy + y^2)$$

$$64z^3 + 1 = (4z + 1)(16z^2 - 4z + 1)$$

6.4 A GENERAL APPROACH TO FACTORING

Step 1 Factor out any common factors.

Step 2 For a binomial, check for the difference of squares, the difference of cubes, or the sum of cubes.

For a trinomial, see if it is a perfect square. If not, factor as in **Section 6.2.**

For more than three terms, try factoring by grouping.

Step 3 Check the factored form by multiplying.

Factor.

$$ak^3 + 2ak^2 - 9ak - 18a$$
$$= a(k^3 + 2k^2 - 9k - 18) \qquad \text{Factor out the common factor.}$$
$$= a[(k^3 + 2k^2) - (9k + 18)] \qquad \text{Factor by grouping.}$$
$$= a[k^2(k + 2) - 9(k + 2)]$$
$$= a[(k + 2)(k^2 - 9)]$$
$$= a(k + 2)(k - 3)(k + 3) \qquad \text{Factor the difference of squares.}$$

6.5 SOLVING EQUATIONS BY FACTORING

Step 1 Rewrite the equation if necessary so that one side is 0.

Step 2 Factor the polynomial.

Solve.

$$2x^2 + 5x = 3$$
$$2x^2 + 5x - 3 = 0 \qquad \text{Standard form}$$
$$(2x - 1)(x + 3) = 0 \qquad \text{Factor.}$$

(continued)

Concepts	Examples
Step 3 Set each factor equal to 0.	$2x - 1 = 0$ or $x + 3 = 0$ Zero-factor property
Step 4 Solve each equation from Step 3.	$2x = 1$ $x = -3$
	$x = \dfrac{1}{2}$
Step 5 Check each solution.	A check verifies that the solution set is $\left\{-3, \tfrac{1}{2}\right\}$.

Answers to Test Your Word Power

1. B; *Example:* $x^2 - 5x - 14 = (x - 7)(x + 2)$

2. D; *Example:* $x^2 + 5x + xy + 5y = x(x + 5) + y(x + 5) = (x + 5)(x + y)$

3. B; *Examples:* $y^2 - 3y + 2 = 0,\ x^2 - 9 = 0,$ $2m^2 = 6m + 8$

4. C; *Example:* Use the zero-factor property to write $(x + 4)(x - 2) = 0$ as $x + 4 = 0$ or $x - 2 = 0$. Then solve each linear equation to find the solution set $\{-4, 2\}$.

Chapter 6 REVIEW EXERCISES

[6.1] *Factor out the greatest common factor.*

1. $12p^2 - 6p$

2. $21x^2 + 35x$

3. $12q^2b + 8qb^2 - 20q^3b^2$

4. $6r^3t - 30r^2t^2 + 18rt^3$

5. $(x + 3)(4x - 1) - (x + 3)(3x + 2)$

6. $(z + 1)(z - 4) + (z + 1)(2z + 3)$

Factor by grouping.

7. $4m + nq + mn + 4q$

8. $x^2 + 5y + 5x + xy$

9. $2m + 6 - am - 3a$

10. $x^2 + 3x - 3y - xy$

[6.2] *Factor completely.*

11. $3p^2 - p - 4$

12. $6k^2 + 11k - 10$

13. $12r^2 - 5r - 3$

14. $10m^2 + 37m + 30$

15. $10k^2 - 11kh + 3h^2$

16. $9x^2 + 4xy - 2y^2$

17. $24x - 2x^2 - 2x^3$

18. $6b^3 - 9b^2 - 15b$

19. $y^4 + 2y^2 - 8$

20. $2k^4 - 5k^2 - 3$

21. $p^2(p + 2)^2 + p(p + 2)^2 - 6(p + 2)^2$

22. $3(r + 5)^2 - 11(r + 5) - 4$

23. *Concept Check* When asked to factor $x^2y^2 - 6x^2 + 5y^2 - 30$, a student gave the following incorrect answer: $x^2(y^2 - 6) + 5(y^2 - 6)$. **WHAT WENT WRONG?** What is the correct answer?

24. If the area of this rectangle is represented by $4p^2 + 3p - 1$, what is the width in terms of p?

$4p - 1$

[6.3] *Factor completely.*

25. $16x^2 - 25$ **26.** $9t^2 - 49$ **27.** $36m^2 - 25n^2$

28. $x^2 + 14x + 49$ **29.** $9k^2 - 12k + 4$ **30.** $r^3 + 27$

31. $125x^3 - 1$ **32.** $m^6 - 1$ **33.** $x^8 - 1$

34. $x^2 + 6x + 9 - 25y^2$ **35.** $(a + b)^3 - (a - b)^3$ **36.** $x^5 - x^3 - 8x^2 + 8$

[6.5] *Solve each equation.*

37. $x^2 - 8x + 16 = 0$ **38.** $(5x + 2)(x + 1) = 0$

39. $p^2 - 5p + 6 = 0$ **40.** $q^2 + 2q = 8$

41. $6z^2 = 5z + 50$ **42.** $6r^2 + 7r = 3$

43. $8k^2 + 14k + 3 = 0$ **44.** $-4m^2 + 36 = 0$

45. $6x^2 + 9x = 0$ **46.** $(2x + 1)(x - 2) = -3$

47. $(r + 2)(r - 2) = (r - 2)(r + 3) - 2$ **48.** $2x^3 - x^2 - 28x = 0$

49. $-t^3 - 3t^2 + 4t + 12 = 0$ **50.** $(r + 2)(5r^2 - 9r - 18) = 0$

Solve each problem.

51. A triangular wall brace has the shape of a right triangle. One of the perpendicular sides is 1 ft longer than twice the other. The area enclosed by the triangle is 10.5 ft². Find the shorter of the perpendicular sides.

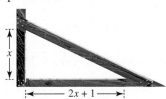

The area is 10.5 ft².

52. A rectangular parking lot has a length 20 ft more than its width. Its area is 2400 ft². What are the dimensions of the lot?

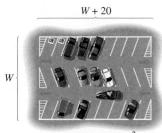

The area is 2400 ft².

A rock is projected directly upward from ground level. After t seconds, its height is given by $f(t) = -16t^2 + 256t$ *(if air resistance is neglected).*

53. When will the rock return to the ground?

54. After how many seconds will it be 240 ft above the ground?

55. Why does the question in Exercise 54 have two answers?

MIXED REVIEW EXERCISES

Factor completely.

56. $30a + am - am^2$ **57.** $16 - 81k^2$ **58.** $8 - a^3$

59. $9x^2 + 13xy - 3y^2$ **60.** $15y^3 + 20y^2$ **61.** $25z^2 - 30zm + 9m^2$

Solve.

62. $5x^2 - 17x = 12$ **63.** $3m^2 - 9m = 0$ **64.** $x^3 - x = 0$

386 **CHAPTER 6** Factoring

65. The length of a rectangular picture frame is 2 in. longer than its width. The area enclosed by the frame is 48 in.^{2}. What is the width?

66. When Europeans arrived in America, many Native Americans of the Northeast lived in *longhouses* that sheltered several related families. The rectangular floor area of a typical Huron longhouse was about 2750 ft^2. The length was 85 ft greater than the width. What were the dimensions of the floor?

Chapter **6** TEST

View the complete solutions to all Chapter Test exercises on the Pass the Test CD.

Factor.

1. $11z^2 - 44z$

2. $10x^2y^5 - 5x^2y^3 - 25x^5y^3$

3. $3x + by + bx + 3y$

4. $-2x^2 - x + 36$

5. $6x^2 + 11x - 35$

6. $4p^2 + 3pq - q^2$

7. $16a^2 + 40ab + 25b^2$

8. $x^2 + 2x + 1 - 4z^2$

9. $a^3 + 2a^2 - ab^2 - 2b^2$

10. $9k^2 - 121j^2$

11. $y^3 - 216$

12. $6k^4 - k^2 - 35$

13. $27x^6 + 1$

14. Explain why $(x^2 + 2y)p + 3(x^2 + 2y)$ is not in factored form. Then factor the polynomial.

15. *Concept Check*   Which one of the following is *not* a factored form of $-x^2 - x + 12$?

A. $(3 - x)(x + 4)$ **B.** $-(x - 3)(x + 4)$

C. $(-x + 3)(x + 4)$ **D.** $(x - 3)(-x + 4)$

Solve each equation.

16. $3x^2 + 8x = -4$

17. $3x^2 - 5x = 0$

18. $5m(m - 1) = 2(1 - m)$

Solve each problem.

19. The area of the rectangle shown is 40 in.^{2}. Find the length and the width of the rectangle.

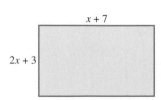

$x + 7$

$2x + 3$

The area is 40 in.^{2}.

20. A ball is projected upward from ground level. After t seconds, its height in feet is a function defined by $f(t) = -16t^2 + 96t$. After how many seconds will it reach a height of 128 ft?

Chapters **1–6** CUMULATIVE REVIEW EXERCISES

Use the properties of real numbers to simplify.

1. $-2(m - 3)$

2. $-(-4m + 3)$

3. $3x^2 - 4x + 4 + 9x - x^2$

Evaluate if $p = -4$, $q = -2$, and $r = 5$.

4. $-3(2q - 3p)$

5. $8r^2 + q^2$

6. $\dfrac{\sqrt{r}}{-p + 2q}$

7. $\dfrac{5p + 6r^2}{p^2 + q - 1}$

Solve.

8. $2z - 5 + 3z = 4 - (z + 2)$

9. $\dfrac{3a - 1}{5} + \dfrac{a + 2}{2} = -\dfrac{3}{10}$

10. $-\dfrac{4}{3}d \geq -5$

11. $3 - 2(m + 3) < 4m$

12. $2k + 4 < 10$ and $3k - 1 > 5$

13. $2k + 4 > 10$ or $3k - 1 < 5$

14. $|5x + 3| - 10 = 3$

15. $|x + 2| < 9$

16. $|2x - 5| \geq 9$

17. $V = lwh$ for h

18. Two planes leave the Dallas–Fort Worth airport at the same time. One travels east at 550 mph, and the other travels west at 500 mph. Assuming no wind, how long will it take for the planes to be 2100 mi apart?

Plane	r	t	d
Eastbound	550	x	
Westbound	500	x	

← Total

19. Graph $4x + 2y = -8$.

20. Find the slope of the line passing through the points $(-4, 8)$ and $(-2, 6)$.

21. What is the slope of the line shown here?

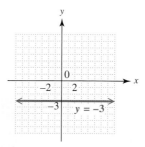

Use the function defined by $f(x) = 2x + 7$ to find each of the following.

22. $f(-4)$

23. The x-intercept of its graph

24. The y-intercept of its graph

Solve each system.

25. $3x - 2y = -7$
$2x + 3y = 17$

26. $2x + 3y - 6z = 5$
$8x - y + 3z = 7$
$3x + 4y - 3z = 7$

Perform the indicated operations. In Exercises 27 and 28, assume that variables represent nonzero real numbers.

27. $(3x^2y^{-1})^{-2}(2x^{-3}y)^{-1}$

28. $\dfrac{5m^{-2}y^3}{3m^{-3}y^{-1}}$

29. $(3x^3 + 4x^2 - 7) - (2x^3 - 8x^2 + 3x)$

30. $(7x + 3y)^2$

31. $(2p + 3)(5p^2 - 4p - 8)$

Factor.

32. $16w^2 + 50wz - 21z^2$

33. $4x^2 - 4x + 1 - y^2$

34. $4y^2 - 36y + 81$

35. $100x^4 - 81$

36. $8p^3 + 27$

Solve.

37. $(p - 1)(2p + 3)(p + 4) = 0$

38. $9q^2 = 6q - 1$

39. A sign is to have the shape of a triangle with a height 3 ft greater than the length of the base. How long should the base be if the area is to be 14 ft²?

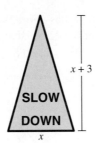

40. A game board has the shape of a rectangle. The longer sides are each 2 in. longer than the distance between them. The area of the board is 288 in.². Find the length of the longer sides and the distance between them.

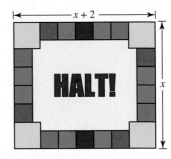

Rational Expressions and Functions

Americans have been car crazy ever since the first automobiles hit the road early in the 20th century. Today there are about 213.5 million vehicles in the United States driving on 3.4 million miles of paved roadways. There is even a museum devoted exclusively to our four-wheeled passion and its influence on our lives and culture. The Museum of Automobile History in Syracuse, New York, features some 200 years of automobile memorabilia, including rare advertising pieces, designer drawings, and Hollywood movie posters. (*Source: Home and Away,* May/June 2002.)

In Exercise 85, Section 7.2, we use a *rational expression* to determine the cost of restoring a vintage automobile.

7.1 Rational Expressions and Functions; Multiplying and Dividing

OBJECTIVE 1 Define rational expressions. In arithmetic, a rational number is the quotient of two integers, with the denominator not 0. In algebra, a **rational expression,** or *algebraic fraction,* is the quotient of two polynomials, again with the denominator not 0. For example,

$$\frac{x}{y}, \quad \frac{-a}{4}, \quad \frac{m+4}{m-2}, \quad \frac{8x^2 - 2x + 5}{4x^2 + 5x}, \quad \text{and} \quad x^5 \left(\text{or } \frac{x^5}{1}\right) \qquad \begin{array}{l}\text{Rational} \\ \text{expressions}\end{array}$$

are all rational expressions. Rational expressions are elements of the set

$$\left\{ \frac{P}{Q} \,\middle|\, P \text{ and } Q \text{ are polynomials, with } Q \neq 0 \right\}.$$

OBJECTIVE 2 Define rational functions and describe their domains. A function that is defined by a quotient of polynomials is called a **rational function** and has the form

$$f(x) = \frac{P(x)}{Q(x)}, \quad \text{where } Q(x) \neq 0.$$

The domain of a rational function consists of all real numbers except those that make $Q(x)$—that is, the denominator—equal to 0. For example, the domain of

$$f(x) = \frac{2}{\underbrace{x-5}_{\text{Cannot equal 0}}}$$

includes all real numbers except 5, because 5 would make the denominator equal to 0.

Figure 1 shows a graph of the function defined by $f(x) = \frac{2}{x-5}$. Notice that the graph does not exist when $x = 5$. It does not intersect the dashed vertical line whose equation is $x = 5$. This line is an *asymptote.* We discuss graphs of rational functions in more detail in **Section 7.4.**

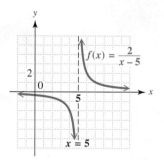

FIGURE 1

EXAMPLE 1 **Finding Numbers That Are Not in the Domains of Rational Functions**

For each rational function, find all numbers that are not in the domain. Then give the domain in set notation.

(a) $f(x) = \dfrac{3}{7x - 14}$

The only values that cannot be used are those that make the denominator 0. To find these values, set the denominator equal to 0 and solve the resulting equation.

$$\begin{aligned} 7x - 14 &= 0 \\ 7x &= 14 \qquad \text{Add 14.} \\ x &= 2 \qquad \text{Divide by 7.} \end{aligned}$$

The number 2 cannot be used as a replacement for x; the domain of f includes all real numbers except 2. This is written $\{x \mid x \neq 2\}$.

(b) $g(x) = \dfrac{3 + x}{x^2 - 4x + 3}$

$$x^2 - 4x + 3 = 0 \qquad \text{Set the denominator equal to 0.}$$

Values that make the denominator 0 must be *excluded*.

$$(x - 1)(x - 3) = 0 \qquad \text{Factor.}$$
$$x - 1 = 0 \quad \text{or} \quad x - 3 = 0 \qquad \text{Zero-factor property}$$
$$x = 1 \quad \text{or} \qquad x = 3 \qquad \text{Solve each equation.}$$

The domain of g includes all real numbers except 1 and 3, written $\{x \mid x \neq 1, 3\}$.

(c) $h(x) = \dfrac{8x + 2}{3}$

The denominator, 3, can never be 0, so the domain of h includes all real numbers, written $(-\infty, \infty)$.

(d) $f(x) = \dfrac{2}{x^2 + 4}$

Setting $x^2 + 4$ equal to 0 leads to $x^2 = -4$. There is no real number whose square is -4. Therefore, any real number can be used, and as in part (c), the domain of f includes all real numbers $(-\infty, \infty)$.

✔ **Now Try Exercises 11, 15, 17, and 19.**

OBJECTIVE 3 Write rational expressions in lowest terms. In arithmetic, we write the fraction $\frac{15}{20}$ in lowest terms by dividing the numerator and denominator by 5 to get $\frac{3}{4}$. We write rational expressions in lowest terms in a similar way, using the **fundamental property of rational numbers.**

Fundamental Property of Rational Numbers

If $\frac{a}{b}$ is a rational number and if c is any nonzero real number, then

$$\frac{a}{b} = \frac{ac}{bc}.$$

That is, the numerator and denominator of a rational number may either be multiplied or divided by the same nonzero number without changing the value of the rational number.

Because $\frac{c}{c}$ is equivalent to 1, the fundamental property is based on the identity property of multiplication.

A rational expression is a quotient of two polynomials. Since the value of a polynomial is a real number for every value of the variable for which it is defined, any statement that applies to rational numbers will also apply to rational expressions.

We use the following steps to write rational expressions in lowest terms.

Writing a Rational Expression in Lowest Terms

Step 1 **Factor** both numerator and denominator to find their greatest common factor (GCF).

Step 2 **Apply the fundamental property.**

EXAMPLE 2 Writing Rational Expressions in Lowest Terms

Write each rational expression in lowest terms.

(a) $\dfrac{(x+5)(x+2)}{(x+2)(x-3)}$

$= \dfrac{(x+5)(x+2)}{(x-3)(x+2)}$ Commutative property

$= \dfrac{x+5}{x-3} \cdot 1$ Fundamental property

$= \dfrac{x+5}{x-3}$ Lowest terms

(b) $\dfrac{a^2 - a - 6}{a^2 + 5a + 6}$

$= \dfrac{(a-3)(a+2)}{(a+3)(a+2)}$ Factor the numerator and the denominator.

$= \dfrac{a-3}{a+3} \cdot 1$ Fundamental property

$= \dfrac{a-3}{a+3}$ Lowest terms

(c) $\dfrac{y^2 - 4}{2y + 4}$

$= \dfrac{(y+2)(y-2)}{2(y+2)}$ Factor; the numerator is a difference of squares.

$= \dfrac{y-2}{2}$ Lowest terms

(d) $\dfrac{x^3 - 27}{x - 3}$

$= \dfrac{(x-3)(x^2 + 3x + 9)}{x - 3}$ Factor the difference of cubes.

$= x^2 + 3x + 9$ Lowest terms

(e) $\dfrac{pr + qr + ps + qs}{pr + qr - ps - qs}$

$= \dfrac{(pr + qr) + (ps + qs)}{(pr + qr) - (ps + qs)}$ Group terms.

$= \dfrac{r(p + q) + s(p + q)}{r(p + q) - s(p + q)}$ Factor within groups.

$= \dfrac{(p + q)(r + s)}{(p + q)(r - s)}$ Factor by grouping.

$= \dfrac{r + s}{r - s}$ Lowest terms

(f) $\dfrac{8 + k}{16}$ Be careful. The numerator cannot be factored.

This expression cannot be simplified further and is in lowest terms.

✔ **Now Try Exercises 27, 31, 35, 39, 43, and 47.**

▶ **CAUTION** Be careful! When you use the fundamental property of rational numbers, *only common factors may be divided.* For example,

$$\frac{y - 2}{2} \neq y \quad \text{and} \quad \frac{y - 2}{2} \neq y - 1$$

because the 2 in $y - 2$ is not a *factor* of the numerator. ***Remember to factor before writing a fraction in lowest terms.*** To see this, replace y with a number and evaluate the fraction. For example, if $y = 5$, then

$$\frac{y - 2}{2} = \frac{5 - 2}{2} = \frac{3}{2}, \quad \text{which does not equal } y \text{ or } y - 1.$$

In the rational expression from Example 2(c), namely,

$$\frac{a^2 - a - 6}{a^2 + 5a + 6}, \quad \text{or} \quad \frac{(a - 3)(a + 2)}{(a + 3)(a + 2)},$$

a can take any value except -3 or -2, since these values make the denominator 0. In the simplified rational expression

$$\frac{a - 3}{a + 3},$$

a cannot equal -3. Because of this,

$$\frac{a^2 - a - 6}{a^2 + 5a + 6} = \frac{a - 3}{a + 3}$$

for all values of a except -3 or -2. From now on, such statements of equality will be made with the understanding that they apply only to those real numbers which make neither denominator equal 0. We will no longer state such restrictions.

EXAMPLE 3 Writing Rational Expressions in Lowest Terms

Write each rational expression in lowest terms.

(a) $\dfrac{m - 3}{3 - m}$

Here, the numerator and denominator are opposites. The given expression can be written in lowest terms by writing the denominator as $-1(m - 3)$, giving

$$\frac{m - 3}{3 - m} = \frac{m - 3}{-1(m - 3)} = \frac{1}{-1} = -1.$$

The numerator could have been rewritten instead to get the same result.

(b) $\dfrac{r^2 - 16}{4 - r}$

$= \dfrac{(r + 4)(r - 4)}{4 - r}$ Factor the difference of squares.

$= \dfrac{(r + 4)(r - 4)}{-1(r - 4)}$ Write $4 - r$ as $-1(r - 4)$.

$= \dfrac{r + 4}{-1}$ Fundamental property

$= -(r + 4)$, or $-r - 4$ Lowest terms

✔ **Now Try Exercises 49 and 51.**

As shown in Examples 3(a) and (b), the quotient $\frac{a}{-a}\,(a \neq 0)$ can be simplified as

$$\frac{a}{-a} = \frac{a}{-1(a)} = \frac{1}{-1} = -1.$$

The following statement summarizes this result.

Quotient of Opposites

In general, if the numerator and the denominator of a rational expression are opposites, then the expression equals -1.

Based on this result, the following are true:

$$\frac{q - 7}{7 - q} = -1 \quad \text{and} \quad \frac{-5a + 2b}{5a - 2b} = -1.$$

Numerator and denominator in each expression are opposites.

However, the expression

$$\frac{r - 2}{r + 2} \quad \text{Numerator and denominator are } not \text{ opposites.}$$

cannot be simplified further.

OBJECTIVE 4 **Multiply rational expressions.** To multiply rational expressions, follow these steps. (In practice, we usually simplify before multiplying.)

Multiplying Rational Expressions

Step 1 **Factor** all numerators and denominators as completely as possible.

Step 2 **Apply the fundamental property.**

Step 3 **Multiply** remaining factors in the numerator and remaining factors in the denominator. Leave the denominator in factored form.

Step 4 **Check** to be sure that the product is in lowest terms.

EXAMPLE 4   **Multiplying Rational Expressions**

Multiply.

(a) $\dfrac{5p-5}{p} \cdot \dfrac{3p^2}{10p-10}$

$\qquad = \dfrac{5(p-1)}{p} \cdot \dfrac{3p \cdot p}{2 \cdot 5(p-1)}$ Factor.

$\qquad = \dfrac{1}{1} \cdot \dfrac{3p}{2}$ Lowest terms

$\qquad = \dfrac{3p}{2}$ Multiply.

(b) $\dfrac{k^2+2k-15}{k^2-4k+3} \cdot \dfrac{k^2-k}{k^2+k-20}$

$\qquad = \dfrac{(k+5)(k-3)}{(k-3)(k-1)} \cdot \dfrac{k(k-1)}{(k+5)(k-4)}$ Factor.

$\qquad = \dfrac{k}{k-4}$ Fundamental property

(c) $(p-4) \cdot \dfrac{3}{5p-20}$

$\qquad = \dfrac{p-4}{1} \cdot \dfrac{3}{5p-20}$ Write $p-4$ as $\frac{p-4}{1}$.

$\qquad = \dfrac{p-4}{1} \cdot \dfrac{3}{5(p-4)}$ Factor.

$\qquad = \dfrac{3}{5}$ Fundamental property

(d) $\dfrac{x^2+2x}{x+1} \cdot \dfrac{x^2-1}{x^3+x^2}$

$\qquad = \dfrac{x(x+2)}{x+1} \cdot \dfrac{(x+1)(x-1)}{x^2(x+1)}$ Factor.

$\qquad = \dfrac{(x+2)(x-1)}{x(x+1)}$ Multiply; lowest terms

(e) $\dfrac{x-6}{x^2-12x+36} \cdot \dfrac{x^2-3x-18}{x^2+7x+12}$

$\qquad = \dfrac{x-6}{(x-6)^2} \cdot \dfrac{(x+3)(x-6)}{(x+3)(x+4)}$ Factor.

$\qquad = \dfrac{1}{x+4}$ Lowest terms

Remember to include
1 in the numerator
when all other factors
are eliminated.

✔ **Now Try Exercises 71, 73, 77, and 83.**

Rational Expression	Reciprocal
3, or $\dfrac{3}{1}$	$\dfrac{1}{3}$
$\dfrac{5}{k}$	$\dfrac{k}{5}$
$\dfrac{m^2 - 9m}{2}$	$\dfrac{2}{m^2 - 9m}$
$\dfrac{0}{4}$	undefined

Reciprocals have a product of 1.

OBJECTIVE 5 **Find reciprocals of rational expressions.** The rational numbers $\frac{a}{b}$ and $\frac{c}{d}$ are reciprocals of each other if they have a product of 1. The **reciprocal** of a rational expression is defined in the same way: Two rational expressions are reciprocals of each other if they have a product of 1. *Recall that 0 has no reciprocal.* The table shows several rational expressions and their reciprocals.

The examples in the table suggest the following procedure.

Finding the Reciprocal

To find the reciprocal of a nonzero rational expression, invert the rational expression.

OBJECTIVE 6 **Divide rational expressions.** Dividing rational expressions is like dividing rational numbers.

Dividing Rational Expressions

To divide two rational expressions, *multiply* the first (the *dividend*) by the reciprocal of the second (the *divisor*).

EXAMPLE 5 Dividing Rational Expressions

Divide.

(a) $\dfrac{2z}{9} \div \dfrac{5z^2}{18}$

$= \dfrac{2z}{9} \cdot \dfrac{18}{5z^2}$ Multiply by the reciprocal of the divisor.

$= \dfrac{2z}{9} \cdot \dfrac{2 \cdot 9}{5z \cdot z}$ Factor.

$= \dfrac{4}{5z}$ Multiply; lowest terms

(b) $\dfrac{m^2pq^3}{mp^4} \div \dfrac{m^5p^2q}{mpq^2}$

$= \dfrac{m^2pq^3}{mp^4} \cdot \dfrac{mpq^2}{m^5p^2q}$ Multiply by the reciprocal.

$= \dfrac{m^3p^2q^5}{m^6p^6q}$ Properties of exponents

$= \dfrac{q^4}{m^3p^4}$ Properties of exponents

(c) $\dfrac{8k - 16}{3k} \div \dfrac{3k - 6}{4k^2}$

$= \dfrac{8k - 16}{3k} \cdot \dfrac{4k^2}{3k - 6}$ Multiply by the reciprocal.

$= \dfrac{8(k - 2)}{3k} \cdot \dfrac{4k \cdot k}{3(k - 2)}$ Factor.

$= \dfrac{32k}{9}$ Multiply; lowest terms

(d) $\dfrac{5m^2 + 17m - 12}{3m^2 + 7m - 20} \div \dfrac{5m^2 + 2m - 3}{15m^2 - 34m + 15}$

$= \dfrac{5m^2 + 17m - 12}{3m^2 + 7m - 20} \cdot \dfrac{15m^2 - 34m + 15}{5m^2 + 2m - 3}$ Definition of division

$= \dfrac{(5m - 3)(m + 4)}{(m + 4)(3m - 5)} \cdot \dfrac{(3m - 5)(5m - 3)}{(5m - 3)(m + 1)}$ Factor.

$= \dfrac{5m - 3}{m + 1}$ Lowest terms

✔ **Now Try Exercises 63, 79, and 91.**

7.1 EXERCISES

Concept Check *Rational expressions often can be written in lowest terms in seemingly different ways. For example,*

$$\frac{y - 3}{-5} \quad and \quad \frac{-y + 3}{5}$$

look different, but we get the second quotient by multiplying the first by -1 *in both the numerator and denominator. To practice recognizing equivalent rational expressions, match the expressions in Exercises 1–6 with their equivalents in choices A–F.*

1. $\dfrac{x - 3}{x + 4}$ **2.** $\dfrac{x + 3}{x - 4}$ **3.** $\dfrac{x - 3}{x - 4}$ **4.** $\dfrac{x + 3}{x + 4}$ **5.** $\dfrac{3 - x}{x + 4}$ **6.** $\dfrac{x + 3}{4 - x}$

A. $\dfrac{-x - 3}{4 - x}$ **B.** $\dfrac{-x - 3}{-x - 4}$ **C.** $\dfrac{3 - x}{-x - 4}$ **D.** $\dfrac{-x + 3}{-x + 4}$ **E.** $\dfrac{x - 3}{-x - 4}$ **F.** $\dfrac{-x - 3}{x - 4}$

▨ **7.** In Example 1(a), we showed that the domain of the rational function defined by $f(x) = \dfrac{3}{7x - 14}$ does not include 2. Explain in your own words why this is so. In general, how do we find the value or values excluded from the domain of a rational function?

▨ **8.** The domain of the rational function defined by $g(x) = \dfrac{x + 1}{x^2 + 3}$ includes all real numbers. Explain.

For each function, find all numbers that are not in the domain. Then give the domain in set notation. See Example 1.

9. $f(x) = \dfrac{x}{x - 7}$

10. $f(x) = \dfrac{x}{x + 3}$

⊙ **11.** $f(x) = \dfrac{6x - 5}{7x + 1}$

12. $f(x) = \dfrac{8x - 3}{2x + 7}$

13. $f(x) = \dfrac{12x + 3}{x}$

14. $f(x) = \dfrac{9x + 8}{x}$

15. $f(x) = \dfrac{3x + 1}{2x^2 + x - 6}$

16. $f(x) = \dfrac{2x + 4}{3x^2 + 11x - 42}$

17. $f(x) = \dfrac{x + 2}{14}$

18. $f(x) = \dfrac{x - 9}{26}$

19. $f(x) = \dfrac{2x^2 - 3x + 4}{3x^2 + 8}$

20. $f(x) = \dfrac{9x^2 - 8x + 3}{4x^2 + 1}$

21. (a) *Concept Check*   Identify the two *terms* in the numerator and the two *terms* in the denominator of the rational expression $\dfrac{x^2 + 4x}{x + 4}$.

(b) Describe the steps you would use to write this rational expression in lowest terms. (*Hint:* It simplifies to x.)

22. *Concept Check*   Which one of the following rational expressions can be simplified?

A. $\dfrac{x^2 + 2}{x^2}$ **B.** $\dfrac{x^2 + 2}{2}$ **C.** $\dfrac{x^2 + y^2}{y^2}$ **D.** $\dfrac{x^2 - 5x}{x}$

23. *Concept Check*   Which one of the following rational expressions is *not* equivalent to $\dfrac{x - 3}{4 - x}$?

A. $\dfrac{3 - x}{x - 4}$ **B.** $\dfrac{x + 3}{4 + x}$ **C.** $-\dfrac{3 - x}{4 - x}$ **D.** $-\dfrac{x - 3}{x - 4}$

24. *Concept Check*   Which two of the following rational expressions equal -1?

A. $\dfrac{2x + 3}{2x - 3}$ **B.** $\dfrac{2x - 3}{3 - 2x}$ **C.** $\dfrac{2x + 3}{3 + 2x}$ **D.** $\dfrac{2x + 3}{-2x - 3}$

Write each rational expression in lowest terms. See Example 2.

25. $\dfrac{x^2(x + 1)}{x(x + 1)}$

26. $\dfrac{y^3(y - 4)}{y^2(y - 4)}$

⊙ **27.** $\dfrac{(x + 4)(x - 3)}{(x + 5)(x + 4)}$

28. $\dfrac{(2x + 7)(x - 1)}{(2x + 3)(2x + 7)}$

29. $\dfrac{4x(x + 3)}{8x^2(x - 3)}$

30. $\dfrac{5y^2(y + 8)}{15y(y - 8)}$

31. $\dfrac{3x + 7}{3}$

32. $\dfrac{4x - 9}{4}$

33. $\dfrac{6m + 18}{7m + 21}$

34. $\dfrac{5r - 20}{3r - 12}$

35. $\dfrac{t^2 - 9}{3t + 9}$

36. $\dfrac{m^2 - 25}{4m - 20}$

37. $\dfrac{2t + 6}{t^2 - 9}$

38. $\dfrac{5s - 25}{s^2 - 25}$

39. $\dfrac{x^2 + 2x - 15}{x^2 + 6x + 5}$

40. $\dfrac{y^2 - 5y - 14}{y^2 + y - 2}$

41. $\dfrac{8x^2 - 10x - 3}{8x^2 - 6x - 9}$

42. $\dfrac{12x^2 - 4x - 5}{8x^2 - 6x - 5}$

43. $\dfrac{a^3 + b^3}{a + b}$

44. $\dfrac{r^3 - s^3}{r - s}$

45. $\dfrac{2c^2 + 2cd - 60d^2}{2c^2 - 12cd + 10d^2}$

46. $\dfrac{3s^2 - 9st - 54t^2}{3s^2 - 6st - 72t^2}$

47. $\dfrac{ac - ad + bc - bd}{ac - ad - bc + bd}$

48. $\dfrac{2xy + 2xw + y + w}{2xy + y - 2xw - w}$

Write each rational expression in lowest terms. See Example 3.

49. $\dfrac{7 - b}{b - 7}$

50. $\dfrac{r - 13}{13 - r}$

51. $\dfrac{x^2 - y^2}{y - x}$

52. $\dfrac{m^2 - n^2}{n - m}$

53. $\dfrac{(a - 3)(x + y)}{(3 - a)(x - y)}$

54. $\dfrac{(8 - p)(x + 2)}{(p - 8)(x - 2)}$

55. $\dfrac{5k - 10}{20 - 10k}$

56. $\dfrac{7x - 21}{63 - 21x}$

57. $\dfrac{a^2 - b^2}{a^2 + b^2}$

58. $\dfrac{p^2 + q^2}{p^2 - q^2}$

59. Explain in a few words how to multiply rational expressions. Give an example.

60. Explain in a few words how to divide rational expressions. Give an example.

Multiply or divide as indicated. See Examples 4 and 5.

61. $\dfrac{x^3}{3y} \cdot \dfrac{9y^2}{x^5}$

62. $\dfrac{a^4}{5b^2} \cdot \dfrac{25b^4}{a^3}$

63. $\dfrac{5a^4b^2}{16a^2b} \div \dfrac{25a^2b}{60a^3b^2}$

64. $\dfrac{s^3t^2}{10s^2t^4} \div \dfrac{8s^4t^2}{5t^6}$

65. $\dfrac{(-3mn)^2 \cdot 64(m^2n)^3}{16m^2n^4(mn^2)^3} \div \dfrac{24(m^2n^2)^4}{(3m^2n^3)^2}$

66. $\dfrac{(-4a^2b^3)^2 \cdot 9(a^2b^4)^2}{(2a^2b^3)^4 \cdot (3a^3b)^2} \div \dfrac{(ab)^4}{(a^2b^3)^2}$

67. $\dfrac{(x + 2)(x + 1)}{(x + 3)(x - 2)} \cdot \dfrac{(x + 3)(x + 4)}{(x + 2)(x + 1)}$

68. $\dfrac{(x + 3)(x - 4)}{(x - 4)(x + 2)} \cdot \dfrac{(x + 5)(x - 6)}{(x + 3)(x - 6)}$

69. $\dfrac{(2x + 3)(x - 4)}{(x + 8)(x - 4)} \div \dfrac{(x - 4)(x + 2)}{(x - 4)(x + 8)}$

70. $\dfrac{(6x + 5)(x - 3)}{(x + 9)(x - 1)} \div \dfrac{(x - 3)(2x + 7)}{(x - 1)(x + 9)}$

71. $\dfrac{4x}{8x + 4} \cdot \dfrac{14x + 7}{6}$

72. $\dfrac{12x - 20}{5x} \cdot \dfrac{6}{9x - 15}$

73. $\dfrac{p^2 - 25}{4p} \cdot \dfrac{2}{5 - p}$

74. $\dfrac{a^2 - 1}{4a} \cdot \dfrac{2}{1 - a}$

75. $(7k + 7) \div \dfrac{4k + 4}{5}$

76. $(8y - 16) \div \dfrac{3y - 6}{10}$

77. $(z^2 - 1) \cdot \dfrac{1}{1 - z}$

78. $(y^2 - 4) \div \dfrac{2 - y}{8y}$

79. $\dfrac{4x - 20}{5x} \div \dfrac{2x - 10}{7x^3}$

80. $\dfrac{k^2 - 4}{3k^2} \div \dfrac{2 - k}{11k}$

81. $\dfrac{12x - 10y}{3x + 2y} \cdot \dfrac{6x + 4y}{10y - 12x}$

82. $\dfrac{9s - 12t}{2s + 2t} \cdot \dfrac{3s + 3t}{4t - 3s}$

83. $\dfrac{x^2 - 25}{x^2 + x - 20} \cdot \dfrac{x^2 + 7x + 12}{x^2 - 2x - 15}$

84. $\dfrac{t^2 - 49}{t^2 + 4t - 21} \cdot \dfrac{t^2 + 8t + 15}{t^2 - 2t - 35}$

85. $\dfrac{a^3 - b^3}{a^2 - b^2} \div \dfrac{2a - 2b}{2a + 2b}$

86. $\dfrac{x^3 + y^3}{2x + 2y} \div \dfrac{x^2 - y^2}{2x - 2y}$

87. $\dfrac{8x^3 - 27}{2x^2 - 18} \cdot \dfrac{2x + 6}{8x^2 + 12x + 18}$

88. $\dfrac{64x^3 + 1}{4x^2 - 100} \cdot \dfrac{4x + 20}{64x^2 - 16x + 4}$

89. $\dfrac{a^3 - 8b^3}{a^2 - ab - 6b^2} \cdot \dfrac{a^2 + ab - 12b^2}{a^2 + 2ab - 8b^2}$

90. $\dfrac{p^3 - 27q^3}{p^2 + pq - 12q^2} \cdot \dfrac{p^2 - 2pq - 24q^2}{p^2 - 5pq - 6q^2}$

91. $\dfrac{6x^2 + 5x - 6}{12x^2 - 11x + 2} \div \dfrac{4x^2 - 12x + 9}{8x^2 - 14x + 3}$

92. $\dfrac{8a^2 - 6a - 9}{6a^2 - 5a - 6} \div \dfrac{4a^2 + 11a + 6}{9a^2 + 12a + 4}$

The next problems are "brain busters." Multiply or divide as indicated.

93. $\dfrac{3k^2 + 17kp + 10p^2}{6k^2 + 13kp - 5p^2} \div \dfrac{6k^2 + kp - 2p^2}{6k^2 - 5kp + p^2}$

94. $\dfrac{16c^2 + 24cd + 9d^2}{16c^2 - 16cd + 3d^2} \div \dfrac{16c^2 - 9d^2}{16c^2 - 24cd + 9d^2}$

95. $\left(\dfrac{6k^2 - 13k - 5}{k^2 + 7k} \div \dfrac{2k - 5}{k^3 + 6k^2 - 7k} \right) \cdot \dfrac{k^2 - 5k + 6}{3k^2 - 8k - 3}$

96. $\left(\dfrac{2x^3 + 3x^2 - 2x}{3x - 15} \div \dfrac{2x^3 - x^2}{x^2 - 3x - 10} \right) \cdot \dfrac{5x^2 - 10x}{3x^2 + 12x + 12}$

97. $\dfrac{a^2(2a + b) + 6a(2a + b) + 5(2a + b)}{3a^2(a + 2b) - 2a(a + 2b) - (a + 2b)} \div \dfrac{a + 1}{a - 1}$

98. $\dfrac{2x^2(x - 3z) - 5x(x - 3z) + 2(x - 3z)}{4x^2(3z - x) - 11x(3z - x) + 6(3z - x)} \div \dfrac{4x + 1}{4x - 3}$

PREVIEW EXERCISES

*Add or subtract as indicated. See **Section 1.2**.*

99. $\dfrac{4}{7} + \dfrac{1}{3} - \dfrac{1}{2}$

100. $\dfrac{9}{10} - \left(-\dfrac{1}{3} \right)$

101. $-\dfrac{3}{4} + \dfrac{1}{12}$

102. $-\dfrac{2}{3} + \left(-\dfrac{4}{7} \right) - \dfrac{1}{8}$

7.2 Adding and Subtracting Rational Expressions

OBJECTIVE 1 **Add and subtract rational expressions with the same denominator.** The following steps, used to add or subtract rational numbers, are also used to add or subtract rational expressions.

Adding or Subtracting Rational Expressions

Step 1 **If the denominators are the same,** add or subtract the numerators. Place the result over the common denominator.

If the denominators are different, first find the least common denominator. Write all rational expressions with this least common denominator, and then add or subtract the numerators. Place the result over the common denominator.

Step 2 **Simplify.** Write all answers in lowest terms.

EXAMPLE 1 Adding and Subtracting Rational Expressions with the Same Denominators

Add or subtract as indicated.

(a) $\dfrac{3y}{5} + \dfrac{x}{5}$

$= \dfrac{3y + x}{5}$ ← Add the numerators.

← Keep the common denominator.

(b) $\dfrac{7}{2r^2} - \dfrac{11}{2r^2}$

$= \dfrac{7 - 11}{2r^2}$ Subtract the numerators; keep the common denominator.

$= \dfrac{-4}{2r^2}$

$= -\dfrac{2}{r^2}$ Lowest terms

(c) $\dfrac{m}{m^2 - p^2} + \dfrac{p}{m^2 - p^2}$

$= \dfrac{m + p}{m^2 - p^2}$ Add the numerators; keep the common denominator.

$= \dfrac{m + p}{(m + p)(m - p)}$ Factor.

$= \dfrac{1}{m - p}$ Lowest terms

Remember to write 1 in the numerator.

402 CHAPTER 7 Rational Expressions and Functions

(d) $\dfrac{4}{x^2 + 2x - 8} + \dfrac{x}{x^2 + 2x - 8}$

$= \dfrac{4 + x}{x^2 + 2x - 8}$ Add.

$= \dfrac{4 + x}{(x - 2)(x + 4)}$ Factor.

$= \dfrac{1}{x - 2}$ Lowest terms

✔ **Now Try Exercises 3, 7, 11, and 13.**

OBJECTIVE 2 **Find a least common denominator.** We add or subtract rational expressions with different denominators by first writing them with a common denominator, usually the **least common denominator (LCD).**

Finding the Least Common Denominator

Step 1 **Factor** each denominator.

Step 2 **Find the least common denominator.** The LCD is the product of all of the different factors from each denominator, with each factor raised to the *greatest* power that occurs in any denominator.

EXAMPLE 2 Finding Least Common Denominators

Assume that the given expressions are denominators of fractions. Find the LCD for each group.

(a) $5xy^2, \quad 2x^3y$

Each denominator is already factored.

$$5xy^2 = 5 \cdot x \cdot y^2$$
$$2x^3y = 2 \cdot x^3 \cdot y$$

Greatest exponent on *x* is 3.

$$LCD = 5 \cdot 2 \cdot x^3 \cdot y^2 \;\leftarrow\; \text{Greatest exponent on } y \text{ is 2.}$$
$$= 10x^3y^2$$

(b) $k - 3, \quad k$

Each denominator is already factored. The LCD, an expression divisible by *both* $k - 3$ and k, is

Don't forget *k*. —— $k(k - 3)$.

It is usually best to leave a least common denominator in factored form.

(c) $y^2 - 2y - 8, \quad y^2 + 3y + 2$

$$\left. \begin{array}{l} y^2 - 2y - 8 = (y - 4)(y + 2) \\ y^2 + 3y + 2 = (y + 2)(y + 1) \end{array} \right\} \text{Factor.}$$

The LCD, divisible by both polynomials, is $(y - 4)(y + 2)(y + 1)$.

(d) $8z - 24, \quad 5z^2 - 15z$

$$\left.\begin{array}{l} 8z - 24 = 8(z - 3) \\ 5z^2 - 15z = 5z(z - 3) \end{array}\right\} \text{ Factor.}$$

The LCD is $8 \cdot 5z \cdot (z - 3) = 40z(z - 3)$.

(e) $m^2 + 5m + 6, \quad m^2 + 4m + 4, \quad 2m^2 + 4m - 6$

$$\left.\begin{array}{l} m^2 + 5m + 6 = (m + 3)(m + 2) \\ m^2 + 4m + 4 = (m + 2)^2 \\ 2m^2 + 4m - 6 = 2(m + 3)(m - 1) \end{array}\right\} \text{ Factor.}$$

The LCD is $2(m + 3)(m + 2)^2(m - 1)$.

✔ **Now Try Exercises 19, 21, 31, and 35.**

OBJECTIVE 3 **Add and subtract rational expressions with different denominators.** Before adding or subtracting two rational expressions, we write each expression with the least common denominator by multiplying its numerator and denominator by the factors needed to get the LCD. This procedure is valid because we are multiplying each rational expression by a form of 1, the identity element for multiplication.

Consider the sum $\frac{7}{15} + \frac{5}{12}$. The LCD for 15 and 12 is 60. Multiply $\frac{7}{15}$ by $\frac{4}{4}$ (a form of 1) and multiply $\frac{5}{12}$ by $\frac{5}{5}$, so that each fraction has denominator 60, and then add the numerators.

$$\begin{aligned} \frac{7}{15} + \frac{5}{12} &= \frac{7 \cdot 4}{15 \cdot 4} + \frac{5 \cdot 5}{12 \cdot 5} \qquad \text{Fundamental property} \\ &= \frac{28}{60} + \frac{25}{60} \\ &= \frac{28 + 25}{60} \qquad \begin{array}{l}\text{Add the numerators; keep}\\ \text{the common denominator.}\end{array} \\ &= \frac{53}{60} \end{aligned}$$

EXAMPLE 3 **Adding and Subtracting Rational Expressions with Different Denominators**

Add or subtract as indicated.

(a) $\dfrac{5}{2p} + \dfrac{3}{8p}$ \qquad The LCD for $2p$ and $8p$ is $8p$.

$$\begin{aligned} &= \frac{5 \cdot 4}{2p \cdot 4} + \frac{3}{8p} \qquad \text{Fundamental property} \\ &= \frac{20}{8p} + \frac{3}{8p} \\ &= \frac{20 + 3}{8p} \qquad \begin{array}{l}\text{Add the numerators; keep}\\ \text{the common denominator.}\end{array} \\ &= \frac{23}{8p} \end{aligned}$$

(b) $\dfrac{6}{r} - \dfrac{5}{r-3}$ The LCD is $r(r-3)$.

$= \dfrac{6(r-3)}{r(r-3)} - \dfrac{r \cdot 5}{r(r-3)}$ Fundamental property

$= \dfrac{6r-18}{r(r-3)} - \dfrac{5r}{r(r-3)}$ Distributive and commutative properties

$= \dfrac{6r-18-5r}{r(r-3)}$ Subtract numerators.

$= \dfrac{r-18}{r(r-3)}$ Combine terms in the numerator.

✔ **Now Try Exercises 39 and 49.**

▶ **CAUTION**   One of the most common sign errors in algebra occurs when one is subtracting a rational expression with two or more terms in the numerator. Remember that in this situation *the subtraction sign must be distributed to every term in the numerator of the fraction that follows it.* Study Example 4 carefully to see how this is done.

EXAMPLE 4   **Using the Distributive Property When Subtracting Rational Expressions**

Subtract.

(a) $\dfrac{7x}{3x+1} - \dfrac{x-2}{3x+1}$

The denominators are the same for both rational expressions. The subtraction sign must be applied to *both* terms in the numerator of the second rational expression.

$$\dfrac{7x}{3x+1} - \dfrac{x-2}{3x+1}$$

Use parentheses to avoid errors.

$= \dfrac{7x-(x-2)}{3x+1}$ Subtract the numerators; keep the common denominator.

Be careful with signs. $= \dfrac{7x-x+2}{3x+1}$ Distributive property

$= \dfrac{6x+2}{3x+1}$ Combine terms in the numerator.

$= \dfrac{2(3x+1)}{3x+1}$ Factor the numerator.

$= 2$ Lowest terms

(b) $\dfrac{1}{q-1} - \dfrac{1}{q+1}$

$$= \dfrac{1(q+1)}{(q-1)(q+1)} - \dfrac{1(q-1)}{(q+1)(q-1)} \qquad \text{The LCD is } (q-1)(q+1);\ \text{fundamental property}$$

$$= \dfrac{(q+1)-(q-1)}{(q-1)(q+1)} \qquad \text{Subtract.}$$

Be careful with signs. $\qquad = \dfrac{q+1-q+1}{(q-1)(q+1)} \qquad \text{Distributive property}$

$$= \dfrac{2}{(q-1)(q+1)} \qquad \text{Combine terms in the numerator.}$$

✔ **Now Try Exercises 53 and 59.**

In some problems, rational expressions to be added or subtracted have denominators that are opposites of each other, such as

$$\dfrac{y}{y-2} + \dfrac{8}{2-y}. \qquad \text{Denominators are opposites.}$$

The next example illustrates how to proceed in such a problem.

EXAMPLE 5 Adding Rational Expressions with Denominators That Are Opposites

Add.

$$\dfrac{y}{y-2} + \dfrac{8}{2-y}$$

$$= \dfrac{y}{y-2} + \dfrac{8(-1)}{(2-y)(-1)} \qquad \text{Multiply the second expression by } \tfrac{-1}{-1}.$$

$$= \dfrac{y}{y-2} + \dfrac{-8}{y-2} \qquad \text{The LCD is } y-2.$$

$$= \dfrac{y-8}{y-2} \qquad \text{Add the numerators.}$$

We could use $2-y$ as the common denominator and rewrite the first expression.

$$\dfrac{y}{y-2} + \dfrac{8}{2-y}$$

$$= \dfrac{y(-1)}{(y-2)(-1)} + \dfrac{8}{2-y}$$

$$= \dfrac{-y+8}{2-y}, \quad \text{or} \quad \dfrac{8-y}{2-y} \qquad \text{This is an equivalent form of the answer.}$$

✔ **Now Try Exercise 55.**

EXAMPLE 6 Adding and Subtracting Three Rational Expressions

Add and subtract as indicated.

$$\frac{3}{x-2} + \frac{5}{x} - \frac{6}{x^2 - 2x}$$

$$= \frac{3}{x-2} + \frac{5}{x} - \frac{6}{x(x-2)} \qquad \text{Factor the third denominator.}$$

$$= \frac{3x}{x(x-2)} + \frac{5(x-2)}{x(x-2)} - \frac{6}{x(x-2)} \qquad \begin{array}{l} \text{The LCD is } x(x-2); \\ \text{fundamental property} \end{array}$$

$$= \frac{3x + 5(x-2) - 6}{x(x-2)} \qquad \text{Add and subtract the numerators.}$$

$$= \frac{3x + 5x - 10 - 6}{x(x-2)} \qquad \text{Distributive property}$$

$$= \frac{8x - 16}{x(x-2)} \qquad \text{Combine terms in the numerator.}$$

$$= \frac{8(x-2)}{x(x-2)} \qquad \text{Factor the numerator.}$$

$$= \frac{8}{x} \qquad \text{Lowest terms}$$

✔ **Now Try Exercise 61.**

EXAMPLE 7 Subtracting Rational Expressions

Subtract.

$$\frac{m+4}{m^2 - 2m - 3} - \frac{2m - 3}{m^2 - 5m + 6}$$

$$= \frac{m+4}{(m-3)(m+1)} - \frac{2m-3}{(m-3)(m-2)} \qquad \begin{array}{l} \text{Factor each} \\ \text{denominator.} \end{array}$$

$$= \frac{(m+4)(m-2)}{(m-3)(m+1)(m-2)} - \frac{(2m-3)(m+1)}{(m-3)(m-2)(m+1)} \qquad \begin{array}{l} \text{The LCD is } (m-3) \cdot \\ (m+1)\,(m-2). \end{array}$$

$$= \frac{(m+4)(m-2) - (2m-3)(m+1)}{(m-3)(m+1)(m-2)} \qquad \text{Subtract.}$$

$$= \frac{m^2 + 2m - 8 - (2m^2 - m - 3)}{(m-3)(m+1)(m-2)} \qquad \begin{array}{l} \text{Note the} \\ \text{careful use of} \\ \text{parentheses.} \end{array} \qquad \begin{array}{l} \text{Multiply in the} \\ \text{numerator.} \end{array}$$

$$\begin{array}{l} \text{Be careful} \\ \text{with signs.} \end{array} = \frac{m^2 + 2m - 8 - 2m^2 + m + 3}{(m-3)(m+1)(m-2)} \qquad \text{Distributive property}$$

$$= \frac{-m^2 + 3m - 5}{(m-3)(m+1)(m-2)} \qquad \begin{array}{l} \text{Combine terms in} \\ \text{the numerator.} \end{array}$$

If we try to factor the numerator, we find that this rational expression is in lowest terms.

✔ **Now Try Exercise 75.**

EXAMPLE 8 Adding Rational Expressions

Add.

$$\frac{5}{x^2 + 10x + 25} + \frac{2}{x^2 + 7x + 10}$$

$$= \frac{5}{(x + 5)^2} + \frac{2}{(x + 5)(x + 2)} \qquad \text{Factor each denominator.}$$

$$= \frac{5(x + 2)}{(x + 5)^2(x + 2)} + \frac{2(x + 5)}{(x + 5)^2(x + 2)} \qquad \begin{array}{l} \text{The LCD is } (x + 5)^2(x + 2); \\ \text{fundamental property} \end{array}$$

$$= \frac{5(x + 2) + 2(x + 5)}{(x + 5)^2(x + 2)} \qquad \text{Add.}$$

$$= \frac{5x + 10 + 2x + 10}{(x + 5)^2(x + 2)} \qquad \text{Distributive property}$$

$$= \frac{7x + 20}{(x + 5)^2(x + 2)} \qquad \text{Combine terms in the numerator.}$$

✔ **Now Try Exercise 83.**

7.2 EXERCISES

Complete solution available on Video Lectures on CD/DVD

Now Try Exercise

Add or subtract as indicated. Write all answers in lowest terms. See Example 1.

1. $\dfrac{7}{t} + \dfrac{2}{t}$

2. $\dfrac{5}{r} + \dfrac{9}{r}$

3. $\dfrac{6x}{7} + \dfrac{y}{7}$

4. $\dfrac{12t}{5} + \dfrac{s}{5}$

5. $\dfrac{11}{5x} - \dfrac{1}{5x}$

6. $\dfrac{7}{4y} - \dfrac{3}{4y}$

7. $\dfrac{9}{4x^3} - \dfrac{17}{4x^3}$

8. $\dfrac{6}{5y^4} - \dfrac{21}{5y^4}$

9. $\dfrac{5x + 4}{6x + 5} + \dfrac{x + 1}{6x + 5}$

10. $\dfrac{6y + 12}{4y + 3} + \dfrac{2y - 6}{4y + 3}$

11. $\dfrac{x^2}{x + 5} - \dfrac{25}{x + 5}$

12. $\dfrac{y^2}{y + 6} - \dfrac{36}{y + 6}$

13. $\dfrac{-3p + 7}{p^2 + 7p + 12} + \dfrac{8p + 13}{p^2 + 7p + 12}$

14. $\dfrac{5x + 6}{x^2 + x - 20} + \dfrac{4 - 3x}{x^2 + x - 20}$

15. $\dfrac{a^3}{a^2 + ab + b^2} - \dfrac{b^3}{a^2 + ab + b^2}$

16. $\dfrac{p^3}{p^2 - pq + q^2} + \dfrac{q^3}{p^2 - pq + q^2}$

🖉 **17.** Write a step-by-step method for adding or subtracting rational expressions that have a common denominator. Illustrate with an example.

🖉 **18.** Write a step-by-step method for adding or subtracting rational expressions that have different denominators. Give an example.

Assume that the expressions given are denominators of fractions. Find the least common denominator (LCD) for each group. See Example 2.

19. $18x^2y^3, \quad 24x^4y^5$

20. $24a^3b^4, \quad 18a^5b^2$

21. $z - 2, \quad z$

22. $k + 3, \quad k$

23. $2y + 8, \quad y + 4$

24. $3r - 21, \quad r - 7$

25. $x^2 - 81$, $\quad x^2 + 18x + 81$

26. $y^2 - 16$, $\quad y^2 - 8y + 16$

27. $m + n$, $\quad m - n$, $\quad m^2 - n^2$

28. $r + s$, $\quad r - s$, $\quad r^2 - s^2$

29. $x^2 - 3x - 4$, $\quad x + x^2$

30. $y^2 - 8y + 12$, $\quad y^2 - 6y$

31. $2t^2 + 7t - 15$, $\quad t^2 + 3t - 10$

32. $s^2 - 3s - 4$, $\quad 3s^2 + s - 2$

33. $2y + 6$, $\quad y^2 - 9$, $\quad y$

34. $9x + 18$, $\quad x^2 - 4$, $\quad x$

35. $2x - 6$, $\quad x^2 - x - 6$, $\quad x^2 + 4x + 4$

36. $3a - 3b$, $\quad a^2 + ab - 2b^2$, $\quad a^2 - 2ab + b^2$

37. *Concept Check* Consider the following *incorrect* work.

$$\frac{x}{x + 2} - \frac{4x - 1}{x + 2}$$

$$= \frac{x - 4x - 1}{x + 2}$$

$$= \frac{-3x - 1}{x + 2}$$

WHAT WENT WRONG? Work the problem correctly.

38. One student added two rational expressions and obtained the answer $\frac{3}{5 - y}$. Another student obtained the answer $\frac{-3}{y - 5}$ for the same problem. Is it possible that both answers are correct? Explain.

Add or subtract as indicated. Write all answers in lowest terms. See Examples 3–6.

39. $\dfrac{8}{t} + \dfrac{7}{3t}$

40. $\dfrac{5}{x} + \dfrac{9}{4x}$

41. $\dfrac{5}{12x^2y} - \dfrac{11}{6xy}$

42. $\dfrac{7}{18a^3b^2} - \dfrac{2}{9ab}$

43. $\dfrac{4}{15a^4b^5} + \dfrac{3}{20a^2b^6}$

44. $\dfrac{5}{12x^5y^2} + \dfrac{5}{18x^4y^5}$

45. $\dfrac{2r}{7p^3q^4} + \dfrac{3s}{14p^4q}$

46. $\dfrac{4t}{9a^8b^7} + \dfrac{5s}{27a^4b^3}$

47. $\dfrac{1}{a^3b^2} - \dfrac{2}{a^4b} + \dfrac{3}{a^5b^7}$

48. $\dfrac{5}{t^4u^7} - \dfrac{3}{t^5u^9} + \dfrac{6}{t^{10}u}$

49. $\dfrac{1}{x - 1} - \dfrac{1}{x}$

50. $\dfrac{3}{x - 3} - \dfrac{1}{x}$

51. $\dfrac{3a}{a + 1} + \dfrac{2a}{a - 3}$

52. $\dfrac{2x}{x + 4} + \dfrac{3x}{x - 7}$

53. $\dfrac{17y + 3}{9y + 7} - \dfrac{-10y - 18}{9y + 7}$

54. $\dfrac{7x + 8}{3x + 2} - \dfrac{x + 4}{3x + 2}$

55. $\dfrac{2}{4 - x} + \dfrac{5}{x - 4}$

56. $\dfrac{3}{2 - t} + \dfrac{1}{t - 2}$

57. $\dfrac{w}{w - z} - \dfrac{z}{z - w}$

58. $\dfrac{a}{a - b} - \dfrac{b}{b - a}$

59. $\dfrac{1}{x + 1} - \dfrac{1}{x - 1}$

60. $\dfrac{-2}{x - 1} + \dfrac{2}{x + 1}$

61. $\dfrac{4x}{x-1} - \dfrac{2}{x+1} - \dfrac{4}{x^2-1}$

62. $\dfrac{4}{x+3} - \dfrac{x}{x-3} - \dfrac{18}{x^2-9}$

63. $\dfrac{15}{y^2+3y} + \dfrac{2}{y} + \dfrac{5}{y+3}$

64. $\dfrac{7}{t-2} - \dfrac{6}{t^2-2t} - \dfrac{3}{t}$

65. $\dfrac{5}{x-2} + \dfrac{1}{x} + \dfrac{2}{x^2-2x}$

66. $\dfrac{5x}{x-3} + \dfrac{2}{x} + \dfrac{6}{x^2-3x}$

67. $\dfrac{3x}{x+1} + \dfrac{4}{x-1} - \dfrac{6}{x^2-1}$

68. $\dfrac{5x}{x+3} + \dfrac{x+2}{x} - \dfrac{6}{x^2+3x}$

69. $\dfrac{4}{x+1} + \dfrac{1}{x^2-x+1} - \dfrac{12}{x^3+1}$

70. $\dfrac{5}{x+2} + \dfrac{2}{x^2-2x+4} - \dfrac{60}{x^3+8}$

71. $\dfrac{2x+4}{x+3} + \dfrac{3}{x} - \dfrac{6}{x^2+3x}$

72. $\dfrac{4x+1}{x+5} - \dfrac{2}{x} + \dfrac{10}{x^2+5x}$

73. $\dfrac{3}{(p-2)^2} - \dfrac{5}{p-2} + 4$

74. $\dfrac{8}{(3r-1)^2} + \dfrac{2}{3r-1} - 6$

*Add or subtract as indicated. Write all answers in lowest terms. See Examples 7 and 8.**

75. $\dfrac{3}{x^2-5x+6} - \dfrac{2}{x^2-4x+4}$

76. $\dfrac{2}{m^2-4m+4} + \dfrac{3}{m^2+m-6}$

77. $\dfrac{5x}{x^2+xy-2y^2} - \dfrac{3x}{x^2+5xy-6y^2}$

78. $\dfrac{6x}{6x^2+5xy-4y^2} - \dfrac{2y}{9x^2-16y^2}$

79. $\dfrac{5x-y}{x^2+xy-2y^2} - \dfrac{3x+2y}{x^2+5xy-6y^2}$

80. $\dfrac{6x+5y}{6x^2+5xy-4y^2} - \dfrac{x+2y}{9x^2-16y^2}$

81. $\dfrac{r+s}{3r^2+2rs-s^2} - \dfrac{s-r}{6r^2-5rs+s^2}$

82. $\dfrac{3y}{y^2+yz-2z^2} + \dfrac{4y-1}{y^2-z^2}$

83. $\dfrac{3}{x^2+4x+4} + \dfrac{7}{x^2+5x+6}$

84. $\dfrac{5}{x^2+6x+9} - \dfrac{2}{x^2+4x+3}$

Work each problem.

85. A **Concours D'elegance** is a competition in which a maximum of 100 points is awarded to a car on the basis of its general attractiveness. The function defined by the rational expression

$$c(x) = \frac{1010}{49(101-x)} - \frac{10}{49}$$

approximates the cost, in thousands of dollars, of restoring a car so that it will win x points.

(a) Simplify the expression for $c(x)$ by performing the indicated subtraction.

(b) Use the simplified expression to determine how much it would cost to win 95 points.

---

*The authors wish to thank Joyce Nemeth of Broward Community College for her suggestions regarding some of these exercises.

86. A **cost–benefit model** expresses the cost of an undertaking in terms of the benefits received. One cost–benefit model gives the cost in thousands of dollars to remove x percent of a certain pollutant as

$$c(x) = \frac{6.7x}{100 - x}.$$

Another model produces the relationship

$$c(x) = \frac{6.5x}{102 - x}.$$

(a) What is the cost found by averaging the two models? (*Hint:* The average of two quantities is half their sum.)

(b) Using the two given models and your answer to part (a), find the cost to the nearest dollar to remove 95% ($x = 95$) of the pollutant.

(c) Average the two costs in part (b) from the given models. What do you notice about this result compared with the cost obtained by using the average of the two models?

RELATING CONCEPTS (EXERCISES 87–92)

FOR INDIVIDUAL OR GROUP WORK

In Example 6, we showed that

$$\frac{3}{x - 2} + \frac{5}{x} - \frac{6}{x^2 - 2x}$$

is equal to $\frac{8}{x}$. *Algebra is, in a sense, a generalized form of arithmetic.* **Work Exercises 87–92 in order,** *to see how the algebra in this example is related to the arithmetic of common fractions.*

87. Perform the following operations, and express your answer in lowest terms.

$$\frac{3}{7} + \frac{5}{9} - \frac{6}{63}$$

88. Substitute 9 for x in the given problem from Example 6. Compare this problem with the one given in Exercise 87. What do you notice?

89. Now substitute 9 for x in the answer given in Example 6. Do your results agree with the result you obtained in Exercise 87?

90. Replace x in the problem from Example 6 with the number of letters in your last name, assuming that this number is not 2. If your last name has two letters, let $x = 3$. Now predict the answer to your problem. Verify that your prediction is correct.

91. Why will $x = 2$ not work for the problem from Example 6?

92. What other value of x is not allowed in the problem from Example 6?

PREVIEW EXERCISES

*Simplify. See **Sections 1.2 and 1.3.***

93. $\dfrac{\dfrac{5}{9} - \dfrac{1}{3}}{\dfrac{2}{3} + \dfrac{1}{6}}$

94. $\dfrac{\dfrac{7}{8} - \dfrac{3}{2}}{-\dfrac{1}{4} - \dfrac{3}{8}}$

95. $\dfrac{2 - \dfrac{1}{4}}{\dfrac{5}{4} + 3}$

96. $\dfrac{\dfrac{4}{3} - 2}{1 - \dfrac{3}{8}}$

7.3 Complex Fractions

A **complex fraction** is an expression having a fraction in the numerator, denominator, or both. Examples of complex fractions include

$$\frac{1 + \dfrac{1}{x}}{2}, \qquad \frac{\dfrac{4}{y}}{6 - \dfrac{3}{y}}, \qquad \text{and} \qquad \frac{\dfrac{m^2 - 9}{m + 1}}{\dfrac{m + 3}{m^2 - 1}}. \qquad \text{Complex fractions}$$

OBJECTIVE 1 Simplify complex fractions by simplifying the numerator and denominator (Method 1). This is the first of two different methods for simplifying complex fractions.

Simplifying a Complex Fraction: Method 1

Step 1 Simplify the numerator and denominator separately.

Step 2 Divide by multiplying the numerator by the reciprocal of the denominator.

Step 3 Simplify the resulting fraction if possible.

In Step 2, we are treating the complex fraction as a quotient of two rational expressions and dividing. Before performing this step, be sure that both numerator and denominator are single fractions.

EXAMPLE 1 Simplifying Complex Fractions by Method 1

Use Method 1 to simplify each complex fraction.

(a) $\dfrac{\dfrac{x + 1}{x}}{\dfrac{x - 1}{2x}}$
Both the numerator and the denominator are already simplified. (Step 1)

$= \dfrac{x + 1}{x} \div \dfrac{x - 1}{2x}$
Write as a division problem.

$= \dfrac{x + 1}{x} \cdot \dfrac{2x}{x - 1}$
Multiply by the reciprocal of $\frac{x - 1}{2x}$. (Step 2)

$= \dfrac{2x(x + 1)}{x(x - 1)}$
Multiply.

$= \dfrac{2(x + 1)}{x - 1}$
Simplify. (Step 3)

(b) $\dfrac{2 + \dfrac{1}{y}}{3 - \dfrac{2}{y}}$ Simplify the numerator and denominator. (Step 1)

$= \dfrac{\dfrac{2y}{y} + \dfrac{1}{y}}{\dfrac{3y}{y} - \dfrac{2}{y}}$

$= \dfrac{\dfrac{2y + 1}{y}}{\dfrac{3y - 2}{y}}$

$= \dfrac{2y + 1}{y} \div \dfrac{3y - 2}{y}$ Write as a division problem.

$= \dfrac{2y + 1}{y} \cdot \dfrac{y}{3y - 2}$ Multiply by the reciprocal of $\frac{3y - 2}{y}$. (Step 2)

$= \dfrac{2y + 1}{3y - 2}$ Multiply and simplify. (Step 3)

✔ **Now Try Exercises 5 and 9.**

OBJECTIVE 2 **Simplify complex fractions by multiplying by a common denominator (Method 2).** The second method for simplifying complex fractions uses the identity property for multiplication.

Simplifying a Complex Fraction: Method 2

Step 1 Multiply the numerator and denominator of the complex fraction by the least common denominator of the fractions in the numerator and the fractions in the denominator of the complex fraction.

Step 2 Simplify the resulting fraction if possible.

EXAMPLE 2 **Simplifying Complex Fractions by Method 2**

Use Method 2 to simplify each complex fraction.

(a) $\dfrac{2 + \dfrac{1}{y}}{3 - \dfrac{2}{y}}$

$= \dfrac{2 + \dfrac{1}{y}}{3 - \dfrac{2}{y}} \cdot 1$ Identity property of multiplication

$$= \frac{\left(2 + \dfrac{1}{y}\right) \cdot y}{\left(3 - \dfrac{2}{y}\right) \cdot y}$$

The LCD of all the fractions is y;
Multiply the numerator and denominator
by y, since $\frac{y}{y} = 1$.

$$= \frac{2 \cdot y + \dfrac{1}{y} \cdot y}{3 \cdot y - \dfrac{2}{y} \cdot y}$$

Distributive property

$$= \frac{2y + 1}{3y - 2}$$

Multiply.

Compare this method of solution with that used in Example 1(b).

(b) $\dfrac{2p + \dfrac{5}{p - 1}}{3p - \dfrac{2}{p}}$

$$= \frac{\left(2p + \dfrac{5}{p - 1}\right) \cdot p(p - 1)}{\left(3p - \dfrac{2}{p}\right) \cdot p(p - 1)}$$

Multiply the numerator and
denominator by the LCD,
$p(p - 1)$.

$$= \frac{2p[p(p - 1)] + \dfrac{5}{p - 1} \cdot p(p - 1)}{3p[p(p - 1)] - \dfrac{2}{p} \cdot p(p - 1)}$$

Distributive property

$$= \frac{2p[p(p - 1)] + 5p}{3p[p(p - 1)] - 2(p - 1)}$$

Multiply.

$$= \frac{2p^3 - 2p^2 + 5p}{3p^3 - 3p^2 - 2p + 2}$$

Multiply again.

This rational expression is in lowest terms.

✔ **Now Try Exercises 9 (using Method 2) and 23.**

OBJECTIVE 3 **Compare the two methods of simplifying complex fractions.** Choosing whether to use Method 1 or Method 2 to simplify a complex fraction is usually a matter of preference. Some students prefer one method over the other, while other students feel comfortable with both methods and rely on practice with many examples to determine which method they will use on a particular problem. In the next example, we illustrate how to simplify a complex fraction by both methods so that you can observe the processes and decide for yourself the pros and cons of each method.

EXAMPLE 3 Simplifying Complex Fractions by Both Methods

Use both Method 1 and Method 2 to simplify each complex fraction.

| **Method 1** | **Method 2** |

(a) $\dfrac{\dfrac{2}{x-3}}{\dfrac{5}{x^2-9}}$

$= \dfrac{\dfrac{2}{x-3}}{\dfrac{5}{(x-3)(x+3)}}$

$= \dfrac{2}{x-3} \div \dfrac{5}{(x-3)(x+3)}$

$= \dfrac{2}{x-3} \cdot \dfrac{(x-3)(x+3)}{5}$

$= \dfrac{2(x+3)}{5}$

(a) $\dfrac{\dfrac{2}{x-3}}{\dfrac{5}{x^2-9}}$

$= \dfrac{\dfrac{2}{x-3} \cdot (x-3)(x+3)}{\dfrac{5}{(x-3)(x+3)} \cdot (x-3)(x+3)}$

$= \dfrac{2(x+3)}{5}$

(b) $\dfrac{\dfrac{1}{x}+\dfrac{1}{y}}{\dfrac{1}{x^2}-\dfrac{1}{y^2}}$

$= \dfrac{\dfrac{y}{xy}+\dfrac{x}{xy}}{\dfrac{y^2}{x^2y^2}-\dfrac{x^2}{x^2y^2}}$

$= \dfrac{\dfrac{y+x}{xy}}{\dfrac{y^2-x^2}{x^2y^2}}$

$= \dfrac{y+x}{xy} \div \dfrac{y^2-x^2}{x^2y^2}$

$= \dfrac{y+x}{xy} \cdot \dfrac{x^2y^2}{(y-x)(y+x)}$

$= \dfrac{xy}{y-x}$

(b) $\dfrac{\dfrac{1}{x}+\dfrac{1}{y}}{\dfrac{1}{x^2}-\dfrac{1}{y^2}}$

$= \dfrac{\left(\dfrac{1}{x}+\dfrac{1}{y}\right) \cdot x^2y^2}{\left(\dfrac{1}{x^2}-\dfrac{1}{y^2}\right) \cdot x^2y^2}$

$= \dfrac{xy^2+x^2y}{y^2-x^2}$

$= \dfrac{xy(y+x)}{(y+x)(y-x)}$

$= \dfrac{xy}{y-x}$

✔ **Now Try Exercises 17 and 19.**

OBJECTIVE ❹ Simplify rational expressions with negative exponents. Rational expressions and complex fractions sometimes involve negative exponents. To simplify such expressions, we begin by rewriting the expressions with only positive exponents.

EXAMPLE 4 Simplifying Rational Expressions with Negative Exponents

Simplify each expression, using only positive exponents in the answer.

(a)
$$\frac{m^{-1} + p^{-2}}{2m^{-2} - p^{-1}}$$

Be careful! The base is m, not $2m$: $2m^{-2} = \dfrac{2}{m^2}$.

$$= \frac{\dfrac{1}{m} + \dfrac{1}{p^2}}{\dfrac{2}{m^2} - \dfrac{1}{p}}$$ Write with positive exponents.

$$= \frac{m^2 p^2 \left(\dfrac{1}{m} + \dfrac{1}{p^2} \right)}{m^2 p^2 \left(\dfrac{2}{m^2} - \dfrac{1}{p} \right)}$$ Simplify by Method 2, multiplying the numerator and denominator by the LCD, $m^2 p^2$.

$$= \frac{m^2 p^2 \cdot \dfrac{1}{m} + m^2 p^2 \cdot \dfrac{1}{p^2}}{m^2 p^2 \cdot \dfrac{2}{m^2} - m^2 p^2 \cdot \dfrac{1}{p}}$$ Distributive property

$$= \frac{mp^2 + m^2}{2p^2 - m^2 p}$$ Lowest terms

(b) $\dfrac{x^{-2} - 2y^{-1}}{y - 2x^2}$

The 2 does *not* go in the denominator of this fraction.

$$= \frac{\dfrac{1}{x^2} - \dfrac{2}{y}}{y - 2x^2}$$ Write with positive exponents.

$$= \frac{\left(\dfrac{1}{x^2} - \dfrac{2}{y} \right) x^2 y}{(y - 2x^2) x^2 y}$$ Use Method 2; multiply by the LCD, $x^2 y$.

$$= \frac{y - 2x^2}{(y - 2x^2) x^2 y}$$ Use the distributive property in the numerator.

$$= \frac{1}{x^2 y}$$ Lowest terms

Remember to write 1 in the numerator.

✔ **Now Try Exercises 35 and 37.**

7.3 EXERCISES

📝 **1.** Explain in your own words the two methods of simplifying complex fractions.

2. Method 2 of simplifying complex fractions says that we can multiply both the numerator and the denominator of the complex fraction by the same nonzero expression. What property of real numbers from **Section 1.4** justifies this method?

Use either method to simplify each complex fraction. See Examples 1–3.

3. $\dfrac{\dfrac{12}{x-1}}{\dfrac{6}{x}}$

4. $\dfrac{\dfrac{24}{t+4}}{\dfrac{6}{t}}$

● **5.** $\dfrac{\dfrac{k+1}{2k}}{\dfrac{3k-1}{4k}}$

6. $\dfrac{\dfrac{1-r}{4r}}{\dfrac{1+r}{8r}}$

7. $\dfrac{\dfrac{4z^2x^4}{9}}{\dfrac{12x^2z^5}{15}}$

8. $\dfrac{\dfrac{3y^2x^3}{8}}{\dfrac{9y^3x^4}{16}}$

● **9.** $\dfrac{6+\dfrac{1}{x}}{7-\dfrac{3}{x}}$

10. $\dfrac{4-\dfrac{1}{p}}{9+\dfrac{5}{p}}$

11. $\dfrac{\dfrac{3}{x}+\dfrac{3}{y}}{\dfrac{3}{x}-\dfrac{3}{y}}$

12. $\dfrac{\dfrac{4}{t}-\dfrac{4}{s}}{\dfrac{4}{t}+\dfrac{4}{s}}$

13. $\dfrac{\dfrac{8x-24y}{10}}{\dfrac{x-3y}{5x}}$

14. $\dfrac{\dfrac{20x-10y}{12y}}{\dfrac{2x-y}{6y^2}}$

15. $\dfrac{\dfrac{x^2-16y^2}{xy}}{\dfrac{1}{y}-\dfrac{4}{x}}$

16. $\dfrac{\dfrac{4t^2-9s^2}{st}}{\dfrac{2}{s}-\dfrac{3}{t}}$

● **17.** $\dfrac{\dfrac{6}{y-4}}{\dfrac{12}{y^2-16}}$

18. $\dfrac{\dfrac{8}{t+7}}{\dfrac{24}{t^2-49}}$

19. $\dfrac{\dfrac{1}{b^2}-\dfrac{1}{a^2}}{\dfrac{1}{b}-\dfrac{1}{a}}$

20. $\dfrac{\dfrac{1}{x^2}-\dfrac{1}{y^2}}{\dfrac{1}{x}+\dfrac{1}{y}}$

21. $\dfrac{x+y}{\dfrac{1}{y}+\dfrac{1}{x}}$

22. $\dfrac{s-r}{\dfrac{1}{r}-\dfrac{1}{s}}$

23. $\dfrac{y-\dfrac{y-3}{3}}{\dfrac{4}{9}+\dfrac{2}{3y}}$

24. $\dfrac{p-\dfrac{p+2}{4}}{\dfrac{3}{4}-\dfrac{5}{2p}}$

25. $\dfrac{\dfrac{x+2}{x}+\dfrac{1}{x+2}}{\dfrac{5}{x}+\dfrac{x}{x+2}}$

26. $\dfrac{\dfrac{y+3}{y}-\dfrac{4}{y-1}}{\dfrac{y}{y-1}+\dfrac{1}{y}}$

RELATING CONCEPTS   (EXERCISES 27–32)

FOR INDIVIDUAL OR GROUP WORK

Simplifying a complex fraction by Method 1 is a good way to review the methods of adding, subtracting, multiplying, and dividing rational expressions. Method 2 gives a good review of the fundamental principle of rational expressions. Refer to the following complex fraction and **work Exercises 27–32 in order.**

$$\frac{\dfrac{4}{m} + \dfrac{m+2}{m-1}}{\dfrac{m+2}{m} - \dfrac{2}{m-1}}$$

27. Add the fractions in the numerator.

28. Subtract as indicated in the denominator.

29. Divide your answer from Exercise 27 by your answer from Exercise 28.

30. Go back to the original complex fraction and find the LCD of all denominators.

31. Multiply the numerator and denominator of the complex fraction by your answer from Exercise 30.

32. Your answers for Exercises 29 and 31 should be the same. Write a paragraph comparing the two methods. Which method do you prefer? Explain why.

Simplify each expression, using only positive exponents in your answer. See Example 4.

33. $\dfrac{1}{x^{-2} + y^{-2}}$ **34.** $\dfrac{1}{p^{-2} - q^{-2}}$ **35.** $\dfrac{x^{-2} + y^{-2}}{x^{-1} + y^{-1}}$

36. $\dfrac{x^{-1} - y^{-1}}{x^{-2} - y^{-2}}$ **37.** $\dfrac{x^{-1} + 2y^{-1}}{2y + 4x}$ **38.** $\dfrac{a^{-2} - 4b^{-2}}{3b - 6a}$

39. (a) Start with the complex fraction $\dfrac{\dfrac{3}{mp} - \dfrac{4}{p} + \dfrac{8}{m}}{2m^{-1} - 3p^{-1}}$, and write it so that there are no negative exponents in your expression.

     (b) Explain why $\dfrac{\dfrac{3}{mp} - \dfrac{4}{p} + \dfrac{8}{m}}{\dfrac{1}{2m} - \dfrac{1}{3p}}$ would *not* be a correct response in part (a).

     (c) Simplify the complex fraction in part (a).

40. Is $\dfrac{m^{-1} + n^{-1}}{m^{-2} + n^{-2}} = \dfrac{m^2 + n^2}{m + n}$ a true statement? Explain why or why not.

PREVIEW EXERCISES

Solve each equation. See Section 2.1.

41. $\dfrac{1}{2}x + \dfrac{1}{4}x = -9$ **42.** $\dfrac{x}{3} - \dfrac{x}{8} = -5$

43. $\dfrac{x-6}{5} = \dfrac{x+4}{10}$ **44.** $1.5x - 0.75x = x - 2$

*For each rational function, find all numbers that are not in the domain. Then give the domain, using set notation. See **Section 7.1**.*

45. $f(x) = \dfrac{-3x + 2}{x - 6}$

46. $f(x) = \dfrac{1}{x^2 - 16}$

47. $f(x) = \dfrac{1}{x}$

48. $f(x) = \dfrac{6}{x^2 + 4}$

7.4 Equations with Rational Expressions and Graphs

OBJECTIVES

1 Determine the domain of the variable in a rational equation.

2 Solve rational equations.

3 Recognize the graph of a rational function.

4 Solve rational equations with a graphing calculator.

At the beginning of this chapter, we defined the domain of a rational expression as the set of all possible values of the variable. Any value that makes the denominator 0 is excluded. A **rational equation** is an equation that contains at least one rational expression with a variable in the denominator.

OBJECTIVE 1 Determine the domain of the variable in a rational equation. The **domain of the variable in a rational equation** is the intersection (overlap) of the domains of the rational expressions in the equation.

EXAMPLE 1 Determining the Domains of the Variables in Rational Equations

Find the domain of the variables in each equation.

(a) $\dfrac{2}{x} - \dfrac{3}{2} = \dfrac{7}{2x}$

The domains of the three rational expressions of the equation $\frac{2}{x} - \frac{3}{2} = \frac{7}{2x}$ are, in order, $\{x \mid x \neq 0\}$, $(-\infty, \infty)$, and $\{x \mid x \neq 0\}$. The intersection of these three domains is all real numbers except 0, which may be written $\{x \mid x \neq 0\}$.

(b) $\dfrac{2}{x - 3} - \dfrac{3}{x + 3} = \dfrac{12}{x^2 - 9}$

The domains of the three expressions are, respectively, $\{x \mid x \neq 3\}$, $\{x \mid x \neq -3\}$, and $\{x \mid x \neq \pm 3\}$. The domain of the variables in the equation is the intersection of the three domains: all real numbers except 3 and -3, written $\{x \mid x \neq \pm 3\}$.

✔ **Now Try Exercises 1 and 5.**

OBJECTIVE 2 Solve rational equations. The easiest way to solve most rational equations is to multiply all terms in the equation by the least common denominator. This step will clear the equation of all denominators, as the next examples show. *We can do this only with equations, not expressions.*

Because the first step in solving a rational equation is to multiply both sides of the equation by a common denominator, it is *necessary* to either check the proposed solutions or verify that they are in the domain.

▶ **CAUTION** When both sides of an equation are multiplied by a *variable* expression, the resulting proposed solutions may not satisfy the original equation. *You must either determine and observe the domain or check all proposed solutions in the original equation. It is wise to do both.*

EXAMPLE 2 Solving a Rational Equation

Solve $\dfrac{2}{x} - \dfrac{3}{2} = \dfrac{7}{2x}$.

The domain, which excludes 0, was found in Example 1(a).

$$2x\left(\frac{2}{x} - \frac{3}{2}\right) = 2x\left(\frac{7}{2x}\right) \qquad \text{Multiply by the LCD, } 2x.$$

$$2x\left(\frac{2}{x}\right) - 2x\left(\frac{3}{2}\right) = 2x\left(\frac{7}{2x}\right) \qquad \text{Distributive property}$$

$$4 - 3x = 7 \qquad \text{Multiply.}$$

$$-3x = 3 \qquad \text{Subtract 4.}$$

$$\text{Proposed solution} \rightarrow x = -1 \qquad \text{Divide by } -3.$$

Check:
$$\frac{2}{x} - \frac{3}{2} = \frac{7}{2x} \qquad \text{Original equation}$$

$$\frac{2}{-1} - \frac{3}{2} = \frac{7}{2(-1)} \qquad ? \quad \text{Let } x = -1.$$

$$-\frac{7}{2} = -\frac{7}{2} \qquad \text{True}$$

The solution set is $\{-1\}$.

✔ **Now Try Exercise 15.**

EXAMPLE 3 Solving a Rational Equation with No Solution

Solve $\dfrac{2}{x-3} - \dfrac{3}{x+3} = \dfrac{12}{x^2-9}$.

Using the result from Example 1(b), we know that the domain excludes 3 and -3. We multiply each side by the LCD, $(x+3)(x-3)$.

$$(x+3)(x-3)\left(\frac{2}{x-3} - \frac{3}{x+3}\right) = (x+3)(x-3)\left(\frac{12}{x^2-9}\right)$$

$$2(x+3) - 3(x-3) = 12 \qquad \text{Distributive property}$$

$$2x + 6 - 3x + 9 = 12 \qquad \text{Distributive property}$$

$$-x + 15 = 12 \qquad \text{Combine like terms.}$$

$$-x = -3 \qquad \text{Subtract 15.}$$

$$\text{Proposed solution} \rightarrow x = 3 \qquad \text{Divide by } -1.$$

Since the proposed solution, 3, is not in the domain, it cannot be an actual solution of the equation. Substituting 3 into the original equation shows why.

$$\frac{2}{x-3} - \frac{3}{x+3} = \frac{12}{x^2-9} \qquad \text{Original equation}$$

$$\frac{2}{3-3} - \frac{3}{3+3} = \frac{12}{3^2-9} \quad ? \quad \text{Let } x = 3.$$

$$\frac{2}{0} - \frac{3}{6} = \frac{12}{0} \quad ?$$

Division by 0 is undefined. The equation has no solution and the solution set is $\emptyset$.

✔ **Now Try Exercise 29.**

EXAMPLE 4 **Solving a Rational Equation**

Solve $\dfrac{3}{p^2+p-2} - \dfrac{1}{p^2-1} = \dfrac{7}{2(p^2+3p+2)}$.

Factor each denominator to find the LCD, $2(p-1)(p+2)(p+1)$. The domain excludes 1, -2, and -1. Multiply each side by the LCD.

$$2(p-1)(p+2)(p+1)\left[\frac{3}{(p+2)(p-1)} - \frac{1}{(p+1)(p-1)}\right]$$

$$= 2(p-1)(p+2)(p+1)\left[\frac{7}{2(p+2)(p+1)}\right]$$

$$2 \cdot 3(p+1) - 2(p+2) = 7(p-1) \qquad \text{Distributive property}$$

$$6p + 6 - 2p - 4 = 7p - 7 \qquad \text{Distributive property}$$

$$4p + 2 = 7p - 7 \qquad \text{Combine like terms.}$$

$$9 = 3p \qquad \text{Subtract } 4p; \text{ add 7.}$$

Proposed solution $\rightarrow\ 3 = p \qquad \text{Divide by 3.}$

Note that 3 is in the domain. Substitute 3 for p in the original equation to check that the solution set is $\{3\}$.

✔ **Now Try Exercise 39.**

EXAMPLE 5 **Solving a Rational Equation That Leads to a Quadratic Equation**

Solve $\dfrac{2}{3x+1} = \dfrac{1}{x} - \dfrac{6x}{3x+1}$.

Since the denominator $3x + 1$ cannot equal 0, $-\frac{1}{3}$ is excluded from the domain, as is 0.

$$x(3x+1)\left(\frac{2}{3x+1}\right) = x(3x+1)\left(\frac{1}{x} - \frac{6x}{3x+1}\right) \qquad \begin{array}{l}\text{Multiply by the LCD,}\\ x(3x+1).\end{array}$$

$$2x = 3x + 1 - 6x^2 \qquad \text{Distributive property}$$

$$6x^2 - x - 1 = 0 \qquad \text{Standard form}$$
$$(3x + 1)(2x - 1) = 0 \qquad \text{Factor.}$$
$$3x + 1 = 0 \quad \text{or} \quad 2x - 1 = 0 \qquad \text{Zero-factor property}$$
$$x = -\frac{1}{3} \quad \text{or} \quad x = \frac{1}{2} \qquad \text{Proposed solutions}$$

Because $-\frac{1}{3}$ is not in the domain of the equation, it is not a solution. Check that the solution set is $\left\{\frac{1}{2}\right\}$.

✔ **Now Try Exercise 35.**

OBJECTIVE 3 Recognize the graph of a rational function. As mentioned in **Section 7.1,** a function defined by a quotient of polynomials is a *rational function*. Because one or more values of x are excluded from the domain of many rational functions, their graphs are often **discontinuous.** That is, there will be one or more breaks in the graph. For example, consider the graph of the simple rational function defined by

$$f(x) = \frac{1}{x}.$$

The domain of this function includes all real numbers except 0. Thus, there will be no point on the graph with $x = 0$. The vertical line with equation $x = 0$ is called a *vertical asymptote* of the graph. If the y-values of a rational function approach ∞ or $-\infty$ as the x-values approach a real number a, the vertical line $x = a$ is called a **vertical asymptote** of the graph of the function.

We show some typical ordered pairs for $f(x) = \frac{1}{x}$ in the table for both negative and positive x-values.

x	-3	-2	-1	-0.5	-0.25	-0.1	0.1	0.25	0.5	1	2	3
y	$-\frac{1}{3}$	$-\frac{1}{2}$	-1	-2	-4	-10	10	4	2	1	$\frac{1}{2}$	$\frac{1}{3}$

Notice that the closer positive values of x are to 0, the larger y is. Similarly, as negative values of x get closer to 0, y-values are negative and get larger in absolute value. Using this observation and the fact that the domain excludes 0, and plotting some of the points just found, produces the graph in Figure 2.

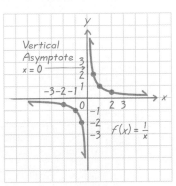

FIGURE 2

The graph of $g(x) = \frac{-2}{x - 3}$ is shown in Figure 3 on the next page. Some ordered pairs that belong to the function are listed in the table.

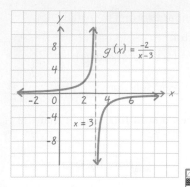

FIGURE 3

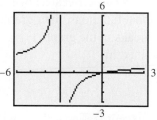

Connected Mode

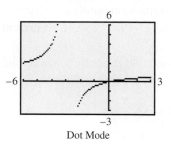

Dot Mode

FIGURE 4

x	−2	−1	0	1	2	2.5	2.75	3.25	3.5	4	5	6	7
y	$\frac{2}{5}$	$\frac{1}{2}$	$\frac{2}{3}$	1	2	4	8	−8	−4	−2	−1	$-\frac{2}{3}$	$-\frac{1}{2}$

There is no point on the graph for $x = 3$, because 3 is excluded from the domain. The dashed line $x = 3$ represents the asymptote and is not part of the graph. As suggested by the points from the table, the graph gets closer to the vertical asymptote $x = 3$ as the x-values get closer to 3.

✔ **Now Try Exercise 45.**

OBJECTIVE 4 **Solve rational equations with a graphing calculator.** Earlier, we solved linear and quadratic equations with a graphing calculator. The procedure is similar with rational equations. Because rational functions often have values of x that are excluded from the domain, a calculator in *connected mode* may show a vertical line on the screen where an asymptote occurs. Using *dot mode* will usually give a more realistic picture.

In Figure 4, we show the graph of

$$g(x) = \frac{x}{x + 3},$$

generated in connected mode and dot mode. If dot mode is used, we must remember that, theoretically, the function is continuous (unbroken) on its domain—in this case, $(-\infty, -3) \cup (-3, \infty)$. As before, the x-intercepts of the graph give the solutions of the equation

$$\frac{x}{x + 3} = 0.$$

From the graph, since the x-intercept is $(0, 0)$, we see that the only solution of this equation is $\{0\}$.

EXAMPLE 6 **Finding Solutions of a Rational Equation from a Graphing Calculator Screen**

Figure 5 shows two views of the graph of

$$f(x) = \frac{1}{x^2} + \frac{1}{x} - \frac{3}{2}.$$

Use the graph to determine the solution set of

$$\frac{1}{x^2} + \frac{1}{x} - \frac{3}{2} = 0.$$

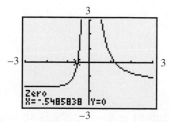 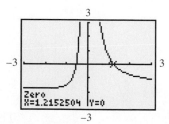

FIGURE 5

Look at the bottom of each calculator screen for the x-values for which $y = 0$: $-.5485838$ and 1.2152504. These are the solutions of the equation. The solution set is $\{-.5485838, 1.2152504\}$.

✔**Now Try Exercise 53.**

7.4 EXERCISES

🔵 *Complete solution available on Video Lectures on CD/DVD*

▨ *Now Try Exercise*

As explained in this section, any values that would cause a denominator to equal 0 must be excluded from the domain and, consequently, as solutions of an equation that has variable expressions in the denominators. **(a)** *Without actually solving the equation, list all possible numbers that would have to be rejected if they appeared as proposed solutions.* **(b)** *Then give the domain, using set notation. See Example 1.*

🔵 **1.** $\dfrac{1}{3x} + \dfrac{1}{2x} = \dfrac{x}{3}$

2. $\dfrac{5}{6x} - \dfrac{8}{2x} = \dfrac{x}{4}$

3. $\dfrac{1}{x + 1} - \dfrac{1}{x - 2} = 0$

4. $\dfrac{3}{x + 4} - \dfrac{2}{x - 9} = 0$

5. $\dfrac{1}{x^2 - 16} - \dfrac{2}{x - 4} = \dfrac{1}{x + 4}$

6. $\dfrac{2}{x^2 - 25} - \dfrac{1}{x + 5} = \dfrac{1}{x - 5}$

7. $\dfrac{2}{x^2 - x} + \dfrac{1}{x + 3} = \dfrac{4}{x - 2}$

8. $\dfrac{3}{x^2 + x} - \dfrac{1}{x + 5} = \dfrac{2}{x - 7}$

9. $\dfrac{6}{4x + 7} - \dfrac{3}{x} = \dfrac{5}{6x - 13}$

10. $\dfrac{4}{3x - 5} + \dfrac{2}{x} = \dfrac{9}{4x + 13}$

11. $\dfrac{3x + 1}{x - 4} = \dfrac{6x + 5}{2x - 7}$

12. $\dfrac{4x - 1}{2x + 3} = \dfrac{12x - 25}{6x - 2}$

▨ **13.** Suppose that in solving $\dfrac{x + 7}{4} - \dfrac{x + 3}{3} = \dfrac{x}{12}$, all of your algebraic steps are correct. Is there a possibility that your proposed solution will have to be rejected? Explain.

14. Consider the equation in Exercise 13.

(a) Solve it. (b) Check your solution, showing all steps.

Solve each equation. See Examples 2–5.

🔵 **15.** $\dfrac{3}{4x} = \dfrac{5}{2x} - \dfrac{7}{4}$

16. $\dfrac{6}{5x} + \dfrac{8}{45} = \dfrac{2}{3x}$

17. $x - \dfrac{24}{x} = -2$

18. $p + \dfrac{15}{p} = -8$

19. $\dfrac{x - 4}{x + 6} = \dfrac{2x + 3}{2x - 1}$

20. $\dfrac{5x - 8}{x + 2} = \dfrac{5x - 1}{x + 3}$

21. $\dfrac{3x + 1}{x - 4} = \dfrac{6x + 5}{2x - 7}$

22. $\dfrac{4x - 1}{2x + 3} = \dfrac{12x - 25}{6x - 2}$

23. $\dfrac{1}{y - 1} + \dfrac{5}{12} = \dfrac{-2}{3y - 3}$

24. $\dfrac{4}{m + 2} - \dfrac{11}{9} = \dfrac{1}{3m + 6}$

25. $\dfrac{7}{6x + 3} - \dfrac{1}{3} = \dfrac{2}{2x + 1}$

26. $\dfrac{3}{4m + 2} = \dfrac{17}{2} - \dfrac{7}{2m + 1}$

27. $\dfrac{3}{k + 2} - \dfrac{2}{k^2 - 4} = \dfrac{1}{k - 2}$

28. $\dfrac{3}{x - 2} + \dfrac{21}{x^2 - 4} = \dfrac{14}{x + 2}$

29. $\dfrac{1}{y + 2} + \dfrac{3}{y + 7} = \dfrac{5}{y^2 + 9y + 14}$

30. $\dfrac{1}{t + 3} + \dfrac{4}{t + 5} = \dfrac{2}{t^2 + 8t + 15}$

31. $\dfrac{9}{x} + \dfrac{4}{6x - 3} = \dfrac{2}{6x - 3}$

32. $\dfrac{5}{n} + \dfrac{4}{6 - 3n} = \dfrac{2n}{6 - 3n}$

33. $\dfrac{6}{w + 3} + \dfrac{-7}{w - 5} = \dfrac{-48}{w^2 - 2w - 15}$

34. $\dfrac{2}{r - 5} + \dfrac{3}{2r + 1} = \dfrac{22}{2r^2 - 9r - 5}$

35. $\dfrac{x}{x - 3} + \dfrac{4}{x + 3} = \dfrac{18}{x^2 - 9}$

36. $\dfrac{2x}{x - 3} + \dfrac{4}{x + 3} = \dfrac{-24}{x^2 - 9}$

37. $\dfrac{1}{x + 4} + \dfrac{x}{x - 4} = \dfrac{-8}{x^2 - 16}$

38. $\dfrac{5}{x - 4} - \dfrac{3}{x - 1} = \dfrac{x^2 - 1}{x^2 - 5x + 4}$

39. $\dfrac{2}{k^2 + k - 6} + \dfrac{1}{k^2 - k - 2} = \dfrac{4}{k^2 + 4k + 3}$

40. $\dfrac{5}{p^2 + 3p + 2} - \dfrac{3}{p^2 - 4} = \dfrac{1}{p^2 - p - 2}$

41. $\dfrac{5x + 14}{x^2 - 9} = \dfrac{-2x^2 - 5x + 2}{x^2 - 9} + \dfrac{2x + 4}{x - 3}$

42. $\dfrac{4x - 7}{4x^2 - 9} = \dfrac{-2x^2 + 5x - 4}{4x^2 - 9} + \dfrac{x + 1}{2x + 3}$

43. *Concept Check* Professor Dan Abbey asked the following question on a test: What is the solution set of $\frac{x + 3}{x + 3} = 1$? Only one student answered it correctly.

(a) What is the solution set?

(b) Many students answered $(-\infty, \infty)$. Why is this not correct?

44. Without actually solving the equation, explain why the solution set must be ∅.

$$\dfrac{1}{x^2 + 2} + \dfrac{2}{x^2 + 1} = -4$$

Graph each rational function. Give the equation of the vertical asymptote. See Figures 2 and 3.

45. $f(x) = \dfrac{2}{x}$

46. $f(x) = \dfrac{3}{x}$

47. $f(x) = \dfrac{1}{x - 2}$

48. $f(x) = \dfrac{1}{x + 2}$

Solve each problem.

49. The average number of vehicles waiting in line to enter a sports arena parking area is modeled by the function defined by

$$w(x) = \dfrac{x^2}{2(1 - x)},$$

where x is a quantity between 0 and 1 known as the **traffic intensity.** (*Source:* Mannering, F., and W. Kilareski, *Principles of Highway Engineering and Traffic Control,* John Wiley and Sons, 1990.) For each traffic intensity, find the average number of vehicles waiting (to the nearest tenth).

(a) 0.1 (b) 0.8 (c) 0.9

(d) What happens to waiting time as traffic intensity increases?

50. The percent of deaths caused by smoking is modeled by the rational function defined by

$$p(x) = \frac{x - 1}{x},$$

where x is the number of times a smoker is more likely to die of lung cancer than a nonsmoker is. This is called the **incidence rate**. (*Source:* Walker, A., *Observation and Inference: An Introduction to the Methods of Epidemiology,* Epidemiology Resources Inc., 1991.) For example, $x = 10$ means that a smoker is 10 times more likely than a nonsmoker to die of lung cancer.

(a) Find $p(x)$ if x is 10.
(b) For what values of x is $p(x) = 80\%$? (*Hint:* Change 80% to a decimal.)
(c) Can the incidence rate equal 0? Explain.

51. The force required to keep a 2000-lb car going 30 mph from skidding on a curve, where r is the radius of the curve in feet, is given by

$$F(r) = \frac{225,000}{r}.$$

(a) What radius must a curve have if a force of 450 lb is needed to keep the car from skidding?
(b) As the radius of the curve is lengthened, how is the force affected?

52. The amount of heating oil produced (in gallons per day) by an oil refinery is modeled by the rational function defined by

$$f(x) = \frac{125,000 - 25x}{125 + 2x},$$

where x is the amount of gasoline produced (in hundreds of gallons per day). Suppose the refinery must produce 300 gal of heating oil per day to meet the needs of its customers.

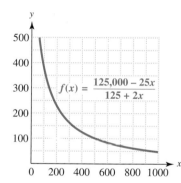

(a) How much gasoline will be produced per day?
(b) The graph of f is shown in the figure. Use it to decide what happens to the amount of gasoline (x) produced as the amount of heating oil (y) produced increases.

TECHNOLOGY INSIGHTS (EXERCISES 53–56)

Two views of the graph of $f(x) = \frac{7}{x^4} - \frac{8}{x^2} + 1$ are shown. Use the graphs to respond to Exercises 53 and 54. See Example 6.

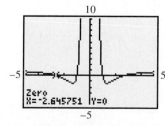
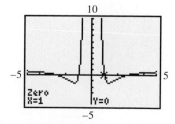

53. How many solutions does the equation $f(x) = 0$ have?

54. Use the display to give the solution set of $f(x) = 0$.

(continued)

In Exercises 55 and 56, use the graph to determine the solution set of the equation
$f(x) = 0$. *All solutions are integers.*

55. $f(x) = \dfrac{x^3 - x^2 - 6x}{x^2 - 1}$

56. $f(x) = \dfrac{-x^3 - x^2 + 2x}{x^2}$

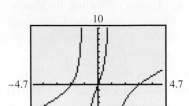

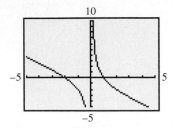

PREVIEW EXERCISES

Solve each formula for the specified variable. See **Section 2.2.**

57. $d = rt$ for t

58. $I = prt$ for r

59. $P = a + b + c$ for c

60. $A = \dfrac{1}{2}h(b + B)$ for B

■ Summary Exercises on Rational Expressions and Equations

A common student error is to confuse an equation, such as $\frac{x}{2} + \frac{x}{3} = -5$, with an expression involving an operation, such as $\frac{x}{2} + \frac{x}{3}$. Look for the equals sign to distinguish between them. Equations are solved for a numerical answer, while problems involving operations result in simplified expressions.

Solving an Equation	**Simplifying an Expression Involving an Operation**
Solve: $\dfrac{x}{2} + \dfrac{x}{3} = -5$.	Add: $\dfrac{x}{2} + \dfrac{x}{3}$.
Multiply each side by the LCD, 6.	Write both fractions with the LCD, 6.
$6\left(\dfrac{x}{2} + \dfrac{x}{3}\right) = 6(-5)$	$\dfrac{x}{2} + \dfrac{x}{3}$
$6\left(\dfrac{x}{2}\right) + 6\left(\dfrac{x}{3}\right) = 6(-5)$	$= \dfrac{x \cdot 3}{2 \cdot 3} + \dfrac{x \cdot 2}{3 \cdot 2}$
$3x + 2x = -30$	$= \dfrac{3x}{6} + \dfrac{2x}{6}$
$5x = -30$	$= \dfrac{3x + 2x}{6}$
$x = -6$	$= \dfrac{5x}{6}$
Check that the solution set is $\{-6\}$.	

Identify each exercise as an equation *or an* expression. *Then simplify the expression or solve the equation, as appropriate.*

1. $\dfrac{x}{2} - \dfrac{x}{4} = 5$

2. $\dfrac{4x - 20}{x^2 - 25} \cdot \dfrac{(x + 5)^2}{10}$

3. $\dfrac{6}{7x} - \dfrac{4}{x}$

4. $\dfrac{\dfrac{1}{x} + \dfrac{1}{y}}{\dfrac{1}{x} - \dfrac{1}{y}}$

5. $\dfrac{5}{7t} = \dfrac{52}{7} - \dfrac{3}{t}$

6. $\dfrac{x - 5}{3} + \dfrac{1}{3} = \dfrac{x - 2}{5}$

7. $\dfrac{7}{6x} + \dfrac{5}{8x}$

8. $\dfrac{4}{x} - \dfrac{8}{x + 1} = 0$

9. $\dfrac{\dfrac{6}{x + 1} - \dfrac{1}{x}}{\dfrac{2}{x} - \dfrac{4}{x + 1}}$

10. $\dfrac{8}{r + 2} - \dfrac{7}{4r + 8}$

11. $\dfrac{x}{x + y} + \dfrac{2y}{x - y}$

12. $\dfrac{3p^2 - 6p}{p + 5} \div \dfrac{p^2 - 4}{8p + 40}$

13. $\dfrac{x - 2}{9} \cdot \dfrac{5}{8 - 4x}$

14. $\dfrac{a - 4}{3} + \dfrac{11}{6} = \dfrac{a + 1}{2}$

15. $\dfrac{b^2 + b - 6}{b^2 + 2b - 8} \cdot \dfrac{b^2 + 8b + 16}{3b + 12}$

16. $\dfrac{10z^2 - 5z}{3z^3 - 6z^2} \div \dfrac{2z^2 + 5z - 3}{z^2 + z - 6}$

17. $\dfrac{5}{x^2 - 2x} - \dfrac{3}{x^2 - 4}$

18. $\dfrac{6}{t + 1} + \dfrac{4}{5t + 5} = \dfrac{34}{15}$

19. $\dfrac{\dfrac{5}{x} - \dfrac{3}{y}}{\dfrac{9x^2 - 25y^2}{x^2y}}$

20. $\dfrac{-2}{a^2 + 2a - 3} - \dfrac{5}{3 - 3a} = \dfrac{4}{3a + 9}$

21. $\dfrac{4y^2 - 13y + 3}{2y^2 - 9y + 9} \div \dfrac{4y^2 + 11y - 3}{6y^2 - 5y - 6}$

22. $\dfrac{8}{3k + 9} - \dfrac{8}{15} = \dfrac{2}{5k + 15}$

23. $\dfrac{3r}{r - 2} = 1 + \dfrac{6}{r - 2}$

24. $\dfrac{6z^2 - 5z - 6}{6z^2 + 5z - 6} \cdot \dfrac{12z^2 - 17z + 6}{12z^2 - z - 6}$

25. $\dfrac{-1}{3 - x} - \dfrac{2}{x - 3}$

26. $\dfrac{\dfrac{t}{4} - \dfrac{1}{t}}{1 + \dfrac{t + 4}{t}}$

27. $\dfrac{2}{y + 1} - \dfrac{3}{y^2 - y - 2} = \dfrac{3}{y - 2}$

28. $\dfrac{7}{2x^2 - 8x} + \dfrac{3}{x^2 - 16}$

29. $\dfrac{3}{y - 3} - \dfrac{3}{y^2 - 5y + 6} = \dfrac{2}{y - 2}$

30. $\dfrac{2k + \dfrac{5}{k - 1}}{3k - \dfrac{2}{k}}$

7.5 Applications of Rational Expressions

OBJECTIVE 1 Find the value of an unknown variable in a formula. Formulas may contain rational expressions, as does $t = \frac{d}{r}$ and $\frac{1}{f} = \frac{1}{p} + \frac{1}{q}$.

EXAMPLE 1 Finding the Value of a Variable in a Formula

In physics, the focal length f of a lens is given by the formula

$$\frac{1}{f} = \frac{1}{p} + \frac{1}{q}.$$

In the formula, p is the distance from the object to the lens and q is the distance from the lens to the image. See Figure 6. Find q if $p = 20$ cm and $f = 10$ cm.

Focal Length of Camera Lens

FIGURE 6

$$\frac{1}{f} = \frac{1}{p} + \frac{1}{q}$$

$$\frac{1}{10} = \frac{1}{20} + \frac{1}{q} \qquad \text{Let } f = 10, p = 20.$$

$$20q \cdot \frac{1}{10} = 20q\left(\frac{1}{20} + \frac{1}{q}\right) \qquad \text{Multiply by the LCD, } 20q.$$

$$2q = q + 20$$

$$q = 20 \qquad \text{Subtract } q.$$

The distance from the lens to the image is 20 cm.

✔ **Now Try Exercise 5.**

OBJECTIVE 2 Solve a formula for a specified variable. Recall that the goal in solving for a specified variable is to isolate it on one side of the equals sign.

EXAMPLE 2 Solving a Formula for a Specified Variable

Solve $\dfrac{1}{f} = \dfrac{1}{p} + \dfrac{1}{q}$ for p.

$$fpq \cdot \frac{1}{f} = fpq\left(\frac{1}{p} + \frac{1}{q}\right) \qquad \text{Multiply by the LCD, } fpq.$$

$$pq = fq + fp \qquad \text{Distributive property}$$

Transform so that the terms with p (the specified variable) are on the same side.

$$pq - fp = fq \qquad \text{Subtract } fp.$$

$$p(q - f) = fq \qquad \text{Factor out } p.$$

This is a key step.

$$p = \frac{fq}{q - f} \qquad \text{Divide by } q - f.$$

✔ **Now Try Exercise 11.**

EXAMPLE 3 Solving a Formula for a Specified Variable

Solve $I = \dfrac{nE}{R + nr}$ for n.

$$(R + nr)I = (R + nr)\dfrac{nE}{R + nr} \qquad \text{Multiply by } R + nr.$$

$$RI + nrI = nE$$

$$RI = nE - nrI \qquad \text{Subtract } nrI.$$

$$RI = n(E - rI) \qquad \text{Factor out } n.$$

$$\dfrac{RI}{E - rI} = n \qquad \text{Divide by } E - rI.$$

✔ **Now Try Exercise 19.**

▶ **CAUTION**   Refer to the steps in Examples 2 and 3 that factor out the desired variable. The variable for which you are solving *must* be a factor on only one side of the equation, so that each side can be divided by the remaining factor in the last step.

We now solve problems that translate into equations with rational expressions, using the six-step problem-solving method from **Chapter 2.**

OBJECTIVE 3 Solve applications by using proportions.   A **ratio** is a comparison of two quantities. The ratio of a to b may be written in any of the following ways:

$$a \text{ to } b, \qquad a:b, \qquad \text{or} \qquad \dfrac{a}{b}. \qquad \text{Ratio of } a \text{ to } b$$

Ratios are usually written as quotients in algebra. A **proportion** is a statement that two ratios are equal, such as

$$\dfrac{a}{b} = \dfrac{c}{d}. \qquad \text{Proportion}$$

EXAMPLE 4 Solving a Proportion

In 2003, about 15 of every 100 Americans had no health insurance. The population at that time was about 288 million. How many million Americans had no health insurance? (*Source:* U.S. Census Bureau.)

Step 1   **Read** the problem.

Step 2   **Assign a variable.** Let $x =$ the number of Americans (in millions) who had no health insurance.

Step 3   **Write an equation.** To get an equation, set up a proportion. The ratio 15 to 100 should equal the ratio x to 288.

$$\dfrac{15}{100} = \dfrac{x}{288} \qquad \text{Write a proportion.}$$

Step 4 **Solve.**

$$\frac{15}{100} = \frac{x}{288}$$

$$28{,}800\left(\frac{15}{100}\right) = 28{,}800\left(\frac{x}{288}\right) \qquad \text{Multiply by a common denominator.}$$

$$4320 = 100x \qquad \text{Simplify.}$$

$$x = 43.2 \qquad \text{Divide by 100.}$$

Step 5 **State the answer.** There were 43.2 million Americans with no health insurance in 2003.

Step 6 **Check** that the ratio of 43.2 million to 288 million equals $\frac{15}{100}$.

✔ **Now Try Exercise 31.**

EXAMPLE 5 Solving a Proportion Involving Rates

Marissa's car uses 10 gal of gasoline to travel 210 mi. She has 5 gal of gasoline in the car, and she wants to know how much more gasoline she will need to drive 640 mi. If we assume the car continues to use gasoline at the same rate, how many more gallons will she need?

Step 1 **Read** the problem.

Step 2 **Assign a variable.** Let $x =$ the additional number of gallons of gas needed.

Step 3 **Write an equation.** To get an equation, set up a proportion.

$$\begin{array}{c} \text{gallons} \rightarrow \\ \text{miles} \rightarrow \end{array} \frac{10}{210} = \frac{5 + x}{640} \begin{array}{c} \leftarrow \text{gallons} \\ \leftarrow \text{miles} \end{array}$$

Step 4 **Solve.** We could multiply both sides by the LCD $10 \cdot 21 \cdot 64$. Instead we use an alternative method that involves *cross products:* For $\frac{a}{b} = \frac{c}{d}$ to be true, then the cross products ad and bc must be equal. Thus,

$$10 \cdot 640 = 210(5 + x) \qquad \text{If } \tfrac{a}{b} = \tfrac{c}{d}, \text{ then } ad = bc.$$

$$6400 = 1050 + 210x \qquad \text{Multiply; distributive property}$$

$$5350 = 210x \qquad \text{Subtract 1050.}$$

$$25.5 \approx x. \qquad \text{Divide by 210; round to the nearest tenth.}$$

Step 5 **State the answer.** Marissa will need about 25.5 more gallons of gas.

Step 6 **Check** the answer in the words of the problem. The 25.5 gal plus the 5 gal equals 30.5 gal.

$$\frac{30.5}{640} \approx 0.048 \qquad \text{and} \qquad \frac{10}{210} \approx 0.048$$

Since the rates are equal, the solution is correct.

✔ **Now Try Exercise 37.**

OBJECTIVE **4** **Solve applications about distance, rate, and time.** The next examples use the distance formula $d = rt$ introduced in **Chapter 2.** A familiar example of a rate is speed, which is the ratio of distance to time, or $r = \frac{d}{t}$.

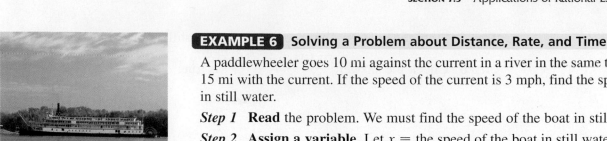

EXAMPLE 6 Solving a Problem about Distance, Rate, and Time

A paddlewheeler goes 10 mi against the current in a river in the same time that it goes 15 mi with the current. If the speed of the current is 3 mph, find the speed of the boat in still water.

Step 1 **Read** the problem. We must find the speed of the boat in still water.

Step 2 **Assign a variable.** Let $x =$ the speed of the boat in still water.

When the boat is traveling *against* the current, the current slows the boat down, and the speed of the boat is the *difference* between its speed in still water and the speed of the current—that is, $(x - 3)$ mph.

When the boat is traveling *with* the current, the current speeds the boat up, and the speed of the boat is the *sum* of its speed in still water and the speed of the current—that is, $(x + 3)$ mph.

Thus, $x - 3 =$ the speed of the boat *against* the current,

and $x + 3 =$ the speed of the boat *with* the current.

Because the time is the same going against the current as with the current, find time in terms of distance and rate (speed) for each situation. Start with the distance formula, $d = rt$, and divide each side by r to get $t = \frac{d}{r}$. Against the current, the distance is 10 mi and the rate is $(x - 3)$ mph, giving

$$t = \frac{d}{r} = \frac{10}{x - 3}. \quad \text{Time \textit{against} the current}$$

With the current, the distance is 15 mi and the rate is $(x + 3)$ mph, so

$$t = \frac{d}{r} = \frac{15}{x + 3}. \quad \text{Time \textit{with} the current}$$

This information is summarized in the following table.

	Distance	Rate	Time	
Against Current	10	$x - 3$	$\dfrac{10}{x - 3}$	Times
With Current	15	$x + 3$	$\dfrac{15}{x + 3}$	are equal.

Step 3 **Write an equation.** Because the times are equal,

$$\frac{10}{x - 3} = \frac{15}{x + 3}.$$

Step 4 **Solve** this equation. The LCD is $(x + 3)(x - 3)$.

$$(x + 3)(x - 3)\left(\frac{10}{x - 3}\right) = (x + 3)(x - 3)\left(\frac{15}{x + 3}\right) \quad \text{Multiply by the LCD.}$$

$$10(x + 3) = 15(x - 3)$$

$$10x + 30 = 15x - 45 \quad \text{Distributive property}$$

$$75 = 5x \quad \text{Subtract } 10x; \text{ add } 45.$$

$$15 = x \quad \text{Divide by 5.}$$

Step 5 **State the answer.** The speed of the boat in still water is 15 mph.

Step 6 **Check** the answer: $\dfrac{10}{15 - 3} = \dfrac{15}{15 + 3}$ is true.

✔ **Now Try Exercise 43.**

EXAMPLE 7 Solving a Problem about Distance, Rate, and Time

At O'Hare International Airport in Chicago, Cheryl and Bill are walking to the gate (at the same speed) to catch their flight to Akron, Ohio. Since Bill wants a window seat, he steps onto the moving sidewalk and continues to walk while Cheryl uses the stationary sidewalk. If the sidewalk moves at 1 m per sec and Bill saves 50 sec covering the 300-m distance, what is their walking speed?

Step 1 **Read** the problem. We must find their walking speed.

Step 2 **Assign a variable.** Let x represent their walking speed in meters per second. Thus, Cheryl travels at x meters per second and Bill travels at $(x + 1)$ meters per second. Since Bill's time is 50 sec less than Cheryl's time, express their times in terms of the known distances and the variable rates. As in Example 6, start with $d = rt$ and divide each side by r to get $t = \frac{d}{r}$. For Cheryl, the distance is 300 m and the rate is x mph. Cheryl's time is

$$t = \frac{d}{r} = \frac{300}{x}.$$

Bill travels 300 m at a rate of $(x + 1)$ mph, so his time is

$$t = \frac{d}{r} = \frac{300}{x + 1}.$$

This information is summarized in the following table.

	Distance	Rate	Time
Cheryl	300	x	$\dfrac{300}{x}$
Bill	300	$x + 1$	$\dfrac{300}{x + 1}$

Step 3 **Write an equation,** using the times from the table.

Bill's time	is	Cheryl's time		less 50 seconds.
$\dfrac{300}{x + 1}$	$=$	$\dfrac{300}{x}$	$-$	50

Step 4 **Solve.**

$$x(x + 1)\left(\frac{300}{x + 1}\right) = x(x + 1)\left(\frac{300}{x} - 50\right) \qquad \text{Multiply by the LCD, } x(x + 1).$$

$$300x = 300(x + 1) - 50x(x + 1) \qquad \text{Multiply; distributive property}$$

$$300x = 300x + 300 - 50x^2 - 50x \qquad \text{Distributive property}$$

$$0 = 50x^2 + 50x - 300 \qquad \text{Standard form}$$
$$0 = x^2 + x - 6 \qquad \text{Divide by 50.}$$
$$0 = (x + 3)(x - 2) \qquad \text{Factor.}$$
$$x + 3 = 0 \quad \text{or} \quad x - 2 = 0 \qquad \text{Zero-factor property}$$
$$x = -3 \quad \text{or} \qquad x = 2 \qquad \text{Solve each equation.}$$

Discard the negative answer, since speed cannot be negative.

Step 5 **State the answer.** Their walking speed is 2 m per sec.

Step 6 **Check** the solution in the words of the original problem.

✔ **Now Try Exercise 49.**

▶ **CAUTION** The method of solving a rational equation by using cross products, introduced in Example 5, cannot be used to solve the equation in Example 7 (as it appears in Step 3). The method can be used only when there is a single rational expression on each side.

OBJECTIVE 5 Solve applications about work rates. Problems about work are closely related to distance problems.

▶ **PROBLEM-SOLVING HINT** If the letters r, t, and A represent the rate at which work is done, the time required, and the amount of work accomplished, respectively, then

$$A = rt.$$

Notice the similarity to the distance formula, $d = rt$.

Amount of work can be measured in terms of jobs accomplished. Thus, if 1 job is completed, $A = 1$, and the formula gives the rate as

$$1 = rt$$
$$r = \frac{1}{t}.$$

To solve a work problem, we begin by using this fact to express all rates of work.

Rate of Work

If a job can be accomplished in t units of time, then the rate of work is

$$\frac{1}{t} \text{ job per unit of time.}$$

See if you can identify the problem-solving steps in the work problem that follows on the next page.

EXAMPLE 8 Solving a Problem about Work

Letitia and Kareem are working on a neighborhood cleanup. Kareem can clean up all the trash in the area in 7 hr, while Letitia can do the same job in 5 hr. How long will it take them if they work together?

Let x = the number of hours it will take the two people working together. Just as we made a table for the distance formula, $d = rt$, we make a table here for $A = rt$, with $A = 1$. Since $A = 1$, the rate for each person will be $\frac{1}{t}$, where t is the time it takes the person to complete the job alone. For example, since Kareem can clean up all the trash in 7 hr, his rate is $\frac{1}{7}$ of the job per hour. Similarly, Letitia's rate is $\frac{1}{5}$ of the job per hour. Fill in the table as shown.

	Rate	Time Working Together	Fractional Part of the Job Done
Kareem	$\frac{1}{7}$	x	$\frac{1}{7}x$
Letitia	$\frac{1}{5}$	x	$\frac{1}{5}x$

Since together they complete 1 job, the sum of the fractional parts accomplished by them should equal 1.

$$\underbrace{\text{Part done by Kareem}}_{\frac{1}{7}x} \quad + \quad \underbrace{\text{part done by Letitia}}_{\frac{1}{5}x} \quad \text{is} \quad \underbrace{\text{1 whole job.}}_{1}$$

$$\frac{1}{7}x \quad + \quad \frac{1}{5}x \quad = \quad 1$$

$$35\left(\frac{1}{7}x + \frac{1}{5}x\right) = 35 \cdot 1 \qquad \text{Multiply by the LCD, 35.}$$

$$5x + 7x = 35 \qquad \text{Distributive property}$$

$$12x = 35 \qquad \text{Combine like terms.}$$

$$x = \frac{35}{12} \qquad \text{Divide by 12.}$$

Working together, Kareem and Letitia can do the job in $\frac{35}{12}$ hr, or 2 hr, 55 min. Check this result in the original problem.

✔ **Now Try Exercise 51.**

There is another way to approach problems about rates of work. For instance, in Example 8, x represents the number of hours it will take the two people working together to complete the entire job. In 1 hr, $\frac{1}{x}$ of the entire job will be completed. In 1 hr, Kareem completes $\frac{1}{7}$ of the job and Letitia completes $\frac{1}{5}$ of the job. The sum of their rates should equal $\frac{1}{x}$. This reasoning gives the equation

$$\frac{1}{7} + \frac{1}{5} = \frac{1}{x}.$$

When each side of this equation is multiplied by $35x$, the result is $5x + 7x = 35$, the same equation we got in Example 8 in the third line from the bottom. Thus, the solution of the equation is the same with either approach.

7.5 EXERCISES

● *Complete solution available on Video Lectures on CD/DVD*

    Now Try Exercise

Concept Check   *Exercises 1–4 present a familiar formula. Give the letter of the choice that is an equivalent form of the formula.*

1. $p = br$ (percent)

   A. $b = \dfrac{p}{r}$ **B.** $r = \dfrac{b}{p}$ **C.** $b = \dfrac{r}{p}$ **D.** $p = \dfrac{r}{b}$

2. $V = LWH$ (geometry)

   A. $H = \dfrac{LW}{V}$ **B.** $L = \dfrac{V}{WH}$ **C.** $L = \dfrac{WH}{V}$ **D.** $W = \dfrac{H}{VL}$

3. $m = \dfrac{F}{a}$ (physics)

   A. $a = mF$ **B.** $F = \dfrac{m}{a}$ **C.** $F = \dfrac{a}{m}$ **D.** $F = ma$

4. $I = \dfrac{E}{R}$ (electricity)

   A. $R = \dfrac{I}{E}$ **B.** $R = IE$ **C.** $E = \dfrac{I}{R}$ **D.** $E = RI$

Solve each problem. See Example 1.

● **5.** In work with electric circuits, the formula

$$\frac{1}{a} = \frac{1}{b} + \frac{1}{c}$$

occurs. Find b if $a = 8$ and $c = 12$.

6. A gas law in chemistry says that

$$\frac{PV}{T} = \frac{pv}{t}.$$

Suppose that $T = 300$, $t = 350$, $V = 9$, $P = 50$, and $v = 8$. Find p.

7. A formula from anthropology says that

$$c = \frac{100b}{L}.$$

Find L if $c = 80$ and $b = 5$.

8. The gravitational force between two masses is given by

$$F = \frac{GMm}{d^2}.$$

Find M if $F = 10$, $G = 6.67 \times 10^{-11}$, $m = 1$, and $d = 3 \times 10^{-6}$.

Solve each formula for the specified variable. See Examples 2 and 3.

9. $F = \dfrac{GMm}{d^2}$ for G (physics)

10. $F = \dfrac{GMm}{d^2}$ for M (physics)

● **11.** $\dfrac{1}{a} = \dfrac{1}{b} + \dfrac{1}{c}$ for a (electricity)

12. $\dfrac{1}{a} = \dfrac{1}{b} + \dfrac{1}{c}$ for b (electricity)

13. $\dfrac{PV}{T} = \dfrac{pv}{t}$ for v (chemistry)

14. $\dfrac{PV}{T} = \dfrac{pv}{t}$ for T (chemistry)

15. $I = \dfrac{nE}{R + nr}$ for r (engineering)

16. $a = \dfrac{V - v}{t}$ for V (physics)

17. $A = \dfrac{1}{2}h(b + B)$ for b (mathematics)

18. $S = \dfrac{n}{2}(a + \ell)d$ for n (mathematics)

19. $\dfrac{E}{e} = \dfrac{R + r}{r}$ for r (engineering)

20. $y = \dfrac{x + z}{a - x}$ for x

21. *Concept Check*   In solving the equation $m = \dfrac{ab}{a - b}$ for a, what is the first step?

22. *Concept Check*   Suppose you are asked to solve the equation

$$rp - rq = p + q$$

for r. What is the first step?

Solve each problem mentally. Use proportions in Exercises 23 and 24.

23. In a mathematics class, 3 of every 4 students are girls. If there are 20 students in the class, how many are girls? How many are boys?

24. In a certain southern state, sales tax on a purchase of $1.50 is $0.12. What is the sales tax on a purchase of $6.00?

25. If Marin can mow her yard in 2 hr, what is her rate (in terms of the proportion of the job per hour)?

26. A van traveling from Atlanta to Detroit averages 50 mph and takes 14 hr to make the trip. How far is it from Atlanta to Detroit?

Use a proportion to solve each problem. See Examples 4 and 5.

27. On a map of the United States, the distance between Seattle and Durango is 4.125 in. The two cities are actually 1238 miles apart. On this same map, what would be the distance between Chicago and El Paso, two cities that are actually 1606 mi apart? (*Source:* Universal Map Atlas 2006.)

28. On a map of the United States, the distance between Reno and Phoenix is 2.5 in. The two cities are actually 768 miles apart. On this same map, what would be the distance between St. Louis and Jacksonville, two cities that are actually 919 mi apart? (*Source:* Universal Map Atlas 2006.)

29. On a world globe, the distance between New York and Cairo, two cities that are actually 5619 mi apart, is 8.5 in. On this same globe, how far apart are Madrid and Rio de Janeiro, two cities that are actually 5045 mi apart? (*Sources:* Author's globe, *The World Almanac and Book of Facts 2006.*)

30. On a world globe, the distance between San Francisco and Melbourne, two cities that are actually 7856 mi apart, is 11.875 in. On this same globe, how far apart are Mexico City and Singapore, two cities that are actually 10,327 mi apart? (*Sources:* Author's globe, *The World Almanac and Book of Facts 2006.*)

31. During the 2001–2002 school year, the average ratio of teachers to students in private elementary schools was approximately 1 to 14. If one such private school had 554 students, how many teachers would be at the school if this ratio was valid for that school? Round your answer to the nearest whole number. (*Source:* U.S. National Center for Education Statistics.)

32. On June 28, 2006, the Boston Red Sox were in first place in the Eastern Division of the American League, having won 47 of their first 75 regular season games. If the team continued to win the same fraction of its games, how many games would the Red Sox win for the complete 162-game season? Round your answer to the nearest whole number. (*Source:* www.mlb.com)

33. To estimate the deer population of a forest preserve, wildlife biologists caught, tagged, and then released 42 deer. A month later, they returned and caught a sample of 75 deer and found that 15 of them were tagged. Based on this experiment, approximately how many deer lived in the forest preserve?

34. Suppose that in the experiment in Exercise 33, when the biologists returned, they found only 5 tagged deer in their sample of 75. What would be the estimate of the deer population?

35. Biologists tagged 500 fish in a lake on January 1. On February 1, they returned and collected a random sample of 400 fish, 8 of which had been previously tagged. On the basis of this experiment, approximately how many fish does the lake have?

36. Suppose that in the experiment of Exercise 35, 10 of the previously tagged fish were collected on February 1. What would be the estimate of the fish population?

🌐 **37.** Bruce Johnston's Shelby Cobra uses 5 gal of gasoline to drive 156 miles. He has 3 gal of gasoline in the car, and he wants to know how much more gasoline he will need to drive 300 mi. If we assume that the car continues to use gasoline at the same rate, how many more gallons will he need?

38. Mike Love's T-bird uses 6 gal of gasoline to drive 141 miles. He has 4 gal of gasoline in the car, and he wants to know how much more gasoline he will need to drive 250 mi. If we assume that the car continues to use gasoline at the same rate, how many more gallons will he need?

In geometry, two triangles with corresponding angle measures equal, called **similar triangles,** *have corresponding sides proportional. For example, in the figure, angle A = angle D, angle B = angle E, and angle C = angle F, so the triangles are similar. Then the following ratios of corresponding sides are equal.*

$$\frac{4}{6} = \frac{6}{9} = \frac{2x+1}{2x+5}$$

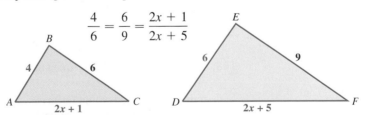

39. Solve for x, using the given proportion to find the lengths of the third sides of the triangles.

40. Suppose the following triangles are similar. Find y and the lengths of the two longest sides of each triangle.

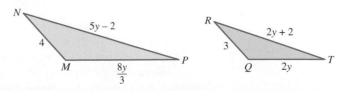

Nurses use proportions to determine the amount of a drug to administer when the dose is measured in milligrams, but the drug is packaged in a diluted form in milliliters. (Source: Hoyles, Celia, Richard Noss, and Stefano Pozzi, "Proportional Reasoning in Nursing Practice," Journal for Research in Mathematics Education, January 2001.) For example, to find the number of milliliters of fluid needed to administer 300 mg *of a drug that comes packaged as* 120 mg *in 2 mL of fluid, a nurse sets up the proportion*

$$\frac{120 \text{ mg}}{2 \text{ mL}} = \frac{300 \text{ mg}}{x \text{ mL}},$$

where x represents the amount to administer in milliliters. Use this method to find the correct dose for each prescription.

41. 120 mg of Amakacine packaged as 100 mg in 2-mL vials

42. 1.5 mg of morphine packaged as 20 mg ampules diluted in 10 mL of fluid

Solve each problem. See Examples 6 and 7.

43. Kellen's boat goes 12 mph. Find the rate of the current of the river if she can go 6 mi upstream in the same amount of time she can go 10 mi downstream.

	Distance	Rate	Time
Downstream	10	12 + x	
Upstream	6	12 − x	

44. Kasey can travel 8 mi upstream in the same time it takes her to go 12 mi downstream. Her boat goes 15 mph in still water. What is the rate of the current?

	Distance	Rate	Time
Downstream			
Upstream			

45. On his drive from Montpelier, Vermont, to Columbia, South Carolina, Dylan Davis averaged 51 mph. If he had been able to average 60 mph, he would have reached his destination 3 hr earlier. What is the driving distance between Montpelier and Columbia?

46. Rebecca Song is a college student who lives in an off-campus apartment. Some days she rides her bike to campus, while other days she walks. When she rides her bike, she gets to her first classroom building 36 min faster than when she walks. If her average walking speed is 3 mph and her average biking speed is 12 mph, how far is it from her apartment to the classroom building?

47. A plane averaged 500 mph on a trip going east, but only 350 mph on the return trip. The total flying time in both directions was 8.5 hr. What was the one-way distance?

48. Leah drove from her apartment to her parents' house for the weekend. Driving to their house on Saturday morning, she was able to average 60 mph because traffic was light. However, returning on Sunday night, she was able to average only 45 mph on the same route, because traffic was heavy. The drive on Sunday took her 1.5 hr longer than the drive on Saturday. What is the distance between Leah's apartment and her parents' house?

49. On the first part of a trip to Carmel traveling on the freeway, Marge averaged 60 mph. On the rest of the trip, which was 10 mi longer than the first part, she averaged 50 mph. Find the total distance to Carmel if the second part of the trip took 30 min more than the first part.

50. While on vacation, Jim and Annie decided to drive all day. During the first part of their trip on the highway, they averaged 60 mph. When they got to Houston, traffic caused them to average only 30 mph. The distance they drove in Houston was 100 mi less than their distance on the highway. What was their total driving distance if they spent 50 min more on the highway than they did in Houston?

Solve each problem. See Example 8.

51. Butch and Peggy want to pick up the mess that their grandson, Grant, has made in his playroom. Butch can do it in 15 min working alone. Peggy, working alone, can clean it in 12 min. How long will it take them if they work together?

	Rate	Time Working Together	Fractional Part of the Job Done
Butch	$\frac{1}{15}$	x	
Peggy	$\frac{1}{12}$	x	

52. Lou can groom Jay Beckenstein's dogs in 8 hr, but it takes his business partner, Janet, only 5 hr to groom the same dogs. How long will it take them to groom Jay's dogs if they work together?

	Rate	Time Working Together	Fractional Part of the Job Done
Lou	$\frac{1}{8}$	x	
Janet	$\frac{1}{5}$	x	

53. Jerry and Kuba are laying a hardwood floor. Working alone, Jerry can do the job in 20 hr. If the two of them work together, they can complete the job in 12 hr. How long would it take Kuba to lay the floor working alone?

54. Mrs. Disher is a high school mathematics teacher. She can grade a set of chapter tests in 5 hr working alone. If her student teacher Mr. Howes helps her, it will take 3 hr to grade the tests. How long would it take Mr. Howes to grade the tests if he worked alone?

55. Dixie can paint a room in 3 hr working alone. Trixie can paint the same room in 6 hr working alone. How long after Dixie starts to paint the room will it be finished if Trixie joins her 1 hr later?

56. Chipper can wash a car in 30 min working alone. Dalie can do the same job in 45 min working alone. How long after Chipper starts to wash the car will it be finished if Dalie joins him 5 min later?

57. If a vat of acid can be filled by an inlet pipe in 10 hr and emptied by an outlet pipe in 20 hr, how long will it take to fill the vat if both pipes are open?

58. A winery has a vat to hold Chardonnay. An inlet pipe can fill the vat in 9 hr, while an outlet pipe can empty it in 12 hr. How long will it take to fill the vat if both the outlet and the inlet pipes are open?

59. Suppose that Hortense and Mort can clean their entire house in 7 hr, while their toddler, Mimi, just by being around, can completely mess it up in only 2 hr. If Hortense and Mort clean the house while Mimi is at her grandma's and then start cleaning up after Mimi the minute she gets home, how long does it take from the time Mimi gets home until the whole place is a shambles?

60. An inlet pipe can fill an artificial lily pond in 60 min, while an outlet pipe can empty it in 80 min. Through an error, both pipes are left open. How long will it take for the pond to fill?

*Find k, given that y = 1 and x = 3. See **Section 2.1.***

61. $y = kx$ **62.** $y = kx^2$ **63.** $y = \dfrac{k}{x}$ **64.** $y = \dfrac{k}{x^2}$

7.6 Variation

OBJECTIVES

1 Write an equation expressing direct variation.

2 Find the constant of variation, and solve direct variation problems.

3 Solve inverse variation problems.

4 Solve joint variation problems.

5 Solve combined variation problems.

$C = 2\pi r$

FIGURE 7

Certain types of functions are common, especially in business and the physical sciences. These are functions in which *y* depends on a multiple of *x* or *y* depends on a number divided by *x*. In such situations, *y* is said to *vary directly as x* (in the first case) or *vary inversely as x* (in the second case). For example, by the distance formula, the distance traveled varies directly as the rate (or speed) and the time. Formulas for area and volume are other familiar examples of *direct variation*.

By contrast, the force required to keep a car from skidding on a curve varies inversely as the radius of the curve. Another example of *inverse variation* is how travel time is inversely proportional to rate or speed.

OBJECTIVE 1 Write an equation expressing direct variation. The circumference of a circle is given by the formula $C = 2\pi r$, where *r* is the radius of the circle. See Figure 7. The circumference is always a constant multiple of the radius. (*C* is always found by multiplying *r* by the constant 2π.) Thus,

As the *radius increases,* the *circumference increases.*

The following is also true:

As the *radius decreases,* the *circumference decreases.*

Because of these relationships, the circumference is said to *vary directly* as the radius.

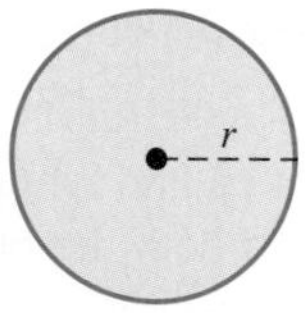

> **Direct Variation**
>
> *y* **varies directly as** *x* if there exists a real number *k* such that
>
> $$y = kx.$$

Also, *y* is said to be **proportional to** *x*. The number *k* is called the **constant of variation.** In direct variation, for $k > 0$, as the value of *x* increases, the value of *y* also increases. Similarly, as *x* decreases, *y* decreases.

OBJECTIVE 2 Find the constant of variation, and solve direct variation problems. The direct variation equation $y = kx$ defines a linear function, where the constant of variation *k* is the slope of the line. For example, we wrote the equation

$$y = 3.20x$$

to describe the cost *y* to buy *x* gallons of gasoline in Example 6 of **Section 3.3.** The cost varies directly as, or is proportional to, the number of gallons of gasoline purchased.

That is, as the number of gallons of gasoline increases, the cost increases; also, as the number of gallons of gasoline decreases, the cost decreases. The constant of variation k is 3.20, the cost of 1 gal of gasoline.

EXAMPLE 1 Finding the Constant of Variation and the Variation Equation

Abby Tanenbaum is paid an hourly wage. One week she worked 43 hr and was paid $795.50. How much does she earn per hour?

Let h represent the number of hours she works and P represent her corresponding pay. Then P varies directly as h, so

$$P = kh.$$

Here, k represents Abby's hourly wage. Since $P = 795.50$ when $h = 43$,

$$795.50 = 43k \qquad \text{Substitute for } P \text{ and } h.$$
$$k = 18.50. \qquad \text{Use a calculator.}$$

Her hourly wage is $18.50, and P and h are related by

$$P = 18.50h.$$

✔ **Now Try Exercise 29.**

EXAMPLE 2 Solving a Direct Variation Problem

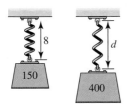

FIGURE 8

Hooke's law for an elastic spring states that the distance a spring stretches is directly proportional to the force applied. If a force of 150 newtons* stretches a certain spring 8 cm, how much will a force of 400 newtons stretch the spring? See Figure 8.

If d is the distance the spring stretches and f is the force applied, then $d = kf$ for some constant k. Since a force of 150 newtons stretches the spring 8 cm, use these values to find k.

$$d = kf \qquad \text{Variation equation}$$
$$8 = k \cdot 150 \qquad \text{Let } d = 8 \text{ and } f = 150.$$
$$k = \frac{8}{150} \qquad \text{Solve for } k.$$
$$k = \frac{4}{75} \qquad \text{Lowest terms}$$

Substitute $\frac{4}{75}$ for k in the variation equation $d = kf$ to get

$$d = \frac{4}{75}f.$$

For a force of 400 newtons,

$$d = \frac{4}{75}(400) = \frac{64}{3}. \qquad \text{Let } f = 400.$$

The spring will stretch $\frac{64}{3}$ cm, or $21\frac{1}{3}$ cm, if a force of 400 newtons is applied.

✔ **Now Try Exercise 31.**

---

*A newton is a unit of measure of force used in physics.

In summary, use the following steps to solve a variation problem.

Solving a Variation Problem

Step 1 Write the variation equation.

Step 2 Substitute the initial values and solve for k.

Step 3 Rewrite the variation equation with the value of k from Step 2.

Step 4 Substitute the remaining values, solve for the unknown, and find the required answer.

The direct variation equation $y = kx$ is a linear equation. However, other kinds of variation involve other types of equations. For example, one variable can be proportional to a power of another variable.

Direct Variation as a Power

y varies directly as the nth power of x if there exists a real number k such that

$$y = kx^n.$$

$A = \pi r^2$

FIGURE 9

An example of direct variation as a power is the formula for the area of a circle, $A = \pi r^2$. See Figure 9. Here, π is the constant of variation, and the area varies directly as the *square* of the radius.

EXAMPLE 3 Solving a Direct Variation Problem

The distance a body falls from rest varies directly as the square of the time it falls (disregarding air resistance). If a skydiver falls 64 ft in 2 sec, how far will she fall in 8 sec?

Step 1 If d represents the distance the skydiver falls and t the time it takes to fall, then d is a function of t, and, for some constant k,

$$d = kt^2.$$

Step 2 To find the value of k, use the fact that the skydiver falls 64 ft in 2 sec.

$$d = kt^2 \qquad \text{Variation equation}$$
$$64 = k(2)^2 \qquad \text{Let } d = 64 \text{ and } t = 2.$$
$$k = 16 \qquad \text{Find } k.$$

Step 3 Using 16 for k, we find that the variation equation is

$$d = 16t^2.$$

Step 4 Now let $t = 8$ to find the number of feet the skydiver will fall in 8 sec.

$$d = 16(8)^2 = 1024 \qquad \text{Let } t = 8.$$

The skydiver will fall 1024 ft in 8 sec.

✔ **Now Try Exercise 35.**

As pressure increases, volume decreases.

FIGURE 10

OBJECTIVE 3 **Solve inverse variation problems.**   In direct variation, where $k > 0$, as x increases, y increases. Similarly, as x decreases, y decreases. Another type of variation is *inverse variation*. With inverse variation, where $k > 0$, as one variable increases, the other variable decreases. For example, in a closed space, volume decreases as pressure increases, as illustrated by a trash compactor. See Figure 10. As the compactor presses down, the pressure on the trash increases; in turn, the trash occupies a smaller space.

Inverse Variation

y **varies inversely as** *x* if there exists a real number k such that

$$y = \frac{k}{x}.$$

Also, *y* **varies inversely as the** *n***th power of** *x* if there exists a real number k such that

$$y = \frac{k}{x^n}.$$

The inverse variation equation also defines a function. Since x is in the denominator, these functions are rational functions. Another example of inverse variation comes from the distance formula. In its usual form, the formula is

$$d = rt.$$

Dividing each side by r gives

$$t = \frac{d}{r}.$$

Here, t (time) varies inversely as r (rate or speed), with d (distance) serving as the constant of variation. For example, if the distance between Chicago and Des Moines is 300 mi, then

$$t = \frac{300}{r},$$

and the values of r and t might be any of the following.

$$\left.\begin{array}{l} r = 50, t = 6 \\ r = 60, t = 5 \\ r = 75, t = 4 \end{array}\right\}\begin{array}{l}\text{As } r \text{ increases,} \\ t \text{ decreases.}\end{array} \qquad \left.\begin{array}{l} r = 30, t = 10 \\ r = 25, t = 12 \\ r = 20, t = 15 \end{array}\right\}\begin{array}{l}\text{As } r \text{ decreases,} \\ t \text{ increases.}\end{array}$$

If we *increase* the rate (speed) at which we drive, time *decreases*. If we *decrease* the rate (speed) at which we drive, time *increases*.

EXAMPLE 4   **Solving an Inverse Variation Problem**

In the manufacture of a certain medical syringe, the cost of producing the syringe varies inversely as the number produced. If 10,000 syringes are produced, the cost is $2 per syringe. Find the cost per syringe of producing 25,000 syringes.

$$\begin{aligned} \text{Let} \quad & x = \text{the number of syringes produced,} \\ \text{and} \quad & c = \text{the cost per syringe.} \end{aligned}$$

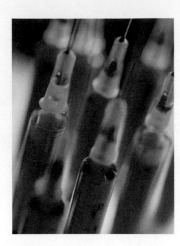

Here, as production increases, cost decreases, and as production decreases, cost increases. Since c varies inversely as x, there is a constant k such that

$$c = \frac{k}{x}.$$

Find k by replacing c with 2 and x with 10,000.

$$2 = \frac{k}{10,000}$$

$$20,000 = k \qquad \text{Multiply by 10,000.}$$

Since $c = \frac{k}{x}$,

$$c = \frac{20,000}{25,000} = 0.80. \qquad \text{Let } k = 20,000 \text{ and } x = 25,000.$$

The cost per syringe to make 25,000 syringes is $0.80.

✔ **Now Try Exercise 37.**

EXAMPLE 5 **Solving an Inverse Variation Problem**

The weight of an object above Earth varies inversely as the square of its distance from the center of Earth. A space shuttle in an elliptical orbit has a maximum distance from the center of Earth (*apogee*) of 6700 mi. Its minimum distance from the center of Earth (*perigee*) is 4090 mi. See Figure 11. If an astronaut in the shuttle weighs 57 lb at its apogee, what does the astronaut weigh at its perigee?

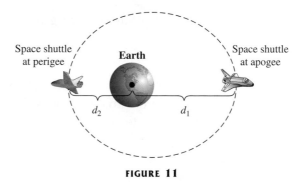

FIGURE 11

If w is the weight and d is the distance from the center of Earth, then

$$w = \frac{k}{d^2}$$

for some constant k. At the apogee, the astronaut weighs 57 lb, and the distance from the center of Earth is 6700 mi. Use these values to find k.

$$57 = \frac{k}{(6700)^2} \qquad \text{Let } w = 57 \text{ and } d = 6700.$$

$$k = 57(6700)^2$$

Then the weight at the perigee with $d = 4090$ mi is

$$w = \frac{57(6700)^2}{(4090)^2} \approx 153 \text{ lb.} \qquad \text{Use a calculator.}$$

✔ **Now Try Exercise 41.**

OBJECTIVE 4 Solve joint variation problems. It is common for one variable to depend on several others. If one variable varies directly as the *product* of several other variables (perhaps raised to powers), the first variable is said to *vary jointly* as the others.

Joint Variation

y **varies jointly as** *x* **and** *z* if there exists a real number *k* such that
$$y = kxz.$$

▶ **CAUTION**   Note that *and* in the expression "*y* varies directly as *x and z*" translates as the product $y = kxz$. The word *and* does not indicate addition here.

EXAMPLE 6 Solving a Joint Variation Problem

The interest on a loan or an investment is given by the formula $I = prt$. Here, for a given principal *p*, the interest earned, *I*, varies jointly as the interest rate *r* and the time *t* the principal is left earning interest. If an investment earns \$100 interest at 5% for 2 yr, how much interest will the same principal earn at 4.5% for 3 yr?

We use the formula $I = prt$, where *p* is the constant of variation because it is the same for both investments. For the first investment,

$$I = prt$$
$$100 = p(0.05)(2) \qquad \text{Let } I = 100, r = 0.05, \text{ and } t = 2.$$
$$100 = 0.1p$$
$$p = 1000. \qquad \text{Divide by 0.1; rewrite.}$$

Now we find *I* when $p = 1000$, $r = 0.045$, and $t = 3$.

$$I = 1000(0.045)(3) = 135 \qquad \text{Let } p = 1000, r = 0.045, \text{ and } t = 3.$$

The interest will be \$135.

✔ **Now Try Exercise 43.**

OBJECTIVE 5 Solve combined variation problems.   There are many combinations of direct and inverse variation. Example 7 shows a typical **combined variation** problem.

EXAMPLE 7 Solving a Combined Variation Problem

Body mass index, or BMI, is used by physicians to assess a person's level of fatness. A BMI from 19 through 25 is considered desirable. BMI varies directly as an individual's weight in pounds and inversely as the square of the individual's height in inches. A person who weighs 118 lb and is 64 in. tall has a BMI of 20. (The BMI is rounded to the nearest whole number.) Find the BMI of a person who weighs 165 lb and is 70 in. tall. (*Source: Washington Post.*)

Let *B* represent the BMI, *w* the weight, and *h* the height. Then

$$B = \frac{kw}{h^2}. \qquad \begin{array}{l} \longleftarrow \text{ BMI varies directly as the weight.} \\ \longleftarrow \text{ BMI varies inversely as the square of the height.} \end{array}$$

To find k, let $B = 20$, $w = 118$, and $h = 64$.

$$20 = \frac{k(118)}{64^2} \qquad B = \frac{kw}{h^2}$$

$$k = \frac{20(64^2)}{118} \qquad \text{Multiply by } 64^2 \text{; divide by 118.}$$

$$k \approx 694 \qquad \text{Use a calculator.}$$

Now find B when $k = 694$, $w = 165$, and $h = 70$.

$$B = \frac{694(165)}{70^2} \approx 23 \qquad \text{Nearest whole number}$$

The person's BMI is 23.

✔ **Now Try Exercise 51.**

7.6 EXERCISES

Concept Check   *Use personal experience or intuition to determine whether the situation suggests* direct *or* inverse *variation.*

1. The number of raffle tickets you buy and your probability of winning that raffle

2. The rate and the distance traveled by a pickup truck in 2 hr

3. The amount of pressure put on the accelerator of a car and the speed of the car

4. The number of days from now until December 25 and the magnitude of the frenzy of Christmas shopping

5. Your age and the probability that you believe in Santa Claus

6. The surface area of a balloon and its diameter

7. The number of days until the end of the baseball season and the number of home runs that Albert Pujols has

8. The amount of gasoline you pump and the amount you pay

Concept Check   *Determine whether each equation represents* direct, inverse, joint, *or* combined *variation.*

9. $y = \dfrac{3}{x}$ 10. $y = \dfrac{8}{x}$ 11. $y = 10x^2$ 12. $y = 2x^3$

13. $y = 3xz^4$ 14. $y = 6x^3z^2$ 15. $y = \dfrac{4x}{wz}$ 16. $y = \dfrac{6x}{st}$

17. *Concept Check*   For $k > 0$, if y varies directly as x, then when x increases, y _____, and when x decreases, y _____.

18. *Concept Check*   For $k > 0$, if y varies inversely as x, then when x increases, y _____, and when x decreases, y _____.

Concept Check   *Solve each problem.*

19. If x varies directly as y, and $x = 9$ when $y = 3$, find x when $y = 12$.

20. If x varies directly as y, and $x = 10$ when $y = 7$, find y when $x = 50$.

21. If a varies directly as the square of b, and $a = 4$ when $b = 3$, find a when $b = 2$.

22. If h varies directly as the square of m, and $h = 15$ when $m = 5$, find h when $m = 7$.

23. If z varies inversely as w, and $z = 10$ when $w = 0.5$, find z when $w = 8$.

24. If t varies inversely as s, and $t = 3$ when $s = 5$, find s when $t = 5$.

25. If m varies inversely as p^2, and $m = 20$ when $p = 2$, find m when $p = 5$.

26. If a varies inversely as b^2, and $a = 48$ when $b = 4$, find a when $b = 7$.

27. p varies jointly as q and r^2, and $p = 200$ when $q = 2$ and $r = 3$. Find p when $q = 5$ and $r = 2$.

28. f varies jointly as g^2 and h, and $f = 50$ when $g = 4$ and $h = 2$. Find f when $g = 3$ and $h = 6$.

Solve each problem involving variation. See Examples 1–7.

29. Ben bought 15 gal of gasoline and paid $43.79. To the nearest tenth of a cent, what is the price of gasoline per gallon?

30. Sara gives horseback rides at Shadow Mountain Ranch. A 2.5-hr ride costs $75.00. What is the price per hour?

31. The weight of an object on Earth is directly proportional to the weight of that same object on the moon. A 200-lb astronaut would weigh 32 lb on the moon. How much would a 50-lb dog weigh on the moon?

32. The pressure exerted by a certain liquid at a given point is directly proportional to the depth of the point beneath the surface of the liquid. The pressure at 30 m is 80 newtons. What pressure is exerted at 50 m?

33. The volume of a can of pork and beans is directly proportional to the height of the can. If the volume of the can is 300 in.^{3} when its height is 10.62 in., find the volume of a can with height 15.92 in.

34. The force required to compress a spring is directly proportional to the change in length of the spring. If a force of 20 newtons is required to compress a certain spring 2 cm, how much force is required to compress the spring from 20 cm to 8 cm?

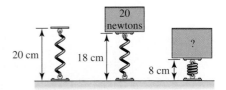

35. For a body falling freely from rest (disregarding air resistance), the distance the body falls varies directly as the square of the time. If an object is dropped from the top of a tower 576 ft high and hits the ground in 6 sec, how far did it fall in the first 4 sec?

36. The amount of water emptied by a pipe varies directly as the square of the diameter of the pipe. For a certain constant water flow, a pipe emptying into a canal will allow 200 gal of water to escape in an hour. The diameter of the pipe is 6 in. How much water would a 12-in. pipe empty into the canal in an hour, assuming the same water flow?

37. Over a specified distance, speed varies inversely with time. If a Dodge Viper on a test track goes a certain distance in one-half minute at 160 mph, what speed is needed to go the same distance in three-fourths minute?

38. For a constant area, the length of a rectangle varies inversely as the width. The length of a rectangle is 27 ft when the width is 10 ft. Find the width of a rectangle with the same area if the length is 18 ft.

39. The frequency of a vibrating string varies inversely as its length. That is, a longer string vibrates fewer times in a second than a shorter string. Suppose a piano string 2 ft long vibrates 250 cycles per sec. What frequency would a string 5 ft long have?

40. The current in a simple electrical circuit varies inversely as the resistance. If the current is 20 amps when the resistance is 5 ohms, find the current when the resistance is 7.5 ohms.

41. The amount of light (measured in foot-candles) produced by a light source varies inversely as the square of the distance from the source. If the illumination produced 1 m from a light source is 768 foot-candles, find the illumination produced 6 m from the same source.

42. The force with which Earth attracts an object above Earth's surface varies inversely as the square of the distance of the object from the center of Earth. If an object 4000 mi from the center of Earth is attracted with a force of 160 lb, find the force of attraction if the object were 6000 mi from the center of Earth.

43. For a given interest rate, simple interest varies jointly as principal and time. If $2000 left in an account for 4 yr earned interest of $280, how much interest would be earned in 6 yr?

44. The collision impact of an automobile varies jointly as its mass and the square of its speed. Suppose a 2000-lb car traveling at 55 mph has a collision impact of 6.1. What is the collision impact of the same car at 65 mph?

45. The weight of a bass varies jointly as its girth and the square of its length. (*Girth* is the distance around the body of the fish.) A prize-winning bass weighed in at 22.7 lb and measured 36 in. long with a 21-in. girth. How much would a bass 28 in. long with an 18-in. girth weigh?

46. The weight of a trout varies jointly as its length and the square of its girth. One angler caught a trout that weighed 10.5 lb and measured 26 in. long with an 18-in. girth. Find the weight of a trout that is 22 in. long with a 15-in. girth.

47. The force needed to keep a car from skidding on a curve varies inversely as the radius of the curve and jointly as the weight of the car and the square of the speed. If 242 lb of force keeps a 2000-lb car from skidding on a curve of radius 500 ft at 30 mph, what force would keep the same car from skidding on a curve of radius 750 ft at 50 mph?

48. The maximum load that a cylindrical column with a circular cross section can hold varies directly as the fourth power of the diameter of the cross section and inversely as the square of the height. A 9-m column 1 m in diameter will support 8 metric tons. How many metric tons can be supported by a column 12 m high and $\frac{2}{3}$ m in diameter?

Load = 8 metric tons

49. The number of long-distance phone calls between two cities during a certain period varies jointly as the populations of the cities, p_1 and p_2, and inversely as the distance between them. If 80,000 calls are made between two cities 400 mi apart, with populations of 70,000 and 100,000, how many calls are made between cities with populations of 50,000 and 75,000 that are 250 mi apart?

50. Natural gas provides 35.8% of U.S. energy. (*Source:* U.S. Energy Department.) The volume of gas varies inversely as the pressure and directly as the temperature. (Temperature must be measured in *kelvins* (K), a unit of measurement used in physics.) If a certain gas occupies a volume of 1.3 L at 300 K and a pressure of 18 newtons, find the volume at 340 K and a pressure of 24 newtons.

51. A body mass index from 27 through 29 carries a slight risk of weight-related health problems, while one of 30 or more indicates a great increase in risk. Use your own height and weight and the information in Example 7 to determine your BMI and whether you are at risk.

52. The maximum load of a horizontal beam that is supported at both ends varies directly as the width and the square of the height and inversely as the length between the supports. A beam 6 m long, 0.1 m wide, and 0.06 m high supports a load of 360 kg. What is the maximum load supported by a beam 16 m long, 0.2 m wide, and 0.08 m high?

53. Explain the difference between inverse variation and direct variation.

54. What is meant by the constant of variation in a direct variation problem? If you were to graph the linear equation $y = kx$ for some nonnegative constant k, what role would k play in the graph?

PREVIEW EXERCISES

*Solve each equation for real solutions. See **Section 6.5**.*

55. $x^2 = 81$ **56.** $x^2 = 625$ **57.** $t^2 = 0.25$

58. $s^2 = 0.16$ **59.** $s^4 = 81$ **60.** $r^4 = 16$

Chapter 7 | Group Activity

IT DEPENDS ON WHAT YOU MEAN BY "AVERAGE"

Objective   Learn when and how to calculate a harmonic mean.

Finding an average seems to be a simple process. Don't we just add the values and divide by the number of values? Well, for rational expressions, it all depends on what you mean by "average."

- To find the average of two fractions, say, $\frac{1}{3}$ and $\frac{3}{4}$, add the two fractions and divide by 2.

$$\dfrac{\dfrac{1}{3} + \dfrac{3}{4}}{2} = \dfrac{\left(\dfrac{1}{3} + \dfrac{3}{4}\right) \cdot 12}{2 \cdot 12} = \dfrac{4 + 9}{24} = \dfrac{13}{24}$$

On a number line, the fraction $\frac{13}{24}$ is the **arithmetic mean,** which is exactly halfway between the fractions $\frac{1}{3}$ and $\frac{3}{4}$.

- Suppose you travel in one direction at 60 mph and return at 30 mph.

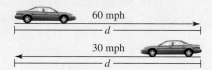

To find your average rate, you have to calculate the total distance divided by the total time. Recall that $d = rt$, so if the distance *each* way is d, then the total distance is $2d$, the time going is $\frac{d}{60}$, and the time returning is $\frac{d}{30}$. Since $r = \frac{d}{t}$,

$$r = \dfrac{2d}{\dfrac{d}{60} + \dfrac{d}{30}} = \dfrac{2d \cdot 60}{\left(\dfrac{d}{60} + \dfrac{d}{30}\right) \cdot 60} = \dfrac{120d}{d + 2d} = \dfrac{120d}{3d} = 40 \text{ mph.}$$

The average rate is 40 mph. This is the *harmonic mean* of 60 and 30. The **harmonic mean** of two numbers a and b is defined as

$$\dfrac{2ab}{a + b}.$$

Note that

$$\dfrac{2 \cdot 60 \cdot 30}{60 + 30} = \dfrac{3600}{90} = 40.$$

- To calculate a batting average, you find the **ratio** of the number of hits to the number of at bats. Suppose a baseball player has 72 hits in 364 at bats. Then his batting average would be $\frac{72}{364} \approx .198$. If the same player gets an additional 3 hits from 8 more at bats during the next week, then his revised batting average would be

$$\frac{72 + 3}{364 + 8} = \frac{75}{372} \approx .202.$$

A carpenter builds wine racks. As a group, find the appropriate "average" quantity for each situation.

A. The carpenter told his helper to cut $\frac{1}{2}$-ft pieces from a dowel. The helper could not find a measuring tape, but he did recall that the distance from the tip of his middle finger to the tip of his thumb was approximately 6 in., so he estimated the lengths. When the carpenter checked his work, he found that the helper had actually cut two pieces that were $\frac{5}{12}$ and $\frac{1}{2}$ ft long. What was the average length of the two pieces?

B. Once the pieces are cut, the carpenter can assemble and finish a wine rack in 2 hr, working alone. His helper takes 4 hr to accomplish the same task, working alone. If the carpenter and the helper work together, what is the average time they take to assemble and finish a wine rack?

C. Of 115 wine racks built, 112 passed a quality control check. What was the acceptance rate? During the next week, the carpenter built 35 additional wine racks, of which 28 were acceptable. What was the revised acceptance rate? Round answers to the nearest thousandth.

Chapter **7** **SUMMARY** *View the Interactive Summary on the Pass the Test CD.*

KEY TERMS

7.1 rational expression
rational function
7.2 least common
denominator (LCD)
7.3 complex fraction

7.4 rational equation
discontinuous
vertical asymptote
domain of the variable
in a rational equation

7.5 ratio
proportion
7.6 vary directly
proportional
constant of variation

vary inversely
vary jointly
combined variation

TEST YOUR WORD POWER

See how well you have learned the vocabulary in this chapter. Answers, with examples, follow the Quick Review.

1. A **rational expression** is
   A. an algebraic expression made up of a term or the sum of a finite number of terms with real coefficients and integer exponents
   B. a polynomial equation of degree 2
   C. an expression with one or more fractions in the numerator, denominator, or both
   D. the quotient of two polynomials with denominator not 0.

2. In a given set of fractions, the **least common denominator** is
   A. the smallest denominator of all the denominators

   B. the smallest expression that is divisible by all the denominators
   C. the largest integer that evenly divides the numerator and denominator of all the fractions
   D. the largest denominator of all the denominators.

3. A **complex fraction** is
   A. an algebraic expression made up of a term or the sum of a finite number of terms with real coefficients and integer exponents
   B. a polynomial equation of degree 2
   C. an expression with one or more fractions in the numerator, denominator, or both

   D. the quotient of two polynomials with denominator not 0.

4. A **ratio**
   A. compares two quantities by using a quotient
   B. says that two quotients are equal
   C. is a product of two quantities
   D. is a difference between two quantities.

5. A **proportion**
   A. compares two quantities by using a quotient
   B. says that two quotients are equal
   C. is a product of two quantities
   D. is a difference between two quantities.

QUICK REVIEW

Concepts	Examples

7.1 RATIONAL EXPRESSIONS AND FUNCTIONS; MULTIPLYING AND DIVIDING

Rational Functions
A function of the form

$$f(x) = \frac{P(x)}{Q(x)}, \quad \text{where } Q(x) \neq 0,$$

is a rational function. Its domain consists of all real numbers except those that make $Q(x) = 0$.

Find the domain.

$$f(x) = \frac{2x + 1}{3x + 6}$$

Solve $3x + 6 = 0$ to find $x = -2$. This is the only real number excluded from the domain. The domain is $\{x \mid x \neq -2\}$.

(continued)

Concepts	Examples

Fundamental Property of Rational Numbers

If $\frac{a}{b}$ is a rational number and if c is any nonzero real number, then

$$\frac{a}{b} = \frac{ac}{bc}.$$

$$\frac{3}{4} = \frac{3 \cdot 5}{4 \cdot 5} = \frac{15}{20}$$

Writing a Rational Expression in Lowest Terms

Step 1 Factor the numerator and the denominator completely.

Step 2 Apply the fundamental property.

Write in lowest terms.

$$\frac{2x + 8}{x^2 - 16}$$

$$= \frac{2(x + 4)}{(x - 4)(x + 4)} \qquad \text{Factor.}$$

$$= \frac{2}{x - 4} \qquad \text{Lowest terms}$$

Multiplying Rational Expressions

Step 1 Factor numerators and denominators.

Step 2 Apply the fundamental property.

Step 3 Multiply the remaining factors in the numerator and in the denominator.

Step 4 Check that the product is in lowest terms.

Multiply.

$$\frac{x^2 + 2x + 1}{x^2 - 1} \cdot \frac{5}{3x + 3}$$

$$= \frac{(x + 1)^2}{(x - 1)(x + 1)} \cdot \frac{5}{3(x + 1)} \qquad \text{Factor.}$$

$$= \frac{5}{3(x - 1)} \qquad \text{Multiply; lowest terms}$$

Dividing Rational Expressions

Multiply the first rational expression by the reciprocal of the second.

Divide.

$$\frac{2x + 5}{x - 3} \div \frac{2x^2 + 3x - 5}{x^2 - 9}$$

$$= \frac{2x + 5}{x - 3} \cdot \frac{x^2 - 9}{2x^2 + 3x - 5} \qquad \text{Multiply by the reciprocal.}$$

$$= \frac{2x + 5}{x - 3} \cdot \frac{(x + 3)(x - 3)}{(2x + 5)(x - 1)} \qquad \text{Factor.}$$

$$= \frac{x + 3}{x - 1} \qquad \text{Multiply; lowest terms}$$

7.2 ADDING AND SUBTRACTING RATIONAL EXPRESSIONS

Adding or Subtracting Rational Expressions

Step 1 If the denominators are the same, add or subtract the numerators. Place the result over the common denominator.

  If the denominators are different, write all rational expressions with the LCD. Then add or subtract the numerators, and place the result over the common denominator.

Step 2 Make sure that the answer is in lowest terms.

Subtract.

$$\frac{1}{x + 6} - \frac{3}{x + 2}$$

$$= \frac{x + 2}{(x + 6)(x + 2)} - \frac{3(x + 6)}{(x + 6)(x + 2)}$$

$$= \frac{x + 2 - 3(x + 6)}{(x + 6)(x + 2)}$$

$$= \frac{x + 2 - 3x - 18}{(x + 6)(x + 2)} \qquad \text{Be careful with signs.}$$

$$= \frac{-2x - 16}{(x + 6)(x + 2)}$$

(continued)

Concepts	Examples

7.3 COMPLEX FRACTIONS

Simplifying a Complex Fraction

Simplify the complex fraction.

Method 1 Simplify the numerator and denominator separately, as much as possible. Then multiply the numerator by the reciprocal of the denominator. Write the answer in lowest terms.

Method 1

$$\frac{\dfrac{1}{x^2} - \dfrac{1}{y^2}}{\dfrac{1}{x} + \dfrac{1}{y}}$$

$$= \frac{\dfrac{y^2}{x^2 y^2} - \dfrac{x^2}{x^2 y^2}}{\dfrac{y}{xy} + \dfrac{x}{xy}}$$

$$= \frac{\dfrac{y^2 - x^2}{x^2 y^2}}{\dfrac{y + x}{xy}}$$

$$= \frac{y^2 - x^2}{x^2 y^2} \div \frac{y + x}{xy}$$

$$= \frac{(y + x)(y - x)}{x^2 y^2} \cdot \frac{xy}{y + x}$$

$$= \frac{y - x}{xy}$$

Method 2 Multiply the numerator and denominator of the complex fraction by the least common denominator of all fractions appearing in the complex fraction. Then simplify the result.

Method 2

$$\frac{\dfrac{1}{x^2} - \dfrac{1}{y^2}}{\dfrac{1}{x} + \dfrac{1}{y}}$$

$$= \frac{x^2 y^2 \left(\dfrac{1}{x^2} - \dfrac{1}{y^2} \right)}{x^2 y^2 \left(\dfrac{1}{x} + \dfrac{1}{y} \right)}$$

$$= \frac{y^2 - x^2}{xy^2 + x^2 y}$$

$$= \frac{(y - x)(y + x)}{xy(y + x)}$$

$$= \frac{y - x}{xy}$$

(continued)

Concepts	Examples

7.4   EQUATIONS WITH RATIONAL EXPRESSIONS AND GRAPHS

Solving an Equation with Rational Expressions
To solve a rational equation, first determine the domain of the variable. Then multiply all the terms in the equation by the least common denominator. Solve the resulting equation. Each proposed solution *must* be checked to see that it is in the domain.

Solve.

$$\frac{3x + 2}{x - 2} + \frac{2}{x(x - 2)} = \frac{-1}{x}$$

Note that 0 and 2 are excluded from the domain.

$$x(3x + 2) + 2 = -(x - 2) \qquad \text{Multiply by } x(x - 2).$$
$$3x^2 + 2x + 2 = -x + 2 \qquad \text{Distributive property}$$
$$3x^2 + 3x = 0 \qquad \text{Add } x; \text{ subtract 2.}$$
$$3x(x + 1) = 0 \qquad \text{Factor.}$$
$$3x = 0 \quad \text{or} \quad x + 1 = 0 \qquad \text{Zero-factor property}$$
$$x = 0 \quad \text{or} \quad x = -1$$

Of the two proposed solutions, 0 must be discarded because it is not in the domain. The solution set is $\{-1\}$.

The graph of a rational function (written in lowest terms) may have one or more breaks. At such points, the graph will approach an asymptote.

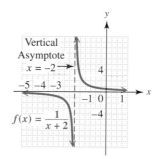

7.5   APPLICATIONS OF RATIONAL EXPRESSIONS

To solve a formula for a particular variable, isolate that variable on one side.

Solve for L.

$$c = \frac{100b}{L}$$
$$cL = 100b \qquad \text{Multiply by } L.$$
$$L = \frac{100b}{c} \qquad \text{Divide by } c.$$

To solve a motion problem, use the formula

$$d = rt$$

or one of its equivalents,

$$t = \frac{d}{r} \quad \text{or} \quad r = \frac{d}{t}.$$

Solve.

A canal has a current of 2 mph. Find the speed of Amy's boat in still water if it goes 11 mi downstream in the same time that it goes 8 mi upstream.
    Let x represent the speed of the boat in still water.

	Distance	Rate	Time
Downstream	11	$x + 2$	$\dfrac{11}{x + 2}$
Upstream	8	$x - 2$	$\dfrac{8}{x - 2}$

(continued)

Concepts	Examples
	Because the times are the same, the equation is

$$\frac{11}{x+2} = \frac{8}{x-2}. \qquad \text{Use } t = \frac{d}{r}.$$

The LCD is $(x+2)(x-2)$.

$$11(x-2) = 8(x+2) \qquad \text{Multiply by the LCD.}$$
$$11x - 22 = 8x + 16 \qquad \text{Distributive property}$$
$$3x = 38 \qquad \text{Subtract } 8x; \text{ add } 22.$$
$$x = 12\frac{2}{3} \qquad \text{Divide by 3.}$$

The speed in still water is $12\frac{2}{3}$ mph.

To solve a work problem, use the fact that if a complete job is done in t units of time, the rate of work is $\frac{1}{t}$ job per unit of time. The amount of work accomplished is $A = rt$, so if one job is accomplished in time t, use the formula

$$1 = rt.$$

7.6 VARIATION

If there is some constant k such that

$y = kx^n$,   then y varies directly as x^n.

$y = \dfrac{k}{x^n}$,   then y varies inversely as x^n.

$y = kxz$,   then y varies jointly as x and z.

The area of a circle varies directly as the square of the radius.

$$A = kr^2 \qquad \text{Here, } k = \pi.$$

Pressure varies inversely as volume.

$$p = \frac{k}{V}$$

For a given principal, interest varies jointly as interest rate and time.

$$I = krt \qquad k \text{ is the given principal.}$$

Answers to Test Your Word Power

1. D; *Examples:* $-\dfrac{3}{4y^2}, \dfrac{5x^3}{x+2}, \dfrac{a+3}{a^2-4a-5}$

2. B; *Example:* The LCD of $\dfrac{1}{x}, \dfrac{2}{3}$, and $\dfrac{5}{x+1}$ is $3x(x+1)$.

3. C; *Examples:* $\dfrac{\frac{2}{3}}{\frac{4}{7}}, \dfrac{x-\frac{1}{x}}{x+\frac{1}{y}}, \dfrac{\frac{2}{a+1}}{a^2-1}$

4. A; *Example:* $\dfrac{7 \text{ in.}}{12 \text{ in.}}$ compares two quantities.

5. B; *Example:* The proportion $\dfrac{2}{3} = \dfrac{8}{12}$ states that the two ratios are equal.

Chapter **7** **REVIEW EXERCISES**

[7.1] *(a) Find all real numbers that are excluded from the domain. (b) Give the domain in set notation.*

1. $f(x) = \dfrac{-7}{3x + 18}$

2. $f(x) = \dfrac{5x + 17}{x^2 - 7x + 10}$

3. $f(x) = \dfrac{9}{x^2 - 18x + 81}$

Write in lowest terms.

4. $\dfrac{12x^2 + 6x}{24x + 12}$

5. $\dfrac{25m^2 - n^2}{25m^2 - 10mn + n^2}$

6. $\dfrac{r - 2}{4 - r^2}$

7. What is meant by the reciprocal of a rational expression?

Multiply or divide. Write the answer in lowest terms.

8. $\dfrac{(2y + 3)^2}{5y} \cdot \dfrac{15y^3}{4y^2 - 9}$

9. $\dfrac{w^2 - 16}{w} \cdot \dfrac{3}{4 - w}$

10. $\dfrac{z^2 - z - 6}{z - 6} \cdot \dfrac{z^2 - 6z}{z^2 + 2z - 15}$

11. $\dfrac{m^3 - n^3}{m^2 - n^2} \div \dfrac{m^2 + mn + n^2}{m + n}$

[7.2] *Assume that each expression is the denominator of a rational expression. Find the least common denominator for each group.*

12. $32b^3, \quad 24b^5$

13. $9r^2, \quad 3r + 1, \quad 9$

14. $6x^2 + 13x - 5, \quad 9x^2 + 9x - 4$

15. $3x - 12, x^2 - 2x - 8, x^2 - 8x + 16$

Add or subtract as indicated.

16. $\dfrac{5}{3x^6y^5} - \dfrac{8}{9x^4y^7}$

17. $\dfrac{5y + 13}{y + 1} - \dfrac{1 - 7y}{y + 1}$

18. $\dfrac{6}{5a + 10} + \dfrac{7}{6a + 12}$

19. $\dfrac{3r}{10r^2 - 3rs - s^2} + \dfrac{2r}{2r^2 + rs - s^2}$

20. *Concept Check* Two students worked the following problem:

$$\text{Add:} \quad \frac{2}{y - x} + \frac{1}{x - y}.$$

One student got a sum of $\dfrac{1}{y - x}$ and the other got a sum of $\dfrac{-1}{x - y}$. Which student(s) got the correct answer?

[7.3] *Simplify each complex fraction.*

21. $\dfrac{\dfrac{3}{t} + 2}{\dfrac{4}{t} - 7}$

22. $\dfrac{\dfrac{2}{m - 3n}}{\dfrac{1}{3n - m}}$

23. $\dfrac{\dfrac{3}{p} - \dfrac{2}{q}}{\dfrac{9q^2 - 4p^2}{qp}}$

24. $\dfrac{x^{-2} - y^{-2}}{x^{-1} - y^{-1}}$

[7.4] *Solve each equation.*

25. $\dfrac{1}{t + 4} + \dfrac{1}{2} = \dfrac{3}{2t + 8}$

26. $\dfrac{-5m}{m + 1} + \dfrac{m}{3m + 3} = \dfrac{56}{6m + 6}$

27. $\dfrac{2}{k-1} - \dfrac{4k+1}{k^2-1} = \dfrac{-1}{k+1}$

28. $\dfrac{5}{x+2} + \dfrac{3}{x+3} = \dfrac{x}{x^2+5x+6}$

29. *Concept Check* After solving the equation

$$\frac{3}{x-3} - \frac{2}{x-2} = \frac{3}{x^2-5x+6},$$

a student got $x = 3$ as her final step. She could not understand why the answer in the back of the book was "$\emptyset$," because she checked her algebra several times and was sure that all her algebraic work was correct. Was she wrong, or was the answer in the back of the book wrong?

30. Explain the difference between the following:

Simplify the expression $\dfrac{4}{x} + \dfrac{1}{2} - \dfrac{1}{3}$; Solve the equation $\dfrac{4}{x} + \dfrac{1}{2} = \dfrac{1}{3}$.

31. *Concept Check* Which graph has a vertical asymptote, and what is its equation?

A. 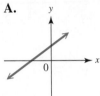 **B.** 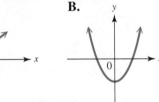 **C.** 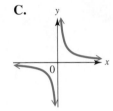 **D.**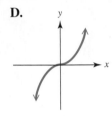

[7.5]

32. The equation from a law from physics is in the form $\frac{1}{A} = \frac{1}{B} + \frac{1}{C}$. Find A if $B = 30$ and $C = 10$.

Solve each formula for the specified variable.

33. $F = \dfrac{GMm}{d^2}$ for m (physics)

34. $\mu = \dfrac{Mv}{M+m}$ for M (electronics)

Solve each problem.

35. An article in *Scientific American* predicts that, in the year 2050, about 23,200 of the 58,000 passenger-km per day in North America will be provided by high-speed trains. If the traffic volume in a typical region of North America is 15,000, how many passenger-kilometers per day will high-speed trains provide there? (*Source:* Schafer, Andreas, and David Victor, "The Past and Future of Global Mobility," *Scientific American,* October 1997.)

36. A river has a current of 4 km per hr. Find the speed of Lynn McTernan's boat in still water if it goes 40 km downstream in the same time that it takes to go 24 km upstream.

	d	r	t
Upstream	24	x − 4	
Downstream	40		

37. A sink can be filled by a cold-water tap in 8 min and by a hot-water tap in 12 min. How long would it take to fill the sink with both taps open?

38. Melena Fenn and Jeff Houck need to sort a pile of bottles at the recycling center. Working alone, Melena could do the entire job in 9 hr, while Jeff could do the entire job in 6 hr. How long will it take them if they work together?

[7.6]

39. *Concept Check*   In which one of the following does y vary inversely as x?

A. $y = 2x$ **B.** $y = \dfrac{x}{3}$ **C.** $y = \dfrac{3}{x}$ **D.** $y = x^2$

Solve each problem.

40. For the subject in a photograph to appear in the same perspective in the photograph as in real life, the viewing distance must be properly related to the amount of enlargement. For a particular camera, the viewing distance varies directly as the amount of enlargement. A picture that is taken with this camera and enlarged 5 times should be viewed from a distance of 250 mm. Suppose a print 8.6 times the size of the negative is made. From what distance should it be viewed?

41. The frequency (number of vibrations per second) of a vibrating guitar string varies inversely as its length. That is, a longer string vibrates fewer times in a second than a shorter string. Suppose a guitar string 0.65 m long vibrates 4.3 times per sec. What frequency would a string 0.5 m long have?

42. The volume of a rectangular box of a given height is proportional to its width and length. A box with width 2 ft and length 4 ft has volume 12 ft^3. Find the volume of a box with the same height, but that is 3 ft wide and 5 ft long.

MIXED REVIEW EXERCISES

Write in lowest terms.

43. $\dfrac{x + 2y}{x^2 - 4y^2}$

44. $\dfrac{x^2 + 2x - 15}{x^2 - x - 6}$

Perform the indicated operations.

45. $\dfrac{2}{m} + \dfrac{5}{3m^2}$

46. $\dfrac{k^2 - 6k + 9}{1 - 216k^3} \cdot \dfrac{6k^2 + 17k - 3}{9 - k^2}$

47. $\dfrac{\dfrac{-3}{x} + \dfrac{x}{2}}{1 + \dfrac{x + 1}{x}}$

48. $\dfrac{9x^2 + 46x + 5}{3x^2 - 2x - 1} \div \dfrac{x^2 + 11x + 30}{x^3 + 5x^2 - 6x}$

49. $\dfrac{\dfrac{3}{x} - 5}{6 + \dfrac{1}{x}}$

50. $\dfrac{9}{3 - x} - \dfrac{2}{x - 3}$

51. $\dfrac{4y + 16}{30} \div \dfrac{2y + 8}{5}$

52. $\dfrac{t^{-2} + s^{-2}}{t^{-1} - s^{-1}}$

53. $\dfrac{4a}{a^2 - ab - 2b^2} - \dfrac{6b - a}{a^2 + 4ab + 3b^2}$

54. $\dfrac{a}{b} + \dfrac{b}{c} + \dfrac{c}{d}$

Solve.

55. $\dfrac{x + 3}{x^2 - 5x + 4} - \dfrac{1}{x} = \dfrac{2}{x^2 - 4x}$

56. $A = \dfrac{Rr}{R + r}$ for r

57. $1 - \dfrac{5}{r} = \dfrac{-4}{r^2}$

58. $\dfrac{3x}{x - 4} + \dfrac{2}{x} = \dfrac{48}{x^2 - 4x}$

59. The strength of a contact lens is given in units called diopters and also in millimeters of arc. As the diopters increase, the millimeters of arc decrease. The rational function defined by

$$f(x) = \frac{337}{x}$$

relates the arc measurement $f(x)$ to the diopter measurement x. (*Source:* Bausch and Lomb.)

(a) What arc measurement will correspond to 40.5-diopter lenses?

(b) A lens with an arc measurement of 7.51 mm will provide what diopter strength?

60. The hot-water tap can fill a tub in 20 min. The cold-water tap takes 15 min to fill the tub. How long would it take to fill the tub with both taps open?

61. At a certain gasoline station, 3 gal of unleaded gasoline cost $4.86. How much would 13 gal of the same gasoline cost?

62. The area of a triangle varies jointly as the lengths of the base and height. A triangle with base 10 ft and height 4 ft has area 20 ft^2. Find the area of a triangle with base 3 ft and height 8 ft.

63. If Dr. Dawson rides his bike to his office, he averages 12 mph. If he drives his car, he averages 36 mph. His time driving is $\frac{1}{4}$ hr less than his time riding his bike. How far is his office from home?

64. A private plane traveled from San Francisco to a secret rendezvous. It averaged 200 mph. On the return trip, the average speed was 300 mph. If the total traveling time was 4 hr, how far from San Francisco was the secret rendezvous?

65. Johnny averages 30 mph when he drives on the old highway to his favorite fishing hole, and he averages 50 mph when most of his route is on the interstate. If both routes are the same length, and he saves 2 hr by traveling on the interstate, how far away is the fishing hole?

Chapter **7** TEST

View the complete solutions to all Chapter Test exercises on the Pass the Test CD.

1. Find all real numbers excluded from the domain of $f(x) = \dfrac{x + 3}{3x^2 + 2x - 8}$. Then give the domain in set notation.

2. Write $\dfrac{6x^2 - 13x - 5}{9x^3 - x}$ in lowest terms.

Multiply or divide.

3. $\dfrac{(x + 3)^2}{4} \cdot \dfrac{6}{2x + 6}$

4. $\dfrac{y^2 - 16}{y^2 - 25} \cdot \dfrac{y^2 + 2y - 15}{y^2 - 7y + 12}$

5. $\dfrac{3 - t}{5} \div \dfrac{t - 3}{10}$

6. $\dfrac{x^2 - 9}{x^3 + 3x^2} \div \dfrac{x^2 + x - 12}{x^3 + 9x^2 + 20x}$

7. Find the least common denominator for the following group of denominators: $t^2 + t - 6$, $t^2 + 3t$, t^2.

Add or subtract as indicated.

8. $\dfrac{7}{6t^2} - \dfrac{1}{3t}$

9. $\dfrac{3}{7a^4b^3} + \dfrac{5}{21a^5b^2}$

10. $\dfrac{9}{x^2 - 6x + 9} + \dfrac{2}{x^2 - 9}$

11. $\dfrac{6}{x + 4} + \dfrac{1}{x + 2} - \dfrac{3x}{x^2 + 6x + 8}$

Simplify each complex fraction.

12. $\dfrac{\dfrac{12}{r + 4}}{\dfrac{11}{6r + 24}}$

13. $\dfrac{\dfrac{1}{a} - \dfrac{1}{b}}{\dfrac{a}{b} - \dfrac{b}{a}}$

14. $\dfrac{2x^{-2} + y^{-2}}{x^{-1} - y^{-1}}$

15. One of the following is an expression to be simplified by algebraic operations, and the other is an equation to be solved. Identify each, and then simplify the one that is an expression and solve the one that is an equation.

(a) $\dfrac{2x}{3} + \dfrac{x}{4} - \dfrac{11}{2}$

(b) $\dfrac{2x}{3} + \dfrac{x}{4} = \dfrac{11}{2}$

Solve each equation.

16. $\dfrac{1}{x} - \dfrac{4}{3x} = \dfrac{1}{x - 2}$

17. $\dfrac{y}{y + 2} - \dfrac{1}{y - 2} = \dfrac{8}{y^2 - 4}$

18. Checking the solution(s) of an equation in earlier chapters verified that the algebraic steps were performed correctly. When an equation includes a term with a variable denominator, what additional reason *requires* that the solutions be checked?

19. Solve for the variable ℓ in this formula from mathematics: $S = \dfrac{n}{2}(a + \ell)$.

20. Graph the function defined by $f(x) = \dfrac{-2}{x + 1}$. Give the equation of its vertical asymptote.

Solve each problem.

21. Wayne can do a job in 9 hr, while Susan can do the same job in 5 hr. How long would it take them to do the job if they worked together?

22. The rate of the current in a stream is 3 mph. Nana's boat can go 36 mi downstream in the same time that it takes to go 24 mi upstream. Find the rate of her boat in still water.

23. Biologists collected a sample of 600 fish from Lake Linda on May 1 and tagged each of them. When they returned on June 1, a new sample of 800 fish was collected and 10 of these had been previously tagged. Use this experiment to determine the approximate fish population of Lake Linda.

24. In biology, the function defined by

$$g(x) = \frac{5x}{2 + x}$$

gives the growth rate of a population for x units of available food. (*Source:* Smith, J. Maynard, *Models in Ecology,* Cambridge University Press, 1974.)

(a) What amount of food (in appropriate units) would produce a growth rate of 3 units of growth per unit of food?

(b) What is the growth rate if no food is available?

25. The current in a simple electrical circuit is inversely proportional to the resistance. If the current is 80 amps when the resistance is 30 ohms, find the current when the resistance is 12 ohms.

26. The force of the wind blowing on a vertical surface varies jointly as the area of the surface and the square of the velocity. If a wind blowing at 40 mph exerts a force of 50 lb on a surface of 500 ft^2, how much force will a wind of 80 mph place on a surface of 2 ft^2?

Chapters 1–7 CUMULATIVE REVIEW EXERCISES

Evaluate if $x = -4$, $y = 3$, and $z = 6$.

1. $|2x| + 3y - z^3$

2. $\dfrac{x(2x - 1)}{3y - z}$

Solve each equation.

3. $7(2x + 3) - 4(2x + 1) = 2(x + 1)$

4. $|6x - 8| - 4 = 0$

5. $ax + by = cx + d$ for x

Solve each inequality.

6. $\dfrac{2}{3}y + \dfrac{5}{12}y \le 20$

7. $|3x + 2| \ge 4$

Solve each problem.

8. Otis Taylor invested some money at 4% interest and twice as much at 3% interest. His interest for the first year was $400. How much did he invest at each rate?

9. A triangle has area 42 m^2. The base is 14 m long. Find the height of the triangle.

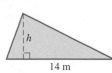

14 m

10. Graph $-4x + 2y = 8$ and give the intercepts.

Find the slope of each line described in Exercises 11 and 12.

11. Through $(-5, 8)$ and $(-1, 2)$

12. Perpendicular to $4x - 3y = 12$

13. Write an equation of the line in Exercise 11. Give the equation in the form $y = mx + b$.

Graph the solution set of each inequality.

14. $2x + 5y > 10$

15. $x - y \geq 3$ and $3x + 4y \leq 12$

Decide whether each relation defined in Exercises 16–18 is a function, and give its domain and range.

16. AVERAGE HOURLY WAGES IN MEXICO

Year	Wage (in U.S. dollars)
1990	1.25
1992	1.61
1994	1.80
1996	1.21
1998	1.94
2000	2.26
2002	2.60

Source: World Almanac and Book of Facts 2006.

17.

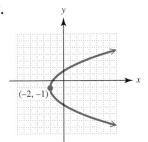

18. $y = -\sqrt{x + 2}$

19. Suppose that $y = f(x)$ and $5x - 3y = 8$.

   (a) Find the equation that defines $f(x)$. That is, $f(x) = $ _____.

   (b) Find $f(1)$.

20. If $f(x) = 3x + 6$, what is $f(x + 3)$?

Solve each system.

21. $4x - y = -7$
       $5x + 2y = 1$

22. $x + y - 2z = -1$
       $2x - y + z = -6$
       $3x + 2y - 3z = -3$

23. $x + 2y + z = 5$
       $x - y + z = 3$
       $2x + 4y + 2z = 11$

24. With traffic taken into account, an automobile can travel 7 km, on the average, in the same time that an airplane can travel 100 km. The average speed of an airplane is 558 km per hr greater than that of an automobile. Find both speeds. (*Source:* Schafer, Andreas, and David Victor, "The Past and Future of Global Mobility," *Scientific American,* October 1997.)

Simplify. Write each answer with only positive exponents. Assume that all variables represent nonzero real numbers.

25. $\left(\dfrac{a^{-3}b^4}{a^2b^{-1}} \right)^{-2}$

26. $\left(\dfrac{m^{-4}n^2}{m^2n^{-3}} \right) \cdot \left(\dfrac{m^5n^{-1}}{m^{-2}n^5} \right)$

Perform the indicated operations.

27. $(3y^2 - 2y + 6) - (-y^2 + 5y + 12)$

28. $-6x^4(x^2 - 3x + 2)$

29. $(4f + 3)(3f - 1)$

30. $(7t^3 + 8)(7t^3 - 8)$

31. $\left(\dfrac{1}{4}x + 5\right)^2$

32. $(3x^3 + 13x^2 - 17x - 7) \div (3x + 1)$

33. For the polynomial functions defined by

$$f(x) = x^2 + 2x - 3, \quad g(x) = 2x^3 - 3x^2 + 4x - 1, \quad \text{and} \quad h(x) = x^2,$$

find **(a)** $(f + g)(x)$, **(b)** $(g - f)(x)$, **(c)** $(f + g)(-1)$, and **(d)** $(f \circ h)(x)$.

Factor each polynomial completely.

34. $2x^2 - 13x - 45$

35. $100t^4 - 25$

36. $8p^3 + 125$

37. Solve the equation $3x^2 + 4x = 7$.

Write each rational expression in lowest terms.

38. $\dfrac{y^2 - 16}{y^2 - 8y + 16}$

39. $\dfrac{8x^2 - 18}{8x^2 + 4x - 12}$

Perform the indicated operations. Express the answer in lowest terms.

40. $\dfrac{2a^2}{a + b} \cdot \dfrac{a - b}{4a}$

41. $\dfrac{x^2 - 9}{2x + 4} \div \dfrac{x^3 - 27}{4}$

42. $\dfrac{x + 4}{x - 2} + \dfrac{2x - 10}{x - 2}$

43. $\dfrac{2x}{2x - 1} + \dfrac{4}{2x + 1} + \dfrac{8}{4x^2 - 1}$

44. Solve the equation $\dfrac{-3x}{x + 1} + \dfrac{4x + 1}{x} = \dfrac{-3}{x^2 + x}$.

45. Solve the formula for q: $\quad \dfrac{1}{f} = \dfrac{1}{p} + \dfrac{1}{q}$.

Solve each problem.

46. Erika Suco can fly her plane 200 mi against the wind in the same time it takes her to fly 300 mi with the wind. The wind blows at 30 mph. Find the speed of her plane in still air.

47. Machine A can complete a certain job in 2 hr. To speed up the work, Machine B, which could complete the job alone in 3 hr, is brought in to help. How long will it take the two machines to complete the job working together?

48. The cost of a pizza varies directly as the square of its radius. If a pizza with a 7-in. radius costs $6.00, how much should a pizza with a 9-in. radius cost?

Roots, Radicals, and Root Functions

Tom Skilling is the chief meteorologist for the *Chicago Tribune*. He writes a column titled "Ask Tom Why," in which readers question him on a variety of topics. In the Saturday, August 17, 2002 issue, reader Ted Fleischaker wrote,

> *I cannot remember the formula to calculate the distance to the horizon. I have a stunning view from my 14th-floor condo, 150 ft above the ground. How far can I see?*

Skilling's answer in Section 8.3, Exercise 125, provides a formula for finding the distance to the horizon. The formula includes a *square root,* one of the topics of this chapter.

8.1 Radical Expressions and Graphs

OBJECTIVE 1 Find roots of numbers. Recall from **Section 1.3** that $6^2 = 36$; that is, 6 *squared* is 36. The opposite (or inverse) of *squaring* a number is taking its *square root.* Thus,

> It is customary to write $\sqrt{}$, rather than $\sqrt[2]{}$.

$$\sqrt{36} = 6, \quad \text{because} \quad 6^2 = 36.$$

We now extend our discussion of roots to *cube roots* $\sqrt[3]{}$, *fourth roots* $\sqrt[4]{}$, and higher roots. In general, $\sqrt[n]{a}$ is a number whose nth power equals a. That is,

$$\sqrt[n]{a} = b \quad \text{means} \quad b^n = a.$$

The number a is the **radicand**, n is the **index** or **order**, and the expression $\sqrt[n]{a}$ is a **radical**.

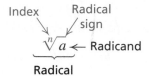

EXAMPLE 1 Simplifying Higher Roots

Simplify.

(a) $\sqrt[3]{64} = 4$, because $4^3 = 64$. **(b)** $\sqrt[3]{125} = 5$, because $5^3 = 125$.

(c) $\sqrt[4]{16} = 2$, because $2^4 = 16$. **(d)** $\sqrt[5]{32} = 2$, because $2^5 = 32$.

(e) $\sqrt[3]{\dfrac{8}{27}} = \dfrac{2}{3}$, because $\left(\dfrac{2}{3}\right)^3 = \dfrac{8}{27}$.

(f) $\sqrt[4]{0.0016} = 0.2$, because $(0.2)^4 = 0.0016$.

✔ **Now Try Exercises 5, 15, 33, and 39.**

OBJECTIVE 2 Find principal roots. If n is even, positive numbers have two nth roots. For example, both 4 and -4 are square roots of 16, and 2 and -2 are fourth roots of 16. In such cases, the notation $\sqrt[n]{a}$ represents the positive root, called the **principal root,** and $-\sqrt[n]{a}$ represents the negative root.

---

nth Root

1. If n is *even* and a is *positive or 0,* then

$$\sqrt[n]{a} \text{ represents the **principal nth root** of } a,$$

and $\qquad -\sqrt[n]{a}$ represents the **negative nth root** of a.

2. If n is *even* and a is *negative,* then

$$\sqrt[n]{a} \text{ is not a real number.}$$

3. If n is *odd,* then

there is exactly one real nth root of a, written $\sqrt[n]{a}$.

If n is even, the two nth roots of a are often written together as $\pm\sqrt[n]{a}$, with $\pm$ read "positive or negative," or "plus or minus."

EXAMPLE 2 Finding Roots

Find each root.

(a) $\sqrt{100} = 10$

Because the radicand, 100, is *positive,* there are two square roots: 10 and -10. We want the principal root, which is 10.

(b) $-\sqrt{100} = -10$

Here, we want the negative square root, -10.

(c) $\sqrt[4]{81} = 3$ Principal 4th root   **(d)** $-\sqrt[4]{81} = -3$ Negative 4th root

Parts (a)–(d) illustrate Case 1 in the preceding box.

(e) $\sqrt[4]{-81}$

The index is *even* and the radicand is *negative,* so $\sqrt[4]{-81}$ is not a real number. This is Case 2 in the preceding box.

(f) $\sqrt[3]{8} = 2$, because $2^3 = 8$.   **(g)** $\sqrt[3]{-8} = -2$, because $(-2)^3 = -8$.

Parts (f) and (g) illustrate Case 3 in the box. The index is *odd,* so each radical represents exactly one nth root (regardless of whether the radicand is positive, negative, or 0).

✔ **Now Try Exercises 13, 17, 21, and 25.**

OBJECTIVE 3 Graph functions defined by radical expressions. A **radical expression** is an algebraic expression that contains radicals.

$$3 - \sqrt{x}, \qquad \sqrt[3]{x}, \qquad \text{and} \qquad \sqrt{2x - 1} \qquad \text{Radical expressions}$$

In earlier chapters, we graphed functions defined by polynomial and rational expressions. Now we examine the graphs of functions defined by the basic radical expressions $f(x) = \sqrt{x}$ and $f(x) = \sqrt[3]{x}$.

Figure 1 shows the graph of the **square root function,** together with a table of selected points. Only nonnegative values can be used for x, so the domain is $[0, \infty)$. Because $\sqrt{x}$ is the principal square root of x, it always has a nonnegative value, so the range is also $[0, \infty)$.

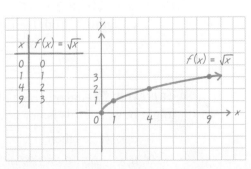

FIGURE 1

Figure 2 shows the graph of the **cube root function** and a table of selected points. Since any real number (positive, negative, or 0) can be used for x in the cube root function, $\sqrt[3]{x}$ can be positive, negative, or 0. Thus, both the domain and the range of the cube root function are $(-\infty, \infty)$.

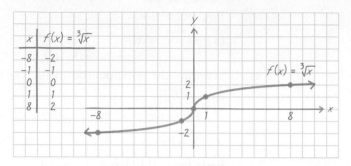

FIGURE 2

EXAMPLE 3 Graphing Functions Defined with Radicals

Graph each function by creating a table of values. Give the domain and range.

(a) $f(x) = \sqrt{x - 3}$

A table of values is given with Figure 3. The x-values were chosen in such a way that the function values are all integers. For the radicand to be nonnegative, we must have $x - 3 \geq 0$, or $x \geq 3$. Therefore, the domain is $[3, \infty)$. Function values are positive or 0, so the range is $[0, \infty)$. The graph is shown in Figure 3.

x	$f(x) = \sqrt{x - 3}$
3	$\sqrt{3 - 3} = 0$
4	$\sqrt{4 - 3} = 1$
7	$\sqrt{7 - 3} = 2$

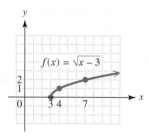

FIGURE 3

x	$f(x) = \sqrt[3]{x} + 2$
-8	$\sqrt[3]{-8} + 2 = 0$
-1	$\sqrt[3]{-1} + 2 = 1$
0	$\sqrt[3]{0} + 2 = 2$
1	$\sqrt[3]{1} + 2 = 3$
8	$\sqrt[3]{8} + 2 = 4$

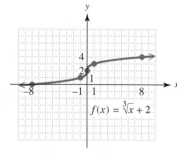

FIGURE 4

(b) $f(x) = \sqrt[3]{x} + 2$

See Figure 4. Both the domain and range are $(-\infty, \infty)$.

✔ **Now Try Exercises 41 and 45.**

OBJECTIVE 4 Find nth roots of nth powers. What does $\sqrt{a^2}$ equal? Your first answer might be a, but this is not necessarily true. For example, consider the following:

If $a = 6$, then $\sqrt{a^2} = \sqrt{6^2} = \sqrt{36} = 6$.

If $a = -6$, then $\sqrt{a^2} = \sqrt{(-6)^2} = \sqrt{36} = 6$. ← Instead of -6, we get 6, the *absolute value* of -6.

Since the symbol $\sqrt{a^2}$ represents the *nonnegative* square root, we express $\sqrt{a^2}$ with absolute value bars as $|a|$ in case a is a negative number.

$\sqrt{a^2}$

For any real number a, $\qquad \sqrt{a^2} = |a|$.

That is, the principal square root of a^2 is the absolute value of a.

EXAMPLE 4 Simplifying Square Roots by Using Absolute Value

Find each square root.

(a) $\sqrt{7^2} = |7| = 7$

(b) $\sqrt{(-7)^2} = |-7| = 7$

(c) $\sqrt{k^2} = |k|$

(d) $\sqrt{(-k)^2} = |-k| = |k|$

✔ **Now Try Exercises 49, 51, 59, and 61.**

We can generalize this idea to any nth root.

$\sqrt[n]{a^n}$

If n is an *even* positive integer, then $\qquad \sqrt[n]{a^n} = |a|$.

If n is an *odd* positive integer, then $\qquad \sqrt[n]{a^n} = a$.

That is, use absolute value when n is even; absolute value is not necessary when n is odd.

EXAMPLE 5 Simplifying Higher Roots by Using Absolute Value

Simplify each root.

(a) $\sqrt[6]{(-3)^6} = |-3| = 3 \qquad$ n is even; use absolute value.

(b) $\sqrt[5]{(-4)^5} = -4 \qquad$ n is odd.

(c) $-\sqrt[4]{(-9)^4} = -|-9| = -9 \qquad$ n is even; use absolute value.

(d) $-\sqrt{m^4} = -|m^2| = -m^2$

No absolute value bars are needed here, because m^2 is nonnegative for any real number value of m.

(e) $\sqrt[3]{a^{12}} = a^4$, because $a^{12} = (a^4)^3$.

(f) $\sqrt[4]{x^{12}} = |x^3|$

We use absolute value bars to guarantee that the result is not negative (because x^3 can be either positive or negative, depending on x). Also, $|x^3|$ can be written as $x^2 \cdot |x|$.

✔ **Now Try Exercises 53, 55, 57, and 65.**

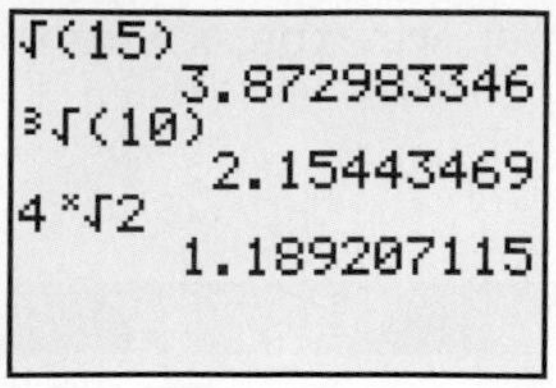

```
√(15)
        3.872983346
³√(10)
        2.15443469
4ˣ√2
        1.189207115
```

(a)

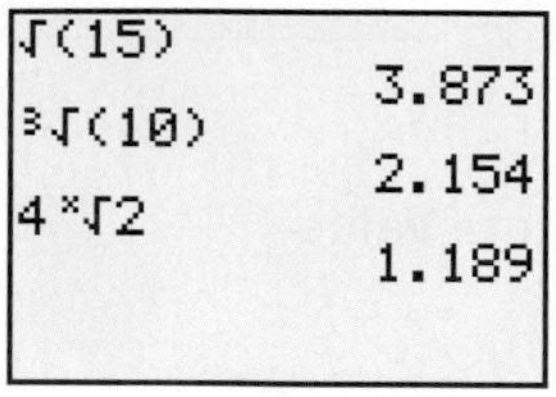

```
√(15)
            3.873
³√(10)
            2.154
4ˣ√2
            1.189
```

(b)

FIGURE 5

OBJECTIVE 5 Use a calculator to find roots. While numbers such as $\sqrt{9}$ and $\sqrt[3]{-8}$ are rational, radicals are often irrational numbers. To find approximations of roots such as $\sqrt{15}$, $\sqrt[3]{10}$, and $\sqrt[4]{2}$, we usually use scientific or graphing calculators. Using a calculator, we find that

$$\sqrt{15} \approx 3.872983346, \quad \sqrt[3]{10} \approx 2.15443469, \quad \text{and} \quad \sqrt[4]{2} \approx 1.189207115,$$

where the symbol $\approx$ means "is approximately equal to." In this book, we usually show approximations rounded to three decimal places. Thus, we would write

$$\sqrt{15} \approx 3.873, \quad \sqrt[3]{10} \approx 2.154, \quad \text{and} \quad \sqrt[4]{2} \approx 1.189.$$

Figure 5 shows how the preceding approximations are displayed on a TI-83/84 Plus graphing calculator. In Figure 5(a), eight or nine decimal places are shown, while in Figure 5(b), the number of decimal places is fixed at three.

There is a simple way to check that a calculator approximation is "in the ball-park." Because 16 is a little larger than 15, $\sqrt{16} = 4$ should be a little larger than $\sqrt{15}$. Thus, 3.873 is reasonable as an approximation for $\sqrt{15}$.

▶ **NOTE** The methods for finding approximations differ among makes and models of calculators. *You should always consult your owner's manual for keystroke instructions.* Be aware that graphing calculators often differ from scientific calculators in the order in which keystrokes are made.

EXAMPLE 6 Finding Approximations for Roots

Use a calculator to verify that each approximation is correct.

(a) $\sqrt{39} \approx 6.245$ **(b)** $-\sqrt{72} \approx -8.485$

(c) $\sqrt[3]{93} \approx 4.531$ **(d)** $\sqrt[4]{39} \approx 2.499$

✔ **Now Try Exercises 69, 73, 75, and 77.**

EXAMPLE 7 Using Roots to Calculate Resonant Frequency

In electronics, the resonant frequency f of a circuit may be found by the formula

$$f = \frac{1}{2\pi\sqrt{LC}},$$

where f is in cycles per second, L is in henrys, and C is in farads. (Henrys and farads are units of measure in electronics.) Find the resonant frequency f if $L = 5 \times 10^{-4}$ henry and $C = 3 \times 10^{-10}$ farad. Give your answer to the nearest thousand.

Find the value of f when $L = 5 \times 10^{-4}$ and $C = 3 \times 10^{-10}$.

$$f = \frac{1}{2\pi\sqrt{LC}} \qquad \text{Given formula}$$

$$= \frac{1}{2\pi\sqrt{(5 \times 10^{-4})(3 \times 10^{-10})}} \qquad \text{Substitute for } L \text{ and } C.$$

$$\approx 411{,}000 \qquad \text{Use a calculator.}$$

The resonant frequency f is approximately 411,000 cycles per sec.

✔ **Now Try Exercise 81.**

8.1 EXERCISES

Match each expression from Column I with the equivalent choice from Column II. Answers may be used more than once. See Examples 1 and 2.

I

1. $-\sqrt{16}$ **2.** $\sqrt{-16}$

3. $\sqrt[3]{-27}$ **4.** $\sqrt[5]{-32}$

● **5.** $\sqrt[4]{81}$ **6.** $\sqrt[3]{8}$

II

A. 3   **B.** -2

C. 2   **D.** -3

E. -4 **F.** Not a real number

Concept Check   *Choose the closest approximation of each square root.*

7. $\sqrt{123.5}$

   A. 9   **B.** 10   **C.** 11   **D.** 12

8. $\sqrt{67.8}$

   A. 7   **B.** 8   **C.** 9   **D.** 10

Refer to the rectangle to answer the questions in Exercises 9 and 10.

9. Which one of the following is the best estimate of its area?

   A. 2500   **B.** 250   **C.** 50   **D.** 100

10. Which one of the following is the best estimate of its perimeter?

   A. 15   **B.** 250   **C.** 100   **D.** 30

11. *Concept Check*   Consider the expression $-\sqrt{-a}$. Decide whether it is *positive, negative, 0,* or *not a real number* if

   (a) $a > 0$,   **(b)** $a < 0$,   **(c)** $a = 0$.

12. *Concept Check*   If n is odd, under what conditions is $\sqrt[n]{a}$

   (a) positive,   **(b)** negative,   **(c)** 0?

Find each root that is a real number. See Examples 1 and 2.

● **13.** $-\sqrt{81}$ **14.** $-\sqrt{121}$ **15.** $\sqrt[3]{216}$ **16.** $\sqrt[3]{343}$

17. $\sqrt[3]{-64}$ **18.** $\sqrt[3]{-125}$ **19.** $-\sqrt[3]{512}$ **20.** $-\sqrt[3]{1000}$

21. $\sqrt[4]{1296}$ **22.** $\sqrt[4]{625}$ **23.** $-\sqrt[4]{16}$ **24.** $-\sqrt[4]{256}$

25. $\sqrt[4]{-625}$ **26.** $\sqrt[4]{-256}$ **27.** $\sqrt[6]{64}$ **28.** $\sqrt[6]{729}$

29. $\sqrt[6]{-32}$ **30.** $\sqrt[8]{-1}$ **31.** $\sqrt{\dfrac{64}{81}}$ **32.** $\sqrt{\dfrac{100}{9}}$

33. $\sqrt[3]{\dfrac{64}{27}}$ **34.** $\sqrt[4]{\dfrac{81}{16}}$ **35.** $-\sqrt[6]{\dfrac{1}{64}}$ **36.** $-\sqrt[5]{\dfrac{1}{32}}$

37. $\sqrt{0.49}$ **38.** $\sqrt{0.81}$ **39.** $\sqrt[3]{0.001}$ **40.** $\sqrt[3]{0.125}$

Graph each function and give its domain and range. See Example 3.

● **41.** $f(x) = \sqrt{x+3}$ **42.** $f(x) = \sqrt{x-5}$ **43.** $f(x) = \sqrt{x}-2$

44. $f(x) = \sqrt{x}+4$ **45.** $f(x) = \sqrt[3]{x}-3$ **46.** $f(x) = \sqrt[3]{x}+1$

47. $f(x) = \sqrt[3]{x-3}$ **48.** $f(x) = \sqrt[3]{x+1}$

Simplify each root. See Examples 4 and 5.

49. $\sqrt{12^2}$ **50.** $\sqrt{19^2}$ **51.** $\sqrt{(-10)^2}$ **52.** $\sqrt{(-13)^2}$

53. $\sqrt[6]{(-2)^6}$ **54.** $\sqrt[6]{(-4)^6}$ **55.** $\sqrt[5]{(-9)^5}$ **56.** $\sqrt[5]{(-8)^5}$

57. $-\sqrt[6]{(-5)^6}$ **58.** $-\sqrt[6]{(-7)^6}$ **59.** $\sqrt{x^2}$ **60.** $-\sqrt{x^2}$

61. $\sqrt{(-z)^2}$ **62.** $\sqrt{(-q)^2}$ **63.** $\sqrt[3]{x^3}$ **64.** $-\sqrt[3]{x^3}$

65. $\sqrt[3]{x^{15}}$ **66.** $\sqrt[3]{m^9}$ **67.** $\sqrt[6]{x^{30}}$ **68.** $\sqrt[4]{k^{20}}$

Find a decimal approximation for each radical. Round the answer to three decimal places. See Example 6.

69. $\sqrt{9483}$ **70.** $\sqrt{6825}$ **71.** $\sqrt{284.361}$ **72.** $\sqrt{846.104}$

73. $-\sqrt{82}$ **74.** $-\sqrt{91}$ **75.** $\sqrt[3]{423}$ **76.** $\sqrt[3]{555}$

77. $\sqrt[4]{100}$ **78.** $\sqrt[4]{250}$ **79.** $\sqrt[5]{23.8}$ **80.** $\sqrt[5]{98.4}$

Solve each problem. See Example 7.

81. Use the formula in Example 7 to calculate the resonant frequency of a circuit to the nearest thousand if $L = 7.237 \times 10^{-5}$ henry and $C = 2.5 \times 10^{-10}$ farad.

82. The threshold weight T for a person is the weight above which the risk of death increases greatly. The threshold weight in pounds for men aged 40–49 is related to height h in inches by the formula

$$h = 12.3\sqrt[3]{T}.$$

What height corresponds to a threshold weight of 216 lb for a 43-year-old man? Round your answer to the nearest inch and then to the nearest tenth of a foot.

83. According to an article in *The World Scanner Report,* the distance D, in miles, to the horizon from an observer's point of view over water or "flat" earth is given by

$$D = \sqrt{2H},$$

where H is the height of the point of view, in feet. If a person whose eyes are 6 ft above ground level is standing at the top of a hill 44 ft above "flat" earth, approximately how far to the horizon will she be able to see?

84. The time for one complete swing of a simple pendulum is

$$t = 2\pi\sqrt{\frac{L}{g}},$$

where t is time in seconds, L is the length of the pendulum in feet, and g, the acceleration due to gravity, is about 32 ft per sec². Find the time of a complete swing of a 2-ft pendulum to the nearest tenth of a second.

85. Heron's formula gives a method of finding the area of a triangle if the lengths of its sides are known. Suppose that a, b, and c are the lengths of the sides. Let s denote one-half of the perimeter of the triangle (called the *semiperimeter*); that is,

$$s = \frac{1}{2}(a + b + c).$$

Then the area of the triangle is

$$A = \sqrt{s(s - a)(s - b)(s - c)}.$$

Find the area of the Bermuda Triangle if the "sides" of this triangle measure approximately 850 mi, 925 mi, and 1300 mi. Give your answer to the nearest thousand square miles.

86. The Vietnam Veterans Memorial in Washington, DC, is in the shape of an unenclosed isosceles triangle with equal sides of length 246.75 ft. If the triangle were enclosed, the third side would have length 438.14 ft. Use Heron's formula from the previous exercise to find the area of this enclosure to the nearest hundred square feet. (*Source:* Information pamphlet obtained at the Vietnam Veterans Memorial.)

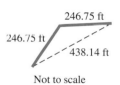

246.75 ft

246.75 ft

438.14 ft

Not to scale

87. The coefficient of self-induction L (in henrys), the energy P stored in an electronic circuit (in joules), and the current I (in amps) are related by the formula

$$I = \sqrt{\frac{2P}{L}}.$$

Find I if $P = 120$ and $L = 80$.

PREVIEW EXERCISES

*Apply the rules for exponents. Write each result with only positive exponents. Assume that all variables represent nonzero real numbers. See **Section 5.1**.*

88. $17^2 \cdot 17^3$

89. $x^5 \cdot x^{-1} \cdot x^{-3}$

90. $(4x^2y^3)(2^3x^5y)$

91. $(13x^0y^5)(13x^4y^3)$

92. $\dfrac{7^0}{7^{-3}}$

93. $\dfrac{5}{5^{-1}}$

94. $\dfrac{11^{-2}}{5^{-1}}$

95. $\left(\dfrac{2}{3}\right)^{-3}$

8.2 Rational Exponents

OBJECTIVE 1 Use exponential notation for *n*th roots. We now look at exponents that are rational numbers of the form $\frac{1}{n}$, or $1/n$, where *n* is a natural number.

Consider the product $(3^{1/2})^2 = 3^{1/2} \cdot 3^{1/2}$. Using the rules of exponents from **Section 5.1,** we can simplify this product as follows:

$$(3^{1/2})^2 = 3^{1/2} \cdot 3^{1/2}$$
$$= 3^{1/2+1/2} \qquad \text{Product rule: } a^m \cdot a^n = a^{m+n}$$
$$= 3^1 \qquad \text{Add exponents.}$$
$$= 3.$$

Also, by definition, $(\sqrt{3})^2 = \sqrt{3} \cdot \sqrt{3} = 3$. Since both $(3^{1/2})^2$ and $(\sqrt{3})^2$ are equal to 3, it seems reasonable to define

$$3^{1/2} = \sqrt{3}.$$

This suggests the following generalization.

$a^{1/n}$

If $\sqrt[n]{a}$ is a real number, then $\qquad a^{1/n} = \sqrt[n]{a}.$

For example,

$$4^{1/2} = \sqrt{4}, \quad 8^{1/3} = \sqrt[3]{8}, \quad \text{and} \quad 16^{1/4} = \sqrt[4]{16}.$$

Notice that the denominator of the rational exponent is the index of the radical.

EXAMPLE 1 Evaluating Exponentials of the Form $a^{1/n}$

Evaluate each exponential.

> The denominator
> is the index.

> The denominator is
> the index.
> $\sqrt{}$ means $\sqrt[2]{}$.

(a) $64^{1/3} = \sqrt[3]{64} = 4$ ⟶ **(b)** $100^{1/2} = \sqrt{100} = 10$

(c) $-256^{1/4} = -\sqrt[4]{256} = -4$

(d) $(-256)^{1/4} = \sqrt[4]{-256}$ is not a real number, because the radicand, -256, is negative and the index is even.

(e) $(-32)^{1/5} = \sqrt[5]{-32} = -2$ **(f)** $\left(\dfrac{1}{8}\right)^{1/3} = \sqrt[3]{\dfrac{1}{8}} = \dfrac{1}{2}$

✔ **Now Try Exercises 11, 13, 19, and 21.**

▶ **CAUTION** Notice the difference between parts (c) and (d) in Example 1. The radical in part (c) is the *negative fourth root of a positive number,* while the radical in part (d) is the *principal fourth root of a negative number, which is not a real number.*

OBJECTIVE 2 Define and use expressions of the form $a^{m/n}$. We know that $8^{1/3} = \sqrt[3]{8}$. How should we define a number like $8^{2/3}$? For past rules of exponents to be valid,

$$8^{2/3} = 8^{(1/3)2} = (8^{1/3})^2.$$

Since $8^{1/3} = \sqrt[3]{8}$,

$$8^{2/3} = \left(\sqrt[3]{8}\right)^2 = 2^2 = 4.$$

Generalizing from this example, we define $a^{m/n}$ as follows.

$a^{m/n}$

If m and n are positive integers with m/n in lowest terms, then

$$a^{m/n} = (a^{1/n})^m,$$

provided that $a^{1/n}$ is a real number. If $a^{1/n}$ is not a real number, then $a^{m/n}$ is not a real number.

EXAMPLE 2 Evaluating Exponentials of the Form $a^{m/n}$

Evaluate each exponential.

Think:
$36^{1/2} = \sqrt{36} = 6$

Think:
$125^{1/3} = \sqrt[3]{125} = 5$

(a) $36^{3/2} = (36^{1/2})^3 = 6^3 = 216$ **(b)** $125^{2/3} = (125^{1/3})^2 = 5^2 = 25$

Be careful.
The base is 4.

(c) $-4^{5/2} = -(4^{5/2}) = -(4^{1/2})^5 = -(2)^5 = -32$

Because the base here is 4, the negative sign is *not* affected by the exponent.

(d) $(-27)^{2/3} = [(-27)^{1/3}]^2 = (-3)^2 = 9$

Notice how the $-$ sign is used in parts (c) and (d). In part (c), we first evaluate the exponential and then find its negative. In part (d), the $-$ sign is part of the base, -27.

(e) $(-100)^{3/2} = [(-100)^{1/2}]^3$, which is not a real number, since $(-100)^{1/2}$, or $\sqrt{-100}$, is not a real number.

✔ **Now Try Exercises 23, 27, and 29.**

When a rational exponent is negative, the earlier interpretation of negative exponents is applied.

$a^{-m/n}$

If $a^{m/n}$ is a real number, then

$$a^{-m/n} = \frac{1}{a^{m/n}} \quad (a \neq 0).$$

EXAMPLE 3 Evaluating Exponentials with Negative Rational Exponents

Evaluate each exponential.

(a) $16^{-3/4} = \dfrac{1}{16^{3/4}} = \dfrac{1}{(16^{1/4})^3} = \dfrac{1}{(\sqrt[4]{16})^3} = \dfrac{1}{2^3} = \dfrac{1}{8}$

> The denominator of 3/4 is the index and the numerator is the exponent.

(b) $25^{-3/2} = \dfrac{1}{25^{3/2}} = \dfrac{1}{(25^{1/2})^3} = \dfrac{1}{(\sqrt{25})^3} = \dfrac{1}{5^3} = \dfrac{1}{125}$

(c) $\left(\dfrac{8}{27}\right)^{-2/3} = \dfrac{1}{\left(\dfrac{8}{27}\right)^{2/3}} = \dfrac{1}{\left(\sqrt[3]{\dfrac{8}{27}}\right)^2} = \dfrac{1}{\left(\dfrac{2}{3}\right)^2} = \dfrac{1}{\dfrac{4}{9}} = \dfrac{9}{4}$

> $\dfrac{1}{\frac{4}{9}} = 1 \div \dfrac{4}{9} = 1 \cdot \dfrac{9}{4}$

We could also use the rule $\left(\dfrac{b}{a}\right)^{-m} = \left(\dfrac{a}{b}\right)^m$ here, as follows:

$$\left(\dfrac{8}{27}\right)^{-2/3} = \left(\dfrac{27}{8}\right)^{2/3} = \left(\sqrt[3]{\dfrac{27}{8}}\right)^2 = \left(\dfrac{3}{2}\right)^2 = \dfrac{9}{4}.$$

> Take the reciprocal only of the base, *not* the exponent.

✔ **Now Try Exercises 31 and 35.**

▶ **CAUTION** Be careful to distinguish between exponential expressions like the following:

$$16^{-1/4} = \dfrac{1}{2}, \qquad -16^{1/4} = -2, \qquad \text{and} \qquad -16^{-1/4} = -\dfrac{1}{2}.$$

A negative exponent does not necessarily lead to a negative result. Negative exponents lead to reciprocals, which may be positive.

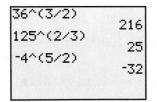

```
36^(3/2)
            216
125^(2/3)
             25
-4^(5/2)
            -32
```

FIGURE 6

The screens in Figures 6 and 7 illustrate how a graphing calculator performs some of the evaluations seen in Examples 2 and 3.

We obtain an alternative definition of $a^{m/n}$ by using the power rule for exponents a little differently than in the earlier definition. If all indicated roots are real numbers,

then $\qquad a^{m/n} = a^{m(1/n)} = (a^m)^{1/n},$

so $\qquad a^{m/n} = (a^m)^{1/n}.$

```
(-27)^(2/3)
              9
16^(-3/4)▶Frac
            1/8
```

FIGURE 7

$a^{m/n}$

If all indicated roots are real numbers, then

$$a^{m/n} = (a^{1/n})^m = (a^m)^{1/n}.$$

We can now evaluate an expression such as $27^{2/3}$ in two ways:

$$27^{2/3} = (27^{1/3})^2 = 3^2 = 9$$

The result is the same.

or $\qquad 27^{2/3} = (27^2)^{1/3} = 729^{1/3} = 9.$

In most cases, it is easier to use $(a^{1/n})^m$.

This rule can also be expressed with radicals as follows.

Radical Form of $a^{m/n}$

If all indicated roots are real numbers, then

$$a^{m/n} = \sqrt[n]{a^m} = \left(\sqrt[n]{a}\right)^m.$$

That is, raise a to the mth power and then take the nth root, or take the nth root of a and then raise to the mth power.

For example,

$$8^{2/3} = \sqrt[3]{8^2} = \sqrt[3]{64} = 4, \qquad \text{and} \qquad 8^{2/3} = \left(\sqrt[3]{8}\right)^2 = 2^2 = 4,$$

so $\qquad 8^{2/3} = \sqrt[3]{8^2} = \left(\sqrt[3]{8}\right)^2.$

OBJECTIVE 3 Convert between radicals and rational exponents. Using the definition of rational exponents, we can simplify many problems involving radicals by converting the radicals to numbers with rational exponents. After simplifying, we convert the answer back to radical form.

EXAMPLE 4 Converting between Rational Exponents and Radicals

Write each exponential as a radical. Assume that all variables represent positive real numbers. Use the definition that takes the root first.

(a) $13^{1/2} = \sqrt{13}$
 (b) $6^{3/4} = \left(\sqrt[4]{6}\right)^3$

(c) $9m^{5/8} = 9\left(\sqrt[8]{m}\right)^5$
 (d) $6x^{2/3} - (4x)^{3/5} = 6\left(\sqrt[3]{x}\right)^2 - \left(\sqrt[5]{4x}\right)^3$

(e) $r^{-2/3} = \dfrac{1}{r^{2/3}} = \dfrac{1}{\left(\sqrt[3]{r}\right)^2}$

(f) $(a^2 + b^2)^{1/2} = \sqrt{a^2 + b^2}$ $\qquad \boxed{\sqrt{a^2 + b^2} \neq a + b}$

In (g)–(i), write each radical as an exponential. Simplify. Assume that all variables represent positive real numbers.

(g) $\sqrt{10} = 10^{1/2}$
 (h) $\sqrt[4]{3^8} = 3^{8/4} = 3^2 = 9$

(i) $\sqrt[6]{z^6} = z$, since z is positive.

✔ **Now Try Exercises 37, 39, 41, 53, and 55.**

▶ **NOTE** In Example 4(i), it was not necessary to use absolute value bars, since the directions specifically stated that the variable represents a positive real number. Because the absolute value of the positive real number z is z itself, the answer is simply z.

OBJECTIVE ④ Use the rules for exponents with rational exponents. The definition of rational exponents allows us to apply the rules for exponents from **Section 5.1.**

Rules for Rational Exponents

Let r and s be rational numbers. For all real numbers a and b for which the indicated expressions exist,

$$a^r \cdot a^s = a^{r+s} \qquad a^{-r} = \frac{1}{a^r} \qquad \frac{a^r}{a^s} = a^{r-s} \qquad \left(\frac{a}{b}\right)^{-r} = \frac{b^r}{a^r}$$

$$(a^r)^s = a^{rs} \qquad (ab)^r = a^r b^r \qquad \left(\frac{a}{b}\right)^r = \frac{a^r}{b^r} \qquad a^{-r} = \left(\frac{1}{a}\right)^r.$$

EXAMPLE 5 Applying Rules for Rational Exponents

Write with only positive exponents. Assume that all variables represent positive real numbers.

(a) $2^{1/2} \cdot 2^{1/4} = 2^{1/2+1/4} = 2^{3/4}$ Product rule

(b) $\dfrac{5^{2/3}}{5^{7/3}} = 5^{2/3-7/3} = 5^{-5/3} = \dfrac{1}{5^{5/3}}$ Quotient rule

(c) $\dfrac{(x^{1/2}y^{2/3})^4}{y} = \dfrac{(x^{1/2})^4(y^{2/3})^4}{y}$ Power rule

$\qquad\qquad = \dfrac{x^2 y^{8/3}}{y^1}$ Power rule

$\qquad\qquad = x^2 y^{8/3-1}$ Quotient rule

$\qquad\qquad = x^2 y^{5/3}$

(d) $\left(\dfrac{x^4 y^{-6}}{x^{-2} y^{1/3}}\right)^{-2/3} = \dfrac{(x^4)^{-2/3}(y^{-6})^{-2/3}}{(x^{-2})^{-2/3}(y^{1/3})^{-2/3}}$

$\qquad\qquad = \dfrac{x^{-8/3} y^4}{x^{4/3} y^{-2/9}}$ Power rule

$\qquad\qquad = x^{-8/3-4/3} y^{4-(-2/9)}$ Quotient rule

$\qquad\qquad = x^{-4} y^{38/9}$ *(Use parentheses to avoid errors.)*

$\qquad\qquad = \dfrac{y^{38/9}}{x^4}$ Definition of negative exponent

The same result is obtained if we simplify within the parentheses first, leading to $(x^6 y^{-19/3})^{-2/3}$. Then apply the power rule. (Show that the result is the same.)

(e) $m^{3/4}(m^{5/4} - m^{1/4}) = m^{3/4} \cdot m^{5/4} - m^{3/4} \cdot m^{1/4}$ Distributive property

$\qquad\qquad = m^{3/4+5/4} - m^{3/4+1/4}$ Product rule

(Do not make the common mistake of multiplying exponents in the first step.)

$\qquad\qquad = m^{8/4} - m^{4/4}$

$\qquad\qquad = m^2 - m$

✔ **Now Try Exercises 61, 63, 71, 81, and 83.**

▶ **CAUTION** Use the rules of exponents in problems like those in Example 5. Do not convert the expressions to radical form.

EXAMPLE 6 Applying Rules for Rational Exponents

Write all radicals as exponentials, and then apply the rules for rational exponents. Leave answers in exponential form. Assume that all variables represent positive real numbers.

(a) $\sqrt[3]{x^2} \cdot \sqrt[4]{x} = x^{2/3} \cdot x^{1/4}$ Convert to rational exponents.

$\qquad\qquad\quad = x^{2/3+1/4}$ Product rule

$\qquad\qquad\quad = x^{8/12+3/12}$ Write exponents with a common denominator.

$\qquad\qquad\quad = x^{11/12}$

(b) $\dfrac{\sqrt{x^3}}{\sqrt[3]{x^2}} = \dfrac{x^{3/2}}{x^{2/3}} = x^{3/2-2/3} = x^{5/6}$ $\quad \dfrac{3}{2} - \dfrac{2}{3} = \dfrac{9}{6} - \dfrac{4}{6} = \dfrac{5}{6}$

(c) $\sqrt{\sqrt[4]{z}} = \sqrt{z^{1/1}} = (z^{1/4})^{1/2} = z^{1/8}$

✔ **Now Try Exercises 89, 91, and 95.**

8.2 EXERCISES

🌐 *Complete solution available on Video Lectures on CD/DVD*

Now Try Exercise

Concept Check Match each expression from Column I with the equivalent choice from Column II.

I		II	
1. $2^{1/2}$	**2.** $(-27)^{1/3}$	**A.** -4	**B.** 8
3. $-16^{1/2}$	**4.** $(-16)^{1/2}$	**C.** $\sqrt{2}$	**D.** $-\sqrt{6}$
5. $(-32)^{1/5}$	**6.** $(-32)^{2/5}$	**E.** -3	**F.** $\sqrt{6}$
7. $4^{3/2}$	**8.** $6^{2/4}$	**G.** 4	**H.** -2
9. $-6^{2/4}$	**10.** $36^{0.5}$	**I.** 6	**J.** Not a real number

Evaluate each exponential. See Examples 1–3.

🌐 **11.** $169^{1/2}$ $\qquad$ **12.** $121^{1/2}$ $\qquad$ **13.** $729^{1/3}$ $\qquad$ **14.** $512^{1/3}$

15. $16^{1/4}$ $\qquad$ **16.** $625^{1/4}$ $\qquad$ **17.** $\left(\dfrac{64}{81}\right)^{1/2}$ $\qquad$ **18.** $\left(\dfrac{8}{27}\right)^{1/3}$

19. $(-27)^{1/3}$ $\qquad$ **20.** $(-32)^{1/5}$ $\qquad$ **21.** $(-144)^{1/2}$ $\qquad$ **22.** $(-36)^{1/2}$

🌐 **23.** $100^{3/2}$ $\qquad$ **24.** $64^{3/2}$ $\qquad$ **25.** $81^{3/4}$ $\qquad$ **26.** $216^{2/3}$

27. $-16^{5/2}$ $\qquad$ **28.** $-32^{3/5}$ $\qquad$ **29.** $(-8)^{4/3}$ $\qquad$ **30.** $(-243)^{2/5}$

🌐 **31.** $32^{-3/5}$ $\qquad$ **32.** $27^{-4/3}$ $\qquad$ **33.** $64^{-3/2}$ $\qquad$ **34.** $81^{-3/2}$

35. $\left(\dfrac{125}{27}\right)^{-2/3}$ $\qquad$ **36.** $\left(\dfrac{64}{125}\right)^{-2/3}$

Write with radicals. Assume that all variables represent positive real numbers. See Example 4.

37. $10^{1/2}$ **38.** $3^{1/2}$ **39.** $8^{3/4}$

40. $7^{2/3}$ **41.** $(9q)^{5/8} - (2x)^{2/3}$ **42.** $(3p)^{3/4} + (4x)^{1/3}$

43. $(2m)^{-3/2}$ **44.** $(5y)^{-3/5}$ **45.** $(2y + x)^{2/3}$

46. $(r + 2z)^{3/2}$ **47.** $(3m^4 + 2k^2)^{-2/3}$ **48.** $(5x^2 + 3z^3)^{-5/6}$

49. Show that, in general, $\sqrt{a^2 + b^2} \neq a + b$ by replacing a with 3 and b with 4.

50. Suppose someone claims that $\sqrt[n]{a^n + b^n}$ must equal $a + b$, since, when $a = 1$ and $b = 0$, a true statement results: $\sqrt[n]{a^n + b^n} = \sqrt[n]{1^n + 0^n} = \sqrt[n]{1^n} = 1 = 1 + 0 = a + b$. Explain why this is faulty reasoning.

Simplify by first converting to rational exponents. Assume that all variables represent positive real numbers. See Example 4.

51. $\sqrt{2^{12}}$ **52.** $\sqrt{5^{10}}$ **53.** $\sqrt[3]{4^9}$ **54.** $\sqrt[4]{6^8}$ **55.** $\sqrt{x^{20}}$

56. $\sqrt{r^{50}}$ **57.** $\sqrt[3]{x} \cdot \sqrt{x}$ **58.** $\sqrt[4]{y} \cdot \sqrt[5]{y^2}$ **59.** $\dfrac{\sqrt[3]{t^4}}{\sqrt[5]{t^4}}$ **60.** $\dfrac{\sqrt[4]{w^3}}{\sqrt[6]{w}}$

Simplify each expression. Write all answers with positive exponents. Assume that all variables represent positive real numbers. See Example 5.

61. $3^{1/2} \cdot 3^{3/2}$ **62.** $6^{4/3} \cdot 6^{2/3}$ **63.** $\dfrac{64^{5/3}}{64^{4/3}}$

64. $\dfrac{125^{7/3}}{125^{5/3}}$ **65.** $y^{7/3} \cdot y^{-4/3}$ **66.** $r^{-8/9} \cdot r^{17/9}$

67. $x^{2/3} \cdot x^{-1/4}$ **68.** $x^{2/5} \cdot x^{-1/3}$ **69.** $\dfrac{k^{1/3}}{k^{2/3} \cdot k^{-1}}$

70. $\dfrac{z^{3/4}}{z^{5/4} \cdot z^{-2}}$ **71.** $\dfrac{(x^{1/4}y^{2/5})^{20}}{x^2}$ **72.** $\dfrac{(r^{1/5}s^{2/3})^{15}}{r^2}$

73. $\dfrac{(x^{2/3})^2}{(x^2)^{7/3}}$ **74.** $\dfrac{(p^3)^{1/4}}{(p^{5/4})^2}$ **75.** $\dfrac{m^{3/4}n^{-1/4}}{(m^2n)^{1/2}}$

76. $\dfrac{(a^2b^5)^{-1/4}}{(a^{-3}b^2)^{1/6}}$ **77.** $\dfrac{p^{1/5}p^{7/10}p^{1/2}}{(p^3)^{-1/5}}$ **78.** $\dfrac{z^{1/3}z^{-2/3}z^{1/6}}{(z^{-1/6})^3}$

79. $\left(\dfrac{b^{-3/2}}{c^{-5/3}}\right)^2 (b^{-1/4}c^{-1/3})^{-1}$ **80.** $\left(\dfrac{m^{-2/3}}{a^{-3/4}}\right)^4 (m^{-3/8}a^{1/4})^{-2}$ **81.** $\left(\dfrac{p^{-1/4}q^{-3/2}}{3^{-1}p^{-2}q^{-2/3}}\right)^{-2}$

82. $\left(\dfrac{2^{-2}w^{-3/4}x^{-5/8}}{w^{3/4}x^{-1/2}}\right)^{-3}$ **83.** $p^{2/3}(p^{1/3} + 2p^{4/3})$ **84.** $z^{5/8}(3z^{5/8} + 5z^{11/8})$

85. $k^{1/4}(k^{3/2} - k^{1/2})$ **86.** $r^{3/5}(r^{1/2} + r^{3/4})$ **87.** $6a^{7/4}(a^{-7/4} + 3a^{-3/4})$

88. $4m^{5/3}(m^{-2/3} - 4m^{-5/3})$

Write with rational exponents, and then apply the properties of exponents. Assume that all radicands represent positive real numbers. Give answers in exponential form. See Example 6.

89. $\sqrt[5]{x^3} \cdot \sqrt[4]{x}$ **90.** $\sqrt[6]{y^5} \cdot \sqrt[3]{y^2}$ **91.** $\dfrac{\sqrt{x^5}}{\sqrt{x^8}}$ **92.** $\dfrac{\sqrt[3]{k^5}}{\sqrt[3]{k^7}}$

93. $\sqrt{y} \cdot \sqrt[3]{yz}$ **94.** $\sqrt[3]{xz} \cdot \sqrt{z}$ **95.** $\sqrt[4]{\sqrt[3]{m}}$ **96.** $\sqrt[3]{\sqrt{k}}$

97. $\sqrt{\sqrt[3]{\sqrt[4]{x}}}$ **98.** $\sqrt[3]{\sqrt[5]{\sqrt{y}}}$

Solve each problem.

99. Meteorologists can determine the duration of a storm by using the function defined by

$$T(D) = 0.07D^{3/2},$$

where D is the diameter of the storm in miles and T is the time in hours. Find the duration of a storm with a diameter of 16 mi. Round your answer to the nearest tenth of an hour.

100. The threshold weight T, in pounds, for a person is the weight above which the risk of death increases greatly. The threshold weight in pounds for men aged 40–49 is related to height h in inches by the function defined by

$$h(T) = (1860.867T)^{1/3}.$$

What height corresponds to a threshold weight of 200 lb for a 46-yr-old man? Round your answer to the nearest inch and then to the nearest tenth of a foot.

*The **windchill factor** is a measure of the cooling effect that the wind has on a person's skin. It calculates the equivalent cooling temperature if there were no wind. The National Weather Service uses the formula*

$$\text{Windchill temperature} = 35.74 + 0.6215T - 35.75V^{4/25} + 0.4275TV^{4/25},$$

where T is the temperature in °F and V is the wind speed in miles per hour, to calculate windchill. The chart gives the windchill factor for various wind speeds and temperatures at which frostbite is a risk, and how quickly it may occur.

Temperature (°F)

Calm	40	30	20	10	0	−10	−20	−30	−40
5	36	25	13	1	−11	−22	−34	−46	−57
10	34	21	9	−4	−16	−28	−41	−53	−66
15	32	19	6	−7	−19	−32	−45	−58	−71
20	30	17	4	−9	−22	−35	−48	−61	−74
25	29	16	3	−11	−24	−37	−51	−64	−78
30	28	15	1	−12	−26	−39	−53	−67	−80
35	28	14	0	−14	−27	−41	−55	−69	−82
40	27	13	−1	−15	−29	−43	−57	−71	−84

(left axis: Wind speed (mph))

Frostbites times: ☐ 30 minutes   ■ 10 minutes   ■ 5 minutes

Source: National Oceanic and Atmospheric Administration, National Weather Service.

Use the formula and a calculator to determine the windchill to the nearest tenth of a degree, given the following conditions. Compare your answers with the appropriate entries in the table.

101. 30°F, 15-mph wind

102. 10°F, 30-mph wind

PREVIEW EXERCISES

*Simplify each pair of expressions, and then compare the results. Use a calculator if necessary. See **Section 8.1**.*

103. $\sqrt{25} \cdot \sqrt{36}, \quad \sqrt{25 \cdot 36}$

104. $\sqrt[3]{8} \cdot \sqrt[3]{27}, \quad \sqrt[3]{8 \cdot 27}$

105. $\dfrac{\sqrt[3]{27}}{\sqrt[3]{729}}, \quad \sqrt[3]{\dfrac{27}{729}}$

106. $\dfrac{\sqrt{16}}{\sqrt{64}}, \quad \sqrt{\dfrac{16}{64}}$

8.3 Simplifying Radical Expressions

OBJECTIVE 1 Use the product rule for radicals. We now develop rules for multiplying and dividing radicals that have the same index. For example, is the product of two nth-root radicals equal to the nth root of the product of the radicands? Are the expressions $\sqrt{36 \cdot 4}$ and $\sqrt{36} \cdot \sqrt{4}$ equal?

$$\sqrt{36 \cdot 4} = \sqrt{144} = 12$$
$$\sqrt{36} \cdot \sqrt{4} = 6 \cdot 2 = 12$$

The result is the same.

This is an example of the **product rule for radicals.**

Product Rule for Radicals

If $\sqrt[n]{a}$ and $\sqrt[n]{b}$ are real numbers and n is a natural number, then

$$\sqrt[n]{a} \cdot \sqrt[n]{b} = \sqrt[n]{ab}.$$

That is, the product of two nth roots is the nth root of the product.

We justify the product rule by using the rules for rational exponents. Since $\sqrt[n]{a} = a^{1/n}$ and $\sqrt[n]{b} = b^{1/n}$,

$$\sqrt[n]{a} \cdot \sqrt[n]{b} = a^{1/n} \cdot b^{1/n} = (ab)^{1/n} = \sqrt[n]{ab}.$$

▶ **CAUTION** *Use the product rule only when the radicals have the same index.*

EXAMPLE 1 Using the Product Rule

Multiply. Assume that all variables represent positive real numbers.

(a) $\sqrt{5} \cdot \sqrt{7} = \sqrt{5 \cdot 7} = \sqrt{35}$ **(b)** $\sqrt{2} \cdot \sqrt{19} = \sqrt{2 \cdot 19} = \sqrt{38}$

(c) $\sqrt{11} \cdot \sqrt{p} = \sqrt{11p}$ **(d)** $\sqrt{7} \cdot \sqrt{11xyz} = \sqrt{77xyz}$

✔ **Now Try Exercises 7, 9, and 11.**

EXAMPLE 2 Using the Product Rule

Multiply. Assume that all variables represent positive real numbers.

Remember to write the index.

(a) $\sqrt[3]{3} \cdot \sqrt[3]{12} = \sqrt[3]{3 \cdot 12} = \sqrt[3]{36}$ **(b)** $\sqrt[4]{8y} \cdot \sqrt[4]{3r^2} = \sqrt[4]{24yr^2}$

(c) $\sqrt[6]{10m^4} \cdot \sqrt[6]{5m} = \sqrt[6]{50m^5}$

(d) $\sqrt[4]{2} \cdot \sqrt[5]{2}$ cannot be simplified by using the product rule for radicals, because the indexes (4 and 5) are different.

✔ **Now Try Exercises 13, 15, 17, and 19.**

OBJECTIVE 2 Use the quotient rule for radicals. The **quotient rule for radicals** is similar to the product rule.

Quotient Rule for Radicals

If $\sqrt[n]{a}$ and $\sqrt[n]{b}$ are real numbers, $b \neq 0$, and n is a natural number, then

$$\sqrt[n]{\frac{a}{b}} = \frac{\sqrt[n]{a}}{\sqrt[n]{b}}.$$

That is, the nth root of a quotient is the quotient of the nth roots.

EXAMPLE 3 Using the Quotient Rule

Simplify. Assume that all variables represent positive real numbers.

(a) $\sqrt{\dfrac{16}{25}} = \dfrac{\sqrt{16}}{\sqrt{25}} = \dfrac{4}{5}$

(b) $\sqrt{\dfrac{7}{36}} = \dfrac{\sqrt{7}}{\sqrt{36}} = \dfrac{\sqrt{7}}{6}$

(c) $\sqrt[3]{-\dfrac{8}{125}} = \sqrt[3]{\dfrac{-8}{125}} = \dfrac{\sqrt[3]{-8}}{\sqrt[3]{125}} = \dfrac{-2}{5} = -\dfrac{2}{5}$ $-\dfrac{a}{b} = \dfrac{-a}{b}$

(d) $\sqrt[3]{\dfrac{7}{216}} = \dfrac{\sqrt[3]{7}}{\sqrt[3]{216}} = \dfrac{\sqrt[3]{7}}{6}$

(e) $\sqrt[5]{\dfrac{x}{32}} = \dfrac{\sqrt[5]{x}}{\sqrt[5]{32}} = \dfrac{\sqrt[5]{x}}{2}$

(f) $-\sqrt[3]{\dfrac{m^6}{125}} = -\dfrac{\sqrt[3]{m^6}}{\sqrt[3]{125}} = -\dfrac{m^2}{5}$

> Think: $\sqrt[3]{m^6} = m^{6/3} = m^2$

> ✔ **Now Try Exercises 23, 25, 31, 33, and 35.**

OBJECTIVE 3 Simplify radicals. We use the product and quotient rules to simplify radicals. A radical is **simplified** if the following four conditions are met.

Conditions for a Simplified Radical

1. The radicand has no factor raised to a power greater than or equal to the index.
2. The radicand has no fractions.
3. No denominator contains a radical.
4. Exponents in the radicand and the index of the radical have no common factor (except 1).

EXAMPLE 4 Simplifying Roots of Numbers

Simplify.

(a) $\sqrt{24}$

Check to see whether 24 is divisible by a perfect square (the square of a natural number), such as 4, 9, ... The largest perfect square that divides into 24 is 4. Write 24 as the product of 4 and 6, and then use the product rule.

$$\sqrt{24} = \sqrt{4 \cdot 6} = \sqrt{4} \cdot \sqrt{6} = 2\sqrt{6}$$

(b) $\sqrt{108}$

As shown on the left, the number 108 is divisible by the perfect square 36. If this is not obvious, try factoring 108 into its prime factors, as shown on the right.

$\sqrt{108} = \sqrt{36 \cdot 3}$	$\sqrt{108} = \sqrt{2^2 \cdot 3^3}$	
$= \sqrt{36} \cdot \sqrt{3}$	$= \sqrt{2^2 \cdot 3^2 \cdot 3}$	
$= 6\sqrt{3}$	$= 2 \cdot 3 \cdot \sqrt{3}$	Product rule; $\sqrt{2^2} = 2, \sqrt{3^2} = 3$
	$= 6\sqrt{3}$	

(c) $\sqrt{10}$

No perfect square (other than 1) divides into 10, so $\sqrt{10}$ cannot be simplified further.

(d) $\sqrt[3]{16}$

The largest perfect *cube* that divides into 16 is 8, so write 16 as $8 \cdot 2$ (or factor 16 into prime factors).

$$\sqrt[3]{16} = \sqrt[3]{8 \cdot 2} = \sqrt[3]{8} \cdot \sqrt[3]{2} = 2\sqrt[3]{2}$$

(e) $-\sqrt[4]{162} = -\sqrt[4]{81 \cdot 2}$ 81 is a perfect 4th power.

> **Remember the negative sign in each line.**

$= -\sqrt[4]{81} \cdot \sqrt[4]{2}$ Product rule

$= -3\sqrt[4]{2}$

✔ **Now Try Exercises 39, 41, 47, 49, and 55.**

▶ **CAUTION** *Be careful with which factors belong outside the radical sign and which belong inside.* Note in Example 4(b) how $2 \cdot 3$ is written outside because $\sqrt{2^2} = 2$ and $\sqrt{3^2} = 3$, while the remaining 3 is left inside the radical.

EXAMPLE 5 Simplifying Radicals Involving Variables

Simplify. Assume that all variables represent positive real numbers.

(a) $\sqrt{16m^3} = \sqrt{16m^2 \cdot m}$ Factor.

$= \sqrt{16m^2} \cdot \sqrt{m}$ Product rule

$= 4m\sqrt{m}$

Absolute value bars are not needed around the m in color because all the variables represent *positive* real numbers.

(b) $\sqrt{200k^7q^8} = \sqrt{10^2 \cdot 2 \cdot (k^3)^2 \cdot k \cdot (q^4)^2}$ Factor.

$= 10k^3q^4\sqrt{2k}$ Remove perfect square factors.

(c) $\sqrt[3]{-8x^4y^5} = \sqrt[3]{(-8x^3y^3)(xy^2)}$ Choose $-8x^3y^3$ as the perfect cube that divides into $-8x^4y^5$.

$= \sqrt[3]{-8x^3y^3} \cdot \sqrt[3]{xy^2}$ Product rule

$= -2xy\sqrt[3]{xy^2}$

(d) $-\sqrt[4]{32y^9} = -\sqrt[4]{(16y^8)(2y)}$ $16y^8$ is the largest 4th power that divides $32y^9$.

$\qquad\qquad = -\sqrt[4]{16y^8} \cdot \sqrt[4]{2y}$ Product rule

$\qquad\qquad = -2y^2\sqrt[4]{2y}$

> ✔ **Now Try Exercises 75, 79, 83, and 87.**

> ▶ **NOTE** From Example 5, we see that if a variable is raised to a power with an exponent divisible by 2, it is a perfect square. If it is raised to a power with an exponent divisible by 3, it is a perfect cube. In general, if it is raised to a power with an exponent divisible by n, it is a perfect nth power.

The conditions for a simplified radical given earlier state that an exponent in the radicand and the index of the radical should have no common factor (except 1). The next example shows how to simplify radicals with such common factors.

EXAMPLE 6 Simplifying Radicals by Using Smaller Indexes

Simplify. Assume that all variables represent positive real numbers.

(a) $\sqrt[9]{5^6}$

We write this radical by using rational exponents and then write the exponent in lowest terms. We then express the answer as a radical.

$$\sqrt[9]{5^6} = (5^6)^{1/9} = 5^{6/9} = 5^{2/3} = \sqrt[3]{5^2}, \quad \text{or} \quad \sqrt[3]{25}$$

(b) $\sqrt[4]{p^2} = (p^2)^{1/4} = p^{2/4} = p^{1/2} = \sqrt{p}$ (Recall the assumption that $p > 0$.)

> ✔ **Now Try Exercises 93 and 97.**

These examples suggest the following rule.

> If m is an integer, n and k are natural numbers, and all indicated roots exist, then
> $$\sqrt[kn]{a^{km}} = \sqrt[n]{a^m}.$$

OBJECTIVE 4 Simplify products and quotients of radicals with different indexes. Since the product and quotient rules for radicals apply only when they have the same index, we multiply and divide radicals with different indexes by using rational exponents.

EXAMPLE 7 Multiplying Radicals with Different Indexes

Simplify $\sqrt{7} \cdot \sqrt[3]{2}$.

Because the different indexes, 2 and 3, have a least common index of 6, use rational exponents to write each radical as a sixth root.

$$\sqrt{7} = 7^{1/2} = 7^{3/6} = \sqrt[6]{7^3} = \sqrt[6]{343}$$
$$\sqrt[3]{2} = 2^{1/3} = 2^{2/6} = \sqrt[6]{2^2} = \sqrt[6]{4}$$

Therefore,

$$\sqrt{7} \cdot \sqrt[3]{2} = \sqrt[6]{343} \cdot \sqrt[6]{4} = \sqrt[6]{1372}. \quad \text{Product rule}$$

> ✔ **Now Try Exercise 99.**

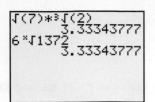

FIGURE 8

Results such as the one in Example 7 can be supported with a calculator, as shown in Figure 8. Notice that the calculator gives the same approximation for the initial product and the final radical that we obtained.

▶ **CAUTION** The computation in Figure 8 is not *proof* that the two expressions are equal. The algebra in Example 7, however, is valid proof of their equality.

OBJECTIVE **5** **Use the Pythagorean formula.** The **Pythagorean formula** relates the lengths of the three sides of a right triangle.

Pythagorean Formula

If c is the length of the longest side of a right triangle and a and b are the lengths of the shorter sides, then

$$c^2 = a^2 + b^2.$$

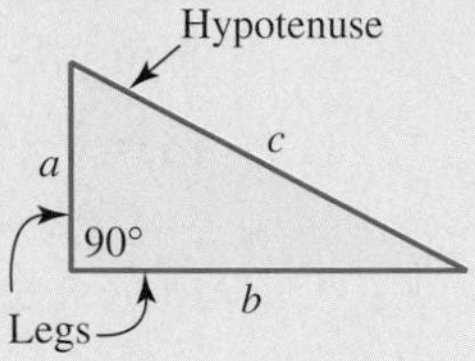

Hypotenuse

The longest side is the **hypotenuse,** and the two shorter sides are the **legs,** of the triangle. The hypotenuse is the side opposite the right angle.

In **Section 9.1** we will see that an equation such as $x^2 = 7$ has two solutions: $\sqrt{7}$ (the principal, or positive, square root of 7) and $-\sqrt{7}$. Similarly, $c^2 = 52$ has two solutions, $\pm\sqrt{52} = \pm2\sqrt{13}$. In applications we often choose only the positive square root, as seen in the example that follows.

EXAMPLE 8 **Using the Pythagorean Formula**

Use the Pythagorean formula to find the length of the hypotenuse of the triangle in Figure 9.

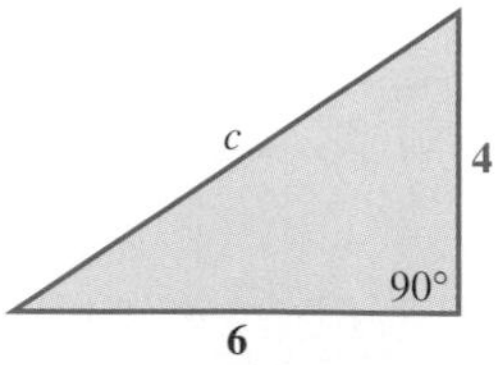

$$c^2 = a^2 + b^2 \qquad \text{Pythagorean formula}$$
$$c^2 = 4^2 + 6^2 \qquad \text{Let } a = 4 \text{ and } b = 6.$$

Substitute carefully.

$$c^2 = 52$$
$$c = \sqrt{52} \qquad \text{Choose the principal root.}$$
$$c = \sqrt{4 \cdot 13} \qquad \text{Factor.}$$
$$c = \sqrt{4} \cdot \sqrt{13} \qquad \text{Product rule}$$
$$c = 2\sqrt{13}$$

FIGURE 9

The length of the hypotenuse is $2\sqrt{13}$.

✔ **Now Try Exercise 105.**

▶ **CAUTION**   When using the formula $c^2 = a^2 + b^2$, be sure that the length of the hypotenuse is substituted for c and that the lengths of the legs are substituted for a and b. Errors often occur because values are substituted incorrectly.

CONNECTIONS

The Pythagorean formula is undoubtedly one of the most widely used and oldest formulas we have. It is very important in trigonometry, which is used in surveying, drafting, engineering, navigation, and many other fields. There is evidence that the Babylonians knew the concept quite well. Although attributed to Pythagoras, it was known to every surveyor from Egypt to China for a thousand years before Pythagoras.

In the 1939 movie *The Wizard of Oz,* the Scarecrow asks the Wizard for a brain. When the Wizard presents him with a diploma granting him a Th.D. (Doctor of Thinkology), the Scarecrow recites the following:

The sum of the square roots of any two sides of an isosceles triangle is equal to the square root of the remaining side. . . . Oh joy! Rapture! I've got a brain.

For Discussion or Writing

Did the Scarecrow recite the Pythagorean formula? (An *isosceles triangle* is a triangle with two equal sides.) Is his statement true? Explain.

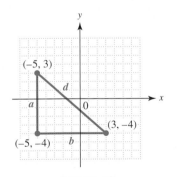

FIGURE 10

OBJECTIVE 6 Use the distance formula.   An important result in algebra is derived from the Pythagorean formula. The *distance formula* allows us to find the distance between two points in the coordinate plane, or the length of the line segment joining those two points. Figure 10 shows the points $(3, -4)$ and $(-5, 3)$. The vertical line through $(-5, 3)$ and the horizontal line through $(3, -4)$ intersect at the point $(-5, -4)$. Thus, the point $(-5, -4)$ becomes the vertex of the right angle in a right triangle. By the Pythagorean formula, the square of the length of the hypotenuse d of the right triangle in Figure 10 is equal to the sum of the squares of the lengths of the two legs a and b:

$$d^2 = a^2 + b^2.$$

The length a is the difference between the y-coordinates of the endpoints. Since the x-coordinate of both points in Figure 10 is -5, the side is vertical, and we can find a by finding the difference between the y-coordinates. We subtract -4 from 3 to get a positive value for a.

$$a = 3 - (-4) = 7$$

Similarly, we find b by subtracting -5 from 3.

$$b = 3 - (-5) = 8$$

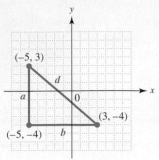

FIGURE 10
(repeated)

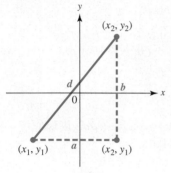

FIGURE 11

Substituting these values into the formula, we obtain

$$d^2 = a^2 + b^2$$
$$d^2 = 7^2 + 8^2 \qquad \text{Let } a = 7 \text{ and } b = 8.$$
$$d^2 = 49 + 64$$
$$d^2 = 113$$
$$d = \sqrt{113}.$$

We choose the principal root, since distance cannot be negative. Therefore, the distance between $(-5, 3)$ and $(3, -4)$ is $\sqrt{113}$.

▶ **NOTE** It is customary to leave the distance in radical form. Do not use a calculator to get an approximation, unless you are specifically directed to do so.

This result can be generalized. Figure 11 shows the two points (x_1, y_1) and (x_2, y_2). Notice that the distance between (x_1, y_1) and (x_2, y_1) is given by

$$a = |x_2 - x_1|,$$

and the distance between (x_2, y_2) and (x_2, y_1) is given by

$$b = |y_2 - y_1|.$$

From the Pythagorean formula,

$$d^2 = a^2 + b^2$$
$$d^2 = (x_2 - x_1)^2 + (y_2 - y_1)^2.$$

Choosing the principal square root gives the **distance formula.**

Distance Formula

The distance between the points (x_1, y_1) and (x_2, y_2) is

$$d = \sqrt{(x_2 - x_1)^2 + (y_2 - y_1)^2}.$$

EXAMPLE 9 Using the Distance Formula

Find the distance between the points $(-3, 5)$ and $(6, 4)$.

Designating the points as (x_1, y_1) and (x_2, y_2) is arbitrary. We choose $(x_1, y_1) = (-3, 5)$ and $(x_2, y_2) = (6, 4)$.

$$d = \sqrt{(x_2 - x_1)^2 + (y_2 - y_1)^2}$$
$$= \sqrt{[6 - (-3)]^2 + (4 - 5)^2} \qquad x_2 = 6, y_2 = 4, x_1 = -3, y_1 = 5$$
$$= \sqrt{9^2 + (-1)^2}$$
$$= \sqrt{82}$$

Start with the *x*-value and *y*-value of the *same* point.

✔ **Now Try Exercise 111.**

8.3 EXERCISES

● *Complete solution available on Video Lectures on CD/DVD*

▨ *Now Try Exercise*

Concept Check Decide whether each statement is true *or* false *by using the product rule explained in this section. Then support your answer by finding a calculator approximation for each expression.*

1. $2\sqrt{12} = \sqrt{48}$

2. $\sqrt{72} = 2\sqrt{18}$

3. $3\sqrt{8} = 2\sqrt{18}$

4. $5\sqrt{72} = 6\sqrt{50}$

5. *Concept Check* Which one of the following is *not* equal to $\sqrt{\frac{1}{2}}$? (Do not use calculator approximations.)

A. $\sqrt{0.5}$ **B.** $\sqrt{\frac{2}{4}}$ **C.** $\sqrt{\frac{3}{6}}$ **D.** $\frac{\sqrt{4}}{\sqrt{16}}$

✎ **6.** Use the π key on your calculator to get a value for π. Now find an approximation for $\sqrt[4]{\frac{2143}{22}}$. Does the result mean that π is actually equal to $\sqrt[4]{\frac{2143}{22}}$? Why or why not?

Multiply, using the product rule. Assume that all variables represent positive real numbers. See Examples 1 and 2.

● **7.** $\sqrt{5} \cdot \sqrt{6}$

8. $\sqrt{10} \cdot \sqrt{3}$

9. $\sqrt{14} \cdot \sqrt{x}$

10. $\sqrt{23} \cdot \sqrt{t}$

11. $\sqrt{14} \cdot \sqrt{3pqr}$

12. $\sqrt{7} \cdot \sqrt{5xt}$

● **13.** $\sqrt[3]{7x} \cdot \sqrt[3]{2y}$

14. $\sqrt[3]{9x} \cdot \sqrt[3]{4y}$

15. $\sqrt[4]{11} \cdot \sqrt[4]{3}$

16. $\sqrt[4]{6} \cdot \sqrt[4]{9}$

17. $\sqrt[4]{2x} \cdot \sqrt[4]{3x^2}$

18. $\sqrt[4]{3y^2} \cdot \sqrt[4]{6y}$

19. $\sqrt[3]{7} \cdot \sqrt[4]{3}$

20. $\sqrt[5]{8} \cdot \sqrt[6]{12}$

✎ **21.** Explain the product rule for radicals in your own words. Give examples.

✎ **22.** Explain the quotient rule for radicals in your own words. Give examples.

Simplify each radical. Assume that all variables represent positive real numbers. See Example 3.

● **23.** $\sqrt{\frac{64}{121}}$

24. $\sqrt{\frac{16}{49}}$

25. $\sqrt{\frac{3}{25}}$

26. $\sqrt{\frac{13}{49}}$

27. $\sqrt{\frac{x}{25}}$

28. $\sqrt{\frac{k}{100}}$

29. $\sqrt{\frac{p^6}{81}}$

30. $\sqrt{\frac{w^{10}}{36}}$

31. $\sqrt[3]{-\frac{27}{64}}$

32. $\sqrt[3]{-\frac{216}{125}}$

33. $\sqrt[3]{\frac{r^2}{8}}$

34. $\sqrt[3]{\frac{t}{125}}$

35. $-\sqrt[4]{\frac{81}{x^4}}$

36. $-\sqrt[4]{\frac{625}{y^4}}$

37. $\sqrt[5]{\frac{1}{x^{15}}}$

38. $\sqrt[5]{\frac{32}{y^{20}}}$

Express each radical in simplified form. See Example 4.

● **39.** $\sqrt{12}$

40. $\sqrt{18}$

41. $\sqrt{288}$

42. $\sqrt{72}$

43. $-\sqrt{32}$

44. $-\sqrt{48}$

45. $-\sqrt{28}$

46. $-\sqrt{24}$

47. $\sqrt{30}$

48. $\sqrt{46}$

49. $\sqrt[3]{128}$

50. $\sqrt[3]{24}$

51. $\sqrt[3]{-16}$

52. $\sqrt[3]{-250}$

53. $\sqrt[3]{40}$

54. $\sqrt[3]{375}$

55. $-\sqrt[4]{512}$

56. $-\sqrt[4]{1250}$

57. $\sqrt[5]{64}$

58. $\sqrt[5]{128}$

59. A student claimed that $\sqrt[3]{14}$ is not in simplified form, since $14 = 8 + 6$, and 8 is a perfect cube. Was his reasoning correct? Why or why not?

60. Explain in your own words why $\sqrt[3]{k^4}$ is not a simplified radical.

Express each radical in simplified form. Assume that all variables represent positive real numbers. See Example 5.

61. $\sqrt{72k^2}$ **62.** $\sqrt{18m^2}$ **63.** $\sqrt{144x^3y^9}$

64. $\sqrt{169s^5t^{10}}$ **65.** $\sqrt{121x^6}$ **66.** $\sqrt{256z^{12}}$

67. $-\sqrt[3]{27t^{12}}$ **68.** $-\sqrt[3]{64y^{18}}$ **69.** $-\sqrt{100m^8z^4}$

70. $-\sqrt{25t^6s^{20}}$ **71.** $-\sqrt[3]{-125a^6b^9c^{12}}$ **72.** $-\sqrt[3]{-216y^{15}x^6z^3}$

73. $\sqrt[4]{\dfrac{1}{16}r^8t^{20}}$ **74.** $\sqrt[4]{\dfrac{81}{256}t^{12}u^8}$ **75.** $\sqrt{50x^3}$ **76.** $\sqrt{300z^3}$

77. $-\sqrt{500r^{11}}$ **78.** $-\sqrt{200p^{13}}$ **79.** $\sqrt{13x^7y^8}$ **80.** $\sqrt{23k^9p^{14}}$

81. $\sqrt[3]{8z^6w^9}$ **82.** $\sqrt[3]{64a^{15}b^{12}}$ **83.** $\sqrt[3]{-16z^5t^7}$ **84.** $\sqrt[3]{-81m^4n^{10}}$

85. $\sqrt[4]{81x^{12}y^{16}}$ **86.** $\sqrt[4]{81t^8u^{28}}$ **87.** $-\sqrt[4]{162r^{15}s^{10}}$ **88.** $-\sqrt[4]{32k^5m^{10}}$

89. $\sqrt{\dfrac{y^{11}}{36}}$ **90.** $\sqrt{\dfrac{v^{13}}{49}}$ **91.** $\sqrt[3]{\dfrac{x^{16}}{27}}$ **92.** $\sqrt[3]{\dfrac{y^{17}}{125}}$

Simplify each radical. Assume that $x \geq 0$. See Example 6.

93. $\sqrt[4]{48^2}$ **94.** $\sqrt[4]{50^2}$ **95.** $\sqrt[4]{2^5}$

96. $\sqrt[6]{8}$ **97.** $\sqrt[10]{x^{25}}$ **98.** $\sqrt[12]{x^{44}}$

Simplify by first writing the radicals as radicals with the same index. Then multiply. Assume that all variables represent positive real numbers. See Example 7.

99. $\sqrt[3]{4} \cdot \sqrt{3}$ **100.** $\sqrt[3]{5} \cdot \sqrt{6}$ **101.** $\sqrt[3]{3} \cdot \sqrt[4]{4}$

102. $\sqrt[5]{7} \cdot \sqrt[7]{5}$ **103.** $\sqrt{x} \cdot \sqrt[3]{x}$ **104.** $\sqrt[3]{y} \cdot \sqrt[4]{y}$

Find the unknown length in each right triangle. Simplify the answer if possible. See Example 8.

105.

106.

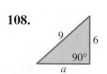

107.

108.

Find the distance between each pair of points. See Example 9.

109. $(6, 13)$ and $(1, 1)$ **110.** $(8, 13)$ and $(2, 5)$

111. $(-6, 5)$ and $(3, -4)$ **112.** $(-1, 5)$ and $(-7, 7)$

113. $(-8, 2)$ and $(-4, 1)$ **114.** $(-1, 2)$ and $(5, 3)$

115. $(4.7, 2.3)$ and $(1.7, -1.7)$

116. $(-2.9, 18.2)$ and $(2.1, 6.2)$

117. $\left(\sqrt{2}, \sqrt{6}\right)$ and $\left(-2\sqrt{2}, 4\sqrt{6}\right)$

118. $\left(\sqrt{7}, 9\sqrt{3}\right)$ and $\left(-\sqrt{7}, 4\sqrt{3}\right)$

119. $(x + y, y)$ and $(x - y, x)$

120. $(c, c - d)$ and $(d, c + d)$

121. *Concept Check* As given in the text, the distance formula is expressed with a radical. Write the distance formula with rational exponents.

122. An alternative form of the distance formula is

$$d = \sqrt{(x_1 - x_2)^2 + (y_1 - y_2)^2}.$$

Compare this with the form given in this section, and explain why the two forms are equivalent.

Find the perimeter of each triangle. $\left(\text{Hint: For Exercise 123, } \sqrt{k} + \sqrt{k} = 2\sqrt{k}.\right)$

123.

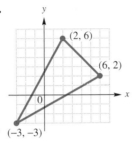

124.

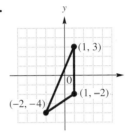

Solve each problem.

125. The following letter appeared in the column "Ask Tom Why," written by Tom Skilling of the *Chicago Tribune*:

> *Dear Tom,*
> *I cannot remember the formula to calculate the distance to the horizon. I have a stunning view from my 14th-floor condo, 150 ft above the ground. How far can I see?*
> *Ted Fleischaker; Indianapolis, Ind.*

Skilling's answer was as follows:

> To find the distance to the horizon in miles, take the square root of the height of your view in feet and multiply that result by 1.224. Your answer will be the number of miles to the horizon. (*Source: Chicago Tribune*, August 17, 2002.)

Assuming that Ted's eyes are 6 ft above the ground, the total height from the ground is $150 + 6 = 156$ ft. To the nearest tenth of a mile, how far can he see to the horizon?

126. The length of the diagonal of a box is given by

$$D = \sqrt{L^2 + W^2 + H^2},$$

where L, W, and H are, respectively, the length, width, and height of the box. Find the length of the diagonal D of a box that is 4 ft long, 2 ft wide, and 3 ft high. Give the exact value, and then round to the nearest tenth of a foot.

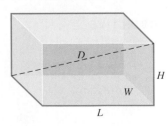

127. A Sanyo color television, model AVM-2755, has a rectangular screen with a 21.7-in. width. Its height is 16 in. What is the measure of the diagonal of the screen, to the nearest tenth of an inch? (*Source:* Actual measurements of the author's television.)

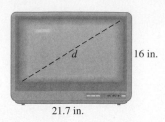

16 in.

21.7 in.

128. A formula from electronics dealing with the impedance of parallel resonant circuits is

$$I = \frac{E}{\sqrt{R^2 + \omega^2 L^2}},$$

where the variables are in appropriate units. Find I if $E = 282$, $R = 100$, $L = 264$, and $\omega = 120\pi$. Give your answer to the nearest thousandth.

129. In the study of sound, one version of the law of tensions is

$$f_1 = f_2 \sqrt{\frac{F_1}{F_2}}.$$

If $F_1 = 300$, $F_2 = 60$, and $f_2 = 260$, find f_1 to the nearest unit.

130. The illumination I, in foot-candles, produced by a light source is related to the distance d, in feet, from the light source by the equation

$$d = \sqrt{\frac{k}{I}},$$

where k is a constant. If $k = 640$, how far from the light source will the illumination be 2 foot-candles? Give the exact value, and then round to the nearest tenth of a foot.

PREVIEW EXERCISES

*Combine like terms. See **Section 5.3.***

131. $13x^4 - 12x^3 + 9x^4 + 2x^3$

132. $-15z^3 - z^2 + 4z^4 + 12z^8$

133. $9q^2 + 2q - 5q - q^2$

134. $7m^5 - 2m^3 + 8m^5 - m^3$

8.4   Adding and Subtracting Radical Expressions

OBJECTIVE

1 Simplify radical expressions involving addition and subtraction.

OBJECTIVE 1 Simplify radical expressions involving addition and subtraction.   An expression such as $4\sqrt{2} + 3\sqrt{2}$ can be simplified by the distributive property.

$4\sqrt{2} + 3\sqrt{2}$
$= (4 + 3)\sqrt{2} = 7\sqrt{2}$

This is similar to simplifying $4x + 3x$ to $7x$.

$2\sqrt{3} - 5\sqrt{3}$
$= (2 - 5)\sqrt{3} = -3\sqrt{3}$

This is similar to simplifying $2x - 5x$ to $-3x$.

▶ **CAUTION** *Only radical expressions with the same index and the same radicand may be combined.* Expressions such as $5\sqrt{3} + 2\sqrt{2}$ or $3\sqrt{3} + 2\sqrt[3]{3}$ cannot be simplified by combining terms.

EXAMPLE 1 Adding and Subtracting Radicals

Add or subtract to simplify each radical expression.

(a) $3\sqrt{24} + \sqrt{54}$

Simplify each radical; then use the distributive property to combine terms.

$$3\sqrt{24} + \sqrt{54}$$
$$= 3\sqrt{4} \cdot \sqrt{6} + \sqrt{9} \cdot \sqrt{6} \qquad \text{Product rule}$$
$$= 3 \cdot 2\sqrt{6} + 3\sqrt{6} \qquad \sqrt{4} = 2; \sqrt{9} = 3$$
$$= 6\sqrt{6} + 3\sqrt{6}$$
$$= 9\sqrt{6} \qquad \text{Combine like terms.}$$

(b) $2\sqrt{20x} - \sqrt{45x}, \quad x \geq 0$

$$= 2\sqrt{4} \cdot \sqrt{5x} - \sqrt{9} \cdot \sqrt{5x} \qquad \text{Product rule}$$
$$= 2 \cdot 2\sqrt{5x} - 3\sqrt{5x}$$
$$= 4\sqrt{5x} - 3\sqrt{5x}$$
$$= \sqrt{5x} \qquad \text{Combine like terms.}$$

(c) $2\sqrt{3} - 4\sqrt{5}$

Here the radicands differ and are already simplified, so $2\sqrt{3} - 4\sqrt{5}$ cannot be simplified further.

✔ **Now Try Exercises 7, 15, and 19.**

▶ **CAUTION** Do not confuse the product rule with combining like terms. *The root of a sum does not equal the sum of the roots.* For example,

$$\sqrt{9 + 16} \neq \sqrt{9} + \sqrt{16}$$

since $\sqrt{9 + 16} = \sqrt{25} = 5$, but $\sqrt{9} + \sqrt{16} = 3 + 4 = 7$.

EXAMPLE 2 Adding and Subtracting Radicals with Higher Indexes

Add or subtract to simplify each radical expression. Assume that all variables represent positive real numbers.

(a) $2\sqrt[3]{16} - 5\sqrt[3]{54}$

$$= 2\sqrt[3]{8 \cdot 2} - 5\sqrt[3]{27 \cdot 2} \qquad \text{Factor.}$$
$$= 2\sqrt[3]{8} \cdot \sqrt[3]{2} - 5\sqrt[3]{27} \cdot \sqrt[3]{2} \qquad \text{Product rule}$$
$$= 2 \cdot 2 \cdot \sqrt[3]{2} - 5 \cdot 3 \cdot \sqrt[3]{2}$$
$$= 4\sqrt[3]{2} - 15\sqrt[3]{2}$$
$$= (4 - 15)\sqrt[3]{2} \qquad \text{Distributive property}$$
$$= -11\sqrt[3]{2} \qquad \text{Combine like terms.}$$

(b)

$$2\sqrt[3]{x^2y} + \sqrt[3]{8x^5y^4}$$

$$= 2\sqrt[3]{x^2y} + \sqrt[3]{(8x^3y^3)x^2y} \qquad \text{Factor.}$$

$$= 2\sqrt[3]{x^2y} + \sqrt[3]{8x^3y^3} \cdot \sqrt[3]{x^2y} \qquad \text{Product rule}$$

$$= 2\sqrt[3]{x^2y} + 2xy\sqrt[3]{x^2y}$$

$$= (2 + 2xy)\sqrt[3]{x^2y} \qquad \text{Distributive property}$$

This result cannot be simplified further.

Be careful! The indexes are different.

(c) $5\sqrt{4x^3} + 3\sqrt[3]{64x^4}$

$$= 5\sqrt{4x^2 \cdot x} + 3\sqrt[3]{64x^3 \cdot x} \qquad \text{Factor.}$$

$$= 5\sqrt{4x^2} \cdot \sqrt{x} + 3\sqrt[3]{64x^3} \cdot \sqrt[3]{x} \qquad \text{Product rule}$$

$$= 5 \cdot 2x\sqrt{x} + 3 \cdot 4x\sqrt[3]{x}$$

$$= 10x\sqrt{x} + 12x\sqrt[3]{x}$$

Remember to write the correct index.

The radicands are both x, but since the indexes are different, this expression cannot be simplified further.

✔ **Now Try Exercises 25, 37, and 41.**

▶ **CAUTION** *Remember to write the index when working with cube roots, fourth roots, and so on.*

EXAMPLE 3 Adding and Subtracting Radicals with Fractions

Perform the indicated operations. Assume that all variables represent positive real numbers.

(a) $2\sqrt{\dfrac{75}{16}} + 4\dfrac{\sqrt{8}}{\sqrt{32}}$

$$= 2\frac{\sqrt{25 \cdot 3}}{\sqrt{16}} + 4\frac{\sqrt{4 \cdot 2}}{\sqrt{16 \cdot 2}} \qquad \text{Quotient rule}$$

$$= 2\left(\frac{5\sqrt{3}}{4}\right) + 4\left(\frac{2\sqrt{2}}{4\sqrt{2}}\right) \qquad \text{Product rule}$$

$$= \frac{5\sqrt{3}}{2} + 2 \qquad \text{Multiply; } \frac{\sqrt{2}}{\sqrt{2}} = 1.$$

$$= \frac{5\sqrt{3}}{2} + \frac{4}{2} \qquad \text{Write with a common denominator.}$$

$$= \frac{5\sqrt{3} + 4}{2} \qquad \text{Add fractions.}$$

(b) $10\sqrt[3]{\dfrac{5}{x^6}} - 3\sqrt[3]{\dfrac{4}{x^9}}$

$= 10\dfrac{\sqrt[3]{5}}{\sqrt[3]{x^6}} - 3\dfrac{\sqrt[3]{4}}{\sqrt[3]{x^9}}$ Quotient rule

$= \dfrac{10\sqrt[3]{5}}{x^2} - \dfrac{3\sqrt[3]{4}}{x^3}$ Simplify denominators.

$= \dfrac{10\sqrt[3]{5} \cdot x}{x^2 \cdot x} - \dfrac{3\sqrt[3]{4}}{x^3}$ Write with a common denominator.

$= \dfrac{10x\sqrt[3]{5} - 3\sqrt[3]{4}}{x^3}$ Subtract fractions.

✔ **Now Try Exercises 51 and 57.**

A calculator can support some of the results obtained in the examples of this section. In Example 1(a), we simplified $3\sqrt{24} + \sqrt{54}$ to obtain $9\sqrt{6}$. The screen in Figure 12(a) shows that the approximations are the same, suggesting that our simplification was correct. Figure 12(b) shows support for the result of Example 2(a): $2\sqrt[3]{16} - 5\sqrt[3]{54} = -11\sqrt[3]{2}$. Figure 12(c) supports the result of Example 3(a).

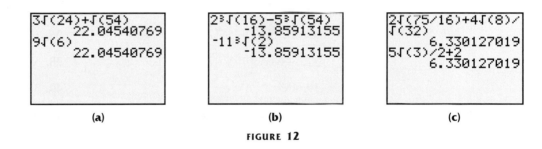

(a)   (b)   (c)

FIGURE 12

8.4 EXERCISES

🌐 *Complete solution available on Video Lectures on CD/DVD*

Now Try Exercise

1. *Concept Check*   Which one of the following sums could be simplified without first simplifying the individual radical expressions?

   A. $\sqrt{50} + \sqrt{32}$ **B.** $3\sqrt{6} + 9\sqrt{6}$ **C.** $\sqrt[3]{32} - \sqrt[3]{108}$ **D.** $\sqrt[5]{6} - \sqrt[5]{192}$

2. *Concept Check*   Let $a = 1$ and let $b = 64$.

   (a) Evaluate $\sqrt{a} + \sqrt{b}$. Then find $\sqrt{a + b}$. Are they equal?

   (b) Evaluate $\sqrt[3]{a} + \sqrt[3]{b}$. Then find $\sqrt[3]{a + b}$. Are they equal?

   (c) Complete the following: In general, $\sqrt[n]{a} + \sqrt[n]{b} \neq$ _____, based on the observations in parts (a) and (b) of this exercise.

3. Even though the root indexes of the terms are not equal, the sum $\sqrt{64} + \sqrt[3]{125} + \sqrt[4]{16}$ can be simplified quite easily. What is this sum? Why can we add these terms so easily?

4. Explain why $28 - 4\sqrt{2}$ is not equal to $24\sqrt{2}$. (This is a common error among algebra students.)

Simplify. Assume that all variables represent positive real numbers. See Examples 1 and 2.

5. $\sqrt{36} - \sqrt{100}$ **6.** $\sqrt{25} - \sqrt{81}$ 🌐 **7.** $-2\sqrt{48} + 3\sqrt{75}$

8. $4\sqrt{32} - 2\sqrt{8}$ **9.** $\sqrt[3]{16} + 4\sqrt[3]{54}$ **10.** $3\sqrt[3]{24} - 2\sqrt[3]{192}$

11. $\sqrt[4]{32} + 3\sqrt[4]{2}$ **12.** $\sqrt[4]{405} - 2\sqrt[4]{5}$

13. $6\sqrt{18} - \sqrt{32} + 2\sqrt{50}$ **14.** $5\sqrt{8} + 3\sqrt{72} - 3\sqrt{50}$

15. $5\sqrt{6} + 2\sqrt{10}$ **16.** $3\sqrt{11} - 5\sqrt{13}$

17. $2\sqrt{5} + 3\sqrt{20} + 4\sqrt{45}$ **18.** $5\sqrt{54} - 2\sqrt{24} - 2\sqrt{96}$

19. $\sqrt{72x} - \sqrt{8x}$ **20.** $4\sqrt{18k} - \sqrt{72k}$

21. $3\sqrt{72m^2} - 5\sqrt{32m^2} - 3\sqrt{18m^2}$ **22.** $9\sqrt{27p^2} - 14\sqrt{108p^2} + 2\sqrt{48p^2}$

23. $2\sqrt[3]{16} - \sqrt[3]{54}$ **24.** $15\sqrt[3]{81} - 4\sqrt[3]{24}$

🌐 **25.** $2\sqrt[3]{27x} - 2\sqrt[3]{8x}$ **26.** $6\sqrt[3]{128m} + 3\sqrt[3]{16m}$

27. $\sqrt[3]{x^2y} - \sqrt[3]{8x^2y}$ **28.** $3\sqrt[3]{x^2y^2} - 2\sqrt[3]{64x^2y^2}$

29. $3x\sqrt[3]{xy^2} - 2\sqrt[3]{8x^4y^2}$ **30.** $6q^2\sqrt[3]{5q} - 2q\sqrt[3]{40q^4}$

31. $5\sqrt[4]{32} + 3\sqrt[4]{162}$ **32.** $2\sqrt[4]{512} + 4\sqrt[4]{32}$

33. $3\sqrt[4]{x^5y} - 2x\sqrt[4]{xy}$ **34.** $2\sqrt[4]{m^9p^6} - 3m^2p\sqrt[4]{mp^2}$

35. $2\sqrt[4]{32a^3} + 5\sqrt[4]{2a^3}$ **36.** $-\sqrt[4]{16r} + 5\sqrt[4]{r}$

37. $\sqrt[3]{64xy^2} + \sqrt[3]{27x^4y^5}$ **38.** $\sqrt[4]{625s^3t} - \sqrt[4]{81s^7t^5}$

39. $4\sqrt[3]{x} - 6\sqrt{x}$ **40.** $3\sqrt{7z} - 9\sqrt[3]{7z}$

41. $2\sqrt[3]{8x^4} + 3\sqrt[4]{16x^5}$ **42.** $3\sqrt[3]{64m^4} + 5\sqrt[4]{81m^5}$

Simplify. Assume that all variables represent positive real numbers. See Example 3.

43. $\sqrt{8} - \dfrac{\sqrt{64}}{\sqrt{16}}$ **44.** $\sqrt{48} - \dfrac{\sqrt{81}}{\sqrt{9}}$ **45.** $\dfrac{2\sqrt{5}}{3} + \dfrac{\sqrt{5}}{6}$

46. $\dfrac{4\sqrt{3}}{3} + \dfrac{2\sqrt{3}}{9}$ **47.** $\sqrt{\dfrac{8}{9}} + \sqrt{\dfrac{18}{36}}$ **48.** $\sqrt{\dfrac{12}{16}} + \sqrt{\dfrac{48}{64}}$

49. $\dfrac{\sqrt{32}}{3} + \dfrac{2\sqrt{2}}{3} - \dfrac{\sqrt{2}}{\sqrt{9}}$ **50.** $\dfrac{\sqrt{27}}{2} - \dfrac{3\sqrt{3}}{2} + \dfrac{\sqrt{3}}{\sqrt{4}}$ 🌐 **51.** $3\sqrt{\dfrac{50}{9}} + 8\dfrac{\sqrt{2}}{\sqrt{8}}$

52. $9\sqrt{\dfrac{48}{25}} - 2\dfrac{\sqrt{2}}{\sqrt{98}}$ **53.** $\sqrt{\dfrac{25}{x^8}} - \sqrt{\dfrac{9}{x^6}}$ **54.** $\sqrt{\dfrac{100}{y^4}} + \sqrt{\dfrac{81}{y^{10}}}$

55. $3\sqrt[3]{\dfrac{m^5}{27}} - 2m\sqrt[3]{\dfrac{m^2}{64}}$ **56.** $2a\sqrt[4]{\dfrac{a}{16}} - 5a\sqrt[4]{\dfrac{a}{81}}$ 🌐 **57.** $3\sqrt[3]{\dfrac{2}{x^6}} - 4\sqrt[3]{\dfrac{5}{x^9}}$

58. $-4\sqrt[3]{\dfrac{4}{t^9}} + 3\sqrt[3]{\dfrac{9}{t^{12}}}$

In Example 1(a), we showed that $3\sqrt{24} + \sqrt{54} = 9\sqrt{6}$. To support this result, we can find a calculator approximation of $3\sqrt{24}$, then find a calculator approximation of $\sqrt{54}$, and add these two approximations. Then we find a calculator approximation of $9\sqrt{6}$. It should correspond to the sum that we just found. (For this example, both approximations are 22.04540769. Due to rounding, there may be a discrepancy in the final digit if you try to duplicate this work.) Follow this procedure to support the statements in Exercises 59 and 60.

59. $3\sqrt{32} - 2\sqrt{8} = 8\sqrt{2}$ **60.** $2\sqrt{40} + 6\sqrt{90} - 3\sqrt{160} = 10\sqrt{10}$

Solve each problem.

61. A rectangular yard has a length of $\sqrt{192}$ m and a width of $\sqrt{48}$ m. Choose the best estimate of its dimensions. Then estimate the perimeter.

    A. 14 m by 7 m   **B.** 5 m by 7 m   **C.** 14 m by 8 m   **D.** 15 m by 8 m

62. If the sides of a triangle are $\sqrt{65}$ in., $\sqrt{35}$ in., and $\sqrt{26}$ in., which one of the following is the best estimate of its perimeter?

    A. 20 in.   **B.** 26 in.   **C.** 19 in.   **D.** 24 in.

Solve each problem. Give answers as simplified radical expressions.

63. Find the perimeter of the triangle.   **64.** Find the perimeter of the rectangle.

$3\sqrt{20}$ in. $2\sqrt{45}$ in.

$\sqrt{75}$ in.

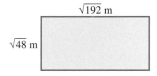

$\sqrt{192}$ m

$\sqrt{48}$ m

65. What is the perimeter of the computer graphic?   **66.** Find the area of the trapezoid.

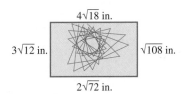

$4\sqrt{18}$ in.

$3\sqrt{12}$ in. $\sqrt{108}$ in.

$2\sqrt{72}$ in.

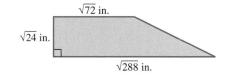

$\sqrt{72}$ in.

$\sqrt{24}$ in.

$\sqrt{288}$ in.

PREVIEW EXERCISES

*Find each product. See **Section 5.4**.*

67. $5xy(2x^2y^3 - 4x)$ **68.** $(3x + 7)(2x - 6)$ **69.** $(a^2 + b)(a^2 - b)$

70. $(2p - 7)^2$ **71.** $(4x^3 + 3)^3$ **72.** $(2 + 3y)(2 - 3y)$

*Write in lowest terms. See **Section 7.1**.*

73. $\dfrac{8x^2 - 10x}{6x^2}$ **74.** $\dfrac{15y^3 - 9y^2}{6y}$

8.5 Multiplying and Dividing Radical Expressions

OBJECTIVE 1 Multiply radicals. We multiply binomial expressions involving radicals by using the FOIL method from **Section 5.4.** For example, we find the product of the binomials $\sqrt{5} + 3$ and $\sqrt{6} + 1$ as follows:

$$\overbrace{(\sqrt{5} + 3)(\sqrt{6} + 1)}^{} = \overbrace{\sqrt{5} \cdot \sqrt{6}}^{\text{First}} + \overbrace{\sqrt{5} \cdot 1}^{\text{Outer}} + \overbrace{3 \cdot \sqrt{6}}^{\text{Inner}} + \overbrace{3 \cdot 1}^{\text{Last}}$$

$$= \sqrt{30} + \sqrt{5} + 3\sqrt{6} + 3.$$

This result cannot be simplified further.

EXAMPLE 1 Multiplying Binomials Involving Radicals

Multiply, using the FOIL method.

(a) $(7 - \sqrt{3})(\sqrt{5} + \sqrt{2}) = \overset{\text{F}}{7\sqrt{5}} + \overset{\text{O}}{7\sqrt{2}} - \overset{\text{I}}{\sqrt{3} \cdot \sqrt{5}} - \overset{\text{L}}{\sqrt{3} \cdot \sqrt{2}}$

$$= 7\sqrt{5} + 7\sqrt{2} - \sqrt{15} - \sqrt{6}$$

(b) $(\sqrt{10} + \sqrt{3})(\sqrt{10} - \sqrt{3})$

$$= \sqrt{10} \cdot \sqrt{10} - \sqrt{10} \cdot \sqrt{3} + \sqrt{10} \cdot \sqrt{3} - \sqrt{3} \cdot \sqrt{3} \quad \text{FOIL}$$

$$= 10 - 3$$

$$= 7$$

The product $(\sqrt{10} + \sqrt{3})(\sqrt{10} - \sqrt{3}) = (\sqrt{10})^2 - (\sqrt{3})^2$ is the difference of squares:

$$(x + y)(x - y) = x^2 - y^2.$$

Here, $x = \sqrt{10}$ and $y = \sqrt{3}$.

(c) $(\sqrt{7} - 3)^2 = (\sqrt{7} - 3)(\sqrt{7} - 3)$

$$= \sqrt{7} \cdot \sqrt{7} - 3\sqrt{7} - 3\sqrt{7} + 3 \cdot 3$$

$$= 7 - 6\sqrt{7} + 9 \qquad \boxed{\text{Be careful! These}}$$

$$= 16 - 6\sqrt{7} \qquad \qquad \text{terms cannot be combined.}$$

(d) $(5 - \sqrt[3]{3})(5 + \sqrt[3]{3}) = 5 \cdot 5 + 5\sqrt[3]{3} - 5\sqrt[3]{3} - \sqrt[3]{3} \cdot \sqrt[3]{3}$

$$= 25 - \sqrt[3]{3^2} \qquad \boxed{\text{Remember to}}$$

$$= 25 - \sqrt[3]{9} \qquad \qquad \text{write the index 3 in each radical.}$$

(e) $(\sqrt{k} + \sqrt{y})(\sqrt{k} - \sqrt{y}) = (\sqrt{k})^2 - (\sqrt{y})^2 \qquad \text{Difference of squares}$

$$= k - y, \quad k \geq 0 \text{ and } y \geq 0$$

✔ **Now Try Exercises 13, 17, 23, 27, and 39.**

▶ **NOTE** In Example 1(c), we could have used the formula for the square of a binomial,

$$(x - y)^2 = x^2 - 2xy + y^2,$$

to obtain the same result:

$$(\sqrt{7} - 3)^2 = (\sqrt{7})^2 - 2(\sqrt{7})(3) + 3^2$$
$$= 7 - 6\sqrt{7} + 9$$
$$= 16 - 6\sqrt{7}.$$

OBJECTIVE 2 Rationalize denominators with one radical term. As defined earlier, a simplified radical expression has no radical in the denominator. The origin of this agreement no doubt occurred before the days of high-speed calculation, when computation was a tedious process performed by hand. To see this, consider the radical expression $\frac{1}{\sqrt{2}}$. To find a decimal approximation by hand, it is necessary to divide 1 by a decimal approximation for $\sqrt{2}$, such as 1.414. It is much easier if the divisor is a whole number. This can be accomplished by multiplying $\frac{1}{\sqrt{2}}$ by 1 in the form $\frac{\sqrt{2}}{\sqrt{2}}$. *Multiplying by 1 in any form does not change the value of the original expression.*

$$\frac{1}{\sqrt{2}} \cdot \frac{\sqrt{2}}{\sqrt{2}} = \frac{\sqrt{2}}{2} \qquad \frac{\sqrt{2}}{\sqrt{2}} = 1$$

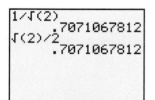

FIGURE 13

Now the computation requires dividing 1.414 by 2 to obtain 0.707, a much easier task.

With current technology, either form of this fraction can be approximated with the same number of keystrokes. See Figure 13, which shows how a calculator gives the same approximation for both forms of the expression.

A common way of "standardizing" the form of a radical expression is to have the denominator contain no radicals. The process of removing radicals from a denominator so that the denominator contains only rational numbers is called **rationalizing the denominator.**

EXAMPLE 2 Rationalizing Denominators with Square Roots

Rationalize each denominator.

(a) $\dfrac{3}{\sqrt{7}}$

Multiply the numerator and denominator by $\sqrt{7}$. This is, in effect, multiplying by 1.

$$\frac{3}{\sqrt{7}} = \frac{3 \cdot \sqrt{7}}{\sqrt{7} \cdot \sqrt{7}} = \frac{3\sqrt{7}}{7}$$

In the denominator, $\sqrt{7} \cdot \sqrt{7} = \sqrt{7 \cdot 7} = \sqrt{49} = 7$. The final denominator is now a rational number.

(b) $\dfrac{5\sqrt{2}}{\sqrt{5}} = \dfrac{5\sqrt{2} \cdot \sqrt{5}}{\sqrt{5} \cdot \sqrt{5}} = \dfrac{5\sqrt{10}}{5} = \sqrt{10}$

(c) $\dfrac{-6}{\sqrt{12}}$

Less work is involved if we simplify the radical in the denominator first.

$$\dfrac{-6}{\sqrt{12}} = \dfrac{-6}{\sqrt{4 \cdot 3}} = \dfrac{-6}{2\sqrt{3}} = \dfrac{-3}{\sqrt{3}}$$

Now we rationalize the denominator.

$$\dfrac{-3}{\sqrt{3}} = \dfrac{-3 \cdot \sqrt{3}}{\sqrt{3} \cdot \sqrt{3}} = \dfrac{-3\sqrt{3}}{3} = -\sqrt{3}$$

✔ **Now Try Exercises 43, 49, and 51.**

EXAMPLE 3 Rationalizing Denominators in Roots of Fractions

Simplify each radical. In part (b), $p > 0$.

(a) $-\sqrt{\dfrac{18}{125}} = -\dfrac{\sqrt{18}}{\sqrt{125}}$　　　Quotient rule

$\qquad\qquad = -\dfrac{\sqrt{9 \cdot 2}}{\sqrt{25 \cdot 5}}$　　　Factor.

$\qquad\qquad = -\dfrac{3\sqrt{2}}{5\sqrt{5}}$　　　Product rule

$\qquad\qquad = -\dfrac{3\sqrt{2} \cdot \sqrt{5}}{5\sqrt{5} \cdot \sqrt{5}}$　　　Multiply by $\dfrac{\sqrt{5}}{\sqrt{5}}$.

$\qquad\qquad = -\dfrac{3\sqrt{10}}{5 \cdot 5}$　　　Product rule

$\qquad\qquad = -\dfrac{3\sqrt{10}}{25}$

(b) $\sqrt{\dfrac{50m^4}{p^5}} = \dfrac{\sqrt{50m^4}}{\sqrt{p^5}}$　　　Quotient rule

$\qquad\qquad = \dfrac{5m^2\sqrt{2}}{p^2\sqrt{p}}$　　　Product rule

$\qquad\qquad = \dfrac{5m^2\sqrt{2} \cdot \sqrt{p}}{p^2\sqrt{p} \cdot \sqrt{p}}$　　　Multiply by $\dfrac{\sqrt{p}}{\sqrt{p}}$.

$\qquad\qquad = \dfrac{5m^2\sqrt{2p}}{p^2 \cdot p}$　　　Product rule

$\qquad\qquad = \dfrac{5m^2\sqrt{2p}}{p^3}$

✔ **Now Try Exercises 55 and 63.**

EXAMPLE 4 Rationalizing Denominators with Cube and Fourth Roots

Simplify.

(a) $\sqrt[3]{\dfrac{27}{16}}$

Use the quotient rule, and simplify the numerator and denominator.

$$\sqrt[3]{\frac{27}{16}} = \frac{\sqrt[3]{27}}{\sqrt[3]{16}} = \frac{3}{\sqrt[3]{8} \cdot \sqrt[3]{2}} = \frac{3}{2\sqrt[3]{2}}$$

To get a rational denominator, multiply the numerator and denominator by a number that will result in a perfect *cube* in the radicand in the denominator. Since $2 \cdot 4 = 8$, a perfect cube, multiply the numerator and denominator by $\sqrt[3]{4}$.

$$\sqrt[3]{\frac{27}{16}} = \frac{3}{2\sqrt[3]{2}} = \frac{3 \cdot \sqrt[3]{4}}{2\sqrt[3]{2} \cdot \sqrt[3]{4}} = \frac{3\sqrt[3]{4}}{2\sqrt[3]{8}} = \frac{3\sqrt[3]{4}}{2 \cdot 2} = \frac{3\sqrt[3]{4}}{4}$$

(b) $\sqrt[4]{\dfrac{5x}{z}} = \dfrac{\sqrt[4]{5x}}{\sqrt[4]{z}} \cdot \dfrac{\sqrt[4]{z^3}}{\sqrt[4]{z^3}} = \dfrac{\sqrt[4]{5xz^3}}{\sqrt[4]{z^4}} = \dfrac{\sqrt[4]{5xz^3}}{z}, \quad x \geq 0, z > 0$$

✔ **Now Try Exercises 71 and 81.**

▶ **CAUTION** It is easy to make mistakes in problems like Example 4(a). A typical error is to multiply the numerator and denominator by $\sqrt[3]{2}$, forgetting that

$$\sqrt[3]{2} \cdot \sqrt[3]{2} \neq 2.$$

You need *three* factors of 2 to obtain 2^3 under the radical.

$$\sqrt[3]{2} \cdot \sqrt[3]{2} \cdot \sqrt[3]{2} = 2$$

OBJECTIVE 3 Rationalize denominators with binomials involving radicals. Recall the special product

$$(x + y)(x - y) = x^2 - y^2.$$

To rationalize a denominator that contains a binomial expression (one that contains exactly two terms) involving radicals, such as

$$\frac{3}{1 + \sqrt{2}},$$

we must use *conjugates*. The conjugate of $1 + \sqrt{2}$ is $1 - \sqrt{2}$. In general, $x + y$ and $x - y$ are **conjugates.**

Rationalizing a Binomial Denominator

Whenever a radical expression has a sum or difference with square root radicals in the denominator, rationalize the denominator by multiplying both the numerator and denominator by the conjugate of the denominator.

For the expression $\frac{3}{1 + \sqrt{2}}$, we rationalize the denominator by multiplying both the numerator and denominator by $1 - \sqrt{2}$, the conjugate of the denominator.

$$\frac{3}{1 + \sqrt{2}} = \frac{3(1 - \sqrt{2})}{(1 + \sqrt{2})(1 - \sqrt{2})}$$

$$= \frac{3(1 - \sqrt{2})}{-1} \qquad (1 + \sqrt{2})(1 - \sqrt{2}) = 1^2 - (\sqrt{2})^2 = 1 - 2 = -1.$$

$$= \frac{3}{-1}(1 - \sqrt{2}) \quad \text{The denominator is now a rational number.}$$

$$= -3(1 - \sqrt{2})$$

$$= -3 + 3\sqrt{2}$$

EXAMPLE 5   Rationalizing Binomial Denominators

Rationalize each denominator.

(a) $\dfrac{5}{4 - \sqrt{3}}$

To rationalize the denominator, multiply both the numerator and denominator by the conjugate of the denominator, $4 + \sqrt{3}$.

$$\frac{5}{4 - \sqrt{3}} = \frac{5(4 + \sqrt{3})}{(4 - \sqrt{3})(4 + \sqrt{3})}$$

$$= \frac{5(4 + \sqrt{3})}{16 - 3}$$

$$= \frac{5(4 + \sqrt{3})}{13}$$

Notice that the numerator is left in factored form. This makes it easier to determine whether the expression is written in lowest terms.

(b) $\dfrac{\sqrt{2} - \sqrt{3}}{\sqrt{5} + \sqrt{3}} = \dfrac{(\sqrt{2} - \sqrt{3})(\sqrt{5} - \sqrt{3})}{(\sqrt{5} + \sqrt{3})(\sqrt{5} - \sqrt{3})}$ Multiply numerator and denominator by $\sqrt{5} - \sqrt{3}$.

$$= \frac{\sqrt{10} - \sqrt{6} - \sqrt{15} + 3}{5 - 3}$$

$$= \frac{\sqrt{10} - \sqrt{6} - \sqrt{15} + 3}{2}$$

(c) $\dfrac{3}{\sqrt{5m} - \sqrt{p}} = \dfrac{3(\sqrt{5m} + \sqrt{p})}{(\sqrt{5m} - \sqrt{p})(\sqrt{5m} + \sqrt{p})}$

$$= \frac{3(\sqrt{5m} + \sqrt{p})}{5m - p}, \quad 5m \neq p, m > 0, p > 0$$

✔ **Now Try Exercises 83, 89, and 93.**

OBJECTIVE 4 Write radical quotients in lowest terms.

EXAMPLE 6 Writing Radical Quotients in Lowest Terms

Write each quotient in lowest terms.

(a) $\dfrac{6 + 2\sqrt{5}}{4}$

Factor the numerator and denominator, and then write in lowest terms.

$$\frac{6 + 2\sqrt{5}}{4} = \frac{2(3 + \sqrt{5})}{2 \cdot 2} = \frac{3 + \sqrt{5}}{2}$$

> Factor first; then divide out the common factor.

Here is an alternative method for writing this quotient in lowest terms:

$$\frac{6 + 2\sqrt{5}}{4} = \frac{6}{4} + \frac{2\sqrt{5}}{4} = \frac{3}{2} + \frac{\sqrt{5}}{2} = \frac{3 + \sqrt{5}}{2}.$$

(b) $\dfrac{5y - \sqrt{8y^2}}{6y} = \dfrac{5y - 2y\sqrt{2}}{6y}, \quad y > 0$ Product rule

$$= \frac{y(5 - 2\sqrt{2})}{6y} \qquad \text{Factor the numerator.}$$

$$= \frac{5 - 2\sqrt{2}}{6} \qquad \text{Divide out the common factor.}$$

✔ **Now Try Exercises 103 and 105.**

▶ **CAUTION** *Be careful to factor before writing a quotient in lowest terms.*

CONNECTIONS

In calculus, it is sometimes desirable to rationalize the *numerator*. For example, to rationalize the numerator of

$$\frac{6 - \sqrt{2}}{4},$$

we multiply the numerator and the denominator by the conjugate of the *numerator*.

$$\frac{6 - \sqrt{2}}{4} = \frac{(6 - \sqrt{2})(6 + \sqrt{2})}{4(6 + \sqrt{2})} = \frac{36 - 2}{4(6 + \sqrt{2})} = \frac{34}{4(6 + \sqrt{2})} = \frac{17}{2(6 + \sqrt{2})}$$

For Discussion or Writing

Rationalize the numerator of each expression, assuming that a and b are non-negative real numbers.

1. $\dfrac{8\sqrt{5} - 1}{6}$ **2.** $\dfrac{3\sqrt{a} + \sqrt{b}}{b}$ **3.** $\dfrac{3\sqrt{a} + \sqrt{b}}{\sqrt{b} - \sqrt{a}}$ $(b \neq a)$

4. Rationalize the denominator of the expression in Exercise 3, and then describe the difference in the procedure you used from what you did in Exercise 3.

8.5 EXERCISES

● *Complete solution available on Video Lectures on CD/DVD*

▨ *Now Try Exercise*

Concept Check   *Match each part of a rule for a special product in Column I with the other part in Column II. Assume that all variables represent positive real numbers.*

I	II
1. $(x + \sqrt{y})(x - \sqrt{y})$	**A.** $x - y$
2. $(\sqrt{x} + y)(\sqrt{x} - y)$	**B.** $x + 2y\sqrt{x} + y^2$
3. $(\sqrt{x} + \sqrt{y})(\sqrt{x} - \sqrt{y})$	**C.** $x - y^2$
4. $(\sqrt{x} + \sqrt{y})^2$	**D.** $x - 2\sqrt{xy} + y$
5. $(\sqrt{x} - \sqrt{y})^2$	**E.** $x^2 - y$
6. $(\sqrt{x} + y)^2$	**F.** $x + 2\sqrt{xy} + y$

Multiply, and then simplify each product. Assume that all variables represent positive real numbers. See Example 1.

7. $\sqrt{6}(3 + \sqrt{2})$ 　　　**8.** $\sqrt{2}(\sqrt{32} - \sqrt{9})$ 　　　**9.** $5(\sqrt{72} - \sqrt{8})$

10. $\sqrt{3}(\sqrt{12} + 2)$ 　　**11.** $(\sqrt{7} + 3)(\sqrt{7} - 3)$ 　　**12.** $(\sqrt{3} - 5)(\sqrt{3} + 5)$

● **13.** $(\sqrt{2} - \sqrt{3})(\sqrt{2} + \sqrt{3})$ 　　　　　**14.** $(\sqrt{7} + \sqrt{3})(\sqrt{7} - \sqrt{3})$

15. $(\sqrt{8} - \sqrt{2})(\sqrt{8} + \sqrt{2})$ 　　　　　**16.** $(\sqrt{20} - \sqrt{5})(\sqrt{20} + \sqrt{5})$

17. $(\sqrt{2} + 1)(\sqrt{3} - 1)$ 　　　　　**18.** $(\sqrt{3} + 3)(\sqrt{5} - 2)$

19. $(\sqrt{11} - \sqrt{7})(\sqrt{2} + \sqrt{5})$ 　　　　　**20.** $(\sqrt{6} + \sqrt{2})(\sqrt{3} + \sqrt{2})$

21. $(2\sqrt{3} + \sqrt{5})(3\sqrt{3} - 2\sqrt{5})$ 　　　　　**22.** $(\sqrt{7} - \sqrt{11})(2\sqrt{7} + 3\sqrt{11})$

23. $(\sqrt{5} + 2)^2$ 　　　　　**24.** $(\sqrt{11} - 1)^2$

25. $(\sqrt{21} - \sqrt{5})^2$ 　　　　　**26.** $(\sqrt{6} - \sqrt{2})^2$

27. $(2 + \sqrt[3]{6})(2 - \sqrt[3]{6})$ 　　　　　**28.** $(\sqrt[3]{3} + 6)(\sqrt[3]{3} - 6)$

29. $(2 + \sqrt[3]{2})(4 - 2\sqrt[3]{2} + \sqrt[3]{4})$ 　　　　　**30.** $(\sqrt[3]{3} - 1)(\sqrt[3]{9} + \sqrt[3]{3} + 1)$

31. $(3\sqrt{x} - \sqrt{5})(2\sqrt{x} + 1)$ 　　　　　**32.** $(4\sqrt{p} + \sqrt{7})(\sqrt{p} - 9)$

33. $(3\sqrt{r} - \sqrt{s})(3\sqrt{r} + \sqrt{s})$ 　　　　　**34.** $(\sqrt{k} + 4\sqrt{m})(\sqrt{k} - 4\sqrt{m})$

35. $(\sqrt[3]{2y} - 5)(4\sqrt[3]{2y} + 1)$ 　　　　　**36.** $(\sqrt[3]{9z} - 2)(5\sqrt[3]{9z} + 7)$

37. $(\sqrt{3x} + 2)(\sqrt{3x} - 2)$ 　　　　　**38.** $(\sqrt{6y} - 4)(\sqrt{6y} + 4)$

39. $(2\sqrt{x} + \sqrt{y})(2\sqrt{x} - \sqrt{y})$ 　　　　　**40.** $(\sqrt{p} + 5\sqrt{s})(\sqrt{p} - 5\sqrt{s})$

41. $\left[(\sqrt{2} + \sqrt{3}) - \sqrt{6}\right]\left[(\sqrt{2} + \sqrt{3}) + \sqrt{6}\right]$

42. $\left[(\sqrt{5} - \sqrt{2}) - \sqrt{3}\right]\left[(\sqrt{5} - \sqrt{2}) + \sqrt{3}\right]$

Rationalize the denominator in each expression. Assume that all variables represent positive real numbers. See Examples 2 and 3.

43. $\dfrac{7}{\sqrt{7}}$ **44.** $\dfrac{11}{\sqrt{11}}$ **45.** $\dfrac{15}{\sqrt{3}}$ **46.** $\dfrac{12}{\sqrt{6}}$ **47.** $\dfrac{\sqrt{3}}{\sqrt{2}}$

48. $\dfrac{\sqrt{7}}{\sqrt{6}}$ **49.** $\dfrac{9\sqrt{3}}{\sqrt{5}}$ **50.** $\dfrac{3\sqrt{2}}{\sqrt{11}}$ **51.** $\dfrac{-6}{\sqrt{18}}$ **52.** $\dfrac{-5}{\sqrt{24}}$

53. $\sqrt{\dfrac{7}{2}}$ **54.** $\sqrt{\dfrac{10}{3}}$ **55.** $-\sqrt{\dfrac{7}{50}}$ **56.** $-\sqrt{\dfrac{13}{75}}$ **57.** $\sqrt{\dfrac{24}{x}}$

58. $\sqrt{\dfrac{52}{y}}$ **59.** $\dfrac{-8\sqrt{3}}{\sqrt{k}}$ **60.** $\dfrac{-4\sqrt{13}}{\sqrt{m}}$ **61.** $-\sqrt{\dfrac{150m^5}{n^3}}$ **62.** $-\sqrt{\dfrac{98r^3}{s^5}}$

63. $\sqrt{\dfrac{288x^7}{y^9}}$ **64.** $\sqrt{\dfrac{242t^9}{u^{11}}}$ **65.** $\dfrac{5\sqrt{2m}}{\sqrt{y^3}}$

66. $\dfrac{2\sqrt{5r}}{\sqrt{m^3}}$ **67.** $-\sqrt{\dfrac{48k^2}{z}}$ **68.** $-\sqrt{\dfrac{75m^3}{p}}$

Simplify. Assume that all variables represent positive real numbers. See Example 4.

69. $\sqrt[3]{\dfrac{2}{3}}$ **70.** $\sqrt[3]{\dfrac{4}{5}}$ **71.** $\sqrt[3]{\dfrac{4}{9}}$ **72.** $\sqrt[3]{\dfrac{5}{16}}$ **73.** $\sqrt[3]{\dfrac{9}{32}}$

74. $\sqrt[3]{\dfrac{10}{9}}$ **75.** $-\sqrt[3]{\dfrac{2p}{r^2}}$ **76.** $-\sqrt[3]{\dfrac{6x}{y^2}}$ **77.** $\sqrt[3]{\dfrac{x^6}{y}}$ **78.** $\sqrt[3]{\dfrac{m^9}{q}}$

79. $\sqrt[4]{\dfrac{16}{x}}$ **80.** $\sqrt[4]{\dfrac{81}{y}}$ **81.** $\sqrt[4]{\dfrac{2y}{z}}$ **82.** $\sqrt[4]{\dfrac{7t}{s^2}}$

Rationalize the denominator in each expression. Assume that all variables represent positive real numbers and no denominators are 0. See Example 5.

83. $\dfrac{3}{4+\sqrt{5}}$ **84.** $\dfrac{4}{3-\sqrt{7}}$ **85.** $\dfrac{\sqrt{8}}{3-\sqrt{2}}$

86. $\dfrac{\sqrt{27}}{2+\sqrt{3}}$ **87.** $\dfrac{2}{3\sqrt{5}+2\sqrt{3}}$ **88.** $\dfrac{-1}{3\sqrt{2}-2\sqrt{7}}$

89. $\dfrac{\sqrt{2}-\sqrt{3}}{\sqrt{6}-\sqrt{5}}$ **90.** $\dfrac{\sqrt{5}+\sqrt{6}}{\sqrt{3}-\sqrt{2}}$ **91.** $\dfrac{m-4}{\sqrt{m}+2}$

92. $\dfrac{r-9}{\sqrt{r}-3}$ **93.** $\dfrac{4}{\sqrt{x}-2\sqrt{y}}$ **94.** $\dfrac{5}{3\sqrt{r}+\sqrt{s}}$

95. $\dfrac{\sqrt{x}-\sqrt{y}}{\sqrt{x}+\sqrt{y}}$ **96.** $\dfrac{\sqrt{a}+\sqrt{b}}{\sqrt{a}-\sqrt{b}}$ **97.** $\dfrac{5\sqrt{k}}{2\sqrt{k}+\sqrt{q}}$

98. $\dfrac{3\sqrt{x}}{\sqrt{x}-2\sqrt{y}}$

Write each expression in lowest terms. Assume that all variables represent positive real numbers. See Example 6.

99. $\dfrac{30 - 20\sqrt{6}}{10}$

100. $\dfrac{24 + 12\sqrt{5}}{12}$

101. $\dfrac{3 - 3\sqrt{5}}{3}$

102. $\dfrac{-5 + 5\sqrt{2}}{5}$

103. $\dfrac{16 - 4\sqrt{8}}{12}$

104. $\dfrac{12 - 9\sqrt{72}}{18}$

105. $\dfrac{6p + \sqrt{24p^3}}{3p}$

106. $\dfrac{11y - \sqrt{242y^5}}{22y}$

Rationalize each denominator. Assume that all radicals represent real numbers and no denominators are 0.

107. $\dfrac{1}{\sqrt{x + y}}$

108. $\dfrac{5}{\sqrt{m - n}}$

109. $\dfrac{p}{\sqrt{p + 2}}$

110. $\dfrac{3q}{\sqrt{5 + q}}$

111. The following expression occurs in a certain standard problem in trigonometry:

$$\frac{1}{\sqrt{2}} \cdot \frac{\sqrt{3}}{2} - \frac{1}{\sqrt{2}} \cdot \frac{1}{2}.$$

Show that it simplifies to $\dfrac{\sqrt{6} - \sqrt{2}}{4}$. Then verify, using a calculator approximation.

112. The following expression occurs in a certain standard problem in trigonometry:

$$\frac{\sqrt{3} + 1}{1 - \sqrt{3}}.$$

Show that it simplifies to $-2 - \sqrt{3}$. Then verify, using a calculator approximation.

Rationalize the numerator in each expression. Assume that all variables represent positive real numbers. (Hint: See the Connections box following Example 6.)

113. $\dfrac{6 - \sqrt{2}}{4}$

114. $\dfrac{8\sqrt{5} - 1}{6}$

115. $\dfrac{3\sqrt{a} + \sqrt{b}}{b}$

116. $\dfrac{\sqrt{p} - 3\sqrt{q}}{4q}$

PREVIEW EXERCISES

*Solve each equation. See **Sections 2.1 and 6.5**.*

117. $-8t + 7 = 4$

118. $3x - 7 = 12$

119. $6x^2 - 7x = 3$

120. $m(15m - 11) = -2$

*Apply the exponent in each expression to give a polynomial as a result. See **Section 5.4**.*

121. $(2x + 5)^2$

122. $(3x - 1)^2$

123. $\left(\sqrt{x^4 + 2x^2 + 5}\right)^2$

124. $\left(\sqrt[3]{4x^2 - 2x + 3}\right)^3$

Determine whether each statement is true *or false for the given value of x. See **Section 1.3**.*

125. $\sqrt{4 - x} = x + 2$ for $x = 0$

126. $\sqrt{4 - x} = x + 2$ for $x = -5$

127. $\sqrt{x^2 - 4x + 9} = x - 1$ for $x = 4$

128. $\sqrt{5x + 6} + \sqrt{3x + 4} = 2$ for $x = 15$

Summary Exercises on Operations with Radicals and Rational Exponents

Recall that a simplified radical satisfies the following conditions.

Conditions for a Simplified Radical

1. The radicand has no factor raised to a power greater than or equal to the index.
2. The radicand has no fractions.
3. No denominator contains a radical.
4. Exponents in the radicand and the index of the radical have no common factor (except 1).

Perform all indicated operations, and express each answer in simplest form with positive exponents. Assume that all variables represent positive real numbers.

1. $6\sqrt{10} - 12\sqrt{10}$

2. $\sqrt{7}\left(\sqrt{7} - \sqrt{2}\right)$

3. $\left(1 - \sqrt{3}\right)\left(2 + \sqrt{6}\right)$

4. $\sqrt{50} - \sqrt{98} + \sqrt{72}$

5. $\left(3\sqrt{5} + 2\sqrt{7}\right)^2$

6. $\dfrac{-3}{\sqrt{6}}$

7. $\dfrac{8}{\sqrt{7} + \sqrt{5}}$

8. $\dfrac{1 - \sqrt{2}}{1 + \sqrt{2}}$

9. $\left(\sqrt{5} + 7\right)\left(\sqrt{5} - 7\right)$

10. $\dfrac{1}{\sqrt{x} - \sqrt{5}}, \quad x \neq 5$

11. $\sqrt[3]{8a^3b^5c^9}$

12. $\dfrac{15}{\sqrt[3]{9}}$

13. $\dfrac{3}{\sqrt{5} + 2}$

14. $\sqrt{\dfrac{3}{5x}}$

15. $\dfrac{16\sqrt{3}}{5\sqrt{12}}$

16. $\dfrac{2\sqrt{25}}{8\sqrt{50}}$

17. $\dfrac{-10}{\sqrt[3]{10}}$

18. $\dfrac{\sqrt{6} + \sqrt{5}}{\sqrt{6} - \sqrt{5}}$

19. $\sqrt{12x} - \sqrt{75x}$

20. $\left(5 - 3\sqrt{3}\right)^2$

21. $\sqrt[3]{\dfrac{13}{81}}$

22. $\dfrac{\sqrt{3} + \sqrt{7}}{\sqrt{6} - \sqrt{5}}$

23. $\dfrac{6}{\sqrt[4]{3}}$

24. $\dfrac{1}{1 - \sqrt[3]{3}}$

25. $\sqrt[3]{\dfrac{x^2y}{x^{-3}y^4}}$

26. $\sqrt{12} - \sqrt{108} - \sqrt[3]{27}$

27. $\dfrac{x^{-2/3}y^{4/5}}{x^{-5/3}y^{-2/5}}$

28. $\left(\dfrac{x^{3/4}y^{2/3}}{x^{1/3}y^{5/8}}\right)^{24}$

29. $\left(125x^3\right)^{-2/3}$

30. $\dfrac{4^{1/2} + 3^{1/2}}{4^{1/2} - 3^{1/2}}$

31. $\sqrt[3]{16x^2} - \sqrt[3]{54x^2} + \sqrt[3]{128x^2}$

32. $\left(1 - \sqrt[3]{3}\right)\left(1 + \sqrt[3]{3} + \sqrt[3]{9}\right)$

33. $\left(\sqrt{74} - \sqrt{73}\right)\left(\sqrt{74} + \sqrt{73}\right)$

34. $-t^2\sqrt[4]{t} + 3\sqrt[4]{t^9} - t\sqrt[4]{t^5}$

35. $\left(3x^{-2/3}y^{1/2}\right)\left(-2x^{5/8}y^{-1/3}\right)$

36. $\left(\sqrt{6} - \sqrt{5}\right)^2\left(\sqrt{6} + \sqrt{5}\right)^2$

Students often have trouble distinguishing between the following two types of problems:

Simplifying a Radical Involving a Square Root	**Solving an Equation Using Square Roots**
Exercise: Simplify $\sqrt{25}$.	*Exercise:* Solve $x^2 = 25$.
Answer: 5	*Answer:* $\{-5, 5\}$
In this situation, $\sqrt{25}$ represents the positive square root of 25, namely 5.	In this situation, $x^2 = 25$ has two solutions, the negative square root of 25 or the positive square root of 25: $-5, 5$. (See Exercise 45.)

In Exercises 37–44, provide the appropriate responses.

37. (a) Simplify $\sqrt{64}$.
   (b) Solve $x^2 = 64$.

38. (a) Simplify $\sqrt{100}$.
   (b) Solve $x^2 = 100$.

39. (a) Solve $x^2 = 16$.
   (b) Simplify $-\sqrt{16}$.

40. (a) Solve $x^2 = 25$.
   (b) Simplify $-\sqrt{25}$.

41. (a) Simplify $-\sqrt{\dfrac{81}{121}}$.
   (b) Solve $x^2 = \dfrac{81}{121}$.

42. (a) Simplify $-\sqrt{\dfrac{49}{100}}$.
   (b) Solve $x^2 = \dfrac{49}{100}$.

43. (a) Solve $x^2 = 0.04$.
   (b) Simplify $\sqrt{0.04}$.

44. (a) Solve $x^2 = 0.09$.
   (b) Simplify $\sqrt{0.09}$.

45. Use the zero-factor property to show that the solution set of $x^2 = 36$ is $\{-6, 6\}$.

8.6 Solving Equations with Radicals

OBJECTIVES

1 Solve radical equations by using the power rule.

2 Solve radical equations that require additional steps.

3 Solve radical equations with indexes greater than 2.

4 Solve radical equations by using a graphing calculator.

5 Use the power rule to solve a formula for a specified variable.

An equation that includes one or more radical expressions with a variable is called a **radical equation.**

$$\sqrt{x-4} = 8, \quad \sqrt{5x+12} = 3\sqrt{2x-1}, \quad \text{and} \quad \sqrt[3]{6+x} = 27 \quad \text{Radical equations}$$

OBJECTIVE 1 Solve radical equations by using the power rule. The equation $x = 1$ has only one solution. Its solution set is $\{1\}$. If we square both sides of this equation, we get $x^2 = 1$. This new equation has *two* solutions: -1 and 1. Notice that the solution of the original equation is also a solution of the squared equation. However, the squared equation has another solution, -1, that is *not* a solution of the original equation. When solving equations with radicals, we use this idea of raising both sides to a power. It is an application of the **power rule.**

Power Rule for Solving an Equation with Radicals

If both sides of an equation are raised to the same power, all solutions of the original equation are also solutions of the new equation.

Read the power rule carefully; it does not say that all solutions of the new equation are solutions of the original equation. They may or may not be. Solutions that do not satisfy the original equation are called **extraneous solutions;** they must be discarded.

▶ **CAUTION** When the power rule is used to solve an equation, *every solution of the new equation must be checked in the original equation.*

EXAMPLE 1 Using the Power Rule

Solve $\sqrt{3x + 4} = 8$.

Use the power rule and square both sides to obtain

$$\left(\sqrt{3x + 4}\right)^2 = 8^2$$
$$3x + 4 = 64$$
$$3x = 60 \qquad \text{Subtract 4.}$$
$$x = 20. \qquad \text{Divide by 3.}$$

Check:

$$\sqrt{3x + 4} = 8 \qquad \text{Original equation}$$
$$\sqrt{3 \cdot 20 + 4} = 8 \quad ? \qquad \text{Let } x = 20.$$
$$\sqrt{64} = 8 \quad ?$$
$$8 - 8 \qquad \text{True}$$

Since 20 satisfies the *original* equation, the solution set is {20}.

✔ **Now Try Exercise 9.**

Use the following steps to solve equations with radicals.

Solving an Equation with Radicals

Step 1 **Isolate the radical.** Make sure that one radical term is alone on one side of the equation.

Step 2 **Apply the power rule.** Raise both sides of the equation to a power that is the same as the index of the radical.

Step 3 **Solve** the resulting equation; if it still contains a radical, repeat Steps 1 and 2.

Step 4 **Check** all proposed solutions in the original equation.

EXAMPLE 2 Using the Power Rule

Solve $\sqrt{5q - 1} + 3 = 0$.

Step 1 To isolate the radical on one side, subtract 3 from each side.

$$\sqrt{5q - 1} = -3$$

Step 2 Now square both sides.

$$\left(\sqrt{5q - 1}\right)^2 = (-3)^2$$

Step 3 $5q - 1 = 9$

                             $5q = 10$ Add 1.

                             $q = 2$ Divide by 5.

Step 4 *Check:* $\sqrt{5q - 1} + 3 = 0$ Original equation

> Be sure to check the
> proposed solution.

$\sqrt{5 \cdot 2 - 1} + 3 = 0$?   Let $q = 2$.

                             $3 + 3 = 0$ False

This false result shows that 2 is *not* a solution of the original equation; it is extraneous. The solution set is $\emptyset$.

✔ **Now Try Exercise 11.**

▶ **NOTE**   We could have determined after Step 1 that the equation in Example 2 has no solution because the expression on the left cannot be negative.

OBJECTIVE 2 Solve radical equations that require additional steps.   The next examples involve finding the square of a binomial. Recall from **Section 5.4** that

$$(x + y)^2 = x^2 + 2xy + y^2.$$

EXAMPLE 3 **Using the Power Rule; Squaring a Binomial**

Solve $\sqrt{4 - x} = x + 2$.

Step 1   The radical is alone on the left side of the equation.

Step 2   Square both sides; the square of $x + 2$ is $(x + 2)^2 = x^2 + 4x + 4$.

$$\left(\sqrt{4 - x}\right)^2 = (x + 2)^2$$

> Remember the
> middle term.

$$4 - x = x^2 + 4x + 4$$

└─ Twice the product of 2 and x

Step 3   The new equation is quadratic, so get 0 on one side.

$0 = x^2 + 5x$ Subtract 4; add x.

$0 = x(x + 5)$ Factor.

> Set *each* factor
> equal to 0.

$x = 0$ or $x + 5 = 0$ Zero-factor property

                             $x = -5$

Step 4   Check each proposed solution in the original equation.

Check: If $x = 0$, then   If $x = -5$, then

$\sqrt{4 - x} = x + 2$ $\sqrt{4 - x} = x + 2$

$\sqrt{4 - 0} = 0 + 2$? $\sqrt{4 - (-5)} = -5 + 2$?

$\sqrt{4} = 2$? $\sqrt{9} = -3$?

$2 = 2$.   True $3 = -3$.   False

The solution set is $\{0\}$. The other proposed solution, -5, is extraneous.

✔ **Now Try Exercise 27.**

EXAMPLE 4 Using the Power Rule; Squaring a Binomial

Solve $\sqrt{x^2 - 4x + 9} = x - 1$.

Squaring both sides gives $(x - 1)^2 = x^2 - 2(x)(1) + 1^2$ on the right.

$$\left(\sqrt{x^2 - 4x + 9}\right)^2 = (x - 1)^2$$

> Remember the middle term.

$$x^2 - 4x + 9 = x^2 - 2x + 1$$

└── Twice the product of x and -1

$$-2x = -8 \qquad \text{Subtract } x^2 \text{ and 9; add } 2x.$$

$$x = 4 \qquad \text{Divide by } -2.$$

Check:

$$\sqrt{x^2 - 4x + 9} = x - 1 \qquad \text{Original equation}$$

$$\sqrt{4^2 - 4 \cdot 4 + 9} = 4 - 1 \qquad ? \quad \text{Let } x = 4.$$

$$3 = 3 \qquad \text{True}$$

The solution set of the original equation is $\{4\}$.

> ✔ Now Try Exercise 29.

EXAMPLE 5 Using the Power Rule; Squaring Twice

Solve $\sqrt{5x + 6} + \sqrt{3x + 4} = 2$.

Start by isolating one radical on one side of the equation. Do this by subtracting $\sqrt{3x + 4}$ from each side.

$$\sqrt{5x + 6} = 2 - \sqrt{3x + 4} \qquad \text{Subtract } \sqrt{3x + 4}.$$

$$\left(\sqrt{5x + 6}\right)^2 = \left(2 - \sqrt{3x + 4}\right)^2 \qquad \text{Square both sides.}$$

$$5x + 6 = 4 - 4\sqrt{3x + 4} + (3x + 4)$$

└── Twice the product of 2 and $-\sqrt{3x + 4}$

> Remember the middle term.

This equation still contains a radical, so isolate the radical term on the right and square both sides again.

$$5x + 6 = 8 + 3x - 4\sqrt{3x + 4}$$

$$2x - 2 = -4\sqrt{3x + 4} \qquad \text{Subtract 8 and } 3x.$$

$$x - 1 = -2\sqrt{3x + 4} \qquad \text{Divide by 2.}$$

> Divide *each* term by 2.

$$(x - 1)^2 = \left(-2\sqrt{3x + 4}\right)^2 \qquad \text{Square both sides again.}$$

$$x^2 - 2x + 1 = (-2)^2\left(\sqrt{3x + 4}\right)^2 \qquad \text{On the right, } (ab)^2 = a^2b^2.$$

$$x^2 - 2x + 1 = 4(3x + 4) \qquad \text{Apply the exponents.}$$

$$x^2 - 2x + 1 = 12x + 16 \qquad \text{Distributive property}$$

$$x^2 - 14x - 15 = 0 \qquad \text{Standard form}$$

$$(x - 15)(x + 1) = 0 \qquad \text{Factor.}$$

$$x - 15 = 0 \quad \text{or} \quad x + 1 = 0 \qquad \text{Zero-factor property}$$

$$x = 15 \quad \text{or} \qquad x = -1$$

Check: If $x = 15$, then

$$\sqrt{5x + 6} + \sqrt{3x + 4} = 2 \qquad \text{Original equation}$$
$$\sqrt{5(15) + 6} + \sqrt{3(15) + 4} = 2 \quad ? \qquad \text{Let } x = 15.$$
$$\sqrt{81} + \sqrt{49} = 2 \quad ?$$
$$9 + 7 = 2 \quad ?$$
$$16 = 2. \qquad \text{False}$$

Thus, 15 is an extraneous solution and must be discarded. Confirm that the proposed solution -1 checks, so the solution set is $\{-1\}$.

✔ **Now Try Exercise 51.**

OBJECTIVE 3 Solve radical equations with indexes greater than 2. The power rule also works for root indexes and powers greater than 2.

EXAMPLE 6 Using the Power Rule for a Root Index Greater than 2

Solve $\sqrt[3]{z + 5} = \sqrt[3]{2z - 6}$.

Raise both sides to the third power.

$$\left(\sqrt[3]{z + 5}\right)^3 = \left(\sqrt[3]{2z - 6}\right)^3$$
$$z + 5 = 2z - 6$$
$$11 = z \qquad \text{Subtract } z; \text{ add 6.}$$

Check:
$$\sqrt[3]{z + 5} = \sqrt[3]{2z - 6} \qquad \text{Original equation}$$
$$\sqrt[3]{11 + 5} = \sqrt[3]{2 \cdot 11 - 6} \quad ? \qquad \text{Let } z = 11.$$
$$\sqrt[3]{16} = \sqrt[3]{16} \qquad \text{True}$$

The solution set is $\{11\}$.

✔ **Now Try Exercise 37.**

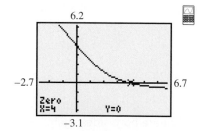

FIGURE 14

OBJECTIVE 4 Solve radical equations by using a graphing calculator. In Example 4, we solved the equation $\sqrt{x^2 - 4x + 9} = x - 1$ by using algebraic methods. If we write this equation with one side equal to 0, we get

$$\sqrt{x^2 - 4x + 9} - x + 1 = 0.$$

Using a graphing calculator to graph the function defined by

$$f(x) = \sqrt{x^2 - 4x + 9} - x + 1,$$

we obtain the graph shown in Figure 14. Notice that its *zero* (*x*-value of the *x*-intercept) is 4, which is the solution we found in Example 4.

In Example 3, we found that the single solution of $\sqrt{4 - x} = x + 2$ is 0. An extraneous solution is -5. If we graph $f(x) = \sqrt{4 - x}$ and $g(x) = x + 2$ in the same window, we find that the *x*-coordinate of the point of intersection of the two graphs is 0, which is the solution of the equation. See Figure 15.

We solved the equation in Example 3 by squaring both sides, obtaining $4 - x = x^2 + 4x + 4$. In Figure 16 on the next page, we show that the two functions defined by $f(x) = 4 - x$ and $g(x) = x^2 + 4x + 4$ have two points of intersection.

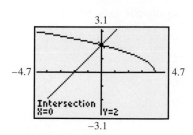

FIGURE 15

The extraneous solution -5 that we found in Example 3 shows up as an x-value of one of these points of intersection. However, our check showed that -5 was not a solution of the *original* equation (before the squaring step). Here we see a graphical interpretation of the extraneous solution.

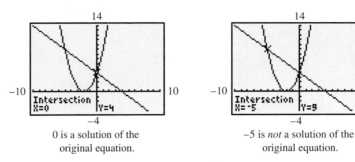

0 is a solution of the original equation.

-5 is *not* a solution of the original equation.

FIGURE 16

✔ **Now Try Exercise 59.**

OBJECTIVE 5 Use the power rule to solve a formula for a specified variable.

EXAMPLE 7 Solving a Formula from Electronics for a Variable

An important property of a radio-frequency transmission line is its **characteristic impedance,** represented by Z and measured in ohms. If L and C are the inductance and capacitance, respectively, per unit of length of the line, then these quantities are related by the formula $Z = \sqrt{\dfrac{L}{C}}$. Solve this formula for C.

$$Z = \sqrt{\frac{L}{C}} \qquad \text{Given formula}$$

Our goal is to isolate C on one side of the equals sign.

$$Z^2 = \frac{L}{C} \qquad \text{Square both sides.}$$

$$CZ^2 = L \qquad \text{Multiply by } C.$$

$$C = \frac{L}{Z^2} \qquad \text{Divide by } Z^2.$$

✔ **Now Try Exercise 65.**

8.6 EXERCISES

🌐 *Complete solution available on Video Lectures on CD/DVD*

▢ *Now Try Exercise*

Concept Check   Check each equation to see if the given value for x is a solution.

1. $\sqrt{3x + 18} = x$

(a) 6   (b) -3

2. $\sqrt{3x - 3} = x - 1$

(a) 1   (b) 4

3. $\sqrt{x + 2} = \sqrt{9x - 2} - 2\sqrt{x - 1}$

(a) 2   (b) 7

4. $\sqrt{8x - 3} = 2x$

(a) $\dfrac{3}{2}$ (b) $\dfrac{1}{2}$

📝 **5.** Is 9 a solution of the equation $\sqrt{x} = -3$? If not, what is the solution of this equation? Explain.

📝 **6.** Before even attempting to solve $\sqrt{3x + 18} = x$, how can you be sure that the equation cannot have a negative solution?

Solve each equation. See Examples 1–4.

7. $\sqrt{r - 2} = 3$ **8.** $\sqrt{q + 1} = 7$ 🌐 **9.** $\sqrt{6k - 1} = 1$

10. $\sqrt{7m - 3} = 5$ 🌐 **11.** $\sqrt{4r + 3} + 1 = 0$ **12.** $\sqrt{5k - 3} + 2 = 0$

13. $\sqrt{3k + 1} - 4 = 0$ **14.** $\sqrt{5z + 1} - 11 = 0$ **15.** $4 - \sqrt{x - 2} = 0$

16. $9 - \sqrt{4k + 1} = 0$ **17.** $\sqrt{9a - 4} = \sqrt{8a + 1}$ **18.** $\sqrt{4p - 2} = \sqrt{3p + 5}$

19. $2\sqrt{x} = \sqrt{3x + 4}$ **20.** $2\sqrt{m} = \sqrt{5m - 16}$ **21.** $3\sqrt{z - 1} = 2\sqrt{2z + 2}$

22. $5\sqrt{4a + 1} = 3\sqrt{10a + 25}$ **23.** $k = \sqrt{k^2 + 4k - 20}$

24. $p = \sqrt{p^2 - 3p + 18}$ **25.** $a = \sqrt{a^2 + 3a + 9}$

26. $z = \sqrt{z^2 - 4z - 8}$ 🌐 **27.** $\sqrt{9 - x} = x + 3$

28. $\sqrt{5 - x} = x + 1$ 🌐 **29.** $\sqrt{k^2 + 2k + 9} = k + 3$

30. $\sqrt{a^2 - 3a + 3} = a - 1$ **31.** $\sqrt{r^2 + 9r + 3} = -r$

32. $\sqrt{p^2 - 15p + 15} = p - 5$ **33.** $\sqrt{z^2 + 12z - 4} + 4 - z = 0$

34. $\sqrt{m^2 + 3m + 12} - m - 2 = 0$

35. *Concept Check* In solving the equation $\sqrt{3x + 4} = 8 - x$, a student wrote the following for her first step:

$$3x + 4 = 64 + x^2.$$

WHAT WENT WRONG? Solve the given equation correctly.

36. *Concept Check* In solving the equation $\sqrt{5x + 6} - \sqrt{x + 3} = 3$, a student wrote the following for his first step:

$$(5x + 6) + (x + 3) = 9.$$

WHAT WENT WRONG? Solve the given equation correctly.

Solve each equation. See Examples 5 and 6.

🌐 **37.** $\sqrt[3]{2x + 5} = \sqrt[3]{6x + 1}$ **38.** $\sqrt[3]{p - 1} = 2$

39. $\sqrt[3]{a^2 + 5a + 1} = \sqrt[3]{a^2 + 4a}$ **40.** $\sqrt[3]{r^2 + 2r + 8} = \sqrt[3]{r^2}$

41. $\sqrt[3]{2m - 1} = \sqrt[3]{m + 13}$ **42.** $\sqrt[3]{2k - 11} - \sqrt[3]{5k + 1} = 0$

43. $\sqrt[4]{a + 8} = \sqrt[4]{2a}$ **44.** $\sqrt[4]{z + 11} = \sqrt[4]{2z + 6}$

45. $\sqrt[3]{x - 8} + 2 = 0$ **46.** $\sqrt[3]{r + 1} + 1 = 0$

47. $\sqrt[4]{2k - 5} + 4 = 0$ **48.** $\sqrt[4]{8z - 3} + 2 = 0$

49. $\sqrt{k + 2} - \sqrt{k - 3} = 1$ **50.** $\sqrt{r + 6} - \sqrt{r - 2} = 2$

🌐 **51.** $\sqrt{2r + 11} - \sqrt{5r + 1} = -1$ **52.** $\sqrt{3x - 2} - \sqrt{x + 3} = 1$

53. $\sqrt{3p + 4} - \sqrt{2p - 4} = 2$ **54.** $\sqrt{4x + 5} - \sqrt{2x + 2} = 1$

55. $\sqrt{3 - 3p} - 3 = \sqrt{3p + 2}$ **56.** $\sqrt{4x + 7} - 4 = \sqrt{4x - 1}$

57. $\sqrt{2\sqrt{x + 11}} = \sqrt{4x + 2}$ **58.** $\sqrt{1 + \sqrt{24 - 10x}} = \sqrt{3x + 5}$

59. Use a graphing calculator to solve $\sqrt{3 - 3x} = 3 + \sqrt{3x + 2}$. What is the domain of $y = \sqrt{3 - 3x} - 3 - \sqrt{3x + 2}$?

60. Use a graphing calculator with a viewing window of $[-1, 4]$ by $[-1, 3]$ to solve $\sqrt{2\sqrt{7x + 2}} = \sqrt{3x + 2}$. What is the domain of $f(x) = \sqrt{2\sqrt{7x + 2}} - \sqrt{3x + 2}$?

For each equation, write the expressions with rational exponents as radical expressions, and then solve, using the procedures explained in this section.

61. $(2x - 9)^{1/2} = 2 + (x - 8)^{1/2}$ **62.** $(3w + 7)^{1/2} = 1 + (w + 2)^{1/2}$

63. $(2w - 1)^{2/3} - w^{1/3} = 0$ **64.** $(x^2 - 2x)^{1/3} - x^{1/3} = 0$

Solve each formula from electricity and radio for the indicated variable. See Example 7. (Source: Cooke, Nelson M., and Joseph B. Orleans, Mathematics Essential to Electricity and Radio, *McGraw-Hill, 1943.)*

65. $V = \sqrt{\dfrac{2K}{m}}$ for K **66.** $V = \sqrt{\dfrac{2K}{m}}$ for m

67. $f = \dfrac{1}{2\pi\sqrt{LC}}$ for L **68.** $r = \sqrt{\dfrac{Mm}{F}}$ for F

A number of useful formulas involve radicals or radical expressions. The formula

$$N = \frac{1}{2\pi}\sqrt{\frac{a}{r}}$$

is used to find the rotational rate N of a space station. Here, a is the acceleration and r represents the radius of the space station, in meters. To find the value of r that will make N simulate the effect of gravity on Earth, the equation must be solved for r, using the required value of N. (Source: Kastner, Bernice, Space Mathematics, *NASA, 1972.)*

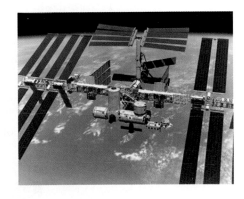

69. Solve the equation for r.

70. (a) Find the value of r that makes $N = 0.063$ rotation per sec if $a = 9.8$ m per sec^2.
     (b) Find the value of r that makes $N = 0.04$ rotation per sec if $a = 9.8$ m per sec^2.

PREVIEW EXERCISES

*Perform the indicated operations. See **Sections 5.2 and 5.4**.*

71. $(5 + 9x) + (-4 - 8x)$ **72.** $(12 + 7y) - (-3 + 2y)$ **73.** $(x + 3)(2x - 5)$

*Simplify each radical. See **Section 8.5**.*

74. $\dfrac{2}{4 + \sqrt{3}}$ **75.** $\dfrac{-7}{5 - \sqrt{2}}$ **76.** $\dfrac{\sqrt{2} + \sqrt{7}}{\sqrt{5} + \sqrt{3}}$

8.7 Complex Numbers

As we saw in **Section 1.1,** the set of real numbers includes many other number sets (the rational numbers, integers, and natural numbers, for example). In this section, a new set of numbers is introduced that includes the set of real numbers, as well as numbers that are even roots of negative numbers, such as $\sqrt{-2}$.

OBJECTIVE 1 Simplify numbers of the form $\sqrt{-b}$, where $b > 0$. The equation $x^2 + 1 = 0$ has no real number solution, since any solution must be a number whose square is -1. In the set of real numbers, all squares are nonnegative numbers because the product of two positive numbers or two negative numbers is positive and $0^2 = 0$. To provide a solution of the equation $x^2 + 1 = 0$, we introduce a new number i.

Imaginary Unit i

The **imaginary unit i** is defined as

$$i = \sqrt{-1}, \quad \text{where} \quad i^2 = -1.$$

That is, i is the principal square root of -1.

This definition of i makes it possible to define any square root of a negative real number as follows.

$\sqrt{-b}$

For any positive real number b, $\quad \sqrt{-b} = i\sqrt{b}.$

EXAMPLE 1 Simplifying Square Roots of Negative Numbers

Write each number as a product of a real number and i.

(a) $\sqrt{-100} = i\sqrt{100} = 10i$ **(b)** $-\sqrt{-36} = -i\sqrt{36} = -6i$

(c) $\sqrt{-2} = i\sqrt{2}$ **(d)** $\sqrt{-54} = i\sqrt{54} = i\sqrt{9 \cdot 6} = 3i\sqrt{6}$

✔ **Now Try Exercises 7, 9, 11, and 13.**

▶ **CAUTION** It is easy to mistake $\sqrt{2}i$ for $\sqrt{2i}$, with the i under the radical. For this reason, we usually write $\sqrt{2}i$ as $i\sqrt{2}$, as in the definition of $\sqrt{-b}$.

When finding a product such as $\sqrt{-4} \cdot \sqrt{-9}$, we cannot use the product rule for radicals because it applies only to *nonnegative* radicands. ***For this reason, we change $\sqrt{-b}$ to the form $i\sqrt{b}$ before performing any multiplications or divisions.***

For example,

$$\sqrt{-4} \cdot \sqrt{-9} = i\sqrt{4} \cdot i\sqrt{9} \qquad \sqrt{-b} = i\sqrt{b}$$

First write all square roots in terms of i.

$$= i \cdot 2 \cdot i \cdot 3$$
$$= 6i^2$$
$$= 6(-1) \qquad \text{Substitute: } i^2 = -1.$$
$$= -6.$$

▶ **CAUTION** Using the product rule for radicals *before* using the definition of $\sqrt{-b}$ gives a *wrong* answer. The preceding example shows that

$$\sqrt{-4} \cdot \sqrt{-9} = -6, \qquad \text{Correct}$$

but

$$\sqrt{-4(-9)} = \sqrt{36} = 6, \qquad \text{Incorrect}$$

so

$$\sqrt{-4} \cdot \sqrt{-9} \neq \sqrt{-4(-9)}.$$

EXAMPLE 2 Multiplying Square Roots of Negative Numbers

Multiply.

(a) $\sqrt{-3} \cdot \sqrt{-7} = i\sqrt{3} \cdot i\sqrt{7} \qquad \sqrt{-b} = i\sqrt{b}$

First write all square roots in terms of i.

$$= i^2\sqrt{3 \cdot 7} \qquad \text{Product rule}$$
$$= (-1)\sqrt{21} \qquad \text{Substitute: } i^2 = -1.$$
$$= -\sqrt{21}$$

(b) $\sqrt{-2} \cdot \sqrt{-8} = i\sqrt{2} \cdot i\sqrt{8} \qquad \sqrt{-b} = i\sqrt{b}$

$$= i^2\sqrt{2 \cdot 8} \qquad \text{Product rule}$$
$$= (-1)\sqrt{16} \qquad i^2 = -1$$
$$= -4$$

(c) $\sqrt{-5} \cdot \sqrt{6} = i\sqrt{5} \cdot \sqrt{6} = i\sqrt{30}$

✔ **Now Try Exercises 15, 17, and 19.**

The methods used to find products also apply to quotients. First write all square roots in terms of i.

EXAMPLE 3 Dividing Square Roots of Negative Numbers

Divide.

(a) $\dfrac{\sqrt{-75}}{\sqrt{-3}} = \dfrac{i\sqrt{75}}{i\sqrt{3}} = \sqrt{\dfrac{75}{3}} = \sqrt{25} = 5$

(b) $\dfrac{\sqrt{-32}}{\sqrt{8}} = \dfrac{i\sqrt{32}}{\sqrt{8}} = i\sqrt{\dfrac{32}{8}} = i\sqrt{4} = 2i$

✔ **Now Try Exercises 21 and 23.**

OBJECTIVE 2 **Recognize complex numbers.** With the imaginary unit i and the real numbers, a new set of numbers, the *complex numbers,* are defined as follows.

Complex Number

If a and b are real numbers, then any number of the form $\boldsymbol{a + bi}$ is called a **complex number.** In the complex number $a + bi$, the number a is called the **real part** and b is called the **imaginary part.***

For a complex number $a + bi$, if $b = 0$, then $a + bi = a$, which is a real number. ***Thus, the set of real numbers is a subset of the set of complex numbers.*** If $a = 0$ and $b \neq 0$, the complex number is said to be a **pure imaginary number.** For example, $3i$ is a pure imaginary number. A number such as $7 + 2i$ is a **nonreal complex number.**

A complex number written in the form $a + bi$ is in **standard form.** In this section, most answers will be given in standard form, but if a or b is 0, we consider answers such as a and bi to be in standard form.

The relationships among the various sets of numbers are shown in Figure 17.

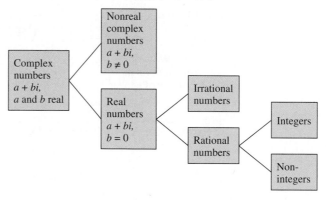

FIGURE 17

OBJECTIVE 3 **Add and subtract complex numbers.** The commutative, associative, and distributive properties for real numbers are also valid for complex numbers. Thus, to add complex numbers, we add their real parts and add their imaginary parts.

EXAMPLE 4 Adding Complex Numbers

Add.

(a) $(2 + 3i) + (6 + 4i) = (2 + 6) + (3 + 4)i$ Properties of real numbers
$$= 8 + 7i$$ Add real parts; add imaginary parts.

(b) $(4 + 2i) + (3 - i) + (-6 + 3i) = [4 + 3 + (-6)] + [2 + (-1) + 3]i$
$$= 1 + 4i$$

✔ **Now Try Exercises 27 and 35.**

---

*Some texts define bi as the imaginary part of the complex number $a + bi$.

We subtract complex numbers by subtracting their real parts and subtracting their imaginary parts.

EXAMPLE 5 Subtracting Complex Numbers

Subtract.

(a) $(6 + 5i) - (3 + 2i) = (6 - 3) + (5 - 2)i$ Properties of real numbers

$\qquad\qquad\qquad\quad = 3 + 3i$ Subtract real parts; subtract imaginary parts.

(b) $(7 - 3i) - (8 - 6i) = (7 - 8) + [-3 - (-6)]i$

$\qquad\qquad\qquad\qquad = -1 + 3i$

(c) $(-9 + 4i) - (-9 + 8i) = (-9 + 9) + (4 - 8)i$

$\qquad\qquad\qquad\qquad\quad = 0 - 4i$

$\qquad\qquad\qquad\qquad\quad = -4i$

✔ **Now Try Exercises 31 and 33.**

OBJECTIVE ④ Multiply complex numbers. We multiply complex numbers as we multiply polynomials. Complex numbers of the form $a + bi$ have the same form as binomials, so when applicable, we multiply two complex numbers in standard form by using the FOIL method for multiplying binomials. (Recall from **Section 5.4** that FOIL stands for *First, Outer, Inner, Last.*)

EXAMPLE 6 Multiplying Complex Numbers

Multiply.

(a) $4i(2 + 3i) = 4i(2) + 4i(3i)$ Distributive property

$\qquad\qquad\quad = 8i + 12i^2$ Multiply.

$\qquad\qquad\quad = 8i + 12(-1)$ Substitute: $i^2 = -1$.

$\qquad\qquad\quad = -12 + 8i$ Standard form

(b) $(3 + 5i)(4 - 2i) = \underbrace{3(4)}_{\text{First}} + \underbrace{3(-2i)}_{\text{Outer}} + \underbrace{5i(4)}_{\text{Inner}} + \underbrace{5i(-2i)}_{\text{Last}}$ (FOIL)

$\qquad\qquad\qquad = 12 - 6i + 20i - 10i^2$

$\qquad\qquad\qquad = 12 + 14i - 10(-1)$ Combine imaginary terms; $i^2 = -1$.

$\qquad\qquad\qquad = 12 + 14i + 10$

$\qquad\qquad\qquad = 22 + 14i$ Combine real terms.

(c) $(2 + 3i)(1 - 5i) = 2(1) + 2(-5i) + 3i(1) + 3i(-5i)$ FOIL

$\qquad\qquad\qquad = 2 - 10i + 3i - 15i^2$

$\qquad\qquad\qquad = 2 - 7i - 15(-1)$ Use parentheses around -1 to avoid errors.

$\qquad\qquad\qquad = 2 - 7i + 15$

$\qquad\qquad\qquad = 17 - 7i$

✔ **Now Try Exercises 45 and 47.**

The two complex numbers $a + bi$ and $a - bi$ are called **complex conjugates,** or simply *conjugates,* of each other. ***The product of a complex number and its conjugate is always a real number,*** as shown here:

$$(a + bi)(a - bi) = a^2 - abi + abi - b^2i^2$$
$$= a^2 - b^2(-1)$$
$$(a + bi)(a - bi) = a^2 + b^2.$$

The product eliminates i.

For example, $(3 + 7i)(3 - 7i) = 3^2 + 7^2 = 9 + 49 = 58$.

OBJECTIVE **5** Divide complex numbers. The quotient of two complex numbers should be a complex number. To write the quotient as a complex number, we need to eliminate i in the denominator. We use conjugates and a process like that for rationalizing a denominator to do this.

EXAMPLE 7 Dividing Complex Numbers

Find each quotient.

(a) $\dfrac{8 + 9i}{5 + 2i}$

Multiply both the numerator and denominator by the conjugate of the denominator. The conjugate of $5 + 2i$ is $5 - 2i$.

$$\frac{8 + 9i}{5 + 2i} = \frac{(8 + 9i)(5 - 2i)}{(5 + 2i)(5 - 2i)} \qquad \tfrac{5 - 2i}{5 - 2i} = 1$$

$$= \frac{40 - 16i + 45i - 18i^2}{5^2 + 2^2} \qquad$$

$$= \frac{58 + 29i}{29} \qquad -18i^2 = -18(-1) = 18;\ \text{combine terms.}$$

Factor first; then divide out the common factor.

$$= \frac{29(2 + i)}{29} \qquad \text{Factor the numerator.}$$

$$= 2 + i \qquad \text{Lowest terms}$$

(b) $\dfrac{1 + i}{i} = \dfrac{(1 + i)(-i)}{i(-i)}$ Multiply numerator and denominator by $-i$, the conjugate of i.

$$= \frac{-i - i^2}{-i^2}$$

$$= \frac{-i - (-1)}{-(-1)} \qquad \text{Substitute: } i^2 = -1.$$

$$= \frac{-i + 1}{1} \qquad$$

Use parentheses to avoid errors.

$$= 1 - i$$

✔ **Now Try Exercises 61 and 67.**

OBJECTIVE 6 Find powers of *i*. Because i^2 is defined to be -1, we can find larger powers of *i*, as shown in the following examples:

$$i^3 = i \cdot i^2 = i(-1) = -i \qquad\qquad i^6 = i^2 \cdot i^4 = (-1) \cdot 1 = -1$$
$$i^4 = i^2 \cdot i^2 = (-1)(-1) = 1 \qquad i^7 = i^3 \cdot i^4 = (-i) \cdot 1 = -i$$
$$i^5 = i \cdot i^4 = i \cdot 1 = i \qquad\qquad i^8 = i^4 \cdot i^4 = 1 \cdot 1 = 1.$$

Notice that the powers of *i* rotate through the four numbers i, -1, $-i$, and 1. Larger powers of *i* can be simplified by using the fact that $i^4 = 1$. For example,

$$i^{75} = (i^4)^{18} \cdot i^3 = 1^{18} \cdot i^3 = 1 \cdot i^3 = i^3 = -i.$$

EXAMPLE 8 Simplifying Powers of *i*

Find each power of *i*.

(a) $i^{12} = (i^4)^3 = 1^3 = 1$

(b) $i^{39} = i^{36} \cdot i^3 = (i^4)^9 \cdot i^3 = 1^9 \cdot (-i) = -i$

(c) $i^{-2} = \dfrac{1}{i^2} = \dfrac{1}{-1} = -1$

(d) $i^{-1} = \dfrac{1}{i} = \dfrac{1(-i)}{i(-i)} = \dfrac{-i}{-i^2} = \dfrac{-i}{-(-1)} = \dfrac{-i}{1} = -i$

✔ **Now Try Exercises 69 and 77.**

8.7 EXERCISES

Concept Check   *Decide whether each expression is equal to* 1, -1, i, *or* $-i$.

1. $\sqrt{-1}$ **2.** $-\sqrt{-1}$ **3.** i^2 **4.** $-i^2$ **5.** $\dfrac{1}{i}$ **6.** $(-i)^2$

Write each number as a product of a real number and i*. Simplify all radical expressions. See Example 1.*

🌐 **7.** $\sqrt{-169}$ **8.** $\sqrt{-225}$ **9.** $-\sqrt{-144}$ **10.** $-\sqrt{-196}$

11. $\sqrt{-5}$ **12.** $\sqrt{-21}$ **13.** $\sqrt{-48}$ **14.** $\sqrt{-96}$

Multiply or divide as indicated. See Examples 2 and 3.

🌐 **15.** $\sqrt{-7} \cdot \sqrt{-15}$ **16.** $\sqrt{-3} \cdot \sqrt{-19}$ **17.** $\sqrt{-4} \cdot \sqrt{-25}$ **18.** $\sqrt{-9} \cdot \sqrt{-81}$

19. $\sqrt{-3} \cdot \sqrt{11}$ **20.** $\sqrt{-10} \cdot \sqrt{2}$ 🌐 **21.** $\dfrac{\sqrt{-300}}{\sqrt{-100}}$ **22.** $\dfrac{\sqrt{-40}}{\sqrt{-10}}$

23. $\dfrac{\sqrt{-75}}{\sqrt{3}}$ **24.** $\dfrac{\sqrt{-160}}{\sqrt{10}}$

✏ **25. (a)** Every real number is a complex number. Explain why this is so.

     (b) Not every complex number is a real number. Give an example, and explain why this statement is true.

✏ **26.** Explain how to add, subtract, multiply, and divide complex numbers. Give examples.

Add or subtract as indicated. Write your answers in the form a + bi. See Examples 4 and 5.

27. $(3 + 2i) + (-4 + 5i)$

28. $(7 + 15i) + (-11 + 14i)$

29. $(5 - i) + (-5 + i)$

30. $(-2 + 6i) + (2 - 6i)$

31. $(4 + i) - (-3 - 2i)$

32. $(9 + i) - (3 + 2i)$

33. $(-3 - 4i) - (-1 - 4i)$

34. $(-2 - 3i) - (-5 - 3i)$

35. $(-4 + 11i) + (-2 - 4i) + (7 + 6i)$

36. $(-1 + i) + (2 + 5i) + (3 + 2i)$

37. $[(7 + 3i) - (4 - 2i)] + (3 + i)$

38. $[(7 + 2i) + (-4 - i)] - (2 + 5i)$

39. *Concept Check*   Fill in the blank with the correct response:

Because $(4 + 2i) - (3 + i) = 1 + i$, using the definition of subtraction, we can check this to find that $(1 + i) + (3 + i) = $ _____ .

40. *Concept Check*   Fill in the blank with the correct response:

Because $\frac{-5}{2 - i} = -2 - i$, using the definition of division, we can check this to find that $(-2 - i)(2 - i) = $ _____ .

Multiply. See Example 6.

41. $(3i)(27i)$

42. $(5i)(125i)$

43. $(-8i)(-2i)$

44. $(-32i)(-2i)$

45. $5i(-6 + 2i)$

46. $3i(4 + 9i)$

47. $(4 + 3i)(1 - 2i)$

48. $(7 - 2i)(3 + i)$

49. $(4 + 5i)^2$

50. $(3 + 2i)^2$

51. $2i(-4 - i)^2$

52. $3i(-3 - i)^2$

53. $(12 + 3i)(12 - 3i)$

54. $(6 + 7i)(6 - 7i)$

55. $(4 + 9i)(4 - 9i)$

56. $(7 + 2i)(7 - 2i)$

57. *Concept Check*   What is the conjugate of $a + bi$?

58. *Concept Check*   If we multiply $a + bi$ by its conjugate, we get _____ , which is always a real number.

Find each quotient. See Example 7.

59. $\dfrac{2}{1 - i}$

60. $\dfrac{29}{5 + 2i}$

61. $\dfrac{-7 + 4i}{3 + 2i}$

62. $\dfrac{-38 - 8i}{7 + 3i}$

63. $\dfrac{8i}{2 + 2i}$

64. $\dfrac{-8i}{1 + i}$

65. $\dfrac{2 - 3i}{2 + 3i}$

66. $\dfrac{-1 + 5i}{3 + 2i}$

67. $\dfrac{3 + i}{i}$

68. $\dfrac{5 - i}{-i}$

Find each power of i. See Example 8.

69. i^{18}

70. i^{26}

71. i^{89}

72. i^{48}

73. i^{38}

74. i^{102}

75. i^{43}

76. i^{83}

77. i^{-5}

78. i^{-17}

79. A student simplified i^{-18} as follows:

$$i^{-18} = i^{-18} \cdot i^{20} = i^{-18+20} = i^2 = -1.$$

Explain the mathematical justification for this correct work.

80. Explain why

$$(46 + 25i)(3 - 6i) \quad \text{and} \quad (46 + 25i)(3 - 6i)i^{12}$$

must be equal. (Do not actually perform the computation.)

Ohm's law for the current I in a circuit with voltage E, resistance R, capacitive reactance X_c, and inductive reactance X_L is

$$I = \frac{E}{R + (X_L - X_c)i}.$$

Use this law to work Exercises 81 and 82.

81. Find I if $E = 2 + 3i$, $R = 5$, $X_L = 4$, and $X_c = 3$.

82. Find E if $I = 1 - i$, $R = 2$, $X_L = 3$, and $X_c = 1$.

*Complex numbers will appear again in this book in **Chapter 9**, when we study quadratic equations. The following exercises examine how a complex number can be a solution of a quadratic equation.*

83. Show that $1 + 5i$ is a solution of

$$x^2 - 2x + 26 = 0.$$

Then show that its conjugate is also a solution.

84. Show that $3 + 2i$ is a solution of

$$x^2 - 6x + 13 = 0.$$

Then show that its conjugate is also a solution.

The next problems are "brain busters." Perform the indicated operations. Give answers in standard form.

85. $\dfrac{3}{2 - i} + \dfrac{5}{1 + i}$

86. $\dfrac{2}{3 + 4i} + \dfrac{4}{1 - i}$

87. $\left(\dfrac{2 + i}{2 - i} + \dfrac{i}{1 + i} \right)i$

88. $\left(\dfrac{4 - i}{1 + i} - \dfrac{2i}{2 + i} \right)4i$

PREVIEW EXERCISES

*Solve each equation. See **Sections 2.1 and 6.5**.*

89. $6x + 13 = 0$

90. $4x - 7 = 0$

91. $x(x + 3) = 40$

92. $2x^2 - 5x - 7 = 0$

93. $5x^2 - 3x = 2$

94. $-6x^2 + 7x = -10$

Chapter **8** | **Group Activity**

SOLAR ELECTRICITY

Objective Apply the Pythagorean formula.

In this activity, you will determine the sizes of frames needed to support solar electric panels on a flat roof.

A. The table that follows gives three different solar modules by Solarex. Have each member of the group choose one of the solar panels.

Model	Watts	Volts	Amps	Size (in inches)	Cost (in dollars)
MSX-77	77	16.9	4.56	44 × 26	475
MSX-83	83	17.1	4.85	44 × 24	490
MSX-60	60	17.1	3.5	44 × 20	382

Source: Solarex table in *Jade Mountain* catalog.

B. To use your solar panel, you must make a wooden frame to support it. The sides of this frame will form a right triangle. The hypotenuse of the triangle will be the width of the solar panel you chose. Make a sketch and use the Pythagorean formula to find the dimensions of the legs for each frame, given the conditions that follow. Round answers to the nearest tenth.

1. The legs have equal length.

2. One leg is twice the length of the other.

3. One leg is three times the length of the other.

C. Compare the different frame sizes for each panel. What factors might determine which of the triangles you would use in your frame?

Chapter 8 SUMMARY

View the Interactive Summary on the Pass the Test CD.

KEY TERMS

8.1 radicand
index (order)
radical
principal root
radical expression

square root function
cube root function
8.3 simplified radical
8.5 rationalizing the
denominator

conjugates
8.6 radical equation
extraneous solution
8.7 complex number
real part

imaginary part
pure imaginary
number
standard form
complex conjugates

NEW SYMBOLS

$\sqrt{}$ radical sign

$\sqrt[n]{a}$ radical; principal nth
root of a

$\pm$ "positive or negative"
or "plus or minus"

$\approx$ is approximately equal
to

$a^{1/n}$ a to the power $\dfrac{1}{n}$

$a^{m/n}$ a to the power $\dfrac{m}{n}$

i imaginary unit

TEST YOUR WORD POWER

See how well you have learned the vocabulary in this chapter. Answers, with examples, follow the Quick Review.

1. A **radicand** is
   A. the index of a radical
   B. the number or expression under the radical sign
   C. the positive root of a number
   D. the radical sign.

2. The **Pythagorean formula** states that, in a right triangle,
   A. the sum of the measures of the angles is 180°
   B. the sum of the lengths of the two shorter sides equals the length of the longest side
   C. the longest side is opposite the right angle
   D. the square of the length of the longest side equals the sum of the squares of the lengths of the two shorter sides.

3. A **hypotenuse** is
   A. either of the two shorter sides of a triangle
   B. the shortest side of a triangle
   C. the side opposite the right angle in a triangle
   D. the longest side in any triangle.

4. **Rationalizing the denominator** is the process of
   A. eliminating fractions from a radical expression
   B. changing the denominator of a fraction from a radical to a rational number
   C. clearing a radical expression of radicals
   D. multiplying radical expressions.

5. An **extraneous solution** is a solution
   A. that does not satisfy the original equation
   B. that makes an equation true
   C. that makes an expression equal 0
   D. that checks in the original equation.

6. A **complex number** is
   A. a real number that includes a complex fraction
   B. a zero multiple of i
   C. a number of the form $a + bi$, where a and b are real numbers
   D. the square root of -1.

QUICK REVIEW

Concepts	Examples

8.1 RADICAL EXPRESSIONS AND GRAPHS

$\sqrt[n]{a} = b$ **means** $b^n = a.$

$\sqrt[n]{a}$ is the principal nth root of $a.$

$\sqrt[n]{a^n} = |a|$ if n is even; $\sqrt[n]{a^n} = a$ if n is odd.

The two square roots of 64 are $\sqrt{64} = 8$ (the principal square root) and $-\sqrt{64} = -8.$

$$\sqrt[4]{(-2)^4} = |-2| = 2 \qquad \sqrt[3]{-27} = -3$$

Functions Defined by Radical Expressions

The square root function defined by $f(x) = \sqrt{x}$ and the cube root function defined by $f(x) = \sqrt[3]{x}$ are two important functions defined by radical expressions.

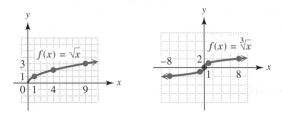

8.2 RATIONAL EXPONENTS

$a^{1/n} = \sqrt[n]{a}$ whenever $\sqrt[n]{a}$ exists.

If m and n are positive integers with $\frac{m}{n}$ in lowest terms, then $a^{m/n} = (a^{1/n})^m$, provided that $a^{1/n}$ is a real number.

All of the usual definitions and rules for exponents are valid for rational exponents.

$$81^{1/2} = \sqrt{81} = 9 \qquad -64^{1/3} = -\sqrt[3]{64} = -4$$
$$8^{5/3} = (8^{1/3})^5 = 2^5 = 32$$
$$5^{-1/2} \cdot 5^{1/4} = 5^{-1/2+1/4} = 5^{-1/4} = \frac{1}{5^{1/4}} \qquad (y^{2/5})^{10} = y^4$$

$$\frac{x^{-1/3}}{x^{-1/2}} = x^{-1/3-(-1/2)} = x^{-1/3+1/2} = x^{1/6}, \quad x > 0$$

8.3 SIMPLIFYING RADICAL EXPRESSIONS

Product and Quotient Rules for Radicals

If $\sqrt[n]{a}$ and $\sqrt[n]{b}$ are real numbers and n is a natural number, then

$$\sqrt[n]{a} \cdot \sqrt[n]{b} = \sqrt[n]{ab}$$

and

$$\sqrt[n]{\frac{a}{b}} = \frac{\sqrt[n]{a}}{\sqrt[n]{b}}, \qquad b \neq 0.$$

$$\sqrt{3} \cdot \sqrt{7} = \sqrt{21}$$
$$\sqrt[5]{x^3 y} \cdot \sqrt[5]{xy^2} = \sqrt[5]{x^4 y^3}$$

$$\frac{\sqrt{x^5}}{\sqrt{x^4}} = \sqrt{\frac{x^5}{x^4}} = \sqrt{x}, \quad x > 0$$

Conditions for a Simplified Radical

1. The radicand has no factor raised to a power greater than or equal to the index.

2. The radicand has no fractions.

3. No denominator contains a radical.

4. Exponents in the radicand and the index of the radical have no common factor (except 1).

$$\sqrt{18} = \sqrt{9 \cdot 2} = 3\sqrt{2}$$
$$\sqrt[3]{54x^5 y^3} = \sqrt[3]{27x^3 y^3 \cdot 2x^2} = 3xy\sqrt[3]{2x^2}$$
$$\sqrt{\frac{7}{4}} = \frac{\sqrt{7}}{\sqrt{4}} = \frac{\sqrt{7}}{2}$$
$$\sqrt[9]{x^3} = x^{3/9} = x^{1/3}, \quad \text{or} \quad \sqrt[3]{x}$$

(continued)

Concepts	Examples

Pythagorean Formula

If c is the length of the longest side of a right triangle and a and b are the lengths of the shorter sides, then $c^2 = a^2 + b^2$. The longest side is the hypotenuse and the two shorter sides are the legs of the triangle. The hypotenuse is opposite the right angle.

Find b for the triangle in the figure.

$$10^2 + b^2 = \left(2\sqrt{61}\right)^2$$
$$b^2 = 4(61) - 100$$
$$b^2 = 144$$
$$b = 12$$

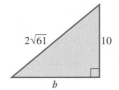

Distance Formula

The distance between (x_1, y_1) and (x_2, y_2) is

$$d = \sqrt{(x_2 - x_1)^2 + (y_2 - y_1)^2}.$$

The distance between $(3, -2)$ and $(-1, 1)$ is

$$\sqrt{(-1 - 3)^2 + [1 - (-2)]^2} = \sqrt{(-4)^2 + 3^2}$$
$$= \sqrt{16 + 9}$$
$$= \sqrt{25}$$
$$= 5.$$

> Start with the *x*-value and *y*-value of the same point.

8.4   ADDING AND SUBTRACTING RADICAL EXPRESSIONS

Only radical expressions with the same index and the same radicand may be combined.

$$3\sqrt{17} + 2\sqrt{17} - 8\sqrt{17} = (3 + 2 - 8)\sqrt{17}$$
$$= -3\sqrt{17}$$

$$\sqrt[3]{2} - \sqrt[3]{250} = \sqrt[3]{2} - 5\sqrt[3]{2} \qquad \sqrt[3]{250} = \sqrt[3]{125 \cdot 2} = 5\sqrt[3]{2}$$
$$= -4\sqrt[3]{2}$$

$$\left.\begin{array}{l}\sqrt{15} + \sqrt{30} \\ \sqrt{3} + \sqrt[3]{9}\end{array}\right\} \text{ cannot be simplified further}$$

8.5   MULTIPLYING AND DIVIDING RADICAL EXPRESSIONS

Multiply binomial radical expressions by using the FOIL method. Special products from **Section 5.4** may apply.

$$\left(\sqrt{2} + \sqrt{7}\right)\left(\sqrt{3} - \sqrt{6}\right)$$
$$= \sqrt{6} - 2\sqrt{3} + \sqrt{21} - \sqrt{42} \qquad \sqrt{12} = 2\sqrt{3}$$

$$\left(\sqrt{5} - \sqrt{10}\right)\left(\sqrt{5} + \sqrt{10}\right) = 5 - 10 = -5$$

$$\left(\sqrt{3} - \sqrt{2}\right)^2 = 3 - 2\sqrt{3} \cdot \sqrt{2} + 2 = 5 - 2\sqrt{6}$$

Rationalize the denominator by multiplying both the numerator and the denominator by the same expression.

$$\frac{\sqrt{7}}{\sqrt{5}} = \frac{\sqrt{7} \cdot \sqrt{5}}{\sqrt{5} \cdot \sqrt{5}} = \frac{\sqrt{35}}{5}$$

$$\frac{\sqrt[3]{2}}{\sqrt[3]{4}} = \frac{\sqrt[3]{2} \cdot \sqrt[3]{2}}{\sqrt[3]{4} \cdot \sqrt[3]{2}} = \frac{\sqrt[3]{4}}{\sqrt[3]{8}} = \frac{\sqrt[3]{4}}{2}$$

$$\frac{4}{\sqrt{5} - \sqrt{2}} = \frac{4\left(\sqrt{5} + \sqrt{2}\right)}{\left(\sqrt{5} - \sqrt{2}\right)\left(\sqrt{5} + \sqrt{2}\right)}$$
$$= \frac{4\left(\sqrt{5} + \sqrt{2}\right)}{5 - 2} = \frac{4\left(\sqrt{5} + \sqrt{2}\right)}{3}$$

$$\frac{5 + 15\sqrt{6}}{10} = \frac{5\left(1 + 3\sqrt{6}\right)}{5 \cdot 2} = \frac{1 + 3\sqrt{6}}{2}$$

> Factor first; then divide out the common factor.

(continued)

Concepts	Examples

8.6 SOLVING EQUATIONS WITH RADICALS

Solving an Equation with Radicals

Step 1 Isolate one radical on one side of the equation.

Step 2 Raise both sides of the equation to a power that is the same as the index of the radical.

Step 3 Solve the resulting equation; if it still contains a radical, repeat Steps 1 and 2.

Step 4 Check all proposed solutions in the *original* equation.

Solve $\sqrt{2x+3} - x = 0$.

$$\sqrt{2x+3} = x \qquad \text{Subtract } x.$$
$$\left(\sqrt{2x+3}\right)^2 = x^2 \qquad \text{Square both sides.}$$
$$2x + 3 = x^2$$
$$x^2 - 2x - 3 = 0 \qquad \text{Standard form}$$
$$(x-3)(x+1) = 0 \qquad \text{Zero-factor property}$$
$$x - 3 = 0 \quad \text{or} \quad x + 1 = 0 \qquad \text{Solve each equation.}$$
$$x = 3 \quad \text{or} \qquad x = -1$$

A check shows that 3 is a solution, but -1 is extraneous. The solution set is $\{3\}$.

8.7 COMPLEX NUMBERS

$i = \sqrt{-1}$, where $i^2 = -1$.

For any positive number b, $\sqrt{-b} = i\sqrt{b}$.

To multiply radicals with negative radicands, first change each factor to the form $i\sqrt{b}$ and then multiply. The same procedure applies to quotients.

$$\sqrt{-25} = i\sqrt{25} = 5i$$

$$\sqrt{-3} \cdot \sqrt{-27} = i\sqrt{3} \cdot i\sqrt{27}$$
$$= i^2\sqrt{81}$$
$$= -1 \cdot 9$$
$$= -9$$

First write all square roots in terms of i.

$$\frac{\sqrt{-18}}{\sqrt{-2}} = \frac{i\sqrt{18}}{i\sqrt{2}} = \sqrt{\frac{18}{2}} = \sqrt{9} = 3$$

Adding and Subtracting Complex Numbers

Add (or subtract) the real parts and add (or subtract) the imaginary parts.

$$(5 + 3i) + (8 - 7i) = 13 - 4i$$
$$(5 + 3i) - (8 - 7i) = -3 + 10i$$

Multiplying Complex Numbers

Multiply complex numbers by using the FOIL method.

$$(2 + i)(5 - 3i) = 10 - 6i + 5i - 3i^2 \qquad \text{FOIL}$$
$$= 10 - i - 3(-1) \qquad i^2 = -1$$
$$= 10 - i + 3$$
$$= 13 - i$$

Dividing Complex Numbers

Divide complex numbers by multiplying the numerator and the denominator by the conjugate of the denominator.

$$\frac{20}{3+i} = \frac{20(3-i)}{(3+i)(3-i)} = \frac{20(3-i)}{9-i^2}$$
$$= \frac{20(3-i)}{10} = 2(3-i) = 6 - 2i$$

Answers to Test Your Word Power

1. B; *Example:* In $\sqrt{3xy}$, $3xy$ is the radicand.

2. D; *Example:* In a right triangle where $a = 6$, $b = 8$, and $c = 10$, $6^2 + 8^2 = 10^2$.

3. C; *Example:* In a right triangle where the sides measure 9, 12, and 15 units, the hypotenuse is the side opposite the right angle, with measure 15 units.

4. B; *Example:* To rationalize the denominator of $\frac{5}{\sqrt{3}+1}$, multiply both the numerator and denominator by $\sqrt{3} - 1$ to get $\frac{5(\sqrt{3}-1)}{2}$.

5. A; *Example:* The proposed solution 2 is extraneous in $\sqrt{5q-1} + 3 = 0$.

6. C; *Examples:* -5 (or $-5 + 0i$), $7i$ (or $0 + 7i$), $\sqrt{2} - 4i$

Chapter **8** REVIEW EXERCISES

[8.1] *Find each root.*

1. $\sqrt{1764}$ **2.** $-\sqrt{289}$ **3.** $\sqrt[3]{216}$

4. $\sqrt[3]{-125}$ **5.** $-\sqrt[3]{27}$ **6.** $\sqrt[5]{-32}$

7. *Concept Check* Under what conditions is $\sqrt[n]{a}$ not a real number?

8. Simplify each radical so that no radicals appear. Assume that x represents any real number.

(a) $\sqrt{x^2}$ **(b)** $-\sqrt{x^2}$ **(c)** $\sqrt[3]{x^3}$

Use a calculator to find a decimal approximation for each number. Give the answer to the nearest thousandth.

9. $-\sqrt{47}$ **10.** $\sqrt[3]{-129}$ **11.** $\sqrt[4]{605}$

12. $500^{-3/4}$ **13.** $-500^{4/3}$ **14.** $-28^{-1/2}$

Graph each function. Give the domain and range.

15. $f(x) = \sqrt{x-1}$ **16.** $f(x) = \sqrt[3]{x} + 4$

17. What is the best estimate of the area of the triangle shown here?

A. 3600 **B.** 30 **C.** 60 **D.** 360

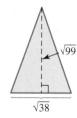

$\sqrt{99}$

$\sqrt{38}$

[8.2]

18. *Concept Check* Fill in the blanks with the correct responses: One way to evaluate $8^{2/3}$ is to first find the _____ root of _____ , which is _____ . Then raise that result to the _____ power, to get an answer of _____ . Therefore, $8^{2/3} =$ _____ .

19. *Concept Check* Which one of the following is a positive number?

A. $(-27)^{2/3}$ **B.** $(-64)^{5/3}$ **C.** $(-100)^{1/2}$ **D.** $(-32)^{1/5}$

20. *Concept Check* If a is a negative number and n is odd, then what must be true about m for $a^{m/n}$ to be

(a) positive **(b)** negative?

21. *Concept Check* If a is negative and n is even, then what can be said about $a^{1/n}$?

Simplify. If the expression does not represent a real number, say so.

22. $49^{1/2}$ **23.** $-121^{1/2}$ **24.** $16^{5/4}$ **25.** $-8^{2/3}$

26. $-\left(\dfrac{36}{25}\right)^{3/2}$ **27.** $\left(-\dfrac{1}{8}\right)^{-5/3}$ **28.** $\left(\dfrac{81}{10,000}\right)^{-3/4}$ **29.** $(-16)^{3/4}$

30. Solve the Pythagorean formula $a^2 + b^2 = c^2$ for b, where $b > 0$.

31. Explain the relationship between the expressions $a^{m/n}$ and $\sqrt[n]{a^m}$. Give an example.

Write each expression as a radical.

32. $(m + 3n)^{1/2}$

33. $(3a + b)^{-5/3}$

Write each expression with a rational exponent.

34. $\sqrt{7^9}$

35. $\sqrt[5]{p^4}$

Use the rules for exponents to simplify each expression. Write the answer with only positive exponents. Assume that all variables represent positive real numbers.

36. $5^{1/4} \cdot 5^{7/4}$

37. $\dfrac{96^{2/3}}{96^{-1/3}}$

38. $\dfrac{(a^{1/3})^4}{a^{2/3}}$

39. $\dfrac{y^{-1/3} \cdot y^{5/6}}{y}$

40. $\left(\dfrac{z^{-1}x^{-3/5}}{2^{-2}z^{-1/2}x}\right)^{-1}$

41. $r^{-1/2}(r + r^{3/2})$

Simplify by first writing each radical in exponential form. Leave the answer in exponential form. Assume that all variables represent positive real numbers.

42. $\sqrt[8]{s^4}$

43. $\sqrt[6]{r^9}$

44. $\dfrac{\sqrt{p^5}}{p^2}$

45. $\sqrt[4]{k^3} \cdot \sqrt{k^3}$

46. $\sqrt[3]{m^5} \cdot \sqrt[3]{m^8}$

47. $\sqrt[4]{\sqrt[3]{z}}$

48. $\sqrt{\sqrt{\sqrt{x}}}$

49. $\sqrt[3]{\sqrt[5]{x}}$

50. $\sqrt{\sqrt[6]{\sqrt[3]{x}}}$

51. By the product rule for exponents, we know that $2^{1/4} \cdot 2^{1/5} = 2^{9/20}$. However, there is no exponent rule for simplifying $3^{1/4} \cdot 2^{1/5}$. Why?

[8.3] *Simplify each radical. Assume that all variables represent positive real numbers.*

52. $\sqrt{6} \cdot \sqrt{11}$

53. $\sqrt{5} \cdot \sqrt{r}$

54. $\sqrt[3]{6} \cdot \sqrt[3]{5}$

55. $\sqrt[4]{7} \cdot \sqrt[4]{3}$

56. $\sqrt{20}$

57. $\sqrt{75}$

58. $-\sqrt{125}$

59. $\sqrt[3]{-108}$

60. $\sqrt{100y^7}$

61. $\sqrt[3]{64p^4q^6}$

62. $\sqrt[3]{108a^8b^5}$

63. $\sqrt[3]{632r^8t^4}$

64. $\sqrt{\dfrac{y^3}{144}}$

65. $\sqrt[3]{\dfrac{m^{15}}{27}}$

66. $\sqrt[3]{\dfrac{r^2}{8}}$

67. $\sqrt[4]{\dfrac{a^9}{81}}$

Simplify each radical expression.

68. $\sqrt[6]{15^3}$

69. $\sqrt[4]{p^6}$

70. $\sqrt[3]{2} \cdot \sqrt[4]{5}$

71. $\sqrt{x} \cdot \sqrt[5]{x}$

72. Find the missing length in the right triangle. Simplify the answer if applicable.

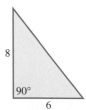

73. Find the distance between the points $(-4, 7)$ and $(10, 6)$.

[8.4] *Perform the indicated operations. Assume that all variables represent positive real numbers.*

74. $2\sqrt{8} - 3\sqrt{50}$

75. $8\sqrt{80} - 3\sqrt{45}$

76. $-\sqrt{27y} + 2\sqrt{75y}$

77. $2\sqrt{54m^3} + 5\sqrt{96m^3}$

78. $3\sqrt[3]{54} + 5\sqrt[3]{16}$

79. $-6\sqrt[4]{32} + \sqrt[4]{512}$

80. $\dfrac{3}{\sqrt{16}} - \dfrac{\sqrt{5}}{2}$

81. $\dfrac{4}{\sqrt{25}} + \dfrac{\sqrt{5}}{4}$

In Exercises 82 and 83, leave answers as simplified radicals.

82. Find the perimeter of a rectangular electronic billboard having sides of lengths shown in the figure.

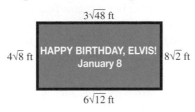

$3\sqrt{48}$ ft

$4\sqrt{8}$ ft HAPPY BIRTHDAY, ELVIS! January 8 $8\sqrt{2}$ ft

$6\sqrt{12}$ ft

83. Find the perimeter of a triangular electronic highway road sign having the dimensions shown in the figure.

All Traffic Must Exit Iowa Highway 64

$\sqrt{108}$ ft $2\sqrt{27}$ ft

$\sqrt{50}$ ft

[8.5] *Multiply.*

84. $\left(\sqrt{3} + 1\right)\left(\sqrt{3} - 2\right)$

85. $\left(\sqrt{7} + \sqrt{5}\right)\left(\sqrt{7} - \sqrt{5}\right)$

86. $\left(3\sqrt{2} + 1\right)\left(2\sqrt{2} - 3\right)$

87. $\left(\sqrt{13} - \sqrt{2}\right)^2$

88. $\left(\sqrt[3]{2} + 3\right)\left(\sqrt[3]{4} - 3\sqrt[3]{2} + 9\right)$

89. $\left(\sqrt[3]{4y} - 1\right)\left(\sqrt[3]{4y} + 3\right)$

90. Use a calculator to show that the answer to Exercise 87, $15 - 2\sqrt{26}$, is not equal to $13\sqrt{26}$.

91. *Concept Check* A friend wants to rationalize the denominator of the fraction $\dfrac{5}{\sqrt[3]{6}}$, and she decides to multiply the numerator and denominator by $\sqrt[3]{6}$. Why will her plan *not* work?

Rationalize each denominator. Assume that all variables represent positive real numbers.

92. $\dfrac{\sqrt{6}}{\sqrt{5}}$

93. $\dfrac{-6\sqrt{3}}{\sqrt{2}}$

94. $\dfrac{3\sqrt{7p}}{\sqrt{y}}$

95. $\sqrt{\dfrac{11}{8}}$

96. $-\sqrt[3]{\dfrac{9}{25}}$

97. $\sqrt[3]{\dfrac{108m^3}{n^5}}$

98. $\dfrac{1}{\sqrt{2} + \sqrt{7}}$

99. $\dfrac{-5}{\sqrt{6} - 3}$

Write in lowest terms.

100. $\dfrac{2 - 2\sqrt{5}}{8}$

101. $\dfrac{4 - 8\sqrt{8}}{12}$

102. $\dfrac{-18 + \sqrt{27}}{6}$

[8.6] *Solve each equation.*

103. $\sqrt{8x + 9} = 5$

104. $\sqrt{2z - 3} - 3 = 0$

105. $\sqrt{3m + 1} - 2 = -3$

106. $\sqrt{7z + 1} = z + 1$

107. $3\sqrt{m} = \sqrt{10m - 9}$

108. $\sqrt{p^2 + 3p + 7} = p + 2$

109. $\sqrt{a + 2} - \sqrt{a - 3} = 1$

110. $\sqrt[3]{5m - 1} = \sqrt[3]{3m - 2}$

111. $\sqrt[3]{2x^2 + 3x - 7} = \sqrt[3]{2x^2 + 4x + 6}$

112. $\sqrt[3]{3y^2 - 4y + 6} = \sqrt[3]{3y^2 - 2y + 8}$

113. $\sqrt[3]{1 - 2k} - \sqrt[3]{-k - 13} = 0$

114. $\sqrt[3]{11 - 2t} - \sqrt[3]{-1 - 5t} = 0$

115. $\sqrt[4]{x - 1} + 2 = 0$

116. $\sqrt[4]{2k + 3} + 1 = 0$

117. $\sqrt[4]{x + 7} = \sqrt[4]{2x}$

118. $\sqrt[4]{x + 8} = \sqrt[4]{3x}$

119. Carpenters stabilize wall frames with a diagonal brace, as shown in the figure. The length of the brace is given by $L = \sqrt{H^2 + W^2}$.

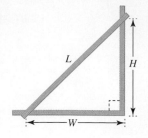

   (a) Solve this formula for H.

   (b) If the bottom of the brace is attached 9 ft from the corner and the brace is 12 ft long, how far up the corner post should it be nailed? Give your answer to the nearest tenth of a foot.

[8.7] *Write each expression as a product of a real number and i.*

120. $\sqrt{-25}$ **121.** $\sqrt{-200}$

122. *Concept Check*   If a is a positive real number, is $-\sqrt{-a}$ a real number?

Perform the indicated operations. Give answers in standard form.

123. $(-2 + 5i) + (-8 - 7i)$ **124.** $(5 + 4i) - (-9 - 3i)$ **125.** $\sqrt{-5} \cdot \sqrt{-7}$

126. $\sqrt{-25} \cdot \sqrt{-81}$ **127.** $\dfrac{\sqrt{-72}}{\sqrt{-8}}$ **128.** $(2 + 3i)(1 - i)$

129. $(6 - 2i)^2$ **130.** $\dfrac{3 - i}{2 + i}$ **131.** $\dfrac{5 + 14i}{2 + 3i}$

Find each power of i.

132. i^{11} **133.** i^{36} **134.** i^{-10} **135.** i^{-8}

MIXED REVIEW EXERCISES

Simplify. Assume that all variables represent positive real numbers.

136. $-\sqrt[4]{256}$ **137.** $1000^{-2/3}$ **138.** $\dfrac{z^{-1/5} \cdot z^{3/10}}{z^{7/10}}$

139. $\sqrt[4]{k^{24}}$ **140.** $\sqrt[3]{54z^9 t^8}$ **141.** $-5\sqrt{18} + 12\sqrt{72}$

142. $\dfrac{-1}{\sqrt{12}}$ **143.** $\sqrt[3]{\dfrac{12}{25}}$ **144.** i^{-1000}

145. $\sqrt{-49}$ **146.** $(4 - 9i) + (-1 + 2i)$ **147.** $\dfrac{\sqrt{50}}{\sqrt{-2}}$

148. $\dfrac{3 + \sqrt{54}}{6}$ **149.** $(3 + 2i)^2$ **150.** $8\sqrt[3]{x^3 y^2} - 2x\sqrt[3]{y^2}$

151. $9\sqrt{5} - 4\sqrt{15}$ **152.** $\left(\sqrt{5} - \sqrt{3}\right)\left(\sqrt{7} + \sqrt{3}\right)$

Solve each equation.

153. $\sqrt{x + 4} = x - 2$ **154.** $\sqrt[3]{2x - 9} = \sqrt[3]{5x + 3}$

155. $\sqrt{6 + 2x} - 1 = \sqrt{7 - 2x}$ **156.** $\sqrt{7x + 11} - 5 = 0$

157. $\sqrt{6x + 2} - \sqrt{5x + 3} = 0$ **158.** $\sqrt{3 + 5x} - \sqrt{x + 11} = 0$

159. $3\sqrt{x} = \sqrt{8x + 9}$ **160.** $6\sqrt{p} = \sqrt{30p + 24}$

161. $\sqrt{11 + 2x} + 1 = \sqrt{5x + 1}$ **162.** $\sqrt{5x + 6} - \sqrt{x + 3} = 3$

Chapter **8** TEST

View the complete solutions to all Chapter Test exercises on the Pass the Test CD.

Evaluate.

1. $-\sqrt{841}$

2. $\sqrt[3]{-512}$

3. $125^{1/3}$

4. *Concept Check* For $\sqrt{146.25}$, which choice gives the best estimate?

    A. 10   **B.** 11   **C.** 12   **D.** 13

Use a calculator to approximate each root to the nearest thousandth.

5. $\sqrt{478}$

6. $\sqrt[3]{-832}$

7. Graph the function defined by $f(x) = \sqrt{x+6}$, and give the domain and range.

Simplify each expression. Assume that all variables represent positive real numbers.

8. $\left(\dfrac{16}{25}\right)^{-3/2}$

9. $(-64)^{-4/3}$

10. $\dfrac{3^{2/5}x^{-1/4}y^{2/5}}{3^{-8/5}x^{7/4}y^{1/10}}$

11. $\left(\dfrac{x^{-4}y^{-6}}{x^{-2}y^3}\right)^{-2/3}$

12. $7^{3/4} \cdot 7^{-1/4}$

13. $\sqrt[3]{a^4} \cdot \sqrt[3]{a^7}$

14. Use the Pythagorean formula to find the exact length of side b in the figure.

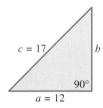

$c = 17$ b $90°$ $a = 12$

15. Find the distance between the points $(-4, 2)$ and $(2, 10)$.

Simplify each expression. Assume that all variables represent positive real numbers.

16. $\sqrt{54x^5y^6}$

17. $\sqrt[4]{32a^7b^{13}}$

18. $\sqrt{2} \cdot \sqrt[3]{5}$ (Express as a radical.)

19. $3\sqrt{20} - 5\sqrt{80} + 4\sqrt{500}$

20. $\sqrt[3]{16t^3s^5} - \sqrt[3]{54t^6s^2}$

21. $\left(7\sqrt{5} + 4\right)\left(2\sqrt{5} - 1\right)$

22. $\left(\sqrt{3} - 2\sqrt{5}\right)^2$

23. $\dfrac{-5}{\sqrt{40}}$

24. $\dfrac{2}{\sqrt[3]{5}}$

25. $\dfrac{-4}{\sqrt{7} + \sqrt{5}}$

26. Write $\dfrac{6 + \sqrt{24}}{2}$ in lowest terms.

27. The following formula is used in physics, relating the velocity V of sound to the temperature T:

$$V = \frac{V_0}{\sqrt{1 - kT}}.$$

    (a) Find an approximation of V to the nearest tenth if $V_0 = 50$, $k = 0.01$, and $T = 30$. Use a calculator.

    (b) Solve the formula for T.

Solve each equation.

28. $\sqrt[3]{5x} = \sqrt[3]{2x - 3}$

29. $x + \sqrt{x + 6} = 9 - x$

30. $\sqrt{x + 4} - \sqrt{1 - x} = -1$

Perform the indicated operations. Give the answers in standard form.

31. $(-2 + 5i) - (3 + 6i) - 7i$

32. $(1 + 5i)(3 + i)$

33. $\dfrac{7 + i}{1 - i}$

34. Simplify i^{37}.

35. *Concept Check* Answer *true* or *false* to each of the following.

   (a) $i^2 = -1$ **(b)** $i = \sqrt{-1}$ **(c)** $i = -1$ **(d)** $\sqrt{-3} = i\sqrt{3}$

Chapters **1–8 CUMULATIVE REVIEW EXERCISES**

Evaluate each expression if $a = -3$, $b = 5$, and $c = -4$.

1. $|2a^2 - 3b + c|$

2. $\dfrac{(a + b)(a + c)}{3b - 6}$

Solve each equation.

3. $3(x + 2) - 4(2x + 3) = -3x + 2$

4. $\dfrac{1}{3}x + \dfrac{1}{4}(x + 8) = x + 7$

5. $0.04x + 0.06(100 - x) = 5.88$

6. $|6x + 7| = 13$

7. $|-2x + 4| = |-2x - 3|$

8. Find the solution set of $-5 - 3(m - 2) < 11 - 2(m + 2)$. Write it in interval notation.

9. Find the measures of the marked angles.

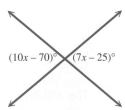

$(10x - 70)°$ $(7x - 25)°$

Solve each problem.

10. A piggy bank has 50 coins, all of which are nickels and quarters. The total value of the money is $8.90. How many of each denomination are there in the bank?

11. How many liters of pure alcohol must be mixed with 40 L of 18% alcohol to obtain a 22% alcohol solution?

12. Graph the equation $4x - 3y = 12$.

13. Find the slope of the line passing through the points $(-4, 6)$ and $(2, -3)$. Then find the equation of the line and write it in the form $y = mx + b$.

14. If $f(x) = 3x - 7$, find $f(-10)$.

15. Solve the system by substitution or elimination.

$$3x - y = 23$$
$$2x + 3y = 8$$

16. Solve the system by matrix methods.

$$x + y + z = 1$$
$$x - y - z = -3$$
$$x + y - z = -1$$

Solve the problem by using a system of equations.

17. In 2006, if you had sent five 2-oz letters and three 3-oz letters by first-class mail, it would have cost you $5.76. Sending three 2-oz letters and five 3-oz letters would have cost $6.24. What was the 2006 postage rate for one 2-oz letter and for one 3-oz letter? (*Source:* U.S. Postal Service.)

Perform the indicated operations.

18. $(3k^3 - 5k^2 + 8k - 2) - (4k^3 + 11k + 7) + (2k^2 - 5k)$

19. $(8x - 7)(x + 3)$

20. $\dfrac{8z^3 - 16z^2 + 24z}{8z^2}$

21. $\dfrac{6y^4 - 3y^3 + 5y^2 + 6y - 9}{2y + 1}$

Factor each polynomial completely.

22. $2p^2 - 5pq + 3q^2$

23. $3k^4 + k^2 - 4$

24. $x^3 + 512$

Solve by factoring.

25. $2x^2 + 11x + 15 = 0$

26. $5t(t - 1) = 2(1 - t)$

27. What is the domain of $f(x) = \dfrac{2}{x^2 - 9}$?

Perform each operation and express the answer in lowest terms.

28. $\dfrac{y^2 + y - 12}{y^3 + 9y^2 + 20y} \div \dfrac{y^2 - 9}{y^3 + 3y^2}$

29. $\dfrac{1}{x + y} + \dfrac{3}{x - y}$

Simplify each complex fraction.

30. $\dfrac{\dfrac{-6}{x - 2}}{\dfrac{8}{3x - 6}}$

31. $\dfrac{\dfrac{1}{a} - \dfrac{1}{b}}{\dfrac{a}{b} - \dfrac{b}{a}}$

32. $\dfrac{x^{-1}}{y - x^{-1}}$

33. Cecily can ride her bike 4 mph faster than her husband, Mike. If Cecily can ride 48 mi in the same time that Mike can ride 24 mi, what are their speeds?

34. Solve the equation $\dfrac{p+1}{p-3} = \dfrac{4}{p-3} + 6$.

Write each expression in simplest form, using only positive exponents. Assume that all variables represent positive real numbers.

35. $27^{-2/3}$

36. $\sqrt{200x^4}$

37. $\sqrt[3]{16x^2y} \cdot \sqrt[3]{3x^3y}$

38. $\sqrt{50} + \sqrt{8}$

39. $\dfrac{1}{\sqrt{10} - \sqrt{8}}$

40. $\left(2\sqrt{x} + \sqrt{y}\right)\left(-3\sqrt{x} - 4\sqrt{y}\right)$

41. Find the distance between the points $(-4, 4)$ and $(-2, 9)$.

42. Solve the equation $\sqrt{3r - 8} = r - 2$.

43. The **fall speed,** in miles per hour, of a vehicle running off the road into a ditch is given by

$$S = \frac{2.74D}{\sqrt{h}},$$

where D is the horizontal distance traveled from the level surface to the bottom of the ditch and h is the height (or depth) of the ditch. What is the fall speed of a vehicle that traveled 32 ft horizontally into a 5-ft-deep ditch?

Perform the indicated operations. Give answers in standard form.

44. $(5 + 7i) - (3 - 2i)$

45. $\dfrac{6 - 2i}{1 - i}$

Quadratic Equations, Inequalities, and Functions

Since 1980, the number of multiple births in the United States has increased 59%, primarily due to greater use of fertility drugs and greater numbers of births to women over age 40. The number of higher-order multiple births—that is, births involving triplets or more—has increased over 400%. One of the most publicized higher-order multiple births occurred November 19, 1997, with the birth of the McCaughey septuplets in Des Moines, Iowa. All seven premature babies survived, a first in medical history. (*Source:* American College of Obstetricians and Gynecologists; *The Gazette,* November 19, 2003.)

In Example 6 of Section 9.5, we determine a *quadratic function* that models the number of higher-order multiple births in the United States.

9.1 The Square Root Property and Completing the Square

We introduced quadratic equations in **Section 6.5.** Recall that a *quadratic equation* is defined as follows.

Quadratic Equation

An equation that can be written in the form
$$ax^2 + bx + c = 0,$$
where a, b, and c are real numbers, with $a \neq 0$, is a **quadratic equation.** The given form is called **standard form.**

A quadratic equation is a *second-degree equation,* that is, an equation with a squared variable term and no terms of greater degree. For example,

$$4x^2 + 4x - 5 = 0 \qquad \text{and} \qquad 3x^2 = 4x - 8$$

are quadratic equations, with the first equation in standard form.

OBJECTIVE 1 Review the zero-factor property. In **Section 6.5** we used factoring and the zero-factor property to solve quadratic equations.

Zero-Factor Property

If two numbers have a product of 0, then at least one of the numbers must be 0. That is, if $ab = 0$, then $a = 0$ or $b = 0$.

We solved a quadratic equation such as $3x^2 - 5x - 28 = 0$ using the zero-factor property as follows.

$$3x^2 - 5x - 28 = 0$$
$$(3x + 7)(x - 4) = 0 \qquad \text{Factor.}$$
$$3x + 7 = 0 \qquad \text{or} \quad x - 4 = 0 \qquad \text{Zero-factor property}$$
$$3x = -7 \qquad \text{or} \qquad x = 4 \qquad \text{Solve each equation.}$$
$$x = -\frac{7}{3}$$

The solution set is $\left\{-\frac{7}{3}, 4\right\}$.

OBJECTIVE 2 Learn the square root property. Although factoring is the simplest way to solve quadratic equations, not every quadratic equation can be solved easily by factoring. In this section and the next, we develop three other methods of solving quadratic equations based on the following property.

Square Root Property

If x and k are complex numbers and $x^2 = k$, then

$$x = \sqrt{k} \quad \text{or} \quad x = -\sqrt{k}.$$

The following steps justify the square root property.

$$x^2 = k$$

$$x^2 - k = 0 \qquad \text{Subtract } k.$$

$$\left(x - \sqrt{k}\right)\left(x + \sqrt{k}\right) = 0 \qquad \text{Factor.}$$

$$x - \sqrt{k} = 0 \quad \text{or} \quad x + \sqrt{k} = 0 \qquad \text{Zero-factor property}$$

$$x = \sqrt{k} \quad \text{or} \quad x = -\sqrt{k} \qquad \text{Solve each equation.}$$

▶ **CAUTION** Remember that if $k \neq 0$, using the square root property always produces *two* square roots, one positive and one negative.

EXAMPLE 1 Using the Square Root Property

Solve each equation.

(a) $x^2 = 5$

By the square root property,

$$x = \sqrt{5} \quad \text{or} \quad x = -\sqrt{5}, \qquad \text{\small Don't forget the negative solution.}$$

and the solution set is $\left\{\sqrt{5}, -\sqrt{5}\right\}$.

(b)
$$4x^2 - 48 = 0$$

$$4x^2 = 48 \qquad \text{Add 48.}$$

$$x^2 = 12 \qquad \text{Divide by 4.}$$

$$x = \sqrt{12} \quad \text{or} \quad x = -\sqrt{12} \qquad \text{Square root property}$$

$$x = 2\sqrt{3} \quad \text{or} \quad x = -2\sqrt{3} \qquad \sqrt{12} = \sqrt{4} \cdot \sqrt{3} = 2\sqrt{3}$$

The solutions are $2\sqrt{3}$ and $-2\sqrt{3}$. Check each in the original equation.

Check: $\qquad\qquad 4x^2 - 48 = 0 \qquad$ Original equation

$$4\left(2\sqrt{3}\right)^2 - 48 = 0 \quad ? \qquad\qquad 4\left(-2\sqrt{3}\right)^2 - 48 = 0 \quad ?$$

$$4(12) - 48 = 0 \quad ? \qquad\qquad 4(12) - 48 = 0 \quad ?$$

$$48 - 48 = 0 \quad ? \qquad\qquad 48 - 48 = 0 \quad ?$$

$$0 = 0 \qquad \text{True} \qquad\qquad\qquad 0 = 0 \qquad \text{True}$$

The solution set is $\left\{2\sqrt{3}, -2\sqrt{3}\right\}$.

✔ **Now Try Exercises 7 and 13.**

▶ **NOTE** Recall that solutions such as those in Example 1 are sometimes abbreviated with the symbol $\pm$ (read "positive or negative," or "plus or minus"). With this symbol the solutions in Example 1 would be written $\pm\sqrt{5}$ and $\pm 2\sqrt{3}$.

EXAMPLE 2 Using the Square Root Property in an Application

Galileo Galilei (1564–1642) developed a formula for freely falling objects described by

$$d = 16t^2,$$

where d is the distance in feet that an object falls (disregarding air resistance) in t seconds, regardless of weight. Galileo dropped objects from the Leaning Tower of Pisa to develop this formula. If the Leaning Tower is about 180 ft tall, use Galileo's formula to determine how long it would take an object dropped from the tower to fall to the ground. (*Source: Microsoft Encarta Encyclopedia.*)

We substitute 180 for d in Galileo's formula.

$$d = 16t^2$$
$$180 = 16t^2 \qquad \text{Let } d = 180.$$
$$11.25 = t^2 \qquad \text{Divide by 16.}$$
$$t = \sqrt{11.25} \quad \text{or} \quad t = -\sqrt{11.25} \qquad \text{Square root property}$$

Since time cannot be negative, we discard the negative solution. In applied problems, we usually prefer approximations to exact values. Using a calculator, $\sqrt{11.25} \approx 3.4$ so $t \approx 3.4$. The object would fall to the ground in about 3.4 sec.

✔ **Now Try Exercise 27.**

OBJECTIVE 3 Solve quadratic equations of the form $(ax + b)^2 = c$ by using the square root property. To solve more complicated equations by using the square root property, such as

$$(x - 5)^2 = 36,$$

substitute $(x - 5)^2$ for x^2 and 36 for k in the square root property to obtain

$$x - 5 = \sqrt{36} \quad \text{or} \quad x - 5 = -\sqrt{36}$$
$$x - 5 = 6 \quad \text{or} \quad x - 5 = -6$$
$$x = 11 \quad \text{or} \quad x = -1.$$

Check:
$$(x - 5)^2 = 36 \qquad \text{Original equation}$$

$$(11 - 5)^2 = 36 \quad ? \qquad\qquad (-1 - 5)^2 = 36 \quad ?$$
$$6^2 = 36 \quad ? \qquad\qquad (-6)^2 = 36 \quad ?$$
$$36 = 36 \quad \text{True} \qquad\qquad 36 = 36 \quad \text{True}$$

Since both solutions satisfy the original equation, the solution set is $\{-1, 11\}$.

EXAMPLE 3 Using the Square Root Property

Solve $(2x - 3)^2 = 18$.

$$2x - 3 = \sqrt{18} \quad \text{or} \quad 2x - 3 = -\sqrt{18} \qquad \text{Square root property}$$
$$2x = 3 + \sqrt{18} \quad \text{or} \quad 2x = 3 - \sqrt{18} \qquad \text{Add 3.}$$
$$x = \frac{3 + \sqrt{18}}{2} \quad \text{or} \quad x = \frac{3 - \sqrt{18}}{2} \qquad \text{Divide by 2.}$$
$$x = \frac{3 + 3\sqrt{2}}{2} \quad \text{or} \quad x = \frac{3 - 3\sqrt{2}}{2} \qquad \sqrt{18} = \sqrt{9 \cdot 2} = 3\sqrt{2}$$

We show the check for the first solution. The check for the other solution is similar.

Check:

$$(2x - 3)^2 = 18 \qquad \text{Original equation}$$

$$\left[2\left(\frac{3 + 3\sqrt{2}}{2}\right) - 3\right]^2 = 18 \quad ? \quad \text{Let } x = \frac{3 + 3\sqrt{2}}{2}.$$

$$\left(3 + 3\sqrt{2} - 3\right)^2 = 18 \quad ? \quad \text{Multiply.}$$

$$\left(3\sqrt{2}\right)^2 = 18 \quad ? \quad \text{Simplify.}$$

$$18 = 18 \qquad \text{True}$$

The solution set is $\left\{ \dfrac{3 + 3\sqrt{2}}{2}, \dfrac{3 - 3\sqrt{2}}{2} \right\}.$

✔ **Now Try Exercise 23.**

OBJECTIVE 4 Solve quadratic equations by completing the square. We can use the square root property to solve *any* quadratic equation by writing it in the form $(x + k)^2 = n$. That is, we must write the left side of the equation as a perfect square trinomial that can be factored as $(x + k)^2$, the square of a binomial, and the right side must be a constant. Rewriting a quadratic equation in this form is called **completing the square.**

Recall that the perfect square trinomial

$$x^2 + 10x + 25$$

can be factored as $(x + 5)^2$. In the trinomial, the coefficient of x (the first-degree term) is 10 and the constant term is 25. Notice that if we take half of 10 and square it, we get the constant term, 25.

$$\overset{\text{Coefficient of } x}{\underset{\downarrow}{}} \qquad \overset{\text{Constant}}{\underset{\downarrow}{}}$$

$$\left[\frac{1}{2}(10)\right]^2 = 5^2 = 25$$

Similarly, in

$$x^2 + 12x + 36, \qquad \left[\frac{1}{2}(12)\right]^2 = 6^2 = 36,$$

and in

$$m^2 - 6m + 9, \qquad \left[\frac{1}{2}(-6)\right]^2 = (-3)^2 = 9.$$

This relationship is true in general and is the idea behind completing the square.

EXAMPLE 4 Solving a Quadratic Equation by Completing the Square

Solve $x^2 + 8x + 10 = 0$.

This quadratic equation cannot be solved by factoring, and it is not in the correct form to solve using the square root property. To solve it by completing the square, we need a perfect square trinomial on the left side of the equation. To get this form, we first subtract 10 from each side.

$$x^2 + 8x + 10 = 0 \qquad \text{Original equation}$$
$$x^2 + 8x = -10 \qquad \text{Subtract 10.}$$

We must add a constant to get a perfect square trinomial on the left.

$$\underbrace{x^2 + 8x + \underline{}}$$

Needs to be a perfect
square trinomial

To find this constant, we apply the ideas preceding this example. Take half the coefficient of the first-degree term and square the result.

$$\left[\frac{1}{2}(8) \right]^2 = 4^2 = 16 \leftarrow \text{Desired constant}$$

We add this constant, 16, to *each* side of the equation.

This is a key step. $\qquad x^2 + 8x + 16 = -10 + 16$

Next we factor the perfect square trinomial on the left and add on the right.

$$(x + 4)^2 = 6$$

We can now solve this equation using the square root property.

$$x + 4 = \sqrt{6} \qquad \text{or} \quad x + 4 = -\sqrt{6}$$
$$x = -4 + \sqrt{6} \quad \text{or} \qquad x = -4 - \sqrt{6}$$

Check: $\qquad\qquad\qquad x^2 + 8x + 10 = 0 \qquad\qquad$ Original equation

Remember the middle term when squaring $-4 + \sqrt{6}$.

$$\left(-4 + \sqrt{6}\right)^2 + 8\left(-4 + \sqrt{6}\right) + 10 = 0 \qquad ? \qquad \text{Let } x = -4 + \sqrt{6}.$$
$$16 - 8\sqrt{6} + 6 - 32 + 8\sqrt{6} + 10 = 0 \qquad ?$$
$$0 = 0 \qquad\qquad \text{True}$$

The check of the other solution is similar. The solution set is

$$\left\{ -4 + \sqrt{6}, -4 - \sqrt{6} \right\}.$$

✔ **Now Try Exercise 49.**

The procedure from Example 4 can be generalized.

Completing the Square

To solve $ax^2 + bx + c = 0 \ (a \neq 0)$ by completing the square, use these steps.

Step 1 **Be sure the second-degree (squared) term has coefficient 1.** If the coefficient of the squared term is 1, proceed to Step 2. If the coefficient of the squared term is not 1 but some other nonzero number a, divide each side of the equation by a.

Step 2 **Write the equation in correct form** so that terms with variables are on one side of the equals sign and the constant is on the other side.

Step 3 **Square half the coefficient of the first-degree (linear) term.**

(continued)

Completing the Square (continued)

Step 4 **Add the square to each side.**

Step 5 **Factor the perfect square trinomial.** One side should now be a perfect square trinomial. Factor it as the square of a binomial. Simplify the other side.

Step 6 **Solve the equation.** Apply the square root property to complete the solution.

▶ **NOTE** Steps 1 and 2 can be done in either order. With some equations, it is more convenient to do Step 2 first.

EXAMPLE 5 Solving a Quadratic Equation with $a = 1$ by Completing the Square

Solve $x^2 + 5x - 1 = 0$.

Follow the steps in the box. Since the coefficient of the squared term is 1, begin with Step 2.

Step 2 $\qquad\qquad x^2 + 5x = 1$ Add 1 to each side.

Step 3 Take half the coefficient of the first-degree term and square the result.

$$\left[\frac{1}{2}(5)\right]^2 = \left(\frac{5}{2}\right)^2 = \frac{25}{4}$$

Step 4 $\qquad x^2 + 5x + \dfrac{25}{4} = 1 + \dfrac{25}{4}$ Add the square to each side of the equation.

Step 5 $\qquad\qquad \left(x + \dfrac{5}{2}\right)^2 = \dfrac{29}{4}$ Factor on the left; add on the right.

Step 6 $x + \dfrac{5}{2} = \sqrt{\dfrac{29}{4}}$ or $x + \dfrac{5}{2} = -\sqrt{\dfrac{29}{4}}$ Square root property

$\qquad x + \dfrac{5}{2} = \dfrac{\sqrt{29}}{2}$ or $x + \dfrac{5}{2} = -\dfrac{\sqrt{29}}{2}$ $\sqrt{\dfrac{a}{b}} = \dfrac{\sqrt{a}}{\sqrt{b}}$

$\qquad\quad x = -\dfrac{5}{2} + \dfrac{\sqrt{29}}{2}$ or $x = -\dfrac{5}{2} - \dfrac{\sqrt{29}}{2}$ Add $-\dfrac{5}{2}$.

$\qquad\quad x = \dfrac{-5 + \sqrt{29}}{2}$ or $x = \dfrac{-5 - \sqrt{29}}{2}$ $\dfrac{a}{c} + \dfrac{b}{c} = \dfrac{a+b}{c}$

Check that the solution set is $\left\{\dfrac{-5 + \sqrt{29}}{2}, \dfrac{-5 - \sqrt{29}}{2}\right\}$.

✔ **Now Try Exercise 51.**

| EXAMPLE 6 | Solving a Quadratic Equation with $a \neq 1$ by Completing the Square |

Solve $2x^2 - 4x - 5 = 0$.

Divide each side by 2 to get 1 as the coefficient of the second-degree term.

$$x^2 - 2x - \frac{5}{2} = 0 \qquad \text{Step 1}$$

$$x^2 - 2x = \frac{5}{2} \qquad \text{Step 2}$$

$$\left[\frac{1}{2}(-2) \right]^2 = (-1)^2 = 1 \qquad \text{Step 3}$$

$$x^2 - 2x + 1 = \frac{5}{2} + 1 \qquad \text{Step 4}$$

$$(x - 1)^2 = \frac{7}{2} \qquad \text{Step 5}$$

$$x - 1 = \sqrt{\frac{7}{2}} \qquad \text{or} \qquad x - 1 = -\sqrt{\frac{7}{2}} \qquad \text{Step 6}$$

$$x = 1 + \sqrt{\frac{7}{2}} \quad \text{or} \qquad x = 1 - \sqrt{\frac{7}{2}} \qquad \text{Add 1.}$$

$$x = 1 + \frac{\sqrt{14}}{2} \quad \text{or} \qquad x = 1 - \frac{\sqrt{14}}{2} \qquad \begin{array}{l}\text{Rationalize} \\ \text{denominators.}\end{array}$$

Add the two terms in each solution as follows.

$$1 + \frac{\sqrt{14}}{2} = \frac{2}{2} + \frac{\sqrt{14}}{2} = \frac{2 + \sqrt{14}}{2} \qquad 1 = \tfrac{2}{2}$$

$$1 - \frac{\sqrt{14}}{2} = \frac{2}{2} - \frac{\sqrt{14}}{2} = \frac{2 - \sqrt{14}}{2}.$$

Check that the solution set is $\left\{ \dfrac{2 + \sqrt{14}}{2}, \dfrac{2 - \sqrt{14}}{2} \right\}$.

✔ **Now Try Exercise 57.**

OBJECTIVE 5 Solve quadratic equations with solutions that are not real numbers. So far, all the equations we have solved using the square root property have had two real solutions. In the equation $x^2 = k$, if $k < 0$, there will be two nonreal complex solutions.

| EXAMPLE 7 | Solving Quadratic Equations with Nonreal Complex Solutions |

Solve each equation.

(a) $x^2 = -15$

$$x = \sqrt{-15} \quad \text{or} \quad x = -\sqrt{-15} \qquad \text{Square root property}$$

$$x = i\sqrt{15} \quad \text{or} \quad x = -i\sqrt{15} \qquad \sqrt{-1} = i$$

The solution set is $\left\{ i\sqrt{15}, -i\sqrt{15} \right\}$.

(b) $(x + 2)^2 = -16$

$$x + 2 = \sqrt{-16} \quad \text{or} \quad x + 2 = -\sqrt{-16} \qquad \text{Square root property}$$
$$x + 2 = 4i \quad \text{or} \quad x + 2 = -4i \qquad \sqrt{-16} = 4i$$
$$x = -2 + 4i \quad \text{or} \quad x = -2 - 4i \qquad \text{Add } -2.$$

The solution set is $\{-2 + 4i, -2 - 4i\}$.

(c) $x^2 + 2x + 7 = 0$

Solve by completing the square.

$$x^2 + 2x = -7 \qquad \text{Subtract 7.}$$
$$x^2 + 2x + 1 = -7 + 1 \qquad \left[\tfrac{1}{2}(2)\right]^2 = 1; \text{ add 1 to each side.}$$
$$(x + 1)^2 = -6 \qquad \text{Factor on the left; add on the right.}$$
$$x + 1 = \pm i\sqrt{6} \qquad \text{Square root property}$$
$$x = -1 \pm i\sqrt{6} \qquad \text{Subtract 1.}$$

The solution set is $\left\{-1 + i\sqrt{6}, -1 - i\sqrt{6}\right\}$.

✔ **Now Try Exercises 65, 67, and 71.**

▶ **NOTE** The procedure for completing the square is also used in other areas of mathematics. For example, we use it in **Section 9.6** when we graph quadratic equations and again in **Chapter 11** when we work with circles.

9.1 EXERCISES

Complete solution available on Video Lectures on CD/DVD

Now Try Exercise

1. *Concept Check* A student was asked to solve the quadratic equation $x^2 = 16$ and did not get full credit for the solution set $\{4\}$. ***WHAT WENT WRONG?*** Give the correct solution set.

2. *Concept Check* Why can't the zero-factor property be used to solve every quadratic equation?

3. Give a one-sentence description or explanation of each phrase.

    (a) Quadratic equation in standard form   **(b)** Zero-factor property
    (c) Square root property

4. *Concept Check* A student tried to solve $x^2 - x - 2 = 5$ as follows.

$$x^2 - x - 2 = 5$$
$$(x - 2)(x + 1) = 5 \qquad \text{Factor.}$$
$$x - 2 = 5 \quad \text{or} \quad x + 1 = 5 \qquad \text{Zero-factor property}$$
$$x = 7 \quad \text{or} \quad x = 4 \qquad \text{Solve each equation.}$$

This method is incorrect. ***WHAT WENT WRONG?***

Use the square root property to solve each equation. See Examples 1 and 3.

5. $x^2 = 81$ **6.** $x^2 = 225$ **7.** $x^2 = 17$

8. $x^2 = 19$ **9.** $x^2 = 32$ **10.** $x^2 = 54$

11. $x^2 - 20 = 0$ **12.** $p^2 - 50 = 0$ **13.** $3n^2 - 72 = 0$

14. $5z^2 - 200 = 0$ **15.** $(x + 2)^2 = 25$ **16.** $(t + 8)^2 = 9$

17. $(x - 4)^2 = 3$ **18.** $(x + 3)^2 = 11$ **19.** $(t + 5)^2 = 48$

20. $(m - 6)^2 = 27$ **21.** $(3x - 1)^2 = 7$ **22.** $(2x + 4)^2 = 10$

23. $(4p + 1)^2 = 24$ **24.** $(5t - 2)^2 = 12$ **25.** $(2 - 5t)^2 = 12$

26. Explain why Exercises 24 and 25 have the same solution set.

Solve Exercises 27 and 28 using Galileo's formula, $d = 16t^2$. Round answers to the nearest tenth. See Example 2.

27. Mount Rushmore National Memorial in South Dakota features a sculpture of four of America's favorite presidents carved into the rim of the mountain, 500 ft above the valley floor. How long would it take a rock dropped from the top of the sculpture to fall to the ground? (*Source: Microsoft Encarta Encyclopedia.*)

28. The Gateway Arch in St. Louis, Missouri, is 630 ft tall. How long would it take an object dropped from the top of it to fall to the ground? (*Source: Home & Away, November/December 2000.*)

29. *Concept Check*   Of the two equations

$$(2x + 1)^2 = 5 \quad \text{and} \quad x^2 + 4x = 12,$$

which one is more suitable for solving by the square root property? Which one is more suitable for solving by completing the square?

30. Why would most students find the equation $x^2 + 4x = 20$ easier to solve by completing the square than the equation $5x^2 + 2x = 3$?

Concept Check   *Decide what number must be added to make each expression a perfect square trinomial. Then factor the trinomial.*

31. $x^2 + 6x + \underline{\hspace{0.8cm}}$ **32.** $x^2 + 14x + \underline{\hspace{0.8cm}}$ **33.** $p^2 - 12p + \underline{\hspace{0.8cm}}$

34. $x^2 + 20x + \underline{\hspace{0.8cm}}$ **35.** $q^2 + 9q + \underline{\hspace{0.8cm}}$ **36.** $t^2 - 13t + \underline{\hspace{0.8cm}}$

37. $x^2 + \dfrac{1}{4}x + \underline{\hspace{0.8cm}}$ **38.** $x^2 + \dfrac{1}{2}x + \underline{\hspace{0.8cm}}$ **39.** $x^2 - 0.8x + \underline{\hspace{0.8cm}}$

40. *Concept Check*   What would be the first step in solving $2x^2 + 8x = 9$ by completing the square?

Determine the number that will complete the square to solve each equation after the constant term has been written on the right side. Do not actually solve. See Examples 4–6.

41. $x^2 + 4x - 2 = 0$ **42.** $t^2 + 2t - 1 = 0$ **43.** $x^2 + 10x + 18 = 0$

44. $x^2 + 8x + 11 = 0$ **45.** $3w^2 - w - 24 = 0$ **46.** $4z^2 - z - 39 = 0$

Solve each equation by completing the square. See Examples 4–6.

47. $x^2 - 2x - 24 = 0$ **48.** $m^2 - 4m - 32 = 0$ 🌐 **49.** $x^2 + 4x - 2 = 0$

50. $t^2 + 2t - 1 = 0$ 🌐 **51.** $x^2 + 7x - 1 = 0$ **52.** $x^2 + 13x - 3 = 0$

53. $3w^2 - w = 24$ **54.** $4z^2 - z = 39$ **55.** $2k^2 + 5k - 2 = 0$

56. $3r^2 + 2r - 2 = 0$ 🌐 **57.** $5x^2 - 10x + 2 = 0$ **58.** $2x^2 - 16x + 25 = 0$

59. $9x^2 - 24x = -13$ **60.** $25n^2 - 20n = 1$

61. $z^2 - \dfrac{4}{3}z = -\dfrac{1}{9}$ **62.** $p^2 - \dfrac{8}{3}p = -1$

63. $0.1x^2 - 0.2x - 0.1 = 0$ **64.** $0.1p^2 - 0.4p + 0.1 = 0$

Find the nonreal complex solutions of each equation. See Example 7.

🌐 **65.** $x^2 = -12$ **66.** $x^2 = -18$ **67.** $(r - 5)^2 = -4$

68. $(t + 6)^2 = -9$ **69.** $(6k - 1)^2 = -8$ **70.** $(4m - 7)^2 = -27$

71. $m^2 + 4m + 13 = 0$ **72.** $t^2 + 6t + 10 = 0$ **73.** $3r^2 + 4r + 4 = 0$

74. $4x^2 + 5x + 5 = 0$ **75.** $-m^2 - 6m - 12 = 0$ **76.** $-k^2 - 5k - 10 = 0$

RELATING CONCEPTS (EXERCISES 77–82)

FOR INDIVIDUAL OR GROUP WORK

The Greeks had a method of completing the square geometrically in which they literally changed a figure into a square. For example, to complete the square for $x^2 + 6x$, we begin with a square of side x, as in the figure on the left. We add three rectangles of width 1 to the right side and the bottom to get a region with area $x^2 + 6x$. To fill in the corner (complete the square), we must add nine 1-by-1 squares as shown in the figure on the right.

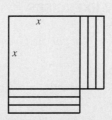 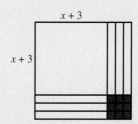

Work Exercises 77–82 in order.

77. What is the area of the original square?

78. What is the area of each strip?

79. What is the total area of the six strips?

80. What is the area of each small square in the corner of the second figure?

81. What is the total area of the small squares?

82. What is the area of the new "complete" square?

The next problems are "brain busters." Solve for x. Assume that a and b represent positive real numbers.

83. $x^2 - b = 0$

84. $x^2 = 4b$

85. $4x^2 = b^2 + 16$

86. $9x^2 - 25a = 0$

87. $(5x - 2b)^2 = 3a$

88. $x^2 - a^2 - 36 = 0$

TECHNOLOGY INSIGHTS (EXERCISES 89–90)

The first two calculator screens show the intersection points of the graph of $y = x^2$ and the graph of the horizontal line $y = 5$. The other screen shows that $\sqrt{5} \approx 2.236068$, so the graphs intersect at $x = -\sqrt{5}$ and $x = \sqrt{5}$. This supports our solution $\pm\sqrt{5}$ for $x^2 = 5$ using the square root property in Example 1(a).

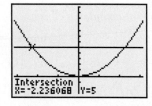 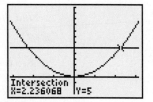

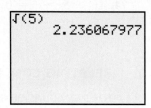

Use the method involving graphs described above to find approximations for the following square roots.

89. $\pm\sqrt{17}$

90. $\pm\sqrt{23}$

PREVIEW EXERCISES

*Evaluate $\sqrt{b^2 - 4ac}$ for the given values of a, b, and c. See **Section 1.3**.*

91. $a = 3, b = 1, c = -1$

92. $a = 4, b = 11, c = -3$

93. $a = 6, b = 7, c = 2$

94. $a = 1, b = -6, c = 9$

*Evaluate $\dfrac{-b + \sqrt{b^2 - 4ac}}{2a}$ for the given values of a, b, and c in each of the following. See **Section 1.3**.*

95. Exercise 91

96. Exercise 92

9.2 The Quadratic Formula

In this section, we complete the square to solve the general quadratic equation

$$ax^2 + bx + c = 0,$$

where a, b, and c are complex numbers and $a \neq 0$. The solution of this general equation gives a formula for finding the solution of any specific quadratic equation.

OBJECTIVE 1 Derive the quadratic formula. To solve $ax^2 + bx + c = 0$ by completing the square (assuming $a > 0$), we follow the steps given in **Section 9.1.**

$$ax^2 + bx + c = 0$$

$$x^2 + \frac{b}{a}x + \frac{c}{a} = 0 \qquad \text{Divide by } a. \text{ (Step 1)}$$

$$x^2 + \frac{b}{a}x = -\frac{c}{a} \qquad \text{Subtract } \tfrac{c}{a}. \text{ (Step 2)}$$

$$\left[\frac{1}{2}\left(\frac{b}{a}\right)\right]^2 = \left(\frac{b}{2a}\right)^2 = \frac{b^2}{4a^2} \qquad \text{(Step 3)}$$

$$x^2 + \frac{b}{a}x + \frac{b^2}{4a^2} = -\frac{c}{a} + \frac{b^2}{4a^2} \qquad \text{Add } \tfrac{b^2}{4a^2} \text{ to each side. (Step 4)}$$

Write the left side as a perfect square and rearrange the right side.

$$\left(x + \frac{b}{2a}\right)^2 = \frac{b^2}{4a^2} + \frac{-c}{a} \qquad \text{(Step 5)}$$

$$\left(x + \frac{b}{2a}\right)^2 = \frac{b^2}{4a^2} + \frac{-4ac}{4a^2} \qquad \text{Write with a common denominator.}$$

$$\left(x + \frac{b}{2a}\right)^2 = \frac{b^2 - 4ac}{4a^2} \qquad \text{Add fractions.}$$

$$x + \frac{b}{2a} = \sqrt{\frac{b^2 - 4ac}{4a^2}} \quad \text{or} \quad x + \frac{b}{2a} = -\sqrt{\frac{b^2 - 4ac}{4a^2}} \qquad \begin{array}{l}\text{Square root}\\ \text{property}\\ \text{(Step 6)}\end{array}$$

Since
$$\sqrt{\frac{b^2 - 4ac}{4a^2}} = \frac{\sqrt{b^2 - 4ac}}{\sqrt{4a^2}} = \frac{\sqrt{b^2 - 4ac}}{2a},$$

the right side of each equation can be expressed as

$$x + \frac{b}{2a} = \frac{\sqrt{b^2 - 4ac}}{2a} \qquad \text{or} \quad x + \frac{b}{2a} = \frac{-\sqrt{b^2 - 4ac}}{2a}$$

$$x = \frac{-b}{2a} + \frac{\sqrt{b^2 - 4ac}}{2a} \qquad \text{or} \quad x = \frac{-b}{2a} - \frac{\sqrt{b^2 - 4ac}}{2a}$$

$$x = \frac{-b + \sqrt{b^2 - 4ac}}{2a} \qquad \text{or} \quad x = \frac{-b - \sqrt{b^2 - 4ac}}{2a}.$$

If $a < 0$, the same two solutions are obtained. The result is the **quadratic formula,** which is abbreviated as follows.

> **Quadratic Formula**
>
> The solutions of the equation $ax^2 + bx + c = 0$ ($a \neq 0$) are given by
> $$x = \frac{-b \pm \sqrt{b^2 - 4ac}}{2a}.$$

▶ **CAUTION** In the quadratic formula, $x = \dfrac{-b \pm \sqrt{b^2 - 4ac}}{2a}$, *the square root is added to or subtracted from the value of $-b$ before dividing by $2a$.*

OBJECTIVE 2 Solve quadratic equations by using the quadratic formula. To use the quadratic formula, first write the given equation in standard form

$$ax^2 + bx + c = 0.$$

Then identify the values of a, b, and c and substitute them into the formula.

EXAMPLE 1 Using the Quadratic Formula (Rational Solutions)

Solve $6x^2 - 5x - 4 = 0$.

Here a, the coefficient of the second-degree term, is 6, while b, the coefficient of the first-degree term, is -5, and the constant c is -4. Substitute these values into the quadratic formula.

$$x = \frac{-b \pm \sqrt{b^2 - 4ac}}{2a} \qquad \text{Quadratic formula}$$

$$x = \frac{-(-5) \pm \sqrt{(-5)^2 - 4(6)(-4)}}{2(6)} \qquad a = 6, b = -5, c = -4$$

Use parentheses and substitute carefully to avoid errors.

$$x = \frac{5 \pm \sqrt{25 + 96}}{12}$$

$$x = \frac{5 \pm \sqrt{121}}{12}$$

$$x = \frac{5 \pm 11}{12}$$

There are two solutions, one from the $+$ sign and one from the $-$ sign.

$$x = \frac{5 + 11}{12} = \frac{16}{12} = \frac{4}{3} \qquad \text{or} \qquad x = \frac{5 - 11}{12} = \frac{-6}{12} = -\frac{1}{2}$$

Check each solution in the original equation. The solution set is $\left\{ -\frac{1}{2}, \frac{4}{3} \right\}$.

✔ **Now Try Exercise 5.**

We could have used factoring to solve the equation in Example 1.

$$6x^2 - 5x - 4 = 0$$

$(3x - 4)(2x + 1) = 0$ Factor.

$3x - 4 = 0$ or $2x + 1 = 0$ Zero-factor property

$3x = 4$ or $2x = -1$ Solve each equation.

$x = \dfrac{4}{3}$ or $x = -\dfrac{1}{2}$ Same solutions as in Example 1

When solving quadratic equations, it is a good idea to try factoring first. If the equation cannot be factored or if factoring is difficult, then use the quadratic formula. Later in this section, we will show a way to determine whether factoring can be used to solve a quadratic equation.

EXAMPLE 2   **Using the Quadratic Formula (Irrational Solutions)**

Solve $4x^2 = 8x - 1$.

Write the equation in standard form as $4x^2 - 8x + 1 = 0$.

$$x = \frac{-b \pm \sqrt{b^2 - 4ac}}{2a}$$ Quadratic formula

$$x = \frac{-(-8) \pm \sqrt{(-8)^2 - 4(4)(1)}}{2(4)}$$ $a = 4,\ b = -8,\ c = 1$

$$= \frac{8 \pm \sqrt{64 - 16}}{8}$$

$$= \frac{8 \pm \sqrt{48}}{8}$$

$$= \frac{8 \pm 4\sqrt{3}}{8}$$ $\sqrt{48} = \sqrt{16} \cdot \sqrt{3} = 4\sqrt{3}$

Factor first; then divide out the common factor.

$$= \frac{4\left(2 \pm \sqrt{3}\right)}{4(2)}$$ Factor.

$$= \frac{2 \pm \sqrt{3}}{2}$$ Lowest terms

The solution set is $\left\{ \dfrac{2 + \sqrt{3}}{2}, \dfrac{2 - \sqrt{3}}{2} \right\}$.

✔ **Now Try Exercise 9.**

▶ **CAUTION**   Remember the following:
1. *Every quadratic equation must be expressed in standard form* $ax^2 + bx + c = 0$ *before we begin to solve it,* whether we use factoring or the quadratic formula.
2. *When writing solutions in lowest terms, be sure to factor first; then divide out the common factor,* as shown in the last two steps in Example 2.

EXAMPLE 3 Using the Quadratic Formula (Nonreal Complex Solutions)

Solve $(9x + 3)(x - 1) = -8$.

Write the equation in standard form.

$$(9x + 3)(x - 1) = -8$$
$$9x^2 - 6x - 3 = -8 \qquad \text{Multiply.}$$
$$9x^2 - 6x + 5 = 0 \qquad \text{Standard form}$$

From the equation $9x^2 - 6x + 5 = 0$, we identify $a = 9$, $b = -6$, and $c = 5$.

$$x = \frac{-(-6) \pm \sqrt{(-6)^2 - 4(9)(5)}}{2(9)} \qquad \begin{array}{l}\text{Substitute into the}\\ \text{quadratic formula.}\end{array}$$

$$= \frac{6 \pm \sqrt{-144}}{18}$$

$$= \frac{6 \pm 12i}{18} \qquad \sqrt{-144} = 12i$$

$$= \frac{6(1 \pm 2i)}{6(3)} \qquad \text{Factor.}$$

$$= \frac{1 \pm 2i}{3} \qquad \text{Lowest terms}$$

The solution set, written in standard form $a + bi$, is $\left\{\frac{1}{3} + \frac{2}{3}i, \frac{1}{3} - \frac{2}{3}i\right\}$.

✔ **Now Try Exercise 37.**

OBJECTIVE ❸ Use the discriminant to determine the number and type of solutions.
The solutions of the quadratic equation $ax^2 + bx + c = 0$ are given by

$$x = \frac{-b \pm \sqrt{b^2 - 4ac}}{2a}. \qquad \longleftarrow \text{Discriminant}$$

If a, b, and c are integers, the type of solutions of a quadratic equation—that is, rational, irrational, or nonreal complex—is determined by the expression under the radical sign, $b^2 - 4ac$. Because it distinguishes among the three types of solutions, $b^2 - 4ac$ is called the *discriminant.* By calculating the discriminant before solving a quadratic equation, we can predict whether the solutions will be rational numbers, irrational numbers, or nonreal complex numbers.

Discriminant

The **discriminant** of $ax^2 + bx + c = 0$ is $\boldsymbol{b^2 - 4ac}$. If a, b, and c are integers, then the number and type of solutions are determined as follows.

Discriminant	*Number and Type of Solutions*
Positive, and the square of an integer	Two rational solutions
Positive, but not the square of an integer	Two irrational solutions
Zero	One rational solution
Negative	Two nonreal complex solutions

Calculating the discriminant can also help you decide whether to solve a quadratic equation by factoring or by using the quadratic formula. *If the discriminant is a perfect square (including 0), then the equation can be solved by factoring. Otherwise, the quadratic formula should be used.*

EXAMPLE 4 Using the Discriminant

Find the discriminant. Use it to predict the number and type of solutions for each equation. Tell whether the equation can be solved by factoring or whether the quadratic formula should be used.

(a) $6x^2 - x - 15 = 0$

We find the discriminant by evaluating $b^2 - 4ac$.

$$b^2 - 4ac = (-1)^2 - 4(6)(-15) \qquad a = 6, b = -1, c = -15$$
$$= 1 + 360$$
$$= 361$$

Use parentheses and substitute carefully.

A calculator shows that $361 = 19^2$, a perfect square. Since a, b, and c are integers and the discriminant is a perfect square, there will be two rational solutions and the equation can be solved by factoring.

(b) $3x^2 - 4x = 5$

Write the equation in standard form as $3x^2 - 4x - 5 = 0$.

$$b^2 - 4ac = (-4)^2 - 4(3)(-5) \qquad a = 3, b = -4, c = -5$$
$$= 16 + 60$$
$$= 76$$

Because 76 is positive but not the square of an integer and a, b, and c are integers, the equation will have two irrational solutions and is best solved using the quadratic formula.

(c) $4x^2 + x + 1 = 0$

Since $a = 4$, $b = 1$, and $c = 1$, the discriminant is

$$b^2 - 4ac = 1^2 - 4(4)(1)$$
$$= 1 - 16$$
$$= -15.$$

Because the discriminant is negative and a, b, and c are integers, this quadratic equation will have two nonreal complex solutions. The quadratic formula should be used to solve it.

(d) $4x^2 + 9 = 12x$

Write the equation as $4x^2 - 12x + 9 = 0$. The discriminant is

$$b^2 - 4ac = (-12)^2 - 4(4)(9) \qquad a = 4, b = -12, c = 9$$
$$= 144 - 144$$
$$= 0.$$

Because the discriminant is 0, the quantity under the radical in the quadratic formula is 0, and there is only one rational solution. Again, the equation can be solved by factoring.

✔ **Now Try Exercises 39, 41, and 43.**

EXAMPLE 5 Using the Discriminant

Find k so that $9x^2 + kx + 4 = 0$ will have only one rational solution.

The equation will have only one rational solution if the discriminant is 0. Since $a = 9$, $b = k$, and $c = 4$, the discriminant is

$$b^2 - 4ac = k^2 - 4(9)(4) = k^2 - 144.$$

Set the discriminant equal to 0 and solve for k.

$$k^2 - 144 = 0$$
$$k^2 = 144 \qquad \text{Subtract 144.}$$
$$k = 12 \quad \text{or} \quad k = -12 \qquad \text{Square root property}$$

The equation will have only one rational solution if $k = 12$ or $k = -12$.

✔ **Now Try Exercise 55.**

9.2 EXERCISES

Concept Check   Answer each question in Exercises 1–4.

1. The Cadillac Bar in Houston, Texas, encourages patrons to write (tasteful) messages on the walls. One person attempted to write the quadratic formula, as shown here.

$$x = \frac{-b\sqrt{b^2 - 4ac}}{2a}$$

Was this correct? If not, correct it.

2. An early version of Microsoft *Word* for Windows included the 1.0 edition of *Equation Editor.* The documentation used the following for the quadratic formula.

$$x = -b \pm \frac{\sqrt{b^2 - 4ac}}{2a}$$

Was this correct? If not, correct it.

3. A student claimed that the equation $2x^2 - 5 = 0$ cannot be solved using the quadratic formula because there is no first-degree x-term. Was the student correct? If not, give the values of a, b, and c.

4. A student attempted to solve $5x^2 - 5x + 1 = 0$ as follows.

$$x = \frac{-(-5) \pm \sqrt{(-5)^2 - 4(5)(1)}}{2(5)} = \frac{5 \pm \sqrt{5}}{10} = \frac{1}{2} \pm \sqrt{5}$$

Solution set: $\left\{ \dfrac{1}{2} \pm \sqrt{5} \right\}$

This is incorrect. **WHAT WENT WRONG?** Give the correct solution set.

Use the quadratic formula to solve each equation. (All solutions for these equations are real numbers.) See Examples 1 and 2.

5. $x^2 - 8x + 15 = 0$ **6.** $x^2 + 3x - 28 = 0$ **7.** $2x^2 + 4x + 1 = 0$

8. $2x^2 + 3x - 1 = 0$ ⊙ **9.** $2x^2 - 2x = 1$ **10.** $9x^2 + 6x = 1$

11. $x^2 + 18 = 10x$ **12.** $x^2 - 4 = 2x$ **13.** $4k^2 + 4k - 1 = 0$

14. $4r^2 - 4r - 19 = 0$ **15.** $2 - 2x = 3x^2$ **16.** $26r - 2 = 3r^2$

17. $\dfrac{x^2}{4} - \dfrac{x}{2} = 1$ **18.** $p^2 + \dfrac{p}{3} = \dfrac{1}{6}$ **19.** $-2t(t + 2) = -3$

20. $-3x(x + 2) = -4$ **21.** $(r - 3)(r + 5) = 2$ **22.** $(k + 1)(k - 7) = 1$

23. $(x + 2)(x - 3) = 1$ **24.** $(x - 5)(x + 2) = 6$ **25.** $p = \dfrac{5(5 - p)}{3(p + 1)}$

26. $x = \dfrac{2(x + 3)}{x + 5}$ **27.** $(2x + 1)^2 = x + 4$ **28.** $(2x - 1)^2 = x + 2$

Use the quadratic formula to solve each equation. (All solutions for these equations are non-real complex numbers.) See Example 3.

29. $x^2 - 3x + 6 = 0$ **30.** $x^2 - 5x + 20 = 0$ **31.** $r^2 - 6r + 14 = 0$

32. $t^2 + 4t + 11 = 0$ **33.** $4x^2 - 4x = -7$ **34.** $9x^2 - 6x = -7$

35. $x(3x + 4) = -2$ **36.** $z(2z + 3) = -2$

37. $(2x - 1)(8x - 4) = -1$ **38.** $(x - 1)(9x - 3) = -2$

Use the discriminant to determine whether the solutions for each equation are
     A. *two rational numbers;*   **B.** *one rational number;*
     C. *two irrational numbers;*   **D.** *two nonreal complex numbers.*

Do not actually solve. See Example 4.

39. $25x^2 + 70x + 49 = 0$ **40.** $4k^2 - 28k + 49 = 0$ **41.** $x^2 + 4x + 2 = 0$

42. $9x^2 - 12x - 1 = 0$ **43.** $3x^2 = 5x + 2$ **44.** $4x^2 = 4x + 3$

45. $3m^2 - 10m + 15 = 0$ **46.** $18x^2 + 60x + 82 = 0$ **47.** $0.5x^2 + 10x + 50 = 0$

48. Using the discriminant, which equations in Exercises 39–47 can be solved by factoring?

Based on your answer in Exercise 48, solve the equation given in each exercise.

49. Exercise 39   **50.** Exercise 40   **51.** Exercise 43   **52.** Exercise 44

53. Find the discriminant for each quadratic equation. Use it to tell whether the equation can be solved by factoring or whether the quadratic formula should be used. Then solve each equation.

     (a) $3k^2 + 13k = -12$ **(b)** $2x^2 + 19 = 14x$

54. *Concept Check*   Is it possible for the solution of a quadratic equation with integer coefficients to include just one irrational number? Why or why not?

Find the value of a, b, or c so that each equation will have exactly one rational solution. See Example 5.

55. $p^2 + bp + 25 = 0$ **56.** $r^2 - br + 49 = 0$ **57.** $am^2 + 8m + 1 = 0$

58. $at^2 + 24t + 16 = 0$ **59.** $9x^2 - 30x + c = 0$ **60.** $4m^2 + 12m + c = 0$

61. One solution of $4x^2 + bx - 3 = 0$ is $-\frac{5}{2}$. Find b and the other solution.

62. One solution of $3x^2 - 7x + c = 0$ is $\frac{1}{3}$. Find c and the other solution.

PREVIEW EXERCISES

Write each expression making the indicated substitution. Then factor the trinomial obtained after the substitution has been made. See **Section 6.4.**

63. $(7z + 3)^2 + 4(7z + 3) - 5$

    Let $7z + 3 = u$.

64. $4\left(\dfrac{1}{2}w + 8\right)^2 - 7\left(\dfrac{1}{2}w + 8\right) - 2$

    Let $\frac{1}{2}w + 8 = u$.

Solve each equation. See **Section 2.1.**

65. $\dfrac{3}{4}x + \dfrac{1}{2}x = -10$

66. $\dfrac{x}{5} + \dfrac{3x}{4} = -19$

Solve each equation. See **Section 8.6.**

67. $\sqrt{2x + 6} = x - 1$

68. $\sqrt{2x + 1} + \sqrt{x + 3} = 0$

9.3   Equations Quadratic in Form

OBJECTIVES

1 Solve an equation with fractions by writing it in quadratic form.

2 Use quadratic equations to solve applied problems.

3 Solve an equation with radicals by writing it in quadratic form.

4 Solve an equation that is quadratic in form by substitution.

We have introduced four methods for solving quadratic equations written in standard form $ax^2 + bx + c = 0$. The following table lists some advantages and disadvantages of each method.

METHODS FOR SOLVING QUADRATIC EQUATIONS

Method	Advantages	Disadvantages
Factoring	This is usually the fastest method.	Not all polynomials are factorable; some factorable polynomials are difficult to factor.
Square root property	This is the simplest method for solving equations of the form $(ax + b)^2 = c$.	Few equations are given in this form.
Completing the square	This method can always be used, although most people prefer the quadratic formula.	It requires more steps than other methods.
Quadratic formula	This method can always be used.	It is more difficult than factoring because of the square root, although calculators can simplify its use.

OBJECTIVE ❶ Solve an equation with fractions by writing it in quadratic form.   A variety of nonquadratic equations can be written in the form of a quadratic equation and solved by using one of the methods in the table. As you solve the equations in this section, try to decide which method is best for each equation.

EXAMPLE 1	Solving an Equation with Fractions that Leads to a Quadratic Equation

Solve $\dfrac{1}{x} + \dfrac{1}{x-1} = \dfrac{7}{12}$.

Clear fractions by multiplying each term by the least common denominator, $12x(x-1)$. (Note that the domain must be restricted to $x \neq 0$, $x \neq 1$.)

$$12x(x-1)\frac{1}{x} + 12x(x-1)\frac{1}{x-1} = 12x(x-1)\frac{7}{12}$$

$$12(x-1) + 12x = 7x(x-1)$$

$12x - 12 + 12x = 7x^2 - 7x$ Distributive property

$24x - 12 = 7x^2 - 7x$ Combine terms.

$7x^2 - 31x + 12 = 0$ Standard form

$(7x - 3)(x - 4) = 0$ Factor.

$7x - 3 = 0$ or $x - 4 = 0$ Zero-factor property

$7x = 3$ or $x = 4$ Solve for x.

$$x = \frac{3}{7}$$

The solution set is $\left\{\frac{3}{7}, 4\right\}$.

✔ **Now Try Exercise 19.**

OBJECTIVE ❷ Use quadratic equations to solve applied problems.   Earlier we solved distance-rate-time (or motion) problems that led to linear equations or rational equations. Now we solve motion problems that lead to quadratic equations. We continue to use the six-step problem-solving method from **Section 2.3.**

EXAMPLE 2 Solving a Motion Problem

A riverboat for tourists averages 12 mph in still water. It takes the boat 1 hr, 4 min to go 6 mi upstream and return. Find the speed of the current. See Figure 1.

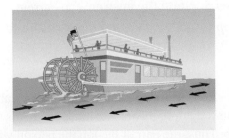

Riverboat, traveling *upstream*—
the current slows it down

FIGURE 1

Step 1 **Read** the problem carefully.

Step 2 **Assign a variable.** Let $x =$ the speed of the current. The current slows down the boat when it is going upstream, so the rate (or speed) of the boat going upstream is its speed in still water less the speed of the current, or $(12 - x)$ mph. Similarly, the current speeds up the boat as it travels downstream, so its speed downstream is $(12 + x)$ mph. Thus,

$$12 - x = \text{the rate upstream;}$$
$$12 + x = \text{the rate downstream.}$$

This information can be used to complete a table. We use the distance formula, $d = rt$, solved for time t, $t = \frac{d}{r}$, to write expressions for t.

	d	r	t
Upstream	6	$12 - x$	$\dfrac{6}{12 - x}$
Downstream	6	$12 + x$	$\dfrac{6}{12 + x}$

Times in hours

Step 3 **Write an equation.** The total time of 1 hr, 4 min can be written as

$$1 + \frac{4}{60} = 1 + \frac{1}{15} = \frac{16}{15} \text{ hr.}$$

Because the time upstream plus the time downstream equals $\frac{16}{15}$ hr,

Time upstream $+$ Time downstream $=$ Total time

$$\frac{6}{12 - x} \quad + \quad \frac{6}{12 + x} \quad = \quad \frac{16}{15}.$$

Step 4 **Solve** the equation. Multiply each side by $15(12 - x)(12 + x)$, the LCD, and solve the resulting quadratic equation.

$$15(12 + x)6 + 15(12 - x)6 = 16(12 - x)(12 + x)$$
$$90(12 + x) + 90(12 - x) = 16(144 - x^2)$$
$$1080 + 90x + 1080 - 90x = 2304 - 16x^2 \qquad \text{Distributive property}$$
$$2160 = 2304 - 16x^2 \qquad \text{Combine terms.}$$
$$16x^2 = 144$$
$$x^2 = 9 \qquad \text{Divide by 16.}$$
$$x = 3 \quad \text{or} \quad x = -3 \qquad \text{Square root property}$$

Step 5 **State the answer.** The current speed cannot be -3, so the answer is 3 mph.

Step 6 **Check** that this value satisfies the original problem.

✔ **Now Try Exercise 33.**

In **Section 7.5** we solved problems about work rates. Recall that a person's work rate is $\frac{1}{t}$ part of the job per hour, where t is the time in hours required to do the complete job. Thus, the part of the job the person will do in x hours is $\frac{1}{t}x$.

EXAMPLE 3   Solving a Work Problem

It takes two carpet layers 4 hr to carpet a room. If each worked alone, one of them could do the job in 1 hr less time than the other. How long would it take each carpet layer to complete the job alone?

Step 1   **Read** the problem again. There will be two answers.

Step 2   **Assign a variable.** Let x = the number of hours for the slower carpet layer to complete the job alone. Then the faster carpet layer could do the entire job in $(x - 1)$ hours. The slower person's rate is $\frac{1}{x}$, and the faster person's rate is $\frac{1}{x-1}$. Together, they do the job in 4 hr. Complete a table.

	Rate	Time Working Together	Fractional Part of the Job Done	
Slower Worker	$\dfrac{1}{x}$	4	$\dfrac{1}{x}(4)$	Sum is 1 whole job.
Faster Worker	$\dfrac{1}{x-1}$	4	$\dfrac{1}{x-1}(4)$	

Step 3   **Write an equation.** The sum of the fractional parts done by the workers should equal 1 (the whole job).

Part done by slower worker   +   Part done by faster worker   =   1 whole job

$$\frac{4}{x} \qquad + \qquad \frac{4}{x-1} \qquad = \qquad 1$$

Step 4   **Solve** the equation from Step 3.

$$x(x-1)\left(\frac{4}{x} + \frac{4}{x-1}\right) = x(x-1)(1) \qquad \text{Multiply by the LCD.}$$

$$4(x-1) + 4x = x(x-1) \qquad \text{Distributive property}$$

$$4x - 4 + 4x = x^2 - x \qquad \text{Distributive property}$$

$$x^2 - 9x + 4 = 0 \qquad \text{Standard form}$$

This equation cannot be solved by factoring, so use the quadratic formula.

$$x = \frac{-(-9) \pm \sqrt{(-9)^2 - 4(1)(4)}}{2(1)} = \frac{9 \pm \sqrt{65}}{2} \qquad a = 1, b = -9, c = 4$$

$$x = \frac{9 + \sqrt{65}}{2} \approx 8.5 \quad \text{or} \quad x = \frac{9 - \sqrt{65}}{2} \approx 0.5 \qquad \text{Use a calculator.}$$

Step 5   **State the answer.** Only the solution 8.5 makes sense in the original problem, because $x - 1 = 0.5 - 1 = -0.5$ cannot represent the time for the faster worker. Thus, the slower worker could do the job in about 8.5 hr and the faster in about $8.5 - 1 = 7.5$ hr.

Step 6   **Check** that these results satisfy the original problem.

✔ **Now Try Exercise 39.**

OBJECTIVE 3 Solve an equation with radicals by writing it in quadratic form.

EXAMPLE 4 Solving Radical Equations That Lead to Quadratic Equations

Solve each equation.

(a) $x = \sqrt{6x - 8}$

This equation is not quadratic. However, squaring both sides of the equation gives a quadratic equation that can be solved by factoring.

$$x^2 = 6x - 8 \qquad \text{Square each side.}$$
$$x^2 - 6x + 8 = 0 \qquad \text{Standard form}$$
$$(x - 4)(x - 2) = 0 \qquad \text{Factor.}$$
$$x - 4 = 0 \quad \text{or} \quad x - 2 = 0 \qquad \text{Zero-factor property}$$
$$x = 4 \quad \text{or} \qquad x = 2 \qquad \text{Proposed solutions}$$

Recall from **Section 8.6** that squaring both sides of an equation can introduce extraneous solutions that do not satisfy the original equation. *All proposed solutions must be checked in the original (not the squared) equation.*

Check: If $x = 4$, then

$$x = \sqrt{6x - 8}$$
$$4 = \sqrt{6(4) - 8} \quad ?$$
$$4 = \sqrt{16} \qquad ?$$
$$4 = 4. \qquad\qquad \text{True}$$

If $x = 2$, then

$$x = \sqrt{6x - 8}$$
$$2 = \sqrt{6(2) - 8} \quad ?$$
$$2 = \sqrt{4} \qquad ?$$
$$2 = 2. \qquad\qquad \text{True}$$

Both solutions check, so the solution set is $\{2, 4\}$.

(b) $\qquad\qquad x + \sqrt{x} = 6$

$$\sqrt{x} = 6 - x \qquad \text{Isolate the radical on one side.}$$
$$x = 36 - 12x + x^2 \qquad \text{Square each side.}$$
$$x^2 - 13x + 36 = 0 \qquad \text{Standard form}$$
$$(x - 4)(x - 9) = 0 \qquad \text{Factor.}$$
$$x - 4 = 0 \quad \text{or} \quad x - 9 = 0 \qquad \text{Zero-factor property}$$
$$x = 4 \quad \text{or} \qquad x = 9 \qquad \text{Proposed solutions}$$

Check both proposed solutions in the *original* equation.

If $x = 4$, then

$$x + \sqrt{x} = 6$$
$$4 + \sqrt{4} = 6 \quad ?$$
$$6 = 6. \qquad \text{True}$$

If $x = 9$, then

$$x + \sqrt{x} = 6$$
$$9 + \sqrt{9} = 6 \quad ?$$
$$12 = 6. \qquad \text{False}$$

Only the solution 4 checks, so the solution set is $\{4\}$.

✔ **Now Try Exercises 43 and 49.**

OBJECTIVE 4 Solve an equation that is quadratic in form by substitution. A nonquadratic equation that can be written in the form $au^2 + bu + c = 0$, for $a \neq 0$ and an algebraic expression u, is called **quadratic in form.**

EXAMPLE 5 Solving Equations That Are Quadratic in Form

Solve each equation.

(a) $x^4 - 13x^2 + 36 = 0$

Because $x^4 = (x^2)^2$, we can write this equation in quadratic form with $u = x^2$ and $u^2 = x^4$. (Any letter except x could be used instead of u.)

$$x^4 - 13x^2 + 36 = 0$$

$$(x^2)^2 - 13x^2 + 36 = 0 \qquad x^4 = (x^2)^2$$

$$u^2 - 13u + 36 = 0 \qquad \text{Let } u = x^2.$$

$$(u - 4)(u - 9) = 0 \qquad \text{Factor.}$$

$$u - 4 = 0 \quad \text{or} \quad u - 9 = 0 \qquad \text{Zero-factor property}$$

Don't stop here. $\qquad u = 4 \quad \text{or} \qquad u = 9 \qquad \text{Solve.}$

$$x^2 = 4 \quad \text{or} \qquad x^2 = 9 \qquad \text{Substitute } x^2 \text{ for } u.$$

$$x = \pm 2 \quad \text{or} \qquad x = \pm 3 \qquad \text{Square root property}$$

The equation $x^4 - 13x^2 + 36 = 0$, a fourth-degree equation, has four solutions.* The solution set is $\{-3, -2, 2, 3\}$. Each solution can be verified by substituting it into the original equation for x.

(b) $$4x^6 + 1 - 5x^3$$

$$4(x^3)^2 + 1 = 5x^3 \qquad x^6 = (x^3)^2$$

$$4u^2 + 1 = 5u \qquad \text{Let } u = x^3.$$

$$4u^2 - 5u + 1 = 0 \qquad \text{Standard form}$$

$$(4u - 1)(u - 1) = 0 \qquad \text{Factor.}$$

$$4u - 1 = 0 \quad \text{or} \quad u - 1 = 0 \qquad \text{Zero-factor property}$$

$$u = \frac{1}{4} \quad \text{or} \qquad u = 1 \qquad \text{Solve.}$$

This is a key step. $\qquad x^3 = \dfrac{1}{4} \quad \text{or} \qquad x^3 = 1 \qquad u = x^3$

From these two equations,

$$x = \sqrt[3]{\frac{1}{4}} = \frac{\sqrt[3]{1}}{\sqrt[3]{4}} = \frac{1}{\sqrt[3]{4}} \cdot \frac{\sqrt[3]{2}}{\sqrt[3]{2}} = \frac{\sqrt[3]{2}}{2} \qquad \text{or} \qquad x = \sqrt[3]{1} = 1.$$

There are nonreal complex solutions for this equation, but finding them involves trigonometry. The real number solution set of $4x^6 + 1 = 5x^3$ is $\left\{ \dfrac{\sqrt[3]{2}}{2}, 1 \right\}$.

---

*In general, an equation in which an nth-degree polynomial equals 0 has n complex solutions, although some of them may be repeated.

(c)
$$x^4 = 6x^2 - 3$$

$$x^4 - 6x^2 + 3 = 0 \qquad \text{Standard form}$$

$$(x^2)^2 - 6x^2 + 3 = 0 \qquad x^4 = (x^2)^2$$

$$u^2 - 6u + 3 = 0 \qquad \text{Let } u = x^2.$$

Since this equation cannot be solved by factoring, use the quadratic formula.

$$u = \frac{-(-6) \pm \sqrt{(-6)^2 - 4(1)(3)}}{2(1)} \qquad a = 1, b = -6, c = 3$$

$$u = \frac{6 \pm \sqrt{24}}{2}$$

$$u = \frac{6 \pm 2\sqrt{6}}{2} \qquad \sqrt{24} = \sqrt{4} \cdot \sqrt{6} = 2\sqrt{6}$$

$$u = \frac{2(3 \pm \sqrt{6})}{2} \qquad \text{Factor.}$$

$$u = 3 \pm \sqrt{6} \qquad \text{Lowest terms}$$

Find both **square roots in each case.** $x^2 = 3 + \sqrt{6} \qquad \text{or} \quad x^2 = 3 - \sqrt{6} \qquad u = x^2$

$$x = \pm\sqrt{3 + \sqrt{6}} \quad \text{or} \quad x = \pm\sqrt{3 - \sqrt{6}}$$

The solution set contains four numbers:

$$\left\{ \sqrt{3 + \sqrt{6}}, -\sqrt{3 + \sqrt{6}}, \sqrt{3 - \sqrt{6}}, -\sqrt{3 - \sqrt{6}} \right\}.$$

✔ **Now Try Exercises 55, 83, and 87.**

▶ **NOTE**   Some students prefer to solve equations like those in Examples 5(a) and (b) by factoring directly. For example,

$$x^4 - 13x^2 + 36 = 0 \qquad \text{Example 5(a) equation}$$

$$(x^2 - 9)(x^2 - 4) = 0 \qquad \text{Factor.}$$

$$(x + 3)(x - 3)(x + 2)(x - 2) = 0. \qquad \text{Factor again.}$$

Using the zero-factor property gives the same solutions obtained in Example 5(a). Equations that cannot be solved by factoring, like that in Example 5(c), must be solved using the method of substitution and the quadratic formula.

EXAMPLE 6   **Solving Equations That Are Quadratic in Form**

Solve each equation.

(a) $2(4x - 3)^2 + 7(4x - 3) + 5 = 0$

Because of the repeated quantity $4x - 3$, this equation is quadratic in form with $u = 4x - 3$.

$$2(4x - 3)^2 + 7(4x - 3) + 5 = 0$$

$$2u^2 + 7u + 5 = 0 \qquad \text{Let } u = 4x - 3.$$

$$(2u + 5)(u + 1) = 0 \qquad \text{Factor.}$$

$$2u + 5 = 0 \quad \text{or} \quad u + 1 = 0 \qquad \text{Zero-factor property}$$

Don't stop here. $\qquad u = -\dfrac{5}{2} \quad \text{or} \qquad u = -1 \qquad \text{Solve for } u.$

$$4x - 3 = -\dfrac{5}{2} \quad \text{or} \quad 4x - 3 = -1 \qquad \text{Substitute } 4x - 3 \text{ for } u.$$

$$4x = \dfrac{1}{2} \quad \text{or} \qquad 4x = 2 \qquad \text{Solve for } x.$$

$$x = \dfrac{1}{8} \quad \text{or} \qquad x = \dfrac{1}{2}$$

Check that the solution set of the original equation is $\left\{\frac{1}{8}, \frac{1}{2}\right\}$.

(b) $2x^{2/3} - 11x^{1/3} + 12 = 0$

Let $x^{1/3} = u$; then $x^{2/3} = (x^{1/3})^2 = u^2$. Substitute into the given equation.

$$2u^2 - 11u + 12 = 0 \qquad \text{Let } x^{1/3} = u; \, x^{2/3} = u^2.$$

$$(2u - 3)(u - 4) = 0 \qquad \text{Factor.}$$

$$2u - 3 = 0 \quad \text{or} \quad u - 4 = 0 \qquad \text{Zero-factor property}$$

$$u = \dfrac{3}{2} \quad \text{or} \qquad u = 4$$

$$x^{1/3} = \dfrac{3}{2} \quad \text{or} \qquad x^{1/3} = 4 \qquad u = x^{1/3}$$

$$(x^{1/3})^3 = \left(\dfrac{3}{2}\right)^3 \quad \text{or} \quad (x^{1/3})^3 = 4^3 \qquad \text{Cube each side.}$$

$$x = \dfrac{27}{8} \quad \text{or} \qquad x = 64$$

Check that the solution set is $\left\{\frac{27}{8}, 64\right\}$.

✔ **Now Try Exercises 63 and 69.**

▶ **CAUTION** A common error when solving problems like those in Examples 5 and 6 is to stop too soon. *Once you have solved for u, remember to substitute and solve for the values of the original variable.*

9.3 EXERCISES

Concept Check   *Refer to the box at the beginning of this section. Decide whether factoring, the square root property, or the quadratic formula is most appropriate for solving each quadratic equation. Do not actually solve the equations.*

1. $(2x + 3)^2 = 4$ **2.** $4x^2 - 3x = 1$ **3.** $x^2 + 5x - 8 = 0$

4. $2x^2 + 3x = 1$ **5.** $3x^2 = 2 - 5x$ **6.** $x^2 = 5$

Concept Check *Write a sentence describing the first step you would take to solve each equation. Do not actually solve.*

7. $\dfrac{14}{x} = x - 5$

8. $\sqrt{1 + x} + x = 5$

9. $(x^2 + x)^2 - 8(x^2 + x) + 12 = 0$

10. $3x = \sqrt{16 - 10x}$

11. *Concept Check* A student solved the equation $x = \sqrt{3x + 4}$ as follows.

$$x = \sqrt{3x + 4}$$
$$x^2 = 3x + 4$$
Square both sides.

$$x^2 - 3x - 4 = 0$$
$$(x - 4)(x + 1) = 0$$
$$x - 4 = 0 \quad \text{or} \quad x + 1 = 0$$
$$x = 4 \quad \text{or} \quad x = -1$$

Solution set: $\{4, -1\}$

This solution set is incorrect. **WHAT WENT WRONG?** Give the correct solution set.

12. *Concept Check* A student solved the following equation as shown.

$$2(x - 1)^2 - 3(x - 1) + 1 = 0$$
$$2u^2 - 3u + 1 = 0$$
Let $u = x - 1$.

$$(2u - 1)(u - 1) = 0$$
$$2u - 1 = 0 \quad \text{or} \quad u - 1 = 0$$
$$u = \dfrac{1}{2} \quad \text{or} \quad u = 1$$

Solution set: $\{\tfrac{1}{2}, 1\}$

This solution set is incorrect. **WHAT WENT WRONG?** Give the correct solution set.

Solve each equation. Check your solutions. See Example 1.

13. $\dfrac{14}{x} = x - 5$

14. $\dfrac{-12}{x} = x + 8$

15. $1 - \dfrac{3}{x} - \dfrac{28}{x^2} = 0$

16. $4 - \dfrac{7}{r} - \dfrac{2}{r^2} = 0$

17. $3 - \dfrac{1}{t} = \dfrac{2}{t^2}$

18. $1 + \dfrac{2}{k} = \dfrac{3}{k^2}$

◐ 19. $\dfrac{1}{x} + \dfrac{2}{x + 2} = \dfrac{17}{35}$

20. $\dfrac{2}{m} + \dfrac{3}{m + 9} = \dfrac{11}{4}$

21. $\dfrac{2}{x + 1} + \dfrac{3}{x + 2} = \dfrac{7}{2}$

22. $\dfrac{4}{3 - p} + \dfrac{2}{5 - p} = \dfrac{26}{15}$

23. $\dfrac{3}{2x} - \dfrac{1}{2(x + 2)} = 1$

24. $\dfrac{4}{3x} - \dfrac{1}{2(x + 1)} = 1$

25. $3 = \dfrac{1}{t + 2} + \dfrac{2}{(t + 2)^2}$

26. $1 + \dfrac{2}{3z + 2} = \dfrac{15}{(3z + 2)^2}$

27. $\dfrac{6}{p} = 2 + \dfrac{p}{p + 1}$

28. $\dfrac{k}{2 - k} + \dfrac{2}{k} = 5$

29. $1 - \dfrac{1}{2x + 1} - \dfrac{1}{(2x + 1)^2} = 0$

30. $1 - \dfrac{1}{3x - 2} - \dfrac{1}{(3x - 2)^2} = 0$

Concept Check *Answer each question.*

31. A boat goes 20 mph in still water, and the rate of the current is t mph.

   (a) What is the rate of the boat when it travels upstream?

   (b) What is the rate of the boat when it travels downstream?

32. If it takes m hours to grade a set of papers, what is the grader's rate (in job per hour)?

Solve each problem. See Examples 2 and 3.

33. On a windy day Eduardo found that he could go 16 mi downstream and then 4 mi back upstream at top speed in a total of 48 min. What was the top speed of Eduardo's boat if the current was 15 mph?

	d	r	t
Upstream	4	x − 15	
Downstream	16		

34. Latarsha flew her plane for 6 hr at a constant speed. She traveled 810 mi with the wind, then turned around and traveled 720 mi against the wind. The wind speed was a constant 15 mph. Find the speed of the plane.

	d	r	t
With Wind	810		
Against Wind	720		

35. In California, the distance from Jackson to Lodi is about 40 mi, as is the distance from Lodi to Manteca. Rico drove from Jackson to Lodi during the rush hour, stopped in Lodi for a high energy drink, and then drove on to Manteca at 10 mph faster. Driving time for the entire trip was 88 min. Find his speed from Jackson to Lodi. (*Source: State Farm Road Atlas.*)

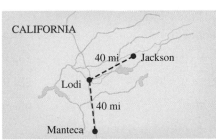

36. In Canada, Medicine Hat and Cranbrook are 300 km apart. Harry rides his Harley 20 km per hr faster than Yoshi rides his Yamaha. Find Harry's average speed if he travels from Cranbrook to Medicine Hat in $1\frac{1}{4}$ hr less time than Yoshi. (*Source: State Farm Road Atlas.*)

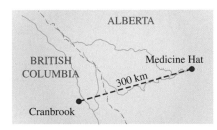

37. Working together, two people can cut a large lawn in 2 hr. One person can do the job alone in 1 hr less time than the other. How long (to the nearest tenth) would it take the faster person to do the job? (*Hint: x is the time of the faster person.*)

	Rate	Time Working Together	Fractional Part of the Job Done
Faster Worker	$\frac{1}{x}$	2	
Slower Worker		2	

38. A janitorial service provides two people to clean an office building. Working together, the two can clean the building in 5 hr. One person is new to the job and would take 2 hr longer than the other person to clean the building alone. How long (to the nearest tenth) would it take the new worker to clean the building alone?

	Rate	Time Working Together	Fractional Part of the Job Done
Faster Worker			
Slower Worker			

39. Rusty and Nancy Brauner are planting flats of spring flowers. Working alone, Rusty would take 2 hr longer than Nancy to plant the flowers. Working together, they do the job in 12 hr. How long would it have taken each person working alone?

40. Jay Beckenstein can work through a stack of invoices in 1 hr less time than Colleen Manley Jones can. Working together they take $1\frac{1}{2}$ hr. How long would it take each person working alone?

41. Two pipes together can fill a large tank in 2 hr. One of the pipes, used alone, takes 3 hr longer than the other to fill the tank. How long would each pipe take to fill the tank alone?

42. A washing machine can be filled in 6 min if both the hot and cold water taps are fully opened. Filling the washer with hot water alone takes 9 min longer than filling it with cold water alone. How long does it take to fill the washer with cold water?

Solve each equation. Check your solutions. See Example 4.

43. $x = \sqrt{7x - 10}$

44. $z = \sqrt{5z - 4}$

45. $2x = \sqrt{11x + 3}$

46. $4x = \sqrt{6x + 1}$

47. $3x = \sqrt{16 - 10x}$

48. $4t = \sqrt{8t + 3}$

49. $k + \sqrt{k} = 12$

50. $p - 2\sqrt{p} = 8$

51. $m = \sqrt{\dfrac{6 - 13m}{5}}$

52. $r = \sqrt{\dfrac{20 - 19r}{6}}$

53. $-x = \sqrt{\dfrac{8 - 2x}{3}}$

54. $-x = \sqrt{\dfrac{3x + 7}{4}}$

Solve each equation. Check your solutions. See Examples 5 and 6.

55. $x^4 - 29x^2 + 100 = 0$

56. $x^4 - 37x^2 + 36 = 0$

57. $4k^4 - 13k^2 + 9 = 0$

58. $9x^4 - 25x^2 + 16 = 0$

59. $x^4 + 48 = 16x^2$

60. $z^4 = 17z^2 - 72$

61. $(x + 3)^2 + 5(x + 3) + 6 = 0$

62. $(k - 4)^2 + (k - 4) - 20 = 0$

63. $3(m + 4)^2 - 8 = 2(m + 4)$

64. $(t + 5)^2 + 6 = 7(t + 5)$

65. $2 + \dfrac{5}{3k - 1} = \dfrac{-2}{(3k - 1)^2}$

66. $3 - \dfrac{7}{2p + 2} = \dfrac{6}{(2p + 2)^2}$

67. $2 - 6(m - 1)^{-2} = (m - 1)^{-1}$

68. $3 - 2(x - 1)^{-1} = (x - 1)^{-2}$

69. $x^{2/3} + x^{1/3} - 2 = 0$

70. $x^{2/3} - 2x^{1/3} - 3 = 0$

71. $r^{2/3} + r^{1/3} - 12 = 0$

72. $3x^{2/3} - x^{1/3} - 24 = 0$

73. $4k^{4/3} - 13k^{2/3} + 9 = 0$

74. $9t^{4/3} - 25t^{2/3} + 16 = 0$

75. $2(1 + \sqrt{r})^2 = 13(1 + \sqrt{r}) - 6$

76. $(k^2 + k)^2 + 12 = 8(k^2 + k)$

77. $2x^4 + x^2 - 3 = 0$

78. $4k^4 + 5k^2 + 1 = 0$

The equations in Exercises 79–88 are not grouped by type. Decide which method of solution applies, and then solve each equation. Give only real solutions. See Examples 1 and 4–6.

79. $12x^4 - 11x^2 + 2 = 0$

80. $\left(x - \dfrac{1}{2}\right)^2 + 5\left(x - \dfrac{1}{2}\right) - 4 = 0$

81. $\sqrt{2x + 3} = 2 + \sqrt{x - 2}$ **82.** $\sqrt{m + 1} = -1 + \sqrt{2m}$

83. $2m^6 + 11m^3 + 5 = 0$ **84.** $8x^6 + 513x^3 + 64 = 0$

85. $6 = 7(2w - 3)^{-1} + 3(2w - 3)^{-2}$ **86.** $m^6 - 10m^3 = -9$

87. $2x^4 - 9x^2 = -2$ **88.** $8x^4 + 1 = 11x^2$

PREVIEW EXERCISES

Solve each equation for the specified variable. See **Section 2.2.**

89. $P = 2L + 2W$ for W **90.** $A = \dfrac{1}{2}bh$ for h

91. $F = \dfrac{9}{5}C + 32$ for C **92.** $S = 2\pi rh + 2\pi r^2$ for h

---

Summary Exercises on Solving Quadratic Equations

*Exercises marked * require knowledge of the complex number system.*

We have introduced four algebraic methods for solving quadratic equations written in the form $ax^2 + bx + c = 0$: **factoring, the square root property, completing the square, and the quadratic formula.** *Refer to the summary box at the beginning of* **Section 9.3** *to review some of the advantages and disadvantages of each method. Then solve each quadratic equation by the method of your choice.*

1. $p^2 = 7$ **2.** $6x^2 - x - 15 = 0$ **3.** $n^2 + 6n + 4 = 0$

4. $(x - 3)^2 = 25$ **5.** $\dfrac{5}{m} + \dfrac{12}{m^2} = 2$ **6.** $3m^2 = 3 - 8m$

7. $2r^2 - 4r + 1 = 0$ ***8.** $x^2 = -12$ **9.** $x\sqrt{2} = \sqrt{5x - 2}$

10. $m^4 - 10m^2 + 9 = 0$ **11.** $(2k + 3)^2 = 8$ **12.** $\dfrac{2}{x} + \dfrac{1}{x - 2} = \dfrac{5}{3}$

13. $t^4 + 14 = 9t^2$ **14.** $8x^2 - 4x = 2$ ***15.** $z^2 + z + 1 = 0$

16. $5x^6 + 2x^3 - 7 = 0$ **17.** $4t^2 - 12t + 9 = 0$ **18.** $x\sqrt{3} = \sqrt{2 - x}$

19. $r^2 - 72 = 0$ **20.** $-3x^2 + 4x = -4$ **21.** $x^2 - 5x - 36 = 0$

22. $w^2 = 169$ ***23.** $3p^2 = 6p - 4$ **24.** $z = \sqrt{\dfrac{5z + 3}{2}}$

*25. $\dfrac{4}{r^2} + 3 = \dfrac{1}{r}$ **26.** $2(3k - 1)^2 + 5(3k - 1) = -2$

9.4 Formulas and Further Applications

OBJECTIVES

1 Solve formulas for variables involving squares and square roots.

2 Solve applied problems using the Pythagorean formula.

3 Solve applied problems using area formulas.

4 Solve applied problems using quadratic functions as models.

OBJECTIVE 1 Solve formulas for variables involving squares and square roots. The methods presented earlier can be used to solve such formulas.

EXAMPLE 1 Solving for Variables Involving Squares or Square Roots

Solve each formula for the given variable. Keep $\pm$ in the answer in part (a).

(a) $w = \dfrac{kFr}{v^2}$ for v

$$w = \frac{kFr}{v^2} \qquad \text{The goal is to isolate } v \text{ on one side.}$$

$$v^2 w = kFr \qquad \text{Multiply by } v^2.$$

$$v^2 = \frac{kFr}{w} \qquad \text{Divide by } w.$$

$$v = \pm\sqrt{\frac{kFr}{w}} \qquad \text{Square root property}$$

$$v = \frac{\pm\sqrt{kFr}}{\sqrt{w}} \cdot \frac{\sqrt{w}}{\sqrt{w}} \qquad \text{Rationalize the denominator.}$$

$$v = \frac{\pm\sqrt{kFrw}}{w}$$

(b) $d = \sqrt{\dfrac{4A}{\pi}}$ for A

$$d = \sqrt{\frac{4A}{\pi}} \qquad \text{The goal is to isolate } A \text{ on one side.}$$

$$d^2 = \frac{4A}{\pi} \qquad \text{Square both sides.}$$

$$\pi d^2 = 4A \qquad \text{Multiply by } \pi.$$

$$\frac{\pi d^2}{4} = A, \quad \text{or} \quad A = \frac{\pi d^2}{4} \qquad \text{Divide by 4.}$$

✔ **Now Try Exercises 9 and 19.**

▶ **NOTE** In many formulas like $v = \frac{\pm\sqrt{kFrw}}{w}$ in Example 1(a), we choose the positive value. In our work here, we will include both positive and negative values.

EXAMPLE 2 Solving for a Variable that Appears in First- and Second-Degree Terms

Solve $s = 2t^2 + kt$ for t.

Since the given equation has terms with t^2 and t, write it in standard form $ax^2 + bx + c = 0$, with t as the variable instead of x.

$$2t^2 + kt - s = 0$$

Now use the quadratic formula with $a = 2$, $b = k$, and $c = -s$.

$$t = \frac{-k \pm \sqrt{k^2 - 4(2)(-s)}}{2(2)} \qquad \text{Solve for } t.$$

$$t = \frac{-k \pm \sqrt{k^2 + 8s}}{4}$$

The solutions are $t = \dfrac{-k + \sqrt{k^2 + 8s}}{4}$ and $t = \dfrac{-k - \sqrt{k^2 + 8s}}{4}$.

✔ **Now Try Exercise 15.**

OBJECTIVE 2 Solve applied problems using the Pythagorean formula. The Pythagorean formula

$$c^2 = a^2 + b^2,$$

illustrated by the figure in the margin, was introduced in **Section 8.3** and is used to solve applications involving right triangles. Such problems often require solving quadratic equations.

Leg a
Hypotenuse
c
$90°$
Leg b
$c^2 = a^2 + b^2$
Pythagorean Formula

EXAMPLE 3 Using the Pythagorean Formula

Two cars left an intersection at the same time, one heading due north, the other due west. Some time later, they were exactly 100 mi apart. The car headed north had gone 20 mi farther than the car headed west. How far had each car traveled?

Step 1 **Read** the problem carefully.

Step 2 **Assign a variable.**

Let $x =$ the distance traveled by the car headed west.

Then $x + 20 =$ the distance traveled by the car headed north.

See Figure 2. The cars are 100 mi apart, so the hypotenuse of the right triangle equals 100.

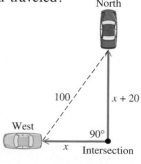

North

West

100

$x + 20$

$90°$

x

Intersection

FIGURE 2

Step 3   **Write an equation.** Use the Pythagorean formula.

$$c^2 = a^2 + b^2$$
$$100^2 = x^2 + (x + 20)^2 \qquad (x + y)^2 = x^2 + 2xy + y^2$$

Step 4   **Solve.**

$$10{,}000 = x^2 + x^2 + 40x + 400 \qquad \text{Square the binomial.}$$
$$0 = 2x^2 + 40x - 9600 \qquad \text{Standard form}$$
$$0 = x^2 + 20x - 4800 \qquad \text{Divide by 2.}$$
$$0 = (x + 80)(x - 60) \qquad \text{Factor.}$$
$$x + 80 = 0 \qquad \text{or} \quad x - 60 = 0 \qquad \text{Zero-factor property}$$
$$x = -80 \quad \text{or} \qquad x = 60$$

Step 5   **State the answer.** Since distance cannot be negative, discard the negative solution. The required distances are 60 mi and $60 + 20 = 80$ mi.

Step 6   **Check.** Since $60^2 + 80^2 = 100^2$, the answer is correct.

✔ **Now Try Exercise 31.**

OBJECTIVE ❸ Solve applied problems using area formulas.

EXAMPLE 4   **Solving an Area Problem**

A rectangular reflecting pool in a park is 20 ft wide and 30 ft long. The park gardener wants to plant a strip of grass of uniform width around the edge of the pool. She has enough seed to cover 336 ft². How wide will the strip be?

Step 1   **Read** the problem carefully.

Step 2   **Assign a variable.** The pool is shown in Figure 3.

Let $x =$ the unknown width of the grass strip.

Then $20 + 2x =$ the width of the large rectangle (the width of the pool plus two grass strips),

and $30 + 2x =$ the length of the large rectangle.

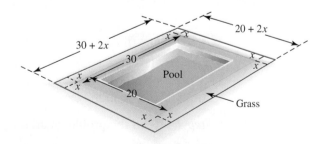

FIGURE 3

Step 3   **Write an equation.** The area of the large rectangle is given by the product of its length and width, $(30 + 2x)(20 + 2x)$. The area of the pool is $30 \cdot 20 = 600$ ft². The area of the large rectangle minus the area of the pool should equal the area of the grass strip. Since the area of the grass strip is to be 336 ft², the equation is $(30 + 2x)(20 + 2x) - 600 = 336$.

$$\begin{array}{ccc}
\text{Area} & \text{Area} & \text{Area} \\
\text{of} & - \quad \text{of} & = \quad \text{of} \\
\text{rectangle} & \text{pool} & \text{grass} \\
\downarrow & \downarrow & \downarrow
\end{array}$$

$$(30 + 2x)(20 + 2x) - 600 = 336$$

Step 4 **Solve.**

$600 + 100x + 4x^2 - 600 = 336$	Multiply.
$4x^2 + 100x - 336 = 0$	Standard form
$x^2 + 25x - 84 = 0$	Divide by 4.
$(x + 28)(x - 3) = 0$	Factor.
$x = -28 \quad \text{or} \quad x = 3$	Zero-factor property

Step 5 **State the answer.** The width cannot be -28 ft, so the grass strip should be 3 ft wide.

Step 6 **Check.** If $x = 3$, then the area of the large rectangle (which includes the grass strip) is

$$(30 + 2 \cdot 3)(20 + 2 \cdot 3) = 36 \cdot 26 = 936 \text{ ft}^2. \qquad \text{Area of pool and strip}$$

The area of the pool is $30 \cdot 20 = 600$ ft². So, the area of the grass strip is $936 - 600 = 336$ ft², as required. The answer is correct.

✔ **Now Try Exercise 37.**

OBJECTIVE 4 **Solve applied problems using quadratic functions as models.** Some applied problems can be modeled by *quadratic functions*, which for real numbers a, b, and c, can be written in the form

$$f(x) = ax^2 + bx + c, \quad \text{with } a \neq 0.$$

EXAMPLE 5 Solving an Applied Problem Using a Quadratic Function

If an object is projected upward from the top of a 144-ft building at 112 ft per sec, its position (in feet above the ground) is given by

$$s(t) = -16t^2 + 112t + 144,$$

where t is time in seconds after it was projected. When does it hit the ground?

When the object hits the ground, its distance above the ground is 0. We must find the value of t that makes $s(t) = 0$.

$0 = -16t^2 + 112t + 144$	Let $s(t) = 0$.
$0 = t^2 - 7t - 9$	Divide by -16.
$t = \dfrac{-(-7) \pm \sqrt{(-7)^2 - 4(1)(-9)}}{2(1)}$	Substitute into the quadratic formula.
$t = \dfrac{7 \pm \sqrt{85}}{2} \approx \dfrac{7 \pm 9.2}{2}$	Use a calculator.

The solutions are $t \approx 8.1$ or $t \approx -1.1$. Time cannot be negative, so we discard the negative solution. The object hits the ground about 8.1 sec after it is projected.

✔ **Now Try Exercise 43.**

| EXAMPLE 6 | Using a Quadratic Function to Model Company Bankruptcy Filings |

The number of publicly traded companies filing for bankruptcy was high in the early 1990s due to an economic recession. The number then declined during the middle 1990s, and in recent years has increased again. The quadratic function defined by

$$f(x) = 3.37x^2 - 28.6x + 133$$

approximates the number of company bankruptcy filings during the years 1990–2001, where x is the number of years since 1990. (*Source:* www.BankruptcyData.com)

(a) Use the model to approximate the number of company bankruptcy filings in 1995. For 1995, $x = 5$, so find $f(5)$.

$$f(5) = 3.37(5)^2 - 28.6(5) + 133 \qquad \text{Let } x = 5.$$
$$= 74.25$$

There were about 74 company bankruptcy filings in 1995.

(b) In what year did company bankruptcy filings reach 150?
Find the value of x that makes $f(x) = 150$.

$$f(x) = 3.37x^2 - 28.6x + 133$$
$$150 = 3.37x^2 - 28.6x + 133 \qquad \text{Let } f(x) = 150.$$
$$0 = 3.37x^2 - 28.6x - 17 \qquad \text{Standard form}$$

Now use $a = 3.37$, $b = -28.6$, and $c = -17$ in the quadratic formula.

$$x = \frac{-(-28.6) \pm \sqrt{(-28.6)^2 - 4(3.37)(-17)}}{2(3.37)}$$

$$x \approx 9.0 \quad \text{or} \quad x \approx -0.56 \qquad \qquad \text{Use a calculator.}$$

The positive solution is $x \approx 9$, so company bankruptcy filings reached 150 in the year $1990 + 9 = 1999$. (Reject the negative solution since the model is not valid for negative values of x.) Note that company bankruptcy filings doubled from about 74 in 1995 to 150 in 1999.

✔ **Now Try Exercises 57 and 59.**

9.4 EXERCISES

Concept Check *Answer each question in Exercises 1–4.*

1. In solving a formula that has the specified variable in the denominator, what is the first step?

2. What is the first step in solving a formula like $gw^2 = 2r$ for w?

3. What is the first step in solving a formula like $gw^2 = kw + 24$ for w?

4. Why is it particularly important to check all proposed solutions to an applied problem against the information in the original problem?

In Exercises 5 and 6, solve for m in terms of the other variables (m > 0).

5.

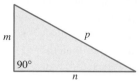

6.

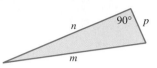

Solve each equation for the indicated variable. (Leave ± in your answers.) See Examples 1 and 2.

7. $d = kt^2$ for t

8. $s = kwd^2$ for d

🌐 **9.** $I = \dfrac{ks}{d^2}$ for d

10. $R = \dfrac{k}{d^2}$ for d

11. $F = \dfrac{kA}{v^2}$ for v

12. $L = \dfrac{kd^4}{h^2}$ for h

13. $V = \dfrac{1}{3}\pi r^2 h$ for r

14. $V = \pi(r^2 + R^2)h$ for r

🌐 **15.** $At^2 + Bt = -C$ for t

16. $S = 2\pi rh + \pi r^2$ for r

17. $D = \sqrt{kh}$ for h

18. $F = \dfrac{k}{\sqrt{d}}$ for d

19. $p = \sqrt{\dfrac{k\ell}{g}}$ for ℓ

20. $p = \sqrt{\dfrac{k\ell}{g}}$ for g

21. $S = 4\pi r^2$ for r

✎ **22.** Refer to Example 2 of this section. Suppose that k and s both represent positive numbers.

    (a) Which one of the two solutions given is positive?

    (b) Which one is negative?   **(c)** How can you tell?

Solve each equation for the indicated variable.

23. $p = \dfrac{E^2 R}{(r + R)^2}$ for R $(E > 0)$

24. $S(6S - t) - t^2$ for S

25. $10p^2c^2 + 7pcr = 12r^2$ for r

26. $S = vt + \dfrac{1}{2}gt^2$ for t

27. $LI^2 + RI + \dfrac{1}{c} = 0$ for I

28. $P = EI - RI^2$ for I

Solve each problem. When appropriate, round answers to the nearest tenth. See Example 3.

29. Find the lengths of the sides of the triangle.

30. Find the lengths of the sides of the triangle.

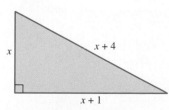

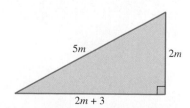

31. Two ships leave port at the same time, one heading due south and the other heading due east. Several hours later, they are 170 mi apart. If the ship traveling south traveled 70 mi farther than the other ship, how many miles did they each travel?

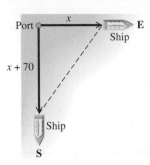

32. Emily McCants is flying a kite that is 30 ft farther above her hand than its horizontal distance from her. The string from her hand to the kite is 150 ft long. How high is the kite?

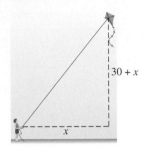

33. A game board is in the shape of a right triangle. The hypotenuse is 2 inches longer than the longer leg, and the longer leg is 1 inch less than twice as long as the shorter leg. How long is each side of the game board?

34. Sedona Levy is planting a vegetable garden in the shape of a right triangle. The longer leg is 3 ft longer than the shorter leg, and the hypotenuse is 3 ft longer than the longer leg. Find the lengths of the three sides of the garden.

35. The diagonal of a rectangular rug measures 26 ft, and the length is 4 ft more than twice the width. Find the length and width of the rug.

36. A 13-ft ladder is leaning against a house. The distance from the bottom of the ladder to the house is 7 ft less than the distance from the top of the ladder to the ground. How far is the bottom of the ladder from the house?

Solve each problem. See Example 4.

37. A club swimming pool is 30 ft wide and 40 ft long. The club members want an exposed aggregate border in a strip of uniform width around the pool. They have enough material for 296 ft². How wide can the strip be?

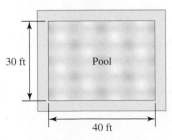

38. A couple wants to buy a rug for a room that is 20 ft long and 15 ft wide. They want to leave an even strip of flooring uncovered around the edges of the room. How wide a strip will they have if they buy a rug with an area of 234 ft²?

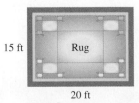

39. A rectangle has a length 2 m less than twice its width. When 5 m are added to the width, the resulting figure is a square with an area of 144 m². Find the dimensions of the original rectangle.

40. Ahmad's backyard measures 20 m by 30 m. He wants to put a flower garden in the middle of the backyard, leaving a strip of grass of uniform width around the flower garden. Ahmad must have 184 m² of grass. Under these conditions, what will the length and width of the garden be?

41. A rectangular piece of sheet metal has a length that is 4 in. less than twice the width. A square piece 2 in. on a side is cut from each corner. The sides are then turned up to form an uncovered box of volume 256 in.³. Find the length and width of the original piece of metal.

42. Another rectangular piece of sheet metal is 2 in. longer than it is wide. A square piece 3 in. on a side is cut from each corner. The sides are then turned up to form an uncovered box of volume 765 in.³. Find the dimensions of the original piece of metal.

Solve each problem. When appropriate, round answers to the nearest tenth. See Example 5.

43. An object is projected directly upward from the ground. After t seconds its distance in feet above the ground is

$$s(t) = 144t - 16t^2.$$

After how many seconds will the object be 128 ft above the ground? (*Hint:* Look for a common factor before solving the equation.)

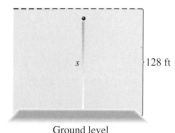

44. When does the object in Exercise 43 strike the ground?

45. A ball is projected upward from the ground. Its distance in feet from the ground in t seconds is given by

$$s(t) = -16t^2 + 128t.$$

At what times will the ball be 213 ft from the ground?

46. A toy rocket is launched from ground level. Its distance in feet from the ground in t seconds is given by

$$s(t) = -16t^2 + 208t.$$

At what times will the rocket be 550 ft from the ground?

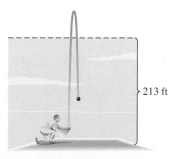

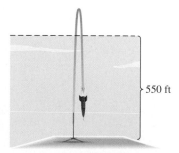

47. The function defined by

$$D(t) = 13t^2 - 100t$$

gives the distance in feet a car going approximately 68 mph will skid in t seconds. Find the time it would take for the car to skid 180 ft.

48. The function given in Exercise 47 becomes $D(t) = 13t^2 - 73t$ for a car going 50 mph. Find the time it takes for this car to skid 218 ft.

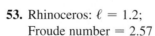 *A rock is projected upward from ground level, and its distance in feet from the ground in t seconds is given by $s(t) = -16t^2 + 160t$. Use algebra and a short explanation to answer Exercises 49 and 50.*

49. After how many seconds does it reach a height of 400 ft? How would you describe in words its position at this height?

50. After how many seconds does it reach a height of 425 ft? How would you interpret the mathematical result here?

Solve each problem using a quadratic equation.

51. If a square piece of cardboard has 3-in. squares cut from its corners and then has the flaps folded up to form an open-top box, the volume of the box is given by the formula $V = 3(x - 6)^2$, where x is the length of each side of the original piece of cardboard in inches. What original length would yield a box with volume 432 in.³?

52. The formula $A = P(1 + r)^2$ gives the amount A in dollars that P dollars will grow to in 2 yr at interest rate r (where r is given as a decimal), using compound interest. What interest rate will cause $2000 to grow to $2142.25 in 2 yr?

William Froude was a 19th century naval architect who used the expression

$$\frac{v^2}{g\ell}$$

*in shipbuilding. This expression, known as the **Froude number**, was also used by R. McNeill Alexander in his research on dinosaurs. (Source: "How Dinosaurs Ran," Scientific American, April 1991.) In Exercises 53 and 54, find the value of v (in meters per second), given $g = 9.8$ m per sec².*

53. Rhinoceros: $\ell = 1.2$;
Froude number = 2.57

54. Triceratops: $\ell = 2.8$;
Froude number = 0.16

Recall that corresponding sides of similar triangles are proportional. Use this fact to find the lengths of the indicated sides of each pair of similar triangles. Check all possible solutions in both triangles. Sides of a triangle cannot be negative (and are not drawn to scale here).

55. Side AC

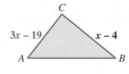

56. Side RQ

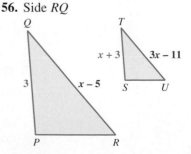

The total number of miles traveled by all motor vehicles in the U.S. for the years 1994–2003 are shown in the bar graph and can be modeled by the quadratic function defined by

$$f(x) = -1.705x^2 + 75.93x + 2351.$$

Here, x = 0 represents 1994, x = 1 represents 1995, and so on. Use the graph and the model to work Exercises 57–60. See Example 6.

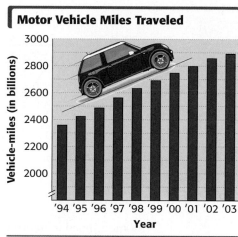

Motor Vehicle Miles Traveled

Source: U.S. Federal Highway Administration.

57. (a) Use the graph to estimate miles traveled in 2000 to the nearest ten billion.
   (b) Use the model to approximate miles traveled in 2000 to the nearest ten billion. How does this result compare to your estimate from part (a)?

58. Based on the model, in what year did miles traveled reach 2600 billion? (Round down to the nearest year.) How does this result compare to the vehicle-miles shown in the graph?

59. Based on the model, in what year did miles traveled reach 2800 billion? (Round down to the nearest year.) How does this result compare to the vehicle-miles shown in the graph?

60. If these data were modeled by a *linear* function defined by $f(x) = ax + b$, would the value of a be positive or negative? Explain.

PREVIEW EXERCISES

*Find each function value. See **Section 3.5.***

61. $f(x) = x^2 + 4x - 3$. Find $f(2)$.

62. $f(x) = 2(x - 3)^2 + 5$. Find $f(3)$.

63. $f(x) = ax^2 + bx + c$. Find $f\left(\dfrac{-b}{2a}\right)$.

*Solve each problem. See **Sections 6.5 and 9.2.***

64. If $f(x) = 8x^2 - 2x - 3$, solve the equation $f(x) = 0$.

65. If $f(x) = (x - 4)^2$, solve the equation $f(x) = 9$.

9.5 Graphs of Quadratic Functions

OBJECTIVES

1. Graph a quadratic function.

2. Graph parabolas with horizontal and vertical shifts.

3. Use the coefficient of x^2 to predict the shape and direction in which a parabola opens.

4. Find a quadratic function to model data.

OBJECTIVE 1 Graph a quadratic function. Figure 4 gives a graph of the simplest *quadratic function,* defined by $y = x^2$.

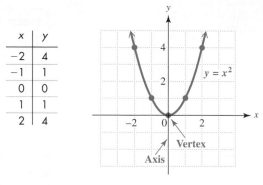

x	y
-2	4
-1	1
0	0
1	1
2	4

FIGURE 4

As mentioned in **Section 5.3**, this graph is called a **parabola.** The point $(0, 0)$, the lowest point on the curve, is the **vertex** of this parabola. The vertical line through the vertex is the **axis** of the parabola, here $x = 0$. A parabola is **symmetric about its axis;** that is, if the graph were folded along the axis, the two portions of the curve would coincide. As Figure 4 suggests, x can be any real number, so the domain of the function defined by $y = x^2$ is $(-\infty, \infty)$. Since y is always nonnegative, the range is $[0, \infty)$.

In **Section 9.4**, we solved applications modeled by quadratic functions. We consider graphs of more general quadratic functions as defined here.

Quadratic Function

A function that can be written in the form

$$f(x) = ax^2 + bx + c$$

for real numbers a, b, and c, with $a \neq 0$, is a **quadratic function.**

The graph of any quadratic function is a parabola with a vertical axis.

▶ **NOTE** We use the variable y and function notation $f(x)$ interchangeably when discussing parabolas. Although we use the letter f most often to name quadratic functions, other letters can be used. We use the capital letter F to distinguish between different parabolas graphed on the same coordinate axes.

Parabolas, which are a type of *conic section* (**Chapter 11**), have many applications. Cross sections of satellite dishes and automobile headlights form parabolas. The cables that are used to support suspension bridges are shaped like parabolas.

OBJECTIVE 2 Graph parabolas with horizontal and vertical shifts. Parabolas need not have their vertices at the origin, as does the graph of $f(x) = x^2$. For example, to graph a parabola of the form

$$F(x) = x^2 + k,$$

start by selecting sample values of x like those that were used to graph $f(x) = x^2$. The corresponding values of $F(x)$ in $F(x) = x^2 + k$ differ by k from those of $f(x) = x^2$. Thus, the graph of $F(x) = x^2 + k$ is *shifted,* or *translated, k* units vertically compared with that of $f(x) = x^2$.

EXAMPLE 1 **Graphing a Parabola with a Vertical Shift**

Graph $F(x) = x^2 - 2$.

This graph has the same shape as that of $f(x) = x^2$, but since k here is -2, the graph is shifted 2 units down, with vertex $(0, -2)$. Every function value is 2 less than the corresponding function value of $f(x) = x^2$. Plotting points on both sides of the vertex gives the graph in Figure 5.

Notice that since the parabola is symmetric about its axis $x = 0$, the plotted points are "mirror images" of each other. Since x can be any real number, the domain is still $(-\infty, \infty)$; the value of y (or $F(x)$) is always greater than or equal to -2, so the range is $[-2, \infty)$. The graph of $f(x) = x^2$ is shown in Figure 5 for comparison.

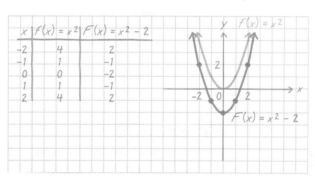

FIGURE 5

✔ **Now Try Exercise 23.**

Vertical Shift

The graph of $F(x) = x^2 + k$ is a parabola with the same shape as the graph of $f(x) = x^2$. The parabola is shifted k units up if $k > 0$, and $|k|$ units down if $k < 0$. The vertex is $(0, k)$.

The graph of the function defined by

$$F(x) = (x - h)^2$$

is also a parabola with the same shape as that of $f(x) = x^2$. Because $(x - h)^2 \geq 0$ for all x, the vertex of $F(x) = (x - h)^2$ is the lowest point on the parabola. The lowest point occurs here when $F(x)$ is 0. To get $F(x)$ equal to 0, let $x = h$ so the vertex of $F(x) = (x - h)^2$ is $(h, 0)$. Thus, the graph of $F(x) = (x - h)^2$ is shifted h units horizontally compared with that of $f(x) = x^2$.

EXAMPLE 2 Graphing a Parabola with a Horizontal Shift

Graph $F(x) = (x - 2)^2$.

If $x = 2$, then $F(x) = 0$, giving the vertex $(2, 0)$. The graph of $F(x) = (x - 2)^2$ has the same shape as that of $f(x) = x^2$ but is shifted 2 units to the right. Plot points on one side of the vertex and use symmetry about the axis $x = 2$ to find corresponding points on the other side. See Figure 6. The domain is $(-\infty, \infty)$; the range is $[0, \infty)$.

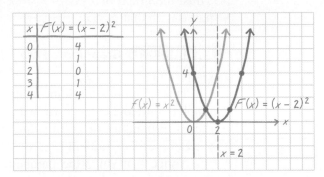

FIGURE 6

✔ **Now Try Exercise 27.**

Horizontal Shift

The graph of $F(x) = (x - h)^2$ is a parabola with the same shape as the graph of $f(x) = x^2$. The parabola is shifted h units horizontally: h units to the right if $h > 0$, and $|h|$ units to the left if $h < 0$. The vertex is $(h, 0)$.

▶ **CAUTION** Errors frequently occur when horizontal shifts are involved. To determine the direction and magnitude of a horizontal shift, find the value that would cause the expression $x - h$ to equal 0.

EXAMPLE 3 Graphing a Parabola with Horizontal and Vertical Shifts

Graph $F(x) = (x + 3)^2 - 2$.

This graph has the same shape as that of $f(x) = x^2$, but is shifted 3 units to the left (since $x + 3 = 0$ if $x = -3$) and 2 units down (because of the -2). See Figure 7. The vertex is $(-3, -2)$, with axis $x = -3$, domain $(-\infty, \infty)$, and range $[-2, \infty)$.

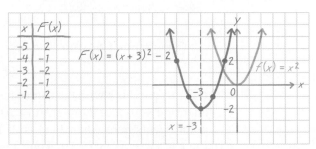

FIGURE 7

✔ **Now Try Exercise 29.**

The characteristics of the graph of a parabola of the form $F(x) = (x - h)^2 + k$ are summarized as follows.

Vertex and Axis of a Parabola

The graph of $F(x) = (x - h)^2 + k$ is a parabola with the same shape as the graph of $f(x) = x^2$, but with vertex (h, k). The axis is the vertical line $x = h$.

OBJECTIVE 3 **Use the coefficient of x^2 to predict the shape and direction in which a parabola opens.** Not all parabolas open up, and not all parabolas have the same shape as the graph of $f(x) = x^2$.

EXAMPLE 4 **Graphing a Parabola That Opens Down**

Graph $f(x) = -\frac{1}{2}x^2$.

This parabola is shown in Figure 8. The coefficient $-\frac{1}{2}$ affects the shape of the graph; the $\frac{1}{2}$ makes the parabola wider $\left(\text{since the values of } \frac{1}{2}x^2 \text{ increase more slowly than those of } x^2\right)$, and the negative sign makes the parabola open down. The graph is not shifted in any direction; the vertex is still $(0, 0)$ and the axis is $x = 0$. Unlike the parabolas graphed in Examples 1–3, the vertex here has the *largest* function value of any point on the graph. The domain is $(-\infty, \infty)$; the range is $(-\infty, 0]$.

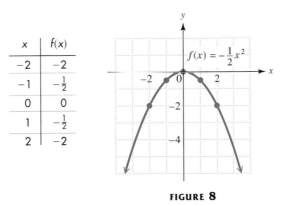

x	$f(x)$
-2	-2
-1	$-\frac{1}{2}$
0	0
1	$-\frac{1}{2}$
2	-2

FIGURE 8

✔ **Now Try Exercise 21.**

Some general principles concerning the graph of $F(x) = a(x - h)^2 + k$ follow.

General Principles

1. The graph of the quadratic function defined by
$$F(x) = a(x - h)^2 + k, \quad a \neq 0,$$
is a parabola with vertex (h, k) and the vertical line $x = h$ as axis.

2. The graph opens up if a is positive and down if a is negative.

3. The graph is wider than that of $f(x) = x^2$ if $0 < |a| < 1$. The graph is narrower than that of $f(x) = x^2$ if $|a| > 1$.

EXAMPLE 5 Using the General Principles to Graph a Parabola

Graph $F(x) = -2(x + 3)^2 + 4$.

   The parabola opens down (because $a < 0$) and is narrower than the graph of $f(x) = x^2$, since $|-2| = 2 > 1$, causing values of $F(x)$ to decrease more quickly than those of $f(x) = -x^2$. This parabola has vertex $(-3, 4)$, as shown in Figure 9. To complete the graph, we plotted the ordered pairs $(-4, 2)$ and, by symmetry, $(-2, 2)$. Symmetry can be used to find additional ordered pairs that satisfy the equation. The domain is $(-\infty, \infty)$; the range is $(-\infty, 4]$.

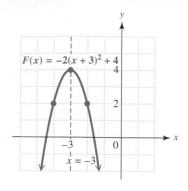

FIGURE 9

✔ **Now Try Exercise 33.**

OBJECTIVE 4 Find a quadratic function to model data.

EXAMPLE 6 Finding a Quadratic Function to Model the Rise in Multiple Births

The number of higher-order multiple births in the United States is rising. Let x represent the number of years since 1970 and y represent the rate of higher-order multiples born per 100,000 births since 1971. The data are shown in the following table.

U.S. HIGHER-ORDER MULTIPLE BIRTHS

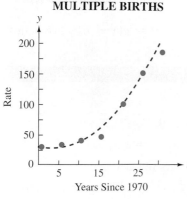

FIGURE 10

Year	x	y
1971	1	29.1
1976	6	35.0
1981	11	40.0
1986	16	47.0
1991	21	100.0
1996	26	152.6
2001	31	185.6

Source: National Center for Health Statistics.

Find a quadratic function that models the data.

   A scatter diagram of the ordered pairs (x, y) is shown in Figure 10. The general shape suggested by the scatter diagram indicates that a parabola should approximate these points, as shown by the dashed curve in Figure 11. The equation for such a parabola would have a positive coefficient for x^2 since the graph opens up. To find a quadratic function of the form

$$y = ax^2 + bx + c$$

that models, or *fits*, these data, we choose three representative ordered pairs and use them to write a system of three equations. Using $(1, 29.1)$, $(11, 40)$, and $(21, 100)$, we substitute the x- and y-values from the ordered pairs into the quadratic form $y = ax^2 + bx + c$ to get the following three equations.

$$a(1)^2 + b(1) + c = 29.1 \qquad \text{or} \qquad a + b + c = 29.1 \qquad (1)$$
$$a(11)^2 + b(11) + c = 40 \qquad \text{or} \qquad 121a + 11b + c = 40 \qquad (2)$$
$$a(21)^2 + b(21) + c = 100 \qquad \text{or} \qquad 441a + 21b + c = 100 \qquad (3)$$

U.S. HIGHER-ORDER MULTIPLE BIRTHS

FIGURE 11

We can find the values of a, b, and c by solving this system of three equations in three variables using the methods of **Section 4.2.** Multiplying equation (1) by -1 and adding the result to equation (2) gives

$$120a + 10b = 10.9. \quad (4)$$

Multiplying equation (2) by -1 and adding the result to equation (3) gives

$$320a + 10b = 60. \quad (5)$$

We eliminate b from this system of two equations in two variables by multiplying equation (4) by -1 and adding the result to equation (5).

$$200a = 49.1$$
$$a = 0.2455 \qquad \text{Use a calculator.}$$

We substitute 0.2455 for a in equation (4) or (5) to find that $b = -1.856$. Substituting the values of a and b into equation (1) gives $c = 30.7105$. Using these values of a, b, and c, our model is defined by

$$y = 0.2455x^2 - 1.856x + 30.7105.$$

✔ **Now Try Exercise 49.**

▶ **NOTE** In Example 6, if we had chosen three different ordered pairs of data, a slightly different model would result. The *quadratic regression* feature on a graphing calculator can also be used to generate the quadratic model that best fits given data. See your owner's manual for details.

9.5 EXERCISES

◉ *Complete solution available on Video Lectures on CD/DVD*

Now Try Exercise

1. *Concept Check* Match each quadratic function with its graph from choices A–D.

(a) $f(x) = (x + 2)^2 - 1$

(b) $f(x) = (x + 2)^2 + 1$

(c) $f(x) = (x - 2)^2 - 1$

(d) $f(x) = (x - 2)^2 + 1$

A.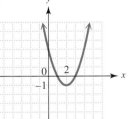

B.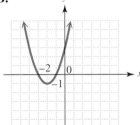

C.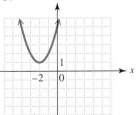

D.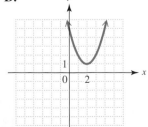

2. *Concept Check* Match each quadratic function with its graph from choices A–D.

(a) $f(x) = -x^2 + 2$

A.
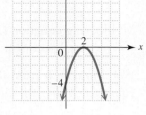

B.
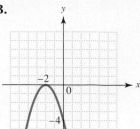

(b) $f(x) = -x^2 - 2$

(c) $f(x) = -(x + 2)^2$

C.

D.
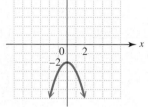

(d) $f(x) = -(x - 2)^2$

Identify the vertex of each parabola. See Examples 1–4.

3. $f(x) = -3x^2$

4. $f(x) = \dfrac{1}{2}x^2$

5. $f(x) = x^2 + 4$

6. $f(x) = x^2 - 4$

7. $f(x) = (x - 1)^2$

8. $f(x) = (x + 3)^2$

9. $f(x) = (x + 3)^2 - 4$

10. $f(x) = (x - 5)^2 - 8$

11. $f(x) = -(x - 5)^2 + 6$

12. (a) How are the graphs of the parabolas in Exercises 9 and 10 shifted compared to the graph of $f(x) = x^2$?

(b) What does the value of a in $F(x) = a(x - h)^2 + k$ tell you about the graph of the function compared to the graph of $f(x) = x^2$?

For each quadratic function, tell whether the graph opens up or down and whether the graph is wider, narrower, or the same shape as the graph of $f(x) = x^2$. See Examples 4 and 5.

13. $f(x) = -\dfrac{2}{5}x^2$

14. $f(x) = -2x^2$

15. $f(x) = 3x^2 + 1$

16. $f(x) = \dfrac{2}{3}x^2 - 4$

17. *Concept Check* For $f(x) = a(x - h)^2 + k$, in what quadrant is the vertex if

(a) $h > 0, k > 0$;　　**(b)** $h > 0, k < 0$;　　**(c)** $h < 0, k > 0$;　　**(d)** $h < 0, k < 0$?

18. *Concept Check* Answer each question.

(a) What is the value of h if the graph of $f(x) = a(x - h)^2 + k$ has vertex on the y-axis?

(b) What is the value of k if the graph of $f(x) = a(x - h)^2 + k$ has vertex on the x-axis?

19. *Concept Check* Match each quadratic function with the description of the parabola that is its graph.

(a) $f(x) = (x - 4)^2 - 2$ **A.** Vertex $(2, -4)$, opens down

(b) $f(x) = (x - 2)^2 - 4$ **B.** Vertex $(2, -4)$, opens up

(c) $f(x) = -(x - 4)^2 - 2$ **C.** Vertex $(4, -2)$, opens down

(d) $f(x) = -(x - 2)^2 - 4$ **D.** Vertex $(4, -2)$, opens up

20. Explain in your own words the meaning of each term.

(a) Vertex of a parabola   (b) Axis of a parabola

Graph each parabola. Plot at least two points as well as the vertex. Give the vertex, axis, domain, and range in Exercises 27–36. See Examples 1–5.

21. $f(x) = -2x^2$

22. $f(x) = \dfrac{1}{3}x^2$

23. $f(x) = x^2 - 1$

24. $f(x) = x^2 + 3$

25. $f(x) = -x^2 + 2$

26. $f(x) = 2x^2 - 2$

27. $f(x) = (x - 4)^2$

28. $f(x) = -2(x + 1)^2$

29. $f(x) = (x + 2)^2 - 1$

30. $f(x) = (x - 1)^2 + 2$

31. $f(x) = 2(x - 2)^2 - 4$

32. $f(x) = -3(x - 2)^2 + 1$

33. $f(x) = -\dfrac{1}{2}(x + 1)^2 + 2$

34. $f(x) = -\dfrac{2}{3}(x + 2)^2 + 1$

35. $f(x) = 2(x - 2)^2 - 3$

36. $f(x) = \dfrac{4}{3}(x - 3)^2 - 2$

RELATING CONCEPTS (EXERCISES 37–42)

FOR INDIVIDUAL OR GROUP WORK

*The procedures that allow the graph of $y = x^2$ to be shifted vertically and horizontally apply to other types of functions. In **Section 3.5** we introduced linear functions of the form $g(x) = ax + b$. Consider the graph of the simplest linear function defined by $g(x) = x$, shown here. **Work Exercises 37–42 in order.***

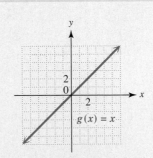

37. Based on the concepts of this section, how does the graph of $F(x) = x^2 + 6$ compare to the graph of $f(x) = x^2$ if a *vertical* shift is considered?

38. Graph the linear function defined by $G(x) = x + 6$.

39. Based on the concepts of **Chapter 3,** how does the graph of $G(x) = x + 6$ compare to the graph of $g(x) = x$ if a vertical shift is considered? (*Hint:* Look at the y-intercept.)

40. Based on the concepts of this section, how does the graph of $F(x) = (x - 6)^2$ compare to the graph of $f(x) = x^2$ if a *horizontal* shift is considered?

41. Graph the linear function $G(x) = x - 6$.

42. Based on the concepts of **Chapter 3,** how does the graph of $G(x) = x - 6$ compare to the graph of $g(x) = x$ if a horizontal shift is considered? (*Hint:* Look at the x-intercept.)

Concept Check In Exercises 43–48, tell whether a linear or quadratic function would be a more appropriate model for each set of graphed data. If linear, tell whether the slope should be positive or negative. If quadratic, tell whether the coefficient a of x^2 should be positive or negative. See Example 6.

43. **AVERAGE DAILY
E-MAIL VOLUME**

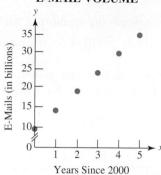

Source: General Accounting Office.

44. **AVERAGE DAILY VOLUME
OF FIRST-CLASS MAIL**

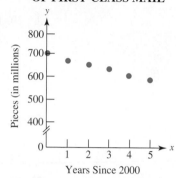

Source: General Accounting Office.

45. **INCREASES IN WHOLESALE
DRUG PRICES**

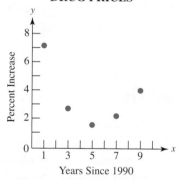

Source: IMS Health, Retail and
Provider Perspective.

46. **U.S. COMMERCIAL
BANK FAILURES**

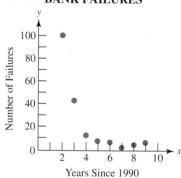

Source: www.ABA.com

47. **CEDAR RAPIDS SCHOOLS—
GENERAL RESERVE FUND**

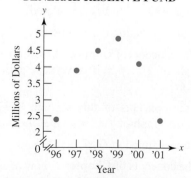

Source: Cedar Rapids School District.

48. **SOCIAL SECURITY ASSETS***

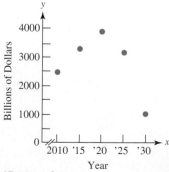

*Projected
Source: Social Security Administration.

Solve each problem. See Example 6.

49. The number of publicly traded companies filing for bankruptcy for selected years between 1990 and 2000 are shown in the table. In the year column, 0 represents 1990, 2 represents 1992, and so on.

Year	Number of Bankruptcies
0	115
2	91
4	70
6	84
8	120
10	176

Source: www.BankruptcyData.com

(a) Use the ordered pairs (year, number of bankruptcies) to make a scatter diagram of the data.

(b) Use the scatter diagram to decide whether a linear or quadratic function would better model the data. If quadratic, should the coefficient a of x^2 be positive or negative?

(c) Use the ordered pairs (0, 115), (4, 70), and (8, 120) to find a quadratic function that models the data. Round the values of a, b, and c in your model to three decimal places, as necessary.

(d) Use your model from part (c) to approximate the number of company bankruptcy filings in 2002. Round your answer to the nearest whole number.

(e) The number of company bankruptcy filings through August 16, 2002 was 129. Based on this, is your estimate from part (d) reasonable? Explain.

50. The percent of U.S. high school students in grades 9–12 who smoke is shown in the table for selected years. In the year column, 1 represents 1991, 3 represents 1993, and so on.

Year	Percent of Students
1	28
3	31
5	35
7	36
9	35
11	29
13	22

Source: National Center for Health Statistics.

(a) Use the ordered pairs (year, percent of students) to make a scatter diagram of the data.

(b) Would a linear or quadratic function better model the data?

(c) Should the coefficient a of x^2 in a quadratic model be positive or negative?

(d) Use the ordered pairs (1, 28), (7, 36), and (11, 29) to find a quadratic function that models the data. Round the values of a, b, and c in your model to the nearest tenth, as necessary.

(e) Use your model from part (d) to approximate the percent of high school students who smoked during 1995 and 2003 to the nearest percent. How well does the model approximate the actual data from the table?

51. In Example 6, we determined that the quadratic function defined by
$$y = 0.2455x^2 - 1.856x + 30.7105$$
modeled the rate (per 100,000) of higher-order multiple births, where x represents the number of years since 1970.

(a) Use this model to approximate the rate of higher-order births in 2002 to the nearest tenth.

(b) The actual rate of higher-order births in 2002 was 184.0. (*Source:* National Center for Health Statistics.) How does the approximation using the model compare to the actual rate for 2002?

52. Should the model from Exercise 51 be used to approximate the rate of higher-order multiple births in years after 2002? Explain.

TECHNOLOGY INSIGHTS (EXERCISES 53–54)

*Recall from **Section 3.3** that the x-value of the x-intercept of the graph of the line $y = mx + b$ is the solution of the linear equation $mx + b = 0$. In the same way, the x-values of the x-intercepts of the graph of the parabola $y = ax^2 + bx + c$ are the real solutions of the quadratic equation $ax^2 + bx + c = 0$.*

In Exercises 53–54, the calculator graphs show the x-values of the x-intercepts of the graph of the polynomial in the equation. Use the graphs to solve each equation.

53. $x^2 - x - 20 = 0$

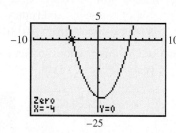

54. $x^2 + 9x + 14 = 0$

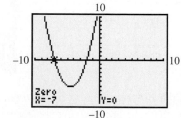

PREVIEW EXERCISES

*Solve each quadratic equation by completing the square. **See Section 9.1.***

55. $x^2 + 6x - 3 = 0$

56. $x^2 + 8x - 4 = 0$

57. $2x^2 - 12x = 5$

58. $3x^2 - 12x = 10$

9.6 More About Parabolas and Their Applications

OBJECTIVE 1 **Find the vertex of a vertical parabola.** When the equation of a parabola is given in the form $f(x) = ax^2 + bx + c$, we need to locate the vertex to sketch an accurate graph. There are two ways to do this:

1. Complete the square as shown in Examples 1 and 2, or

2. Use a formula derived by completing the square, as shown in Example 3.

EXAMPLE 1 **Completing the Square to Find the Vertex**

Find the vertex of the graph of $f(x) = x^2 - 4x + 5$.

We need to express $x^2 - 4x + 5$ in the form $(x - h)^2 + k$ by completing the square on $x^2 - 4x$, as in **Section 9.1.** The process is slightly different here because we want to keep $f(x)$ alone on one side of the equation. Instead of adding the appropriate number to each side, we *add and subtract* it on the right. This is equivalent to adding 0.

$$f(x) = x^2 - 4x + 5$$
$$= (x^2 - 4x \quad\quad) + 5 \qquad \text{Group the variable terms.}$$

$$\left[\frac{1}{2}(-4)\right]^2 = (-2)^2 = 4$$

$$= (x^2 - 4x + 4 - 4) + 5 \qquad \text{Add and subtract 4.}$$
$$= (x^2 - 4x + 4) - 4 + 5 \qquad \text{Bring } -4 \text{ outside the parentheses.}$$
$$f(x) = (x - 2)^2 + 1 \qquad\qquad \text{Factor; combine terms.}$$

The vertex of this parabola is $(2, 1)$.

✔ **Now Try Exercise 5.**

EXAMPLE 2 **Completing the Square to Find the Vertex When $a \neq 1$**

Find the vertex of the graph of $f(x) = -3x^2 + 6x - 1$.

We must complete the square on $-3x^2 + 6x$. Because the x^2-term has a coefficient other than 1, we factor that coefficient out of the first two terms.

$$f(x) = -3x^2 + 6x - 1$$
$$= -3(x^2 - 2x) - 1 \qquad\qquad \text{Factor out } -3.$$

$$\left[\frac{1}{2}(-2)\right]^2 = (-1)^2 = 1$$

$$= -3(x^2 - 2x + 1 - 1) - 1 \qquad \begin{array}{l}\text{Add and subtract 1} \\ \text{within the parentheses.}\end{array}$$

Bring -1 outside the parentheses; *be sure to multiply it by -3.*

$$= -3(x^2 - 2x + 1) + (-3)(-1) - 1 \qquad \text{Distributive property}$$
$$= -3(x^2 - 2x + 1) + 3 - 1 \qquad \text{This is a key step.}$$
$$f(x) = -3(x - 1)^2 + 2 \qquad\qquad \text{Factor; combine terms.}$$

The vertex is $(1, 2)$.

✔ **Now Try Exercise 7.**

To derive a formula for the vertex of the graph of the quadratic function defined by $f(x) = ax^2 + bx + c \ (a \neq 0)$, complete the square.

$$f(x) = ax^2 + bx + c \qquad \qquad \text{Standard form}$$

$$= a\left(x^2 + \frac{b}{a}x\right) + c \qquad \qquad \begin{array}{l}\text{Factor } a \text{ from the}\\ \text{first two terms.}\end{array}$$

$$\left[\frac{1}{2}\left(\frac{b}{a}\right)\right]^2 = \left(\frac{b}{2a}\right)^2 = \frac{b^2}{4a^2}$$

$$= a\left(x^2 + \frac{b}{a}x + \frac{b^2}{4a^2} - \frac{b^2}{4a^2}\right) + c \qquad \qquad \text{Add and subtract } \tfrac{b^2}{4a^2}.$$

$$= a\left(x^2 + \frac{b}{a}x + \frac{b^2}{4a^2}\right) + a\left(-\frac{b^2}{4a^2}\right) + c \qquad \text{Distributive property}$$

$$= a\left(x^2 + \frac{b}{a}x + \frac{b^2}{4a^2}\right) - \frac{b^2}{4a} + c$$

$$= a\left(x + \frac{b}{2a}\right)^2 + \frac{4ac - b^2}{4a} \qquad \qquad \text{Factor; combine terms.}$$

$$f(x) = a\left[x - \left(\frac{-b}{2a}\right)\right]^2 + \underbrace{\frac{4ac - b^2}{4a}}_{} \qquad \qquad f(x) = (x - h)^2 + k$$
$$\underbrace{\phantom{a\left[x - \left(\frac{-b}{2a}\right)\right]}}_{h} \qquad \underbrace{\phantom{\frac{4ac-b^2}{4a}}}_{k}$$

Thus, the vertex (h, k) can be expressed in terms of a, b, and c. However, it is not necessary to remember this expression for k, since it can be found by replacing x with $\frac{-b}{2a}$. Using function notation, if $y = f(x)$, then the y-value of the vertex is $f\left(\frac{-b}{2a}\right)$.

Vertex Formula

The graph of the quadratic function defined by $f(x) = ax^2 + bx + c \ (a \neq 0)$ has vertex

$$\left(\frac{-b}{2a}, f\left(\frac{-b}{2a}\right)\right),$$

and the axis of the parabola is the line

$$x = \frac{-b}{2a}.$$

EXAMPLE 3 Using the Formula to Find the Vertex

Use the vertex formula to find the vertex of the graph of $f(x) = x^2 - x - 6$.

For this function, $a = 1$, $b = -1$, and $c = -6$. The x-coordinate of the vertex of the parabola is given by

$$\frac{-b}{2a} = \frac{-(-1)}{2(1)} = \frac{1}{2}.$$

The y-coordinate is $f\left(\frac{-b}{2a}\right) = f\left(\frac{1}{2}\right)$.

$$f\left(\frac{1}{2}\right) = \left(\frac{1}{2}\right)^2 - \frac{1}{2} - 6 = \frac{1}{4} - \frac{1}{2} - 6 = -\frac{25}{4}$$

The vertex is $\left(\frac{1}{2}, -\frac{25}{4}\right)$.

✔ **Now Try Exercise 9.**

OBJECTIVE 2 Graph a quadratic function. We give a general approach for graphing any quadratic function here.

Graphing a Quadratic Function $y = f(x)$

Step 1 **Determine whether the graph opens up or down.** If $a > 0$, the parabola opens up; if $a < 0$, it opens down.

Step 2 **Find the vertex.** Use either the vertex formula or completing the square.

Step 3 **Find any intercepts.** To find the x-intercepts (if any), solve $f(x) = 0$. To find the y-intercept, evaluate $f(0)$.

Step 4 **Complete the graph.** Plot the points found so far. Find and plot additional points as needed, using symmetry about the axis.

EXAMPLE 4 **Using the Steps to Graph a Quadratic Function**

Graph the quadratic function defined by

$$f(x) = x^2 - x - 6.$$

Step 1 From the equation, $a = 1$, so the graph of the function opens up.

Step 2 The vertex, $\left(\frac{1}{2}, -\frac{25}{4}\right)$, was found in Example 3 by using the vertex formula.

Step 3 Find any intercepts. Since the vertex, $\left(\frac{1}{2}, -\frac{25}{4}\right)$, is in quadrant IV and the graph opens up, there will be two x-intercepts. To find them, let $f(x) = 0$ and solve.

$$\begin{aligned}
f(x) &= x^2 - x - 6 & \\
0 &= x^2 - x - 6 & \text{Let } f(x) = 0. \\
0 &= (x - 3)(x + 2) & \text{Factor.} \\
x - 3 = 0 \quad &\text{or} \quad x + 2 = 0 & \text{Zero-factor property} \\
x = 3 \quad &\text{or} \qquad x = -2 &
\end{aligned}$$

The x-intercepts are $(3, 0)$ and $(-2, 0)$.

Now find the y-intercept by evaluating $f(0)$.

$$\begin{aligned}
f(x) &= x^2 - x - 6 & \\
f(0) &= 0^2 - 0 - 6 & \text{Let } x = 0. \\
f(0) &= -6 &
\end{aligned}$$

The y-intercept is $(0, -6)$.

Step 4 Plot the points found so far and additional points as needed using symmetry about the axis, $x = \frac{1}{2}$. The graph is shown in Figure 12. The domain is $(-\infty, \infty)$, and the range is $\left[-\frac{25}{4}, \infty\right)$.

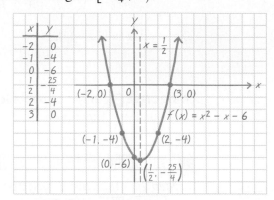

FIGURE 12

✔ **Now Try Exercise 23.**

OBJECTIVE 3 Use the discriminant to find the number of *x*-intercepts of a parabola with a vertical axis. Recall from **Section 9.2** that $b^2 - 4ac$ is called the *discriminant* of the quadratic equation $ax^2 + bx + c = 0$ and that we can use it to determine the number of real solutions of a quadratic equation.

In a similar way, we can use the discriminant of a quadratic *function* to determine the number of *x*-intercepts of its graph. See Figure 13. If the discriminant is positive, the parabola will have two *x*-intercepts. If the discriminant is 0, there will be only one *x*-intercept, and it will be the vertex of the parabola. If the discriminant is negative, the graph will have no *x*-intercepts.

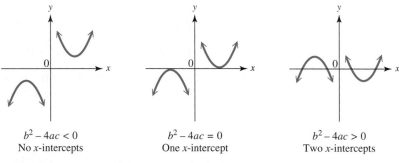

| $b^2 - 4ac < 0$ No *x*-intercepts | $b^2 - 4ac = 0$ One *x*-intercept | $b^2 - 4ac > 0$ Two *x*-intercepts |

FIGURE 13

EXAMPLE 5 **Using the Discriminant to Determine the Number of *x*-Intercepts**

Find the discriminant and use it to determine the number of *x*-intercepts of the graph of each quadratic function.

(a) $f(x) = 2x^2 + 3x - 5$

The discriminant is $b^2 - 4ac$. Here $a = 2$, $b = 3$, and $c = -5$, so

$$b^2 - 4ac = 9 - 4(2)(-5) = 49.$$

Since the discriminant is positive, the parabola has two *x*-intercepts.

(b) $f(x) = -3x^2 - 1$

Here, $a = -3$, $b = 0$, and $c = -1$. The discriminant is

$$b^2 - 4ac = 0 - 4(-3)(-1) = -12.$$

The discriminant is negative, so the graph has no x-intercepts.

(c) $f(x) = 9x^2 + 6x + 1$

Here, $a = 9$, $b = 6$, and $c = 1$. The discriminant is

$$b^2 - 4ac = 36 - 4(9)(1) = 0.$$

The parabola has only one x-intercept (its vertex).

✔ **Now Try Exercises 11 and 13.**

OBJECTIVE ④ Use quadratic functions to solve problems involving maximum or minimum value. The vertex of the graph of a quadratic function is either the highest or the lowest point on the parabola. The y-value of the vertex gives the maximum or minimum value of y, while the x-value tells where that maximum or minimum occurs.

▶ **PROBLEM-SOLVING HINT** In many applied problems we must find the greatest or least value of some quantity. When we can express that quantity in terms of a quadratic function, the value of k in the vertex (h, k) gives that optimum value.

EXAMPLE 6 Finding the Maximum Area of a Rectangular Region

A farmer has 120 ft of fencing. He wants to put a fence around a rectangular field next to a building. (See Figure 14.) Find the maximum area he can enclose, and the dimensions of the field when the area is maximized.

FIGURE 14

Let $x =$ the width of the field. Then

$x + x + \text{length} = 120$	Sum of the sides is 120 ft.
$2x + \text{length} = 120$	Combine terms.
$\text{length} = 120 - 2x.$	Subtract $2x$.

The area is given by the product of the width and length, so

$$A(x) = x(120 - 2x)$$
$$= 120x - 2x^2.$$

To determine the maximum area, find the vertex of the parabola given by $A(x) = 120x - 2x^2$ using the vertex formula. Writing the equation in standard form as $A(x) = -2x^2 + 120x$ gives $a = -2$, $b = 120$, and $c = 0$, so

$$h = \frac{-b}{2a} = \frac{-120}{2(-2)} = \frac{-120}{-4} = 30;$$

$$A(30) = -2(30)^2 + 120(30) = -2(900) + 3600 = 1800.$$

The graph is a parabola that opens down, and its vertex is (30, 1800). Thus, the maximum area will be 1800 ft². This area will occur if x, the width of the field, is 30 ft and the length is $120 - 2(30) = 60$ ft.

✔ **Now Try Exercise 35.**

▶ **CAUTION** *Be careful when interpreting the meanings of the coordinates of the vertex.* The first coordinate, x, gives the value for which the *function value* is a maximum or a minimum. Be sure to read the problem carefully to determine whether you are asked to find the value of the independent variable, the function value, or both.

EXAMPLE 7 Finding the Maximum Height Attained by a Projectile

If air resistance is neglected, a projectile on Earth shot straight upward with an initial velocity of 40 m per sec will be at a height s in meters given by

$$s(t) = -4.9t^2 + 40t,$$

where t is the number of seconds elapsed after projection. After how many seconds will it reach its maximum height, and what is this maximum height?

For this function, $a = -4.9$, $b = 40$, and $c = 0$. Use the vertex formula.

$$h = \frac{-b}{2a} = \frac{-40}{2(-4.9)} \approx 4.1 \qquad \text{Use a calculator.}$$

This indicates that the maximum height is attained at 4.1 sec. To find this maximum height, calculate $s(4.1)$.

$$s(4.1) = -4.9(4.1)^2 + 40(4.1)$$
$$\approx 81.6 \qquad \text{Use a calculator.}$$

The projectile will attain a maximum height of approximately 81.6 m.

✔ **Now Try Exercise 37.**

OBJECTIVE 5 Graph parabolas with horizontal axes. If x and y are interchanged in the equation $y = ax^2 + bx + c$, the equation becomes $x = ay^2 + by + c$. Because of the interchange of the roles of x and y, these parabolas are horizontal (with horizontal lines as axes), compared with the vertical ones graphed previously.

> ### Graph of a Parabola with Horizontal Axis
>
> The graph of
> $$x = ay^2 + by + c \quad \text{or} \quad x = a(y - k)^2 + h$$
> is a parabola with vertex (h, k) and the horizontal line $y = k$ as axis. The graph opens to the right if $a > 0$ and to the left if $a < 0$.

EXAMPLE 8 Graphing a Parabola with Horizontal Axis

Graph $x = (y - 2)^2 - 3$. Give the vertex, axis, domain, and range.

This graph has its vertex at $(-3, 2)$, since the roles of x and y are interchanged. It opens to the right because $a = 1 > 0$, and has the same shape as $y = x^2$. Plotting a few additional points gives the graph shown in Figure 15. Note that the graph is symmetric about its axis, $y = 2$. The domain is $[-3, \infty)$, and the range is $(-\infty, \infty)$.

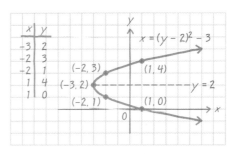

x	y
-3	2
-2	3
-2	1
1	4
1	0

FIGURE 15

✔ **Now Try Exercise 27.**

When a quadratic equation is given in the form $x = ay^2 + by + c$, completing the square on y allows us to find the vertex.

EXAMPLE 9 Completing the Square to Graph a Parabola with Horizontal Axis

Graph $x = -2y^2 + 4y - 3$. Give the vertex, axis, domain, and range of the relation.

$$x = -2y^2 + 4y - 3$$
$$= -2(y^2 - 2y) - 3 \qquad \text{Factor out } -2.$$
$$= -2(y^2 - 2y + 1 - 1) - 3 \qquad \text{Complete the square within the parentheses; add and subtract 1.}$$
$$= -2(y^2 - 2y + 1) + (-2)(-1) - 3 \qquad \text{Distributive property}$$

Be careful here.

$$x = -2(y - 1)^2 - 1 \qquad \text{Factor; simplify.}$$

Because of the negative coefficient (-2) in $x = -2(y - 1)^2 - 1$, the graph opens to the left (the negative x-direction). The graph is narrower than the graph of $y = x^2$ because $|-2| > 1$.

As shown in Figure 16, the vertex is $(-1, 1)$ and the axis is $y = 1$. The domain is $(-\infty, -1]$, and the range is $(-\infty, \infty)$.

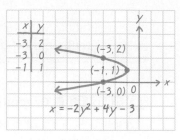

FIGURE 16

✔ **Now Try Exercise 31.**

▶ **CAUTION** Only quadratic equations solved for y (whose graphs are vertical parabolas) are examples of functions. ***The horizontal parabolas in Examples 8 and 9 are not graphs of functions,*** because they do not satisfy the vertical line test.

In summary, the graphs of parabolas studied in **Sections 9.5 and 9.6** fall into the following categories.

GRAPHS OF PARABOLAS

Equation	Graph
$y = ax^2 + bx + c$ or $y = a(x - h)^2 + k$	![graphs] (h, k) $a > 0$ (h, k) $a < 0$ These graphs represent functions.
$x = ay^2 + by + c$ or $x = a(y - k)^2 + h$	![graphs] (h, k) $a > 0$ (h, k) $a < 0$ These graphs are not graphs of functions.

9.6 EXERCISES

Complete solution available on Video Lectures on CD/DVD

Now Try Exercise

Concept Check   In Exercises 1–4, answer each question.

1. How can you determine just by looking at the equation of a parabola whether it has a vertical or a horizontal axis?

2. Why can't the graph of a quadratic function be a parabola with a horizontal axis?

3. How can you determine the number of x-intercepts of the graph of a quadratic function without graphing the function?

4. If the vertex of the graph of a quadratic function is $(1, -3)$, and the graph opens down, how many x-intercepts does the graph have?

Find the vertex of each parabola. See Examples 1–3.

5. $f(x) = x^2 + 8x + 10$

6. $f(x) = x^2 + 10x + 23$

7. $f(x) = -2x^2 + 4x - 5$

8. $f(x) = -3x^2 + 12x - 8$

9. $f(x) = x^2 + x - 7$

10. $f(x) = x^2 - x + 5$

Find the vertex of each parabola. For each equation, decide whether the graph opens up, down, to the left, or to the right, and whether it is wider, narrower, or the same shape as the graph of $y = x^2$. If it is a parabola with vertical axis, find the discriminant and use it to determine the number of x-intercepts. See Examples 1–3, 5, 8, and 9.

11. $f(x) = 2x^2 + 4x + 5$

12. $f(x) = 3x^2 - 6x + 4$

13. $f(x) = -x^2 + 5x + 3$

14. $x = -y^2 + 7y + 2$

15. $x = \dfrac{1}{3}y^2 + 6y + 24$

16. $x = \dfrac{1}{2}y^2 + 10y - 5$

Concept Check   Use the concepts of this section to match each equation in Exercises 17–22 with its graph in A–F.

17. $y = 2x^2 + 4x - 3$

18. $y = -x^2 + 3x + 5$

19. $y = -\dfrac{1}{2}x^2 - x + 1$

20. $x = y^2 + 6y + 3$

21. $x = -y^2 - 2y + 4$

22. $x = 3y^2 + 6y + 5$

A.

B.
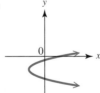

C.
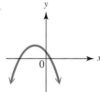

D.
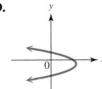

E.
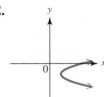

F.
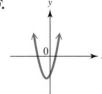

Graph each parabola. (Use the results of Exercises 5–8 to help graph the parabolas in Exercises 23–26.) Give the vertex, axis, domain, and range. See Examples 4, 8, and 9.

23. $f(x) = x^2 + 8x + 10$

24. $f(x) = x^2 + 10x + 23$

25. $f(x) = -2x^2 + 4x - 5$

26. $f(x) = -3x^2 + 12x - 8$

27. $x = (y + 2)^2 + 1$

28. $x = (y + 3)^2 - 2$

29. $x = -\dfrac{1}{5}y^2 + 2y - 4$

30. $x = -\dfrac{1}{2}y^2 - 4y - 6$

31. $x = 3y^2 + 12y + 5$

32. $x = 4y^2 + 16y + 11$

Solve each problem. See Examples 6 and 7.

33. Find the pair of numbers whose sum is 40 and whose product is a maximum. (*Hint:* Let x and $40 - x$ represent the two numbers.)

34. Find the pair of numbers whose sum is 60 and whose product is a maximum.

35. Morgan's Department Store wants to construct a rectangular parking lot on land bordered on one side by a highway. It has 280 ft of fencing that is to be used to fence off the other three sides. What should be the dimensions of the lot if the enclosed area is to be a maximum? What is the maximum area?

36. Emmylou has 100 ft of fencing material to enclose a rectangular exercise run for her dog. One side of the run will border her house, so she will only need to fence three sides. What dimensions will give the enclosure the maximum area? What is the maximum area?

37. If an object on Earth is projected upward with an initial velocity of 32 ft per sec, then its height after t seconds is given by

$$s(t) = -16t^2 + 32t.$$

Find the maximum height attained by the object and the number of seconds it takes to hit the ground.

38. A projectile on Earth is fired straight upward so that its distance (in feet) above the ground t seconds after firing is given by

$$s(t) = -16t^2 + 400t.$$

Find the maximum height it reaches and the number of seconds it takes to reach that height.

39. After experimentation, two Pacific Institute physics students find that when a bottle of California wine is shaken several times, held upright, and uncorked, its cork travels according to the function defined by

$$s(t) = -16t^2 + 64t + 3,$$

where s is its height in feet above the ground t seconds after being released. After how many seconds will it reach its maximum height? What is the maximum height?

40. Professor Bernstein has found that the number of students attending his intermediate algebra class is approximated by

$$S(x) = -x^2 + 20x + 80,$$

where x is the number of hours that the Campus Center is open daily. Find the number of hours that the center should be open so that the number of students attending class is a maximum. What is this maximum number of students?

41. The percent of births in the U.S. to teenage mothers in the years 1990–2002 can be modeled by the quadratic function defined by

$$f(x) = -0.0334x^2 + 0.2351x + 12.79,$$

where $x = 0$ represents 1990, $x = 1$ represents 1991, and so on. (*Source:* U.S. National Center for Health Statistics.)

(a) Since the coefficient of x^2 in the model is negative, the graph of this quadratic function is a parabola that opens down. Will the y-value of the vertex of this graph be a maximum or a minimum?

(b) In what year during this period was the percent of births in the U.S. to teenage mothers a maximum? (Round down to the nearest year.) Use the actual y-value of the vertex, to the nearest tenth, to find this percent.

42. The number of tickets sold (in millions) to the top 50 rock concerts in the years 1998–2002 is modeled by the quadratic function defined by

$$f(x) = -0.386x^2 + 1.28x + 11.3,$$

where $x = 0$ represents 1998, $x = 1$ represents 1999, and so on. (*Source:* Pollstar Online.)

(a) Since the coefficient of x^2 in the model is negative, the graph of this quadratic function is a parabola that opens down. Will the y-value of the vertex of this graph be a maximum or a minimum?

(b) In what year was the maximum number of tickets sold? (Round down to the nearest year.) Use the actual x-value of the vertex, to the nearest tenth, to find this number.

43. The graph shows how Social Security trust fund assets are expected to change as the number of retirees receiving benefits increases.

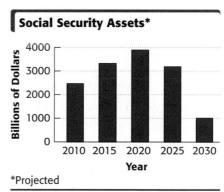

Social Security Assets*

*Projected

Source: Social Security Administration.

The graph suggests that a quadratic function would be a good fit to the data. The data are approximated by the function defined by

$$f(x) = -20.57x^2 + 758.9x - 3140.$$

In the model, $x = 10$ represents 2010, $x = 15$ represents 2015, and so on, and $f(x)$ is in billions of dollars.

(a) *Concept Check* How could you have predicted this quadratic model would have a negative coefficient for x^2, based only on the graph shown?

(b) Algebraically determine the vertex of the graph, with coordinates to four significant digits.

(c) Interpret the answer to part (b) as it applies to this application.

44. The graph shows the performance of investment portfolios with different mixtures of U.S. and foreign investments over a 25-yr period.

(a) Is this the graph of a function? Explain.

(b) What investment mixture shown on the graph appears to represent the vertex? What relative amount of risk does this point represent? What return on investment does it provide?

(c) Which point on the graph represents the riskiest investment mixture? What return on investment does it provide?

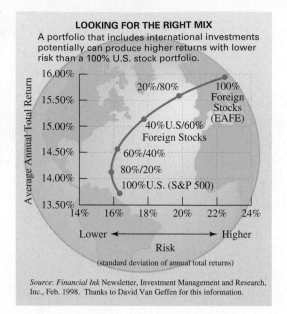

LOOKING FOR THE RIGHT MIX

A portfolio that includes international investments potentially can produce higher returns with lower risk than a 100% U.S. stock portfolio.

Source: *Financial Ink* Newsletter, Investment Management and Research, Inc., Feb. 1998. Thanks to David Van Geffen for this information.

45. A charter flight charges a fare of $200 per person, plus $4 per person for each unsold seat on the plane. If the plane holds 100 passengers and if x represents the number of unsold seats, find the following.

(a) A function defined by $R(x)$ that describes the total revenue received for the flight (*Hint:* Multiply the number of people flying, $100 - x$, by the price per ticket, $200 + 4x$.)

(b) The graph of the function from part (a)

(c) The number of unsold seats that will produce the maximum revenue

(d) The maximum revenue

46. For a trip to a resort, a charter bus company charges a fare of $48 per person, plus $2 per person for each unsold seat on the bus. If the bus has 42 seats and x represents the number of unsold seats, find the following.

(a) A function defined by $R(x)$ that describes the total revenue from the trip (*Hint:* Multiply the total number riding, $42 - x$, by the price per ticket, $48 + 2x$.)

(b) The graph of the function from part (a)

(c) The number of unsold seats that produces the maximum revenue

(d) The maximum revenue

TECHNOLOGY INSIGHTS (EXERCISES 47–50)

Graphing calculators are capable of determining the coordinates of "peaks" and "valleys" of graphs. In the case of quadratic functions, these peaks and valleys are the vertices and are called maximum and minimum points. For example, the vertex of the graph of $f(x) = -x^2 - 6x - 13$ is $(-3, -4)$, as indicated in the display at the bottom of the screen. In this case, the vertex is a maximum point.

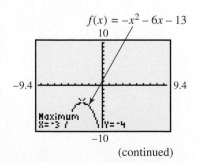

(continued)

In Exercises 47–50, match the function with its calculator graph in A–D by determining the vertex and using the display at the bottom of the screen.

47. $f(x) = x^2 - 8x + 18$

48. $f(x) = x^2 + 8x + 18$

49. $f(x) = x^2 - 8x + 14$

50. $f(x) = x^2 + 8x + 14$

A.

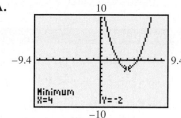

B.

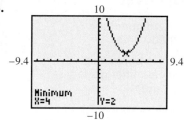

C.

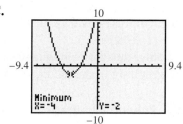

D.

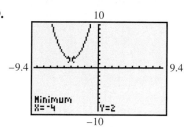

In the following exercise, the distance formula is used to develop the equation of a parabola.

51. A parabola can be defined as the set of all points in a plane equally distant from a given point and a given line not containing the point. (The point is called the **focus** and the line is called the **directrix.**) See the figure.

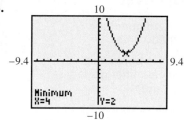

  (a) Suppose (x, y) is to be on the parabola. Suppose the directrix has equation $x = -p$. Find the distance between (x, y) and the directrix. (The distance from a point to a line is the length of the perpendicular from the point to the line.)

  (b) If $x = -p$ is the equation of the directrix, why should the focus have coordinates $(p, 0)$? (*Hint:* See the figure.)

  (c) Find an expression for the distance from (x, y) to $(p, 0)$.

  (d) Find an equation for the parabola in the figure. (*Hint:* Use the results of parts (a) and (c) and the fact that (x, y) is equally distant from the focus and the directrix.)

52. Use the equation derived in Exercise 51 to find an equation for a parabola with focus $(3, 0)$ and directrix with equation $x = -3$.

PREVIEW EXERCISES

Graph each interval on a number line. See **Section 2.5.**

53. $[1, 5]$

54. $(-6, 1]$

55. $(-\infty, 1] \cup [5, \infty)$

Solve each inequality. See **Section 2.5.**

56. $3 - x \le 5$

57. $-2x + 1 < 4$

58. $-\dfrac{1}{2}x - 3 > 5$

9.7 Quadratic and Rational Inequalities

Now we combine the methods of solving linear inequalities with the methods of solving quadratic equations to solve *quadratic inequalities.*

Quadratic Inequality

A **quadratic inequality** can be written in the form
$$ax^2 + bx + c < 0 \qquad \text{or} \qquad ax^2 + bx + c > 0,$$
where a, b, and c are real numbers, with $a \neq 0$.

As before, the symbols $<$ and $>$ may be replaced with $\leq$ and $\geq$.

OBJECTIVE 1 Solve quadratic inequalities. One method for solving a quadratic inequality is by graphing the related quadratic function.

EXAMPLE 1 Solving Quadratic Inequalities by Graphing

Solve each inequality.

(a) $x^2 - x - 12 > 0$

To solve the inequality, we graph the related quadratic function defined by $f(x) = x^2 - x - 12$. We are particularly interested in the *x*-intercepts, which are found as in **Section 9.6** by letting $f(x) = 0$ and solving the quadratic equation

$$x^2 - x - 12 = 0.$$
$$(x - 4)(x + 3) = 0 \qquad \text{Factor.}$$
$$x - 4 = 0 \quad \text{or} \quad x + 3 = 0 \qquad \text{Zero-factor property}$$
$$x = 4 \quad \text{or} \qquad x = -3$$

Thus, the *x*-intercepts are $(4, 0)$, and $(-3, 0)$. The graph, which opens up since the coefficient of x^2 is positive, is shown in Figure 17(a). Notice from this graph that *x*-values less than -3 or greater than 4 result in *y*-values *greater than* 0. Thus, the solution set of $x^2 - x - 12 > 0$, written in interval notation, is $(-\infty, -3) \cup (4, \infty)$.

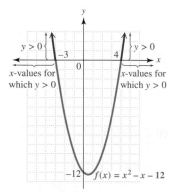

The graph is *above* the *x*-axis for
$(-\infty, -3) \cup (4, \infty)$.

(a)

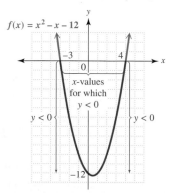

The graph is *below* the *x*-axis for
$(-3, 4)$.

(b)

FIGURE 17

(b) $x^2 - x - 12 < 0$

We want values of y that are *less than* 0. Referring to Figure 17(b), we notice from the graph that x-values between -3 and 4 result in y-values less than 0. Thus, the solution set of $x^2 - x - 12 < 0$, written in interval notation, is $(-3, 4)$.

✔ **Now Try Exercise 1.**

▶ **NOTE** If the inequalities in Example 1 had used $\geq$ and $\leq$, the solution sets would have included the x-values of the intercepts and been written in interval notation as $(-\infty, -3] \cup [4, \infty)$ for Example 1(a) and $[-3, 4]$ for Example 1(b).

In Example 1, we used graphing to divide the x-axis into intervals. Then using the graphs in Figure 17, we determined which x-values resulted in y-values that were either greater than or less than 0. Another method for solving a quadratic inequality uses these basic ideas without actually graphing the related quadratic function.

EXAMPLE 2 Solving a Quadratic Inequality Using Test Numbers

Solve $x^2 - x - 12 > 0$.

Solve the quadratic equation $x^2 - x - 12 = 0$ by factoring, as in Example 1(a).

$$(x - 4)(x + 3) = 0$$
$$x - 4 = 0 \quad \text{or} \quad x + 3 = 0$$
$$x = 4 \quad \text{or} \quad x = -3$$

The numbers 4 and -3 divide a number line into the three intervals shown in Figure 18. *Be careful to put the lesser number on the left.* (Notice the similarity between Figure 18 and the x-axis with intercepts $(-3, 0)$ and $(4, 0)$ in Figure 17(a).)

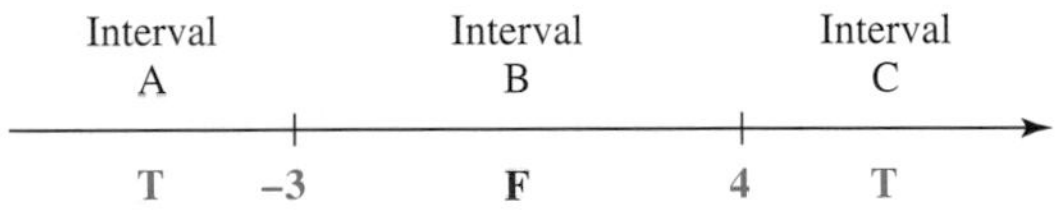

FIGURE 18

The numbers 4 and -3 are the only numbers that make the expression $x^2 - x - 12$ equal to 0. All other numbers make the expression either positive or negative. The sign of the expression can change from positive to negative or from negative to positive only at a number that makes it 0. Therefore, if one number in an interval satisfies the inequality, then all the numbers in that interval will satisfy the inequality.

To see if the numbers in Interval A satisfy the inequality, choose any number from Interval A in Figure 18 (that is, any number less than -3). Substitute this test number for x in the original inequality $x^2 - x - 12 > 0$. If the result is *true,* then all numbers in Interval A satisfy the inequality.

Try -5 from Interval A. Substitute -5 for x.

$$x^2 - x - 12 > 0 \qquad \text{Original inequality}$$
$$(-5)^2 - (-5) - 12 > 0 \qquad ?$$
$$25 + 5 - 12 > 0 \qquad ?$$
$$18 > 0 \qquad \text{True}$$

Use parentheses to avoid sign errors.

Because -5 from Interval A satisfies the inequality, all numbers from Interval A are solutions.

Now try 0 from Interval B. If $x = 0$, then

$$0^2 - 0 - 12 > 0 \qquad ?$$

$$-12 > 0. \qquad \text{False}$$

The numbers in Interval B are *not* solutions. Verify that the test number 5 satisfies the inequality, so the numbers in Interval C are also solutions.

Based on these results (shown by the colored letters in Figure 18), the solution set includes the numbers in Intervals A and C, as shown on the graph in Figure 19. The solution set is written in interval notation as

$$(-\infty, -3) \cup (4, \infty).$$

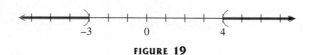

FIGURE 19

This agrees with the solution set we found by graphing the related quadratic function in Example 1(a).

✔ **Now Try Exercise 11.**

In summary, follow these steps to solve a quadratic inequality.

Solving a Quadratic Inequality

Step 1   **Write the inequality as an equation and solve it.**

Step 2   **Use the solutions from Step 1 to determine intervals.** Graph the numbers found in Step 1 on a number line. These numbers divide the number line into intervals.

Step 3   **Find the intervals that satisfy the inequality.** Substitute a test number from each interval into the original inequality to determine the intervals that satisfy the inequality. All numbers in those intervals are in the solution set. A graph of the solution set will usually look like one of these. (Square brackets might be used instead of parentheses.)

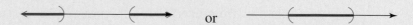

Step 4   **Consider the endpoints separately.** The numbers from Step 1 are included in the solution set if the inequality symbol is $\leq$ or $\geq$; they are not included if it is $<$ or $>$.

Special cases of quadratic inequalities may occur, as in the next example.

EXAMPLE 3   Solving Special Cases

Solve each inequality.

(a) $(2x - 3)^2 > -1$

Because $(2x - 3)^2$ is never negative, it is always greater than -1. Thus, the solution for $(2x - 3)^2 > -1$ is the set of all real numbers, $(-\infty, \infty)$.

(b) $(2x - 3)^2 < -1$

Using the same reasoning as in part (a), there is no solution for this inequality. The solution set is $\emptyset$.

✔ **Now Try Exercises 25 and 27.**

OBJECTIVE 2 **Solve polynomial inequalities of degree 3 or greater.**   Higher-degree polynomial inequalities that can be factored are solved in the same way as quadratic inequalities.

EXAMPLE 4   Solving a Third-Degree Polynomial Inequality

Solve $(x - 1)(x + 2)(x - 4) \leq 0$.

This is a *cubic* (third-degree) inequality rather than a quadratic inequality, but it can be solved using the method shown in the box by extending the zero-factor property to more than two factors. Begin by setting the factored polynomial *equal* to 0 and solving the equation. (Step 1)

$$(x - 1)(x + 2)(x - 4) = 0$$

$$x - 1 = 0 \quad \text{or} \quad x + 2 = 0 \quad \text{or} \quad x - 4 = 0$$

$$x = 1 \quad \text{or} \quad x = -2 \quad \text{or} \quad x = 4$$

Locate the numbers -2, 1, and 4 on a number line, as in Figure 20, to determine the Intervals A, B, C, and D. (Step 2)

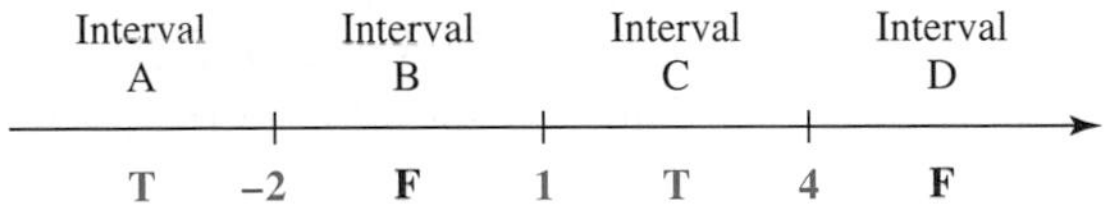

Interval A		Interval B		Interval C		Interval D
T	-2	F	1	T	4	F

FIGURE 20

Substitute a test number from each interval in the *original* inequality to determine which intervals satisfy the inequality. (Step 3) Use a table to organize this information.

Interval	Test Number	Test of Inequality	True or False?
A	-3	$-28 \leq 0$	**T**
B	0	$8 \leq 0$	**F**
C	2	$-8 \leq 0$	**T**
D	5	$28 \leq 0$	**F**

Verify the information given in the table and graphed in Figure 21 on the next page.

The numbers in Intervals A and C are in the solution set, which is written in interval notation as

$$(-\infty, -2] \cup [1, 4].$$

The three endpoints are included since the inequality symbol is $\leq$. (Step 4)

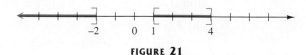

FIGURE 21

✔ **Now Try Exercise 29.**

OBJECTIVE 3 Solve rational inequalities. Inequalities that involve rational expressions, called **rational inequalities,** are solved similarly using the following steps.

Solving a Rational Inequality

Step 1 **Write the inequality** so that 0 is on one side and there is a single fraction on the other side.

Step 2 **Determine the numbers that make the numerator or denominator equal to 0.**

Step 3 **Divide a number line into intervals.** Use the numbers from Step 2.

Step 4 **Find the intervals that satisfy the inequality.** Test a number from each interval by substituting it into the *original* inequality.

Step 5 **Consider the endpoints separately.** Exclude any values that make the denominator 0.

▶ **CAUTION** As indicated in Step 5, any number that makes the denominator 0 *must* be excluded from the solution set.

EXAMPLE 5 Solving a Rational Inequality

Solve $\dfrac{-1}{x-3} > 1$.

Write the inequality so that 0 is on one side. (Step 1)

$$\frac{-1}{x-3} - 1 > 0 \qquad \text{Subtract 1.}$$

$$\frac{-1}{x-3} - \frac{x-3}{x-3} > 0 \qquad \text{Use } x-3 \text{ as the common denominator.}$$

Be careful with signs. $\dfrac{-1-x+3}{x-3} > 0 \qquad$ Write the left side as a single fraction.

$$\frac{-x+2}{x-3} > 0 \qquad \text{Combine terms.}$$

The sign of the rational expression $\frac{-x+2}{x-3}$ will change from positive to negative or negative to positive only at those numbers that make the numerator or denominator 0. The number 2 makes the numerator 0, and 3 makes the denominator 0. (Step 2) These two numbers, 2 and 3, divide a number line into three intervals. See Figure 22. (Step 3)

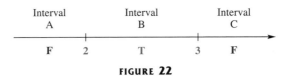

FIGURE 22

Testing a number from each interval in the *original* inequality, $\frac{-1}{x-3} > 1$, gives the results shown in the table. (Step 4)

Interval	Test Number	Test of Inequality	True or False?
A	0	$\frac{1}{3} > 1$	**F**
B	2.5	$2 > 1$	**T**
C	4	$-1 > 1$	**F**

The solution set is the interval $(2, 3)$. This interval does not include 3 since it would make the denominator of the original equality 0; 2 is not included either since the inequality symbol is $>$. (Step 5) A graph of the solution set is given in Figure 23.

FIGURE 23

☑ **Now Try Exercise 37.**

EXAMPLE 6   Solving a Rational Inequality

Solve $\dfrac{x-2}{x+2} \le 2$.

Write the inequality so that 0 is on one side. (Step 1)

$$\frac{x-2}{x+2} - 2 \le 0 \qquad \text{Subtract 2.}$$

$$\frac{x-2}{x+2} - \frac{2(x+2)}{x+2} \le 0 \qquad \text{Use } x+2 \text{ as the common denominator.}$$

$$\frac{x-2-2x-4}{x+2} \le 0 \qquad \text{Write as a single fraction.}$$

Be careful with signs.

$$\frac{-x-6}{x+2} \le 0 \qquad \text{Combine terms.}$$

The number -6 makes the numerator 0, and -2 makes the denominator 0. (Step 2) These two numbers determine three intervals. (Step 3) Test one number from each interval (Step 4) to see that the solution set is

$$(-\infty, -6] \cup (-2, \infty).$$

The number -6 satisfies the original inequality, but -2 cannot be used as a solution since it makes the denominator 0. (Step 5) A graph of the solution set is shown in Figure 24.

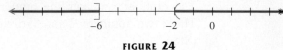

FIGURE 24

✔ **Now Try Exercise 43.**

9.7 EXERCISES

🌐 *Complete solution available on Video Lectures on CD/DVD*

▨ *Now Try Exercise*

In Example 1, we determined the solution sets of the quadratic inequalities $x^2 - x - 12 > 0$ and $x^2 - x - 12 < 0$ by graphing $f(x) = x^2 - x - 12$. The x-intercepts of this graph indicated the solutions of the equation $x^2 - x - 12 = 0$. The x-values of the points on the graph that were above the x-axis formed the solution set of $x^2 - x - 12 > 0$, and the x-values of the points on the graph that were below the x-axis formed the solution set of $x^2 - x - 12 < 0$.

In Exercises 1–4, the graph of a quadratic function f is given. Use the graph to find the solution set of each equation or inequality. See Example 1.

🌐 **1. (a)** $x^2 - 4x + 3 = 0$
  (b) $x^2 - 4x + 3 > 0$
  (c) $x^2 - 4x + 3 < 0$

2. (a) $3x^2 + 10x - 8 = 0$
  (b) $3x^2 + 10x - 8 \geq 0$
  (c) $3x^2 + 10x - 8 < 0$

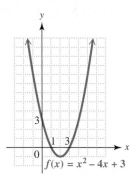

3. (a) $-2x^2 - x + 15 = 0$
  (b) $-2x^2 - x + 15 \geq 0$
  (c) $-2x^2 - x + 15 \leq 0$

4. (a) $-x^2 + 3x + 10 = 0$
  (b) $-x^2 + 3x + 10 \geq 0$
  (c) $-x^2 + 3x + 10 \leq 0$

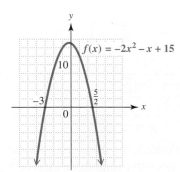

▨ **5.** Explain how to determine whether to include or exclude endpoints when solving a quadratic or greater-degree inequality.

6. *Concept Check* The solution set of the inequality $x^2 + x - 12 < 0$ is the interval $(-4, 3)$. Without actually performing any work, give the solution set of the inequality $x^2 + x - 12 \geq 0$.

Solve each inequality and graph the solution set. See Example 2. (Hint: In Exercises 23 and 24, use the quadratic formula.)

7. $(x + 1)(x - 5) > 0$ **8.** $(x + 6)(x - 2) > 0$ **9.** $(x + 4)(x - 6) < 0$

10. $(x + 4)(x - 8) < 0$ **11.** $x^2 - 4x + 3 \geq 0$ **12.** $x^2 - 3x - 10 \geq 0$

13. $10x^2 + 9x \geq 9$ **14.** $3x^2 + 10x \geq 8$ **15.** $4x^2 - 9 \leq 0$

16. $9x^2 - 25 \leq 0$ **17.** $6x^2 + x \geq 1$ **18.** $4x^2 + 7x \geq -3$

19. $z^2 - 4z \geq 0$ **20.** $x^2 + 2x < 0$ **21.** $3k^2 - 5k \leq 0$

22. $2z^2 + 3z > 0$ **23.** $x^2 - 6x + 6 \geq 0$ **24.** $3k^2 - 6k + 2 \leq 0$

Solve each inequality. See Example 3.

25. $(4 - 3x)^2 \geq -2$ **26.** $(6x + 7)^2 \geq -1$

27. $(3x + 5)^2 \leq -4$ **28.** $(8x + 5)^2 \leq -5$

Solve each inequality and graph the solution set. See Example 4.

29. $(x - 1)(x - 2)(x - 4) < 0$ **30.** $(2x + 1)(3x - 2)(4x + 7) < 0$

31. $(x - 4)(2x + 3)(3x - 1) \geq 0$ **32.** $(x + 2)(4x - 3)(2x + 7) \geq 0$

Solve each inequality and graph the solution set. See Examples 5 and 6.

33. $\dfrac{x - 1}{x - 4} > 0$ **34.** $\dfrac{x + 1}{x - 5} > 0$ **35.** $\dfrac{2x + 3}{x - 5} \leq 0$ **36.** $\dfrac{3x + 7}{x - 3} \leq 0$

37. $\dfrac{8}{x - 2} \geq 2$ **38.** $\dfrac{20}{x - 1} \geq 1$ **39.** $\dfrac{3}{2x - 1} < 2$ **40.** $\dfrac{6}{x - 1} < 1$

41. $\dfrac{x - 3}{x + 2} \geq 2$ **42.** $\dfrac{m + 4}{m + 5} \geq 2$ **43.** $\dfrac{x - 8}{x - 4} < 3$ **44.** $\dfrac{2t - 3}{t + 1} > 4$

45. $\dfrac{4k}{2k - 1} < k$ **46.** $\dfrac{r}{r + 2} < 2r$ **47.** $\dfrac{2x - 3}{x^2 + 1} \geq 0$ **48.** $\dfrac{9x - 8}{4x^2 + 25} < 0$

49. $\dfrac{(3x - 5)^2}{x + 2} > 0$ **50.** $\dfrac{(5x - 3)^2}{2x + 1} \leq 0$

PREVIEW EXERCISES

*Give the domain and the range of each function. See **Section 3.5**.*

51. $\left\{ \left(-3, \dfrac{1}{8}\right), \left(-2, \dfrac{1}{4}\right), \left(-1, \dfrac{1}{2}\right), (0, 1), (1, 2), (2, 4), (3, 8) \right\}$

52. $f(x) = x^2$

*Decide whether each graph is that of a function. See **Section 3.5**.*

53.

54.

Chapter **9** | **Group Activity**

FINDING THE PATH OF A COMET

Objective   Find and graph an equation of a parabola with a given focus.

The orbit that a comet takes as it approaches the sun depends on its velocity, as well as other factors. If the velocity of a comet equals escape velocity—that is, it is going just fast enough to get away from the sun—then its orbit will be parabolic. (Other possible orbits are *hyperbolic* or *elliptical*. These figures are discussed in **Chapter 11.**) The sun is at the focus of the parabola. The vertex is the point where the comet is closest to the sun.

A. Refer to **Section 9.6,** Exercise 51, and make a sketch of the comet and the sun.

1. Place the vertex of the parabola at the origin, with the focus on the *y*-axis and a horizontal directrix.

2. Assume that the comet is 0.75 astronomical unit* from the sun at its closest point.

B. What are the coordinates of the focus?

C. What is the equation of the directrix?

D. Using the information from **Section 9.6,** Exercise 51 as a guide, find an equation of the parabola.

E. Graph the parabola (as a vertical parabola). Include the focus and directrix on your graph.

---

*One astronomical unit (AU) is the average distance between Earth and the sun.

Chapter **9** SUMMARY *View the Interactive Summary on the Pass the Test CD.*

KEY TERMS

9.1 quadratic equation
9.2 quadratic formula
discriminant

9.3 quadratic in form
9.5 parabola
vertex

axis
quadratic function

9.7 quadratic inequality
rational inequality

TEST YOUR WORD POWER

See how well you have learned the vocabulary in this chapter. Answers, with examples, follow the Quick Review.

1. The **quadratic formula** is
   A. a formula to find the number of solutions of a quadratic equation
   B. a formula to find the type of solutions of a quadratic equation
   C. the standard form of a quadratic equation
   D. a general formula for solving any quadratic equation.

2. A **quadratic function** is a function that can be written in the form
   A. $f(x) = mx + b$ for real numbers m and b
   B. $f(x) = \frac{P(x)}{Q(x)}$, where $Q(x) \neq 0$
   C. $f(x) = ax^2 + bx + c$ for real numbers a, b, and c ($a \neq 0$)
   D. $f(x) = \sqrt{x}$ for $x \geq 0$.

3. A **parabola** is the graph of
   A. any equation in two variables
   B. a linear equation
   C. an equation of degree 3
   D. a quadratic equation in two variables.

4. The **vertex** of a parabola is
   A. the point where the graph intersects the y-axis
   B. the point where the graph intersects the x-axis
   C. the lowest point on a parabola that opens up or the highest point on a parabola that opens down
   D. the origin.

5. The **axis** of a parabola is
   A. either the x-axis or the y-axis

   B. the vertical line (of a vertical parabola) or the horizontal line (of a horizontal parabola) through the vertex
   C. the lowest or highest point on the graph of a parabola
   D. a line through the origin.

6. A parabola is **symmetric about its axis** since
   A. its graph is near the axis
   B. its graph is identical on each side of the axis
   C. its graph looks different on each side of the axis
   D. its graph intersects the axis.

QUICK REVIEW

Concepts	Examples

9.1 THE SQUARE ROOT PROPERTY AND COMPLETING THE SQUARE

Square Root Property

If x and k are complex numbers and $x^2 = k$, then

$$x = \sqrt{k} \quad \text{or} \quad x = -\sqrt{k}.$$

Completing the Square

To solve $ax^2 + bx + c = 0$ ($a \neq 0$):

Step 1 If $a \neq 1$, divide each side by a.

Step 2 Write the equation with the variable terms on one side and the constant on the other.

Step 3 Take half the coefficient of x and square it.

Solve $(x - 1)^2 = 8$.

$$x - 1 = \sqrt{8} \quad \text{or} \quad x - 1 = -\sqrt{8}$$
$$x = 1 + 2\sqrt{2} \quad \text{or} \quad x = 1 - 2\sqrt{2}$$

The solution set is $\left\{1 + 2\sqrt{2}, 1 - 2\sqrt{2}\right\}$.

Solve $2x^2 - 4x - 18 = 0$.

$$x^2 - 2x - 9 = 0 \qquad \text{Divide by 2.}$$
$$x^2 - 2x = 9 \qquad \text{Add 9.}$$

$$\left[\tfrac{1}{2}(-2)\right]^2 = (-1)^2 = 1$$

(continued)

Concepts	Examples

Step 4 Add the square to each side.

Step 5 Factor the perfect square trinomial, and write it as the square of a binomial. Simplify the other side.

Step 6 Use the square root property to complete the solution.

$$x^2 - 2x + 1 = 9 + 1 \quad \text{Add 1.}$$
$$(x - 1)^2 = 10 \quad \text{Factor.}$$
$$x - 1 = \sqrt{10} \quad \text{or} \quad x - 1 = -\sqrt{10} \qquad \text{Square root}$$
$$x = 1 + \sqrt{10} \quad \text{or} \qquad x = 1 - \sqrt{10} \qquad \text{property}$$

The solution set is $\{1 + \sqrt{10}, 1 - \sqrt{10}\}$.

9.2 THE QUADRATIC FORMULA

Quadratic Formula

The solutions of $ax^2 + bx + c = 0$ $(a \neq 0)$ are given by

$$x = \frac{-b \pm \sqrt{b^2 - 4ac}}{2a}.$$

Solve $3x^2 + 5x + 2 = 0$.

$$x = \frac{-5 \pm \sqrt{5^2 - 4(3)(2)}}{2(3)} = \frac{-5 \pm 1}{6}$$

$$x = \frac{-5 + 1}{6} = -\frac{2}{3} \quad \text{or} \quad x = \frac{-5 - 1}{6} = -1$$

The solution set is $\{-1, -\frac{2}{3}\}$.

The Discriminant

If a, b, and c are integers, then the discriminant, $b^2 - 4ac$, of $ax^2 + bx + c = 0$ determines the number and type of solutions as follows.

Discriminant	Number and Type of Solutions
Positive, the square of an integer	Two rational solutions
Positive, not the square of an integer	Two irrational solutions
Zero	One rational solution
Negative	Two nonreal complex solutions

For $x^2 + 3x - 10 = 0$, the discriminant is

$$3^2 - 4(1)(-10) = 49. \qquad \text{Two rational solutions}$$

For $4x^2 + x + 1 = 0$, the discriminant is

$$1^2 - 4(4)(1) = -15. \qquad \text{Two nonreal complex solutions}$$

9.3 EQUATIONS QUADRATIC IN FORM

A nonquadratic equation that can be written in the form

$$au^2 + bu + c = 0,$$

for $a \neq 0$ and an algebraic expression u, is called quadratic in form. Substitute u for the expression, solve for u, and then solve for the variable in the expression.

Solve $3(x + 5)^2 + 7(x + 5) + 2 = 0$.

$$3u^2 + 7u + 2 = 0 \qquad \text{Let } u = x + 5.$$
$$(3u + 1)(u + 2) = 0 \qquad \text{Factor.}$$

$$u = -\frac{1}{3} \quad \text{or} \qquad u = -2$$

$$x + 5 = -\frac{1}{3} \quad \text{or} \quad x + 5 = -2 \qquad x + 5 = u$$

$$x = -\frac{16}{3} \quad \text{or} \qquad x = -7$$

The solution set is $\{-7, -\frac{16}{3}\}$.

(continued)

Concepts	Examples

9.4 FORMULAS AND FURTHER APPLICATIONS

To solve a formula for a squared variable, proceed as follows.

(a) If the variable appears only to the second power: Isolate the squared variable on one side of the equation, and then use the square root property.

Solve $A = \dfrac{2mp}{r^2}$ for r.

$r^2 A = 2mp$ Multiply by r^2.

$r^2 = \dfrac{2mp}{A}$ Divide by A.

$r = \pm \sqrt{\dfrac{2mp}{A}}$ Square root property

$r = \dfrac{\pm\sqrt{2mpA}}{A}$ Rationalize the denominator.

(b) If the variable appears to the first and second powers: Write the equation in standard form, and then use the quadratic formula.

Solve $x^2 + rx = t$ for x.

$x^2 + rx - t = 0$ Standard form

$x = \dfrac{-r \pm \sqrt{r^2 - 4(1)(-t)}}{2(1)}$ $a = 1,\ b = r,\ c = -t$

$x = \dfrac{-r \pm \sqrt{r^2 + 4t}}{2}$

9.5 GRAPHS OF QUADRATIC FUNCTIONS

1. The graph of the quadratic function defined by $F(x) = a(x - h)^2 + k,\ a \neq 0$, is a parabola with vertex at (h, k) and the vertical line $x = h$ as axis.

2. The graph opens up if a is positive and down if a is negative.

3. The graph is wider than the graph of $f(x) = x^2$ if $0 < |a| < 1$ and narrower if $|a| > 1$.

Graph $f(x) = -(x + 3)^2 + 1$.

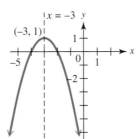

The graph opens down since $a < 0$. It is shifted 3 units left and 1 unit up, so the vertex is $(-3, 1)$, with axis $x = -3$. The domain is $(-\infty, \infty)$; the range is $(-\infty, 1]$.

9.6 MORE ABOUT PARABOLAS AND THEIR APPLICATIONS

The vertex of the graph of $f(x) = ax^2 + bx + c$, $a \neq 0$, may be found by completing the square.

The vertex has coordinates $\left(\dfrac{-b}{2a},\ f\left(\dfrac{-b}{2a} \right) \right)$.

Graphing a Quadratic Function

Step 1   Determine whether the graph opens up or down.

Step 2   Find the vertex.

Step 3   Find the x-intercepts (if any). Find the y-intercept.

Step 4   Find and plot additional points as needed.

Graph $f(x) = x^2 + 4x + 3$.

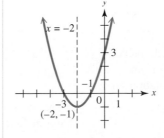

The graph opens up since $a > 0$. The vertex is $(-2, -1)$. The solutions of $x^2 + 4x + 3 = 0$ are -1 and -3, so the x-intercepts are $(-1, 0)$ and $(-3, 0)$. Since $f(0) = 3$, the y-intercept is $(0, 3)$. The domain is $(-\infty, \infty)$; the range is $[-1, \infty)$.

(continued)

Concepts	Examples

Horizontal Parabolas

The graph of $x = ay^2 + by + c$ is a horizontal parabola, opening to the right if $a > 0$ or to the left if $a < 0$. Horizontal parabolas do not represent functions.

Graph $x = 2y^2 + 6y + 5$.

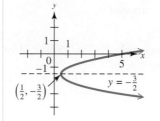

The graph opens to the right since $a > 0$. The vertex is $\left(\frac{1}{2}, -\frac{3}{2}\right)$. The axis is $y = -\frac{3}{2}$. The domain is $\left[\frac{1}{2}, \infty\right)$; the range is $(-\infty, \infty)$.

9.7 QUADRATIC AND RATIONAL INEQUALITIES

Solving a Quadratic (or Higher-Degree Polynomial) Inequality

Step 1 Write the inequality as an equation and solve.

Step 2 Use the numbers found in Step 1 to divide a number line into intervals.

Step 3 Substitute a test number from each interval into the original inequality to determine the intervals that belong to the solution set.

Step 4 Consider the endpoints separately.

Solve $2x^2 + 5x + 2 < 0$.

$$2x^2 + 5x + 2 = 0$$
$$(2x + 1)(x + 2) = 0$$
$$x = -\tfrac{1}{2} \quad \text{or} \quad x = -2$$

Intervals: $(-\infty, -2)$, $\left(-2, -\frac{1}{2}\right), \left(-\frac{1}{2}, \infty\right)$

Test values: $-3, -1, 0$

$x = -3$ makes the original inequality false; $x = -1$ makes it true; $x = 0$ makes it false. Choose the interval(s) which yield(s) a true statement. The solution set is the interval $\left(-2, -\frac{1}{2}\right)$.

Solving a Rational Inequality

Step 1 Write the inequality so that 0 is on one side and there is a single fraction on the other side.

Solve $\dfrac{x}{x + 2} \geq 4$.

$$\frac{x}{x + 2} - 4 \geq 0 \qquad \text{Subtract 4.}$$

$$\frac{x}{x + 2} - \frac{4(x + 2)}{x + 2} \geq 0 \qquad \begin{array}{l}\text{Write with a}\\ \text{common denominator.}\end{array}$$

$$\frac{-3x - 8}{x + 2} \geq 0 \qquad \text{Subtract fractions.}$$

Step 2 Determine the numbers that make the numerator or denominator 0.

Step 3 Use the numbers from Step 2 to divide a number line into intervals.

Step 4 Substitute a test number from each interval into the original inequality to determine the intervals that belong to the solution set.

Step 5 Consider the endpoints separately.

$-\frac{8}{3}$ makes the numerator 0; -2 makes the denominator 0.

-4 from A makes the original inequality false; $-\frac{7}{3}$ from B makes it true; 0 from C makes it false.

The solution set is the interval $\left[-\frac{8}{3}, -2\right)$. The endpoint -2 is not included since it makes the denominator 0.

Answers to Test Your Word Power

1. D; *Example:* The solutions of $ax^2 + bx + c = 0$ $(a \neq 0)$ are given by $x = \dfrac{-b \pm \sqrt{b^2 - 4ac}}{2a}$.

2. C; *Examples:* $f(x) = x^2 - 2$, $f(x) = (x + 4)^2 + 1$, $f(x) = x^2 - 4x + 5$

3. D; *Examples:* See the figures in the Quick Review for **Sections 9.5 and 9.6.**

4. C; *Example:* The graph of $y = (x + 3)^2$ has vertex $(-3, 0)$, which is the lowest point on the graph.

5. B; *Example:* The axis of $y = (x + 3)^2$ is the vertical line $x = -3$.

6. B; *Example:* Since the graph of $y = (x + 3)^2$ is symmetric about its axis $x = -3$, the points $(-2, 1)$ and $(-4, 1)$ are on the graph.

Chapter 9 REVIEW EXERCISES

*Exercises marked * require knowledge of the complex number system.*

[9.1] *Solve each equation by using the square root property or completing the square.*

1. $t^2 = 121$

2. $p^2 = 3$

3. $(2x + 5)^2 = 100$

***4.** $(3k - 2)^2 = -25$

5. $x^2 + 4x = 15$

6. $2m^2 - 3m = -1$

7. *Concept Check*   A student gave the following "solution" to the equation $x^2 = 12$.

$$x^2 = 12$$
$$x = \sqrt{12} \qquad \text{Square root property}$$
$$x = 2\sqrt{3}$$

Solution set: $\left\{ 2\sqrt{3} \right\}$

The answer is not correct. ***WHAT WENT WRONG?*** Give the correct solution set.

8. Navy Pier Center in Chicago, Illinois, features a 150-ft tall Ferris wheel. Use Galileo's formula $d = 16t^2$ to find how long it would take an MP-3 player dropped from the top of the Ferris wheel to fall to the ground. Round your answer to the nearest tenth of a second. (*Source: Microsoft Encarta Encyclopedia.*)

[9.2] *Use the discriminant to predict whether the solutions to each equation are*

A. *two rational numbers;*   **B.** *one rational number;*
C. *two irrational numbers;*   **D.** *two nonreal complex numbers.*

9. $x^2 + 5x + 2 = 0$

10. $4t^2 = 3 - 4t$

11. $4x^2 = 6x - 8$

12. $9z^2 + 30z + 25 = 0$

Solve each equation by using the quadratic formula.

13. $2x^2 + x - 21 = 0$ **14.** $k^2 + 5k = 7$ **15.** $(t + 3)(t - 4) = -2$

*16. $2x^2 + 3x + 4 = 0$ ***17.** $3p^2 = 2(2p - 1)$ **18.** $m(2m - 7) = 3m^2 + 3$

[9.3] *Solve each equation. Check your solutions.*

19. $\dfrac{15}{x} = 2x - 1$

20. $\dfrac{1}{n} + \dfrac{2}{n + 1} = 2$

21. $-2r = \sqrt{\dfrac{48 - 20r}{2}}$

22. $8(3x + 5)^2 + 2(3x + 5) - 1 = 0$

23. $2x^{2/3} - x^{1/3} - 28 = 0$

24. $p^4 - 10p^2 + 9 = 0$

Solve each problem. Round answers to the nearest tenth, as necessary.

25. Phong paddled his canoe 20 mi upstream, then paddled back. If the speed of the current was 3 mph and the total trip took 7 hr, what was Phong's speed?

26. Maureen O'Connor drove 8 mi to pick up her friend Laurie, and then drove 11 mi to a mall at a speed 15 mph faster. If Maureen's total travel time was 24 min, what was her speed on the trip to pick up Laurie?

27. An old machine processes a batch of checks in 1 hr more time than a new one. How long would it take the old machine to process a batch of checks that the two machines together process in 2 hr?

28. Greg Tobin can process a stack of invoices 1 hr faster than Carter Fenton can. Working together, they take 1.5 hr. How long would it take each person working alone?

[9.4] *Solve each formula for the indicated variable. (Give answers with ±.)*

29. $k = \dfrac{rF}{wv^2}$ for v

30. $p = \sqrt{\dfrac{yz}{6}}$ for y

31. $mt^2 = 3mt + 6$ for t

Solve each problem. Round answers to the nearest tenth, as necessary.

32. A large machine requires a part in the shape of a right triangle with a hypotenuse 9 ft less than twice the length of the longer leg. The shorter leg must be $\frac{3}{4}$ the length of the longer leg. Find the lengths of the three sides of the part.

33. A square has an area of 256 cm². If the same amount is removed from one dimension and added to the other, the resulting rectangle has an area 16 cm² less. Find the dimensions of the rectangle.

34. Nancy wants to buy a mat for a photograph that measures 14 in. by 20 in. She wants to have an even border around the picture when it is mounted on the mat. If the area of the mat she chooses is 352 in.², how wide will the border be?

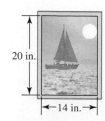

35. A searchlight moves horizontally back and forth along a wall with the distance of the light from a starting point at t minutes given by the quadratic function defined by

$$f(t) = 100t^2 - 300t.$$

How long will it take before the light returns to the starting point?

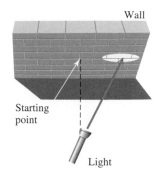

Wall

Starting point

Light

36. Lewis Tower, built in Philadelphia, Pennsylvania, in 1929, is 400 ft high. Suppose that a ball is projected upward from the top of the Tower, and its position in feet above the ground is given by the quadratic function defined by

$$f(t) = -16t^2 + 45t + 400,$$

where t is the number of seconds elapsed. How long will it take for the ball to reach a height of 200 ft above the ground? (*Source: World Almanac and Book of Facts 2006.*)

37. The Alberta Stock Exchange in Calgary, Alberta, is 407 ft high. Suppose that a ball is projected upward from the top of the Exchange, and its position in feet above the ground is given by the quadratic function defined by

$$s(t) = -16t^2 + 75t + 407,$$

where t is the number of seconds elapsed. How long will it take for the ball to reach a height of 450 ft above the ground? (*Source: World Almanac and Book of Facts 2006.*)

38. The manager of a restaurant has determined that the demand for frozen yogurt is $\frac{25}{p}$ units per day, where p is the price (in dollars) per unit. The supply is $70p + 15$ units per day. Find the price at which supply and demand are equal.

39. Use the formula $A = P(1 + r)^2$ to find the interest rate r at which a principal P of $10,000 will increase to $10,920.25 in 2 yr.

40. The number of e-mail boxes in North America (in millions) for the years 1995 through 2001 are shown in the graph and can be modeled by the quadratic function defined by

$$f(x) = 3.29x^2 - 10.4x + 21.6.$$

In the model, $x = 5$ represents 1995, $x = 10$ represents 2000, and so on.

(a) Use the model to approximate the number of e-mail boxes in 2001 to the nearest whole number. How does this result compare to the number shown in the graph?

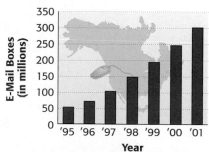

Growth of E-mail Boxes in North America

Source: IDC research.

(b) Based on the model, in what year did the number of e-mail boxes reach 200 million? (Round down to the nearest year.) How does this result compare to the number shown in the graph?

[9.5–9.6] *Identify the vertex of each parabola.*

41. $y = 6 - 2x^2$ **42.** $f(x) = -(x - 1)^2$ **43.** $f(x) = (x - 3)^2 + 7$

44. $y = -3x^2 + 4x - 2$ **45.** $x = (y - 3)^2 - 4$

46. *Concept Check* If the discriminant of a quadratic function is negative, what do you know about the graph of the function?

Graph each parabola. Give the vertex, axis, domain, and range.

47. $y = 2(x - 2)^2 - 3$ **48.** $f(x) = -2x^2 + 8x - 5$

49. $x = 2(y + 3)^2 - 4$ **50.** $x = -\dfrac{1}{2}y^2 + 6y - 14$

Solve each problem.

51. Consumer spending for home video games in dollars per person per year is given in the table. Let $x = 0$ represent 1995, $x = 1$ represent 1996, and so on.

CONSUMER SPENDING FOR
HOME VIDEO GAMES

Year	Dollars
1995	10.54
1996	11.47
1997	16.45
1998	18.49
1999	24.45
2000	25.89
2001	26.89
2002	29.59
2003	29.78

Source: Communications Industry Forecast, (annual).

(a) Use the data for 1996, 1999, and 2003 in the quadratic form $ax^2 + bx + c = y$ to write a system of three equations.

(b) Solve the system from part (a) to find a quadratic function f that models the data.

(c) Use the model found in part (b) to approximate consumer spending for home video games in 2000 to the nearest cent. How does your answer compare to the actual data from the table?

52. The height (in feet) of a projectile t seconds after being fired from Earth into the air is given by

$$f(t) = -16t^2 + 160t.$$

Find the number of seconds required for the projectile to reach maximum height. What is the maximum height?

53. Find the length and width of a rectangle having a perimeter of 200 m if the area is to be a maximum. What is the maximum area?

[9.7] *Solve each inequality and graph the solution set.*

54. $(x - 4)(2x + 3) > 0$ **55.** $x^2 + x \le 12$

56. $(x + 2)(x - 3)(x + 5) \leq 0$

57. $(4m + 3)^2 \leq -4$

58. $\dfrac{6}{2z - 1} < 2$

59. $\dfrac{3t + 4}{t - 2} \leq 1$

MIXED REVIEW EXERCISES

Solve.

60. $V = r^2 + R^2 h$ for R

*61. $3t^2 - 6t = -4$

*62. $x^4 - 1 = 0$

63. $(x^2 - 2x)^2 = 11(x^2 - 2x) - 24$

64. $(r - 1)(2r + 3)(r + 6) < 0$

65. $2x - \sqrt{x} = 6$

66. $(3k + 11)^2 = 7$

67. $S = \dfrac{Id^2}{k}$ for d

68. $(8k - 7)^2 \geq -1$

69. $6 + \dfrac{15}{s^2} = -\dfrac{19}{s}$

70. $x^4 - 8x^2 = -1$

71. $\dfrac{-2}{x + 5} \leq -5$

Concept Check   *Match each equation with the figure that most closely resembles its graph.*

72. $g(x) = x^2 - 5$

73. $h(x) = -x^2 + 4$

74. $F(x) = (x - 1)^2$

75. $G(x) = (x + 1)^2$

76. $H(x) = (x - 1)^2 + 1$

77. $K(x) = (x + 1)^2 + 1$

A.
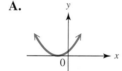

B.
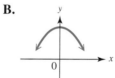

C.
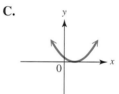

D.
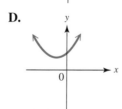

E.
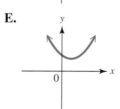

F.
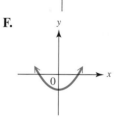

78. Graph $f(x) = 4x^2 + 4x - 2$. Give the vertex, axis, domain, and range.

79. Nuclear power consumption in the United States in quadrillions of Btu (British thermal units) from 1970 through 2000 can be modeled by the quadratic function defined by

$$f(x) = -0.001x^2 + 0.295x + 0.227,$$

where $x = 0$ represents 1970, $x = 5$ represents 1975, and so on. (*Source:* U.S. Energy Information Administration.)

(a) Use the model to approximate nuclear consumption in 2002.

(b) Actual nuclear consumption in 2002 was 8.14 quadrillion Btu. How does this compare to the amount found in part (a) using the model?

(c) Based on the model, in what year will nuclear energy consumption in the U.S. reach 10 quadrillon Btu? (Round down to the nearest year.)

80. In 4 hr, Kerrie can go 15 mi upriver and come back. The speed of the current is 5 mph. Find the speed of the boat in still water.

81. Refer to Exercise 36. Suppose that a wire is attached to the top of Lewis Tower and pulled tight. It is attached to the ground 100 ft from the base of the building. How long is the wire?

82. Two pieces of a large wooden puzzle fit together to form a rectangle with length 1 cm less than twice the width. The diagonal, where the two pieces meet, is 2.5 cm in length. Find the length and width of the rectangle.

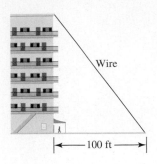

Chapter **9** TEST

*Problems marked * require knowledge of the complex number system.*

Solve each equation by using the square root property or completing the square.

1. $t^2 = 54$ **2.** $(7x + 3)^2 = 25$ **3.** $2x^2 + 4x = 8$

Solve by using the quadratic formula.

4. $2x^2 - 3x - 1 = 0$ ***5.** $3t^2 - 4t - -5$ **6.** $3x = \sqrt{\dfrac{9x + 2}{2}}$

***7.** *Concept Check*   If k is a negative number, then which one of the following equations will have two nonreal complex solutions?

   A. $x^2 = 4k$ **B.** $x^2 = -4k$ **C.** $(x + 2)^2 = -k$ **D.** $x^2 + k = 0$

8. What is the discriminant for $2x^2 - 8x - 3 = 0$? How many and what type of solutions does this equation have? (Do not actually solve.)

Solve by any method.

9. $3 - \dfrac{16}{x} - \dfrac{12}{x^2} = 0$ **10.** $4x^2 + 7x - 3 = 0$

11. $9x^4 + 4 = 37x^2$ **12.** $12 = (2n + 1)^2 + (2n + 1)$

13. Solve $S = 4\pi r^2$ for r. (Leave $\pm$ in your answer.)

Solve each problem.

14. Andrew and Kent do word processing. For a certain prospectus, Kent can prepare it 2 hr faster than Andrew can. If they work together, they can do the entire prospectus in 5 hr. How long will it take each of them working alone to prepare the prospectus? Round your answers to the nearest tenth of an hour.

15. Abby Tanenbaum paddled her canoe 10 mi upstream and then paddled back to her starting point. If the rate of the current was 3 mph and the entire trip took $3\frac{1}{2}$ hr, what was Abby's rate?

16. Tyler McGinnis has a pool 24 ft long and 10 ft wide. He wants to construct a concrete walk around the pool. If he plans for the walk to be of uniform width and cover 152 ft², what will the width of the walk be?

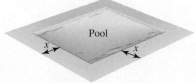

17. At a point 30 m from the base of a tower, the distance to the top of the tower is 2 m more than twice the height of the tower. Find the height of the tower.

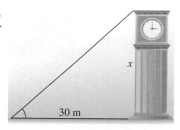

18. *Concept Check* Which one of the following figures most closely resembles the graph of $f(x) = a(x - h)^2 + k$ if $a < 0$, $h > 0$, and $k < 0$?

A.

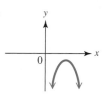

B.

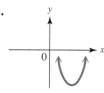

C.

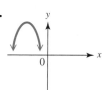

D.

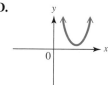

Graph each parabola. Identify the vertex, axis, domain, and range.

19. $f(x) = \dfrac{1}{2}x^2 - 2$ **20.** $f(x) = -x^2 + 4x - 1$ **21.** $x = -(y - 2)^2 + 2$

Solve each problem.

22. The percent increase for in-state tuition at Iowa public universities during the years 1992 through 2002 can be modeled by the quadratic function defined by

$$f(x) = 0.156x^2 - 2.05x + 10.2,$$

where $x = 2$ represents 1992, $x = 3$ represents 1993, and so on. (*Source:* Iowa Board of Regents.)

   (a) Based on this model, by what percent (to the nearest tenth) did tuition increase in 2001?

   (b) In what year was the minimum tuition increase? (Round down to the nearest year.) To the nearest tenth, by what percent did tuition increase that year?

23. Palo Alto College is planning to construct a rectangular parking lot on land bordered on one side by a highway. The plan is to use 640 ft of fencing to fence off the other three sides. What should the dimensions of the lot be if the enclosed area is to be a maximum?

Solve, and graph each solution set.

24. $2x^2 + 7x > 15$ **25.** $\dfrac{5}{t - 4} \le 1$

Chapters 1–9 CUMULATIVE REVIEW EXERCISES

1. Let $S = \left\{-\frac{7}{3}, -2, -\sqrt{3}, 0, 0.7, \sqrt{12}, \sqrt{-8}, 7, \frac{32}{3}\right\}$. List the elements of S that are elements of each set.

    (a) Integers   **(b)** Rational numbers   **(c)** Real numbers   **(d)** Complex numbers

Simplify each expression.

2. $|-3| + 8 - |-9| - (-7 + 3)$ **3.** $2(-3)^2 + (-8)(-5) + (-17)$

Solve each equation.

4. $7 - (4 + 3t) + 2t = -6(t - 2) - 5$ **5.** $|6x - 9| = |-4x + 2|$

6. $2x = \sqrt{\dfrac{5x + 2}{3}}$ **7.** $\dfrac{3}{x - 3} - \dfrac{2}{x - 2} = \dfrac{3}{x^2 - 5x + 6}$

8. $(r - 5)(2r + 3) = 1$ **9.** $x^4 - 5x^2 + 4 = 0$

Solve each inequality.

10. $-2x + 4 \le -x + 3$ **11.** $|3x - 7| \le 1$

12. $x^2 - 4x + 3 < 0$ **13.** $\dfrac{3}{p + 2} > 1$

Graph each relation. Tell whether or not y can be expressed as a function f of x, and if so, give its domain and range, and write using function notation.

14. $4x - 5y = 15$ **15.** $4x - 5y < 15$ **16.** $y = -2(x - 1)^2 + 3$

17. Find the slope and intercepts of the line with equation $-2x + 7y = 16$.

18. Write an equation for the specified line. Express each equation in slope-intercept form.

    (a) Through $(2, -3)$ and parallel to the line with equation $5x + 2y = 6$
    (b) Through $(-4, 1)$ and perpendicular to the line with equation $5x + 2y = 6$

19. Sales of NASCAR-licensed merchandise for recent years are given in the table and can be modeled by a linear equation. Let $x = 0$ represent 1995, $x = 2$ represent 1997, and so on.

Year	Sales (in millions of dollars)
1995	600
1997	800
1999	1130
2001	1340
2002	1400

Source: NASCAR.

    (a) Use the ordered pairs $(0, 600)$ and $(6, 1340)$ to write a linear equation that models these data. Round the slope to the nearest whole number.
    (b) Use your model to approximate sales of NASCAR-licensed merchandise in 2002. How does it compare to the actual value from the table?

20. Does the relation $x = 5$ define y as a function of x? Explain.

21. For the function defined by $f(x) = 2(x - 1)^2 - 5$, find each of the following.

   (a) $f(-2)$ **(b)** the domain and range

Solve each system of equations.

22. $2x - 4y = 10$
$9x + 3y = 3$

23. $\begin{aligned} x + y + 2z &= 3 \\ -x + y + z &= -5 \\ 2x + 3y - z &= -8 \end{aligned}$

24. The merger in 2000 of America Online and Time Warner was at that time the largest in U.S. history. The two companies had combined sales of $34.2 billion. Sales for AOL were $0.3 billion less than 4 times the sales for Time Warner. What were sales for each company? (*Source:* Company reports.)

   (a) Write a system of equations to solve the problem.
   (b) Solve the problem.

Write with positive exponents only. Assume that variables represent positive real numbers.

25. $\left(\dfrac{x^{-3}y^2}{x^5y^{-2}} \right)^{-1}$

26. $\dfrac{(4x^{-2})^2(2y^3)}{8x^{-3}y^5}$

Perform the indicated operations.

27. $(7x + 4)(2x - 3)$

28. $\left(\dfrac{2}{3}t + 9 \right)^2$

29. $(3t^3 + 5t^2 - 8t + 7) - (6t^3 + 4t - 8)$

30. Divide $4x^3 + 2x^2 - x + 26$ by $x + 2$.

Factor completely.

31. $16x - x^3$

32. $24m^2 + 2m - 15$

33. $8x^3 + 27y^3$

34. $9x^2 - 30xy + 25y^2$

Perform the indicated operations or simplify the complex fraction, and express each answer in lowest terms. Assume denominators are nonzero.

35. $\dfrac{x^2 - 3x - 10}{x^2 + 3x + 2} \cdot \dfrac{x^2 - 2x - 3}{x^2 + 2x - 15}$

36. $\dfrac{3}{2 - k} - \dfrac{5}{k} + \dfrac{6}{k^2 - 2k}$

37. $\dfrac{\dfrac{r}{s} - \dfrac{s}{r}}{\dfrac{r}{s} + 1}$

38. $\dfrac{1 - x^{-2}y^2}{x^{-1} - x^{-2}y}$

Simplify each radical expression.

39. $\sqrt[3]{\dfrac{27}{16}}$

40. $\dfrac{2}{\sqrt{7} - \sqrt{5}}$

Solve each problem.

41. Tri rode his bicycle for 12 mi and then walked an additional 8 mi. The total time for the trip was 5 hr. If his rate while walking was 10 mph less than his rate while riding, what was each rate?

42. Two cars left an intersection at the same time, one heading due south and the other due east. Later they were exactly 95 mi apart. The car heading east had gone 38 mi less than twice as far as the car heading south. How far had each car traveled?

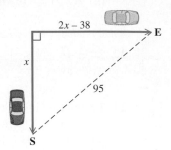

At Automated Teller Machines (ATMs) people often find themselves doing what they have done for years when faced with a soft drink machine that won't respond: They talk to it. According to one report, the following are percentages of people in the United States, the United Kingdom (UK), and Germany who talk to ATMs and what they say.

	United States	UK	Germany
Thanking the ATM	22%	24%	14%
Cursing the ATM	31%	41%	53%
Telling the ATM to Hurry Up	47%	36%	33%

Source: BMRB International for NCR.

In a random sample of 4000 people, how many would there be in each category?

43. People in the United States who curse the ATM

44. People in the UK who thank the ATM

45. People in Germany who tell the ATM to hurry up

46. How many more German cursers would there be than United States thankers?

Inverse, Exponential, and Logarithmic Functions

In 2001, Apple Computer Inc., introduced the iPod. Since then, the company has sold over 40 million of the popular music players, in spite of warnings by experts that listening to the devices at high volumes may put people at increased risk of hearing loss. In 2006, a federal class-action lawsuit was filed against the company, accusing it of not taking adequate steps to protect the hearing of iPod users. As a result, Apple issued a software update that allows listeners to set maximum volume limits on some of the newer iPod models. (*Source: Sacramento Bee, USA Today.*)

In Example 4 of Section 10.5, we use a *logarithmic function* to calculate the volume level, in *decibels,* of an iPod.

10.1 Inverse Functions

OBJECTIVES

1 Decide whether a function is one-to-one and, if it is, find its inverse.

2 Use the horizontal line test to determine whether a function is one-to-one.

3 Find the equation of the inverse of a function.

4 Graph f^{-1} from the graph of f.

5 Use a graphing calculator to graph inverse functions.

In this chapter we will study two important types of functions, *exponential* and *logarithmic*. These functions are related in a special way: They are *inverses* of one another. We begin by discussing inverse functions in general.

OBJECTIVE 1 Decide whether a function is one-to-one and, if it is, find its inverse. Suppose we define the function

$$G = \{(-2, 2), (-1, 1), (0, 0), (1, 3), (2, 5)\}.$$

We can form another set of ordered pairs from G by interchanging the x- and y-values of each pair in G. We can call this set F, so

$$F = \{(2, -2), (1, -1), (0, 0), (3, 1), (5, 2)\}.$$

To show that these two sets are related, F is called the *inverse* of G. For a function f to have an inverse, f must be a *one-to-one function*.

One-to-One Function

In a **one-to-one function,** each x-value corresponds to only one y-value, and each y-value corresponds to only one x-value.

The function shown in Figure 1(a) is not one-to-one because the y-value 7 corresponds to *two* x-values, 2 and 3. That is, the ordered pairs (2, 7) and (3, 7) both belong to the function. The function in Figure 1(b) is one-to-one.

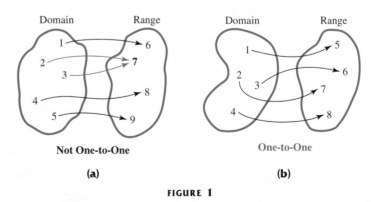

Not One-to-One

(a)

One-to-One

(b)

FIGURE 1

The *inverse* of any one-to-one function f is found by interchanging the components of the ordered pairs of f. The inverse of f is written f^{-1}. Read f^{-1} as **"the inverse of f"** or **"f-inverse."**

▶ **CAUTION** The symbol $f^{-1}(x)$ does *not* represent $\dfrac{1}{f(x)}$.

The definition of the inverse of a function follows.

Inverse of a Function

The **inverse** of a one-to-one function f, written f^{-1}, is the set of all ordered pairs of the form (y, x), where (x, y) belongs to f. Since the inverse is formed by interchanging x and y, the domain of f becomes the range of f^{-1} and the range of f becomes the domain of f^{-1}.

For inverses f and f^{-1}, it follows that

$$(f \circ f^{-1})(x) = x \quad \text{and} \quad (f^{-1} \circ f)(x) = x.$$

EXAMPLE 1 Finding the Inverses of One-to-One Functions

Decide whether each function is one-to-one. If it is, find the inverse.

(a) $F = \{(-2, 1), (-1, 0), (0, 1), (1, 2), (2, 2)\}$

Each x-value in F corresponds to just one y-value. However, the y-value 1 corresponds to two x-values, -2 and 0. Also, the y-value 2 corresponds to both 1 and 2. Because some y-values correspond to more than one x-value, F is not one-to-one and does not have an inverse.

(b) $G = \{(3, 1), (0, 2), (2, 3), (4, 0)\}$

Every x-value in G corresponds to only one y-value, and every y-value corresponds to only one x-value, so G is a one-to-one function. The inverse function is found by interchanging the x- and y-values in each ordered pair.

$$G^{-1} = \{(1, 3), (2, 0), (3, 2), (0, 4)\}$$

Notice how the domain and range of G become the range and domain, respectively, of G^{-1}.

(c) The Pollutant Standard Index (PSI) is an indicator of air quality. If the PSI exceeds 100 on a particular day, then that day is classified as unhealthy. The table shows the number of unhealthy days in Chicago for the years 1991–2002, based on new standards set in 1998.

Year	Number of Unhealthy Days	Year	Number of Unhealthy Days
1991	24	1997	10
1992	5	1998	12
1993	4	1999	19
1994	13	2000	2
1995	24	2001	22
1996	7	2002	21

Source: U.S. Environmental Protection Agency, Office of Air Quality Planning and Standards.

Let f be the function defined in the table on the previous page, with the years forming the domain and the numbers of unhealthy days forming the range. Then f is not one-to-one, because in two different years (1991 and 1995), the number of unhealthy days was the same, 24.

✔ **Now Try Exercises 1, 9, and 11.**

OBJECTIVE ② Use the horizontal line test to determine whether a function is one-to-one. It may be difficult to decide whether a function is one-to-one just by looking at the equation that defines the function. However, by graphing the function and observing the graph, we can use the *horizontal line test* to tell whether the function is one-to-one.

Horizontal Line Test

A function is one-to-one if every horizontal line intersects the graph of the function at most once.

The horizontal line test follows from the definition of a one-to-one function. Any two points that lie on the same horizontal line have the same y-coordinate. No two ordered pairs that belong to a one-to-one function may have the same y-coordinate, and therefore no horizontal line will intersect the graph of a one-to-one function more than once.

EXAMPLE 2 Using the Horizontal Line Test

Use the horizontal line test to determine whether each graph is the graph of a one-to-one function.

(a)

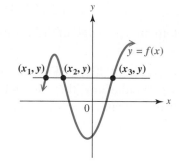

(b)

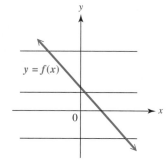

Because a horizontal line intersects the graph in more than one point (actually three points), the function is not one-to-one.

Every horizontal line will intersect the graph in exactly one point. This function is one-to-one.

✔ **Now Try Exercise 7.**

OBJECTIVE ③ Find the equation of the inverse of a function. The inverse of a one-to-one function is found by interchanging the x- and y-values of each of its ordered pairs. The equation of the inverse of a function defined by $y = f(x)$ is found in the same way.

> **Finding the Equation of the Inverse of $y = f(x)$**
>
> For a one-to-one function f defined by an equation $y = f(x)$, find the defining equation of the inverse as follows.
>
> **Step 1** Interchange x and y.
>
> **Step 2** Solve for y.
>
> **Step 3** Replace y with $f^{-1}(x)$.

EXAMPLE 3 Finding Equations of Inverses

Decide whether each equation defines a one-to-one function. If so, find the equation that defines the inverse.

(a) $f(x) = 2x + 5$

    The graph of $y = 2x + 5$ is a nonvertical line, so by the horizontal line test, f is a one-to-one function. To find the inverse, let $y = f(x)$ so that

$$y = 2x + 5$$
$$x = 2y + 5 \qquad \text{Interchange } x \text{ and } y. \text{ (Step 1)}$$
$$2y = x - 5 \qquad \text{Solve for } y. \text{ (Step 2)}$$
$$y = \frac{x - 5}{2}$$
$$f^{-1}(x) = \frac{x - 5}{2}, \qquad \text{Replace } y \text{ with } f^{-1}(x). \text{ (Step 3)}$$

which can be written

$$f^{-1}(x) = \frac{x}{2} - \frac{5}{2}, \quad \text{or} \quad f^{-1}(x) = \frac{1}{2}x - \frac{5}{2}. \qquad \frac{a - b}{c} = \frac{a}{c} - \frac{b}{c}$$

Thus, f^{-1} is a linear function. In the function defined by $y = 2x + 5$, the value of y is found by starting with a value of x, multiplying by 2, and adding 5. The equation $f^{-1}(x) = \frac{x - 5}{2}$ for the inverse has us *subtract* 5, and then *divide* by 2. This shows how an inverse is used to "undo" what a function does to the variable x.

(b) $y = x^2 + 2$

    This equation has a vertical parabola as its graph, so some horizontal lines will intersect the graph at two points. For example, both $x = 3$ and $x = -3$ correspond to $y = 11$. Because of the x^2-term, there are many pairs of x-values that correspond to the same y-value. This means that the function defined by $y = x^2 + 2$ is not one-to-one and does not have an inverse.

    If this is not noticed, then following the steps for finding the equation of an inverse leads to

$$y = x^2 + 2$$
$$x = y^2 + 2 \qquad \text{Interchange } x \text{ and } y.$$
$$y^2 = x - 2 \qquad \text{Solve for } y.$$
$$y = \pm\sqrt{x - 2}. \qquad \text{Square root property}$$

The last step shows that there are two y-values for each choice of $x > 2$, so the given function is not one-to-one and does not have an inverse.

(c) $f(x) = (x - 2)^3$

Refer to **Section 5.3** to see that the graphs of cubing functions are one-to-one.

$$y = (x - 2)^3 \qquad \text{Replace } f(x) \text{ with } y.$$

$$x = (y - 2)^3 \qquad \text{Interchange } x \text{ and } y.$$

$$\sqrt[3]{x} = \sqrt[3]{(y - 2)^3} \qquad \text{Take the cube root on each side.}$$

$$\sqrt[3]{x} = y - 2$$

$$y = \sqrt[3]{x} + 2 \qquad \text{Solve for } y.$$

$$f^{-1}(x) = \sqrt[3]{x} + 2 \qquad \text{Replace } y \text{ with } f^{-1}(x).$$

✔ **Now Try Exercises 13, 17, and 19.**

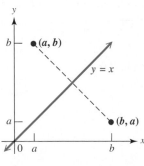

FIGURE 2

OBJECTIVE 4 Graph f^{-1} from the graph of f. One way to graph the inverse of a function f whose equation is known is to find some ordered pairs that belong to f, interchange x and y to get ordered pairs that belong to f^{-1}, plot those points, and sketch the graph of f^{-1} through the points. A simpler way is to select points on the graph of f and use symmetry to find corresponding points on the graph of f^{-1}.

For example, suppose the point (a, b) shown in Figure 2 belongs to a one-to-one function f. Then the point (b, a) belongs to f^{-1}. The line segment connecting (a, b) and (b, a) is perpendicular to, and cut in half by, the line $y = x$. The points (a, b) and (b, a) are "mirror images" of each other with respect to $y = x$. For this reason *we can find the graph of f^{-1} from the graph of f by locating the mirror image of each point in f with respect to the line $y = x$.*

EXAMPLE 4 Graphing the Inverse

Graph the inverses of the functions f (shown in blue) in Figure 3.

In Figure 3 the graphs of two functions f are shown in blue. Their inverses are shown in red. In each case, the graph of f^{-1} is a reflection of the graph of f with respect to the line $y = x$.

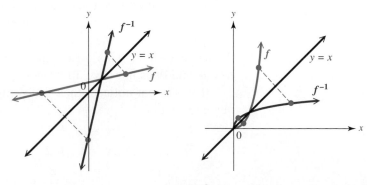

FIGURE 3

✔ **Now Try Exercises 25 and 29.**

OBJECTIVE 5 Use a graphing calculator to graph inverse functions. We have described how inverses of one-to-one functions may be determined algebraically. We also explained how the graph of a one-to-one function f compares to the graph of its inverse f^{-1}: It is a reflection of the graph of f^{-1} across the line $y = x$.

In Example 3 we showed that the inverse of the one-to-one function defined by $f(x) = 2x + 5$ is given by $f^{-1}(x) = \frac{x - 5}{2}$. If we use a square viewing window of a graphing calculator and graph $y_1 = f(x) = 2x + 5$, $y_2 = f^{-1}(x) = \frac{x - 5}{2}$, and $y_3 = x$, we can see how this reflection appears on the screen. See Figure 4.

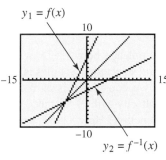

FIGURE 4

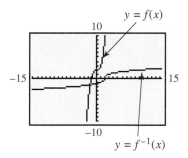

FIGURE 5

Some graphing calculators have the capability to "draw" the inverse of a function. Figure 5 shows the graphs of $f(x) = x^3 + 2$ and its inverse in a square viewing window.

✔ **Now Try Exercise 43.**

10.1 EXERCISES

● *Complete solution available on Video Lectures on CD/DVD*

Now Try Exercise

✎ *In Exercises 1–4, write a few sentences of explanation. See Example 1.*

● **1.** A new study found that the trans fat content in fast-food products varied widely around the world, based on the type of frying oil used, as shown in the table.

If the set of countries is the domain and the set of trans fat percentages is the range of the function consisting of the eight pairs listed, is it a one-to-one function? Why or why not?

Country	Percentage of Trans Fat in McDonald's Chicken
Scotland	14
France	11
United States	11
Peru	9
Hungary	8
Poland	8
Russia	5
Denmark	1

Source: New England Journal of Medicine.

2. The table shows the number of uncontrolled hazardous waste sites in 2004 that require further investigation to determine whether remedies are needed under the Superfund program. The eight states listed are ranked in the top ten in the United States.

If this correspondence is considered to be a function that pairs each state with its number of uncontrolled waste sites, is it one-to-one? If not, explain why.

State	Number of Sites
New Jersey	114
Pennsylvania	96
California	95
New York	91
Michigan	69
Florida	52
Illinois	47
Washington	47

Source: U.S. Environmental Protection Agency.

3. The road mileage between Denver, Colorado, and several selected U.S. cities is shown in the table. If we consider this as a function that pairs each city with a distance, is it a one-to-one function? How could we change the answer to this question by adding 1 mile to one of the distances shown?

City	Distance to Denver (in miles)
Atlanta	1398
Dallas	781
Indianapolis	1058
Kansas City, MO	600
Los Angeles	1059
San Francisco	1235

4. Suppose you consider the set of ordered pairs (x, y) such that x represents a person in your mathematics class and y represents that person's mother. Explain how this function might not be a one-to-one function.

In Exercises 5–8, choose the correct response from the given list.

5. *Concept Check*   If a function is made up of ordered pairs in such a way that the same y-value appears in a correspondence with two different x-values, then

A. the function is one-to-one
B. the function is not one-to-one
C. its graph does not pass the vertical line test
D. it has an inverse function associated with it.

6. Which equation defines a one-to-one function? Explain why the others are not, using specific examples.

A. $f(x) = x$ **B.** $f(x) = x^2$ **C.** $f(x) = |x|$ **D.** $f(x) = -x^2 + 2x - 1$

7. Only one of the graphs illustrates a one-to-one function. Which one is it? (See Example 2.)

A. 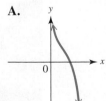   **B.** 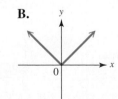   **C.** 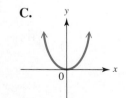   **D.**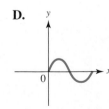

8. *Concept Check*   If a function f is one-to-one and the point (p, q) lies on the graph of f, then which point *must* lie on the graph of f^{-1}?

A. $(-p, q)$ **B.** $(-q, -p)$ **C.** $(p, -q)$ **D.** (q, p)

If the function is one-to-one, find its inverse. See Examples 1–3.

9. $\{(3, 6), (2, 10), (5, 12)\}$

10. $\left\{(-1, 3), (0, 5), (5, 0), \left(7, -\dfrac{1}{2}\right)\right\}$

11. $\{(-1, 3), (2, 7), (4, 3), (5, 8)\}$

12. $\{(-8, 6), (-4, 3), (0, 6), (5, 10)\}$

13. $f(x) = 2x + 4$

14. $f(x) = 3x + 1$

15. $g(x) = \sqrt{x - 3}, \quad x \geq 3$

16. $g(x) = \sqrt{x + 2}, \quad x \geq -2$

17. $f(x) = 3x^2 + 2$

18. $f(x) = -4x^2 - 1$

19. $f(x) = x^3 - 4$

20. $f(x) = x^3 - 3$

Let $f(x) = 2^x$. We will see in the next section that this function is one-to-one. Find each value, always working part (a) before part (b).

21. **(a)** $f(3)$ **(b)** $f^{-1}(8)$

22. **(a)** $f(4)$ **(b)** $f^{-1}(16)$

23. **(a)** $f(0)$ **(b)** $f^{-1}(1)$

24. **(a)** $f(-2)$ **(b)** $f^{-1}\left(\dfrac{1}{4}\right)$

*The graphs of some functions are given in Exercises 25–30. **(a)** Use the horizontal line test to determine whether the function is one-to-one. **(b)** If the function is one-to-one, then graph the inverse of the function. (Remember that if f is one-to-one and (a, b) is on the graph of f, then (b, a) is on the graph of f^{-1}.) See Example 4.*

25.

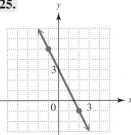

26.

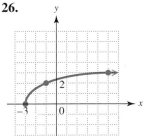

27.

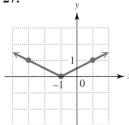

28.

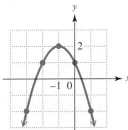

29.

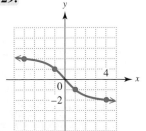

30.

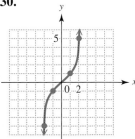

Each function defined in Exercises 31–38 is a one-to-one function. Graph the function as a solid line (or curve) and then graph its inverse on the same set of axes as a dashed line (or curve). In Exercises 35–38 you are given a table to complete so that graphing the function will be easier. See Example 4.

31. $f(x) = 2x - 1$ **32.** $f(x) = 2x + 3$ **33.** $g(x) = -4x$ **34.** $g(x) = -2x$

35. $f(x) = \sqrt{x}$,
$x \geq 0$

x	f(x)
0	
1	
4	

36. $f(x) = -\sqrt{x}$,
$x \geq 0$

x	f(x)
0	
1	
4	

37. $f(x) = x^3 - 2$

x	f(x)
−1	
0	
1	
2	

38. $f(x) = x^3 + 3$

x	f(x)
−2	
−1	
0	
1	

RELATING CONCEPTS (EXERCISES 39–42)

FOR INDIVIDUAL OR GROUP WORK

Inverse functions are used by government agencies and other businesses to send and receive coded information. The functions they use are usually very complicated. A simple example might use the function defined by $f(x) = 2x + 5$. (Note that it is one-to-one.) Suppose that each letter of the alphabet is assigned a numerical value according to its position, as follows:

A	1	G	7	L	12	Q	17
B	2	H	8	M	13	R	18
C	3	I	9	N	14	S	19
D	4	J	10	O	15	T	20
E	5	K	11	P	16	U	21
F	6						

V	22
W	23
X	24
Y	25
Z	26

This is an Enigma machine, used by the Germans in World War II to send coded messages.

Using the function, the word ALGEBRA *would be encoded as*

$$7 \quad 29 \quad 19 \quad 15 \quad 9 \quad 41 \quad 7,$$

because

$$f(A) = f(1) = 2(1) + 5 = 7, \quad f(L) = f(12) = 2(12) + 5 = 29,$$

and so on. The message would then be decoded by using the inverse of f, defined by $f^{-1}(x) = \frac{x-5}{2}$ $\left(or\ f^{-1}(x) = \frac{1}{2}x - \frac{5}{2}\right)$. For example,

$$f^{-1}(7) = \frac{7-5}{2} = 1 = A, \quad f^{-1}(29) = \frac{29-5}{2} = 12 = L,$$

and so on. **Work Exercises 39–42 in order.**

39. Suppose that you are an agent for a detective agency. Today's function for your code is defined by $f(x) = 4x - 5$. Find the rule for f^{-1} algebraically.

40. You receive the following coded message today. (Read across from left to right.)

47 95 23 67 −1 59 27 31 51 23 7 −1 43 7 79 43 −1 75 55 67

31 71 75 27 15 23 67 15 −1 75 15 71 75 75 27 31 51

23 71 31 51 7 15 71 43 31 7 15 11 3 67 15 −1 11

Use the letter/number assignment described earlier to decode the message.

41. Why is a one-to-one function essential in this encoding/decoding process?

42. Use $f(x) = x^3 + 4$ to encode your name, using the letter/number assignment described earlier.

Each function defined is one-to-one. Find the inverse algebraically, and then graph both the function and its inverse on the same graphing calculator screen. Use a square viewing window. See Objective 5.

43. $f(x) = 2x - 7$ **44.** $f(x) = -3x + 2$

45. $f(x) = x^3 + 5$ **46.** $f(x) = \sqrt[3]{x + 2}$

Some graphing calculators have the capability to draw the "inverse" of a function even if the function is not one-to-one; therefore, the inverse is not technically a function, but it is a relation. For example, the graphs of $y = x^2$ and $x = y^2$ are shown in the screen using a square viewing window.

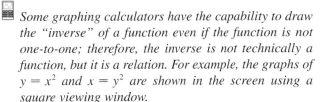

 Read your instruction manual to see if your model has this capability. If so, draw both Y_1 and its inverse in the same square window.

47. $Y_1 = X^2 + 3X + 4$ **48.** $Y_1 = X^3 - 9X$

PREVIEW EXERCISES

If $f(x) = 4^x$, find each value indicated. In Exercises 53 and 54, use a calculator, and give answers to the nearest hundredth. See Section 3.5.

49. $f(3)$ **50.** $f(0)$ **51.** $f\left(\dfrac{1}{2}\right)$

52. $f\left(-\dfrac{1}{2}\right)$ **53.** $f(2.73)$ **54.** $f(1.68)$

10.2   Exponential Functions

OBJECTIVES

1 Define an exponential function.

2 Graph an exponential function.

3 Solve exponential equations of the form $a^x = a^k$ for x.

4 Use exponential functions in applications involving growth or decay.

OBJECTIVE 1 Define an exponential function.   In **Section 8.2** we showed how to evaluate 2^x for rational values of x. For example,

$$2^3 = 8, \qquad 2^{-1} = \frac{1}{2}, \qquad 2^{1/2} = \sqrt{2}, \qquad \text{and} \qquad 2^{3/4} = \sqrt[4]{2^3} = \sqrt[4]{8}.$$

In more advanced courses it is shown that 2^x exists for all real number values of x, both rational and irrational. (Later in this chapter, we will see how to approximate the value of 2^x for irrational x.) The following definition of an exponential function assumes that a^x exists for all real numbers x.

> **Exponential Function**
>
> For $a > 0$, $a \neq 1$, and all real numbers x,
>
> $$f(x) = a^x$$
>
> defines the **exponential function with base a.**

▶ **NOTE** *The two restrictions on a in the definition of an exponential function $f(x) = a^x$ are important.*

1. The restriction $a > 0$ is necessary so that the function can be defined for all real numbers x. For example, letting a be negative ($a = -2$, for instance) and letting $x = \frac{1}{2}$ would give the expression $(-2)^{1/2}$, which is not real.

2. The restriction $a \neq 1$ is necessary because 1 raised to any power is equal to 1, and the function would then be the linear function defined by $f(x) = 1$.

OBJECTIVE 2 Graph an exponential function. We graph an exponential function by finding several ordered pairs that belong to the function, plotting these points, and connecting them with a smooth curve.

EXAMPLE 1 Graphing an Exponential Function with $a > 1$

Graph $f(x) = 2^x$.

Choose some values of x, and find the corresponding values of $f(x)$. Plotting these points and drawing a smooth curve through them gives the darker graph shown in Figure 6. This graph is typical of the graphs of exponential functions of the form $F(x) = a^x$, where $a > 1$. *The larger the value of a, the faster the graph rises.* Compare the lighter graph of $F(x) = 5^x$ with the graph of $f(x) = 2^x$ in Figure 6.

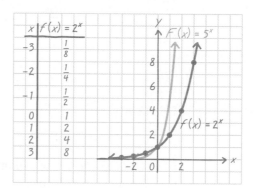

x	$f(x) = 2^x$
-3	$\frac{1}{8}$
-2	$\frac{1}{4}$
-1	$\frac{1}{2}$
0	1
1	2
2	4
3	8

FIGURE 6

By the vertical line test, the graphs in Figure 6 represent functions. As these graphs suggest, the domain of an exponential function includes all real numbers. Because y is always positive, the range is $(0, \infty)$. Figure 6 also shows an important characteristic of exponential functions where $a > 1$: *As x gets larger, y increases at a faster and faster rate.*

✔ **Now Try Exercise 5.**

▶ **CAUTION** The graph of an exponential function *approaches* the x-axis, but does ***not*** touch it.

EXAMPLE 2 Graphing an Exponential Function with $0 < a < 1$

Graph $g(x) = \left(\dfrac{1}{2}\right)^x$.

Again, find some points on the graph. The graph, shown in Figure 7, is very similar to that of $f(x) = 2^x$ (Figure 6) with the same domain and range, except that here *as x gets larger, y decreases.* This graph is typical of the graph of a function of the form $F(x) = a^x$, where $0 < a < 1$.

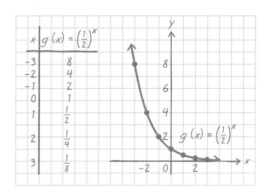

FIGURE 7

✔ **Now Try Exercise 7.**

Based on Examples 1 and 2, we make the following generalizations about the graphs of exponential functions of the form $F(x) = a^x$.

Graph of $F(x) = a^x$

1. The graph contains the point $(0, 1)$.

2. When $a > 1$, the graph will *rise* from left to right. (See Figure 6.) When $0 < a < 1$, the graph will *fall* from left to right. (See Figure 7.) In both cases, the graph goes from the second quadrant to the first.

3. The graph will approach the x-axis, but never touch it. (Recall from **Section 7.4** that such a line is called an *asymptote.*)

4. The domain is $(-\infty, \infty)$, and the range is $(0, \infty)$.

EXAMPLE 3 Graphing a More Complicated Exponential Function

Graph $f(x) = 3^{2x-4}$.

Find some ordered pairs.

$$\text{If } x = 0, \text{ then } y = 3^{2(0)-4} = 3^{-4} = \frac{1}{81}.$$

$$\text{If } x = 2, \text{ then } y = 3^{2(2)-4} = 3^0 = 1.$$

These ordered pairs, $\left(0, \frac{1}{81}\right)$ and $(2, 1)$, along with the other ordered pairs shown in the table, lead to the graph in Figure 8. The graph is similar to the graph of $f(x) = 3^x$ except that it is shifted to the right and rises more rapidly.

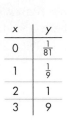

x	y
0	$\frac{1}{81}$
1	$\frac{1}{9}$
2	1
3	9

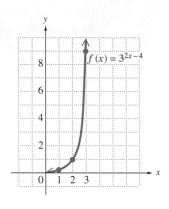

$f(x) = 3^{2x-4}$

FIGURE 8

✔ **Now Try Exercise 11.**

OBJECTIVE 3 **Solve exponential equations of the form $a^x = a^k$ for x.**   Until this chapter, we have solved only equations that had the variable as a base, like $x^2 = 8$; all exponents have been constants. An **exponential equation** is an equation that has a variable in an exponent, such as

$$9^x = 27.$$

By the horizontal line test, the exponential function defined by $F(x) = a^x$ is a one-to-one function, so we can use the following property to solve many exponential equations.

Property for Solving an Exponential Equation

For $a > 0$ and $a \neq 1$,   if $a^x = a^y$ then $x = y.$

This property would not necessarily be true if $a = 1$.

To solve an exponential equation using this property, follow these steps.

Solving an Exponential Equation

Step 1   **Each side must have the same base.** If the two sides of the equation do not have the same base, express each as a power of the same base if possible.

Step 2   **Simplify exponents** if necessary, using the rules of exponents.

Step 3   **Set exponents equal** using the property given in this section.

Step 4   **Solve** the equation obtained in Step 3.

▶ **NOTE** These steps cannot be applied to an exponential equation like
$$3^x = 12$$
because Step 1 cannot easily be done. A method for solving such equations is given in **Section 10.6.**

EXAMPLE 4 **Solving an Exponential Equation**

Solve the equation $9^x = 27$.

$$9^x = 27$$

$$(3^2)^x = 3^3 \qquad \text{Write with the same base;}$$
$$\qquad\qquad\qquad 9 = 3^2 \text{ and } 27 = 3^3. \text{ (Step 1)}$$

$$3^{2x} = 3^3 \qquad \text{Power rule for exponents (Step 2)}$$

$$2x = 3 \qquad \text{If } a^x = a^y, \text{ then } x = y. \text{ (Step 3)}$$

$$x = \frac{3}{2} \qquad \text{Solve for } x. \text{ (Step 4)}$$

Check that the solution set is $\left\{\frac{3}{2}\right\}$ by substituting $\frac{3}{2}$ for x:

$$9^x = 9^{3/2} = \left(9^{1/2}\right)^3 = 3^3 = 27, \quad \text{as required.}$$

✔ **Now Try Exercise 17.**

EXAMPLE 5 **Solving Exponential Equations**

Solve each equation.

(a) $\quad 4^{3x-1} = 16^{x+2}$

> Be careful multiplying the exponents.

$$4^{3x-1} = (4^2)^{x+2} \qquad \text{Write with the same base; } 16 = 4^2.$$

$$4^{3x-1} = 4^{2x+4} \qquad \text{Power rule for exponents}$$

$$3x - 1 = 2x + 4 \qquad \text{Set exponents equal.}$$

$$x = 5 \qquad \text{Subtract } 2x; \text{ add 1.}$$

Verify that the solution set is $\{5\}$.

(b) $6^x = \dfrac{1}{216}$

$$6^x = \frac{1}{6^3} \qquad 216 = 6^3$$

$$6^x = 6^{-3} \qquad \text{Write with the same base; } \frac{1}{6^3} = 6^{-3}.$$

$$x = -3 \qquad \text{Set exponents equal.}$$

Check that the solution set is $\{-3\}$ by substituting -3 for x:

$$6^x = 6^{-3} = \frac{1}{6^3} = \frac{1}{216}, \quad \text{as required.}$$

(c) $\left(\dfrac{2}{3}\right)^x = \dfrac{9}{4}$

$\left(\dfrac{2}{3}\right)^x = \left(\dfrac{4}{9}\right)^{-1}$ $\qquad \dfrac{9}{4} = \left(\dfrac{4}{9}\right)^{-1}$

$\left(\dfrac{2}{3}\right)^x = \left[\left(\dfrac{2}{3}\right)^2\right]^{-1}$ Write with the same base.

$\left(\dfrac{2}{3}\right)^x = \left(\dfrac{2}{3}\right)^{-2}$ Power rule for exponents

$x = -2$ Set exponents equal.

Check that the solution set is $\{-2\}$.

✔ **Now Try Exercises 19, 21, and 25.**

OBJECTIVE ➍ Use exponential functions in applications involving growth or decay.

EXAMPLE 6 Solving an Application Involving Exponential Growth

The graph in Figure 9 shows the concentration of carbon dioxide (in parts per million) in the air. This concentration is increasing exponentially.

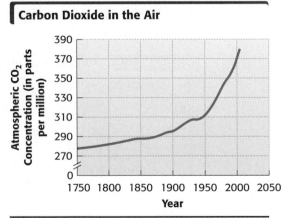

Carbon Dioxide in the Air

Source: Sacramento Bee; National Oceanic and Atmospheric Administration.

FIGURE 9

The data are approximated by the function defined by $f(x) = 266(1.001)^x$, where x is the number of years since 1750. Use this function and a calculator to approximate the concentration of carbon dioxide in parts per million for each year.

(a) 1900

Because x represents the number of years since 1750, $x = 1900 - 1750 = 150$. Thus, evaluate $f(150)$.

$$f(x) = 266(1.001)^x$$
$$f(150) = 266(1.001)^{150} \qquad \text{Let } x = 150.$$
$$\approx 309 \text{ parts per million} \qquad \text{Use a calculator.}$$

(b) 1950

$$f(200) = 266(1.001)^{200} \qquad\qquad x = 1950 - 1750 = 200$$
$$\approx 325 \text{ parts per million}$$

✔ **Now Try Exercise 37.**

EXAMPLE 7 **Applying an Exponential Decay Function**

The atmospheric pressure (in millibars) at a given altitude x, in meters, can be approximated by the function defined by

$$f(x) = 1038(1.000134)^{-x},$$

for values of x between 0 and 10,000. Because the base is greater than 1 and the coefficient of x in the exponent is negative, the function values decrease as x increases. This means that as the altitude increases, the atmospheric pressure decreases. (*Source:* Miller, A. and J. Thompson, *Elements of Meteorology,* Fourth Edition, Charles E. Merrill Publishing Company, 1993.)

(a) According to this function, what is the pressure at ground level?

At ground level, $x = 0$, so

$$f(0) = 1038(1.000134)^{-0} = 1038(1) = 1038.$$

The pressure is 1038 millibars.

(b) What is the pressure at 5000 m?

$$f(5000) = 1038(1.000134)^{-5000}$$
$$\approx 531 \qquad\qquad \text{Use a calculator.}$$

The pressure is approximately 531 millibars.

✔ **Now Try Exercise 39.**

10.2 EXERCISES

Concept Check   Choose the correct response in Exercises 1–4.

1. Which point lies on the graph of $f(x) = 2^x$?

   A. $(1, 0)$ **B.** $(2, 1)$ **C.** $(0, 1)$ **D.** $\left(\sqrt{2}, \dfrac{1}{2}\right)$

2. Which statement is true?

   A. The y-intercept of the graph of $f(x) = 10^x$ is $(0, 10)$.
   B. For any $a > 1$, the graph of $f(x) = a^x$ falls from left to right.
   C. The point $\left(\frac{1}{2}, \sqrt{5}\right)$ lies on the graph of $f(x) = 5^x$.
   D. The graph of $y = 4^x$ rises at a faster rate than the graph of $y = 10^x$.

3. The asymptote of the graph of $F(x) = a^x$

    A. is the x-axis   **B.** is the y-axis

    C. has equation $x = 1$ **D.** has equation $y = 1$.

4. Which equation is graphed here?

    A. $y = 1000\left(\dfrac{1}{2}\right)^{0.3x}$ **B.** $y = 1000\left(\dfrac{1}{2}\right)^{x}$

    C. $y = 1000(2)^{0.3x}$ **D.** $y = 1000^{x}$

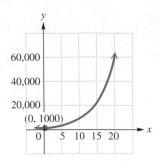

Graph each exponential function. See Examples 1–3.

  5. $f(x) = 3^x$ **6.** $f(x) = 5^x$ **7.** $g(x) = \left(\dfrac{1}{3}\right)^x$

  8. $g(x) = \left(\dfrac{1}{5}\right)^x$ **9.** $y = 4^{-x}$ **10.** $y = 6^{-x}$

  11. $y = 2^{2x-2}$ **12.** $y = 2^{2x+1}$

13. *Concept Check*

    (a) For an exponential function defined by $f(x) = a^x$, if $a > 1$, the graph _____

                                                           (rises/falls)

        from left to right. If $0 < a < 1$, the graph _____ from left to right.
                                           (rises/falls)

    (b) Based on your answers in part (a), make a conjecture (an educated guess) concerning whether an exponential function defined by $f(x) = a^x$ is one-to-one. Then decide whether it has an inverse based on the concepts of **Section 10.1.**

14. In your own words, describe the characteristics of the graph of an exponential function. Use the exponential function defined by $f(x) = 3^x$ (Exercise 5) and the words *asymptote*, *domain*, and *range* in your explanation.

Solve each equation. See Examples 4 and 5.

  15. $6^x = 36$ **16.** $8^x = 64$ **17.** $100^x = 1000$

  18. $8^x = 4$ **19.** $16^{2x+1} = 64^{x+3}$ **20.** $9^{2x-8} = 27^{x-4}$

  21. $5^x = \dfrac{1}{125}$ **22.** $3^x = \dfrac{1}{81}$ **23.** $5^x = 0.2$

  24. $10^x = 0.1$ **25.** $\left(\dfrac{3}{2}\right)^x = \dfrac{8}{27}$ **26.** $\left(\dfrac{4}{3}\right)^x = \dfrac{27}{64}$

Use the exponential key of a calculator to find an approximation to the nearest thousandth.

  27. $12^{2.6}$ **28.** $13^{1.8}$ **29.** $0.5^{3.921}$

  30. $0.6^{4.917}$ **31.** $2.718^{2.5}$ **32.** $2.718^{-3.1}$

The graph shown here accompanied the article "Is Our World Warming?" which appeared in National Geographic. *It shows projected temperature increases using two graphs: one an exponential-type curve, and the other linear. From the graph, approximate the increase* **(a)** *for the exponential curve and* **(b)** *for the linear graph for each year.*

33. 2000

34. 2010

35. 2020

36. 2040

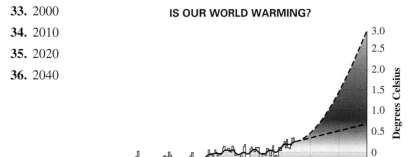

IS OUR WORLD WARMING?

Degrees Celsius

Year

Graph, "Zero Equals Average Global Temperature for the Period 1950–1979." Dale D. Glasgow, © National Geographic Society. Reprinted by permission.

Solve each problem. See Examples 6 and 7.

37. Based on figures from 1970 through 2002, the worldwide carbon monoxide emissions in thousands of tons are approximated by the exponential function defined by

$$f(x) = 220{,}717(1.0217)^{-x},$$

where $x = 0$ corresponds to 1970, $x = 5$ corresponds to 1975, and so on. (*Source:* U.S. Environmental Protection Agency.)

(a) Use this model to approximate the emissions in 1970.

(b) Use this model to approximate the emissions in 1995.

(c) In 2002, the actual amount of emissions was 112,049 million tons. How does this compare to the number that the model provides?

38. Based on figures from 1980 through 2003, the municipal solid waste generated in millions of tons can be approximated by the exponential function defined by

$$f(x) = 158.64(1.0189)^{x},$$

where $x = 0$ corresponds to 1980, $x = 5$ corresponds to 1985, and so on. (*Source:* U.S. Environmental Protection Agency.)

(a) Use the model to approximate the number of tons of this waste in 1980.

(b) Use the model to approximate the number of tons of this waste in 1995.

(c) In 2003, the actual number of millions of tons of this waste was 236.2. How does this compare to the number that the model provides?

39. A small business estimates that the value $V(t)$ of a copy machine is decreasing according to the function defined by

$$V(t) = 5000(2)^{-0.15t},$$

where t is the number of years that have elapsed since the machine was purchased, and $V(t)$ is in dollars.

(a) What was the original value of the machine?

(b) What is the value of the machine 5 yr after purchase? Give your answer to the nearest dollar.

(c) What is the value of the machine 10 yr after purchase? Give your answer to the nearest dollar.

(d) Graph the function.

40. The amount of radioactive material in an ore sample is given by the function defined by

$$A(t) = 100(3.2)^{-0.5t},$$

where $A(t)$ is the amount present, in grams, of the sample t months after the initial measurement.

(a) How much was present at the initial measurement? (*Hint:* $t = 0$.)

(b) How much was present 2 months later?

(c) How much was present 10 months later?

(d) Graph the function.

41. Refer to the function in Exercise 39. When will the value of the machine be $2500? (*Hint:* Let $V(t) = 2500$, divide both sides by 5000, and use the method of Example 4.)

42. Refer to the function in Exercise 39. When will the value of the machine be $1250?

> **PREVIEW EXERCISES**

Determine what number would have to be placed in each box for the statement to be true. See Section 5.1.

43. $2^{\square} = 16$ **44.** $2^{\square} = \dfrac{1}{16}$ **45.** $2^{\square} = 1$ **46.** $2^{\square} = \sqrt{2}$

10.3 Logarithmic Functions

OBJECTIVES

1 Define a logarithm.

2 Convert between exponential and logarithmic forms.

3 Solve logarithmic equations of the form $\log_a b = k$ for a, b, or k.

4 Define and graph logarithmic functions.

5 Use logarithmic functions in applications involving growth or decay.

The graph of $y = 2^x$ is the curve shown in blue in Figure 10. Because $y = 2^x$ defines a one-to-one function, it has an inverse. Interchanging x and y gives

$$x = 2^y, \quad \text{the inverse of} \quad y = 2^x.$$

As we saw in **Section 10.1,** the graph of the inverse is found by reflecting the graph of $y = 2^x$ about the line $y = x$. The graph of $x = 2^y$ is shown as a red curve in Figure 10.

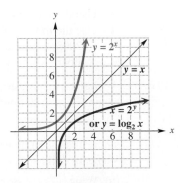

FIGURE 10

OBJECTIVE ❶ Define a logarithm. We cannot solve the equation $x = 2^y$ for the dependent variable y with the methods presented up to now. The following definition is used to solve $x = 2^y$ for y.

Logarithm

For all positive numbers a, with $a \neq 1$, and all positive numbers x,

$$y = \log_a x \quad \text{means the same as} \quad x = a^y.$$

This key statement should be memorized. The abbreviation **log** is used for the word **logarithm**. Read $\log_a x$ as "**the logarithm of x to the base a**" or "**the base a logarithm of x.**" To remember the location of the base and the exponent in each form, refer to the following diagrams.

Exponent ↓
Logarithmic form: $y = \log_a x$
↑ Base

Exponent ↓
Exponential form: $x = a^y$
↑ Base

In working with logarithmic form and exponential form, remember the following.

Meaning of $\log_a x$

A logarithm is an exponent. The expression $\log_a x$ represents the exponent to which the base a must be raised to obtain x.

OBJECTIVE ❷ Convert between exponential and logarithmic forms. We can use the definition of logarithm to write exponential statements in logarithmic form and logarithmic statements in exponential form. The following table shows several pairs of equivalent statements.

Exponential Form	Logarithmic Form
$3^2 = 9$	$\log_3 9 = 2$
$\left(\frac{1}{5}\right)^{-2} = 25$	$\log_{1/5} 25 = -2$
$10^5 = 100,000$	$\log_{10} 100,000 = 5$
$4^{-3} = \frac{1}{64}$	$\log_4 \frac{1}{64} = -3$

✔ **Now Try Exercises 3 and 11.**

OBJECTIVE ❸ Solve logarithmic equations of the form $\log_a b = k$ for a, b, or k. A **logarithmic equation** is an equation with a logarithm in at least one term. We solve logarithmic equations of the form $\log_a b = k$ for any of the three variables by first writing the equation in exponential form.

EXAMPLE 1 Solving Logarithmic Equations

Solve each equation.

(a) $\log_4 x = -2$

By the definition of logarithm, $\log_4 x = -2$ is equivalent to $x = 4^{-2}$. Solve this exponential equation.

$$x = 4^{-2} = \frac{1}{16}$$

The solution set is $\left\{\frac{1}{16}\right\}$.

(b) $\qquad \log_{1/2}(3x + 1) = 2$

$$3x + 1 = \left(\frac{1}{2}\right)^2 \qquad \text{Write in exponential form.}$$

$$3x + 1 = \frac{1}{4} \qquad \text{Apply the exponent.}$$

$$12x + 4 = 1 \qquad \text{Multiply each term by 4.}$$

$$12x = -3 \qquad \text{Subtract 4.}$$

$$x = -\frac{1}{4} \qquad \text{Divide by 12.}$$

The solution set is $\left\{-\frac{1}{4}\right\}$.

(c) $\qquad \log_x 3 = 2$

$$x^2 = 3 \qquad \text{Write in exponential form.}$$

$$x = \pm\sqrt{3} \qquad \text{Take square roots.}$$

Only the *principal* square root satisfies the equation since the base must be a positive number. The solution set is $\left\{\sqrt{3}\right\}$.

(d) $\qquad \log_{49} \sqrt[3]{7} = x$

$$49^x = \sqrt[3]{7} \qquad \text{Write in exponential form.}$$

$$(7^2)^x = 7^{1/3} \qquad \text{Write with the same base.}$$

$$7^{2x} = 7^{1/3} \qquad \text{Power rule for exponents}$$

$$2x = \frac{1}{3} \qquad \text{Set exponents equal.}$$

$$x = \frac{1}{6} \qquad \text{Divide by 2.}$$

The solution set is $\left\{\frac{1}{6}\right\}$.

✔ **Now Try Exercises 21, 25, 37, and 39.**

For any real number b, we know that $b^1 = b$ and for $b \neq 0$, $b^0 = 1$. Writing these two statements in logarithmic form gives the following two properties of logarithms.

Properties of Logarithms

For any positive real number b, with $b \neq 1$,

$$\log_b b = 1 \quad \text{and} \quad \log_b 1 = 0.$$

EXAMPLE 2 Using Properties of Logarithms

Use the preceding two properties of logarithms to evaluate each logarithm.

(a) $\log_7 7 = 1$ **(b)** $\log_{\sqrt{2}} \sqrt{2} = 1$

(c) $\log_9 1 = 0$ **(d)** $\log_{0.2} 1 = 0$

✔ **Now Try Exercise 19.**

OBJECTIVE 4 Define and graph logarithmic functions. Now we define the logarithmic function with base a.

Logarithmic Function

If a and x are positive numbers, with $a \neq 1$, then

$$G(x) = \log_a x$$

defines the **logarithmic function with base a.**

EXAMPLE 3 Graphing a Logarithmic Function with $a > 1$

Graph $f(x) = \log_2 x$.

By writing $y = f(x) = \log_2 x$ in exponential form as $x = 2^y$, we can identify ordered pairs that satisfy the equation. It is easier to choose values for y and find the corresponding values of x. Plotting the points in the table of ordered pairs and connecting them with a smooth curve gives the graph in Figure 11. This graph is typical of logarithmic functions with base $a > 1$.

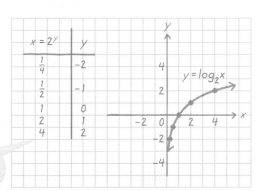

Be careful to write the x- and y-values in the correct order.

FIGURE 11

As the table and graph suggest, x is always positive, so the domain of a logarithmic function is $(0, \infty)$. The range includes all real numbers, $(-\infty, \infty)$.

✔ **Now Try Exercise 41.**

EXAMPLE 4 Graphing a Logarithmic Function with $0 < a < 1$

Graph $g(x) = \log_{1/2} x$.

We write $y = g(x) = \log_{1/2} x$ in exponential form as $x = \left(\frac{1}{2}\right)^y$, then choose values for y and find the corresponding values of x. See the table of ordered pairs.

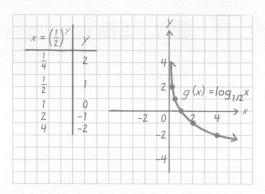

$x = \left(\frac{1}{2}\right)^y$	y
$\frac{1}{4}$	2
$\frac{1}{2}$	1
1	0
2	-1
4	-2

FIGURE 12

Plotting these points and connecting them with a smooth curve gives the graph in Figure 12. This graph, which is similar to that of $f(x) = \log_2 x$ (Figure 11) with the same domain and range, is typical of logarithmic functions with $0 < a < 1$.

✔ **Now Try Exercise 43.**

Based on the graphs of the functions defined by $y = \log_2 x$ in Figure 11 and $y = \log_{1/2} x$ in Figure 12, we make the following generalizations about the graphs of logarithmic functions of the form $G(x) = \log_a x$.

Graph of $G(x) = \log_a x$

1. The graph contains the point $(1, 0)$.

2. When $a > 1$, the graph will *rise* from left to right, from the fourth quadrant to the first. (See Figure 11.) When $0 < a < 1$, the graph will *fall* from left to right, from the first quadrant to the fourth. (See Figure 12.)

3. The graph will approach the y-axis, but never touch it. (The y-axis is an asymptote.)

4. The domain is $(0, \infty)$, and the range is $(-\infty, \infty)$.

Compare these generalizations to the similar ones for exponential functions found in **Section 10.2.**

OBJECTIVE 5 Use logarithmic functions in applications involving growth or decay. Logarithmic functions, like exponential functions, can be applied to growth or decay of real-world phenomena.

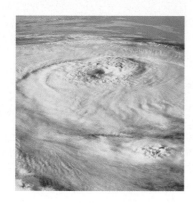

EXAMPLE 5 Solving an Application of a Logarithmic Function

The function defined by

$$f(x) = 27 + 1.105 \log_{10}(x + 1)$$

approximates the barometric pressure in inches of mercury at a distance of x miles from the eye of a typical hurricane. (*Source:* Miller, A. and R. Anthes, *Meteorology,* Fifth Edition, Charles E. Merrill Publishing Company, 1985.)

(a) Approximate the pressure 9 mi from the eye of the hurricane.

Let $x = 9$, and find $f(9)$.

$$
\begin{aligned}
f(9) &= 27 + 1.105 \log_{10}(9 + 1) &&\text{Let } x = 9. \\
&= 27 + 1.105 \log_{10} 10 &&\text{Add inside parentheses.} \\
&= 27 + 1.105(1) &&\log_{10} 10 = 1 \\
&= 28.105
\end{aligned}
$$

The pressure 9 mi from the eye of the hurricane is 28.105 in.

(b) Approximate the pressure 99 mi from the eye of the hurricane.

$$
\begin{aligned}
f(99) &= 27 + 1.105 \log_{10}(99 + 1) &&\text{Let } x = 99. \\
&= 27 + 1.105 \log_{10} 100 &&\text{Add inside parentheses.} \\
&= 27 + 1.105(2) &&\log_{10} 100 = 2 \\
&= 29.21
\end{aligned}
$$

The pressure 99 mi from the eye of the hurricane is 29.21 in.

✔ **Now Try Exercise 53.**

10.3 EXERCISES

1. *Concept Check* Match the logarithmic equation in Column I with the corresponding exponential equation from Column II.

I	II
(a) $\log_{1/3} 3 = -1$	**A.** $8^{1/3} = \sqrt[3]{8}$
(b) $\log_5 1 = 0$	**B.** $\left(\dfrac{1}{3}\right)^{-1} = 3$
(c) $\log_2 \sqrt{2} = \dfrac{1}{2}$	**C.** $4^1 = 4$
(d) $\log_{10} 1000 = 3$	**D.** $2^{1/2} = \sqrt{2}$
(e) $\log_8 \sqrt[3]{8} = \dfrac{1}{3}$	**E.** $5^0 = 1$
(f) $\log_4 4 = 1$	**F.** $10^3 = 1000$

2. *Concept Check* Use the definition of logarithm to match the logarithm in Column I with its value in Column II. (*Example:* $\log_3 9$ is equal to 2 because 2 is the exponent to which 3 must be raised in order to obtain 9.)

I	II
(a) $\log_4 16$	**A.** -2
(b) $\log_3 81$	**B.** -1
(c) $\log_3\left(\dfrac{1}{3}\right)$	**C.** 2
(d) $\log_{10} 0.01$	**D.** 0
(e) $\log_5 \sqrt{5}$	**E.** $\dfrac{1}{2}$
(f) $\log_{13} 1$	**F.** 4

Write in logarithmic form. See the table in Objective 2.

3. $4^5 = 1024$

4. $3^6 = 729$

5. $\left(\dfrac{1}{2}\right)^{-3} = 8$

6. $\left(\dfrac{1}{6}\right)^{-3} = 216$

7. $10^{-3} = 0.001$

8. $36^{1/2} = 6$

9. $\sqrt[4]{625} = 5$

10. $\sqrt[3]{343} = 7$

Write in exponential form. See the table in Objective 2.

11. $\log_4 64 = 3$

12. $\log_2 512 = 9$

13. $\log_{10} \dfrac{1}{10,000} = -4$

14. $\log_{100} 100 = 1$

15. $\log_6 1 = 0$

16. $\log_\pi 1 = 0$

17. $\log_9 3 = \dfrac{1}{2}$

18. $\log_{64} 2 = \dfrac{1}{6}$

19. Match each logarithm in Column I with its value in Column II. See Example 2.

I	II
(a) $\log_8 8$	**A.** -1
(b) $\log_{16} 1$	**B.** 0
(c) $\log_{0.3} 1$	**C.** 1
(d) $\log_{\sqrt{7}} \sqrt{7}$	**D.** 0.1

20. When a student asked his teacher to explain how to evaluate $\log_9 3$ without showing any work, his teacher told him, "Think radically." Explain what the teacher meant by this hint.

Solve each equation. See Examples 1 and 2.

21. $x = \log_{27} 3$

22. $x = \log_{125} 5$

23. $\log_x 9 = \dfrac{1}{2}$

24. $\log_x 5 = \dfrac{1}{2}$

25. $\log_x 125 = -3$

26. $\log_x 64 = -6$

27. $\log_{12} x = 0$

28. $\log_4 x = 0$

29. $\log_x x = 1$

30. $\log_x 1 = 0$

31. $\log_x \dfrac{1}{25} = -2$

32. $\log_x \dfrac{1}{10} = -1$

33. $\log_8 32 = x$

34. $\log_{81} 27 = x$

35. $\log_\pi \pi^4 = x$

36. $\log_{\sqrt{2}} \sqrt{2^9} = x$

37. $\log_6 \sqrt{216} = x$

38. $\log_4 \sqrt{64} = x$

39. $\log_4(2x + 4) = 3$

40. $\log_3(2x + 7) = 4$

*If the point (p, q) is on the graph of $f(x) = a^x$ (for $a > 0$ and $a \neq 1$), then the point (q, p) is on the graph of $f^{-1}(x) = \log_a x$. Use this fact, and refer to the graphs required in Exercises 5–8 in **Section 10.2** to graph each logarithmic function. See Examples 3 and 4.*

41. $y = \log_3 x$

42. $y = \log_5 x$

43. $y = \log_{1/3} x$

44. $y = \log_{1/5} x$

45. Explain why 1 is not allowed as a base for a logarithmic function.

46. Compare the summary of facts about the graph of $F(x) = a^x$ in **Section 10.2** with the similar summary of facts about the graph of $G(x) = \log_a x$ in this section. Make a list of the facts that reinforce the concept that F and G are inverse functions.

47. *Concept Check* The domain of $F(x) = a^x$ is $(-\infty, \infty)$, while the range is $(0, \infty)$. Therefore, since $G(x) = \log_a x$ defines the inverse of F, the domain of G is _____, while the range of G is _____ .

48. *Concept Check* The graphs of both $F(x) = 3^x$ and $G(x) = \log_3 x$ rise from left to right. Which one rises at a faster rate?

Concept Check *Use the graph at the right to predict the value of $f(t)$ for the given value of t.*

49. $t = 0$

50. $t = 10$

51. $t = 60$

52. Show that the points determined in Exercises 49–51 lie on the graph of $f(t) = 8 \log_5(2t + 5)$.

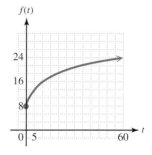

Solve each problem. See Example 5.

53. According to selected figures from 1981 through 2003, the number of billion cubic feet of natural gas gross withdrawals from crude oil wells in the United States can be approximated by the function defined by

$$f(x) = 3800 + 585 \log_2 x,$$

where $x = 1$ corresponds to 1981, $x = 2$ to 1982, and so on. (*Source:* Energy Information Administration.) Use this function to approximate the number of cubic feet withdrawn in each of the following years.

(a) 1982   **(b)** 1988   **(c)** 1996

54. According to selected figures from the last two decades of the 20th century, the number of trillion cubic feet of dry natural gas consumed worldwide can be approximated by the function defined by

$$f(x) = 51.47 + 6.044 \log_2 x,$$

where $x = 1$ corresponds to 1980, $x = 2$ to 1981, and so on. (*Source:* Energy Information Administration.) Use the function to approximate consumption in each year.

(a) 1980   **(b)** 1987   **(c)** 1995

55. Sales (in thousands of units) of a new product are approximated by the function defined by

$$S(t) = 100 + 30 \log_3(2t + 1),$$

where t is the number of years after the product is introduced.

(a) What were the sales after 1 yr?
(b) What were the sales after 13 yr?
(c) Graph $y = S(t)$.

56. A study showed that the number of mice in an old abandoned house was approximated by the function defined by

$$M(t) = 6 \log_4(2t + 4),$$

where t is measured in months and $t = 0$ corresponds to January 1998. Find the number of mice in the house in

(a) January 1998   **(b)** July 1998   **(c)** July 2000.
(d) Graph the function.

*In the United States, the intensity of an earthquake is rated using the **Richter scale.** The Richter scale rating of an earthquake of intensity x is given by*

$$R = \log_{10} \frac{x}{x_0},$$

where x_0 is the intensity of an earthquake of a certain (small) size. The figure here shows Richter scale ratings for major Southern California earthquakes since 1930. As the figure indicates, earthquakes "come in bunches."

Major Southern California Earthquakes
(with magnitudes greater than 4.7)

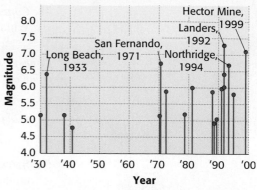

Source: Caltech; U.S. Geological Survey.

57. The 1994 Northridge earthquake had a Richter scale rating of 6.7; the 1992 Landers earthquake had a rating of 7.3. How much more powerful was the Landers earthquake than the Northridge earthquake?

58. Compare the smallest rated earthquake in the figure (at 4.8) with the Landers quake. How much more powerful was the Landers quake?

*As mentioned in **Section 10.1,** some graphing calculators have the capability of drawing the inverse of a function. For example, the two screens that follow show the graphs of $f(x) = 2^x$ and $g(x) = \log_2 x$. The graph of g was obtained by drawing the graph of f^{-1}, since $g(x) = f^{-1}(x)$. (Compare to Figure 10 in this section.)*

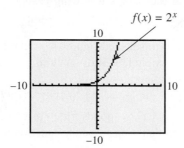

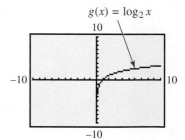

Use a graphing calculator with the capability of drawing the inverse of a function to draw the graph of each logarithmic function. Use the standard viewing window.

59. $g(x) = \log_3 x$
(Compare to Exercise 41.)

60. $g(x) = \log_5 x$
(Compare to Exercise 42.)

61. $g(x) = \log_{1/3} x$
(Compare to Exercise 43.)

62. $g(x) = \log_{1/5} x$
(Compare to Exercise 44.)

PREVIEW EXERCISES

*Simplify each expression. Write answers using only positive exponents. See **Section 5.1.***

63. $4^7 \cdot 4^2$

64. $3^{-3} \cdot 3^{16}$

65. $\dfrac{5^{-3}}{5^8}$

66. $\dfrac{7^8}{7^{-4}}$

67. $(9^3)^{-2}$

68. $(6x^5)^5$

10.4 Properties of Logarithms

OBJECTIVES

1 Use the product rule for logarithms.

2 Use the quotient rule for logarithms.

3 Use the power rule for logarithms.

4 Use properties to write alternative forms of logarithmic expressions.

Logarithms have been used as an aid to numerical calculation for several hundred years. Today the widespread use of calculators has made the use of logarithms for calculation obsolete. However, logarithms are still very important in applications and in further work in mathematics.

OBJECTIVE 1 Use the product rule for logarithms. One way in which logarithms simplify problems is by changing a problem of multiplication into one of addition. We know that $\log_2 4 = 2$, $\log_2 8 = 3$, and $\log_2 32 = 5$. Since $2 + 3 = 5$,

$$\log_2 32 = \log_2 4 + \log_2 8$$
$$\log_2 (4 \cdot 8) = \log_2 4 + \log_2 8.$$

This is true in general.

Product Rule for Logarithms

If x, y, and b are positive real numbers, where $b \neq 1$, then

$$\log_b xy = \log_b x + \log_b y.$$

That is, the logarithm of a product is the sum of the logarithms of the factors.

▶ **NOTE** The word statement of the product rule can be restated by replacing "logarithm" with "exponent." The rule then becomes the familiar rule for multiplying exponential expressions: The *exponent* of a product is the sum of the *exponents* of the factors.

To prove this rule, let $m = \log_b x$ and $n = \log_b y$, and recall that

$$\log_b x = m \quad \text{means} \quad b^m = x.$$
$$\log_b y = n \quad \text{means} \quad b^n = y.$$

Now consider the product xy.

$xy = b^m \cdot b^n$	Substitute.
$xy = b^{m+n}$	Product rule for exponents
$\log_b xy = m + n$	Convert to logarithmic form.
$\log_b xy = \log_b x + \log_b y$	Substitute.

The last statement is the result we wished to prove.

EXAMPLE 1 Using the Product Rule

Use the product rule to rewrite each logarithm. Assume $x > 0$.

(a) $\log_5(6 \cdot 9) = \log_5 6 + \log_5 9$

(b) $\log_7 8 + \log_7 12 = \log_7(8 \cdot 12)$

$$= \log_7 96$$

(c) $\log_3(3x) = \log_3 3 + \log_3 x$ Product rule

        $= 1 + \log_3 x$ $\log_3 3 = 1$

(d) $\log_4 x^3 = \log_4(x \cdot x \cdot x)$ $x^3 = x \cdot x \cdot x$

        $= \log_4 x + \log_4 x + \log_4 x$ Product rule

        $= 3 \log_4 x$

✔ **Now Try Exercises 7 and 21.**

OBJECTIVE 2 **Use the quotient rule for logarithms.** The rule for division is similar to the rule for multiplication.

Quotient Rule for Logarithms

If x, y, and b are positive real numbers, where $b \neq 1$, then

$$\log_b \frac{x}{y} = \log_b x - \log_b y.$$

That is, the logarithm of a quotient is the difference between the logarithm of the numerator and the logarithm of the denominator.

The proof of this rule is very similar to the proof of the product rule.

EXAMPLE 2 **Using the Quotient Rule**

Use the quotient rule to rewrite each logarithm. Assume $x > 0$.

(a) $\log_4 \dfrac{7}{9} = \log_4 7 - \log_4 9$

(b) $\log_5 6 - \log_5 x = \log_5 \dfrac{6}{x}$

(c) $\log_3 \dfrac{27}{5} = \log_3 27 - \log_3 5$

          $= 3 - \log_3 5$ $\log_3 27 = 3$

✔ **Now Try Exercises 9 and 23.**

▶ **CAUTION** *There is no property of logarithms to rewrite the logarithm of a sum or difference.* For example, we *cannot* write $\log_b(x + y)$ in terms of $\log_b x$ and $\log_b y$. Also,

$$\log_b \frac{x}{y} \neq \frac{\log_b x}{\log_b y}.$$

OBJECTIVE 3 Use the power rule for logarithms.   An exponential expression such as 2^3 means $2 \cdot 2 \cdot 2$; the base is used as a factor 3 times. Similarly, the product rule can be extended to rewrite the logarithm of a power as the product of the exponent and the logarithm of the base. For example, by the product rule for logarithms,

$$\begin{aligned} \log_5 2^3 &= \log_5(2 \cdot 2 \cdot 2) \\ &= \log_5 2 + \log_5 2 + \log_5 2 \\ &= 3 \log_5 2. \end{aligned} \qquad \begin{aligned} \log_2 7^4 &= \log_2(7 \cdot 7 \cdot 7 \cdot 7) \\ &= \log_2 7 + \log_2 7 + \log_2 7 + \log_2 7 \\ &= 4 \log_2 7. \end{aligned}$$

Furthermore, we saw in Example 1(d) that $\log_4 x^3 = 3 \log_4 x$. These examples suggest the following rule.

Power Rule for Logarithms

If x and b are positive real numbers, where $b \neq 1$, and if r is any real number, then

$$\log_b x^r = r \log_b x.$$

That is, the logarithm of a number to a power equals the exponent times the logarithm of the number.

As further examples of this rule,

$$\log_b m^5 = 5 \log_b m \qquad \text{and} \qquad \log_3 5^4 = 4 \log_3 5.$$

To prove the power rule, let $\log_b x = m$.

$$\begin{aligned} \log_b x &= m \\ b^m &= x && \text{Convert to exponential form.} \\ (b^m)^r &= x^r && \text{Raise to the power } r. \\ b^{mr} &= x^r && \text{Power rule for exponents} \\ \log_b x^r &= mr && \text{Convert to logarithmic form.} \\ \log_b x^r &= rm && \text{Commutative property} \\ \log_b x^r &= r \log_b x && m = \log_b x \end{aligned}$$

This is the statement to be proved.

As a special case of the power rule, let $r = \frac{1}{p}$, so

$$\log_b \sqrt[p]{x} = \log_b x^{1/p} = \frac{1}{p} \log_b x.$$

For example, using this result, with $x > 0$,

$$\log_b \sqrt[5]{x} = \log_b x^{1/5} = \frac{1}{5} \log_b x \qquad \text{and} \qquad \log_b \sqrt[3]{x^4} = \log_b x^{4/3} = \frac{4}{3} \log_b x.$$

Another special case is

$$\log_b \frac{1}{x} = \log_b x^{-1} = -\log_b x.$$

▶ **NOTE**   For a review of rational exponents, refer to **Section 8.2.**

EXAMPLE 3 Using the Power Rule

Use the power rule to rewrite each logarithm. Assume $b > 0$, $x > 0$, and $b \neq 1$.

(a) $\log_5 4^2 = 2 \log_5 4$ **(b)** $\log_b x^5 = 5 \log_b x$

(c) $\log_b \sqrt{7} = \log_b 7^{1/2}$ $\sqrt{x} = x^{1/2}$ **(d)** $\log_2 \sqrt[5]{x^2} = \log_2 x^{2/5}$ $\sqrt[5]{x^2} = x^{2/5}$

$\qquad\qquad = \dfrac{1}{2} \log_b 7$ Power rule $\qquad\qquad\qquad = \dfrac{2}{5} \log_2 x$ Power rule

✔ **Now Try Exercise 11.**

      Two special properties involving both exponential and logarithmic expressions come directly from the fact that logarithmic and exponential functions are inverses of each other.

Special Properties

If $b > 0$ and $b \neq 1$, then

$$b^{\log_b x} = x, \quad x > 0 \qquad \text{and} \qquad \log_b b^x = x.$$

To prove the first statement, let $y = \log_b x$.

$$y = \log_b x$$
$$b^y = x \qquad \text{Convert to exponential form.}$$
$$b^{\log_b x} = x \qquad \text{Replace } y \text{ with } \log_b x.$$

The proof of the second statement is similar.

EXAMPLE 4 Using the Special Properties

Find each value.

(a) $\log_5 5^4 = 4$, since $\log_b b^x = x$.   **(b)** $\log_3 9 = \log_3 3^2 = 2$

(c) $4^{\log_4 10} = 10$

✔ **Now Try Exercises 3 and 5.**

      Here is a summary of the properties of logarithms.

Properties of Logarithms

If x, y, and b are positive real numbers, where $b \neq 1$, and r is any real number, then

Product Rule $\log_b xy = \log_b x + \log_b y$

Quotient Rule $\log_b \dfrac{x}{y} = \log_b x - \log_b y$

Power Rule $\log_b x^r = r \log_b x$

Special Properties $b^{\log_b x} = x \quad \text{and} \quad \log_b b^x = x.$

OBJECTIVE 4 Use properties to write alternative forms of logarithmic expressions.

EXAMPLE 5 Writing Logarithms in Alternative Forms

Use the properties of logarithms to rewrite each expression if possible. Assume that all variables represent positive real numbers.

(a) $\log_4 4x^3$

$$= \log_4 4 + \log_4 x^3 \qquad \text{Product rule}$$

$$= 1 + 3 \log_4 x \qquad \log_4 4 = 1; \text{ power rule}$$

(b) $\log_7 \sqrt{\dfrac{m}{n}}$

$$= \log_7 \left(\frac{m}{n} \right)^{1/2} \qquad \text{Write the radical expression with a rational exponent.}$$

$$= \frac{1}{2} \log_7 \frac{m}{n} \qquad \text{Power rule}$$

$$= \frac{1}{2} (\log_7 m - \log_7 n) \qquad \text{Quotient rule}$$

(c) $\log_5 \dfrac{a^2}{bc}$

$$= \log_5 a^2 - \log_5 bc \qquad \text{Quotient rule}$$

$$= 2 \log_5 a - \log_5 bc \qquad \text{Power rule}$$

$$= 2 \log_5 a - (\log_5 b + \log_5 c) \qquad \text{Product rule}$$

$$= 2 \log_5 a - \log_5 b - \log_5 c \qquad \text{Use parentheses to avoid errors.}$$

(d) $4 \log_b m - \log_b n$

$$= \log_b m^4 - \log_b n \qquad \text{Power rule}$$

$$= \log_b \frac{m^4}{n} \qquad \text{Quotient rule}$$

(e) $\log_b(x + 1) + \log_b(2x + 1) - \dfrac{2}{3} \log_b x$

$$= \log_b(x + 1) + \log_b(2x + 1) - \log_b x^{2/3} \qquad \text{Power rule}$$

$$= \log_b \frac{(x + 1)(2x + 1)}{x^{2/3}} \qquad \text{Product and quotient rules}$$

$$= \log_b \frac{2x^2 + 3x + 1}{x^{2/3}} \qquad \text{Multiply in the numerator.}$$

(f) $\log_8(2p + 3r)$ cannot be rewritten using the properties of logarithms. There is no property of logarithms to rewrite the logarithm of a sum.

✔ **Now Try Exercises 13, 15, 27, and 31.**

In the next example, we use numerical values for $\log_2 5$ and $\log_2 3$. While we use the equals sign to give these values, they are actually just approximations since most logarithms of this type are irrational numbers. We use $=$ with the understanding that the values are correct to four decimal places.

EXAMPLE 6 Using the Properties of Logarithms with Numerical Values

Given that $\log_2 5 = 2.3219$ and $\log_2 3 = 1.5850$, evaluate the following.

(a) $\log_2 15 = \log_2(3 \cdot 5)$

$\quad\quad\quad = \log_2 3 + \log_2 5$ Product rule

$\quad\quad\quad = 1.5850 + 2.3219$ Substitute the given values.

$\quad\quad\quad = 3.9069$ Add.

(b) $\log_2 0.6 = \log_2 \dfrac{3}{5}$ $0.6 = \frac{6}{10} = \frac{3}{5}$

$\quad\quad\quad = \log_2 3 - \log_2 5$ Quotient rule

$\quad\quad\quad = 1.5850 - 2.3219$ Substitute the given values.

$\quad\quad\quad = -0.7369$ Subtract.

(c) $\log_2 27 = \log_2 3^3$

$\quad\quad\quad = 3 \log_2 3$ Power rule

$\quad\quad\quad = 3(1.5850)$ Substitute the given value.

$\quad\quad\quad = 4.7550$ Multiply.

✔ **Now Try Exercises 33, 35, and 43.**

EXAMPLE 7 Deciding Whether Statements about Logarithms Are True

Decide whether each statement is *true* or *false*.

(a) $\log_2 8 - \log_2 4 = \log_2 4$

Evaluate both sides.

Left side: $\log_2 8 - \log_2 4 = \log_2 2^3 - \log_2 2^2 = 3 - 2 = 1$

Right side: $\log_2 4 = \log_2 2^2 = 2$

The statement is false because $1 \neq 2$.

(b) $\log_3(\log_2 8) = \dfrac{\log_7 49}{\log_8 64}$

Evaluate both sides.

Left side: $\log_3(\log_2 8) = \log_3 3 = 1$

Right side: $\dfrac{\log_7 49}{\log_8 64} = \dfrac{\log_7 7^2}{\log_8 8^2} = \dfrac{2}{2} = 1$

The statement is true because $1 = 1$.

✔ **Now Try Exercises 45 and 51.**

Napier's Rods
Source: IBM Corporate Archives.

CONNECTIONS

Long before the days of calculators and computers, the search for making calculations easier was an ongoing process. Machines built by Charles Babbage and Blaise Pascal, a system of "rods" used by John Napier, and slide rules were the forerunners of today's electronic marvels. The invention of logarithms by John Napier in the sixteenth century was a great breakthrough in the search for easier methods of calculation.

Since logarithms are exponents, their properties allowed users of tables of common logarithms to multiply by adding, divide by subtracting, raise to powers by multiplying, and take roots by dividing. Although logarithms are no longer used for computations, they play an important part in higher mathematics.

For Discussion or Writing

1. To multiply 458.3 by 294.6 using logarithms, we add $\log_{10} 458.3$ and $\log_{10} 294.6$, and then find 10 to this power. Perform this multiplication using the $\boxed{\log x}$ key* and the $\boxed{10^x}$ key on your calculator. Check your answer by multiplying directly with your calculator.

2. Try division, raising to a power, and taking a root by this method.

---

*In this text, the notation log x is used to mean $\log_{10} x$. This is also the meaning of the log key on calculators.

10.4 EXERCISES

Complete solution available on Video Lectures on CD/DVD

Now Try Exercise

Use the indicated rule of logarithms to complete each equation. See Examples 1–4.

1. $\log_{10}(3 \cdot 4) =$ _____   (product rule)

2. $\log_{10} \dfrac{3}{4} =$ _____   (quotient rule)

3. $3^{\log_3 4} =$ _____   (special property)

4. $\log_{10} 3^4 =$ _____   (power rule)

5. $\log_3 3^4 =$ _____   (special property)

6. Evaluate $\log_2(8 + 8)$. Then evaluate $\log_2 8 + \log_2 8$. Are the results the same? How could you change the operation in the first expression to make the two expressions equal?

Use the properties of logarithms to express each logarithm as a sum or difference of logarithms, or as a single number if possible. Assume that all variables represent positive real numbers. See Examples 1–5.

7. $\log_7(4 \cdot 5)$ 8. $\log_8(9 \cdot 11)$ 9. $\log_5 \dfrac{8}{3}$

10. $\log_3 \dfrac{7}{5}$ 11. $\log_4 6^2$ 12. $\log_5 7^4$

13. $\log_3 \dfrac{\sqrt[3]{4}}{x^2 y}$ **14.** $\log_7 \dfrac{\sqrt[3]{13}}{pq^2}$ **15.** $\log_3 \sqrt{\dfrac{xy}{5}}$

16. $\log_6 \sqrt{\dfrac{pq}{7}}$ **17.** $\log_2 \dfrac{\sqrt[3]{x} \cdot \sqrt[5]{y}}{r^2}$ **18.** $\log_4 \dfrac{\sqrt[4]{z} \cdot \sqrt[5]{w}}{s^2}$

19. A student erroneously wrote $\log_a(x + y) = \log_a x + \log_a y$. When his teacher explained that this was indeed wrong, the student claimed that he had used the distributive property. Write a few sentences explaining why the distributive property does not apply in this case.

20. Write a few sentences explaining how the rules for multiplying and dividing powers of the same base are similar to the rules for finding logarithms of products and quotients.

Use the properties of logarithms to write each expression as a single logarithm. Assume that all variables are defined in such a way that the variable expressions are positive, and bases are positive numbers not equal to 1. See Examples 1–5.

21. $\log_b x + \log_b y$ **22.** $\log_b 2 + \log_b z$

23. $\log_a m - \log_a n$ **24.** $\log_b x - \log_b y$

25. $(\log_a r - \log_a s) + 3 \log_a t$ **26.** $(\log_a p - \log_a q) + 2 \log_a r$

27. $3 \log_a 5 - 4 \log_a 3$ **28.** $3 \log_a 5 + \dfrac{1}{2} \log_a 9$

29. $\log_{10}(x + 3) + \log_{10}(x - 3)$ **30.** $\log_{10}(y + 4) + \log_{10}(y - 4)$

31. $3 \log_p x + \dfrac{1}{2} \log_p y - \dfrac{3}{2} \log_p z - 3 \log_p a$

32. $\dfrac{1}{3} \log_b x + \dfrac{2}{3} \log_b y - \dfrac{3}{4} \log_b s - \dfrac{2}{3} \log_b t$

To four decimal places, the values of $\log_{10} 2$ and $\log_{10} 9$ are

$$\log_{10} 2 = 0.3010 \qquad \log_{10} 9 = 0.9542.$$

Evaluate each logarithm by applying the appropriate rule or rules from this section. DO NOT USE A CALCULATOR. See Example 6.

33. $\log_{10} 18$ **34.** $\log_{10} \dfrac{9}{2}$ **35.** $\log_{10} \dfrac{2}{9}$ **36.** $\log_{10} 4$

37. $\log_{10} 36$ **38.** $\log_{10} 162$ **39.** $\log_{10} 3$ **40.** $\log_{10} \sqrt[5]{2}$

41. $\log_{10} \sqrt[4]{9}$ **42.** $\log_{10} \dfrac{1}{9}$ **43.** $\log_{10} 9^5$ **44.** $\log_{10} 2^{19}$

Decide whether each statement is true *or* false. *See Example 7.*

45. $\log_2(8 + 32) = \log_2 8 + \log_2 32$ **46.** $\log_2(64 - 16) = \log_2 64 - \log_2 16$

47. $\log_3 7 + \log_3 7^{-1} = 0$ **48.** $\log_9 14 - \log_{14} 9 = 0$

49. $\log_6 60 - \log_6 10 = 1$ **50.** $\log_3 8 + \log_3 \dfrac{1}{8} = 0$

51. $\dfrac{\log_{10} 7}{\log_{10} 14} = \dfrac{1}{2}$ **52.** $\dfrac{\log_{10} 10}{\log_{10} 100} = \dfrac{1}{10}$

53. *Concept Check* Refer to the Note following the word statement of the product rule for logarithms in this section. Now, state the quotient rule in words, replacing "logarithm" with "exponent."

54. Explain why the statement for the power rule for logarithms requires that x be a positive real number.

55. *Concept Check* Refer to Example 7(a). Change the left side of the equation using the quotient rule so that the statement becomes true, and simplify.

56. *Concept Check* Consider the following "proof" that $\log_2 16$ does not exist.

$$\log_2 16 = \log_2(-4)(-4)$$
$$= \log_2(-4) + \log_2(-4)$$

Since the logarithm of a negative number is not defined, the final step cannot be evaluated, and so $\log_2 16$ does not exist. *WHAT WENT WRONG?*

PREVIEW EXERCISES

Write each exponential statement in logarithmic form. See ***Section 10.3.***

57. $10^4 = 10,000$ **58.** $10^{1/2} = \sqrt{10}$ **59.** $10^{-2} = 0.01$

Write each logarithmic statement in exponential form. See ***Section 10.3.***

60. $\log_{10} 0.001 = -3$ **61.** $\log_{10} 1 = 0$ **62.** $\log_{10} \sqrt[3]{10} = \dfrac{1}{3}$

10.5 Common and Natural Logarithms

OBJECTIVES

1 Evaluate common logarithms using a calculator.

2 Use common logarithms in applications.

3 Evaluate natural logarithms using a calculator.

4 Use natural logarithms in applications.

5 Use the change-of-base rule.

As mentioned earlier, logarithms are important in many applications of mathematics to everyday problems, particularly in biology, engineering, economics, and social science. In this section we find numerical approximations for logarithms. Traditionally, base 10 logarithms were used most often because our number system is base 10. Logarithms to base 10 are called **common logarithms,** and $\log_{10} x$ is abbreviated as simply **log x,** where the base is understood to be 10.

OBJECTIVE 1 **Evaluate common logarithms using a calculator.** We use calculators to evaluate common logarithms. In the next example we give the results of evaluating some common logarithms using a calculator with a $\boxed{\text{LOG}}$ key. (This may be a second function key on some calculators.) For simple scientific calculators, just enter the number, then press the $\boxed{\text{LOG}}$ key. For graphing calculators, these steps are reversed. We give most approximations for logarithms to four decimal places.

EXAMPLE 1 Evaluating Common Logarithms

Evaluate each logarithm using a calculator.

(a) $\log 327.1 \approx 2.5147$ **(b)** $\log 437{,}000 \approx 5.6405$

(c) $\log 0.0615 \approx -1.2111$

Figure 13 shows how a graphing calculator displays these common logarithms to four decimal places.

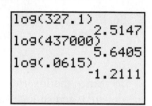

```
log(327.1)
           2.5147
log(437000)
           5.6405
log(.0615)
          -1.2111
```

FIGURE 13

In part (c), notice that $\log 0.0615 \approx -1.2111$, a negative result. *The common logarithm of a number between 0 and 1 is always negative* because the logarithm is the exponent on 10 that produces the number. In this case, we have

$$10^{-1.2111} \approx 0.0615.$$

If the exponent (the logarithm) were positive, the result would be greater than 1 because $10^0 = 1$. The graph in Figure 14 illustrates these concepts.

✔ **Now Try Exercises 7, 9, and 11.**

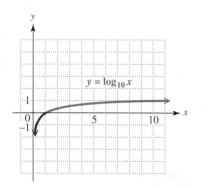

FIGURE 14

OBJECTIVE 2 Use common logarithms in applications. In chemistry, pH is a measure of the acidity or alkalinity of a solution; pure water, for example, has pH 7. In general, acids have pH numbers less than 7, and alkaline solutions have pH values greater than 7, as shown in Figure 15. The **pH** of a solution is defined as

$$\mathbf{pH} = -\log[\mathbf{H_3O^+}],$$

where $[H_3O^+]$ is the hydronium ion concentration in moles per liter. It is customary to round pH values to the nearest tenth.

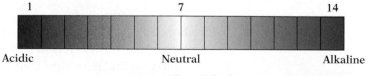

Acidic   Neutral   Alkaline

FIGURE 15   pH Scale

EXAMPLE 2 Using pH in an Application

Wetlands are classified as *bogs, fens, marshes,* and *swamps,* on the basis of pH values. A pH value between 6.0 and 7.5, such as that of Summerby Swamp in Michigan's Hiawatha National Forest, indicates that the wetland is a "rich fen." When the pH is between 4.0 and 6.0, the wetland is a "poor fen," and if the pH falls to 3.0 or less, it is a "bog." (*Source:* Mohlenbrock, R., "Summerby Swamp, Michigan," *Natural History,* March 1994.)

Suppose that the hydronium ion concentration of a sample of water from a wetland is 6.3×10^{-3}. How would this wetland be classified?

$$\begin{aligned}
\text{pH} &= -\log(6.3 \times 10^{-3}) && \text{Definition of pH} \\
&= -(\log 6.3 + \log 10^{-3}) && \text{Product rule} \\
&= -[0.7993 - 3(1)] && \text{Use a calculator to find log 6.3.} \\
&= -0.7993 + 3 && \text{Distributive property} \\
&\approx 2.2
\end{aligned}$$

Since the pH is less than 3.0, the wetland is a bog.

✔ **Now Try Exercise 29.**

EXAMPLE 3 Finding Hydronium Ion Concentration

Find the hydronium ion concentration of drinking water with pH 6.5.

$$\begin{aligned}
\text{pH} &= -\log[\text{H}_3\text{O}^+] \\
6.5 &= -\log[\text{H}_3\text{O}^+] && \text{Let pH} = 6.5. \\
\log[\text{H}_3\text{O}^+] &= -6.5 && \text{Multiply by } -1.
\end{aligned}$$

Solve for $[\text{H}_3\text{O}^+]$ by writing the equation in exponential form using base 10.

$$\begin{aligned}
[\text{H}_3\text{O}^+] &= 10^{-6.5} \\
[\text{H}_3\text{O}^+] &\approx 3.2 \times 10^{-7} && \text{Use a calculator.}
\end{aligned}$$

✔ **Now Try Exercise 35.**

Decibel Level	Example
60	Normal conversation
90	Rush hour traffic, lawn mower
100	Garbage truck, chain saw, pneumatic drill
120	Rock concert, thunderclap
140	Gunshot blast, jet engine
180	Rocket launching pad

Source: Deafness Research Foundation.

The loudness of sound is measured in a unit called a **decibel,** abbreviated **dB.** To measure with this unit, we first assign an intensity of I_0 to a very faint sound, called the **threshold sound.** If a particular sound has intensity I, then the decibel level of this louder sound is

$$D = 10 \log\left(\frac{I}{I_0}\right).$$

The table in the margin gives average decibel levels for some common sounds. Any sound over 85 dB exceeds what hearing experts consider safe. Permanent hearing damage can be suffered at levels above 150 dB.

EXAMPLE 4 Measuring the Loudness of Sound

If music delivered through the earphones of an iPod has intensity I of $3.162 \times 10^{11} I_0$, find the average decibel level. (*Source: Sacramento Bee.*)

$$\begin{aligned}
D &= 10 \log\left(\frac{I}{I_0}\right) \\
D &= 10 \log\left(\frac{3.162 \times 10^{11} I_0}{I_0}\right) && \text{Substitute the given value for } I. \\
D &= 10 \log(3.162 \times 10^{11}) \\
D &\approx 115 && \text{Use a calculator.}
\end{aligned}$$

✔ **Now Try Exercise 39.**

Leonhard Euler (1707–1783)

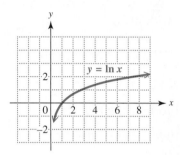

FIGURE 16

OBJECTIVE 3 Evaluate natural logarithms using a calculator. The most important logarithms used in applications are **natural logarithms,** which have as base the number *e*. The number *e* is a fundamental number in our universe. For this reason *e*, like π, is called a **universal constant.** The letter *e* is used to honor Leonhard Euler, who published extensive results on the number in 1748. Since it is an irrational number, its decimal expansion never terminates and never repeats.

The first few digits of the decimal value of *e* are **2.718281828.** On a calculator, use $\boxed{e^x}$ or the two keys $\boxed{\text{INV}}$ and $\boxed{\ln x}$ to approximate powers of *e*, such as

$$e^2 \approx 7.389056099, \quad e^3 \approx 20.08553692, \quad \text{and} \quad e^{0.6} \approx 1.8221188.$$

Logarithms to base *e* are called natural logarithms because they occur in natural situations that involve growth or decay. The base *e* logarithm of *x* is written **ln x** (read "el en *x*"). The graph of $y = \ln x$ is given in Figure 16.

A calculator key labeled $\boxed{\ln x}$ is used to evaluate natural logarithms. If your scientific calculator has an $\boxed{e^x}$ key, but not a key labeled $\boxed{\ln x}$, find a natural logarithm by entering the number and pressing the $\boxed{\text{INV}}$ key and then the $\boxed{e^x}$ key. This works because $y = e^x$ defines the inverse function of $y = \ln x$ (or $y = \log_e x$).

EXAMPLE 5 Evaluating Natural Logarithms

Evaluate each logarithm using a calculator.

(a) $\ln 0.5841 \approx -0.5377$

As with common logarithms, *a number between 0 and 1 has a negative natural logarithm.*

(b) $\ln 192.7 \approx 5.2611$ **(c)** $\ln 10.84 \approx 2.3832$

See Figure 17.

```
ln(.5841)
            -.5377
ln(192.7)
            5.2611
ln(10.84)
            2.3832
```

FIGURE 17

✔ **Now Try Exercises 15, 17, and 19.**

OBJECTIVE 4 Use natural logarithms in applications.

EXAMPLE 6 Applying a Natural Logarithmic Function

The altitude in meters that corresponds to an atmospheric pressure of *x* millibars is given by the logarithmic function defined by

$$f(x) = 51,600 - 7457 \ln x.$$

(*Source:* Miller, A. and J. Thompson, *Elements of Meteorology,* Fourth Edition, Charles E. Merrill Publishing Company, 1993.) Use this function to find the altitude when atmospheric pressure is 400 millibars.

Let $x = 400$ and substitute in the expression for $f(x)$.

$$f(400) = 51,600 - 7457 \ln 400$$
$$\approx 6900$$

Atmospheric pressure is 400 millibars at approximately 6900 m.

✔ **Now Try Exercise 41.**

▶ **NOTE** In Example 6, the final answer was obtained using a calculator *without* rounding the intermediate values. In general, it is best to wait until the final step to round the answer; otherwise, a buildup of round-off error may cause the final answer to have an incorrect final decimal place digit.

OBJECTIVE 5 Use the change-of-base rule. We have used a calculator to approximate the values of common logarithms (base 10) and natural logarithms (base *e*). However, some applications involve logarithms to other bases. For example, the percentage of women who had a baby in the last year and returned to work is given by

$$f(x) = 38.83 + 4.208 \log_2 x,$$

for year *x* since 1980. (*Source:* U.S. Census Bureau.) To use this function, we need to find a base 2 logarithm. The following rule is used to convert logarithms from one base to another.

Change-of-Base Rule

If $a > 0$, $a \neq 1$, $b > 0$, $b \neq 1$, and $x > 0$, then

$$\log_a x = \frac{\log_b x}{\log_b a}.$$

▶ **NOTE** Any positive number other than 1 can be used for base *b* in the change-of-base rule, but usually the only practical bases are *e* and 10 because calculators give logarithms only for these two bases.

To derive the change-of-base rule, let $\log_a x = m$.

$$\log_a x = m$$
$$a^m = x \quad \text{Change to exponential form.}$$

Since logarithmic functions are one-to-one, if all variables are positive and if $x = y$, then $\log_b x = \log_b y$.

$$\log_b(a^m) = \log_b x$$
$$m \log_b a = \log_b x \quad \text{Power rule}$$
$$(\log_a x)(\log_b a) = \log_b x \quad \text{Substitute for } m.$$
$$\log_a x = \frac{\log_b x}{\log_b a} \quad \text{Divide by } \log_b a.$$

The last step gives the change-of-base rule.

EXAMPLE 7 Using the Change-of-Base Rule

Find $\log_5 12$.

Use common logarithms and the change-of-base rule.

$$\log_5 12 = \frac{\log 12}{\log 5} \approx 1.5440 \qquad \text{Use a calculator.}$$

✔ **Now Try Exercise 47.**

▶ **NOTE** Either common or natural logarithms can be used when applying the change-of-base rule. Verify that the same value is found in Example 7 if natural logarithms are used.

EXAMPLE 8 Using the Change-of-Base Rule in an Application

Use natural logarithms in the change-of-base rule and the function

$$f(x) = 38.83 + 4.208 \log_2 x$$

(given earlier) to find the percent of women who returned to work after having a baby in 1995. In the equation, $x = 0$ represents 1980.

$$f(15) = 38.83 + 4.208 \log_2 15 \qquad 1995 - 1980 = 15, \text{ so let } x = 15.$$

$$= 38.83 + 4.208 \left(\frac{\ln 15}{\ln 2} \right) \qquad \text{Change-of-base rule}$$

$$\approx 55.3\% \qquad \text{Use a calculator.}$$

This is very close to the actual value of 55%.

✔ **Now Try Exercise 59.**

CONNECTIONS

As previously mentioned, the number $e \approx 2.718281828$ is a fundamental number in our universe. If there are intelligent beings elsewhere, they too will have to use e to do higher mathematics. The properties of e are used extensively in calculus and in higher mathematics. In **Section 10.6** we see how it applies to growth and decay in the physical world.

For Discussion or Writing

The value of e can be expressed as

$$e = 1 + \frac{1}{1} + \frac{1}{1 \cdot 2} + \frac{1}{1 \cdot 2 \cdot 3} + \frac{1}{1 \cdot 2 \cdot 3 \cdot 4} + \cdots.$$

Approximate e using two terms of this expression, then three terms, four terms, five terms, and six terms. How close is the approximation to the value of e given above with six terms? Does this infinite sum approach the value of e very quickly?

10.5 EXERCISES

Concept Check   Choose the correct response in Exercises 1–4.

1. What is the base in the expression log x?

    A. e **B.** 1   **C.** 10   **D.** x

2. What is the base in the expression ln x?

    A. e **B.** 1   **C.** 10   **D.** x

3. Since $10^0 = 1$ and $10^1 = 10$, between what two consecutive integers is the value of log 5.6?

    A. 5 and 6   **B.** 10 and 11   **C.** 0 and 1   **D.** -1 and 0

4. Since $e^1 \approx 2.718$ and $e^2 \approx 7.389$, between what two consecutive integers is the value of ln 5.6?

    A. 5 and 6   **B.** 2 and 3   **C.** 1 and 2   **D.** 0 and 1

5. *Concept Check*   Without using a calculator, give the value of log $10^{19.2}$.

6. *Concept Check*   Without using a calculator, give the value of ln $e^{\sqrt{2}}$.

You will need a calculator for the remaining exercises in this set.

Find each logarithm. Give approximations to four decimal places. See Examples 1 and 5.

⊙ **7.** log 43   **8.** log 98   **9.** log 328.4

10. log 457.2   **11.** log 0.0326   **12.** log 0.1741

13. log(4.76×10^9) **14.** log(2.13×10^4) ⊙ **15.** ln 7.84

16. ln 8.32   **17.** ln 0.0556   **18.** ln 0.0217

19. ln 388.1   **20.** ln 942.6   **21.** ln$(8.59 \times e^2)$

22. ln$(7.46 \times e^3)$ **23.** ln 10   **24.** log e

25. Use your calculator to find approximations of the following logarithms:

    (a) log 356.8   **(b)** log 35.68   **(c)** log 3.568.

    ▨ **(d)** Observe your answers and make a conjecture concerning the decimal values of the common logarithms of numbers greater than 1 that have the same digits.

26. Let k represent the number of letters in your last name.

    (a) Use your calculator to find log k.

    (b) Raise 10 to the power indicated by the number in part (a). What is your result?

    ▨ **(c)** Use the concepts of **Section 10.1** to explain why you obtained the answer you found in part (b). Would it matter what number you used for k to observe the same result?

▨ **27.** Try to find log(-1) using a calculator. (If you have a graphing calculator, it should be in real number mode.) What happens? Explain.

Suppose that water from a wetland area is sampled and found to have the given hydronium ion concentration. Is the wetland a rich fen, *a* poor fen, *or a* bog? *See Example 2.*

28. 2.5×10^{-5} ⊙ **29.** 2.5×10^{-2} **30.** 2.5×10^{-7}

Find the pH *of the substance with the given hydronium ion concentration. See Example 2.*

31. Ammonia, 2.5×10^{-12}

32. Sodium bicarbonate, 4.0×10^{-9}

33. Grapes, 5.0×10^{-5}

34. Tuna, 1.3×10^{-6}

Find the hydronium ion concentration of the substance with the given pH. *See Example 3.*

🌐 **35.** Human blood plasma, 7.4

36. Human gastric contents, 2.0

37. Spinach, 5.4

38. Bananas, 4.6

Solve each problem. See Examples 4 and 6.

🌐 **39.** Consumers can now enjoy movies at home in elaborate home-theater systems. Find the average decibel level

$$D = 10 \log \left(\frac{I}{I_0} \right)$$

for each popular movie with the given intensity I.

(a) *Spider-Man 2;* $5.012 \times 10^{10} I_0$

(b) *Finding Nemo;* $10^{10} I_0$

(c) *Saving Private Ryan;* $6{,}310{,}000{,}000 \, I_0$

40. The time t in years for an amount increasing at a rate of r (in decimal form) to double is given by

$$t(r) = \frac{\ln 2}{\ln(1 + r)}.$$

This is called **doubling time.** Find the doubling time to the nearest tenth for an investment at each interest rate.

(a) 2% (or 0.02)   (b) 5% (or 0.05)   (c) 8% (or 0.08)

🌐 **41.** The number of years, $N(r)$, since two independently evolving languages split off from a common ancestral language is approximated by

$$N(r) = -5000 \ln r,$$

where r is the percent of words (in decimal form) from the ancestral language common to both languages now. Find the number of years (to the nearest hundred years) since the split for each percent of common words.

(a) 85% (or 0.85)   (b) 35% (or 0.35)   (c) 10% (or 0.10)

42. The concentration of a drug injected into the bloodstream decreases with time. The intervals of time T when the drug should be administered are given by

$$T = \frac{1}{k} \ln \frac{C_2}{C_1},$$

where k is a constant determined by the drug in use, C_2 is the concentration at which the drug is harmful, and C_1 is the concentration below which the drug is ineffective. (*Source:* Horelick, Brindell and Sinan Koont, "Applications of Calculus to Medicine: Prescribing Safe and Effective Dosage," *UMAP Module 202, 1977.*) Thus, if $T = 4$, the drug should be administered every 4 hr. For a certain drug, $k = \frac{1}{3}$, $C_2 = 5$, and $C_1 = 2$. How often should the drug be administered? (*Hint:* Round down.)

43. The growth of outpatient surgeries as a percent of total surgeries at hospitals is approximated by

$$f(x) = -1317 + 304 \ln x,$$

where x is the number of years since 1900. (*Source:* American Hospital Association.)

(a) What does this function predict for the percent of outpatient surgeries in 1998?

(b) When did outpatient surgeries reach 50%? (*Hint:* Substitute for y, then write the equation in exponential form to solve it.)

44. In the central Sierra Nevada of California, the percent of moisture that falls as snow rather than rain is approximated reasonably well by

$$f(x) = 86.3 \ln x - 680,$$

where x is the altitude in feet.

(a) What percent of the moisture at 5000 ft falls as snow?

(b) What percent at 7500 ft falls as snow?

45. The **cost-benefit equation**

$$T = -0.642 - 189 \ln(1 - p)$$

describes the approximate tax T, in dollars per ton, that would result in a $p\%$ (in decimal form) reduction in carbon dioxide emissions.

(a) What tax will reduce emissions 25%?

✐ **(b)** Explain why the equation is not valid for $p = 0$ or $p = 1$.

46. The age in years of a female blue whale of length L in feet is approximated by

$$t = -2.57 \ln\left(\frac{87 - L}{63}\right).$$

(a) How old is a female blue whale that measures 80 ft?

✐ **(b)** The equation that defines t has domain $24 < L < 87$. Explain why.

Use the change-of-base rule (with either common or natural logarithms) to find each logarithm to four decimal places. See Example 7.

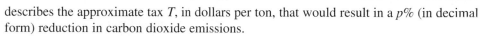

47. $\log_3 12$ **48.** $\log_4 18$ **49.** $\log_5 3$

50. $\log_7 4$ **51.** $\log_3 \sqrt{2}$ **52.** $\log_6 \sqrt[3]{5}$

53. $\log_\pi e$ **54.** $\log_\pi 10$ **55.** $\log_e 12$

✐ **56.** Explain why the answer to Exercise 55 is the same one that you get when you use a calculator to approximate ln 12.

57. Let m be the number of letters in your first name, and let n be the number of letters in your last name.

✐ **(a)** In your own words, explain what $\log_m n$ means.

(b) Use your calculator to find $\log_m n$.

(c) Raise m to the power indicated by the number found in part (b). What is your result?

58. To solve the equation $5^x = 7$, we must find the exponent to which 5 must be raised in order to obtain 7: this is $\log_5 7$.

(a) Use the change-of-base rule and your calculator to find $\log_5 7$.

(b) Raise 5 to the number you found in part (a). What is your result?

(c) Using as many decimal places as your calculator gives, write the solution set of $5^x = 7$. (Equations of this type will be studied in more detail in **Section 10.6.**)

Solve each application of a logarithmic function (from Exercises 53 and 54 of Section 10.3). See Example 8.

59. According to selected figures from 1981 through 2003, the number of billion cubic feet of natural gas gross withdrawals from crude oil wells in the United States can be approximated by the function defined by

$$f(x) = 3800 + 585 \log_2 x,$$

where $x = 1$ represents 1981, $x = 2$ represents 1982, and so on. (*Source:* Energy Information Administration.) Use this function to approximate the number of cubic feet withdrawn in 2003.

60. According to selected figures from the last two decades of the 20th century, the number of trillion cubic feet of dry natural gas consumed worldwide can be approximated by the function defined by

$$f(x) = 51.47 + 6.044 \log_2 x,$$

where $x = 1$ represents 1980, $x = 2$ represents 1981, and so on. (*Source:* Energy Information Administration.) Use this function to approximate consumption in 2003.

TECHNOLOGY INSIGHTS (EXERCISES 61–64)

Because graphing calculators are equipped with log *x and* ln *x keys, it is possible to graph the functions defined by* $f(x) = \log x$ *and* $g(x) = \ln x$ *directly, as shown in the figures that follow.*

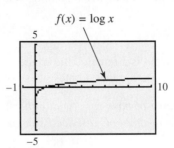

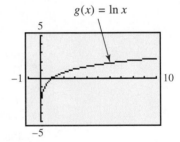

To graph functions defined by logarithms to bases other than 10 or e, however, we must use the change-of-base rule. For example, to graph $y = \log_2 x$*, we may enter* Y_1 *as* $\dfrac{\log X}{\log 2}$ *or* $\dfrac{\ln X}{\ln 2}$*. This is shown in the figure at the right. (Compare it to the figure in Exercises 59–62 of Section 10.3, where it was drawn using the fact that* $y = \log_2 x$ *is the inverse of* $y = 2^x$*.)*

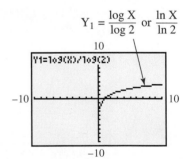

Use the change-of-base rule to graph each logarithmic function with a graphing calculator. Use a viewing window with Xmin = −1, Xmax = 10, Ymin = −5, and Ymax = 5.

61. $g(x) = \log_3 x$ **62.** $g(x) = \log_5 x$ **63.** $g(x) = \log_{1/3} x$ **64.** $g(x) = \log_{1/5} x$

PREVIEW EXERCISES

*Solve each equation. See **Sections 10.2 and 10.3.***

65. $4^{2x} = 8^{3x+1}$ **66.** $2^{5x} = \left(\dfrac{1}{16}\right)^{x+3}$ **67.** $\log_3(x + 4) = 2$

68. $\log_x 64 = 2$ **69.** $\log_{1/2} 8 = x$ **70.** $\log_a 1 = 0$

*Write as a single logarithm. See **Section 10.4.***

71. $\log(x + 2) + \log(x - 3)$ **72.** $\log_4(x + 4) - 2 \log_4(3x + 1)$

10.6   Exponential and Logarithmic Equations; Further Applications

OBJECTIVES

1 Solve equations involving variables in the exponents.

2 Solve equations involving logarithms.

3 Solve applications of compound interest.

4 Solve applications involving base e exponential growth and decay.

5 Use a graphing calculator to solve exponential and logarithmic equations.

We solved exponential and logarithmic equations in **Sections 10.2 and 10.3.** General methods for solving these equations depend on the following properties.

> **Properties for Solving Exponential and Logarithmic Equations**
>
> For all real numbers $b > 0$, $b \neq 1$, and any real numbers x and y:
>
> **1.** If $x = y$, then $b^x = b^y$.
>
> **2.** If $b^x = b^y$, then $x = y$.
>
> **3.** If $x = y$, and $x > 0, y > 0$, then $\log_b x = \log_b y$.
>
> **4.** If $x > 0, y > 0$, and $\log_b x = \log_b y$, then $x = y$.

We used Property 2 to solve exponential equations in **Section 10.2.**

OBJECTIVE 1 Solve equations involving variables in the exponents. The first two examples illustrate the method for solving exponential equations using Property 3.

EXAMPLE 1 Solving an Exponential Equation

Solve $3^x = 12$.

$$3^x = 12$$

$$\log 3^x = \log 12 \qquad \text{Property 3 (common logs)}$$

$$x \log 3 = \log 12 \qquad \text{Power rule}$$

Exact solution $\longrightarrow x = \dfrac{\log 12}{\log 3} \qquad$ Divide by log 3.

Decimal approximation $\longrightarrow x \approx 2.262 \qquad$ Use a calculator.

The solution set is $\{2.262\}$. Check with a calculator that $3^{2.262} \approx 12$.

✔ **Now Try Exercise 5.**

▶ **CAUTION** Be careful: $\frac{\log 12}{\log 3}$ is *not* equal to log 4. Check to see that log 4 $\approx$ 0.6021, but $\frac{\log 12}{\log 3} \approx 2.262$.

When an exponential equation has e as the base, as in the next example, it is easiest to use base e logarithms.

EXAMPLE 2 Solving an Exponential Equation with Base e

Solve $e^{0.003x} = 40$.

$$\ln e^{0.003x} = \ln 40 \qquad \text{Property 3 (natural logs)}$$

$$0.003x \ln e = \ln 40 \qquad \text{Power rule}$$

$$0.003x = \ln 40 \qquad \ln e = \ln e^1 = 1$$

$$x = \frac{\ln 40}{0.003} \qquad \text{Divide by 0.003.}$$

$$x \approx 1230 \qquad \text{Use a calculator.}$$

The solution set is $\{1230\}$. Check that $e^{0.003(1230)} \approx 40$.

✔ **Now Try Exercise 15.**

General Method for Solving an Exponential Equation

Take logarithms to the same base on both sides and then use the power rule of logarithms or the special property $\log_b b^x = x$. (See Examples 1 and 2.)

As a special case, if both sides can be written as exponentials with the same base, do so, and set the exponents equal. (See **Section 10.2.**)

OBJECTIVE ❷ Solve equations involving logarithms. The properties of logarithms from **Section 10.4** are useful here, as is using the definition of a logarithm to change the equation to exponential form.

EXAMPLE 3 Solving a Logarithmic Equation

Solve $\log_2(x + 5)^3 = 4$. Give the exact solution.

$$\log_2(x + 5)^3 = 4$$

$$(x + 5)^3 = 2^4 \qquad \text{Convert to exponential form.}$$

$$(x + 5)^3 = 16$$

$$x + 5 = \sqrt[3]{16} \qquad \text{Take the cube root on each side.}$$

$$x = -5 + \sqrt[3]{16} \qquad \text{Subtract 5.}$$

$$x = -5 + 2\sqrt[3]{2} \qquad \sqrt[3]{16} = \sqrt[3]{8 \cdot 2} = \sqrt[3]{8} \cdot \sqrt[3]{2} = 2\sqrt[3]{2}$$

Check:
$$\log_2(x + 5)^3 = 4 \qquad \text{Original equation}$$

$$\log_2(-5 + 2\sqrt[3]{2} + 5)^3 = 4 \qquad ? \qquad \text{Let } x = -5 + 2\sqrt[3]{2}.$$

$$\log_2(2\sqrt[3]{2})^3 = 4 \qquad ? \qquad \text{Work inside the parentheses.}$$

$$\log_2 16 = 4 \qquad ? \qquad (2\sqrt[3]{2})^3 = 2^3(\sqrt[3]{2})^3 = 8 \cdot 2 = 16$$

$$2^4 = 16 \qquad ? \qquad \text{Write in exponential form.}$$

$$16 = 16 \qquad \text{True}$$

A true statement results, so the solution set is $\left\{-5 + 2\sqrt[3]{2}\right\}$.

✔ **Now Try Exercise 29.**

▶ **CAUTION** Recall that the domain of $y = \log_b x$ is $(0, \infty)$. For this reason, *always check that the solution of an equation with logarithms yields only logarithms of positive numbers in the original equation.*

EXAMPLE 4 Solving a Logarithmic Equation

Solve $\log_2(x + 1) - \log_2 x = \log_2 7$.

$$\log_2(x + 1) - \log_2 x = \log_2 7$$

> **Transform the left side to an expresssion with only *one* logarithm.**

$$\log_2 \frac{x + 1}{x} = \log_2 7 \qquad \text{Quotient rule}$$

$$\frac{x + 1}{x} = 7 \qquad \text{Property 4}$$

$$x + 1 = 7x \qquad \text{Multiply by } x.$$

$$1 = 6x \qquad \text{Subtract } x.$$

$$\frac{1}{6} = x \qquad \text{Divide by 6.}$$

Since we cannot take the logarithm of a *nonpositive* number, both $x + 1$ and x must be positive here. If $x = \frac{1}{6}$, then this condition is satisfied.

Check: $\log_2(x + 1) - \log_2 x = \log_2 7 \qquad \text{Original equation}$

$$\log_2\left(\frac{1}{6} + 1\right) - \log_2 \frac{1}{6} = \log_2 7 \qquad ? \qquad \text{Let } x = \frac{1}{6}.$$

$$\log_2 \frac{7}{6} - \log_2 \frac{1}{6} = \log_2 7 \qquad ?$$

$$\log_2 \frac{\frac{7}{6}}{\frac{1}{6}} = \log_2 7 \qquad ? \qquad \text{Quotient rule}$$

> $\frac{\frac{7}{6}}{\frac{1}{6}} = \frac{7}{6} \div \frac{1}{6} = \frac{7}{6} \cdot \frac{6}{1} = 7$

$$\log_2 7 = \log_2 7 \qquad \text{True}$$

A true statement results, so the solution set is $\left\{\frac{1}{6}\right\}$.

✔ **Now Try Exercise 35.**

EXAMPLE 5 Solving a Logarithmic Equation

Solve $\log x + \log(x - 21) = 2$.

$$\log x + \log(x - 21) = 2$$

$$\log x(x - 21) = 2 \qquad \text{Product rule}$$

The base is 10. $\qquad x(x - 21) = 10^2 \qquad$ Write in exponential form.

$$x^2 - 21x = 100 \qquad \text{Distributive property; multiply.}$$

$$x^2 - 21x - 100 = 0 \qquad \text{Standard form}$$

$$(x - 25)(x + 4) = 0 \qquad \text{Factor.}$$

$$x - 25 = 0 \quad \text{or} \quad x + 4 = 0 \qquad \text{Zero-factor property}$$

$$x = 25 \quad \text{or} \qquad x = -4 \qquad \text{Solve each equation.}$$

The value -4 must be rejected as a solution since it leads to the logarithm of a negative number in the original equation:

$$\log(-4) + \log(-4 - 21) = 2. \qquad \text{The left side is undefined.}$$

Check that the only solution is 25, so the solution set is $\{25\}$.

✔ **Now Try Exercise 39.**

▶ **CAUTION** *Do not reject a potential solution just because it is nonpositive. Reject any value that leads to the logarithm of a nonpositive number.*

In summary, we use the following steps to solve a logarithmic equation.

Solving a Logarithmic Equation

Step 1 **Transform the equation so that a single logarithm appears on one side.** Use the product rule or quotient rule of logarithms to do this.

Step 2 **(a) Use Property 4.** If $\log_b x = \log_b y$, then $x = y$. (See Example 4.)

(b) Write the equation in exponential form. If $\log_b x = k$, then $x = b^k$. (See Examples 3 and 5.)

OBJECTIVE 3 Solve applications of compound interest. So far in this book, we have solved simple interest problems using the formula $I = prt$. In most cases, interest paid or charged is *compound interest* (interest paid on both principal and interest). The formula for compound interest is an application of exponential functions.

Compound Interest Formula (for a Finite Number of Periods)

If a principal of P dollars is deposited at an annual rate of interest r compounded (paid) n times per year, then the account will contain

$$A = P\left(1 + \frac{r}{n}\right)^{nt}$$

dollars after t years. (In this formula, r is expressed as a decimal.)

EXAMPLE 6 Solving a Compound Interest Problem for *A*

How much money will there be in an account at the end of 5 yr if $1000 is deposited at 6% compounded quarterly? (Assume no withdrawals are made.)

Because interest is compounded quarterly, $n = 4$. The other given values are $P = 1000$, $r = 0.06$ (because $6\% = 0.06$), and $t = 5$.

$$A = P\left(1 + \frac{r}{n}\right)^{nt}$$

$$A = 1000\left(1 + \frac{0.06}{4}\right)^{4 \cdot 5} \qquad \text{Substitute the given values.}$$

$$A = 1000(1.015)^{20}$$

$$A = 1346.86 \qquad \text{Use a calculator; round to the nearest cent.}$$

The account will contain $1346.86. (The actual amount of interest earned is $1346.86 - \$1000 = \$346.86$. Why?)

✔ **Now Try Exercise 45(a).**

EXAMPLE 7 Solving a Compound Interest Problem for *t*

Suppose inflation is averaging 3% per year. How many years will it take for prices to double?

We want the number of years t for $1 to grow to $2 at a rate of 3% per year. In the compound interest formula, we let $A = 2$, $P = 1$, $r = 0.03$, and $n = 1$.

$$2 = 1\left(1 + \frac{0.03}{1}\right)^{1t} \qquad \text{Substitute in the compound interest formula.}$$

$$2 = (1.03)^t \qquad \text{Simplify.}$$

$$\log 2 = \log(1.03)^t \qquad \text{Property 3}$$

$$\log 2 = t \log(1.03) \qquad \text{Power rule}$$

$$t = \frac{\log 2}{\log 1.03} \qquad \text{Divide by log 1.03.}$$

$$t \approx 23.45 \qquad \text{Use a calculator.}$$

Prices will double in about 23 yr. (This is called the **doubling time** of the money.) To check, verify that $1.03^{23.45} \approx 2$.

✔ **Now Try Exercise 45(b).**

Interest can be compounded annually, semiannually, quarterly, daily, and so on. The number of compounding periods can get larger and larger. If the value of n is allowed to approach infinity, we have an example of *continuous compounding*. However, the compound interest formula above cannot be used for continuous compounding since there is no finite value for n. The formula for continuous compounding is an example of exponential growth involving the number e.

> ### Continuous Compound Interest Formula
>
> If a principal of P dollars is deposited at an annual rate of interest r compounded continuously for t years, the final amount on deposit is
> $$A = Pe^{rt}.$$

EXAMPLE 8 Solving a Continuous Compound Interest Problem

In Example 6 we found that $1000 invested for 5 yr at 6% interest compounded quarterly would grow to $1346.86.

(a) How much would this same investment grow to if interest were compounded continuously?

$$
\begin{aligned}
A &= Pe^{rt} && \text{Formula for continuous compounding} \\
&= 1000e^{0.06(5)} && \text{Let } P = 1000, r = 0.06, \text{ and } t = 5. \\
&= 1000e^{0.30} \\
&= 1349.86 && \text{Use a calculator; round to the nearest cent.}
\end{aligned}
$$

Continuous compounding would cause the investment to grow to $1349.86. This is $3.00 more than the amount the investment grew to in Example 6, when interest was compounded quarterly.

(b) How long would it take for the initial investment to double its original amount?

   We must find the value of t that will cause A to be $2(\$1000) = \$2000$. Once again, we use the continuous compounding formula with $P = 1000$ and $r = 0.06$.

$$
\begin{aligned}
A &= Pe^{rt} \\
2000 &= 1000e^{0.06t} && \text{Let } A = 2P = 2000. \\
2 &= e^{0.06t} && \text{Divide by 1000.} \\
\ln 2 &= \ln e^{0.06t} && \text{Take natural logarithms.} \\
\ln 2 &= 0.06t && \ln e^k = k \\
t &= \frac{\ln 2}{0.06} && \text{Divide by 0.06.} \\
t &\approx 11.55 && \text{Use a calculator.}
\end{aligned}
$$

It would take about 11.55 yr for the original investment to double.

✔ **Now Try Exercise 47.**

OBJECTIVE 4 Solve applications involving base e exponential growth and decay. One of the most common applications of exponential functions depends on the fact that in many situations involving growth or decay of a population, the amount or number of some quantity present at time t can be closely approximated by

$$y = y_0 e^{kt},$$

where y_0 is the amount or number present at time $t = 0$, k is a constant, and e is the base of natural logarithms.

EXAMPLE 9 Applying a Base e Exponential Function

Hybrid cars, powered by a combination of a battery and gasoline, emit less pollution and get considerably better gas mileage than conventional cars of comparable size. The first hybrid car came on the U.S. market in 1999. Since then, sales of hybrid cars have been increasing rapidly. The number of hybrid cars sold in the United States over the period 2000–2005 can be modeled by the function defined by

$$f(x) = 10,014e^{0.5761x},$$

where $x = 0$ represents 2000, $x = 1$ represents 2001, and so on. (*Source:* www.hybridcars.com) Use this function to approximate the number of hybrid cars sold in 2002.

Here, $x = 2$ represents 2002. Evaluate $f(2)$ using a calculator.

$$f(2) = 10,014e^{0.5761(2)} \approx 31,696$$

In 2002, about 31,696 hybrid cars were sold in the United States.

✔ **Now Try Exercise 55.**

EXAMPLE 10 Solving an Application Involving Exponential Decay

Carbon 14 is a radioactive form of carbon that is found in all living plants and animals. After a plant or animal dies, the radioactive carbon 14 disintegrates according to the function defined by

$$y = y_0 e^{-0.000121t},$$

where t is time in years, y is the amount of the sample at time t, and y_0 is the initial amount present at $t = 0$.

(a) If an initial sample contains $y_0 = 10$ g of carbon 14, how many grams will be present after 3000 yr?

Let $y_0 = 10$ and $t = 3000$ in the formula, and use a calculator.

$$y = 10e^{-0.000121(3000)} \approx 6.96 \text{ g}$$

(b) How long would it take for the initial sample to decay to half of its original amount? (This is called the **half-life.**)

Let $y = \frac{1}{2}(10) = 5$, and solve for t.

$$5 = 10e^{-0.000121t} \qquad \text{Substitute.}$$

$$\frac{1}{2} = e^{-0.000121t} \qquad \text{Divide by 10.}$$

$$\ln \frac{1}{2} = -0.000121t \qquad \text{Take natural logarithms; } \ln e^k = k.$$

$$t = \frac{\ln \frac{1}{2}}{-0.000121} \qquad \text{Divide by } -0.000121.$$

$$t \approx 5728 \qquad \text{Use a calculator.}$$

The half-life is just over 5700 yr.

✔ **Now Try Exercise 59.**

OBJECTIVE 5 Use a graphing calculator to solve exponential and logarithmic equations. Recall that the x-intercepts of the graph of a function f correspond to the real solutions of the equation $f(x) = 0$. In Example 1, we solved the equation $3^x = 12$ algebraically using rules for logarithms and found the solution set to be $\{2.262\}$. This can be supported graphically by showing that the x-intercept of the graph of the function defined by $y = 3^x - 12$ corresponds to this solution. See Figure 18.

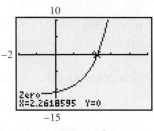

FIGURE 18

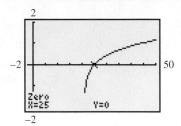

FIGURE 19

In Example 5, we solved $\log x + \log(x - 21) = 2$ to find the solution set $\{25\}$. (We rejected the apparent solution -4 since it led to the logarithm of a negative number.) Figure 19 shows that the x-intercept of the graph of the function defined by $y = \log x + \log(x - 21) - 2$ supports this result.

10.6 EXERCISES

Complete solution available on Video Lectures on CD/DVD

Now Try Exercise

RELATING CONCEPTS (EXERCISES 1–4)

FOR INDIVIDUAL OR GROUP WORK

*In **Section 10.2** we solved an equation such as $5^x = 125$ by writing each side as a power of the same base, setting exponents equal, and then solving the resulting equation as follows.*

$$5^x = 125 \qquad \text{Original equation}$$
$$5^x = 5^3 \qquad 125 = 5^3$$
$$x = 3 \qquad \text{Set exponents equal.}$$

Solution set: $\{3\}$

The method described in this section can also be used to solve this equation. **Work Exercises 1–4 in order,** *to see how this is done.*

1. Take common logarithms on both sides, and write this equation.

2. Apply the power rule for logarithms on the left.

3. Write the equation so that x is alone on the left.

4. Use a calculator to find the decimal form of the solution. What is the solution set?

Many of the problems in the remaining exercises require a scientific calculator.

Solve each equation. Give solutions to three decimal places. See Example 1.

5. $7^x = 5$

6. $4^x = 3$

7. $9^{-x+2} = 13$

8. $6^{-t+1} = 22$

9. $3^{2x} = 14$

10. $5^{0.3x} = 11$

11. $2^{x+3} = 5^x$

12. $6^{m+3} = 4^m$

13. $2^{x+3} = 3^{x-4}$

Solve each equation. Use natural logarithms. When appropriate, give solutions to three decimal places. See Example 2.

14. $e^{0.006x} = 30$

15. $e^{0.012x} = 23$

16. $e^{-0.103x} = 7$

17. $e^{-0.205x} = 9$

18. $\ln e^x = 4$

19. $\ln e^{3x} = 9$

20. $\ln e^{0.04x} = \sqrt{3}$

21. $\ln e^{0.45x} = \sqrt{7}$

22. $\ln e^{2x} = \pi$

23. Solve one of the equations in Exercises 14–17 using common logarithms rather than natural logarithms. (You should get the same solution.) Explain why using natural logarithms is a better choice.

24. *Concept Check* If you were asked to solve

$$10^{0.0025x} = 75,$$

would natural or common logarithms be a better choice? Why?

Solve each equation. Give the exact solution. See Example 3.

25. $\log_3(6x + 5) = 2$

26. $\log_5(12x - 8) = 3$

27. $\log_2(2x - 1) = 5$

28. $\log_6(4x + 2) = 2$

29. $\log_7(x + 1)^3 = 2$

30. $\log_4(x - 3)^3 = 4$

31. *Concept Check* Suppose that in solving a logarithmic equation having the term $\log(x - 3)$ you obtain an apparent solution of 2. All algebraic work is correct. Why must you reject 2 as a solution of the equation?

32. *Concept Check* Suppose that in solving a logarithmic equation having the term $\log(3 - x)$ you obtain an apparent solution of -4. All algebraic work is correct. Should you reject -4 as a solution of the equation? Why or why not?

Solve each equation. Give exact solutions. See Examples 4 and 5.

33. $\log(6x + 1) = \log 3$

34. $\log(7 - x) = \log 12$

35. $\log_5(3t + 2) - \log_5 t = \log_5 4$

36. $\log_2(x + 5) - \log_2(x - 1) = \log_2 3$

37. $\log 4x - \log(x - 3) = \log 2$

38. $\log(-x) + \log 3 = \log(2x - 15)$

39. $\log_2 x + \log_2(x - 7) = 3$

40. $\log(2x - 1) + \log 10x = \log 10$

41. $\log 5x - \log(2x - 1) = \log 4$

42. $\log_3 x + \log_3(2x + 5) = 1$

43. $\log_2 x + \log_2(x - 6) = 4$

44. $\log_2 x + \log_2(x + 4) = 5$

Solve each problem. See Examples 6–8.

45. (a) How much money will there be in an account at the end of 6 yr if $2000 is deposited at 4% compounded quarterly? (Assume no withdrawals are made.)

(b) To one decimal place, how long will it take for the account to grow to $3000?

46. (a) How much money will there be in an account at the end of 7 yr if $3000 is deposited at 3.5% compounded quarterly? (Assume no withdrawals are made.)
   (b) To one decimal place, when will the account grow to $5000?

47. (a) What will be the amount A in an account with initial principal $4000 if interest is compounded continuously at an annual rate of 3.5% for 6 yr?
   (b) To one decimal place, how long will it take for the initial amount to double?

48. Refer to Exercise 46. Does the money grow to a larger value under those conditions, or when invested for 7 yr at 3% compounded continuously?

49. Find the amount of money in an account after 12 yr if $5000 is deposited at 7% annual interest compounded as follows.

   (a) Annually   **(b)** Semiannually   **(c)** Quarterly
   (d) Daily (Use $n = 365$.)   **(e)** Continuously

50. How much money will be in an account at the end of 8 yr if $4500 is deposited at 6% annual interest compounded as follows?

   (a) Annually   **(b)** Semiannually   **(c)** Quarterly
   (d) Daily (Use $n = 365$.)   **(e)** Continuously

51. How much money must be deposited today to amount to $1850 in 40 yr at 6.5% compounded continuously?

52. How much money must be deposited today to amount to $1000 in 10 yr at 5% compounded continuously?

Solve each problem. See Examples 9 and 10.

53. The total volume in millions of tons of materials recovered from municipal solid waste collections in the United States during the period 1980–2003 can be approximated by the function defined by

$$f(x) = 15.80e^{0.0708x},$$

where $x = 0$ corresponds to 1980, $x = 1$ to 1981, and so on. Approximate the volume recovered each year. (*Source:* Environmental Protection Agency.)

   (a) 1980   **(b)** 1985   **(c)** 1995   **(d)** 2003

54. Worldwide emissions in millions of metric tons of the greenhouse gas carbon dioxide from fossil fuel consumption during the period 1980–2003 can be modeled by the function defined by

$$f(x) = 18{,}315e^{0.01338x},$$

where $x = 0$ corresponds to 1980, $x = 1$ to 1981, and so on. Approximate the emissions for each year. (*Source:* U.S. Department of Energy.)

   (a) 1980   **(b)** 1995   **(c)** 2000   **(d)** 2003

55. Consumer expenditures on all types of books in the United States for the years 1995–2004 can be modeled by the function defined by

$$B(x) = 27{,}190e^{0.0448x},$$

where $x = 0$ represents 1995, $x = 1$ represents 1996, and so on, and $B(x)$ is in millions of dollars. Approximate consumer expenditures for 2004. (*Source:* Book Industry Study Group.)

56. Based on selected figures obtained during the years 1980–2003, the total number of bachelor's degrees earned in the United States can be modeled by the function defined by

$$D(x) = 918{,}030e^{0.0154x},$$

where $x = 0$ corresponds to 1980, $x = 10$ corresponds to 1990, and so on. Approximate the number of bachelor's degrees earned in 2003. (*Source:* U.S. National Center for Education Statistics.)

57. Suppose that the amount, in grams, of plutonium 241 present in a given sample is determined by the function defined by

$$A(t) = 2.00e^{-0.053t},$$

where t is measured in years. Find the amount present in the sample after the given number of years.

(a) 4   **(b)** 10   **(c)** 20   **(d)** What was the initial amount present?

58. Suppose that the amount, in grams, of radium 226 present in a given sample is determined by the function defined by

$$A(t) = 3.25e^{-0.00043t},$$

where t is measured in years. Find the amount present in the sample after the given number of years.

(a) 20   **(b)** 100   **(c)** 500   **(d)** What was the initial amount present?

59. A sample of 400 g of lead 210 decays to polonium 210 according to the function defined by

$$A(t) = 400e^{-0.032t},$$

where t is time in years.

(a) How much lead will be left in the sample after 25 yr?
(b) How long will it take the initial sample to decay to half of its original amount?

60. The concentration of a drug in a person's system decreases according to the function defined by

$$C(t) = 2e^{-0.125t},$$

where $C(t)$ is in appropriate units, and t is in hours.

(a) How much of the drug will be in the system after 1 hr?
(b) Find the time that it will take for the concentration to be half of its original amount.

61. Refer to Exercise 53. Assuming that the function continued to apply past 2003, in what year could we have expected the volume of materials recovered to have reached 100 million tons? (*Source:* Environmental Protection Agency.)

62. Refer to Exercise 54. Assuming that the function continued to apply past 2003, in what year can we expect worldwide carbon dioxide emissions from fossil fuel consumption to reach 28,000 million metric tons? (*Source:* U.S. Department of Energy.)

63. The number of ants in an anthill grows according to the function defined by

$$f(t) = 300e^{0.4t},$$

where t is time measured in days. Find the time it will take for the number of ants to double.

TECHNOLOGY INSIGHTS (EXERCISES 64–66)

64. The function defined by $A(x) = 3.25e^{-0.00043x}$, with $x = t$, described in Exercise 58, is graphed on the screen at the right. Interpret the meanings of X and Y in the display at the bottom of the screen in the context of Exercise 58.

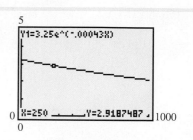

65. The screen shows a table of selected values for the function defined by $Y_1 = \left(1 + \frac{1}{X}\right)^X$.

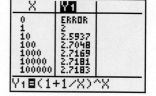

(a) Why is there an error message for X = 0?

(b) What number does the function value seem to approach as X takes on larger and larger values?

(c) Use a calculator to evaluate this function for X = 1,000,000. What value do you get? Now evaluate $e = e^1$. How close are these two values?

(d) Make a conjecture: As the values of x approach infinity, the value of $\left(1 + \frac{1}{x}\right)^x$ approaches _____.

66. Here is another property of logarithms: For $b > 0$, $x > 0$, $b \neq 1$, $x \neq 1$,

$$\log_b x = \frac{1}{\log_x b}.$$

Now observe the calculator screen.

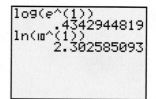

(a) Without using a calculator, give a decimal representation for $\frac{1}{0.4342944819}$. Then support your answer using the reciprocal key of your calculator.

(b) Without using a calculator, give a decimal representation for $\frac{1}{2.302585093}$. Then support your answer using the reciprocal key of your calculator.

PREVIEW EXERCISES

*Graph each function. See **Section 9.5**.*

67. $f(x) = 2x^2$

68. $f(x) = x^2 - 1$

69. $f(x) = (x + 1)^2$

70. $f(x) = (x - 1)^2 + 2$

Chapter **10** | **Group Activity**

HOW MUCH SPACE DO WE NEED?

Objective   Calculate using natural logarithms and exponential equations.

The formula for exponential population growth is $P(t) = P_0 e^{kt}$, where $P(t)$ is population after t years, P_0 is initial population, k is annual growth rate, and t is number of years elapsed.

A. If Earth's population will double in 30 yr at the current growth rate, what is this growth rate? (Express your answer as a percent.)

B. Earth has a total surface area of approximately 5.1×10^{14} m^2. Seventy percent of this surface area is rock, ice, sand, and open ocean. Another 8% of the total surface area, made up of tundra, lakes and streams, continental shelves, algae beds and reefs, and estuaries, is unfit for living space. The remaining area is suitable for growing food and for living space. Determine the surface area available for growing food and for living space.

C. Suppose that each person needs 100 m^2 of Earth's surface for living space. Use 5.5×10^9 for Earth's population. If none of the surface area available for living space is used for food, how long will it take for the livable surface of Earth to be covered with people? (Use the growth rate from part A and the surface area from part B.)

D. Measure a space that is 1 m^2 in area. Discuss with your partner whether or not you would want to be packed this closely together on Earth.

E. Now suppose that for each person 100 m^2 of Earth's surface is needed for living space and growing food.

   1. Using the population and growth rate from part C, how long will it take to fill Earth with people? (Use the surface area from part B.)

   2. Does 100 m^2 per person for living space and growing food seem reasonable? Consider the following:

     • How much space do you think it takes to raise animals for food? To grow grains, nuts, fruits, and vegetables?

     • Would food grow as well in deserts, mountains, or jungles?

     • Would there be any space left for shopping malls, movie theaters, concert halls, factories, office buildings, or parking lots?

   3. Write a paragraph summarizing your results and your discussion.

Chapter **10** **SUMMARY** *View the Interactive Summary on the Pass the Test CD.*

KEY TERMS

10.1 one-to-one function
inverse of a function

10.2 exponential function
asymptote
exponential equation

10.3 logarithm
logarithmic equation
logarithmic function
with base *a*

10.5 common logarithm
natural logarithm
universal constant

NEW SYMBOLS

$f^{-1}(x)$ the inverse of $f(x)$

$\log_a x$ the logarithm of x
to the base a

$\log x$ common (base 10)
logarithm of x

$\ln x$ natural (base e)
logarithm of x

e a constant,
approximately
2.7182818

TEST YOUR WORD POWER

See how well you have learned the vocabulary in this chapter. Answers, with examples, follow the Quick Review.

1. In a **one-to-one function**
  A. each x-value corresponds to only one y-value
  B. each x-value corresponds to one or more y-values
  C. each x-value is the same as each y-value
  D. each x-value corresponds to only one y-value and each y-value corresponds to only one x-value.

2. If f is a one-to-one function, then the **inverse** of f is
  A. the set of all solutions of f
  B. the set of all ordered pairs formed by interchanging the coordinates of the ordered pairs of f
  C. the set of all ordered pairs that are the opposite (negative) of the coordinates of the ordered pairs of f

  D. an equation involving an exponential expression.

3. An **exponential function** is a function defined by an expression of the form
  A. $f(x) = ax^2 + bx + c$ for real numbers a, b, c ($a \neq 0$)
  B. $f(x) = \log_a x$ for positive numbers a and x ($a \neq 1$)
  C. $f(x) = a^x$ for all real numbers x ($a > 0, a \neq 1$)
  D. $f(x) = \sqrt{x}$ for $x \geq 0$.

4. An **asymptote** is
  A. a line that a graph intersects just once
  B. a line that the graph of a function more and more closely approaches as the x-values increase or decrease

  C. the x-axis or y-axis
  D. a line about which a graph is symmetric.

5. A **logarithm** is
  A. an exponent
  B. a base
  C. an equation
  D. a polynomial.

6. A **logarithmic function** is a function that is defined by an expression of the form
  A. $f(x) = ax^2 + bx + c$ for real numbers a, b, c ($a \neq 0$)
  B. $f(x) = \log_a x$ for positive numbers a and x ($a \neq 1$)
  C. $f(x) = a^x$ for all real numbers x ($a > 0, a \neq 1$)
  D. $f(x) = \sqrt{x}$ for $x \geq 0$.

QUICK REVIEW

Concepts	Examples

10.1 INVERSE FUNCTIONS

Horizontal Line Test
A function is one-to-one if every horizontal line intersects the graph of the function at most once.

Find f^{-1} if $f(x) = 2x - 3$.

The graph of f is a straight line, so f is one-to-one by the horizontal line test.

Inverse Functions
For a one-to-one function f defined by an equation $y = f(x)$, the equation that defines the inverse function f^{-1} is found by interchanging x and y, solving for y, and replacing y with $f^{-1}(x)$.

To find $f^{-1}(x)$, interchange x and y in the equation $y = 2x - 3$.
$$x = 2y - 3$$
Solve for y to get
$$y = \frac{x + 3}{2}.$$

Therefore,
$$f^{-1}(x) = \frac{x + 3}{2}, \text{ or } f^{-1}(x) = \frac{1}{2}x + \frac{3}{2}.$$

In general, the graph of f^{-1} is the mirror image of the graph of f with respect to the line $y = x$.

The graphs of a function f and its inverse f^{-1} are shown here.

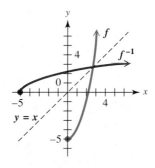

10.2 EXPONENTIAL FUNCTIONS

For $a > 0$, $a \neq 1$, $F(x) = a^x$ defines the exponential function with base a.

Graph of $F(x) = a^x$
1. The graph contains the point $(0, 1)$.
2. When $a > 1$, the graph rises from left to right. When $0 < a < 1$, the graph falls from left to right.
3. The x-axis is an asymptote.
4. The domain is $(-\infty, \infty)$; the range is $(0, \infty)$.

$F(x) = 3^x$ defines the exponential function with base 3.

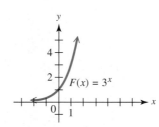

10.3 LOGARITHMIC FUNCTIONS

$y = \log_a x$ means $x = a^y$.

For $b > 0$, $b \neq 1$, $\log_b b = 1$ and $\log_b 1 = 0$.

$y = \log_2 x$ means $x = 2^y$.
$$\log_3 3 = 1 \qquad \log_5 1 = 0$$

(continued)

Concepts	Examples

For $a > 0$, $a \neq 1$, $x > 0$, $\boldsymbol{G(x) = \log_a x}$ defines the logarithmic function with base a.

$G(x) = \log_3 x$ defines the logarithmic function with base 3.

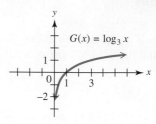

Graph of $\boldsymbol{G(x) = \log_a x}$

1. The graph contains the point $(1, 0)$.
2. When $a > 1$, the graph rises from left to right. When $0 < a < 1$, the graph falls from left to right.
3. The y-axis is an asymptote.
4. The domain is $(0, \infty)$; the range is $(-\infty, \infty)$.

10.4 PROPERTIES OF LOGARITHMS

Product Rule $\log_a xy = \log_a x + \log_a y$

Quotient Rule $\log_a \dfrac{x}{y} = \log_a x - \log_a y$

Power Rule $\log_a x^r = r \log_a x$

Special Properties

$$b^{\log_b x} = x \quad \text{and} \quad \log_b b^x = x$$

$$\log_2 3m = \log_2 3 + \log_2 m$$

$$\log_5 \frac{9}{4} = \log_5 9 - \log_5 4$$

$$\log_{10} 2^3 = 3 \log_{10} 2$$

$$6^{\log_6 10} = 10 \qquad \log_3 3^4 = 4$$

10.5 COMMON AND NATURAL LOGARITHMS

Common logarithms (base 10) are used in applications such as pH, sound level, and intensity of an earthquake. Use the $\boxed{\text{LOG}}$ key of a calculator to evaluate common logarithms.

Use the formula $\text{pH} = -\log[\text{H}_3\text{O}^+]$ to find the pH (to one decimal place) of grapes with hydronium ion concentration 5.0×10^{-5}.

$$
\begin{aligned}
\text{pH} &= -\log(5.0 \times 10^{-5}) & &\text{Substitute.} \\
&= -(\log 5.0 + \log 10^{-5}) & &\text{Property of logarithms} \\
&\approx 4.3 & &\text{Evaluate.}
\end{aligned}
$$

Natural logarithms (base e) are often used in applications of growth and decay, such as time for money invested to double, decay of chemical compounds, and biological growth. Use the $\boxed{\ln x}$ key or both the $\boxed{\text{INV}}$ and $\boxed{e^x}$ keys to evaluate natural logarithms.

Use the formula for doubling time (in years) $t(r) = \dfrac{\ln 2}{\ln(1 + r)}$ to find the doubling time to the nearest tenth at an interest rate of 4%.

$$
\begin{aligned}
t(0.04) &= \frac{\ln 2}{\ln(1 + 0.04)} & &\text{Substitute.} \\
&\approx 17.7 & &\text{Evaluate.}
\end{aligned}
$$

The doubling time is about 17.7 yr.

Change-of-Base Rule

If $a > 0$, $a \neq 1$, $b > 0$, $b \neq 1$, $x > 0$, then

$$\log_a x = \frac{\log_b x}{\log_b a}.$$

$$\log_3 17 = \frac{\ln 17}{\ln 3} = \frac{\log 17}{\log 3} \approx 2.5789$$

(continued)

Concepts	Examples

10.6 EXPONENTIAL AND LOGARITHMIC EQUATIONS; FURTHER APPLICATIONS

To solve exponential equations, use these properties ($b > 0$, $b \neq 1$).

1. If $b^x = b^y$, then $x = y$.

Solve.

$$2^{3x} = 2^5$$
$$3x = 5 \quad \text{Set exponents equal.}$$
$$x = \frac{5}{3} \quad \text{Divide by 3.}$$

The solution set is $\left\{\frac{5}{3}\right\}$.

2. If $x = y$, $x > 0$, $y > 0$, then $\log_b x = \log_b y$.

Solve.

$$5^m = 8$$
$$\log 5^m = \log 8 \quad \text{Take common logarithms.}$$
$$m \log 5 = \log 8 \quad \text{Power rule}$$
$$m = \frac{\log 8}{\log 5} \approx 1.2920 \quad \text{Divide by log 5.}$$

The solution set is $\{1.2920\}$.

To solve logarithmic equations, use these properties, where $b > 0$, $b \neq 1$, $x > 0$, $y > 0$. First use the properties of **Section 10.4,** if necessary, to write the equation in the proper form.

1. If $\log_b x = \log_b y$, then $x = y$.

Solve.

$$\log_3 2x = \log_3(x + 1)$$
$$2x = x + 1$$
$$x = 1 \quad \text{Subtract } x.$$

This value checks, so the solution set is $\{1\}$.

2. If $\log_b x = y$, then $b^y = x$.

Solve.

$$\log_2(3a - 1) = 4$$
$$3a - 1 = 2^4 \quad \text{Exponential form}$$
$$3a - 1 = 16 \quad \text{Apply the exponent.}$$
$$3a = 17 \quad \text{Add 1.}$$
$$a = \frac{17}{3} \quad \text{Divide by 3.}$$

The solution set is $\left\{\frac{17}{3}\right\}$.

Answers to Test Your Word Power

1. D; *Example:* The function $f = \{(0, 2), (1, -1), (3, 5), (-2, 3)\}$ is one-to-one.

2. B; *Example:* The inverse of the one-to-one function f defined in Answer 1 is $f^{-1} = \{(2, 0), (-1, 1), (5, 3), (3, -2)\}$.

3. C; *Examples:* $f(x) = 4^x$, $g(x) = \left(\frac{1}{2}\right)^x$, $h(x) = 2^{-x+3}$

4. B; *Example:* The graph of $F(x) = 2^x$ has the x-axis ($y = 0$) as an asymptote.

5. A; *Example:* $\log_a x$ is the exponent to which a must be raised to obtain x; $\log_3 9 = 2$ since $3^2 = 9$.

6. B; *Examples:* $y = \log_3 x$, $y = \log_{1/3} x$

Chapter **10** REVIEW EXERCISES

[10.1] *Determine whether each graph is the graph of a one-to-one function.*

1.

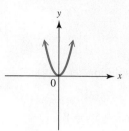

2.

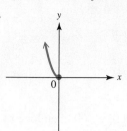

3. The table lists caffeine amounts in several popular 12-oz sodas. If the set of sodas is the domain and the set of caffeine amounts is the range of the function consisting of the six pairs listed, is it a one-to-one function? Why or why not?

Soda	Caffeine (mg)
Mountain Dew	55
Diet Coke	45
Dr. Pepper	41
Sunkist Orange Soda	41
Diet Pepsi-Cola	36
Coca-Cola Classic	34

Source: National Soft Drink Association.

Determine whether each function is one-to-one. If it is, find its inverse.

4. $f(x) = -3x + 7$ **5.** $f(x) = \sqrt[3]{6x - 4}$ **6.** $f(x) = -x^2 + 3$

Each function graphed is one-to-one. Graph its inverse.

7.

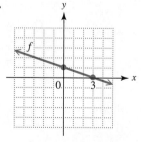

8.

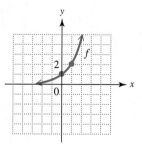

[10.2] *Graph each function.*

9. $f(x) = 3^x$ **10.** $f(x) = \left(\dfrac{1}{3}\right)^x$ **11.** $y = 3^{x+1}$ **12.** $y = 2^{2x+3}$

Solve each equation.

13. $4^{3x} = 8^{x+4}$ **14.** $\left(\dfrac{1}{27}\right)^{x-1} = 9^{2x}$

15. The gross wastes generated in plastics, in millions of tons, from 1980 through 2003 can be approximated by the exponential function defined by

$$W(x) = 7.77(1.059)^x,$$

where $x = 0$ corresponds to 1980, $x = 5$ to 1985, and so on. Use this function to approximate the plastic waste amounts for each year. (*Source:* U.S. Environmental Protection Agency.)

(**a**) 1985   (**b**) 1995   (**c**) 2000

[10.3] *Graph each function.*

16. $g(x) = \log_3 x$ (*Hint:* See Exercise 9.)   **17.** $g(x) = \log_{1/3} x$ (*Hint:* See Exercise 10.)

Solve each equation.

18. $\log_8 64 = x$ **19.** $\log_2 \sqrt{8} = x$ **20.** $\log_x \left(\dfrac{1}{49} \right) = -2$

21. $\log_4 x = \dfrac{3}{2}$ **22.** $\log_k 4 = 1$ **23.** $\log_b b^2 = 2$

 24. In your own words, explain the meaning of $\log_b a$.

25. *Concept Check* Based on the meaning of $\log_b a$, what is the simplest form of $b^{\log_b a}$?

26. A company has found that total sales, in thousands of dollars, are given by the function defined by

$$S(x) = 100 \log_2(x + 2),$$

where x is the number of weeks after a major advertising campaign was introduced.

(**a**) What were the total sales 6 weeks after the campaign was introduced?

(**b**) Graph the function.

[10.4] *Apply the properties of logarithms to express each logarithm as a sum or difference of logarithms. Assume that all variables represent positive real numbers.*

27. $\log_2 3xy^2$ **28.** $\log_4 \dfrac{\sqrt{x} \cdot w^2}{z}$

Apply the properties of logarithms to write each expression as a single logarithm. Assume that all variables represent positive real numbers, $b \neq 1$.

29. $\log_b 3 + \log_b x - 2 \log_b y$ **30.** $\log_3(x + 7) - \log_3(4x + 6)$

[10.5] *Evaluate each logarithm. Give approximations to four decimal places.*

31. $\log 28.9$ **32.** $\log 0.257$ **33.** $\ln 28.9$ **34.** $\ln 0.257$

Use the change-of-base rule (with either common or natural logarithms) to find each logarithm. Give approximations to four decimal places.

35. $\log_{16} 13$ **36.** $\log_4 12$

Use the formula $\text{pH} = -\log[\text{H}_3\text{O}^+]$ *to find the* pH *of each substance with the given hydronium ion concentration.*

37. Milk, 4.0×10^{-7}

38. Crackers, 3.8×10^{-9}

39. If orange juice has pH 4.6, what is its hydronium ion concentration?

40. Suppose the quantity, measured in grams, of a radioactive substance present at time t is given by

$$Q(t) = 500e^{-0.05t},$$

where t is measured in days. Find the quantity present at the following times.

(a) $t = 0$ **(b)** $t = 4$

41. Section 10.5, Exercise 40 introduced the doubling function defined by

$$t(r) = \frac{\ln 2}{\ln(1 + r)},$$

that gives the number of years required to double your money when it is invested at interest rate r (in decimal form) compounded annually. How long does it take to double your money at each rate? Round answers to the nearest year.

(a) 4% **(b)** 6% **(c)** 10% **(d)** 12%

✎ **(e)** Compare each answer in parts (a)–(d) with these numbers:

$$\frac{72}{4}, \frac{72}{6}, \frac{72}{10}, \frac{72}{12}.$$

What do you find?

[10.6] *Solve each equation. Give solutions to three decimal places.*

42. $3^x = 9.42$ **43.** $2^{x-1} = 15$ **44.** $e^{0.06x} = 3$

Solve each equation. Give exact solutions.

45. $\log_3(9x + 8) = 2$ **46.** $\log_5(y + 6)^3 = 2$

47. $\log_3(p + 2) - \log_3 p = \log_3 2$ **48.** $\log(2x + 3) = 1 + \log x$

49. $\log_4 x + \log_4(8 - x) = 2$ **50.** $\log_2 x + \log_2(x + 15) = \log_2 16$

51. *Concept Check* Consider the following "solution" of the equation $\log x^2 = 2$. *WHAT WENT WRONG?* Give the correct solution set.

$\log x^2 = 2$	Original equation
$2 \log x = 2$	Power rule for logarithms
$\log x = 1$	Divide both sides by 2.
$x = 10^1$	Write in exponential form.
$x = 10$	$10^1 = 10$

Solution set: $\{10\}$

Solve each problem. Use a calculator as necessary.

52. If \$20,000 is deposited at 7% annual interest compounded quarterly, how much will be in the account after 5 yr, assuming no withdrawals are made?

53. How much will \$10,000 compounded continuously at 6% annual interest amount to in 3 yr?

54. Which is a better plan?

> *Plan A:*   Invest $1000 at 4% compounded quarterly for 3 yr
> *Plan B:*   Invest $1000 at 3.9% compounded monthly for 3 yr

55. What is the half-life of the radioactive substance described in Exercise 40?

56. A machine purchased for business use **depreciates,** or loses value, over a period of years. The value of the machine at the end of its useful life is called its **scrap value.** By one method of depreciation (where it is assumed a constant percentage of the value depreciates annually), the scrap value, S, is given by

$$S = C(1 - r)^n,$$

where C is the original cost, n is the useful life in years, and r is the constant percent of depreciation.

(a) Find the scrap value of a machine costing $30,000, having a useful life of 12 yr and a constant annual rate of depreciation of 15%.

(b) A machine has a "half-life" of 6 yr. Find the constant annual rate of depreciation.

57. Recall from Exercise 41 in **Section 10.5** that the number of years, $N(r)$, since two independently evolving languages split off from a common ancestral language is approximated by

$$N(r) = -5000 \ln r,$$

where r is the percent of words from the ancestral language common to both languages now. Find r if the split occurred 2000 yr ago.

58. *Concept Check* Which one of the following is *not* equal to the solution of $7^x = 23$?

   A. $\dfrac{\log 23}{\log 7}$ **B.** $\dfrac{\ln 23}{\ln 7}$ **C.** $\log_7 23$ **D.** $\log_{23} 7$

MIXED REVIEW EXERCISES

Evaluate.

59. $\log_2 128$ **60.** $5^{\log_5 36}$ **61.** $e^{\ln 4}$

62. $10^{\log e}$ **63.** $\log_3 3^{-5}$ **64.** $\ln e^{5.4}$

Solve.

65. $\log_3(x + 9) = 4$ **66.** $\ln e^x = 3$ **67.** $\log_x \dfrac{1}{81} = 2$

68. $27^x = 81$ **69.** $2^{2x-3} = 8$ **70.** $5^{x+2} = 25^{2x+1}$

71. $\log_3(x + 1) - \log_3 x = 2$ **72.** $\log(3x - 1) = \log 10$ **73.** $\ln(x^2 + 3x + 4) = \ln 2$

74. Consider the logarithmic equation

$$\log(2x + 3) = \log x + 1.$$

(a) Solve the equation using properties of logarithms.

(b) If $Y_1 = \log(2X + 3)$ and $Y_2 = \log X + 1$, then the graph of $Y_1 - Y_2$ looks like that shown. Explain how the display at the bottom of the screen confirms the solution set found in part (a).

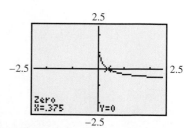

75. Based on selected figures from 1980 through 2003, the fractional part of the generation of municipal solid waste recovered can be approximated by the function defined by

$$R(x) = 10.001e^{0.0521x},$$

where $x = 0$ corresponds to 1980, $x = 10$ to 1990, and so on. Based on this model, what *percent* of municipal solid waste was recovered in 2000? (*Source:* Environmental Protection Agency.)

76. One measure of the diversity of the species in an ecological community is the **index of diversity,** the logarithmic expression

$$-(p_1 \ln p_1 + p_2 \ln p_2 + \cdots + p_n \ln p_n),$$

where $p_1, p_2, \ldots, p_n$ are the proportions of a sample belonging to each of n species in the sample. (*Source:* Ludwig, John and James Reynolds, *Statistical Ecology: A Primer on Methods and Computing,* New York, John Wiley and Sons, 1988.) Find the index of diversity to three decimal places if a sample of 100 from a community produces the following numbers.

(a) 90 of one species, 10 of another   **(b)** 60 of one species, 40 of another

Chapter **10** TEST

View the complete solutions to all Chapter Test exercises on the Pass the Test CD.

1. Decide whether each function is one-to-one.

(a) $f(x) = x^2 + 9$ **(b)**

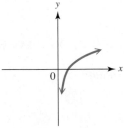

2. Find $f^{-1}(x)$ for the one-to-one function defined by $f(x) = \sqrt[3]{x + 7}$.

3. Graph the inverse of f, given the graph of f at the right.

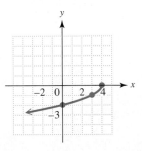

Graph each function.

4. $f(x) = 6^x$ **5.** $g(x) = \log_6 x$

6. Explain how the graph of the function in Exercise 5 can be obtained from the graph of the function in Exercise 4.

Solve each equation. Give the exact solution.

7. $5^x = \dfrac{1}{625}$

8. $2^{3x-7} = 8^{2x+2}$

9. A recent report predicts that the U.S. Hispanic population will increase from 26.7 million in 1995 to 96.5 million in 2050. (*Source:* U.S. Census Bureau.) Assuming an exponential growth pattern, the population is approximated by

$$f(x) = 26.7e^{0.023x},$$

where x represents the number of years since 1995. Use this function to estimate the Hispanic population in each year.

(a) 2010   **(b)** 2015

10. Write in logarithmic form: $4^{-2} = 0.0625$.

11. Write in exponential form: $\log_7 49 = 2$.

Solve each equation.

12. $\log_{1/2} x = -5$

13. $x = \log_9 3$

14. $\log_x 16 = 4$

15. *Concept Check* Fill in the blanks with the correct responses: The value of $\log_2 32$ is _____. This means that if we raise _____ to the _____ power, the result is _____.

Use properties of logarithms to write each expression as a sum or difference of logarithms. Assume that variables represent positive real numbers.

16. $\log_3 x^2 y$

17. $\log_5\left(\dfrac{\sqrt{x}}{yz}\right)$

Use properties of logarithms to write each expression as a single logarithm. Assume that variables represent positive real numbers, $b \neq 1$.

18. $3 \log_b s - \log_b t$

19. $\dfrac{1}{4} \log_b r + 2 \log_b s - \dfrac{2}{3} \log_b t$

20. Use a calculator to approximate each logarithm to four decimal places.

(a) $\log 23.1$ **(b)** $\ln 0.82$

21. Use the change-of-base rule to express $\log_3 19$

(a) in terms of common logarithms   **(b)** in terms of natural logarithms
(c) correct to four decimal places.

22. Solve $3^x = 78$, giving the correct solution to four decimal places.

23. Solve $\log_8(x + 5) + \log_8(x - 2) = 1$.

24. Suppose that $10,000 is invested at 4.5% annual interest, compounded quarterly. How much will be in the account in 5 yr if no money is withdrawn?

25. Suppose that $15,000 is invested at 5% annual interest, compounded continuously.

(a) How much will be in the account in 5 yr if no money is withdrawn?
(b) How long will it take for the initial principal to double?

Chapters 1–10 CUMULATIVE REVIEW EXERCISES

Let $S = \left\{ -\frac{9}{4}, -2, -\sqrt{2}, 0, 0.6, \sqrt{11}, \sqrt{-8}, 6, \frac{30}{3} \right\}$. List the elements of S that are members of each set.

1. Integers

2. Rational numbers

3. Irrational numbers

4. Real numbers

Simplify each expression.

5. $|-8| + 6 - |-2| - (-6 + 2)$

6. $-12 - |-3| - 7 - |-5|$

7. $2(-5) + (-8)(4) - (-3)$

Solve each equation or inequality.

8. $7 - (3 + 4a) + 2a = -5(a - 1) - 3$

9. $2m + 2 \le 5m - 1$

10. $|2x - 5| = 9$

11. $|3p| - 4 = 12$

12. $|3k - 8| \le 1$

13. $|4m + 2| > 10$

Graph.

14. $5x + 2y = 10$

15. $-4x + y \le 5$

16. The graph indicates that the number of U.S. travelers to international countries dropped from 61,327 thousand in 2000 to 56,250 thousand in 2003.

(a) Does the graph represent a function?

(b) What is the slope to the nearest integer of the line in the graph? Interpret the slope in the context of U.S. travelers to international countries.

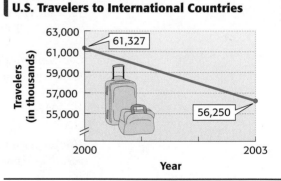

U.S. Travelers to International Countries

Source: U.S. Department of Commerce.

17. Find an equation of the line through $(5, -1)$ and parallel to the line with equation $3x - 4y = 12$. Write the equation in slope-intercept form.

Solve each system.

18. $5x - 3y = 14$
$2x + 5y = 18$

19. $2x - 7y = 8$
$4x - 14y = 3$

20. $x + 2y + 3z = 11$
$3x - y + z = 8$
$2x + 2y - 3z = -12$

21. Candy worth \$1.00 per lb is to be mixed with 10 lb of candy worth \$1.96 per lb to get a mixture that will be sold for \$1.60 per lb. How many pounds of the \$1.00 candy should be used?

Number of Pounds	Price per Pound	Value
x	\$1.00	$1x$
	\$1.60	

Perform the indicated operations.

22. $(2p + 3)(3p - 1)$ **23.** $(4k - 3)^2$

24. $(3m^3 + 2m^2 - 5m) - (8m^3 + 2m - 4)$

25. Divide $6t^4 + 17t^3 - 4t^2 + 9t + 4$ by $3t + 1$.

Factor.

26. $8x + x^3$ **27.** $24y^2 - 7y - 6$ **28.** $5z^3 - 19z^2 - 4z$

29. $16a^2 - 25b^4$ **30.** $8c^3 + d^3$ **31.** $16r^2 + 56rq + 49q^2$

Perform the indicated operations.

32. $\dfrac{(5p^3)^4(-3p^7)}{2p^2(4p^4)}$ **33.** $\dfrac{x^2 - 9}{x^2 + 7x + 12} \div \dfrac{x - 3}{x + 5}$

34. $\dfrac{2}{k + 3} - \dfrac{5}{k - 2}$ **35.** $\dfrac{3}{p^2 - 4p} - \dfrac{4}{p^2 + 2p}$

Simplify.

36. $\sqrt{288}$ **37.** $2\sqrt{32} - 5\sqrt{98}$

38. Solve $\sqrt{2x + 1} - \sqrt{x} = 1$.   **39.** Multiply $(5 + 4i)(5 - 4i)$.

Solve each equation or inequality.

40. $3x^2 - x - 1 = 0$ **41.** $k^2 + 2k - 8 > 0$ **42.** $x^4 - 5x^2 + 4 = 0$

43. Find two numbers whose sum is 300 and whose product is a maximum.

44. Graph $f(x) = \dfrac{1}{3}(x - 1)^2 + 2$.   **45.** Graph $f(x) = 2^x$.

46. Solve $5^{x+3} = \left(\dfrac{1}{25}\right)^{3x+2}$.   **47.** Graph $f(x) = \log_3 x$.

48. Given that $\log_2 9 = 3.1699$, what is the value of $\log_2 81$?

49. Rewrite the following using the product, quotient, and power properties of logarithms.

$$\log \frac{x^3\sqrt{y}}{z}$$

50. Let the number of bacteria present in a certain culture be given by

$$B(t) = 25{,}000e^{0.2t},$$

where t is time measured in hours, and $t = 0$ corresponds to noon. Find, to the nearest hundred, the number of bacteria present at

(a) noon   **(b)** 1 P.M.   **(c)** 2 P.M.
(d) When will the population double?

Nonlinear Functions, Conic Sections, and Nonlinear Systems

In this chapter, we study a group of curves known as *conic sections*. One conic section, the *ellipse*, has a special reflecting property responsible for "whispering galleries." In a whispering gallery, a person whispering at a certain point in the room can be heard clearly at another point across the room.

The Old House Chamber of the U.S. Capitol, now called Statuary Hall, is a whispering gallery. History has it that John Quincy Adams, whose desk was positioned at exactly the right point beneath the ellipsoidal ceiling, often pretended to sleep there as he listened to political opponents whispering strategies across the room. (*Source: We, the People, The Story of the United States Capitol*, 1991.)

In Section 11.2, we investigate ellipses.

11.1 Additional Graphs of Functions

In earlier chapters we introduced the function defined by $f(x) = x^2$, sometimes called the **squaring function.** This is an important elementary function in algebra.

OBJECTIVE 1 Recognize the graphs of the elementary functions defined by $|x|$, $\frac{1}{x}$, and $\sqrt{x}$, and graph their translations. The elementary function defined by $f(x) = |x|$ is called the **absolute value function.** Its graph, along with a table of selected ordered pairs, is shown in Figure 1. Its domain is $(-\infty, \infty)$, and its range is $[0, \infty)$.

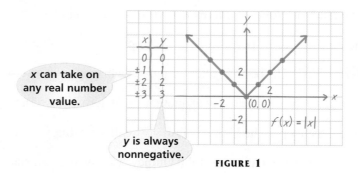

x can take on any real number value.

y is always nonnegative.

FIGURE 1

The **reciprocal function,** defined by $f(x) = \frac{1}{x}$, is a rational function. Rational functions were introduced in **Section 7.1.** The graph of this function is shown in Figure 2. Since x can never equal 0 for this function, as x gets closer and closer to 0, $\frac{1}{x}$ approaches either ∞ or $-\infty$. Also, $\frac{1}{x}$ can never equal 0, and as x approaches ∞ or $-\infty$, $\frac{1}{x}$ approaches 0. The axes are called **asymptotes** for the function. For the reciprocal function, the domain and the range are both $(-\infty, 0) \cup (0, \infty)$.

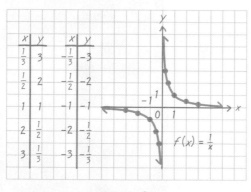

FIGURE 2

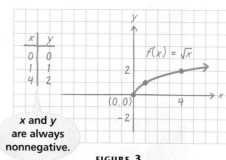

x and *y* are always nonnegative.

FIGURE 3

The **square root function,** defined by $f(x) = \sqrt{x}$, was introduced in **Section 8.1.** Its graph is shown in Figure 3. Since we restrict function values to be real numbers, x cannot take on negative values. Thus, the domain of the square root function is $[0, \infty)$. Because the principal square root is always nonnegative, the range is also $[0, \infty)$.

Just as the graph of $f(x) = x^2$ can be shifted, or translated, as we saw in **Section 9.5,** so can the graphs of these other elementary functions.

EXAMPLE 1 Applying a Horizontal Shift

Graph $f(x) = |x - 2|$. Give the domain and range.

The graph of $y = (x - 2)^2$ is obtained by shifting the graph of $y = x^2$ two units to the right. In a similar manner, the graph of $f(x) = |x - 2|$ is found by shifting the graph of $y = |x|$ two units to the right, as shown in Figure 4. The table of ordered pairs accompanying the graph supports this, as you can see by comparing it to the table with Figure 1. The domain of this function is $(-\infty, \infty)$, and its range is $[0, \infty)$.

✔ **Now Try Exercise 9.**

x	y
0	2
1	1
2	0
3	1
4	2

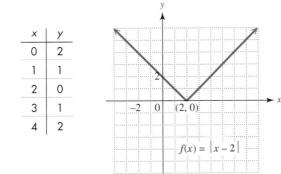

FIGURE 4

x	y
$\frac{1}{3}$	6
$\frac{1}{2}$	5
1	4
2	3.5

x	y
$-\frac{1}{3}$	0
$-\frac{1}{2}$	1
-1	2
-2	2.5

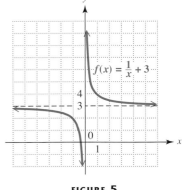

FIGURE 5

EXAMPLE 2 Applying a Vertical Shift

Graph $f(x) = \dfrac{1}{x} + 3$. Give the domain and range.

The graph of this function is found by shifting the graph of $y = \frac{1}{x}$ three units up. See Figure 5. The domain is $(-\infty, 0) \cup (0, \infty)$, and the range is $(-\infty, 3) \cup (3, \infty)$.

✔ **Now Try Exercise 11.**

EXAMPLE 3 Applying Both Horizontal and Vertical Shifts

Graph $f(x) = \sqrt{x + 1} - 4$. Give the domain and range.

The graph of $y = (x + 1)^2 - 4$ is obtained by shifting the graph of $y = x^2$ one unit to the left and four units down. Following this pattern here, we shift the graph of $y = \sqrt{x}$ one unit to the left and four units down to get the graph of $f(x) = \sqrt{x + 1} - 4$. See Figure 6. The domain is $[-1, \infty)$, and the range is $[-4, \infty)$.

x	y
-1	-4
0	-3
3	-2

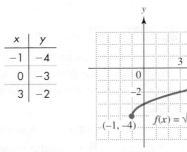

FIGURE 6

✔ **Now Try Exercise 17.**

OBJECTIVE 2 Recognize and graph step functions.   The **greatest integer function,** usually written $f(x) = [\![x]\!]$, is defined as follows.

> $[\![x]\!]$ **denotes the greatest integer that is less than or equal to x.**

For example,

$$[\![8]\!] = 8, \quad [\![7.45]\!] = 7, \quad [\![\pi]\!] = 3, \quad [\![-1]\!] = -1, \quad [\![-2.6]\!] = -3,$$

and so on.

✔ **Now Try Exercises 21 and 25.**

EXAMPLE 4 Graphing the Greatest Integer Function

Graph $f(x) = [\![x]\!]$. Give the domain and range.
    For $[\![x]\!]$,

$$\text{if } -1 \le x < 0, \quad \text{then} \quad [\![x]\!] = -1;$$
$$\text{if } \quad 0 \le x < 1, \quad \text{then} \quad [\![x]\!] = 0;$$
$$\text{if } \quad 1 \le x < 2, \quad \text{then} \quad [\![x]\!] = 1,$$

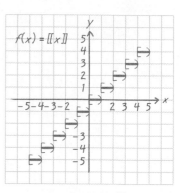

and so on. Thus, the graph, as shown in Figure 7, consists of a series of horizontal line segments. In each one, the left endpoint is included and the right endpoint is excluded. These segments continue infinitely following this pattern to the left and right. Since x can take any real number value, the domain is $(-\infty, \infty)$. The range is the set of integers $\{\ldots, -4, -3, -2, -1, 0, 1, 2, 3, 4, \ldots\}$. The appearance of the graph is the reason that this function is called a **step function.**

FIGURE 7

The graph of a step function also may be shifted. For example, the graph of $h(x) = [\![x - 2]\!]$ is the same as the graph of $f(x) = [\![x]\!]$ shifted two units to the right. Similarly, the graph of $g(x) = [\![x]\!] + 2$ is the graph of $f(x)$ shifted two units up.

✔ **Now Try Exercise 27.**

EXAMPLE 5 Applying a Greatest Integer Function

An overnight delivery service charges $25 for a package weighing up to 2 lb. For each additional pound or fraction of a pound there is an additional charge of $3. Let $D(x)$ represent the cost to send a package weighing x pounds. Graph $D(x)$ for x in the interval $(0, 6]$.

$$\text{For } x \text{ in the interval } (0, 2], \quad y = 25.$$
$$\text{For } x \text{ in the interval } (2, 3], \quad y = 25 + 3 = 28.$$
$$\text{For } x \text{ in the interval } (3, 4], \quad y = 28 + 3 = 31, \text{ and so on.}$$

The graph, which is that of a step function, is shown in Figure 8.

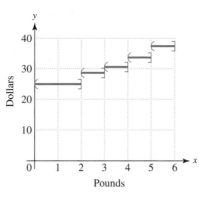

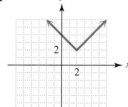

FIGURE 8

✔ Now Try Exercise 29.

11.1 EXERCISES

Now Try Exercise

Concept Check Fill in each blank with the correct response.

1. For the reciprocal function defined by $f(x) = \frac{1}{x}$, _____ is the only real number not in the domain.

2. The range of the square root function, given by $f(x) = \sqrt{x}$, is _____.

3. The lowest point on the graph of $f(x) = |x|$ has coordinates (_____, _____).

4. The range of $f(x) = [\![x]\!]$, the greatest integer function, is _____.

Concept Check Without actually plotting points, match each function defined by the absolute value expression with its graph.

5. $f(x) = |x - 2| + 2$ **A.**

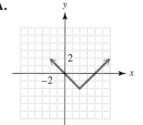

6. $f(x) = |x + 2| + 2$

B.

7. $f(x) = |x - 2| - 2$ **C.**

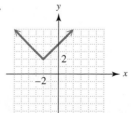

8. $f(x) = |x + 2| - 2$

D.

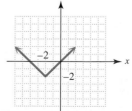

Graph each function. Give the domain and range. See Examples 1–3.

9. $f(x) = |x + 1|$

10. $f(x) = |x - 1|$

11. $f(x) = \dfrac{1}{x} + 1$

12. $f(x) = \dfrac{1}{x} - 1$

13. $f(x) = \sqrt{x - 2}$

14. $f(x) = \sqrt{x + 5}$

15. $f(x) = \dfrac{1}{x - 2}$

16. $f(x) = \dfrac{1}{x + 2}$

17. $f(x) = \sqrt{x + 3} - 3$

Evaulate each expression. See Objective 2.

18. $[\![18]\!]$

19. $[\![3]\!]$

20. $[\![8.7]\!]$

21. $\left[\!\!\left[\dfrac{1}{2}\right]\!\!\right]$

22. $[\![-5]\!]$

23. $[\![-14]\!]$

24. $[\![-6.9]\!]$

25. $[\![-10.1]\!]$

26. Explain how the graph of $f(x) = \frac{1}{x-3} + 2$ is obtained from the graph of $g(x) = \frac{1}{x}$.

Graph each step function. See Examples 4 and 5.

27. $f(x) = [\![x - 3]\!]$

28. $g(x) = [\![x + 2]\!]$

29. Assume that postage rates are 39¢ for the first ounce, plus 24¢ for each additional ounce, and that each letter carries one 39¢ stamp and as many 24¢ stamps as necessary. Graph the function defined by $y = p(x) =$ the number of stamps on a letter weighing x ounces. Use the interval $(0, 5]$.

30. The cost of parking a car at an airport hourly parking lot is $3 for the first half-hour and $2 for each additional half-hour or fraction thereof. Graph the function defined by $f(x) =$ the cost of parking a car for x hours. Use the interval $(0, 2]$.

PREVIEW EXERCISES

*Find the distance between each pair of points. See **Section 8.3**.*

31. $(2, -1)$ and $(4, 3)$

32. $(5, 6)$ and $(-2, -3)$

33. (x, y) and $(-2, 5)$

34. (x, y) and (h, k)

11.2 The Circle and the Ellipse

OBJECTIVES

1 Find an equation of a circle given the center and radius.

2 Determine the center and radius of a circle given its equation.

3 Recognize an equation of an ellipse.

4 Graph ellipses.

5 Graph circles and ellipses using a graphing calculator.

When an infinite cone is intersected by a plane, the resulting figure is called a **conic section.** The parabola is one example of a conic section; circles, ellipses, and hyperbolas may also result. See Figure 9.

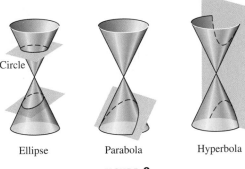

Circle

Ellipse   Parabola   Hyperbola

FIGURE 9

OBJECTIVE 1 Find an equation of a circle given the center and radius. A **circle** is the set of all points in a plane that lie a fixed distance from a fixed point. The fixed point is called the **center,** and the fixed distance is called the **radius.** We use the distance formula from **Section 8.3** to find an equation of a circle.

EXAMPLE 1 Finding an Equation of a Circle and Graphing It

Find an equation of the circle with radius 3 and center at $(0, 0)$, and graph it.

If the point (x, y) is on the circle, then the distance from (x, y) to the center $(0, 0)$ is 3.

$$\sqrt{(x_2 - x_1)^2 + (y_2 - y_1)^2} = d \qquad \text{Distance formula}$$
$$\sqrt{(x - 0)^2 + (y - 0)^2} = 3 \qquad \text{Let } x_1 = 0, y_1 = 0, \text{ and } d = 3.$$
$$x^2 + y^2 = 9 \qquad \text{Square each side.}$$

An equation of this circle is $x^2 + y^2 = 9$. The graph is shown in Figure 10.

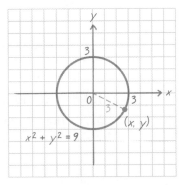

FIGURE 10

✔ **Now Try Exercise 1.**

A circle may not be centered at the origin, as seen in the next example.

EXAMPLE 2 Finding an Equation of a Circle and Graphing It

Find an equation of the circle with center at $(4, -3)$ and radius 5, and graph it.

$$\sqrt{(x-4)^2 + [y-(-3)]^2} = 5 \qquad \text{Distance formula}$$
$$(x-4)^2 + (y+3)^2 = 25 \qquad \text{Square each side.}$$

To graph the circle, plot the center $(4, -3)$, then move 5 units right, left, up, and down from the center, plotting points as you move. Draw a smooth curve through these four points, sketching one quarter of the circle at a time. The graph of this circle is shown in Figure 11.

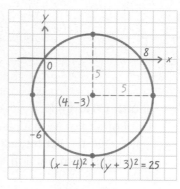

FIGURE 11

✔ **Now Try Exercise 7.**

Examples 1 and 2 suggest the form of an equation of a circle with radius r and center at (h, k). If (x, y) is a point on the circle, then the distance from the center (h, k) to the point (x, y) is r. By the distance formula,

$$\sqrt{(x-h)^2 + (y-k)^2} = r.$$

Squaring both sides gives the **center-radius form** of the equation of a circle.

Equation of a Circle (Center-Radius Form)

$$(x-h)^2 + (y-k)^2 = r^2$$

is an equation of the circle with radius r and center at (h, k).

EXAMPLE 3 Using the Center-Radius Form of the Equation of a Circle

Find an equation of the circle with center at $(-1, 2)$ and radius $\sqrt{7}$.

Use the center-radius form, with $h = -1$, $k = 2$, and $r = \sqrt{7}$.

$$(x-h)^2 + (y-k)^2 = r^2$$
$$[x-(-1)]^2 + (y-2)^2 = \left(\sqrt{7}\right)^2$$

Pay attention to signs here. $(x+1)^2 + (y-2)^2 = 7$

✔ **Now Try Exercise 9.**

OBJECTIVE 2 Determine the center and radius of a circle given its equation. In the equation found in Example 2, multiplying out $(x - 4)^2$ and $(y + 3)^2$ gives

$$(x - 4)^2 + (y + 3)^2 = 25$$
$$x^2 - 8x + 16 + y^2 + 6y + 9 = 25$$
$$x^2 + y^2 - 8x + 6y = 0.$$

This general form suggests that an equation with both x^2- and y^2-terms with equal coefficients may represent a circle. The next example shows how to tell, by completing the square. This procedure was introduced in **Section 9.1.**

EXAMPLE 4 **Completing the Square to Find the Center and Radius**

Find the center and radius of the circle $x^2 + y^2 + 2x + 6y - 15 = 0$, and graph it.

Since the equation has x^2- and y^2-terms with equal coefficients, its graph might be that of a circle. To find the center and radius, complete the squares on x and y.

$$x^2 + y^2 + 2x + 6y = 15 \qquad \text{Transform so that the constant is on the right.}$$

$$(x^2 + 2x \qquad) + (y^2 + 6y \qquad) = 15 \qquad \text{Write in anticipation of completing the square.}$$

$$\left[\frac{1}{2}(2)\right]^2 = 1 \qquad \left[\frac{1}{2}(6)\right]^2 = 9 \qquad \text{Square half the coefficient of each middle term.}$$

Add 1 and 9 on *both* sides of the equation.

$$(x^2 + 2x + 1) + (y^2 + 6y + 9) = 15 + 1 + 9 \qquad \text{Complete the squares on both } x \text{ and } y.$$

$$(x + 1)^2 + (y + 3)^2 = 25 \qquad \text{Factor on the left; add on the right.}$$

$$[x - (-1)]^2 + [y - (-3)]^2 = 5^2 \qquad \text{Center-radius form}$$

The final equation

$$[x - (-1)]^2 + [y - (-3)]^2 = 5^2$$

or

$$(x + 1)^2 + (y + 3)^2 = 5^2$$

shows that the graph is a circle with center at $(-1, -3)$ and radius 5. The graph is shown in Figure 12.

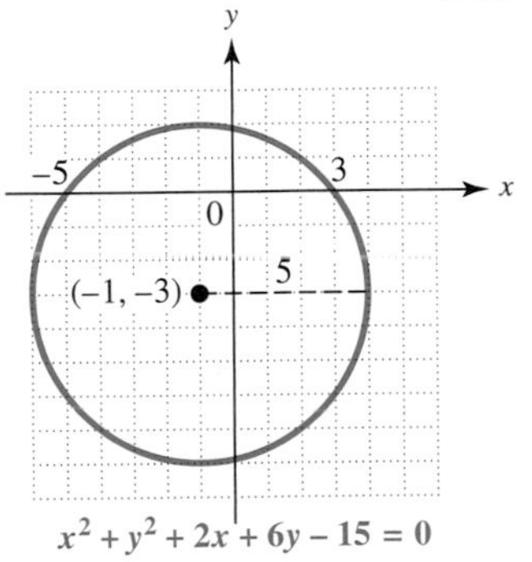

✔ Now Try Exercise 11.

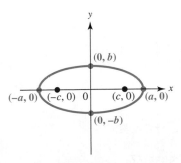

$x^2 + y^2 + 2x + 6y - 15 = 0$

FIGURE 12

▶ **NOTE** If the procedure of Example 4 leads to an equation of the form $(x - h)^2 + (y - k)^2 = 0$, then the graph is the single point (h, k). If the constant on the right side is negative, then the equation has no graph.

OBJECTIVE 3 Recognize an equation of an ellipse. An **ellipse** is the set of all points in a plane the *sum* of whose distances from two fixed points is constant. These fixed points are called **foci** (singular: *focus*). Figure 13 shows an ellipse whose foci are $(c, 0)$ and $(-c, 0)$, with x-intercepts $(a, 0)$ and $(-a, 0)$ and y-intercepts $(0, b)$ and $(0, -b)$. It is shown in more advanced courses that $c^2 = a^2 - b^2$ for an ellipse of this type. The origin is the **center** of the ellipse.

An ellipse has the following equation.

FIGURE 13

Equation of an Ellipse

The ellipse whose x-intercepts are $(a, 0)$ and $(-a, 0)$ and whose y-intercepts are $(0, b)$ and $(0, -b)$ has an equation of the form

$$\frac{x^2}{a^2} + \frac{y^2}{b^2} = 1.$$

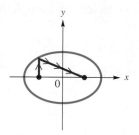

Reflecting property
of an ellipse

FIGURE 14

FIGURE 15

▶ **NOTE** A circle is a special case of an ellipse, where $a^2 = b^2$.

When a ray of light or sound emanating from one focus of an ellipse bounces off the ellipse, it passes through the other focus. See Figure 14. As mentioned in the chapter introduction, this reflecting property is responsible for whispering galleries. John Quincy Adams was able to listen in on his opponents' conversations because his desk was positioned at one of the foci beneath the ellipsoidal ceiling and his opponents were located across the room at the other focus.

The paths of Earth and other planets around the sun are approximately ellipses; the sun is at one focus and a point in space is at the other.

Elliptical bicycle gears are designed to respond to the legs' natural strengths and weaknesses. At the top and bottom of the powerstroke, where the legs have the least leverage, the gear offers little resistance, but as the gear rotates, the resistance increases. This allows the legs to apply more power where it is most naturally available. See Figure 15.

OBJECTIVE 4 Graph ellipses.

EXAMPLE 5 Graphing Ellipses

Graph each ellipse.

(a) $\dfrac{x^2}{49} + \dfrac{y^2}{36} = 1$

Here, $a^2 = 49$, so $a = 7$, and the x-intercepts for this ellipse are $(7, 0)$ and $(-7, 0)$. Similarly, $b^2 = 36$, so $b = 6$, and the y-intercepts are $(0, 6)$ and $(0, -6)$. Plotting the intercepts and sketching the ellipse through them gives the graph in Figure 16.

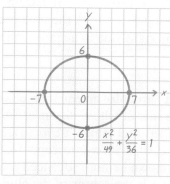

FIGURE 16

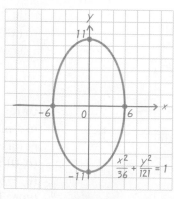

FIGURE 17

(b) $\dfrac{x^2}{36} + \dfrac{y^2}{121} = 1$

The x-intercepts for this ellipse are $(6, 0)$ and $(-6, 0)$, and the y-intercepts are $(0, 11)$ and $(0, -11)$. Join these with the smooth curve of an ellipse. The graph has been sketched in Figure 17 on the preceding page.

✔ **Now Try Exercises 29 and 35.**

As with the graphs of functions and circles, the graph of an ellipse may be shifted horizontally and vertically, as in the next example.

EXAMPLE 6 **Graphing an Ellipse Shifted Horizontally and Vertically**

Graph $\dfrac{(x - 2)^2}{25} + \dfrac{(y + 3)^2}{49} = 1.$

Just as $(x - 2)^2$ and $(y + 3)^2$ would indicate that the center of a circle would be $(2, -3)$, so it is with this ellipse. Figure 18 shows that the graph goes through the four points $(2, 4), (7, -3), (2, -10),$ and $(-3, -3)$. The x-values of these points are found by adding $\pm a = \pm 5$ to 2, and the y-values come from adding $\pm b = \pm 7$ to -3.

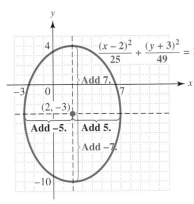

FIGURE 18

✔ **Now Try Exercise 37.**

OBJECTIVE 5 Graph circles and ellipses using a graphing calculator. The only conic section whose graph is a function is the parabola with vertical axis having equation $f(x) = ax^2 + bx + c$. Therefore, a graphing calculator in function mode cannot directly graph a circle or an ellipse. We must first solve the equation for y, getting two functions y_1 and y_2. The union of these two graphs is the graph of the entire figure.

For example, to graph $(x + 3)^2 + (y + 2)^2 = 25$, begin by solving for y.

$$(x + 3)^2 + (y + 2)^2 = 25$$
$$(y + 2)^2 = 25 - (x + 3)^2 \qquad \text{Subtract } (x + 3)^2.$$
$$y + 2 = \pm\sqrt{25 - (x + 3)^2} \qquad \text{Take square roots.}$$
Remember both roots. $\qquad y = -2 \pm \sqrt{25 - (x + 3)^2} \qquad \text{Subtract 2.}$

The two functions to be graphed are

$$y_1 = -2 + \sqrt{25 - (x + 3)^2} \qquad \text{and} \qquad y_2 = -2 - \sqrt{25 - (x + 3)^2}.$$

To get an undistorted screen, a **square viewing window** must be used. (Refer to your instruction manual for details.) See Figure 19. The two semicircles seem to be disconnected. This is because the graphs are nearly vertical at those points, and the calculator cannot show a true picture of the behavior there.

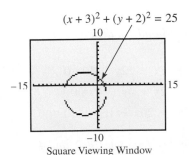

$(x + 3)^2 + (y + 2)^2 = 25$

Square Viewing Window

FIGURE 19

✔ **Now Try Exercise 43.**

> ▶ **NOTE** *The graphs in this section are not graphs of functions.* As mentioned, the only conic section whose graph is a function is the vertical parabola with equation $f(x) = ax^2 + bx + c$.

11.2 EXERCISES

Now Try Exercise

◑ **1.** See Example 1. Consider the circle whose equation is $x^2 + y^2 = 25$.

    (a) What are the coordinates of its center?   **(b)** What is its radius?

    (c) Sketch its graph.

✎ **2.** Why does a set of points defined by a circle *not* satisfy the definition of a function?

Concept Check   *Match each equation with the correct graph.*

3. $(x - 3)^2 + (y - 2)^2 = 25$

4. $(x - 3)^2 + (y + 2)^2 = 25$

5. $(x + 3)^2 + (y - 2)^2 = 25$

6. $(x + 3)^2 + (y + 2)^2 = 25$

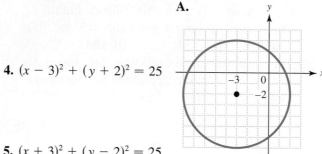

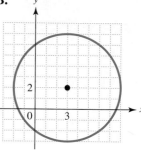

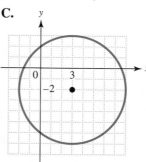

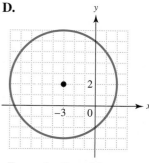

Find the equation of a circle satisfying the given conditions. See Examples 2 and 3.

◑ **7.** Center: $(-4, 3)$; radius: 2   **8.** Center: $(5, -2)$; radius: 4

◑ **9.** Center: $(-8, -5)$; radius: $\sqrt{5}$ **10.** Center: $(-12, 13)$; radius: $\sqrt{7}$

Find the center and radius of each circle. (Hint: In Exercises 15 and 16, divide each side by a common factor.) See Example 4.

◑ **11.** $x^2 + y^2 + 4x + 6y + 9 = 0$ **12.** $x^2 + y^2 - 8x - 12y + 3 = 0$

13. $x^2 + y^2 + 10x - 14y - 7 = 0$ **14.** $x^2 + y^2 - 2x + 4y - 4 = 0$

15. $3x^2 + 3y^2 - 12x - 24y + 12 = 0$ **16.** $2x^2 + 2y^2 + 20x + 16y + 10 = 0$

Graph each circle. Identify the center if it is not at the origin. See Examples 1, 2, and 4.

17. $x^2 + y^2 = 9$ **18.** $x^2 + y^2 = 4$ **19.** $2y^2 = 10 - 2x^2$ **20.** $3x^2 = 48 - 3y^2$

21. $(x + 3)^2 + (y - 2)^2 = 9$

22. $(x - 1)^2 + (y + 3)^2 = 16$

23. $x^2 + y^2 - 4x - 6y + 9 = 0$

24. $x^2 + y^2 + 8x + 2y - 8 = 0$

25. $x^2 + y^2 + 6x - 6y + 9 = 0$

26. $x^2 + y^2 - 4x + 10y + 20 = 0$

27. A circle can be drawn on a piece of posterboard by fastening one end of a string with a thumbtack, pulling the string taut with a pencil, and tracing a curve, as shown in the figure. Explain why this method works.

28. An ellipse can be drawn on a piece of posterboard by fastening two ends of a length of string with thumbtacks, pulling the string taut with a pencil, and tracing a curve, as shown in the figure. Explain why this method works.

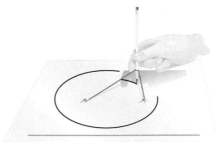

Graph each ellipse. See Examples 5 and 6.

29. $\dfrac{x^2}{9} + \dfrac{y^2}{25} = 1$

30. $\dfrac{x^2}{9} + \dfrac{y^2}{16} = 1$

31. $\dfrac{x^2}{36} = 1 - \dfrac{y^2}{16}$

32. $\dfrac{x^2}{9} = 1 - \dfrac{y^2}{4}$

33. $\dfrac{y^2}{25} = 1 - \dfrac{x^2}{49}$

34. $\dfrac{y^2}{9} = 1 - \dfrac{x^2}{16}$

35. $\dfrac{x^2}{16} + \dfrac{y^2}{4} = 1$

36. $\dfrac{x^2}{49} + \dfrac{y^2}{81} = 1$

37. $\dfrac{(x + 1)^2}{64} + \dfrac{(y - 2)^2}{49} = 1$

38. $\dfrac{(x - 4)^2}{9} + \dfrac{(y + 2)^2}{4} = 1$

39. $\dfrac{(x - 2)^2}{16} + \dfrac{(y - 1)^2}{9} = 1$

40. $\dfrac{(x + 3)^2}{25} + \dfrac{(y + 2)^2}{36} = 1$

41. Explain why a set of ordered pairs whose graph forms an ellipse does not satisfy the definition of a function.

42. (a) How many points are there on the graph of $(x - 4)^2 + (y - 1)^2 = 0$? Explain.

(b) How many points are there on the graph of $(x - 4)^2 + (y - 1)^2 = -1$? Explain.

TECHNOLOGY INSIGHTS (EXERCISES 43 AND 44)

43. The circle shown in the calculator graph was created using function mode, with a square viewing window. It is the graph of $(x + 2)^2 + (y - 4)^2 = 16$. What are the two functions y_1 and y_2 that were used to obtain this graph?

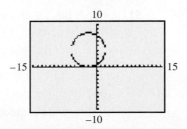

(continued)

44. The ellipse shown in the calculator graph was graphed using function mode, with a square viewing window. It is the graph of

$$\frac{x^2}{4} + \frac{y^2}{9} = 1.$$

What are the two functions y_1 and y_2 that were used to obtain this graph?

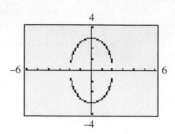

 Use a graphing calculator in function mode to graph each circle or ellipse. Use a square viewing window. See Objective 5.

45. $x^2 + y^2 = 36$

46. $(x - 2)^2 + y^2 = 49$

47. $\dfrac{x^2}{16} + \dfrac{y^2}{4} = 1$

48. $\dfrac{(x - 3)^2}{25} + \dfrac{y^2}{9} = 1$

*A **lithotripter** is a machine used to crush kidney stones using shock waves. The patient is placed in an elliptical tub with the kidney stone at one focus of the ellipse. A beam is projected from the other focus to the tub, so that it reflects to hit the kidney stone. See the figure.*

49. Suppose a lithotripter is based on the ellipse with equation

$$\frac{x^2}{36} + \frac{y^2}{9} = 1.$$

How far from the center of the ellipse must the kidney stone and the source of the beam be placed?

50. Rework Exercise 49 if the equation of the ellipse is

$$9x^2 + 4y^2 = 36.$$

(*Hint:* Write the equation in fraction form by dividing each term by 36, and use $c^2 = b^2 - a^2$, since $b > a$ here.)

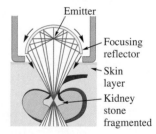

The top of an ellipse is illustrated in this depiction of how a lithotripter crushes kidney stones.

Source: Adapted drawing of an ellipse in illustration of a lithotripter. The American Medical Association, *Encyclopedia of Medicine*, 1989.

Solve each problem.

51. An arch has the shape of half an ellipse. The equation of the ellipse is

$$100x^2 + 324y^2 = 32{,}400,$$

where x and y are in meters.

(a) How high is the center of the arch?

(b) How wide is the arch across the bottom?

NOT TO SCALE

52. A one-way street passes under an overpass, which is in the form of the top half of an ellipse, as shown in the figure. Suppose that a truck 12 ft wide passes directly under the overpass. What is the maximum possible height of this truck?

NOT TO SCALE

In Exercises 53 and 54, see Figure 13 and use the fact that $c^2 = a^2 - b^2$ where $a^2 > b^2$.

53. The orbit of Mars is an ellipse with the sun at one focus. For x and y in millions of miles, the equation of the orbit is

$$\frac{x^2}{141.7^2} + \frac{y^2}{141.1^2} = 1.$$

(*Source:* Kaler, James B., *Astronomy!*, Addison-Wesley, 1997.)

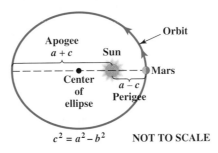

$c^2 = a^2 - b^2$ NOT TO SCALE

(a) Find the greatest distance (the **apogee**) from Mars to the sun.
(b) Find the smallest distance (the **perigee**) from Mars to the sun.

54. The orbit of Venus around the sun (one of the foci) is an ellipse with equation

$$\frac{x^2}{5013} + \frac{y^2}{4970} = 1,$$

where x and y are measured in millions of miles. (*Source:* Kaler, James B., *Astronomy!*, Addison-Wesley, 1997.)

(a) Find the greatest distance between Venus and the sun.
(b) Find the smallest distance between Venus and the sun.

PREVIEW EXERCISES

*For Exercises 55–57, see **Section 3.1**.*

55. Plot the points $(3, 4)$, $(-3, 4)$, $(3, -4)$, and $(-3, -4)$.

56. Sketch the graphs of $y = \frac{4}{3}x$ and $y = -\frac{4}{3}x$ on the same axes.

57. Find the x- and y-intercepts of the graph of $4x + 3y = 12$.

58. Solve the equation $x^2 = 121$. See **Section 9.1**.

11.3 The Hyperbola and Functions Defined by Radicals

OBJECTIVE 1 Recognize the equation of a hyperbola. A **hyperbola** is the set of all points in a plane such that the absolute value of the *difference* of the distances from two fixed points (called *foci*) is constant. Figure 20 shows a hyperbola; using the distance formula and the definition above, we can show that this hyperbola has equation

$$\frac{x^2}{16} - \frac{y^2}{12} = 1.$$

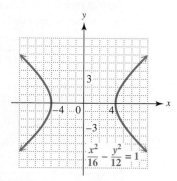

FIGURE 20

To graph hyperbolas centered at the origin, we need to find their intercepts. For the hyperbola in Figure 20, we proceed as follows.

x-Intercepts	*y*-Intercepts
Let $y = 0$.	Let $x = 0$.

x-Intercepts

Let $y = 0$.

$$\frac{x^2}{16} - \frac{0^2}{12} = 1 \qquad \text{Let } y = 0.$$

$$\frac{x^2}{16} = 1$$

$$x^2 = 16 \qquad \text{Multiply by 16.}$$

$$x = \pm 4$$

The x-intercepts are $(4, 0)$ and $(-4, 0)$.

y-Intercepts

Let $x = 0$.

$$\frac{0^2}{16} - \frac{y^2}{12} = 1 \qquad \text{Let } x = 0.$$

$$-\frac{y^2}{12} = 1$$

$$y^2 = -12 \qquad \text{Multiply by } -12.$$

Because there are no *real* solutions to the equation $y^2 = -12$, the graph has no y-intercepts.

The graph of $\dfrac{x^2}{16} - \dfrac{y^2}{12} = 1$ has no y-intercepts. On the other hand, the hyperbola in Figure 21 has no x-intercepts. Its equation is

$$\frac{y^2}{25} - \frac{x^2}{9} = 1,$$

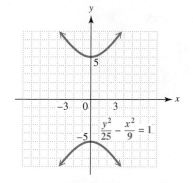

FIGURE 21

with y-intercepts $(0, 5)$ and $(0, -5)$.

Equations of Hyperbolas

A hyperbola with x-intercepts $(a, 0)$ and $(-a, 0)$ has an equation of the form

$$\frac{x^2}{a^2} - \frac{y^2}{b^2} = 1,$$

and a hyperbola with y-intercepts $(0, b)$ and $(0, -b)$ has an equation of the form

$$\frac{y^2}{b^2} - \frac{x^2}{a^2} = 1.$$

OBJECTIVE 2 Graph hyperbolas by using asymptotes. The two branches of the graph of a hyperbola approach a pair of intersecting straight lines, which are its asymptotes. See Figure 22 on the next page. The asymptotes are useful for sketching the graph of the hyperbola.

Asymptotes of Hyperbolas

The extended diagonals of the rectangle with vertices (corners) at the points (a, b), $(-a, b)$, $(-a, -b)$, and $(a, -b)$ are the **asymptotes** of the hyperbolas

$$\frac{x^2}{a^2} - \frac{y^2}{b^2} = 1 \qquad \text{and} \qquad \frac{y^2}{b^2} - \frac{x^2}{a^2} = 1.$$

This rectangle is called the **fundamental rectangle.** Using the methods of **Chapter 3,** we could show that the equations of these asymptotes are

$$y = \frac{b}{a}x \qquad \text{and} \qquad y = -\frac{b}{a}x.$$

To graph hyperbolas, follow these steps.

Graphing a Hyperbola

Step 1 **Find the intercepts.** Locate the intercepts at $(a, 0)$ and $(-a, 0)$ if the x^2-term has a positive coefficient, or at $(0, b)$ and $(0, -b)$ if the y^2-term has a positive coefficient.

Step 2 **Find the fundamental rectangle.** Locate the vertices of the fundamental rectangle at (a, b), $(-a, b)$, $(-a, -b)$, and $(a, -b)$.

Step 3 **Sketch the asymptotes.** The extended diagonals of the rectangle are the asymptotes of the hyperbola, and they have equations $y = \pm\frac{b}{a}x$.

Step 4 **Draw the graph.** Sketch each branch of the hyperbola through an intercept and approaching (but not touching) the asymptotes.

EXAMPLE 1 Graphing a Horizontal Hyperbola

Graph $\dfrac{x^2}{16} - \dfrac{y^2}{25} = 1$.

Step 1 Here $a = 4$ and $b = 5$. The x-intercepts are $(4, 0)$ and $(-4, 0)$.

Step 2 The four points $(4, 5)$, $(-4, 5)$, $(-4, -5)$, and $(4, -5)$ are the vertices of the fundamental rectangle, as shown in Figure 22 on the next page.

Steps 3 The equations of the asymptotes are $y = \pm\frac{5}{4}x$, and the hyperbola approaches
and 4 these lines as x and y get larger and larger in absolute value.

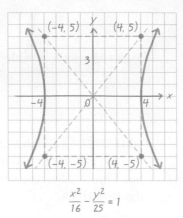

$$\frac{x^2}{16} - \frac{y^2}{25} = 1$$

FIGURE 22

✔ **Now Try Exercise 7.**

▶ **CAUTION** When sketching the graph of a hyperbola, be sure that the branches do not touch the asymptotes.

EXAMPLE 2 Graphing a Vertical Hyperbola

Graph $\dfrac{y^2}{49} - \dfrac{x^2}{16} = 1$.

This hyperbola has y-intercepts $(0, 7)$ and $(0, -7)$. The asymptotes are the extended diagonals of the rectangle with vertices at $(4, 7)$, $(-4, 7)$, $(-4, -7)$, and $(4, -7)$. Their equations are $y = \pm\frac{7}{4}x$. See Figure 23.

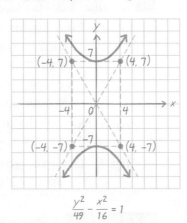

$$\frac{y^2}{49} - \frac{x^2}{16} = 1$$

FIGURE 23

✔ **Now Try Exercise 9.**

As with circles and ellipses, hyperbolas are graphed with a graphing calculator by first writing the equations of two functions whose union is equivalent to the equation of the hyperbola. A square window gives a truer shape for hyperbolas, too.

SUMMARY OF CONIC SECTIONS

Equation	Graph	Description	Identification
$y = ax^2 + bx + c$ **or** $y = a(x - h)^2 + k$	 Parabola	It opens up if $a > 0$, down if $a < 0$. The vertex is (h, k).	It has an x^2-term. y is not squared.
$x = ay^2 + by + c$ **or** $x = a(y - k)^2 + h$	 Parabola	It opens to the right if $a > 0$, to the left if $a < 0$. The vertex is (h, k).	It has a y^2-term. x is not squared.
$(x - h)^2 + (y - k)^2 = r^2$	 Circle	The center is (h, k), and the radius is r.	x^2- and y^2-terms have the same positive coefficient.
$\dfrac{x^2}{a^2} + \dfrac{y^2}{b^2} = 1$	 Ellipse	The x-intercepts are $(a, 0)$ and $(-a, 0)$. The y-intercepts are $(0, b)$ and $(0, -b)$.	x^2- and y^2-terms have different positive coefficients.
$\dfrac{x^2}{a^2} - \dfrac{y^2}{b^2} = 1$	 Hyperbola	The x-intercepts are $(a, 0)$ and $(-a, 0)$. The asymptotes are found from (a, b), $(a, -b)$, $(-a, -b)$, and $(-a, b)$.	x^2 has a positive coefficient. y^2 has a negative coefficient.
$\dfrac{y^2}{b^2} - \dfrac{x^2}{a^2} = 1$	 Hyperbola	The y-intercepts are $(0, b)$ and $(0, -b)$. The asymptotes are found from (a, b), $(a, -b)$, $(-a, -b)$, and $(-a, b)$.	y^2 has a positive coefficient. x^2 has a negative coefficient.

OBJECTIVE ③ Identify conic sections by their equations. Rewriting a second-degree equation in one of the forms given for ellipses, hyperbolas, circles, or parabolas makes it possible to determine when the graph is one of these.

EXAMPLE 3 Identifying the Graphs of Equations

Identify the graph of each equation.

(a) $9x^2 = 108 + 12y^2$

Both variables are squared, so the graph is either an ellipse or a hyperbola. (This situation also occurs for a circle, which is a special case of an ellipse.) To see whether the graph is an ellipse or a hyperbola, rewrite the equation so that the x^2- and y^2-terms are on one side of the equation and 1 is on the other.

$$9x^2 - 12y^2 = 108 \quad \text{Subtract } 12y^2.$$

$$\frac{x^2}{12} - \frac{y^2}{9} = 1 \quad \text{Divide by 108.}$$

Because of the minus sign, the graph of this equation is a hyperbola.

(b) $x^2 = y - 3$

Only one of the two variables, x, is squared, so this is the vertical parabola $y = x^2 + 3$.

(c) $x^2 = 9 - y^2$

Write the variable terms on the same side of the equation.

$$x^2 + y^2 = 9 \quad \text{Add } y^2.$$

The graph of this equation is a circle with center at the origin and radius 3.

✔ **Now Try Exercises 17, 19, and 21.**

OBJECTIVE ④ Graph certain square root functions. Recall that no vertical line will intersect the graph of a function in more than one point. Thus, horizontal parabolas, all circles and ellipses, and most hyperbolas discussed in this chapter are examples of graphs that do not satisfy the conditions of a function. However, by considering only a part of the graph of each of these we have the graph of a function, as seen in Figure 24.

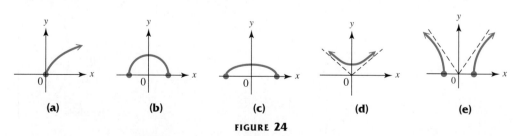

FIGURE 24

In parts (a), (b), (c), and (d) of Figure 24, the top portion of a conic section is shown (parabola, circle, ellipse, and hyperbola, respectively). In part (e), the top two portions of a hyperbola are shown. In each case, the graph is that of a function since the graph satisfies the conditions of the vertical line test.

In **Sections 8.1 and 11.1** we observed the square root function defined by $f(x) = \sqrt{x}$. To find equations for the types of graphs shown in Figure 24, we extend its definition.

> ### Square Root Function
>
> For an algebraic expression u, with $u \geq 0$, a function of the form
>
> $$f(x) = \sqrt{u}$$
>
> is called a **square root function.**

EXAMPLE 4 Graphing a Semicircle

Graph $f(x) = \sqrt{25 - x^2}$. Give the domain and range.

Replace $f(x)$ with y and square both sides to get the equation

$$y^2 = 25 - x^2 \quad \text{or} \quad x^2 + y^2 = 25.$$

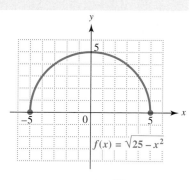

FIGURE 25

This is the graph of a circle with center at $(0, 0)$ and radius 5. Since $f(x)$, or y, represents a principal square root in the original equation, $f(x)$ must be nonnegative. This restricts the graph to the upper half of the circle, as shown in Figure 25. Use the graph and the vertical line test to verify that it is indeed a function. The domain is $[-5, 5]$, and the range is $[0, 5]$.

✔ **Now Try Exercise 25.**

EXAMPLE 5 Graphing a Portion of an Ellipse

Graph $\dfrac{y}{6} = -\sqrt{1 - \dfrac{x^2}{16}}$. Give the domain and range.

Square both sides to get an equation whose form is known.

$$\frac{y^2}{36} = 1 - \frac{x^2}{16}$$

$$\frac{x^2}{16} + \frac{y^2}{36} = 1 \qquad \text{Add } \tfrac{x^2}{16}.$$

This is the equation of an ellipse with x-intercepts $(4, 0)$ and $(-4, 0)$ and y-intercepts $(0, 6)$ and $(0, -6)$. Since $\frac{y}{6}$ equals a negative square root in the original equation, y must be nonpositive, restricting the graph to the lower half of the ellipse, as shown in Figure 26. Verify that this is the graph of a function, using the vertical line test. The domain is $[-4, 4]$, and the range is $[-6, 0]$.

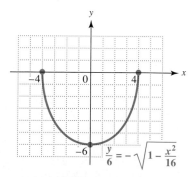

FIGURE 26

✔ **Now Try Exercise 27.**

Root functions, since they are functions, can be entered and graphed directly with a graphing calculator.

11.3 EXERCISES

 Complete solution available on Video Lectures on CD/DVD

Now Try Exercise

Concept Check *Based on the discussions of ellipses in the previous section and of hyperbolas in this section, match each equation with its graph.*

1. $\dfrac{x^2}{25} + \dfrac{y^2}{9} = 1$

A.

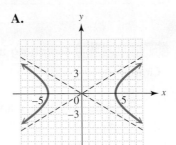

B.

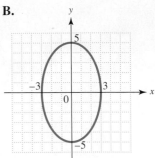

2. $\dfrac{x^2}{9} + \dfrac{y^2}{25} = 1$

3. $\dfrac{x^2}{9} - \dfrac{y^2}{25} = 1$

C.

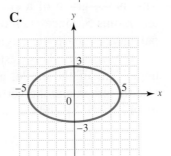

D.

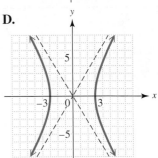

4. $\dfrac{x^2}{25} - \dfrac{y^2}{9} = 1$

✎ **5.** Write an explanation of how you can tell from the equation whether the branches of a hyperbola open up and down or left and right.

✎ **6.** Describe how the fundamental rectangle is used to sketch a hyperbola.

Graph each hyperbola. See Examples 1 and 2.

7. $\dfrac{x^2}{16} - \dfrac{y^2}{9} = 1$

8. $\dfrac{y^2}{9} - \dfrac{x^2}{9} = 1$

9. $\dfrac{y^2}{4} - \dfrac{x^2}{25} = 1$

10. $\dfrac{x^2}{49} - \dfrac{y^2}{16} = 1$

11. $\dfrac{x^2}{25} - \dfrac{y^2}{36} = 1$

12. $\dfrac{y^2}{9} - \dfrac{x^2}{4} = 1$

13. $\dfrac{y^2}{16} - \dfrac{x^2}{16} = 1$

14. $\dfrac{x^2}{25} - \dfrac{y^2}{9} = 1$

Identify the graph of each equation as a parabola, circle, ellipse, *or* hyperbola, *and then sketch the graph. See Example 3.*

15. $x^2 - y^2 = 16$

16. $x^2 + y^2 = 16$

17. $4x^2 + y^2 = 16$

18. $y^2 = 36 - x^2$

19. $x^2 - 2y = 0$

20. $9x^2 + 25y^2 = 225$

21. $9x^2 = 144 + 16y^2$

22. $x^2 + 9y^2 = 9$

23. $y^2 = 4 + x^2$

✎ **24.** State in your own words the major difference between the definitions of *ellipse* and *hyperbola*.

Graph each function defined by a radical expression. Give the domain and range. See Examples 4 and 5.

25. $f(x) = \sqrt{16 - x^2}$ **26.** $f(x) = \sqrt{9 - x^2}$ **27.** $f(x) = -\sqrt{36 - x^2}$

28. $f(x) = -\sqrt{25 - x^2}$ **29.** $\dfrac{y}{3} = \sqrt{1 + \dfrac{x^2}{9}}$ **30.** $y = \sqrt{\dfrac{x + 4}{2}}$

In **Section 11.2,** Example 6, we saw that the center of an ellipse may be shifted away from the origin. The same process applies to hyperbolas. For example, the hyperbola shown at the right,

$$\frac{(x + 5)^2}{4} - \frac{(y - 2)^2}{9} = 1,$$

has the same graph as $\dfrac{x^2}{4} - \dfrac{y^2}{9} = 1$, but it is centered at $(-5, 2)$. Graph each hyperbola with center shifted away from the origin.

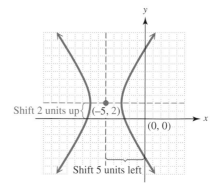

31. $\dfrac{(x - 2)^2}{4} - \dfrac{(y + 1)^2}{9} = 1$ **32.** $\dfrac{(x + 3)^2}{16} - \dfrac{(y - 2)^2}{25} = 1$

33. $\dfrac{y^2}{36} - \dfrac{(x - 2)^2}{49} = 1$ **34.** $\dfrac{(y - 5)^2}{9} - \dfrac{x^2}{25} = 1$

Solve each problem.

35. Two buildings in a sports complex are shaped and positioned like a portion of the branches of the hyperbola with equation

$$400x^2 - 625y^2 = 250,000,$$

where x and y are in meters.

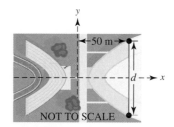

(a) How far apart are the buildings at their closest point?

(b) Find the distance d in the figure.

36. In rugby, after a *try* (similar to a touchdown in American football) the scoring team attempts a kick for extra points. The ball must be kicked from directly behind the point where the try was scored. The kicker can choose the distance but cannot move the ball sideways. It can be shown that the kicker's best choice is on the hyperbola with equation

$$\frac{x^2}{g^2} - \frac{y^2}{g^2} = 1,$$

where $2g$ is the distance between the goal posts. Since the hyperbola approaches its asymptotes, it is easier for the kicker to estimate points on the asymptotes instead of on the hyperbola. What are the asymptotes of this hyperbola? Why is it relatively easy to estimate them? (*Source:* Isaksen, Daniel C., "How to Kick a Field Goal," *The College Mathematics Journal*, September 1996.)

TECHNOLOGY INSIGHTS (EXERCISES 37 AND 38)

37. The hyperbola shown in the figure was graphed in function mode, with a square viewing window. It is the graph of $\frac{x^2}{9} - y^2 = 1$. What are the two functions y_1 and y_2 that were used to obtain this graph?

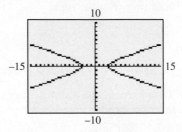

38. Repeat Exercise 37 for the graph of $\frac{y^2}{9} - x^2 = 1$, shown in the figure.

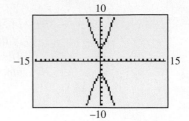

Use a graphing calculator in function mode to graph each hyperbola. Use a square viewing window.

39. $\dfrac{x^2}{25} - \dfrac{y^2}{49} = 1$ **40.** $\dfrac{x^2}{4} - \dfrac{y^2}{16} = 1$ **41.** $y^2 - 9x^2 = 9$ **42.** $y^2 - 9x^2 = 36$

PREVIEW EXERCISES

*Solve each system by substitution. See **Section 4.1.***

43. $2x + y = 13$
$\quad y = 3x + 3$

44. $3x + y = 4$
$\quad x = y - 8$

45. $9x + 2y = 10$
$\quad x - y = -5$

46. $5x + 2y = 15$
$\quad x - y = 3$

*Solve each equation. See **Section 9.3.***

47. $2x^4 - 5x^2 - 3 = 0$

48. $3x^4 + 26x^2 + 35 = 0$

49. $x^4 - 7x^2 + 12 = 0$

50. $x^4 - 14x^2 + 45 = 0$

11.4 Nonlinear Systems of Equations

An equation in which some terms have more than one variable or a variable of degree 2 or greater is called a **nonlinear equation.** A **nonlinear system of equations** includes at least one nonlinear equation.

When solving a nonlinear system, it helps to visualize the types of graphs of the equations of the system to determine the possible number of points of intersection. For example, if a system includes two equations where the graph of one is a parabola and the graph of the other is a line, then there may be zero, one, or two points of intersection, as illustrated in Figure 27.

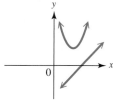
No points of intersection

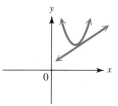
One point of intersection

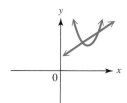
Two points of intersection

FIGURE 27

A similar situation exists for a system consisting of a circle and a line. See Figure 28.

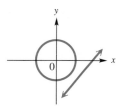
No points of intersection

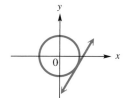
One point of intersection

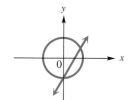
Two points of intersection

FIGURE 28

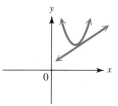

This system has four solutions, since there are four points of intersection.

FIGURE 29

If a system consists of two second-degree equations, then there may be zero, one, two, three, or four solutions. Figure 29 shows a case where a system consisting of a circle and a parabola has four solutions, all made up of ordered pairs of real numbers. (Exercises 7–14 are designed for you to visualize similar situations.)

OBJECTIVE 1 Solve a nonlinear system by substitution. We solve nonlinear systems by the elimination method, the substitution method, or a combination of the two. The substitution method **(Section 4.1)** is usually appropriate when one equation is linear.

EXAMPLE 1 Solving a Nonlinear System by Substitution

Solve the system

$$x^2 + y^2 = 9 \quad \text{(1)}$$
$$2x - y = 3. \quad \text{(2)}$$

The graph of (1) is a circle and the graph of (2) is a line. Recall from Figure 28 that the graphs could intersect in zero, one, or two points.

Solve the linear equation for one of the two variables and then substitute the resulting expression into the nonlinear equation to obtain an equation in one variable.

$$2x - y = 3 \qquad (2)$$
$$y = 2x - 3 \qquad (3)$$

Substitute $2x - 3$ for y in equation (1).

$$
\begin{aligned}
x^2 + y^2 &= 9 & &(1) \\
x^2 + (2x - 3)^2 &= 9 & &\text{Let } y = 2x - 3. \\
x^2 + 4x^2 - 12x + 9 &= 9 & &\text{Square } 2x - 3. \\
5x^2 - 12x &= 0 & &\text{Combine like terms; subtract 9.} \\
x(5x - 12) &= 0 & &\text{Factor; GCF is } x.
\end{aligned}
$$

$$x = 0 \quad \text{or} \quad 5x - 12 = 0 \qquad \text{Zero-factor property}$$

> **Set *both* factors equal to 0.**

$$x = \frac{12}{5}$$

Let $x = 0$ in equation (3) to get $y = -3$. If $x = \frac{12}{5}$, then $y = \frac{9}{5}$. The solution set of the system is $\left\{(0, -3), \left(\frac{12}{5}, \frac{9}{5}\right)\right\}$. The graph in Figure 30 confirms the two points of intersection.

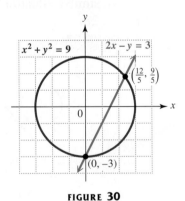

FIGURE 30

> ✔ **Now Try Exercise 19.**

EXAMPLE 2 Solving a Nonlinear System by Substitution

Solve the system

$$6x - y = 5 \qquad (1)$$
$$xy = 4. \qquad (2)$$

The graph of (1) is a line. We have not specifically mentioned equations like (2); however, it can be shown by plotting points that its graph is a hyperbola. Visualizing a line and a hyperbola indicates that there may be zero, one, or two points of intersection. Since neither equation has a squared term, we can solve either equation for one of the variables and then substitute the result into the other equation. Solving $xy = 4$ for x gives $x = \frac{4}{y}$. We substitute $\frac{4}{y}$ for x in equation (1).

$$6\left(\frac{4}{y}\right) - y = 5 \qquad \text{Let } x = \tfrac{4}{y} \text{ in equation (1).}$$

$$\frac{24}{y} - y = 5 \qquad \text{Multiply.}$$

$$24 - y^2 = 5y \qquad \text{Multiply by } y,\ y \neq 0.$$

$$y^2 + 5y - 24 = 0 \qquad \text{Standard form}$$

$$(y - 3)(y + 8) = 0 \qquad \text{Factor.}$$

$$y = 3 \quad \text{or} \quad y = -8 \qquad \text{Zero-factor property}$$

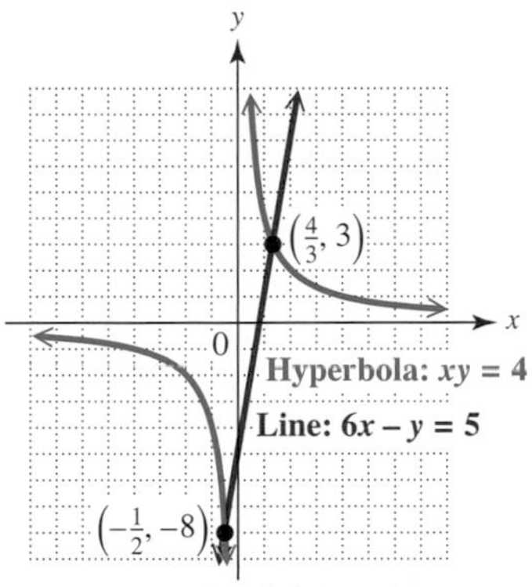

We substitute these results into $x = \frac{4}{y}$ to obtain the corresponding values of x.

$$\text{If } y = 3, \text{ then } x = \frac{4}{3}. \qquad \text{If } y = -8, \text{ then } x = -\frac{1}{2}.$$

The solution set of the system is $\left\{\left(\tfrac{4}{3}, 3\right), \left(-\tfrac{1}{2}, -8\right)\right\}$. The graph in Figure 31 shows these two points of intersection.

FIGURE 31

✔ **Now Try Exercise 21.**

OBJECTIVE 2 Use the elimination method to solve a system with two second-degree equations. The elimination method (**Section 4.1**) is often used when both equations are second degree.

EXAMPLE 3 Solving a Nonlinear System by Elimination

Solve the system

$$x^2 + y^2 = 9 \qquad (1)$$
$$2x^2 - y^2 = -6. \qquad (2)$$

The graph of (1) is a circle, while the graph of (2) is a hyperbola. By analyzing the possibilities we conclude that there may be zero, one, two, three, or four points of intersection. Adding the two equations will eliminate y, leaving an equation that can be solved for x.

$$\begin{array}{rl}
x^2 + y^2 = & 9 \\
\underline{2x^2 - y^2 = -6} & \\
3x^2 \quad\ \ = & 3 \qquad \text{Add.} \\
x^2 = & 1 \qquad \text{Divide by 3.} \\
x = 1 \quad \text{or} \quad x = -1 & \qquad \text{Square root property}
\end{array}$$

Each value of x gives corresponding values for y when substituted into one of the original equations. Using equation (1) gives the following.

If $x = 1$, then

$$1^2 + y^2 = 9$$
$$y^2 = 8$$
$$y = \sqrt{8} \quad \text{or} \quad y = -\sqrt{8}$$
$$y = 2\sqrt{2} \quad \text{or} \quad y = -2\sqrt{2}.$$

If $x = -1$, then

$$(-1)^2 + y^2 = 9$$
$$y^2 = 8$$
$$y = 2\sqrt{2} \quad \text{or} \quad y = -2\sqrt{2}.$$

The solution set is $\{(1, 2\sqrt{2}), (1, -2\sqrt{2}), (-1, 2\sqrt{2}), (-1, -2\sqrt{2})\}$. Figure 32 shows the four points of intersection.

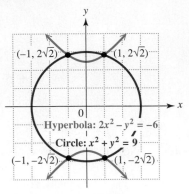

FIGURE 32

✔ **Now Try Exercise 35.**

OBJECTIVE 3 Solve a system that requires a combination of methods. Solving a system of second-degree equations may require a combination of methods.

EXAMPLE 4 Solving a Nonlinear System by a Combination of Methods

Solve the system

$$x^2 + 2xy - y^2 = 7 \quad (1)$$
$$x^2 - y^2 = 3. \quad (2)$$

While we have not graphed equations like (1), its graph is a hyperbola. The graph of (2) is also a hyperbola. Two hyperbolas may have zero, one, two, three, or four points of intersection. We use the elimination method here in combination with the substitution method. We begin by eliminating the squared terms by multiplying each side of equation (2) by -1 and then adding the result to equation (1).

$$\begin{array}{rl} x^2 + 2xy - y^2 = & 7 \quad (1) \\ -x^2 \qquad\;\; + y^2 = & -3 \\ \hline 2xy \qquad = & 4 \quad \text{Add.} \end{array}$$

Next, we solve $2xy = 4$ for y. (Either variable would do.)

$$2xy = 4$$
$$y = \frac{2}{x} \quad \text{Divide by 2x.} \quad (3)$$

Now, we substitute $y = \frac{2}{x}$ into one of the original equations. It is easier to do this with equation (2).

$$x^2 - y^2 = 3 \quad (2)$$
$$x^2 - \left(\frac{2}{x}\right)^2 = 3 \quad \text{Let } y = \frac{2}{x}.$$
$$x^2 - \frac{4}{x^2} = 3 \quad \text{Square } \frac{2}{x}.$$

$$x^4 - 4 = 3x^2 \quad \text{Multiply by } x^2, x \neq 0.$$

$$x^4 - 3x^2 - 4 = 0 \quad \text{Subtract } 3x^2.$$

$$(x^2 - 4)(x^2 + 1) = 0 \quad \text{Factor.}$$

$$x^2 - 4 = 0 \quad \text{or} \quad x^2 + 1 = 0$$

$$x^2 = 4 \quad \text{or} \quad x^2 = -1$$

$$x = 2 \quad \text{or} \quad x = -2 \quad \text{or} \quad x = i \quad \text{or} \quad x = -i$$

Substituting these four values into $y = \frac{2}{x}$ (equation (3)) gives the corresponding values for y.

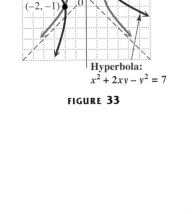

Hyperbola: $x^2 - y^2 = 3$

(2, 1)

(−2, −1)

Hyperbola: $x^2 + 2xy - y^2 = 7$

FIGURE 33

If $x = 2$, then $y = \frac{2}{2} = 1$.

If $x = -2$, then $y = \frac{2}{-2} = -1$.

If $x = i$, then $y = \frac{2}{i} = \frac{2}{i} \cdot \frac{-i}{-i} = -2i$.

If $x = -i$, then $y = \frac{2}{-i} = \frac{2}{-i} \cdot \frac{i}{i} = 2i$.

> **Multiply by the complex conjugate of the denominator.**

Note that if we substitute the x-values we found into equation (1) or (2) instead of into equation (3), we get extraneous solutions. ***It is always wise to check all solutions in both of the given equations.*** There are four ordered pairs in the solution set, two with real values and two with pure imaginary values. The solution set is

$$\{(2, 1), (-2, -1), (i, -2i), (-i, 2i)\}.$$

The graph of the system, shown in Figure 33, shows only the two real intersection points because the graph is in the real number plane. In general, if solutions contain nonreal complex numbers as components, they do not appear on the graph.

✔ **Now Try Exercise 41.**

▶ **NOTE** In the examples of this section, we analyzed the possible number of points of intersection of the graphs in each system. However, in Examples 2 and 4, we worked with equations whose graphs had not been studied. Keep in mind that it is not essential to visualize the number of points of intersection in order to solve the system. Furthermore, as in Example 4, there are sometimes nonreal complex solutions to nonlinear systems that do not appear as points of intersection in the real plane. Visualizing the geometry of the graphs is only an aid to solving these systems.

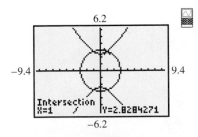

FIGURE 34

OBJECTIVE 4 Use a graphing calculator to solve a nonlinear system. If the equations in a nonlinear system can be solved for y, then we can graph the equations of the system with a graphing calculator and use the capabilities of the calculator to identify all intersection points. For instance, the two equations in Example 3 would require graphing the four separate functions

$$Y_1 = \sqrt{9 - X^2}, \quad Y_2 = -\sqrt{9 - X^2}, \quad Y_3 = \sqrt{2X^2 + 6}, \quad \text{and} \quad Y_4 = -\sqrt{2X^2 + 6}.$$

Figure 34 indicates the coordinates of one of the points of intersection.

11.4 EXERCISES

✐ **1.** Write an explanation of the steps you would use to solve the system

$$x^2 + y^2 = 25$$
$$y = x - 1$$

by the substitution method. Why would the elimination method not be appropriate for this system?

✐ **2.** Write an explanation of the steps you would use to solve the system

$$x^2 + y^2 = 12$$
$$x^2 - y^2 = 13$$

by the elimination method.

Concept Check *Each sketch represents the graphs of a pair of equations in a system. How many points are in each solution set?*

3.

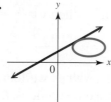

4.

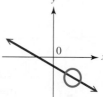

5.

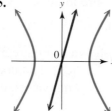

6.

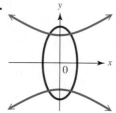

Concept Check *Suppose that a nonlinear system is composed of equations whose graphs are those described, and the number of points of intersection of the two graphs is as given. Make a sketch satisfying these conditions. (There may be more than one way to do this.)*

7. A line and a circle; no points

8. A line and a circle; one point

9. A line and a hyperbola; one point

10. A line and an ellipse; no points

11. A circle and an ellipse; four points

12. A parabola and an ellipse; one point

13. A parabola and an ellipse; four points

14. A parabola and a hyperbola; two points

Solve each system by the substitution method. See Examples 1 and 2.

15. $y = 4x^2 - x$
$\quad\ y = x$

16. $y = x^2 + 6x$
$\quad\ 3y = 12x$

17. $y = x^2 + 6x + 9$
$\quad\ x + y = 3$

18. $y = x^2 + 8x + 16$
$\quad\ x - y = -4$

⊙ **19.** $x^2 + y^2 = 2$
$\quad\ 2x + y = 1$

20. $2x^2 + 4y^2 = 4$
$\quad\ x = 4y$

⊙ **21.** $xy = 4$
$\quad\ 3x + 2y = -10$

22. $xy = -5$
$\quad\ 2x + y = 3$

23. $xy = -3$
$x + y = -2$

24. $xy = 12$
$x + y = 8$

25. $y = 3x^2 + 6x$
$y = x^2 - x - 6$

26. $y = 2x^2 + 1$
$y = 5x^2 + 2x - 7$

27. $2x^2 - y^2 = 6$
$y = x^2 - 3$

28. $x^2 + y^2 = 4$
$y = x^2 - 2$

29. $x^2 - xy + y^2 = 0$
$x - 2y = 1$

30. $x^2 - 3x + y^2 = 4$
$2x - y = 3$

Solve each system by the elimination method or a combination of the elimination and substitution methods. See Examples 3 and 4.

31. $3x^2 + 2y^2 = 12$
$x^2 + 2y^2 = 4$

32. $5x^2 - 2y^2 = -13$
$3x^2 + 4y^2 = 39$

33. $2x^2 + 3y^2 = 6$
$x^2 + 3y^2 = 3$

34. $6x^2 + y^2 = 9$
$3x^2 + 4y^2 = 36$

35. $2x^2 + y^2 = 28$
$4x^2 - 5y^2 = 28$

36. $x^2 + 6y^2 = 9$
$4x^2 + 3y^2 = 36$

37. $2x^2 = 8 - 2y^2$
$3x^2 = 24 - 4y^2$

38. $5x^2 = 20 - 5y^2$
$2y^2 = 2 - x^2$

39. $x^2 + xy + y^2 = 15$
$x^2 + y^2 = 10$

40. $2x^2 + 3xy + 2y^2 = 21$
$x^2 + y^2 = 6$

41. $3x^2 + 2xy - 3y^2 = 5$
$-x^2 - 3xy + y^2 = 3$

42. $-2x^2 + 7xy - 3y^2 = 4$
$2x^2 - 3xy + 3y^2 = 4$

Use a graphing calculator to solve each system. Then confirm your answer algebraically.

43. $xy = -6$
$x + y = -1$

44. $y = 2x^2 + 4x$
$y = -x^2 - 1$

Solve each problem by using a nonlinear system.

45. The area of a rectangular rug is 84 ft² and its perimeter is 38 ft. Find the length and width of the rug.

46. Find the length and width of a rectangular room whose perimeter is 50 m and whose area is 100 m².

47. A company has found that the price p (in dollars) of its scientific calculator is related to the supply x (in thousands) by the equation $px = 16$. The price is related to the demand x (in thousands) for the calculator by the equation $p = 10x + 12$. The **equilibrium price** is the value of p where demand equals supply. Find the equilibrium price and the supply/demand at that price by solving a system of equations. (*Hint:* Demand, price, and supply must all be positive.)

48. The calculator company in Exercise 47 has also determined that the cost y to make x (thousand) calculators is $y = 4x^2 + 36x + 20$, while the revenue y from the sale of x (thousand) calculators is $36x^2 - 3y = 0$. Find the **break-even point,** where cost equals revenue, by solving a system of equations.

PREVIEW EXERCISES

*Graph each inequality. See **Section 3.4.***

49. $2x - y \le 4$ **50.** $-x + 3y > 9$ **51.** $-5x + 3y \le 15$ **52.** $2x \le y$

11.5 Second-Degree Inequalities and Systems of Inequalities

OBJECTIVES

1. Graph second-degree inequalities.

2. Graph the solution set of a system of inequalities.

OBJECTIVE 1 Graph second-degree inequalities. The linear inequality $3x + 2y \le 5$ is graphed by first graphing the boundary line $3x + 2y = 5$. *Second-degree inequalities* are graphed in the same way. A **second-degree inequality** is an inequality with at least one variable of degree 2 and no variable with degree greater than 2. An example is $x^2 + y^2 \le 36$. The boundary of the inequality $x^2 + y^2 \le 36$ is the graph of the equation $x^2 + y^2 = 36$, a circle with radius 6 and center at the origin, as shown in Figure 35.

The inequality $x^2 + y^2 \le 36$ will include either the points outside the circle or the points inside the circle, as well as the boundary. We decide which region to shade by substituting any test point not on the circle, such as $(0, 0)$, into the original inequality. Since $0^2 + 0^2 \le 36$ is a true statement, the original inequality includes the points inside the circle, the shaded region in Figure 35, and the boundary.

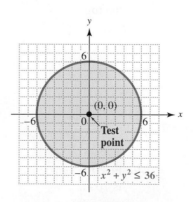

FIGURE 35

EXAMPLE 1 Graphing a Second-Degree Inequality

Graph $y < -2(x - 4)^2 - 3$.

The boundary, $y = -2(x - 4)^2 - 3$, is a parabola that opens down with vertex at $(4, -3)$. Using $(0, 0)$ as a test point gives

$$0 < -2(0 - 4)^2 - 3 \quad ?$$
$$0 < -32 - 3 \quad ?$$
$$0 < -35. \quad \text{False}$$

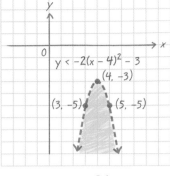

Because the final inequality is a false statement, the points in the region containing $(0, 0)$ do not satisfy the inequality. Figure 36 shows the final graph; the parabola is drawn as a dashed curve since the points of the parabola itself do not satisfy the inequality, and the region inside (or below) the parabola is shaded.

FIGURE 36

✔ Now Try Exercise 11.

▶ **NOTE** Since the substitution is easy, the origin is the test point of choice unless the graph actually passes through $(0, 0)$.

EXAMPLE 2 Graphing a Second-Degree Inequality

Graph $16y^2 \leq 144 + 9x^2$.

$$16y^2 - 9x^2 \leq 144 \quad \text{Subtract } 9x^2.$$

$$\frac{y^2}{9} - \frac{x^2}{16} \leq 1 \quad \text{Divide by 144.}$$

This form shows that the boundary is the hyperbola given by

$$\frac{y^2}{9} - \frac{x^2}{16} = 1.$$

Since the graph is a vertical hyperbola, the desired region will be either the region between the branches or the regions above the top branch and below the bottom branch. Choose $(0, 0)$ as a test point. Substituting into the original inequality leads to $0 \leq 144$, a true statement, so the region between the branches containing $(0, 0)$ is shaded, as shown in Figure 37.

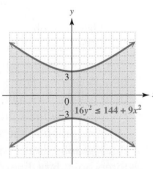

FIGURE 37

✔ Now Try Exercise 17.

OBJECTIVE ② Graph the solution set of a system of inequalities. If two or more inequalities are considered at the same time, we have a **system of inequalities.** To find the solution set of the system, we find the intersection of the graphs (solution sets) of the inequalities in the system.

EXAMPLE 3 Graphing a System of Two Inequalities

Graph the solution set of the system

$$2x + 3y > 6$$
$$x^2 + y^2 < 16.$$

Begin by graphing the solution set of $2x + 3y > 6$. The boundary line is the graph of $2x + 3y = 6$ and is a dashed line because of the symbol $>$. The test point $(0, 0)$ leads to a false statement in the inequality $2x + 3y > 6$, so shade the region above the line, as shown in Figure 38. The graph of $x^2 + y^2 < 16$ is the interior of a dashed circle centered at the origin with radius 4. This is shown in Figure 39.

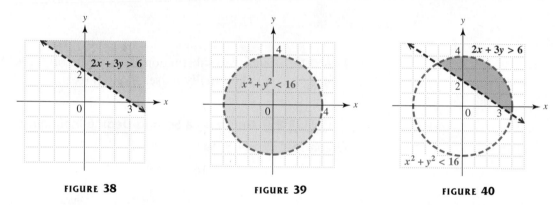

FIGURE 38   **FIGURE 39**   **FIGURE 40**

To get the graph of the solution set of the system, determine the intersection of the graphs of the two inequalities. The overlapping region in Figure 40 is the solution set.

✔ **Now Try Exercise 29.**

EXAMPLE 4 Graphing a Linear System with Three Inequalities

Graph the solution set of the system

$$x + y < 1$$
$$y \leq 2x + 3$$
$$y \geq -2.$$

Graph each inequality separately, on the same axes. The graph of $x + y < 1$ consists of all points that lie below the dashed line $x + y = 1$. The graph of $y \leq 2x + 3$ is the region that lies below the solid line $y = 2x + 3$. Finally, the graph of $y \geq -2$ is the region above the solid horizontal line $y = -2$.

The graph of the system, the intersection of these three graphs, is the triangular region enclosed by the three boundary lines in Figure 41, including two of its boundaries.

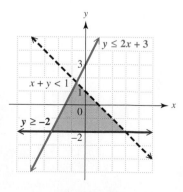

FIGURE 41

✔ **Now Try Exercise 31.**

EXAMPLE 5 Graphing a System with Three Inequalities

Graph the solution set of the system

$$y \geq x^2 - 2x + 1$$
$$2x^2 + y^2 > 4$$
$$y < 4.$$

The graph of $y = x^2 - 2x + 1$ is a parabola with vertex at $(1, 0)$. Those points above (or in the interior of) the parabola satisfy the condition $y > x^2 - 2x + 1$. Thus, points on the parabola or in the interior are the solution set of $y \geq x^2 - 2x + 1$. The graph of the equation $2x^2 + y^2 = 4$ is an ellipse. We draw it as a dashed curve. To satisfy the inequality $2x^2 + y^2 > 4$, a point must lie outside the ellipse. The graph of $y < 4$ includes all points below the dashed line $y = 4$. Finally, the graph of the system is the shaded region in Figure 42, which lies outside the ellipse, inside or on the boundary of the parabola, and below the line $y = 4$.

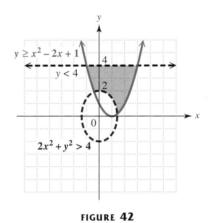

FIGURE 42

✔ **Now Try Exercise 33.**

11.5 EXERCISES

Now Try Exercise

☑ **1.** Explain how to graph the solution set of a nonlinear inequality.

☑ **2.** Explain how to graph the solution set of a system of nonlinear inequalities.

3. *Concept Check* Which one of the following is a description of the graph of the solution set of the following system?

$$x^2 + y^2 < 25$$
$$y > -2$$

A. All points outside the circle $x^2 + y^2 = 25$ and above the line $y = -2$
B. All points outside the circle $x^2 + y^2 = 25$ and below the line $y = -2$
C. All points inside the circle $x^2 + y^2 = 25$ and above the line $y = -2$
D. All points inside the circle $x^2 + y^2 = 25$ and below the line $y = -2$

4. *Concept Check*   Fill in each blank with the appropriate response. The graph of the system

$$y > x^2 + 1$$

$$\frac{x^2}{9} + \frac{y^2}{4} > 1$$

$$y < 5$$

consists of all points _____ the parabola $y = x^2 + 1$, _____ the
                              (above/below)   (inside/outside)

ellipse $\dfrac{x^2}{9} + \dfrac{y^2}{4} = 1$, and _____ the line $y = 5$.
                                          (above/below)

Concept Check   Match each nonlinear inequality with its graph.

5. $y \geq x^2 + 4$ **6.** $y \leq x^2 + 4$ **7.** $y < x^2 + 4$ **8.** $y > x^2 + 4$

A. 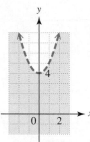   **B.** 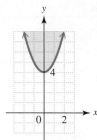   **C.** 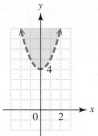   **D.**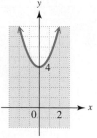

Graph each nonlinear inequality. See Examples 1 and 2.

9. $y^2 > 4 + x^2$ **10.** $y^2 \leq 4 - 2x^2$ **11.** $y \geq x^2 - 2$

12. $x^2 \leq 16 - y^2$ **13.** $2y^2 \geq 8 - x^2$ **14.** $x^2 \leq 16 + 4y^2$

15. $y \leq x^2 + 4x + 2$ **16.** $9x^2 < 16y^2 - 144$

17. $9x^2 > 16y^2 + 144$ **18.** $4y^2 \leq 36 - 9x^2$

19. $x^2 - 4 \geq -4y^2$ **20.** $x \geq y^2 - 8y + 14$

21. $x \leq -y^2 + 6y - 7$ **22.** $y^2 - 16x^2 \leq 16$

Graph each system of inequalities. See Examples 3–5.

23. $2x + 5y < 10$ **24.** $3x - y > -6$ **25.** $5x - 3y \leq 15$
       $x - 2y < 4$ $4x + 3y > 12$ $4x + y \geq 4$

26. $4x - 3y \leq 0$ **27.** $x \leq 5$ **28.** $x \geq -2$
       $x + y \leq 5$ $y \leq 4$ $y \leq 4$

29. $y > x^2 - 4$ **30.** $x^2 - y^2 \geq 9$ **31.** $x^2 + y^2 \geq 4$
       $y < -x^2 + 3$ $\dfrac{x^2}{16} + \dfrac{y^2}{9} \leq 1$ $x + y \leq 5$
                                                                   $x \geq 0$
                                                                   $y \geq 0$

32. $y^2 - x^2 \geq 4$ **33.** $y \leq -x^2$ **34.** $y < x^2$
       $-5 \leq y \leq 5$ $y \geq x - 3$ $y > -2$
                                     $y \leq -1$ $x + y < 3$
                                     $x < 1$ $3x - 2y > -6$

For each nonlinear inequality in Exercises 35–42, a restriction is placed on one or both variables. For example, the graph of

$$x^2 + y^2 \le 4, \quad x \ge 0$$

would be as shown in the figure. Only the right half of the interior of the circle and its boundary is shaded, because of the restriction that x must be nonnegative. Graph each nonlinear inequality with the given restrictions.

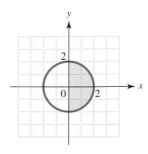

35. $x^2 + y^2 > 36, \quad x \ge 0$

36. $4x^2 + 25y^2 < 100, \quad y < 0$

37. $x < y^2 - 3, \quad x < 0$

38. $x^2 - y^2 < 4, \quad x < 0$

39. $4x^2 - y^2 > 16, \quad x < 0$

40. $x^2 + y^2 > 4, \quad y < 0$

41. $x^2 + 4y^2 \ge 1, \quad x \ge 0, y \ge 0$

42. $2x^2 - 32y^2 \le 8, \quad x \le 0, y \ge 0$

 Use the shading feature of a graphing calculator to graph each system.

43. $y \ge x - 3$
     $y \le -x + 4$

44. $y \ge -x^2 + 5$
     $y \le x^2 - 3$

45. $y < x^2 + 4x + 4$
     $y > -3$

46. $y > (x - 4)^2 - 3$
     $y < 5$

PREVIEW EXERCISES

*Evaluate each expression for (**a**) $n = 1$, (**b**) $n = 2$, (**c**) $n = 3$, and (**d**) $n = 4$. See Section 1.3.*

47. $\dfrac{n + 5}{n}$

48. $\dfrac{n - 1}{n + 1}$

49. $n^2 - n$

50. $n(n - 3)$

Chapter **11** **Group Activity**

FINDING THE PATHS OF NATURAL SATELLITES

Objective Write and graph equations of ellipses from given data.

The moon, which orbits Earth, and Halley's comet, which orbits the sun, are both natural satellites. In **Section 11.2,** you solved problems where you were given equations of ellipses for the orbits of planets and were asked to find apogees (greatest distance from the sun) and perigees (smallest distance from the sun). This activity reverses the process; that is, given apogees and perigees you must find equations of ellipses.

A. Have each student choose a natural satellite from the table. Predict the shape of the orbital ellipse for your satellite.

Natural Satellite	Apogee	Perigee
Moon	406.7 thousand km from Earth	356.4 thousand km from Earth
Halley's comet	35 astronomical units* from the sun	0.6 astronomical unit* from the sun

Source: World Book Encyclopedia.
*One astronomical unit (AU) is the distance from Earth to the sun.

B. For your satellite, do the following.

1. Find values for a, b, and c. Note that apogee $= a + c$, perigee $= a - c$, and $c^2 = a^2 - b^2$.

2. Write the equation of the ellipse in the form $\dfrac{x^2}{a^2} + \dfrac{y^2}{b^2} = 1$.

3. Rewrite the equation so it can be graphed on a graphing calculator. (See **Section 11.2,** Objective 5.)

4. Graph your equation on a graphing calculator. Adjust the window setting in order to see the entire graph. Once the window is set correctly, get a square window to see the true shape of the ellipse.

C. Compare your graph with your partner's graph.

1. Do the graphs reflect the shapes you predicted in part A?

2. What window was used to graph each ellipse?

Chapter **11** SUMMARY

View the Interactive Summary on the Pass the Test CD.

KEY TERMS

11.1 asymptotes
greatest integer
function
step function
composition

11.2 conic section
circle
center
radius
center-radius form
ellipse
foci

11.3 hyperbola
asymptotes of a
hyperbola
fundamental
rectangle
square root function

11.4 nonlinear equation
nonlinear system of
equations
11.5 second-degree
inequality
system of
inequalities

TEST YOUR WORD POWER

See how well you have learned the vocabulary in this chapter. Answers, with examples, follow the Quick Review.

1. **Conic sections** are
   A. graphs of first-degree equations
   B. the result of two or more intersecting planes
   C. graphs of first-degree inequalities
   D. figures that result from the intersection of an infinite cone with a plane.

2. A **circle** is the set of all points in a plane
   A. such that the absolute value of the difference of the distances from two fixed points is constant
   B. that lie a fixed distance from a fixed point
   C. the sum of whose distances from two fixed points is constant
   D. that make up the graph of any second-degree equation.

3. An **ellipse** is the set of all points in a plane
   A. such that the absolute value of the difference of the distances from two fixed points is constant
   B. that lie a fixed distance from a fixed point
   C. the sum of whose distances from two fixed points is constant
   D. that make up the graph of any second-degree equation.

4. A **hyperbola** is the set of all points in a plane
   A. such that the absolute value of the difference of the distances from two fixed points is constant
   B. that lie a fixed distance from a fixed point
   C. the sum of whose distances from two fixed points is constant
   D. that make up the graph of any second-degree equation.

5. A **nonlinear equation** is an equation
   A. in which some terms have more than one variable or a variable of degree 2 or greater
   B. in which the terms have only one variable
   C. of degree 1
   D. of a linear function.

6. A **nonlinear system of equations** is a system
   A. with at least one linear equation
   B. with two or more inequalities
   C. with at least one nonlinear equation
   D. with at least two linear equations.

QUICK REVIEW

Concepts	Examples

11.1 ADDITIONAL GRAPHS OF FUNCTIONS

Other Functions

In addition to the squaring function, some other important elementary functions in algebra are the absolute value function, defined by $f(x) = |x|$; the reciprocal function, defined by $f(x) = \frac{1}{x}$; and the square root function, defined by $f(x) = \sqrt{x}$. Their graphs can be translated, as shown in the examples at the right.

Step functions, such as the greatest integer function, defined by $f(x) = [\![x]\!]$, are useful in applications.

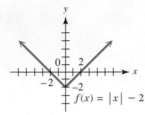

$f(x) = |x| - 2$

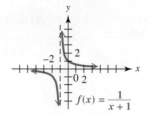

$f(x) = \frac{1}{x + 1}$

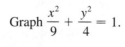

$f(x) = \sqrt{x - 2} + 1$

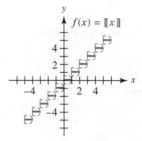

$f(x) = [\![x]\!]$

11.2 THE CIRCLE AND THE ELLIPSE

Circle

The circle with radius r and center at (h, k) has an equation of the form

$$(x - h)^2 + (y - k)^2 = r^2.$$

The circle with equation $(x + 2)^2 + (y - 3)^2 = 25$, which can be written $[x - (-2)]^2 + (y - 3)^2 = 5^2$, has center $(-2, 3)$ and radius 5.

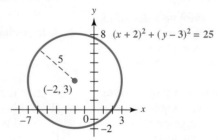

$(x + 2)^2 + (y - 3)^2 = 25$

Ellipse

The ellipse whose x-intercepts are $(a, 0)$ and $(-a, 0)$ and whose y-intercepts are $(0, b)$ and $(0, -b)$ has an equation of the form

$$\frac{x^2}{a^2} + \frac{y^2}{b^2} = 1.$$

Graph $\dfrac{x^2}{9} + \dfrac{y^2}{4} = 1.$

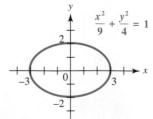

$\dfrac{x^2}{9} + \dfrac{y^2}{4} = 1$

(continued)

Concepts	Examples

11.3 THE HYPERBOLA AND FUNCTIONS DEFINED BY RADICALS

Hyperbola

A hyperbola with x-intercepts $(a, 0)$ and $(-a, 0)$ has an equation of the form

$$\frac{x^2}{a^2} - \frac{y^2}{b^2} = 1,$$

and a hyperbola with y-intercepts $(0, b)$ and $(0, -b)$ has an equation of the form

$$\frac{y^2}{b^2} - \frac{x^2}{a^2} = 1.$$

Graph $\dfrac{x^2}{4} - \dfrac{y^2}{4} = 1$.

The graph has x-intercepts $(2, 0)$ and $(-2, 0)$.

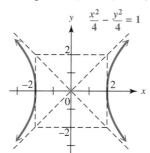

The extended diagonals of the fundamental rectangle with vertices at the points (a, b), $(-a, b)$, $(-a, -b)$, and $(a, -b)$ are the asymptotes of these hyperbolas.

The fundamental rectangle has vertices at $(2, 2)$, $(-2, 2)$, $(-2, -2)$, and $(2, -2)$.

Graphing a Square Root Function

To graph a square root function defined by

$$f(x) = \sqrt{u}$$

for an algebraic expression u, with $u \geq 0$, square both sides so that the equation can be easily recognized. Then graph only the part indicated by the original equation.

Graph $y = -\sqrt{4 - x^2}$.

Square both sides and rearrange terms to get

$$x^2 + y^2 = 4.$$

This equation has a circle as its graph. However, graph only the lower half of the circle, since the original equation indicates that y cannot be positive.

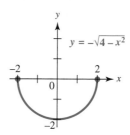

11.4 NONLINEAR SYSTEMS OF EQUATIONS

Solving a Nonlinear System

A nonlinear system can be solved by the substitution method, the elimination method, or a combination of the two.

Solve the system

$$\begin{array}{ll} x^2 + 2xy - y^2 = 14 & (1) \\ x^2 - y^2 = -16. & (2) \end{array}$$

Multiply equation (2) by -1 and use elimination.

$$\begin{array}{rcl} x^2 + 2xy - y^2 &=& 14 \\ -x^2 + y^2 &=& 16 \\ \hline 2xy &=& 30 \\ xy &=& 15 \end{array}$$

(continued)

Concepts	Examples

Solve $xy = 15$ for y to obtain $y = \frac{15}{x}$, and substitute into equation (2).

$$x^2 - y^2 = -16 \quad (2)$$

$$x^2 - \left(\frac{15}{x}\right)^2 = -16$$

$$x^2 - \frac{225}{x^2} = -16$$

$x^4 + 16x^2 - 225 = 0$ Multiply by x^2; add $16x^2$.

$(x^2 - 9)(x^2 + 25) = 0$ Factor.

$x = \pm 3$ or $x = \pm 5i$ Zero-factor property

Find corresponding y-values to get the solution set

$$\{(3, 5), (-3, -5), (5i, -3i), (-5i, 3i)\}.$$

11.5 SECOND-DEGREE INEQUALITIES AND SYSTEMS OF INEQUALITIES

Graphing a Second-Degree Inequality
To graph a second-degree inequality, graph the corresponding equation as a boundary and use test points to determine which region(s) form the solution set. Shade the appropriate region(s).

Graphing a System of Inequalities
The solution set of a system of inequalities is the intersection of the solution sets of the individual inequalities.

Graph $y \geq x^2 - 2x + 3$.

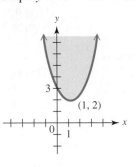

(1, 2)

Graph the solution set of the system

$$3x - 5y > -15$$
$$x^2 + y^2 \leq 25.$$

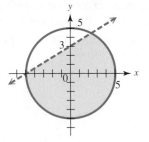

Answers to Test Your Word Power

1. D; *Example:* Parabolas, circles, ellipses, and hyperbolas are conic sections.

2. B; *Example:* See the graph of $x^2 + y^2 = 9$ in Figure 10 of **Section 11.2.**

3. C; *Example:* See the graph of $\frac{x^2}{49} + \frac{y^2}{36} = 1$ in Figure 16 of **Section 11.2.**

4. A; *Example:* See the graph of $\frac{x^2}{16} - \frac{y^2}{12} = 1$ in Figure 20 of **Section 11.3.**

5. A; *Examples:* $y = x^2 + 8x + 16$, $xy = 5$, $2x^2 - y^2 = 6$

6. C; *Example:* $x^2 + y^2 = 2$ $2x + y = 1$

Chapter **11** REVIEW EXERCISES

[11.1] *Graph each function.*

1. $f(x) = |x + 4|$ **2.** $f(x) = \dfrac{1}{x - 4}$ **3.** $f(x) = \sqrt{x} + 3$ **4.** $f(x) = [\![x]\!] - 2$

[11.2] *Write an equation for each circle.*

5. Center $(-2, 4)$, $r = 3$ **6.** Center $(-1, -3)$, $r = 5$ **7.** Center $(4, 2)$, $r = 6$

Find the center and radius of each circle.

8. $x^2 + y^2 + 6x - 4y - 3 = 0$ **9.** $x^2 + y^2 - 8x - 2y + 13 = 0$

10. $2x^2 + 2y^2 + 4x + 20y = -34$ **11.** $4x^2 + 4y^2 - 24x + 16y = 48$

Graph each equation.

12. $x^2 + y^2 = 16$ **13.** $\dfrac{x^2}{16} + \dfrac{y^2}{9} = 1$ **14.** $\dfrac{x^2}{49} + \dfrac{y^2}{25} = 1$

15. A satellite is in an elliptical orbit around Earth with perigee altitude of 160 km and apogee altitude of 16,000 km. See the figure. (*Source:* Kastner, Bernice, *Space Mathematics,* NASA.) Find the equation of the ellipse.

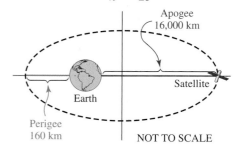

16. (a) The Roman Colosseum is an ellipse with $a = 310$ ft and $b = \frac{513}{2}$ ft. Find the distance between the foci of this ellipse.

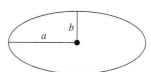

(b) A formula for the approximate circumference of an ellipse is

$$C \approx 2\pi\sqrt{\dfrac{a^2 + b^2}{2}},$$

where a and b are the lengths given in part (a). Use this formula to find the approximate circumference of the Roman Colosseum.

[11.3] *Graph each equation.*

17. $\dfrac{x^2}{16} - \dfrac{y^2}{25} = 1$ **18.** $\dfrac{y^2}{25} - \dfrac{x^2}{4} = 1$ **19.** $f(x) = -\sqrt{16 - x^2}$

Identify the graph of each equation as a parabola, circle, ellipse, *or* hyperbola.

20. $x^2 + y^2 = 64$ **21.** $y = 2x^2 - 3$ **22.** $y^2 = 2x^2 - 8$

23. $y^2 = 8 - 2x^2$ **24.** $x = y^2 + 4$ **25.** $x^2 - y^2 = 64$

26. Ships and planes often use a location-finding system called LORAN. With this system, a radio transmitter at M sends out a series of pulses. (See the figure.) When each pulse is received at transmitter S, it then sends out a pulse. A ship at P receives pulses from both M and S. A receiver on the ship measures the difference in the arrival times of the pulses. A special map gives hyperbolas that correspond to the differences in arrival times (which give the distances d_1 and d_2 in the figure.) The ship can then be located as lying on a branch of a particular hyperbola.

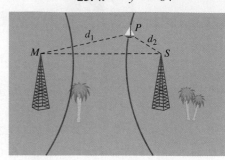

Suppose $d_1 = 80$ mi and $d_2 = 30$ mi, and the distance between transmitters M and S is 100 mi. Use the definition to find an equation of the hyperbola the ship is located on.

[11.4] *Solve each system.*

27. $2y = 3x - x^2$
     $x + 2y = -12$

28. $y + 1 = x^2 + 2x$
     $y + 2x = 4$

29. $x^2 + 3y^2 = 28$
     $y - x = -2$

30. $xy = 8$
     $x - 2y = 6$

31. $x^2 + y^2 = 6$
     $x^2 - 2y^2 = -6$

32. $3x^2 - 2y^2 = 12$
     $x^2 + 4y^2 = 18$

33. *Concept Check* How many solutions are possible for a system of two equations whose graphs are a circle and a line?

34. *Concept Check* How many solutions are possible for a system of two equations whose graphs are a parabola and a hyperbola?

[11.5] *Graph each inequality.*

35. $9x^2 \geq 16y^2 + 144$ **36.** $4x^2 + y^2 \geq 16$ **37.** $y < -(x + 2)^2 + 1$

Graph each system of inequalities.

38. $2x + 5y \leq 10$
     $3x - y \leq 6$

39. $|x| \leq 2$
     $|y| > 1$
     $4x^2 + 9y^2 \leq 36$

40. $9x^2 \leq 4y^2 + 36$
     $x^2 + y^2 \leq 16$

MIXED REVIEW EXERCISES

Graph.

41. $\dfrac{x^2}{64} + \dfrac{y^2}{25} = 1$ **42.** $\dfrac{y^2}{4} - 1 = \dfrac{x^2}{9}$ **43.** $x^2 + y^2 = 25$ **44.** $x^2 + 9y^2 = 9$

45. $x^2 - 9y^2 = 9$ **46.** $f(x) = \sqrt{4 - x}$ **47.** $3x + 2y \geq 0$ **48.** $4y > 3x - 12$
                                                      $y \leq 4$ $x^2 < 16 - y^2$
                                                      $x \leq 4$

Chapter **11** TEST

View the complete solutions to all Chapter Test exercises on the Pass the Test CD.

1. *Concept Check*   Match each function with its graph from choices A–D.

(a) $f(x) = \sqrt{x} - 2$ **A.**

(b) $f(x) = \sqrt{x + 2}$

(c) $f(x) = \sqrt{x} + 2$ **C.**

(d) $f(x) = \sqrt{x} - 2$

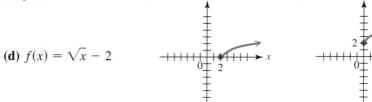

2. Sketch the graph of $f(x) = |x - 3| + 4$.

3. Find the center and radius of the circle whose equation is $(x - 2)^2 + (y + 3)^2 = 16$. Sketch the graph.

4. Find the center and radius of the circle whose equation is $x^2 + y^2 + 8x - 2y = 8$.

Graph.

5. $f(x) = \sqrt{9 - x^2}$

6. $4x^2 + 9y^2 = 36$

7. $16y^2 - 4x^2 = 64$

8. $\dfrac{y}{2} = -\sqrt{1 - \dfrac{x^2}{9}}$

Identify the graph of each equation as a parabola, hyperbola, ellipse, *or* circle.

9. $6x^2 + 4y^2 = 12$

10. $16x^2 = 144 + 9y^2$

11. $4y^2 + 4x = 9$

Solve each nonlinear system.

12. $2x - y = 9$
$xy = 5$

13. $x - 4 = 3y$
$x^2 + y^2 = 8$

14. $x^2 + y^2 = 25$
$x^2 - 2y^2 = 16$

15. Graph the inequality $y < x^2 - 2$.

16. Graph the system $\begin{aligned} x^2 + 25y^2 &\le 25 \\ x^2 + y^2 &\le 9. \end{aligned}$

Chapters **1–11** CUMULATIVE REVIEW EXERCISES

1. Simplify $-10 + |-5| - |3| + 4$.

Solve.

2. $4 - (2x + 3) + x = 5x - 3$

3. $-4k + 7 \geq 6k + 1$

4. $|5m| - 6 = 14$

5. $|2p - 5| > 15$

6. Find the slope of the line through $(2, 5)$ and $(-4, 1)$.

7. Find the equation of the line through the point $(-3, -2)$ and perpendicular to the graph of $2x - 3y = 7$.

Solve each system.

8. $3x - y = 12$
$2x + 3y = -3$

9. $x + y - 2z = 9$
$2x + y + z = 7$
$3x - y - z = 13$

10. $xy = -5$
$2x + y = 3$

Solve each problem.

11. Al and Bev traveled from their apartment to a picnic 20 mi away. Al traveled on his bike while Bev, who left later, took her car. Al's average speed was half of Bev's average speed. The trip took Al $\frac{1}{2}$ hr longer than Bev. What was Bev's average speed?

12. The president of InstaTune, a chain of franchised automobile tune-up shops, reports that people who buy a franchise and open a shop pay a weekly fee (in dollars) to company headquarters, according to the linear function defined by

$$f(x) = 0.07x + 135,$$

where $f(x)$ is the fee and x is the total amount of money taken in during the week by the shop. Find the weekly fee if \$2000 is taken in for the week. (*Source: Business Week.*)

Perform the indicated operations.

13. $(5y - 3)^2$

14. $(2r + 7)(6r - 1)$

15. $\dfrac{8x^4 - 4x^3 + 2x^2 + 13x + 8}{2x + 1}$

Factor.

16. $12x^2 - 7x - 10$

17. $2y^4 + 5y^2 - 3$

18. $z^4 - 1$

19. $a^3 - 27b^3$

Perform the indicated operations.

20. $\dfrac{5x - 15}{24} \cdot \dfrac{64}{3x - 9}$

21. $\dfrac{y^2 - 4}{y^2 - y - 6} \div \dfrac{y^2 - 2y}{y - 1}$

22. $\dfrac{5}{c + 5} - \dfrac{2}{c + 3}$

23. $\dfrac{p}{p^2 + p} + \dfrac{1}{p^2 + p}$

Solve.

24. Kareem and Jamal want to clean their office. Kareem can do the job alone in 3 hr, while Jamal can do it alone in 2 hr. How long will it take them if they work together?

Simplify. Assume all variables represent positive real numbers.

25. $\left(\dfrac{4}{3}\right)^{-1}$

26. $\dfrac{(2a)^{-2}a^4}{a^{-3}}$

27. $4\sqrt[3]{16} - 2\sqrt[3]{54}$

28. $\dfrac{3\sqrt{5x}}{\sqrt{2x}}$

29. $\dfrac{5 + 3i}{2 - i}$

Solve.

30. $2\sqrt{k} = \sqrt{5k + 3}$

31. $10q^2 + 13q = 3$

32. $(4x - 1)^2 = 8$

33. $3k^2 - 3k - 2 = 0$

34. $2(x^2 - 3)^2 - 5(x^2 - 3) = 12$

35. $F = \dfrac{kwv^2}{r}$ for v

36. If $f(x) = x^3 + 4$, find $f^{-1}(x)$.

37. Evaluate $3^{\log_3 4}$.

38. Evaluate $e^{\ln 7}$.

39. Use properties of logarithms to write $2 \log(3x + 7) - \log 4$ as a single logarithm.

40. Solve $\log(x + 2) + \log(x - 1) = 1$.

41. If \$10,000 is invested at 5% for 4 yr, how much will there be in the account if interest is compounded

    (a) quarterly;   **(b)** continuously?

The bar graph shows on-line U.S. retail sales (in billions of dollars). A reasonable model for sales y in billions of dollars is the exponential function defined by

$$y = 1.38(1.65)^x,$$

where x is the number of years since 1995.

42. Use the model to estimate sales in 2000. (*Hint:* Let $x = 5$.)

43. Use the model to predict sales in 2003.

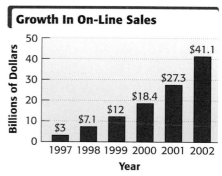

Growth In On-Line Sales

Source: Jupiter Communications.

44. Give the domain and range of the function defined by $f(x) = |x - 3|$.

Graph.

45. $f(x) = -3x + 5$

46. $f(x) = -2(x - 1)^2 + 3$

47. $\dfrac{x^2}{25} + \dfrac{y^2}{16} \le 1$

48. $f(x) = \sqrt{x - 2}$

49. $\dfrac{x^2}{4} - \dfrac{y^2}{16} = 1$

50. $f(x) = 3^x$

Sequences and Series

The male honeybce hatches from an unfertilized egg, while the female hatches from a fertilized one. The "family tree" of a male honeybee is shown below, where M represents male and F represents female. Starting with the male honeybee at the top, and counting the number of bees in each generation, we obtain the following numbers in the order shown:

$$1, 1, 2, 3, 5, 8.$$

Notice the pattern. After the first two terms (1 and 1), each successive term is obtained by adding the two previous terms. This se quence of numbers is called the *Fibonacci sequence*.

In this chapter, we study *sequences* and sums of terms of sequences, known as *series*.

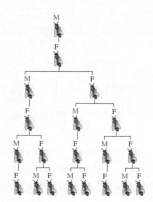

12.1 Sequences and Series

A **sequence** is a function whose domain is the set of natural numbers. Intuitively, a sequence is a list of numbers in which the order of their appearance is important. For instance, the interest portions of monthly payments made to pay off an automobile or home loan form a sequence.

In the Palace of the Alhambra, residence of the Moorish rulers of Granada, Spain, the Sultana's quarters feature an interesting architectural pattern: There are 2 matched marble slabs inlaid in the floor, 4 walls, an octagon (8-sided) ceiling, 16 windows, 32 arches, and so on. If this pattern is continued indefinitely, the set of numbers forms an *infinite sequence.*

Infinite Sequence

An **infinite sequence** is a function with the set of positive integers as the domain.

OBJECTIVE 1 Find the terms of a sequence given the general term. For any positive integer n, the function value (y-value) of a sequence is written as a_n (read "a sub-n") instead of $a(n)$ or $f(n)$. The function values $a_1, a_2, a_3, \ldots$, written in order, are the **terms** of the sequence, with a_1 the first term, a_2 the second term, and so on. The expression a_n, which defines the sequence, is called the **general term** of the sequence.

In the Palace of the Alhambra example, the first five terms of the sequence are

$$a_1 = 2, \quad a_2 = 4, \quad a_3 = 8, \quad a_4 = 16, \quad \text{and} \quad a_5 = 32.$$

The general term for this sequence is $a_n = 2^n$.

EXAMPLE 1 Writing the Terms of Sequences from the General Term

Given an infinite sequence with $a_n = n + \frac{1}{n}$, find the following.

(a) The second term of the sequence

$$a_2 = 2 + \frac{1}{2} = \frac{5}{2} \qquad \text{Replace } n \text{ with 2.}$$

(b) $a_{10} = 10 + \frac{1}{10} = \frac{101}{10}$ **(c)** $a_{12} = 12 + \frac{1}{12} = \frac{145}{12}$

✔ **Now Try Exercises 1 and 11.**

Graphing calculators can be used to generate and graph sequences, as shown in Figure 1. The calculator must be in dot mode, so that the discrete points on the graph are not connected. *Remember that the domain of a sequence consists only of natural numbers.*

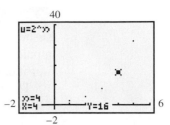

```
seq(2^N,N,1,5)
    {2 4 8 16 32}
```

The first five terms of
the sequence $a_n = 2^n$

(a)

The first five terms of $a_n = 2^n$ are graphed
here. The display indicates that the fourth
term is 16; that is, $a_4 = 2^4 = 16$.

(b)

FIGURE 1

OBJECTIVE ② Find the general term of a sequence. Sometimes we need to find a
general term to fit the first few terms of a given sequence. There are no rules for finding the general term of a sequence from the first few terms. In fact, it is possible to
give more than one general term, all of which produce the same first three or four
terms. However, in many examples, the terms may suggest a general term.

EXAMPLE 2 Finding the General Term of a Sequence

Find an expression for the general term a_n of the sequence

$$5, 10, 15, 20, 25, \ldots.$$

The first term is 5(1), the second is 5(2), and so on. By inspection,

$$a_n = 5n$$

will produce the given first five terms.

✔ **Now Try Exercise 17.**

▶ **CAUTION** One problem with using just a few terms to suggest a general
term, as in Example 2, is that there may be more than one general term that
gives the same first few terms.

OBJECTIVE ③ Use sequences to solve applied problems. Practical problems often involve *finite sequences.*

Finite Sequence

A **finite sequence** has a domain that includes only the first n positive integers.

For example, if $n = 5$, the domain is $\{1, 2, 3, 4, 5\}$ and the sequence has five terms.

EXAMPLE 3 Using a Sequence in an Application

Keshon borrows $5000 and agrees to pay $500 monthly, plus interest of 1% on the unpaid balance from the beginning of the first month. Find the payments for the first four months and the remaining debt at the end of that period.

The payments and remaining balances are calculated as follows:

First month	Payment:	$500 + 0.01(5000) = 550$ dollars
	Balance:	$5000 - 500 = 4500$ dollars
Second month	Payment:	$500 + 0.01(4500) = 545$ dollars
	Balance:	$5000 - 2 \cdot 500 = 4000$ dollars
Third month	Payment:	$500 + 0.01(4000) = 540$ dollars
	Balance:	$5000 - 3 \cdot 500 = 3500$ dollars
Fourth month	Payment:	$500 + 0.01(3500) = 535$ dollars
	Balance:	$5000 - 4 \cdot 500 = 3000$ dollars.

The payments for the first four months, in dollars, are

$$550, \ 545, \ 540, \ 535$$

and the remaining debt at the end of the period is 3000 dollars.

✔ **Now Try Exercise 21.**

OBJECTIVE 4 Use summation notation to evaluate a series. By adding the terms of a sequence, we obtain a *series.*

Series

The indicated sum of the terms of a sequence is called a **series.**

For example, if we consider the sum of the payments listed in Example 3, namely,

$$550 + 545 + 540 + 535,$$

we obtain a series that represents the total amount of payments for the first four months.

Since a sequence can be finite or infinite, there are finite or infinite series. One type of infinite series is discussed in **Section 12.3,** and the binomial theorem discussed in **Section 12.4** defines an important finite series. In this section, we discuss only finite series.

We use a compact notation, called **summation notation,** to write a series from the general term of the corresponding sequence. For example, the sum of the first six terms of the sequence with general term $a_n = 3n + 2$ is written with the Greek letter Σ **(sigma)** as

$$\sum_{i=1}^{6} (3i + 2).$$

We read this as "the sum from $i = 1$ to 6 of $3i + 2$." To find this sum, we replace the letter i in $3i + 2$ with 1, 2, 3, 4, 5, and 6, as follows:

$$\sum_{i=1}^{6} (3i + 2) = (3 \cdot 1 + 2) + (3 \cdot 2 + 2) + (3 \cdot 3 + 2)$$

Replace i with
1, 2, 3, 4, 5, 6.

Multiply;
then add.

$$+ (3 \cdot 4 + 2) + (3 \cdot 5 + 2) + (3 \cdot 6 + 2)$$

$$= 5 + 8 + 11 + 14 + 17 + 20$$

$$= 75.$$

The letter i is called the **index of summation.**

▶ **CAUTION**   This use of i has no connection with the complex number i.

EXAMPLE 4   **Evaluating Series Written in Summation Notation**

Write out the terms and evaluate each series.

(a) $\displaystyle\sum_{i=1}^{5} (i - 4) = (1 - 4) + (2 - 4) + (3 - 4) + (4 - 4) + (5 - 4)$

$$= -3 - 2 - 1 + 0 + 1$$

$$= -5$$

(b) $\displaystyle\sum_{i=3}^{7} 3i^2 = 3(3)^2 + 3(4)^2 + 3(5)^2 + 3(6)^2 + 3(7)^2$

$$= 27 + 48 + 75 + 108 + 147$$

$$= 405$$

✔ **Now Try Exercises 25 and 27.**

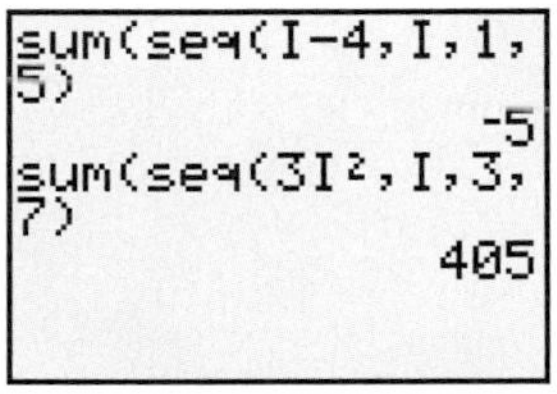

```
sum(seq(I-4,I,1,
5)
                -5
sum(seq(3I²,I,3,
7)
               405
```

FIGURE 2

Figure 2 shows how a graphing calculator can be used in Example 4.

OBJECTIVE 5   **Write a series with summation notation.**   In Example 4, we started with summation notation and wrote each series using + signs. It is possible to go in the other direction; that is, given a series, we can write it with summation notation, by observing a pattern in the terms and writing the general term accordingly.

EXAMPLE 5   **Writing Series with Summation Notation**

Write each sum with summation notation.

(a) $2 + 5 + 8 + 11$

First, find a general term a_n that will give these four terms for a_1, a_2, a_3, and a_4, respectively. Inspection (and trial and error) shows that $3i - 1$ will work, since

$$3(1) - 1 = 2$$
$$3(2) - 1 = 5$$
$$3(3) - 1 = 8$$
$$3(4) - 1 = 11.$$

(Remember, there may be other expressions that also work; the four terms may be the first terms of more than one sequence.) Since i ranges from 1 to 4,

$$2 + 5 + 8 + 11 = \sum_{i=1}^{4} (3i - 1).$$

(b) $8 + 27 + 64 + 125 + 216$

These numbers are the cubes of 2, 3, 4, 5, and 6, so the general term is i^3.

$$8 + 27 + 64 + 125 + 216 = \sum_{i=2}^{6} i^3$$

✔ **Now Try Exercises 37 and 41.**

OBJECTIVE 6 Find the arithmetic mean (average) of a group of numbers.

Arithmetic Mean or Average

The **arithmetic mean,** or **average,** of a group of numbers is symbolized $\bar{x}$ and is found by dividing the sum of the numbers by the number of numbers. That is,

$$\bar{x} = \frac{\sum_{i=1}^{n} x_i}{n}.$$

The values of x_i represent the individual numbers in the group, and n represents the number of numbers.

EXAMPLE 6 Finding the Arithmetic Mean or Average

The following table shows the number of companies listed on the New York Stock Exchange for each year during the period from 1998 through 2004. What was the average number of listings per year for this 7-yr period?

Year	Number of Listings
1998	3114
1999	3025
2000	2862
2001	2798
2002	2783
2003	2750
2004	2768

Source: New York Stock Exchange.

Let $x_1 = 3114$, $x_2 = 3025$, and so on. There are 7 numbers in the group, so $n = 7$.

$$\bar{x} = \frac{\sum_{i=1}^{7} x_i}{7}$$

$$= \frac{3114 + 3025 + 2862 + 2798 + 2783 + 2750 + 2768}{7}$$

$$= 2871 \quad \text{(rounded to the nearest unit)}$$

The average number of listings per year for this 7-yr period was 2871.

✔ **Now Try Exercise 51.**

12.1 EXERCISES

⊙ *Complete solution available on Video Lectures on CD/DVD*

▢ *Now Try Exercise*

Write out the first five terms of each sequence. See Example 1.

⊙ **1.** $a_n = n + 1$

2. $a_n = n - 4$

3. $a_n = \dfrac{n + 3}{n}$

4. $a_n = \dfrac{n + 2}{n + 1}$

5. $a_n = 3^n$

6. $a_n = 1^{n-1}$

7. $a_n = \dfrac{1}{n^2}$

8. $a_n = \dfrac{n^2}{n + 1}$

9. $a_n = (-1)^n$

10. $a_n = (-1)^{2n-1}$

Find the indicated term for each sequence. See Example 1.

11. $a_n = -9n + 2; \; a_8$

12. $a_n = 3n - 7; \; a_{12}$

13. $a_n = \dfrac{3n + 7}{2n - 5}; \; a_{14}$

14. $a_n = \dfrac{5n - 9}{3n + 8}; \; a_{16}$

15. $a_n = (n + 1)(2n + 3); \; a_8$

16. $a_n = (5n - 2)(3n + 1); \; a_{10}$

Find a general term a_n for the given terms of each sequence. See Example 2.

⊙ **17.** $4, 8, 12, 16, \ldots$

18. $-10, -20, -30, -40, \ldots$

19. $\dfrac{1}{3}, \dfrac{1}{9}, \dfrac{1}{27}, \dfrac{1}{81}, \ldots$

20. $\dfrac{1}{2}, \dfrac{2}{3}, \dfrac{3}{4}, \dfrac{4}{5}, \ldots$

Solve each applied problem by writing the first few terms of a sequence. See Example 3.

⊙ **21.** Anne borrows $1000 and agrees to pay $100 plus interest of 1% on the unpaid balance each month. Find the payments for the first six months and the remaining debt at the end of that period.

22. Larissa Perez is offered a new modeling job with a salary of $20,000 + 2500n$ dollars per year at the end of the *n*th year. Write a sequence showing her salary at the end of each of the first 5 yr. If she continues in this way, what will her salary be at the end of the tenth year?

23. Suppose that an automobile loses $\frac{1}{5}$ of its value each year; that is, at the end of any given year, the value is $\frac{4}{5}$ of the value at the beginning of that year. If a car costs $20,000 new, what is its value at the end of 5 yr?

24. A certain car loses $\frac{1}{2}$ of its value each year. If this car cost $40,000 new, what is its value at the end of 6 yr?

Write out each series and evaluate it. See Example 4.

⊙ **25.** $\displaystyle\sum_{i=1}^{5} (i + 3)$

26. $\displaystyle\sum_{i=1}^{6} (i + 9)$

27. $\displaystyle\sum_{i=1}^{3} (i^2 + 2)$

28. $\displaystyle\sum_{i=1}^{4} i(i + 3)$

29. $\displaystyle\sum_{i=1}^{6} (-1)^i$

30. $\displaystyle\sum_{i=1}^{5} (-1)^i \cdot i$

31. $\displaystyle\sum_{i=3}^{7} (i - 3)(i + 2)$

32. $\displaystyle\sum_{i=2}^{6} \dfrac{i^2 + 1}{2}$

Write out the terms of each series.

33. $\displaystyle\sum_{i=1}^{5} 2x \cdot i$

34. $\displaystyle\sum_{i=1}^{6} x^i$

35. $\displaystyle\sum_{i=1}^{5} i \cdot x^i$

36. $\displaystyle\sum_{i=2}^{6} \dfrac{x + i}{x - i}$

Write each series with summation notation. See Example 5.

37. $3 + 4 + 5 + 6 + 7$

38. $1 + 4 + 9 + 16$

39. $\dfrac{1}{2} + \dfrac{1}{3} + \dfrac{1}{4} + \dfrac{1}{5} + \dfrac{1}{6}$

40. $-1 + 2 - 3 + 4 - 5 + 6$

41. $1 + 4 + 9 + 16 + 25$

42. $1 + 16 + 81 + 256$

43. Suppose that f is the function defined by $f(x) = 2x + 4$ with domain all real numbers. Suppose that an infinite sequence is defined by $a_n = 2n + 4$. Discuss the similarities and differences between the function and the sequence. Give examples using each.

44. What is wrong with the following?

For the sequence defined by $a_n = 2n + 4$, find $a_{1/2}$.

45. Explain the basic difference between a sequence and a series.

46. Evaluate $\displaystyle\sum_{i=1}^{3} 5i$ and $5 \displaystyle\sum_{i=1}^{3} i$. Notice that the sums are the same. Explain how the distributive property demonstrates that the two sums are equal.

Find the arithmetic mean for each collection of numbers. See Example 6.

47. 8, 11, 14, 9, 3, 6, 8

48. 10, 12, 8, 19, 23

49. 5, 9, 8, 2, 4, 7, 3, 2

50. 2, 1, 4, 8, 3, 7

Solve each problem. See Example 6.

51. The number of mutual funds available to investors each year during the period 2000 through 2004 is given in the table.

Year	Number of Funds Available
2000	8155
2001	8305
2002	8244
2003	8126
2004	8044

Source: Investment Company Institute.

To the nearest whole number, what was the average number of funds available per year during the given period?

52. The total assets of mutual funds, in billions of dollars, for each year during the period 2000 through 2004 are shown in the table. To the nearest unit (in billions of dollars), what were the average assets per year during this period?

Year	Assets (in billions of dollars)
2000	6965
2001	6975
2002	6390
2003	7414
2004	8107

Source: Investment Company Institute.

PREVIEW EXERCISES

*Find the values of a and d by solving each system. See **Section 4.1.***

53. $a + 3d = 12$
$a + 8d = 22$

54. $a + 7d = 12$
$a + 2d = 7$

*Evaluate $a + (n - 1)d$ for the given values. See **Section 1.3.***

55. $a = -2, n = 5, d = 3$

56. $a = \dfrac{1}{2}, n = 10, d = \dfrac{3}{2}$

12.2 Arithmetic Sequences

OBJECTIVES

1 Find the common difference of an arithmetic sequence.

2 Find the general term of an arithmetic sequence.

3 Use an arithmetic sequence in an application.

4 Find any specified term or the number of terms of an arithmetic sequence.

5 Find the sum of a specified number of terms of an arithmetic sequence.

OBJECTIVE 1 Find the common difference of an arithmetic sequence. In this section, we introduce a special type of sequence that has many applications.

Arithmetic Sequence

A sequence in which each term after the first differs from the preceding term by a constant amount is called an **arithmetic sequence,** or **arithmetic progression.**

For example, the sequence

$$6, 11, 16, 21, 26, \ldots \quad \text{Arithmetic sequence}$$

is an arithmetic sequence, since the difference between any two adjacent terms is always 5. The number 5 is called the **common difference** of the arithmetic sequence. The common difference, d, is found by subtracting any pair of terms a_n and a_{n+1}.

$$d = a_{n+1} - a_n \quad \text{Common difference}$$

EXAMPLE 1 Finding the Common Difference

Find d for the arithmetic sequence

$$-11, -4, 3, 10, 17, 24, \ldots.$$

Since the sequence is arithmetic, d is the difference between any two adjacent terms. Choosing the terms 10 and 17 gives

$$d = 17 - 10$$
$$= 7.$$

The terms -11 and -4 would give $d = -4 - (-11) = 7$, the same result.

✔ **Now Try Exercise 7.**

| **EXAMPLE 2** | **Writing the Terms of a Sequence from the First Term and the Common Difference** |

Write the first five terms of the arithmetic sequence with first term 3 and common difference -2.

The second term is found by adding -2 to the first term 3, getting 1. For the next term, add -2 to 1, and so on. The first five terms are

$$3, 1, -1, -3, -5.$$

✔ **Now Try Exercise 9.**

OBJECTIVE 2 Find the general term of an arithmetic sequence. Generalizing from Example 2, if we know the first term a_1 and the common difference d of an arithmetic sequence, then the sequence is completely defined as

$$a_1, \quad a_2 = a_1 + d, \quad a_3 = a_1 + 2d, \quad a_4 = a_1 + 3d, \ldots.$$

Writing the terms of the sequence in this way suggests the following rule.

General Term of an Arithmetic Sequence

The general term of an arithmetic sequence with first term a_1 and common difference d is

$$a_n = a_1 + (n - 1)d.$$

Since $a_n = a_1 + (n - 1)d = dn + (a_1 - d)$ is a linear function in n, any linear expression of the form $kn + c$, where k and c are real numbers, defines an arithmetic sequence.

EXAMPLE 3 **Finding the General Term of an Arithmetic Sequence**

Find the general term of the arithmetic sequence

$$-9, -6, -3, 0, 3, 6, \ldots.$$

Then use the general term to find a_{20}.

The first term is $a_1 = -9$.

$$d = -3 - (-6) = 3. \qquad \text{Let } d = a_3 - a_2.$$

Now find a_n.

$$\begin{aligned}
a_n &= a_1 + (n - 1)d && \text{Formula for } a_n \\
&= -9 + (n - 1)(3) && \text{Let } a_1 = -9, d = 3. \\
&= -9 + 3n - 3 && \text{Distributive property} \\
a_n &= 3n - 12 && \text{Combine terms.}
\end{aligned}$$

The general term is $a_n = 3n - 12$.

$$a_{20} = 3(20) - 12 = 60 - 12 = 48 \qquad \text{Let } n = 20.$$

✔ **Now Try Exercise 13.**

OBJECTIVE 3 Use an arithmetic sequence in an application.

EXAMPLE 4 Applying an Arithmetic Sequence

Howie Sorkin's uncle decides to start a fund for Howie's education. He makes an initial contribution of $3000 and deposits an additional $500 each month. Thus, after one month the fund will have $3000 + $500 = $3500. How much will it have after 24 months? (Disregard any interest.)

Using an arithmetic sequence, the fund will contain

$$a_n = 3000 + 500n \text{ dollars}$$

after n months. To find the amount in the fund after 24 months, find a_{24}.

$$
\begin{aligned}
a_{24} &= 3000 + 500(24) && \text{Let } n = 24. \\
&= 3000 + 12{,}000 && \text{Multiply.} \\
&= 15{,}000 && \text{Add.}
\end{aligned}
$$

The account will contain $15,000 (disregarding interest) after 24 months.

✔ **Now Try Exercise 43.**

OBJECTIVE 4 Find any specified term or the number of terms of an arithmetic sequence. The formula for the general term of an arithmetic sequence has four variables: a_n, a_1, n, and d. If we know any three of these, the formula can be used to find the value of the fourth variable.

EXAMPLE 5 Finding Specified Terms

Find the indicated term for each arithmetic sequence.

(a) $a_1 = -6$, $d = 12$; a_{15}

We use the formula $a_n = a_1 + (n - 1)d$.

$$
\begin{aligned}
a_{15} &= a_1 + (15 - 1)d && \text{Let } n = 15. \\
&= -6 + 14(12) && \text{Let } a_1 = -6, d = 12. \\
&= 162
\end{aligned}
$$

(b) $a_5 = 2$ and $a_{11} = -10$; a_{17}

Any term can be found if a_1 and d are known. Use the formula for a_n.

$$
\begin{array}{ll}
a_5 = a_1 + (5 - 1)d & a_{11} = a_1 + (11 - 1)d \\
a_5 = a_1 + 4d & a_{11} = a_1 + 10d \\
2 = a_1 + 4d \quad a_5 = 2 & -10 = a_1 + 10d \quad a_{11} = -10
\end{array}
$$

This gives a system of two equations in two variables, a_1 and d. Solve the system for d by adding -1 times one equation to the other equation to eliminate a_1.

$$
\begin{aligned}
-10 &= a_1 + 10d \\
\underline{-2 = -a_1 - 4d} && \text{Multiply } 2 = a_1 + 4d \text{ by } -1. \\
-12 &= 6d && \text{Add.} \\
-2 &= d && \text{Divide by 6.}
\end{aligned}
$$

Now find a_1 by substituting -2 for d into either equation.

$$-10 = a_1 + 10(-2) \qquad \text{Let } d = -2.$$
$$-10 = a_1 - 20$$
$$10 = a_1$$

Use the formula for a_n to find a_{17}.

$$a_{17} = a_1 + (17 - 1)d \qquad \text{Let } n = 17.$$
$$= a_1 + 16d$$
$$= 10 + 16(-2) \qquad \text{Let } a_1 = 10, d = -2.$$
$$= -22$$

Multiply; then add.

✔ **Now Try Exercises 19 and 23.**

EXAMPLE 6 **Finding the Number of Terms in a Sequence**

Find the number of terms in the arithmetic sequence

$$-8, -2, 4, 10, \ldots, 52.$$

Let n represent the number of terms in the sequence. Since $a_n = 52$, $a_1 = -8$, and $d = -2 - (-8) = 6$, use the formula for a_n to find n.

$$a_n = a_1 + (n - 1)d \qquad \text{Formula for } a_n$$
$$52 = -8 + (n - 1)6 \qquad \text{Let } a_n = 52, a_1 = -8, d = 6.$$
$$52 = -8 + 6n - 6 \qquad \text{Distributive property}$$
$$66 = 6n \qquad \text{Simplify.}$$
$$n = 11 \qquad \text{Divide by 6.}$$

The sequence has 11 terms.

✔ **Now Try Exercise 25.**

OBJECTIVE 5 **Find the sum of a specified number of terms of an arithmetic sequence.**
To find a formula for the sum S_n of the first n terms of an arithmetic sequence, we can write out the terms as

$$S_n = a_1 + (a_1 + d) + (a_1 + 2d) + \cdots + [a_1 + (n - 1)d].$$

This same sum can be written in reverse as

$$S_n = a_n + (a_n - d) + (a_n - 2d) + \cdots + [a_n - (n - 1)d].$$

Now add the corresponding terms of these two expressions for S_n to get

$$2S_n = (a_1 + a_n) + (a_1 + a_n) + (a_1 + a_n) + \cdots + (a_1 + a_n).$$

The right-hand side of this expression contains n terms, each equal to $a_1 + a_n$, so

$$2S_n = n(a_1 + a_n)$$

$$S_n = \frac{n}{2}(a_1 + a_n).$$

EXAMPLE 7 Finding the Sum of the First n Terms of an Arithmetic Sequence

Find the sum of the first five terms of the arithmetic sequence in which $a_n = 2n - 5$.

We can use the formula $S_n = \frac{n}{2}(a_1 + a_n)$ to find the sum of the first five terms. Here, $n = 5$, $a_1 = 2(1) - 5 = -3$, and $a_5 = 2(5) - 5 = 5$. From the formula,

$$S_5 = \frac{5}{2}(-3 + 5) = \frac{5}{2}(2) = 5.$$

✔ **Now Try Exercise 35.**

It is sometimes useful to express the sum S_n of an arithmetic sequence in terms of a_1 and d, the quantities that define the sequence. We can do this as follows: Since

$$S_n = \frac{n}{2}(a_1 + a_n) \qquad \text{and} \qquad a_n = a_1 + (n - 1)d,$$

by substituting the expression for a_n into the expression for S_n we obtain

$$S_n = \frac{n}{2}(a_1 + [a_1 + (n - 1)d])$$

$$S_n = \frac{n}{2}[2a_1 + (n - 1)d].$$

The summary box gives both of the alternative forms that may be used to find the sum of the first n terms of an arithmetic sequence.

Sum of the First n Terms of an Arithmetic Sequence

The sum of the first n terms of the arithmetic sequence with first term a_1, nth term a_n, and common difference d is

$$S_n = \frac{n}{2}(a_1 + a_n) \qquad \text{or} \qquad S_n = \frac{n}{2}[2a_1 + (n - 1)d].$$

EXAMPLE 8 Finding the Sum of the First n Terms of an Arithmetic Sequence

Find the sum of the first eight terms of the arithmetic sequence having first term 3 and common difference -2.

Since the known values, $a_1 = 3$, $d = -2$, and $n = 8$, appear in the second formula for S_n, we use it.

$$S_n = \frac{n}{2}[2a_1 + (n - 1)d]$$

$$S_8 = \frac{8}{2}[2(3) + (8 - 1)(-2)] \qquad \text{Let } a_1 = 3, d = -2, n = 8.$$

$$= 4[6 - 14] \qquad \text{Work inside the brackets.}$$

$$= -32$$

✔ **Now Try Exercise 31.**

As mentioned earlier, linear expressions of the form $kn + c$, where k and c are real numbers, define an arithmetic sequence. For example, the sequences defined by $a_n = 2n + 5$ and $a_n = n - 3$ are arithmetic sequences. For this reason,

$$\sum_{i=1}^{n} (ki + c)$$

represents the sum of the first n terms of an arithmetic sequence having first term $a_1 = k(1) + c = k + c$ and general term $a_n = k(n) + c = kn + c$. We can find this sum with the first formula for S_n, as shown in the next example.

EXAMPLE 9 Using S_n to Evaluate a Summation

Find $\sum_{i=1}^{12} (2i - 1)$.

This is the sum of the first 12 terms of the arithmetic sequence having $a_n = 2n - 1$. This sum, S_{12}, is found with the first formula for S_n,

$$S_n = \frac{n}{2}(a_1 + a_n).$$

Here, $n = 12$, $a_1 = 2(1) - 1 = 1$, and $a_{12} = 2(12) - 1 = 23$. Substitute these values into the formula to find that

$$S_{12} = \frac{12}{2}(1 + 23) = 6(24) = 144.$$

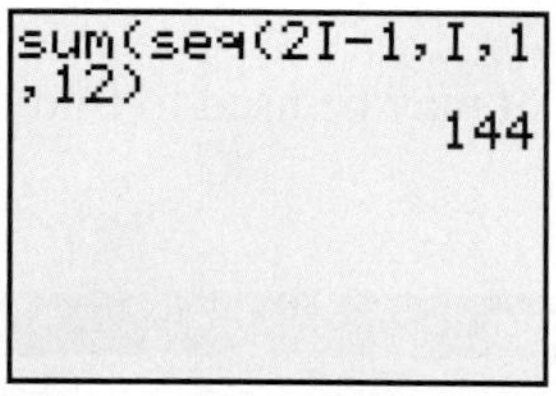

✔ **Now Try Exercise 37.**

```
sum(seq(2I-1,I,1
,12)
           144
```

FIGURE 3

Figure 3 shows how a graphing calculator supports the result of Example 9.

12.2 EXERCISES

⊕ *Complete solution available on Video Lectures on CD/DVD*

Now Try Exercise

✎ **1.** Using several examples, explain the meaning of *arithmetic sequence.*

✎ **2.** Can any two terms of an arithmetic sequence be used to find the common difference? Explain.

If the given sequence is arithmetic, find the common difference d. If the sequence is not arithmetic, say so. See Example 1.

3. $1, 2, 3, 4, 5, \ldots$

4. $2, 5, 8, 11, \ldots$

5. $2, -4, 6, -8, 10, -12, \ldots$

6. $-6, -10, -14, -18, \ldots$

7. $-10, -5, 0, 5, 10, \ldots$

8. $1, 2, 4, 7, 11, 16, \ldots$

Write the first five terms of each arithmetic sequence. See Example 2.

9. $a_1 = 5, d = 4$

10. $a_1 = 6, d = 7$

11. $a_1 = -2, d = -4$

12. $a_1 = -3, d = -5$

Use the formula for a_n to find the general term of each arithmetic sequence. See Example 3.

13. $a_1 = 2, d = 5$

14. $a_1 = 5, d = -3$

15. $3, \dfrac{15}{4}, \dfrac{9}{2}, \dfrac{21}{4}, \dots$

16. $4, 14, 24, \dots$

17. $-3, 0, 3, \dots$

18. $-10, -5, 0, 5, 10, \dots$

Find the indicated term for each arithmetic sequence. See Examples 2 and 5.

19. $a_1 = 4, d = 3; a_{25}$

20. $a_1 = 1, d = -\dfrac{1}{2}; a_{12}$

21. $2, 4, 6, \dots; a_{24}$

22. $1, 5, 9, \dots; a_{50}$

23. $a_{12} = -45, a_{10} = -37; a_1$

24. $a_{10} = -2, a_{15} = -8; a_3$

Find the number of terms in each arithmetic sequence. See Example 6.

25. $3, 5, 7, \dots, 33$

26. $2, \dfrac{3}{2}, 1, \dfrac{1}{2}, \dots, -5$

27. $\dfrac{3}{4}, 3, \dfrac{21}{4}, \dots, 12$

28. $4, 1, -2, \dots, -32$

29. *Concept Check* In the formula for S_n, what does n represent?

30. Explain when you would use each of the two formulas for S_n.

Find S_6 for each arithmetic sequence. See Examples 7 and 8.

31. $a_1 = 6, d = 3$

32. $a_1 = 5, d = 4$

33. $a_1 = 7, d = -3$

34. $a_1 = -5, d = -4$

35. $a_n = 4 + 3n$

36. $a_n = 9 + 5n$

Use a formula for S_n to evaluate each series. See Example 9.

37. $\displaystyle\sum_{i=1}^{10} (8i - 5)$

38. $\displaystyle\sum_{i=1}^{17} (i - 1)$

39. $\displaystyle\sum_{i=1}^{20} (2i - 5)$

40. $\displaystyle\sum_{i=1}^{10} \left(\dfrac{1}{2}i - 1\right)$

41. $\displaystyle\sum_{i=1}^{250} i$

42. $\displaystyle\sum_{i=1}^{2000} i$

Solve each problem. (Hint: Immediately after reading the problem, determine whether you need to find a specific term of a sequence or the sum of the terms of a sequence.) See Example 4.

43. Nancy Bondy's aunt has promised to deposit $1 in her account on the first day of her birthday month, $2 on the second day, $3 on the third day, and so on for 30 days. How much will this amount to over the entire month?

44. Repeat Exercise 43, but assume that the deposits are $2, $4, $6, and so on, and that the month is February of a leap year.

45. Suppose that Randy Morgan is offered a job at $1600 per month with a guaranteed increase of $50 every six months for 5 yr. What will his salary be at the end of that time?

46. Repeat Exercise 45, but assume that the starting salary is $2000 per month and the guaranteed increase is $100 every four months for 3 yr.

47. A seating section in a theater-in-the-round has 20 seats in the first row, 22 in the second row, 24 in the third row, and so on for 25 rows. How many seats are there in the last row? How many seats are there in the section?

48. José Valdivielso has started on a fitness program. He plans to jog 10 min per day for the first week and then add 10 min per day each week until he is jogging an hour each day. In which week will this occur? What is the total number of minutes he will run during the first four weeks?

49. A child builds with blocks, placing 35 blocks in the first row, 31 in the second row, 27 in the third row, and so on. Continuing this pattern, can she end with a row containing exactly 1 block? If not, how many blocks will the last row contain? How many rows can she build this way?

50. A stack of firewood has 28 pieces on the bottom, 24 on top of those, then 20, and so on. If there are 108 pieces of wood, how many rows are there? (*Hint: n ≤ 7.*)

PREVIEW EXERCISES

*Evaluate ar^n for the given values of a, r, and n. See **Section 5.1.***

51. $a = 2, r = 3, n = 2$

52. $a = 3, r = 2, n = 4$

53. $a = 4, r = \dfrac{1}{2}, n = 3$

54. $a = 5, r = \dfrac{1}{4}, n = 2$

12.3 Geometric Sequences

In an arithmetic sequence, each term after the first is found by *adding* a fixed number to the previous term. A *geometric sequence* is defined as follows.

Geometric Sequence

A **geometric sequence,** or **geometric progression,** is a sequence in which each term after the first is a constant nonzero multiple of the preceding term.

OBJECTIVE 1 Find the common ratio of a geometric sequence. We find the constant multiplier, called the **common ratio,** by dividing any term after the first by the preceding term. That is, the common ratio is

$$r = \frac{a_{n+1}}{a_n}. \quad \text{Common ratio}$$

For example,

$$2, 6, 18, 54, 162, \dots \quad \text{Geometric sequence}$$

is a geometric sequence in which the first term, a_1, is 2 and the common ratio is

$$r = \frac{6}{2} = \frac{18}{6} = \frac{54}{18} = \frac{162}{54} = 3.$$

EXAMPLE 1 Finding the Common Ratio

Find r for the geometric sequence

$$15, \frac{15}{2}, \frac{15}{4}, \frac{15}{8}, \dots.$$

To find r, choose any two successive terms and divide the second one by the first. Choosing the second and third terms of the sequence gives

$$r = \frac{a_3}{a_2} = \frac{15}{4} \div \frac{15}{2} = \frac{15}{4} \cdot \frac{2}{15} = \frac{1}{2}.$$

Any other two successive terms could have been used to find r. Additional terms of the sequence can be found by multiplying each successive term by $\frac{1}{2}$.

✔ **Now Try Exercise 3.**

OBJECTIVE 2 Find the general term of a geometric sequence. The general term a_n of a geometric sequence $a_1, a_2, a_3, \dots$ is expressed in terms of a_1 and r by writing the first few terms as

$$a_1, \quad a_2 = a_1 r, \quad a_3 = a_1 r^2, \quad a_4 = a_1 r^3, \dots,$$

which suggests the next rule.

General Term of a Geometric Sequence

The general term of the geometric sequence with first term a_1 and common ratio r is

$$a_n = a_1 r^{n-1}.$$

▶ **CAUTION** In finding $a_1 r^{n-1}$, be careful to use the correct order of operations. The value of r^{n-1} must be found first. Then multiply the result by a_1.

EXAMPLE 2 Finding the General Term

Find the general term of the sequence in Example 1.
   The first term is $a_1 = 15$ and the common ratio is $r = \frac{1}{2}$.

$$a_n = a_1 r^{n-1} = 15\left(\frac{1}{2}\right)^{n-1} \qquad \text{Substitute into the formula for } a_n.$$

Notice that it is not possible to simplify further, because the exponent must be applied before the multiplication can be done.

✔ **Now Try Exercise 11.**

OBJECTIVE 3 Find any specified term of a geometric sequence. We can use the formula for the general term to find any particular term.

EXAMPLE 3 Finding Specified Terms

Find the indicated term for each geometric sequence.

(a) $a_1 = 4$, $r = -3$; a_6
   Use the formula for the general term.

$$a_n = a_1 r^{n-1}$$
$$a_6 = a_1 \cdot r^{6-1} \qquad \text{Let } n = 6.$$
$$\text{Evaluate } (-3)^5; \qquad = 4 \cdot (-3)^5 \qquad \text{Let } a_1 = 4, r = -3.$$
$$\text{then multiply.} \qquad = -972$$

(b) $\dfrac{3}{4}, \dfrac{3}{8}, \dfrac{3}{16}, \dots ; a_7$
   Here, $r = \frac{1}{2}$, $a_1 = \frac{3}{4}$, and $n = 7$.

$$a_7 = \frac{3}{4} \cdot \left(\frac{1}{2}\right)^6 = \frac{3}{4} \cdot \frac{1}{64} = \frac{3}{256}$$

✔ **Now Try Exercises 17 and 19.**

EXAMPLE 4 Writing the Terms of a Sequence

Write the first five terms of the geometric sequence whose first term is 5 and whose common ratio is $\frac{1}{2}$.

$$a_1 = 5, \quad a_2 = 5\left(\frac{1}{2}\right) = \frac{5}{2}, \quad a_3 = 5\left(\frac{1}{2}\right)^2 = \frac{5}{4}, \qquad \text{Use } a_n = a_1 r^{n-1}.$$

$$a_4 = 5\left(\frac{1}{2}\right)^3 = \frac{5}{8}, \quad a_5 = 5\left(\frac{1}{2}\right)^4 = \frac{5}{16}$$

✔ **Now Try Exercise 23.**

OBJECTIVE 4 Find the sum of a specified number of terms of a geometric sequence. It is convenient to have a formula for the sum S_n of the first n terms of a geometric sequence. We can develop a formula by first writing out S_n.

$$S_n = a_1 + a_1 r + a_1 r^2 + a_1 r^3 + \cdots + a_1 r^{n-1}$$

Next, we multiply both sides by r.

$$r S_n = a_1 r + a_1 r^2 + a_1 r^3 + a_1 r^4 + \cdots + a_1 r^n$$

We subtract the first result from the second.

$$r S_n - S_n = (a_1 r - a_1) + (a_1 r^2 - a_1 r) + (a_1 r^3 - a_1 r^2)$$
$$\qquad\qquad + (a_1 r^4 - a_1 r^3) + \cdots + (a_1 r^n - a_1 r^{n-1})$$

$$r S_n - S_n = (a_1 r - a_1 r) + (a_1 r^2 - a_1 r^2) \qquad \text{Commutative, associative,}$$
$$\qquad\qquad + (a_1 r^3 - a_1 r^3) + \cdots + (a_1 r^n - a_1) \qquad \text{and distributive properties}$$

$$S_n(r - 1) = a_1 r^n - a_1$$

$$S_n = \frac{a_1 r^n - a_1}{r - 1} \qquad\qquad\qquad \text{Divide by } r - 1, \text{ where } r \neq 1.$$

$$S_n = \frac{a_1(r^n - 1)}{r - 1} \qquad\qquad\qquad \text{Factor out } a_1.$$

Sum of the First n Terms of a Geometric Sequence

The sum of the first n terms of the geometric sequence with first term a_1 and common ratio r is

$$S_n = \frac{a_1(r^n - 1)}{r - 1} \quad (r \neq 1).$$

If $r = 1$, then $S_n = a_1 + a_1 + a_1 + \cdots + a_1 = na_1$.

Multiplying the formula for S_n by $\frac{-1}{-1}$ gives an alternative form.

$$S_n = \frac{a_1(r^n - 1)}{r - 1} \cdot \frac{-1}{-1} = \frac{a_1(1 - r^n)}{1 - r} \qquad \text{Alternative form}$$

EXAMPLE 5 Finding the Sum of the First n Terms of a Geometric Sequence

Find the sum of the first six terms of the geometric sequence with first term -2 and common ratio 3.

$$S_n = \frac{a_1(r^n - 1)}{r - 1} \qquad \text{Formula for } S_n$$

$$S_6 = \frac{-2(3^6 - 1)}{3 - 1} \qquad \text{Let } n = 6, a_1 = -2, r = 3.$$

$$= \frac{-2(729 - 1)}{2} \qquad \text{Evaluate } 3^6.$$

$$= -728$$

✔ **Now Try Exercise 27.**

A series of the form

$$\sum_{i=1}^{n} a \cdot b^i$$

represents the sum of the first n terms of a geometric sequence having first term $a_1 = a \cdot b^1 = ab$ and common ratio b. The next example illustrates this form.

EXAMPLE 6 Using the Formula for S_n to Find a Summation

Find $\sum_{i=1}^{4} 3 \cdot 2^i$.

Since the series is in the form $\sum_{i=1}^{n} a \cdot b^i$, it represents the sum of the first n terms of the geometric sequence with $a_1 = a \cdot b^1$ and $r = b$.

$$S_n = \frac{a_1(r^n - 1)}{r - 1} \qquad \text{Formula for } S_n$$

$$S_4 = \frac{6(2^4 - 1)}{2 - 1} \qquad \text{Let } n = 4, a_1 = 6, r = 2.$$

$$= \frac{6(16 - 1)}{1} \qquad \text{Evaluate } 2^4.$$

$$= 90$$

✔ **Now Try Exercise 31.**

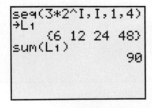

FIGURE 4

Figure 4 shows how a graphing calculator can store the terms in a list and then find the sum of these terms. The figure supports the result of Example 6.

OBJECTIVE 5 Apply the formula for the future value of an ordinary annuity. A sequence of equal payments made over equal periods is called an **annuity.** If the payments are made at the end of the period, and if the frequency of payments is the same as the frequency of compounding, the annuity is called an **ordinary annuity.** The time between payments is the **payment period,** and the time from the beginning of the first payment period to the end of the last is called the **term of the annuity.** The **future value of the annuity,** the final sum on deposit, is defined as the sum of the compound amounts of all the payments, compounded to the end of the term.

For example, suppose $1500 is deposited at the end of the year for the next 6 yr in an account paying 8% per yr compounded annually. To find the future value of this annuity, look separately at each of the $1500 payments. The first of these payments will produce a compound amount of

$$1500(1 + 0.08)^5 = 1500(1.08)^5.$$

Use 5 as the exponent instead of 6, since the money is deposited at the *end* of the first year and earns interest for only 5 yr. The second payment of $1500 will produce the compound amount $1500(1.08)^4$. Continuing in this way and finding the sum of all the terms gives

$$1500(1.08)^5 + 1500(1.08)^4 + 1500(1.08)^3 + 1500(1.08)^2 + 1500(1.08)^1 + 1500.$$

(The last payment earns no interest at all.) Reading in reverse order, we see that this expression is the sum of the first six terms of a geometric sequence with $a_1 = 1500$, $r = 1.08$, and $n = 6$. Therefore, the sum is

$$\frac{a_1(r^n - 1)}{r - 1} = \frac{1500[(1.08)^6 - 1]}{1.08 - 1} = 11{,}003.89, \quad \text{or} \quad \$11{,}003.89.$$

We state the following formula without proof.

Future Value of an Ordinary Annuity

The future value of an ordinary annuity is

$$S = R\left[\frac{(1 + i)^n - 1}{i}\right],$$

where S is the future value,

R is the payment at the end of each period,

i is the interest rate per period, and

n is the number of periods.

EXAMPLE 7   Applying the Formula for the Future Value of an Annuity

(a) Rocky Rhodes is an athlete who believes that his playing career will last 7 yr. To prepare for his future, he deposits $22,000 at the end of each year for 7 yr in an account paying 6% compounded annually. How much will he have on deposit after 7 yr?

Rocky's payments form an ordinary annuity with $R = 22{,}000$, $n = 7$, and $i = 0.06$. The future value of this annuity (from the formula) is

$$S = 22{,}000\left[\frac{(1.06)^7 - 1}{0.06}\right] = 184{,}664.43, \quad \text{or} \quad \$184{,}664.43. \qquad \text{Use a calculator.}$$

(b) Judy Zahrndt has decided to deposit $200 at the end of each month in an account that pays interest of 7.2% compounded monthly for retirement in 20 yr. How much will be in the account at that time?

Because the interest is compounded monthly, $i = \frac{0.072}{12}$. Also, $R = 200$ and $n = 12(20)$. The future value is

$$S = 200 \left[\frac{\left(1 + \frac{0.072}{12} \right)^{12(20)} - 1}{\frac{0.072}{12}} \right] = 106{,}752.47, \quad \text{or} \quad \$106{,}752.47.$$

✔ **Now Try Exercise 35.**

OBJECTIVE 6 **Find the sum of an infinite number of terms of certain geometric sequences.** Consider an infinite geometric sequence such as

$$\frac{1}{3}, \frac{1}{6}, \frac{1}{12}, \frac{1}{24}, \frac{1}{48}, \ldots.$$

Can the sum of the terms of such a sequence be found somehow? The sum of the first two terms is

$$S_2 = \frac{1}{3} + \frac{1}{6} = \frac{1}{2} = 0.5.$$

In a similar manner,

$$S_3 = S_2 + \frac{1}{12} = \frac{1}{2} + \frac{1}{12} = \frac{7}{12} \approx 0.583, \quad S_4 = S_3 + \frac{1}{24} = \frac{7}{12} + \frac{1}{24} = \frac{15}{24} = 0.625,$$

$$S_5 = \frac{31}{48} \approx 0.64583, \quad S_6 = \frac{21}{32} = 0.65625, \quad S_7 = \frac{127}{192} \approx 0.6614583.$$

Each term of the geometric sequence is smaller than the preceding one, so each additional term is contributing less and less to the sum. In decimal form (to the nearest thousandth), the first 7 terms and the 10th term are given in the table.

Term	a_1	a_2	a_3	a_4	a_5	a_6	a_7	a_{10}
Value	0.333	0.167	0.083	0.042	0.021	0.010	0.005	0.001

As the table suggests, the value of a term gets closer and closer to 0 as the number of the term increases. To express this idea, we say that as n increases without bound (written $n \to \infty$), the limit of the term a_n is 0, written

$$\lim_{n \to \infty} a_n = 0.$$

A number that can be defined as the sum of an infinite number of terms of a geometric sequence can be found by starting with the expression for the sum of a finite number of terms:

$$S_n = \frac{a_1(r^n - 1)}{r - 1}.$$

If $|r| < 1$, then as n increases without bound, the value of r^n gets closer and closer to 0. For example, in the infinite sequence just discussed, $r = \frac{1}{2} = 0.5$. The following table shows how $r^n = (0.5)^n$, given to the nearest thousandth, gets smaller as n increases.

n	1	2	3	4	5	6	7	10
r^n	0.5	0.25	0.125	0.063	0.031	0.016	0.008	0.001

As r^n approaches 0, $r^n - 1$ approaches $0 - 1 = -1$, and S_n approaches the quotient $\dfrac{-a_1}{r - 1}$.

$$\lim_{r^n \to 0} S_n = \lim_{r^n \to 0} \frac{a_1(r^n - 1)}{r - 1} = \frac{a_1(0 - 1)}{r - 1} = \frac{-a_1}{r - 1} = \frac{a_1}{1 - r}$$

This limit is defined to be the sum of the infinite geometric sequence

$$a_1 + a_1 r + a_1 r^2 + a_1 r^3 + \cdots = \frac{a_1}{1 - r}, \quad \text{if } |r| < 1.$$

What happens if $|r| > 1$? For example, suppose the sequence is

$$6, 12, 24, \ldots, 3(2)^n, \ldots.$$

In this kind of sequence, as n increases, the value of r^n also increases and so does the sum S_n. Since each new term adds a larger and larger amount to the sum, there is no limit to the value of S_n, and the sum S_n does not exist. A similar situation exists if $r = 1$.

The sum of the terms of an infinite geometric sequence is defined as follows.

Sum of the Terms of an Infinite Geometric Sequence

The sum S of the terms of an infinite geometric sequence with first term a_1 and common ratio r, where $|r| < 1$, is

$$S = \frac{a_1}{1 - r}.$$

If $|r| \geq 1$, then the sum does not exist.

EXAMPLE 8 Finding the Sum of the Terms of an Infinite Geometric Sequence

Find the sum of the terms of the infinite geometric sequence with $a_1 = 3$ and $r = -\frac{1}{3}$.

From the preceding rule, the sum is

$$S = \frac{a_1}{1 - r} = \frac{3}{1 - \left(-\frac{1}{3}\right)} = \frac{3}{\frac{4}{3}} = 3 \div \frac{4}{3} = \frac{3}{1} \cdot \frac{3}{4} = \frac{9}{4}.$$

✔ **Now Try Exercise 39.**

In summation notation, the sum of an infinite geometric sequence is written as

$$\sum_{i=1}^{\infty} a_i.$$

For instance, the sum in Example 8 would be written

$$\sum_{i=1}^{\infty} 3\left(-\frac{1}{3}\right)^{i-1}.$$

EXAMPLE 9 **Finding the Sum of the Terms of an Infinite Geometric Series**

Find $\sum_{i=1}^{\infty} \left(\frac{1}{2}\right)^i$.

This is the infinite geometric series

$$\frac{1}{2} + \frac{1}{4} + \frac{1}{8} + \cdots,$$

with $a_1 = \frac{1}{2}$ and $r = \frac{1}{2}$. Since $|r| < 1$, we find the sum as follows:

$$S = \frac{a_1}{1 - r} = \frac{\frac{1}{2}}{1 - \frac{1}{2}} = \frac{\frac{1}{2}}{\frac{1}{2}} = 1.$$

✔**Now Try Exercise 43.**

12.3 EXERCISES

✐ **1.** Using several examples, explain the meaning of *geometric sequence*.

✐ **2.** Explain why the sequence 5, 5, 5, 5, . . . can be considered either arithmetic or geometric.

If the given sequence is geometric, find the common ratio r. If the sequence is not geometric, say so. See Example 1.

● **3.** 4, 8, 16, 32, . . .   **4.** 5, 15, 45, 135, . . .   **5.** $\frac{1}{3}, \frac{2}{3}, \frac{3}{3}, \frac{4}{3}, \frac{5}{3}, \ldots$

6. $\frac{1}{3}, \frac{2}{3}, \frac{4}{3}, \frac{8}{3}, \ldots$ **7.** 1, −3, 9, −27, 81, . . .   **8.** 1, −3, 7, −11, . . .

9. 1, $-\frac{1}{2}, \frac{1}{4}, -\frac{1}{8}, \frac{1}{16}, \ldots$ **10.** $\frac{2}{3}, \frac{2}{15}, \frac{2}{75}, \frac{2}{375}, \ldots$

Find a general term for each geometric sequence. See Example 2.

● **11.** 5, 10, . . .   **12.** −2, −6, . . .   **13.** $\frac{1}{9}, \frac{1}{3}, \ldots$

14. $-3, \frac{3}{2}, \ldots$ **15.** 10, −2, . . .   **16.** −4, 8, . . .

Find the indicated term for each geometric sequence. See Example 3.

17. $a_1 = 2, r = 5; a_{10}$

18. $a_1 = -1, r = 3; a_{15}$

19. $\dfrac{1}{2}, \dfrac{1}{6}, \dfrac{1}{18}, \ldots ; a_{12}$

20. $\dfrac{2}{3}, -\dfrac{1}{3}, \dfrac{1}{6}, \ldots ; a_{18}$

21. $a_3 = \dfrac{1}{2}, a_7 = \dfrac{1}{32}; a_{25}$

22. $a_5 = 48, a_8 = -384; a_{10}$

Write the first five terms of each geometric sequence. See Example 4.

23. $a_1 = 2, r = 3$

24. $a_1 = 4, r = 2$

25. $a_1 = 5, r = -\dfrac{1}{5}$

26. $a_1 = 6, r = -\dfrac{1}{3}$

Use the formula for S_n to find the sum of the terms of each geometric sequence. See Examples 5 and 6. In Exercises 29–34, give the answer to the nearest thousandth.

27. $\dfrac{1}{3}, \dfrac{1}{9}, \dfrac{1}{27}, \dfrac{1}{81}, \dfrac{1}{243}$

28. $\dfrac{4}{3}, \dfrac{8}{3}, \dfrac{16}{3}, \dfrac{32}{3}, \dfrac{64}{3}, \dfrac{128}{3}$

29. $-\dfrac{4}{3}, -\dfrac{4}{9}, -\dfrac{4}{27}, -\dfrac{4}{81}, -\dfrac{4}{243}, -\dfrac{4}{729}$

30. $\dfrac{5}{16}, -\dfrac{5}{32}, \dfrac{5}{64}, -\dfrac{5}{128}, \dfrac{5}{256}$

31. $\displaystyle\sum_{i=1}^{7} 4\left(\dfrac{2}{5}\right)^i$

32. $\displaystyle\sum_{i=1}^{8} 5\left(\dfrac{2}{3}\right)^i$

33. $\displaystyle\sum_{i=1}^{10} (-2)\left(\dfrac{3}{5}\right)^i$

34. $\displaystyle\sum_{i=1}^{6} (-2)\left(-\dfrac{1}{2}\right)^i$

Solve each problem involving an ordinary annuity. See Example 7.

35. A father opened a savings account for his daughter on the day she was born, depositing $1000. Each year on her birthday he deposits another $1000, making the last deposit on her 21st birthday. If the account pays 6.5% interest compounded annually, how much is in the account at the end of the day on the daughter's 21st birthday?

36. Sam Smith puts $1000 in a retirement account at the end of each quarter $\left(\frac{1}{4} \text{ of a year}\right)$ for 15 yr. If the account pays 4% annual interest compounded quarterly, how much will be in the account at that time?

37. At the end of each quarter, a 50-year-old woman puts $1200 in a retirement account that pays 7% interest compounded quarterly. When she reaches age 60, she withdraws the entire amount and places it in a mutual fund that pays 9% interest compounded monthly. From then on, she deposits $300 in the mutual fund at the end of each month. How much is in the account when she reaches age 65?

38. John Stealy deposits $10,000 at the beginning of each year for 12 yr in an account paying 5% compounded annually. He then puts the total amount on deposit in another account paying 6% compounded semiannually for another 9 yr. Find the final amount on deposit after the entire 21-yr period.

Find the sum, if it exists, of the terms of each infinite geometric sequence. See Examples 8 and 9.

39. $a_1 = 6, r = \dfrac{1}{3}$

40. $a_1 = 10, r = \dfrac{1}{5}$

41. $a_1 = 1000, r = -\dfrac{1}{10}$

42. $a_1 = 8500, r = \dfrac{3}{5}$

43. $\displaystyle\sum_{i=1}^{\infty} \dfrac{9}{8}\left(-\dfrac{2}{3}\right)^i$

44. $\displaystyle\sum_{i=1}^{\infty} \dfrac{3}{5}\left(\dfrac{5}{6}\right)^i$

45. $\displaystyle\sum_{i=1}^{\infty} \dfrac{12}{5}\left(\dfrac{5}{4}\right)^i$

46. $\displaystyle\sum_{i=1}^{\infty} \left(-\dfrac{16}{3}\right)\left(-\dfrac{9}{8}\right)^i$

Solve each application. (Hint: Immediately after reading the problem, determine whether you need to find a specific term of a sequence or the sum of the terms of a sequence.)

47. When dropped from a certain height, a ball rebounds $\frac{3}{5}$ of the original height. How high will the ball rebound after the fourth bounce if it was dropped from a height of 10 ft?

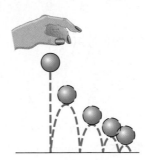

48. A fully wound yo-yo has a string 40 in. long. It is allowed to drop, and on its first rebound it returns to a height 15 in. lower than its original height. Assuming that this "rebound ratio" remains constant until the yo-yo comes to rest, how far does it travel on its third trip up the string?

49. A particular substance decays in such a way that it loses half its weight each day. In how many days will 256 g of the substance be reduced to 32 g? How much of the substance is left after 10 days?

50. A tracer dye is injected into a system with an ingestion and an excretion. After 1 hr, $\frac{2}{3}$ of the dye is left. At the end of the second hour, $\frac{2}{3}$ of the remaining dye is left, and so on. If one unit of the dye is injected, how much is left after 6 hr?

51. In a certain community, the consumption of electricity has increased about 6% per yr.

(a) If a community uses 1.1 billion units of electricity now, how much will it use 5 yr from now?

(b) Find the number of years it will take for the consumption to double.

52. Suppose the community in Exercise 51 reduces its increase in consumption to 2% per yr.

    (a) How much will it use 5 yr from now?

    (b) Find the number of years it will take for the consumption to double.

53. A machine depreciates by $\frac{1}{4}$ of its value each year. If it cost $50,000 new, what is its value after 8 yr?

54. Refer to Exercise 48. Theoretically, how far does the yo-yo travel before coming to rest?

RELATING CONCEPTS (EXERCISES 55–60)

FOR INDIVIDUAL OR GROUP WORK

*In **Chapter 1**, we learned that any repeating decimal is a rational number; that is, it can be expressed as a quotient of integers. Thus, the repeating decimal*

$$0.99999\ldots,$$

*with an endless string of 9s, must be a rational number. **Work Exercises 55–60 in order**, to discover the surprising simplest form of this rational number.*

55. Use long division or your previous experience to write a repeating decimal representation for $\frac{1}{3}$.

56. Use long division or your previous experience to write a repeating decimal representation for $\frac{2}{3}$.

57. Because $\frac{1}{3} + \frac{2}{3} = 1$, the sum of the decimal representations in Exercises 55 and 56 must also equal 1. Line up the decimals in the usual vertical method for addition, and obtain the repeating decimal result. The value of this decimal is exactly 1.

58. The repeating decimal $0.99999\ldots$ can be written as the sum of the terms of a geometric sequence with $a_1 = 0.9$ and $r = 0.1$:

$$0.99999\ldots = 0.9 + 0.9(0.1) + 0.9(0.1)^2 + 0.9(0.1)^3 + 0.9(0.1)^4 + 0.9(0.1)^5 + \cdots.$$

    Since $|0.1| < 1$, this sum can be found from the formula $S = \dfrac{a_1}{1 - r}$. Use this formula to support the result you found another way in Exercises 55–57.

59. Which one of the following is true, based on your results in Exercises 57 and 58?

    A. $0.99999\ldots < 1$ **B.** $0.99999\ldots = 1$ **C.** $0.99999\ldots \approx 1$

60. Show that $0.49999\ldots = \frac{1}{2}$.

PREVIEW EXERCISES

*Multiply. See **Section 5.4**.*

61. $(3x + 2y)^2$ **62.** $(4x - 3y)^2$

63. $(a - b)^3$ **64.** $(x + y)^4$

12.4 The Binomial Theorem

OBJECTIVES

1 Expand a binomial raised to a power.

2 Find any specified term of the expansion of a binomial.

OBJECTIVE 1 Expand a binomial raised to a power. Writing out the binomial expression $(x + y)^n$ for nonnegative integer values of n gives a family of expressions that is important in the study of mathematics and its applications. For example,

$$(x + y)^0 = 1,$$
$$(x + y)^1 = x + y,$$
$$(x + y)^2 = x^2 + 2xy + y^2,$$
$$(x + y)^3 = x^3 + 3x^2y + 3xy^2 + y^3,$$
$$(x + y)^4 = x^4 + 4x^3y + 6x^2y^2 + 4xy^3 + y^4,$$
$$(x + y)^5 = x^5 + 5x^4y + 10x^3y^2 + 10x^2y^3 + 5xy^4 + y^5.$$

Inspection shows that these expansions follow a pattern. By identifying the pattern, we can write a general expansion for $(x + y)^n$.

First, if n is a positive integer, each expansion after $(x + y)^0$ begins with x raised to the same power to which the binomial is raised. That is, the expansion of $(x + y)^1$ has a first term of x^1, the expansion of $(x + y)^2$ has a first term of x^2, the expansion of $(x + y)^3$ has a first term of x^3, and so on. Also, the last term in each expansion is y to this same power, so the expansion of $(x + y)^n$ should begin with the term x^n and end with the term y^n.

The exponents on x decrease by 1 in each term after the first, while the exponents on y, beginning with y in the second term, increase by 1 in each succeeding term. Thus, the *variables* in the expansion of $(x + y)^n$ have the following pattern:

$$x^n, \quad x^{n-1}y, \quad x^{n-2}y^2, \quad x^{n-3}y^3, \ldots, xy^{n-1}, \quad y^n.$$

This pattern suggests that the sum of the exponents on x and y in each term is n. For example, in the third term shown, the variable part is $x^{n-2}y^2$ and the sum of the exponents, $n - 2$ and 2, is n.

Now examine the pattern for the *coefficients* of the terms of the preceding expansions. Writing the coefficients alone in a triangular pattern gives **Pascal's triangle,** named in honor of the 17th-century mathematician Blaise Pascal, one of the first to use it extensively.

Blaise Pascal (1623–1662)

Pascal's Triangle

```
                  1
               1     1
            1     2     1
         1     3     3     1
      1     4     6     4     1
   1     5    10    10     5     1      and so on
```

Arranging the coefficients in this way shows that each number in the triangle is the sum of the two numbers just above it (one to the right and one to the left). For example, in the fifth row from the top, 1 is the sum of 1 (the only number above it), 4 is the sum of 1 and 3, 6 is the sum of 3 and 3, and so on.

To obtain the coefficients for $(x + y)^6$, we attach the seventh row to the table by adding pairs of numbers from the sixth row.

1   6   15   20   15   6   1   Seventh row

We then use these coefficients to expand $(x + y)^6$ as

$$(x + y)^6 = x^6 + 6x^5y + 15x^4y^2 + 20x^3y^3 + 15x^2y^4 + 6xy^5 + y^6.$$

Although it is possible to use Pascal's triangle to find the coefficients in $(x + y)^n$ for any positive integer value of n, it is impractical for large values of n. A more efficient way to determine these coefficients uses a notational shorthand with the symbol **$n!$** (read **"n factorial"**), defined as follows.

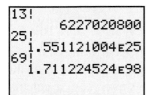

(a)

---

n Factorial ($n!$)

For any positive integer n,

$$n(n - 1)(n - 2)(n - 3) \cdots (2)(1) = n!.$$

For example,

$$3! = 3 \cdot 2 \cdot 1 = 6 \qquad \text{and} \qquad 5! = 5 \cdot 4 \cdot 3 \cdot 2 \cdot 1 = 120.$$

From the definition of n factorial, $n[(n - 1)!] = n!$. If $n = 1$, then $1(0!) = 1! = 1$. Because of this, $0!$ is defined as

$$0! = 1.$$

> ✔ **Now Try Exercise 1.**

```
13!          6227020800
25!
    1.551121004E25
69!
    1.711224524E98
```

A graphing calculator with a 10-digit display will give the exact value of $n!$ for $n \leq 13$ and approximate values of $n!$ for $14 \leq n \leq 69$.

(b)

FIGURE 5

Scientific and graphing calculators can compute factorials. The three factorial expressions just discussed are shown in Figure 5(a). Figure 5(b) shows some larger factorials.

EXAMPLE 1 Evaluating Expressions Involving Factorials

Find the value of each expression.

(a) $\dfrac{5!}{4!1!} = \dfrac{5 \cdot 4 \cdot 3 \cdot 2 \cdot 1}{(4 \cdot 3 \cdot 2 \cdot 1)(1)} = 5$

(b) $\dfrac{5!}{3!2!} = \dfrac{5 \cdot 4 \cdot 3 \cdot 2 \cdot 1}{(3 \cdot 2 \cdot 1)(2 \cdot 1)} = \dfrac{5 \cdot 4}{2 \cdot 1} = 10$

(c) $\dfrac{6!}{3!3!} = \dfrac{6 \cdot 5 \cdot 4 \cdot 3 \cdot 2 \cdot 1}{(3 \cdot 2 \cdot 1)(3 \cdot 2 \cdot 1)} = \dfrac{6 \cdot 5 \cdot 4}{3 \cdot 2 \cdot 1} = 20$

(d) $\dfrac{4!}{4!0!} = \dfrac{4 \cdot 3 \cdot 2 \cdot 1}{(4 \cdot 3 \cdot 2 \cdot 1)(1)} = 1$

> ✔ **Now Try Exercises 3 and 7.**

Now look again at the coefficients of the expansion

$$(x + y)^5 = x^5 + 5x^4y + 10x^3y^2 + 10x^2y^3 + 5xy^4 + y^5.$$

The coefficient of the second term is 5, and the exponents on the variables in that term are 4 and 1. From Example 1(a), $\frac{5!}{4!1!} = 5$. The coefficient of the third term is 10, and the exponents are 3 and 2. From Example 1(b), $\frac{5!}{3!2!} = 10$. Similar results are true for the remaining terms. The first term can be written as $1x^5y^0$, and the last term can be written as $1x^0y^5$. Then the coefficient of the first term should be $\frac{5!}{5!0!} = 1$, and the coefficient of the last term would be $\frac{5!}{0!5!} = 1$. Generalizing, we see that the coefficient of a term in $(x + y)^n$ in which the variable part is x^ry^{n-r} will be

$$\frac{n!}{r!(n - r)!}.$$

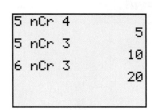

FIGURE 6

▶ **NOTE** The denominator factorials in the coefficient of a term are the same as the exponents on the variables in that term.

The expression $\frac{n!}{r!(n - r)!}$ is often represented by the symbol $_nC_r$. This notation comes from the fact that if we choose *combinations* of n things taken r at a time, the result is given by that expression. A graphing calculator can evaluate the expression for particular values of n and r. Figure 6 shows how a calculator evaluates $_5C_4$, $_5C_3$, and $_6C_3$. Compare these results with parts (a), (b), and (c) of Example 1.

✔ **Now Try Exercise 5.**

We now state the **binomial theorem,** or the **general binomial expansion.**

Binomial Theorem

For any positive integer n,

$$(x + y)^n = x^n + \frac{n!}{(n - 1)!1!}x^{n-1}y + \frac{n!}{(n - 2)!2!}x^{n-2}y^2$$

$$+ \frac{n!}{(n - 3)!3!}x^{n-3}y^3 + \cdots + \frac{n!}{1!(n - 1)!}xy^{n-1} + y^n.$$

The binomial theorem can be written in summation notation as

$$(x + y)^n = \sum_{i=0}^{n} \frac{n!}{(n - i)!i!}x^{n-i}y^i.$$

▶ **NOTE** The letter i is used instead of r in $(x + y)^n = \displaystyle\sum_{i=0}^{n} \frac{n!}{(n - i)!i!} x^{n-i}y^i$ because we are using summation notation. It is not the imaginary unit i.

EXAMPLE 2 Using the Binomial Theorem

Expand $(2m + 3)^4$.

$$(2m + 3)^4 = (2m)^4 + \frac{4!}{3!1!}(2m)^3(3) + \frac{4!}{2!2!}(2m)^2(3)^2 + \frac{4!}{1!3!}(2m)(3)^3 + 3^4$$

$$= 16m^4 + 4(8m^3)(3) + 6(4m^2)(9) + 4(2m)(27) + 81$$

Remember:
$(ab)^m = a^m b^m.$

$$= 16m^4 + 96m^3 + 216m^2 + 216m + 81$$

✔ **Now Try Exercise 17.**

EXAMPLE 3 Using the Binomial Theorem

Expand $\left(a - \dfrac{b}{2} \right)^5$.

$$\left(a - \frac{b}{2} \right)^5 = a^5 + \frac{5!}{4!1!}a^4\left(-\frac{b}{2} \right) + \frac{5!}{3!2!}a^3\left(-\frac{b}{2} \right)^2 + \frac{5!}{2!3!}a^2\left(-\frac{b}{2} \right)^3$$

$$+ \frac{5!}{1!4!}a\left(-\frac{b}{2} \right)^4 + \left(-\frac{b}{2} \right)^5$$

$$= a^5 + 5a^4\left(-\frac{b}{2} \right) + 10a^3\left(\frac{b^2}{4} \right) + 10a^2\left(-\frac{b^3}{8} \right)$$

$$+ 5a\left(\frac{b^4}{16} \right) + \left(-\frac{b^5}{32} \right)$$

Notice that signs alternate positive and negative.

$$= a^5 - \frac{5}{2}a^4b + \frac{5}{2}a^3b^2 - \frac{5}{4}a^2b^3 + \frac{5}{16}ab^4 - \frac{1}{32}b^5$$

✔ **Now Try Exercise 19.**

▶ **CAUTION** When the binomial is the *difference* of two terms, as in Example 3, the signs of the terms in the expansion will alternate. Those terms with odd exponents on the second variable expression $\left(-\frac{b}{2} \text{ in Example 3} \right)$ will be negative, while those with even exponents on the second variable expression will be positive.

OBJECTIVE 2 Find any specified term of the expansion of a binomial. Any single term of a binomial expansion can be determined without writing out the whole expansion. For example, if $n \geq 10$, then the 10th term of $(x + y)^n$ has y raised to the ninth power (since y has the power of 1 in the second term, the power of 2 in the third term, and so on). Since the exponents on x and y in any term must have a sum of n, the exponent on x in the 10th term is $n - 9$. The quantities 9 and $n - 9$ determine the factorials in the denominator of the coefficient. Thus,

$$\frac{n!}{(n-9)!9!}x^{n-9}y^9$$

is the 10th term of $(x + y)^n$. A generalization of this idea follows.

---

rth Term of the Binomial Expansion

If $n \geq r - 1$, then the rth term of the expansion of $(x + y)^n$ is

$$\frac{n!}{[n-(r-1)]!(r-1)!}x^{n-(r-1)}y^{r-1}.$$

In this general expression, remember to start with the exponent on y, which is 1 less than the term number r. Then subtract that exponent from n to get the exponent on x: $n - (r - 1)$. The two exponents are then used as the factorials in the denominator of the coefficient.

EXAMPLE 4 Finding a Single Term of a Binomial Expansion

Find the fourth term of $(a + 2b)^{10}$.

In the fourth term, $2b$ has an exponent of $4 - 1 = 3$ and a has an exponent of $10 - 3 = 7$. The fourth term is

$$\frac{10!}{7!3!}(a^7)(2b)^3 = \frac{10 \cdot 9 \cdot 8}{3 \cdot 2 \cdot 1}(a^7)(8b^3)$$
$$= 120a^7(8b^3)$$
$$= 960a^7b^3.$$

✔ **Now Try Exercise 29.**

12.4 EXERCISES

⊙ *Complete solution available on Video Lectures on CD/DVD*

Now Try Exercise

Evaluate each expression. See Example 1.

⊙ **1.** $6!$

2. $4!$

⊙ **3.** $\dfrac{6!}{4!2!}$

4. $\dfrac{7!}{3!4!}$

⊙ **5.** $_6C_2$

6. $_7C_4$

7. $\dfrac{4!}{0!4!}$

8. $\dfrac{5!}{5!0!}$

9. $4! \cdot 5$ **10.** $6! \cdot 7$ **11.** $_{13}C_{11}$ **12.** $_{13}C_2$

Use the binomial theorem to expand each expression. See Examples 2 and 3.

13. $(m + n)^4$ **14.** $(x + r)^5$ **15.** $(a - b)^5$ **16.** $(p - q)^4$

17. $(2x + 3)^3$ **18.** $\left(\dfrac{x}{3} + 2y\right)^5$ **19.** $\left(\dfrac{x}{2} - y\right)^4$ **20.** $(x^2 + 1)^4$

21. $(mx - n^2)^3$ **22.** $(2p^2 - q^2)^3$

Write the first four terms of each binomial expansion.

23. $(r + 2s)^{12}$ **24.** $(m - n)^{20}$ **25.** $(3x - y)^{14}$

26. $(2p + 3q)^{11}$ **27.** $(t^2 + u^2)^{10}$ **28.** $(x^2 - y^2)^{15}$

Find the indicated term of each binomial expansion. See Example 4.

29. $(2m + n)^{10}$; fourth term   **30.** $(a - 3b)^{12}$; fifth term

31. $\left(x + \dfrac{y}{2}\right)^8$; seventh term   **32.** $(3p - 2q)^{15}$; eighth term

33. $(k - 1)^9$; third term   **34.** $(-4 - s)^{11}$; fourth term

35. The middle term of $(x^2 - 2y)^6$ **36.** The middle term of $(m^3 + 3)^8$

37. The term with x^9y^4 in $(3x^3 - 4y^2)^5$ **38.** The term with x^{10} in $\left(x^3 - \dfrac{2}{x}\right)^6$

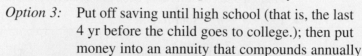

Chapter **12**   **Group Activity**

INVESTING FOR THE FUTURE

Objective   Calculate compound interest; understand the effects of monthly, quarterly, and annual compounding.

In this chapter, you have seen many different types of financing and investing options, including loans, mutual funds, savings accounts, and annuities. In this activity, you will analyze annuities with different periods of compound interest.

Consider a family that wants to save money for college for a 10-year-old child. The family is considering three different savings options.

Option 1:   An annuity that compounds quarterly

Option 2:   An annuity that compounds monthly

Option 3:   Put off saving until high school (that is, the last 4 yr before the child goes to college.); then put money into an annuity that compounds annually

A. Have each member of your group calculate total savings for one of the three options. Use the following formula, where S is the future value of the money, R is the payment amount made each *period, i* is the annual interest rate divided by the number of periods per year, and n is the total number of compounding periods, along with the specific information given below for each option:

$$S = R\left[\frac{(1 + i)^n - 1}{i}\right].$$

Option 1:   The family plans to save $300 per quarter (3 months), the interest rate is 5%, and the annuity compounds quarterly. Savings will be for 8 yr.

Option 2:   The family plans to save $100 per month, the interest rate is 5%, and the annuity compounds monthly. Again, savings will be for 8 yr.

Option 3:   The family plans to wait and save only the last 4 yr. The family will save $2400 per yr, the interest rate is 5%, and the annuity compounds annually.

B. Compare your answers.

1. How much is being invested, according to each option?

2. Which option resulted in the largest amount of savings?

3. Explain why this option produced more savings.

4. What other considerations might be involved in deciding how to save?

Chapter **12** SUMMARY

View the Interactive Summary on the Pass the Test CD.

KEY TERMS

12.1 sequence
infinite sequence
terms of a sequence
general term
finite sequence
series
summation notation

index of summation
arithmetic mean
(average)
12.2 arithmetic sequence
(arithmetic
progression)
common difference

12.3 geometric sequence
(geometric
progression)
common ratio
annuity
ordinary annuity
payment period

future value of an
annuity
term of an annuity
12.4 Pascal's triangle
binomial theorem
(general binomial
expansion)

NEW SYMBOLS

a_n nth term of a
sequence

$\sum_{i=1}^{n} a_i$ summation notation

S_n sum of first n
terms of a
sequence

$\lim_{n \to \infty} a_n$ limit of a_n as n
gets larger and
larger

$\sum_{i=1}^{\infty} a_i$ sum of an infinite
number of terms

$n!$ n factorial

$_nC_r$ binomial coefficient
(combinations of n
things taken r at a
time)

TEST YOUR WORD POWER

See how well you have learned the vocabulary in this chapter. Answers, with examples, follow the Quick Review.

1. An **infinite sequence** is
   A. the values of a function
   B. a function whose domain is the
      set of natural numbers
   C. the sum of the terms of a function
   D. the average of a group of
      numbers.

2. A **series** is
   A. the sum of the terms of a
      sequence
   B. the product of the terms of a
      sequence
   C. the average of the terms of a
      sequence
   D. the function values of a sequence.

3. An **arithmetic sequence** is a
   sequence in which
   A. each term after the first is a
      constant multiple of the preceding
      term

B. the numbers are written in a
   triangular array
C. the terms are added
D. each term after the first differs
   from the preceding term by a
   common amount.

4. A **geometric sequence** is a sequence
   in which
   A. each term after the first is a
      constant multiple of the preceding
      term
   B. the numbers are written in a
      triangular array
   C. the terms are multiplied
   D. each term after the first differs
      from the preceding term by a
      common amount.

5. The **common difference** is
   A. the average of the terms in a
      sequence

B. the constant multiplier in a
   geometric sequence
C. the difference between any two
   adjacent terms in an arithmetic
   sequence
D. the sum of the terms of an
   arithmetic sequence.

6. The **common ratio** is
   A. the average of the terms in a
      sequence
   B. the constant multiplier in a
      geometric sequence
   C. the difference between any two
      adjacent terms in an arithmetic
      sequence
   D. the product of the terms of a
      geometric sequence.

QUICK REVIEW

Concepts	Examples

12.1 SEQUENCES AND SERIES

Sequence

General Term a_n

Series

$1, \dfrac{1}{2}, \dfrac{1}{3}, \dfrac{1}{4}, \ldots, \dfrac{1}{n}$ has general term $\dfrac{1}{n}$.

The corresponding series is the *sum*

$$1 + \frac{1}{2} + \frac{1}{3} + \frac{1}{4} + \cdots + \frac{1}{n}.$$

12.2 ARITHMETIC SEQUENCES

Assume that a_1 is the first term, a_n is the nth term, and d is the common difference.

Common Difference

$$d = a_{n+1} - a_n$$

nth Term

$$a_n = a_1 + (n - 1)d$$

Sum of the First n Terms

$$S_n = \frac{n}{2}(a_1 + a_n)$$

or $\qquad S_n = \dfrac{n}{2}[2a_1 + (n - 1)]$

The arithmetic sequence 2, 5, 8, 11, ... has $a_1 = 2$.

$$d = 5 - 2 = 3 \qquad \text{Use } a_2 - a_1.$$

(Any two successive terms could have been used.)

Suppose that $n = 10$. Then the 10th term is

$$a_{10} = 2 + (10 - 1)3$$
$$= 2 + 9 \cdot 3 = 29.$$

The sum of the first 10 terms is

$$S_{10} = \frac{10}{2}(2 + a_{10})$$
$$= 5(2 + 29) = 5(31) = 155$$

or $\qquad S_{10} = \dfrac{10}{2}[2(2) + (10 - 1)3]$

$$= 5(4 + 9 \cdot 3)$$
$$= 5(4 + 27) = 5(31) = 155.$$

12.3 GEOMETRIC SEQUENCES

Assume that a_1 is the first term, a_n is the nth term, and r is the common ratio.

Common Ratio

$$r = \frac{a_{n+1}}{a_n}$$

nth Term

$$a_n = a_1 r^{n-1}$$

The geometric sequence 1, 2, 4, 8, ... has $a_1 = 1$.

$$r = \frac{8}{4} = 2 \qquad \text{Use } \frac{a_4}{a_3}.$$

(Any two successive terms could have been used.)

Suppose that $n = 6$. Then the sixth term is

$$a_6 = (1)(2)^{6-1} = 1(2)^5 = 32.$$

(continued)

Concepts	Examples

Sum of the First n Terms

$$S_n = \frac{a_1(1 - r^n)}{1 - r} \quad \text{or} \quad S_n = \frac{a_1(r^n - 1)}{r - 1} \quad (r \neq 1)$$

The sum of the first six terms is

$$S_6 = \frac{1(2^6 - 1)}{2 - 1} = \frac{64 - 1}{1} = 63.$$

Future Value of an Ordinary Annuity

$$S = R\left[\frac{(1 + i)^n - 1}{i}\right],$$

where S is the future value, R is the payment at the end of each period, i is the interest rate per period, and n is the number of periods.

If \$5800 is deposited into an ordinary annuity at the end of each quarter for 4 yr and interest is earned at 6.4% compounded quarterly, then

$$R = \$5800, \quad i = \frac{0.064}{4} = 0.016, \quad n = 4(4) = 16,$$

and

$$S = 5800\left[\frac{(1 + 0.016)^{16} - 1}{0.016}\right] = \$104{,}812.44.$$

Sum of the Terms of an Infinite Geometric Sequence with $|r| < 1$

$$S = \frac{a_1}{1 - r}$$

The sum S of the terms of an infinite geometric sequence with $a_1 = 1$ and $r = \frac{1}{2}$ is

$$S = \frac{1}{1 - \frac{1}{2}} = \frac{1}{\frac{1}{2}} = 2.$$

12.4 THE BINOMIAL THEOREM

Factorials
For any positive integer n,

$$n(n - 1)(n - 2) \cdots (2)(1) = n!$$
$$0! = 1.$$

$$4! = 4 \cdot 3 \cdot 2 \cdot 1 = 24$$

Binomial Coefficient

$$_nC_r = \frac{n!}{r!(n - r)!}$$

$$_5C_3 = \frac{5!}{3!(5 - 3)!} = \frac{5!}{3!2!}$$
$$= \frac{5 \cdot 4 \cdot 3 \cdot 2 \cdot 1}{3 \cdot 2 \cdot 1 \cdot 2 \cdot 1} = 10$$

General Binomial Expansion
For any positive integer n,

$$(x + y)^n = x^n + \frac{n!}{(n - 1)!1!}x^{n-1}y$$
$$+ \frac{n!}{(n - 2)!2!}x^{n-2}y^2$$
$$+ \frac{n!}{(n - 3)!3!}x^{n-3}y^3 + \cdots$$
$$+ \frac{n!}{1!(n - 1)!}xy^{n-1} + y^n.$$

$$(2m + 3)^4 = (2m)^4 + \frac{4!}{3!1!}(2m)^3(3) + \frac{4!}{2!2!}(2m)^2(3)^2$$
$$+ \frac{4!}{1!3!}(2m)(3)^3 + 3^4$$
$$= 2^4m^4 + 4(2)^3m^3(3) + 6(2)^2m^2(9)$$
$$+ 4(2m)(27) + 81$$
$$= 16m^4 + 12(8)m^3 + 54(4)m^2 + 216m + 81$$
$$= 16m^4 + 96m^3 + 216m^2 + 216m + 81$$

(continued)

Concepts	Examples
rth Term of the Binomial Expansion of $(x + y)^n$ $$\frac{n!}{[n - (r - 1)]!(r - 1)!}x^{n-(r-1)}y^{r-1}$$	The eighth term of $(a - 2b)^{10}$ is $$\frac{10!}{3!7!}a^3(-2b)^7 = \frac{10 \cdot 9 \cdot 8}{3 \cdot 2 \cdot 1}a^3(-2)^7b^7 \qquad x = a, y = -2b$$ $$= 120(-128)a^3b^7 \qquad n = 10, r = 8$$ $$= -15{,}360a^3b^7.$$

Answers to Test Your Word Power

1. B; *Example:* The ordered list of numbers 3, 6, 9, 12, 15, . . . is an infinite sequence.

2. A; *Example:* $3 + 6 + 9 + 12 + 15$, written in summation notation as $\sum\limits_{i=1}^{5} 3i$, is a series.

3. D; *Example:* The sequence $-3, 2, 7, 12,$ $17, . . .$ is arithmetic.

4. A; *Example:* The sequence 1, 4, 16, 64, 256, . . . is geometric.

5. C; *Example:* The common difference of the arithmetic sequence in Answer 3 is 5, since $2 - (-3) = 5, 7 - 2 = 5,$ $12 - 7 = 5$, and so on.

6. B; *Example:* The common ratio of the geometric sequence in Answer 4 is 4, since $\frac{4}{1} = \frac{16}{4} = \frac{64}{16} = \frac{256}{64} = 4.$

Chapter **12** REVIEW EXERCISES

[12.1] *Write out the first four terms of each sequence.*

1. $a_n = 2n - 3$

2. $a_n = \dfrac{n - 1}{n}$

3. $a_n = n^2$

4. $a_n = \left(\dfrac{1}{2}\right)^n$

5. $a_n = (n + 1)(n - 1)$

Write each series as a sum of terms.

6. $\sum\limits_{i=1}^{5} i^2 x$

7. $\sum\limits_{i=1}^{6} (i + 1)x^i$

Evaluate each series.

8. $\sum\limits_{i=1}^{4} (i + 2)$

9. $\sum\limits_{i=1}^{6} 2^i$

10. $\sum\limits_{i=4}^{7} \dfrac{i}{i + 1}$

11. Find the arithmetic mean, or average, of the mutual fund's retirement assets for the years 1999 through 2003 shown in the table.

Year	Assets (in billions of dollars)
1999	2535
2000	2478
2001	2342
2002	2078
2003	2662

Source: Investment Company Institute.

[12.2–12.3] *Decide whether each sequence is* arithmetic, geometric, *or* neither. *If the sequence is arithmetic, find the common difference d. If it is geometric, find the common ratio r.*

12. 2, 5, 8, 11, . . .   **13.** −6, −2, 2, 6, 10, . . .   **14.** $\dfrac{2}{3}, -\dfrac{1}{3}, \dfrac{1}{6}, -\dfrac{1}{12}, \ldots$

15. −1, 1, −1, 1, −1, . . .   **16.** 64, 32, 8, $\dfrac{1}{2}$, . . .

17. 64, 32, 16, 8, . . .   **18.** 10, 8, 6, 4, . . .

[12.2] *Find the indicated term of each arithmetic sequence.*

19. $a_1 = -2, d = 5; a_{16}$ **20.** $a_6 = 12, a_8 = 18; a_{25}$

Find the general term of each arithmetic sequence.

21. $a_1 = -4, d = -5$ **22.** 6, 3, 0, −3, . . .

Find the number of terms in each arithmetic sequence.

23. 7, 10, 13, . . . , 49   **24.** 5, 1, −3, . . . , −79

Find S_8 for each arithmetic sequence.

25. $a_1 = -2, d = 6$ **26.** $a_n = -2 + 5n$

[12.3] *Find the general term for each geometric sequence.*

27. −1, −4, . . .   **28.** $\dfrac{2}{3}, \dfrac{2}{15}, \ldots$

Find the indicated term for each geometric sequence.

29. 2, −6, 18, . . . ; a_{11} **30.** $a_3 = 20, a_5 = 80; a_{10}$

Find each sum if it exists.

31. $\displaystyle\sum_{i=1}^{5} \left(\frac{1}{4}\right)^i$ **32.** $\displaystyle\sum_{i=1}^{8} \frac{3}{4}(-1)^i$ **33.** $\displaystyle\sum_{i=1}^{\infty} 4\left(\frac{1}{5}\right)^i$ **34.** $\displaystyle\sum_{i=1}^{\infty} 2(3)^i$

[12.4] *Use the binomial theorem to expand each binomial.*

35. $(2p - q)^5$ **36.** $(x^2 + 3y)^4$ **37.** $\left(\sqrt{m} + \sqrt{n}\right)^4$

38. Write the fourth term of the expansion of $(3a + 2b)^{19}$.

39. Write the 23rd term of the expansion of $(-2k + 3)^{25}$.

MIXED REVIEW EXERCISES

Find the indicated term and S_{10} for each sequence.

40. a_{40}: arithmetic; 1, 7, 13, . . .   **41.** a_{10}: geometric; −3, 6, −12, . . .

42. a_9: geometric; $a_1 = 1, r = -3$ **43.** a_{15}: arithmetic; $a_1 = 4, d = 3$

Find the general term for each arithmetic or geometric sequence.

44. 2, 7, 12, . . .   **45.** 2, 8, 32, . . .

46. 27, 9, 3, . . .   **47.** 12, 9, 6, . . .

Solve each problem.

48. When Mary's sled goes down the hill near her home, she covers 3 ft in the first second; then, for each second after that, she goes 4 ft more than in the preceding second. If the distance she covers going down is 210 ft, how long does it take her to reach the bottom?

49. An ordinary annuity is set up so that $672 is deposited at the end of each quarter for 7 yr. The money earns 6% annual interest compounded quarterly. What is the future value of the annuity?

50. The school population in Middleton has been dropping 3% per yr. The current population is 50,000. If this trend continues, what will the population be in 6 yr?

51. A pump removes $\frac{1}{2}$ of the liquid in a container with each stroke. What fraction of the liquid is left in the container after seven strokes?

52. Consider the repeating decimal number 0.55555 . . . .

   (a) Write it as the sum of the terms of an infinite geometric sequence.
   (b) What is r for this sequence?
   (c) Find this infinite sum if it exists, and write it as a common fraction in lowest terms.

53. Can the sum of the terms of the infinite geometric sequence defined by $a_n = 5(2)^n$ be found? Explain.

54. Can any two terms of a geometric sequence be used to find the common ratio? Explain.

Chapter **12** TEST

View the complete solutions to all Chapter Test exercises on the Pass the Test CD.

Write the first five terms of each sequence described.

1. $a_n = (-1)^n + 1$ **2.** Arithmetic, with $a_1 = 4$ and $d = 2$

3. Geometric, with $a_4 = 6$ and $r = \frac{1}{2}$

Find a_4 for each sequence described.

4. Arithmetic, with $a_1 = 6$ and $d = -2$ **5.** Geometric, with $a_5 = 16$ and $a_7 = 9$

Find S_5 for each sequence described.

6. Arithmetic, with $a_2 = 12$ and $a_3 = 15$ **7.** Geometric, with $a_5 = 4$ and $a_7 = 1$

8. The numbers of commercial banks in the United States for the years 2000 through 2004 are given in the table. What was the average number of banks per year for that period? Round to the nearest unit.

Year	Number
2000	71,911
2001	72,458
2002	73,527
2003	74,638
2004	76,579

Source: U.S. Federal Deposit Insurance Corporation.

9. If $4000 is deposited in an ordinary annuity at the end of each quarter for 7 yr and earns 6% interest compounded quarterly, how much will be in the account at the end of this term?

10. *Concept Check* Under what conditions does an infinite geometric series have a sum?

Find each sum that exists.

11. $\displaystyle\sum_{i=1}^{5} (2i + 8)$

12. $\displaystyle\sum_{i=1}^{6} (3i - 5)$

13. $\displaystyle\sum_{i=1}^{500} i$

14. $\displaystyle\sum_{i=1}^{3} \frac{1}{2}(4^i)$

15. $\displaystyle\sum_{i=1}^{\infty} \left(\frac{1}{4}\right)^i$

16. $\displaystyle\sum_{i-1}^{\infty} 6\left(\frac{3}{2}\right)^i$

Evaluate.

17. $8!$

18. $0!$

19. $\dfrac{6!}{4!2!}$

20. $_{12}C_{10}$

21. Expand $(3k - 5)^4$.

22. Write the fifth term of $\left(2x - \dfrac{y}{3}\right)^{12}$.

Solve each problem.

23. Cheryl bought a new sewing machine for $300. She agreed to pay $20 per month for 15 months, plus interest of 1% each month, on the unpaid balance. Find the total cost of the machine.

24. During the summer months, the population of a certain insect colony triples each week. If there are 20 insects in the colony at the end of the first week in July, how many are present by the end of September? (Assume exactly four weeks in a month.)

Chapters **1–12** **CUMULATIVE REVIEW EXERCISES**

This set of exercises may be considered a final examination for the course.

Simplify each expression.

1. $|-7| + 6 - |-10| - (-8 + 3)$

2. $-15 - |-4| - 10 - |-6|$

3. $4(-6) + (-8)(5) - (-9)$

Let $P = \left\{-\frac{8}{3}, 10, 0, \sqrt{13}, -\sqrt{3}, \frac{45}{15}, \sqrt{-7}, 0.82, -3\right\}$. List the elements of P that are members of each set.

4. Integers

5. Rational numbers

6. Irrational numbers

7. Real numbers

Solve each equation or inequality.

8. $9 - (5 + 3a) + 5a = -4(a - 3) - 7$

9. $7m + 18 \leq 9m - 2$

10. $|4x - 3| = 21$

11. $\dfrac{x + 3}{12} - \dfrac{x - 3}{6} = 0$

12. $2x > 8$ or $-3x > 9$

13. $|2m - 5| \geq 11$

14. Find the slope of the line through $(4, -5)$ and $(-12, -17)$.

15. Find the standard form of the equation of the line through $(-2, 10)$ and parallel to the line with equation $3x + y = 7$.

Graph.

16. $x - 3y = 6$

17. $4x - y < 4$

18. Consider the set of ordered pairs

$$\{(-3, 2), (-2, 6), (0, 4), (1, 2), (2, 6)\}.$$

(a) Is this a function?
(b) What is its domain?
(c) What is its range?

Solve each system of equations by the method indicated.

19. $2x + 5y = -19$
        $-3x + 2y = -19$ (Elimination)

20. $y = 5x + 3$
       $2x + 3y = -8$ (Substitution)

21. $x + 2y + \ \ z = 8$
        $2x - \ \ y + 3z = 15$
        $-x + 3y - 3z = -11$ (Row operations)

22. Nuts worth $3 per lb are to be mixed with 8 lb of nuts worth $4.25 per lb to obtain a mixture that will be sold for $4 per lb. How many pounds of the $3 nuts should be used?

Perform the indicated operations.

23. $(4p + 2)(5p - 3)$

24. $(3k - 7)^2$

25. $(2m^3 - 3m^2 + 8m) - (7m^3 + 5m - 8)$

26. Divide $6t^4 + 5t^3 - 18t^2 + 14t - 1$ by $3t - 2$.

Factor.

27. $7x + x^3$

28. $14y^2 + 13y - 12$

29. $6z^3 + 5z^2 - 4z$

30. $49a^4 - 9b^2$

31. $c^3 + 27d^3$

32. $64r^2 + 48rq + 9q^2$

Solve each equation or inequality.

33. $2x^2 + x = 10$

34. $k^2 - k - 6 \leq 0$

Simplify.

35. $\left(\dfrac{2}{3}\right)^{-2}$

36. $\dfrac{(3p^2)^3(-2p^6)}{4p^3(5p^7)}$

37. What is the domain of the rational function defined by $f(x) = \dfrac{2}{x^2 - 81}$?

Simplify.

38. $\dfrac{x^2 - 16}{x^2 + 2x - 8} \div \dfrac{x - 4}{x + 7}$

39. $\dfrac{5}{p^2 + 3p} - \dfrac{2}{p^2 - 4p}$

Solve.

40. $\dfrac{4}{x - 3} - \dfrac{6}{x + 3} = \dfrac{24}{x^2 - 9}$

41. $6x^2 + 5x = 8$

42. $\sqrt{3x - 2} = x$

Work each problem.

43. Simplify $5\sqrt{72} - 4\sqrt{50}$.

44. Multiply $(8 + 3i)(8 - 3i)$.

45. Find $f^{-1}(x)$ if $f(x) = 9x + 5$.

46. Graph $g(x) = \left(\dfrac{1}{3}\right)^x$.

47. Solve $3^{2x-1} = 81$.

48. Graph $y = \log_{1/3} x$.

49. Solve $\log_8 x + \log_8(x + 2) = 1$.

Graph.

50. $f(x) = 2(x - 2)^2 - 3$ **51.** $\dfrac{x^2}{9} + \dfrac{y^2}{25} = 1$ **52.** $x^2 - y^2 = 9$

53. Solve the system $\begin{array}{l} xy = -5 \\ 2x + y = 3. \end{array}$

54. Find the equation of a circle with center at $(-5, 12)$ and radius 9.

55. Write the first five terms of the sequence defined by $a_n = 5n - 12$.

56. Find each sum.

  (a) The sum of the first six terms of the arithmetic sequence with $a_1 = 8$ and $d = 2$

  (b) The sum of the geometric series $15 - 6 + \frac{12}{5} - \frac{24}{25} + \cdots$

57. Find the sum: $\displaystyle\sum_{i=1}^{4} 3i$.

58. Evaluate $9!$.

59. Use the binomial theorem to expand $(2a - 1)^5$.

60. What is the fourth term in the expansion of $\left(3x^4 - \frac{1}{2}y^2\right)^5$?

Appendix A

An Introduction to Calculators

There is little doubt that the appearance of handheld calculators more than three decades ago and the later development of scientific and graphing calculators have changed the methods of learning and studying mathematics forever. For example, computations with tables of logarithms and slide rules made up an important part of mathematics courses prior to 1970. Today, with the widespread availability of calculators, these topics are studied only for their historical significance.

Calculators come in a large array of different types, sizes, and prices. *For the course for which this textbook is intended, the most appropriate type is the scientific calculator,* which costs $10–$20.

In this introduction, we explain some of the features of scientific and graphing calculators. However, remember that calculators vary among manufacturers and models and that, while the methods explained here apply to many of them, they may not apply to your specific calculator. *This introduction is only a guide and is not intended to take the place of your owner's manual.* Always refer to the manual whenever you need an explanation of how to perform a particular operation.

Scientific Calculators

Scientific calculators are capable of much more than the typical four-function calculator that you might use for balancing your checkbook. Most scientific calculators use *algebraic logic.* (Models sold by Texas Instruments, Sharp, Casio, and Radio Shack, for example, use algebraic logic.) A notable exception is Hewlett-Packard, a company whose calculators use *Reverse Polish Notation* (RPN). In this introduction, we explain the use of calculators with algebraic logic.

Arithmetic Operations To perform an operation of arithmetic, simply enter the first number, press the operation key $\oplus$, $\ominus$, $\otimes$, or $\oslash$, enter the second number, and then press the $\equiv$ key. For example, to add 4 and 3, use the following keystrokes.

Change Sign Key The key marked $+/-$ allows you to change the sign of a display. This is particularly useful when you wish to enter a negative number. For example, to enter -3, use the following keystrokes.

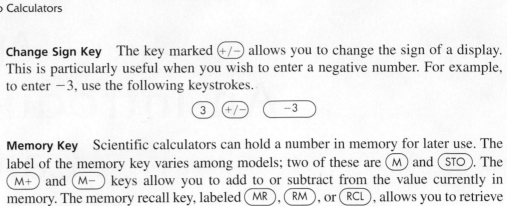

Memory Key Scientific calculators can hold a number in memory for later use. The label of the memory key varies among models; two of these are (M) and (STO). The (M+) and (M−) keys allow you to add to or subtract from the value currently in memory. The memory recall key, labeled (MR), (RM), or (RCL), allows you to retrieve the value stored in memory.

Suppose that you wish to store the number 5 in memory. Enter 5, and then press the key for memory. You can then perform other calculations. When you need to retrieve the 5, press the key for memory recall.

If a calculator has a constant memory feature, the value in memory will be retained even after the power is turned off. Some advanced calculators have more than one memory. Read the owner's manual for your model to see exactly how memory is activated.

Clearing/Clear Entry Keys The key (C) or (CE) allows you to clear the display or clear the last entry entered into the display. In some models, pressing the (C) key once will clear the last entry, while pressing it twice will clear the entire operation in progress.

Second Function Key This key, usually marked (2nd), is used in conjunction with another key to activate a function that is printed *above* an operation key (and not on the key itself). For example, suppose you wish to find the square of a number, and the squaring function (explained in more detail later) is printed above another key. You would need to press (2nd) before the desired squaring function can be activated.

Square Root Key Pressing $(\sqrt{})$ or $(\sqrt{x})$ will give the square root (or an approximation of the square root) of the number in the display. On some scientific calculators, the square root key is pressed *before* entering the number, while other calculators use the opposite order. Experiment with your calculator to see which method it uses. For example, to find the square root of 36, use the following keystrokes.

$(\sqrt{})$ (3) (6) (6)   or   (3) (6) $(\sqrt{})$ (6)

The square root of 2 is an example of an irrational number (**Chapter 8**). The calculator will give an approximation of its value, since the decimal for $\sqrt{2}$ never terminates and never repeats. The number of digits shown will vary among models. To find an approximation for $\sqrt{2}$, use the following keystrokes.

$(\sqrt{})$ (2) (1.4142136)   or   (2) $(\sqrt{})$ (1.4142136)

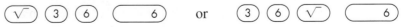

An approximation for $\sqrt{2}$

Squaring Key The (x^2) key allows you to square the entry in the display. For example, to square 35.7, use the following keystrokes.

(3) (5) (.) (7) (x^2) (1274.49)

The squaring key and the square root key are often found together, with one of them being a second function (that is, activated by the second function key previously described).

Reciprocal Key   The key marked $\boxed{1/x}$ is the reciprocal key. (When two numbers have a product of 1, they are called *reciprocals*. See **Chapter 1.**) Suppose that you wish to find the reciprocal of 5. Use the following keystrokes.

$$\boxed{5} \quad \boxed{1/x} \quad \boxed{\qquad 0.2}$$

Inverse Key   Some calculators have an inverse key, marked $\boxed{\text{INV}}$. Inverse operations are operations that "undo" each other. For example, the operations of squaring and taking the square root are inverse operations. The use of the $\boxed{\text{INV}}$ key varies among different models of calculators, so read your owner's manual carefully.

Exponential Key   The key marked $\boxed{x^y}$ or $\boxed{y^x}$ allows you to raise a number to a power. For example, if you wish to raise 4 to the fifth power (that is, find 4^5, as explained in **Chapter 1**), use the following keystrokes.

$$\boxed{4} \quad \boxed{x^y} \quad \boxed{5} \quad \boxed{=} \quad \boxed{\qquad 1024}$$

Root Key   Some calculators have a key specifically marked $\boxed{\sqrt[x]{x}}$ or $\boxed{\sqrt[x]{y}}$; with others, the operation of taking roots is accomplished by using the inverse key in conjunction with the exponential key. Suppose, for example, your calculator is of the latter type and you wish to find the fifth root of 1024. Use the following keystrokes.

$$\boxed{1} \quad \boxed{0} \quad \boxed{2} \quad \boxed{4} \quad \boxed{\text{INV}} \quad \boxed{x^y} \quad \boxed{5} \quad \boxed{=} \quad \boxed{\qquad 4}$$

Notice how this "undoes" the operation explained in the discussion of the exponential key.

Pi Key   The number π is an important number in mathematics. It occurs, for example, in the area and circumference formulas for a circle. One popular model gives the following display when the $\boxed{\pi}$ key is pressed. (Because π is irrational, the display shows only an approximation.)

$$\boxed{3.1415927} \qquad \text{An approximation for } \pi$$

Methods of Display   When decimal approximations are shown on scientific calculators, they are either *truncated* or *rounded*. To see how a particular model is programmed, evaluate 1/18 as an example. If the display shows 0.0555555 (last digit 5), the calculator truncates the display. If the display shows 0.0555556 (last digit 6), the calculator rounds the display.

When very large or very small numbers are obtained as answers, scientific calculators often express these numbers in scientific notation (**Chapter 5**). For example, if you multiply 6,265,804 by 8,980,591, the display might look like this:

$$\boxed{5.6270623 \ 13}$$

The 13 at the far right means that the number on the left is multiplied by 10^{13}. This means that the decimal point must be moved 13 places to the right if the answer is to be expressed in its usual form. Even then, the value obtained will only be an approximation: 56,270,623,000,000.

Graphing Calculators

While you are not expected to have a graphing calculator to study from this book, we include the following as background information and reference should your course or future courses require the use of graphing calculators.

Basic Features   In addition to possessing the typical keys found on scientific calculators, graphing calculators have keys that can be used to create graphs, make tables, analyze data, and change settings. One of the major differences between graphing and scientific calculators is that a graphing calculator has a larger viewing screen with graphing capabilities. The following screens illustrate the graphs of $Y = X$ and $Y = X^2$.

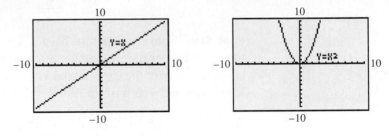

If you look closely at the screens, you will see that the graphs appear to be jagged rather than smooth. The reason for this is that graphing calculators have much lower resolution than computer screens. Because of this, graphs generated by graphing calculators must be interpreted carefully.

Editing Input   The screen of a graphing calculator can display several lines of text at a time. This feature allows you to view both previous and current expressions. If an incorrect expression is entered, an error message is displayed. The erroneous expression can be viewed and corrected by using various editing keys, much like a word-processing program. You do not need to enter the entire expression again. Many graphing calculators can also recall past expressions for editing or updating. The screen on the left shows how two expressions are evaluated. The final line is entered incorrectly, and the resulting error message is shown in the screen on the right.

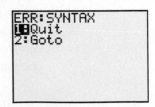

Order of Operations   Arithmetic operations on graphing calculators are usually entered as they are written in mathematical expressions. For example, to evaluate $\sqrt{36}$ you would first press the square root key and then enter 36. See the left screen below. The order of operations on a graphing calculator is also important, and current models assist the user by inserting parentheses when typical errors might occur. The open parenthesis that follows the square root symbol is automatically entered by the calculator so that an expression such as $\sqrt{2 \times 8}$ will not be calculated incorrectly as $\sqrt{2} \times 8$. Compare the two entries and their results in the screen on the right.

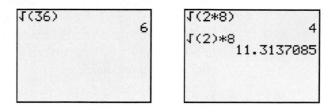

Viewing Windows   The viewing window for a graphing calculator is similar to the viewfinder in a camera. A camera usually cannot take a photograph of an entire view of a scene. The camera must be centered on some object and can capture only a portion of the available scenery. A camera with a zoom lens can photograph different views of the same scene by zooming in and out. Graphing calculators have similar capabilities. The *xy*-coordinate plane is infinite. The calculator screen can show only a finite, rectangular region in the plane, and it must be specified before the graph can be drawn. This is done by setting both minimum and maximum values for the *x*- and *y*-axes. The scale (distance between tick marks) is usually specified as well. Determining an appropriate viewing window for a graph is often a challenge, and many times it will take a few attempts before a satisfactory window is found.

  The screen on the left shows a standard viewing window, and the graph of $Y = 2X + 1$ is shown on the right. Using a different window would give a different view of the line.

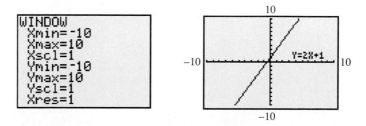

Locating Points on a Graph: Tracing and Tables   Graphing calculators allow you to trace along the graph of an equation and display the coordinates of points on the graph. For example, the screen on the left at the top of the next page indicates that the point (2, 5) lies on the graph of $Y = 2X + 1$. Tables for equations can also be displayed. The screen on the right on the next page shows a partial table for this same equation. Note the middle of the screen, which indicates that when X = 2, Y = 5.

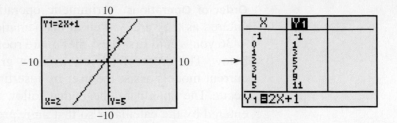

Additional Features There are many features of graphing calculators that go far beyond the scope of this book. These calculators can be programmed, much like computers. Many of them can solve equations at the stroke of a key, analyze statistical data, and perform symbolic algebraic manipulations. Calculators also provide the opportunity to ask "What if . . . ?" more easily. Values in algebraic expressions can be altered and conjectures tested quickly.

Final Comments Despite the power of today's calculators, they cannot replace human thought. ***In the entire problem-solving process, your brain is the most important component.*** Calculators are only tools, and like any tool, they must be used appropriately in order to enhance our ability to understand mathematics. Mathematical insight may often be the quickest and easiest way to solve a problem; a calculator may be neither needed nor appropriate. By applying mathematical concepts, you can make the decision whether to use a calculator.

Appendix B

Determinants and Cramer's Rule

OBJECTIVES

1 Evaluate 2 × 2 determinants.

2 Use expansion by minors to evaluate 3 × 3 determinants.

3 Understand the derivation of Cramer's rule.

4 Apply Cramer's rule to solve linear systems.

Recall from **Section 4.4** that an ordered array of numbers within square brackets is called a *matrix* (plural *matrices*). Matrices are named according to the number of rows and columns they contain. A *square matrix* has the same number of rows and columns.

$$
\text{Rows} \begin{bmatrix} 2 & 3 & 5 \\ 7 & 1 & 2 \end{bmatrix} \begin{array}{l} 2 \times 3 \\ \text{matrix} \end{array} \qquad \begin{bmatrix} -1 & 0 \\ 1 & -2 \end{bmatrix} \begin{array}{l} 2 \times 2 \\ \text{square matrix} \end{array}
$$

Columns

Associated with every *square matrix* is a real number called the **determinant** of the matrix. A determinant is symbolized by the entries of the matrix placed between two vertical lines, such as

$$
\begin{vmatrix} 2 & 3 \\ 7 & 1 \end{vmatrix} \begin{array}{l} 2 \times 2 \\ \text{determinant} \end{array} \qquad \begin{vmatrix} 7 & 4 & 3 \\ 0 & 1 & 5 \\ 6 & 0 & 1 \end{vmatrix}. \begin{array}{l} 3 \times 3 \\ \text{determinant} \end{array}
$$

Like matrices, determinants are named according to the number of rows and columns they contain.

OBJECTIVE 1 Evaluate 2 × 2 determinants. As mentioned above, the value of a determinant is a *real number.* We use the following rule to evaluate a 2×2 determinant.

Value of a 2 × 2 Determinant

$$
\begin{vmatrix} a & b \\ c & d \end{vmatrix} = ad - bc
$$

EXAMPLE 1 Evaluating a 2 × 2 Determinant

Evaluate the determinant.

$$
\begin{vmatrix} -1 & -3 \\ 4 & -2 \end{vmatrix}
$$

Here $a = -1$, $b = -3$, $c = 4$, and $d = -2$, so

$$\begin{vmatrix} -1 & -3 \\ 4 & -2 \end{vmatrix} = -1(-2) - (-3)4 = 2 + 12 = 14.$$

✔ **Now Try Exercise 3.**

A 3×3 determinant is evaluated in a similar way.

Value of a 3 × 3 Determinant

$$\begin{vmatrix} a_1 & b_1 & c_1 \\ a_2 & b_2 & c_2 \\ a_3 & b_3 & c_3 \end{vmatrix} = (a_1 b_2 c_3 + b_1 c_2 a_3 + c_1 a_2 b_3) - (a_3 b_2 c_1 + b_3 c_2 a_1 + c_3 a_2 b_1)$$

To calculate a 3×3 determinant, we rearrange terms using the distributive property.

$$\begin{vmatrix} a_1 & b_1 & c_1 \\ a_2 & b_2 & c_2 \\ a_3 & b_3 & c_3 \end{vmatrix} = a_1(b_2 c_3 - b_3 c_2) - a_2(b_1 c_3 - b_3 c_1) + a_3(b_1 c_2 - b_2 c_1) \qquad (1)$$

Each quantity in parentheses represents a 2×2 determinant that is the part of the 3×3 determinant remaining when the row and column of the multiplier are eliminated, as shown below.

$a_1(b_2 c_3 - b_3 c_2)$ $\begin{vmatrix} a_1 & b_1 & c_1 \\ a_2 & b_2 & c_2 \\ a_3 & b_3 & c_3 \end{vmatrix}$ Eliminate the 1st row and 1st column.

$a_2(b_1 c_3 - b_3 c_1)$ $\begin{vmatrix} a_1 & b_1 & c_1 \\ a_2 & b_2 & c_2 \\ a_3 & b_3 & c_3 \end{vmatrix}$ Eliminate the 2nd row and 1st column.

$a_3(b_1 c_2 - b_2 c_1)$ $\begin{vmatrix} a_1 & b_1 & c_1 \\ a_2 & b_2 & c_2 \\ a_3 & b_3 & c_3 \end{vmatrix}$ Eliminate the 3rd row and 1st column.

These 2×2 determinants are called **minors** of the elements in the 3×3 determinant. In the determinant above, the minors of a_1, a_2, and a_3 are, respectively,

$$\begin{vmatrix} b_2 & c_2 \\ b_3 & c_3 \end{vmatrix}, \qquad \begin{vmatrix} b_1 & c_1 \\ b_3 & c_3 \end{vmatrix}, \qquad \text{and} \qquad \begin{vmatrix} b_1 & c_1 \\ b_2 & c_2 \end{vmatrix}. \qquad \text{Minors}$$

OBJECTIVE 2 **Use expansion by minors to evaluate 3 × 3 determinants.** We evaluate a 3×3 determinant by multiplying each element in the first column by its minor and combining the products as indicated in equation (1). This procedure is called **expansion of the determinant by minors** about the first column.

EXAMPLE 2 Evaluating a 3 × 3 Determinant

Evaluate the determinant using expansion by minors about the first column.

$$\begin{vmatrix} 1 & 3 & -2 \\ -1 & -2 & -3 \\ 1 & 1 & 2 \end{vmatrix}$$

In this determinant, $a_1 = 1$, $a_2 = -1$, and $a_3 = 1$. Multiply each of these numbers by its minor, and combine the three terms using the definition. Notice that the second term in the definition is *subtracted*.

$$\begin{vmatrix} 1 & 3 & -2 \\ -1 & -2 & -3 \\ 1 & 1 & 2 \end{vmatrix} = 1\begin{vmatrix} -2 & -3 \\ 1 & 2 \end{vmatrix} - (-1)\begin{vmatrix} 3 & -2 \\ 1 & 2 \end{vmatrix} + 1\begin{vmatrix} 3 & -2 \\ -2 & -3 \end{vmatrix}$$

Use parentheses and brackets to avoid errors.

$$= 1[-2(2) - (-3)1] + 1[3(2) - (-2)1]$$
$$+ 1[3(-3) - (-2)(-2)]$$
$$= 1(-1) + 1(8) + 1(-13)$$
$$= -1 + 8 - 13$$
$$= -6$$

✔ **Now Try Exercise 9.**

To obtain equation (1), we could have rearranged terms in the definition of the determinant and used the distributive property to factor out the three elements of the second or third column or of any of the three rows. Therefore, expanding by minors about any row or any column results in the same value for a 3 × 3 determinant. To determine the correct signs for the terms of other expansions, the following **array of signs** is helpful.

Array of Signs for a 3 × 3 Determinant

$$\begin{array}{ccc} + & - & + \\ - & + & - \\ + & - & + \end{array}$$

The signs alternate for each row and column beginning with a + in the first row, first column position. For example, if the expansion is to be about the second column, the first term would have a minus sign associated with it, the second term a plus sign, and the third term a minus sign.

EXAMPLE 3 Evaluating a 3 × 3 Determinant

Evaluate the determinant of Example 2 using expansion by minors about the second column.

$$\begin{vmatrix} 1 & 3 & -2 \\ -1 & -2 & -3 \\ 1 & 1 & 2 \end{vmatrix} = -3 \begin{vmatrix} -1 & -3 \\ 1 & 2 \end{vmatrix} + (-2) \begin{vmatrix} 1 & -2 \\ 1 & 2 \end{vmatrix} - 1 \begin{vmatrix} 1 & -2 \\ -1 & -3 \end{vmatrix}$$

$$= -3(1) - 2(4) - 1(-5)$$
$$= -3 - 8 + 5$$
$$= -6 \qquad \text{The result is the same as in Example 2.}$$

✔ **Now Try Exercise 15.**

OBJECTIVE 3 **Understand the derivation of Cramer's rule.** We can use determinants to solve a system of equations of the form

$$a_1 x + b_1 y = c_1 \qquad (1)$$
$$a_2 x + b_2 y = c_2. \qquad (2)$$

The result will be a formula that can be used to solve any system of two equations with two variables. To get this general solution, we eliminate y and solve for x by first multiplying each side of equation (1) by b_2 and each side of equation (2) by $-b_1$. Then we add these results and solve for x.

$$a_1 b_2 x + b_1 b_2 y = c_1 b_2 \qquad \text{Multiply equation (1) by } b_2.$$
$$\underline{-a_2 b_1 x - b_1 b_2 y = -c_2 b_1} \qquad \text{Multiply equation (2) by } -b_1.$$
$$(a_1 b_2 - a_2 b_1) x = c_1 b_2 - c_2 b_1 \qquad \text{Add.}$$
$$x = \frac{c_1 b_2 - c_2 b_1}{a_1 b_2 - a_2 b_1} \qquad \text{(if } a_1 b_2 - a_2 b_1 \neq 0)$$

To solve for y, we multiply each side of equation (1) by $-a_2$ and each side of equation (2) by a_1 and add.

$$-a_1 a_2 x - a_2 b_1 y = -a_2 c_1 \qquad \text{Multiply equation (1) by } -a_2.$$
$$\underline{a_1 a_2 x + a_1 b_2 y = a_1 c_2} \qquad \text{Multiply equation (2) by } a_1.$$
$$(a_1 b_2 - a_2 b_1) y = a_1 c_2 - a_2 c_1 \qquad \text{Add.}$$
$$y = \frac{a_1 c_2 - a_2 c_1}{a_1 b_2 - a_2 b_1} \qquad \text{(if } a_1 b_2 - a_2 b_1 \neq 0)$$

We can write both numerators and the common denominator of these values for x and y as determinants because

$$a_1 c_2 - a_2 c_1 = \begin{vmatrix} a_1 & c_1 \\ a_2 & c_2 \end{vmatrix}, \qquad c_1 b_2 - c_2 b_1 = \begin{vmatrix} c_1 & b_1 \\ c_2 & b_2 \end{vmatrix}, \quad \text{and} \quad a_1 b_2 - a_2 b_1 = \begin{vmatrix} a_1 & b_1 \\ a_2 & b_2 \end{vmatrix}.$$

Using these results, the solutions for x and y become

$$x = \frac{\begin{vmatrix} c_1 & b_1 \\ c_2 & b_2 \end{vmatrix}}{\begin{vmatrix} a_1 & b_1 \\ a_2 & b_2 \end{vmatrix}} \qquad \text{and} \qquad y = \frac{\begin{vmatrix} a_1 & c_1 \\ a_2 & c_2 \end{vmatrix}}{\begin{vmatrix} a_1 & b_1 \\ a_2 & b_2 \end{vmatrix}}, \qquad \begin{vmatrix} a_1 & b_1 \\ a_2 & b_2 \end{vmatrix} \neq 0.$$

For convenience, we denote the three determinants in the solution as

$$\begin{vmatrix} a_1 & b_1 \\ a_2 & b_2 \end{vmatrix} = D, \qquad \begin{vmatrix} c_1 & b_1 \\ c_2 & b_2 \end{vmatrix} = D_x, \qquad \text{and} \qquad \begin{vmatrix} a_1 & c_1 \\ a_2 & c_2 \end{vmatrix} = D_y.$$

Notice that the elements of D are the four coefficients of the variables in the given system; the elements of D_x are obtained by replacing the coefficients of x by the respective constants; the elements of D_y are obtained by replacing the coefficients of y by the respective constants.

These results are summarized as **Cramer's rule.**

Cramer's Rule for 2 × 2 Systems

Given the system $\begin{array}{l} a_1x + b_1y = c_1 \\ a_2x + b_2y = c_2 \end{array}$ with $a_1b_2 - a_2b_1 = D \neq 0,$

then

$$x = \frac{\begin{vmatrix} c_1 & b_1 \\ c_2 & b_2 \end{vmatrix}}{\begin{vmatrix} a_1 & b_1 \\ a_2 & b_2 \end{vmatrix}} = \frac{D_x}{D} \qquad \text{and} \qquad y = \frac{\begin{vmatrix} a_1 & c_1 \\ a_2 & c_2 \end{vmatrix}}{\begin{vmatrix} a_1 & b_1 \\ a_2 & b_2 \end{vmatrix}} = \frac{D_y}{D}.$$

OBJECTIVE 4 Apply Cramer's rule to solve linear systems. To use Cramer's rule to solve a system of equations, we find the three determinants, D, D_x, and D_y, and then write the necessary quotients for x and y.

▶ **CAUTION** As indicated in the box, *Cramer's rule does not apply if $D = a_1b_2 - a_2b_1 = 0$.* When $D = 0$, the system is inconsistent or has dependent equations. For this reason, it is a good idea to evaluate D first.

EXAMPLE 4 Using Cramer's Rule to Solve a 2 × 2 System

Use Cramer's rule to solve the system

$$5x + 7y = -1$$
$$6x + 8y = 1.$$

By Cramer's rule, $x = \frac{D_x}{D}$ and $y = \frac{D_y}{D}$. If $D \neq 0$, then we find D_x and D_y.

$$D = \begin{vmatrix} 5 & 7 \\ 6 & 8 \end{vmatrix} = 5(8) - 7(6) = -2$$

$$D_x = \begin{vmatrix} -1 & 7 \\ 1 & 8 \end{vmatrix} = -1(8) - 7(1) = -15$$

$$D_y = \begin{vmatrix} 5 & -1 \\ 6 & 1 \end{vmatrix} = 5(1) - (-1)6 = 11$$

From Cramer's rule,

$$x = \frac{D_x}{D} = \frac{-15}{-2} = \frac{15}{2} \quad \text{and} \quad y = \frac{D_y}{D} = \frac{11}{-2} = -\frac{11}{2}.$$

The solution set is $\left\{\left(\frac{15}{2}, -\frac{11}{2}\right)\right\}$, as can be verified by checking in the given system.

✔ **Now Try Exercise 23.**

We can extend Cramer's rule to systems of three equations with three variables.

Cramer's Rule for 3 × 3 Systems

Given the system

$$a_1x + b_1y + c_1z = d_1$$
$$a_2x + b_2y + c_2z = d_2$$
$$a_3x + b_3y + c_3z = d_3$$

with

$$D_x = \begin{vmatrix} d_1 & b_1 & c_1 \\ d_2 & b_2 & c_2 \\ d_3 & b_3 & c_3 \end{vmatrix}, \quad D_y = \begin{vmatrix} a_1 & d_1 & c_1 \\ a_2 & d_2 & c_2 \\ a_3 & d_3 & c_3 \end{vmatrix},$$

$$D_z = \begin{vmatrix} a_1 & b_1 & d_1 \\ a_2 & b_2 & d_2 \\ a_3 & b_3 & d_3 \end{vmatrix}, \quad D = \begin{vmatrix} a_1 & b_1 & c_1 \\ a_2 & b_2 & c_2 \\ a_3 & b_3 & c_3 \end{vmatrix} \neq 0,$$

then

$$x = \frac{D_x}{D}, \quad y = \frac{D_y}{D}, \quad \text{and} \quad z = \frac{D_z}{D}.$$

EXAMPLE 5 Using Cramer's Rule to Solve a 3 × 3 System

Use Cramer's rule to solve the system

$$x + y - z + 2 = 0$$
$$2x - y + z + 5 = 0$$
$$x - 2y + 3z - 4 = 0.$$

To use Cramer's rule, we first rewrite the system in the form

$$x + y - z = -2$$
$$2x - y + z = -5$$
$$x - 2y + 3z = 4.$$

We expand by minors about row 1 to find D.

$$D = \begin{vmatrix} 1 & 1 & -1 \\ 2 & -1 & 1 \\ 1 & -2 & 3 \end{vmatrix}$$

$$= 1 \begin{vmatrix} -1 & 1 \\ -2 & 3 \end{vmatrix} - 1 \begin{vmatrix} 2 & 1 \\ 1 & 3 \end{vmatrix} + (-1) \begin{vmatrix} 2 & -1 \\ 1 & -2 \end{vmatrix}$$

$$= 1(-1) - 1(5) - 1(-3)$$

$$= -3$$

Expanding D_x by minors about row 1 gives

$$D_x = \begin{vmatrix} -2 & 1 & -1 \\ -5 & -1 & 1 \\ 4 & -2 & 3 \end{vmatrix}$$

$$= -2 \begin{vmatrix} -1 & 1 \\ -2 & 3 \end{vmatrix} - 1 \begin{vmatrix} -5 & 1 \\ 4 & 3 \end{vmatrix} + (-1) \begin{vmatrix} -5 & -1 \\ 4 & -2 \end{vmatrix}$$

$$= -2(-1) - 1(-19) - 1(14)$$

$$= 7.$$

Verify that $D_y = -22$ and $D_z = -21$. Thus,

$$x = \frac{D_x}{D} = \frac{7}{-3} = -\frac{7}{3}, \qquad y = \frac{D_y}{D} = \frac{-22}{-3} = \frac{22}{3}, \qquad z = \frac{D_z}{D} = \frac{-21}{-3} = 7.$$

Check that the solution set is $\left\{ \left(-\frac{7}{3}, \frac{22}{3}, 7 \right) \right\}$.

✔ **Now Try Exercise 29.**

EXAMPLE 6 **Determining When Cramer's Rule Does Not Apply**

Use Cramer's rule, if possible, to solve the system

$$2x - 3y + 4z = 8$$
$$6x - 9y + 12z = 24$$
$$x + 2y - 3z = 5.$$

First, find D.

$$D = \begin{vmatrix} 2 & -3 & 4 \\ 6 & -9 & 12 \\ 1 & 2 & -3 \end{vmatrix}$$

$$= 2 \begin{vmatrix} -9 & 12 \\ 2 & -3 \end{vmatrix} - 6 \begin{vmatrix} -3 & 4 \\ 2 & -3 \end{vmatrix} + 1 \begin{vmatrix} -3 & 4 \\ -9 & 12 \end{vmatrix}$$

$$= 2(3) - 6(1) + 1(0)$$

$$= 0$$

Since $D = 0$ here, Cramer's rule does not apply and we must use another method to solve the system. Multiplying each side of the first equation by 3 shows that the first two equations have the same solution set, so this system has dependent equations and an infinite solution set.

✔ **Now Try Exercise 31.**

Cramer's rule can be extended to 4×4 or larger systems. See a standard college algebra text for details.

APPENDIX B EXERCISES

Now Try Exercise

1. *Concept Check* Decide whether each statement is *true* or *false*. If a statement is false, explain why.

 (a) A matrix is an array of numbers, while a determinant is a single number.

 (b) A square matrix has the same number of rows as columns.

 (c) The determinant $\begin{vmatrix} a & b \\ c & d \end{vmatrix}$ is equal to $ad + bc$.

 (d) The value of $\begin{vmatrix} 0 & 0 \\ x & y \end{vmatrix}$ is 0 for any replacements for x and y.

2. *Concept Check* Which one of the following is the expression for the determinant

$$\begin{vmatrix} -2 & -3 \\ 4 & -6 \end{vmatrix}?$$

 A. $-2(-6) + (-3)4$ **B.** $-2(-6) - 3(4)$
 C. $-3(4) - (-2)(-6)$ **D.** $-2(-6) - (-3)4$

Evaluate each determinant. See Example 1.

3. $\begin{vmatrix} -2 & 5 \\ -1 & 4 \end{vmatrix}$

4. $\begin{vmatrix} 3 & -6 \\ 2 & -2 \end{vmatrix}$

5. $\begin{vmatrix} 1 & -2 \\ 7 & 0 \end{vmatrix}$

6. $\begin{vmatrix} -5 & -1 \\ 1 & 0 \end{vmatrix}$

7. $\begin{vmatrix} 0 & 4 \\ 0 & 4 \end{vmatrix}$

8. $\begin{vmatrix} 8 & -3 \\ 0 & 0 \end{vmatrix}$

Evaluate each determinant by expansion by minors about the first column. See Example 2.

9. $\begin{vmatrix} -1 & 2 & 4 \\ -3 & -2 & -3 \\ 2 & -1 & 5 \end{vmatrix}$

10. $\begin{vmatrix} 2 & -3 & -5 \\ 1 & 2 & 2 \\ 5 & 3 & -1 \end{vmatrix}$

11. $\begin{vmatrix} 1 & 0 & -2 \\ 0 & 2 & 3 \\ 1 & 0 & 5 \end{vmatrix}$

12. $\begin{vmatrix} 2 & -1 & 0 \\ 0 & -1 & 1 \\ 1 & 2 & 0 \end{vmatrix}$

13. Explain in your own words how to evaluate a 2×2 determinant. Illustrate with an example.

Evaluate each determinant by expansion by minors about any row or column. (Hint: The work is easier if you choose a row or a column with 0s.) See Example 3.

14. $\begin{vmatrix} 4 & 4 & 2 \\ 1 & -1 & -2 \\ 1 & 0 & 2 \end{vmatrix}$

15. $\begin{vmatrix} 3 & -1 & 2 \\ 1 & 5 & -2 \\ 0 & 2 & 0 \end{vmatrix}$

16. $\begin{vmatrix} 3 & 5 & -2 \\ 1 & -4 & 1 \\ 3 & 1 & -2 \end{vmatrix}$

17. $\begin{vmatrix} 0 & 0 & 3 \\ 4 & 0 & -2 \\ 2 & -1 & 3 \end{vmatrix}$

18. $\begin{vmatrix} 3 & 0 & -2 \\ 1 & -4 & 1 \\ 3 & 1 & -2 \end{vmatrix}$

19. $\begin{vmatrix} 1 & 1 & 2 \\ 5 & 5 & 7 \\ 3 & 3 & 1 \end{vmatrix}$

20. *Concept Check*   Consider the system

$$4x + 3y - 2z = 1$$
$$7x - 4y + 3z = 2$$
$$-2x + y - 8z = 0.$$

Match each determinant in parts (a)–(d) with its correct representation from choices A–D.

(a) D **(b)** D_x **(c)** D_y **(d)** D_z

A. $\begin{vmatrix} 1 & 3 & -2 \\ 2 & -4 & 3 \\ 0 & 1 & -8 \end{vmatrix}$ **B.** $\begin{vmatrix} 4 & 3 & 1 \\ 7 & -4 & 2 \\ -2 & 1 & 0 \end{vmatrix}$ **C.** $\begin{vmatrix} 4 & 1 & -2 \\ 7 & 2 & 3 \\ -2 & 0 & -8 \end{vmatrix}$ **D.** $\begin{vmatrix} 4 & 3 & -2 \\ 7 & -4 & 3 \\ -2 & 1 & -8 \end{vmatrix}$

21. For the system

$$x + 3y - 6z = 7$$
$$2x - y + z = 1$$
$$x + 2y + 2z = -1,$$

$D = -43$, $D_x = -43$, $D_y = 0$, and $D_z = 43$. What is the solution set of the system?

Use Cramer's rule to solve each linear system in two variables. See Example 4.

22. $3x + 5y = -5$
$-2x + 3y = 16$

23. $5x + 2y = -3$
$4x - 3y = -30$

24. $8x + 3y = 1$
$6x - 5y = 2$

25. $3x - y = 9$
$2x + 5y = 8$

26. $2x + 3y = 4$
$5x + 6y = 7$

27. $4x + 5y = 6$
$7x + 8y = 9$

Use Cramer's rule where applicable to solve each linear system in three variables. See Examples 5 and 6.

28. $2x + 3y + 2z = 15$
$x - y + 2z = 5$
$x + 2y - 6z = -26$

29. $x - y + 6z = 19$
$3x + 3y - z = 1$
$x + 9y + 2z = -19$

30. $2x - 3y + 4z = 8$
$6x - 9y + 12z = 24$
$-4x + 6y - 8z = -16$

31. $7x + y - z = 4$
$2x - 3y + z = 2$
$-6x + 9y - 3z = -6$

32. $3x + 5z = 0$
$2x + 3y = 1$
$-y + 2z = -11$

33. $-x + 2y = 4$
$3x + y = -5$
$2x + z = -1$

34. $x - 3y = 13$
$2y + z = 5$
$-x + z = -7$

35. $-5x - y = -10$
$3x + 2y + z = -3$
$-y - 2z = -13$

36. Under what conditions can a system *not* be solved using Cramer's rule?

Solve each equation by finding an expression for the determinant on the left, and then solving using the methods of **Chapter 2.**

37. $\begin{vmatrix} 4 & x \\ 2 & 3 \end{vmatrix} = 8$

38. $\begin{vmatrix} -2 & 10 \\ x & 6 \end{vmatrix} = 0$

39. $\begin{vmatrix} x & 4 \\ x & -3 \end{vmatrix} = 0$

40. $\begin{vmatrix} 5 & 3 \\ x & x \end{vmatrix} = 20$

Appendix C
Synthetic Division

OBJECTIVES

1 Use synthetic division to divide by a polynomial of the form $x - k$.

2 Use the remainder theorem to evaluate a polynomial.

3 Decide whether a given number is a solution of an equation.

OBJECTIVE 1 Use synthetic division to divide by a polynomial of the form $x - k$.
Often, when one polynomial is divided by a second, the second polynomial has the form $x - k$, where the coefficient of the x term is 1. To see a shortcut for these divisions, look first below left, where the division of $3x^3 - 2x + 5$ by $x - 3$ is shown. Notice that we inserted 0 for the missing x^2-term.

$$
\begin{array}{r}
3x^2 + 9x + 25 \\
x - 3\overline{)3x^3 + 0x^2 - 2x + 5} \\
\underline{3x^3 - 9x^2} \\
9x^2 - 2x \\
\underline{9x^2 - 27x} \\
25x + 5 \\
\underline{25x - 75} \\
80
\end{array}
$$

$$
\begin{array}{r}
3 \quad 9 \quad 25 \\
1 - 3\overline{)3 \quad\; 0 \quad -2 \quad\; 5} \\
3 \quad -9 \\
\overline{9 \quad -2} \\
9 \quad -27 \\
\overline{25 \quad\; 5} \\
25 \quad -75 \\
\overline{80}
\end{array}
$$

On the right, exactly the same division is shown written without the variables. This is why it is *essential* to use 0 as a placeholder in synthetic division. All the numbers in color on the right are repetitions of the numbers directly above them, so we omit them, as shown on the left below.

$$
\begin{array}{r}
3 \quad 9 \quad 25 \\
1 - 3\overline{)3 \quad\; 0 \quad -2 \quad\; 5} \\
\underline{-9} \\
9 \quad -2 \\
\underline{-27} \\
25 \quad 5 \\
\underline{-75} \\
80
\end{array}
$$

$$
\begin{array}{r}
3 \quad 9 \quad 25 \\
1 - 3\overline{)3 \quad\; 0 \quad -2 \quad\; 5} \\
\underline{-9} \\
9 \\
\underline{-27} \\
25 \\
\underline{-75} \\
80
\end{array}
$$

The numbers in color on the left are again repetitions of the numbers directly above them; they too are omitted, as shown on the right above.

Now we can condense the problem. If we bring the 3 in the dividend down to the beginning of the bottom row, the top row can be omitted, since it duplicates the bottom row.

$$
\begin{array}{r}
1 - 3\overline{)3 \quad\; 0 \quad -2 \quad\; 5} \\
\underline{-9 \quad -27 \quad -75} \\
3 \quad 9 \quad\; 25 \quad\; 80
\end{array}
$$

Finally, we omit the 1 at the upper left. Also, to simplify the arithmetic, we replace subtraction in the second row by addition. To compensate for this, we change the -3 at the upper left to its additive inverse, 3.

$$
\text{Additive inverse} \longrightarrow 3\overline{)\begin{array}{cccc} 3 & 0 & -2 & 5 \\ & 9 & 27 & 75 \\ \hline 3 & 9 & 25 & 80 \end{array}} \leftarrow \text{Signs changed} \\ \leftarrow \text{Remainder}
$$

The quotient is read from the bottom row.

$$3x^2 + 9x + 25 + \dfrac{80}{x-3}$$

The first three numbers in the bottom row are the coefficients of the quotient polynomial with degree 1 less than the degree of the dividend. The last number gives the remainder.

Synthetic Division

This shortcut procedure is called **synthetic division.** It is used only when dividing a polynomial by a binomial of the form $x - k$.

EXAMPLE 1 Using Synthetic Division

Use synthetic division to divide $5x^2 + 16x + 15$ by $x + 2$.

We change $x + 2$ into the form $x - k$ by writing it as

$$x + 2 = x - (-2),$$

where $k = -2$. Now write the coefficients of $5x^2 + 16x + 15$, placing -2 to the left.

$$x + 2 \text{ leads to } -2. \longrightarrow -2\overline{)\begin{array}{ccc} 5 & 16 & 15 \end{array}} \leftarrow \text{Coefficients}$$

Bring down the 5, and multiply: $-2 \cdot 5 = -10$.

$$
-2\overline{)\begin{array}{ccc} 5 & 16 & 15 \\ & -10 & \\ \hline 5 & & \end{array}}
$$

Add 16 and -10, getting 6, and multiply 6 and -2 to get -12.

$$
-2\overline{)\begin{array}{ccc} 5 & 16 & 15 \\ & -10 & -12 \\ \hline 5 & 6 & \end{array}}
$$

Add 15 and -12, getting 3.

$$
-2\overline{)\begin{array}{ccc} 5 & 16 & 15 \\ & -10 & -12 \\ \hline 5 & 6 & 3 \end{array}} \leftarrow \text{Remainder}
$$

The result is read from the bottom row.

$$\dfrac{5x^2 + 16x + 15}{x + 2} = 5x + 6 + \dfrac{3}{x + 2}$$

✔ **Now Try Exercise 7.**

EXAMPLE 2   Using Synthetic Division with a Missing Term

Use synthetic division to find $(-4x^5 + x^4 + 6x^3 + 2x^2 + 50) \div (x - 2)$.

Use the steps given above, first inserting a 0 for the missing x-term.

$$
\begin{array}{r}
2)\overline{-4 \quad\ 1 \quad\ \ 6 \quad\ \ \ 2 \quad\ \ \ 0 \quad\ \ 50} \\
-8 \ \ -14 \ \ -16 \ \ -28 \ \ -56 \\
\hline
-4 \ \ -7 \ \ \ -8 \ \ -14 \ \ -28 \ \ \ -6
\end{array}
$$

Read the result from the bottom row.

$$
\frac{-4x^5 + x^4 + 6x^3 + 2x^2 + 50}{x - 2} = -4x^4 - 7x^3 - 8x^2 - 14x - 28 + \frac{-6}{x - 2}
$$

✔ **Now Try Exercise 13.**

OBJECTIVE 2   Use the remainder theorem to evaluate a polynomial.   We can use synthetic division to evaluate polynomials. For example, in the synthetic division of Example 2, where the polynomial was divided by $x - 2$, the remainder was -6.

Replacing x in the polynomial with 2 gives

$$
\begin{aligned}
-4x^5 + x^4 + 6x^3 + 2x^2 + 50 &= -4 \cdot 2^5 + 2^4 + 6 \cdot 2^3 + 2 \cdot 2^2 + 50 \\
&= -4 \cdot 32 + 16 + 6 \cdot 8 + 2 \cdot 4 + 50 \\
&= -128 + 16 + 48 + 8 + 50 \\
&= -6,
\end{aligned}
$$

the same number as the remainder; that is, dividing by $x - 2$ produced a remainder equal to the result when x is replaced with 2. This always happens, as the following **remainder theorem** states.

Remainder Theorem

If the polynomial $P(x)$ is divided by $x - k$, then the remainder is equal to $P(k)$.

This result is proved in more advanced courses.

EXAMPLE 3   Using the Remainder Theorem

Let $P(x) = 2x^3 - 5x^2 - 3x + 11$. Find $P(-2)$.

Use the remainder theorem; divide $P(x)$ by $x - (-2)$.

$$
\text{Value of } k \rightarrow \quad -2)\overline{\begin{array}{rrrr} 2 & -5 & -3 & 11 \end{array}}
$$
$$
\begin{array}{rrrr} & -4 & 18 & -30 \\ \hline 2 & -9 & 15 & -19 \end{array} \leftarrow \text{Remainder}
$$

Thus, $P(-2) = -19$.

✔ **Now Try Exercise 21.**

OBJECTIVE 3   Decide whether a given number is a solution of an equation.   We can also use the remainder theorem to show that a given number is a solution of an equation.

EXAMPLE 4 Using the Remainder Theorem

Show that -5 is a solution of the equation

$$2x^4 + 12x^3 + 6x^2 - 5x + 75 = 0.$$

One way to show that -5 is a solution is by substituting -5 for x in the equation. However, an easier way is to use synthetic division and the remainder theorem.

$$
\begin{array}{r}
\text{Proposed solution} \rightarrow \ -5 \overline{)\ 2 \quad\ \ 12 \quad\ \ \ 6 \quad\ -5 \quad\ \ \ 75\ } \\
\underline{\qquad\quad -10 \ \ -10 \quad\ 20 \ \ -75} \\
2 \quad\ \ \ 2 \quad -4 \quad\ \ 15 \qquad 0 \ \leftarrow \text{Remainder}
\end{array}
$$

Since the remainder is 0, the polynomial has value 0 when $k = -5$, and so -5 is a solution of the given equation.

✔ **Now Try Exercises 29 and 31.**

The synthetic division in Example 4 shows that $x - (-5)$ divides the polynomial with 0 remainder. Thus $x - (-5) = x + 5$ is a *factor* of the polynomial and

$$2x^4 + 12x^3 + 6x^2 - 5x + 75 = (x + 5)(2x^3 + 2x^2 - 4x + 15).$$

The second factor is the quotient polynomial found in the last row of the synthetic division.

CONNECTIONS

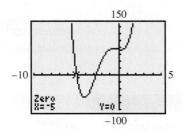

The procedure in Example 4 is exactly how we use a graphing calculator to find real solutions of an equation by determining the x-intercepts of the graph. The screen shows the graph of $P(x) = 2x^4 + 12x^3 + 6x^2 - 5x + 75$ and shows that one value of x that makes $P(x) = 0$ is -5. This agrees with our result in Example 4.

For Discussion or Writing

The graph of $P(x) = x^3 + 5x^2 + 3x - 4$ has one x-intercept that is an integer. Find that integer, and then use synthetic division to show that it makes $P(x) = 0$.

APPENDIX C EXERCISES

Now Try Exercise

📝 **1.** What is the purpose of synthetic division?

2. *Concept Check* What type of polynomial divisors may be used with synthetic division?

Use synthetic division to find each quotient. See Examples 1 and 2.

3. $\dfrac{x^2 - 6x + 5}{x - 1}$

4. $\dfrac{x^2 - 4x - 21}{x + 3}$

5. $\dfrac{4m^2 + 19m - 5}{m + 5}$

6. $\dfrac{3k^2 - 5k - 12}{k - 3}$

7. $\dfrac{2a^2 + 8a + 13}{a + 2}$

8. $\dfrac{4y^2 - 5y - 20}{y - 4}$

9. $(p^2 - 3p + 5) \div (p + 1)$

10. $(z^2 + 4z - 6) \div (z - 5)$

11. $\dfrac{4a^3 - 3a^2 + 2a - 3}{a - 1}$

12. $\dfrac{5p^3 - 6p^2 + 3p + 14}{p + 1}$

13. $(x^5 - 2x^3 + 3x^2 - 4x - 2) \div (x - 2)$

14. $(2y^5 - 5y^4 - 3y^2 - 6y - 23) \div (y - 3)$

15. $(-4r^6 - 3r^5 - 3r^4 + 5r^3 - 6r^2 + 3r + 3) \div (r - 1)$

16. $(2t^6 - 3t^5 + 2t^4 - 5t^3 + 6t^2 - 3t - 2) \div (t - 2)$

17. $(-3y^5 + 2y^4 - 5y^3 - 6y^2 - 1) \div (y + 2)$

18. $(m^6 + 2m^4 - 5m + 11) \div (m - 2)$

19. $\dfrac{y^3 + 1}{y - 1}$

20. $\dfrac{z^4 + 81}{z - 3}$

Use the remainder theorem to find $P(k)$. See Example 3.

21. $P(x) = 2x^3 - 4x^2 + 5x - 3; k = 2$

22. $P(y) = y^3 + 3y^2 - y + 5; k = -1$

23. $P(r) = -r^3 - 5r^2 - 4r - 2; k = -4$

24. $P(z) = -z^3 + 5z^2 - 3z + 4; k = 3$

25. $P(y) = 2y^3 - 4y^2 + 5y - 33; k = 3$

26. $P(x) = x^3 - 3x^2 + 4x - 4; k = 2$

27. Explain why a 0 remainder in synthetic division of $P(x)$ by $x - k$ indicates that k is a solution of the equation $P(x) = 0$.

28. Explain why it is important to insert 0s as placeholders for missing terms before performing synthetic division.

Use synthetic division to decide whether the given number is a solution of the equation. See Example 4.

29. $x^3 - 2x^2 - 3x + 10 = 0; x = -2$

30. $x^3 - 3x^2 - x + 10 = 0; x = -2$

31. $m^4 + 2m^3 - 3m^2 + 8m - 8 = 0; m = -2$

32. $r^4 - r^3 - 6r^2 + 5r + 10 = 0; r = -2$

33. $3a^3 + 2a^2 - 2a + 11 = 0; a = -2$

34. $3z^3 + 10z^2 + 3z - 9 = 0; z = -2$

35. $2x^3 - x^2 - 13x + 24 = 0; x = -3$

36. $5p^3 + 22p^2 + p - 28 = 0; p = -4$

RELATING CONCEPTS (EXERCISES 37–41)

FOR INDIVIDUAL OR GROUP WORK

*In **Section 5.5** we saw the close connection between polynomial division and writing a quotient of polynomials in lowest terms after factoring the numerator. Now we can show a connection between dividing one polynomial by another and factoring the first polynomial. Let $P(x) = 2x^2 + 5x - 12$. **Work Exercises 37–41 in order.***

37. Factor $P(x)$.

38. Solve $P(x) = 0$.

39. Find $P(-4)$ and $P\left(\frac{3}{2}\right)$.

40. Complete the following sentence. If $P(a) = 0$, then $x -$ _____ is a factor of $P(x)$.

41. Use the conclusion in Exercise 40 to decide whether $x - 3$ is a factor of $Q(x) = 3x^3 - 4x^2 - 17x + 6$. Factor $Q(x)$ completely.

TECHNOLOGY INSIGHTS (EXERCISES 42–45)

Use the graph to determine a solution of each equation.

42. $2x^3 + 12x^2 + 24x + 16 = 0$

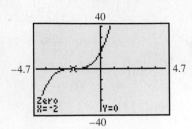

43. $x^3 - x^2 - 21x + 45 = 0$

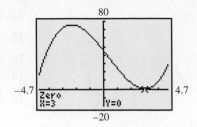

44. $x^3 + 3x^2 - 10x - 24 = 0$

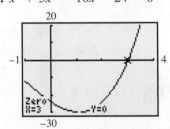

45. $x^3 + 3x^2 - 13x - 15 = 0$

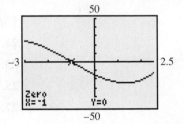

Answers to Selected Exercises

In this section, we provide the answers that we think most students will obtain when they work the exercises using the methods explained in the text. If your answer does not look exactly like the one given here, it is not necessarily wrong. In many cases, there are equivalent forms of the answer that are correct. For example, if the answer section shows $\frac{3}{4}$ and your answer is 0.75, you have obtained the right answer, but written it in a different (yet equivalent) form. Unless the directions specify otherwise, 0.75 is just as valid an answer as $\frac{3}{4}$.

In general, if your answer does not agree with the one given in the text, see whether it can be transformed into the other form. If it can, then it is the correct answer. If you still have doubts, talk with your instructor. You might also want to obtain a copy of the *Student's Solutions Manual* that goes with this book. Your college bookstore either has this manual or can order it for you.

CHAPTER 1   Review of the Real Number System

Section 1.1   (pages 12–15)

1. $\{1, 2, 3, 4, 5\}$ **3.** $\{5, 6, 7, 8, \ldots\}$ **5.** $\{\ldots, -1, 0, 1, 2, 3, 4\}$ **7.** $\{10, 12, 14, 16, \ldots\}$ **9.** $\emptyset$ **11.** $\{-4, 4\}$

In Exercises 13 and 15, we give one possible answer.   **13.** $\{x \mid x \text{ is an even natural number less than or equal to } 8\}$

15. $\{x \mid x \text{ is a multiple of 4 greater than } 0\}$ **17.** yes   **19.**   **21.**

23. (a) $8, 13, \dfrac{75}{5}$ (or 15)   **(b)** $0, 8, 13, \dfrac{75}{5}$ **(c)** $-9, 0, 8, 13, \dfrac{75}{5}$ **(d)** $-9, -0.7, 0, \dfrac{6}{7}, 4.\overline{6}, 8, \dfrac{21}{2}, 13, \dfrac{75}{5}$ **(e)** $-\sqrt{6}, \sqrt{7}$

(f) All are real numbers.   **25.** False; some are whole numbers, but negative integers are not.   **27.** False; no irrational number is an integer.   **29.** true   **31.** true   **33.** true   **35. (a)** A   **(b)** A   **(c)** B   **(d)** B   **37. (a)** -6 **(b)** 6   **39. (a)** 12

(b) 12   **41. (a)** $-\dfrac{6}{5}$ **(b)** $\dfrac{6}{5}$ **43.** 8   **45.** $\dfrac{3}{2}$ **47.** -5 **49.** -2 **51.** -4.5 **53.** 5   **55.** 6   **57.** 0

59. (a) Las Vegas; the population increased by 11.5%. **(b)** Chicago; the population decreased by 1.2%.   **61.** Pacific Ocean, Indian Ocean, Caribbean Sea, South China Sea, Gulf of California   **63.** true   **65.** true   **67.** false   **69.** true   **71.** true

73. $2 < 6$ **75.** $4 > -9$ **77.** $-10 < -5$ **79.** $x > 0$ **81.** $7 > y$ **83.** $5 \geq 5$ **85.** $3t - 4 \leq 10$ **87.** $5x + 3 \neq 0$

89. $-3 < t < 5$ **91.** $-3 \leq 3x < 4$ **93.** $-6 < 10$; true   **95.** $10 \geq 10$; true   **97.** $-3 \geq -3$; true

99. $-8 > -6$; false   **101.** $(-1, \infty)$ **103.** $(-\infty, 6]$

105. $(0, 3.5)$ **107.** $[2, 7]$ **109.** $(-4, 3]$

111. $(0, 3]$ **113.** Iowa (IA), Ohio (OH), Pennsylvania (PA), Indiana (IN)   **115.** $x < y$

117. The *natural numbers* are the numbers with which we count. Some examples are 1, 2, and 3. The *whole numbers* are formed by including 0 with the natural numbers, such as 0, 2, and 17. The *integers* are formed by including the negatives of the natural numbers with the whole numbers, such as -1, -2, and -3. The *rational numbers,* such as $\frac{1}{2}$, 0.75, and $-\frac{3}{4}$, are formed by quotients of integers. The *irrational numbers* include positive or negative numbers that are not rational, such as π, $\sqrt{2}$, and $-\sqrt{5}$. The *real numbers* include all positive numbers, negative numbers, and 0, such as $-\pi$, 0, and $\sqrt{2}$.

Section 1.2   (pages 22–26)
1. the numbers are additive inverses; $4 + (-4) = 0$ **3.** negative; $-7 + (-21) = -28$ **5.** the positive number has larger absolute value; $15 + (-2) = 13$ **7.** the number with smaller absolute value is subtracted from the one with larger absolute value; $-15 - (-3) = -12$ **9.** negative; $-5(15) = -75$ **11.** -19 **13.** 9   **15.** $-\dfrac{19}{12}$ **17.** -1.85 **19.** -11

21. 21   **23.** -13 **25.** -10.18 **27.** $\dfrac{67}{30}$ **29.** 14   **31.** -5 **33.** -6 **35.** -11 **37.** 16   **39.** -4 **41.** 8

43. 3.018   **45.** $-\dfrac{7}{4}$ **47.** $-\dfrac{7}{8}$ **49.** 1   **51.** 6   **53.** $\dfrac{13}{2}$, or $6\dfrac{1}{2}$ **55.** It is true for multiplication (and division). It is false for addition and for subtraction when the number to be subtracted has the smaller absolute value. A more precise statement is, "The product or quotient of two negative numbers is positive."   **57.** -35 **59.** 40   **61.** 2   **63.** -12 **65.** $\dfrac{6}{5}$

67. 1   **69.** 5.88   **71.** -10.676 **73.** -7 **75.** 6   **77.** -4 **79.** 0   **81.** undefined   **83.** $\dfrac{25}{102}$ **85.** $-\dfrac{9}{13}$

87. -2.1 **89.** 10,000   **91.** $\dfrac{17}{18}$ **93.** $\dfrac{17}{36}$ **95.** $-\dfrac{19}{24}$ **97.** $-\dfrac{22}{45}$ **99.** $-\dfrac{2}{15}$ **101.** $\dfrac{3}{5}$ **103.** $-\dfrac{35}{27}$ **105.** $-\dfrac{4}{9}$

107. -12.351 **109.** -15.876 **111.** -4.14 **113.** 4800   **115.** 51.495   **117.** $112°F$ **119.** $30.13
121. (a) $466.02   **(b)** $190.68   **123. (a)** $-$475 thousand   **(b)** $262 thousand   **(c)** $-$83 thousand
125. (a) 2000: $129 billion; 2010: $206 billion; 2020: $74 billion; 2030: $-$501 billion   **(b)** The cost of Social Security will exceed revenue in 2030 by $501 billion.

Section 1.3   (pages 32–35)
1. false; $-4^6 = -(4^6)$ **3.** true   **5.** true   **7.** true   **9.** false; the base is 3.   **11. (a)** 64   **(b)** -64 **(c)** 64   **(d)** -64
13. 10^4 **15.** $\left(\dfrac{3}{4}\right)^5$ **17.** $(-9)^3$ **19.** z^7 **21.** 16   **23.** 0.021952   **25.** $\dfrac{1}{125}$ **27.** $\dfrac{256}{625}$ **29.** -125 **31.** 256

33. -729 **35.** -4096 **37.** 9   **39.** 13   **41.** -20 **43.** $\dfrac{10}{11}$ **45.** -0.7 **47.** not a real number   **49. (a)** B   **(b)** C
(c) A   **51.** not a real number   **53.** 24   **55.** 15   **57.** 55   **59.** -91 **61.** -8 **63.** -48 **65.** -2 **67.** -79

69. -10 **71.** 2   **73.** -2 **75.** undefined   **77.** -7 **79.** -1 **81.** 17   **83.** -96 **85.** $-\dfrac{15}{238}$ **87.** 8

89. $-\dfrac{5}{16}$ **91.** -2.75 **93.** $-\dfrac{3}{16}$ **95.** $1572   **97.** $3296   **99.** 0.035   **101. (a)** $16.7 billion   **(b)** $27 billion
(c) $35.6 billion   **(d)** The amount spent on pets more than doubled from 1994 to 2005.

Section 1.4   (pages 41–42)
1. B   **3.** A   **5.** product; 0   **7.** grouping   **9.** like   **11.** $2m + 2p$ **13.** $-12x + 12y$ **15.** $8k$ **17.** $-2r$
19. cannot be simplified   **21.** $8a$ **23.** $-2d + f$ **25.** $-6y + 3$ **27.** $p + 11$ **29.** $-2k + 15$ **31.** $-3m + 2$
33. -1 **35.** $2p + 7$ **37.** $-6z - 39$ **39.** $(5 + 8)x = 13x$ **41.** $(5 \cdot 9)r = 45r$ **43.** $9y + 5x$ **45.** 7   **47.** 0
49. $8(-4) + 8x = -32 + 8x$ **51.** 0   **53.** Answers will vary. One example is washing your face and brushing your teeth;

one example is putting on your socks and putting on your shoes. **55.** 1900 **57.** 75 **59.** 431 **61.** associative property
62. associative property **63.** commutative property **64.** associative property **65.** distributive property
66. arithmetic facts **67.** No. One example is $7 + (5 \cdot 3) = (7 + 5)(7 + 3)$, which is false.

Chapter 1 Review Exercises (pages 47–50)

1. (number line showing $\frac{9}{4}$ marked, points at $-4, -2, 0, 2, 4$) **3.** 16 **5.** 5 **7.** $-9, -\sqrt{4}$ (or -2), $0, \frac{12}{3}$ (or 4) **9.** All are real numbers except $\sqrt{-9}$.

11. $\{0, 1, 2, 3\}$ **13.** false **15.** Chrysler; 13.7% **17.** false **19.** $(-\infty, -5)$ (number line with open circle at -5) **21.** $\frac{41}{24}$ **23.** -3

25. -39 **27.** $\frac{23}{20}$ **29.** $37.5 million **31.** $\frac{2}{3}$ **33.** 3.21 **35.** 10,000 **37.** -125 **39.** 20 **41.** -0.9 **43.** -4

45. -2 **47.** -30 **49.** (a) 25 (b) Answers will vary. **51.** $-4z$ **53.** $4p$ **55.** $6r + 18$ **57.** $-p - 3q$ **59.** 0
61. $(2 + 3)x = 5x$ **63.** $(2 \cdot 4)x = 8x$ **65.** 0 **67.** 7 **69.** $51,671 million; negative **71.** $76,521 million; negative
73. 25 **75.** 9 **77.** -5 **79.** -6.16 **81.** 2 **83.** not a real number **85.** -11.408 **87.** $-6x + 4$ **89.** (a) -116

(b) $-\frac{9}{4}$

Chapter 1 Test (pages 50–51)

[1.1] 1. (number line with points 0.75, $\frac{5}{3}$, 6.3 marked; $-2, 0, 2, 4, 6$) **2.** $0, 3, \sqrt{25}$ (or 5), $\frac{24}{2}$ (or 12) **3.** $-1, 0, 3, \sqrt{25}$ (or 5), $\frac{24}{2}$ (or 12) **4.** $-1, -0.5, 0, 3,$

$\sqrt{25}$ (or 5), $7.5, \frac{24}{2}$ (or 12) **5.** All are real numbers except $\sqrt{-4}$. **6.** $(-\infty, -3)$ (number line with open circle at -3)

7. $(-4, 2]$ (number line from -4 to 2) **[1.2] 8.** 0 **[1.3] 9.** -26 **10.** 19 **11.** 1 **12.** $\frac{16}{7}$ **13.** $\frac{11}{23}$ **[1.2] 14.** 50,395 ft

15. 37,486 ft **16.** 1345 ft **[1.3] 17.** 14 **18.** -15 **19.** not a real number **20.** (a) a must be positive. (b) a must

be negative. (c) a must be 0. **21.** $-\frac{6}{23}$ **[1.4] 22.** $10k - 10$ **23.** It changes the sign of each term. The simplified form

is $7r + 2$. **24.** B **25.** D **26.** A **27.** F **28.** C **29.** C **30.** E

CHAPTER 2 Linear Equations, Inequalities, and Applications

Section 2.1 (pages 60–62)
1. A and C **3.** Both sides are evaluated as 30, so 6 is a solution. **5.** (a) equation (b) expression (c) equation
(d) expression **7.** The solution contains a sign error when the distributive property was applied. The left side of the second
line of the solution should be $8x - 4x + 6$. The correct solution is 1. **9.** A conditional equation has one solution, an identity
has infinitely many solutions, and a contradiction has no solution. **11.** $\{-1\}$ **13.** $\{-4\}$ **15.** $\{-7\}$ **17.** $\{0\}$ **19.** $\{4\}$

21. $\left\{-\frac{7}{8}\right\}$ **23.** $\emptyset$; contradiction **25.** $\left\{-\frac{5}{3}\right\}$ **27.** $\left\{-\frac{1}{2}\right\}$ **29.** $\{2\}$ **31.** $\{-2\}$ **33.** $\{$all real numbers$\}$; identity

35. $\{-1\}$ **37.** $\{7\}$ **39.** $\{2\}$ **41.** $\{$all real numbers$\}$; identity **43.** $\left\{\frac{3}{2}\right\}$ **45.** 12 **47.** (a) 10^2, or 100 (b) 10^3, or 1000

49. $\left\{-\dfrac{18}{5}\right\}$ **51.** $\left\{-\dfrac{5}{6}\right\}$ **53.** $\{6\}$ **55.** $\{4\}$ **57.** $\{3\}$ **59.** $\{3\}$ **61.** $\{0\}$ **63.** $\{2000\}$ **65.** $\{25\}$ **67.** $\{40\}$

69. $\{3\}$ **71.** 36   **73.** 72   **75.** 50   **77.** $\dfrac{39}{2}$, or $19\dfrac{1}{2}$

Section 2.2 (pages 67–72)

1. (a) $3x = 5x + 8$ **(b)** $ct = bt + k$ **2. (a)** $3x - 5x = 8$ **(b)** $ct - bt = k$ **3. (a)** $-2x = 8$; distributive property

(b) $t(c - b) = k$; distributive property   **4. (a)** $x = -4$ **(b)** $t = \dfrac{k}{c - b}$ **5.** $c \neq b$; if $c = b$, the denominator is 0.

6. To solve an equation for a particular variable, such as t in the second equation, go through the same steps as you would in

solving for x in the first equation. Treat all other variables as constants.   **7.** $r = \dfrac{I}{pt}$ **9.** $L = \dfrac{P - 2W}{2}$, or $L = \dfrac{P}{2} - W$

11. (a) $W = \dfrac{V}{LH}$ **(b)** $H = \dfrac{V}{LW}$ **13.** $r = \dfrac{C}{2\pi}$ **15. (a)** $h = \dfrac{2A}{b + B}$ **(b)** $B = \dfrac{2A}{h} - b$, or $B = \dfrac{2A - hb}{h}$

17. $C = \dfrac{5}{9}(F - 32)$ **19.** D   **21.** $r = \dfrac{-2k - 3y}{a - 1}$, or $r = \dfrac{2k + 3y}{1 - a}$ **23.** $y = \dfrac{-x}{w - 3}$, or $y = \dfrac{x}{3 - w}$ **25.** 3.699 hr

27. 52 mph   **29.** 104°F   **31.** 230 m   **33.** radius: 240 in.; diameter: 480 in.   **35.** 2 in.   **37.** 75% water; 25% alcohol

39. 3%   **41.** $10.51   **43.** $45.66   **45.** .679, .519, .407, .200   **47.** $82,304   **49.** $33,890   **51.** 1500   **53.** 18,900

55. $\{12\}$ **57.** $\left\{\dfrac{1}{2}\right\}$ **59.** $\{-6\}$ **61.** -3 **63.** 6

Connections **(page 81)**   Our Steps 1, 3, 4, and 6 correspond to Polya's steps.

Section 2.3 (pages 81–87)

1. (a) $x + 12$ **(b)** $12 > x$ **3. (a)** $x - 4$ **(b)** $4 < x$ **5.** D   **7.** $2x - 13$ **9.** $12 + 3x$ **11.** $8(x - 12)$ **13.** $\dfrac{3x}{7}$

15. $x + 6 = -31$; -37 **17.** $x - (-4x) = x + 9$; $\dfrac{9}{4}$ **19.** $12 - \dfrac{2}{3}x = 10$; 3   **21.** expression   **23.** equation

25. expression   **27.** *Step 1:* the number of patents each university secured; *Step 2:* patents that Stanford secured;
Step 3: x; $x - 38$; *Step 4:* 134; *Step 5:* 134; 96; *Step 6:* 38; MIT patents; 96; 230   **29.** width: 165 ft; length: 265 ft
31. 24.34 in. by 29.88 in.   **33.** 850 mi; 925 mi; 1300 mi   **35.** Exxon Mobil: $340 billion; Wal-Mart: $316 billion
37. Eiffel Tower: 984 ft; Leaning Tower: 180 ft   **39.** Bush: 286 votes; Kerry: 251 votes   **41.** 45.2%   **43.** $5481
45. $35.67   **47.** $225   **49.** $4000 at 3%; $8000 at 4%   **51.** $1800 at 5%; $3200 at 6.5%   **53.** $13,500   **55.** 5 L
57. 4 L   **59.** 1 gal   **61.** 150 lb   **63.** We cannot expect the final mixture to be worth more than either of the ingredients.
65. (a) $800 - x$ **(b)** $800 - y$ **66. (a)** $0.05x$; $0.10(800 - x)$ **(b)** $0.05y$; $0.10(800 - y)$
67. (a) $0.05x + 0.10(800 - x) = 800(0.0875)$ **(b)** $0.05y + 0.10(800 - y) = 800(0.0875)$ **68. (a)** $200 at 5%; $600 at 10%
(b) 200 L of 5% acid; 600 L of 10% acid   **69.** The processes are the same. The amounts of money in Problem A correspond
to the amounts of solution in Problem B.   **71.** 32   **73.** 19   **75.** 5

Section 2.4 (pages 92–96)

1. $4.50   **3.** 60 mph   **5.** The problem asks for the *distance* to the workplace. To find this distance, we must multiply the rate,

10 mph, by the time, $\dfrac{3}{4}$ hr.   **7.** No, the answers must be whole numbers because they represent the number of coins.

9. 17 pennies; 17 dimes; 10 quarters   **11.** 23 loonies; 14 toonies   **13.** 28 $10 coins; 13 $20 coins   **15.** 872 adult tickets

17. 8.08 m per sec   **19.** 8.40 m per sec   **21.** $2\frac{1}{2}$ hr   **23.** 7:50 P.M.   **25.** 45 mph   **27.** $\frac{1}{2}$ hr   **29.** 60°, 60°, 60°

31. 40°, 45°, 95°   **33.** 40°, 80°   **34.** 120°   **35.** The sum is equal to the measure of the angle found in Exercise 34.

36. The sum of the measures of angles ① and ② is equal to the measure of angle ③.   **37.** Both measure 122°.

39. 64°, 26°   **41.** 19, 20, 21   **43.** 61 yr old   **45.** ⟶ (4)   **47.** ⟶ (−2, 6)

49. ⟶ (−4, 9)

Summary Exercises on Solving Applied Problems (pages 97–98)

1. length: 8 in.; width: 5 in.   **2.** length: 60 m; width: 30 m   **3.** $50.94   **4.** $425   **5.** $800 at 4%; $1600 at 5%

6. $12,000 at 3%; $14,000 at 4%   **7.** *Frasier:* 31; *The Simpsons:* 20   **8.** *Titanic:* $600.8 million; *Star Wars:* $461 million

9. 5 hr   **10.** Negussie: 11.88 mph; Ndereba: 10.78 mph   **11.** $13\frac{1}{3}$ L   **12.** $53\frac{1}{3}$ kg   **13.** fives: 84; tens: 42

14. 1650 tickets at $8; 810 tickets at $6   **15.** 20°, 30°, 130°   **16.** 107°, 73°   **17.** 31, 32, 33   **18.** 9 and 11

19. 6 in., 12 in., 16 in.   **20.** 23 in.

Section 2.5   (pages 107–112)

1. D   **3.** B   **5.** F   **7. (a)** $131 \le s \le 155$ **(b)** $s > 155$ **(c)** $9 \le x \le 12$ **(d)** $x > 18$ **9.** Since $4 > 0$, the student should not have reversed the direction of the inequality symbol when dividing by 4. We reverse the symbol only when multiplying or dividing by a *negative* number. The solution set is $[-16, \infty)$.   **11.** $[16, \infty)$ ⟶ (16)

13. $(7, \infty)$ ⟶ (7)   **15.** $(-\infty, -4)$ ⟵ (−4)   **17.** $(-\infty, -40]$ ⟵ (−40)

19. $(-\infty, 4]$ ⟵ (4)   **21.** $(-\infty, -10]$ ⟵ (−10)   **23.** $(-\infty, 14)$ ⟵ (14)

25. $\left(-\infty, -\dfrac{15}{2}\right)$ ⟵ (−15/2)   **27.** $\left[\dfrac{1}{2}, \infty\right)$ ⟶ (1/2)   **29.** $[2, \infty)$ ⟶ (2)

31. $(3, \infty)$ ⟶ (3)   **33.** $(-\infty, 4)$ ⟵ (4)   **35.** $\left(-\infty, \dfrac{23}{6}\right]$ ⟵ (23/6)

37. $\left(-\infty, \dfrac{76}{11}\right)$ ⟵ (76/11)   **39.** $(-\infty, \infty)$ ⟵⟶ (0)   **41.** ∅   **43.** $\{-9\}$ ⟶ (−9)

44. $(-9, \infty)$ ⟶ (−9)   **45.** $(-\infty, -9)$ ⟵ (−9)   **46.** the set of all real numbers ⟵⟶ (−9)

47. $(-\infty, -3)$ **49.** $(1, 11)$ ⟶ (1, 11)   **51.** $[-14, 10]$ ⟶ (−14, 10)   **53.** $[-5, 6]$ ⟶ (−5, 6)

55. $\left[-\dfrac{14}{3}, 2\right]$ ⟶ (−14/3, 2)   **57.** $\left[-\dfrac{1}{2}, \dfrac{35}{2}\right]$ ⟶ (−1/2, 35/2)   **59.** $\left(-\dfrac{1}{3}, \dfrac{1}{9}\right]$ ⟶ (−1/3, 1/9)

61. all numbers between −2 and 2—that is, $(-2, 2)$ **63.** all numbers greater than or equal to 3—that is, $[3, \infty)$

65. all numbers greater than or equal to −9—that is, $[-9, \infty)$ **67.** from about 2:30 P.M. to 6:00 P.M.   **69.** about 84°F–91°F

71. at least 80   **73.** 26 months   **75.** 26 DVDs   **77. (a)** 140 to 184 lb **(b)** Answers will vary.

79. (a) **(b)** **(c)** All numbers greater than 4 and less than 5 (that is, $4 < x < 5$) belong to both sets. **81. (a)** **(b)** **(c)** All real numbers belong to either one or both of these sets.

Section 2.6 (pages 119–121)

1. true **3.** false; the union is $(-\infty, 8) \cup (8, \infty)$. **5.** false; the intersection is $\emptyset$. **7.** $\{1, 3, 5\}$, or B **9.** $\{4\}$, or D **11.** $\emptyset$
13. $\{1, 2, 3, 4, 5, 6\}$, or A **15.** **17.** **19.** $(-3, 2)$

21. $(-\infty, 2]$ **23.** $\emptyset$ **25.** $[5, 9]$ **27.** $(-3, -1)$

29. $(-\infty, 4]$ **31.** **33.** **35.** $(-\infty, 8]$

37. $[-2, \infty)$ **39.** $(-\infty, \infty)$ **41.** $(-\infty, -5) \cup (5, \infty)$

43. $(-\infty, -1) \cup (2, \infty)$ **45.** $(-\infty, \infty)$ **47.** $[-4, -1]$ **49.** $[-9, -6]$

51. $(-\infty, 3)$ **53.** $[3, 9)$ **55.** intersection; $(-5, -1)$ **57.** union; $(-\infty, 4)$

59. union; $(-\infty, 0] \cup [2, \infty)$ **61.** intersection; $[4, 12]$ **63.** {Tuition and fees}

65. {Tuition and fees, Dormitory charges} **67.** Maria, Joe **68.** none of them **69.** none of them **70.** Luigi, Than
71. Maria, Joe **72.** all of them **73.** $[-6, \infty)$ **75.** $(-3, 2)$ **77.** -21

Connections **(page 128)** The filled carton may contain between 30.4 and 33.6 oz, inclusive.

Section 2.7 (pages 128–131)

1. E; C; D; B; A **3. (a)** one **(b)** two **(c)** none **5.** $\{-12, 12\}$ **7.** $\{-5, 5\}$ **9.** $\{-6, 12\}$ **11.** $\{-5, 6\}$

13. $\left\{-3, \dfrac{11}{2}\right\}$ **15.** $\left\{-\dfrac{19}{2}, \dfrac{9}{2}\right\}$ **17.** $\{-10, -2\}$ **19.** $\left\{-\dfrac{32}{3}, 8\right\}$ **21.** $(-\infty, -3) \cup (3, \infty)$

23. $(-\infty, -4] \cup [4, \infty)$ **25.** $(-\infty, -25] \cup [15, \infty)$

27. $(-\infty, -12) \cup (8, \infty)$ **29.** $(-\infty, -2) \cup (8, \infty)$

31. $\left(-\infty, -\dfrac{9}{5}\right] \cup [3, \infty)$ **33. (a)** **(b)**

35. $[-3, 3]$ **37.** $(-4, 4)$ **39.** $[-25, 15]$

41. $[-12, 8]$ **43.** $[-2, 8]$ **45.** $\left[-\dfrac{9}{5}, 3\right]$

47. $(-\infty, -5) \cup (13, \infty)$ **49.** $(-\infty, -25) \cup (15, \infty)$

51. $\{-6, -1\}$ **53.** $\left[-\dfrac{10}{3}, 4\right]$ **55.** $\left[-\dfrac{7}{6}, -\dfrac{5}{6}\right]$

57. $(-\infty, -3] \cup [4, \infty)$ **59.** $\{-5, 1\}$ **61.** $\{3, 9\}$ **63.** $\{-5, 5\}$ **65.** $\{-5, -3\}$

67. $(-\infty, -3) \cup (2, \infty)$ **69.** $[-10, 0]$ **71.** $\{-1, 3\}$ **73.** $\left\{-3, \dfrac{5}{3}\right\}$ **75.** $\left\{-\dfrac{1}{3}, -\dfrac{1}{15}\right\}$ **77.** $\left\{-\dfrac{5}{4}\right\}$

79. $(-\infty, \infty)$ **81.** $\emptyset$ **83.** $\left\{-\dfrac{1}{4}\right\}$ **85.** $\emptyset$ **87.** $(-\infty, \infty)$ **89.** $\left\{-\dfrac{3}{7}\right\}$ **91.** $\left\{\dfrac{2}{5}\right\}$ **93.** $(-\infty, \infty)$ **95.** $\emptyset$

97. $|x - 1000| \le 100$; $900 \le x \le 1100$ **99.** 475.6 ft **100.** 1201 Walnut, Fidelity Bank and Trust Building, Kansas City Power and Light, City Hall, Hyatt Regency Crown Center **101.** City Center Square, Commerce Tower, Federal Office Building, 1201 Walnut, City Hall, Fidelity Bank and Trust Building, Kansas City Power and Light, Hyatt Regency Crown Center **102. (a)** $|x - 475.6| \ge 75$ **(b)** $x \ge 550.6$ or $x \le 400.6$ **(c)** Town Pavilion, One Kansas City Place **(d)** It makes sense because it includes all buildings *not* listed earlier. **103. (a)** 12 **(b)** 4 **105. (a)** 0 **(b)** $\dfrac{36}{5}$

Summary Exercises on Solving Linear and Absolute Value Equations and Inequalities (page 132)

1. $\{12\}$ **2.** $\{-5, 7\}$ **3.** $\{7\}$ **4.** $\left\{-\dfrac{2}{5}\right\}$ **5.** $\emptyset$ **6.** $(-\infty, -1]$ **7.** $\left[-\dfrac{2}{3}, \infty\right)$ **8.** $\{-1\}$ **9.** $\{-3\}$

10. $\left\{1, \dfrac{11}{3}\right\}$ **11.** $(-\infty, 5]$ **12.** $(-\infty, \infty)$ **13.** $\{2\}$ **14.** $(-\infty, -8] \cup [8, \infty)$ **15.** $\emptyset$ **16.** $(-\infty, \infty)$

17. $(-5.5, 5.5)$ **18.** $\left\{\dfrac{13}{3}\right\}$ **19.** $\left\{-\dfrac{96}{5}\right\}$ **20.** $(-\infty, 32]$ **21.** $(-\infty, -24)$ **22.** $\left\{\dfrac{3}{8}\right\}$ **23.** $\left\{\dfrac{7}{2}\right\}$ **24.** $(-6, 8)$

25. $(-\infty, \infty)$ **26.** $(-\infty, 5)$ **27.** $(-\infty, -4) \cup (7, \infty)$ **28.** $\{24\}$ **29.** $\left\{-\dfrac{1}{5}\right\}$ **30.** $\left(-\infty, -\dfrac{5}{2}\right]$ **31.** $\left[-\dfrac{1}{3}, 3\right]$

32. $[1, 7]$ **33.** $\left\{-\dfrac{1}{6}, 2\right\}$ **34.** $\{-3\}$ **35.** $(-\infty, -1] \cup \left[\dfrac{5}{3}, \infty\right)$ **36.** $\left[\dfrac{3}{4}, \dfrac{15}{8}\right]$ **37.** $\left\{-\dfrac{5}{2}\right\}$ **38.** $\{60\}$

39. $\left[-\dfrac{9}{2}, \dfrac{15}{2}\right]$ **40.** $(1, 9)$ **41.** $(-\infty, \infty)$ **42.** $\left\{\dfrac{1}{3}, 9\right\}$ **43.** $(-\infty, \infty)$ **44.** $\left\{-\dfrac{10}{9}\right\}$ **45.** $\{-2\}$ **46.** $\emptyset$

47. $(-\infty, -1) \cup (2, \infty)$ **48.** $[-3, -2]$

Chapter 2 Review Exercises (pages 138–143)

1. $\left\{-\dfrac{9}{5}\right\}$ **3.** $\left\{-\dfrac{7}{5}\right\}$ **5.** $(-\infty, \infty)$; identity **7.** $\{0\}$; conditional **9.** $b = \dfrac{2A - Bh}{h}$, or $b = \dfrac{2A}{h} - B$

11. $x = \dfrac{4}{3}(P + 12)$, or $x = \dfrac{4}{3}P + 16$ **13.** 6 ft **15.** 6.5% **17.** \$525 billion **19.** $9 - \dfrac{1}{3}x$ **21.** length: 13 m; width: 8 m **23.** 12 kg **25.** 10 L **27.** 15 dimes; 8 quarters **29.** A **31.** 2.2 hr **33.** 1 hr **35.** 40°, 45°, 95°

37. $(-9, \infty)$ **39.** $\left(\dfrac{3}{2}, \infty\right)$ **41.** $[3, 5)$ **43.** 38 m or less **45.** any score greater than or equal to 61 **47.** $\{a, c\}$

49. $\{a, c, e, f, g\}$ **51.** $(6, 9)$ **53.** $(-\infty, -3] \cup (5, \infty)$ **55.** $\emptyset$ **57.** $(-3, 4)$

59. $(4, \infty)$ **61.** $\{-7, 7\}$ **63.** $\left\{-\dfrac{1}{3}, 5\right\}$ **65.** $\{0, 7\}$ **67.** $\left\{-\dfrac{3}{4}, \dfrac{1}{2}\right\}$ **69.** $(-14, 14)$ **71.** $[-3, -2]$

73. $(-2, \infty)$ **75.** $[-2, 3)$ **77.** $(-\infty, \infty)$ **79.** 10 ft **81.** $\left\{-\dfrac{7}{3}, 1\right\}$ **83.** $[-16, 10]$ **85.** $\left(-3, \dfrac{7}{2}\right)$ **87.** 80°

89. $\left(-\infty, -\dfrac{13}{5}\right) \cup (3, \infty)$ **91.** 5 L **93.** $\{30\}$ **95.** $\left\{1, \dfrac{11}{3}\right\}$ **97. (a)** $\emptyset$ **(b)** $(-\infty, \infty)$ **(c)** $\emptyset$ **99.** ⟵++++++++++[++⟶ -2 0 7

Chapter 2 Test (pages 143–144)

[2.1] **1.** $\{-19\}$ **2.** $\{5\}$ **3.** $(-\infty, \infty)$ **4. (a)** $\emptyset$; contradiction **(b)** $(-\infty, \infty)$; identity **(c)** $\{0\}$; conditional equation

[2.2] **5.** $v = \dfrac{S + 16t^2}{t}$ **6.** $r = \dfrac{-2 - 6t}{a - 3}$, or $r = \dfrac{2 + 6t}{3 - a}$ [2.3, 2.4] **7.** 3.173 hr **8.** 6.25% **9.** 73.7%

10. \$8000 at 3%; \$20,000 at 5% **11.** faster car: 60 mph; slower car: 45 mph **12.** $40°, 40°, 100°$

[2.5] **13.** $[1, \infty)$ ⟶——[⟶ 1 **14.** $(-\infty, 28)$ ⟵)——⟶ 28 **15.** $[-3, 3]$ ——[——]⟶ -3 3 **16.** C **17.** 82

18. $[500, \infty)$ [2.6] **19. (a)** $\{1, 5\}$ **(b)** $\{1, 2, 5, 7, 9, 12\}$ **20.** $[2, 9)$ **21.** $(-\infty, 3) \cup [6, \infty)$ [2.7] **22.** $\left[-\dfrac{5}{2}, 1\right]$

23. $\left(-\infty, -\dfrac{7}{6}\right) \cup \left(\dfrac{17}{6}, \infty\right)$ **24.** $\emptyset$ **25.** $\left(\dfrac{1}{3}, \dfrac{7}{3}\right)$ **26.** $\left\{-\dfrac{5}{3}, 3\right\}$ **27.** $\left\{-\dfrac{5}{7}, \dfrac{11}{3}\right\}$ **28. (a)** $\emptyset$ **(b)** $(-\infty, \infty)$ **(c)** $\emptyset$

Chapters 1–2 Cumulative Review Exercises (pages 145–146)

[1.1] **1.** 9, 6 **2.** 0, 9, 6 **3.** $-8, 0, 9, 6$ **4.** $-8, -\dfrac{2}{3}, 0, \dfrac{4}{5}, 9, 6$ **5.** $-\sqrt{6}$ **6.** All are real numbers. [1.2] **7.** $-\dfrac{22}{21}$

8. 8 [1.3] **9.** 8 **10.** 0 **11.** -243 **12.** $\dfrac{216}{343}$ **13.** $-\dfrac{8}{27}$ **14.** -4096 **15.** $\sqrt{-36}$ is not a real number.

16. $\dfrac{4 + 4}{4 - 4}$ is undefined. **17.** -16 **18.** 184 **19.** $\dfrac{27}{16}$ [1.4] **20.** $-20r + 17$ **21.** $13k + 42$

22. commutative property **23.** distributive property [2.1] **24.** $\{5\}$ **25.** $\{30\}$ **26.** $\{15\}$ [2.2] **27.** $b = P - a - c$

[2.1] **28.** $\emptyset$ **29.** $\{$all real numbers$\}$, or $(-\infty, \infty)$ [2.5] **30.** $[-14, \infty)$ ⟶——[——⟶ -14 **31.** $\left[\dfrac{5}{3}, 3\right)$ ——[——)⟶ $\frac{5}{3}$ 3

[2.6] **32.** $(-\infty, 0) \cup (2, \infty)$ ⟵)———(⟶ 0 2 [2.7] **33.** $\left(-\infty, -\dfrac{1}{7}\right] \cup [1, \infty)$ ⟵]———[⟶ $-\frac{1}{7}$ 1

[2.3–2.5] **34.** \$5000 at 5%; \$7000 at 6% **35.** $6\dfrac{1}{3}$ g **36.** 74 or greater **37.** 2 L **38.** 9 pennies, 12 nickels, 8 quarters

[2.2] **39.** 44 mg [2.3] **40. (a)** 155 **(b)** 9.6%

CHAPTER 3 Graphs, Linear Equations, and Functions

Section 3.1 (pages 156–160)

1. (a) x represents the year; y represents the revenue in billions of dollars. **(b)** about \$1850 billion **(c)** $(2002, 1850)$

(d) In 2000, federal tax revenues were about \$2030 billion. **3.** origin **5.** y; x; x; y **7.** two **9. (a)** I **(b)** III **(c)** II

(d) IV **(e)** none **(f)** none **11. (a)** I or III **(b)** II or IV **(c)** II or IV **(d)** I or III **13.–21.**

23. (a) $-4; -3; -2; -1; 0$ **(b)**   **25. (a)** $-3; 3; 2; -1$ **(b)**

27. (a) $\dfrac{5}{2}; 5; \dfrac{3}{2}; 1$ **(b)**   **29. (a)** $-4; 5; -\dfrac{12}{5}; \dfrac{5}{4}$ **(b)**

31. (a) $3; 1; -1; -3$ **(b)**   **33. (a)** 1 **(b)** 2 **(c)** For every increase in x by 1 unit, y increases by 2 units.

35. Choose a value *other than* 0 for either x or y. For example, if $x = -5$, then $y = 4$.

37. $(6, 0); (0, 4)$ **39.** $(6, 0); (0, -2)$ 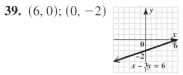 **41.** $\left(\dfrac{21}{2}, 0\right); \left(0, -\dfrac{7}{3}\right)$

43. none; $(0, 5)$ **45.** $(2, 0)$; none   **47.** $(-4, 0)$; none

49. $(0, 0); (0, 0)$ **51.** $(0, 0); (0, 0)$ **53.** $(0, 0); (0, 0)$ **55.** $(-5, -1)$

57. $\left(\dfrac{9}{2}, -\dfrac{3}{2}\right)$ **59.** $\left(0, \dfrac{11}{2}\right)$ **61.** $(2.1, 0.9)$ **63.** $(1, 1)$ **65.** $\left(-\dfrac{5}{12}, \dfrac{5}{28}\right)$ **67.** $Q(11, -4)$ **69.** $Q\left(\dfrac{8}{3}, \dfrac{9}{5}\right)$

71. 57,225 thousand (or 57,225,000) **73.** B **75.** The screen on the right is more useful because it shows the intercepts.

77. 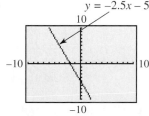 **79.** 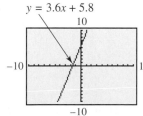 **81.** 2 **83.** $\dfrac{5}{2}$ **85.** 0

Connections **(page 163)** **1.** 60 **2. (a)** $\dfrac{13}{7}$ **(b)** $\dfrac{1}{20}$ **(c)** -2

Section 3.2 (pages 170–175)

1. A, B, and D **3. (a)** C **(b)** A **(c)** D **(d)** B **5.** 0 **7.** $-\dfrac{1}{3}$ **9.** -4 **11.** 2 **13.** $\dfrac{5}{2}$ **15.** 0 **17.** undefined

19. undefined **21.** 8 **23.** $\dfrac{5}{6}$ **25.** 0 **27.** -1 **29.** 6 **31.** -3 **33.** $-\dfrac{5}{2}$ **35.** undefined **37.** B **39.** A

41. $-\dfrac{1}{2}$ 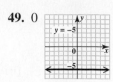 **43.** $\dfrac{5}{2}$ **45.** 4 **47.** undefined **49.** 0

51. 0 **53.** **55.** **57.** **59.** **61.**

63. $-\dfrac{4}{9};\dfrac{9}{4}$ **65.** parallel **67.** perpendicular **69.** neither **71.** parallel **73.** neither **75.** perpendicular

77. $-\$4000$ per yr; The value of the machine is decreasing $\$4000$ each year during these years. **79.** 0% per yr (or no change); The percent of pay raise is not changing—it is 3% each year during these years. **81.** 19.5 ft

83. (a) 23,431 per yr; 18,897 per yr; 12,392 per yr; 17,955 per yr; 23,418 per yr **(b)** no; no **85. (a)** $\$18.25$ billion per yr **(b)** The positive slope means that personal spending on recreation in the United States *increased* by an average of $\$18.25$ billion each year. **87.** $-\$69$ per yr; the price decreased an average of $\$69$ each year from 1997 to 2002.

89. Since the slopes of both pairs of opposite sides are equal, the figure is a parallelogram. **91.** A is y_1 and B is y_2. **93.** $\dfrac{1}{3}$

94. $\dfrac{1}{3}$ **95.** $\dfrac{1}{3}$ **96.** $\dfrac{1}{3}=\dfrac{1}{3}=\dfrac{1}{3}$ is true. **97.** collinear **98.** not collinear **99.** $y=-\dfrac{3}{2}x+4$ **101.** $y=4x+14$

103. $5x-3y=-17$ **105.** $x+2y=-4$

Section 3.3 (pages 186–191)

1. A **3.** A **5.** $3x+y=10$ **7.** A **9.** C **11.** H **13.** B **15.** $y=5x+15$ **17.** $y=-\dfrac{2}{3}x+\dfrac{4}{5}$

19. $y=\dfrac{2}{5}x+5$ **21.** $y=\dfrac{2}{3}x+1$ **23. (a)** $y=x+4$ **(b)** 1 **(c)** $(0,4)$ **(d)**

25. (a) $y=-\dfrac{6}{5}x+6$ **(b)** $-\dfrac{6}{5}$ **(c)** $(0,6)$ **(d)**   **27. (a)** $y=\dfrac{4}{5}x-4$ **(b)** $\dfrac{4}{5}$ **(c)** $(0,-4)$ **(d)**

29. (a) $y=-\dfrac{1}{2}x-2$ **(b)** $-\dfrac{1}{2}$ **(c)** $(0,-2)$ **(d)**   **31. (a)** $y=\dfrac{4}{3}x+4$ **(b)** $\dfrac{4}{3}$ **(c)** $(0,4)$ **(d)**

33. (a) $2x+y=18$ **(b)** $y=-2x+18$ **35. (a)** $3x+4y=10$ **(b)** $y=-\dfrac{3}{4}x+\dfrac{5}{2}$

37. (a) $x - 2y = -13$ **(b)** $y = \dfrac{1}{2}x + \dfrac{13}{2}$ **39. (a)** $4x - y = 12$ **(b)** $y = 4x - 12$ **41. (a)** $7x - 5y = -20$

(b) $y = 1.4x + 4$ **43.** $y = 5$ **45.** $x = 9$ **47.** $x = 0.5$ **49.** $y = 8$ **51. (a)** $2x - y = 2$ **(b)** $y = 2x - 2$

53. (a) $x + 2y = 8$ **(b)** $y = -\dfrac{1}{2}x + 4$ **55. (a)** $2x - 13y = -6$ **(b)** $y = \dfrac{2}{13}x + \dfrac{6}{13}$ **57. (a)** $y = 5$ **(b)** $y = 5$

59. (a) $x = 7$ **(b)** not possible **61. (a)** $y = -3$ **(b)** $y = -3$ **63. (a)** $y = 3x - 19$ **(b)** $3x - y = 19$

65. (a) $y = \dfrac{1}{2}x - 1$ **(b)** $x - 2y = 2$ **67. (a)** $y = -\dfrac{1}{2}x + 9$ **(b)** $x + 2y = 18$ **69. (a)** $y = 7$ **(b)** $y = 7$

71. $y = 45x$; $(0, 0), (5, 225), (10, 450)$ **73.** $y = 3.01x$; $(0,0), (5, 15.05), (10, 30.10)$ **75. (a)** $y = 39x + 99$ **(b)** $(5, 294)$; the cost of a 5-month membership is \$294. **(c)** \$567 **77. (a)** $y = 35x + 44.95$ **(b)** $(5, 219.95)$; the cost of the plan for 5 months is \$219.95. **(c)** \$464.95 **79. (a)** $y = 6x + 30$ **(b)** $(5, 60)$; it costs \$60 to rent the saw for 5 days. **(c)** 18 days **81. (a)** $y = -5.6x + 91$; the percent of households accessing the Internet by dial-up is decreasing 5.6% per year. **(b)** 57% **83. (a)** $y = 906.5x + 17,860.5$ **(b)** about \$26,019; it is slightly lower than the actual value. **85. (a)** $y = -3x + 9$ **(b)** 3 **(c)** $\{3\}$ **87. (a)** $y = 4x + 2$ **(b)** -0.5 **(c)** $\{-0.5\}$ **89.** D **91.** 32; 212 **92.** $(0, 32)$ and $(100, 212)$ **93.** $\dfrac{9}{5}$

94. $F = \dfrac{9}{5}C + 32$ **95.** $C = \dfrac{5}{9}(F - 32)$ **96.** When the Celsius temperature is 50°, the Fahrenheit temperature is 122°.

97. $(-\infty, 2)$ **99.** $\left(-\infty, -\dfrac{4}{3}\right]$

Summary Exercises on Slopes and Equations of Lines (page 192)

1. $-\dfrac{3}{5}$ **2.** $-\dfrac{4}{7}$ **3.** 2 **4.** $\dfrac{5}{2}$ **5.** -0.25 **6.** 2 **7. (a)** $y = -\dfrac{5}{6}x + \dfrac{13}{3}$ **(b)** $5x + 6y = 26$ **8. (a)** $y = 3x + 11$

(b) $3x - y = -11$ **9. (a)** $y = -\dfrac{5}{2}x$ **(b)** $5x + 2y = 0$ **10. (a)** $y = -8$ **(b)** $y = -8$ **11. (a)** $y = -\dfrac{7}{9}$ **(b)** $9y = -7$

12. (a) $y = 8x - 27$ **(b)** $8x - y = 27$ **13. (a)** $y = \dfrac{2}{3}x + \dfrac{14}{3}$ **(b)** $2x - 3y = -14$ **14. (a)** $y = 2x - 10$

(b) $2x - y = 10$ **15. (a)** $y = -\dfrac{7}{5}x + \dfrac{32}{5}$ **(b)** $7x + 5y = 32$ **16. (a)** $y = -2x + \dfrac{7}{6}$ **(b)** $12x + 6y = 7$

17. (a) $y = \dfrac{1}{5}x + \dfrac{17}{210}$ **(b)** $42x - 210y = -17$ **18. (a)** $y = -\dfrac{3}{4}x + \dfrac{53}{42}$ **(b)** $63x + 84y = 106$ **19. (a)** $y = 2x + 0.9$

(b) $20x - 10y = -9$ **20. (a)** $y = 0.5x + 0.5$ **(b)** $x - 2y = -1$ **21.** B **22.** D **23.** A **24.** C **25.** E

Section 3.4 (pages 197–199)

1. solid; below **3.** dashed; above **5.** The graph of $Ax + By = C$ divides the plane into two regions. In one of the regions, the ordered pairs satisfy $Ax + By < C$; in the other, they satisfy $Ax + By > C$. **7.** **9.**

11. 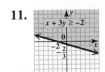 **13.** 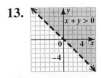 **15.** **17.** **19.** **21.**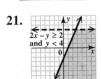

23. 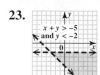   **25.** $-3 < x < 3$ 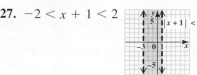   **27.** $-2 < x + 1 < 2$ 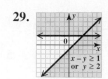   **29.**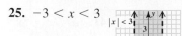

31. 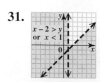   **33.**    **35.** C   **37.** A   **39. (a)** $\{-4\}$ **(b)** $(-\infty, -4)$ **(c)** $(-4, \infty)$

41. (a) $\{3.5\}$ **(b)** $(3.5, \infty)$ **(c)** $(-\infty, 3.5)$ *We include a calculator graph and supporting explanation only with the answer to Exercise 43.* **43. (a)** $\{-0.6\}$ **(b)** $(-0.6, \infty)$ **(c)** $(-\infty, -0.6)$ The graph of $y_1 = 5x + 3$ has x-intercept $(-0.6, 0)$, supporting the result of part (a). The graph of y_1 lies *above* the x-axis for values of x *greater than* -0.6, supporting the result of part (b). The graph of y_1 lies *below* the x-axis for values of x *less than* -0.6, supporting the result of part (c).

$y_1 = 5x + 3$

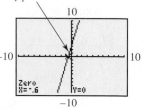

45. (a) $\{-1.2\}$ **(b)** $(-\infty, -1.2]$ **(c)** $[-1.2, \infty)$ **47.** $x \le 200, x \ge 100, y \ge 3000$ **48.**

49. $C = 50x + 100y$ **50.** Some examples are $(100, 5000)$, $(150, 3000)$, and $(150, 5000)$. The corner points are $(100, 3000)$ and $(200, 3000)$.   **51.** The least value occurs when $x = 100$ and $y = 3000$.   **52.** The company should use 100 workers and manufacture 3000 units to achieve the least possible cost.   **53.** -9 **55.** 1   **57.** $y = \dfrac{3}{7}x - \dfrac{8}{7}$ **59.** $y = \dfrac{1}{8}x - \dfrac{5}{4}$

Section 3.5   (pages 210–215)

1. We give one of many possible answers here. A function is a set of ordered pairs in which each first component corresponds to exactly one second component. For example, $\{(0, 1), (1, 2), (2, 3), (3, 4), \ldots\}$ is a function.   **3.** independent variable
5. function   **7.** not a function   **9.** function   **11.** not a function; domain: $\{0, 1, 2\}$; range: $\{-4, -1, 0, 1, 4\}$ **13.** function; domain: $\{2, 3, 5, 11, 17\}$; range: $\{1, 7, 20\}$ **15.** not a function; domain: $\{1\}$; range: $\{5, 2, -1, -4\}$ **17.** function; domain: $(-\infty, \infty)$; range: $(-\infty, \infty)$ **19.** function; domain: $(-\infty, \infty)$; range: $(-\infty, 4]$ **21.** not a function; domain: $[-4, 4]$; range: $[-3, 3]$ **23.** function; domain: $(-\infty, \infty)$ **25.** not a function; domain: $[0, \infty)$ **27.** function; domain: $(-\infty, \infty)$ **29.** not a function; domain: $(-\infty, \infty)$ **31.** function; domain: $[0, \infty)$ **33.** function; domain: $(-\infty, 0) \cup (0, \infty)$
35. function; domain: $\left[-\dfrac{1}{2}, \infty \right)$ **37.** function; domain: $(-\infty, 4) \cup (4, \infty)$ **39.** B   **41.** 4   **43.** -11 **45.** 3
47. 2.75   **49.** $-3p + 4$ **51.** $3x + 4$ **53.** $-3x - 2$ **55.** $-\pi^2 + 4\pi + 1$ **57.** $-3x - 3h + 4$ **59.** -9
61. (a) 2 **(b)** 3   **63. (a)** 15 **(b)** 10   **65. (a)** 3 **(b)** -3 **67. (a)** $f(x) = -\dfrac{1}{3}x + 4$ **(b)** 3   **69. (a)** $f(x) = 3 - 2x^2$
(b) -15 **71. (a)** $f(x) = \dfrac{4}{3}x - \dfrac{8}{3}$ **(b)** $\dfrac{4}{3}$ **73.** line; -2; $-2x + 4$; -2; 3; -2

75. domain: $(-\infty, \infty)$; range: $(-\infty, \infty)$

77. domain: $(-\infty, \infty)$; range: $(-\infty, \infty)$

79. domain: $(-\infty, \infty)$; range: $(-\infty, \infty)$

81. domain: $(-\infty, \infty)$; range: $\{-4\}$

83. domain: $(-\infty, \infty)$; range: $\{0\}$

85. (a) \$8.25 **(b)** 3 is the value of the independent variable, which represents a package weight of 3 lb; $f(3)$ is the value of the dependent variable, representing the cost to mail a 3-lb package. **(c)** \$13.75; $f(5) = 13.75$ **87. (a)** 194.53 cm **(b)** 177.29 cm **(c)** 177.41 cm **(d)** 163.65 cm   **89. (a)** $[0, 100]$; $[0, 3000]$ **(b)** 25 hr; 25 hr **(c)** 2000 gal **(d)** $f(0) = 0$; the pool is empty at time 0. **(e)** $f(25) = 3000$; after 25 hr, there are 3000 gal of water in the pool.   **91.** $f(3) = 7$ **93.** $y = 4$ **95.** $y = 4$ **97.** $\{6\}$

Chapter 3 Review Exercises   (pages 219–223)

1.

x	y
0	5
$\dfrac{10}{3}$	0
2	2
$\dfrac{14}{3}$	-2

3. $(3, 0)$; $(0, -4)$

5. $(10, 0)$; $(0, 4)$

7. If both coordinates are positive, the point lies in quadrant I. If the first coordinate is negative and the second is positive, the point lies in quadrant II. To lie in quadrant III, the point must have both coordinates negative. To lie in quadrant IV, the first coordinate must be positive and the second must be negative.   **9.** $-\dfrac{1}{2}$ **11.** $\dfrac{3}{4}$ **13.** $\dfrac{2}{3}$ **15.** undefined   **17.** -1

19. negative   **21.** 0   **23.** 12 ft   **25. (a)** $y = -\dfrac{1}{3}x - 1$ **(b)** $x + 3y = -3$ **27. (a)** $y = -\dfrac{4}{3}x + \dfrac{29}{3}$

(b) $4x + 3y = 29$ **29. (a)** not possible **(b)** $x = 2$ **31. (a)** $y = \dfrac{7}{5}x + \dfrac{16}{5}$ **(b)** $7x - 5y = -16$ **33. (a)** $y = 4x - 29$

(b) $4x - y = 29$ **35. (a)** $y = 57x + 159$; \$843 **(b)** $y = 47x + 159$; \$723   **37.**   **39.**   **41.** D

43. domain: $\{9, 11, 4, 17, 25\}$; range: $\{32, 47, 69, 14\}$; function   **45.** domain: $(-\infty, 0]$; range: $(-\infty, \infty)$; not a function

47. not a function; domain: $(-\infty, \infty)$ **49.** function; domain: $\left[-\dfrac{7}{4}, \infty\right)$ **51.** function; domain: $(-\infty, 6) \cup (6, \infty)$

53. -6 **55.** -8 **57. (a)** yes **(b)** domain: $\{1943, 1953, 1963, 1973, 1983, 1993, 2003\}$; range: $\{63.3, 68.8, 69.9, 71.4, 74.6, 75.5, 77.6\}$ **(c)** Answers will vary. Two possible answers are $(1943, 63.3)$ and $(1953, 68.8)$.

57. (d) 71.4; in 1973, life expectancy at birth was 71.4 yr. **(e)** 1993 **59.** C **61.** Because it falls from left to right, the slope is negative. **62.** $-\dfrac{3}{2}$ **63.** $-\dfrac{3}{2}; \dfrac{2}{3}$ **64.** $\left(\dfrac{7}{3}, 0\right)$ **65.** $\left(0, \dfrac{7}{2}\right)$ **66.** $f(x) = -\dfrac{3}{2}x + \dfrac{7}{2}$ **67.** $f(8) = -\dfrac{17}{2}$

68. $x = \dfrac{23}{3}$ **69.** **70.** $\left\{\dfrac{7}{3}\right\}$ **71.** $\left(\dfrac{7}{3}, \infty\right)$ **72.** $\left(-\infty, \dfrac{7}{3}\right)$

Chapter 3 Test (pages 223–224)

[3.1] **1.** $-\dfrac{10}{3}; -2; 0$ [3.1, 3.3] **2.** $\left(\dfrac{20}{3}, 0\right); (0, -10)$ **3.** none; $(0, 5)$

4. $(2, 0)$; none [3.2] **5.** $\dfrac{1}{2}$ **6.** It is a vertical line. **7.** perpendicular **8.** neither

9. -1200 farms per yr; the number of farms decreased, on the average, by about 1200 each year from 1980 to 2005.

[3.3] **10. (a)** $y = -5x + 19$ **(b)** $5x + y = 19$ **11. (a)** $y = 14$ **(b)** $y = 14$ **12. (a)** $y = -\dfrac{3}{5}x - \dfrac{11}{5}$

(b) $3x + 5y = -11$ **13. (a)** $y = -\dfrac{1}{2}x - \dfrac{3}{2}$ **(b)** $x + 2y = -3$ **14. (a)** $y = -\dfrac{1}{2}x + 2$ **(b)** $x + 2y = 4$

15. (a) not possible **(b)** $x = 5$ **16.** B **17. (a)** $y = 1784.17x + 13,939.17$ **(b)** \$29,997; it is slightly less than the actual value. [3.4] **18.** **19.** [3.5] **20.** D **21.** D **22.** domain: $[0, \infty)$; range: $(-\infty, \infty)$

23. domain: $\{0, -2, 4\}$; range: $\{1, 3, 8\}$ **24. (a)** 0 **(b)** $-a^2 + 2a - 1$ **25.** domain: $(-\infty, \infty)$; range: $(-\infty, \infty)$

Chapters 1–3 Cumulative Review Exercises (pages 225–226)

[1.1] **1.** always true **2.** always true **3.** never true **4.** sometimes true; for example, $3 + (-3) = 0$, but

$3 + (-1) = 2 \neq 0$. [1.2] **5.** 4 [1.3] **6.** 0.64 **7.** not a real number [1.2] **8.** $\dfrac{8}{5}$ [1.4] **9.** $4m - 3$

10. $2x^2 + 5x + 4$ [1.3] **11.** $-\dfrac{19}{2}$ [1.1] **12.** $(-3, 5]$ **13.** no [1.3] **14.** -39 **15.** 62 **16.** undefined

[2.1] **17.** $\left\{\dfrac{7}{6}\right\}$ **18.** $\{-1\}$ [2.2] **19.** $h = \dfrac{3V}{\pi r^2}$ **20.** $-58°F$ [2.4] **21.** 6 in. **22.** 2 hr

[2.5] **23.** $\left(-3, \dfrac{7}{2}\right)$ **24.** $(-\infty, 1]$ [2.6] **25.** $(6, 8)$

26. $(-\infty, -2] \cup (7, \infty)$ <img_line>   [2.7] **27.** $\{0, 7\}$ **28.** $(-\infty, \infty)$ [2.5] **29.** The union of the three solution sets

is $(-\infty, \infty)$.   [3.1] **30.** $-3; 4; -\dfrac{3}{2}$ **31.** x-intercept: $(4, 0)$; y-intercept: $\left(0, \dfrac{12}{5}\right)$ [3.2] **32.** (a) $-\dfrac{6}{5}$ (b) $\dfrac{5}{6}$

[3.4] **33.** <img_graph>   [3.3] **34.** (a) $y = -\dfrac{3}{4}x - 1$ (b) $3x + 4y = -4$ **35.** (a) $y = -2$ (b) $y = -2$

36. (a) $y = -\dfrac{4}{3}x + \dfrac{7}{3}$ (b) $4x + 3y = 7$ **37.** (a) not possible (b) $x = 4$ [3.5] **38.** domain: $\{14, 91, 75, 23\}$;

range: $\{9, 70, 56, 5\}$; not a function; 75 in the domain is paired with two different values, 70 and 56, in the range.

39. (a) domain: $(-\infty, \infty)$; range: $(-\infty, \infty)$ (b) 22   [3.2] **40.** 10.571 thousand (or 10,571) per year; the number of motor

scooters sold in the United States increased by an average of 10,571 per year from 1997 to 2004.

CHAPTER 4   Systems of Linear Equations

Section 4.1   (pages 237–243)

1. $3; -6$ **3.** $\emptyset$ **5.** 0   **7.** D; The ordered pair solution must be in quadrant IV, since that is where the graphs of the

equations intersect.   **9.** (a) B (b) C (c) A (d) D   **11.** yes   **13.** no   **15.** $\{(-2, -3)\}$ <img_graph>   **17.** $\{(1, 2)\}$

19. $\{(2, 3)\}$ **21.** $\left\{\left(\dfrac{22}{9}, \dfrac{22}{3}\right)\right\}$ **23.** $\{(5, 4)\}$ **25.** $\{(1, 3)\}$ **27.** $\left\{\left(-5, -\dfrac{10}{3}\right)\right\}$ **29.** $\{(2, 6)\}$

31. $\{(x, y) \mid 2x - y = 0\}$; dependent equations   **33.** $\emptyset$; inconsistent system   **35.** $\{(2, -4)\}$ **37.** $\{(3, -1)\}$ **39.** $\{(2, -3)\}$

41. $\{(x, y) \mid 7x + 2y = 6\}$; dependent equations   **43.** $\left\{\left(\dfrac{3}{2}, -\dfrac{3}{2}\right)\right\}$ **45.** $\emptyset$; inconsistent system   **47.** $\{(0, 0)\}$

49. $\{(0, -4)\}$ **51.** $\{(2, -4)\}$ **53.** $y = -\dfrac{3}{7}x + \dfrac{4}{7}; y = -\dfrac{3}{7}x + \dfrac{3}{14}$; no solution   **55.** Both are $y = -\dfrac{2}{3}x + \dfrac{1}{3}$;

infinitely many solutions   **57.** (a) Use substitution, since the second equation is solved for y. (b) Use elimination, since the

coefficients of the y-terms are opposites. (c) Use elimination, since the equations are in standard form with no coefficients of 1

or -1. Solving by substitution would involve fractions.   **59.** $\{(-3, 2)\}$ **61.** $\{(-4, 6)\}$ **63.** $\{(x, y) \mid 4x - y = -2\}$

65. $\left\{\left(1, \dfrac{1}{2}\right)\right\}$ **67.** $(3, -4)$ **69.** A   **71.** (a) $\{(5, 5)\}$ (b)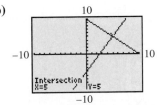

73. (a) $\{(0, -2)\}$ **(b)**

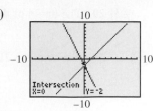

75. (a) \$4 **(b)** 300 half-gallons **(c)** supply: 200 half-gallons; demand: 400 half-gallons **77. (a)** 1994–1996 **(b)** 1997; CBS; 16% **(c)** 1998: 16% share; 1999: 16% share **(d)** 1994; CBS and NBC; $(1994, 18)$ **(e)** Viewership has generally declined during these years.

79. 2000, 2001, 2002, first half of 2003 **81.** $(3.6, 10.4)$ (Values may vary slightly based on the method of solution used.)

83. $\{(2, 4)\}$ **85.** $\left\{\left(\frac{1}{2}, 2\right)\right\}$ **87.** $\left(\frac{c}{a}, 0\right)$ **89.** $\left\{\left(-\frac{1}{a}, -5\right)\right\}$ **91.** $\{(1, 3)\}$ **92.** $f(x) = -3x + 6$; linear

93. $g(x) = \frac{2}{3}x + \frac{7}{3}$; linear **94.** one; 1; 3; 1; 3; 1; 3 **95.** $8x - 12y + 4z = 20$ **97.** 4 **99.** -3

Section 4.2 (pages 249–251)

1. B **3.** $\{(3, 2, 1)\}$ **5.** $\{(1, 4, -3)\}$ **7.** $\{(0, 2, -5)\}$ **9.** $\{(1, 0, 3)\}$ **11.** $\{(-12, 18, 0)\}$ **13.** $\left\{\left(1, \frac{3}{10}, \frac{2}{5}\right)\right\}$

15. $\left\{\left(-\frac{7}{3}, \frac{22}{3}, 7\right)\right\}$ **17.** $\{(4, 5, 3)\}$ **19.** $\{(2, 2, 2)\}$ **21.** $\left\{\left(\frac{8}{3}, \frac{2}{3}, 3\right)\right\}$ **23.** $\{(-1, 0, 0)\}$ **25.** $\{(-3, 5, -6)\}$

27. $\emptyset$; inconsistent system **29.** $\{(x, y, z) \mid x - y + 4z = 8\}$; dependent equations **31.** $\{(3, 0, 2)\}$

33. $\{(x, y, z) \mid 2x + y - z = 6\}$; dependent equations **35.** $\{(0, 0, 0)\}$ **37.** $\emptyset$; inconsistent system **39.** $\{(2, 1, 5, 3)\}$

41. $\{(-2, 0, 1, 4)\}$ **43.** 100 in., 103 in., 120 in. **45.** $-4, 8, 12$

Section 4.3 (pages 259–265)

1. wins: 99; losses: 63 **3.** length: 78 ft; width: 36 ft **5.** ExxonMobil: \$340 billion; Wal-Mart: \$316 billion **7.** $x = 40$ and $y = 50$, so the angles measure 40° and 50°. **9.** NHL: \$247.32; NBA: \$267.37 **11.** Regular Roast Beef: \$2.39; Large Roast Beef: \$4.19 **13. (a)** 6 oz **(b)** 15 oz **(c)** 24 oz **(d)** 30 oz **15.** \$1.29x **17.** 6 gal of 25%; 14 gal of 35% **19.** pure acid: 6 L; 10% acid: 48 L **21.** nuts: 14 kg; cereal: 16 kg **23.** \$1000 at 2%; \$2000 at 4% **25. (a)** $(10 - x)$ mph **(b)** $(10 + x)$ mph **27.** train: 60 km per hr; plane: 160 km per hr **29.** boat: 21 mph; current: 3 mph **31.** \$0.75-per-lb candy: 5.22 lb; \$1.25-per-lb candy: 3.78 lb **33.** general admission: 76; with student ID: 108 **35.** 8 for a citron; 5 for a wood apple **37.** $x + y + z = 180$; angle measures: 70°, 30°, 80° **39.** first: 20°; second: 70°; third: 90° **41.** shortest: 12 cm; middle: 25 cm; longest: 33 cm **43.** gold: 35; silver: 39; bronze: 29 **45.** \$14 tickets: 300; \$20 tickets: 225; \$50 tickets: 60 **47.** bookstore A: 140; bookstore B: 280; bookstore C: 380 **49.** first chemical: 50 kg; second chemical: 400 kg; third chemical: 300 kg **51.** wins: 46; losses: 25; ties: 11 **53. (a)** 6 **(b)** $-\frac{1}{6}$

55. (a) $-\frac{7}{8}$ **(b)** $\frac{8}{7}$

Section 4.4 (pages 271–273)

1. (a) $0, 5, -3$ **(b)** $1, -3, 8$ **(c)** yes; the number of rows is the same as the number of columns (three). **(d)** $\begin{bmatrix} 1 & 4 & 8 \\ 0 & 5 & -3 \\ -2 & 3 & 1 \end{bmatrix}$

(e) $\begin{bmatrix} 1 & -\frac{3}{2} & -\frac{1}{2} \\ 0 & 5 & -3 \\ 1 & 4 & 8 \end{bmatrix}$ **(f)** $\begin{bmatrix} 1 & 15 & 25 \\ 0 & 5 & -3 \\ 1 & 4 & 8 \end{bmatrix}$ **3.** $\begin{bmatrix} 1 & 2 & | & 11 \\ 2 & -1 & | & -3 \end{bmatrix}$; $\begin{bmatrix} 1 & 2 & | & 11 \\ 0 & -5 & | & -25 \end{bmatrix}$; $\begin{bmatrix} 1 & 2 & | & 11 \\ 0 & 1 & | & 5 \end{bmatrix}$; $x + 2y = 11, y = 5$;

$\{(1, 5)\}$ **5.** $\{(4, 1)\}$ **7.** $\{(1, 1)\}$ **9.** $\{(-1, 4)\}$ **11.** $\emptyset$ **13.** $\{(x, y) \mid 2x + y = 4\}$

15. $\begin{bmatrix} 1 & 1 & -1 & | & -3 \\ 0 & -1 & 3 & | & 10 \\ 0 & -6 & 7 & | & 38 \end{bmatrix}; \begin{bmatrix} 1 & 1 & -1 & | & -3 \\ 0 & 1 & -3 & | & -10 \\ 0 & -6 & 7 & | & 38 \end{bmatrix}; \begin{bmatrix} 1 & 1 & -1 & | & -3 \\ 0 & 1 & -3 & | & -10 \\ 0 & 0 & -11 & | & -22 \end{bmatrix}; \begin{bmatrix} 1 & 1 & -1 & | & -3 \\ 0 & 1 & -3 & | & -10 \\ 0 & 0 & 1 & | & 2 \end{bmatrix}; x + y - z = -3,$

$y - 3z = -10, z = 2; \{(3, -4, 2)\}$ **17.** $\{(4, 0, 1)\}$ **19.** $\{(-1, 23, 16)\}$ **21.** $\{(3, 2, -4)\}$

23. $\{(x, y, z) \mid x - 2y + z = 4\}$ **25.** $\emptyset$ **27.** $\{(1, 1)\}$ **29.** $\{(-1, 2, 1)\}$ **31.** $\{(1, 7, -4)\}$ **33.** 64   **35.** 625

37. $\dfrac{81}{256}$

Chapter 4 Review Exercises   (pages 278–281)

1. (a) 1980 and 1985 **(b)** just less than 500,000   **3.** D   **5.** $\left\{\left(-\dfrac{8}{9}, -\dfrac{4}{3}\right)\right\}$ **7.** $\{(2, 4)\}$ **9.** $\{(0, 1)\}$ **11.** $\{(-6, 3)\}$

13. $\emptyset$; inconsistent system   **15.** $\{(1, -5, 3)\}$ **17.** $\emptyset$; inconsistent system   **19.** Boston Red Sox: $40.77;
Chicago Cubs: $28.45   **21.** $2-per-lb nuts: 30 lb; $1-per-lb candy: 70 lb   **23.** $40,000 at 10%; $100,000 at 6%;
$140,000 at 5%   **25.** Mantle: 54; Maris: 61; Blanchard: 21   **27.** $\{(-1, 5)\}$ **29.** $\{(1, 2, -1)\}$ **31.** $\{(12, 9)\}$
33. $\{(3, -1)\}$ **35.** $\{(0, 4)\}$ **37.** 20 L

Chapter 4 Test   (pages 281–282)

[4.1] **1. (a)** Houston, Phoenix, Dallas **(b)** Philadelphia **(c)** Dallas, Phoenix, Philadelphia, Houston

2. (a) 2010; 1.45 million **(b)** $(2025, 2.8)$ **3.** $\{(6, 1)\}$ **4.** $\{(6, -4)\}$ **5.** $\left\{\left(-\dfrac{9}{4}, \dfrac{5}{4}\right)\right\}$

6. $\{(x, y) \mid 12x - 5y = 8\}$; dependent equations   **7.** $\{(3, 3)\}$ **8.** $\{(0, -2)\}$ **9.** $\emptyset$; inconsistent system

[4.2] **10.** $\left\{\left(-\dfrac{2}{3}, \dfrac{4}{5}, 0\right)\right\}$ **11.** $\{(3, -2, 1)\}$ [4.3] **12.** *Ocean's Eleven:* $183.4 million; *Runaway Bride:* $152.1 million

13. 45 mph, 75 mph   **14.** 20% solution: 4 L; 50% solution: 8 L   **15.** AC adaptor: $8; rechargeable flashlight: $15

16. Orange Pekoe: 60 oz; Irish Breakfast: 30 oz; Earl Grey: 10 oz   [4.4] **17.** $\left\{\left(\dfrac{2}{5}, \dfrac{7}{5}\right)\right\}$ **18.** $\{(-1, 2, 3)\}$

Chapters 1–4 Cumulative Review Exercises   (pages 283–284)

[1.2] **1.** $-\dfrac{23}{20}$ **2.** $-\dfrac{2}{9}$ [1.3] **3.** 81   **4.** -81 **5.** -81 **6.** 0.7   **7.** -0.7 **8.** It is not a real number.   **9.** -199

10. 455   [1.4] **11.** commutative property   [2.1] **12.** $\left\{-\dfrac{15}{4}\right\}$ [2.7] **13.** $\left\{\dfrac{2}{3}, 2\right\}$ [2.2] **14.** $x = \dfrac{d - by}{a - c}$, or $x = \dfrac{by - d}{c - a}$

[2.1] **15.** $\{11\}$ [2.5] **16.** $\left(-\infty, \dfrac{240}{13}\right]$ [2.7] **17.** $\left[-2, \dfrac{2}{3}\right]$ **18.** $(-\infty, \infty)$ [2.2] **19.** 2010; 1813; 62.8%; 57.2%

[2.3, 2.4] **20.** 6 m   **21.** pennies: 35; nickels: 29; dimes: 30   **22.** 46°, 46°, 88°   [3.1] **23.** $y = 6$ **24.** $x = 4$

[3.2] **25.** $-\dfrac{4}{3}$ **26.** $\dfrac{3}{4}$ [3.3] **27.** $4x + 3y = 10$ [3.2] **28.** 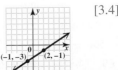   [3.4] **29.**

[3.5] **30. (a)** -6 **(b)** $a^2 + 3a - 6$ [4.1, 4.4] **31.** $\{(3, -3)\}$ **32.** $\{(x, y) \mid x - 3y = 7\}$ [4.2, 4.4] **33.** $\{(5, 3, 2)\}$
[4.3] **34.** oranges: 5 lb; apples: 1 lb   **35.** Tickle Me Elmo: \$27.63; T.M.X.: \$40.00   **36.** small: \$1.50; large: \$2.50
37. peanuts: \$2 per lb; cashews: \$4 per lb   [4.1] **38. (a)** $x = 8$, or 800 items; \$3000 **(b)** about \$400

CHAPTER 5   Exponents, Polynomials, and Polynomial Functions

Section 5.1   (pages 296–301)

1. incorrect; $(ab)^2 = a^2b^2$ **3.** incorrect; $\left(\dfrac{4}{a}\right)^3 = \dfrac{4^3}{a^3}$ **5.** The product rule says that when exponential expressions with like
bases are multiplied, the base stays the same and the exponents are added. For example, $x^5 \cdot x^6 = x^{11}$.   **7.** 13^{12} **9.** x^{17}
11. $-27w^8$ **13.** $18x^3y^8$ **15.** The product rule does not apply.   **17. (a)** B **(b)** C **(c)** B **(d)** C   **19.** 1   **21.** -1
23. 1   **25.** 2   **27.** 0   **29.** -2 **31. (a)** B **(b)** D **(c)** B **(d)** D   **33.** $\dfrac{1}{5^4}$, or $\dfrac{1}{625}$ **35.** $\dfrac{1}{8}$ **37.** $\dfrac{1}{16x^2}$ **39.** $\dfrac{4}{x^2}$
41. $-\dfrac{1}{a^3}$ **43.** $\dfrac{1}{a^4}$ **45.** $\dfrac{11}{30}$ **47.** $-\dfrac{5}{24}$ **49.** When n is even, the expressions are opposites. When n is odd, they are equal.
51. 16   **53.** $\dfrac{27}{4}$ **55.** $\dfrac{27}{8}$ **57.** $\dfrac{25}{16}$ **59. (a)** B **(b)** D **(c)** D **(d)** B   **61.** The quotient rule says that when exponential
expressions with like bases are divided, the base stays the same and the exponents are subtracted. For example, $\dfrac{x^8}{x^5} = x^3$.
63. 4^2, or 16   **65.** x^4 **67.** $\dfrac{1}{r^3}$ **69.** 6^6 **71.** $\dfrac{1}{6^{10}}$ **73.** 7^2, or 49   **75.** r^3 **77.** The quotient rule does not apply.
79. x^{18} **81.** $\dfrac{27}{125}$ **83.** $64t^3$ **85.** $-216x^6$ **87.** $-\dfrac{64m^6}{t^3}$ **89.** $\dfrac{1}{3}$ **91.** $\dfrac{1}{a^5}$ **93.** $\dfrac{1}{k^2}$ **95.** $-4r^6$ **97.** $\dfrac{625}{a^{10}}$ **99.** $\dfrac{z^4}{x^3}$
101. $-14k^3$ **103.** $\dfrac{p^4}{5}$ **105.** $\dfrac{1}{2pq}$ **107.** $\dfrac{4}{a^2}$ **109.** $\dfrac{1}{6y^{13}}$ **111.** $\dfrac{4k^5}{m^2}$ **113.** $\dfrac{4k^{17}}{125}$ **115.** $\dfrac{2k^5}{3}$ **117.** $\dfrac{8}{3pq^{10}}$ **119.** $\dfrac{y^9}{8}$
121. $-\dfrac{3}{32m^8p^4}$ **123.** $\dfrac{2}{3y^4}$ **125.** $\dfrac{3p^8}{16q^{14}}$ **127.** 5.3×10^2 **129.** 8.3×10^{-1} **131.** 6.92×10^{-6} **133.** -3.85×10^4
135. 72,000   **137.** 0.00254   **139.** $-60,000$ **141.** 0.000012   **143.** 0.06   **145.** 0.0000025   **147.** 200,000
149. 3000   **151.** $\$1 \times 10^9$; $\$1 \times 10^{12}$; $\$2.128 \times 10^{12}$; 1.44419×10^5 **153. (a)** 2.814×10^8 **(b)** $\$1 \times 10^{12}$ (or $\$10^{12}$)
(c) \$3554   **155.** 300 sec   **157.** approximately 5.87×10^{12} mi   **159. (a)** 20,000 hr **(b)** 833 days   **161.** 7.5×10^9
163. 4×10^{17} **165.** $9x$ **167.** $4 - 11z$ **169.** $2x - 2$

Section 5.2   (pages 305–307)

1. $2x^3 - 3x^2 + x + 4$ **3.** $p^7 - 8p^5 + 4p^3$ **5.** $3m^4 - m^3 + 5m^2 + 10$ **7.** 7; 1   **9.** -15; 2   **11.** 1; 4   **13.** $\dfrac{1}{6}$; 1
15. -1; 6   **17.** monomial; 0   **19.** binomial; 1   **21.** binomial; 8   **23.** trinomial; 3   **25.** none of these; 5   **27.** A
29. $8z^4$ **31.** $7m^3$ **33.** $5x$ **35.** already simplified   **37.** $-t + 13s$ **39.** $8k^2 + 2k - 7$ **41.** $-2n^4 - n^3 + n^2$
43. $-2ab^2 + 20a^2b$ **45.** $3m + 11$ **47.** $-p - 4$ **49.** A *monomial* (or *term*) is a numeral, a variable, or a product of
numerals and variables raised to positive integer powers. Some examples of monomials are 6, x, and $-4x^2y^3$. A *binomial* is a
sum or difference of exactly two terms, such as $x^2 + y^2$ and $x^2 - y^2$. A *trinomial* consists of exactly three terms, such as
$x^2 - 3x + 8$. These are all examples of *polynomials*.   **51.** $8x^2 + x - 2$ **53.** $-t^4 + 2t^2 - t + 5$
55. $5y^3 - 3y^2 + 5y + 1$ **57.** $r + 13$ **59.** $-2a^2 - 2a - 7$ **61.** $-3z^5 + z^2 + 7z$ **63.** $12p - 4$
65. $-9p^2 + 11p - 9$ **67.** $5a + 18$ **69.** $14m^2 - 13m + 6$ **71.** $13z^2 + 10z - 3$ **73.** $10y^3 - 7y^2 + 5y + 8$

75. $-5a^4 - 6a^3 + 9a^2 - 11$ **77.** $3y^2 - 4y + 2$ **79.** $-4m^2 + 4n^2 - 7n$ **81.** $y^4 - 4y^2 - 4$ **83.** $10z^2 - 16z$
85. (a) -2 **(b)** 7 **87. (a)** -9 **(b)** 0 **89.** function **91.** function

Section 5.3 (pages 315–317)

1. (a) -10 **(b)** 8 **3. (a)** 8 **(b)** 2 **5. (a)** 8 **(b)** 74 **7. (a)** -11 **(b)** 4 **9. (a)** 15,163 **(b)** 17,401 **(c)** 19,656
11. (a) 9 million **(b)** 61 million **(c)** 118 million **13. (a)** $8x - 3$ **(b)** $2x - 17$ **15. (a)** $-x^2 + 12x - 12$

(b) $9x^2 + 4x + 6$ **17.** $x^2 + 2x - 9$ **19.** 6 **21.** $x^2 - x - 6$ **23.** 6 **25.** -33 **27.** 0 **29.** $-\dfrac{9}{4}$ **31.** $-\dfrac{9}{2}$

33. For example, let $f(x) = 2x^3 + 3x^2 + x + 4$ and $g(x) = 2x^4 + 3x^3 - 9x^2 + 2x - 4$. For these functions, $(f - g)(x) = -2x^4 - x^3 + 12x^2 - x + 8$, and $(g - f)(x) = 2x^4 + x^3 - 12x^2 + x - 8$. Because the two differences are not equal, subtraction of polynomial functions is not commutative. **35.** 16 **37.** 83 **39.** 13 **41.** $(2x + 3)^2 + 4$

43. $(x + 5)^2 + 4$ **45.** $2x + 8$ **47.** $\dfrac{137}{4}$ **49.** 8 **51.** $(f \circ g)(x) = 63{,}360x$; it computes the number of inches in

x miles. **53.** $(A \circ r)(t) = 4\pi t^2$; this is the area of the circular layer as a function of time.

55. 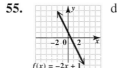 domain: $(-\infty, \infty)$; range: $(-\infty, \infty)$ **57.** domain: $(-\infty, \infty)$; range: $(-\infty, 0]$

59. domain: $(-\infty, \infty)$; range: $(-\infty, \infty)$ **61.** $12m^5$ **63.** $-6a^3b^9$ **65.** $60x^3y^4$

Section 5.4 (pages 324–326)

1. C **3.** D **5.** $-24m^5$ **7.** $-28x^7y^4$ **9.** $-6x^2 + 15x$ **11.** $-2q^3 - 3q^4$ **13.** $18k^4 + 12k^3 + 6k^2$
15. $6m^3 + m^2 - 14m - 3$ **17.** $m^3 - 3m^2 - 40m$ **19.** $24z^3 - 20z^2 - 16z$ **21.** $4x^5 - 4x^4 - 24x^3$ **23.** $6y^2 + y - 12$
25. $-2b^3 + 2b^2 + 18b + 12$ **27.** $25m^2 - 9n^2$ **29.** $8z^4 - 14z^3 + 17z^2 + 20z - 3$ **31.** $6p^4 + p^3 + 4p^2 - 27p - 6$
33. $m^2 - 3m - 40$ **35.** $12k^2 + k - 6$ **37.** $3z^2 + zw - 4w^2$ **39.** $12c^2 + 16cd - 3d^2$ **41.** $0.1x^2 + 0.63x - 0.13$

43. $3w^2 - \dfrac{23}{4}wz - \dfrac{1}{2}z^2$ **45.** The product of two binomials is the sum of the product of the first terms, the product of the

outer terms, the product of the inner terms, and the product of the last terms. **47.** $4p^2 - 9$ **49.** $25m^2 - 1$ **51.** $9a^2 - 4c^2$

53. $16x^2 - \dfrac{4}{9}$ **55.** $16m^2 - 49n^4$ **57.** $75y^7 - 12y$ **59.** $y^2 - 10y + 25$ **61.** $4p^2 + 28p + 49$

63. $16n^2 + 24nm + 9m^2$ **65.** $k^2 - \dfrac{10}{7}kp + \dfrac{25}{49}p^2$ **67.** $0.04x^2 - 0.56xy + 1.96y^2$ **69.** Write 101 as $100 + 1$ and 99 as

$100 - 1$. Then $101 \cdot 99 = (100 + 1)(100 - 1) = 100^2 - 1^2 = 10{,}000 - 1 = 9999$.
71. $25x^2 + 10x + 1 + 60xy + 12y + 36y^2$ **73.** $4a^2 + 4ab + b^2 - 12a - 6b + 9$ **75.** $4a^2 + 4ab + b^2 - 9$
77. $4h^2 - 4hk + k^2 - j^2$ **79.** $y^3 + 6y^2 + 12y + 8$ **81.** $125r^3 - 75r^2s + 15rs^2 - s^3$ **83.** $q^4 - 8q^3 + 24q^2 - 32q + 16$
85. $6a^3 + 7a^2b + 4ab^2 + b^3$ **87.** $4z^4 - 17z^3x + 12z^2x^2 - 6zx^3 + x^4$ **89.** $m^4 - 4m^2p^2 + 4mp^3 - p^4$

91. $a^4b - 7a^2b^3 - 6ab^4$ **93.** 49; 25; $49 \neq 25$ **95.** 2401; 337; $2401 \neq 337$ **97.** $\dfrac{9}{2}x^2 - 2y^2$ **99.** $15x^2 - 2x - 24$

101. $a - b$ **102.** $A = s^2$; $(a - b)^2$ **103.** $(a - b)b$, or $ab - b^2$; $2ab - 2b^2$ **104.** b^2 **105.** a^2; a

106. $a^2 - (2ab - 2b^2) - b^2 = a^2 - 2ab + b^2$ **107. (a)** They must be equal to each other. **(b)** $(a - b)^2 = a^2 - 2ab + b^2$; this equation reinforces the special product for the square of a binomial difference. **108.**

a	Area: a^2	Area: ab
b	Area: ab	Area: b^2
	a	b

The large square is made up of two smaller squares and two congruent rectangles. The sum of the areas is $a^2 + 2ab + b^2$. Since $(a + b)^2$ must represent the same quantity, they must be equal. Thus $(a + b)^2 = a^2 + 2ab + b^2$. **109.** $10x^2 - 2x$ **111.** $2x^2 - x - 3$
113. $8x^3 - 27$ **115.** $2x^3 - 18x$ **117.** -20 **119.** $2x^2 - 6x$ **121.** 36 **123.** $\dfrac{35}{4}$ **125.** $\dfrac{1859}{64}$ **127.** $2p^4$

129. $\dfrac{-4b^6}{3a^2}$ **131.** $-8a^2 + a + 4$

Section 5.5 (pages 331–333)

1. quotient; exponents **3.** descending **5.** $3x^3 - 2x^2 + 1$ **7.** $3y + 4 - \dfrac{5}{y}$ **9.** $3m + 5 + \dfrac{6}{m}$ **11.** $n - \dfrac{3n^2}{2m} + 2$

13. $\dfrac{2y}{x} + \dfrac{3}{4} + \dfrac{3w}{x}$ **15.** $r^2 - 7r + 6$ **17.** $y - 4$ **19.** $q + 8$ **21.** $t + 5$ **23.** $p - 4 + \dfrac{44}{p + 6}$ **25.** $m^2 + 2m - 1$

27. $m^2 + m + 3$ **29.** $z^2 + 3$ **31.** $x^2 + 2x - 3 + \dfrac{6}{4x + 1}$ **33.** $2x - 5 + \dfrac{-4x + 5}{3x^2 - 2x + 4}$ **35.** $x^2 + x + 3$

37. $3x^2 + 6x + 11 + \dfrac{26}{x - 2}$ **39.** $2k^2 + 3k - 1$ **41.** $2y^2 + 2$ **43.** $x^2 - 4x + 2 + \dfrac{9x - 4}{x^2 + 3}$

45. $p^2 + \dfrac{5}{2}p + 2 + \dfrac{-1}{2p + 2}$ **47.** $\dfrac{3}{2}a - 10 + \dfrac{77}{2a + 6}$ **49.** $p^2 + p + 1$ **51.** $\dfrac{2}{3}x - 1$ **53.** $\dfrac{3}{4}a - 2 + \dfrac{1}{4a + 3}$

55. $2p + 7$ **57.** -13; -13; they are the same, which suggests that when $P(x)$ is divided by $x - r$, the result is $P(r)$. Here,

$r = -1$. **59.** $5x - 1$; 0 **61.** $2x - 3$; -1 **63.** $4x^2 + 6x + 9$; $\dfrac{3}{2}$ **65.** $\dfrac{x^2 - 9}{2x}$, $x \neq 0$ **67.** $-\dfrac{5}{4}$

69. $\dfrac{x - 3}{2x}$, $x \neq 0$ **71.** 0 **73.** $-\dfrac{35}{4}$ **75.** $\dfrac{7}{2}$ **77.** $9(6 + r^2)$ **79.** $7(2x - 3z)$ **81.** $(3x + 4)(x + 1)$ **83.** $18z^7 w^9$

85. $\dfrac{12p^3}{5q^3}$

Chapter 5 Review Exercises (pages 338–341)

1. 64 **3.** -125 **5.** $\dfrac{81}{16}$ **7.** $\dfrac{11}{30}$ **9.** 0 **11.** x^8 **13.** $\dfrac{1}{z^{15}}$ **15.** $\dfrac{r^{17}}{9}$ **17.** $\dfrac{1}{96m^7}$ **19.** $-12x^2 y^8$ **21.** $\dfrac{10p^8}{q^7}$

23. In $(-6)^0$, the base is -6 and the expression simplifies to 1. In -6^0, the base is 6 and the expression simplifies to -1.
25. yes **27.** For example, let $x = 2$ and $y = 3$. Then $(x^2 + y^2)^2 = (2^2 + 3^2)^2 = 169$; $x^4 + y^4 = 2^4 + 3^4 = 97 \neq 169$.
29. 7.65×10^{-8} **31.** 2.814×10^8; 5.0454×10^4; 1×10^2 **33.** 0.0058 **35.** 1.5×10^3; 1500 **37.** 2.7×10^{-2}; 0.027
39. (a) 5.449×10^3 **(b)** 63 mi^2 **41.** -1 **43.** 504 **45. (a)** $9m^7 + 14m^6$ **(b)** binomial **(c)** 7 **47. (a)** $-7q^5 r^3$
(b) monomial **(c)** 8 **49.** $-x^2 - 3x + 1$ **51.** $6a^3 - 4a^2 - 16a + 15$ **53.** $12x^2 + 8x + 5$ **55. (a)** $5x^2 - x + 5$
(b) $-5x^2 + 5x + 1$ **(c)** 11 **(d)** -9 **57. (a)** 15.9 million **(b)** 31.6 million **(c)** 27.244 million **59.**

61. $-12k^3 - 42k$ **63.** $6w^2 - 13wt + 6t^2$ **65.** $3q^3 - 13q^2 - 14q + 20$ **67.** $36r^4 - 1$ **69.** $16m^2 + 24m + 9$

71. $y^2 - 3y + \dfrac{5}{4}$ **73.** $p^2 + 6p + 9 + \dfrac{54}{2p - 3}$ **75. (a)** A **(b)** G **(c)** C **(d)** C **(e)** A **(f)** E **(g)** B **(h)** H **(i)** F

77. $\dfrac{1}{125}$ **79.** $21p^9 + 7p^8 + 14p^7$ **81.** $-\dfrac{1}{5z^9}$ **83.** $8x + 1 + \dfrac{5}{x - 3}$ **85.** $9m^2 - 30mn + 25n^2 - p^2$

87. $-3k^2 + 4k - 7$

Chapter 5 Test (pages 341–342)

[5.1] **1. (a)** C **(b)** A **(c)** D **(d)** A **(e)** E **(f)** F **(g)** B **(h)** G **(i)** C **2.** $\dfrac{4x^7}{9y^{10}}$ **3.** $\dfrac{6}{r^{14}}$ **4.** $\dfrac{16}{9p^{10}q^{28}}$ **5.** $\dfrac{16}{x^6y^{16}}$

6. 0.00000091 **7.** 3×10^{-4}; 0.0003 [5.3] **8. (a)** -18 **(b)** $-2x^2 + 12x - 9$ **(c)** $-2x^2 - 2x - 3$ **(d)** -7

9. (a) 23 **(b)** $3x^2 + 11$ **(c)** $9x^2 + 30x + 27$ **10.**   **11.**   **12. (a)** 616 thousand

(b) 740 thousand **(c)** 854 thousand [5.2] **13.** $x^3 - 2x^2 - 10x - 13$ [5.4] **14.** $10x^2 - x - 3$

15. $6m^3 - 7m^2 - 30m + 25$ **16.** $36x^2 - y^2$ **17.** $9k^2 + 6kq + q^2$ **18.** $4y^2 - 9z^2 + 6zx - x^2$ [5.5] **19.** $4p - 8 + \dfrac{6}{p}$

20. $x^2 + 4x + 4$ [5.4] **21. (a)** $x^3 + 4x^2 + 5x + 2$ **(b)** 0 [5.5] **22. (a)** $x + 2, \ x \neq -1$ **(b)** 0

Chapters 1–5 Cumulative Review Exercises   (pages 343–344)

[1.1] **1.** A, B, C, D, F **2.** B, C, D, F **3.** D, F **4.** C, D, F **5.** E, F **6.** D, F [1.3] **7.** 32 **8.** $-\dfrac{1}{72}$ **9.** 0

[2.1] **10.** $\{-65\}$ **11.** $(-\infty, \infty)$ [2.2] **12.** $t = \dfrac{A - p}{pr}$ [2.5] **13.** $(-\infty, 6)$ [2.7] **14.** $\left\{-\dfrac{1}{3}, 1\right\}$

15. $\left(-\infty, -\dfrac{8}{3}\right] \cup [2, \infty)$ **16.** $\emptyset$ [2.3, 2.4] **17.** 32%; 390; 270; 10% **18.** 15°, 35°, 130° [3.2] **19.** $-\dfrac{4}{3}$ **20.** 0

[3.3] **21. (a)** $y = -4x + 15$ **(b)** $4x + y = 15$ **22. (a)** $y = 4x$ **(b)** $4x - y = 0$ [3.1] **23.**

[3.4] **24.** **25.** [3.2, 3.4] **26. (a)** 3222 per yr; the number of twin births increased an average of

3222 per yr. **(b)** $y = 3222x + 96{,}445$ **(c)** about 138,331 [3.5] **27.** domain: $\{-4, -1, 2, 5\}$; range: $\{-2, 0, 2\}$; function

28. -9 [4.1] **29.** $\{(3, 2)\}$ **30.** $\emptyset$ [4.2] **31.** $\{(1, 0, -1)\}$ [4.3] **32.** length: 42 ft; width: 30 ft **33.** 15% solution: 6 L;

30% solution: 3 L [5.1] **34.** $\dfrac{8m^9n^3}{p^6}$ **35.** $\dfrac{y^7}{x^{13}z^2}$ **36.** $\dfrac{m^6}{8n^9}$ **37.** $\dfrac{3}{10}$ [5.2] **38.** $2x^2 - 4x + 38$

[5.4] **39.** $15x^2 + 7xy - 2y^2$ **40.** $64m^2 - 25n^2$ **41.** $x^3 + 8y^3$ [5.5] **42.** $4xy^4 - 2y + \dfrac{1}{x^2y}$ **43.** $m^2 - 2m + 3$

[5.3] **44.** -7

CHAPTER 6   Factoring

Section 6.1   (pages 350–351)

1. $12(m - 5)$ **3.** $4(1 + 5z)$ **5.** cannot be factored   **7.** $8k(k^2 + 3)$ **9.** $-2p^2q^4(2p + q)$ **11.** $7x^3(3x^2 + 5x - 2)$
13. $2t^3(5t^2 - 4t - 2)$ **15.** $5ac(3ac^2 - 5c + 1)$ **17.** $16zn^3(zn^3 + 4n^4 - 2z^2)$ **19.** $7ab(2a^2b + a - 3a^4b^2 + 6b^3)$
21. $(m - 4)(2m + 5)$ **23.** $11(2z - 1)$ **25.** $(2 - x)^2(1 + 2x)$ **27.** $(3 - x)(6 + 2x - x^2)$ **29.** $20z(2z + 1)(3z + 4)$
31. $5(m + p)^2(m + p - 2 - 3m^2 - 6mp - 3p^2)$ **33.** $r(-r^2 + 3r + 5); -r(r^2 - 3r - 5)$ **35.** $12s^4(-s + 4);$
$-12s^4(s - 4)$ **37.** $2x^2(-x^3 + 3x + 2); -2x^2(x^3 - 3x - 2)$ **39.** $(m + q)(x + y)$ **41.** $(5m + n)(2 + k)$
43. $(2 - q)(2 - 3p)$ **45.** $(p + q)(p - 4z)$ **47.** $(2x + 3)(y + 1)$ **49.** $(m + 4)(m^2 - 6)$ **51.** $(a^2 + b^2)(-3a + 2b)$
53. $(y - 2)(x - 2)$ **55.** $(3y - 2)(3y^3 - 4)$ **57.** $(1 - a)(1 - b)$ **59.** $m^{-5}(3 + m^2)$ **61.** $p^{-3}(3 + 2p)$
63. The directions said that the student was to factor the polynomial *completely*. The completely factored form is $4xy^3(xy^2 - 2)$.
65. C   **67.** $k^2 + 6k - 7$ **69.** $20y^2 + 14y - 12$ **71.** $25x^2 - 4t^2$ **73.** $6y^6 + y^3 - 12$ **75.** $15t^3 - 110t^2 - 80t$

Section 6.2   (pages 358–359)

1. D   **3.** B   **5.** $(y - 3)(y + 10)$ **7.** $(p + 8)(p + 7)$ **9.** prime   **11.** $(a + 5b)(a - 7b)$ **13.** prime
15. $(xy + 9)(xy + 2)$ **17.** $-(6m - 5)(m + 3)$ **19.** $(5x - 6)(2x + 3)$ **21.** $(4k + 3)(5k + 8)$
23. $(3a - 2b)(5a - 4b)$ **25.** $(6m - 5)^2$ **27.** prime   **29.** $(2xz - 1)(3xz + 4)$ **31.** $3(4x + 5)(2x + 1)$
33. $-5(a + 6)(3a - 4)$ **35.** $-11x(x - 6)(x - 4)$ **37.** $2xy^3(x - 12y)^2$ **39.** $6a(a - 3)(a + 5)$
41. $13y(y + 4)(y - 1)$ **43.** $3p(2p - 1)^2$ **45.** She did not factor the polynomial *completely*. The factor $(4x + 10)$ can be
factored further into $2(2x + 5)$, giving the final form as $2(2x + 5)(x - 2)$.   **47.** $(6p^3 - r)(2p^3 - 5r)$
49. $(5k + 4)(2k + 1)$ **51.** $(3m + 3p + 5)(m + p - 4)$ **53.** $(a + b)^2(a - 3b)(a + 2b)$ **55.** $(p + q)^2(p + 3q)$
57. $(z - x)^2(z + 2x)$ **59.** $(p^2 - 8)(p^2 - 2)$ **61.** $(2x^2 + 3)(x^2 - 6)$ **63.** $(4x^2 + 3)(4x^2 + 1)$ **65.** $9x^2 - 25$
67. $p^2 + 6pq + 9q^2$ **69.** $y^3 + 27$

Section 6.3   (pages 364–366)

1. A, D   **3.** B, C   **5.** The sum of two squares can be factored only if the binomial has a common factor.
7. $(p + 4)(p - 4)$ **9.** $(5x + 2)(5x - 2)$ **11.** $2(3a + 7b)(3a - 7b)$ **13.** $4(4m^2 + y^2)(2m + y)(2m - y)$
15. $(y + z + 9)(y + z - 9)$ **17.** $(4 + x + 3y)(4 - x - 3y)$ **19.** $(p^2 + 16)(p + 4)(p - 4)$ **21.** $(k - 3)^2$
23. $(2z + w)^2$ **25.** $(4m - 1 + n)(4m - 1 - n)$ **27.** $(2r - 3 + s)(2r - 3 - s)$ **29.** $(x + y - 1)(x - y + 1)$
31. $2(7m + 3n)^2$ **33.** $(p + q + 1)^2$ **35.** $(a - b + 4)^2$ **37.** $(x - 3)(x^2 + 3x + 9)$ **39.** $(t - 6)(t^2 + 6t + 36)$
41. $(x + 4)(x^2 - 4x + 16)$ **43.** $(10 + y)(100 - 10y + y^2)$ **45.** $(2x + 1)(4x^2 - 2x + 1)$
47. $(5x - 6)(25x^2 + 30x + 36)$ **49.** $(x - 2y)(x^2 + 2xy + 4y^2)$ **51.** $(4g - 3h)(16g^2 + 12gh + 9h^2)$
53. $(7p + 5q)(49p^2 - 35pq + 25q^2)$ **55.** $3(2n + 3p)(4n^2 - 6np + 9p^2)$
57. $(y + z + 4)(y^2 + 2yz + z^2 - 4y - 4z + 16)$ **59.** $(m^2 - 5)(m^4 + 5m^2 + 25)$
61. $(10x^3 - 3)(100x^6 + 30x^3 + 9)$ **63.** $(5y^2 + z)(25y^4 - 5y^2z + z^2)$ **64.** $(x^3 - y^3)(x^3 + y^3); (x - y)(x^2 + xy + y^2) \cdot$
$(x + y)(x^2 - xy + y^2)$ **65.** $(x^2 + xy + y^2)(x^2 - xy + y^2)$ **66.** $(x^2 - y^2)(x^4 + x^2y^2 + y^4); (x - y)(x + y)(x^4 + x^2y^2 + y^4)$
67. $x^4 + x^2y^2 + y^4$ **68.** The product must equal $x^4 + x^2y^2 + y^4$. Multiply $(x^2 + xy + y^2)(x^2 - xy + y^2)$ to verify this.
69. Start by factoring as a difference of squares.   **71.** $(5p + 2q)(25p^2 - 10pq + 4q^2 + 5p - 2q)$
73. $(3a - 4b)(9a^2 + 12ab + 16b^2 + 5)$ **75.** $(t - 3)(2t + 1)(4t^2 - 2t + 1)$
77. $(8m - 9n)(8m + 9n - 64m^2 - 72mn - 81n^2)$ **79.** $(2x + y)(a - b)$ **81.** $(p + 7)(p - 3)$

Section 6.4 (pages 368–369)

1. $(10a + 3b)(10a - 3b)$ **3.** $3p^2(p - 6)(p + 5)$ **5.** $3pq(a + 6b)(a - 5b)$ **7.** prime   **9.** $(6b + 1)(b - 3)$
11. $(x - 10)(x^2 + 10x + 100)$ **13.** $(p + 2)(4 + m)$ **15.** $9m(m - 5 + 2m^2)$ **17.** $2(3m - 10)(9m^2 + 30m + 100)$
19. $(3m - 5n)^2$ **21.** $(k - 9)(q + r)$ **23.** $16z^2x(zx - 2)$ **25.** $(x + 7)(x - 5)$ **27.** $(x - 5)(x + 5)(x^2 + 25)$
29. $(p + 1)(p^2 - p + 1)$ **31.** $(8m + 25)(8m - 25)$ **33.** $6z(2z^2 - z + 3)$ **35.** $16(4b + 5c)(4b - 5c)$
37. $8(5z + 4)(25z^2 - 20z + 16)$ **39.** $(5r - s)(2r + 5s)$ **41.** $4pq(2p + q)(3p + 5q)$ **43.** $3(4k^2 + 9)(2k + 3)(2k - 3)$
45. $(m - n)(m^2 + mn + n^2 + m + n)$ **47.** $(x - 2m - n)(x + 2m + n)$ **49.** $6p^3(3p^2 - 4 + 2p^3)$ **51.** $2(x + 4)(x - 5)$
53. $8mn$ **55.** $2(5p + 9)(5p - 9)$ **57.** $4rx(3m^2 + mn + 10n^2)$ **59.** $(7a - 4b)(3a + b)$ **61.** prime
63. $(p + 8q - 5)^2$ **65.** $(7m^2 + 1)(3m^2 - 5)$ **67.** $(2r - t)(r^2 - rt + 19t^2)$ **69.** $(x + 3)(x^2 + 1)(x + 1)(x - 1)$
71. $(m + n - 5)(m - n + 1)$ **73.** $\left\{-\dfrac{2}{3}\right\}$ **75.** $\{0\}$ **77.** $\{-10\}$

Section 6.5 (pages 376–380)

1. First rewrite the equation so that one side is 0. Factor the other side and set each factor equal to 0. The solutions of these
linear equations are solutions of the quadratic equation.   **3.** $\{-10, 5\}$ **5.** $\left\{-\dfrac{8}{3}, \dfrac{5}{2}\right\}$ **7.** $\{-2, 5\}$ **9.** $\{-6, -3\}$
11. $\left\{-\dfrac{1}{2}, 4\right\}$ **13.** $\left\{-\dfrac{1}{3}, \dfrac{4}{5}\right\}$ **15.** $\{-3, 4\}$ **17.** $\left\{-5, -\dfrac{1}{5}\right\}$ **19.** $\{-4, 0\}$ **21.** $\{0, 6\}$ **23.** $\{-2, 2\}$
25. $\{-3, 3\}$ **27.** $\{3\}$ **29.** $\left\{-\dfrac{4}{3}\right\}$ **31.** $\{-4, 2\}$ **33.** $\left\{-\dfrac{1}{2}, 6\right\}$ **35.** $\{1, 6\}$ **37.** $\left\{-\dfrac{1}{2}, 0, 5\right\}$ **39.** $\{-1, 0, 3\}$
41. $\left\{-\dfrac{4}{3}, 0, \dfrac{4}{3}\right\}$ **43.** $\left\{-\dfrac{5}{2}, -1, 1\right\}$ **45.** $\{-3, 3, 6\}$ **47.** By dividing each side by a variable expression, she "lost"
the solution 0. The solution set is $\left\{-\dfrac{4}{3}, 0, \dfrac{4}{3}\right\}$.   **49.** $\left\{-\dfrac{1}{2}, 6\right\}$ **51.** $\left\{-\dfrac{2}{3}, \dfrac{4}{15}\right\}$ **53.** $\left\{-\dfrac{3}{2}, \dfrac{1}{2}\right\}$ **55.** width: 16 ft;
length: 20 ft   **57.** base: 12 ft; height: 5 ft   **59.** 50 ft by 100 ft   **61.** -6 and -5 or 5 and 6   **63.** length: 15 in.;
width: 9 in.   **65.** 5 sec   **67.** $6\dfrac{1}{4}$ sec   **69.** $\{-0.5, 4\}$ **71.** $\{-2, 5\}$ **73.** $4p$ **75.** $-\dfrac{3}{4m^4n^3}$ **77.** $\dfrac{36}{75}$

Chapter 6 Review Exercises (pages 384–386)

1. $6p(2p - 1)$ **3.** $4qb(3q + 2b - 5q^2b)$ **5.** $(x + 3)(x - 3)$ **7.** $(m + q)(4 + n)$ **9.** $(m + 3)(2 - a)$
11. $(3p - 4)(p + 1)$ **13.** $(3r + 1)(4r - 3)$ **15.** $(2k - h)(5k - 3h)$ **17.** $2x(4 + x)(3 - x)$ **19.** $(y^2 + 4)(y^2 - 2)$
21. $(p + 2)^2(p + 3)(p - 2)$ **23.** It is not factored because there are two terms: $x^2(y^2 - 6)$ and $5(y^2 - 6)$. The correct
answer is $(y^2 - 6)(x^2 + 5)$.   **25.** $(4x + 5)(4x - 5)$ **27.** $(6m - 5n)(6m + 5n)$ **29.** $(3k - 2)^2$
31. $(5x - 1)(25x^2 + 5x + 1)$ **33.** $(x^4 + 1)(x^2 + 1)(x + 1)(x - 1)$ **35.** $2b(3a^2 + b^2)$
37. $\{4\}$ **39.** $\{2, 3\}$ **41.** $\left\{-\dfrac{5}{2}, \dfrac{10}{3}\right\}$ **43.** $\left\{-\dfrac{3}{2}, -\dfrac{1}{4}\right\}$ **45.** $\left\{-\dfrac{3}{2}, 0\right\}$ **47.** $\{4\}$ **49.** $\{-3, -2, 2\}$ **51.** 3 ft
53. after 16 sec   **55.** The rock reaches a height of 240 ft once on its way up and once on its way down.
57. $(4 + 9k)(4 - 9k)$ **59.** prime   **61.** $(5z - 3m)^2$ **63.** $\{0, 3\}$ **65.** 6 in.

Chapter 6 Test (page 386)

[6.1–6.4] **1.** $11z(z - 4)$ **2.** $5x^2y^3(2y^2 - 1 - 5x^3)$ **3.** $(x + y)(3 + b)$ **4.** $-(2x + 9)(x - 4)$ **5.** $(3x - 5)(2x + 7)$
6. $(4p - q)(p + q)$ **7.** $(4a + 5b)^2$ **8.** $(x + 1 + 2z)(x + 1 - 2z)$ **9.** $(a + b)(a - b)(a + 2)$

10. $(3k + 11j)(3k - 11j)$ **11.** $(y - 6)(y^2 + 6y + 36)$ **12.** $(2k^2 - 5)(3k^2 + 7)$ **13.** $(3x^2 + 1)(9x^4 - 3x^2 + 1)$

[6.1] **14.** It is not in factored form because there are two terms: $(x^2 + 2y)p$ and $3(x^2 + 2y)$. The common factor is $x^2 + 2y$, and

the factored form is $(x^2 + 2y)(p + 3)$.   [6.2] **15.** D   [6.5] **16.** $\left\{-2, -\dfrac{2}{3}\right\}$ **17.** $\left\{0, \dfrac{5}{3}\right\}$ **18.** $\left\{-\dfrac{2}{5}, 1\right\}$

19. length: 8 in.; width: 5 in.   **20.** 2 sec and 4 sec

Chapters 1–6 Cumulative Review Exercises (pages 387–388)

[1.4] **1.** $-2m + 6$ **2.** $4m - 3$ **3.** $2x^2 + 5x + 4$ [1.3] **4.** -24 **5.** 204   **6.** undefined   **7.** 10   [2.1] **8.** $\left\{\dfrac{7}{6}\right\}$

9. $\{-1\}$ [2.5] **10.** $\left(-\infty, \dfrac{15}{4}\right]$ **11.** $\left(-\dfrac{1}{2}, \infty\right)$ [2.6] **12.** $(2, 3)$ **13.** $(-\infty, 2) \cup (3, \infty)$ [2.7] **14.** $\left\{-\dfrac{16}{5}, 2\right\}$

15. $(-11, 7)$ **16.** $(-\infty, -2] \cup [7, \infty)$ [2.2] **17.** $h = \dfrac{V}{lw}$ [2.4] **18.** 2 hr   [3.1] **19.**

[3.2] **20.** -1

21. 0   [3.5] **22.** -1 [3.1] **23.** $\left(-\dfrac{7}{2}, 0\right)$ **24.** $(0, 7)$ [4.1] **25.** $\{(1, 5)\}$ [4.2] **26.** $\{(1, 1, 0)\}$ [5.1] **27.** $\dfrac{y}{18x}$

28. $\dfrac{5my^4}{3}$ [5.2] **29.** $x^3 + 12x^2 - 3x - 7$ [5.4] **30.** $49x^2 + 42xy + 9y^2$ **31.** $10p^3 + 7p^2 - 28p - 24$

[6.1–6.4] **32.** $(2w + 7z)(8w - 3z)$ **33.** $(2x - 1 + y)(2x - 1 - y)$ **34.** $(2y - 9)^2$ **35.** $(10x^2 + 9)(10x^2 - 9)$

36. $(2p + 3)(4p^2 - 6p + 9)$ [6.5] **37.** $\left\{-4, -\dfrac{3}{2}, 1\right\}$ **38.** $\left\{\dfrac{1}{3}\right\}$ **39.** 4 ft   **40.** longer sides: 18 in.;

distance between: 16 in.

CHAPTER 7   Rational Expressions and Functions

Section 7.1   (pages 397–400)

1. C   **3.** D   **5.** E   **7.** Replacing x with 2 makes the denominator 0 and the value of the expression undefined. To find the
values excluded from the domain, set the denominator equal to 0 and solve the equation. All solutions of the equation are

excluded from the domain.   **9.** $7; \{x \mid x \neq 7\}$ **11.** $-\dfrac{1}{7}; \left\{x \mid x \neq -\dfrac{1}{7}\right\}$ **13.** $0; \{x \mid x \neq 0\}$

15. $-2, \dfrac{3}{2}; \left\{x \mid x \neq -2, \dfrac{3}{2}\right\}$ **17.** none; $(-\infty, \infty)$ **19.** none; $(-\infty, \infty)$ **21. (a)** numerator: $x^2, 4x$; denominator: $x, 4$

(b) First factor the numerator, getting $x(x + 4)$. Then divide the numerator and denominator by the common factor $x + 4$ to get

$\dfrac{x}{1}$, or x.   **23.** B   **25.** x **27.** $\dfrac{x - 3}{x + 5}$ **29.** $\dfrac{x + 3}{2x(x - 3)}$ **31.** It is already in lowest terms.   **33.** $\dfrac{6}{7}$ **35.** $\dfrac{t - 3}{3}$

37. $\dfrac{2}{t - 3}$ **39.** $\dfrac{x - 3}{x + 1}$ **41.** $\dfrac{4x + 1}{4x + 3}$ **43.** $a^2 - ab + b^2$ **45.** $\dfrac{c + 6d}{c - d}$ **47.** $\dfrac{a + b}{a - b}$ **49.** -1

In Exercises 51 and 53, there are other acceptable ways to express each answer.   **51.** $-(x + y)$ **53.** $-\dfrac{x + y}{x - y}$

55. $-\dfrac{1}{2}$ **57.** It is already in lowest terms.

59. Multiply the numerators, multiply the denominators, and factor each numerator and denominator. (Factoring can be performed first.) Divide the numerator and denominator by any common factors to write the rational expression in lowest terms.

For example, $\dfrac{6r - 5s}{3r + 2s} \cdot \dfrac{6r + 4s}{5s - 6r} = \dfrac{(6r - 5s)(6r + 4s)}{(3r + 2s)(5s - 6r)} = \dfrac{(6r - 5s)2(3r + 2s)}{(3r + 2s)(-1)(6r - 5s)} = \dfrac{2}{-1} = -2.$ **61.** $\dfrac{3y}{x^2}$

63. $\dfrac{3a^3b^2}{4}$ **65.** $\dfrac{27}{2mn^7}$ **67.** $\dfrac{x + 4}{x - 2}$ **69.** $\dfrac{2x + 3}{x + 2}$ **71.** $\dfrac{7x}{6}$ **73.** $-\dfrac{p + 5}{2p}$ (There are other ways.) **75.** $\dfrac{35}{4}$

77. $-(z + 1)$, or $-z - 1$ **79.** $\dfrac{14x^2}{5}$ **81.** -2 **83.** $\dfrac{x + 4}{x - 4}$ **85.** $\dfrac{a^2 + ab + b^2}{a - b}$ **87.** $\dfrac{2x - 3}{2(x - 3)}$ **89.** $\dfrac{a^2 + 2ab + 4b^2}{a + 2b}$

91. $\dfrac{2x + 3}{2x - 3}$ **93.** $\dfrac{k + 5p}{2k + 5p}$ **95.** $(k - 1)(k - 2)$ **97.** $\dfrac{(a + 5)(2a + b)}{(3a + 1)(a + 2b)}$ **99.** $\dfrac{17}{42}$ **101.** $-\dfrac{2}{3}$

Section 7.2 (pages 407–410)

1. $\dfrac{9}{t}$ **3.** $\dfrac{6x + y}{7}$ **5.** $\dfrac{2}{x}$ **7.** $-\dfrac{2}{x^3}$ **9.** 1 **11.** $x - 5$ **13.** $\dfrac{5}{p + 3}$ **15.** $a - b$ **17.** First add or subtract the numerators. Then place the result over the common denominator. Write the answer in lowest terms. We give one example:

$\dfrac{5}{x} - \dfrac{3x + 1}{x} = \dfrac{5 - (3x + 1)}{x} = \dfrac{5 - 3x - 1}{x} = \dfrac{4 - 3x}{x}.$ **19.** $72x^4y^5$ **21.** $z(z - 2)$ **23.** $2(y + 4)$

25. $(x + 9)^2(x - 9)$ **27.** $(m + n)(m - n)$ **29.** $x(x - 4)(x + 1)$ **31.** $(t + 5)(t - 2)(2t - 3)$ **33.** $2y(y + 3)(y - 3)$

35. $2(x + 2)^2(x - 3)$ **37.** The expression $\dfrac{x - 4x - 1}{x + 2}$ is incorrect. The third term in the numerator should be $+1$, since the

$-$ sign should be distributed over both $4x$ and -1. The answer should be $\dfrac{-3x + 1}{x + 2}$. **39.** $\dfrac{31}{3t}$ **41.** $\dfrac{5 - 22x}{12x^2y}$

43. $\dfrac{16b + 9a^2}{60a^4b^6}$ **45.** $\dfrac{4pr + 3sq^3}{14p^4q^4}$ **47.** $\dfrac{a^2b^5 - 2ab^6 + 3}{a^5b^7}$ **49.** $\dfrac{1}{x(x - 1)}$ **51.** $\dfrac{5a^2 - 7a}{(a + 1)(a - 3)}$ **53.** 3

55. $\dfrac{3}{x - 4}$, or $\dfrac{-3}{4 - x}$ **57.** $\dfrac{w + z}{w - z}$, or $\dfrac{-w - z}{z - w}$ **59.** $\dfrac{-2}{(x + 1)(x - 1)}$ **61.** $\dfrac{2(2x - 1)}{x - 1}$ **63.** $\dfrac{7}{y}$ **65.** $\dfrac{6}{x - 2}$ **67.** $\dfrac{3x - 2}{x - 1}$

69. $\dfrac{4x - 7}{x^2 - x + 1}$ **71.** $\dfrac{2x + 1}{x}$ **73.** $\dfrac{4p^2 - 21p + 29}{(p - 2)^2}$ **75.** $\dfrac{x}{(x - 2)^2(x - 3)}$ **77.** $\dfrac{2x(x + 12y)}{(x + 2y)(x - y)(x + 6y)}$

79. $\dfrac{2x^2 + 21xy - 10y^2}{(x + 2y)(x - y)(x + 6y)}$ **81.** $\dfrac{3r - 2s}{(2r - s)(3r - s)}$ **83.** $\dfrac{10x + 23}{(x + 2)^2(x + 3)}$ **85.** **(a)** $c(x) = \dfrac{10x}{49(101 - x)}$

(b) approximately 3.23 thousand dollars **87.** $\dfrac{8}{9}$ **88.** $\dfrac{3}{7} + \dfrac{5}{9} - \dfrac{6}{63}$; They are the same. **89.** $\dfrac{8}{9}$; yes **90.** Answers

will vary. Suppose the name is Gore, so that $x = 4$. The problem is $\dfrac{3}{2} + \dfrac{5}{4} - \dfrac{6}{8}$. The predicted answer is $\dfrac{8}{4} = 2$, which is

correct. **91.** It causes $\dfrac{3}{x - 2}$ and $\dfrac{6}{x^2 - 2x}$ to be undefined, since 0 appears in the denominators. **92.** 0 **93.** $\dfrac{4}{15}$ **95.** $\dfrac{7}{17}$

Section 7.3 (pages 416–418)

1. *Method 1:* Begin by simplifying the numerator to a single fraction. Then simplify the denominator to a single fraction. Write as a division problem, and multiply by the reciprocal of the denominator. Simplify the result if possible. *Method 2:* Find the LCD of all fractions in the complex fraction. Multiply the numerator and denominator of the complex fraction by this LCD.

Simplify the result if possible. **3.** $\dfrac{2x}{x - 1}$ **5.** $\dfrac{2(k + 1)}{3k - 1}$ **7.** $\dfrac{5x^2}{9z^3}$ **9.** $\dfrac{6x + 1}{7x - 3}$ **11.** $\dfrac{y + x}{y - x}$ **13.** $4x$ **15.** $x + 4y$

17. $\dfrac{y+4}{2}$ **19.** $\dfrac{a+b}{ab}$ **21.** xy **23.** $\dfrac{3y}{2}$ **25.** $\dfrac{x^2+5x+4}{x^2+5x+10}$ **27.** $\dfrac{m^2+6m-4}{m(m-1)}$ **28.** $\dfrac{m^2-m-2}{m(m-1)}$

29. $\dfrac{m^2+6m-4}{m^2-m-2}$ **30.** $m(m-1)$ **31.** $\dfrac{m^2+6m-4}{m^2-m-2}$ **32.** Answers will vary. **33.** $\dfrac{x^2y^2}{y^2+x^2}$ **35.** $\dfrac{y^2+x^2}{xy^2+x^2y}$, or

$\dfrac{y^2+x^2}{xy(y+x)}$ **37.** $\dfrac{1}{2xy}$ **39. (a)** $\dfrac{\dfrac{3}{mp}-\dfrac{4}{p}+\dfrac{8}{m}}{\dfrac{2}{m}-\dfrac{3}{p}}$ **(b)** In the denominator, $2m^{-1}=\dfrac{2}{m}$, not $\dfrac{1}{2m}$, and $3p^{-1}=\dfrac{3}{p}$, not $\dfrac{1}{3p}$.

(c) $\dfrac{3-4m+8p}{2p-3m}$ **41.** $\{-12\}$ **43.** $\{16\}$ **45.** 6; $\{x\mid x\neq 6\}$ **47.** 0; $\{x\mid x\neq 0\}$

Section 7.4 (pages 423–426)

1. (a) 0 **(b)** $\{x\mid x\neq 0\}$ **3. (a)** $-1, 2$ **(b)** $\{x\mid x\neq -1, 2\}$ **5. (a)** $-4, 4$ **(b)** $\{x\mid x\neq \pm 4\}$ **7. (a)** $0, 1, -3, 2$

(b) $\{x\mid x\neq 0, 1, -3, 2\}$ **9. (a)** $-\dfrac{7}{4}, 0, \dfrac{13}{6}$ **(b)** $\left\{x\mid x\neq -\dfrac{7}{4}, 0, \dfrac{13}{6}\right\}$ **11. (a)** $4, \dfrac{7}{2}$ **(b)** $\left\{x\mid x\neq 4, \dfrac{7}{2}\right\}$ **13.** No, there

is no possibility that the proposed solution will be rejected, because there are no variables in the denominators in the original

equation. **15.** $\{1\}$ **17.** $\{-6, 4\}$ **19.** $\left\{-\dfrac{7}{12}\right\}$ **21.** $\emptyset$ **23.** $\{-3\}$ **25.** $\{0\}$ **27.** $\{5\}$ **29.** $\emptyset$ **31.** $\left\{\dfrac{27}{56}\right\}$

33. $\emptyset$ **35.** $\{-10\}$ **37.** $\{-1\}$ **39.** $\{13\}$ **41.** $\{x\mid x\neq \pm 3\}$ **43. (a)** $\{x\mid x\neq -3\}$ **(b)** -3 is not in the domain.

45. 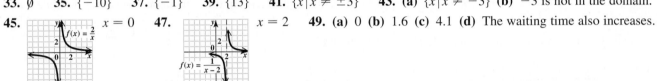 $x=0$ **47.** $x=2$ **49. (a)** 0 **(b)** 1.6 **(c)** 4.1 **(d)** The waiting time also increases.

51. (a) 500 ft **(b)** It decreases. **53.** four **55.** $\{-2, 0, 3\}$ **57.** $t=\dfrac{d}{r}$ **59.** $c=P-a-b$

Summary Exercises on Rational Expressions and Equations (page 427)

1. equation; $\{20\}$ **2.** expression; $\dfrac{2(x+5)}{5}$ **3.** expression; $-\dfrac{22}{7x}$ **4.** expression; $\dfrac{y+x}{y-x}$ **5.** equation; $\left\{\dfrac{1}{2}\right\}$

6. equation; $\{7\}$ **7.** expression; $\dfrac{43}{24x}$ **8.** equation; $\{1\}$ **9.** expression; $\dfrac{5x-1}{-2x+2}$, or $\dfrac{5x-1}{-2(x-1)}$

10. expression; $\dfrac{25}{4(r+2)}$ **11.** expression; $\dfrac{x^2+xy+2y^2}{(x+y)(x-y)}$ **12.** expression; $\dfrac{24p}{p+2}$ **13.** expression; $-\dfrac{5}{36}$

14. equation; $\{0\}$ **15.** expression; $\dfrac{b+3}{3}$ **16.** expression; $\dfrac{5}{3z}$ **17.** expression; $\dfrac{2x+10}{x(x-2)(x+2)}$

18. equation; $\{2\}$ **19.** expression; $\dfrac{-x}{3x+5y}$ **20.** equation; $\{-13\}$ **21.** expression; $\dfrac{3y+2}{y+3}$ **22.** equation; $\left\{\dfrac{5}{4}\right\}$

23. equation; $\emptyset$ **24.** expression; $\dfrac{2z-3}{2z+3}$ **25.** expression; $\dfrac{-1}{x-3}$, or $\dfrac{1}{3-x}$ **26.** expression; $\dfrac{t-2}{8}$

27. equation; $\{-10\}$ **28.** expression; $\dfrac{13x+28}{2x(x+4)(x-4)}$ **29.** equation; $\emptyset$ **30.** expression; $\dfrac{k(2k^2-2k+5)}{(k-1)(3k^2-2)}$

Section 7.5   (pages 435–440)

1. A   **3.** D   **5.** 24   **7.** $\dfrac{25}{4}$ **9.** $G = \dfrac{Fd^2}{Mm}$ **11.** $a = \dfrac{bc}{c+b}$ **13.** $v = \dfrac{PVt}{pT}$ **15.** $r = \dfrac{nE - IR}{In}$, or $r = \dfrac{IR - nE}{-In}$

17. $b = \dfrac{2A}{h} - B$, or $b = \dfrac{2A - hB}{h}$ **19.** $r = \dfrac{eR}{E - e}$ **21.** Multiply each side by $a - b$.   **23.** 15 girls, 5 boys

25. $\dfrac{1}{2}$ job per hr   **27.** 5.351 in.   **29.** 7.6 in.   **31.** 40 teachers   **33.** 210 deer   **35.** 25,000 fish   **37.** 6.6 more gallons

39. $x = \dfrac{7}{2}$; $AC = 8$; $DF = 12$ **41.** 2.4 mL   **43.** 3 mph   **45.** 1020 mi   **47.** 1750 mi   **49.** 190 mi   **51.** $6\dfrac{2}{3}$ min

53. 30 hr   **55.** $2\dfrac{1}{3}$ hr   **57.** 20 hr   **59.** $2\dfrac{4}{5}$ hr   **61.** $\dfrac{1}{3}$ **63.** 3

Section 7.6   (pages 446–449)

1. direct   **3.** direct   **5.** inverse   **7.** inverse   **9.** inverse   **11.** direct   **13.** joint   **15.** combined   **17.** increases;

decreases   **19.** 36   **21.** $\dfrac{16}{9}$ **23.** 0.625   **25.** $\dfrac{16}{5}$ **27.** $222\dfrac{2}{9}$ **29.** \$2.919, or \$2.91$\dfrac{9}{10}$ **31.** 8 lb   **33.** about 450 in.^{3}

35. 256 ft   **37.** $106\dfrac{2}{3}$ mph   **39.** 100 cycles per sec   **41.** $21\dfrac{1}{3}$ foot-candles   **43.** \$420   **45.** 11.8 lb   **47.** 448.1 lb

49. approximately 68,600 calls   **51.** Answers will vary.   **53.** If y varies inversely as x, then x is in the denominator; however, if y varies directly as x, then x is in the numerator. If $k > 0$, then, with inverse variation, as x increases, y decreases. With direct variation, y increases as x increases.   **55.** $\{\pm 9\}$ **57.** $\{\pm 0.5\}$ **59.** $\{\pm 3\}$

Chapter 7 Review Exercises   (pages 457–460)

1. (a) -6 **(b)** $\{x \mid x \ne -6\}$ **3. (a)** 9 **(b)** $\{x \mid x \ne 9\}$ **5.** $\dfrac{5m + n}{5m - n}$ **7.** The reciprocal of a rational expression is another rational expression such that the two rational expressions have a product of 1.   **9.** $\dfrac{-3(w + 4)}{w}$ **11.** 1   **13.** $9r^2(3r + 1)$

15. $3(x - 4)^2(x + 2)$ **17.** 12   **19.** $\dfrac{13r^2 + 5rs}{(5r + s)(2r - s)(r + s)}$ **21.** $\dfrac{3 + 2t}{4 - 7t}$ **23.** $\dfrac{1}{3q + 2p}$ **25.** $\{-3\}$ **27.** $\{0\}$

29. Although her algebra was correct, 3 is not a solution because it is not in the domain of the variable. Thus, $\emptyset$ is correct.

31. C; $x = 0$ **33.** $m = \dfrac{Fd^2}{GM}$ **35.** 6000 passenger-km per day   **37.** $4\dfrac{4}{5}$ min   **39.** C   **41.** 5.59 vibrations per sec

43. $\dfrac{1}{x - 2y}$ **45.** $\dfrac{6m + 5}{3m^2}$ **47.** $\dfrac{x^2 - 6}{2(2x + 1)}$ **49.** $\dfrac{3 - 5x}{6x + 1}$ **51.** $\dfrac{1}{3}$ **53.** $\dfrac{5a^2 + 4ab + 12b^2}{(a + 3b)(a - 2b)(a + b)}$ **55.** $\left\{\dfrac{1}{3}\right\}$

57. $\{1, 4\}$ **59. (a)** 8.32 **(b)** 44.9   **61.** \$21.06   **63.** $4\dfrac{1}{2}$ mi   **65.** 150 mi

Chapter 7 Test   (pages 460–462)

[7.1] **1.** $-2, \dfrac{4}{3}$; $\left\{x \mid x \ne -2, \dfrac{4}{3}\right\}$ **2.** $\dfrac{2x - 5}{x(3x - 1)}$ **3.** $\dfrac{3(x + 3)}{4}$ **4.** $\dfrac{y + 4}{y - 5}$ **5.** -2 **6.** $\dfrac{x + 5}{x}$

[7.2] **7.** $t^2(t + 3)(t - 2)$ **8.** $\dfrac{7 - 2t}{6t^2}$ **9.** $\dfrac{9a + 5b}{21a^5b^3}$ **10.** $\dfrac{11x + 21}{(x - 3)^2(x + 3)}$ **11.** $\dfrac{4}{x + 2}$ [7.3] **12.** $\dfrac{72}{11}$ **13.** $-\dfrac{1}{a + b}$

14. $\dfrac{2y^2 + x^2}{xy(y - x)}$ **[7.4] 15. (a)** expression; $\dfrac{11(x - 6)}{12}$ **(b)** equation; $\{6\}$ **16.** $\left\{\dfrac{1}{2}\right\}$ **17.** $\{5\}$ **18.** A solution cannot

make a denominator 0.   **19.** $\ell = \dfrac{2S}{n} - a$, or $\ell = \dfrac{2S - na}{n}$ **20.** 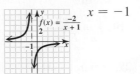 $x = -1$ **[7.5] 21.** $3\dfrac{3}{14}$ hr   **22.** 15 mph

23. 48,000 fish   **24. (a)** 3 units **(b)** 0   **[7.6] 25.** 200 amps   **26.** 0.8 lb

Chapters 1–7 Cumulative Review Exercises   (pages 462–464)

[1.3] 1. -199 **2.** 12   **[2.1] 3.** $\left\{-\dfrac{15}{4}\right\}$ **[2.7] 4.** $\left\{\dfrac{2}{3}, 2\right\}$ **[2.3] 5.** $x = \dfrac{d - by}{a - c}$, or $x = \dfrac{by - d}{c - a}$ **[2.5] 6.** $\left(-\infty, \dfrac{240}{13}\right]$

[2.7] 7. $(-\infty, -2] \cup \left[\dfrac{2}{3}, \infty\right)$ **[2.3] 8.** \$4000 at 4%; \$8000 at 3%   **9.** 6 m   **[3.1] 10.** x-intercept: $(-2, 0)$;

y-intercept: $(0, 4)$ **[3.2] 11.** $-\dfrac{3}{2}$ **12.** $-\dfrac{3}{4}$ **[3.3] 13.** $y = -\dfrac{3}{2}x + \dfrac{1}{2}$ **[3.4] 14.**

15.   **[3.5] 16.** function; domain: $\{1990, 1992, 1994, 1996, 1998, 2000, 2002\}$;

range: $\{1.25, 1.61, 1.80, 1.21, 1.94, 2.26, 2.60\}$

17. not a function; domain: $[-2, \infty)$; range: $(-\infty, \infty)$ **18.** function; domain: $[-2, \infty)$; range: $(-\infty, 0]$

19. (a) $\dfrac{5x - 8}{3}$, or $\dfrac{5}{3}x - \dfrac{8}{3}$ **(b)** -1 **20.** $3x + 15$ **[4.1, 4.4] 21.** $\{(-1, 3)\}$ **[4.2, 4.4] 22.** $\{(-2, 3, 1)\}$ **23.** $\emptyset$

[4.3] 24. automobile: 42 km per hr; airplane: 600 km per hr   **[5.1] 25.** $\dfrac{a^{10}}{b^{10}}$ **26.** $\dfrac{m}{n}$ **[5.2] 27.** $4y^2 - 7y - 6$

[5.4] 28. $-6x^6 + 18x^5 - 12x^4$ **29.** $12f^2 + 5f - 3$ **30.** $49t^6 - 64$ **31.** $\dfrac{1}{16}x^2 + \dfrac{5}{2}x + 25$ **[5.5] 32.** $x^2 + 4x - 7$

[5.3] 33. (a) $2x^3 - 2x^2 + 6x - 4$ **(b)** $2x^3 - 4x^2 + 2x + 2$ **(c)** -14 **(d)** $x^4 + 2x^2 - 3$ **[6.1] 34.** $(2x + 5)(x - 9)$

[6.2] 35. $25(2t^2 + 1)(2t^2 - 1)$ **36.** $(2p + 5)(4p^2 - 10p + 25)$ **[6.5] 37.** $\left\{-\dfrac{7}{3}, 1\right\}$ **[7.1] 38.** $\dfrac{y + 4}{y - 4}$ **39.** $\dfrac{2x - 3}{2(x - 1)}$

40. $\dfrac{a(a - b)}{2(a + b)}$ **41.** $\dfrac{2(x + 3)}{(x + 2)(x^2 + 3x + 9)}$ **[7.2] 42.** 3   **43.** $\dfrac{2(x + 2)}{2x - 1}$ **[7.4] 44.** $\{-4\}$ **[7.5] 45.** $q = \dfrac{fp}{p - f}$,

or $q = \dfrac{-fp}{f - p}$ **46.** 150 mph   **47.** $\dfrac{6}{5}$ hr   **[7.6] 48.** \$9.92

CHAPTER 8   Roots, Radicals, and Root Functions

Section 8.1   (pages 471–473)

1. E   **3.** D   **5.** A   **7.** C   **9.** C   **11.** (a) It is not a real number. (b) negative (c) 0   **13.** -9 **15.** 6   **17.** -4

19. -8 **21.** 6   **23.** -2 **25.** It is not a real number.   **27.** 2   **29.** It is not a real number.   **31.** $\dfrac{8}{9}$ **33.** $\dfrac{4}{3}$

35. $-\dfrac{1}{2}$ **37.** 0.7   **39.** 0.1   *In Exercises 41–47, we give the domain and then the range.*

41. $[-3, \infty)$; $[0, \infty)$ 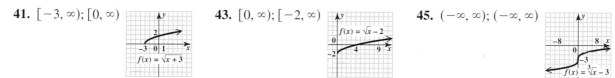   **43.** $[0, \infty)$; $[-2, \infty)$ **45.** $(-\infty, \infty)$; $(-\infty, \infty)$

47. $(-\infty, \infty)$; $(-\infty, \infty)$ 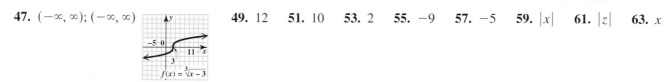   **49.** 12   **51.** 10   **53.** 2   **55.** -9 **57.** -5 **59.** $|x|$ **61.** $|z|$ **63.** x

65. x^5 **67.** $|x|^5$ (or $|x^5|$)   **69.** 97.381   **71.** 16.863   **73.** -9.055 **75.** 7.507   **77.** 3.162   **79.** 1.885

81. 1,183,000 cycles per sec   **83.** 10 mi   **85.** 392,000 mi^2 **87.** 1.732 amps   **89.** x^1, or x **91.** $13^2x^4y^8$, or $169x^4y^8$

93. 5^2, or 25   **95.** $\dfrac{3^3}{2^3}$, or $\dfrac{27}{8}$

Section 8.2   (pages 479–481)

1. C   **3.** A   **5.** H   **7.** B   **9.** D   **11.** 13   **13.** 9   **15.** 2   **17.** $\dfrac{8}{9}$ **19.** -3 **21.** It is not a real number.

23. 1000   **25.** 27   **27.** -1024 **29.** 16   **31.** $\dfrac{1}{8}$ **33.** $\dfrac{1}{512}$ **35.** $\dfrac{9}{25}$ **37.** $\sqrt{10}$ **39.** $\left(\sqrt[4]{8}\right)^3$

41. $\left(\sqrt[8]{9q}\right)^5 - \left(\sqrt[3]{2x}\right)^2$ **43.** $\dfrac{1}{\left(\sqrt{2m}\right)^3}$ **45.** $\left(\sqrt[3]{2y + x}\right)^2$ **47.** $\dfrac{1}{\left(\sqrt[3]{3m^4 + 2k^2}\right)^2}$ **49.** $\sqrt{a^2 + b^2} = \sqrt{3^2 + 4^2} = 5$;

$a + b = 3 + 4 = 7$; $5 \neq 7$ **51.** 64   **53.** 64   **55.** x^{10} **57.** $\sqrt[6]{x^5}$ **59.** $\sqrt[15]{t^8}$ **61.** 9   **63.** 4   **65.** y **67.** $x^{5/12}$

69. $k^{2/3}$ **71.** x^3y^8 **73.** $\dfrac{1}{x^{10/3}}$ **75.** $\dfrac{1}{m^{1/4}n^{3/4}}$ **77.** p^2 **79.** $\dfrac{c^{11/3}}{b^{11/4}}$ **81.** $\dfrac{q^{5/3}}{9p^{7/2}}$ **83.** $p + 2p^2$ **85.** $k^{7/4} - k^{3/4}$

87. $6 + 18a$ **89.** $x^{17/20}$ **91.** $\dfrac{1}{x^{3/2}}$ **93.** $y^{5/6}z^{1/3}$ **95.** $m^{1/12}$ **97.** $x^{1/24}$ **99.** 4.5 hr   **101.** 19.0°; the table gives 19°.

103. 30; 30; they are the same.   **105.** $\dfrac{1}{3}$; $\dfrac{1}{3}$; they are the same.

Connections   **(page 487)**   no; no; answers will vary.

Section 8.3   (pages 489–492)

1. true; both are equal to $4\sqrt{3}$ and approximately 6.92820323.   **3.** true; both are equal to $6\sqrt{2}$ and approximately 8.485281374.

5. D   **7.** $\sqrt{30}$ **9.** $\sqrt{14x}$ **11.** $\sqrt{42pqr}$ **13.** $\sqrt[3]{14xy}$ **15.** $\sqrt[4]{33}$ **17.** $\sqrt[4]{6x^3}$ **19.** This expression cannot be

simplified by the product rule.   **21.** To multiply two radical expressions with the same index, multiply the radicands and keep

the index. For example, $\sqrt[3]{3} \cdot \sqrt[3]{5} = \sqrt[3]{15}$.   **23.** $\dfrac{8}{11}$ **25.** $\dfrac{\sqrt{3}}{5}$ **27.** $\dfrac{\sqrt{x}}{5}$ **29.** $\dfrac{p^3}{9}$ **31.** $-\dfrac{3}{4}$ **33.** $\dfrac{\sqrt[3]{r^2}}{2}$ **35.** $-\dfrac{3}{x}$

37. $\dfrac{1}{x^3}$ **39.** $2\sqrt{3}$ **41.** $12\sqrt{2}$ **43.** $-4\sqrt{2}$ **45.** $-2\sqrt{7}$ **47.** This radical cannot be simplified further.   **49.** $4\sqrt[3]{2}$

51. $-2\sqrt[3]{2}$ **53.** $2\sqrt[3]{5}$ **55.** $-4\sqrt[4]{2}$ **57.** $2\sqrt[5]{2}$ **59.** His reasoning was incorrect. Here, 8 is a term, not a factor.

61. $6k\sqrt{2}$ **63.** $12xy^4\sqrt{xy}$ **65.** $11x^3$ **67.** $-3t^4$ **69.** $-10m^4z^2$ **71.** $5a^2b^3c^4$ **73.** $\dfrac{1}{2}r^2t^5$ **75.** $5x\sqrt{2x}$

77. $-10r^5\sqrt{5r}$ **79.** $x^3y^4\sqrt{13x}$ **81.** $2z^2w^3$ **83.** $-2zt^2\sqrt[3]{2z^2t}$ **85.** $3x^3y^4$ **87.** $-3r^3s^2\sqrt[4]{2r^3s^2}$ **89.** $\dfrac{y^5\sqrt{y}}{6}$

91. $\dfrac{x^5\sqrt[3]{x}}{3}$ **93.** $4\sqrt{3}$ **95.** $\sqrt{5}$ **97.** $x^2\sqrt{x}$ **99.** $\sqrt[6]{432}$ **101.** $\sqrt[12]{6912}$ **103.** $\sqrt[6]{x^5}$ **105.** 5   **107.** $8\sqrt{2}$

109. 13   **111.** $9\sqrt{2}$ **113.** $\sqrt{17}$ **115.** 5   **117.** $6\sqrt{2}$ **119.** $\sqrt{5y^2 - 2xy + x^2}$

121. $d = [(x_2 - x_1)^2 + (y_2 - y_1)^2]^{1/2}$ **123.** $2\sqrt{106} + 4\sqrt{2}$ **125.** 15.3 mi   **127.** 27.0 in.   **129.** 581

131. $22x^4 - 10x^3$ **133.** $8q^2 - 3q$

Section 8.4   (pages 495–497)

1. B   **3.** 15; Each radical expression simplifies to a whole number.   **5.** -4 **7.** $7\sqrt{3}$ **9.** $14\sqrt[3]{2}$ **11.** $5\sqrt[4]{2}$

13. $24\sqrt{2}$ **15.** The expression cannot be simplified further.   **17.** $20\sqrt{5}$ **19.** $4\sqrt{2x}$ **21.** $-11m\sqrt{2}$ **23.** $\sqrt[3]{2}$

25. $2\sqrt[3]{x}$ **27.** $-\sqrt[3]{x^2y}$ **29.** $-x\sqrt[3]{xy^2}$ **31.** $19\sqrt[4]{2}$ **33.** $x\sqrt[4]{xy}$ **35.** $9\sqrt[4]{2a^3}$ **37.** $(4 + 3xy)\sqrt[3]{xy^2}$

39. The expression cannot be simplified further.   **41.** $4x\sqrt[3]{x} + 6x\sqrt[4]{x}$ **43.** $2\sqrt{2} - 2$ **45.** $\dfrac{5\sqrt{5}}{6}$ **47.** $\dfrac{7\sqrt{2}}{6}$ **49.** $\dfrac{5\sqrt{2}}{3}$

51. $5\sqrt{2} + 4$ **53.** $\dfrac{5 - 3x}{x^4}$ **55.** $\dfrac{m\sqrt[3]{m^2}}{2}$ **57.** $\dfrac{3x\sqrt{2} - 4\sqrt[3]{5}}{x^3}$ **59.** Both are approximately 11.3137085.   **61.** A; 42 m

63. $\left(12\sqrt{5} + 5\sqrt{3}\right)$ in.   **65.** $\left(24\sqrt{2} + 12\sqrt{3}\right)$ in.   **67.** $10x^3y^4 - 20x^2y$ **69.** $a^4 - b^2$

71. $64x^9 + 144x^6 + 108x^3 + 27$ **73.** $\dfrac{4x - 5}{3x}$

Connections   (page 503)   **1.** $\dfrac{319}{6(8\sqrt{5} + 1)}$ **2.** $\dfrac{9a - b}{b(3\sqrt{a} - \sqrt{b})}$ **3.** $\dfrac{9a - b}{(\sqrt{b} - \sqrt{a})(3\sqrt{a} - \sqrt{b})}$

4. $\dfrac{(3\sqrt{a} + \sqrt{b})(\sqrt{b} + \sqrt{a})}{b - a}$; Instead of multiplying by the conjugate of the numerator, we use the conjugate of the denominator.

Section 8.5   (pages 504–506)

1. E   **3.** A   **5.** D   **7.** $3\sqrt{6} + 2\sqrt{3}$ **9.** $20\sqrt{2}$ **11.** -2 **13.** -1 **15.** 6   **17.** $\sqrt{6} - \sqrt{2} + \sqrt{3} - 1$

19. $\sqrt{22} + \sqrt{55} - \sqrt{14} - \sqrt{35}$ **21.** $8 - \sqrt{15}$ **23.** $9 + 4\sqrt{5}$ **25.** $26 - 2\sqrt{105}$ **27.** $4 - \sqrt[3]{36}$ **29.** 10

31. $6x + 3\sqrt{x} - 2\sqrt{5x} - \sqrt{5}$ **33.** $9r - s$ **35.** $4\sqrt[3]{4y^2} - 19\sqrt[3]{2y} - 5$ **37.** $3x - 4$ **39.** $4x - y$ **41.** $2\sqrt{6} - 1$

43. $\sqrt{7}$ **45.** $5\sqrt{3}$ **47.** $\dfrac{\sqrt{6}}{2}$ **49.** $\dfrac{9\sqrt{15}}{5}$ **51.** $-\sqrt{2}$ **53.** $\dfrac{\sqrt{14}}{2}$ **55.** $-\dfrac{\sqrt{14}}{10}$ **57.** $\dfrac{2\sqrt{6x}}{x}$ **59.** $\dfrac{-8\sqrt{3k}}{k}$

61. $\dfrac{-5m^2\sqrt{6mn}}{n^2}$ **63.** $\dfrac{12x^3\sqrt{2xy}}{y^5}$ **65.** $\dfrac{5\sqrt{2my}}{y^2}$ **67.** $-\dfrac{4k\sqrt{3z}}{z}$ **69.** $\dfrac{\sqrt[3]{18}}{3}$ **71.** $\dfrac{\sqrt[3]{12}}{3}$ **73.** $\dfrac{\sqrt[3]{18}}{4}$ **75.** $-\dfrac{\sqrt[3]{2pr}}{r}$

77. $\dfrac{x^2\sqrt[3]{y^2}}{y}$ **79.** $\dfrac{2\sqrt[4]{x^3}}{x}$ **81.** $\dfrac{\sqrt[4]{2yz^3}}{z}$ **83.** $\dfrac{3(4 - \sqrt{5})}{11}$ **85.** $\dfrac{6\sqrt{2} + 4}{7}$ **87.** $\dfrac{2(3\sqrt{5} - 2\sqrt{3})}{33}$

89. $2\sqrt{3} + \sqrt{10} - 3\sqrt{2} - \sqrt{15}$ **91.** $\sqrt{m} - 2$ **93.** $\dfrac{4(\sqrt{x} + 2\sqrt{y})}{x - 4y}$ **95.** $\dfrac{x - 2\sqrt{xy} + y}{x - y}$ **97.** $\dfrac{5\sqrt{k}(2\sqrt{k} - \sqrt{q})}{4k - q}$

99. $3 - 2\sqrt{6}$ **101.** $1 - \sqrt{5}$ **103.** $\dfrac{4 - 2\sqrt{2}}{3}$ **105.** $\dfrac{6 + 2\sqrt{6p}}{3}$ **107.** $\dfrac{\sqrt{x + y}}{x + y}$ **109.** $\dfrac{p\sqrt{p + 2}}{p + 2}$

111. Each expression is approximately equal to 0.2588190451.   **113.** $\dfrac{17}{2(6 + \sqrt{2})}$ **115.** $\dfrac{9a - b}{b(3\sqrt{a} - \sqrt{b})}$ **117.** $\left\{\dfrac{3}{8}\right\}$

119. $\left\{-\dfrac{1}{3}, \dfrac{3}{2}\right\}$ **121.** $4x^2 + 20x + 25$ **123.** $x^4 + 2x^2 + 5$ **125.** true   **127.** true

Summary Exercises on Operations with Radicals and Rational Exponents
(pages 507–508)

1. $-6\sqrt{10}$ **2.** $7 - \sqrt{14}$ **3.** $2 + \sqrt{6} - 2\sqrt{3} - 3\sqrt{2}$ **4.** $4\sqrt{2}$ **5.** $73 + 12\sqrt{35}$ **6.** $\dfrac{-\sqrt{6}}{2}$ **7.** $4\left(\sqrt{7} - \sqrt{5}\right)$

8. $-3 + 2\sqrt{2}$ **9.** -44 **10.** $\dfrac{\sqrt{x} + \sqrt{5}}{x - 5}$ **11.** $2abc^3\sqrt[3]{b^2}$ **12.** $5\sqrt[3]{3}$ **13.** $3\left(\sqrt{5} - 2\right)$ **14.** $\dfrac{\sqrt{15x}}{5x}$ **15.** $\dfrac{8}{5}$

16. $\dfrac{\sqrt{2}}{8}$ **17.** $-\sqrt[3]{100}$ **18.** $11 + 2\sqrt{30}$ **19.** $-3\sqrt{3x}$ **20.** $52 - 30\sqrt{3}$ **21.** $\dfrac{\sqrt[3]{117}}{9}$

22. $3\sqrt{2} + \sqrt{15} + \sqrt{42} + \sqrt{35}$ **23.** $2\sqrt[4]{27}$ **24.** $\dfrac{1 + \sqrt[3]{3} + \sqrt[3]{9}}{-2}$ **25.** $\dfrac{x\sqrt[3]{x^2}}{y}$ **26.** $-4\sqrt{3} - 3$ **27.** $xy^{6/5}$

28. $x^{10}y$ **29.** $\dfrac{1}{25x^2}$ **30.** $7 + 4 \cdot 3^{1/2}$, or $7 + 4\sqrt{3}$ **31.** $3\sqrt[3]{2x^2}$ **32.** -2 **33.** 1 **34.** $t^2\sqrt[4]{t}$ **35.** $\dfrac{-6y^{1/6}}{x^{1/24}}$ **36.** 1

37. (a) 8 (b) $\{-8, 8\}$ **38.** (a) 10 (b) $\{-10, 10\}$ **39.** (a) $\{-4, 4\}$ (b) -4 **40.** (a) $\{-5, 5\}$ (b) -5 **41.** (a) $-\dfrac{9}{11}$

(b) $\left\{-\dfrac{9}{11}, \dfrac{9}{11}\right\}$ **42.** (a) $-\dfrac{7}{10}$ (b) $\left\{-\dfrac{7}{10}, \dfrac{7}{10}\right\}$ **43.** (a) $\{-0.2, 0.2\}$ (b) 0.2 **44.** (a) $\{-0.3, 0.3\}$ (b) 0.3

45.
$$x^2 = 36$$
$$x^2 - 36 = 0$$
$$(x + 6)(x - 6) = 0$$
$$x + 6 = 0 \quad \text{or} \quad x - 6 = 0$$
$$x = -6 \quad \text{or} \quad x = 6$$
Solution set: $\{-6, 6\}$

Section 8.6   (pages 513–515)

1. (a) yes (b) no   **3.** (a) yes (b) no   **5.** no; there is no solution. The radical expression, which is positive, cannot equal a negative number.   **7.** $\{11\}$ **9.** $\left\{\dfrac{1}{3}\right\}$ **11.** $\emptyset$ **13.** $\{5\}$ **15.** $\{18\}$ **17.** $\{5\}$ **19.** $\{4\}$ **21.** $\{17\}$ **23.** $\{5\}$ **25.** $\emptyset$

27. $\{0\}$ **29.** $\{0\}$ **31.** $\left\{-\dfrac{1}{3}\right\}$ **33.** $\emptyset$ **35.** You cannot just square each term. The right side should be $(8 - x)^2 = 64 - 16x + x^2$. The correct first step is $3x + 4 = 64 - 16x + x^2$, and the solution set is $\{4\}$.   **37.** $\{1\}$

39. $\{-1\}$ **41.** $\{14\}$ **43.** $\{8\}$ **45.** $\{0\}$ **47.** $\emptyset$ **49.** $\{7\}$ **51.** $\{7\}$ **53.** $\{4, 20\}$ **55.** $\emptyset$ **57.** $\left\{\dfrac{5}{4}\right\}$

59. $\emptyset$; domain: $\left[-\dfrac{2}{3}, 1\right]$ **61.** $\{9, 17\}$ **63.** $\left\{\dfrac{1}{4}, 1\right\}$ **65.** $K = \dfrac{V^2m}{2}$ **67.** $L = \dfrac{1}{4\pi^2f^2C}$ **69.** $r = \dfrac{a}{4\pi^2N^2}$ **71.** $1 + x$

73. $2x^2 + x - 15$ **75.** $\dfrac{-7(5 + \sqrt{2})}{23}$

Section 8.7 (pages 521–523)

1. i **3.** -1 **5.** $-i$ **7.** $13i$ **9.** $-12i$ **11.** $i\sqrt{5}$ **13.** $4i\sqrt{3}$ **15.** $-\sqrt{105}$ **17.** -10 **19.** $i\sqrt{33}$ **21.** $\sqrt{3}$ **23.** $5i$ **25. (a)** Any real number a can be written as $a + 0i$, a complex number with imaginary part 0. **(b)** A complex number such as $2 + 3i$, with nonzero imaginary part, is not real. **27.** $-1 + 7i$ **29.** 0 **31.** $7 + 3i$ **33.** -2 **35.** $1 + 13i$ **37.** $6 + 6i$ **39.** $4 + 2i$ **41.** -81 **43.** -16 **45.** $-10 - 30i$ **47.** $10 - 5i$ **49.** $-9 + 40i$ **51.** $-16 + 30i$ **53.** 153 **55.** 97 **57.** $a - bi$ **59.** $1 + i$ **61.** $-1 + 2i$ **63.** $2 + 2i$ **65.** $-\dfrac{5}{13} - \dfrac{12}{13}i$ **67.** $1 - 3i$ **69.** -1 **71.** i **73.** -1 **75.** $-i$ **77.** $-i$ **79.** Since $i^{20} = (i^4)^5 = 1^5 = 1$, the student multiplied by 1, which is justified by the identity property for multiplication. **81.** $\dfrac{1}{2} + \dfrac{1}{2}i$ **83.** Substitute both $1 + 5i$ and $1 - 5i$ for x, and show that the result is $0 = 0$ in each case. **85.** $\dfrac{37}{10} - \dfrac{19}{10}i$ **87.** $-\dfrac{13}{10} + \dfrac{11}{10}i$ **89.** $\left\{ -\dfrac{13}{6} \right\}$ **91.** $\{-8, 5\}$ **93.** $\left\{ -\dfrac{2}{5}, 1 \right\}$

Chapter 8 Review Exercises (pages 529–532)

1. 42 **3.** 6 **5.** -3 **7.** $\sqrt[n]{a}$ is not a real number if n is even and a is negative. **9.** -6.856 **11.** 4.960 **13.** -3968.503 **15.** domain: $[1, \infty)$; range: $[0, \infty)$ **17.** B **19.** A **21.** It is not a real number.

23. -11 **25.** -4 **27.** -32 **29.** It is not a real number. **31.** The radical $\sqrt[n]{a^m}$ is equivalent to $a^{m/n}$. For example, $\sqrt[3]{8^2} = \sqrt[3]{64} = 4$, and $8^{2/3} = (8^{1/3})^2 = 2^2 = 4$. **33.** $\dfrac{1}{(\sqrt[3]{3a + b})^5}$, or $\dfrac{1}{\sqrt[3]{(3a + b)^5}}$ **35.** $p^{4/5}$ **37.** 96 **39.** $\dfrac{1}{y^{1/2}}$ **41.** $r^{1/2} + r$ **43.** $r^{3/2}$ **45.** $k^{9/4}$ **47.** $z^{1/12}$ **49.** $x^{1/15}$ **51.** The product rule for exponents applies only if the bases are the same. **53.** $\sqrt{5r}$ **55.** $\sqrt[4]{21}$ **57.** $5\sqrt{3}$ **59.** $-3\sqrt[3]{4}$ **61.** $4pq^2\sqrt[3]{p}$ **63.** $2r^2t\sqrt[3]{79r^2t}$ **65.** $\dfrac{m^5}{3}$ **67.** $\dfrac{a^2\sqrt[4]{a}}{3}$ **69.** $p\sqrt{p}$ **71.** $\sqrt[10]{x^7}$ **73.** $\sqrt{197}$ **75.** $23\sqrt{5}$ **77.** $26m\sqrt{6m}$ **79.** $-8\sqrt[4]{2}$ **81.** $\dfrac{16 + 5\sqrt{5}}{20}$ **83.** $\left(12\sqrt{3} + 5\sqrt{2}\right)$ ft **85.** 2 **87.** $15 - 2\sqrt{26}$ **89.** $2\sqrt[3]{2y^2} + 2\sqrt[3]{4y} - 3$ **91.** The denominator would become $\sqrt[3]{6^2} = \sqrt[3]{36}$, which is not rational. **93.** $-3\sqrt{6}$ **95.** $\dfrac{\sqrt{22}}{4}$ **97.** $\dfrac{3m\sqrt[3]{4n}}{n^2}$ **99.** $\dfrac{5(\sqrt{6} + 3)}{3}$ **101.** $\dfrac{1 - 4\sqrt{2}}{3}$ **103.** $\{2\}$ **105.** $\varnothing$ **107.** $\{9\}$ **109.** $\{7\}$ **111.** $\{-13\}$ **113.** $\{14\}$ **115.** $\varnothing$ **117.** $\{7\}$ **119. (a)** $H = \sqrt{L^2 - W^2}$ **(b)** 7.9 ft **121.** $10i\sqrt{2}$ **123.** $-10 - 2i$ **125.** $-\sqrt{35}$ **127.** 3 **129.** $32 - 24i$ **131.** $4 + i$ **133.** 1 **135.** 1 **137.** $\dfrac{1}{100}$ **139.** k^6 **141.** $57\sqrt{2}$ **143.** $\dfrac{\sqrt[3]{60}}{5}$ **145.** $7i$ **147.** $-5i$ **149.** $5 + 12i$ **151.** The expression cannot be simplified further. **153.** $\{5\}$ **155.** $\left\{ \dfrac{3}{2} \right\}$ **157.** $\{1\}$ **159.** $\{9\}$ **161.** $\{7\}$

Chapter 8 Test   (pages 533–534)

[8.1] **1.** -29 **2.** -8 [8.2] **3.** 5   [8.1] **4.** C   **5.** 21.863   **6.** -9.405

7. domain: $[-6, \infty)$; range: $[0, \infty)$ [8.2] **8.** $\dfrac{125}{64}$ **9.** $\dfrac{1}{256}$ **10.** $\dfrac{9y^{3/10}}{x^2}$ **11.** $x^{4/3}y^6$ **12.** $7^{1/2}$, or $\sqrt{7}$

[8.3] **13.** $a^3\sqrt[3]{a^2}$, or $a^{11/3}$ **14.** $\sqrt{145}$ **15.** 10   **16.** $3x^2y^3\sqrt{6x}$ **17.** $2ab^3\sqrt[4]{2a^3b}$ **18.** $\sqrt[6]{200}$ [8.4] **19.** $26\sqrt{5}$

20. $(2ts - 3t^2)\sqrt[3]{2s^2}$ [8.5] **21.** $66 + \sqrt{5}$ **22.** $23 - 4\sqrt{15}$ **23.** $-\dfrac{\sqrt{10}}{4}$ **24.** $\dfrac{2\sqrt[3]{25}}{5}$ **25.** $-2\left(\sqrt{7} - \sqrt{5}\right)$

26. $3 + \sqrt{6}$ [8.6] **27.** **(a)** 59.8 **(b)** $T = \dfrac{V_0^2 - V^2}{-V^2k}$, or $T = \dfrac{V^2 - V_0^2}{V^2k}$ **28.** $\{-1\}$ **29.** $\{3\}$ **30.** $\{-3\}$

[8.7] **31.** $-5 - 8i$ **32.** $-2 + 16i$ **33.** $3 + 4i$ **34.** i **35.** **(a)** true **(b)** true **(c)** false **(d)** true

Chapters 1–8 Cumulative Review Exercises   (pages 534–536)

[1.3] **1.** 1   **2.** $-\dfrac{14}{9}$ [2.1] **3.** $\{-4\}$ **4.** $\{-12\}$ **5.** $\{6\}$ [2.7] **6.** $\left\{-\dfrac{10}{3}, 1\right\}$ **7.** $\left\{\dfrac{1}{4}\right\}$ [2.5] **8.** $(-6, \infty)$

[2.2] **9.** Both angles measure $80°$.   [2.3] **10.** 18 nickels; 32 quarters   **11.** $2\dfrac{2}{39}$ L   [3.1] **12.**

[3.2, 3.3] **13.** $-\dfrac{3}{2}$; $y = -\dfrac{3}{2}x$ [3.5] **14.** -37 [4.1] **15.** $\{(7, -2)\}$ [4.4] **16.** $\{(-1, 1, 1)\}$ [4.3] **17.** 2-oz letter: \$0.63;

3-oz letter: \$0.87   [5.2] **18.** $-k^3 - 3k^2 - 8k - 9$ [5.4] **19.** $8x^2 + 17x - 21$ [5.5] **20.** $z - 2 + \dfrac{3}{z}$

21. $3y^3 - 3y^2 + 4y + 1 + \dfrac{-10}{2y + 1}$ [6.2] **22.** $(2p - 3q)(p - q)$ [6.3] **23.** $(3k^2 + 4)(k - 1)(k + 1)$

24. $(x + 8)(x^2 - 8x + 64)$ [6.5] **25.** $\left\{-3, -\dfrac{5}{2}\right\}$ **26.** $\left\{-\dfrac{2}{5}, 1\right\}$ [7.1] **27.** $\{x \mid x \neq \pm 3\}$ **28.** $\dfrac{y}{y + 5}$

[7.2] **29.** $\dfrac{4x + 2y}{(x + y)(x - y)}$ [7.3] **30.** $-\dfrac{9}{4}$ **31.** $\dfrac{-1}{a + b}$ **32.** $\dfrac{1}{xy - 1}$ [7.6] **33.** Cecily: 8 mph; Mike: 4 mph

[7.5] **34.** $\emptyset$ [8.2] **35.** $\dfrac{1}{9}$ [8.3] **36.** $10x^2\sqrt{2}$ **37.** $2x\sqrt[3]{6x^2y^2}$ [8.4] **38.** $7\sqrt{2}$ [8.5] **39.** $\dfrac{\sqrt{10} + 2\sqrt{2}}{2}$

40. $-6x - 11\sqrt{xy} - 4y$ [8.3] **41.** $\sqrt{29}$ [8.6] **42.** $\{3, 4\}$ [8.1] **43.** 39.2 mph   [8.7] **44.** $2 + 9i$ **45.** $4 + 2i$

CHAPTER 9 Quadratic Equations, Inequalities, and Functions

Section 9.1   (pages 545–548)

1. The equation is also true for $x = -4$. The solution set is $\{-4, 4\}$.   **3.** **(a)** A quadratic equation in standard form has a second-degree polynomial in decreasing powers equal to 0. **(b)** The zero-factor property states that if a product equals 0, then at least one of the factors equals 0. **(c)** The square root property states that if the square of a quantity equals a number, then the quantity equals the positive or negative square root of the number.   **5.** $\{9, -9\}$ **7.** $\left\{\sqrt{17}, -\sqrt{17}\right\}$ **9.** $\left\{4\sqrt{2}, -4\sqrt{2}\right\}$

11. $\left\{2\sqrt{5}, -2\sqrt{5}\right\}$ **13.** $\left\{2\sqrt{6}, -2\sqrt{6}\right\}$ **15.** $\{-7, 3\}$ **17.** $\left\{4 + \sqrt{3}, 4 - \sqrt{3}\right\}$ **19.** $\left\{-5 + 4\sqrt{3}, -5 - 4\sqrt{3}\right\}$

21. $\left\{\dfrac{1 + \sqrt{7}}{3}, \dfrac{1 - \sqrt{7}}{3}\right\}$ **23.** $\left\{\dfrac{-1 + 2\sqrt{6}}{4}, \dfrac{-1 - 2\sqrt{6}}{4}\right\}$ **25.** $\left\{\dfrac{2 + 2\sqrt{3}}{5}, \dfrac{2 - 2\sqrt{3}}{5}\right\}$ **27.** 5.6 sec

29. $(2x + 1)^2 = 5$ is more suitable for solving by the square root property, while $x^2 + 4x = 12$ is more suitable for solving by

completing the square.   **31.** $9; (x + 3)^2$ **33.** $36; (p - 6)^2$ **35.** $\dfrac{81}{4}; \left(q + \dfrac{9}{2}\right)^2$ **37.** $\dfrac{1}{64}; \left(x + \dfrac{1}{8}\right)^2$

39. $0.16; (x - 0.4)^2$ **41.** 4   **43.** 25   **45.** $\dfrac{1}{36}$ **47.** $\{-4, 6\}$ **49.** $\left\{-2 + \sqrt{6}, -2 - \sqrt{6}\right\}$

51. $\left\{\dfrac{-7 + \sqrt{53}}{2}, \dfrac{-7 - \sqrt{53}}{2}\right\}$ **53.** $\left\{-\dfrac{8}{3}, 3\right\}$ **55.** $\left\{\dfrac{-5 + \sqrt{41}}{4}, \dfrac{-5 - \sqrt{41}}{4}\right\}$ **57.** $\left\{\dfrac{5 + \sqrt{15}}{5}, \dfrac{5 - \sqrt{15}}{5}\right\}$

59. $\left\{\dfrac{4 + \sqrt{3}}{3}, \dfrac{4 - \sqrt{3}}{3}\right\}$ **61.** $\left\{\dfrac{2 + \sqrt{3}}{3}, \dfrac{2 - \sqrt{3}}{3}\right\}$ **63.** $\left\{1 + \sqrt{2}, 1 - \sqrt{2}\right\}$ **65.** $\left\{2i\sqrt{3}, -2i\sqrt{3}\right\}$

67. $\{5 + 2i, 5 - 2i\}$ **69.** $\left\{\dfrac{1}{6} + \dfrac{\sqrt{2}}{3}i, \dfrac{1}{6} - \dfrac{\sqrt{2}}{3}i\right\}$ **71.** $\{-2 + 3i, -2 - 3i\}$ **73.** $\left\{-\dfrac{2}{3} + \dfrac{2\sqrt{2}}{3}i, -\dfrac{2}{3} - \dfrac{2\sqrt{2}}{3}i\right\}$

75. $\left\{-3 + i\sqrt{3}, -3 - i\sqrt{3}\right\}$ **77.** x^2 **78.** x **79.** $6x$ **80.** 1   **81.** 9   **82.** $(x + 3)^2$, or $x^2 + 6x + 9$ **83.** $\left\{\sqrt{b}, -\sqrt{b}\right\}$

85. $\left\{\dfrac{\sqrt{b^2 + 16}}{2}, -\dfrac{\sqrt{b^2 + 16}}{2}\right\}$ **87.** $\left\{\dfrac{2b + \sqrt{3a}}{5}, \dfrac{2b - \sqrt{3a}}{5}\right\}$ **89.** ± 4.1231056 **91.** $\sqrt{13}$ **93.** 1   **95.** $\dfrac{-1 + \sqrt{13}}{6}$

Section 9.2   (pages 554–556)

1. The patron forgot the $\pm$ sign in the numerator. The correct formula is $x = \dfrac{-b \pm \sqrt{b^2 - 4ac}}{2a}$.   **3.** No, the quadratic

formula can be used to solve *any* quadratic equation. Here, the quadratic formula can be used with $a = 2$, $b = 0$,

and $c = -5$.   **5.** $\{3, 5\}$ **7.** $\left\{\dfrac{-2 + \sqrt{2}}{2}, \dfrac{-2 - \sqrt{2}}{2}\right\}$ **9.** $\left\{\dfrac{1 + \sqrt{3}}{2}, \dfrac{1 - \sqrt{3}}{2}\right\}$ **11.** $\left\{5 + \sqrt{7}, 5 - \sqrt{7}\right\}$

13. $\left\{\dfrac{-1 + \sqrt{2}}{2}, \dfrac{-1 - \sqrt{2}}{2}\right\}$ **15.** $\left\{\dfrac{-1 + \sqrt{7}}{3}, \dfrac{-1 - \sqrt{7}}{3}\right\}$ **17.** $\left\{1 + \sqrt{5}, 1 - \sqrt{5}\right\}$

19. $\left\{\dfrac{-2 + \sqrt{10}}{2}, \dfrac{-2 - \sqrt{10}}{2}\right\}$ **21.** $\left\{-1 + 3\sqrt{2}, -1 - 3\sqrt{2}\right\}$ **23.** $\left\{\dfrac{1 + \sqrt{29}}{2}, \dfrac{1 - \sqrt{29}}{2}\right\}$

25. $\left\{\dfrac{-4 + \sqrt{91}}{3}, \dfrac{-4 - \sqrt{91}}{3}\right\}$ **27.** $\left\{\dfrac{-3 + \sqrt{57}}{8}, \dfrac{-3 - \sqrt{57}}{8}\right\}$ **29.** $\left\{\dfrac{3}{2} + \dfrac{\sqrt{15}}{2}i, \dfrac{3}{2} - \dfrac{\sqrt{15}}{2}i\right\}$

31. $\left\{3 + i\sqrt{5}, 3 - i\sqrt{5}\right\}$ **33.** $\left\{\dfrac{1}{2} + \dfrac{\sqrt{6}}{2}i, \dfrac{1}{2} - \dfrac{\sqrt{6}}{2}i\right\}$ **35.** $\left\{-\dfrac{2}{3} + \dfrac{\sqrt{2}}{3}i, -\dfrac{2}{3} - \dfrac{\sqrt{2}}{3}i\right\}$

37. $\left\{\dfrac{1}{2} + \dfrac{1}{4}i, \dfrac{1}{2} - \dfrac{1}{4}i\right\}$ **39.** B   **41.** C   **43.** A   **45.** D   **47.** B   **49.** $\left\{-\dfrac{7}{5}\right\}$ **51.** $\left\{-\dfrac{1}{3}, 2\right\}$

53. (a) Discriminant is 25, or 5^2; solve by factoring; $\left\{-3, -\dfrac{4}{3}\right\}$ **(b)** Discriminant is 44; use the quadratic formula;

$\left\{\dfrac{7 + \sqrt{11}}{2}, \dfrac{7 - \sqrt{11}}{2}\right\}$ **55.** -10 or 10   **57.** 16   **59.** 25   **61.** $b = \dfrac{44}{5}; \dfrac{3}{10}$ **63.** $u^2 + 4u - 5; (u + 5)(u - 1)$

65. $\{-8\}$ **67.** $\{5\}$

Section 9.3 (pages 563–567)

1. square root property   **3.** quadratic formula   **5.** factoring   **7.** Multiply by the LCD, x.   **9.** Substitute a variable for $x^2 + x$.   **11.** The proposed solution -1 does not check. The solution set is $\{4\}$.   **13.** $\{-2, 7\}$ **15.** $\{-4, 7\}$

17. $\left\{-\dfrac{2}{3}, 1\right\}$ **19.** $\left\{-\dfrac{14}{17}, 5\right\}$ **21.** $\left\{-\dfrac{11}{7}, 0\right\}$ **23.** $\left\{\dfrac{-1 + \sqrt{13}}{2}, \dfrac{-1 - \sqrt{13}}{2}\right\}$ **25.** $\left\{-\dfrac{8}{3}, -1\right\}$

27. $\left\{\dfrac{2 + \sqrt{22}}{3}, \dfrac{2 - \sqrt{22}}{3}\right\}$ **29.** $\left\{\dfrac{-1 + \sqrt{5}}{4}, \dfrac{-1 - \sqrt{5}}{4}\right\}$ **31. (a)** $(20 - t)$ mph **(b)** $(20 + t)$ mph   **33.** 25 mph

35. 50 mph   **37.** 3.6 hr   **39.** Rusty: 25.0 hr; Nancy: 23.0 hr   **41.** 3 hr; 6 hr   **43.** $\{2, 5\}$ **45.** $\{3\}$ **47.** $\left\{\dfrac{8}{9}\right\}$

49. $\{9\}$ **51.** $\left\{\dfrac{2}{5}\right\}$ **53.** $\{-2\}$ **55.** $\{-5, -2, 2, 5\}$ **57.** $\left\{-\dfrac{3}{2}, -1, 1, \dfrac{3}{2}\right\}$ **59.** $\left\{-2\sqrt{3}, -2, 2, 2\sqrt{3}\right\}$

61. $\{-6, -5\}$ **63.** $\left\{-\dfrac{16}{3}, -2\right\}$ **65.** $\left\{-\dfrac{1}{3}, \dfrac{1}{6}\right\}$ **67.** $\left\{-\dfrac{1}{2}, 3\right\}$ **69.** $\{-8, 1\}$ **71.** $\{-64, 27\}$

73. $\left\{-\dfrac{27}{8}, -1, 1, \dfrac{27}{8}\right\}$ **75.** $\{25\}$ **77.** $\left\{-1, 1, -\dfrac{\sqrt{6}}{2}i, \dfrac{\sqrt{6}}{2}i\right\}$ **79.** $\left\{-\dfrac{\sqrt{6}}{3}, -\dfrac{1}{2}, \dfrac{1}{2}, \dfrac{\sqrt{6}}{3}\right\}$ **81.** $\{3, 11\}$

83. $\left\{-\sqrt[3]{5}, -\dfrac{\sqrt[3]{4}}{2}\right\}$ **85.** $\left\{\dfrac{4}{3}, \dfrac{9}{4}\right\}$ **87.** $\left\{\dfrac{\sqrt{9 + \sqrt{65}}}{2}, -\dfrac{\sqrt{9 + \sqrt{65}}}{2}, \dfrac{\sqrt{9 - \sqrt{65}}}{2}, -\dfrac{\sqrt{9 - \sqrt{65}}}{2}\right\}$

89. $W = \dfrac{P - 2L}{2}$, or $W = \dfrac{P}{2} - L$ **91.** $C = \dfrac{5}{9}(F - 32)$

Summary Exercises on Solving Quadratic Equations (pages 567–568)

1. $\left\{\sqrt{7}, -\sqrt{7}\right\}$ **2.** $\left\{-\dfrac{3}{2}, \dfrac{5}{3}\right\}$ **3.** $\left\{-3 + \sqrt{5}, -3 - \sqrt{5}\right\}$ **4.** $\{-2, 8\}$ **5.** $\left\{-\dfrac{3}{2}, 4\right\}$ **6.** $\left\{-3, \dfrac{1}{3}\right\}$

7. $\left\{\dfrac{2 + \sqrt{2}}{2}, \dfrac{2 - \sqrt{2}}{2}\right\}$ **8.** $\left\{2i\sqrt{3}, -2i\sqrt{3}\right\}$ **9.** $\left\{\dfrac{1}{2}, 2\right\}$ **10.** $\{-3, -1, 1, 3\}$ **11.** $\left\{\dfrac{-3 + 2\sqrt{2}}{2}, \dfrac{-3 - 2\sqrt{2}}{2}\right\}$

12. $\left\{\dfrac{4}{5}, 3\right\}$ **13.** $\left\{-\sqrt{7}, -\sqrt{2}, \sqrt{2}, \sqrt{7}\right\}$ **14.** $\left\{\dfrac{1 + \sqrt{5}}{4}, \dfrac{1 - \sqrt{5}}{4}\right\}$ **15.** $\left\{-\dfrac{1}{2} + \dfrac{\sqrt{3}}{2}i, -\dfrac{1}{2} - \dfrac{\sqrt{3}}{2}i\right\}$

16. $\left\{-\dfrac{\sqrt[3]{175}}{5}, 1\right\}$ **17.** $\left\{\dfrac{3}{2}\right\}$ **18.** $\left\{\dfrac{2}{3}\right\}$ **19.** $\left\{6\sqrt{2}, -6\sqrt{2}\right\}$ **20.** $\left\{-\dfrac{2}{3}, 2\right\}$ **21.** $\{-4, 9\}$ **22.** $\{13, -13\}$

23. $\left\{1 + \dfrac{\sqrt{3}}{3}i, 1 - \dfrac{\sqrt{3}}{3}i\right\}$ **24.** $\{3\}$ **25.** $\left\{\dfrac{1}{6} + \dfrac{\sqrt{47}}{6}i, \dfrac{1}{6} - \dfrac{\sqrt{47}}{6}i\right\}$ **26.** $\left\{-\dfrac{1}{3}, \dfrac{1}{6}\right\}$

Section 9.4 (pages 572–577)

1. Find a common denominator, and then multiply both sides by the common denominator.   **3.** Write it in standard form (with 0 on one side, in decreasing powers of w).   **5.** $m = \sqrt{p^2 - n^2}$ **7.** $t = \dfrac{\pm\sqrt{dk}}{k}$ **9.** $d = \dfrac{\pm\sqrt{skI}}{I}$ **11.** $v = \dfrac{\pm\sqrt{kAF}}{F}$

13. $r = \dfrac{\pm\sqrt{3\pi Vh}}{\pi h}$ **15.** $t = \dfrac{-B \pm \sqrt{B^2 - 4AC}}{2A}$ **17.** $h = \dfrac{D^2}{k}$ **19.** $\ell = \dfrac{p^2 g}{k}$ **21.** $r = \dfrac{\pm\sqrt{S\pi}}{2\pi}$

23. $R = \dfrac{E^2 - 2pr \pm E\sqrt{E^2 - 4pr}}{2p}$ **25.** $r = \dfrac{5pc}{4}$ or $r = -\dfrac{2pc}{3}$ **27.** $I = \dfrac{-cR \pm \sqrt{c^2R^2 - 4cL}}{2cL}$ **29.** 7.9, 8.9, 11.9

31. eastbound ship: 80 mi; southbound ship: 150 mi   **33.** 8 in., 15 in., 17 in.   **35.** length: 24 ft; width: 10 ft   **37.** 2 ft

39. 7 m by 12 m   **41.** 20 in. by 12 in.   **43.** 1 sec and 8 sec   **45.** 2.4 sec and 5.6 sec   **47.** 9.2 sec   **49.** It reaches its *maximum* height at 5 sec because this is the only time it reaches 400 ft.   **51.** 18 in.   **53.** 5.5 m per sec   **55.** 5 or 14

57. (a) 2750 billion **(b)** 2750 billion; They are the same.   **59.** 2001; The graph indicates that vehicle-miles reached 2800

in 2001.   **61.** 9   **63.** $\dfrac{-b^2 + 4ac}{4a}$ **65.** $\{1, 7\}$

Section 9.5 (pages 583–588)

1. (a) B **(b)** C **(c)** A **(d)** D   **3.** (0, 0)   **5.** (0, 4)   **7.** (1, 0)   **9.** (−3, −4)   **11.** (5, 6)   **13.** down; wider

15. up; narrower   **17. (a)** I **(b)** IV **(c)** II **(d)** III   **19. (a)** D **(b)** B   **(c)** C **(d)** A

21.   **23.**   **25.**   **27.** vertex: (4, 0);
axis: $x = 4$;
domain: $(-\infty, \infty)$;
range: $[0, \infty)$

29. vertex: (−2, −1);
axis: $x = -2$;
domain: $(-\infty, \infty)$;
range: $[-1, \infty)$

31. vertex: (2, −4);
axis: $x = 2$;
domain: $(-\infty, \infty)$;
range: $[-4, \infty)$

33. vertex: (−1, 2);
axis: $x = -1$;
domain: $(-\infty, \infty)$;
range: $(-\infty, 2]$

35. vertex: (2, −3);
axis: $x = 2$;
domain: $(-\infty, \infty)$;
range: $[-3, \infty)$

37. It is shifted 6 units up.   **38.**    **39.** It is shifted 6 units up.   **40.** It is shifted 6 units to the right.

41.   **42.** It is shifted 6 units to the right.   **43.** linear; positive   **45.** quadratic; positive

47. quadratic; negative   **49. (a)** 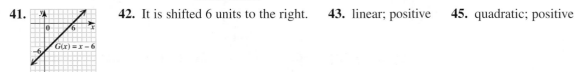   **(b)** quadratic; positive **(c)** $y = 2.969x^2 - 23.125x + 115$ **(d)** 265
(e) No. About 16 companies filed for bankruptcy each month, so at this rate, filings for 2002 would be about 192. The approximation from the model seems high.

51. (a) 222.7 (per 100,000) **(b)** The approximation using the model is high.   **53.** $\{-4, 5\}$ **55.** $\{-3 + 2\sqrt{3}, -3 - 2\sqrt{3}\}$

57. $\left\{\dfrac{6 + \sqrt{46}}{2}, \dfrac{6 - \sqrt{46}}{2}\right\}$

Section 9.6 (pages 597–601)

1. If x is squared, it has a vertical axis; if y is squared, it has a horizontal axis.   **3.** Use the discriminant of the function. If it is positive, there are two x-intercepts. If it is 0, there is one x-intercept (at the vertex), and if it is negative, there is no x-intercept.

5. $(-4, -6)$ **7.** $(1, -3)$ **9.** $\left(-\dfrac{1}{2}, -\dfrac{29}{4}\right)$ **11.** $(-1, 3)$; up; narrower; no x-intercepts   **13.** $\left(\dfrac{5}{2}, \dfrac{37}{4}\right)$; down; same;

two x-intercepts   **15.** $(-3, -9)$; to the right; wider   **17.** F   **19.** C   **21.** D

23. vertex: $(-4, -6)$;
axis: $x = -4$;
domain: $(-\infty, \infty)$;
range: $[-6, \infty)$

25. vertex: $(1, -3)$;
axis: $x = 1$;
domain: $(-\infty, \infty)$;
range: $(-\infty, -3]$

27. vertex: $(1, -2)$;
axis: $y = -2$;
domain: $[1, \infty)$;
range: $(-\infty, \infty)$

29. vertex: $(1, 5)$;
axis: $y = 5$;
domain: $(-\infty, 1]$;
range: $(-\infty, \infty)$

31. vertex: $(-7, -2)$;
axis: $y = -2$;
domain: $[-7, \infty)$;
range: $(-\infty, \infty)$

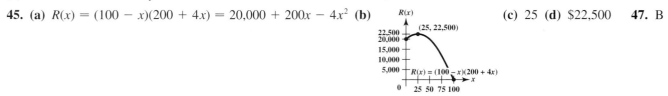

33. 20 and 20   **35.** 140 ft by 70 ft; 9800 ft^2 **37.** 16 ft; 2 sec   **39.** 2 sec; 67 ft   **41.** (a) maximum (b) 1993; 13.2%

43. (a) The coefficient of x^2 is negative because a parabola that models the data must open down. (b) $(18.45, 3860)$

(c) In 2018 Social Security assets will reach their maximum value of $3860 billion.

45. (a) $R(x) = (100 - x)(200 + 4x) = 20,000 + 200x - 4x^2$ (b)   (c) 25 (d) $22,500   **47.** B

49. A   **51.** (a) $|x + p|$ (b) The distance from the focus to the origin should equal the distance from the directrix to the origin.

(c) $\sqrt{(x - p)^2 + y^2}$ (d) $y^2 = 4px$ **53.**   **55.**   **57.** $\left(-\dfrac{3}{2}, \infty\right)$

Section 9.7 (pages 608–609)

1. (a) $\{1, 3\}$ (b) $(-\infty, 1) \cup (3, \infty)$ (c) $(1, 3)$ **3.** (a) $\left\{-3, \dfrac{5}{2}\right\}$ (b) $\left[-3, \dfrac{5}{2}\right]$ (c) $(-\infty, -3] \cup \left[\dfrac{5}{2}, \infty\right)$

5. Include the endpoints if the symbol is $\geq$ or $\leq$. Exclude the endpoints if the symbol is $>$ or $<$.

7. $(-\infty, -1) \cup (5, \infty)$ **9.** $(-4, 6)$ **11.** $(-\infty, 1] \cup [3, \infty)$

13. $\left(-\infty, -\dfrac{3}{2}\right] \cup \left[\dfrac{3}{5}, \infty\right)$ **15.** $\left[-\dfrac{3}{2}, \dfrac{3}{2}\right]$

17. $\left(-\infty, -\dfrac{1}{2}\right] \cup \left[\dfrac{1}{3}, \infty\right)$ **19.** $(-\infty, 0] \cup [4, \infty)$ **21.** $\left[0, \dfrac{5}{3}\right]$

23. $\left(-\infty, 3 - \sqrt{3}\right] \cup \left[3 + \sqrt{3}, \infty\right)$ **25.** $(-\infty, \infty)$ **27.** Ø   **29.** $(-\infty, 1) \cup (2, 4)$

31. $\left[-\dfrac{3}{2}, \dfrac{1}{3}\right] \cup [4, \infty)$ **33.** $(-\infty, 1) \cup (4, \infty)$ **35.** $\left[-\dfrac{3}{2}, 5\right)$

37. $(2, 6]$ 　**39.** $\left(-\infty, \dfrac{1}{2}\right) \cup \left(\dfrac{5}{4}, \infty\right)$ 　**41.** $[-7, -2)$

43. $(-\infty, 2) \cup (4, \infty)$ 　**45.** $\left(0, \dfrac{1}{2}\right) \cup \left(\dfrac{5}{2}, \infty\right)$ 　**47.** $\left[\dfrac{3}{2}, \infty\right)$

49. $\left(-2, \dfrac{5}{3}\right) \cup \left(\dfrac{5}{3}, \infty\right)$ 　**51.** domain: $\{-3, -2, -1, 0, 1, 2, 3\}$; range: $\left\{\dfrac{1}{8}, \dfrac{1}{4}, \dfrac{1}{2}, 1, 2, 4, 8\right\}$

53. function

Chapter 9 Review Exercises　(pages 615–620)

1. $\{-11, 11\}$　**3.** $\left\{-\dfrac{15}{2}, \dfrac{5}{2}\right\}$　**5.** $\{-2 + \sqrt{19}, -2 - \sqrt{19}\}$　**7.** By the square root property, the first step should be

$x = \sqrt{12}$ or $x = -\sqrt{12}$. The solution set is $\{-2\sqrt{3}, 2\sqrt{3}\}$.　**9.** C　**11.** D　**13.** $\left\{-\dfrac{7}{2}, 3\right\}$

15. $\left\{\dfrac{1 + \sqrt{41}}{2}, \dfrac{1 - \sqrt{41}}{2}\right\}$　**17.** $\left\{\dfrac{2}{3} + \dfrac{\sqrt{2}}{3}i, \dfrac{2}{3} - \dfrac{\sqrt{2}}{3}i\right\}$　**19.** $\left\{-\dfrac{5}{2}, 3\right\}$　**21.** $\{-4\}$　**23.** $\left\{-\dfrac{343}{8}, 64\right\}$

25. 7 mph　**27.** 4.6 hr　**29.** $v = \dfrac{\pm\sqrt{rFkw}}{kw}$　**31.** $t = \dfrac{3m \pm \sqrt{9m^2 + 24m}}{2m}$　**33.** 12 cm by 20 cm

35. 3 min　**37.** 0.7 sec and 4.0 sec　**39.** 4.5%　**41.** $(0, 6)$　**43.** $(3, 7)$　**45.** $(-4, 3)$

47. vertex: $(2, -3)$;
axis: $x = 2$;
domain: $(-\infty, \infty)$;
range: $[-3, \infty)$

49. vertex: $(-4, -3)$;
axis: $y = -3$;
domain: $[-4, \infty)$;
range: $(-\infty, \infty)$

51. **(a)** $a + b + c = 11.47$;
$16a + 4b + c = 24.45$;
$64a + 8b + c = 29.78$
(b) $f(x) = -0.428x^2 + 6.47x + 5.43$
(c) \$27.08; the result using the model is slightly high.

53. length: 50 m; width: 50 m; maximum area: 2500 m^2　**55.** $[-4, 3]$ 　**57.** $\emptyset$

59. $[-3, 2)$ 　**61.** $\left\{1 + \dfrac{\sqrt{3}}{3}i, 1 - \dfrac{\sqrt{3}}{3}i\right\}$　**63.** $\{-2, -1, 3, 4\}$　**65.** $\{4\}$　**67.** $d = \dfrac{\pm\sqrt{SkI}}{I}$

69. $\left\{-\dfrac{5}{3}, -\dfrac{3}{2}\right\}$　**71.** $\left(-5, -\dfrac{23}{5}\right]$　**73.** B　**75.** A　**77.** D　**79.** **(a)** 8.64 quadrillion Btu **(b)** The result using the

model is high. **(c)** 2008　**81.** 412.3 ft

Chapter 9 Test　(pages 620–621)

[9.1] **1.** $\{3\sqrt{6}, -3\sqrt{6}\}$　**2.** $\left\{-\dfrac{8}{7}, \dfrac{2}{7}\right\}$　**3.** $\{-1 + \sqrt{5}, -1 - \sqrt{5}\}$　[9.2] **4.** $\left\{\dfrac{3 + \sqrt{17}}{4}, \dfrac{3 - \sqrt{17}}{4}\right\}$

5. $\left\{\dfrac{2}{3} + \dfrac{\sqrt{11}}{3}i, \dfrac{2}{3} - \dfrac{\sqrt{11}}{3}i\right\}$　**6.** $\left\{\dfrac{2}{3}\right\}$　[9.1] **7.** A　[9.2] **8.** discriminant: 88; There are two irrational solutions.

[9.1–9.3] **9.** $\left\{-\dfrac{2}{3}, 6\right\}$ **10.** $\left\{\dfrac{-7 + \sqrt{97}}{8}, \dfrac{-7 - \sqrt{97}}{8}\right\}$ **11.** $\left\{-2, -\dfrac{1}{3}, \dfrac{1}{3}, 2\right\}$ **12.** $\left\{-\dfrac{5}{2}, 1\right\}$

[9.4] **13.** $r = \dfrac{\pm\sqrt{\pi S}}{2\pi}$ [9.3] **14.** Andrew: 11.1 hr; Kent: 9.1 hr   **15.** 7 mph   [9.4] **16.** 2 ft   **17.** 16 m   [9.5] **18.** A

19. vertex: $(0, -2)$; axis: $x = 0$;   [9.6] **20.** vertex: $(2, 3)$; axis: $x = 2$;   **21.** vertex: $(2, 2)$; axis: $y = 2$;
domain: $(-\infty, \infty)$; range: $[-2, \infty)$ domain: $(-\infty, \infty)$; range: $(-\infty, 3]$ domain: $(-\infty, 2]$; range: $(-\infty, \infty)$

22. (a) 6.5% **(b)** 1996; 3.5%   **23.** 160 ft by 320 ft   [9.7] **24.** $(-\infty, -5) \cup \left(\dfrac{3}{2}, \infty\right)$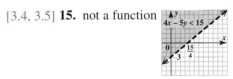

25. $(-\infty, 4) \cup [9, \infty)$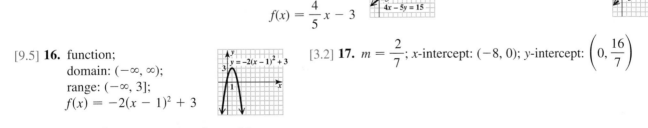

Chapters 1–9 Cumulative Review Exercises   (pages 622–624)

[1.1, 8.7] **1. (a)** $-2, 0, 7$ **(b)** $-\dfrac{7}{3}, -2, 0, 0.7, 7, \dfrac{32}{3}$ **(c)** All are real except $\sqrt{-8}$. **(d)** All are complex numbers.

[1.1–1.3] **2.** 6   **3.** 41   [2.1] **4.** $\left\{\dfrac{4}{5}\right\}$ [2.7] **5.** $\left\{\dfrac{11}{10}, \dfrac{7}{2}\right\}$ [8.6] **6.** $\left\{\dfrac{2}{3}\right\}$ [7.4] **7.** $\emptyset$

[9.1, 9.2] **8.** $\left\{\dfrac{7 + \sqrt{177}}{4}, \dfrac{7 - \sqrt{177}}{4}\right\}$ [9.3] **9.** $\{-2, -1, 1, 2\}$ [2.5] **10.** $[1, \infty)$ [2.7] **11.** $\left[2, \dfrac{8}{3}\right]$

[9.7] **12.** $(1, 3)$ **13.** $(-2, 1)$ [3.1, 3.5] **14.** function;   [3.4, 3.5] **15.** not a function
domain: $(-\infty, \infty)$;
range: $(-\infty, \infty)$;
$f(x) = \dfrac{4}{5}x - 3$

[9.5] **16.** function;   [3.2] **17.** $m = \dfrac{2}{7}$; x-intercept: $(-8, 0)$; y-intercept: $\left(0, \dfrac{16}{7}\right)$
domain: $(-\infty, \infty)$;
range: $(-\infty, 3]$;
$f(x) = -2(x - 1)^2 + 3$

18. (a) $y = -\dfrac{5}{2}x + 2$ **(b)** $y = \dfrac{2}{5}x + \dfrac{13}{5}$ **19. (a)** $y = 123x + 600$ **(b)** \$1461 million; the result using the model is a bit high.   [3.5] **20.** No. The graph is a vertical line, which is not the graph of a function by the vertical line test. Also, the only domain value, 5, can have infinitely many range values paired with it.   **21. (a)** 13 **(b)** domain: $(-\infty, \infty)$; range: $[-5, \infty)$ [4.1] **22.** $\{(1, -2)\}$ [4.2] **23.** $\{(3, -4, 2)\}$ [4.3] **24. (a)** $x + y = 34.2$; $x = 4y - 0.3$ **(b)** AOL: \$27.3 billion;

Time Warner: \$6.9 billion   [5.1] **25.** $\dfrac{x^8}{y^4}$ **26.** $\dfrac{4}{xy^2}$ [5.4] **27.** $14x^2 - 13x - 12$ **28.** $\dfrac{4}{9}t^2 + 12t + 81$

[5.2] **29.** $-3t^3 + 5t^2 - 12t + 15$ [5.5] **30.** $4x^2 - 6x + 11 + \dfrac{4}{x + 2}$ [6.1–6.3] **31.** $x(4 + x)(4 - x)$

32. $(4m - 3)(6m + 5)$ **33.** $(2x + 3y)(4x^2 - 6xy + 9y^2)$ **34.** $(3x - 5y)^2$ [7.1] **35.** $\dfrac{x - 5}{x + 5}$ [7.2] **36.** $-\dfrac{8}{k}$

[7.3] **37.** $\dfrac{r - s}{r}$ **38.** $x + y$ [8.1] **39.** $\dfrac{3\sqrt[3]{4}}{4}$ [8.4] **40.** $\sqrt{7} + \sqrt{5}$ [7.5] **41.** biking: 12 mph; walking: 2 mph

[9.4] **42.** southbound car: 57 mi; eastbound car: 76 mi [2.3] **43.** 1240 **44.** 960 **45.** 1320 **46.** 1240

CHAPTER 10 Inverse, Exponential, and Logarithmic Functions

Section 10.1 (pages 631–635)

1. This function is not one-to-one because both France and the United States are paired with the same trans fat percentage, 11. Also both Hungary and Poland are paired with the same trans fat percentage, 8. **3.** Yes. By adding 1 to 1058 two distances would be the same, so the function would not be one-to-one. **5.** B **7.** A **9.** $\{(6, 3), (10, 2), (12, 5)\}$ **11.** not one-to-one

13. $f^{-1}(x) = \dfrac{x - 4}{2}$, or $f^{-1}(x) = \dfrac{1}{2}x - 2$ **15.** $g^{-1}(x) = x^2 + 3$, $x \geq 0$ **17.** not one-to-one **19.** $f^{-1}(x) = \sqrt[3]{x + 4}$

21. (a) 8 **(b)** 3 **23. (a)** 1 **(b)** 0 **25. (a)** one-to-one **(b)** **27. (a)** not one-to-one

29. (a) one-to-one **(b)** **31.** **33.** **35.**

x	$f(x)$
0	0
1	1
4	2

37.

x	$f(x)$
-1	-3
0	-2
1	-1
2	6

39. $f^{-1}(x) = \dfrac{x + 5}{4}$, or $f^{-1}(x) = \dfrac{1}{4}x + \dfrac{5}{4}$ **40.** My graphing calculator is the greatest thing since sliced bread.

41. If the function were not one-to-one, there would be ambiguity in some of the characters, as they could represent more than one letter. **42.** Answers will vary. For example, Jane Doe is 1004 5 2748 129 68 3379 129.

43. $f^{-1}(x) = \dfrac{x + 7}{2}$, or $f^{-1}(x) = \dfrac{1}{2}x + \dfrac{7}{2}$ **45.** $f^{-1}(x) = \sqrt[3]{x - 5}$ **47.**

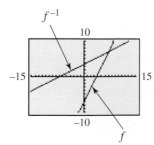

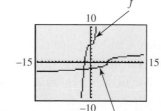

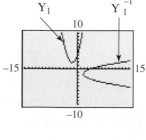

49. 64 **51.** 2 **53.** 44.02

Section 10.2 (pages 641–644)

1. C **3.** A **5.** **7.** **9.** **11.** 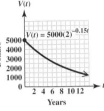 **13. (a)** rises; falls

(b) It is one-to-one and thus has an inverse. **15.** {2} **17.** $\left\{\dfrac{3}{2}\right\}$ **19.** {7} **21.** {−3} **23.** {−1}

25. {−3} **27.** 639.545 **29.** 0.066 **31.** 12.179 **33. (a)** 0.5°C **(b)** 0.35°C **35. (a)** 1.6°C **(b)** 0.5°C
37. (a) 220,717 thousand tons **(b)** 129,048 thousand tons **(c)** It is slightly greater than what the model provides
(111,042 thousand tons). **39. (a)** \$5000 **(b)** \$2973 **(c)** \$1768 **(d)**

41. 6.67 yr after it was purchased **43.** 4 **45.** 0

Section 10.3 (pages 649–652)

1. (a) B **(b)** E **(c)** D **(d)** F **(e)** A **(f)** C **3.** $\log_4 1024 = 5$ **5.** $\log_{1/2} 8 = -3$ **7.** $\log_{10} 0.001 = -3$

9. $\log_{625} 5 = \dfrac{1}{4}$ **11.** $4^3 = 64$ **13.** $10^{-4} = \dfrac{1}{10,000}$ **15.** $6^0 = 1$ **17.** $9^{1/2} = 3$ **19. (a)** C **(b)** B **(c)** B **(d)** C

21. $\left\{\dfrac{1}{3}\right\}$ **23.** {81} **25.** $\left\{\dfrac{1}{5}\right\}$ **27.** {1} **29.** $\{x \mid x > 0, x \neq 1\}$ **31.** {5} **33.** $\left\{\dfrac{5}{3}\right\}$ **35.** {4} **37.** $\left\{\dfrac{3}{2}\right\}$

39. {30} **41.** 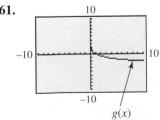 **43.** **45.** Every power of 1 is equal to 1, and thus it cannot be used as a base.

47. $(0, \infty); (-\infty, \infty)$ **49.** 8 **51.** 24 **53. (a)** 4385 billion ft³ **(b)** 5555 billion ft³ **(c)** 6140 billion ft³

55. (a) 130 thousand units **(b)** 190 thousand units **(c)** **57.** about 4 times as powerful

59. 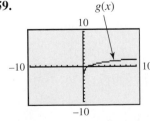 **61.** **63.** 4^9 **65.** $\dfrac{1}{5^{11}}$ **67.** $\dfrac{1}{9^6}$

Connections **(page 659)** **1.** $\log_{10} 458.3 \approx 2.661149857$ **2.** Answers will vary.

$\underline{+ \ \log_{10} 294.6 \approx 2.469232743}$

≈ 5.130382600

$10^{5.130382600} \approx 135{,}015.18$

A calculator gives

$(458.3)(294.6) = 135{,}015.18.$

Section 10.4 (pages 659–661)

1. $\log_{10} 3 + \log_{10} 4$ **3.** 4 **5.** 4 **7.** $\log_7 4 + \log_7 5$ **9.** $\log_5 8 - \log_5 3$ **11.** $2 \log_4 6$

13. $\dfrac{1}{3} \log_3 4 - 2 \log_3 x - \log_3 y$ **15.** $\dfrac{1}{2} \log_3 x + \dfrac{1}{2} \log_3 y - \dfrac{1}{2} \log_3 5$ **17.** $\dfrac{1}{3} \log_2 x + \dfrac{1}{5} \log_2 y - 2 \log_2 r$

19. The distributive property tells us that the *product* $a(x + y)$ equals the sum $ax + ay$. In the notation $\log_a(x + y)$, the parentheses do not indicate multiplication. They indicate that $x + y$ is the result of raising a to some power. **21.** $\log_b xy$

23. $\log_a \dfrac{m}{n}$ **25.** $\log_a \dfrac{rt^3}{s}$ **27.** $\log_a \dfrac{125}{81}$ **29.** $\log_{10}(x^2 - 9)$ **31.** $\log_p \dfrac{x^3 y^{1/2}}{z^{3/2} a^3}$ **33.** 1.2552 **35.** -0.6532

37. 1.5562 **39.** 0.4771 **41.** 0.2386 **43.** 4.7710 **45.** false **47.** true **49.** true **51.** false **53.** The exponent of a quotient is the difference between the exponent of the numerator and the exponent of the denominator.

55. $\log_2 8 - \log_2 4 = \log_2 \dfrac{8}{4} = \log_2 2 = 1$ **57.** $\log_{10} 10{,}000 = 4$ **59.** $\log_{10} 0.01 = -2$ **61.** $10^0 = 1$

Connections **(page 666)** 2; 2.5; $2.\overline{6}$; $2.708\overline{3}$; $2.71\overline{6}$; The difference is 0.0016151618. It approaches e fairly quickly.

Section 10.5 (pages 667–671)

1. C **3.** C **5.** 19.2 **7.** 1.6335 **9.** 2.5164 **11.** -1.4868 **13.** 9.6776 **15.** 2.0592 **17.** -2.8896 **19.** 5.9613
21. 4.1506 **23.** 2.3026 **25.** (a) 2.552424846 (b) 1.552424846 (c) 0.552424846 (d) The whole number parts will vary but the decimal parts are the same. **27.** An error message appears, because we cannot find the common logarithm of a negative number. **29.** bog **31.** 11.6 **33.** 4.3 **35.** 4.0×10^{-8} **37.** 4.0×10^{-6} **39.** (a) 107 dB (b) 100 dB (c) 98 dB
41. (a) 800 yr (b) 5200 yr (c) 11,500 yr **43.** (a) 77% (b) 1989 **45.** (a) \$54 per ton (b) If $p = 0$, then $\ln(1 - p) = \ln 1 = 0$, so T would be negative. If $p = 1$, then $\ln(1 - p) = \ln 0$, but the domain of $\ln x$ is $(0, \infty)$. **47.** 2.2619
49. 0.6826 **51.** 0.3155 **53.** 0.8736 **55.** 2.4849 **57.** Answers will vary. Suppose the name is Jeffery Cole, with $m = 7$ and $n = 4$. (a) $\log_7 4$ is the exponent to which 7 must be raised to obtain 4. (b) 0.7124143742 (c) 4 **59.** 6446 billion ft³

61.

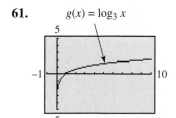

$g(x) = \log_3 x$

63.

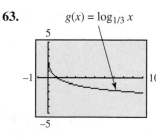

$g(x) = \log_{1/3} x$

65. $\left\{ -\dfrac{3}{5} \right\}$ **67.** $\{5\}$ **69.** $\{-3\}$

71. $\log(x + 2)(x - 3)$, or $\log(x^2 - x - 6)$

Section 10.6 (pages 678–682)

1. $\log 5^x = \log 125$ **2.** $x \log 5 = \log 125$ **3.** $x = \dfrac{\log 125}{\log 5}$ **4.** $\dfrac{\log 125}{\log 5} = 3; \{3\}$ **5.** $\{0.827\}$ **7.** $\{0.833\}$

9. $\{1.201\}$ **11.** $\{2.269\}$ **13.** $\{15.967\}$ **15.** $\{261.291\}$ **17.** $\{-10.718\}$ **19.** $\{3\}$ **21.** $\{5.879\}$

23. Natural logarithms are a better choice because e is the base. **25.** $\left\{ \dfrac{2}{3} \right\}$ **27.** $\left\{ \dfrac{33}{2} \right\}$ **29.** $\left\{ -1 + \sqrt[3]{49} \right\}$

31. 2 cannot be a solution because $\log(2 - 3) = \log(-1)$, and -1 is not in the domain of $\log x$.   **33.** $\left\{\dfrac{1}{3}\right\}$ **35.** $\{2\}$

37. $\emptyset$ **39.** $\{8\}$ **41.** $\left\{\dfrac{4}{3}\right\}$ **43.** $\{8\}$ **45. (a)** \$2539.47 **(b)** 10.2 yr   **47. (a)** \$4934.71 **(b)** 19.8 yr

49. (a) \$11,260.96 **(b)** \$11,416.64 **(c)** \$11,497.99 **(d)** \$11,580.90 **(e)** \$11,581.83   **51.** \$137.41   **53. (a)** 15.8 million tons
(b) 22.5 million tons **(c)** 45.7 million tons **(d)** 80.5 million tons   **55.** \$40,693 million   **57. (a)** 1.62 g **(b)** 1.18 g
(c) 0.69 g **(d)** 2.00 g   **59. (a)** 179.73 g **(b)** 21.66 yr   **61.** 2006   **63.** 1.733 days   **65. (a)** The expression $\dfrac{1}{x}$ in the
base cannot be evaluated since division by 0 is not defined. **(b)** e **(c)** 2.718280469; 2.718281828; They differ in the sixth
decimal place. **(d)** e **67.**   **69.**

Chapter 10 Review Exercises   (pages 688–692)

1. not one-to-one   **3.** This function is not one-to-one because two sodas in the list have 41 mg of caffeine.

5. $f^{-1}(x) = \dfrac{x^3 + 4}{6}$ **7.**   **9.**   **11.**   **13.** $\{4\}$ **15. (a)** 10.3 million tons

(b) 18.4 million tons **(c)** 24.5 million tons   **17.**   **19.** $\left\{\dfrac{3}{2}\right\}$ **21.** $\{8\}$ **23.** $\{b \mid b > 0, b \neq 1\}$ **25.** a

27. $\log_2 3 + \log_2 x + 2\log_2 y$ **29.** $\log_b \dfrac{3x}{y^2}$ **31.** 1.4609   **33.** 3.3638   **35.** 0.9251   **37.** 6.4   **39.** 2.5×10^{-5}

41. (a) 18 yr **(b)** 12 yr **(c)** 7 yr **(d)** 6 yr **(e)** Each comparison shows approximately the same number. For example, in

part (a) the doubling time is 18 yr (rounded) and $\dfrac{72}{4} = 18$. Thus, the formula $t = \dfrac{72}{100r}$ (called the *rule of 72)* is an excellent

approximation of the doubling time formula. (It is used by bankers for that purpose.)   **43.** $\{4.907\}$ **45.** $\left\{\dfrac{1}{9}\right\}$ **47.** $\{2\}$

49. $\{4\}$ **51.** When the power rule was applied in the second step, the domain was changed from $\{x \mid x \neq 0\}$ to $\{x \mid x > 0\}$.
The valid solution -10 was "lost." The solution set is $\{\pm 10\}$.   **53.** \$11,972.17   **55.** about 13.9 days   **57.** about 67%

59. 7   **61.** 4   **63.** -5 **65.** $\{72\}$ **67.** $\left\{\dfrac{1}{9}\right\}$ **69.** $\{3\}$ **71.** $\left\{\dfrac{1}{8}\right\}$ **73.** $\{-2, -1\}$ **75.** about 28.35%

Chapter 10 Test   (pages 692–693)

[10.1] **1. (a)** not one-to-one **(b)** one-to-one   **2.** $f^{-1}(x) = x^3 - 7$ **3.**   [10.2] **4.**

[10.3] **5.**

[10.1–10.3] **6.** Once the graph of $f(x) = 6^x$ is sketched, interchange the x- and y-values of its ordered pairs. The resulting points will be on the graph of $g(x) = \log_6 x$ since f and g are inverses.

[10.2] **7.** $\{-4\}$ **8.** $\left\{-\dfrac{13}{3}\right\}$ [10.5] **9. (a)** 37.7 million **(b)** 42.3 million   [10.3] **10.** $\log_4 0.0625 = -2$ **11.** $7^2 = 49$

12. $\{32\}$ **13.** $\left\{\dfrac{1}{2}\right\}$ **14.** $\{2\}$ **15.** 5; 2; 5th; 32   [10.4] **16.** $2\log_3 x + \log_3 y$ **17.** $\dfrac{1}{2}\log_5 x - \log_5 y - \log_5 z$

18. $\log_b \dfrac{s^3}{t}$ **19.** $\log_b \dfrac{r^{1/4}s^2}{t^{2/3}}$ [10.5] **20. (a)** 1.3636 **(b)** -0.1985 **21. (a)** $\dfrac{\log 19}{\log 3}$ **(b)** $\dfrac{\ln 19}{\ln 3}$ **(c)** 2.6801

[10.6] **22.** $\{3.9656\}$ **23.** $\{3\}$ **24.** \$12,507.51   **25. (a)** \$19,260.38 **(b)** approximately 13.9 yr

Chapters 1–10 Cumulative Review Exercises   (pages 694–695)

[1.1] **1.** $-2, 0, 6, \dfrac{30}{3}$ (or 10)   **2.** $-\dfrac{9}{4}, -2, 0, 0.6, 6, \dfrac{30}{3}$ (or 10)   **3.** $-\sqrt{2}, \sqrt{11}$ **4.** $-\dfrac{9}{4}, -2, -\sqrt{2}, 0, 0.6, \sqrt{11}, 6,$

$\dfrac{30}{3}$ (or 10)   [1.2, 1.3] **5.** 16   **6.** -27 **7.** -39 [2.1] **8.** $\left\{-\dfrac{2}{3}\right\}$ [2.5] **9.** $[1, \infty)$ [2.7] **10.** $\{-2, 7\}$

11. $\left\{\pm\dfrac{16}{3}\right\}$ **12.** $\left[\dfrac{7}{3}, 3\right]$ **13.** $(-\infty, -3) \cup (2, \infty)$ [3.1] **14.**

[3.4] **15.**

[3.2, 3.5] **16. (a)** yes **(b)** -1692; The number of U.S. travelers to international countries decreased by approximately 1692 thousand per year during 2000–2003.   [3.3] **17.** $y = \dfrac{3}{4}x - \dfrac{19}{4}$ [4.1, 4.4] **18.** $\{(4, 2)\}$ **19.** $\varnothing$

[4.2, 4.4] **20.** $\{(1, -1, 4)\}$ [4.3] **21.** 6 lb   [5.4] **22.** $6p^2 + 7p - 3$ **23.** $16k^2 - 24k + 9$

[5.2] **24.** $-5m^3 + 2m^2 - 7m + 4$ [5.5] **25.** $2t^3 + 5t^2 - 3t + 4$ [6.1] **26.** $x(8 + x^2)$ [6.2] **27.** $(3y - 2)(8y + 3)$

28. $z(5z + 1)(z - 4)$ [6.3] **29.** $(4a + 5b^2)(4a - 5b^2)$ **30.** $(2c + d)(4c^2 - 2cd + d^2)$ **31.** $(4r + 7q)^2$

[5.1] **32.** $-\dfrac{1875p^{13}}{8}$ [7.1] **33.** $\dfrac{x + 5}{x + 4}$ [7.2] **34.** $\dfrac{-3k - 19}{(k + 3)(k - 2)}$ **35.** $\dfrac{22 - p}{p(p - 4)(p + 2)}$ [8.3] **36.** $12\sqrt{2}$

[8.4] **37.** $-27\sqrt{2}$ [8.6] **38.** $\{0, 4\}$ [8.7] **39.** 41   [9.1, 9.2] **40.** $\left\{\dfrac{1 \pm \sqrt{13}}{6}\right\}$ [9.7] **41.** $(-\infty, -4) \cup (2, \infty)$

[9.3] **42.** $\{\pm 1, \pm 2\}$ [9.6] **43.** 150 and 150   [9.5] **44.** $f(x) = \frac{1}{3}(x - 1)^2 + 2$

[10.2] **45.**

46. $\{-1\}$

[10.3] **47.**

[10.4] **48.** 6.3398   **49.** $3\log x + \dfrac{1}{2}\log y - \log z$ [10.6] **50. (a)** 25,000 **(b)** 30,500

(c) 37,300 **(d)** in about 3.5 hr, or at about 3:30 P.M.

CHAPTER 11 Nonlinear Functions, Conic Sections, and Nonlinear Systems

Section 11.1   (pages 701–702)

1. 0   **3.** 0, 0   **5.** B   **7.** A   **9.** domain: $(-\infty, \infty)$;
range: $[0, \infty)$

11. domain: $(-\infty, 0) \cup (0, \infty)$;
range: $(-\infty, 1) \cup (1, \infty)$

13. domain: $[2, \infty)$;
range: $[0, \infty)$

15. domain: $(-\infty, 2) \cup (2, \infty)$;
range: $(-\infty, 0) \cup (0, \infty)$

17. domain: $[-3, \infty)$;
range: $[-3, \infty)$

19. 3   **21.** 0   **23.** -14 **25.** -11

27.

29.

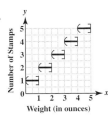

31. $2\sqrt{5}$ **33.** $\sqrt{(x + 2)^2 + (y - 5)^2}$

Section 11.2   (pages 708–711)

1. **(a)** $(0, 0)$ **(b)** 5 **(c)**

3. B   **5.** D   **7.** $(x + 4)^2 + (y - 3)^2 = 4$ **9.** $(x + 8)^2 + (y + 5)^2 = 5$

11. center: $(-2, -3)$; $r = 2$ **13.** center: $(-5, 7)$; $r = 9$ **15.** center: $(2, 4)$; $r = 4$

17.

19.

21. center: $(-3, 2)$

23. center: $(2, 3)$

25. center: $(-3, 3)$

27. The thumbtack acts as the center and the length of the string acts as the radius.

29.

31.

33.

35.

37.

39.

41. By the vertical line test the set is not a function, because a vertical line may intersect the graph of an ellipse in two points.

43. $y_1 = 4 + \sqrt{16 - (x + 2)^2}$, $y_2 = 4 - \sqrt{16 - (x + 2)^2}$

45.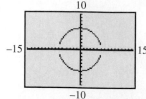

47.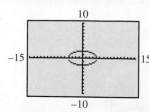

49. $3\sqrt{3}$ units   **51. (a)** 10 m **(b)** 36 m

53. (a) 154.7 million mi **(b)** 128.7 million mi (Answers are rounded.)   **55.**   **57.** (3, 0); (0, 4)

Section 11.3   (pages 718–720)

1. C   **3.** D   **5.** When written in one of the forms given in the box titled "Equations of Hyperbolas" in this section, it will open up and down if the $-$ sign precedes the x^2-term; it will open left and right if the $-$ sign precedes the y^2-term.

7. $\frac{x^2}{16} - \frac{y^2}{9} = 1$

9. $\frac{y^2}{4} - \frac{x^2}{25} = 1$

11. $\frac{x^2}{25} - \frac{y^2}{36} = 1$

13. $\frac{y^2}{16} - \frac{x^2}{16} = 1$

15. hyperbola 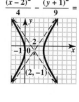 $x^2 - y^2 = 16$

17. ellipse $4x^2 + y^2 = 16$

19. parabola $x^2 - 2y = 0$

21. hyperbola $9x^2 = 144 + 16y^2$

23. hyperbola $y^2 = 4 + x^2$

25. domain: $[-4, 4]$; range: $[0, 4]$

 $f(x) = \sqrt{16 - x^2}$

27. domain: $[-6, 6]$; range: $[-6, 0]$

$f(x) = -\sqrt{36 - x^2}$

29. domain: $(-\infty, \infty)$; range: $[3, \infty)$

 $\frac{y}{3} = \sqrt{1 + \frac{x^2}{9}}$

31. $\frac{(x - 2)^2}{4} - \frac{(y + 1)^2}{9} = 1$

33. $\frac{y^2}{36} - \frac{(x - 2)^2}{49} = 1$

35. (a) 50 m **(b)** 69.3 m

37. $y_1 = \sqrt{\frac{x^2}{9} - 1}$, $y_2 = -\sqrt{\frac{x^2}{9} - 1}$

39.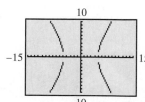

41.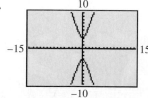

43. $\{(2, 9)\}$ **45.** $\{(0, 5)\}$ **47.** $\left\{ -\sqrt{3}, \sqrt{3}, -\frac{\sqrt{2}}{2}i, \frac{\sqrt{2}}{2}i \right\}$ **49.** $\left\{ -2, -\sqrt{3}, 2, \sqrt{3} \right\}$

Section 11.4 (pages 726–728)

1. Substitute $x - 1$ for y in the first equation. Then solve for x. Find the corresponding y-values by substituting back into $y = x - 1$. In the first equation, both variables are squared and in the second, both variables are to the first power, so the elimination method is not appropriate. **3.** one **5.** none

7. **9.** 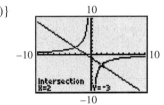 **11.** **13.** 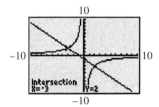 **15.** $\left\{(0, 0), \left(\dfrac{1}{2}, \dfrac{1}{2}\right)\right\}$ **17.** $\{(-6, 9), (-1, 4)\}$

19. $\left\{\left(-\dfrac{1}{5}, \dfrac{7}{5}\right), (1, -1)\right\}$ **21.** $\left\{(-2, -2), \left(-\dfrac{4}{3}, -3\right)\right\}$ **23.** $\{(-3, 1), (1, -3)\}$ **25.** $\left\{\left(-\dfrac{3}{2}, -\dfrac{9}{4}\right), (-2, 0)\right\}$

27. $\left\{\left(-\sqrt{3}, 0\right), \left(\sqrt{3}, 0\right), \left(-\sqrt{5}, 2\right), \left(\sqrt{5}, 2\right)\right\}$ **29.** $\left\{\left(\dfrac{\sqrt{3}}{3}i, -\dfrac{1}{2} + \dfrac{\sqrt{3}}{6}i\right), \left(-\dfrac{\sqrt{3}}{3}i, -\dfrac{1}{2} - \dfrac{\sqrt{3}}{6}i\right)\right\}$

31. $\{(-2, 0), (2, 0)\}$ **33.** $\left\{\left(\sqrt{3}, 0\right), \left(-\sqrt{3}, 0\right)\right\}$ **35.** $\left\{\left(-2\sqrt{3}, -2\right), \left(-2\sqrt{3}, 2\right), \left(2\sqrt{3}, -2\right), \left(2\sqrt{3}, 2\right)\right\}$

37. $\left\{\left(-2i\sqrt{2}, -2\sqrt{3}\right), \left(-2i\sqrt{2}, 2\sqrt{3}\right), \left(2i\sqrt{2}, -2\sqrt{3}\right), \left(2i\sqrt{2}, 2\sqrt{3}\right)\right\}$ **39.** $\left\{\left(-\sqrt{5}, -\sqrt{5}\right), \left(\sqrt{5}, \sqrt{5}\right)\right\}$

41. $\{(i, 2i), (-i, -2i), (2, -1), (-2, 1)\}$ **43.** $\{(2, -3), (-3, 2)\}$

45. length: 12 ft; width: 7 ft **47.** \$20; $\dfrac{4}{5}$ thousand or 800 calculators **49.** **51.**

Section 11.5 (pages 731–733)

1. Answers will vary. **3.** C **5.** B **7.** A **9.** **11.** **13.** **15.**

17. **19.** **21.** **23.** **25.** **27.**

29. $y > x^2 - 4$
$y < -x^2 + 3$
 31. **33.** $y \le -x^2$
$y \ge x - 3$
$y \le -1$
$x < 1$
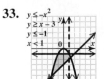 **35.** **37.** 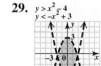 **39.**

41. **43.** 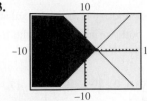 **45.** 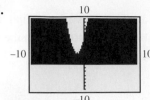 **47.** (a) 6 (b) $\frac{7}{2}$ (c) $\frac{8}{3}$ (d) $\frac{9}{4}$

49. (a) 0 (b) 2 (c) 6 (d) 12

Chapter 11 Review Exercises (pages 739–740)

1. 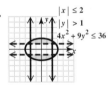 **3.** **5.** $(x+2)^2+(y-4)^2=9$ **7.** $(x-4)^2+(y-2)^2=36$ **9.** center: (4, 1); $r=2$

11. center: (3, −2); $r=5$ **13.** **15.** $\dfrac{x^2}{65{,}286{,}400}+\dfrac{y^2}{2{,}560{,}000}=1$ **17.** **19.**

21. parabola **23.** ellipse **25.** hyperbola **27.** $\{(6,-9),(-2,-5)\}$ **29.** $\{(4,2),(-1,-3)\}$

31. $\left\{\left(-\sqrt{2},2\right),\left(-\sqrt{2},-2\right),\left(\sqrt{2},-2\right),\left(\sqrt{2},2\right)\right\}$ **33.** 0, 1, or 2 **35.** **37.**

39. **41.** **43.** **45.** **47.**

Chapter 11 Test (page 741)

[11.1] **1.** (a) C (b) A (c) D (d) B **2.** [11.2] **3.** center: (2, −3); radius: 4

4. center: (−4, 1); radius: 5 [11.3] **5.** [11.2] **6.** [11.3] **7.** **8.**

9. ellipse **10.** hyperbola **11.** parabola [11.4] **12.** $\left\{\left(-\frac{1}{2},-10\right),(5,1)\right\}$ **13.** $\left\{(-2,-2),\left(\frac{14}{5},-\frac{2}{5}\right)\right\}$

14. $\left\{\left(-\sqrt{22}, -\sqrt{3}\right), \left(-\sqrt{22}, \sqrt{3}\right), \left(\sqrt{22}, -\sqrt{3}\right), \left(\sqrt{22}, \sqrt{3}\right)\right\}$ [11.5] **15.**   **16.**

Chapters 1–11 Cumulative Review Exercises   (pages 742–743)

[1.1, 1.2] **1.** -4 [2.1] **2.** $\left\{\dfrac{2}{3}\right\}$ [2.5] **3.** $\left(-\infty, \dfrac{3}{5}\right]$ [2.7] **4.** $\{-4, 4\}$ **5.** $(-\infty, -5) \cup (10, \infty)$ [3.2] **6.** $\dfrac{2}{3}$

[3.3] **7.** $3x + 2y = -13$ [4.1] **8.** $\{(3, -3)\}$ [4.2] **9.** $\{(4, 1, -2)\}$ [11.4] **10.** $\left\{(-1, 5), \left(\dfrac{5}{2}, -2\right)\right\}$ [4.3] **11.** 40 mph

[3.5] **12.** \$275   [5.4] **13.** $25y^2 - 30y + 9$ **14.** $12r^2 + 40r - 7$ [5.5] **15.** $4x^3 - 4x^2 + 3x + 5 + \dfrac{3}{2x + 1}$

[6.2] **16.** $(3x + 2)(4x - 5)$ **17.** $(2y^2 - 1)(y^2 + 3)$ [6.3] **18.** $(z^2 + 1)(z + 1)(z - 1)$ **19.** $(a - 3b)(a^2 + 3ab + 9b^2)$

[7.1] **20.** $\dfrac{40}{9}$ **21.** $\dfrac{y - 1}{y(y - 3)}$ [7.2] **22.** $\dfrac{3c + 5}{(c + 5)(c + 3)}$ **23.** $\dfrac{1}{p}$ [7.5] **24.** $1\dfrac{1}{5}$ hr   [5.1] **25.** $\dfrac{3}{4}$ **26.** $\dfrac{a^5}{4}$ [8.4] **27.** $2\sqrt[3]{2}$

[8.5] **28.** $\dfrac{3\sqrt{10}}{2}$ [8.7] **29.** $\dfrac{7}{5} + \dfrac{11}{5}i$ [8.6] **30.** $\emptyset$ [6.5] **31.** $\left\{\dfrac{1}{5}, -\dfrac{3}{2}\right\}$ [9.1, 9.2] **32.** $\left\{\dfrac{1 + 2\sqrt{2}}{4}, \dfrac{1 - 2\sqrt{2}}{4}\right\}$

33. $\left\{\dfrac{3 + \sqrt{33}}{6}, \dfrac{3 - \sqrt{33}}{6}\right\}$ [9.3] **34.** $\left\{-\dfrac{\sqrt{6}}{2}, \dfrac{\sqrt{6}}{2}, -\sqrt{7}, \sqrt{7}\right\}$ [9.4] **35.** $v = \dfrac{\pm\sqrt{rFkw}}{kw}$ [10.1] **36.** $f^{-1}(x) = \sqrt[3]{x - 4}$

[10.4] **37.** 4   [10.5] **38.** 7   [10.4] **39.** $\log \dfrac{(3x + 7)^2}{4}$ [10.6] **40.** $\{3\}$ **41.** **(a)** \$12,198.90 **(b)** \$12,214.03

42. \$16.9 billion   **43.** \$75.8 billion   [11.1] **44.** domain: $(-\infty, \infty)$; range: $[0, \infty)$

[3.5] **45.**   [9.5] **46.**   [11.5] **47.**   [11.1] **48.**

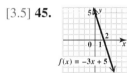 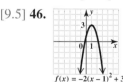 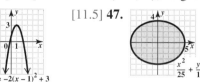 

[11.3] **49.**   [10.2] **50.**

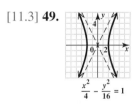 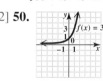

CHAPTER 12 Sequences and Series

Section 12.1   (pages 751–753)

1. $2, 3, 4, 5, 6$ **3.** $4, \dfrac{5}{2}, 2, \dfrac{7}{4}, \dfrac{8}{5}$ **5.** $3, 9, 27, 81, 243$ **7.** $1, \dfrac{1}{4}, \dfrac{1}{9}, \dfrac{1}{16}, \dfrac{1}{25}$ **9.** $-1, 1, -1, 1, -1$ **11.** -70 **13.** $\dfrac{49}{23}$

15. 171 **17.** $4n$ **19.** $\dfrac{1}{3^n}$ **21.** \$110, \$109, \$108, \$107, \$106, \$105; \$400   **23.** \$6554   **25.** $4 + 5 + 6 + 7 + 8 = 30$

27. $3 + 6 + 11 = 20$ **29.** $-1 + 1 - 1 + 1 - 1 + 1 = 0$ **31.** $0 + 6 + 14 + 24 + 36 = 80$

33. $2x + 4x + 6x + 8x + 10x$ **35.** $x + 2x^2 + 3x^3 + 4x^4 + 5x^5$ *Answers may vary for Exercises 37–41.*

37. $\sum_{i=1}^{5} (i + 2)$ **39.** $\sum_{i=1}^{5} \frac{1}{i + 1}$ **41.** $\sum_{i=1}^{5} i^2$ **43.** The similarities are that both are defined by the same linear expression and

that points satisfying both lie in a straight line. The difference is that the domain of f consists of all real numbers, but the domain of the sequence is $\{1, 2, 3, \ldots\}$. An example of a similarity is that $f(1) = 6$ and $a_1 = 6$. An example of a difference is that

$f\left(\frac{3}{2}\right) = 7$ but $a_{3/2}$ is not defined.   **45.** A sequence is a list of terms in a specific order, while a series is the indicated sum of

the terms of a sequence.   **47.** $\frac{59}{7}$ **49.** 5   **51.** 8175   **53.** $a = 6, d = 2$ **55.** 10

Section 12.2   (pages 758–760)

1. An arithmetic sequence is a sequence (list) of numbers in a specific order such that there is a common difference between any two successive terms. For example, the sequence 1, 5, 9, 13, . . . is arithmetic with difference $d = 5 - 1 = 9 - 5 = 13 - 9 = 4$. As another example, 2, -1, -4, -7, . . . is an arithmetic sequence with $d = -3$.   **3.** $d = 1$ **5.** not arithmetic

7. $d = 5$ **9.** 5, 9, 13, 17, 21   **11.** $-2, -6, -10, -14, -18$ **13.** $a_n = 5n - 3$ **15.** $a_n = \frac{3}{4}n + \frac{9}{4}$

17. $a_n = 3n - 6$ **19.** 76   **21.** 48   **23.** -1 **25.** 16   **27.** 6   **29.** n represents the number of terms.   **31.** 81

33. -3 **35.** 87   **37.** 390   **39.** 320   **41.** 31,375   **43.** \$465   **45.** \$2100 per month   **47.** 68; 1100   **49.** no; 3; 9

51. 18   **53.** $\frac{1}{2}$

Section 12.3   (pages 768–771)

1. A geometric sequence is an ordered list of numbers such that each term after the first is obtained by multiplying the previous term by a constant, r, called the common ratio. For example, if the first term is 3 and $r = 4$, then the sequence is 3, 12, 48, 192, ... If the first term is 2 and $r = -1$, then the sequence is 2, -2, 2, -2, ...   **3.** $r = 2$ **5.** not geometric

7. $r = -3$ **9.** $r = -\frac{1}{2}$ *There are alternative forms of the answers in Exercises 11–15.*   **11.** $a_n = 5(2)^{n-1}$

13. $a_n = \frac{3^{n-1}}{9}$ **15.** $a_n = 10\left(-\frac{1}{5}\right)^{n-1}$ **17.** $2(5)^9 = 3,906,250$ **19.** $\frac{1}{2}\left(\frac{1}{3}\right)^{11}$ **21.** $2\left(\frac{1}{2}\right)^{24} = \frac{1}{2^{23}}$

23. 2, 6, 18, 54, 162   **25.** 5, -1, $\frac{1}{5}$, $-\frac{1}{25}$, $\frac{1}{125}$ **27.** $\frac{121}{243}$ **29.** -1.997 **31.** 2.662   **33.** -2.982 **35.** \$46,101.64

37. \$130,159.72   **39.** 9   **41.** $\frac{10,000}{11}$ **43.** $-\frac{9}{20}$ **45.** The sum does not exist.   **47.** $10\left(\frac{3}{5}\right)^4 \approx 1.3$ ft

49. 3 days; $\frac{1}{4}$ g   **51.** (a) $1.1(1.06)^5 \approx 1.5$ billion units **(b)** approximately 12 yr   **53.** $\$50,000\left(\frac{3}{4}\right)^8 \approx \$5000$

55. 0.33333 . . .   **56.** 0.66666 . . .   **57.** 0.99999 . . .   **58.** $\frac{a_1}{1 - r} = \frac{0.9}{1 - 0.1} = \frac{0.9}{0.9} = 1$; therefore, 0.99999 . . . = 1

59. B   **60.** 0.49999 . . . = $0.4 + 0.09999 \ldots = \frac{4}{10} + \frac{1}{10}(0.9999 \ldots) = \frac{4}{10} + \frac{1}{10}(1) = \frac{5}{10} = \frac{1}{2}$ **61.** $9x^2 + 12xy + 4y^2$

63. $a^3 - 3a^2b + 3ab^2 - b^3$

Section 12.4 (pages 776–777)

1. 720   **3.** 15   **5.** 15   **7.** 1   **9.** 120   **11.** 78   **13.** $m^4 + 4m^3n + 6m^2n^2 + 4mn^3 + n^4$ **15.** $a^5 - 5a^4b + 10a^3b^2 -$

$10a^2b^3 + 5ab^4 - b^5$ **17.** $8x^3 + 36x^2 + 54x + 27$ **19.** $\dfrac{x^4}{16} - \dfrac{x^3y}{2} + \dfrac{3x^2y^2}{2} - 2xy^3 + y^4$ **21.** $m^3x^3 - 3m^2n^2x^2 + 3mn^4x - n^6$

23. $r^{12} + 24r^{11}s + 264r^{10}s^2 + 1760r^9s^3$ **25.** $3^{14}x^{14} - 14(3^{13})x^{13}y + 91(3^{12})x^{12}y^2 - 364(3^{11})x^{11}y^3$ **27.** $t^{20} + 10t^{18}u^2 +$

$45t^{16}u^4 + 120t^{14}u^6$ **29.** $120(2^7)m^7n^3$ **31.** $\dfrac{7x^2y^6}{16}$ **33.** $36k^7$ **35.** $-160x^6y^3$ **37.** $4320x^9y^4$

Chapter 12 Review Exercises   (pages 782–784)

1. $-1, 1, 3, 5$ **3.** $1, 4, 9, 16$ **5.** $0, 3, 8, 15$ **7.** $2x + 3x^2 + 4x^3 + 5x^4 + 6x^5 + 7x^6$ **9.** 126

11. 2419 billion dollars   **13.** arithmetic; $d = 4$ **15.** geometric; $r = -1$ **17.** geometric; $r = \dfrac{1}{2}$ **19.** 73

21. $a_n = -5n + 1$ **23.** 15   **25.** 152   **27.** $a_n = -1(4)^{n-1}$ **29.** $2(-3)^{10} = 118,098$ **31.** $\dfrac{341}{1024}$

33. 1   **35.** $32p^5 - 80p^4q + 80p^3q^2 - 40p^2q^3 + 10pq^4 - q^5$ **37.** $m^2 + 4m\sqrt{mn} + 6mn + 4n\sqrt{mn} + n^2$

39. $-18,400(3)^{22}k^3$ **41.** $a_{10} = 1536; S_{10} = 1023$ **43.** $a_{15} = 38; S_{10} = 95$ **45.** $a_n = 2(4)^{n-1}$ **47.** $a_n = -3n + 15$

49. \$23,171.55   **51.** $\dfrac{1}{128}$ **53.** No, the sum cannot be found, because $r = 2$ and this value of r does not satisfy $|r| < 1$.

Chapter 12 Test   (pages 784–785)

[12.1] **1.** $0, 2, 0, 2, 0$ [12.2] **2.** $4, 6, 8, 10, 12$ [12.3] **3.** $48, 24, 12, 6, 3$ [12.2] **4.** 0   [12.3] **5.** $\dfrac{64}{3}$ or $-\dfrac{64}{3}$

[12.2] **6.** 75   [12.3] **7.** 124 or 44   [12.1] **8.** 73,823   [12.3] **9.** \$137,925.91   **10.** It has a sum if $|r| < 1$.

[12.2] **11.** 70   **12.** 33   **13.** 125,250   [12.3] **14.** 42   **15.** $\dfrac{1}{3}$ **16.** The sum does not exist.   [12.4] **17.** 40,320

18. 1   **19.** 15   **20.** 66   **21.** $81k^4 - 540k^3 + 1350k^2 - 1500k + 625$ **22.** $\dfrac{14,080x^8y^4}{9}$ [12.1] **23.** \$324

[12.3] **24.** $20(3^{11}) = 3,542,940$

Chapters 1–12 Cumulative Review Exercises   (pages 785–787)

[1.2, 1.3] **1.** 8   **2.** -35 **3.** -55 [1.1] **4.** $10, 0, \dfrac{45}{15}$ (or 3), -3 **5.** $-\dfrac{8}{3}, 10, 0, \dfrac{45}{15}$ (or 3), 0.82, -3 **6.** $\sqrt{13}, -\sqrt{3}$

7. All are real except $\sqrt{-7}$.   [2.1] **8.** $\left\{\dfrac{1}{6}\right\}$ [2.5] **9.** $[10, \infty)$ [2.7] **10.** $\left\{-\dfrac{9}{2}, 6\right\}$ [2.1] **11.** $\{9\}$

[2.6] **12.** $(-\infty, -3) \cup (4, \infty)$ [2.7] **13.** $(-\infty, -3] \cup [8, \infty)$ [3.2] **14.** $\dfrac{3}{4}$ [3.3] **15.** $3x + y = 4$ [3.1] **16.**

[3.4] **17.**   [3.5] **18. (a)** yes **(b)** $\{-3, -2, 0, 1, 2\}$ **(c)** $\{2, 6, 4\}$ [4.1] **19.** $\{(3, -5)\}$ **20.** $\{(-1, -2)\}$

[4.4] **21.** $\{(2, 1, 4)\}$ [4.3] **22.** 2 lb [5.4] **23.** $20p^2 - 2p - 6$ **24.** $9k^2 - 42k + 49$

[5.2] **25.** $-5m^3 - 3m^2 + 3m + 8$ [5.5] **26.** $2t^3 + 3t^2 - 4t + 2 + \dfrac{3}{3t - 2}$ [6.1] **27.** $x(7 + x^2)$

[6.2] **28.** $(7y - 4)(2y + 3)$ **29.** $z(3z + 4)(2z - 1)$ [6.3] **30.** $(7a^2 + 3b)(7a^2 - 3b)$ **31.** $(c + 3d)(c^2 - 3cd + 9d^2)$

32. $(8r + 3q)^2$ [6.5] **33.** $\left\{-\dfrac{5}{2}, 2\right\}$ [9.7] **34.** $[-2, 3]$ [5.1] **35.** $\dfrac{9}{4}$ **36.** $-\dfrac{27p^2}{10}$ [7.1] **37.** $\{x \mid x \neq -9, 9\}$

38. $\dfrac{x + 7}{x - 2}$ [7.2] **39.** $\dfrac{3p - 26}{p(p + 3)(p - 4)}$ [7.4] **40.** $\emptyset$ [9.2] **41.** $\left\{\dfrac{-5 + \sqrt{217}}{12}, \dfrac{-5 - \sqrt{217}}{12}\right\}$ [8.6] **42.** $\{1, 2\}$

[8.4] **43.** $10\sqrt{2}$ [8.7] **44.** 73 [10.1] **45.** $f^{-1}(x) = \dfrac{x - 5}{9}$, or $f^{-1}(x) = \dfrac{1}{9}x - \dfrac{5}{9}$ [10.2] **46.** 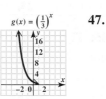 **47.** $\left\{\dfrac{5}{2}\right\}$

[10.3] **48.**   [10.6] **49.** $\{2\}$ [9.5] **50.**   [11.2] **51.**   [11.3] **52.**

[11.4] **53.** $\left\{(-1, 5), \left(\dfrac{5}{2}, -2\right)\right\}$ [11.2] **54.** $(x + 5)^2 + (y - 12)^2 = 81$ [12.1] **55.** $-7, -2, 3, 8, 13$

[12.2, 12.3] **56. (a)** 78 **(b)** $\dfrac{75}{7}$ [12.2] **57.** 30 [12.4] **58.** 362,880 **59.** $32a^5 - 80a^4 + 80a^3 - 40a^2 + 10a - 1$

60. $-\dfrac{45x^8y^6}{4}$

APPENDIXES

Appendix B (pages 802–804)

1. (a) true **(b)** true **(c)** false; the determinant equals $ad - bc$. **(d)** true **3.** -3 **5.** 14 **7.** 0 **9.** 59 **11.** 14
13. Multiply the upper left and lower right entries. Then multiply the upper right and lower left entries. Subtract the

second product from the first to obtain the determinant. For example, $\begin{vmatrix} 4 & 2 \\ 7 & 1 \end{vmatrix} = 4 \cdot 1 - 2 \cdot 7 = 4 - 14 = -10.$ **15.** 16

17. -12 **19.** 0 **21.** $\{(1, 0, -1)\}$ **23.** $\{(-3, 6)\}$ **25.** $\left\{\left(\dfrac{53}{17}, \dfrac{6}{17}\right)\right\}$ **27.** $\{(-1, 2)\}$ **29.** $\{(4, -3, 2)\}$

31. Cramer's rule does not apply. **33.** $\{(-2, 1, 3)\}$ **35.** $\left\{\left(\dfrac{49}{9}, -\dfrac{155}{9}, \dfrac{136}{9}\right)\right\}$ **37.** $\{2\}$ **39.** $\{0\}$

Connections **(page 808)** -4

Appendix C (pages 808–810)

1. Synthetic division provides a quick, easy way to divide a polynomial by a binomial of the form $x - k$. **3.** $x - 5$

5. $4m - 1$ **7.** $2a + 4 + \dfrac{5}{a + 2}$ **9.** $p - 4 + \dfrac{9}{p + 1}$ **11.** $4a^2 + a + 3$ **13.** $x^4 + 2x^3 + 2x^2 + 7x + 10 + \dfrac{18}{x - 2}$

15. $-4r^5 - 7r^4 - 10r^3 - 5r^2 - 11r - 8 + \dfrac{-5}{r - 1}$ **17.** $-3y^4 + 8y^3 - 21y^2 + 36y - 72 + \dfrac{143}{y + 2}$

19. $y^2 + y + 1 + \dfrac{2}{y - 1}$ **21.** 7 **23.** -2 **25.** 0 **27.** By the remainder theorem, a 0 remainder means that $P(k) = 0$;

that is, k is a number that makes $P(x) = 0$. **29.** yes **31.** no **33.** no **35.** yes **37.** $(2x - 3)(x + 4)$

38. $\left\{ -4, \dfrac{3}{2} \right\}$ **39.** $P(-4) = 0, P\left(\dfrac{3}{2} \right) = 0$ **40.** a **41.** Yes, $x - 3$ is a factor. $Q(x) = (x - 3)(3x - 1)(x + 2)$

43. 3 **45.** -1

Glossary

A

absolute value The absolute value of a number is the distance between 0 and the number on a number line. (Section 1.1)

absolute value equation An absolute value equation is an equation that involves the absolute value of a variable expression. (Section 2.7)

absolute value function The function defined by $f(x) = |x|$ with a graph that includes portions of two lines is called the absolute value function. (Section 11.1)

absolute value inequality An absolute value inequality is an inequality that involves the absolute value of a variable expression. (Section 2.7)

addition property of equality The addition property of equality states that the same number can be added to (or subtracted from) both sides of an equation to obtain an equivalent equation. (Section 2.1)

addition property of inequality The addition property of inequality states that the same number can be added to (or subtracted from) both sides of an inequality without changing the solution set. (Section 2.5)

additive inverse (negative, opposite) Two numbers that are the same distance from 0 on a number line but on opposite sides of 0 are called additive inverses. (Section 1.1)

algebraic expression Any collection of numbers or variables joined by the basic operations of addition, subtraction, multiplication, or division (except by 0), or the operations of raising to powers or taking roots, formed according to the rules of algebra, is called an algebraic expression. (Sections 1.3, 5.2)

annuity An annuity is a sequence of equal payments made at equal periods of time. (Section 12.3)

arithmetic mean (average) The arithmetic mean of a group of numbers is the sum of all the numbers divided by the number of numbers. (Section 12.1)

arithmetic sequence (arithmetic progression) An arithmetic sequence is a sequence in which each term after the first differs from the preceding term by a constant amount. (Section 12.2)

array of signs An array of signs is used when evaluating a determinant using expansion by minors. The signs alternate for each row and column, beginning with $+$ in the first row, first column position. (Appendix B)

associative property of addition The associative property of addition states that the way in which numbers being added are grouped does not change the sum. (Section 1.4)

associative property of multiplication The associative property of multiplication states that the way in which numbers being multiplied are grouped does not change the product. (Section 1.4)

asymptote A line that a graph more and more closely approaches as the graph gets farther away from the origin is called an asymptote of the graph. (Section 7.4)

asymptotes of a hyperbola The two intersecting straight lines that the branches of a hyperbola approach are called asymptotes of the hyperbola. (Section 11.3)

augmented matrix An augmented matrix is a matrix that has a vertical bar that separates the columns of the matrix into two groups. (Section 4.4)

axis (axis of symmetry) The axis of a parabola is the vertical or horizontal line through the vertex of the parabola. (Section 9.5)

B

base The base is the number that is a repeated factor when written with an exponent. (Sections 1.3, 5.1)

binomial A binomial is a polynomial with exactly two terms. (Section 5.2)

binomial theorem (general binomial expansion) The binomial theorem is a formula used to expand a binomial raised to a power. (Section 12.4)

boundary line In the graph of a linear inequality, the boundary line separates the region that satisfies the inequality from the region that does not satisfy the inequality. (Sections 3.4, 11.5)

C

center of a circle The fixed point that is a fixed distance from all the points that form a circle is the center of the circle. (Section 11.2)

center of an ellipse The center of an ellipse is the fixed point located exactly halfway between the two foci. (Section 11.2)

center-radius form of the equation of a circle The center-radius form of the equation of a circle with center (h, k) and radius r is $(x - h)^2 + (y - k)^2 = r^2$. (Section 11.2)

circle A circle is the set of all points in a plane that lie a fixed distance from a fixed point. (Section 11.2)

coefficient (numerical coefficient) A coefficient is the numerical factor of a term. (Sections 1.4, 5.2)

column of a matrix A column of a matrix is a group of elements that are read vertically. (Section 4.4)

combined variation If a problem involves a combination of direct and inverse variation, then it is called a combined variation problem. (Section 7.6)

combining like terms Combining like terms is a method of adding or subtracting like terms by using the properties of real numbers. (Section 1.4)

common difference The common difference d is the difference between any two adjacent terms of an arithmetic sequence. (Section 12.2)

common logarithm   A common logarithm is a logarithm to base 10. (Section 10.5)

common ratio   A common ratio r is the constant multiplier between adjacent terms in a geometric sequence. (Section 12.3)

commutative property of addition   The commutative property of addition states that the order of numbers in an addition problem can be changed without changing the sum. (Section 1.4)

commutative property of multiplication   The commutative property of multiplication states that the product in a multiplication problem remains the same regardless of the order of the factors. (Section 1.4)

complementary angles (complements)   Complementary angles are angles whose measures have a sum of 90°. (Section 2.4 Exercises)

completing the square   The process of adding to a binomial the number that makes it a perfect square trinomial is called completing the square. (Section 9.1)

complex conjugates   The complex conjugate of $a + bi$ is $a - bi$. (Section 8.7)

complex fraction   A complex fraction is an expression with one or more fractions in the numerator, denominator, or both. (Section 7.3)

complex number   A complex number is any number that can be written in the form $a + bi$, where a and b are real numbers. (Section 8.7)

composite function   A function in which some quantity depends on a variable that, in turn, depends on another variable is called a composite function. (Section 5.3)

composition of functions   Replacing a variable with an algebraic expression is called composition of functions. (Section 5.3)

compound inequality   A compound inequality consists of two inequalities linked by a connective word such as *and* or *or*. (Section 2.6)

conditional equation   A conditional equation is true for some replacements of the variable and false for others. (Section 2.1)

conic section   When a plane intersects an infinite cone at different angles, the figures formed by the intersections are called conic sections. (Section 11.2)

conjugate   The conjugate of $a + b$ is $a - b$. (Section 8.5)

consecutive integers   Two integers that differ by one are called consecutive integers. (Section 2.4 Exercises)

consistent system   A system of equations with a solution is called a consistent system. (Section 4.1)

constant function   A linear function of the form $f(x) = b$, where b is a constant, is called a constant function. (Section 3.5)

constant of variation   In the variation equations $y = kx$, or $y = \frac{k}{x}$, or $y = kxz$, the number k is called the constant of variation. (Section 7.6)

contradiction   A contradiction is an equation that is never true. It has no solution. (Section 2.1)

coordinate on a number line   Each number on a number line is called the coordinate of the point that it labels. (Section 1.1)

coordinates of a point   The numbers in an ordered pair are called the coordinates of the corresponding point in the plane. (Section 3.1)

Cramer's rule   Cramer's rule uses determinants to solve systems of linear equations. (Appendix B)

cube root function   The function defined by $f(x) = \sqrt[3]{x}$ is called the cube root function. (Section 8.1)

cubing function   The polynomial function defined by $f(x) = x^3$ is called the cubing function. (Section 5.3)

D

degree of a polynomial   The degree of a polynomial is the greatest degree of any of the terms in the polynomial. (Section 5.2)

degree of a term   The degree of a term is the sum of the exponents on the variables in the term. (Section 5.2)

dependent equations   Equations of a system that have the same graph (because they are different forms of the same equation) are called dependent equations. (Section 4.1)

dependent variable   In an equation relating x and y, if the value of the variable y depends on the variable x, then y is called the dependent variable. (Section 3.5)

descending powers   A polynomial in one variable is written in descending powers of the variable if the degree of the terms of the

polynomial decreases from left to right. (Section 5.2)

determinant   Associated with every square matrix is a real number called the determinant of the matrix, symbolized by the entries of the matrix placed between two vertical lines. (Appendix B)

difference   The answer to a subtraction problem is called the difference. (Section 1.2)

difference of cubes   The difference of cubes, $x^3 - y^3$, can be factored as $x^3 - y^3 = (x - y)(x^2 + xy + y^2)$. (Section 6.3)

difference of squares   The difference of squares, $x^2 - y^2$, can be factored as the product of the sum and difference of two terms, or $x^2 - y^2 = (x + y)(x - y)$. (Section 6.3)

direct variation y varies directly as x if there exists a real number k such that $y = kx$. (Section 7.6)

discriminant   The discriminant is the quantity under the radical, $b^2 - 4ac$, in the quadratic formula. (Section 9.2)

distributive property   For any real numbers a, b, and c, the distributive property states that $a(b + c) = ab + ac$ and $(b + c)a = ba + ca$. (Section 1.4)

distance   The distance between two points on a number line is the absolute value of the difference between the two numbers. (Section 1.2)

domain   The set of all first components (x-values) in the ordered pairs of a relation is the domain. (Section 3.5)

domain of a rational equation   The domain of a rational equation is the intersection (overlap) of the domains of the rational expressions in the equation. (Section 7.4)

E

element of a matrix   The numbers in a matrix are called the elements of the matrix. (Section 4.4)

elements (members)   Elements are the objects that belong to a set. (Section 1.1)

elimination method   The elimination method is an algebraic method used to solve a system of equations in which the equations of the system are combined so that one or more variables is eliminated. (Section 4.1)

ellipse An ellipse is the set of all points in a plane the sum of whose distances from two fixed points is constant. (Section 11.2)

empty set (null set) The empty set, denoted by { } or $\emptyset$, is the set containing no elements. (Section 1.1)

equation An equation is a statement that two algebraic expressions are equal. (Section 1.1)

equivalent equations Equivalent equations are equations that have the same solution set. (Section 2.1)

equivalent inequalities Equivalent inequalities are inequalities that have the same solution set. (Section 2.5)

expansion by minors A method of evaluating a 3×3 or larger determinant is called expansion by minors. (Appendix B)

exponent (power) An exponent is a number that indicates how many times a factor is repeated. (Sections 1.3, 5.1)

exponential equation An exponential equation is an equation that has a variable as an exponent. (Section 10.2)

exponential expression A number or letter (variable) written with an exponent is an exponential expression. (Section 1.3)

exponential function An exponential function is a function defined by an expression of the form $f(x) = a^x$, where $a > 0$ and $a \neq 1$ for all real numbers x. (Section 10.2)

extraneous solution A solution to a new equation that does not satisfy the original equation is called an extraneous solution. (Section 8.6)

F

factor A factor of a given number is any number that divides evenly (without remainder) into the given number. (Section 1.3)

factoring Writing a polynomial as the product of two or more simpler polynomials is called factoring. (Section 6.1)

factoring by grouping Factoring by grouping is a method of grouping the terms of a polynomial in such a way that the polynomial can be factored even though the greatest common factor of its individual terms is 1. (Section 6.1)

factoring out the greatest common factor Factoring out the greatest common factor is

the process of using the distributive property to write a polynomial as a product of the greatest common factor and a simpler polynomial. (Section 6.1)

finite sequence A finite sequence has a domain that includes only the first n positive integers. (Section 12.1)

first-degree equation A first-degree (linear) equation has no term with the variable to a power other than 1. (Section 2.1)

foci (singular, **focus**) Foci are fixed points used to determine the points that form a parabola, an ellipse, or a hyperbola. (Sections 11.2, 11.3)

FOIL FOIL is a method for multiplying two binomials $(A + B)(C + D)$. Multiply First terms AC, Outer terms AD, Inner terms BC, and Last terms BD. Then combine like terms. (Section 5.4)

formula A formula is an equation in which variables are used to describe a relationship. (Section 2.2)

function A function is a set of ordered pairs (relation) in which each value of the first component x corresponds to exactly one value of the second component y. (Section 3.5)

function notation Function notation $f(x)$ represents the value of the function at x, that is, the y-value that corresponds to x. (Section 3.5)

fundamental rectangle The asymptotes of a hyperbola are the extended diagonals of its fundamental rectangle, with corners at the points (a, b), $(-a, b)$, $(-a, -b)$, and $(a, -b)$. (Section 11.3)

future value of an annuity The future value of an annuity is the sum of the compound amounts of all the payments, compounded to the end of the term. (Section 12.3)

G

general term of a sequence The expression a_n, which defines a sequence, is called the general term of the sequence. (Section 12.1)

geometric sequence (geometric progression) A geometric sequence is a sequence in which each term after the first is a constant multiple of the preceding term. (Section 12.3)

graph of a number The point on a number line that corresponds to a number is its graph. (Section 1.1)

graph of an equation The graph of an equation is the set of all points that correspond to all of the ordered pairs that satisfy the equation. (Section 3.1)

graph of a relation The graph of a relation is the graph of its ordered pairs. (Section 3.5)

greatest common factor (GCF) The greatest common factor of a list of integers is the largest common factor of those integers. The greatest common factor of a polynomial is the largest term that is a factor of all terms in the polynomial. (Section 6.1)

greatest integer function The function defined by $f(x) = [\![x]\!]$, where the symbol $[\![x]\!]$ is used to represent the greatest integer less than or equal to x, is called the greatest integer function. (Section 11.1)

H

horizontal line test The horizontal line test states that a function is one-to-one if every horizontal line intersects the graph of the function at most once. (Section 10.1)

hyperbola A hyperbola is the set of all points in a plane such that the absolute value of the difference of the distances from two fixed points is constant. (Section 11.3)

hypotenuse The hypotenuse is the longest side in a right triangle. It is the side opposite the right angle. (Section 8.3)

I

identity An identity is an equation that is true for all replacements of the variable. It has an infinite number of solutions. (Section 2.1)

identity element for addition Since adding 0 to a number does not change the number, 0 is called the identity element for addition. (Section 1.4)

identity element for multiplication Since multiplying a number by 1 does not change the number, 1 is called the identity element for multiplication. (Section 1.4)

identity function The simplest polynomial function is the identity function, defined by $f(x) = x$. (Section 5.3)

identity property The identity properties state that the sum of 0 and any number equals the number, and the product of 1 and any number equals the number. (Section 1.4)

imaginary part The imaginary part of a complex number $a + bi$ is b. (Section 8.7)

imaginary unit The symbol i is called the imaginary unit. (Section 8.7)

inconsistent system An inconsistent system of equations is a system with no solution. (Section 4.1)

independent equations Equations of a system that have different graphs are called independent equations. (Section 4.1)

independent variable In an equation relating x and y, if the value of the variable y depends on the variable x, then x is called the independent variable. (Section 3.5)

index (order) In a radical of the form $\sqrt[n]{a}$, n is called the index or order. (Section 8.1)

index of summation When using summation notation, $\sum_{i=1}^{n} f(i)$, the letter i is called the index of summation. (Section 12.1)

inequality An inequality is a statement that two expressions are not equal. (Section 1.1)

infinite sequence An infinite sequence is a function with the set of positive integers as the domain. (Section 12.1)

integers The set of integers is $\{\ldots, -3, -2, -1, 0, 1, 2, 3, \ldots\}$. (Section 1.1)

intersection The intersection of two sets A and B, written $A \cap B$, is the set of elements that belong to both A and B. (Section 2.6)

interval An interval is a portion of a number line. (Section 1.1)

interval notation Interval notation is a simplified notation that uses parentheses ( ) and/or brackets [ ] to describe an interval on a number line. (Section 1.1)

inverse of a function f If f is a one-to-one function, then the inverse of f is the set of all ordered pairs of the form (y, x) where (x, y) belongs to f. (Section 10.1)

inverse property The inverse properties state that a number added to its opposite is 0, and a number multiplied by its reciprocal is 1. (Section 1.4)

inverse variation y varies inversely as x if there exists a real number k such that $y = \frac{k}{x}$. (Section 7.6)

irrational numbers Irrational numbers cannot be written as the quotient of two integers but can be represented by points on the number line. (Section 1.1)

J

joint variation y varies jointly as x and z if there exists a real number k such that $y = kxz$. (Section 7.6)

L

least common denominator (LCD) Given several denominators, the smallest expression that is divisible by all the denominators is called the least common denominator. (Section 7.2)

legs of a right triangle The two shorter sides of a right triangle are called the legs. (Section 8.3)

like terms Terms with exactly the same variables raised to exactly the same powers are called like terms. (Sections 1.4, 5.2)

linear equation in one variable A linear equation in one variable can be written in the form $Ax + B = C$, where A, B, and C are real numbers, with $A \neq 0$. (Section 2.1)

linear equation in two variables A linear equation in two variables is an equation that can be written in the form $Ax + By = C$, where A, B, and C are real numbers and A and B are not both 0. (Section 3.1)

linear function A function defined by an equation of the form $f(x) = ax + b$, for real numbers a and b, is a linear function. The value of a is the slope m of the graph of the function. (Section 3.5)

linear inequality in one variable A linear inequality in one variable can be written in the form $Ax + B < C$ or $Ax + B > C$ (or with $\leq$ or $\geq$), where A, B, and C are real numbers, with $A \neq 0$. (Section 2.5)

linear inequality in two variables A linear inequality in two variables can be written in the form $Ax + By < C$ or $Ax + By > C$ (or with $\leq$ or $\geq$), where A, B, and C are real numbers and A and B are not both 0. (Section 3.4)

linear system (system of linear equations) Two or more linear equations form a linear system. (Section 4.1)

logarithm A logarithm is an exponent; $\log_a x$ is the exponent on the base a that gives the number x. (Section 10.3)

logarithmic equation A logarithmic equation is an equation with a logarithm in at least one term. (Section 10.3)

logarithmic function with base a If a and x are positive numbers with $a \neq 1$, then $f(x) = \log_a x$ defines the logarithmic function with base a. (Section 10.3)

lowest terms A fraction is in lowest terms when there are no common factors in the numerator and denominator (except 1). (Section 7.1)

M

mathematical model In a real-world problem, a mathematical model is one or more equations (or inequalities) that describe the situation. (Section 2.2)

matrix (plural, **matrices**) A matrix is a rectangular array of numbers, consisting of horizontal rows and vertical columns. (Section 4.4, Appendix B)

minors The minor of an element in a 3×3 determinant is the 2×2 determinant remaining when a row and a column of the 3×3 determinant are eliminated. (Appendix B)

monomial A monomial is a polynomial with only one term. (Section 5.2)

multiplication property of equality The multiplication property of equality states that the same nonzero number can be multiplied by (or divided into) both sides of an equation to obtain an equivalent equation. (Section 2.1)

multiplication property of inequality The multiplication property of inequality states that both sides of an inequality may be multiplied (or divided) by a positive number without changing the direction of the inequality symbol. Multiplying (or dividing) by a negative number reverses the inequality symbol. (Section 2.5)

multiplication property of 0 The multiplication property of 0 states that the product of any real number and 0 is 0. (Section 1.4)

multiplicative inverse (reciprocal) The multiplicative inverse of a nonzero real number a is $\frac{1}{a}$. (Section 1.2)

N

n-factorial ($n!$) For any positive integer n, $n(n-1)(n-2)(n-3)\cdots(2)(1) = n!$. (Section 12.4)

natural logarithm A natural logarithm is a logarithm to base e. (Section 10.5)

natural numbers (counting numbers) The set of natural numbers includes the numbers used for counting: $\{1, 2, 3, 4, \ldots\}$. (Section 1.1)

negative of a polynomial The negative of a polynomial is that polynomial with the sign of every term changed. (Section 5.2)

nonlinear equation A nonlinear equation is an equation in which some terms have more than one variable or a variable of degree 2 or higher. (Section 11.4)

nonlinear system of equations A nonlinear system of equations is a system that includes at least one nonlinear equation. (Section 11.4)

nonlinear system of inequalities A nonlinear system of inequalities is two or more inequalities to be considered at the same time, at least one of which is nonlinear. (Section 11.5)

number line A number line is a line with a scale that is used to show how numbers relate to each other. (Section 1.1)

numerical coefficient The numerical factor in a term is its numerical coefficient. (Sections 1.4, 5.2)

O

one-to-one function A one-to-one function is a function in which each x-value corresponds to only one y-value and each y-value corresponds to just one x-value. (Section 10.1)

ordered pair An ordered pair is a pair of numbers written within parentheses in which the order of the numbers is important. (Section 3.1)

ordered triple A solution of an equation in three variables, written (x, y, z), is called an ordered triple. (Section 4.2)

ordinary annuity An ordinary annuity is an annuity in which the payments are made at the end of each time period and the frequency of payments is the same as the frequency of compounding. (Section 12.3)

origin The point at which the x-axis and y-axis of a rectangular coordinate system intersect is called the origin. (Section 3.1)

P

parabola The graph of a second-degree (quadratic) equation in two variables is called a parabola. (Section 9.5)

parallel lines Parallel lines are two lines in the same plane that never intersect. (Section 3.2)

Pascal's triangle Pascal's triangle is a triangular array of numbers that is helpful in expanding binomials. (Section 12.4)

payment period In an annuity, the time between payments is called the payment period. (Section 12.3)

percent Percent, written with the sign %, means "per one hundred." (Section 2.2)

perfect square trinomial A perfect square trinomial is a trinomial that can be factored as the square of a binomial. (Section 6.3)

perpendicular lines Perpendicular lines are two lines that intersect to form a right ($90°$) angle. (Section 3.2)

plot To plot an ordered pair is to locate it on a rectangular coordinate system. (Section 3.1)

point-slope form A linear equation is written in point-slope form if it is in the form $y - y_1 = m(x - x_1)$, where m is the slope of the line and (x_1, y_1) is a point on the line. (Section 3.3)

polynomial A polynomial is a term or a finite sum of terms in which all coefficients are real, all variables have whole number exponents, and no variables appear in denominators. (Section 5.2)

polynomial function A function defined by a polynomial in one variable, consisting of one or more terms, is called a polynomial function. (Section 5.3)

polynomial in x A polynomial containing only the variable x is called a polynomial in x. (Section 5.2)

prime polynomial A prime polynomial is a polynomial that cannot be factored using only integer coefficients. (Section 6.1)

principal root (principal nth root) For even indexes, the symbols $\sqrt{}$, $\sqrt[4]{}$, $\sqrt[6]{}, \ldots, \sqrt[n]{}$, are used for nonnegative roots, which are called principal roots. (Section 8.1)

product The answer to a multiplication problem is called the product. (Section 1.2)

product of the sum and difference of two terms The product of the sum and difference of two terms is the difference of the squares of the terms: $(x + y)(x - y) = x^2 - y^2$. (Section 5.4)

proportion A proportion is a statement that two ratios are equal. (Section 7.5)

proportional If y varies directly as x and there exists some number (constant) k such that $y = kx$, then y is said to be proportional to x. (Section 7.6)

pure imaginary number A complex number $a + bi$ with $a = 0$ and $b \neq 0$ is called a pure imaginary number. (Section 8.7)

Pythagorean formula The Pythagorean formula states that the square of the length of the hypotenuse of a right triangle equals the sum of the squares of the lengths of the two legs. (Section 8.3)

Q

quadrant A quadrant is one of the four regions in the plane determined by a rectangular coordinate system. (Section 3.1)

quadratic equation A quadratic equation is an equation that can be written in the form $ax^2 + bx + c = 0$, where a, b, and c are real numbers, with $a \neq 0$. (Sections 6.5, 9.1)

quadratic formula The quadratic formula is a general formula used to solve any quadratic equation. (Section 9.2)

quadratic function A function defined by an equation of the form $f(x) = ax^2 + bx + c$, for real numbers a, b, and c, with $a \neq 0$, is a quadratic function. (Section 9.5)

quadratic inequality A quadratic inequality can be written in the form $ax^2 + bx + c < 0$ or $ax^2 + bx + c > 0$ (or with $\leq$ or $\geq$), where a, b, and c are real numbers, with $a \neq 0$. (Section 9.7)

quadratic in form A nonquadratic equation that is written in the form $au^2 + bu + c = 0$, for $a \neq 0$ and an algebraic expression u, is called quadratic in form. (Section 9.3)

quotient The answer to a division problem is called the quotient. (Section 1.2)

R

radical A radical sign with a radicand is called a radical. (Section 8.1)

radical equation An equation that includes one or more radical expressions with a variable is called a radical equation. (Section 8.6)

radical expression A radical expression is an algebraic expression that contains radicals. (Section 8.1)

radical sign The symbol $\sqrt{}$ is called a radical sign. (Section 1.3)

radicand The number or expression under a radical sign is called the radicand. (Section 8.1)

radius The radius of a circle is the fixed distance between the center and any point on the circle. (Section 11.2)

range The set of all second components (y-values) in the ordered pairs of a relation is the range. (Section 3.5)

ratio A ratio is a comparison of two quantities. (Section 7.5)

rational expression The quotient of two polynomials with denominator not 0 is called a rational expression, or algebraic fraction. (Section 7.1)

rational function A function that is defined by a quotient of polynomials is called a rational function. (Section 7.1)

rational inequality An inequality that involves fractions is called a rational inequality. (Section 9.7)

rational numbers Rational numbers can be written as the quotient of two integers, with denominator not 0. (Section 1.1)

rationalizing the denominator The process of removing radicals from a denominator so that the denominator contains only rational numbers is called rationalizing the denominator. (Section 8.5)

real numbers Real numbers include all numbers that can be represented by points on the number line, that is, all rational and irrational numbers. (Section 1.1)

real part The real part of a complex number $a + bi$ is a. (Section 8.7)

reciprocal Pairs of numbers whose product is 1 are called reciprocals of each other. (Sections 1.2, 7.1)

reciprocal function The reciprocal function is defined by $f(x) = \frac{1}{x}$. (Section 11.1)

rectangular (Cartesian) coordinate system The x-axis and y-axis placed at a right angle at their zero points form a rectangular coordinate system, also called the Cartesian coordinate system. (Section 3.1)

relation A relation is a set of ordered pairs. (Section 3.5)

rise Rise is the vertical change between two points on a line, that is, the change in y-values. (Section 3.2)

row echelon form If a matrix is written with 1s on the diagonal from upper left to lower right and 0s below the 1s, it is said to be in row echelon form. (Section 4.4)

row of a matrix A row of a matrix is a group of elements that are read horizontally. (Section 4.4)

row operations Row operations are operations on a matrix that produce equivalent matrices leading to systems that have the same solutions as the original system of equations. (Section 4.4)

run Run is the horizontal change between two points on a line, that is, the change in x-values. (Section 3.2)

S

scientific notation A number is written in scientific notation when it is expressed in the form $a \times 10^n$, where $1 \leq |a| < 10$ and n is an integer. (Section 5.1)

second-degree inequality A second-degree inequality is an inequality with at least one variable of degree 2 and no variable with degree greater than 2. (Section 11.5)

sequence A sequence is a function whose domain is the set of natural numbers. (Section 12.1)

series The indicated sum of the terms of a sequence is called a series. (Section 12.1)

set A set is a collection of objects. (Section 1.1)

set-builder notation Set-builder notation is used to describe a set of numbers without actually having to list all of the elements. (Section 1.1)

signed numbers Signed numbers are numbers that can be written with a positive or negative sign. (Section 1.1)

simplified radical A simplified radical meets four conditions:

1. The radicand has no factor raised to a power greater than or equal to the index.

2. The radicand has no fractions.

3. No denominator contains a radical.

4. Exponents in the radicand and the index of the radical have no common factor (except 1).

(Section 8.3)

slope The ratio of the change in y to the change in x along a line is called the slope of the line. (Section 3.2)

slope-intercept form A linear equation is written in slope-intercept form if it is in the form $y = mx + b$, where m is the slope and $(0, b)$ is the y-intercept. (Section 3.3)

solution of an equation A solution of an equation is any replacement for the variable that makes the equation true. (Section 2.1)

solution set The solution set of an equation is the set of all solutions of the equation. (Section 2.1)

solution set of a linear system The solution set of a linear system of equations includes all ordered pairs that satisfy all the equations of the system at the same time. (Section 4.1)

solution set of a system of linear inequalities The solution set of a system of linear inequalities includes all ordered pairs that make all inequalities of the system true at the same time. (Section 11.5)

square matrix A square matrix is a matrix that has the same number of rows as columns. (Section 4.4, Appendix B)

square of a binomial The square of a binomial is the sum of the square of the first term, twice the product of the two terms, and the square of the last term. That is,

$(x + y)^2 = x^2 + 2xy + y^2$ or $(x - y)^2 = x^2 - 2xy + y^2$. (Section 5.4)

square root The opposite of squaring a number is called taking its square root; that is, a number b is a square root of a if $b^2 = a$. (Section 1.3)

square root function The function defined by $f(x) = \sqrt{x}$, with $x \geq 0$, is called the square root function. (Sections 8.1, 11.3)

square root property The square root property states that if $x^2 = k$, then $x = \sqrt{k}$ or $x = -\sqrt{k}$. (Section 9.1)

squaring function The polynomial function defined by $f(x) = x^2$ is called the squaring function. (Section 5.3)

standard form of a complex number The standard form of a complex number is $a + bi$. (Section 8.7)

standard form of a linear equation A linear equation in two variables written in the form $Ax + By = C$, where A, B, and C are integers with no common factor (except 1) and $A \geq 0$, is in standard form. (Sections 3.1, 3.3)

standard form of a quadratic equation A quadratic equation written in the form $ax^2 + bx + c = 0$, where a, b, and c are real numbers with $a \neq 0$, is in standard form. (Sections 6.5, 9.1)

step function A function with a graph that looks like a series of steps is called a step function. (Section 11.1)

substitution method The substitution method is an algebraic method for solving a system of equations in which one equation is solved for one of the variables and the result is substituted in the other equation. (Section 4.1)

sum The answer to an addition problem is called the sum. (Section 1.2)

sum of cubes The sum of cubes, $x^3 + y^3$, can be factored as $x^3 + y^3 = (x + y) \cdot (x^2 - xy + y^2)$. (Section 6.3)

summation (sigma) notation Summation notation is a compact way of writing a series using the general term of the corresponding sequence. (Section 12.1)

supplementary angles (supplements) Supplementary angles are angles whose measures have a sum of 180°. (Section 2.4 Exercises)

synthetic division Synthetic division is a shortcut procedure for dividing a polynomial by a binomial of the form $x - k$. (Appendix C)

system of equations A system of equations consists of two or more equations to be solved at the same time. (Section 4.1)

system of inequalities A system of inequalities consists of two or more inequalities to be solved at the same time. (Section 11.5)

T

term A term is a number, a variable, or the product or quotient of a number and one or more variables raised to powers. (Sections 1.4, 5.2)

term of an annuity The time from the beginning of the first payment period to the end of the last period is called the term of an annuity. (Section 12.3)

terms of a sequence The function values written in order are called the terms of the sequence. (Section 12.1)

trinomial A trinomial is a polynomial with exactly three terms. (Section 5.2)

U

union The union of two sets A and B, written $A \cup B$, is the set of elements that belong to either A or B (or both). (Section 2.6)

universal constant The number e is called a universal constant because of its importance in many areas of mathematics. (Section 10.5)

V

variable A variable is a symbol, usually a letter, used to represent an unknown number. (Section 1.1)

vary directly (is directly proportional to) y varies directly as x if there exists a real number (constant) k such that $y = kx$. (Section 7.6)

vary inversely y varies inversely as x if there exists a real number (constant) k such that $y = \frac{k}{x}$. (Section 7.6)

vary jointly If one variable varies as the product of several other variables (sometimes raised to powers), then the first variable is said to vary jointly as the others. (Section 7.6)

vertex The point on a parabola that has the smallest y-value (if the parabola opens up) or the largest y-value (if the parabola opens down) is called the vertex of the parabola. (Section 9.5)

vertical asymptote A vertical line that a graph approaches, but never touches or intersects, is called a vertical asymptote. (Section 7.4)

vertical line test The vertical line test states that any vertical line drawn through the graph of a function must intersect the graph in at most one point. (Section 3.5)

W

whole numbers The set of whole numbers is $\{0, 1, 2, 3, 4, \dots\}$. (Section 1.1)

X

x-axis The horizontal number line in a rectangular coordinate system is called the x-axis. (Section 3.1)

x-intercept A point where a graph intersects the x-axis is called an x-intercept. (Section 3.1)

Y

y-axis The vertical number line in a rectangular coordinate system is called the y-axis. (Section 3.1)

y-intercept A point where a graph intersects the y-axis is called the y-intercept. (Section 3.1)

Z

zero-factor property The zero-factor property states that if two numbers have a product of 0, then at least one of the numbers must be 0. (Sections 6.5, 9.1)

INDEX

Formulas

Figure	Formulas	Illustration
Square	Perimeter: $P = 4s$ Area: $A = s^2$	
Rectangle	Perimeter: $P = 2L + 2W$ Area: $A = LW$	
Triangle	Perimeter: $P = a + b + c$ Area: $A = \dfrac{1}{2}bh$	
Parallelogram	Perimeter: $P = 2a + 2b$ Area: $A = bh$	
Trapezoid	Perimeter: $P = a + b + c + B$ Area: $A = \dfrac{1}{2}h(b + B)$	
Circle	Diameter: $d = 2r$ Circumference: $C = 2\pi r$ $C = \pi d$ Area: $A = \pi r^2$	